REPRODUCTIVE MEDICINE

Molecular, Cellular and Genetic Fundamentals

REPRODUCTIVE MEDICINE

Molecular, Cellular and Genetic Fundamentals

EDITOR-IN-CHIEF

Bart C.J.M. Fauser

Center of Reproductive Medicine, Erasmus Medical Center, Rotterdam, The Netherlands

SECTION EDITORS

Philippe Bouchard
Hôpital Saint-Antoine
Paris, France

Aaron J.W. Hsueh
Stanford University
School of Medicine
Stanford, CA, USA

Anthony J. Rutherford
Leeds General Infirmary
Leeds, UK

Joe Leigh Simpson
Baylor College of Medicine
Houston, TX, USA

Jerome F. Strauss III
University of Pennsylvania
School of Medicine
Philadelphia, PA, USA

André Van Steirteghem
Center for Reproductive Medicine
Vrije Universiteit Brussels
Brussels, Belgium

The Parthenon Publishing Group
International Publishers in Medicine, Science & Technology

A CRC PRESS COMPANY
BOCA RATON LONDON NEW YORK WASHINGTON, D.C.

Library of Congress Cataloging-in-Publication Data
Reproductive medicine : molecular, cellular, and genetic fundamentals / edited by B.C.J.M. Fauser ... [et al.]. – 2nd ed.
p. ; cm.
Includes bibliographical references and index.
ISBN 1-84214-019-1 (alk. paper)
1. Human reproductive technology. 2. Reproductive health. I. Fauser, B.C.J.M. (Bart C.J.M.)
[DNLM: 1. Reproductive Medicine. 2. Molecular Biology. WQ 205 R428753 2002]
RG133.5.R463 2002
616.6'9206–dc21 2002025335

British Library Cataloguing in Publication Data

Data available on request

ISBN 1-84214-019-1

Published in the USA by
The Parthenon Publishing Group
345 Park Avenue South, 10th Floor
New York
NY 10010
USA

Published in the UK by
The Parthenon Publishing Group
23–25 Blades Court
Deodar Road
London SW15 2NU
UK

First published in 2003

Typeset by The Parthenon Publishing Group
Printed and bound by Butler & Tanner Ltd., Frome and London, UK

Contents

List of the Editors

Bart C.J.M. Fauser
Center of Reproductive Medicine
Erasmus Medical Center
Dr Molewaterplein 40
3015 GD Rotterdam
The Netherlands

Philippe Bouchard
Hôpital Saint-Antoine
Pavillon de l'Horloge
184 rue du Faubourg Saint-Antoine
Paris 75571, Cedex 12
France

Aaron J.W. Hsueh
Department of Obstetrics and Gynecology
Division of Reproductive Biology
Stanford University School of Medicine
300 Pasteur Drive, Room A 344
Stanford
California 94305-5317
USA

Anthony J. Rutherford
Clarendon Wing
Leeds General Infirmary
Belmont Grove
Leeds LS2 9NS
UK

Joe Leigh Simpson
Department of Obstetrics and Gynecology
Baylor College of Medicine
6550 Fannin Suite 901A
Houston
Texas 77030
USA

Jerome F. Strauss III
University of Pennsylvania School of Medicine
778 Clinical Research Building
415 Curie Boulevard
Philadelphia
Pennsylvania 19104-6142
USA

André Van Steirteghem
Center for Reproductive Medicine
University Hospital
Vrije Universiteit Brussels
Laarbeeklaan 107
1090 Brussels
Belgium

List of the Contributors

Willy M. Baarends
Department of Reproduction and Development
Erasmus Medical Center
Dr Molewaterplein 50
3015 GE Rotterdam
The Netherlands

Chrinne Belville
Unité de Recherches sur l'Endocrinologie du Développement
INSERM U. 293
Ecole Normale Supérieure
Département de Biologie
1 rue Maurice Arnoux
92120 Montrouge
France

Irving Boime
Department of Pharmacology
Washington University Medical School
60 South Euclid
St. Louis
Missouri 63110
USA

Jeff Boyd
Department of Surgery
Memorial Sloan-Kettering Cancer Center
Box 201, 1275 York Avenue
New York
New York 10021
USA

Albert O. Brinkmann
Department of Endocrinology and Reproduction
Erasmus Medical Center
PO Box 1738
3000 DR Rotterdam
The Netherlands

Serdar E. Bulun
Department of Obstetrics and Gynecology and Molecular Genetics
University of Illinois at Chicago
820 S. Wood Street, M/C 808
Chicago
Illinois 60612-7313
USA

Kathleen H. Burns
Department of Pathology, Room S 217
Baylor College of Medicine
1 Baylor Plaza
Houston
Texas 77030
USA

Nathalie Chabbert-Buffet
Hôpital Saint-Antoine
Pavillon de l'Horloge
184 rue du Faubourg Saint-Antoine
Paris 75571, Cedex 12
France

Stacey C. Chapman
Department of Neurobiology and Physiology
Northwestern University
O.T. Hogan 4-150
2153 N Campus Dr
Evanston
Illinois 60208
USA

Bin Chen
Breast Cancer Research Program
Robert H. Lurie Cancer Center
Northwestern University Medical School
303 East Chicago Avenue
8258 Olson
Chicago
Illinois 60611
USA

Sophie Christin-Maitre
Service d'Endocrinologie
Hôpital Saint-Antoine
184 rue du Faubourg Saint-Antoine
Paris 75012
France

Hugh Clarke
Division of Reproductive Biology
Department of Obstetrics and Gynecology
McGill University
Women's Pavilion (F3.38)
Royal Victoria Hospital
687 Pine Avenue West
Montreal
Quebec H3A 1A1
Canada

Gerard S. Conway
Cobbold Laboratories
The Middlesex Hospital
Mortimer Street
London W1A 8AA
UK

Francesco J. DeMayo
Department of Molecular and Cellular Biology
Baylor College of Medicine
1 Baylor Plaza
Houston
Texas 77030
USA

Guido de Wert
Institute for Bioethics
PO Box 616
6200 MD Maastricht
The Netherlands

Linda C. Giudice
Division of Reproductive Endocrinology and Infertility
Department of Obstetrics and Gynecology
Room HH333, Stanford University Medical Center
School of Medicine
Palo Alto
California 94305-5317
USA

Roger Gosden
Division of Reproductive Biology
Department of Obstetrics and Gynecology
McGill University
Women's Pavilion (F3.38)
Royal Victoria Hospital
687 Pine Avenue West
Montreal
Quebec H3A 1A1
Canada

J. Anton Grootegoed
Erasmus Medical Center
Department of Reproduction and Development
Dr Molewaterplein 50
3015 GE Rotterdam
The Netherlands

Laura Hewitson
Department of Obstetrics, Gynecology and Reproductive Sciences
Pittsburgh Development Center (West Coast Division)
40 SW 147th Place
Beaverton
Oregon 97006
USA

Ilpo T. Huhtaniemi
Department of Physiology
University of Turku
Kiinamyllinkatu 10
FIN-20520 Turku
Finland

Joan S. Hunt
Department of Anatomy and Cell Biology
University of Kansas Medical Center
3901 Rainbow Boulevard
Kansas City
Kansas 66160-7400
USA

Albina Jablonka-Shariff
Department of Pharmacology
Washington University Medical School
60 South Euclid
St. Louis
Missouri 63110
USA

V. Craig Jordan
Breast Cancer Research Program
Robert H. Lurie Cancer Center
Northwestern University Medical School
303 East Chicago Avenue
8258 Olson
Chicago
Illinois 60611
USA

Marianne Jørgensen
Department of Growth and Reproduction
Section GR-5064
Rigshospitalet
Blegdamsvej 9
2100 Copenhagen
Denmark

Nathalie Josso
Unité de Recherches sur l'Endocrinologie du Développement
INSERM U. 293
Ecole Normale Supérieure
Département de Biologie
1 rue Maurice Arnoux
92120 Montrouge
France

Stephen Kennedy
Nuffield Department of Obstetrics and Gynaecology
University of Oxford
John Radcliffe Hospital
Oxford OX3 9DU
UK

Peter S. Klein
University of Pennsylvania School of Medicine
364 Clinical Research Building
415 Curie Boulevard
Philadelphia
Pennsylvania 19104-6148
USA

Keith E. Latham
The Fels Institute for Cancer Research and Molecular Biology
Temple University School of Medicine
3307 North Broad Street, Room 302
Philadelphia
Pennsylvania 19140
USA

Henrik Leffers
Department of Growth and Reproduction
Section GR-5064
Rigshospitalet
Blegdamsvej 9
2100 Copenhagen
Denmark

Richard S. Legro
Department of Obstetrics and Gynecology
Hershey Medical Center
PO Box 850
Hershey
Pennsylvania 17033
USA

Inge Liebaers
Center of Medical Genetics
University Hospital
Vrije Universiteit Brussels
Laarbeeklaan 101
1090 Brussels
Belgium

Willy Lissens
Center for Medical Genetics
University Hospital
Vrije Universiteit Brussels
Laarbeeklaan 101
1090 Brussels
Belgium

Martin M. Matzuk
Department of Pathology, Room S 217
Baylor College of Medicine
1 Baylor Plaza
Houston
Texas 77030
USA

Synthia H. Mellon
Department of Obstetrics, Gynecology and Reproductive Sciences
The Metabolic Research Unit
University of California at San Francisco
San Francisco
California 94143
USA

Robert P. Millar
MCR Human Reproductive Sciences Unit
37 Chalmers Street
Edinburgh
EH3 9ET
UK

David Miller
Academic Unit of Paediatrics, Obstetrics and Gynaecology
University of Leeds
Clarendon Wing
Leeds General Infirmary
Belmont Grove
Leeds LS2 9NS
UK

Walter L. Miller
Department of Pediatrics
University of California at San Francisco
Building MR-4/Room 209
San Francisco
California 94143
USA

Linda R. Nelson
Department of Obstetrics and Gynecology
University of Illinois at Chicago
820 S. Wood Street, M/C 808
Chicago
Illinois 60612-7313
USA

Maria I. New
Department of Pediatrics
Pediatric Endocrinology
New York Presbyterian Hospital
Weill Cornell Campus
525 East 68th Street, Room M.622
New York
New York 10021
USA

Wiebe Olijve
NV Organon
PO Box 20
5340 BH Oss
The Netherlands

Sandra Timm Pearce
Breast Cancer Research Program
Robert H. Lurie Cancer Center
Northwestern University Medical School
303 East Chicago Avenue
8258 Olson
Chicago
Illinois 60611
USA

Jean-Yves Picard
Unité de Recherches sur l'Endocrinologie du Développement
INSERM U. 293
Ecole Normale Supérieure
Département de Biologie
1 rue Maurice Arnoux
92120 Montrouge
France

G. Praveen Raju
University of Pennsylvania School of Medicine
364 Clinical Research Building
415 Curie Boulevard
Philadelphia
Pennsylvania 19104-6148
USA

Richard M. Schultz
Department of Biology
University of Pennsylvania
415 South University Avenue
Philadelphia
Pennsylvania 19104-6018
USA

Karen Sermon
Center for Medical Genetics
University Hospital
Vrije Universiteit Brussels
Laarbeeklaan 101
1090 Brussels
Belgium

Gerald Schatten
Department of Obstetrics, Gynecology and Reproductive Sciences
Pittsburgh Development Center (West Coast Division)
40 SW 147th Place
Beaverton
Oregon 97006
USA

Margaret A. Shupnik
Department of Medicine
Division of Endocrinology
University of Virginia Medical Center
PO Box 578
Charlottesville
Virginia 22902-0578
USA

Calvin Simerly
Department of Obstetrics, Gynecology and Reproductive Sciences
Pittsburgh Development Center (West Coast Division)
40 SW 147th Place
Beaverton
Oregon 97006
USA

Niels E. Skakkebaek
Department of Growth and Reproduction
Section GR-5064
Rigshospitalet
Blegdamsvej 9
2100 Copenhagen
Denmark

Stephen Smith
Department of Obstetrics and Gynaecology
University of Cambridge
The Rosie Hospital
Robinson Way
Cambridge CB2 2SW
UK

Change Tan
University of Pennsylvania School of Medicine
364 Clinical Research Building
415 Curie Boulevard
Philadelphia
Pennsylvania 19104-6148
USA

Manuel Tena-Sempere
Department of Physiology
University of Turku
Kiinamyllinkatu 10
FIN-20520 Turku
Finland

Jonathan L. Tilly
Vincent Center for Reproductive Biology
Massachusetts General Hospital
VBK137C-GYN
55 Fruit Street
Boston
Massachusetts 02114
USA

Debra J. Wolgemuth
Department of Genetics and Development
Columbia University College of Physicians and Surgeons
1613 Black Building
630 West 168th Street
New York
New York 10032-3702
USA

Teresa K. Woodruff
Department of Neurobiology and Physiology
Northwestern University
O.T. Hogan 4-150
2153 N Campus Dr
Evanston
Illinois 60208
USA

Editors' Introduction

'Molecular biology has inescapably entered the arena of clinical reproductive medicine'. The introduction to the first edition of this book commenced with this sentence. Since then, molecular biology has become established in clinical reproductive medicine. We are beginning to witness the impact of the complete sequencing of the human genome, new possibilities for large-scale DNA sequencing of patients and novel algorithms to analyze the huge amount of information obtained. We are moving from conventional hypothesis-driven research to entirely novel paradigms, where technical limitations are the only restricting factors. We are in the midst of a transition from the classical single-gene approach to the study of interactions between multiple genes and environment in complex disease states. We are moving from genomics to proteomics, providing us with novel information regarding expressed proteins in a given cell under given conditions.

Owing to the success of the first edition of this book – introduced in early 1999 under the title *Molecular Biology in Reproductive Medicine* – we were all motivated to renew our efforts to cover this extremely fascinating and rapidly moving field. A new section entitled 'molecular pharmacology' has been added, separate editors have been allocated to each section and the total number of chapters has increased from 23 to 36. The aim remains to keep each paper concise with adherence to a pre-set format in order to render the book as reader-friendly as this complex material allows.

Modern clinicians are confronted with a rapidly increasing number of papers using novel molecular and genetic approaches, even in clinical journals. The aim of this book remains to provide the reader – in a comprehensive, but reader-friendly fashion – with improved understanding of complex mechanisms regulating hormone synthesis, reception and signal transduction, allowing to better comprehend new concepts developed in reproductive sciences. This new edition illustrates the immense progress that has been made in less than 4 years. These advances have already resulted in clinical applications in *in vitro* fertilization and the production of modified hormone molecules for therapeutic application. However, what we witness today may just represent the first pivotal steps in revolutionary changes in our approaches to the prevention, diagnosis and treatment of reproductive disorders.

The editors would like to express their gratitude to the authors and the publisher for their efforts. We hope that the reader will benefit from our concerted activities and that this volume will continue to meet an important need in advancing genetic and molecular reproductive medicine.

May 2002
Rotterdam, The Netherlands

Bart C.J.M. Fauser, MD, PhD
Professor of Reproductive Medicine

Also on behalf of the section editors

Foreword

The first edition of this book, *Molecular Biology in Reproductive Medicine*, was published in 1999. The underlying concept, formulated by the editors, was to integrate molecular/cellular biology and clinical sciences in reproductive medicine. This second edition contains 36 chapters written by noted experts in the field who have skillfully applied the most up-to-date research findings to the arena of reproductive health and disorders.

The elucidation of protein molecules and their respective genes has led to the identification of amino acid sequences of hormone receptors, gene expressions and post-receptor transcriptions. These are crucial to the investigation of molecular mechanisms regulating the reproductive processes.

In March 2001, the sequence of the complete human genome was published. This unprecedented achievement, together with the development and refinement of analytical technologies, are the major revelations of biomedicine. Indeed, we are entering a new era of bioinformatics with vast potential to identify the genes and gene mutations causing human diseases relevant to endocrinology and reproduction. At the same time, molecular pharmacology, gene therapy and related fields will undoubtedly progress rapidly with the aim of overcoming gene-related defects in reproductive failures. In recognition of these advances, the theme for the annual meeting of the Endocrine Society in San Francisco in June 2002 was 'The Impact of the Human Genome on Endocrinology'. Health-care professionals and those in the relevant areas of practice have been made aware of new means of manipulating the male and female gametes for the enhancement of fertility. There can be no doubt that in the near future these technologies will expand rapidly and will enrich the lives of patients. I have personally experienced the joy parenthood can bring, as I became a grandfather just two and a half years ago.

The chapters presented in the second edition of this book have been extensively revised and updated, both in scope and in depth. The book is replete with information concerning molecular and cellular communications involving the interplay of the brain–pituitary system and their target, the gonads, in maintaining the functional integrity of the immensely complex reproductive system and in dissecting their malfunctions. Reproductive scientists, clinicians and students alike will benefit by this new text.

This book, therefore, is notable both for the information of the rapidly expanding molecular reproductive medicine that has been achieved, and for the mapping of vast frontiers yet to be explored.

Samuel S.C. Yen, MD, DSc
Wallace R. Persons Professor Emeritus in Reproductive Medicine

Department of Reproductive Medicine
University of California San Diego
La Jolla, California, USA

Molecular biology

Aaron J.W. Hsueh

As we enter the twenty-first century, reproductive medicine is being heavily influenced by recent breakthroughs in molecular biology, and genetic and genomic research. This section deals with the basic concepts and tools used by modern biologists to advance our present knowledge of reproductive medicine. The first two chapters describe the essential components of the reproductive organs as well as the basic techniques of molecular biology used to decipher the genetic basis of reproductive physiology and pathophysiology. In addition, the third chapter relates to the impact of the recent genomic revolution in reproductive medicine. The sequence of the complete human genome was published in March 2001, and the information is freely accessible on the Internet. We are indeed in the midst of a major revolution of biomedicine. Armed with a basic knowledge of molecular biology and genomics, readers will be able to appreciate the remaining chapters of this book detailing the cutting-edge science on specific topics of molecular reproduction. More importantly, the readers can apply these molecular biology and genomic concepts and tools for the advancement of their chosen area of basic or clinical research. It is the intent of the editors and authors of this book to facilitate reproductive research for a better understanding of the genetic and environmental etiologies of reproductive disorders, thus leading to more accurate diagnoses and effective therapies.

Chapter 1

Introduction to molecular biology

Margaret A. Shupnik

INTRODUCTION

Molecular biology techniques have revolutionized the way in which reproductive biologists approach basic and clinical questions. They enable us to understand many of the mechanisms underlying reproductive processes and pathological states, and hold the possibility for development of novel therapies. Understanding the power of these methods requires a basic knowledge of information flow inside cells, the structure and organization of genetic material (DNA and RNA) into functional units and the regulation of gene expression. The merger of this information with powerful selection techniques, permitting the identification of a single molecule from a large population, provides the framework to answer important physiological and clinical questions and to design specific diagnostic tests. No one publication may cover all potential interest areas, but many general and specialized texts on molecular biology are available,[1–4] along with comprehensive publications that provide protocols for specific molecular techniques.[5,6] Technology evolves rapidly, and emerging techniques and modifications of established procedures appear regularly in the literature. For the investigator, a critical step in approaching a molecular process in reproduction is defining the specific question and the techniques best suited to answer it unequivocally.

GENE ORGANIZATION

Chemical nature of DNA and RNA

All the information that any organism or cell needs for biological specificity is encoded in the nucleic acids, DNA (deoxyribonucleic acid) and RNA (ribonucleic acid). DNA is found almost exclusively in the nucleus in chromosomes, while RNA is primarily cytoplasmic. Both consist of nucleotides, containng five-carbon sugar molecules linked to a triphosphate group on carbon 5 and to nitrogen-containing bases on carbon 1. In DNA the sugar is deoxyribose, and the four bases are adenine (A), thymine (T), guanosine (G) and cytosine (C), while in RNA the sugar is ribose and the T is replaced by uracil (U). A and G are purines while T, C and U are pyrimidine bases. Both RNA and DNA are polymers of nucleotides, in which a diphosphate bond is formed between the triphosphate on carbon 5 of one sugar and the hydroxyl group on carbon 3 of the following sugar. This structure gives both polarity and order to the molecule, and is designated by the order of nucleotide bases. The sequence is always written and read beginning with the nucleotide with the carbon 5 phosphate (the 5' end) to the nucleotide with the remaining hydroxyl group on carbon 3 (the 3' end); the longer the molecule, the more specific the sequence.

DNA is a double-stranded helical structure stabilized by hydrogen bonds on the inside of the helix between the purine and pyrimidine bases emerging from the sugar–phosphate DNA backbone. This 'base-pairing' occurs only between G and C (three hydrogen bonds) and A and T or U (two hydrogen bonds). The higher is the number of hydrogen bonds (more bases or more G–C pairs), the more stable is the structure. DNA strands are oriented in a specific manner, with the upper strand oriented 5' to 3' and the lower strand oriented in the opposite direction. The two DNA strands are thus complementary to one another, and if the sequence of one strand is known, that of the other strand is defined. Complementary nucleic acid structure is critical in understanding DNA replication and RNA transcription, and is exploited to prepare tools to detect and identify specific DNA and RNA sequences.

Gene structure and information flow

A gene is the chief functional unit of DNA, and contains the information for transcription into either structural (i.e. ribosomal or transfer) RNA or messenger RNA (mRNA) to be translated into proteins. Every cell in each tissue of the body contains the same DNA, except for gametes – eggs and sperm – which contain only half of the genetic material. Researchers can thus use any cell to investigate individual genes; for example, it is standard to collect blood cells to search for genetic mutations. However, the biological attributes of any given cell are defined by the specific mRNAs and proteins and can differ quite widely.

Genes from higher organisms have a similar organization (Figure 1), in which information for

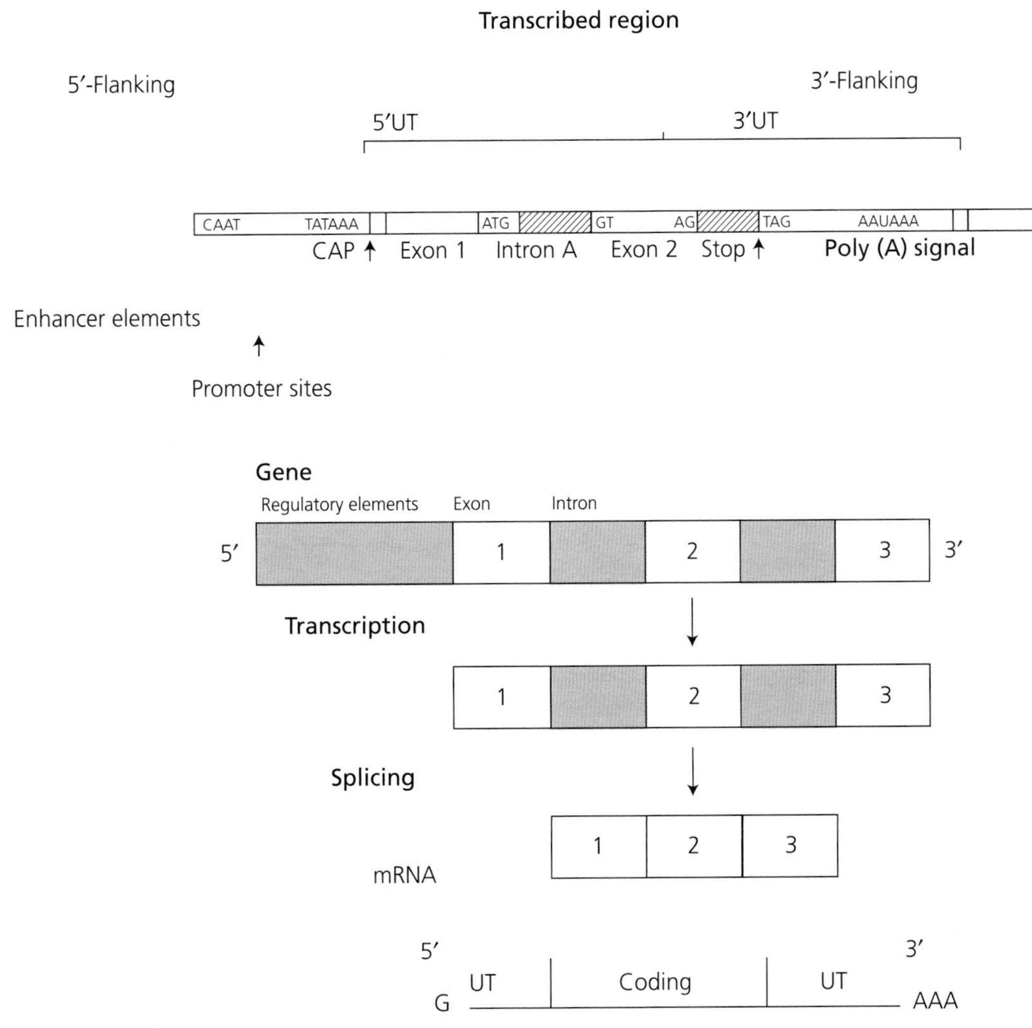

Figure 1 Eukaryotic gene structure and its relationship to mRNA structure. A gene consists of exons that contain sequences in mature mRNA (open bars), and intervening sequences called introns. The sites of transcription initiation and completion are encoded in the gene, as well as binding sites (regulatory elements) for factors that regulate transcription. After transcription, RNA is processed by splicing out the introns at defined sequences, and adding a 7-methyl G residue to the 5' end and a string of As (a poly(A) tail) to the 3' end. This processed mRNA is transported to the cytoplasm for translation. UT, untranslated regions of mRNA

the amino acid coding sequence of a protein is contained in gene units called exons. Exons are interspersed with non-coding areas called introns. There is great variation among genes in terms of the number or length of introns and exons.[7,8] Intron/exon boundaries can sometimes follow functional protein domains. For example, transcription factors have several domains including DNA-binding regions and other 'activation' domains that stimulate transcription, and these are often in separate exons. Complicated genes with alternative exons may be mixed and matched to form alternative mRNAs and proteins.[9,10]

Genes are transcribed from 5' to 3' from the promoter, the site of transcription initiation by RNA polymerase, which is determined by the location of the TATA box. This sequence helps to align RNA polymerase with the DNA by interactions with at least seven protein transcription initiation factors (TFs), which bind in an ordered fashion.

The rate of transcription initiation is determined by CAAT box binding proteins, which bring RNA polymerase to the start site, probably by DNA bending, to assemble the transcriptional machinery.

Additional gene regions, usually located 5' to the TATA box, known as enhancers, silencers and hormone response elements play critical roles in regulating transcription. These elements have characteristic DNA sequences and bind specific proteins that influence the rate of RNA transcription,[1–4] thus determining tissue-specific expression, or physiological regulation of a gene. Expression of these transcription factors may be limited or widespread. For example, CREB (cyclic adenosine monophosphate (cAMP) responsive element binding protein) is expressed in most cells, binds to the sequence TGACGTCA and responds to increases in intracellular cAMP.[11] Estrogen receptor α (ERα), on the other hand, binds to the sequence AGGTCA NNN TGACCT, with 'N' denoting any base, and is expressed in reproductive tissues, brain, bone and liver. DNA enhancers and regulatory elements can be active in either orientation (5' to 3', or 3' to 5'), and influence transcription even from several thousand bases away. A model for DNA enhancer–transcription factor action involves indirect interaction with the transcription initiation complex mediated by interactions with coactivators and integrator proteins. Several coactivators such as steroid receptor coactivator-1 (SRC-1) interact with steroid receptors to increase gene transcription.[12,13] Additional 'integrator proteins' such as CBP (CREB binding protein) bind to steroid receptors, receptor coactivators and many other transcription factors.[11] CBP and some coactivator proteins have intrinsic histone acetylase activity, and acetylate histones on chromatin and modify overall chromosomal structure. This modification is believed to result in increased accessibility of the promoter to RNA polymerase, and an increased rate of transcription. Conversely, proteins such as steroid receptor corepressors can recruit proteins such as Sin3 and enzymes with histone deactylase activity, which remove acetylase groups from histones on chromatin and result in a more compact structure and decreased transcription.[13,14]

RNA is transcribed in the 5' to 3' direction from the DNA template, including all exon (translated and untranslated) and intron sequences. RNA is single stranded and identical to one strand of the DNA called the 'sense' or coding strand. Gene sequences are written as the sense DNA strand, from 5' to 3', and are collinear with the mRNA and translated protein. The nuclear RNA precursor is processed into the cytoplasmic mature mRNA by splicing to remove the introns and join the exons together, and by addition of a 5' 'cap' (a 7-methyl guanosine pyrophosphate), and a 3' poly(A) tail, which is exploited by oligo(dT) affinity columns to purify mRNA.[5] The transcription initiation site and 'stop' site are encoded in the gene, and RNA molecules are therefore of discrete sizes. Proteins are synthesized on the ribosomes by translation of the mRNA from 5' to 3', and for the polypeptide chain from N-terminus to C-terminus. Physiologically relevant modulation steps may occur along any step of this pathway. However, much of the hormonal or developmental control occurs at the level of gene transcription, and attention is currently focused on the measurement of specific mRNA levels and of gene transcription regulation.

RECOMBINANT DNA TECHNOLOGY

Biochemical characteristics of DNA and RNA such as their complementarity and specificity enable the investigator to detect, measure and characterize these molecules and to determine their most interesting characteristic – the biological roles of the proteins they represent. Most general procedures rely on the use of DNA rather than RNA, or convert RNA to a complementary DNA (cDNA) form. This is because DNA is more stable, and there are more biological and biochemical tools to manipulate, amplify and characterize DNA.

Restriction enzymes

While RNA molecules are of discrete, short lengths (10^3 bases), eukaryotic genes lie within chromosomes containing up to millions of bases. Fortunately, a group of bacterial enzymes known as restriction endonucleases cleave DNA internally and can cut any DNA into discrete fragments that can be mapped and manipulated. Their most important and distinctive feature is that they cleave DNA in a sequence-specific, and therefore predictable, manner into defined pieces. Most restriction enzyme sites are defined by 4–6 base pairs, but some can be much longer; the longer the site, the more rare its occurrence in the genome. The enzyme cuts on both strands of the DNA within a restriction site. Cleavage can be in the middle of the sequence, resulting in DNA ends that are blunt (Sma I: 5'GGG/CCC 3'), or asymmetrically, to form cohesive or complementary ends (Hind III: 5' A/AGCTT 3'). Cleavage sites with cohesive ends are particularly useful, as the single stranded ends can reassociate by base pairing. Thus, one specific Hind III site can reassociate with any other Hind III site. This permits DNA fragments to be organized and manipulated, and is integral to the process of preparing recombinant DNA and cloning.

Nucleic acid hybridization

Because most DNA and RNA molecules are very long, are present at low levels and have similar overall charges per unit length, they are difficult to detect in the complicated population of molecules found in most tissues by standard biochemical separation procedures. However, each molecule has a characteristic sequence that can be exploited for detection, based on the complementarity of nucleic acids. Any single stranded DNA or RNA will associate with a complementary RNA or DNA owing to the formation of hydrogen bonds between A–T and G–C pairs. The procedure by which a labelled DNA or RNA, or probe, is used to detect its specific complementary molecule is known as hybridization. Nucleic acids can associate, or anneal, in solution, but many detection techniques are conducted with the labelled probe in solution and the DNA or RNA to be screened on a solid support such as nitrocellulose or a nylon membrane. The procedure in which DNAs are separated by size via gel electrophoresis and transferred to a membrane for hybridization is called Southern blotting, and for RNA is termed Northern blotting.[15] Double-stranded DNA must be disrupted or denatured prior to hybridization by heat or chemicals. Binding of the probe is determined by the stringency (temperature, salt and solvent) of hybridization and washing conditions, and the stability of the nucleic acid hybrids. This is the direct result of the number of G–C versus A–T pairs, and the number of continuous perfect matches between sequences. Thus, a 'mismatch' in the middle of a sequence will be more disruptive than one at the end of a probe, even if the overall percentage agreement is the same. The chemical nature of the probe is also important, as RNA–RNA hybrids are more stable than RNA–DNA hybrids, which are more stable than DNA–DNA hybrids. High-stringency conditions (high temperature, low salt, polar solvent) will allow only perfectly matched probes to bind, while low-stringency conditions (lower temperature, higher salt) allow some mismatches. Several DNAs encoding molecules important in reproductive processes, such as steroidogenic factor-1,[16,17] and ERα,[18,19] were first identified in one species (rodent or human), and the probes or information used to clone homologs under low-stringency conditions from other species.

Several different types of probe are available and useful in specific circumstances. Double-stranded complementary DNAs (usually several hundred bases in length) are most readily available, and are commonly used for screening under both high- and low-stringency conditions. Oligonucleotides (usually 17–40 bases) are easily synthesized, and are valuable to distinguish precise portions of a molecule, or if only a small amount of DNA or amino acid sequence is known. Single-stranded RNA probes are synthesized from cDNA cloned into commercial constructs containing viral promoter sites.[20] Depending on insert orientation, the addition of viral polymerase can synthesize RNA in either sense (same sequence as the translatable mRNA) or antisense (complementary and thus hybridizes to mRNA) directions. RNA probes are very sensitive, and used to quantitate rare mRNAs in solution by RNAse protection analysis, or in tissue sections by *in situ* hybridization.

Polymerase chain reaction technology

The most sensitive current technology to detect and analyze nucleic acid sequences exploits the polymerase chain reaction (PCR).[21] PCR methods can be used with extremely small amounts of material – a few cultured cells, a portion of a biopsy – by exploiting the properties of specific heat-sensitive DNA polymerases (*Taq* polymerase) and protocols that amplify the original target sequence. This permits use in automated synthesis schemes (Figure 2) for multiple rounds of DNA amplification.

PCR techniques require knowledge of the DNA sequence to be amplified to design two short oligonucleotides (usually 17–20 bases) called primers, one complementary to each of the DNA strands, and oriented with the 3' ends of the primers facing each other. Each primer will serve as the 5' end for DNA synthesis of one amplified strand, and the number of bases between the primers determines the length of the amplified sequence. PCR reactions consist of multiple cycles of three basic steps, including DNA denaturation to separate the strands serving as templates, primer annealing and DNA synthesis to amplify the target sequence. In the next cycle, the newly synthesized DNA also serves as a template for amplification, and a single molecule of double-stranded DNA results in one million molecules after 20 amplification cycles. PCR has been invaluable for screening human DNA for specific gene mutations,[22] because it eliminates the need to prepare and screen genomic libraries. PCR requires DNA templates, but a modification of this technique, called reverse transcription-PCR (RT-PCR), uses RNA as the initial template, by synthesis of complementary DNA (cDNA) from RNA using primers complementary to the RNA and the enzyme reverse transcriptase (Figure 2). When an oligo(dT) primer is used, mRNA sequences are preferentially transcribed from total RNA. RT-PCR is extremely sensitive, and has been used to detect, quantitate and clone mRNAs.[23–25]

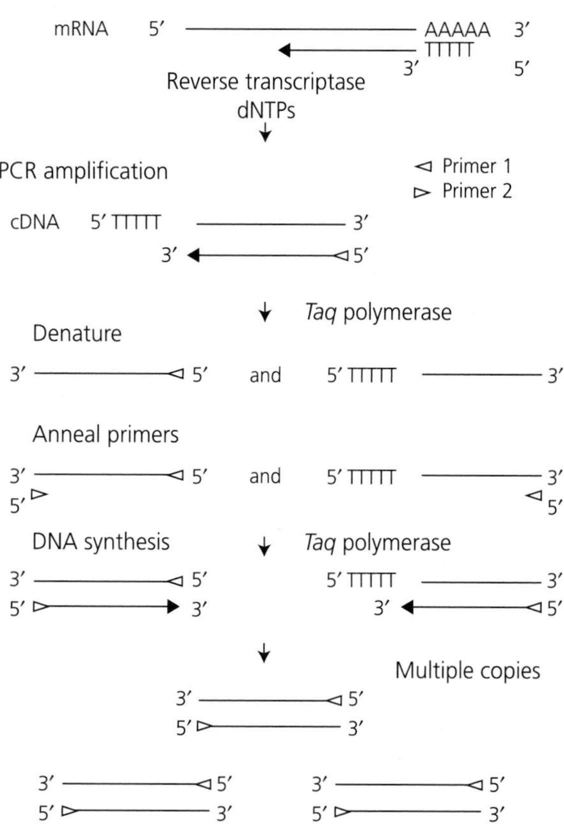

Figure 2 The polymerase chain reaction (PCR) and the reverse transcriptase-PCR analysis of mRNA sequences. In the first step, mRNA is used as a template by the enzyme reverse transcriptase to produce a complementary DNA. Specific sequences are then amplified in the presence of defined primer sequences. The primers are complementary to both strands of the resulting DNA and oriented such that the 3' ends of the primers face each other. Multiple rounds of DNA denaturation, primer annealing and *Taq* polymerase activity result in many copies of DNA sequences bounded by the two primers. dNTPs, all four deoxynucleotide triphosphates (dATP, dTTP, dGTP, dCTP)

DNA cloning principles

Determination of gene structure allows us to understand gene regulation and function and the basis of genetic defects, and to develop strategies to screen for and treat these conditions. DNA cloning relies on recombinant DNA technology to prepare and screen rare events in large populations, and to replicate DNA reproducibly in bacterial host cells. There are three components to cloning, including the desired DNA to be cloned, the vector and the host cell.

The DNA to be cloned is the insert DNA. The DNA vector serves as a vehicle for the cloned DNA, and has features that permit it to be sorted and amplified. The most common cloning vehicle is a bacterial plasmid, but bacteriophage, viruses or yeast artificial chromosomes (YACs) may also be used for large DNA inserts, including portions of chromosomes.[26] Plasmids are closed circular DNAs that replicate independently of bacteria chromosomes, and are engineered to contain genes that confer resistance to antibiotics. Under appropriate salt and temperature conditions, plasmids can enter bacterial hosts. If the bacteria are in vast excess, only one plasmid enters each bacterium. When the bacteria are grown on agar plates with antibiotics, only the bacteria containing plasmids will form colonies, which can be screened to identify the particular DNA insert.

DNA libraries, representing every DNA molecule in the population, may be genomic, or cDNA prepared from mRNAs of a particular cell. Genomic DNA is prepared for cloning by cleaving the DNA with restriction enzymes, typically rare-frequency cutters. Fragments are mixed with a vector cut with the same enzyme(s), and the cohesive ends of the vector and genomic DNA anneal. If only a single enzyme is used, fragments anneal in either orientation (i.e. both 5' to 3', and 3' to 5'). Another enzyme, DNA ligase, joins the sequences, and a complete plasmid is formed. RNA cannot be cloned directly, but must be converted to DNA by reverse transcriptase. The RNA is removed, a second strand of DNA is synthesized and enzyme sites are added with synthetic linkers. The cDNA can then be treated like genomic DNA.

Libraries

A genomic library represents every gene, and can be prepared from any cell or tissue; circulating human lymphocytes are often used to prepare human genomic DNA. Each DNA sequence will be represented at a frequency comparable to the gene number in the genome (generally 2, or some small number), and most genes are represented at roughly equivalent levels. A genomic clone will contain introns and exons, and may be much larger than predicted by the size of the mRNA or proteins. A genomic library is the only source for isolation and identification of the promoter and regulatory regions of a gene that modulate transcription.

A cDNA library, in contrast, only contains sequences of the mRNAs being actively transcribed, and will thus be tissue- or even cell-specific. In addition, the representation of a particular sequence will correlate with the abundance of the particular mRNA. For example, human placenta from the first trimester will contain many copies of mRNA encoding human chorionic gonadotropin-β, while third-trimester placenta will contain very little, and tissues such as the pituitary or breast none at all. Clearly, choice of appropriate cDNA library is critical, although even less abundant molecules can be identified by powerful screening techniques.

Screening approaches

Most common methods use nucleic acid hybridization with any of the previously mentioned probes. If a portion of protein sequence is known, oligonucleotides can be synthesized using relevant codons, keeping in mind that some amino acids have several different codons, and redundant probes with varied codons may be required. Specialized probes may be made by RT-PCR to exploit the considerable homology between similar proteins and mRNAs from different species, or among several proteins of a given type. For example, the DNA-binding regions of the nuclear receptors have high homology among all family members. Investigators using primers homologous to the estrogen receptor cloned molecules from prostate RNA, and found a new form of the receptor called ERβ.[27]

Other screening options use a cDNA expression library to measure some attribute of the specific protein. An expression library is prepared with a vector containing a promoter region next to the cDNA cloning site.[18] Thus, the cDNA, if inserted in phase and in the correct orientation, is transcribed from the promoter and the mRNA is translated into protein. Colonies can be screened with antibodies,[19] or for other specific biological activities such as ligand or DNA binding. One elegant strategy for cloning the receptor for the gonadal peptide activin used a cDNA library introduced into eukaryotic cultured cells, which processed the protein and inserted it into their membrane, thus allowing screening for activin binding.[28] Several transcription factors have been cloned on the basis of their ability to bind specific DNA sequences, which were used as probes to screen expression cDNA libraries.[29] Additional specialized screening approaches including yeast or mammalian two-hybrid, discussed below, are variations on these methods.

DNA sequencing

After screening, clones are characterized by DNA sequencing. Most investigators use the Sanger method,[30] which relies on premature termination of DNA chains newly synthesized by DNA polymerase. This enzyme synthesizes DNA in the 5' to 3' direction, requires a short complementary DNA primer (usually 15–18 bases long) to begin synthesis and adds a nucleotide triphosphate (dNTP) to the growing DNA chain. The technique uses radioactive nucleotide precursors mixed with individual dideoxynucleotides, which have a hydrogen, instead of a carboxyl group, at carbon 3. When the dideoxynucleotide is incorporated in a random and rare event, no additional nucleotides can be added at carbon 3, and the chain is terminated. A population of DNA molecules, terminated at different residues and thus of different lengths, is obtained. Four separate reactions – one for each nucleotide – are performed, the DNA products are subjected to electrophoresis and the sequence is read directly from the autoradiogram of the gel (Figure 3).

Once DNA sequence information is obtained, it can be compared with other DNA or translated protein sequences, using computer databases to predict translated protein sequences in any phase from both possible coding strands, align DNA sequences and highlight areas of homology, or perform other tailored manipulations. GenBank is part of the International Nucleotide Sequence Database Collaboration, and contains millions of sequence records from the DNA Data Bank of Japan, the European Molecular Biology Laboratory and the United States National Center for Biotechnology Information (NCBI). Databases including the entire human and mouse genome can be accessed via the University of California at Santa Cruz (http://genome.ucsc.edu/), NCBI (http://www.ncbi.nlm.nih.gov/genome/guide/ and click on 'MapViewer' and the European Bioinformatics Institute (EBI) (http://www.ensembl.org/). Investigators can compare sequence information of genes and proteins from several species, to scan sequences for specific motifs at the protein (enzymatic active sites, DNA-binding domains) or gene (transcription factor-binding sites, exon/intron boundaries) level, to map chromosomal localization and proximity of specific genes and to identify full gene sequences from partial clones. Additional data repositories include short sequences from several different cDNA libraries generated from randomly selected cDNA clones, called Expressed Sequence Tags (ESTs). Many ESTs are derived from currently uncharacterized genes, and are a starting point for the identification and comparison of novel RNAs.

Once a gene or cDNA is cloned, it can be amplified perpetually, and serves as a renewable resource for that gene, mRNA or protein. Several current therapies, including insulin treatment of diabetes mellitus, rely on the production of large quantities of the human hormone synthesized from cloned cDNAs. Clones for genes and cDNAs can also be mutated to determine which gene regions are critical for regulated transcription, or to learn structure–function relationships of encoded proteins.[31]

TRANSCRIPTIONAL AND POST-TRANSCRIPTIONAL REGULATION

The biological functions of any tissue or cell are defined by the identity and amount of the proteins synthesized in the cells, and are a direct reflection of the transcriptional activation of specific genes. Some structural proteins, such as collagen, turn over very slowly and are synthesized continuously. However, most proteins involved in reproduction have short half-lives (minutes to hours), are often highly regulated and may vary greatly in quantity during the reproductive cycle. The rate of protein

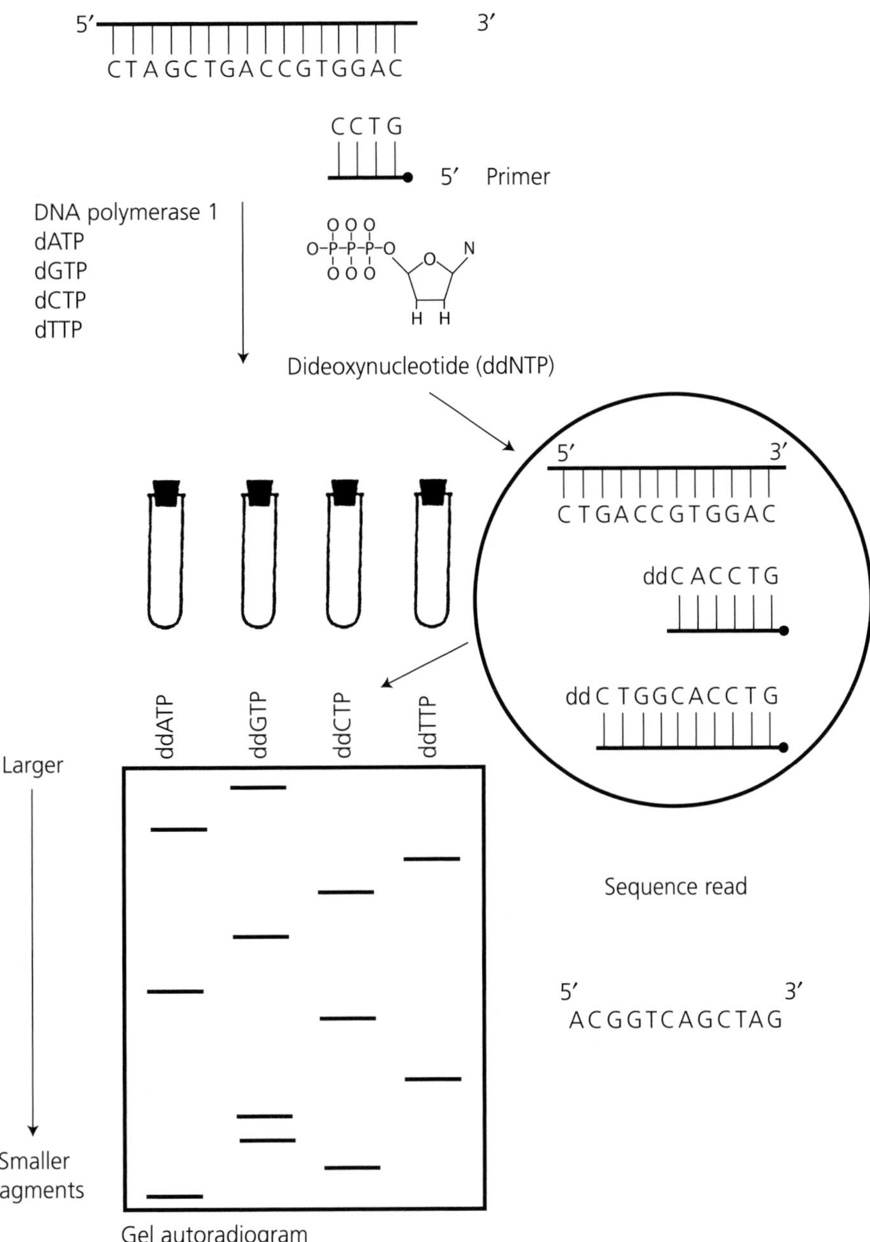

Figure 3 Sanger DNA sequencing protocol and representation of an autoradiogram of the resulting sequencing gel. The primer (usually 15–18 bases, but shorter here for clarity) binds to the denatured DNA, and the sequence is extended by the action of DNA polymerase adding deoxynucleotide triphosphates (dNTPs). Either the primer or one of the precursors is labelled. When a ddNTP is inserted, the chain is terminated, because the carbon 3 has a hydrogen and cannot form a phosphodiester bond. A population of terminated chains, as shown in the circular inset, is generated for each base. The sequence is read from 5' to 3' and represents the extending primer strand

synthesis, usually directly proportional to the levels of mRNA encoding the protein, has profound effects on cellular physiology. Steady-state levels of mRNA are regulated by the rate of mRNA synthesis, or transcription, and mRNA turnover. Although changes in mRNA half-life have physiological consequences, as in activin stabilization of follicle stimulating hormone-β (FSHβ) mRNA,[32] changes in transcription appear to play a larger general role in governing cellular mRNA levels. In addition to gene expression, specific mRNA detection procedures provide information on mRNA localization to specific cell and tissue types, and identification of variant mRNAs.

mRNA quantitation: hybridization

The detection of any mRNA depends on specific tools to measure its unique nucleic acid sequence or the protein it encodes. Translatable mRNA can be measured in commercially available systems prepared from wheat germ or rabbit reticulocyte lysate, containing the cellular ribosomes and other factors to synthesize protein. The technique is indirect, and requires a tool such as a specific antibody to detect the protein.[18] The simplest way to detect mRNA is the blot hybridization method (Northern blotting). RNA is denatured, separated on an agarose gel and transferred to a membrane that is hybridized to a radiolabelled probe.[20,25] The position and intensity of the hybridizing band or bands indicate the size and relative abundance of the mRNAs. Band intensitites are quantitated by autoradiography and are normalized for some RNA such as actin or ribosomal RNA, which is not expected to vary. This is the only method that permits mRNA size determination, and heterologous probes can be used. The technique is helpful in scanning many tissues for mRNA expression,[19] but is relatively insensitive and may not detect rare mRNAs.

Solution hybridization: RNAse protection

Solution hybridization methods are very sensitive, and provide information on RNA structure and abundance. The population of RNA to be assayed is incubated in solution with a radiolabelled complementary RNA present in excess. After the nucleic acids form hybrids, the mixture is treated with RNAse, which digests all single-stranded RNA. Double-stranded RNA is protected from degradation; the protected probe is displayed on a gel and quantitated by autoradiography. The assay requires homologous probes, because potential 'mismatches' during hybridization will be points of RNAse digestion. Specific probes will distinguish between closely related mRNAs, such as splice variants that encode isoforms of the same peptide, or mRNAs with different 5' ends transcribed from different promoters.[33]

In situ hybridization

Often, investigators require knowledge of mRNA expression in complex reproductive tissues containing many cell types, and *in situ* hybridization can be used to localize expression of the mRNA. Tissue is cryopreserved or treated with a fixative, and sectioned into thin slices that preserve anatomical detail. The slices are then hybridized to a radiolabelled or colorimetric tagged probe, which binds to cells that have expressed the mRNA.[34,35] Both types of probe may be used to colocalize two mRNAs to the same cell.[34] An elegant example of this technique was provided by investigators studying rat ovarian inhibin expression, which was regulated during the estrous cycle. Because the ovary is a heterogeneous tissue, consisting of follicles in many different stages of growth and atresia and corpora lutea of different ages, the physiological site of this expression was unclear. *In situ* hybridization studies[35] demonstrated inhibin mRNA expression in developing follicles, which increased in Graafian follicles, then precipitously dropped in mature preovulatory follicles after gonadotropin stimulation. This confirmed the follicular site of inhibin synthesis, and suggested an important developmental role for the peptide in follicular maturation.

Reverse transcriptase polymerase chain reaction

The most sensitive method for detecting any mRNA sequence is RT-PCR, which requires some knowledge of the sequence to be measured.

Sequences are amplified exponentially, and reach a plateau as reagents (primers and nucleotides) are depleted; different mRNAs will stay in the exponential phase of amplification for different numbers of cycles. To address this issue, investigators have used low numbers of amplification cycles, normalization to other transcripts and comparison with internal standards.[23,24] In the latter case, competitive PCR protocols in which standard curves of modified competing templates are added to the PCR reaction have been used successfully to calculate the number of mRNA molecules in a given sample.

Microarray technology

One of the most powerful new technologies to screen thousands of genes simultaneously for changes in expression is microarray or 'chip' technology. In this approach, up to 30 000 oligonucleotide or genomic DNA probes are spotted on a solid support, and probed with dye-tagged cDNAs prepared from RNA provided by the investigator.[36] Relative hybridization intensities between two related cDNA–RNA preparations – control and steroid- or thyroid hormone-treated, for example – can be compared, as well as the identity of expressed genes in a given sample.[37] To this point, most chips have been available from commercial companies that also sell software to analyze the data and compare hybridization profiles. To some extent, current analysis is limited by cost, the genes chosen for inclusion on the chips, the degree of reproducibility between RNA–cDNA samples and hybridization assays, and the degree of difference that can be scored between samples. However, as more investigators and institutions establish these techniques and groups of investigators collaborate to fashion custom microarrays, this potentially powerful technique will become more standardized and available.

mRNA transcription

mRNA synthesis

The mechanism of mRNA regulation by physiological signals may be investigated at the level of gene transcription. Rates of mRNA synthesis may be determined directly in the cell of interest by measuring the incorporation of radioactive nucleotides into newly synthesized mRNA (transcription run-off or run-on assays) before hybridization to unlabelled DNA[38] or by detection of mRNA precursors prior to splicing events by quantitative competitive RT-PCR.[39] Transcription run-off assays can be difficult to perform, particularly if the mRNA is not synthesized at a high rate. However, it directly measures the transcription rate of the endogenous gene in normal tissue, and can quantitate transcriptional responses to physiologically relevant changes. Alternatively, RT-PCR can quantitate specific unprocessed (containing both introns and exons) primary transcripts by using one primer in an exon with the other in an adjacent intron.[39] Both methods can be the first step in understanding the regulated gene expression in the tissue of interest. However, they cannot provide information on specific gene regulatory regions or mechanisms of regulation, which might be obtained with cell-based systems such as transient transfection analysis.

Transfection analysis

One of the most powerful and versatile molecular techniques is transfection analysis, in which exogenous DNA is introduced and expressed in cultured cells. These methods can help to define protein structure–function relationships and regulatory regions of gene promoters, and are the first step in the more sophisticated techniques of gene targeting. Plasmid DNA is introduced into cultured cells by one of several methods (lipid vesicles, calcium phosphate, electroporation).[5,6] More than one type of plasmid (cotransfection) may be used, but one construct must contain a reporter gene fused to a promoter sequence that can be transcribed in the transfected cell. The reporter construct encodes a protein or enzyme that is not usually present in mammalian cells, such as chloramphenicol acetyltransferase (CAT)[32] or luciferase,[40] and serves as a marker for transfection and a measure of promoter activation. Gene regulatory elements are usually located 5' to the RNA polymerase binding site (Figure 1), in the promoter. To characterize a specific element (for example, an estrogen response element), the promoter is inserted in a reporter construct, and the plasmid transfected into cells containing the estrogen receptor. Reporter gene activity reflects promoter activity, and increased enzyme activity with estrogen treatment indicates that the promoter contains an estrogen-responsive element. This element can be localized with a series of promoter deletion/mutation constructs. If the cells are receptor-negative, it is possible to cotransfect an estrogen receptor driven by another promoter. Using the latter cotransfection protocol, a defined estrogen-responsive reporter construct and various individual receptor expression constructs, specific properties of the estrogen receptor protein may be defined.

Although transfection studies are powerful and often easy to perform, they are artificial, and may not always accurately reflect the physiological system. Much of the success of this procedure depends on the availability of a cell culture system or cell line that can be transfected, and which reflects the physiological background of the gene to be tested. Unfortunately, cell lines are not available for all reproductive tissues, particularly those that retain the appropriate differentiated characteristics. In some cases, dispersed cell cultures from primary tissues may be used for transfection.[41,42] Transfection studies can also help in interpreting data from pathophysiological states by transfecting a mutant gene to test its function directly, or help form the framework for future sophisticated studies of gene manipulation in transgenic animals.

DNA–protein interactions

Once an enhancer or response element has been identified, attention can be concentrated on the transcription factors binding to these regions, as they ultimately control gene activity. Although many proteins have consensus DNA-binding motifs, the presence of a similar DNA sequence in a promoter is not sufficient to determine whether a given transcription factor is involved in the cell type of interest. Among the first techniques to evaluate DNA–protein interactions was DNAase footprinting.[3–5,43] DNA (typically 300–500 bp) is labelled at one end and incubated with either nuclear protein extracts or purified proteins. The DNA is then treated with DNAse I to cleave at every DNA base that is not protected by protein. Digested DNA is then run on a sequencing gel with a sequencing run of the same DNA. DNA protected by proteins will not be represented on the gel (specific bands will be absent), and will form a 'footprint' that can be localized by comparison with the sequencing reaction. Hundreds of bases can be screened in one run, but binding of rare proteins or proteins that bind with low affinity may not be seen, as a background of unprotected and digested DNA will obliterate a weak footprint.

The most common assay used to evaluate DNA–protein interactions is the electrophoretic mobility shift assay (EMSA), also referred to as gel or band-shift studies.[42–44] EMSA (Figure 4) can detect DNA–protein interactions of even low affinity, and requires knowledge of the regions of DNA likely to bind transcriptionally relevant proteins. Double-stranded oligonucleotides, typically 18–50 bp, are labelled and mixed with proteins in solution, then displayed on a non-denaturing acrylamide gel. Unbound DNA will migrate to the bottom of the gel. DNA–protein complexes are retarded, resulting in an array of bands. Antibodies to individual proteins can be added to determine which bands contain those proteins, and may result in a 'supershift' band of slower mobility (lane 3) or elimination of a specific band, depending on the specific antibody and the protein epitope bound. Addition of unlabelled oligonucleotides can be used to compete for binding to specific proteins (lane 4), and mutated oligonucleotides can determine specific DNA bases important for complex formation. Investigators typically examine mutations that eliminate regulated transcriptional activity in transfection assays, and compare these with the loss of specific protein binding in EMSA.[42–44] EMSA is a relatively rapid way to scan proteins binding to a given sequence, and is used to verify binding of proteins from cDNA clones screened by DNA enhancer methods.

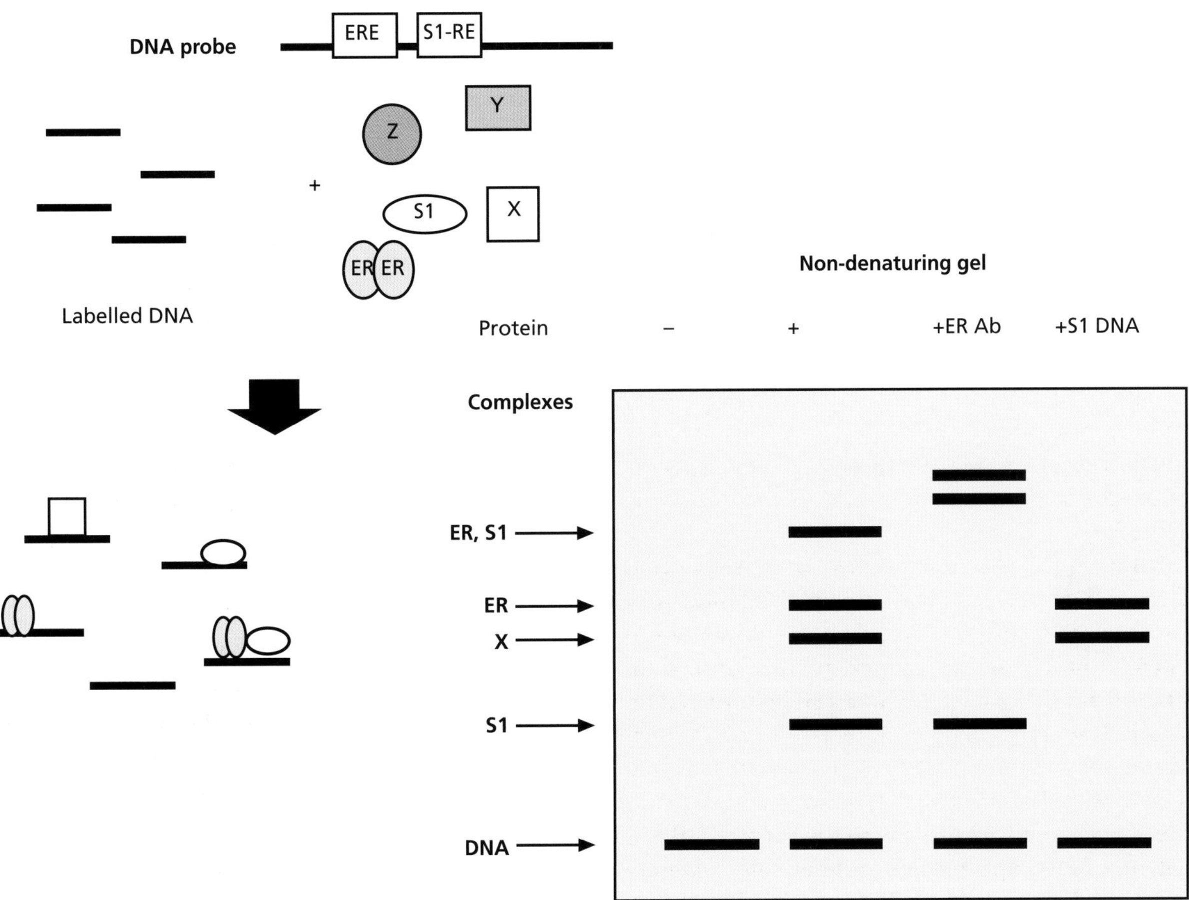

Figure 4 Electrophoretic mobility shift assay. An oligonucleotide of a gene region containing known binding sites for the estrogen receptor (estrogen-response element, ERE) and the protein S1 (S1-RE) is incubated with nuclear proteins from uterine cells. DNA–protein complexes are formed in solution, including DNA bound to the estrogen receptor (ER) and S1 alone and in combination, and an unknown protein 'X'. Individual reactions include DNA probe alone (lane 1), or DNA plus nuclear proteins (lanes 2–4). The reaction in lane 3 also includes an antibody (Ab) to ER, while the reaction in lane 4 contains an unlabelled competitor oligonucleotide for a consensus DNA-binding site for S1. The reactions are then subjected to electrophoresis on a non-denaturing gel. Unbound DNA migrates to the bottom of the gel, and the various complexes, shown by arrows, have slower mobility. Addition of ER antibody 'supershifts' complexes containing bound ER, and the addition of unlabelled competitor DNA binding to S1 eliminates the bands containing bound S1

Two-hybrid cloning

In addition to DNA binding, transcription factors participate in important protein–protein interactions that modify their activity. This includes the formation of heterodimers, such as the participation of c-*jun* and c-*fos* in the activated protein-1 (AP-1) complex, and the interaction of nuclear receptors with coactivator and corepressor proteins.[12,13] Several proteins of this type have been isolated by two-hybrid screening of a specialized expression library in yeast or mammalian cells (Figure 5). This approach is based on the principle that transcription factors consist of at least two functional and often separable protein domains. One domain binds sequence-specific DNA elements, and the second mediates transcriptional activation. Neither domain alone is capable of activating transcription, but if the domains are brought together by protein–protein interactions, transcriptional activation of a target gene occurs.[45–49]

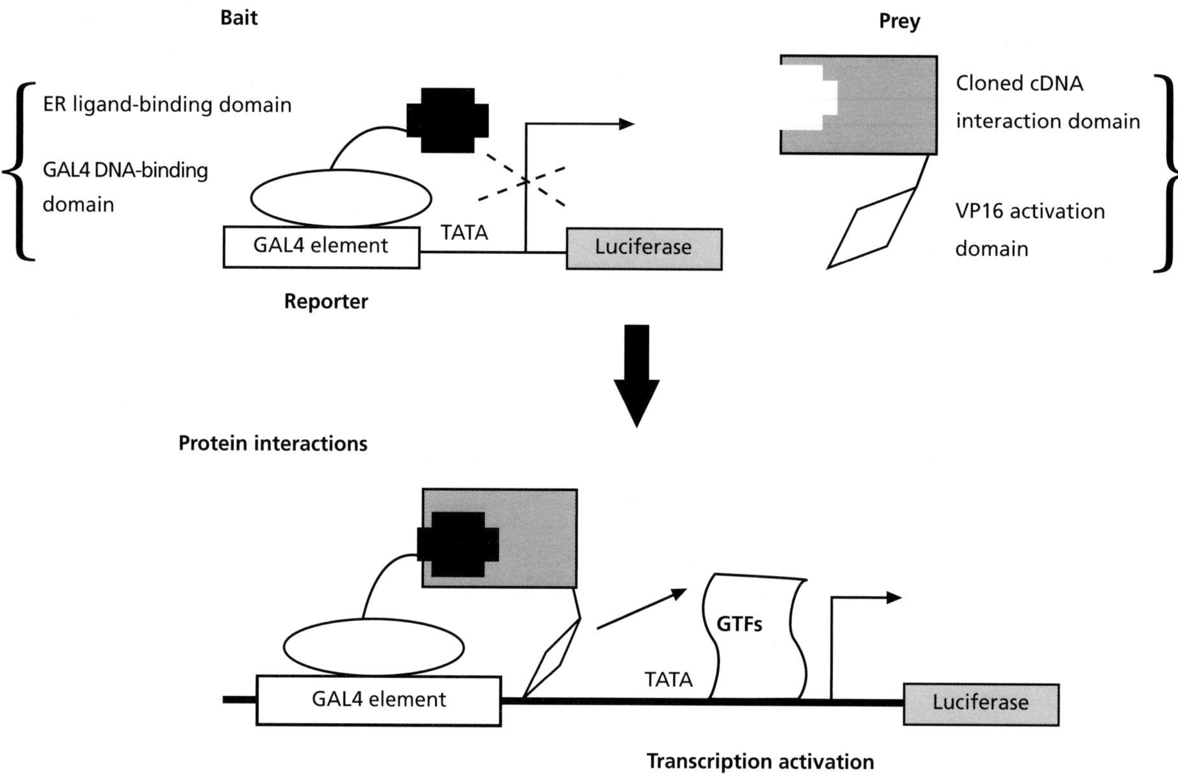

Figure 5 Mammalian two-hybrid screening assay to detect interactions between two proteins. In this experiment, the investigator is looking for uterine proteins that bind to the ligand-binding domain of the estrogen receptor (ER). The 'bait' protein is a fusion protein consisting of the ligand-binding domain (the 'interaction domain') of ER fused to a GAL4 DNA-binding domain. This protein can bind to the GAL4-response element luciferase construct, but will not activate transcription because it does not have an activation function. An expression library is engineered with cDNAs from uterine RNA fused to the activation domain of VP16. These proteins are the 'prey', and also cannot activate transcription from the reporter, because they cannot bind to DNA. When the uterine cDNA encodes a protein that can interact with the ER, it forms a stable complex and brings the VP16 activation domain to the promoter region, where it can interact with integrator proteins and general transcription factors (GTFs) to stimulate transcription

The system requires a reporter gene construct, a 'bait' expression construct that binds to the reporter and is used to screen for protein–protein interactions, and a 'prey' expression cDNA library to be screened (Figure 5). The reporter construct is similar to those used for transfection analysis, and consists typically of a luciferase (mammalian cells) or *lacZ* (yeast) enzyme reporter under the transcriptional control of a promoter containing sequence-specific binding sites for the DNA-binding domain-containing bait protein. DNA binding domains of the yeast transcriptional activators GAL4 and LexA are typically used for this purpose. Transcription requires interaction of the GAL4 DNA binding domain with the activation domain of another factor, such as the yeast VP16 activation domain protein; however, the two domains will not interact directly themselves, and require proximity dictated by the interaction of fusion proteins. Several excellent reviews[45–47] and many commercial kits and libraries are available to help perform these procedures. Mammalian cell systems are used for transient transfections. Yeast-based screening procedures use yeast stably transfected with a *lacZ* reporter and a second reporter to permit selection on nutrient-restricted media.

The bait construct shown in Figure 5 expresses a fusion protein with the DNA-binding domain of GAL4 fused to the ligand-binding domain of ERα. The prey cDNA expression library to be screened for proteins that interact with the bait expresses fusion proteins with the interaction domain of VP16 linked to cDNAs from uterus. If one of these fusion proteins binds to the ligand-binding region of ERα, the VP16 interaction domain will be brought into proximity with the GAL4 DNA-binding domain, the resulting complex will initiate transcription from the promoter and luciferase activity will be measured. Success relies upon low background, and thus low intrinsic, activation of the bait protein. In this example, the ER fusion protein has low activity without ligand, and screening the library in the presence and absence of estrogen will identify proteins that bind preferentially in the presence of ligand and influence receptor activity in a physiologically relevant manner.[12,49] Two-hybrid approaches have been used to identify nuclear receptor coactivators that bind preferentially in the presence of activating ligand,[12,13] and corepressors that bind preferentially in the absence of ligand or in the presence of receptor antagonist.[48,49]. The procedures are very versatile, and can be used to clone new proteins, and to demonstrate functional interactions between known proteins, with variations on these methods constantly under development.

SUMMARY

Molecular biology tools can be powerful means of detecting specific molecules and dissecting specific problems or questions in reproduction. Studies with the estrogen receptor, illustrated in several figures in this chapter, represent one of many examples in which a rational molecular approach, coupled with an understanding of the physiological system and biological implications, can help to advance the clinical and basic knowledge of a given system. With rapidly advancing technology, applications are limited only by the imagination, and the ability and judgement in posing the biological question.

REFERENCES

1. Watson JD, Alberts BA, Bray D, Lewis J. Raff M, Roberts K. *Molecular Biology of the Cell*. New York: Garland Publishers, 1994.
2. Darnell J, Lodish H, Baltimore D. *Molecular Cell Biology*, 3rd edn. New York: Scientific American Books, 1995.
3. Weintraub B. *Molecular Endocrinology: Basic Concepts and Clinical Correlates*. New York: Raven Press, 1995.
4. Jameson JL. *Principles of Molecular Medicine*. Totowa, NJ: Humana Press, 1998.
5. Sambrook J, Russell DW. *Molecular Cloning: a Laboratory Manual*. Cold Spring Harbor, NY: Cold Spring Harbor Laboratory Press, 2001.
6. Ansubel FM, Brent R, Kingston RE *et al. Current Protocols in Molecular Biology*. New York: John Wiley & Sons, 1995.
7. Gharib SD, Wierman ME, Shupnik MA, Chin WW. Molecular biology of the pituitary gonadotropins. *Endocr Rev* 1990 11:177–199
8. Ponglikitmongkol M, Green S, Chambon P. Genomic organization of the human oestrogen receptor gene. *EMBO J* 1988 11: 3385–3388.
9. Sakano H, Rogers JH, Huppi K *et al.* Domains and the hinge region of an immunoglobulin heavy chain are encoded in separate DNA segments. *Nature* 1979 277: 627–633.
10. Mittsuhashi T, Tennyson GE, Nikodem V. Alternative splicing generates messages encoding rat c-erbA proteins that do not bind thyroid hormone. *Proc Natl Acad Sci USA* 1988 85: 5804–5808.
11. Kwok RPS, Lundblad JR, Chevia JC *et al.* Nuclear protein CBP is a coactivator for the transcription factor CREB. *Nature* 1994 370: 223–226.
12. Onate SA, Tsai BY, Tsai M-J, O'Malley BW. Sequence and characterization of a coactivator for the steroid receptor superfamily. *Science* 1995 270: 1354–1357.
13. Horwitz KB, Jackson TA, Bain DL, Richer JK, Takimoto GS, Tung L. Nuclear receptor coactivators and corepressors. *Mol Endocrinol* 1996 10: 1167–1177.
14. Nagy L, Kao H-Y, Chakravarti D *et al.* Nuclear receptor repression mediated by a complex containing SMRT, mSin3a, and histone deacetylase. *Cell* 1997 89: 373–380.

15. Thomas PS. Hybridization of denatured RNA and small DNA fragments transferred to nitrocellulose. *Proc Natl Acad Sci USA* 1980 77: 5201–5205.

16. Luo X, Ikeda Y, Parker KL. A cell-specific nuclear receptor is essential for adrenal and gonadal development and sexual differentiation. *Cell* 1994 77: 481–490.

17. Wong M, Ramayya MS, Chrousos GP, Driggers PH, Parker KL. Cloning and sequence analysis of the human gene encoding steroidogenic factor 1. *J Mol Endocrinol* 1996 17: 139–147.

18. Walter P, Green S, Green G *et al.* Cloning of the human estrogen receptor complementary DNA. *Proc Natl Acad Sci USA* 1985 82: 7889–7893.

19. White R, Lees JA, Needham M, Ham J, Parker M. Structural organization and expression of the mouse estrogen receptor. *Mol Endocrinol* 1987 1: 735–744.

20. Tabor S, Richardson CC. A bacteriophage T7 RNA promoter system for controlled exclusive expression of specific genes. *Proc Natl Acad Sci USA* 1985 82: 1074–1078.

21. McPherson MJ, Quirke P, Taylor GR, eds. *PCR: a Practical Approach.* Oxford: Oxford University Press, 1991.

22. Weinstein LS, Schenker A, Gejman PV, Merino MJ, Friedman E, Spiegel AM. Activation mutations of the stimulatory G protein in the McCune–Albright syndrome. *N Engl J Med* 1991 325: 1688–1695.

23. Gilliland G, Perrin S, Blanchard K, Bunn HF. Analysis of cytokine mRNA and DNA: detection and quantitation by competitive polymerase chain reaction. *Proc Natl Acad Sci USA* 1990 87: 2725–2729.

24. Park C-Y, Mayo KE. Transient expression of progesterone receptor messenger RNA in ovarian granulosa cells after the preovulatory luteinizing hormone surge. *Mol Endocrinol* 1991 5: 967–978.

25. Friend KE, Ang LW, Shupnik MA. Estrogen regulates the expression of several different isoforms of estrogen receptor mRNA in rat pituitary. *Proc Natl Acad Sci USA* 1995 92: 4367–4371.

26. Boehm CD, Kazazian HH Jr. The molecular basis of genetic disease. *Curr Opin Biotech* 1990 1: 180–187.

27. Kuiper GJM, Enmark E, Pelto-Huikko M, Nilsson S, Gustaffson J-A. Cloning of a novel estrogen receptor expressed in rat prostate. *Proc Natl Acad Sci USA* 1996 93: 5925–5930.

28. Matthews LS, Vale WW. Expression cloning of an activin receptor, a predicted transmembrane serine kinase. *Cell* 1991 65: 973–982.

29. Lipkin SM, Naar AM, Kalla KA, Sack RA, Rosenfeld MG. Identification of a novel zinc finger protein binding a conserved element critical for Pit-1 dependent growth hormone gene expression. *Genes Dev* 1993 7: 1674–1687.

30. Sanger F, Nicklen S, Coulson AR. DNA sequencing with chain-terminating inhibitors. *Proc Natl Acad Sci USA* 1977 74: 5463–5467.

31. Attardi B, Winters SJ. Decay of follicle-stimulating hormone-β messenger RNA in the presence of transcriptional inhibitors and/or inhibin, activin or follistatin. *Mol Endocrinol* 1993 7: 668–680.

32. Ekena K, Weis E, Katzenellenbogen JA, Katzenellenbogen BS. Different residues in the human estrogen receptor are involved in recognition of structurally diverse estrogens and antiestrogens. *J Biol Chem* 1997 272: 5069–5075.

33. Friend KE, Resnick EM, Ang LW, Shupnik MA. Specific modulation of steroid receptor isoforms in rat pituitary throughout the estrous cycle and in response to steroid hormones. *Mol Cell Endocrinol* 1997 131: 147–155.

34. Marks DL, Wiemann JN, Burton KA, Lent KL, Clifton DK, Steiner RA. Simultaneous visualization of two cellular mRNA species in individual neurons by use of a new double *in situ* hybridization method. *Mol Cell Neurol* 1992 3: 395–405.

35. Woodruff TK, D'Angostino J, Schwartz NB, Mayo KE. Dynamic changes in inhibin mRNAs in rat ovarian follicles during the reproductive cycle. *Science* 1988 239: 1296–1299.

36. Schena M, Heller RA, Theriault TP, Konrad K, Lachenmeier E, Davis RW. Microarrays: biotechnology's discovery platform for functional genomics. *Trends Biotechnol* 1998 16: 301–306.

37. Feng X, Jiang Y, Meltzer P, Yen PM. Thyroid hormone regulation of hepatic genes *in vivo* detected by complementary cDNA microarray. *Mol Endocrinol* 2000 14: 947–955.

38. Shupnik MA. Effects of gonadotropin-releasing hormone on rat gonadotropin gene transcription *in vitro*: requirement for pulsatile administration of luteinizing hormone-β gene stimulation. *Mol Endocrinol* 1990 4: 1444–1450.

39. Dalkin AC, Burger LL, Aylor KW *et al.* Regulation of gonadotropin subunit gene transcription by gonadotropin-releasing hormone: measurement of primary transcript ribonucleic acids by quantitative reverse transcription-polymerase chain reaction assays. *Endocrinology* 2001 142: 139–146.
40. de Wet JR, Wood KV, De Luca M, Helinski DR, Subramani S. Firefly luciferase gene: structure and expression in mammalian cells. *Mol Cell Biol* 1987 7: 725–737.
41. Gasic S, Bodenburg Y, Nagamani M, Green A, Urban RJ. Troglitazone inhibits progesterone production in porcine granulosa cells. *Endocrinology* 1998 139: 4962–4966.
42. Weck J, Anderson AC, Jenkins S, Fallest PC, Shupnik MA. Divergent and composite gonadotropin-releasing hormone-responsive elements in the rat luteinizing hormone subunit genes. *Mol Endocrinol* 2000 14: 472–485.
43. Petz LN, Nardulli AM. Sp1 binding sites and an estrogen response element half-site are involved in the regulation of the human progesterone receptor A promoter. *Mol Endocrinol* 2000 14: 972–985.
44. Belsham DD, Mellon PL. Transcription factors Oct-1 and C/EBPβ are involved in the glutamate/nitric oxide/cyclic guanosine-5'-monophosphate-mediated repression of gonadotropin-releasing hormone expression. *Mol Endocrinol* 2000 14: 212–228.
45. Allen JB, Wallberg MW, Edwards MC, Elledge SJ. Finding prospective partners in the library: the yeast two-hybrid system and phage display find a match. *Trends Biol Sci* 1995 20: 511–516.
46. McNabb DS, Guarente L. Genetic and biochemical probes for protein–protein interactions. *Curr Opin Biotechnol* 1996 7: 554–559.
47. Young KH. Yeast two-hybrid: so many interactions, (in) so little time. *Biol Reprod* 1998 58: 302–311.
48. Chen JD, Evans RM. A transcriptional co-repressor that interacts with nuclear hormone receptors. *Nature* 1995 377: 454–457.
49. Montano MM, Ekena K, Delage-Mourroux R, Chang W, Martini P, Katzenellenbogen BS. An estrogen receptor-selective coregulator that potentiates the effectiveness of antiestrogens and represses the activity of estrogens. *Proc Natl Acad Sci USA* 1999 96: 6947–6952.

Chapter 2

Functional genomics

Aaron J.W. Hsueh

INTRODUCTION

The biomedical sciences are entering a new era at the beginning of the 21st century as a result of revolutionary advances in genomics, transcriptomics, proteomics, computational biology, and web-based text and sequence data integration. The major paradigm shift started in the last decade,[1] and is exemplified by the abundance of biological information due to the introduction of high-throughput tools to allow a global perspective on cell and organ functions. Instead of the traditional single gene approach, we are now gaining an understanding of the entire genome of diverse organisms as well as gene families and signalling pathways, leading to a comprehensive elucidation of physiology and pathophysiology.

Beginning in the early 1990s, major efforts to sequence the entire genome of human and lower organisms provided detailed information on entire human genome sequences, as well as the genomic sequences for more than 60 model organisms. In the summer of 2000, the Human Genome Project announced the complete sequencing of the entire human genome. This achievement has been compared with the completion of the periodic table at the end of the 19th century. Accompanying this unprecedented availability of sequence information is a major advance in computational power. Now known as Moore's law, an observation of Gordon Moore was that, starting in 1965, the number of transistors in a computer chip doubles every 1.5 years. This unprecedented increase in computational power revolutionized daily life, and allowed biomedical scientists to access massive amounts of data using a personal computer via diverse web sites. Advances in large-scale DNA sequencing, coupled with the design of new algorithms to analyze these data, allowed the opportunity of understanding complete genomes of human and lower organisms. Taking advantage of the extensive use of the World Wide Web, the massive DNA sequences were immediately available to individual investigators throughout the world.

In addition to large-scale automated DNA sequencing, other high throughput tools of modern biology include microarrays, mass spectrometry and sophisticated imaging systems. Recent advances in DNA and protein arrays have provided the basis for analyzing the expression of hundreds of thousands of gene products in different physiological and pathophysiological conditions.[2] This approach enables the development of transcriptomics for the global analysis of all gene transcripts in different tissues under different physiological and pathological conditions. Future progress in protein arrays could further permit the analysis of protein expression profiles using similar approaches. Major advances in mass spectrometry allow the sequencing and identification of peptides and proteins for matching with the genomic data. Together with efforts to investigate the post-translational processing and modification of primary transcripts, the next decade will herald the era of proteomics. In addition, sophisticated computer tools have been developed to elucidate the three-dimensional structures of proteins, and the introduction of new imaging systems with high resolution allows scientists and physicians to view the inner workings of cell organelles and organ systems in a non-invasive manner. Starting with advances in genomics that investigate the entire genomes of human and lower species, functional genomics is the science of understanding the function of all human genes.

It is becoming clear that every discipline of biology will feel the impact of functional genomics and bioinformatics. As indicated by Freeman Dyson, 'The effect of concept-driven revolution is to explain old things in new ways. The effect of tool-driven revolution is to discover new things that have to be explained.' Biomedical research is moving from the hypothesis-driven research of single genes by individual laboratories, to new paradigms based on high-throughput tools and computational methods. Genomics will plot the passage from DNA to genes to chromosomes. Transcriptomics will provide information on all mRNAs that are transcribed in the cell. Proteomics will tell us which proteins are expressed in each cell under specific conditions, as well as the structure of each domain of the individual proteins and the dynamic changes in the post-translational modification of proteins.

GENOMICS

Each genome is a script written in a four-letter alphabet: A, T, C and G. Continuous strings of DNA constitute the individual chromosomes, whereas the total number of chromosomes in a given species represents the entire genome. Within each chromosome, individual genes intersperse among the large non-coding and repeat sequences. The human genome has three billion (10^9) base pairs of DNA, with 1.5% representing exon sequences that encode approximately 35 000 genes. Recent expansion of genome projects for diverse species provides evolutionary perspectives on the origin of genes in man, and allows the grouping of these genes into families. As indicated by Max Delbruck in 1949, 'Any living cell carries with it the experience of a billion years of evolution.' Based on the comparison of human genes with those from lower species, it is now possible to trace the origin of all human genes, thus enhancing the understanding of their physiological roles. Analyses of human genes have been based on several model organisms in which the entire genomic sequence is known and in which extensive experiments have been performed. Table 1 lists the genome size (in base pairs, bp), number of genes, number of haploid chromosomes and number of cells for four model organisms including man. It is becoming clear that analysis of genomes for the unicellular yeast, *Saccharomyces cerevisiae*, and the invertebrates, nematode (*Caenorhabditis elegans*) and fly (*Drosophila melanogaster*), provides invaluable information for the understanding of the human genome.

Evolution of chromosomes, intron–exon regions and regulatory sequences

Sequencing of entire genomes of multiple species and the alignment of all genes in individual chromosomes allowed the investigation of the evolution of chromosomes. As shown in Table 1, the chromosome number varies greatly in divergent species and does not provide useful information about the evolution of chromosomes. However, investigation of chromosome evolution in closely related species clearly indicated that chromosomes undergo duplication, fusion, translocation and inversion during evolution. Investigation of the genes located in different chromosomes leads to the identification of syntenic regions of chromosomes in different vertebrates. In these regions, partial or complete conservation of gene order during chromosome evolution is evident.

For decades, genetic mapping of gene positions in chromosomes has been performed based on family history in patients or based on cross-breeding in animals. The linkage analysis is based on the frequency of meiosis recombination between genetic markers, such as microsatellite DNAs, to estimate the position of genes with a unique phenotype. Recently, understanding of gene positions in chromosomes has also been assisted by physical mapping based on fluorescent *in situ* hybridization (FISH), restriction enzyme length polymorphism and sequence tagged sites. With the availability of complete genomes of an increasing number of vertebrate species, genomic analysis of syntenic regions is becoming a valuable tool for understanding human gene localization.

Table 1 Genome size (base pairs, bp), number of genes and haploid chromosomes, and number of cells for four model organisms

	*Yeast** *(S. cerevisiae)*	*Nematode†* *(C. elegans)*	*Fly†* *(D. melanogaster)*	*Human‡* *(H. sapiens)*
Genome size (bp)	10^7	10^8	1.4×10^8	3×10^9
Genes (*n*)	6000	19 000	13 000	31 000
Chromosome number (haploid) (*n*)	16	4	4	23
Cells (*n*)	1	1×10^3	1×10^6	$1 \times 10^{12-14}$

*Eukaryotic unicellular; †invertebrate multicellular; ‡vertebrate multicellular

Recent studies have demonstrated the important role of the growth differentiation factor (GDF)-9, a member of the transforming growth factor-β (TGF-β) family, in ovarian follicle development. Mutant mice without this oocyte hormone showed an arrest of ovarian follicles at the secondary stage,[3] whereas *in vivo* treatment with GDF-9 enhances the progression of early follicles.[4] Furthermore, another oocyte gene with a high homology to GDF-9, named GDF-9B or bone morphogenetic protein (BMP)-15, was also found to stimulate granulosa cell proliferation[5] and mapped to human chromosome Xp11.2–11.4. In sheep, a naturally occurring X-linked mutation was identified that causes an increased ovulation rate and twin and triplet births in heterozygotes, but primary ovarian failure in homozygotes. The genetic locus of this Inverdale sheep mutation was found to be syntenic to the human Xp region, and a point mutation in the GDF-9B/BMP-15 gene was found in the Inverdale sheep. Thus, this oocyte-expressed gene is essential for female fertility, and natural mutations of this oocyte-derived factor can cause both an increased ovulation rate and infertility phenotypes in a dosage-sensitive manner.[6] It is becoming clear that future analysis of chromosome evolution will facilitate the identification of other important reproductive genes based on their unique chromosome location.

In addition to studies of syntenic chromosome regions, analysis of linked genes in the human chromosome could also provide important information on gene location. Kallmann's syndrome is exemplified by hypogonadotropic hypogonadism and an inability to smell as the result of a defect in the migration of olfactory and gonadotropin-releasing hormone (GnRH) neurons. Although this disease has been associated with both X-linked and autosomal inheritance, identification of patients with a contiguous gene syndrome, showing both infertility and skin lesions of scalp, ears and neck, allowed the localization of the KAL1 gene to Xp22.3. In this region, the KAL1 gene is situated adjacent to the steroid sulfatase gene responsible for X-linked ichthyosis exhibiting the characteristic skin lesion. Thus, microdeletion of neighboring chromosomal regions containing linked genes could reveal the location of important disease genes following analysis of gene location in the finished human genome.

Although the entire human genome has been sequenced, we have only limited knowledge about the encoded genes. Algorithms have been formulated to extract the genomic sequence for the identification of open reading frames (ORFs) that could encode for expressed genes. Identification of ORFs is needed before the eventual assignment of gene function. The search criteria are based on the location of the initiation codon (ATG), termination codon (TAA, TAG and TGA), promoter sequences, and intron/exon prediction as well as sequence matching with known gene transcripts. The present gene identification algorithms are estimated to predict 70% of genes, and this is a challenging area for genomic research. Thus, we now have a map for all the super-highways of the human genome, but only limited knowledge of the entrances, exits and interchanges.

Because more than 98% of the human genome does not have an ORF to encode proteins, the function of most of these DNA sequences is still unclear. Sequencing of the entire genome allows the opportunity to elucidate the function of the so-called 'junk' DNA as well as the promoter, enhancer and silencer regions that are important for the regulation of the expression of genes. Promoters are usually located at the 5'-end adjacent to the coding region, whereas enhancers and silencers are located further away and are important for the regulation of the expression of individual genes. Recently, insulator sequences have been identified that mark the boundaries between genes by limiting the range of action of enhancers and silencers.[7] The insulators are believed to be important for forming loops of DNA, leading to a three-dimensional chromatin structure. Studies of the evolution of DNA binding sites, as well as different transcription factors, are becoming feasible through the alignment of essential promoter regions based on phylogenetic conservation between human and mouse orthologs[8] as well as the analysis of regulatory elements following analysis of multiple co-expressed genes. Studies using yeast two-hybrid assays also allow a global analysis of the DNA sequences in the promoter, essential for binding diverse transcriptional factors. Of particular interest are future studies on the 'junk' DNA that accounts for most of the human genome. Fifty per cent of these sequences are transposable elements and remnants of ancient pathogen DNAs. Their exact role in human evolution remains to be unravelled.

Evolution of human genes

Human genes can be studied on the basis of their evolutionary history. Orthologous genes are those that did not significantly change in either structure or function during evolution. It has been estimated that 40% and 60% of human genes responsible for genetic diseases have orthologs in yeast and fly, respectively. Genes belonging to this group are usually involved in cell division or basic metabolism. Orthologs found in yeast and fly represent the conservation of the gene structure and function for 1.5 and 1.0 billion years, respectively. Genes in a given species with a common evolutionary origin are called paralogs; they belong to the same family and usually have similar structural motifs but could be responsible for related but non-identical physiological functions. By comparing genomes from different phylogenies, it has been hypothesized that, during early chordate evolution, before Cambrian explosion and during the early Devonian period, two entire genome duplications took place,[9,10] leading to the generation of multiple mammalian paralogs, and the opportunity to develop complex regulatory mechanisms and functions. Thus, most human genes belong to a family as the result of gene duplication and domain shuffling. Gene duplication and domain rearrangement allow for the amplification and diversification of existing domains, recruitment of existing domains for new functions, and the development of new domains through domain combination and shuffling.

Comparative genomic analysis allowed the understanding of the evolution of hormones essential for metabolism. Only one human insulin-like ortholog has been found in the nematode, and this gene was shown to determine the life span of the worm. In contrast, multiple insulin paralogs are present in the human genome, including insulin for carbohydrate metabolism, insulin-like growth factor-I (IGF-I) and IGF-II for organ and body growth, relaxin and Leydig cell relaxin for reproductive tract functions, as well as EPIL (early placenta insulin-like factor) and RIFs (relaxin-insulin-like factors) with unknown functions.[11] Sequence comparison indicated that all these genes encode secreted proteins consisting of B and A subunits connected by a long C domain peptide. These human paralogs all have highly conserved cysteine residues essential for maintaining a similar secondary structure.

Genomic studies of the presence of genes with sequence similarities to human gonadotropin and thyrotropin receptors also allowed understanding of the evolution of these genes. All these genes encode proteins with a large N-terminal extracellular region important for ligand binding, followed by a seven-transmembrane region known to be essential for G protein coupling. Although many orthologs have been found in lower species together with four novel paralogs in mammals,[12,13] the ligands for these receptors are unknown. They were named LGRs, reflecting the presence of leucine-rich repeats in the extracellular region and a G protein-coupling transmembrane region. Based on the structural comparison of LGRs from diverse species, and the evolutionary relationship of these animals, a putative evolutionary tree for the LGR family of proteins can be proposed (Figure 1). The primitive LGR gene probably replicated before the emergence of the cnidarian (sea anemone) to form three LGRs (LGRA, LGRB and LGRC), each with homologs in modern vertebrates. Based on their structural similarity and putative evolutionary origins, mammalian LGRs can be divided into three subgroups: LGRA (luteinizing hormone (LH), follicle stimulating hormone (FSH) and thyroid stimulating hormone (TSH) receptors), LGRB (LGR4, -5 and -6) and LGRC (LGR7 and -8). In the fly there are at least three LGRs, and the first two LGRs are closely related to the LGRA subgroup, whereas the third one is homologous to the primitive LGRB. Because only one LGR with similarities to the LGRA subgroup could be identified in *C. elegans*, it is likely that a gene loss occurred during evolution.

An understanding of the evolutionary origin of diverse human genes is useful for functional

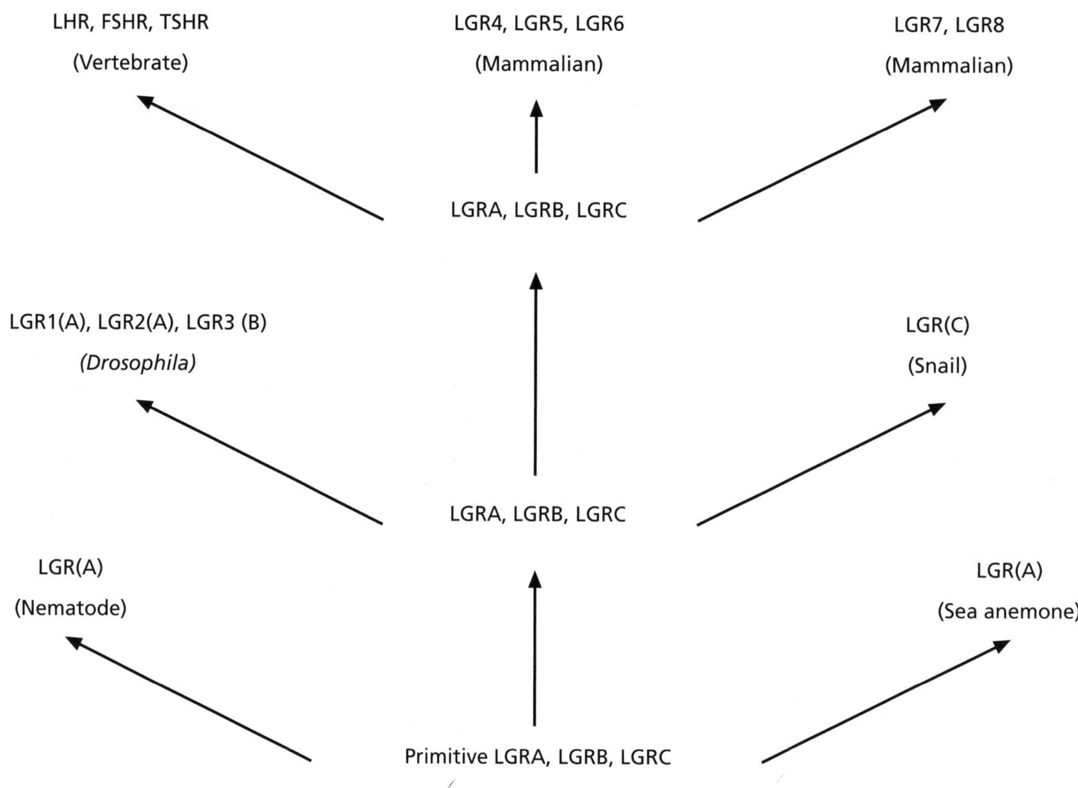

Figure 1 Evolution of leucine-rich repeat-containing, G protein-coupled receptor (LGR) family proteins. The LGRs are probably derived from three primitive receptors present before the development of all animals. In the completely sequenced genome of the *Caenorhabditis elegans*, there is only one subtype (A) of LGR present whereas the fly genome contains two subtypes. In mammals, subtype A evolved into gonadotropin and thyrotropin receptors, subtype B evolved into the orphan LGR4–6, and subtype C evolved into LGR7 and -8. LHR, luteinizing hormone receptor; FSHR, follicle stimulating hormone receptor; TSHR, thyroid stimulating hormone receptor

genomic research. In addition to investigations into the evolution of gene families, attempts have also been made to trace the evolution of signalling pathways and cellular gene circuits that are important for different biological functions. With the expansion of genomic sequencing to multiple vertebrate species, elucidation of gene pathways is becoming easier.

Gene mutations and polymorphism

The out-of-Africa hypothesis regards all modern human populations as descended from a small group of about 100 000 individuals that migrated from Africa less than 200 000 years ago and replaced archaic populations.[14] Owing to common origins, modern men share common changes in their genome. Base pair changes that lead to minimal alterations in the property of encoded amino acids will not significantly affect the overall function of a given protein; however, more drastic changes in the properties of encoded amino acids could lead to disease phenotypes (disease mutations). Depending on the severity of the disease, most mutations are eliminated from the human population. Because of beneficial side-effects under unique environmental conditions, some disease mutations are conserved in restricted geographical regions. In contrast, variations in nucleotide sequences that are associated with silent mutations owing to the use of degenerate codons, or the manifestation of minimal phenotypes, are tolerated during evolution and inherited over generations. These single nucleotide polymorphisms (SNPs) represent minor alterations in DNA that occur with varying frequencies in different ethnic populations, and are the central focus of pharmacogenomic studies.[15] It has been estimated that there are 300 000 potential SNPs in the human genome.[16,17] Polymorphisms of individual or multiple genes could account for variations in phenotypes.

Gene mutations can be broadly divided into two subtypes: gain-of-function mutations and loss-of-function mutations. The former is usually associated with nucleotide changes in one of the two alleles, leading to altered phenotypes. For example, point mutation of the LH receptor gene found in patients with familial male precocious puberty is present in only one allele, corresponding to the transmembrane region of this signalling protein.[18] In contrast, the loss-of-function mutations are present in lower frequency in the population, and are usually associated with mutations in both alleles of the affected gene due to homozygous mutations, or compound heterozygous mutations found in two different locations of the same gene. Inactivating mutations of the human LH receptor lead to Leydig cell hypoplasia, a form of male pseudohermaphroditism resulting from the failure of fetal testicular Leydig cell differentiation.[19] For genes located in the X chromosome, haploinsufficiency is also found in male patients. These loss-of-function mutations are associated with changes in the only allele located at the single X chromosome in males. Familial premature ovarian failure has recently been found to be associated with haploinsufficiency of the forkhead transcription factor FOXL2 gene, which resides in chromosome 3.[20] In these patients, only one allele of the FOXL2 gene is mutated. However, the normal function of the remaining allele is probably disrupted by a balanced chromosomal translocation adjacent to the FOXL2 gene.

Polymorphisms of individual genes have been shown to account for the differences in an individual's susceptibility to various diseases as well as their responsiveness to drug treatment. For example, polymorphisms in the coding region of the chemokine receptor CCR-5 gene, and in the promoter region of its ligand RANTES, are associated with differences in human immunodeficiency virus (HIV) resistance. Patients with the defective CCR-5 receptor, the co-receptor for the viral infection of immune cells, or patients who exhibit overproduction of RANTES, can be exposed to HIV but with delayed or minimal disease symptoms.[21,22] In both cases, the viral particles failed to infect cells because they could not enter the cells through the defective receptor, or because the functional receptor was already occupied by high levels of the endogenous ligand. Likewise, polymorphism of the β-adrenergic receptor is associated with altered sensitivity to β-agonists in asthmatics.[23] Studies on the polymorphic human cytochrome P450 enzyme genes further demonstrated that SNP is associated with abolished, qualitatively altered or enhanced drug metabolism.[24]

For reproductive genes, two linked polymorphisms of the human FSH receptor gene have been found (680 Asn/Ser and 307 Thr/Ala). Of interest, the observed polymorphism is associated with the differential FSH responsiveness of gonadal cells, because the amount of FSH required for ovarian stimulation in patients is different. Lower doses of FSH are needed for patients with 680 Asn/Asn alleles than for those with 680 Ser/Ser alleles.[25] In addition, polymorphism of the human LHβ gene is associated with two linked amino acid changes (8 Trp/Arg and 15 Ile/Thr), and the latter introduces a new glycosylation signal, leading to alterations in immunoassay results.[26] The carrier frequency for these linked polymorphisms varies in different populations (28% in Finns, 7.5% in North American Hispanics, 4.5% in Singaporean Chinese and 3% in East Indians). Although the exact physiological consequences of the LHβ polymorphism are still unclear, variable pubertal onset in boys has been reported.

Recent advances in DNA microarray technology allow a sequencing-by-hybridization approach, based on array analysis of overlapping oligomers corresponding to the entire sequence of a known gene. This method allows detection of mutations or polymorphisms of individual genes.[27] Mutations or polymorphisms for the cancer suppressor gene p53 have been determined by hybridization of patient's genomic DNA, thus demonstrating the feasibility for future analysis of polymorphisms of different candidate genes of pathological and pharmacological interest. Because the genome of each individual needs to be sequenced only once, it is likely that future patient care will involve the complete sequencing of the entire genome for each patient.

TRANSCRIPTOMICS

With the exception of the haploid germ cells, almost every cell in the body contains the same diploid genetic material. Although the so-called housekeeping genes are expressed ubiquitously, most genes are expressed in a cell type-, time- and developmental stage-specific manner. Genomic advances are incomplete without concomitant progress in transcriptomics, which deals with the analysis of the transcripts of genes in given cell types at selective developmental stages, or under different pharmacological and pathological conditions. In addition to changes in the qualitative and quantitative expression pattern of the transcripts, the nature of the transcript is also under regulation, and each primary transcript can lead to the formation of different mRNAs following differential splicing events.

Concomitant with the large-scale sequencing of genomic DNA, major efforts have also been made to sequence cDNA libraries from different tissues under different physiological or pathological states. To facilitate cDNA sequencing, most of the cDNA was analyzed as 400–600-bp fragments, called expressed sequence tags (ESTs). Although attempts have also been made to sequence full-length cDNAs, the majority of ESTs in the GenBank represents 3'-end sequences that could easily be derived from a reverse transcriptase reaction following priming of the poly(A) tail in mRNAs from diverse tissues and cell types. With the sequencing of cDNA libraries for reproductive tissues under different physiological states, future investigations will focus on the digital analysis of gene expression patterns. Recent studies have shown that previously unidentified genes associated with steroid synthesis, prostate cancer, insulin synthesis and neurotransmitter processing could be discovered by using a 'guilt-by-association' approach based on digital analysis of co-ordinated expression of genes in multiple cDNA libraries.[28]

Because many genes are expressed in multiple tissues, one can also survey the expression of genes by identifying short unique tags of cDNAs in a given tissue of interest. This approach takes advantage of the expression of most genes in multiple tissues, and is a short cut to gene discovery. The serial analysis of gene expression (SAGE) method (Figure 2) allows the quantitative and simultaneous analysis of a large number of transcripts. Using this strategy, short diagnostic sequence tags can be isolated from multiple cDNAs in tissues of interest, linked and cloned. Sequencing of hundreds of tags can be performed with relative ease to reveal the gene expression pattern based on searches of the SAGE tags database containing unique tags for diverse cDNAs.[29] The molecular phenotype of the human oocyte was recently analyzed using the SAGE approach.[30] Consecutive application of the polymerase chain reaction and SAGE was used to generate a catalog of approximately 50 000 SAGE tags from human oocytes. Matches for known genes were identified and included surface receptors, second-messenger systems, and cytoskeletal, apoptotic and secreted proteins. Many of these proteins were not previously known to be expressed in mammalian oocytes.

Tissue-, time- and development stage-specific pattern of mRNA expression

With the availability of the sequences of many human cDNAs and predicted ORFs, it is important to analyze the transcript levels of multiple genes using a simple method. DNA microarrays, also known as hybridization arrays, gene chips and/or high-density oligonucleotide arrays, bring gene expression analysis to a genomic scale by permitting investigators simultaneously to examine changes in the expression of literally thousands of genes. Gene-specific sequences (probes) are immobilized on a solid-state matrix (nylon membrane, glass microscope slide or silicon/ceramic chip). These sequences are then queried with labelled copies of nucleic acids from biological samples (targets). The greater is the expression of a gene, the greater is the amount of labelled target and,

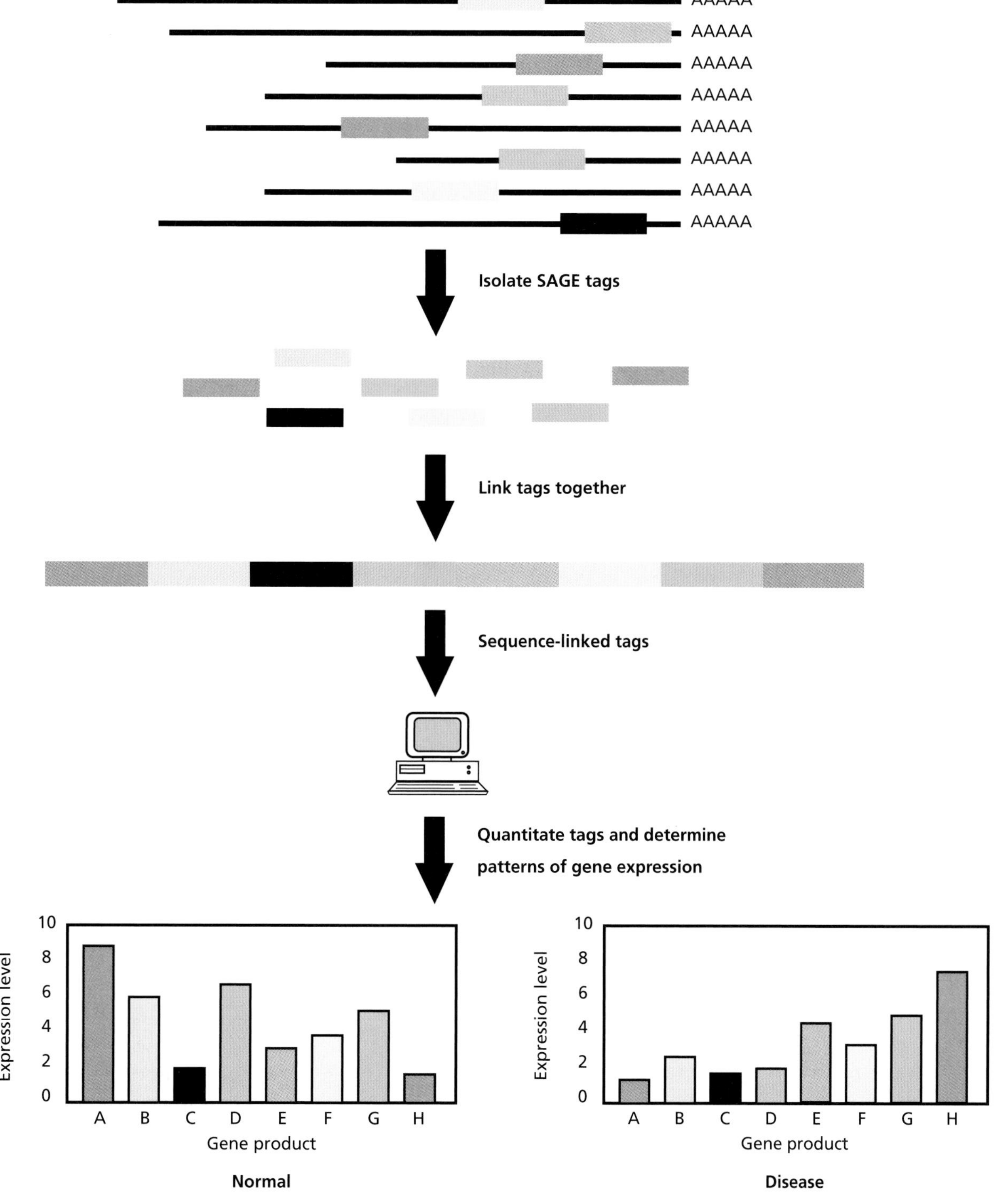

Figure 2 Serial analysis of gene expression (SAGE). Messenger RNAs were first reverse transcribed into cDNAs. Using specific restriction enzymes, unique DNA fragments (SAGE tags), representing the signature of individual cDNAs, were isolated. Following ligation of SAGE tags into long DNA strands, the linked tags were sequenced to determine the pattern of gene expression and for semi-quantitative estimation of mRNA expression pattern

hence, the greater is the output signal. Recent advances in DNA arrays allowed the survey of the expression pattern of the entire yeast genome.[31] Likewise, a genome-wide expression profiling of mid-gestation placenta and embryo, using a 15 000 mouse developmental cDNA microarray, was reported.[32] The DNA array approach has also been used to monitor changes in gene expression in response to drug treatments, as well as for molecular phenotyping of diseased tissues. Owing to the complexities of most diseased tissues, future molecular profiling of individual patient specimens will employ DNA microarrays, together with laser-guided microdissections of tissue sections followed by bioinformatic analysis.[33] Accompanying the major expansion of data acquisition using microarrays, new hierarchical clustering methods have been developed to analyze these results. It is anticipated that an improved version of the DNA microarray will become a routine diagnostic and therapeutic tool for molecular genotyping to augment or replace the present reliance on histological analyses.

Alternative splicing leads to multiple transcripts

Although entire human genome sequences are now available, the understanding of their relationship to the encoded proteins is far from complete. The primary transcript of an individual gene is processed in the nucleus before it is transported to the ribosomes as messenger RNAs to direct protein synthesis. The total number of genes increased only 60% from the simple nematode with 1000 cells, to man with 10^{12-14} cells (Table 1). Because mammalian proteins are needed to carry out sophisticated higher-level brain and other physiological processes, the lack of a major increase in gene number is partially explained by the extensive alternative processing of individual genes. Evolutionary analysis indicated that cellular functions are carried out by similar signalling modules within protein clusters, and the alternative use of these functional motifs encoded by one or more exons provides an economic means for higher organisms to perform complicated tasks with a limited number of genes.

Processing of the primary transcript into mRNA requires intervening sequences to be excised in a process mediated by the mRNA splicing machinery (Figure 3). Exons can be spliced into the mRNA or skipped. Introns that are normally excised can be retained in the mRNA. Unique sequences flanking the exons are called splicing sites. However, the position of the 5' or 3' splicing site can shift to allow the generation of longer or shorter exons. In addition, selective use of the transcriptional start site, or polyadenylation site, allows the generation of multiple mRNAs from a single gene. Prediction of intron–exon junctions and the splicing sites is still an emerging field and the exact basis of differential RNA splicing is not entirely known. Although consensus sequences have been proposed for the splicing junction, the functional splice sites do not always match the consensus sequence. Recent advances in the comparison of rodent and human genomes have facilitated the prediction of potential transcripts because splicing sites are likely to be conserved. However, investigation of the combinatorial use of exons still remains a major challenge for bioinformaticians.

Alignment of ESTs with genomic sequences allowed an estimation of gene splicing, and more

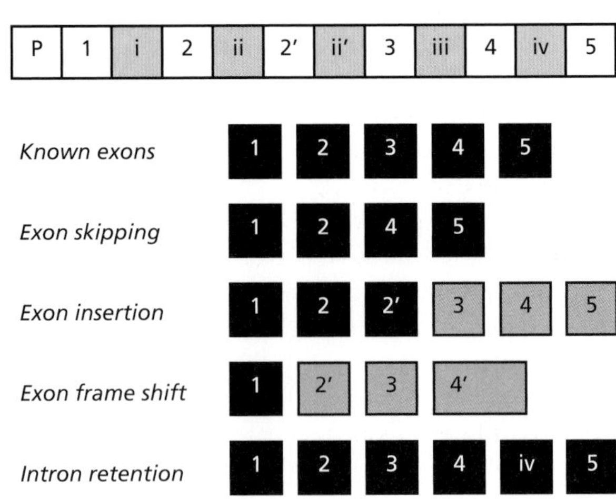

Figure 3 Alternative splicing of genes to derive diverse transcripts. A prototype gene consists of the promoter (P) region followed by exons (1–5) and introns (i–iv). The transcript may consist of all exons (1–5) or other alternative splicing transcripts as the result of exon skipping or insertion. Some transcripts use different open reading frames as the result of a frame shift at the splicing junction. In some cases, the intron could be retained

than 35% of human genes have been found to encode splicing variants. Using this approach, splicing variants were identified for presomatotropin, the presomatotropin-related gene and sex hormone-binding globulin.[34] Furthermore, it is common for individual genes to have a dozen variants. Some splicing variants are derived from the skipping and jumping of exons of a given gene, and most of the alternative splices of coding regions generate additional protein domains rather than alternating domains.[34] For selective genes (such as connexin, *n*-cadherin and calcium-activated potassium channels), more than 100 splicing variants have been found.[35]

Several hormones are derived from the same gene following alternative splicing. Classic examples include calcitonin and calcitonin gene-related proteins (CGRPs)[36] as well as the pro-opiomelanocortin group of hormones (β-endorphin, β-lipotropin, α-melanocyte stimulating hormone (α-MSH) and adrenocorticotropic hormone (ACTH), etc.).[37–39] In these cases, a single gene can encode hormones of diverse functions following alternative splicing. Recent studies on the ESTs of Mcl-1, an antiapoptotic Bcl-2 family protein in the ovary, indicated the presence of a splicing variant as a result of skipping one exon. The protein encoded by this variant, named Mcl-1 short, has proapoptotic activity and antagonizes the survival action of the known Mcl-1 protein by forming non-covalent dimers.[40] This study demonstrated the importance of the alternative splicing mechanism in generating proteins with diametrically opposing actions under different physiological conditions. With the rapid increases in ESTs, and the availability of complete genome sequences, future characterization of gene splicing events and intron–exon arrangement will undoubtedly provide a new understanding of the complexity of the transcriptome of different tissues.

PROTEOMICS

Proteins are the main workhorse of the cellular machinery. Genomic advances set the stage to elucidate the function of all human proteins encoded by individual genes. In addition to analysis of the transcript expression patterns at the RNA level, studies of protein expression patterns based on protein arrays and mass spectrometry are important. Because proteins are further altered by proteolytic processing and post-translational modification, the qualitative nature of expressed proteins is a major focus of proteomics. The other challenges for proteomics include the elucidation of cellular pathways based on protein-interacting characteristics, and prediction of protein functions using protein structure analysis. The combination of genomics, transcriptomics and proteomics could allow eventual linking of genotypes to phenotypes.

Tissue-, development- and regulation-specific pattern of protein expression

As described for the analysis of mRNA levels for different genes, it is even more important to study the tissue-, time- and development-specific pattern of protein expression. Owing to the complicated nature of proteins, tools for proteome analysis are still in an early stage of development. Recently, a high-precision robot designed to manufacture DNA microarrays was used to spot proteins onto chemically derivitized glass slides.[41] The proteins that attached covalently to the slide surface retained their ability to interact specifically with other proteins, or with small molecules in solution. In the future, protein microarrays containing different binding proteins could be used to screen for ligands, receptors and intracellular signalling mediators, as well as for identifying the protein targets of small molecules. Protein chips based on antibodies to survey large numbers of expressed proteins will also become feasible.

Mass spectrometry of peptide fragments and integration with genomic data

Before protein arrays become a reality, studies of protein expression relied on recent advances in the analysis of peptide fragments based on mass spectrometry. Recent advances in two-dimensional gel electrophoresis, automated mass spectrometry and other protein separation tools have provided data on the sequences of peptide fragments under different physiological and pathological conditions.

Advances of the genomic projects facilitated the progression of the proteomic projects, because the origins of small fragments of the proteins could easily be traced back to the entire gene encoding the protein of interest through comparison of nucleotide and deduced amino acid sequences in the genome of a given species. Peptide mass mapping is the most commonly employed mass spectrometric approach for protein identification. The basis of the method is the matching of experimentally determined peptide masses with peptide masses calculated for each entry in a sequence database, based on the specificity of the enzyme used to generate the experimental data.

Modern mass spectrometers can provide sufficient information to allow unique recognition of protein fragments, as well as detection of secondary modifications such as phosphorylation and glycosylation. Mass spectrometry is based on the determination of the exact mass of molecules. To measure the mass of proteins, the test material must be charged (ionized) and desolvated (dried). The two most successful mechanisms for ionization of peptides and proteins are matrix-assisted laser desorption ionization (MALDI) and electrospray ionization (ESI). In MALDI, the analyte of interest is embedded in a matrix that is dried and then volatilized in a vacuum under ultraviolet laser irradiation. Typically, the mass analyzer coupled with MALDI is a time-of-flight (TOF) mass analyzer that simply measures the elapsed time from acceleration of the charged (ionized) molecules through a field-free drift region. For ESI, the analyte is sprayed from a fine needle at high voltage towards the inlet of the mass spectrometer. Typically, the spray is from a reversed-phase high-pressure liquid chromatography (HPLC) column that is similar to a microinjection needle. During this process, the droplets containing analyte are dried and gain charge (ionize). The ions formed during this process are directed into the mass analyzer.

Using the mass spectrometer approach, the mass of peptide fragments can be determined. Protein identification is accelerated by matching EST databases with peptide sequences. Recently, intact *Escherichia coli* ribosomes, containing a mixture of proteins, were analyzed using a mass spectrometer. Peaks in the mass spectra could be assigned to individual ribosomal proteins and to non-covalent complexes of up to five component proteins.[42] In the future, an improved high-throughput and large-scale approach to identify proteins will advance the studies of protein expression and the nature of proteins in complexes.[43]

Proteolytic processing of proteins

In addition to the splicing variants encoded by a single gene, many large proteins are cleaved by proteolysis into multiple functional units. These protease cutting events may remove segments from one or both ends of the pre-peptide, resulting in a shortened form of the protein, or they may cut the polypeptides into a number of different segments, each with a unique function. Many hormones have been shown to undergo extensive proteolytic processing in a tissue-specific manner. For example, inhibins, activins, GDF-9 and several other related TGF-β family members all undergo proteolytic cleavage to remove the pre-hormone sequence that is believed to be important for the folding of the hormones before secretion as the mature form with biological activity.[44] There are many convertase enzymes involved in the proteolytic processing of hormones in addition to other proteases essential for the cleavage of diverse proproteins. Future studies on the tissue-specific expression and function of different processing enzymes, together with advanced computational analysis, are needed to elucidate the generation of final protein products.

Post-translational protein modifications

In addition to the 20 amino acids encoded by the genetic code, this repertoire is increased dramatically by post-translational processing. Individual amino acids in the polypeptides are often modified by the attachment of new chemical groups. The most common modifications are glycosylation and phosphorylation. In addition, amidation, palmitoylation, acetylation, methylation, ubiquitination, prenylation, adenosine diphosphate (ADP)-ribosylation and nitrosylation occur. Many reproductive hormones, including FSH, LH, and human chorionic gonadotropin (hCG), are glycoproteins with sugar side-chains attached to the amino group of asparagine (N-linked) or to the hydroxyl group of serine or threonine (O-linked). Variations in the carbohydrate contents of gonadotropins affect their circulating half-life as well as biological activities. In contrast, the pre-protein of the decapeptide gonadotropin releasing hormone is followed by a Gly–Lys–Arg sequence, leading to enzymatic cleavage of the decapeptide from its precursor and amidation of the carboxy-terminal of this peptide to form the more stable mature hormone. For multiple intracellular proteins, their phosphorylation status is tightly regulated to allow rapid control of their activation and/or intracellular location.

Protein–protein interactions and cellular circuits

It is becoming clear that the majority of cellular functions are mediated through direct protein–protein interactions. In addition to the traditional approach of determining protein–protein interactions through co-precipitation, the yeast two-hybrid method has been used to investigate interactions of intracellular proteins. To study the genes involved in ovarian follicle atresia, diverse known and novel genes have been found to constitute a defined intracellular apoptosis pathway.[45] To apply this technology for the generation of a human protein linkage map, attempts have been made to construct two-hybrid cDNA libraries that cover the entire human genome. With a homologous recombination-mediated approach, modular human EST-derived yeast two-hybrid libraries can be made to allow the eventual elucidation of cellular circuits.[46]

Evolution of protein structures

All the information needed to specify a protein's three-dimensional structure is contained within its amino acid sequences. Major computational efforts have been made to predict the spontaneous self-assembly of protein molecules with a large degree of freedom into a unique, three-dimensional, biologically active structure. It is becoming clear that only a limited number of motifs have evolved to perform the multiple tasks usually carried out by complex proteins. A fingerprint is a group of conserved motifs used to characterize a protein family. Usually, the motifs do not overlap, but are separated along a sequence, although they may be contiguous in three-dimensional space. Fingerprints can encode protein folds and functionalities more powerfully, and with more flexibility, than can single motifs, their full potency deriving from the mutual context afforded by motif neighbors. The crystal structures of hundreds of proteins have been analyzed. Because protein orthologs retain their three-dimensional structure whereas paralogs in the same protein family are likely to have similar structures, the structure and function of diverse proteins can be predicted. The *ab initio* prediction of protein structures and protein-folding mechanisms is a major area of proteomics.[47]

BIOINFORMATIC INTEGRATION OF SEQUENCE- AND TEXT-BASED DATA

Despite the overabundance of DNA, mRNA and protein data, the development of modern computational tools allows their analysis using sophisticated algorithms. In addition, most of these data are interchangeable via web-based communication devices. The global analysis of genomic, transcriptomic and proteomic information is incomplete without integration with the biomedical knowledge already available in the literature derived from traditional approaches. A major challenge for biomedicine in the coming decade is the development of bioinformatic tools to integrate sequence- and text-based data.

In silico analysis of genes based on their sequences

To allow efficient comparison of DNA or protein sequences, a variety of paired sequence comparison programs using different scoring matrices have been developed for aligning individual sequences with a catalogued sequence database. The basic local alignment search tool (BLAST) used by the National Center for Biotechnology Information (NCBI; http://www.ncbi.nlm.nih.gov/) and other related programs have become the essential tools for gene sequence analysis and for the deduction of the functions of encoded proteins.[48–50] In the GenBank, sequences are subdivided into different divisions based on their origins (Figure 4). This includes non-redundant (NR) sequences representing individual genes, ESTs consisting of fragments of cDNA derived from mRNAs, high-throughput genome sequences (HTGSs) containing genomic sequences, and others. In addition to the Entrez web site, the Unigene site (http://www.ncbi.nlm.nih.gov/UniGene) lists all genomic information dealing with a single gene (clustering of NR and ESTs), whereas the SNP site (http://www.ncbi.nlm.nih.gov/SNP/index.html) lists polymorphisms of individual human genes.

In the Entrez web site of the NCBI, one can compare a given query sequence with diverse nucleotide and amino acid sequences of millions of genes in the GenBank. The BLAST uses computer algorithms to align different nucleotide and amino acid sequences from diverse species (Table 2). Based on this analysis, one can trace the evolutionary origin of different genes based on nucleotide sequences and the proteins they encode to reveal the family relationship of a subgroup of genes and their conserved domains. Furthermore, it is possible to identify splicing variants of a given gene through comparison of multiple EST sequences. In the future, SNPs found in different ethnic groups can also be used to predict an individual's disease susceptibility and drug responses.

Text-based bioinformatics

With the explosion of genomic information, one of the major tasks facing the scientist today is the integration of sequence data with the vast and growing

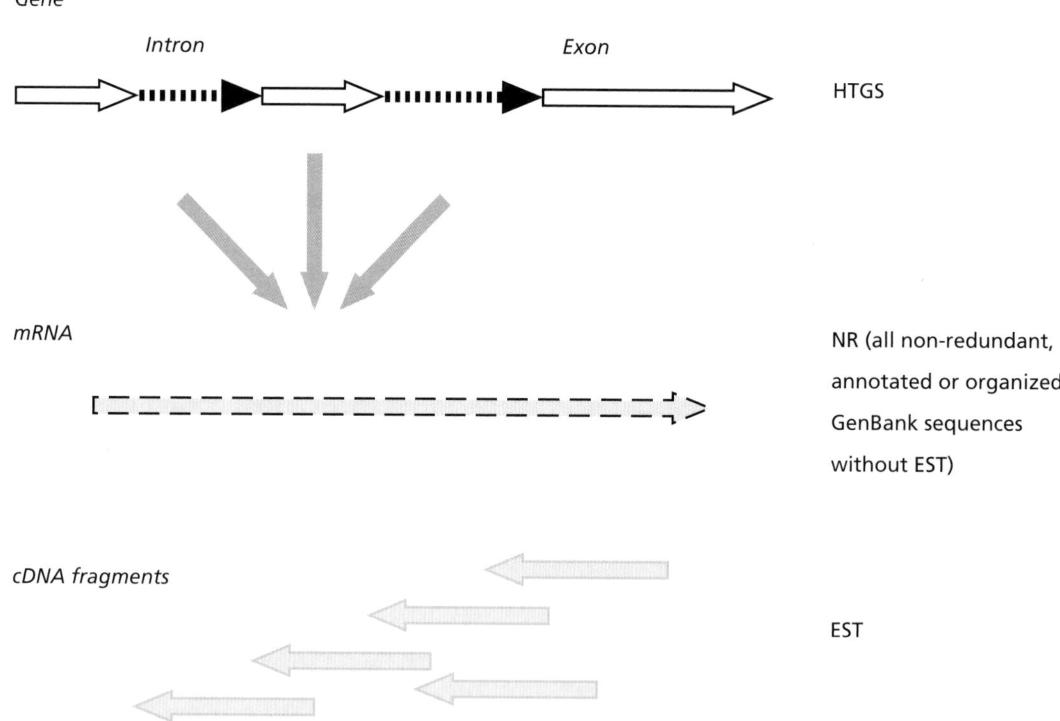

Figure 4 Origin of sequences in the GenBank. Nucleotide and amino acid sequences deposited in the GenBank are derived from the genomic sequences, full-length mRNA or fragments of cDNA. HTGS, high-throughput genomic sequencing; NR, non-redundant; EST, expressed sequence tag

Table 2 Basic local alignment search tool (BLAST). GenBank sequences are analyzed by using different nucleotide and amino acid sequences as queries to search for different databases

blastp	Amino acid query vs. protein sequence database
blastn	Nucleotide query vs. nucleotide sequence database
blastx	Nucleotide query translated in all reading frames vs. protein sequence database
tblastn	Protein query vs. nucleotide sequence database translated in all reading frames
tblastx	Six-frame translations of a nucleotide query vs. six-frame translations of a nucleotide sequence database

body of literature based on functional analyses of genes. The implementation of a user-friendly interface for easy query of diverse databases is necessary for either novice or experienced investigators, to sort out the expanding knowledge base of gene sequence and functions. In the popular PubMed site (http://www4.ncbi.nlm.nih.gov/entrez/query.fcgi), simple keyword-based searches allow the retrieval of published literature based mainly on original biomedical publications. In addition, the Online Mendelian Inheritance in Men (OMIM) database (http://www.ncbi.nlm.nih.gov/Omim/) provides a searchable database of human genes and genetic disorders. Each gene entry contains textual information and hypertext links to the GenBank database and PubMed, allowing quick reference checks for most known genes. Multiple online databases for specific gene groups or tissues have also been developed. For example, the Ovarian Kaleidoscope database (OKdb) (http://ovary.stanford.edu/) provides information regarding the biological function, expression pattern and regulation of genes expressed in the ovary.[51] It also serves as a gateway to other online information resources offering data about nucleotide and amino acid sequences, chromosomal localization and mutation phenotypes, and biomedical publications relevant to ovarian research. For all these databases, gene sequence and functional information are interlinked. Continuing development of databases and portal sites focusing on specific biomedical areas would allow easy access to updated and organized discipline-specific information and perhaps make the analysis of gene sequences a part of routine literature searches. Although only abstracts are available for most earlier publications, the new trend is to allow online access to full texts of original publications (for example, *Proceedings of the National Academy of Sciences of the United States of America* at http://www.pnas.org/).

THE NEW ENCYCLOPEDIA OF LIFE: THE POST-GENOMIC ERA OF REPRODUCTIVE MEDICINE

In the post-genomic era, scientists are shifting from the reductionistic approach to a reconstructionistic one. We are reading the new encyclopedia of life from a global rather than a gene-by-gene perspective. With the integration of existing biomedical knowledge with genomic data, new research will include knowledge of common motifs in all human proteins, transcriptional profiling of all genes, the identification of intracellular pathways and computer modelling of cell function, as well as the analysis of intercellular communications for the understanding of organ physiology and pathophysiology.

For reproductive medicine, comprehensive genome-based health care will become a reality. Diagnosis of diseases will include DNA array analyses of individual polymorphisms in patients, molecular phenotyping of diseased tissues, and protein array testing of multiple hormones in serum or intracellular proteins in specific organs. Prenatal diagnosis will also involve multiple genes or the entire genome instead of the present analysis of chromosome morphology and testing of selected genes.

The present treatment of reproductive diseases has already included the use of recombinant proteins such as FSH, but facilitated drug discovery in the post-genomic era will allow the development of many new drugs. DNA and protein arrays of diseased tissues will become tools to evaluate whether a particular gene expression profile can predict successful treatment outcomes. The ultimate development of safe gene therapy for both somatic and germ cells will change the concept of disease management. These advances will be accompanied by the widespread use of online patient files and remote diagnosis and surgery.

The advent of predictive genetic testing for a broad range of conditions will deal with susceptibility to diseases, for example, the prediction of ovarian cancer development based on heterozygosity mutations of BRCA and other tumor suppressor genes. A major emphasis will be on preventive medicine and the careful monitoring of the effect of environmental factors such as cigarette smoking, ultraviolet light exposure, vitamins, diet, etc. on reproductive tissues that could contribute to disease phenotypes. With the explosion of genomic information, society also has to face the ethical and biomedical implications of genetic interventions, and the corresponding effective legislation.

REFERENCES

1. Gilbert W. Towards a paradigm shift in biology. *Nature* 1991 349: 99
2. Lockhart DJ, Winzeler EA. Genomics, gene expression and DNA arrays. *Nature* 2000 405: 827–836.
3. Dong J, Albertini DF, Nishimori K, Kumar TR, Lu N, Matzuk MM. Growth differentiation factor-9 is required during early ovarian folliculogenesis. *Nature* 1996 383: 531–535.
4. Vitt UA, McGee EA, Hayashi M, Hsueh AJ. *In vivo* treatment with GDF-9 stimulates primordial and primary follicle progression and theca cell marker CYP17 in ovaries of immature rats. *Endocrinology* 2000 141: 3814–3820.
5. Otsuka F, Yamamoto S, Erickson GF, Shimasaki S. Bone morphogenetic protein-15 inhibits follicle-stimulating hormone (FSH) action by suppressing FSH receptor expression. *J Biol Chem* 2001 349: 2.
6. Galloway SM, McNatty KP, Cambridge LM *et al.* Mutations in an oocyte-derived growth factor gene (BMP15) cause increased ovulation rate and infertility in a dosage-sensitive manner. *Nat Genet* 2000 25: 279–283.
7. Bell AC, Felsenfeld G. Stopped at the border: boundaries and insulators. *Curr Opin Genet Dev* 1999 9: 191–198.
8. Wasserman WW, Palumbo M, Thompson W, Fickett JW, Lawrence CE. Human–mouse genome comparisons to locate regulatory sites. *Nat Genet* 2000 26: 225–228.
9. Meyer A, Schartl M. Gene and genome duplications in vertebrates: the one-to-four (-to-eight in fish) rule and the evolution of novel gene functions. *Curr Opin Cell Biol* 1999 11: 699–704.
10. Pebusque MJ, Coulier F, Birnbaum D, Pontarotti P. Ancient large-scale genome duplications: phylogenetic and linkage analyses shed light on chordate genome evolution. *Mol Biol Evol* 1998 15: 1145–1159.
11. Hsu SY. Cloning of two novel mammalian paralogs of relaxin/insulin family proteins and their expression in testis and kidney. *Mol Endocrinol* 1999 13: 2163–2174.
12. Hsu SY, Kudo M, Chen T, Nakabayashi K *et al.* The three subfamilies of leucine-rich repeat-containing G protein-coupled receptors (LGR): identification of LGR6 and LGR7 and the signaling mechanism for LGR7. *Mol Endocrinol* 2000 4: 1257–1271.
13. Hsu SY, Liang SG, Hsueh AJ. Characterization of two LGR genes homologous to gonadotropin and thyrotropin receptors with extracellular leucine-rich repeats and a G protein-coupled, seven-transmembrane region. *Mol Endocrinol* 1998 12: 1830–1845.
14. Stringer CB, Andrews P. Genetic and fossil evidence for the origin of modern humans. *Science* 1988 239: 1263–1268.
15. Wang DG, Fan JB, Siao CJ *et al.* Large-scale identification, mapping, and genotyping of single-nucleotide polymorphisms in the human genome. *Science* 1998 280: 1077–1082.
16. Smigielski EM, Sirotkin K, Ward M, Sherry ST. dbSNP: a database of single nucleotide polymorphisms. *Nucleic Acids Res* 2000 28: 352–355.
17. Evans WE, Relling MV. Pharmacogenomics: translating functional genomics into rational therapeutics. *Science* 1999 286: 487–491.
18. Laue L, Chan WY, Hsueh AJ *et al.* Genetic heterogeneity of constitutively activating mutations of the human luteinizing hormone receptor in familial male-limited precocious puberty. *Proc Natl Acad Sci USA* 1995 92: 1906–1910.
19. Laue LL, Wu SM, Kudo M *et al.* Compound heterozygous mutations of the luteinizing hormone receptor gene in Leydig cell hypoplasia. *Mol Endocrinol* 1996 10: 987–997.

20. Crisponi L, Deiana M, Loi A *et al.* The putative forkhead transcription factor FOXL2 is mutated in blepharophimosis/ptosis/epicanthus inversus syndrome. *Nat Genet* 2001 27: 159–166.
21. Michael NL, Louie LG, Rohrbaugh AL *et al.* The role of CCR5 and CCR2 polymorphisms in HIV-1 transmission and disease progression. *Nat Med* 1997 3: 1160–1162.
22. Liu H, Chao D, Nakayama EE *et al.* Polymorphism in RANTES chemokine promoter affects HIV-1 disease progression. *Proc Natl Acad Sci USA* 1999 96: 4581–4585.
23. Buscher R, Herrmann V, Insel PA. Human adrenoceptor polymorphisms: evolving recognition of clinical importance. *Trends Pharmacol Sci* 1999 20: 94–99.
24. Ingelman-Sundberg M, Oscarson M, McLellan RA. Polymorphic human cytochrome P450 enzymes: an opportunity for individualized drug treatment. *Trends Pharmacol Sci* 1999 20: 342–349.
25. Perez Mayorga M, Gromoll J, Behre HM, Gassner C, Nieschlag E, Simoni M. Ovarian response to follicle-stimulating hormone (FSH) stimulation depends on the FSH receptor genotype. *J Clin Endocrinol Metab* 2000 85: 3365–3369.
26. Huhtaniemi I, Jiang M, Nilsson C, Pettersson K. Mutations and polymorphisms in gonadotropin genes. *Mol Cell Endocrinol* 1999 151: 89–94.
27. Ahrendt SA, Halachmi S, Chow JT *et al.* Rapid p53 sequence analysis in primary lung cancer using an oligonucleotide probe array. *Proc Natl Acad Sci USA* 1999 96: 7382–7387.
28. Walker MG, Volkmuth W, Sprinzak E, Hodgson D, Klingler T. Prediction of gene function by genome-scale expression analysis: prostate cancer-associated genes. *Genome Res* 1999 9: 1198–1203.
29. Velculescu VE, Zhang L, Vogelstein B, Kinzler KW. Serial analysis of gene expression. *Science* 1995 70: 484–487.
30. Neilson L, Andalibi A, Kang D *et al.* Molecular phenotype of the human oocyte by PCR-SAGE. *Genomics* 2000 63: 13–24.
31. Brown PO, Botstein D. Exploring the new world of the genome with DNA microarrays. *Nat Genet* 1999 21: 33–37.
32. Tanaka TS, Jaradat SA, Lim MK *et al.* Genome-wide expression profiling of mid-gestation placenta and embryo using a 15 000 mouse developmental cDNA microarray. *Proc Natl Acad Sci USA* 2000 97: 9127–9132.
33. Osin P, Shipley J, Lu YJ, Crook T, Gusterson BA. Experimental pathology and breast cancer genetics: new technologies. *Recent Res Cancer Res* 1998 152: 35–48.
34. Mironov AA, Fickett JW, Gelfand MS. Frequent alternative splicing of human genes. *Genome Res* 1999 9: 1288–1293.
35. Black DL. Protein diversity from alternative splicing: a challenge for bioinformatics and post-genome biology. *Cell* 2000 103: 367–370.
36. Amara SG, Jonas V, Rosenfeld MG, Ong ES, Evans RM. Alternative RNA processing in calcitonin gene expression generates mRNAs encoding different polypeptide products. *Nature* 1982 298: 240–244.
37. Herbert E, Birnberg N, Civelli O, Lissitzky JC, Uhler M, Durrin L. Regulation of genetic expression of pro-opiomelanocortin in pituitary and extrapituitary tissues of mouse and rat. *Adv Biochem Psychopharmacol* 1982 33: 9–18.
38. Herbert E, Uhler M. Biosynthesis of polyprotein precursors to regulatory peptides. *Cell* 1982 30: 1–2.
39. Chretien M, Seidah NG. Chemistry and biosynthesis of pro-opiomelanocortin. ACTH, MSHs, endorphins and their related peptides. *Mol Cell Biochem* 1981 34: 101–127.
40. Bae J, Leo CP, Hsu SY, Hsueh AJ. MCL-1S, a splicing variant of the antiapoptotic BCL-2 family member MCL-1, encodes a proapoptotic protein possessing only the BH3 domain. *J Biol Chem* 2000 275: 25255–25261.
41. MacBeath G, Schreiber SL. Printing proteins as microarrays for high-throughput function determination. *Science* 2000 289: 1760–1763.
42. Benjamin DR, Robinson CV, Hendrick JP, Hartl FU, Dobson CM. Mass spectrometry of ribosomes and ribosomal subunits. *Proc Natl Acad Sci USA* 1998 95: 7391–7395.
43. Yates JR. Mass spectrometry. From genomics to proteomics. *Trends Genet* 2000 16: 5–8.
44. Massague J, Cheifetz S, Laiho M, Ralph DA, Weis FM, Zentella A. Transforming growth factor-beta. *Cancer Surv* 1992 12: 81–103.
45. Hsu SY, Hsueh AJ. Tissue-specific Bcl-2 protein partners in apoptosis: an ovarian paradigm. *Physiol Rev* 2000 80: 593–614.

46. Hua SB, Luo Y, Qiu M, Chan E, Zhou H, Zhu L. Construction of a modular yeast two-hybrid cDNA library from human EST clones for the human genome protein linkage map. *Gene* 1998 215: 143–152.
47. Baker D. A surprising simplicity to protein folding. *Nature* 2000 405: 39–42.
48. Altschul SF, Gish W, Miller W, Myers EW, Lipman DJ. Basic local alignment search tool. *J Mol Biol* 1990 215: 403–410.
49. Altschul SF, Lipman DJ. Protein database searches for multiple alignments. *Proc Natl Acad Sci USA* 1990 87: 5509–5513.
50. Pearson WR, Lipman DJ. Improved tools for biological sequence comparison. *Proc Natl Acad Sci USA* 1988 85: 2444–2448.
51. Leo CP, Vitt UA, Hsueh AJ. The Ovarian Kaleidoscope database: an online resource for the ovarian research community. *Endocrinology* 2000 141: 3052–3054.

Chapter 3

Molecular genetics

Margaret A. Shupnik

INTRODUCTION

Remarkable advances in DNA diagnostics have been expedited by the development of the polymerase chain reaction (PCR), the ability to isolate DNA from many different sources, and effective screening techniques to scan genomic DNA for point mutations. The first physical map of the human genome based upon 15 000 sequence tagged sites (STSs) distributed over all the human chromosomes was completed in 1995,[1] and the entire human genome sequence will soon be available. The continued development of technology for mapping genes and mutations will ultimately result in automated DNA diagnosis for the practicing clinician. This chapter is designed to acquaint the reader with current techniques of analyzing genes and certain prototype mutations in the reproductive sciences.

TECHNIQUES OF DNA ANALYSIS

Polymerase chain reaction

Almost all of the diagnostic techniques discussed in this chapter use genomic DNA isolated from nucleated blood cells as the starting material. Genomic DNA isolation is followed by amplification of specific nucleic acid segments by PCR.[2] Genes relevant to the reproductive sciences are usually present as single-copy DNA, for which the sensitivity of PCR is critical. This approach requires sufficient knowledge of the gene to design appropriate oligonucleotide (17–30 bp) primers, each complementary to one of the original DNA strands, to either the left (5') or the right (3') side of the region defined for amplification. The use of heat-stable *Taq* DNA polymerase permits automated rounds of primer annealing, DNA synthesis and DNA in which newly synthesized DNA strands become templates for subsequent rounds of amplification. After 30 cycles (4–6 h), a single copy of DNA is amplified to one million copies. A striking example of the power of the technique was provided by investigators who, starting with a single blastomere from each of 25 human embryos, amplified a 149-bp segment of DNA from the Yq11–12 chromosome region and could correctly sex 15 of the embryos.[3]

The ability to amplify rapidly a single region of a gene in a diploid cell has opened the way for preimplantation embryonic diagnosis, and for any disease for which the molecular defect is known at the molecular level. PCR generates sufficient product to analyze restriction fragment length polymorphism (RFLPs), or to detect point mutations with allele-specific oligonucleotide (ASO) probes or heteroduplex analysis.[4] Final detection assays usually rely on hybridization of the amplified product to specific synthetic DNA probes representing a portion of the amplified sequences. PCR offers the advantage of increased signal intensity for gene analysis with any subsequent detection system, and sufficient quantities of DNA are made for direct nucleotide sequencing. An RNA sequence can be similarly amplified, even from a single cell, but a DNA copy of it (cDNA) must be initially synthesized by using reverse transcriptase before the PCR is begun.[5] Given the reported sensitivities of PCR (detection of less than ten copies of a genetic sequence), contamination can be catastrophic and must be contained for clinical diagnostic platforms by requiring minimal pipetting steps with barrier pipette tips to prevent transfer of aerosol.

Multiplex genomic analysis

Innovative PCR modifications include the ability to amplify and screen multiple segments of a gene for deleted regions, a technique known as multiplex genomic DNA amplification. This permits the detection of deletions over megabase regions in a gene by PCR. Multiple primers homologous to sequences flanking a number of exons are used to amplify five or six different segments of the region with PCR. Lack of signal for any of these segments flanked by the primers would indicate a deletion, and is especially helpful for X- or Y-chromosome-localized genes in males, as they are hemizygous, i.e. there is only one copy of the gene to analyze.[6] A homozygous identical mutation on both alleles of genes on other chromosomes will also result in a single amplified product or simple pattern for each gene region, while a heterozygous mutation on only one chromosome allele will result in two amplified products for the given gene region, or a decreased signal with one set of primers, indicating a deletion. In this case, additional amplification products might be required to confirm such a deletion.

Primer-extension preamplification

One of the important by-products of PCR technology, having important implications for preimplantation diagnosis, is primer-extension preamplification (PEP). PEP was developed to duplicate the entire genome from limited amounts of DNA, such as that derived from a single haploid cell, and involves repeated primer extensions using a mixture of 15-base random oligonucleotides to ensure amplification of segments throughout the genome. Through PEP, it is possible to perform multiple genotyping experiments on DNA from a single sperm, oocyte or blastomere cell. PEP is a valuable adjunct to single-cell diagnosis,[7] and affords a simple and rapid means to increase signal/noise ratios in many types of DNA analysis.

Analysis of extracted and amplified DNA

Southern blot

One of the earliest and most basic methods to characterize genomic DNA is Southern blotting, which examines the restriction enzyme digest pattern of a gene or chromosome region by using specific DNA probes. Because restriction enzymes cleave DNA at defined sequence sites, the digestion pattern of a given gene into discretely sized DNA fragments is predictable. A few micrograms of digested DNA are then separated by size on an agarose gel, with the smaller fragments migrating towards the bottom of the gel (Figure 1). Fragments of 500–30 000 bp are readily resolved, and pulse-field or field-inversion gel electrophoresis can extend

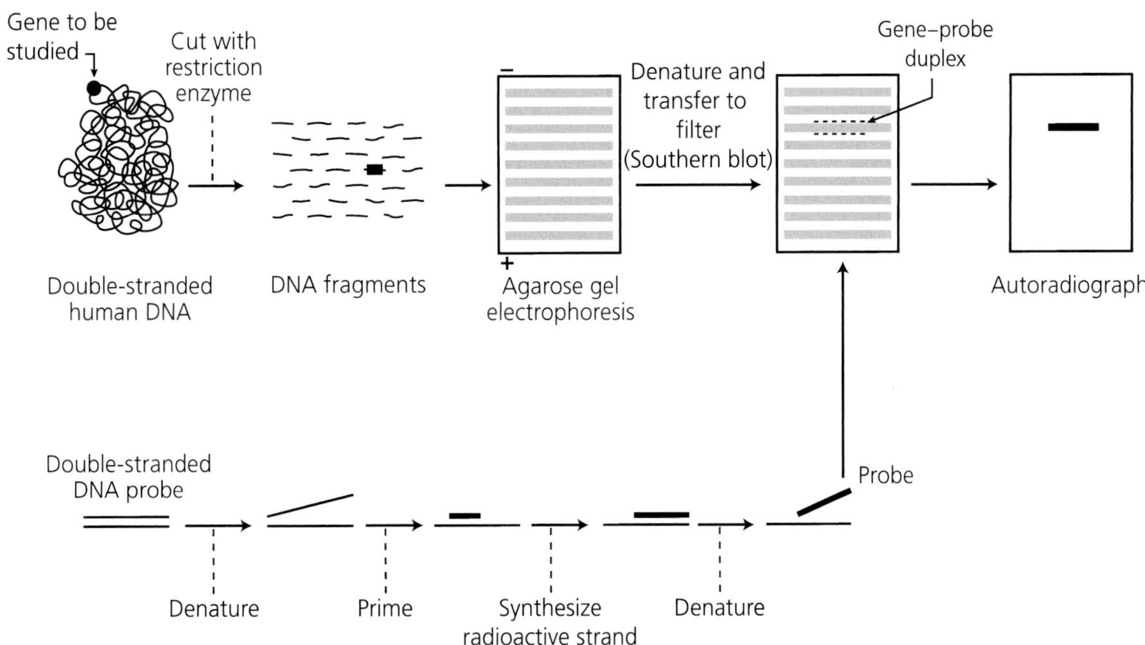

Figure 1 Technique of Southern blotting. Extracted DNA is exposed to an appropriate restriction enzyme and the resultant fragments are subjected to electrophoresis on an agarose gel. The denatured DNA fragments are transferred and permanently fixed to a nitrocellulose filter. The filter is placed in a solution of an appropriate probe of labelled DNA or RNA for hybridization to occur. Hybridization with the radioactively labelled probe is detected by autoradiography. Reprinted with permission from reference 8

the upper limits of resolution to several megabases or fragments greater than 50 000 bp. After denaturation with alkali to allow for hybridization, fragments are transferred to a nitrocellulose or nylon membrane without their pattern being disturbed. This membrane is called a Southern blot after its inventor,[9] and similar techniques can be used to transfer RNA (Northern blotting), or proteins (Western blotting). Specific DNA sequences are then analyzed by hybridization to defined labelled probes. The probe for the gene of interest will hybridize to its complementary strand immobilized on the membrane. After the membrane is washed and exposed to film, the autoradiograph will reveal precise fragment(s) of DNA corresponding to the probe being utilized. If DNA being analyzed has been mutated, there can be gain or loss of a restriction enzyme site, and resulting changes in DNA fragment size are diagnostic of a change in the nucleotide sequence of the gene or its flanking regions. The technique is limited, however, in detecting only mutations that change the size of restriction fragments.

Allele-specific oligonucleotide probes

In most cases, gene mutations do not alter restriction enzyme sites, and it becomes necessary to have alternative approaches for detection of DNA mutations. As more genes are sequenced and mutations recognized, direct approaches to diagnosis can be utilized. Allele-specific oligonucleotide (ASO) probing depends on the development of two DNA probes, one corresponding to the normal gene and one complementary to the known mutant sequence. In practice, two probes of 18–30 bases each are used, with the known mutation typically in the middle of the mutant probe. Because optimal hybridization temperature is dependent on the number of G–C pairs and the number of continuous bases with complete identity, this placement of the mutated base provides maximum disruption of homology with the wild type. Under stringent hybridization conditions of high temperature and low salt, only probes with complete identity will bind. An individual with two normal genes will show hybridization only to the wild-type probe, while DNA from a heterozygous carrier will hybridize to both the normal and the mutant probes. Genes that are known to have many different mutations, or that may have uncharacterized mutations, are not amenable to this approach. For reverse ASO screening, wild-type and mutant DNAs are spotted to filters and labelled amplified DNA is used as a probe. ASO screening is suitable for screening many patient samples for only one or a few possible missense mutations. Reverse ASO is better suited to screening one patient at a time for many different possible missense base-pair substitutions. ASO screening can be valuable in reproductive medicine to screen closely related genes. For example, there are six genes for human chorionic gonadotropin-β (hCGβ) and one for luteinizing hormone-β (LHβ). This poses a problem at the diagnostic level, but they can be distinguished by hybridization to defined oligonucleotides that are completely specific for each gene.[10,11] The final sequencing of the human genome will facilitate the continued development of synthetic oligonucleotides to probe for gene mutations.

Dot blot (slot blot)

The use of standard Southern blotting as described for gene analysis is specific, but requires restriction digestion of the DNA sample and electrophoresis before hybridization. A more rapid method is called 'slot' or 'dot' blotting of DNA, in which DNA or RNA, denatured in high salt, is spotted directly onto nitrocellulose or nylon membranes, baked in a vacuum apparatus and hybridized to labelled probes. The signal may be non-specific because appropriate band size and background hybridization is not determined, but, once optimal conditions of probe design and hybridization are established, the technique is rapid and may be used to screen multiple samples.

Fluorescent in situ *hybridization of chromosomes*

In situ hybridization methods for DNA and RNA analysis can be performed on chromosomal preparations, frozen sections, cytological specimens or formalin-fixed paraffin-embedded sections. Its utility results from the ability to localize relevant nucleic acid sequences within larger structures, thus linking biochemistry with cytogenetics or histochemistry. At the DNA level, genes can be localized to specific chromosomes by radioactive (autoradiography) or fluorochrome-labelled probes (fluorescent *in situ* hybridization, FISH), with FISH analysis currently more common. Probes are placed directly on tissue or cells treated with a protease to allow probe access to the nuclear DNA. The probe is then detected with avidin–biotin enzyme reactions, fluorescence or autoradiography. FISH analysis is particularly powerful in localizing genes and alleles to specific chromosomes, and detecting events such as chromosomal mutations and rearrangements. For example, FISH has been used successfully to track X-chromosome inactivation and to track parental origin of X chromosomes with specific mutations.[12,13]

Restriction fragment length polymorphisms

The earliest technique to follow the transmission of genetic pathology was the use of genomic DNA restriction fragment length polymorphisms (RFLPs). The first step in this type of indirect gene analysis depends on the identification of a DNA fragment that is so closely linked to the gene of interest that they are invariably transmitted together during meiosis. The tight linkage makes the segregating fragment a clear signpost for the presence of the gene itself, and could only be disrupted by a cross-over between chromosomal homolog during meiosis that might transfer the marker or RFLP. In general, the smaller the physical distance between the gene of interest and RFLP, the less is the likelihood that cross-over or recombination might occur. In order to use this DNA marker to follow the gene, it is necessary that the restriction fragment generated is polymorphic, or at least dimorphic, in the family under consideration. This variation in the structure of DNA is seen particularly in regions that do not encode proteins or are not involved in important regulatory functions. These variations between sequences at the same loci on two homologous chromosomes can be detected by different restriction digest patterns and are referred to as RFLPs.

Figure 2 shows a hypothetical example of the use of this technique for disease diagnosis. The family has three daughters. The eldest is a 22-year-old diagnosed with Kallmann's syndrome due to an absence of gonadotropin releasing hormone (GnRH). The two younger children are 5 and 7 years old and the parents would like to know whether either or both of their younger girls will have Kallman's syndrome.

The gene for GnRH has been mapped to the short arm of the number 8 chromosome in humans.[14,15] All family members have identical Southern blots. Because the entire nucleotide sequence encoding the normal human GnRH gene and its flanking sequences has been determined, the synthesis of an ASO probe for the normal gene is possible. However, the different reported types and numbers of mutations affecting this gene hinder the development of a single ASO mutant probe. This lack of knowledge concerning the molecular pathology of the gene under study limits the first two approaches for analysis. However, RFLP analysis does not require any presumptive information concerning gene sequence or pathology. Diagnosis, instead, is based upon a neighboring DNA marker that can be used to follow the inheritance of a putative gene (GnRH) as it segregates through a family. The gene for plasminogen tissue activator (PLAT or tPA) is closely linked to the GnRH gene on chromosome 8.[16] Non-coding sequences in (introns) and around (flanking) the gene for PLAT are polymorphic, i.e. have differences in nucleotide sequences that are transmitted in Mendelian fashion, and can be used to distinguish individual genes within a family. These DNA polymorphisms or markers are called alleles. In the family under study, the DNAs of both parents are cut with a restriction enzyme and probed with the linked *PLAT* gene. On a Southern blot, DNA from each parent has two bands indicating two polymorphisms or alleles, one at 10 kb and a lower molecular weight 5-kb band; they are heterozygotes for this gene. In order to be homozygous for GnRH deficiency, the affected child must have received one copy of the mutant gene from each parent, but the identity of the 'mutant' band requires further analysis. This linkage or 'phase' can be determined by analyzing the DNA of an

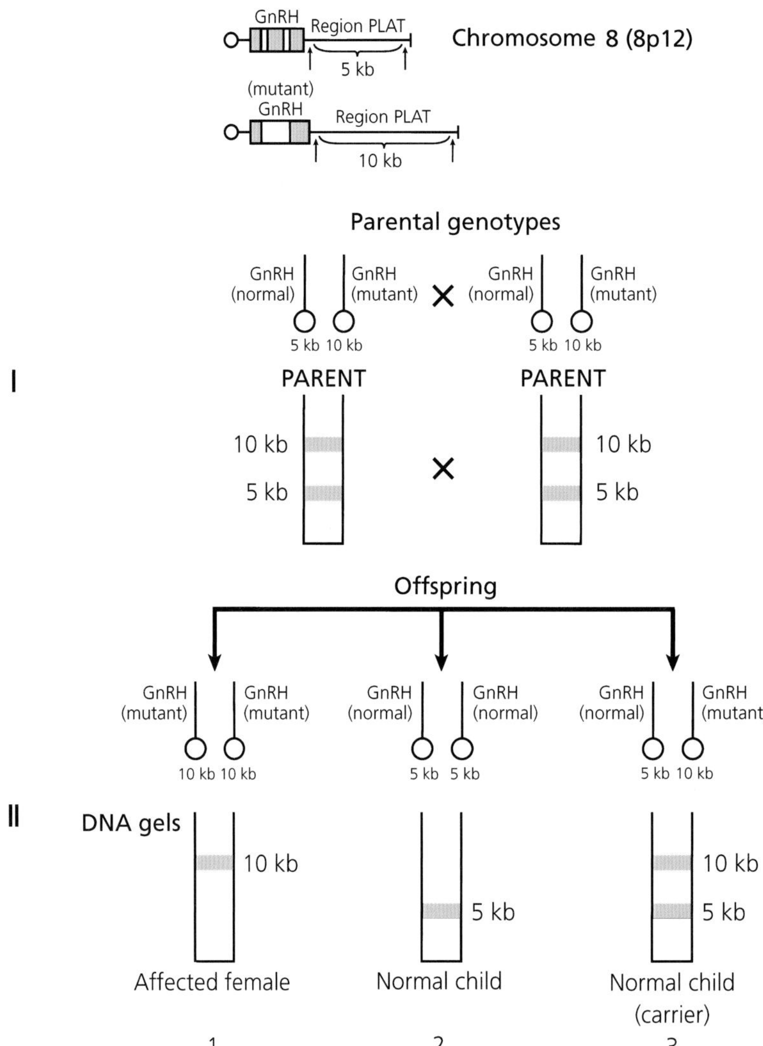

Figure 2 Hypothetical illustration of the use of a restriction fragment length polymorphism to follow the segregation of a mutant allele, showing the pedigree of a family with three female offspring. The eldest daughter (II-1) has hypogonadotropic hypogonadism due to a deficiency of gonadotropic releasing hormone (GnRH). The parents wish to know whether their two prepubertal daughters (II-2 and II-3) will be similarly affected. The gene for GnRH is mapped to 8p, but is not polymorphic. A closely linked gene for plasminogen activator tissue (*PLAT*) is dimorphic, demonstrating two different alleles, one 10 kb and the other 5 kb. Extracted DNAs from the parents probed with *PLAT* demonstrate that they are heterozygous for these two different alleles. The affected daughter (II-1) is homozygous for the 10-kb band, indicating that the mutant GnRH gene is segregating in this family with this *PLAT* allele. One younger daughter (II-2) is homozygous for the 5-kb band (normal) and the other child (II-3) is a heterozygous carrier, demonstrating both bands, similar to the parents. Reprinted with permission from reference 8

affected individual in the family. Southern blotting analysis of the DNA of the affected adult female using the tPA probe reveals only the 10-kb band, and indicates that the mutant phenotype is segregating with the 10-kb band contributed by both parents. This can be inferred without any knowledge of the precise mutation affecting the GnRH gene or even regulatory regions of the gene. The diagnosis and phase are further corroborated by similar analysis of the DNAs of the other two children. One child is homozygous for the 5-kb allele whereas the other has received a 10-kb allele from one parent and a 5-kb allele from the other parent. The former child is normal phenotypically and genotypically, whereas the latter is phenotypically normal but a carrier like her parents. Thus, the use of RFLPs for diagnosis requires the DNA of at least one affected individual in the pedigree.

By following an RFLP that co-segregates with a mutant phenotype, it should be possible to recognize recessive carriers, the non-penetrant dominant carrier and delayed-onset mutant phenotypes such as Kallmann's syndrome, which generally do not express themselves until adolescence. Genetic linkage analysis has relied heavily on this type of information, but genetic dissection of complex traits can be challenging.[17–20] The complete sequencing of the human genome provides the raw data to define all expressed genes, but functional genetic linkage studies will continue to play a critical role in determining the importance of these genes in human disease.

Microsatellite regions

The overall advantage of the technique of RFLP is the likelihood that a gene can be linked with one of its flanking DNA polymorphisms. In addition to RFLPs, useful genetic markers include microsatellites or simple sequence repeats (SSRPs), and single nucleotide polymorphisms (SNPs). As well as direct sequence polymorphisms, variability in the lengths of repetitive DNA sequences that are flanked by a restriction enzyme – variable number of tandem repeat (VNTR) sequences or minisatellites – can also be viewed as a RFLP. VNTR or minisatellite repeats have a core repeating unit of 10–100 bp, and each unit may contain slight variations in length and base sequence.

Microsatellites, or SSRPs, are composed of mono-, di-, tri- and tetranucleotide repeats in the genome that are random in their distribution, occur about once per 1000 bp and can be scored by PCR. They have rapidly expanded the numbers of DNA markers available, and the ultimate goal is to map the human genome with overlapping markers that are no greater distance apart than 1 centiMorgan: the distance between a gene and another marker gene that results in recombination in 1% of meioses. SSRPs can easily be analyzed in single cells (e.g. sperm, polar body, blastomere), and have been used effectively to map several autosomal dominant endocrine cancer syndromes including multiple endocrine neoplasia type IIA and type IIB, and familial medullary thyroid carcinoma-1.[21,22] In the past, predictive testing for the inheritance of mutant alleles in individuals at risk for these disorders was limited by the availability of highly informative and closely linked markers, which has now been provided by this technology (Figure 3). At a practical level, development of these high-density microsatellite-based maps will help in the mapping of genes causing less common genetic disorders, allowing for prenatal diagnosis and eventual gene cloning. A high-resolution map of this sort is also required for identifying the genetic components of complex common diseases such as heart disease or behavioral disorders.

Single nucleotide polymorphisms (SNPs) are positions in the human genome at which two alternative bases occur at an appreciable frequency (> 1%) in the population. They therefore include the base variations defined as RFLPs but also include much larger numbers of changes that are detectable by DNA sequencing, but do not fortuitously belong to a restriction enzyme cleavage site. SNPs are the most prevalent variation in the human genome, constituting approximately 80% of all known polymorphisms, and are estimated to occur once every 1000 bp. Because they are biallelic, SNPs are less informative than microsatellites. However, because of their frequency and possible scoring by automated methods, they may become powerful tools in genetic studies.

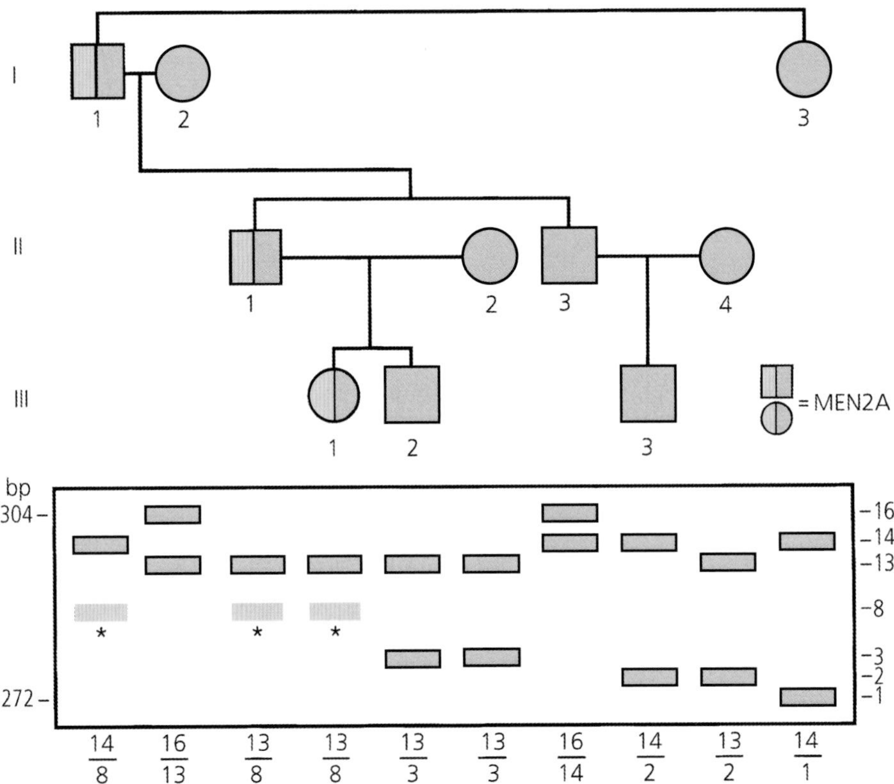

Figure 3 Illustration of the use of microsatellite DNA markers or simple sequence repeat polymorphisms (SSRPs) to follow the segregation of the disease multiple endocrinopathy type IIA (MEN2A). The gene responsible for MEN2A was mapped to the proximal part of the long arm of chromosome 10 (10q). A locus on 10q close to the centromere was found to contain 16 different microsatellite repeats. These dinucleotide repeats ranged in size from 272 to 304 bp. A single individual at a given genetic locus carries only two variations of these polymorphic simple sequence repeats. This figure illustrates a family with MEN2A in which three family members are affected (I-1, II-1, III-1). For simplicity, the different alleles at this locus are numbered numerically from 1 to 16, corresponding to the size of the alleles (272–304 bp). It is clear that the three individuals within the family who have MEN2A and medullary carcinoma of the thyroid all have the allele numbered 8. More recently, the specific gene causing this disorder has been isolated by positional cloning. The gene called *ret* is a proto-oncogene, encodes a tyrosine kinase-type receptor and maps close to this polymorphic locus on 10q. Current studies rely on the direct detection of activating mutations in the *ret* proto-oncogene. The principal value of microsatellite loci is the high degree of polymorphisms associated with these simple sequence repeats (tri-, di- or tetranucleotides). Modified with permission from reference 22

Techniques to screen for mutations in genomic DNA

Southern blotting permitted detection of gross gene deletions, insertions and rearrangements, but small alterations could be detected only if they were in restriction sites. Identification of previously characterized mutations was possible by selective oligonucleotide hybridization to genomic DNA. During the past decade, rapid and effective methods to identify unknown small deletion or point mutations over thousands of DNA bases were established, and can be categorized into either scanning methods to determine whether a mutation occurs, or diagnostic methods once a mutation has been categorized.[23] Most methods use genomic material amplified by PCR over the entire coding and non-coding regions. The ultimate proof that a mutation disrupts function can come only from *in vitro* expression studies, but inspection of the mutation (premature stop codon, mutation of a phosphorylation site) may give a clear indication of its likely effect.

Single-stranded conformation polymorphisms

The single-stranded conformation polymorphism (SSCP) method is the simplest and most widely used technique to detect sequence differences between DNA fragments, and is based on the observation that, owing to secondary structure, the migration of short, single-stranded fragments in non-denaturing gels is a function of their length and sequence. A single base change can result in altered structure and changed mobility in the gel. A comparison of denatured, amplified genomic DNA to be tested with 'wild-type' DNA can identify even single base variations in sequence, which can be confirmed directly (Figure 4). The shifts in mobility are decreasingly apparent in fragments of increasing size, and the technique favors the scanning of PCR products that are 100–250 bp in length. In Figure 5, the PCR product of the conserved region of the mammalian sex-related Y (*SRY*) gene has been denatured and electrophoresed on a non-denaturing gel to compare with DNA from seven subjects with 46,XY gonadal dysgenesis (Swyer's syndrome). The sample in lane 3 migrates differently, and sequencing revealed a missense mutation in the conserved region of the *SRY* gene. PCR–SSCP is well suited to identifying the presence, but not the precise identity, of new mutations, and has obvious value in the genetic analysis of families in which members carry a mutation that has not been previously characterized.

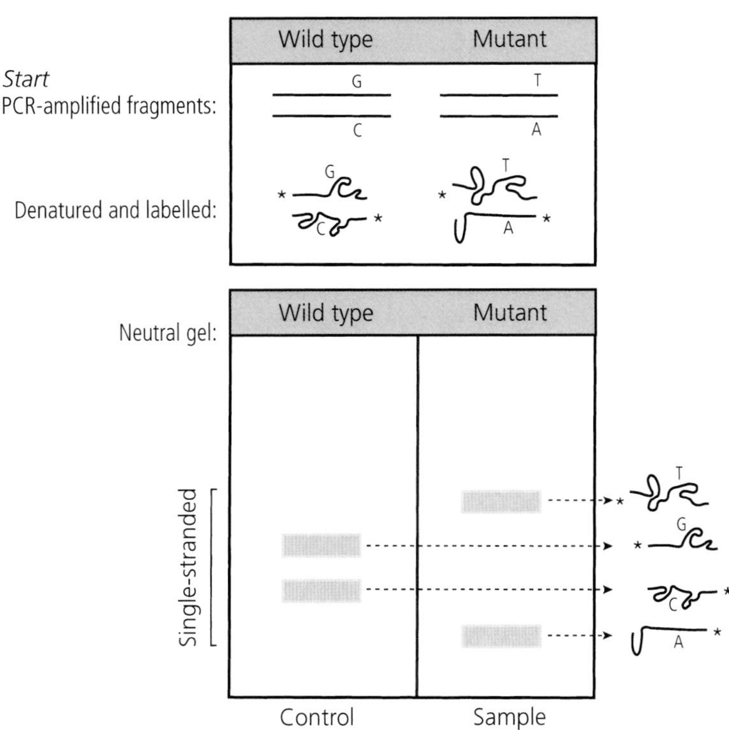

Figure 4 Illustration of the technique of detecting previously unidentified mutations using the technique of single-stranded conformation polymorphism (SSCP). The gene of interest in the control (wild type) and sample (mutant) DNA is first amplified by the polymerase chain reaction (PCR). The PCR products are denatured in order to convert them into single-stranded DNA. The single-stranded DNAs are run on a non-denaturing gel. The single strands will form conformations consistent with their specific base-pair sequence. The assumption of these specific 'snap back' configurations will alter the migration of the mutant strand in the gel in comparison with the control. Samples that have migration patterns different from those of the control are then sequenced to identify specific mutations. Modified with permission from reference 23

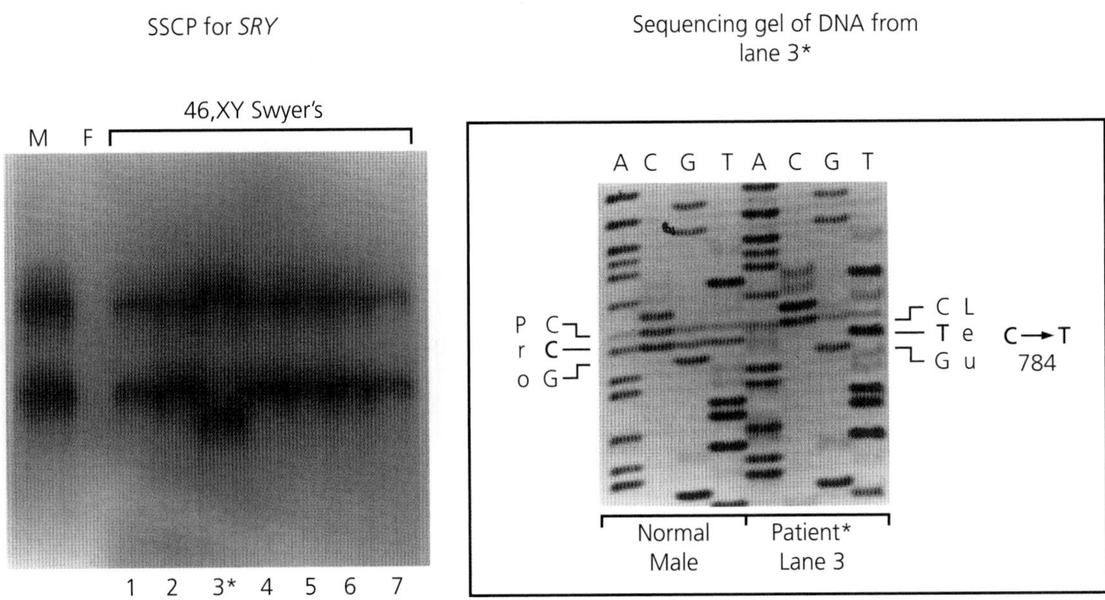

Figure 5 Use of single-stranded conformation polymorphisms (SSCPs) to identify previously unidentified mutations in the sex-related Y gene (*SRY*). The *SRY* gene in the DNAs of controls (male (M) and female (F)) and subjects with Swyer's syndrome ($n = 7$) was amplified by the polymerase chain reaction and use of specific primers for the conserved region of *SRY*. The amplified products were denatured and run on a non-denaturing gel. The DNA of one of the 46,XY sex-reversed females* (Swyer's syndrome; lane 3) has a distinctive migration pattern compared with that of the control male. Sequencing of the DNA from the subject in lane 3 revealed a single missense mutation (C→T), changing a proline to a leucine and inactivating the *SRY* gene

Heteroduplex method

This technique is virtually identical to SSCP in concept, but separates double-stranded heteroduplex molecules on non-denaturing gels. Particular mismatches and their location in the double-stranded fragment distort the usual shape and alter the mobility of the fragment, and a heteroduplex with a single mismatch will have a different mobility from the corresponding duplex without the mismatch. Heteroduplex analysis probably detects 80% of the mutations and is very easy to perform. It has been suggested that a combination of SSCP and heteroduplex analysis might detect close to 100% of mutations, and the same PCR-amplified fragment could be prepared for electrophoresis under both denaturing and non-denaturing conditions.

Denaturing gradient gel electrophoresis

Denaturing gradient gel electrophoresis (DGGE) is based upon the electrophoretic mobility of a double-stranded DNA molecule through linearly increasing concentrations of a denaturing agent such as formamide and urea, or increasing temperature gradients. A DNA duplex with a mismatched pair denatures more rapidly than a perfectly matched duplex, and will result in decreased gel migration and a higher apparent molecular weight. The mismatched duplex can then be eluted and sequenced to characterize the specific mutation. The PCR–DGGE technique is extremely powerful when applied to the detection of heterozygous nucleotide variants. DGGE has high sensitivity with DNA fragments up to 500 bp, and will detect approximately 50–99% of single base changes, depending on gel optimization.[23]

Microarray analysis

Although DNA sequencing is the ultimate technique for detecting and identifying mutations, it is currently not practical for routine diagnosis. One solution to this technical challenge may be provided by DNA chip or microarray technology, in which a large array of DNA probes (up to 10 000 different sequences), representing oligonucleotides corresponding to the gene under investigation, are bound directly to wafer-thin glass or silicon chips. These arrays are hybridized to single-stranded DNA from a patient sample. Mismatching sequences, indicating mutations, will not bind, and can be identified automatically by advanced software. This approach is covered extensively in Chapter 2.

Comparative genomic hybridization

The development of fluorochrome-labelled probes, complemented by confocal laser microscopy, has allowed the identification of specific DNA sequences in resting interphase cells with resolution intervals as small as 150 kb. Deletions, duplications or amplifications of DNA can be seen by variations in the intensity of the fluorochrome signal. Comparative genomic hybridization (CGH) is a novel outgrowth of this technology in which total tumor DNA and normal genomic reference DNA are labelled with different fluorochromes and simultaneously hybridized to normal metaphase spreads. The relative amounts of tumor and reference DNA bound at a given chromosomal locus are dependent on the relative abundance of those sequences in the two DNA samples, and can be quantitated by measurement of the ratio of specific fluorescence.[24] Amplification of a specific oncogene (for example, c-*myc*) in the tumor DNA will produce an elevated tumor fluorescence ratio, whereas deletions of anti-oncogenes (*p53*, *Rb*, etc.) or chromosome loss will cause a reduced ratio. The ability to survey the whole genome in a single hybridization is a distinct advantage over allelic loss studies by RFLPs that target only one locus at a time.

Representational difference analysis

One of the most challenging areas in DNA diagnostics is the identification of sequence differences between individuals or, as in the case of cancer, differences between tumor and genomic DNA in the same individual. This type of comparative analysis involves whole genomes, cut with restriction enzymes, and would include the identification of polymorphic variation between individuals. One technique for cloning the differences between complex genomes is representational difference analysis (RDA).[25] RDA uses PCR to enrich for differences that are identified initially by subtractive hybridization. Using, as an example, the cloning of chromosome Y-specific DNA probes, an excess of denatured DNA from a female (driver DNA) is allowed to hybridize with restriction enzyme-cleaved male DNA (tester DNA), linked to special DNA adapters at their 5' ends. Y chromosome-specific DNA that remains after the hybridization reaction can be amplified at an exponential rate by the use of primers that are specific for these adapters, then cloned into plasmids for characterization. Using this modification, one is able to enrich for the unique sequences in the target DNA to a magnitude of 10^6, compared with only 100 or 1000 times with standard subtraction hybridization.

Genomic mismatch scanning

Genomic mismatch scanning (GMS) is also a technique that compares whole genomes, but, rather than trying to identify differences, the strategy is to identify regions that are shared by individuals with a diseased phenotype.[26] GMS also involves DNA hybridization, and begins by cutting the two genomes with restriction enzymes to produce DNA fragments of varying size. One sample is treated so that methyl groups are added at intervals along the DNA fragments. The two samples are melted to produce single-stranded DNA, then mixed together and allowed to hybridize. The result is a mixture of double-stranded DNA fragments, some containing two methylated strands, others with two unmethylated strands, and some containing one strand from each. The next chemical step eliminates those duplexes that are derived from only one sample, and the last step eliminates those strands from different samples that contain mismatches. This leaves only those hybrids in which each strand is derived from a different sample, and the strands from the different samples are identical for the two genomes being compared. With this technique, it would be possible to compare the genomes of two individuals with polycystic ovary syndrome, from the same or different families, in order to identify common regions that might be inherited from a distant ancestor. Both RDA and GMS offer the prospect of comparing whole genomes for similarities or differences in one single analysis.

RNA AND PROTEIN ANALYSIS

mRNA

mRNA hybridization

DNA is technically easier to analyze than RNA and protein, given that sequence specific restriction enzyme analysis and automated sequencing are available for DNA and not RNA, and that it is possible to amplify nucleic acid sequences directly by PCR techniques. However, genes typically under clinical analysis encode for mRNAs that are translated into proteins, and there can be value in measuring either or both of these molecules in specific situations if material from primary tissue is available. *In situ* hybridization for mRNA has been particularly effective in examining complicated tissues such as the ovary. For example, expression of StAR (steroidogenic acute regulatory protein, required for transport of cholesterol into the mitochondria), aromatase and P450scc mRNA is limited only to those granulosa cells confined to ovulatory follicles, and thus, only those cells have the capacity to synthesize estrogens.[27] Quantitation of specific mRNAs may easily be performed by hybridization to DNA or RNA probes with the techniques of Northern blotting or RNAase protection, as discussed in Chapter 1. Alternatively, the detection of mRNA in clinical samples by reverse transcriptase (RT)-PCR is extremely sensitive and easily adaptable to typical amounts of biopsy material or even from single cells.[27,28] Finally, the development of microarray or DNA chip technology, discussed in Chapter 2, is beginning to allow rapid scanning of RNA expression profiles between RNA samples of known genes or ESTs, and will undoubtedly be a useful tool to focus attention on groups of regulated genes.

Differential display assay for mRNAs

Differential display analysis (DDA) is a procedure for quantitative detection of differentially expressed genes,[29] and permits simultaneous identification of genes that are up- or down-regulated under different conditions. The procedure is based on the PCR amplification of mRNAs of cells or tissues, using a short 5' arbitrary primer, an oligo(dT) 3' primer and radiolabelled DNA nucleotides. Differential gene expression is visualized by autoradiography after electrophoretic separation (Figure 6). The cDNA bands are then recovered from the gel and cloned for sequencing and further characterization. DDA has several technical advantages over differential hybridization, including rapidity, the requirement for only

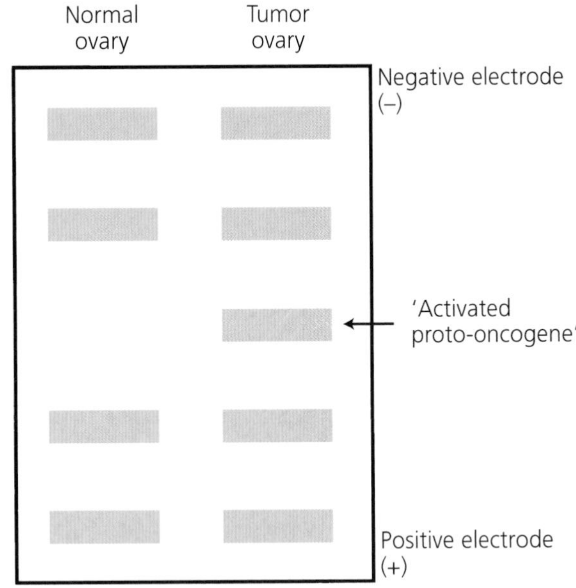

Figure 6 Illustration of the principle of differential display of eukaryotic mRNA. Differential display of mRNAs enables one to isolate those genes that are differentially expressed in different tissues (normal vs. tumor) and under different conditions (e.g. stimulated vs. non-stimulated). Arbitrary primers are designed to amplify the cDNAs after reverse transcription. The different mRNA subpopulations are resolved on a DNA sequencing gel

small amounts of starting material (1 μg of mRNA) and simultaneous detection of both groups of differentially expressed genes, i.e. both stimulated and suppressed genes. This method has important uses in reproductive endocrinology, cancer and developmental biology, and is usually followed by confirmatory RNA studies such as Northern analysis or quantitative RT-PCR.

Protein

Additional information can be obtained by measuring proteins. In some pathophysiological cases, notably comparisons between tumor and normal tissues from the same patients, protein expression of specific molecules such as steroid receptors, oncogene products and enzymes can differ dramatically, although mRNA levels are comparable. Furthermore, the expression of specific variant proteins can differ among cell types and between normal and cancerous tissue. For example, constitutively active variant proteins of the estrogen receptor α are found in endometrial cancer, but not in normal endometrium.[28]

An additional reason for protein analysis is that protein modifications, including glycosylation, phosphorylation and acetylation play a critical role in determining protein activity, and variation in these modifications can have significant impact on reproduction. For example, glycosylation status of the gonadotropins helps to determine their biological activity and clearance in the plasma. One of the most common protein modifications is that of phosphorylation on serine, threonine or tyrosine by specific enzymes called kinases. In the reproductive system, several enzymes and receptors are phosphorylated, but the degree of phosphorylation can alter activity. For example, P450c17 is the single enzyme catalyzing both 17α-hydroxylase and 17,20-lyase activity in the adrenal and ovary. Serine hyperphosphorylation of P450c17 increases the enzyme's 17,20-lyase activity, thereby favoring androgen production, and is believed to contribute to polycystic ovarian disease.[30] These changes in modified proteins would not be scored by DNA or RNA testing.

Examination of proteins relies on electrophoresis and immunological tools. Amino acids differ widely in terms of charge, and proteins can be distinguished on the basis of either size or overall charge. A very powerful technique, two-dimensional polyacrylamide gel electrophoresis, examines both of these properties and can be invaluable in distinguishing individual species of modified proteins. Proteins are first separated by electrophoresis in tube gels containing urea and charged ampholines, which will separate proteins on the basis of charge; this is the first dimension. Gel rods are then incubated with reducing agent and detergent (sodium dodecylsulfate, SDS), and placed on top of slab denaturing SDS–polyacrylamide gels to separate the proteins on the basis of size: the second dimension. After electrophoresis, proteins are blotted and identified with antibodies. The resulting patterns for a phosphorylated protein might show several species with the same apparent molecular weight (from top to bottom in the gel), but several different species from left to right, representing unphosphorylated and phosphorylated forms. This technique was used recently to demonstrate changes in the phosphorylation state of StAR protein after adrenocorticotropic hormone (ACTH) treatment, which is postulated to contribute to changes in StAR import into the mitochondria and subsequent changes in steroidogenesis.[31] An exciting new approach to protein analysis combines high-pressure liquid chromatography with mass spectrometry to detect protein modifications in very small amounts (picomoles or less) of material and may soon replace slower, less sensitive methods.[32]

FUTURE PERSPECTIVES

The identification and mapping of genes as a result of the Human Genome Project, the rapid detection of unknown mutations in human DNA and the identification of differences between genomes require the continued development of many of the techniques mentioned in this chapter. Many different strategies are being developed to make direct sequencing of DNA from the amplified PCR product rapid, accurate and efficient, including the immense power of microarray technology. These new, evolving techniques are helping us to understand the biochemical basis for disease and the etiology of fundamental developmental aberrations, and applications of these techniques should allow more widespread preclinical DNA diagnosis. In many cases, there is still a large hiatus between diagnosis and therapy. In the future, maneuvers such as gene therapy, antisense technology or use of ribozymes may become available to replace a defective gene or to control the expression of inappropriate genes in defined tissues – a challenging proposition that is just beginning to be addressed clinically.

REFERENCES

1. Hudson TJ, Stein LD, Gerety SS *et al.* An STS based map of the human genome. *Science* 1995 270: 1945–1954.
2. Saiki RK, Scharf S, Faloona F *et al.* Enzymatic amplification of β-globin genomic sequences and restriction site analysis for diagnosis of sickle cell anemia. *Science* 1985 230: 1350–1354.
3. Handyside ASH, Pattinson JK, Penketh RJA *et al.* Biopsy of human preimplantation embryos and sexing by DNA amplification. *Lancet* 1989 1: 347–349.
4. King D, Wall JR. Identification of specific gene sequences in preimplantation embryos by genomic amplification: detection of a transgene. *Mol Reprod Dev* 1988 1: 57–62.
5. Skynner MJ, Sim JA, Herbison AE. Detection of estrogen receptor alpha and beta messenger ribonucleic acids in adult gonadotropin-releasing hormone neuron. *Endocrinology* 1999 140: 5195–5201.
6. Chamberlain JS, Gibbs RA, Ranier JE, Nguyen PN, Caskey CT. Deletion screening of the Duchenne muscular dystrophy locus via multiple DNA amplification. *Nucleic Acids Res* 1958 16: 11141–11156.
7. Zhang L, Cui X, Schmitt K *et al.* Whole genome amplification from a single cell: implications for genetic analysis. *Proc Natl Acad Sci USA* 1992 89: 5847–5851.
8. McDonough PG. Molecular biology in reproductive endocrinology. In Yen SCC, Jaffe RB, eds. *Reproductive Endocrinology*, 3rd edn. Philadelphia: WB Saunders, 1991: 25–64.
9. Southern EM. Detection of specific sequences among DNA fragments separated by gel electrophoresis. *J Mol Biol* 1975 98: 503–517.
10. Policastro PF, Daniels-McQueen S, Carle G, Boime I. A map of the hCGβ-LHβ gene cluster. *J Biol Chem* 1986 261: 5907–5916.
11. Juliet C, Weil D, Couillin P *et al.* The beta chorionic gonadotropin–beta luteinizing gene cluster maps to human chromosome 19. *Hum Genet* 1984 67: 174–177.
12. Lee JT, Strauss WM, Sausman JA, Jaenisch R. A 450 kb transgene displays properties of the mammalian X-inactivation center. *Cell* 1996 86: 83–84.
13. Lee JT. Disruption of imprinted X inactivation by parent-of-origin effects at *Tsix*. *Cell* 2000 103: 17–27.
14. Yang-Feng TL, Seeburg PH, Francke U. Human luteinizing hormone-releasing hormone gene (LHRH) is located on short arm of chromosome 8 (region 8p11.2–p21). *Somat Cell Mol Genet* 1986 12: 95–100.
15. Hayflick JS, Adelman JP, Seeburg PH. The complete nucleotide sequence of the human gonadotropin-releasing hormone gene. *Nucleic Acids Res* 1989 17: 6403–6404.
16. Yang-Feng TL, Opdenakker G, Volckaert G, Francke U. Human tissue-type plasminogen activator gene is located near chromosomal breakpoint in myeloproliferative disorders. *Am J Hum Genet* 1986 36: 79–87.
17. Kruglyak L. What is significant in whole-genome linkage disequilibrium studies? *Am J Hum Genet* 1997 61: 810–812.
18. Lander E, Kruglyak L. Genetic dissection of complex traits: guidelines for interpreting and reporting linkage results. *Nat Genet* 1995 11: 241–247.
19. Morton NE. Significance levels in human inheritance. *Am J Hum Genet* 1998 62: 690–697.
20. Elston RC. Methods of linkage analysis – and the assumptions underlying them. *Am J Hum Genet* 1998 63:931–934.
21. Hubert R, Weber JL, Schmitt K, Zhang L, Arnheim N. A new source of polymorphic DNA markers for sperm typing: analysis of microsatellite repeats in single cells. *Am J Hum Genet* 1992 51: 985–991.
22. Howe JR, Lairmore TC, Mishra SK *et al.* Improved predictive test for MEN2, using

flanking dinucleotide repeats and RFLPs. *Am J Hum Genet* 1992 51: 1430–1442.

23. Prosser J. Detecting single-base mutations. *Trends Biotechnol* 1993 11: 238–246.
24. Kallioniemi A, Kallioniemi OP, Sudar D *et al.* Comparative genomic hybridization for molecular cytogenetic analysis of solid tumors. *Science* 1992 258: 818–821.
25. Lisitsyn NA, Lisitsyn NM, Wigler MH. Cloning the difference between complex genomes. *Science* 1993 259: 946–951.
26. Brown PO. Genome scanning methods. *Curr Opin Genet Dev* 1994 4: 366–373
27. Orly J, Stocco DM. The role of steroidogenic acute regulatory (StAR) protein in ovarian function. *Horm Metab Res* 1999 31: 389–398.
28. Jazaeri O, Shupnik MA, Jazaeri A, Rice LW. Estrogen receptor mRNA and protein isoforms expressed in human endometria and endometrial cancer: selective expression of an exon 5 deletion variant in cancer tissue. *Gynecol Oncol* 1999 74: 38–47.
29. Liang P, Pardee AB. Differential display of eukaryotic messenger RNA by means of the polymerase chain reaction. *Science* 1992 257: 967–971.
30. Zhang L, Rodriguez H, Ohno S, Miller WL. Serine phosphorylation of human P450c17 increases 17,20 lyase activity: implications for adrenarche and for the polycystic ovary syndrome. *Proc Natl Acad Sci USA* 1995 92: 10619–10623.
31. Lehroux J-G, Hakles DB, Fleury A, Briere N, Martel D, Ducharme L. The *in vivo* effects of adrenocorticotropin and sodium restriction on the formation of different species of steroidogenic acute regulatory protein rat adrenal. *Endocrinology* 2000 140: 5154–5164.
32. Strahl BD, Briggs SD, Brame CJ *et al.* Methylation of histone H4 at arginine 3 occurs *in vivo* and is mediated by the nuclear receptor coactivator PRMT1. *Curr Biol* 2001 11:996–1000.

Cell biology

Jerome F. Strauss III

To understand important events in reproduction, including germ cell replication and maturation, the proliferation of the endometrium, the growth and differentiation of the embryo and extra-embryonic tissues as well as neoplasia, biologists and clinicians in the field of reproductive medicine need to appreciate advances in cell biology related to tissue homeostasis. The proteins that govern the cell cycle, including cyclins, cyclin-dependent kinases, regulating phosphatases and kinase inhibitors, have been identified and placed into an ordered sequence of reactions. The fidelity of the process is closely monitored at specific check points. Damage to DNA during mitosis or meiosis can be repaired by proteins and enzymes that maintain the integrity of the genome. Alternatively, cells with damaged DNA are eliminated by a process of programmed cell death. Mutations in genes encoding DNA repair proteins are associated with familial cancers and cancer progression. Programmed cell death is also central to the process of follicular atresia. Granulosa cells as well as oocytes undergo apoptosis, as do male germ cells. Programmed cell death, or the lack of it, respectively eliminates abnormal cells or allows transformed cells to flourish. A number of death-promoting proteins have been identified, some of which have selective patterns of expression in reproductive tissues. Counterbalancing these are survival factors. The balance of activity of these proteins determines whether an execution pathway leading to cell demise is initiated. The components of this pathway, as well as the way in which external signals from growth factors and cytokines influence cell death or survival, have been a major focus of investigation during the past decade. Chapter 4 describes the cell biology of the cell cycle while Chapter 5 reviews basic concepts of programmed cell death. Chapter 6 reviews the fundamentals of carcinogenesis including perturbation of the cell cycle and mechanisms that sustain genome stability.

Tissue homeostasis through angiogenesis and programmed cell death are prominent features of reproductive tissues. The ovary, the endometrium and the placenta undergo the most dramatic changes in blood vessel formation and regression described to date. The processes of vasculogenesis and angiogenesis are complex and involve proteins that promote proliferation, branching and stabilization of endothelial cells as well as vascular permeability. The growth of reproductive tract cancers also depends on angiogenesis, and strategies to suppress blood vessel formation figure prominently in preclinical and early-phase clinical chemotherapy strategies. If successful, such approaches might also be employed in the treatment of endometriosis. The repertoire and regulation of angiogenesis proteins and their cognate receptors are described in Chapter 7.

Hyperplasia, neoplasia and tumor progression in certain tissues are profoundly influenced by sex steroid hormones through effects on the cell cycle, genome integrity and cell survival. Chapter 8 reviews the influence of sex steroid hormones on cell transformation and tumor progression, integrating the concepts described in Chapters 4, 5 and 6.

The immune system, innate and adaptative, is also critical for tissue homeostasis. It removes cells recognized as non-self, including malignant cells, but can be modulated through novel mechanisms during pregnancy to allow acceptance of a fetal semi-allograft. Chapter 9 describes the components of the innate and acquired immune system, the process of antigen presentation, the important pro-inflammatory and anti-inflammatory cytokine networks and the ways in which hormones and the placenta help the fetus to evade the maternal immune response. Abnormalities in this still incompletely understood system may lead to early as well as repetitive miscarriage or abnormalities in placentation.

Cell and molecular biological methods have been married in transgenic technology in which genes are introduced into fertilized oocytes or embryonic stem cells, creating animals with altered expression or activity of genes, in some cases in a cell-specific as well as regulated fashion. The over-expression, mis-expression as well as ablation of genes, have informed us about the roles of ligands, including hormones and cytokines, their respective receptors, and downstream signalling molecules and target transcription factors, in reproductive processes. This technology has also created valuable models of human disease or provided the phenotypic information that has pointed clinical geneticists to human disease genes. Chapter 10 reviews transgenic technology and the animal models created with it that are relevant to reproductive medicine.

Chapter 4

The cell cycle and differentiation in the reproductive system

Debra J. Wolgemuth

INTRODUCTION

The cell cycle consists of a series of events required for successful cellular duplication. In most cells, the two major phases are the S phase, when DNA synthesis takes place, and the mitotic or M phase, when DNA in the form of chromosomes is divided into two new cells. Between these phases are the 'gap' or G phases, G_1 and G_2. The transition between these stages is regulated by a series of check-points, of which the restriction point at late G_1 in mammalian cells (known as START in yeast) is particularly important (Figure 1).

This chapter highlights basic aspects of cell cycle control, set within the context of differentiation and function of the reproductive system during key developmental stages. These stages include differentiation of the germ line, early embryogenesis and function of the reproductive system in the adult. The focus will be on mammalian development, notably in the mouse model, in which the combination of genetics and gene targeting approaches has provided insight into the function of specific genes in cell cycle control. The emphasis will be on the control of the transition states, particularly from G_1 to S and G_2 to M phases, rather than on the control of specific processes within the phases, such as DNA replication during S phase. For recent reviews on the biochemical functions of cyclins and cyclin-dependent kinases (CDKs), the reader is referred to articles by Pines,[1] Fisher[2] and Pestell;[3] for replication to Stillman;[4] for check-points and repair to Elledge[5] and Clarke and Gimenez-Abian;[6] and for degradation of cell cycle regulators to Yew.[7]

REVIEW OF THE COMPONENTS OF CELL CYCLE MACHINERY

An orderly progression through the phases of the cell cycle, G_1, S, G_2 and M, is essential for normal development and differentiation. The components of this regulatory process include a complex array of kinases, phosphatases and regulatory proteins such as the cyclins, and an equally complex array of substrates, including nuclear and cytoplasmic structures involved in the mechanics of cell

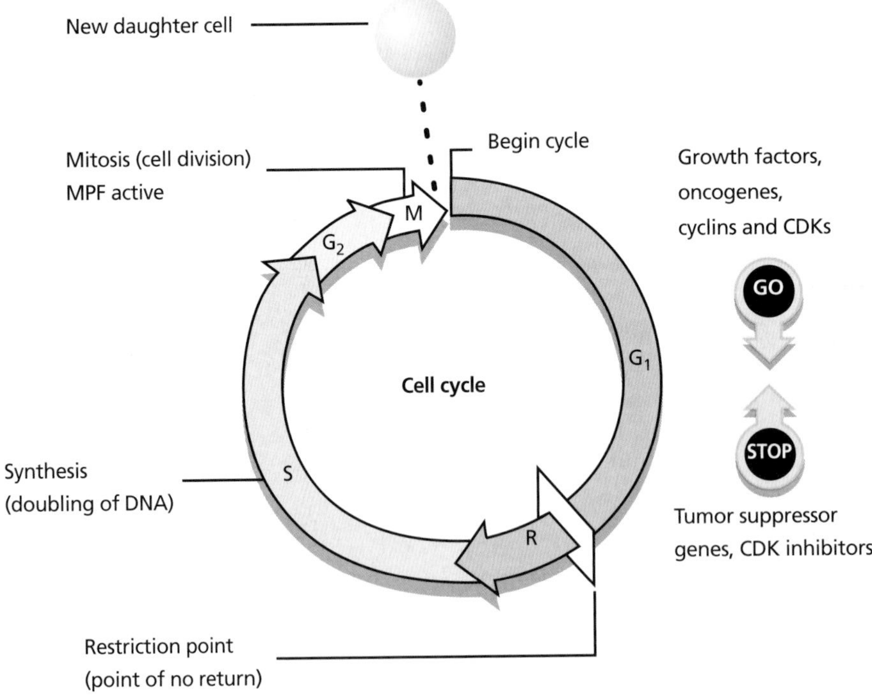

Figure 1 Schematic diagram outlining the stages of the mammalian mitotic cell cycle. Two key transition stages are noted, the G_1 restriction point (R) and the G_2/M commitment to mitosis. MPF, maturation promoting factor; CDK, cyclin-dependent kinase

division. A key player is maturation promoting factor or M phase promoting factor (MPF). MPF was originally identified biochemically as an activity found in unfertilized frog eggs capable of inducing germinal vesicle breakdown and resumption of meiosis when injected into immature oocytes.[8] MPF is composed of a regulatory subunit, cyclin B, and p34^{cdc2} (CDK1), the catalytic serine/threonine protein kinase (Figure 2). The components of MPF were also identified by genetic studies, initially in yeast and subsequently in cultured mammalian cell lines.[9]

Periodic activation of the different cyclin–CDK complexes is, in large part, responsible for the characteristic sequence of cell cycle progression, including mitosis itself, but also for DNA synthesis, chromatin assembly and alterations in subcellular structures such as microtubules and membranes. How these complexes control this vast array of functions during development and differentiation is not understood, primarily because so few *in vivo* physiological substrates have been identified.

Cyclins and cyclin-dependent kinases

Cyclins were originally identified as proteins that were degraded at each mitotic division.[10] Cyclins are highly conserved, and are found in a variety of organisms ranging from yeast to man, based on amino acid homology and functional complementation of yeast cell cycle mutants. Common to cyclins is the 100 amino acid-long consensus 'cyclin box'. The cyclin box has been shown to be involved in binding to its catalytic partner, the cyclin-dependent kinase (CDK). The cyclins have been divided into at least nine classes in higher vertebrates, cyclins A–I (Table 1), based on their amino acid similarity and the timing of their appearance during the cell cycle (reviewed in references 1, 11, 12). The complexity of the function of the mammalian cyclins is underscored by the existence not only of multiple classes of cyclins but also of multiple members of the A-, B- and D-type cyclin families.

The CDKs are defined as protein kinases that must bind to a cyclin to be active. The CDKs share certain structural similarities, among which the characteristic PSTAIRE motif (or variants thereof) is probably the best recognized. Structural studies have revealed that this motif is exposed on the surface of the enzyme, and some amino acids are part of the active site[13] (reviewed in reference 1). Mutations in this highly conserved PSTAIRE region alter the ability of cyclins to bind to their

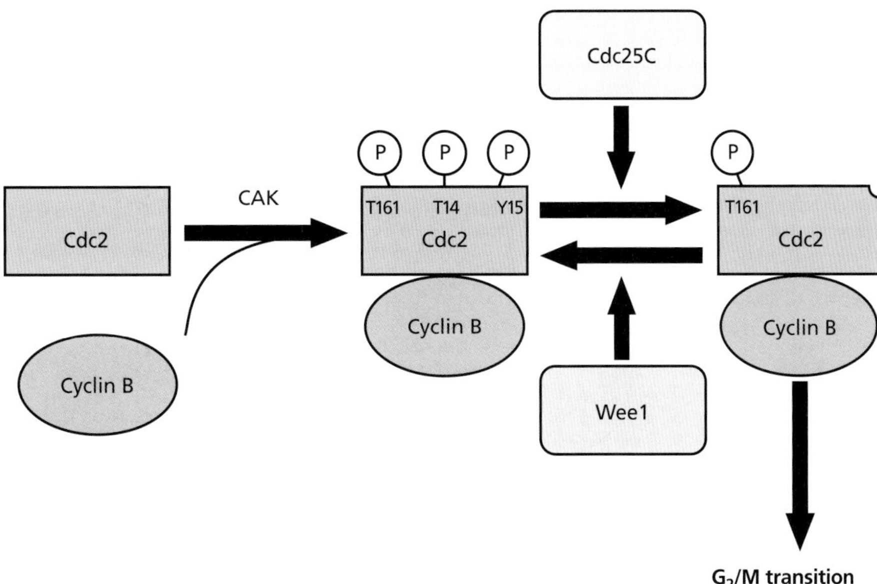

Figure 2 Key components of the G_2/M transition machinery are indicated, along with several of the regulatory kinases and phosphatases. CAK, cyclin-dependent kinase activating kinase; P, phosphorylation

Table 1 Mitotic cell cycle stage-specific activity of cyclins/cyclin-dependent kinases (CDKs)

Cyclin	*CDK partner*	*Cell cycle phase*
A2	CDK1, CDK2, other?	S, G_2/M
B1, B2, B3	CDK1, CDK2, other?	G_2/M
C	CDK8	multiple phases
D1, D2, D3	CDK4, CDK6	G_1
E	CDK2	G_1/S
F	?	G_2
G1, G2	CDK5	G_1, G_2/M?
H	CDK7	G_2
I	?	multiple phases

CDK partners. Antibodies against the PSTAIRE region recognize the CDKs only in their monomeric form, not when complexed with other proteins. A second region in the CDKs that appears to be involved in cyclin binding includes a threonine residue that is phosphorylated in all active protein kinases.

CDK activities are controlled by several different mechanisms, including binding to a specific cyclin partner, state of phosphorylation and interaction with CDK inhibitors. At least two mechanisms are, in turn, involved in the timing of cyclin expression: regulation at the level of transcription[14] and an ubiquitin-mediated degradation.[15] The destruction of the cyclin component of the cyclin–CDK complex irreversibly inactivates it. The degradation appears to involve either the anaphase promoting complex or the Skp1p–Cdc53p/cullin–F-box (SCF) protein pathways, both of which lead to degradation by the 26S proteasome.[16,17] In addition to activating its CDK partner in the kinase complex, recent studies point increasingly to a role for cyclins in determining substrate specificity.[18,19]

Activating and inhibitory phosphatases and kinases

Activation or inhibition of the cyclin–CDK complex is also controlled by its state of phosphorylation (Figure 2). CDKs are typically phosphorylated on two inhibitory and one activating residue. The inhibitory sites in mammalian CDK1 are Thr14 and Tyr15, which are phosphorylated by the Myt1 and Wee1 kinases, respectively (reviewed in references 1, 20, 21). These residues remain phosphorylated until removed by the dual-specificity phosphatase, Cdc25. Mammals appear to have three members of the Cdc25 family, Cdc25A, Cdc25B and Cdc25C.

A third common and highly conserved phosphorylation occurs on Thr-167 (in *Schizosaccharomyces pombe* CDK1) and equivalent sites of Cdk1 in other species. These residues are phosphorylated by an enzyme activity known as CDK-activating kinase (CAK), which is itself composed of a cyclin–CDK complex (reviewed in reference 22). Mammalian CAK is made up of a regulatory subunit, cyclin H, a catalytic subunit, CDK7, and a third protein believed to be an assembly factor sometimes designated as MAK1 (ménage à trois kinase). CDK7 (also known as p40^{MO15}) was originally identified as a negative regulator of oocyte maturation in *Xenopus* oocytes, and subsequently in a wide variety of organisms from yeast to human. An interesting feature of cyclin H is that, at least in mitotic cells, it appears to be expressed throughout the cell cycle, rather than undergoing the characteristic, tightly regulated pattern of synthesis and degradation of the other cyclins.

Cyclin-dependent kinase inhibitors

In mammals and other higher eukaryotes, there are two classes of relatively small (16–57 kDa), tightly binding proteins that function as CDK inhibitors (CKIs) (reviewed in reference 23). They include the CIP–KIP family, consisting of p21^{CIP1}/WAF1, p27^{KIP1} and p57^{KIP2}, and the INK family, consisting of p16^{INK4a}, p15^{INK4b}, p18^{INK4c} and p19^{INK4d}. The CKIs appear to exhibit their inhibitory function by binding in a stoichiometric manner, usually to cyclin–CDK complexes but sometimes to non-complexed CDKs. The two mammalian CDIs, p16^{INK4a} and p15^{INK4b} bind CDK4 and CDK6, perhaps competing for binding with the D-type cyclins (reviewed in reference 1). p15^{INK4b} protein appears to be induced in response to transforming growth factor-β (TGF-β), as is another CKI, p27Kip1. Some data suggest that p27 may exhibit its inactivating capacity by preventing the phosphorylation of a key residue in CDK4 by the CDK-activating kinase CAK.[24] p27 exhibits homology to the p21 CKI, which forms a complex with proliferating cell nuclear antigen (PCNA) and several cyclin–CDK complexes, including cyclins A2, D1 and E.

Function of cell cycle regulators in mitosis versus meiosis

Given the high level of conservation of these key molecules across a widely diverse range of organisms, it is likely that they will be important for both mitosis and meiosis in mammals, as they are in the yeasts in which they have been best characterized. However, it is clear that there are control points and 'check-points' in the cell cycles of higher eukaryotes that do not exist in simpler organisms, and which may differ depending on the developmental and tissue origin of the cells. For example, the oocytes of mammals exhibit unique cell cycle control points; the signals to enter meiosis are different from those found in yeast, there is an arrest in diplotene of meiosis, and the cell pauses again in metaphase II, awaiting fertilization to complete meiosis (Figure 3). The spermatocyte shares some of these control points, but, strikingly, does not normally arrest in either diplotene or at metaphase of meiosis II. It is therefore likely that there will be genes uniquely involved in these regulatory check-points which will be distinct from the yeast genes studied to date.

CELL CYCLE REGULATION IN GAMETES

Control of the cell cycle during mammalian gametogenesis

Mammalian gametogenesis provides a unique system in which to study cell cycle regulation. Understanding the genetic program controlling the mitotic and meiotic divisions of the germ line will furthermore provide insight into understanding infertility and new directions for contraception. Male and female germ cells have stages of cell cycle regulation that are similar between the two lineages, including a mitotic proliferative stage, entry into meiosis, completion of a reductive division and entry into a quiescent state awaiting signals at fertilization. However, the timing of these events and indeed even the stage of development at which these events occurs differ in the two sexes, as reviewed in reference 25. For example, female germ cells enter meiosis during fetal development, whereas this is a postnatal event in the male. Once the male germ cell has entered meiosis, the process continues without interruption until the haploid spermatozoon is produced some weeks later. In contrast, the oocyte is arrested in the diplotene stage of meiotic prophase I, where it can remain for months or years. Following a growth period, the oocyte resumes meiosis, only to be arrested at a second point, at metaphase II. Fertilization then triggers the completion of meiosis and extrusion of the second polar body. Yet another curiosity of mammalian germ cells is that the two haploid pronuclei commence DNA synthesis independently after fertilization, prior to a fusion event. These and other critical control points in the specialized cell cycles of mammalian germ cells are depicted in Figure 3.

Although the role of the known cell cycle regulators in mammalian germ cell development is only beginning to be investigated, observations from our laboratory and others have shown that various cyclins and CDKs exhibit distinct patterns of expression in the male and female germ cell lineages, as well as in the somatic compartments of the gonads (reviewed in reference 25). These studies have revealed specific patterns of cell-type and cell cycle-stage expression *in vivo* of several of the cyclins that had not been detected previously in cultured cell systems.

Common to both germ lines is the need for MPF activity at the G_2/M transition. MPF activity, characterized biochemically by histone H1 kinase assays, was shown to be low in immature mammalian oocytes, to peak at metaphase of meiosis I and II, and to be low again immediately following the meiotic divisions.[27,28] MPF activity has been detected in pachytene spermatocytes as

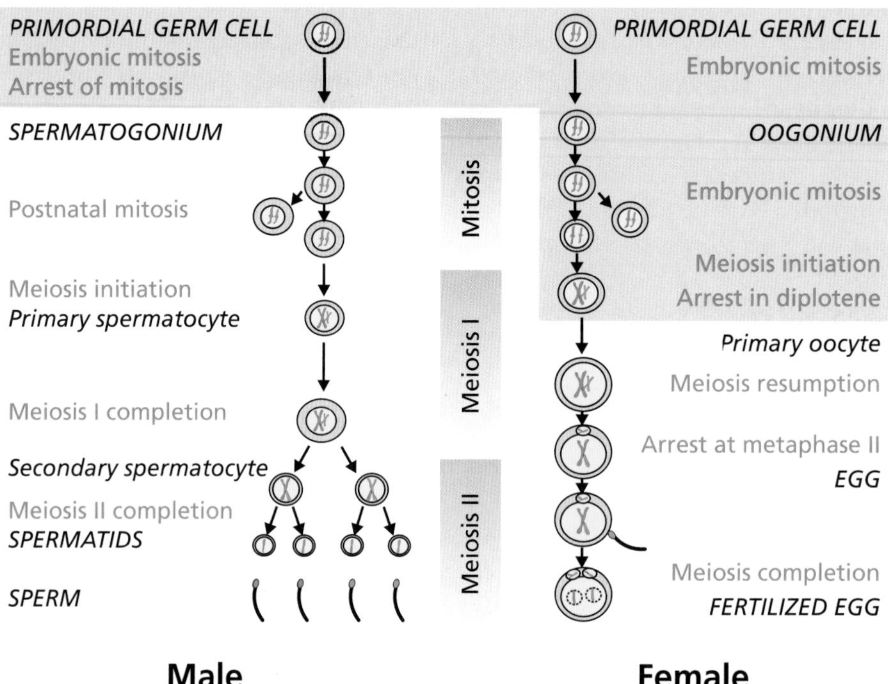

Figure 3 Comparison of the key mitotic and meiotic regulatory stages of mammalian gametogenesis is outlined in this diagram, modified from references 25 and 26. The highlighted parts of the flowchart indicate those developmental stages that occur during embryogenesis in most mammals

well, and is believed to be responsible for entry into the first meiotic division by spermatocytes.[29,30]

Example of an essential role in meiosis for a cell cycle regulator: cyclin A1

Of the mammalian cyclins identified to date, cyclin A has remained somewhat enigmatic, in part because there are no homologs in the yeasts. Our laboratory has shown that there are two distinct cyclin A genes in mammals, one of which, *Ccna1*, is testis-specific and restricted to the germ line.[31,32] Two A-type cyclin genes have now also been documented in human and frog.[33,34] Mouse cyclin A1 and A2 share 44% identity at the amino acid level; however, they share a much higher level of identity (84%) within the two highly conserved cyclin boxes.[31,32] A major issue of interest is the extent to which the two genes are functionally redundant or distinct.

In the mouse, *Ccna1* mRNA and protein are present in late pachytene to diplotene spermatocytes,[32] but not at significant levels during the second division of meiosis.[35] In contrast, the originally described mammalian cyclin A gene, *Ccna2*, was expressed in both spermatogonia and preleptotene spermatocytes, suggesting functions in mitotic and pre-meiotic cell cycles.[31,35] Furthermore, both somatic and germ cells of the adult ovary express *Ccna2* mRNA and protein (Liao and colleagues, unpublished data). The different expression patterns suggest distinct functions for *Ccna1* and *Ccna2* in the somatic and germinal lineages, which also differ between the male and female.

Ccna2 is expressed ubiquitously in cultured cells, and in a broad variety of tissues in the adult mouse and during embryogenesis.[31,32] Perhaps not surprisingly then, targeted mutagenesis of murine *Ccna2* resulted in early embryonic lethality, apparently around the peri-implantation stage.[36] This embryonic lethality has obviated understanding of the role of cyclin A2 in other aspects of mammalian development, including the germ line. In contrast, the strikingly restricted expression of *Ccna1* led to the hypothesis that its primary site of function is in the male germ line, specifically at the first meiotic division.

To test this hypothesis and to begin to address possible redundancy of the two A-type cyclin genes, cyclin A1-deficient mice were generated by targeted mutagenesis of the *Ccna1* gene.[37] *Ccna1* –/– males were sterile owing to a block of spermatogenesis before the first meiotic division, whereas females were normal. The meiotic arrest in *Ccna1* –/– males was associated with increased germ cell apoptosis and desynapsis abnormalities. There was a striking reduction in the activation of MPF kinase at the end of meiotic prophase, although both CDK1 and cyclin B proteins were present. Cyclin A1 is therefore essential for spermatocyte passage into the first meiotic division in male mice, a function that cannot be complemented by the concurrently expressed B-type cyclins. It is clear that cyclin A2 cannot substitute for cyclin A1, which was not unexpected, given their strikingly different patterns of expression. It is also clear that there is a sex-specific difference in the requirement for cyclin A1, as no abnormalities were observed during oogenesis and the females are fully fertile.

CELL CYCLE CONTROL DURING EMBRYOGENESIS

Proper spatial development of a multicellular organism requires exquisite integration of cellular proliferation, differentiation and migration. Key to this process is having the appropriate number of cells available for when the differentiative event will occur. This can be achieved by striking a balance between controlled cell proliferation and cell death; in both processes, cell cycle regulation plays an important role. The unique properties of the early embryonic cell divisions have been the focus of much investigation in a variety of model systems, and each system has its own peculiar characteristics (reviewed by reference 38). For example, the length of the cycle varies drastically, from more than 20 h in mammals to minutes in *Drosophila*, as does the number of cell divisions preceding the activation of the embryonic genome. Common to early embryos of various species are first, the use of some maternal products in the initial mitotic cell division(s), and second, cell divisions without an accompanying increase in mass of the zygote. That is, cell growth and cell division are uncoupled. In the mouse, no increased mass is detected until the blastocyst stage. Finally, studies in *Xenopus* and *Drosophila* suggest that the early embryonic divisions may exhibit modes of cell cycle regulation distinct from the typical somatic mitotic divisions of ensuing embryogenesis and adult life. The timing of the cell cycle in early frog embryos appears to evolve from regulation by accumulation of mitotic cyclins (A1, B1, B2 and E) to mechanisms involving periodic G_1 cyclin expression and inhibitory tyrosine phosphorylation of CDK1.[39] Furthermore, studies in *Drosophila* suggest that, as maternal cell cycle factors are depleted, they are replaced by differentially expressed zygotic factors that can alter the mode of regulation. Cyclin B and E and Cdc25 have been shown to act as limiting regulators in flies in specific cell types and at particular developmental stages (reviewed in reference 38).

The early mouse embryo

The first mitotic cell cycle that follows fertilization is unusually long in the mouse embryo, 18–20 h in duration (reviewed in reference 40). It also appears that this first cell cycle is completed using maternal mRNA and proteins, but clearly the zygotic genome is active by the two-cell stage.[41] The subsequent divisions have a much shortened cell cycle length, in the range of 10–12 h. After the 32-cell stage in the mouse, diversification of cell cycles is apparent, with increased asynchrony among the individual blastomeres and the beginning of endoreplicative cycles in the forming trophectoderm. During the egg cylinder stage, cells in the epiblast are dividing rapidly. The timing of these cycles has been examined in rat embryos and suggests that cells in the primitive streak have the fastest cycle, as short as 3 h.[42] It has been suggested that this rapid division in mammalian embryos resembles the rapid division that precedes the well-characterized mid-blastula transition of *Xenopus* and other organisms. While the cell cycles are driven by zygotic products in mammalian embryos, compared with the predominantly maternal products of the frog oocyte, the net result is the same: providing an adequate number of cells for the patterning and differentiative events to follow.

Examples of essential roles for cell cycle regulators in early embryos

Cyclin B1

There are at least three B-type cyclins identified and studied at the functional level. The mammalian cyclin B1 and B2 show little similarity in the first 100 amino acids and approximately 57% in the remaining 300 residues.[43–45]. Work in cultured cells indicated that the two B cyclins are co-expressed in the majority of dividing cells, although their subcellular localization is distinct. However, cyclin B1 and B2 were shown to have quite distinct patterns of expression during mouse spermatogenesis[44] and, in frog oocytes, cyclin B2 protein is present in unactivated oocytes while B1 is not detected until oocytes are induced to undergo meiotic maturation.[46,47] Brandeis and colleagues[48] therefore set out to test the hypothesis that mouse cyclin B1 and B2 had distinct roles that might be revealed by perturbation of their function in *in vivo* physiological systems by gene targeting. Cyclin B1 proved to be an essential gene in that its deletion resulted in early embryonic lethality. To date, the stage at which development is interrupted has not been reported; however, it is noted that analysis of embryonic-day-10 embryos did not reveal the presence of any homozygous null cyclin B1 fetuses, suggesting a much earlier block.[48] In contrast, homozygous cyclin B2 null animals were viable and fertile, with no obvious defects other than a somewhat reduced body size.[48] It is noted that the litter sizes of the homozygous mating pairs were consistently smaller than those of their heterozygous littermate pairs, raising the possible role of cyclin B2 in optimal fertility in mammals.

Cyclin A2

As noted above, mammalian cyclin A2 is believed to function at both G_1/S and G_2/M transitions of the cell cycle and to be almost ubiquitously expressed. To address the role of cyclin A2 in embryonic cell cycles, targeted mutagenesis was performed. No viable homozygous animals null for cyclin A2 were detected.[36] Homozygous embryos could be detected at embryonic day 3.5 at the expected Mendelian ratio, but, as embryogenesis proceeded, so did the loss of cyclin A2 null embryos until, at embryonic-day 8.5, no null embryos were observed. Subsequent analysis revealed that the embryos complete preimplantation development and implant, but die shortly thereafter.[49] The pool of cyclin A2 mRNA present in the oocyte persists until the second mitotic cycle, but is undectable in the null embryos after that. The ability of embryos to develop from the four-cell to implantation stage in the lack of detectable cyclin A2 shows that it is dispensable for these early embryonic divisions, where the blastomeres are dividing but not growing. However, it is essential for subsequent development.

Other cell cycle regulators essential for normal early development

Gene targeting in the mouse model system is providing clues as to other cell cycle components that are essential for early embryonic cell cycle control, as well as uncovering redundancies in the system. For example, Bub3 is a conserved component of the mitotic spindle assembly complex. Mice that are homozygous null for *Bub3* are also embryonic lethal; in this case, the embryos appear normal up to embryonic-day 3.5, but appear to accumulate mitotic errors from day 4.5 to 6.5 in the form of micronuclei, chromatin bridges and irregularly shaped nuclei.[50] A critical role in early mouse development has also been ascribed to a quite different cell cycle regulator, a protein involved in the ubiquitination and degradation of the cyclins, Cul1. Mutations in the *Cul1* gene, which encodes a component of the above-mentioned SCF complex, result in embryonic lethality at embryonic-day 6.5, before the onset of gastrulation, and accumulate large amounts of cyclin E protein.[51,52] In contrast, the extensive information on the importance of cyclin D1 in mitotic cell cycle progression would have predicted an essential role for this cyclin in early development. Although loss of cyclin D1 does lead to some abnormalities, offspring are born in the predicted Mendelian ratios.[52,53]

CELL CYCLE CONTROL IN ADULT REPRODUCTIVE SYSTEMS

Tissues undergoing hormonally modulated cyclic changes

While many adult tissues have only very low levels of cell turnover and replenishment, several parts of the male and female reproductive systems undergo quite striking remodelling on a regular basis. In the female, the cyclic changes in the development of ovarian follicles and the lining of the uterus and vagina are marked by stages of intense cellular proliferation, differentiation and degradation. While clearly these are hormonally modulated processes, cell cycle regulators must be downstream of these signals. The mammary gland and uterus also undergo highly specialized proliferation and differentiation during pregnancy. In the male, as discussed above, the process of spermatogenesis throughout adult life is exquisitely sensitive to alterations in cell division and differentiation.

Examples of essential roles for cell cycle regulators in adult reproductive function

Cyclin D1

Although mutation of cyclin D1 yielded viable homozygous mice, the mice were small and exhibited symptoms of neurological defects. The most striking phenotypes were defects in retinal development in the eye and in mammary gland development.[53,54] The mammary gland appeared to be relatively normal during postnatal development, but, during pregnancy, a marked reduction in acinar differentiation was observed. In adult mutant females, the mammary epithelium does not undergo the robust proliferative changes associated with pregnancy, even though there are normal levels of ovarian steroids present. Cyclin D1 may thus be involved with steroid-induced proliferation in the adult mammary epithelium, but is clearly not essential for the development of most tissues and organs. Whether most organs compensate for the loss of cyclin D1 by using other members of the D-type cyclin (or other cyclins), or by activating alternative pathways for stimulating cell proliferation, is still not known. An additional intriguing property of cyclin D1 function in the mammary epithelium is its function in potentiating transcription of estrogen receptor-regulated genes. Cyclin D1 has been shown to activate estrogen receptor-regulated genes in the absence of the estrogen ligand (and also the absence of CDK).[55]

Cyclin D2

Gene targeting of cyclin D2 revealed that it is a follicle-stimulating hormone (FSH)-responsive gene involved in ovarian and testicular cell proliferation.[54] Homozygous cyclin D2-deficient mice were born at the expected Mendelian ratio and were indistinguishable phenotypically from their heterozygous or non-mutant littermates. However, cyclin D2-deficient females were sterile because of a defect in granulosa cell proliferation in response to FSH, and the males, although fertile, displayed hypoplastic testes. The numbers of ovarian follicles and oocytes were normal in mutant females, but the number of granulosa cells was markedly decreased. The oocytes were apparently unaffected in these animals, since oocytes removed from the follicles and subjected to *in vitro* maturation and fertilization produced viable blastocysts.[54]

The CKI p27

p27-deficient mice exhibit hyperplasia in multiple organs and marked disorganization of cellular patterns with variable penetrance.[56,57] Of particular interest to this discussion is the observation of female infertility in p27 –/– mice, comprising a combination of abnormal estrous cycles, a poor rate of ovulation and the failure to produce estrogen for implantation. The ovarian defect is correlated with an abnormally prolonged cellular proliferation of the developing corpus luteum. A particularly intriguing recent study focused on the apparently opposing functions of cyclin D1 and p27.[58] The experimental approach was to generate mice that were doubly mutant for both p27 and cyclin D1. The double mutants restored growth rates to near wild-type levels and correction of some but not all of the individual mutant traits. For cyclin D1, body weight, retinal hypoplasia and male aggressiveness were ameliorated; for p27, body weight, retinal hyperplasia and implantation were corrected. However, p27 –/– defects in aberrant estrous cycles, luteal cell proliferation and susceptibility to pituitary tumors were still present.

PERSPECTIVES ON THE ROLE OF CELL CYCLE REGULATORS IN HUMAN REPRODUCTION

The examples cited above provide clear evidence for diverse roles for cell cycle genes in mammalian, and, very probably, human reproduction. Infertility is a common clinical problem, with an estimated 10% of couples being affected. Once systemic complications are ruled out, genetic mutations should be considered as possible culprits. The search for genes involved in reproduction has taken two routes: the first, starting with a phenotype and searching for the responsible gene; the second, mutating genes and looking for resulting sterility. The second strategy is represented by the examples cited above, in which known cell cycle regulators were mutated and the effects on gametogenesis and development, etc. were subsequently observed. Although no cell cycle genes *per se* have been identified by the first (reverse genetics) approach, interesting DNA repair and genome stability genes, such as the *Ataxia telangiectasia mutated* (*ATM*) and the *BRCA2* genes, play roles in germ cell differentiation (reviewed in reference 59). A 'combination of the two approaches' is under way in our laboratory and others, in which cell cycle genes known to affect germ cell development in mice (for example cyclin A1) are screened for mutations in DNA from infertile patients with a phenotype similar to that observed in the mouse knock-out model (arrest in meiosis, in this case).

The cellular specificity of the male sterile phenotype resulting from mutation in the cyclin A1 gene, and the possibility that other such molecules may be discovered, suggest that such proteins may be considered as targets for contraceptive intervention. The prediction would be that small molecules capable of crossing the blood–testis barrier could affect either the production of cyclin A1, by interfering specifically with its transcription, or its activity, possibly by interfering with its ability to interact with its CDK partner.

Finally, one of the more intriguing considerations of cell cycle control has emerged from efforts to use a somatic, as opposed to gametic, nucleus to support embryonic development. Two very provocative but concise reviews of recent attempts to clone mammalian embryos may be found in articles by Campbell and colleagues[60] and Wakayama and Yanagamachi.[61] These discussions outline some of the challenges for understanding the very high failure of apparently 'fertilized' and activated embryos to develop into viable fetuses and liveborns. They can be read bearing in mind the underlying question of whether there is a finely tuned cell cycle regulator required for synchrony between the nuclear and cytoplasmic compartments of the embryo.

ACKNOWLEDGEMENTS

This work was supported in part by a grant from the National Institutes of Health (NIH), HD34915. The help of Erika Laurion in various aspects of the preparation of this chapter is gratefully acknowledged.

REFERENCES

1. Pines J. Cyclins and cyclin-dependent kinases: a biochemical view. *Biochem J* 1995 308: 697–711.
2. Fisher RP. CDKs and cyclins in transition(s). *Curr Opin Genet Dev* 1997 7: 32–38.
3. Pestell RG, Albanese C, Reutens AT, Segall JE, Lee RJ, Arnold A. The cyclins and cyclin-dependent kinase inhibitors in hormonal regulation of

proliferation and differentiation. *Endocr Rev* 1999 20: 501–534.

4. Stillman B. Cell cycle control of DNA replication. *Science* 1996 274: 1659–1664.
5. Elledge SJ. Cell cycle checkpoints: preventing an identity crisis. *Science* 1996 274: 1664–1672.
6. Clarke DJ, Gimenez-Abian JF. Checkpoints controlling mitosis. *Bioessays* 2000 22: 351–363.
7. Yew PR. Ubiquitin-mediated proteolysis of vertebrate G_1- and S-phase regulators. *J Cell Physiol* 2001 187: 1–10.
8. Masui Y, Markert CL. Cytoplasmic control of nuclear behavior during meiotic maturation of frog oocytes. *J Exp Zool* 1971 177: 129–145.
9. Nurse P. Universal control mechanism regulating onset of M-phase. *Nature* 1990 344: 503–508.
10. Hunt T. Cyclins and their partners: from a simple idea to complicated reality. *Semin Cell Biol* 1991 2: 213–222.
11. Sherr CJ. Mammalian G_1 cyclins. *Cell* 1993 73: 1059–1065.
12. Roberts JM. Evolving ideas about cyclins. *Cell* 1999 98: 129–132.
13. De Bondt HL, Rosenblatt J, Jancarik J, Jones HD, Morgan DO, Kim SH. Crystal structure of cyclin-dependent kinase 2. *Nature* 1993 363: 595–602.
14. King RW, Deshaies RJ, Peters JM, Kirschner MW. How proteolysis drives the cell cycle. *Science* 1996 274: 1652–1659.
15. Koch C, Nasmyth K. Cell cycle regulated transcription in yeast. *Curr Opin Cell Biol* 1994 6: 451–459.
16. Patton EE, Willems AR, Tyers M. Combinatorial control in ubiquitin-dependent proteolysis: don't Skp the F-box hypothesis. *Trends Genet* 1998 14: 236–243.
17. Townsley FM, Ruderman JV. Proteolytic ratchets that control progression through mitosis. *Trends Cell Biol* 1998 8: 238–244.
18. Peeper DS, Parker LL, Ewen ME *et al.* A- and B-type cyclins differentially modulate substrate specificity of cyclin-CDK complexes. *EMBO J* 1993 12: 1947–1954.
19. Schulman BA, Lindstrom DL, Harlow E. Substrate recruitment to cyclin-dependent kinase 2 by a multipurpose docking site on cyclin A. *Proc Natl Acad Sci USA* 1998 95: 10453–10458.
20. Coleman TR, Dunphy WG. Cdc2 regulatory factors. *Curr Opin Cell Biol* 1994 6: 877–882.
21. Solomon MJ, Kaldis P. Regulation of CDKs by phosphorylation. *Results Probl Cell Differ* 1998 22: 79–109.
22. Kaldis P. The CDK-activating kinase (CAK): from yeast to mammals. *Cell Mol Life Sci* 1999 55: 284–296.
23. Sherr CJ, Roberts JM. Inhibitors of mammalian G_1 cyclin-dependent kinases. *Genes Dev* 1995 9: 1149–1163.
24. Kato JY, Matsuoka M, Strom DK, Sherr CJ. Regulation of cyclin D-dependent kinase 4 (CDK4) by CDK4-activating kinase. *Mol Cell Biol* 1994 14: 2713–2721.
25. Wolgemuth DJ, Rhee K, Wu S, Ravnik SE. Genetic control of mitosis, meiosis and cellular differentiation during mammalian spermatogenesis. *Reprod Fertil Dev* 1995 7: 669–683.
26. Alberts B, Bray D, Lewis J, Raff M, Roberts K, Watson J. *Molecular Biology of the Cell.* New York: Garland Publishing, 1983: 27.
27. Choi T, Aoki F, Mori M, Yamashita M, Nagahama Y, Kohmoto K. Activation of p34cdc2 protein kinase activity in meiotic and mitotic cell cycles in mouse oocytes and embryos. *Development* 1991 113: 789–795.
28. Fulka J Jr, Jung T, Moor RM. The fall of biological maturation promoting factor (MPF) and histone H1 kinase activity during anaphase and telophase in mouse oocytes. *Mol Reprod Dev* 1992 32: 378–382.
29. Chapman DL, Wolgemuth DJ. Expression of proliferating cell nuclear antigen in the mouse germ line and surrounding somatic cells suggests both proliferation-dependent and -independent modes of function. *Int J Dev Biol* 1994 38: 491–497.
30. Wiltshire T, Park C, Caldwell KA, Handel MA. Induced premature G_2/M-phase transition in pachytene spermatocytes includes events unique to meiosis. *Dev Biol* 1995 169: 557–567.
31. Ravnik SE, Wolgemuth DJ. The developmentally restricted pattern of expression in the male germ line of a murine cyclin A, cyclin A2, suggests roles in both mitotic and meiotic cell cycles. *Dev Biol* 1996 173: 69–78.
32. Sweeney C, Murphy M, Kubelka M *et al.* A distinct cyclin A is expressed in germ cells in the mouse. *Development* 1996 122: 53–64.
33. Howe JA, Howell M, Hunt T, Newport JW. Identification of a developmental timer regulat-

ing the stability of embryonic cyclin A and a new somatic A-type cyclin at gastrulation. *Genes Dev* 1995 9: 1164–1176.

34. Yang R, Morosetti R, Koeffler HP. Characterization of a second human cyclin A that is highly expressed in testis and in several leukemic cell lines. *Cancer Res* 1997 57: 913–920.
35. Ravnik SE, Wolgemuth DJ. Regulation of meiosis during mammalian spermatogenesis: the A-type cyclins and their associated cyclin-dependent kinases are differentially expressed in the germ-cell lineage. *Dev Biol* 1999 207: 408–418.
36. Murphy M, Stinnakre MG, Senamaud-Beaufort C *et al.* Delayed early embryonic lethality following disruption of the murine cyclin A2 gene [Published erratum appears in *Nat Genet* 1999 23: 481]. *Nat Genet* 1997 15: 83–86.
37. Liu D, Matzuk MM, Sung WK, Guo Q, Wang P, Wolgemuth DJ. Cyclin A1 is required for meiosis in the male mouse. *Nat Genet* 1998 20: 377–380.
38. Edgar B. Diversification of cell cycle controls in developing embryos. *Curr Opin Cell Biol* 1995 7: 815–824.
39. Hartley RS, Rempel RE, Maller JL. *In vivo* regulation of the early embryonic cell cycle in *Xenopus*. *Dev Biol* 1996 173: 408–419.
40. McConnell J. Molecular basis of cell cycle control in early mouse embryos. *Int Rev Cytol* 1991 129: 75–90.
41. Schultz RM. Regulation of zygotic gene activation in the mouse. *Bioessays* 1993 15: 531–538.
42. MacAuley A, Werb Z, Mirkes PE. Characterization of the unusually rapid cell cycles during rat gastrulation. *Development* 1993 117: 873–883.
43. Chapman DL, Wolgemuth DJ. Identification of a mouse B-type cyclin which exhibits developmentally regulated expression in the germ line. *Mol Reprod Dev* 1992 33: 259–269.
44. Chapman DL, Wolgemuth DJ. Isolation of the murine cyclin B2 cDNA and characterization of the lineage and temporal specificity of expression of the B1 and B2 cyclins during oogenesis, spermatogenesis and early embryogenesis. *Development* 1993 118: 229–240.
45. Minshull J, Pines J, Golsteyn R *et al.* The role of cyclin synthesis, modification and destruction in the control of cell division. *J Cell Sci Suppl* 1989 12: 77–97.
46. Kobayashi H, Minshull J, Ford C, Golsteyn R, Poon R, Hunt T. On the synthesis and destruction of A- and B-type cyclins during oogenesis and meiotic maturation in *Xenopus laevis*. *J Cell Biol* 1991 114: 755–765.
47. Gautier J, Minshull J, Lohka M, Glotzer M, Hunt T, Maller JL. Cyclin is a component of maturation-promoting factor from *Xenopus*. *Cell* 1990 60: 487–494.
48. Brandeis M, Rosewell I, Carrington M *et al.* Cyclin B2-null mice develop normally and are fertile whereas cyclin B1-null mice die *in utero*. *Proc Natl Acad Sci USA* 1998 95: 4344–4349.
49. Winston N, Bourgain-Guglielmetti F, Ciemerych MA *et al.* Early development of mouse embryos null mutant for the cyclin A2 gene occurs in the absence of maternally derived cyclin A2 gene products. *Dev Biol* 2000 223: 139–153.
50. Kalitsis P, Earle E, Fowler KJ, Choo KH. *Bub3* gene disruption in mice reveals essential mitotic spindle checkpoint function during early embryogenesis. *Genes Dev* 2000 14: 2277–2282.
51. Dealy MJ, Nguyen KV, Lo J *et al.* Loss of Cul1 results in early embryonic lethality and dysregulation of cyclin E. *Nat Genet* 1999 23: 245–248.
52. Wang Y, Penfold S, Tang X *et al.* Deletion of the Cul1 gene in mice causes arrest in early embryogenesis and accumulation of cyclin E. *Curr Biol* 1999 9: 1191–1194.
53. Fantl V, Stamp G, Andrews A, Rosewell I, Dickson C. Mice lacking cyclin D1 are small and show defects in eye and mammary gland development. *Genes Dev* 1995 9: 2364–2372.
54. Sicinski P, Donaher JL, Parker SB *et al.* Cyclin D1 provides a link between development and oncogenesis in the retina and breast. *Cell* 1995 82: 621–630.
55. Zwijsen RM, Wientjens E, Klompmaker R, van der Sman J, Bernards R, Michalides RJ. CDK-independent activation of estrogen receptor by cyclin D1. *Cell* 1997 88: 405–415.
56. Kiyokawa H, Kineman RD, Manova-Todorova KO *et al.* Enhanced growth of mice lacking the cyclin-dependent kinase inhibitor function of p27 (KIP1). *Cell* 1996 85: 721–732.
57. Nakayama K, Ishida N, Shirane M *et al.* Mice lacking p27 (KIP1) display increased body size, multiple organ hyperplasia, retinal dysplasia, and pituitary tumors. *Cell* 1996 85: 707–720.
58. Tong W, Pollard JW. Genetic evidence for the interactions of cyclin D1 and p27 (KIP1) in mice. *Mol Cell Biol* 2001 21: 1319–1328.

59. Venables JP, Cooke HJ. Lessons from knockout and transgenic mice for infertility in men. *J Endocrinol Invest* 2000 23: 584–591.
60. Campbell KH, Loi P, Otaegui PJ, Wilmut I. Cell cycle co-ordination in embryo cloning by nuclear transfer. *Rev Reprod* 1996 1: 40–46.
61. Wakayama T, Yanagimachi R. Cloning the laboratory mouse. *Semin Cell Dev Biol* 1999 10: 253–258.

Apoptosis

Jonathan L. Tilly

INTRODUCTION

As of the end of January 2002, there were over 90 000 citations in the PubMed (National Center for Biotechnology and Information, National Library of Medicine) scientific literature database that were retrieved using 'cell death' as the key word, nearly 42 000 using 'programmed cell death' as the key word and over 52 000 using 'apoptosis' as the key word. While these numbers may not appear overly impressive at first glance, one has to consider that, during the time period prior to and including 1980, only a few more than 12 000 were retrieved using 'cell death' as the key word (over seven-fold increase), 29 using 'programmed cell death' as the key word (over 1400-fold increase) and 40 using 'apoptosis' as the key word (1300-fold increase). That so many papers have been published within the past two decades dealing with cell death is even more striking when one considers that reports of 'naturally occurring' cell death can be found in the scientific literature as far back as the 1880s, and in scattered research publications throughout the first five to six decades of the 20th century.[1,2]

Two of the earliest studies are particularly relevant to the subject matter of this chapter. In 1885, Walther Flemming described a process of 'chromatolysis' as being responsible for granulosa cell loss during ovarian follicular atresia[3] and for degeneration of testicular germ cell populations.[4] Apart from the fact that these studies were conducted over a century ago, the work of Flemming is most noteworthy because he documented beautifully the morphological features of apoptotic cells almost nine decades before the term 'apoptosis' was first coined.[5] This chapter explores how the application of powerful molecular biological approaches to study cell death in the female and male reproductive systems has allowed scientists to extend dramatically the pioneering work of Flemming. In addition, several examples of how apoptosis-based therapeutic approaches could be used to prevent normal and premature menopause, as well as to enhance fertility, are discussed.

BACKGROUND

Defining and detecting apoptosis

Kerr, Wyllie and Currie first proposed the use of the term 'apoptosis' (from the Greek words 'apo' and 'ptosis', meaning 'to fall away from') in 1972 to describe a relatively conserved set of morphological features observed in a wide variety of cell types during 'physiological' episodes of cell death in mammals.[5] Importantly, this study unequivocally established that developmental and homeostatic cell deaths controlled by the body can be easily distinguished from 'accidental' cell deaths initiated by a noxious insult. From the work of Kerr and colleagues,[5,6] apoptosis is generally characterized at the light and electron microscopic levels by separation of the cell from its neighboring cells and/or its extracellular matrix (for epithelial cell lineages, this step has been referred to as 'anoikis', a Greek term for 'homelessness'[7]), a loss of cell volume (cytoplasmic shrinkage), chromatin condensation and margination along the nuclear envelope (nuclear pyknosis), and the final budding and fragmentation of the cell into plasma membrane-bound vesicles termed apoptotic bodies (Figure 1).[5,6] The exact temporal sequence of events may differ, depending on cell lineage, but the ultimate fate of the dead cell is phagocytosis of the corpse

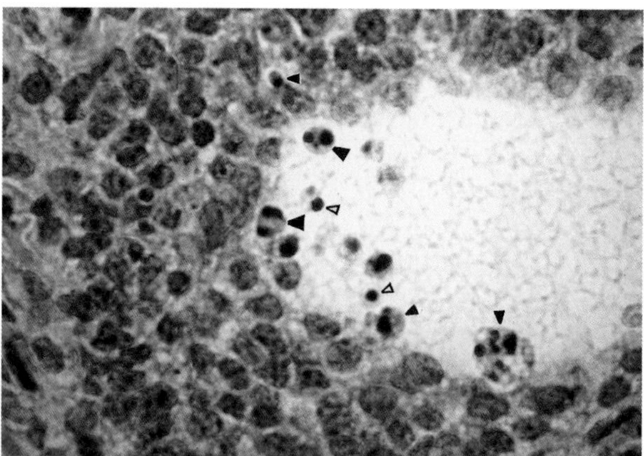

Figure 1 Hematoxylin–eosin staining of a mouse ovarian antral follicle in the early stages of atresia, showing the presence of apoptotic granulosa cells (darkened arrowheads) and apoptotic bodies (open arrowheads) in the periantral region of the follicle. Note the cytoplasmic condensation and nuclear pyknosis characteristic of cells undergoing apoptosis. Reproduced from reference 8, with permission

fragments by resident macrophages or neighboring epithelial cells. Although this rapid and efficient removal of apoptotic bodies by phagocytosis makes detection of apoptosis *in vivo* extremely difficult in most cases,[9] phagocytosis prevents these cellular remnants from undergoing secondary necrosis in the extracellular space, an event that would eventually lead to an inflammatory response. This latter feature further segregates apoptosis from accidental cell death since widespread tissue destruction and inflammation are key components of pathological tissue injury.[1,9,10]

Apoptotic cells also exhibit a number of characteristic biochemical changes, one of the most well known being the fragmentation of nuclear chromatin into a discrete pattern of mono- and oligonucleosomal units (180–200-bp multiples) that appear as the 'rungs of a ladder' following conventional agarose gel electrophoresis (Figure 2 lower panel).[11] Although widely used as a marker for the detection of apoptosis, internucleosomal cleavage of DNA is not required for apoptotic cell death in some systems.[15–17] Importantly, however, cleavage of DNA into higher-molecular-weight structures termed rosettes (300 kbp) and loops (50 kbp) precedes internucleosomal fragmentation during apoptosis, and appears to demarcate all apoptotic cell deaths (Figure 2 upper panel). Thus, caution should be exercised in the interpretation of data derived solely from the analysis of low-molecular-weight DNA ladders since the absence of internucleosomal DNA cleavage does not always confirm the absence of apoptosis. Nonetheless, the development of histochemical techniques for labelling both high- and low-molecular-weight DNA fragments in fixed tissue sections or cultured cells has greatly facilitated *in situ* detection of apoptotic cell death in tissues composed of multiple cell lineages (Figure 3a and b).[18,19] In addition, technologies that permit fluorescence microscopy-based visualization of chromatin condensation in cells undergoing apoptosis *in vivo* or *in vitro* (using DNA-binding dyes such as Hoechst 33442 or 4',6-diamidino-2-phenylindole (DAPI); Figure 3c),[24] or DNA cleavage in single cells (using the microelectrophoresis-based comet assay; Figure 3d–g),[25] have been widely used to detect apoptosis, regardless of the extent of DNA fragmentation.

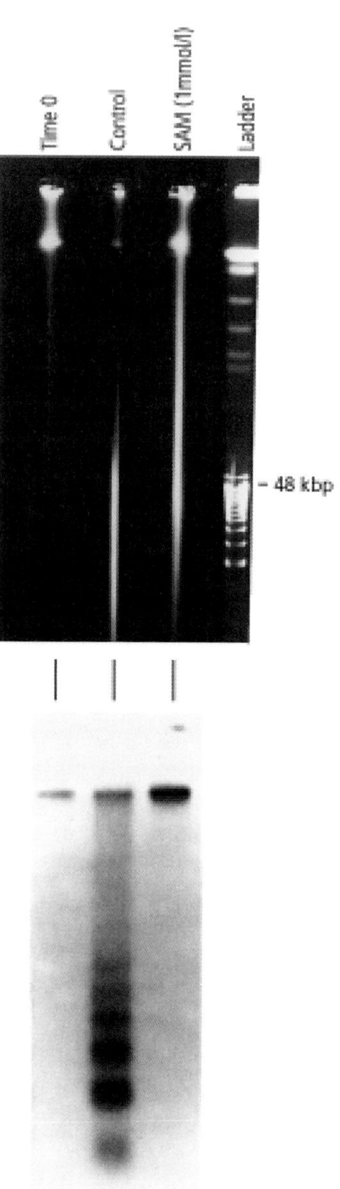

Figure 2 Biochemical analysis of DNA cleavage associated with apoptosis, as detected by pulsed-field gel electrophoresis (upper panel; Ladder, size standards) and conventional agarose gel electrophoresis (lower panel), to assess high- and low-molecular-weight DNA fragments, respectively. Healthy antral follicles isolated from gonadotropin-primed immature female rats[12,13] were analyzed immediately (Time 0) or were incubated *in vitro* without tropic support in the absence (Control) or presence of the nuclease inhibitor, sodium aurothiomalate (SAM). Note that SAM prevents the formation of the 'DNA ladder' but is incapable of suppressing high-molecular-weight DNA cleavage, consistent with the inability of SAM to prevent morphological aspects of apoptosis in granulosa cells.[14] Reproduced from reference 14, with permission

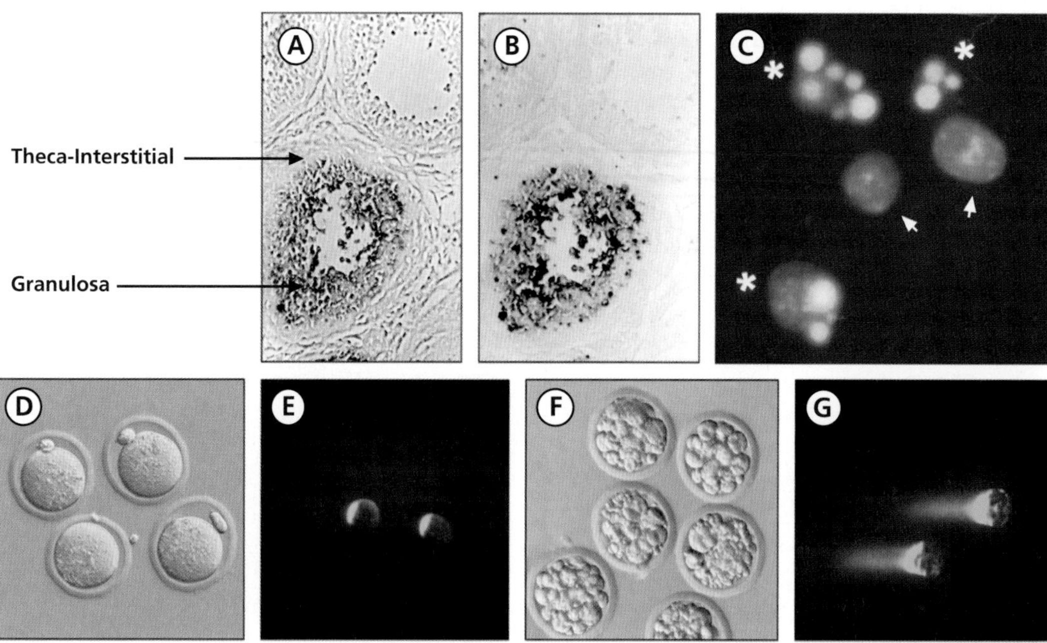

Figure 3 Alternative methods to detect DNA cleavage associated with apoptosis. (a, b) *In situ* labelling of DNA 3′-ends in fixed ovarian tissue sections prepared from a prepubertal female rat, localizing the occurrence of DNA cleavage to granulosa cells of atretic early antral follicles. (a) A phase-contrast photomicrograph to highlight tissue architecture; (b) the same field of view under bright-field microscopy (darkly stained cells are positive for DNA cleavage). For details, see reference 19. (c) Fluorescence microscopic analysis of nuclear collapse using the DNA-binding dye, 4′,6-diamidino-2-phenylindole (DAPI), in mouse granulosa cells induced to undergo apoptosis by serum starvation *in vitro*.[20,21] Compared with nuclei of non-apoptotic cells that appear spherical with lightly fluorescent and evenly distributed chromatin (arrowheads), nuclei of apoptotic cells show condensation and highly fluorescent punctate dots of clumped chromatin (asterisks). Reproduced from reference 20, with permission. (d–g) Analysis of DNA cleavage in single cells using the comet assay. In mouse oocytes induced to undergo apoptosis by doxorubicin exposure *in vitro* (f),[22] a plume of DNA fragments resembling the tail of a comet can be observed migrating away from the clump of intact DNA which does not exit the cell (g). By comparison, vehicle-treated, non-apoptotic oocytes (d) show only a single clump of intact DNA (e). Reproduced from reference 23, with permission

Other 'forms' of cell death

One of the most misunderstood and confusing issues raised when evaluating the cell death literature stems from the numerous terms used to describe the process of cellular elimination. In addition to apoptosis, 'programmed cell death' (PCD) and 'autophagocytosis' (or 'autophagic vacuolization') are the most prevalent, along with the widely misused catch-all term for accidental or pathological cell death, 'necrosis'. Programmed cell death was set forth in the literature by Lockshin and Williams in 1964, and was originally used to describe the developmental death of intersegmental muscle cells during metamorphosis in silkmoths.[26] Shortly thereafter, PCD was most often used in the context of identifying developmental cell deaths, often thought to be intrinsically controlled by a predefined genetic sequence of events.[1] The emergence of a new term, apoptosis, to describe the characteristic morphology of cells dying by 'natural' means was both a blessing and a curse, since it became unclear what the exact relationship was between apoptosis and PCD.[1,27,28] Majno and Joris published one of the clearest descriptions of the terminology problems faced by cell death researchers, and concluded from a number of observations that the two terms, in their truest sense, are not necessarily interchangeable.[1]

In contrast to the confusion over apoptosis and PCD, autophagocytosis is clearly a distinct form of cell death but is nonetheless just as physiological.[29,30] The primary hallmark of a cell dying via autophagocytosis is the presence of numerous large vacuoles in the cytoplasm. There are relatively few documented instances of this form of cell death in the body, although increasing awareness that not all physiological cell deaths display the characteristic features of apoptosis has prompted scientists to examine tissues more carefully for these 'non-apoptotic' deaths. Of direct relevance to reproductive biology, autophagocytosis has been identified in cells of the human regressing corpus luteum[31] and in amniotic epithelial cells of the rat term placenta.[32] Moreover, there is clear biochemical overlap between apoptosis and autophagocytosis, at least in the rat placenta, since internucleosomal DNA cleavage, a predominant feature of apoptosis, occurs in vacuolated amniotic cells during their demise.[32,33] Consequently, assessment of the occurrence and regulation of cell death in various organs of the male and female reproductive systems should be made with care, since important features of tissue turnover may be missed if all of the criteria for identifying apoptotic cells are strictly adhered to as the end-points. Finally, a comment should also be made regarding the use of the term 'necrosis' as a descriptor for 'unnatural', pathological or accidental cell deaths. In its truest sense, necrosis is the end-result of all cell deaths, that being the final and irreversible dissolution of a dead cell. Thus, as proposed by Majno and Joris[1] and by Trump and Berezesky,[34] use of the term 'necrosis' to identify accidental cell deaths should be discouraged.

The core machinery of apoptosis

A great deal of what is known about the control of cell death in mammals has been derived from or supported by comparisons with data obtained from genetic studies of PCD in the nematode, *Caenorhabditis elegans*, and in the fruit fly, *Drosophila melanogaster* (Figure 4).[35–37] For example, in the *C. elegans* hermaphrodite, 131 of the 1090 somatic cells generated during development undergo PCD, and these deaths are dependent upon the actions of at least four key genes: *egl-1*, *ced-9*, *ced-3* and *ced-4*. The *egl-1* gene product triggers cell death by antagonizing the antiapoptotic functions of the CED-9 protein. By comparison, the CED-3 and CED-4 proteins are required for PCD, since inactivation of either gene results in survival of the cells that normally undergo death during *C. elegans* development. Importantly, orthologs of *egl-1/ced-9*, *ced-4* and *ced-3* have been identified in *Drosophila* and in vertebrates, providing evidence of a cell death pathway that has been remarkably conserved through evolution (Figure 4).[35–37]

Prior to the cloning of the *ced-9* gene, a gene in vertebrates termed *bcl-2* was shown to encode a protein with biological activity similar to that of the CED-9 protein, namely a repression of cell death.[38,40,47] Once *ced-9* was isolated, alignment of this gene with *bcl-2* revealed significant structural and sequence similarity, in addition to common function. Further support for the hypothesis that *ced-9* and *bcl-2* are evolutionary orthologs comes from studies demonstrating that forced expression of human Bcl-2 protein in nematodes provides a partial restoration of cellular survival.[42] Although only one other CED-9-related protein (i.e. EGL-1) has been identified in *C. elegans* to date,[43] over the past several years a large family of Bcl-2-related proteins has been characterized in vertebrate species.[38,40,41] These proteins, which interact to form hetero- and homodimers, can be segregated into two classes based on function: those that delay or inhibit apoptosis (e.g. Bcl-2, Bcl-x_L, Mcl-1, A1/Bfl-1, Bcl-w, NR-13) and those that facilitate or induce apoptosis (e.g. Bcl-x_S, Bax, Bak, Bad, Hrk/DP5, Bid, Bik/Nbk, Blk, Bim, Bok/Mtd, Noxa, Nix, Bnip3, PUMA, Bcl-rambo, Bcl-G_L, Bcl-G_S). A common structural feature of these proteins is the Bcl-2 homology (BH) domain. This motif has been used to segregate further the members of this family into those that are 'BH3-domain only' (i.e. Bad, Bid, Bim, Hrk/DP5, Bik/Nbk, Blk, Bnip3, Nix, Noxa, PUMA, Bcl-G_S) and those that are 'multi-domain' (i.e. contain at least two, and up to four, BH domains; all others). Finally, while many roles for Bcl-2 family members have been proposed, their principal function is probably to alter mitochondrial integrity and/or regulate release of 'apoptogenic' factors, such as cytochrome *c*, apoptosis-inducing factor (AIF), Smac/DIABLO and endonuclease-G from mitochondria into the cytoplasm (Figure 5).[39–41,44–46]

Indeed, it is through this activity that Bcl-2 family members are thought to serve as critical determinants of whether or not the final

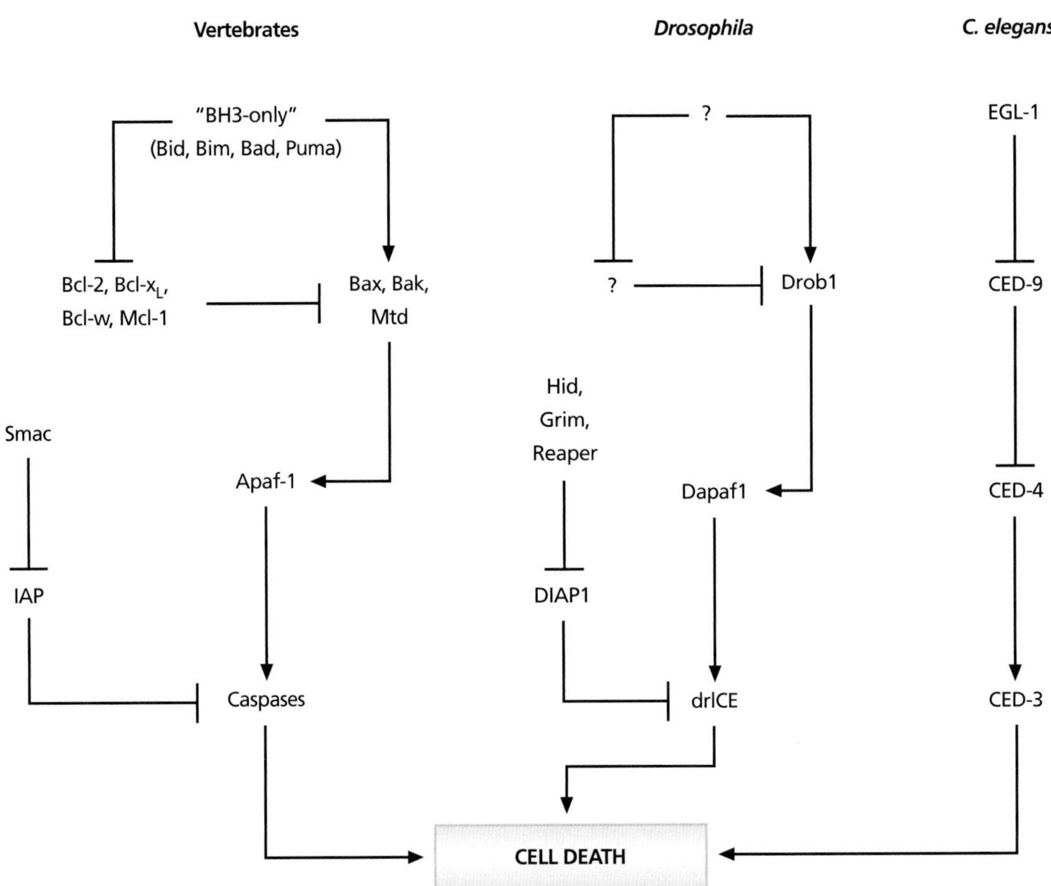

Figure 4 Conservation of the core cell death machinery through evolution. A number of genetic studies have established a remarkably conserved framework of orthologous proteins that regulate apoptosis in vertebrates, *Drosophila melanogaster* and *Caenorhabditis elegans*. There are three principal steps, the first of which involves Bcl-2 family members in vertebrates (including pro-apoptotic BH3 domain-only, as well as pro- and antiapoptotic multidomain, family members; see 'The core machinery of apoptosis' for details and references). Corresponding proteins in *Drosophila* and *C. elegans* are referred to as Drob1 (also named dBorg-1, Debcl or DBok) and EGL-1/CED-9, respectively. These proteins, through various mechanisms, function ultimately as modulators of a molecular 'bridge' protein termed Apaf-1 in vertebrates, Dapaf-1 (also referred to as HAC-1 or Dark) in *Drosophila*, or CED-4 in *C. elegans*, which activates proteolytic degradation of the cell. This final step is carried out by a family of enzymes referred to as caspases in vertebrates, drICE (and other related enzymes including Dronc, Dcp-1, Dcp-2/Dredd and Decay) in *Drosophila*, or CED-3 in *C. elegans*. Caspase activation and/or activity is held in check by the inhibitor-of-apoptosis proteins (IAP) in vertebrates or DIAP1 in *Drosophila*. The function of IAP can, in turn, be suppressed by a group of structurally related proteins referred to as Smac (also named DIABLO) in vertebrates, or Hid, Grim and Reaper in *Drosophila*.[35–39]

executioners of apoptosis, a family of CED-3 orthologs in vertebrates termed caspases (cysteine aspartic acid-specific proteases; see below), are recruited into action. Parallel studies published in 1997 demonstrated that Bcl-2 directly blocks mitochondrial cytochrome c release and the subsequent activation of caspases known to occur downstream of Bcl-2 function.[48,49] This observation has since been extended to a second antiapoptotic Bcl-2 family member, Bcl-x$_L$.[50] By contrast, overexpression of Bax has been shown to cause a rapid destabilization of mitochondria and release of cytochrome *c*, leading to caspase activation.[51,52] These findings have since been reinforced by data

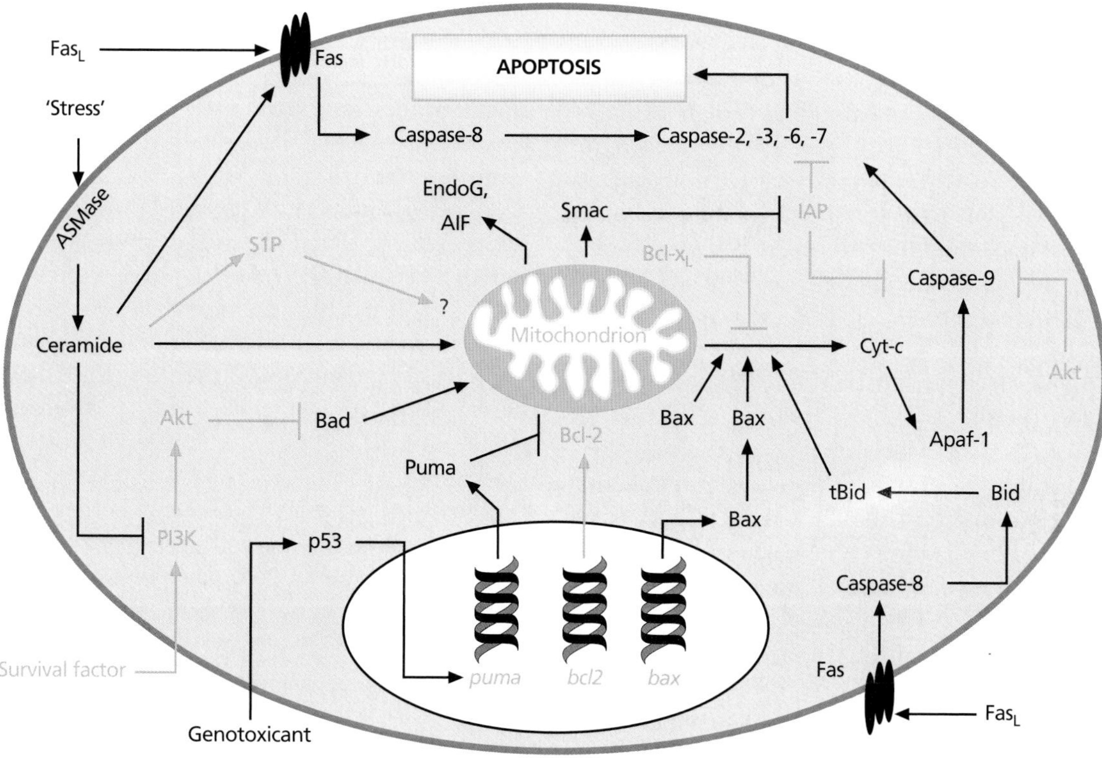

Figure 5 Schematic diagram depicting interactions between key components of the apoptotic cell death program in vertebrates (see 'The core machinery of apoptosis' for further details and references). A central check-point in the cell death pathway is believed to exist at the level of the mitochondrion, where Bcl-2 family members, such as Bcl-2, Bcl-x$_L$, Bax and Bak, localize and regulate release of cytochrome *c* (cyt-c) from this organelle. Cytoplasmic accumulation of cytochrome *c* promotes a conformational change in Apaf-1, leading to procaspase-9 interaction and activation. Caspase-9 then activates proforms of downstream ('death-effector') caspases, such as caspases-2, -3, -6 and -7. These enzymes coordinate proteolytic cleavage events that disrupt cellular homeostasis and produce the morphological and biochemical characteristics of apoptotic cells. Input at the level of the mitochondrion can come from many stimuli, including both survival factors coupled through receptors to phosphatidylinositol-3′-kinase (PI3K) and Akt, as well as death factors such as Fas ligand (Fas$_L$). Of note, the generation of active caspase-8 following Fas ligation can kill cells either by directly activating death-effector caspases (upper left) or by cleaving Bid to promote cytochrome *c* release and death-effector caspase activation (lower right). In addition, numerous external stresses that generate ceramide via acid sphingomyelinase (ASMase) can trigger apoptosis, which in some cases is counterbalanced by the antiapoptotic ceramide metabolite, sphingosine-1-phosphate (S1P). Finally, genotoxic damage is thought to use, among other things, p53 to signal transcriptional changes in cell death gene expression that may ultimately modulate mitochondrial cytochrome *c* release followed by death-effector caspase activation. As discussed in Figure 4, the inhibitor-of-apoptosis proteins (IAP) tightly regulate caspase activation, and in turn IAP function is suppressed by mitochondrial-derived Smac (or DIABLO). Other proapoptotic factors released from mitochondria include apoptosis-inducing factor (AIF) and endonuclease-G (EndoG), both of which drive chromatin degradation. Black text and arrows signify proapoptotic pathways, whereas blue text and arrows demarcate antiapoptotic pathways. Reproduced with modifications from reference 47, with permission

showing that recombinant Bax protein directly triggers cytochrome *c* release from isolated mitochondria, and that this event is abrogated by recombinant Bcl-x_L.[50]

However, the exact role of cytochrome *c* in caspase activation and apoptosis remained elusive until the purification and cloning of the first CED-4 ortholog in vertebrate species.[53] This protein, termed Apaf-1, was isolated as part of a trimeric protein complex that could reconstitute caspase activation and apoptotic events in cell-free assays (hence, the original names ascribed to these proteins were apoptotic protease-activating factors or Apaf-1, -2 and -3). The other two proteins of this complex were identified as cytochrome *c* (Apaf-2) and caspase-9 (Apaf-3).[53,54] Using a number of biochemical approaches, Wang and co-workers methodically dissected the actions and interactions of each component of the Apaf trimer in caspase activation. Their results, combined with additional data from other laboratories, have served as the basis for the following model of apoptosis execution. Release of cytochrome *c* from mitochondria, an event directly regulated by Bcl-2 family members, produces a conformational change in Apaf-1, allowing for heterodimeric interaction of the protein with procaspase-9. The resulting interaction generates procaspase-9 oligomers, leading to auto- or transcatalytic processing of the precursor enzyme to fully functional caspase-9. Once activated, caspase-9 then initiates a proteolytic cascade by sequential activation of death-effector caspases (-2, -3, -6 and -7; see below).[55] This model therefore predicts that, like CED-4 in *C. elegans*,[35] Apaf-1 indeed couples Bcl-2 protein family function (regulation of cytochrome *c* release) to caspase function (processing of procaspase-9 and activation of death-effector caspases) in the apoptotic cell death program in vertebrates. The recent knockout of the *caspase-9* and *apaf1* genes in mice supports this hypothesis in that the vast majority of null fetuses die prior to birth owing to an absence of critical apoptotic events during development.[56–58]

Finally, cloning of the *ced-3* gene in the nematode revealed its similarity to the cytokine-processing cysteine protease in vertebrates, interleukin (IL)-1β-converting enzyme or ICE.[59] These findings suggested that PCD in *C. elegans* was dependent upon the activity of a protease(s), a hypothesis later confirmed by experiments demonstrating that CED-3 is in fact a cysteine protease with cleavage site specificity at aspartic acid residues.[60] Moreover, a conserved active site sequence surrounding the catalytically important cysteine residue (QACXG), along with the specificity of these enzymes to cleave at aspartic acid residues, are fundamental structural and functional features that have permitted the subsequent characterization of at least 15 CED-3 orthologs in vertebrates to date.[61–64] It has also become clear that, like CED-3, caspases are instrumental to the completion of apoptosis in vertebrates through their selective destruction of key structural and functional proteins in the cell destined for deletion. Included among the many targets for caspase attack are proforms of caspases, cytoskeletal proteins, signal transduction molecules, DNA repair enzymes, RNA splicing components, cell cycle regulatory proteins, nuclease activity-modulating factors and nuclear matrix proteins.[62–64]

Connecting to the core: sphingolipids and phosphatidylinositol 3'-kinase/Akt as signals

Ceramide is a sphingolipid second messenger generated in cells by either membrane hydrolysis, via the actions of a sphingomyelinase, or *de novo* synthesis under conditions of inadequate growth factor support or other stresses known to induce apoptosis. Since the initial discovery of the sphingomyelin pathway over 13 years ago, numerous studies have been published on the potential role of ceramide in signalling cell death.[65,66] Interestingly, it is now known that ceramide can also be metabolized via ceramidase to sphingosine, which is then phosphorylated by sphingosine kinase to generate sphingosine-1-phosphate (S1P).[67] In some cell types, S1P can effectively counterbalance apoptosis induced by membrane-permeable ceramide analogs or external stressors known to work through elevations in intracellular ceramide levels.[67,68] Therefore, a rheostat model has been proposed in which cell fate is controlled by shifts in the balance between ceramide and S1P levels.[67]

From a mechanistic standpoint, ceramide may lead to the induction of apoptosis by at least three mechanisms. The first involves antagonism of phosphatidylinositol 3'-kinase (PI3K) and Akt, enzymes that constitute a prominent signal transduction pathway which couples growth factor receptor activation to cellular survival.[69–71] Evidence for this comes from studies that have both overexpressed and inhibited these enzymes in many cell types. Moreover, Akt has been reported to inactivate, via direct phosphorylation, the pro-apoptotic Bcl-2 family member, Bad,[72] and the proform of the apoptotic protease, caspase-9,[73] providing important evidence for direct communication between the PI3K/Akt pathway and key cell death regulatory molecules. Keeping in mind that the PI3K/Akt pathway thus serves a critical anti-apoptotic function in cells, several groups reported that ceramide directly suppresses both PI3K and Akt activities.[74–76] Furthermore, overexpression of Akt can suppress ceramide generation and protect cells from ceramide-induced apoptosis.[77] These data, when viewed as a whole, support the existence of a complex signalling network involving cross-talk between ceramide and PI3K/Akt, the outcome of which, namely cell survival or death, is dependent upon which pathway predominates (Figure 5).

In addition to altering the activity of the PI3K/Akt pathway, ceramide has also been reported to facilitate Bax-induced destabilization of isolated mitochondria[78] and to form 'pores' in lipid bilayers.[79] Accordingly, ceramide may induce apoptosis, at least in some cells, via a more direct route involving the induction of mitochondrial permeability transition and/or release of apoptogenic factors from mitochondria. Finally, ceramide has also recently been shown to be involved in, if not needed for, the 'capping' mechanism required for 'death receptors', such as Fas (see section on 'Death receptors and death domains' below), to become activated and signal apoptosis.[80] Collectively, these reports indicate that ceramide can act at multiple levels in the apoptosis signalling pathway to coordinate and transduce extracellular signals, ultimately leading to the execution phase of apoptosis.

Despite these insights into how ceramide may function in apoptosis, comparatively little is known of the mechanisms of action of S1P. One recent study has shown that S1P-mediated protection of human T cells from apoptosis is associated with reduced cellular levels of Bax protein.[81] In addition, S1P has been reported to inhibit caspase activation in Jurkat T lymphocytes exposed to various proapoptotic stimuli, such as death receptor activation (see below) or ceramide.[82] Unfortunately, in this latter study it was not determined whether S1P directly suppresses caspase activation or, alternatively, whether caspase activation is suppressed owing to inhibitory effects of S1P on more upstream components of the cell death machinery. Nevertheless, these data support a role for S1P in affecting at least two central components of the core apoptosis pathway: Bax and caspases.

Death receptors and death domains

Before concluding this section, a few comments should be made regarding the concept of 'death receptors', such as the receptors for Fas ligand (Fas_L) and tumor necrosis factor-α (TNF-α), and intracellular proteins recruited to these receptors upon ligand activation that signal for apoptosis. Efforts to understand the mechanisms by which extracellular signals actively initiate apoptosis have become a focal point of research in a number of laboratories. For example, a family of proteins, including TNF-α, Fas_L (also known as CD95 or APO-1), lymphotoxin, CD40 ligand, CD30 ligand, CD27 ligand, 4-1BB ligand and TNF-related apoptosis-inducing ligand (TRAIL), have been intensively studied in the context of apoptosis induction.[83,84] All of these ligands bind with specific receptors that, in most cases, share an important structural feature: an approximate 80-amino-acid segment in the cytoplasmic region of the receptor, termed the death domain, that is believed to be critical for initiating the cell death signal.[83,84]

Using Fas_L or Fas-activating antibodies as a stimulus for apoptosis, a number of investigations have provided evidence for the assembly of an intracellular death-inducing signalling complex (DISC) initiated by ligand–receptor interaction and subsequent receptor trimerization.[84] Association of the three death domains of the activated receptor trimer recruits an adapter protein, termed FADD or MORT1, that also possesses a death domain (for Fas interaction) as well as a death effector domain (for apoptosis signalling). Interaction of the death-domain of Fas with that of FADD is then thought to recruit a third protein to complete the DISC, that being the proform of caspase-8 (originally termed FLICE or MACH), through interactions involving the death-effector domains present in both FADD and the protease. Recruitment of procaspase-8 to the DISC then probably triggers autoprocessing of the enzyme, thus serving as a catalyst for the activation of downstream effector caspases that dismantle the cell during apoptosis (Figure 5).[84]

The precise intracellular mechanism set in motion by Fas ligation that leads to apoptosis remains somewhat controversial, however. Two pathways have been described for Fas-induced apoptosis in cells, referred to as type-I and type-II cells, that appear to be dependent upon the cell type examined and the stimulus used to activate the receptor (i.e. natural ligand versus receptor-activating antibody).[84,85] In type-I cells, relatively large amounts of procaspase-8 become activated at the DISC, producing a situation in which the levels of active caspase-8 are sufficient to process directly, and thus activate, downstream effector caspases (see 'The core machinery of apoptosis' above), which then rapidly execute apoptosis. In such a case, the need for mitochondrial involvement is bypassed. By comparison, in type-II cells, relatively low levels of active caspase-8 are generated owing to limited DISC assembly following Fas activation, and thus effector caspase activation cannot be propagated directly by the enzyme. In these cells, an amplification loop for caspase activation is recruited through caspase-8-mediated cleavage of Bid, a pro-apoptotic Bcl-2 family member (Figure 5).[86,87] Once cleaved, truncated Bid undergoes myristoylation,[88] facilitating its insertion into mitochondrial membranes to promote, in conjunction with Bax and Bak, release of the various apoptogenic factors that lead to effector caspase activation and nuclear disintegration.[89] Therefore, unlike type-I cells, mitochondrial involvement is a necessity for Fas-induced apoptosis to occur in type-II cells.[84]

EMERGING CONCEPTS IN REPRODUCTIVE MEDICINE

Overview

Cell death occurs in most, if not all, tissues of the female and male reproductive systems at some point during fetal development or postnatal life. Since a thorough discussion of each of these models cannot be accomplished within the scope of this chapter, three specific examples of how manipulating apoptosis may influence reproductive function are briefly overviewed below.

Postponement of menopause

Once the stockpile of quiescent primordial follicles is endowed in the ovaries shortly after birth, current evidence indicates that, unlike spermatogenesis in males, this pool of germ cells is not renewable. As such, mammalian females are provided with a finite number of follicle-enclosed oocytes that sustain reproductive function until the point of ovarian senescence (such as the menopause in humans and some primates), at which time this pool of germ cells has been almost entirely depleted.[90,91] In humans, it is estimated that between 100 and 1000 follicles remain in the ovaries at the time of the menopause. This is striking when one considers that the human female is born with up to 500 000 non-atretic follicles per ovary,[92] and that only 400 or so of these follicles will survive, and develop to the preovulatory stage to release an oocyte into the oviduct for possible fertilization. Consequently, the vast majority of female germ cells in the postnatal ovaries are depleted through the degenerative process of follicular atresia.[93,94]

The incidence or rate of atresia of immature (i.e. primordial and primary) follicles is clearly a significant event in determining the reproductive life span in females. Early studies by Baker estimated that up to one-half of the quiescent follicle pool endowed in the human ovaries at birth can be atretic.[92] These seminal studies also revealed that atresia of immature follicles in the human ovary is apparently initiated in the oocyte,[92] findings that have also been observed in laboratory animal models.[95] Importantly, the morphological features of oocyte death during atresia of primordial and primary follicles resemble many aspects of apoptosis in somatic cells.[92,95] Assuming, then, that postnatal loss of immature oocytes via follicular atresia is apoptotic in nature, could the process be targeted for manipulation as a means to prolong ovarian function into advanced age? Recent studies with Bax-deficient female mice indeed tested, and ultimately validated, this exciting possibility (Figure 6).

As discussed earlier (see 'The core machinery of apoptosis' above), Bax is a key proapoptotic member of the Bcl-2 family of cell death regulators. In the original report describing the effects of disrupting the *bax* gene in mice, Knudson and colleagues reported the presence of 'a marked accumulation of unusual atretic follicles' characterized by 'atrophic granulosa cells that presumably failed to undergo apoptosis'.[96] However, a subsequent in-depth assessment of oocyte numbers and atresia rates in the mutant females throughout postnatal life revealed another startling outcome of Bax deficiency, that being a dramatic extension of functional ovarian life span into very advanced chronological age.[95] This phenotype was determined to be a result of attenuated rates of atresia in the primordial and primary follicle pools during postnatal life. When taken with the evidence of a cell-autonomous role for Bax in killing oocytes,[23,97] it is probably safe to conclude that the reduced incidence of immature follicle atresia responsible for prolonging ovarian lifespan in *bax* mutant females is due to enhanced oocyte survival. As such, we must at least consider that oocyte-preserving strategies, developed for whatever reasons (see also

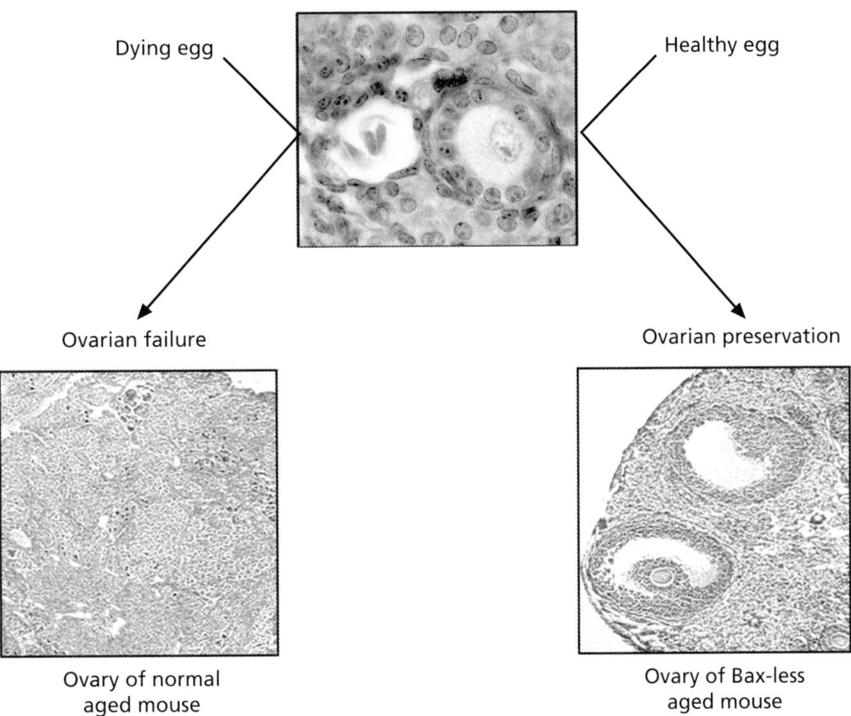

Figure 6 Schema representing the impact of Bax deficiency on postnatal primordial and primary oocyte survival in female mice, leading to a dramatic prolongation of ovarian function into advanced age. Compiled and reproduced from reference 95, with permission

below), could be used to slow the natural rate of oocyte depletion and thus delay the menopause in women.

Preservation of ovarian function and fertility in female cancer patients

Another paradigm of oocyte death with tremendous clinical significance concerns the pathological loss of germ cells in young girls and reproductive-age women treated for cancer with radiation or chemotherapy.[98–100] Although the mechanisms responsible for oocyte destruction in female cancer patients are not yet fully understood, observations from several recent studies have begun to provide some important insights. The first key finding was that mouse oocytes cultured *in vitro* in the presence of therapeutic levels of the anticancer drug, doxorubicin, rapidly undergo apoptosis rather than a pathological form of cell death (Figure 3d–g).[22,23,97] As a consequence, a potential new therapeutic approach to combat premature ovarian failure in female cancer patients was conceived: target key components of the oocyte death pathway needed for apoptosis to occur.[101] Indeed, using a combination of inhibitor-based approaches and genetically manipulated mice, it was reported that several discrete signalling molecules, including ceramide, Bax and caspases (specifically, caspase-2), are required for the induction of apoptosis in oocytes exposed to this chemotherapeutic drug *in vitro*.[23,97,102] Furthermore, Bax-deficient female mice were shown to be resistant to the ovotoxic effects of doxorubicin treatment *in vivo*.[97]

Despite the mechanistic insight provided by these investigations into how oocytes respond to anticancer therapies, a major hurdle remained before clinical application could be considered a realistic possibility: therapeutic methods designed to protect the female germline from pathological insults had to be technically feasible for eventual human use. Accordingly, two approaches were taken, one based on gene therapy and the other on small-molecule therapy. In the first study, the promoter of an oocyte-specific gene (*zona pellucida-3* or *ZP3*) was used to direct overexpression of the Bax antagonist, Bcl-2, in transgenic mice.[103] As anticipated, expression of the transgene product was detected only in oocytes. More importantly, the excess Bcl-2 protein in oocytes both prevented follicular atresia and conveyed resistance to doxorubicin-induced apoptosis.[102]

The second approach, probably even more amenable to therapeutic development, was based on previous work showing that S1P (see 'Connecting to the core: sphingolipids and phosphatidylinositol 3'-kinase/Akt as signals' above) could suppress apoptosis in oocytes exposed to chemotherapy *in vitro*.[97] Assuming that S1P would function similarly *in vivo*, Morita and colleagues then tested *in vivo* the efficacy of S1P in preventing radiotherapy-induced ovarian failure.[23] Results from this study showed that the massive oocyte destruction caused by the exposure of young adult female mice to ionizing radiation could be completely prevented by prior treatment with S1P delivered into the bursal cavity surrounding each ovary. Furthermore, the 'protected' oocytes remained fully competent for maturation, fertilization and preimplantation embryonic development,[23] and preliminary results from *in vivo* mating trials support the theory that S1P preserves a normal level of fertility in irradiated females.[104] Therefore, strategies such as S1P therapy represent promising new directions to pursue in our efforts to finally protect the ovaries in young girls and reproductive-age women from the ravages of anticancer therapy.[105] This knowledge may also prove useful for manipulating the numbers of primordial follicles present in the ovaries under normal conditions (see above), as well as for combating oocyte loss resulting from exposure of females to other pathological insults such as environmental toxicants.[106,107]

Impact of apoptosis on spermatogenesis and testicular homeostasis

Although it has been long recognized that a significant loss of male germ cells occurs during the process of spermatogenesis,[108] only in recent years has the term apoptosis been used to describe the naturally occurring death of spermatogonia (types A2, A3 and A4), spermatocytes and spermatids in the testis.[109,110] As in the female, however, it is at present unknown precisely why so many germ cells undergo apoptosis in the male. One of the most plausible theories suggests that apoptosis serves as a homeostatic mechanism for maintaining an appropriate number of germ cells that can be adequately supported and matured by the surrounding Sertoli cells. Whatever the case, under physiological conditions there appear to be at least three primary factors that determine the susceptibility of male germ cells to apoptosis: the developmental stage of the testis, the specific stage of the seminiferous epithelial cycle and the serum levels of gonadotropins.[109]

In male rats, the survival of developing germ cells clearly depends on gonadotropins, since hypophysectomy or treatment with a gonadotropin-releasing hormone (GnRH) antagonist results in a rapid induction of germ cell apoptosis.[111] Of note, the effects of hypophysectomy on apoptosis induction in the seminiferous epithelium of juvenile male rats could be overcome by testosterone replacement,[111] suggesting that the loss of gonadotropins leads indirectly to male germ cell apoptosis via a decline in androgen biosynthesis. This model of androgen deprivation may have important clinical significance in the case of young boys treated for cryptorchidism by human chorionic gonadotropin (hCG) administration, since the termination of therapy would lead to an abrupt loss of circulating androgens stimulated by hCG treatment. Indeed, a retrospective analysis of human testicular biopsies collected from cryptorchid boys treated with hCG has indicated the occurrence of excessive spermatogonial apoptosis post-therapy, leading to reduced testicular volumes and disrupted reproductive function in adult life.[112]

The intratesticular Fas–Fas_L system (see 'Death receptors and death domains' above) appears to play a prominent role in mediating germ cell death, at least in rodent models, under both normal and pathophysiological (for example, exposure to environmental toxicants) conditions.[113–115] The robust expression of Fas_L by Sertoli cells, coupled with the presence of receptors for this cytokine in some but not all germ cells, has been suggested as a mechanism to explain how Sertoli cells may directly limit the size of the testis germ cell population by a 'negative selection process' involving destruction of Fas-positive spermatocytes.[113] Mechanistically, studies of testicular dysfunction in mutant mice lacking various apoptosis regulators have identified several critical modulators of male germ cell development and death.[116] For example, Bax-deficient male mice are infertile owing to an accumulation of atypical premeiotic germ cells and multinucleated giant cells, and a complete absence of haploid sperm.[96] These data suggest that Bax function is necessary for ensuring the normal death of early germ cell populations in the adult testis, and that excessive survival of mitotic germ cells in Bax-deficient mice overwhelms the capacity of the seminiferous epithelium to support any postmeiotic germ cells.

Evidence of a central role for Bax in controlling male germ cell death also comes from studies of impaired germ cell development in mice with hypomorphic expression of *bcl-x*.[117] In these mice, the males lack spermatogonia and are sterile owing to increased rates of germ cell death during fetal testicular development, phenotypes that are reversed by simultaneous loss of Bax function. Disruption of the gene encoding the antiapoptotic Bcl-2 family member, Bcl-w, results in spermatogenic arrest with a progressive and almost complete degeneration of the adult testes with age.[118–120] Interestingly, expression of the *bcl-w* gene in the testis is essentially confined to Sertoli cells,[120] suggesting that the germ cell depletion observed in these mutants is actually a consequence of defective Sertoli cell function. Of final note, male mice lacking Apaf-1, the CED-4 ortholog that couples Bcl-2 family function to caspase activation (see 'The core machinery of apoptosis' above), exhibit infertility due to excessive apoptosis of spermatogonia leading to a lack of mature sperm production.[121] These findings, along with the male sterility reported to occur in transgenic mice that overexpress antiapoptotic Bcl-2 family members in the testis,[122,123] collectively underscore how precisely apoptosis must be regulated in the testis to maintain spermatogenesis and male fertility.

SUMMARY AND FUTURE PERSPECTIVES

Since the work of Flemming in the 1880s,[3,4] the importance of apoptosis to development and function (normal or otherwise) of the female and male reproductive systems has been unequivocally established. In addition to its role as a quality-control mechanism for ensuring oocyte and sperm quality, apoptosis appears to lie at the heart of several major pathologies associated with gonadal failure and infertility. Indeed, the examples provided herein highlight the potential value of developing new therapies to target apoptosis in germ cells as a means, among other things, to preserve fertility in female cancer patients and possibly prolong the natural life span of the female gonads. This latter possibility is particularly intriguing since the health risks faced by post-menopausal women (such as osteoporosis, cardiovascular disease, neurological disorders) could possibly be alleviated by allowing their ovaries to sustain normal function into advanced age. In addition, the prospects that the fertile life span in women could be increased for at least a few years beyond its current time of cessation cannot be ignored.

It should also be stressed that the clinical impact of apoptosis research will certainly expand outside the application to germ cells. For example, even if a competent viable egg and a competent viable sperm are successfully produced for fertilization, pregnancy rates remain less than 50% without or with the use of assisted reproductive technologies. Is apoptosis involved in other aspects related to infertility? The answer to this question appears to be 'yes', as exemplified by the following three possibilities offered in closing. The first involves pre-implantation embryo fragmentation, a process identified to occur in several species, including humans, via apoptosis.[124–126] Accordingly, methods to improve embryo quality during development *in vitro* prior to intrauterine transfer may be aided by the future incorporation of antiapoptotic agents designed to suppress blastomere fragmentation. Importantly, these new 'agents' may not only surface in the form of hormones or drugs, but also as 'organelle' therapies, such as mitochondrial microinjection.[127,128] Indeed, the ability of ooplasmic transfer to boost the developmental potential of a recipient egg after fertilization[129] may be due to effects of the mitochondria present in the donor egg cytoplasm that become incorporated, and subsequently replicate, in the newly conceived embryo.[128,130]

The remaining two possibilities involve recurrent pregnancy loss due to inappropriate activation of cell death in either the corpus luteum (second possibility) or the endometrium (third possibility). In the normal menstrual cycle, the corpus luteum formed following ovulation exists for a predetermined length of time, and then regresses via apoptosis in the absence of hCG produced by the conceptus.[131] The resultant loss of steroidogenic support to the endometrium then triggers apoptosis in this tissue, leading to menses.[132,133] In the context of pregnancy, these two paradigms of apoptosis are therefore interrelated, since sustained function of the corpus luteum (i.e. progesterone secretion) is required to prevent menses, allowing for the establishment of pregnancy as well as its maintenance during the first 8 weeks of gestation in humans. While more work is needed to test these possibilities rigorously, and to develop further those examples provided earlier (see 'Emerging concepts in reproductive medicine' above), the data in hand strongly support that apoptosis will serve as an important therapeutic target in our efforts to enhance human reproductive health and performance in the new millennium.

ACKNOWLEDGEMENTS

The author would like to thank the past and present members of his laboratory for their dedicated research efforts, as well as the many collaborators and colleagues whose assistance was invaluable for the completion of numerous studies described herein. Work conducted by the author was supported by the National Institutes of Health (R01–AG12279, R01–ES08430, R01–HD34226), the Department of Defense–US Army Medical Research and Material Command's Office of Congressionally Directed Medical Research (DAMD17–00–1–0567), the Steven and Michele Kirsch Foundation, and Vincent Memorial Research Funds.

REFERENCES

1. Majno G, Joris I. Apoptosis, oncosis, and necrosis. An overview of cell death. *Am J Pathol* 1995 146: 3–15.
2. Lockshin R. The early modern period in cell death. *Cell Death Differ* 1997 4: 347–351.
3. Flemming W. Über die bildung von richtungsfiguren in säugethiereiern beim untergang Graaf'scher follikel. *Archiv Anat Entwickelungsgeschichte (Archiv Anat Physiol)* 1885: 221–244.
4. Flemming W. Neue beiträge zur kenntniss der zelle: I. Die kerntheilung bei den spermatocyten von salamandra maculosa. *Archiv Mikrosk Anat Entw Mach* 1887 29: 389–463.
5. Kerr JFR, Wyllie AH, Currie AR. Apoptosis: a basic biological phenomenon with wide-ranging implications in tissue kinetics. *Br J Cancer* 1972 26: 239–257.
6. Kerr JFR, Winterford CM, Harmon BV. Morphological criteria for identifying apoptosis. In Celis JE, ed. *Cell Biology: a Laboratory Handbook*. San Diego: Academic Press, 1994: 319–329.
7. Frisch SM, Francis H. Disruption of epithelial cell–matrix interactions induces apoptosis. *J Cell Biol* 1994 124: 619–626.
8. Tilly JL, Ratts VS. Biological and clinical importance of ovarian cell death. *Contemp Obstet Gynecol* 1996 41: 59–86.
9. Bursch W, Kleine L, Tenniswood M. The biochemistry of cell death via apoptosis. *Biochem Cell Biol* 1990 68: 1071–1074.
10. Wyllie AH, Kerr JFR, Currie AR. Cell death: the significance of apoptosis. *Int Rev Cytol* 1980 68: 251–306.
11. Wyllie AH. Glucocorticoid-induced thymocyte apoptosis is associated with endogenous endonuclease activation. *Nature* 1980 284: 555–556.
12. Tilly JL, Billig H, Kowalski KI, Hsueh AJW. Epidermal growth factor and basic fibroblast growth factor suppress the spontaneous onset of apoptosis in cultured rat ovarian granulosa cells and follicles by a tyrosine kinase-dependent mechanism. *Mol Endocrinol* 1992 6:1942–1950.
13. Tilly JL, Tilly KI, Kenton ML, Johnson AL. Expression of members of the *bcl-2* gene family in the immature rat ovary: equine chorionic gonadotropin-mediated inhibition of granulosa cell apoptosis is associated with decreased *bax* and constitutive *bcl-2* and *bcl-x*$_{long}$ messenger RNA levels. *Endocrinology* 1995 136: 232–241.
14. Trbovich AM, Hughes FM Jr, Perez GI *et al.* High and low molecular weight DNA cleavage in ovarian granulosa cells: characterization and protease modulation in intact cells and in cell-free nuclear autodigestion assays. *Cell Death Differ* 1998 5: 38–49.
15. Walker PR, Sikorska M. Endonuclease activities, chromatin structure, and DNA degradation in apoptosis. *Biochem Cell Biol* 1994 72: 615–623.
16. Cohen GM, Sun X-M, Snowden RT, Dinsdale D, Skilleter DN. Key morphological features of apoptosis may occur in the absence of internucleosomal DNA fragmentation. *Biochem J* 1992 286: 331–334.
17. Oberhammer F, Wilson JW, Dive C *et al.* Apoptotic death in epithelial cells: cleavage of DNA to 300 and/or 50 kb fragments prior to or in the absence of internucleosomal fragmentation. *EMBO J* 1993 12: 3679–3684.
18. Gavrieli Y, Sherman Y, Ben-Sasson SA. Identification of programmed cell death *in situ* by specific labeling of nuclear DNA fragmentation. *J Cell Biol* 1992 119: 493–501.
19. Tilly JL. Use of the terminal transferase DNA labeling reaction for the biochemical and *in situ* analysis of apoptosis. In Celis JE, ed. *Cell Biology: a Laboratory Handbook*. San Diego: Academic Press, 1994: 331–337.
20. Robles R, Tao X-J, Trbovich AM *et al.* Localization, regulation and possible consequences of apoptotic protease-activating factor-1 (Apaf-1) expression in granulosa cells of the mouse ovary. *Endocrinology* 1999 140: 2641–2644.
21. Matikainen T, Perez GI, Zheng TS *et al.* *Caspase-3* gene knockout defines cell lineage specificity for programmed cell death signaling in the ovary. *Endocrinology* 2001 142: 2468–2480.
22. Perez GI, Tao X-J, Tilly JL. Fragmentation and death (aka. apoptosis) of ovulated oocytes. *Mol Hum Reprod* 1999 5: 414–420.
23. Morita Y, Perez GI, Paris F *et al.* Oocyte apoptosis is suppressed by disruption of the *acid sphingomyelinase* gene or by sphingosine-1-phosphate therapy. *Nat Med* 2000 6: 1109–1114.
24. Collins JA, Schandl CA, Young KK, Vesely J, Willingham MC. Major DNA fragmentation is a

late event in apoptosis. *J Histochem Cytochem* 1997 45: 923–934.

25. Fairbairn DW, Olive PL, O'Neill KL. The comet assay: a comprehensive overview. *Mutat Res* 1995 339: 37–59.
26. Lockshin RA, Williams CM. Programmed cell death. II. Endocrine potentiation of the breakdown of the intersegmental muscles of silkmoths. *J Insect Physiol* 1964 10: 643–649.
27. Vaux DL. Toward an understanding of the molecular mechanisms of physiological cell death. *Proc Natl Acad Sci USA* 1993 90: 786–789.
28. Zakeri Z, Bursch W, Tenniswood M, Lockshin RA. Cell death: programmed, apoptosis, necrosis, or other? *Cell Death Differ* 1995 2: 87–96.
29. Ericcson JLE. Mechanism of cellular autophagy. In Dingle JT, Fell HB, eds. *Lysosomes in Biology and Pathology*. Amsterdam–London: North-Holland Publication Company, 1969: 345–394.
30. Bursh W. The autophagosomal-lysosomal compartment in programmed cell death. *Cell Death Differ* 2001 8: 545–548.
31. Quatacker JR. Formation of autophagic vacuoles during human corpus luteum involution. *Z Zellforschung Mikroskop Anat* 1971 122: 479–487.
32. Paavola LG, Furth EE, Delgado V *et al.* Striking changes in the structure and organization of rat fetal membranes precedes parturition. *Biol Reprod* 1995 53: 321–338.
33. Lei H, Furth EE, Kalluri R *et al.* A program of cell death and extracellular matrix degradation is activated in the amnion prior to the onset of labor. *J Clin Invest* 1996 98: 1971–1978.
34. Trump BF, Berezesky IK. The reactions of cells to lethal injury: oncosis and necrosis – the role of calcium. In Lockshin RA, Zakeri Z, Tilly JL, eds. *When Cells Die. A Comprehensive Evaluation of Apoptosis and Programmed Cell Death*. New York: Wiley-Liss, 1998: 57–96.
35. Liu QA, Hengartner MO. The molecular mechanism of programmed cell death in *C. elegans*. *Ann NY Acad Sci* 1999 887: 92–104.
36. Tittel JN, Steller H. A comparison of programmed cell death between species. *Genome Biol* 2000 1: 3.1–3.6.
37. Chen P, Abrams JM. *Drosophila* apoptosis and Bcl-2 genes: outliers fly in. *J Cell Biol* 2000 148: 625–627.
38. Gross A, McDonnell JM, Korsmeyer SJ. BCL-2 family members and the mitochondria in apoptosis. *Genes Dev* 1999 13: 1899–1911.
39. Shi Y. A structural view of mitochondria-mediated apoptosis. *Nat Struct Biol* 2001 8: 394–401.
40. Antonsson B, Martinou JC. The Bcl-2 protein family. *Exp Cell Res* 2000 256: 50–57.
41. Reed JC. Mechanisms of apoptosis. *Am J Pathol* 2000 157: 1415–1430.
42. Vaux DL, Weissman IL, Kim SK. Prevention of programmed cell death in *Caenorhabditis elegans* by human bcl-2. *Science* 1992 258: 1955–1957.
43. Conradt B, Horvitz HR. The *C. elegans* protein EGL-1 is required for programmed cell death and interacts with the Bcl-2-like protein CED-9. *Cell* 1998 93: 519–529.
44. Green DR, Reed JC. Mitochondria and apoptosis. *Science* 1998 281: 1309–1312.
45. Kroemer G, Reed JC. Mitochondrial control of cell death. *Nat Med* 2000 6: 513–519.
46. Lily Y, Luo X, Wang X. Endonuclease G is an apoptotic Dnase when released from mitochondria. *Nature* 2001 412: 95–99.
47. Pru JK, Tilly JL. Programmed cell death in the ovary: insights and future prospects using genetic technologies. *Mol Endocrinol* 2001 15: 845–853.
48. Yang J, Liu X, Bhalla K *et al.* Prevention of apoptosis by Bcl-2: release of cytochrome *c* blocked. *Science* 1997 275: 1129–1132.
49. Kluck RM, Bossy-Wetzel E, Green DR, Newmeyer DD. The release of cytochrome *c* from mitochondria: a primary site for Bcl-2 regulation of apoptosis. *Science* 1997 275: 1132–1136.
50. Jürgensmeier JM, Xie Z, Deveraux Q, Ellerby L, Bredesen D, Reed JC. Bax directly induces cytochrome *c* release from isolated mitochondria. *Proc Natl Acad Sci USA* 1998 95: 4997–5002.
51. Xiang J, Chao DT, Korsmeyer SJ. BAX-induced cell death may not require interleukin-1β-converting enzyme-like proteases. *Proc Natl Acad Sci USA* 1996 93: 14559–14563.
52. Marzo I, Brenner C, Zamzami N *et al.* Bax and adenine nucleotide translocator cooperate in the mitochondrial control of apoptosis. *Science* 1998 281: 2027–2031.
53. Zou H, Henzel WJ, Liu X, Lutschg A, Wang X. Apaf-1, a human protein homologous to *C. elegans* CED-4, participates in cytochrome *c*-dependent activation of caspase-3. *Cell* 1997 90: 405–413.

54. Li P, Nijhawan D, Budihardjo I *et al.* Cytochrome *c* and dATP-dependent formation of Apaf-1/caspase-9 complex initiates an apoptotic protease cascade. *Cell* 1997 91: 479–489.
55. Budihardjo I, Oliver H, Lutter M, Luo X, Wang, X. Biochemical pathways of caspase activation during apoptosis. *Ann Rev Cell Dev Biol* 1999 15: 269–290.
56. Kuida K, Haydar TF, Kuan C-Y *et al.* Reduced apoptosis and cytochrome *c*-mediated caspase activation in mice lacking caspase-9. *Cell* 1998 94: 325–337.
57. Cecconi F, Alvarez-Bolado G, Meyer BI, Roth KA, Gruss P. Apaf1 (CED-4 homolog) regulates programmed cell death in mammalian development. *Cell* 1998 94: 727–737.
58. Yoshida H, Kong Y-Y, Yoshida R *et al.* Apaf1 is required for mitochondrial pathways of apoptosis and brain development. *Cell* 1998 94: 739–750.
59. Yuan J, Shaham S, Ledoux S, Ellis H, Horvitz HR. The *C. elegans* cell death gene *ced-3* encodes a protein similar to mammalian interleukin-1β-converting enzyme. *Cell* 1993 75: 641–652.
60. Xue D, Shaham S, Horvitz HR. The *Caenorhabditis elegans* cell-death protein CED-3 is a cysteine protease with substrate specificities similar to those of the human CPP32 protease. *Genes Dev* 1996 10: 1073–1083.
61. Alnemri ES, Livingston DJ, Nicholson DW *et al.* Human ICE/CED-3 protease nomenclature. *Cell* 1996 87: 171.
62. Casciola-Rosen L, Nicholson DW, Chong T *et al.* Apopain/CPP32 cleaves proteins that are essential for cellular repair: a fundamental principle of apoptotic death. *J Exp Med* 1996 183: 1957–1964.
63. Adrain C, Martin SJ. The mitochondrial apoptosome: a killer unleashed by the cytochrome seas. *Trends Biochem Sci* 2001 26: 390–397.
64. Leist M, Jäättela M. Four deaths and a funeral: from caspases to alternative mechanisms. *Nat Rev Mol Cell Biol* 2001 2: 1–10.
65. Hannun YA. Functions of ceramide in coordinating cellular responses to stress. *Science* 1996 274: 1855–1859.
66. Kolesnick RN, Kronke M. Regulation of ceramide production and apoptosis. *Annu Rev Physiol* 1998 60: 643–665.
67. Spiegel S. Sphingosine 1-phosphate: a prototype of a new class of second messengers. *J Leukoc Biol* 1999 65: 341–44.
68. Cuvillier O, Pirianov G, Kleuser B *et al.* Suppression of ceramide-mediated programmed cell death by sphingosine-1-phosphate. *Nature* 1996 381: 800–803.
69. Kulik G, Klippel A, Weber MJ. Antiapoptotic signalling by the insulin-like growth factor I receptor, phosphatidylinositol 3-kinase, and Akt. *Mol Cell Biol* 1997 17: 1595–1606.
70. Kennedy SG, Wagner AJ, Conzen SD, *et al.* The PI 3-kinase/Akt signaling pathway delivers an anti-apoptotic signal. *Genes Dev* 1997 11: 701–713.
71. Franke TF, Kaplan DR, Cantley LC. PI3K; downstream AKTion blocks apoptosis. *Cell* 1997 88: 435–437.
72. Datta SR, Dudek H, Tao X *et al.* Akt phosphorylation of BAD couples survival signals to the cell-intrinsic death machinery. *Cell* 1997 91: 231–241.
73. Cardone MH, Roy N, Stennicke HR *et al.* Regulation of cell death protease caspase-9 by phosphorylation. *Science* 1998 282: 1318–1321.
74. Zhou H, Summers SA, Birnbaum MJ, Pittman RN. Inhibition of Akt kinase by cell-permeable ceramide and its implications for ceramide-induced apoptosis. *J Biol Chem* 1998 273: 16568–16575.
75. Zundel W, Giaccia A. Inhibition of the anti-apoptotic PI(3)K/Akt/Bad pathway by stress. *Genes Dev* 1998 12: 1941–1946.
76. Summers SA, Garza LA, Zhou H, Birnbaum MJ. Regulation of insulin-stimulated glucose transporter GLUT4 translocation and Akt kinase activity by ceramide. *Mol Cell Biol* 1998 18: 5457–5464.
77. Goswami R, Kilkus J, Dawson SA, Dawson G. Overexpression of Akt (protein kinase B) confers protection against apoptosis and prevents formation of ceramide in response to pro-apoptotic stimuli. *J Neurosci Res* 1999 57: 884–893.
78. Pastorino JG, Tafani M, Rothman RJ *et al.* Functional consequences of the sustained or transient activation by Bax of the mitochondrial permeability transition pore. *J Biol Chem* 1999 274: 31734–31739.
79. Siskind LJ, Colombini M. The lipids C2- and C16-ceramide form large stable channels: impli-

cations for apoptosis. *J Biol Chem* 2000 275: 38640–38644.

80. Cremesti A, Paris F, Grassme H *et al.* Ceramide enables Fas to cap and kill. *J Biol Chem* 2001 276: 23954–23961.
81. Goetzl EJ, Kong Y, Mei B. Lysophosphatidic acid and sphingosine-1-phosphate protection of T cells from apoptosis in association with suppression of Bax. *J Immunol* 1999 162: 2049–2056.
82. Cuvillier O, Rosenthal DS, Smulson ME, Spiegel S. Sphingosine-1-phosphate inhibits activation of caspases that cleave poly(ADP-ribose) polymerase and lamins during Fas- and ceramide-mediated apoptosis in Jurkat T lymphocytes. *J Biol Chem* 1998 273: 2910–2916.
83. Tewari M, Dixit VM. Recent advances in tumor necrosis factor and CD40 signaling. *Curr Opin Genet Dev* 1996 6: 39–44.
84. Krammer PH. CD95's deadly mission in the immune system. *Nature* 2000 407: 789–795.
85. Scaffidi C, Fulda S, Srinivasan A *et al.* Two CD95 (APO-1/Fas) signaling pathways. *EMBO J* 1998 17: 1675–1687.
86. Luo X, Budihardo I, Zou H, Slaughter C, Wang X. Bid, a Bcl2 interacting protein, mediates cytochrome *c* release from mitochondria in response to activation of cell surface death receptors. *Cell* 1998 94: 481–490.
87. Li H, Zhu H, Xu C-J, Yuan J. Cleavage of BID by caspase 8 mediates the mitochondrial damage in the Fas pathway of apoptosis. *Cell* 1998 94: 491–501.
88. Zha J, Weiler S, Oh KJ, Wei MC, Korsmeyer SJ. Posttranslational N-myristoylation of BID as a molecular switch for targeting mitochondria and apoptosis. *Science* 2000 290: 1761–1765.
89. Korsmeyer SJ, Wei MC, Saito M *et al.* Pro-apoptotic cascade activates BID, which oligomerizes BAK or BAX into pores that result in the release of cytochrome *c*. *Cell Death Differ* 2000 7: 1166–1173.
90. Richardson SJ, Senikas V, Nelson JF. Follicular depletion during the menopausal transition: evidence for accelerated loss and ultimate exhaustion. *J Clin Endocrinol Metab* 1987 65: 1231–1237.
91. Gosden RG, Laing SC, Felicio LS, Nelson JF, Finch CE. Imminent oocyte exhaustion and reduced follicular recruitment mark the transition to acyclicity in aging C57BL/6J mice. *Biol Reprod* 1983 28: 255–260.
92. Baker TG. A quantitative and cytological study of germ cells in human ovaries. *Proc R Soc London (B)* 1963 158: 417–433.
93. Gougeon A. Regulation of ovarian follicular development in primates: facts and hypotheses. *Endocr Rev* 1996 17: 121–155.
94. Tilly JL. Commuting the death sentence: how oocytes strive to survive. *Nat Rev Mol Cell Biol* 2001 2: 838–848.
95. Perez GI, Robles R, Knudson CM, Flaws JA, Korsmeyer SJ, Tilly JL. Prolongation of ovarian lifespan into advanced chronological age by *Bax*-deficiency. *Nat Genet* 1999 21: 200–203.
96. Knudson CM, Tung KSK, Tourtellotte WG, Brown GAJ, Korsmeyer SJ. Bax-deficient mice with lymphoid hyperplasia and male germ cell death. *Science* 1995 270: 96–99.
97. Perez GI, Knudson CM, Leykin L, Korsmeyer SJ, Tilly JL. Apoptosis-associated signaling pathways are required for chemotherapy-mediated female germ cell destruction. *Nat Med* 1997 3: 1228–1332.
98. Waxman J. Chemotherapy and the adult gonad: a review. *J R Soc Med* 1983 76: 144–148.
99. Ried HL, Jaffe N. Radiation-induced changes in long-term survivors of childhood cancer after treatment with radiation therapy. *Semin Roentgenol* 1994 29: 6–14.
100. Blumenfeld Z, Avivi I, Ritter M, Rowe JM. Preservation of fertility and ovarian function and minimizing chemotherapy-induced gonadotoxicity in young women. *J Soc Gyn Invest* 1999 6: 229–239.
101. Reynolds T. Cell death genes may hold clues to preserving fertility after chemotherapy. *J Natl Cancer Inst* 1999 91: 664–666.
102. Bergeron L, Perez GI, Mcdonald G *et al.* Defects in regulation of apoptosis in caspase-2-deficient mice. *Genes Dev* 1998 12: 1304–1314.
103. Morita Y, Perez GI, Maravei DV, Tilly KI, Tilly JL. Targeted expression of Bcl-2 in mouse oocytes inhibits ovarian follicle atresia and prevents spontaneous and chemotherapy-induced oocyte apoptosis *in vitro*. *Mol Endocrinol* 1999 13: 841–850.
104. Paris F, Perez GI, Kolesnick RN, Tilly JL. Suppression of oocyte apoptosis by sphingosine-1-phosphate therapy *in vivo* preserves fertility following radiotherapy. Presented at the *83rd*

Annual Meeting of the Endocrine Society, Denver, CO, June 2001: 374.
105. Tilly JL. Emerging technologies to control oocyte apoptosis are finally treading on fertile ground. In *The Scientific World* (www.thescientificworld.com) 2001 1: 181–183.
106. Tilly JL. Apoptosis in female reproductive toxicology. In Roberts R, ed. *Apoptosis in Toxicology*. London: Taylor & Francis, 2000: 95–116.
107. Matikainen T, Perez GI, Jurisicova A *et al.* Aromatic hydrocarbon receptor-driven *Bax* gene expression is required for premature ovarian failure caused by biohazardous environmental chemicals. *Nat Genet* 2001 28: 355–360.
108. Russell LD, Clermont Y. Degeneration of germ cells in normal, hypophysectomized and hormone-treated hypophysectomized rats. *Anat Rec* 1977 187: 347–366.
109. Dunkel L, Hirvonen V, Erkkilä K. Clinical aspects of male germ cell apoptosis during testis development and spermatogenesis. *Cell Death Differ* 1997 4: 171–179.
110. Print CG, Loveland KL. Germ cell suicide: new insights into apoptosis during spermatogenesis. *Bioessays* 2000 22: 423–430.
111. Tapanainen J, Tilly JL, Vihko KK, Hsueh AJW. Hormonal control of apoptotic cell death in the testis: gonadotropins and androgens as testicular cell survival factors. *Mol Endocrinol* 1993 7: 643–650.
112. Dunkel L, Taskinen S, Hovatta O, Tilly JL, Wikström S. Germ cell apoptosis after treatment of cryptorchidism with human chorionic gonadotropin is associated with impaired reproductive function in the adult. *J Clin Invest* 1997 100: 2341–2346.
113. Lee J, Richburg JH, Younkin SC, Boekelheide K. The Fas system is a key regulator of germ cell apoptosis in the testis. *Endocrinology* 1997 138: 2081–2088.
114. Lysiak JJ, Turner SD, Turner TT. Molecular pathway of germ cell apoptosis following ischemia/reperfusion of the rat testis. *Biol Reprod* 2000 63: 1465–1472.
115. Yu X, Kubota H, Wang R *et al.* Involvement of Bcl-2 family genes and Fas signaling system in primary and secondary male germ cell apoptosis induced by 2-bromopropane in rat. *Toxicol Appl Pharmacol* 2001 174: 35–48.
116. Ranger AM, Malynn BA, Korsmeyer SJ. Mouse models of cell death. *Nat Genet* 2001 28: 113–118.
117. Rucker EB, Dierisseau P, Wagner KU *et al.* Bcl-x and Bax regulate mouse primordial germ cell survival and apoptosis during embryogenesis. *Mol Endocrinol* 2000 14: 1038–1052.
118. Ross A, Waymire KG, Moss JE *et al.* Testicular degeneration in *Bclw*-deficient mice. *Nat Genet* 1998 18: 251–255.
119. Print CG, Loveland KL, Gibson L *et al.* Apoptosis regulator Bclw is essential for spermatogenesis but appears otherwise redundant. *Proc Natl Acad Sci USA* 1998 95: 12424–12431.
120. Russel LD, Warren J, Debeljuk L *et al.* Spermatogenesis in *Bclw*-deficient mice. *Biol Reprod* 2001 65: 318–322.
121. Honarpour N, Du C, Richardson JA *et al.* Adult Apaf-1-deficient mice exhibit male infertility. *Dev Biol* 2000 218: 248–258.
122. Furuchi T, Masuko K, Nishimune Y, Obinata M, Matsui Y. Inhibition of testicular germ cell apoptosis and differentiation in mice misexpressing Bcl-2 in spermatogonia. *Development* 1996 122: 1703–1709.
123. Rodriguez I, Ody C, Araki K, Garcia I, Vassalli P. An early and massive wave of germinal cell apoptosis is required for development of functional spermatogenesis. *EMBO J* 1997 16: 2262–2270.
124. Jurisicova A, Varmuza S, Casper RF. Programmed cell death and human embryo fragmentation. *Mol Hum Reprod* 1996 2: 93–98.
125. Jurisicova A, Varmuza S, Casper RF. Developmental consequences of programmed cell death during mammalian embryo development. In Tilly JL, Strauss JF, Tenniswood MP, eds. *Cell Death in Reproductive Physiology*. New York: Springer-Verlag, 1997: 32–47.
126. Levy R, Benchaib M, Cordonier H, Souchier C, Guerin J. Annexin-V labelling and terminal transferase-mediated DNA end-labeling (TUNEL) assay in human arrested embryos. *Mol Hum Reprod* 1998 4: 775–783.
127. Perez GI, Trbovich A M, Gosden RG, Tilly JL. Mitochondria and the death of oocytes. *Nature* 2000 403: 500–501.
128. Jurisicova A, Acton BM, Perez GI *et al.* Mitochondrial function and behaviour in mammalian oocytes and during preimplantation

embryo development. *J Soc Gynecol Invest* 2001 8 (Suppl): 49A.

129. Cohen J, Scott R, Alikani M *et al.* Ooplasmic transfer in mature human oocytes. *Mol Hum Reprod* 1998 4: 269–280.
130. Barritt JA, Brenner CA, Malter HE, Cohen J. Mitochondria in human offspring derived from ooplasmic transplantation. *Hum Reprod* 2001 16: 513–516.
131. Rueda BR, Hoyer PB, Hamernik DL, Tilly JL. Potential regulators of physiological cell death in the corpus luteum. In Tilly JL, Strauss JF, Tenniswood MP, eds. *Cell Death in Reproductive Physiology*. New York: Springer-Verlag, 1997: 161–181.
132. Hopwood D, Levison DA. Atrophy and apoptosis in the cyclical human endometrium. *J Pathol* 1995 119: 159–166.
133. Tao X-J, Tilly KI, Maravei DV *et al.* Differential expression of members of the *bcl-2* gene family in proliferative and secretory human endometrium: glandular epithelial cell apoptosis is associated with increased expression of bax. *J Clin Endocrinol Metab* 1997 82: 2738–2746.

Chapter 6

Molecular carcinogenesis

Jeff Boyd

INTRODUCTION

The question of how cancer arises remains one of the most fundamental and complex problems in all of human biology. Understanding cancer will ultimately require an understanding of 'what makes a cell a cell'.[1] Historically, many theoretical models have received temporary favor for their efforts to address empirically the problem of cancer etiology, including those founded upon the action of environmental agents, chemical carcinogens, viruses, somatic chromosomal abnormalities and congenital predisposition. We now know that all of these paradigms are correct by virtue of their convergence into the genetic paradigm: cancer is the result of an accumulation of mutations in genes that govern the tumor phenotype.[2]

Since genetic mutations are the central etiological factor in tumorigenesis, a chapter on molecular carcinogenesis must include the basic principles of cancer molecular genetics, including evidence for the multistep, multigenic basis of tumorigenesis, and a summary of our current state of knowledge regarding the genes involved in this process. Detailed discussions of the molecular features of the most common malignancies of the female reproductive tract are presented, followed by a summary of the enormous recent progress made in defining the genetic basis of hereditary gynecological cancers. Molecular carcinogenesis is intimately linked to perturbations in cell cycle regulation, and an overview of the enormous progress recently made in this area is also given.

PRINCIPLES OF CANCER MOLECULAR GENETICS

All cancers are genetic in origin, in the sense that the driving force of tumor development is genetic mutation. A given tumor may arise through the accumulation of mutations that are exclusively somatic in origin, or through the inheritance of a mutation(s) through the germline, followed by the acquisition of additional somatic mutations. These two genetic scenarios distinguish what are colloquially referred to as 'sporadic' and 'hereditary'-cancers, respectively. While the neoplastic phenotype is partially derived from epigenetic alterations in gene expression, the sequential mutation of cancer-related genes, and their subsequent selection and accumulation in a clonal population of cells, are the factors determining whether a tumor develops and the time required for its development and progression. The data to support this multistep, multigenic paradigm are extensive,[3–6] but perhaps the most compelling evidence is that the age-specific incidence rates for most human epithelial tumors increase at roughly the fourth to eighth power of elapsed time, suggesting that a series of four to eight genetic alterations are rate-limiting for cancer development.[7,8]

Genetic alterations in cancer cells have thus far been described in two major families of genes, oncogenes[9] and tumor suppressor genes.[10] Proteins encoded by oncogenes may generally be viewed as stimulatory and those encoded by tumor suppressor genes as inhibitory to the neoplastic phenotype; mutational activation of proto-oncogenes to oncogenes and mutational inactivation of tumor suppressor genes must both occur for cancer development to take place. Proto-oncogene mutations are nearly always somatic; two known exceptions involve the *ret* and *met* proto-oncogenes, mutations of which may be inherited through the germline, predisposing to multiple endocrine neoplasia type 2[11] and papillary renal carcinoma,[12] respectively. Tumor suppressor gene mutations may be inherited or acquired somatically. Other than the above-noted exceptions, all hereditary cancer syndromes for which predisposing genes have been identified are linked to tumor suppressor genes.

Oncogenes

Oncogenes result from gain-of-function mutations in their normal cellular counterpart proto-oncogenes, the normal function of which is to drive cell proliferation in the appropriate contexts. Activated oncogenes behave in a dominant fashion at the cellular level; that is, cell proliferation or development of the neoplastic phenotype is stimulated following the mutation of only one allele. This class of genes was originally discovered through studies of the mechanism of retroviral tumorigenesis,[13] which involves viral transduction of the vertebrate proto-oncogene and re-integration into the host genome under the transcriptional control of viral promoters, such that expression is constitutive and thus oncogenic. The most common mechanisms for mutational activation of human proto-oncogenes are gene amplification, typically resulting in overexpression of an otherwise normal protein product, point mutation, generally leading to constitutive activation of a mutant form of the protein product, and chromosomal translocation, which usually results in juxtaposition of the oncogene with the promoter region of a constitutively expressed gene, thus resulting in overexpression of the oncogene-encoded protein. This last mechanism is most common in hematopoietic malignancies, while the first two are more common in solid cancers. The oncogenes most relevant to human solid malignancies, their mechanism of activation and biochemical function, and the tumor types most often affected by each are summarized in Table 1.

Tumor suppressor genes

The protein products of tumor suppressor genes normally function to inhibit cell proliferation, and are inactivated through loss-of-function mutations. Knudson's two-hit model established the paradigm for tumor suppressor gene recessivity at the cellular level, wherein both alleles must be inactivated in order to exert a phenotypic effect on tumorigenesis.[14] The most common mutations observed in tumor suppressor genes are point mutations, either missense or non-sense, microdeletions or insertions of one or several nucleotides causing frameshifts, large deletions and, rarely, translocations. A mutation in one allele, whether germline or somatic, is then revealed following somatic inactivation of the homologous wild-type allele. In theory, the same spectrum of mutational events could contribute to inactivation of the second allele, but what is typically observed in tumors is homozygosity or hemizygosity for the first mutation, indicating loss of the wild-type allele. As originally demonstrated for the retinoblastoma-susceptibility gene,[15] loss of the second allele may occur through mitotic non-disjunction or recombination mechanisms, or large deletions. This so-called 'loss of heterozygosity' (LOH) has become recognized as the hallmark of tumor suppressor gene inactivation at a particular genomic locus. Table 2 summarizes the known tumor suppressor genes, their chromosomal locations and suspected biochemical

Table 1 Summary of representative oncogenes mutated in human solid cancers

Gene	*Chromosomal location*	*Function*	*Mutation*	*Tumor*
ras (K-, H-, N-)	12p12, 11p15, 1p13	membrane-associated GTPase; signal transduction	point mutation (codon 12, 13 or 61)	many
ERBB-2	17q12	transmembrane tyrosine kinase receptor	gene amplification	breast, ovary, endometrium
myc (C-, N-, L-)	8q24, 2p24, 1p34	transcription factor	gene amplification	C: many; N: neuroblastoma; L: lung
EGFR	7p12	transmembrane tyrosine kinase receptor	gene amplification	glioma
MDM2	12q14	p53-binding protein	gene amplification	sarcoma
CCND1 (cyclin D1)	11q13	cell cycle regulator	gene amplification	many
ret	10q11	transmembrane tyrosine kinase receptor	point mutation	endocrine
met	7q31	transmembrane tyrosine kinase receptor	point mutation	renal

EGFR, epidermal growth factor receptor; GTPase, guanosine triphosphatase

Table 2 Summary of representative tumor suppressor genes mutated in human solid cancers

Gene	*Chromosomal location*	*Function*	*Tumors*	
			Hereditary	*Sporadic*
Rb1	13q14	cell cycle regulator	retinoblastoma, osteosarcoma	retinoblastoma, sarcoma, bladder, breast, lung
WT1	11p13	transcription factor	Wilms' tumor	Wilms' tumor
p53	17p13	transcription factor; regulator of cell cycle, apoptosis	Li–Fraumeni syndrome	many
APC	5q21–q22	signal transduction	familial adenomatous polyposis	colorectal, gastric
VHL	3p26–p25	transcriptional elongation	von Hippel–Lindau syndrome	renal
MSH2, MLH1, PMS2, MSH6	2p16, 3p21, 7p22, 2p15	DNA mismatch repair	hereditary non-polyposis colorectal cancer syndrome	colorectal, endometrial, gastric
BRCA1	17q21	transcription factor, DNA repair	breast, ovary	ovary (rare)
BRCA2	13q12	transcription factor, DNA repair	breast, ovary, others	ovary (rare)
NF1	17q11	negative regulator of *ras*	neurofibromatosis	none
DPC4	18q21	TGF-β signalling pathway	none	pancreatic
CDKN2 (p16)	9p21	negative regulator of cyclin D	melanoma	many
PTEN (MMAC1)	10q24	phosphatase	Cowden disease	many
LKB1 (STK11)	19p13	serine–threonine protein kinase	Peutz–Jeghers syndrome	none

APC, adenomatous polyposis coli; TGF-β, transforming growth factor-β

functions, and the hereditary and sporadic tumors with which they are most commonly associated.

Multistep tumorigenesis

A human cancer represents the end-point of a long and complex process involving multiple changes in genotype and phenotype. Human solid tumors are monoclonal in nature; every cell in a given malignancy may be shown to have arisen from a single progenitor cell. As proposed by Nowell,[16] the process by means of which a cell and its offspring sustain and accumulate multiple mutations, with the stepwise selection of variant sublines, is known as clonal evolution or clonal expansion (Figure 1). A long-term goal in studying the molecular genetics of a particular tumor type is to catalog the specific genes that are affected by mutations and the relative order in which they are affected, and, ultimately, to use this molecular blueprint to improve methods of diagnosis, prognostication

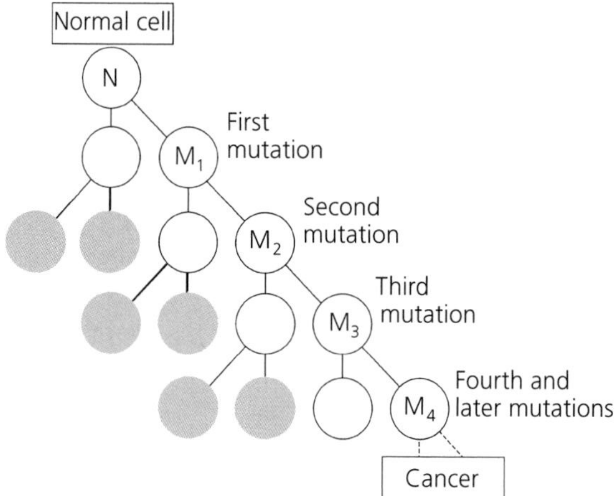

Figure 1 Model of clonal evolution in neoplasia. Following the initiating mutation in a normal cell, stepwise genetic mutations and selective pressures result in a cancer consisting of a clonal population of cells all derived from the original progenitor cell. Each critical mutation in the evolving tumor may be viewed as having provided a selective advantage leading to clonal expansion

and treatment. This task will undoubtedly prove difficult, however, as a defining characteristic of cancer is genetic instability.[17] There are multiple types of such instability, operative at both the chromosomal and the molecular levels. Distinguishing the genetic mutations that are simply the byproducts of genetic instability from those that are critical to the neoplastic phenotype or, indeed, responsible for increasing genetic instability of one form or another is among the greatest challenges to be faced in cancer research.

The greatest progress in this context has clearly been achieved for colorectal cancer, and a model has been proposed that applies molecular detail for this particular cancer type to the general paradigm of multistep tumorigenesis and clonal evolution. In addition to the recent demonstration that most colon cancer cell lines are affected by one of two types of genetic instability,[18] specific molecular genetic alterations have been shown to occur at discrete stages of neoplastic progression in the colon: for example, mutation of the adenomatous polyposis coli (*APC*) tumor suppressor gene at a very early stage of hyperproliferation, mutation of the K-*ras* oncogene in the progression of early to intermediate adenoma and mutation of the *Tp53* tumor suppressor gene in the progression of late adenoma to carcinoma.[19] Several features of colorectal cancer facilitate this type of characterization, including the well-defined histopathological progression of normal colonic epithelium to cancer and the accessibility of the various premalignant lesions for molecular analysis, as well as the occurrence of some of these genetic mutations in unusually large fractions of all colorectal tumors. The model is limited in applicability to other cancer types, however, as non-malignant precursor lesions for many solid tumor types (for example, ovarian cancer) are not readily detectable, and few molecular genetic changes have been described that occur in major fractions of other cancer types.

MOLECULAR GENETICS OF COMMON GYNECOLOGICAL CANCERS

The greatest progress in defining critical molecular genetic alterations in gynecological cancers has been achieved for the most commonly occurring malignancies, epithelial ovarian carcinoma and carcinomas of the uterine endometrium and cervix. The degree to which *ras* and *ERBB-2* oncogenes, the *Tp53* tumor suppressor gene and microsatellite instability affect the sporadic manifestations of these cancers has been reasonably well established. In addition, allelotype analyses have pointed to the involvement of additional tumor suppressor genes for each tumor type. Research into the molecular biology and genetics of gynecological cancers has provided important insights pertaining to the clinical distinction of two types of endometrial carcinoma, the biological behavior of various types of ovarian tumors and the mechanism through which human papillomaviruses initiate tumorigenesis in the cervix.

Endometrial carcinoma

Epidemiological observations support the existence of two forms of endometrial carcinoma, one related to estrogen and most common in North America and Western Europe, and the other unrelated to estrogen and occurring similarly throughout the world.[20] Additional lines of clinical and pathological evidence have led to the development of two classifications for endometrial carcinoma, type I (estrogen-related) and type II (non-estrogen-related).[21–23] Type I tumors, in addition to their relationship to estrogen, occur in relatively younger, perimenopausal women, are frequently associated with endometrial hyperplasia, are of low grade and minimal myometrial invasion, exhibit a stable clinical course and are associated with a good prognosis. In contrast, type II tumors occur in relatively older, postmenopausal women with an absence of estrogen-related risk history, are not preceded or accompanied by hyperplasia, are of high grade and deep myometrial invasion, exhibit a progressive clinical course and are associated with a poor prognosis. Histologically, type I tumors are typically well-differentiated endometrioid adenocarcinomas and, less commonly, secretory, ciliated, villoglandular and squamous variants, whereas type II tumors include adenosquamous, papillary serous and clear cell carcinomas.[22] Presumably, there is a molecular genetic basis for the existence of these two categories of tumor, and any attempt to address the molecular etiology of endometrial carcinoma must be founded on the recognition of this distinction (Table 3).

Although endometrial carcinoma is the most common gynecological malignancy encountered, the molecular characterization of this cancer remains relatively limited. Among published studies examining abnormalities in oncogene structure or expression, several of the genes most commonly altered in human solid tumors have also been implicated in carcinomas of the endometrium, including *ras* and *ERBB-2*. However, it should be emphasized that no single oncogene has been found to be altered in more than approximately 20% of the endometrial cancers studied. Activating point mutation of a member of the *ras* gene family is, in general, the most commonly found oncogene aberration in human cancers. Several studies have confirmed that a *ras* mutation, predominantly in codon 12 of the K-*ras* gene, is present in 10–30% of endometrial cancers.[24–27] In those tumors that contain *ras* mutations, this event appears to occur early in the neoplastic process, as the incidence of mutant *ras* is the same in endometrial hyperplasia as that found in carcinoma. Attempts to correlate *ras* mutation with clinical outcome have produced conflicting data,[24,25] possibly reflecting an age-dependence of this phenomenon.[28]

Table 3 Two types of endometrial carcinoma

Type I	*Type II*
Clinicopathological features	
Hyperestrogenism	No estrogen risk factors
Younger age	Older age
Endometrioid histology	Non-endometrioid histology
Associated with hyperplasia	No hyperplasia
Low grade	High grade
Good prognosis	Poor prognosis
Genetic features	
Diploid	Aneuploid
Low loss of heterozygosity	High loss of heterozygosity
K-*ras*	K-*ras*
Microsatellite instability	*Tp53*
PTEN	*ERBB-2*

Involvement of the *ERBB-2* (also known as HER-2/*neu*) oncogene in tumorigenesis is through overexpression, with or without gene amplification. It is clear that 10–15% of endometrial cancers display overexpression of *ERBB-2* protein, compared with normal endometrial epithelium, as quantitated by immunohistochemistry.[29,30] Some studies have also documented *ERBB-2* gene amplification in endometrial cancers, although in only a subset of those tumors showing overexpression.[31–33] Overexpression of *ERBB-2* appears to be confined to a subset of high-grade and/or advanced-stage tumors. Correlation of expression with clinical outcome has been less conclusive, although the trend has been towards a positive correlation between overexpression and worsening of the prognosis.

Mutation of the *Tp53* tumor suppressor gene is the single most common genetic abnormality currently recognized in human cancers. Relative overexpression of the p53 protein is frequently observed in conjunction with many of the common missense mutations to which the gene is subject; thus, immunohistochemical analysis of p53 expression is frequently used as an end-point in many human tumor studies. Through the analysis of overexpression or mutation, or both, the presence of *Tp53* mutations in endometrial carcinomas has now been well established, although the incidence is clearly limited to a subset (10–30%) of all tumors studied.[26,34–37] Overexpression and/or mutation are associated with poor prognostic features, such as high grade, advanced stage, non-endometrioid histology and disease recurrence.[35,38–40] Several studies have also focused on papillary serous endometrial carcinomas, a majority of which display *Tp53* mutations or overexpression.[41–45] In a study of over 100 premalignant endometrial hyperplasia specimens, *Tp53* mutations were uniformly absent.[46]

As discussed below, all cancers associated with germline mutations in a mismatch repair gene, occurring within the context of the hereditary non-polyposis colorectal cancer (HNPCC) syndrome, display somatic genetic instability of microsatellite repeat sequences. Endometrial carcinoma is one of several tumor types in which a fraction of sporadic, non-hereditary forms of the cancer also display microsatellite instability. This phenomenon is evident in approximately 20% of all endometrial carcinomas.[47–50] While it was assumed that these cases probably result from somatic mutations in the same mismatch repair genes that cause HNPCC, it appears that only a fraction of sporadic endometrial cancers with microsatellite instability have acquired mutations in *hMSH2*, *hMLH1* or *hPMS2*.[51–54] Microsatellite instability in endometrial carcinoma instead seems to result from transcriptional silencing of the *MLH1* promoter through hypermethylation, and loss of MLH1 expression.[53,55] Correlative analyses of the clinicopathological features of these tumors suggest that microsatellite instability occurs in diploid tumors of endometrioid histology with good prognosis,[47,56] and is present in atypical hyperplasia lesions associated with adenocarcinoma.[57,58] Conversely, microsatellite instability is not seen in papillary serous carcinomas of the endometrium.[59,60]

Most recently, a novel tumor suppressor gene responsible for the hereditary cancer syndrome Cowden's disease was cloned and characterized.[61–63] Named *PTEN* or *MMAC1*, this gene encodes a tyrosine phosphatase with additional significant homology to the cytoskeletal proteins tensin and auxilin. Although endometrial cancer is not recognized as a component of Cowden's syndrome, sporadic endometrial carcinomas frequently exhibit LOH in a region of chromosome 10q with reasonably close proximity to the *PTEN* gene.[50,64] Mutation analyses of *PTEN* in endometrial carcinomas indicate that this gene is somatically inactivated in 30–50% of all such tumors,[65–67] representing the most frequent molecular genetic alteration in endometrial cancers yet defined. Interestingly, there is a strong correlation between the presence of microsatellite instability and *PTEN* mutation.[65,66]

Clues to the involvement of additional tumor suppressor genes in endometrial carcinogenesis have emerged from 'allelotyping' studies, in which LOH is quantitated throughout the genome. The principle of this type of analysis is discussed above. At least five studies have published partial or complete allelotyping data for endometrial cancers, allowing for several conclusions.[34,50,68–70] In general, LOH in endometrial carcinoma occurs at a relatively low frequency, compared with other common solid tumor types. Sites at which frequent LOH has been observed include chromosomes 3p, 6p, 8p, 9q, 10q, 14q, 16q, 17p and 18q. Chromosome arms for which the underlying tumor suppressor gene defects have been established with reasonable certainty are 17p and 10q, LOH at which generally correlates with mutation of *Tp53* and *PTEN*, respectively. The only anonymous chromosomal region for which a clinical correlation has been noted is 14q, LOH at which is correlated with a poor prognosis.[69]

In conclusion, a relatively limited body of data on the molecular genetics of endometrial carcinoma suggests that mutations of K-*ras* codon 12, microsatellite instability and *PTEN* mutations may be associated with type I tumors, whereas amplification or overexpression of *ERBB-2*, mutation of *Tp53* and LOH at chromosome 14q are features of some type II cancers (Table 3). Further studies are clearly warranted to define the molecular bases of these two types of malignancies.

Epithelial ovarian carcinoma

As the most lethal of all gynecological cancers, epithelial carcinoma of the ovary has been the focus of intense research. Several features of this cancer type, however, present problems relating to its study. Foremost is the tendency for clinical presentation at an advanced stage, and the absence of a well-defined premalignant lesion associated with the progression to invasive cancer. It is now well accepted that benign and borderline (or low malignant potential) ovarian tumors are generally not precursor lesions for invasive ovarian carcinomas, but rather represent distinct biological entities. The molecular genetic characterization of borderline and invasive cancers supports this distinction. Another common feature of ovarian cancers is the diffuse nature of intrapelvic disease evident at diagnosis, causing confusion as to the primary site of tumor origin, or even the clonal nature of the cancer. Molecular studies have supported the clinicopathological impression that papillary serous tumors of the peritoneum may be multifocal in origin, while serous tumors arising in the ovary are monofocal in origin.[71,72]

As for endometrial carcinoma, significant literature exists pertaining to the status of *ras*, *ERBB-2*, and *Tp53* genes in ovarian carcinoma. Activating point mutations of K-*ras*, primarily at codon 12, are found in 30–50% of both borderline and invasive cancers of mucinous histology, but at a very low frequency in serous tumors.[73–75] Gene amplification and overexpression of the *ERBB-2* gene are observed in 20–30% of invasive cancers, and have been associated with a poorer prognosis in some but not all studies.[76–80] Mutations of the *Tp53* gene accompanied by allelic deletions on chromosome 17p are found in 10–15% of early-stage and 40–50% of advanced-stage ovarian cancers,[74,81–84] but not in benign or borderline tumors.[85,86] Overexpression of the p53 protein correlates with shorter survival time in advanced-stage disease.[87–89] In contrast to endometrial cancer, microsatellite instability is very rare in sporadic ovarian carcinomas, and is not likely to play a significant role in this tumor type.[90–92] Interestingly, somatic mutations of the *PTEN* gene may occur frequently in ovarian cancers of endometrioid histology,[93] but not in the more common serous subtype.[65,94] Similarly, the gene encoding β-catenin is also mutated in endometrioid but not serous ovarian cancers,[95] suggesting that this histological type of ovarian cancer may have a distinct etiological mechanism.

Chromosomal instability and allelic deletions are common in ovarian carcinoma, and allelotyping data implicate many, if not most, chromosome arms as sites of potential tumor suppressor gene mutations.[90,96–98] Chromosome 17 is subject to LOH in 50–80% of all ovarian cancers. High-frequency deletion on the short arm is often associated with *p53* mutation, but there is probably an additional 17p tumor suppressor locus as well.[92] The common LOH observed on 17q is not associated with somatic mutation of the *BRCA1* gene,[99] and there are likely to be one or more novel tumor suppressor genes distal to this locus.[100,101] Other frequently affected chromosomal sites include 6q,[97,102,103] 11p,[102–105] 13q,[97,98,106–108] 14q,[109] 18q,[97,110] 22q[97,111] and Xp.[90,98] The recently developed technique of comparative genomic hybridization also allows for the detection of increases in copy number, which may reflect the location of amplified oncogenes. Increases in copy number on chromosomes 3q, 8q and 20q appear to be common in ovarian cancers.[91] Newly described genes that are amplified to a variable and as yet uncertain extent in ovarian carcinomas include *AKT2* on chromosome 19q,[112] *AIB1* on chromosome 20q[113] and the *PIK3CA* gene on chromosome 3p.[114]

Cervical carcinoma

Squamous cell carcinomas constitute the majority of invasive cervical cancers, with much of the remaining fraction attributable to adenocarcinomas of various types, and there is now considerable evidence that all of these tumors arise from cervical intraepithelial neoplasia (CIN), and that human papilloma virus (HPV) is critically involved in their initiation.[115] Consequently, a large body of research has elucidated in great detail the molecular biology of HPV function in tumorigenesis. It should be emphasized, however, that cervical cancers undoubtedly conform to the paradigm of multistep, multigenic carcinogenesis, and that additional oncogenes and tumor suppressor genes are certain to be involved in their etiology.

The more than 60 known types of HPV may generally be categorized into one of two groups, low-risk types associated with benign lesions, and oncogenic, high-risk types associated with invasive cancers. In cervical cancers and derived cell lines, the HPV DNA is usually integrated into the host genome, and there appears to be selection for the integrity of the region encoding two proteins, E6 and E7, which are regularly expressed in cancers.[116] The mechanism through which E6 and E7 contribute to cellular transformation involves physical interaction with two tumor suppressor proteins that play critical roles in mammalian cell cycle regulation, p53 and the retinoblastoma gene product (Rb). The E6 protein of high-risk HPV types binds to p53 with a higher affinity than E6 of low-risk types, and promotes p53 degradation.[117,118] Similarly, E7 binds and inactivates the Rb protein.[119,120] The combined effects of p53 and Rb inactivation in infected cells are likely to provide a proliferative advantage that is conducive to further genetic alterations, leading to cancer.

That only a small fraction of those infected by HPV eventually develop cervical cancer is further evidence that HPV alone is insufficient for malignant transformation.

The prevalence of *ras* oncogene mutations in cervical cancers has been inadequately surveyed. The absence of published data, however, suggests that such mutations are probably rare. Activating point mutations of codon 12 of the H-*ras* gene were found in 24% of advanced-stage squamous carcinomas,[121] and K-*ras* codon 12 mutations were identified in 13% of invasive endocervical adenocarcinomas.[122] Amplification and overexpression of the *ERBB-2* oncogene are present in only small fractions of squamous carcinomas and adenocarcinomas of the cervix,[123–125] and a correlation with poor prognosis exists for adenocarcinomas.[123,125] A thorough analysis of cervical cancers for *Tp53* status indicates that mutations are rare, and that there is no correlation with HPV status.[126–131] In contrast to other gynecological cancers, the c-*myc* oncogene may play a significant role in cervical carcinogenesis, with amplification and/or rearrangement of c-*myc* observed in 30–50% of cervical carcinomas.[132,133] Microsatellite instability appears to be very rare in cervical carcinoma.[134] Allelotyping analyses implicate potential tumor suppressor genes for cervical carcinoma on chromosomes 3p, 4q, 5p, 6p, 11q and 18q,[135–139] and deletions on chromosome 3p appear to represent a frequent and early event in cervical tumorigenesis.[140]

HEREDITARY GYNECOLOGICAL CANCERS

It has long been recognized that a family history of cancer confers one of the greatest risks of all known factors, and endometrial carcinoma and epithelial ovarian carcinoma both occur as components of cancer predisposition syndromes. As defined above, hereditary cancers are those associated with the inheritance of a mutant genetic allele through the germline, conferring predisposition to the development of one or more cancer types. Extraordinary progress has recently been made in identifying the molecular basis for essentially all of the autosomal dominant, highly penetrant manifestations of endometrial and ovarian carcinoma. It should be stressed that there are likely to exist many other gene variants that confer predisposition to one or another cancer type, with lower penetrance or through interactions with other susceptibility loci, but much less is currently known regarding these types of genes. Thus, it is likely that a much larger fraction of all cancers will eventually be regarded as 'hereditary' in nature, but, for the purposes of this discussion, hereditary cancer refers to those associated with dominant, highly penetrant loci.

Virtually all hereditary endometrial cancers, probably fewer than 5% of all cases, occur within the context of the HNPCC syndrome. The existence of a 'site-specific' endometrial cancer syndrome has been postulated,[141–143] but no genetic data exist to support the occurrence of such a phenomenon, which probably represents a rare manifestation of HNPCC. Indeed, recent evidence implicates the *MSH6* mismatch repair gene in some endometrial cancer families.[144] Hereditary ovarian cancers, about 10% of all cases, occur primarily within the context of the breast and ovarian cancer syndrome and to a lesser extent as part of the HNPCC syndrome. A 'site-specific' form of hereditary ovarian cancer has also been described,[145–147] but, similarly, this phenomenon appears to represent a variant of the breast and ovarian cancer syndrome, as the same genetic locus (*BRCA1*) is responsible for both.[147]

Hereditary non-polyposis colorectal cancer syndrome

Known previously as the 'cancer family syndrome' and 'Lynch syndromes I and II' (depending upon the absence or presence of extracolonic malignancies, respectively), HNPCC is an autosomal dominant genetic syndrome defined clinically by the so-called Amsterdam criteria:

(1) Colorectal cancer in at least three relatives, one of whom is a first-degree relative of the other two;

(2) Disease in at least two successive generations;

(3) Diagnosis under the age of 50 in at least one patient.[148]

In addition to cancers of the colon, HNPCC family members are at increased risk for malignancies of the endometrium, other gastrointestinal sites, the upper urological tract and the ovary.[149] The risk of endometrial cancer by age 70 in these families appears to be approximately 20%, compared with 3% in the general population.[150] Limited data on the risk of ovarian cancer in these families indicate a 3.5-fold increase in the number of cases observed over that expected in the general population;[149] HNPCC accounts for approximately 10–15% of all familial ovarian cancer cases.[151] Significant heterogeneity in ovarian cancer frequency is seen among HNPCC families,[149] suggestive of genetic heterogeneity. Consistent with this observation are linkage data indicative of several genetic loci that contribute to the HNPCC phenotype.[152–155] The cloning and characterization of the genes responsible for HNPCC have provided significant insights into the etiology of HNPCC-associated tumors and the potential for genetic screening for this disorder.

The HNPCC syndrome arises from an inherited defect in any one of several known genes: *hMSH2* (chromosome 2p), *hMLH1* (chromosome 3p), *hPMS2* (chromosome 7p) or *hMSH6* (chromosome 2p).[156–158] Although mutations have been described in five genes, the great majority of HNPCC kindreds appear to be linked to either *hMSH2* or *hMLH1*, and more than 90% of all reported mutations affect one of these two genes.[159] The proteins encoded by these genes participate in the same DNA mismatch repair (MMR) pathway, and loss-of-function mutations are associated with genetic instability in the tumors of affected family members.[156,160] The MMR genes appear to function as classical tumor suppressors, insofar as the wild-type allele inherited from the unaffected parent is lost or mutated somatically in HNPCC-linked tumors.[161–163]

The genetic instability phenotype associated with defective MMR genes is most readily observed through somatic length alterations in simple repeat sequences, for example (CA)*n*, located throughout the genome and known as 'microsatellites'. Replication errors in these repeat sequences are probably common, and their inefficient repair results in the 'microsatellite instability' phenotype. Since the discovery of mutant MMR genes and the corresponding microsatellite instability, a large number of studies have documented microsatellite instability in many sporadic tumor types, including those not associated with the HNPCC syndrome.[164,165] While mutations of the MMR genes have been readily identified in many HNPCC kindreds, somatic MMR gene mutations in sporadic tumors with the microsatellite instability phenotype are not commonly detected.[51,163] Thus, it is possible that this type of genetic instability arises in sporadic tumors through a different mechanism than in HNPCC-associated cancers. Hypermethylation of the *hMLH1* promoter, resulting in down-regulation of its expression, is one likely mechanism, and has been observed in sporadic colorectal, gastric and endometrial cancers with microsatellite instability.[53,55,166–169]

It is not clear how microsatellite instability *per se* contributes to tumorigenesis in the ovary or in other organs affected by the HNPCC syndrome. Microsatellites exist throughout the genome in predominantly non-coding regions of DNA. Simple repeat sequences are known to occur in the coding regions of genes, however, and their somatic mutation may result in loss of function for genes critical to the regulation of proliferation, invasion or metastasis. Examples include genes encoding the transforming growth factor-β (TGF-β) receptor type II,[170,171] the regulator of apoptosis, Bax,[172] the insulin-like growth factor II receptor[173] and caspase-5,[174] all of which contain homopolymeric microsatellite repeats that are mutated in one or another tumor type with microsatellite instability.

Breast and ovarian cancer syndrome

The breast and ovarian cancer syndrome accounts for 85–90% of all familial ovarian cancer cases.[146,151] The probable genetic relationship of these two malignancies in a hereditary context has been demonstrated in population-based, case–control epidemiological studies.[175–177] Families with a total of five or more breast or ovarian cancers in first- or second-degree relatives have been suggested to qualify as having the breast and ovarian cancer syndrome;[178] alternatively, families that contain at least three cases of early-onset (before age 60 years) breast or ovarian cancer have been similarly classified.[179] Following the original report of genetic linkage of early-onset breast cancer families to the *BRCA1* locus,[180] some breast and ovarian cancer families were shown to demonstrate linkage to *BRCA1* as well.[181] This finding has been extended, such that it is now clear that most (76–92%) breast and ovarian cancer families are linked to *BRCA1*.[178,179] The variable estimates of linkage are the probable result of genetic heterogeneity. Lower estimates are obtained if all families, including those with cases of male breast cancer, are considered, while higher estimates (approaching 100%) are obtained if families with cases of male breast cancer or fewer than two cases of ovarian cancer are excluded.[179] Most of the breast and ovarian cancer families not linked to *BRCA1* are linked to the *BRCA2* locus on chromosome 13q12–13, especially those with cases of male breast cancer.[182,183] The incidence of ovarian cancer, compared with breast cancer, appears to be lower in *BRCA2*-linked families, however, raising questions pertaining to the penetrance of this gene for ovarian cancer.[184,185]

BRCA1

As discussed above, genetic linkage analyses implicated a gene on chromosome 17q12–21, named *BRCA1*, as responsible for most cases of hereditary ovarian cancer occurring in the context of the breast and ovarian cancer syndrome. The discovery of a candidate *BRCA1* gene[186] was confirmed by several subsequent studies describing the segregation of inactivating mutations in this gene with disease phenotype in numerous breast and ovarian cancer families.[187–189] Deleterious germline mutations of *BRCA1* confer a relatively high lifetime risk of developing ovarian cancer, consistent with the observation that most unselected cases of ovarian cancer found to be associated with germline *BRCA1* mutations occur in women with remarkable medical or family histories, such as an early age of diagnosis, a previous diagnosis of breast cancer, or relatives with breast or ovarian cancer.[99,190–192] These studies of unselected series of ovarian cancers also indicate that germline mutations in *BRCA1* are associated with approximately 5% of all ovarian cancers.

The *BRCA1* gene consists of 22 coding exons distributed over approximately 100 kb of genomic DNA on chromosome 17q21.[186] The 7.8-kb mRNA transcript is expressed most abundantly in the testis and thymus, and at lower levels in the breast and ovary, and encodes a 1863-amino-acid residue protein.[186] Mutations of *BRCA1* described to date are located throughout the gene, with little evidence of clustering; about 80% of the mutations are loss-of-function non-sense or frameshift alterations,[193] consistent with the classification of *BRCA1* as a tumor suppressor gene. Other studies have shown that allelic deletions (as detected by loss of heterozygosity) affecting the 17q21 region in *BRCA1*-linked breast or ovarian cancers invariably involve the wild-type chromosome, as would be expected if *BRCA1* were behaving as a typical tumor suppressor gene.[194,195] Two specific founder mutations, 185delAG and 5382insC, are present in approximately 1.0% and 0.1% of the Ashkenazi Jewish population, respectively.[196,197] Relatively high frequencies of other specific *BRCA1* founder mutations have also been described in Dutch, Belgian, Scandinavian and French-Canadian breast and ovarian cancer families.[198–201]

The data linking *BRCA1* to most hereditary ovarian cancers are now unequivocal. It was believed that this gene would be found to play a major role in sporadic ovarian carcinomas as well. This speculation centered around the observation that loss of heterozygosity, the genetic hallmark of tumor suppressor gene inactivation, is observed on chromosome 17q in up to 80% of sporadic ovarian carcinomas.[90,96–98,202,203] Analyses of several series of unselected ovarian carcinomas, most of which would be sporadic, suggest that somatic mutations of *BRCA1* are rare, however.[99,190,192,204,205] This finding is supported by fine-deletion mapping studies, which implicate one and perhaps two regions distal to *BRCA1* as harboring an additional tumor suppressor gene(s).[100,101,206] These data suggest that mutational inactivation of *BRCA1* is necessary at a relatively early stage of development in order to contribute to ovarian tumorigenesis. Under this hypothesis, somatic mutation of *BRCA1* in the adult ovarian epithelium seldom provides a significant selective advantage in a developing malignancy, and thus rarely leads to clonal selection and manifestation as a somatic mutation in the ovarian cancer.

BRCA2

Similar approaches were used to localize and clone the *BRCA2* cancer susceptibility gene. Genetic linkage analysis of high-risk breast cancer families that were unlinked to *BRCA1* revealed a locus at chromosome 13q12–13, to which some of these families were linked.[182] A candidate gene from this region of chromosome 13 was determined to represent *BRCA2*, based on the presence of germline-inactivating mutations that segregated with disease in linked families.[207] Confirmation of this finding was provided by additional studies in which inactivating mutations were identified in breast cancer families, especially those with male breast cancer.[208–210] Unlike *BRCA1*, inherited mutations in *BRCA2* appear to confer a substantially lower risk of ovarian cancer, compared with breast cancer, as inferred from tumor incidence rates in linked families,[207,208,211] and from penetrance analysis studies, which suggest that the lifetime risk of ovarian cancer in *BRCA2* mutation carriers is in the range of 16–27%.[212–214] Furthermore, an analysis of unselected ovarian cancer cases suggests that germline-inactivating mutations of *BRCA2* may be found in patients with late-onset disease, and no remarkable medical or family histories with regard to breast or ovarian cancer.[108,215,216] Taken together, these findings are consistent with the hypothesis that inherited mutations in *BRCA2* contribute to ovarian cancer with a lower penetrance than *BRCA1*, and that the fraction of all ovarian cancers resulting from hereditary predisposition may be higher than previously suspected, based on estimates derived from the study of cancer-prone families segregating highly penetrant alleles.

The *BRCA2* gene consists of 26 coding exons distributed over approximately 70 kb of genomic DNA, encoding a transcript of 11–12 kb.[208] As for BRCA1, the BRCA2 mRNA is most highly expressed in the testis and thymus, with lower levels in the breast and ovary.[208] In addition to tissue-specific expression profiles, *BRCA1* and *BRCA2* share a number of additional structural and functional similarities. Both are unusually large genes in terms of the number of exons and size of the encoded message, both have a large exon 11 that contains approximately half of the entire coding region, both contain translation start sites in exon 2, and both are relatively A/T-rich. Mutations of *BRCA2* reported to date are, as for *BRCA1*, dispersed throughout the gene, with little evidence for hotspots or clustering.[193] The great majority of mutations are frameshift in nature, with microdeletions being most common; microinsertions, non-sense and missense mutations occur rarely. Loss of heterozygosity at the *BRCA2* locus in tumors from linked individuals invariably involves the wild-type allele, consistent with the classification of *BRCA2* as a tumor suppressor gene.[217,218] Some mutations have been observed in multiple unrelated families, suggesting that a subset of *BRCA2* mutations may also occur relatively frequently. A single mutation in *BRCA2*, 6174delT, is found in approximately 1.4% of the Ashkenazi Jewish population;[196,219,220] this mutation together with those in *BRCA1* are thus present in approximately 1 in 40 Ashkenazi Jewish individuals. As might be expected, a relatively large fraction of all ovarian cancer cases in Ashkenazi Jews, approximately 40%, are associated with a germline mutation in *BRCA1* or *BRCA2*.[221] Relatively high frequencies of other specific *BRCA2* founder mutations have also been described in Scandinavian, Yemenite Jewish and French-Canadian breast and ovarian cancer families.[200,201,222,223]

As was the case for *BRCA1*, it was believed that somatic mutations of *BRCA2* might be involved in a significant fraction of sporadic ovarian cancers, based on the high frequency of allelic loss observed on chromosome 13q in these tumors.[97,98,224] Analysis of a large sample of unselected ovarian cancers indicates that, although loss of heterozygosity which includes the *BRCA2* locus is seen in over half of the tumors, somatic mutations of the gene are rare,[108,215] but probably more common than for *BRCA1*. These studies further indicate that approximately 3–4% of all ovarian cancers are associated with germline mutations in *BRCA2*.

Function of BRCA proteins

The precise functions of the BRCA proteins remain to be determined in detail.[225,226] Published studies to date provide evidence supporting multiple discrete functions, related primarily to transcriptional activation and DNA repair, not improbable given the large size of both proteins (Figure 2). For BRCA1, multiple specific antibodies detect a 220- kDa protein in the nucleus of cultured cells and normal tissues.[227–230] Several

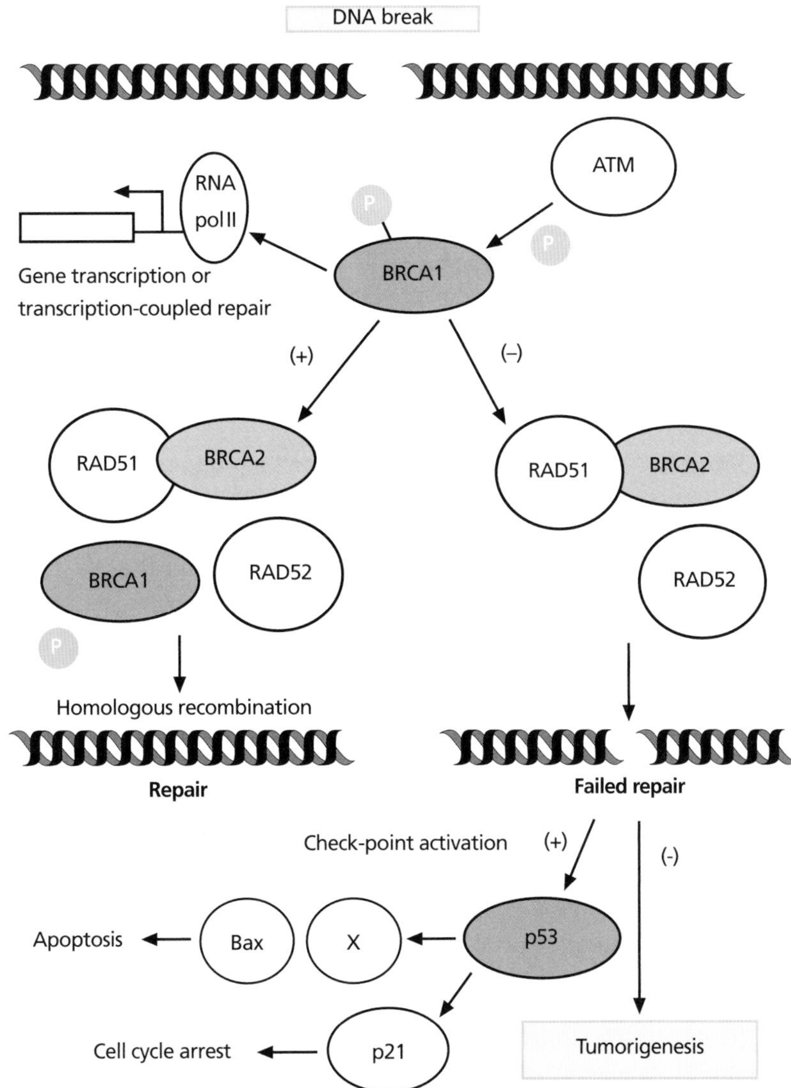

Figure 2 Working model for the function of the breast and ovarian cancer-linked BRCA proteins. Following DNA damage, for example a double-strand break, the ATM protein phosphorylates (P)BRCA1. Phosphorylated BRCA1 may regulate gene transcription or transcription-coupled DNA repair. In co-operation with a BRCA2–RAD51 complex and with other RAD52-related proteins, phosphorylated BRCA1 mediates double-strand break repair through homologous recombination. In the absence of BRCA1 (or BRCA2), repair does not occur, leading to activation of the p53-mediated cell cycle checkpoint(s). The proliferation of cells containing damaged DNA is blocked either through the Bax-mediated induction of apoptosis, or through the p21-mediated induction of cell cycle arrest. In the absence of functional p53, cells with damaged DNA continue to proliferate, leading to tumorigenesis. Hereditary ovarian cancers associated with *BRCA* mutations have sustained inactivating mutations in *BRCA1* or *BRCA2* and *Tp53*. RNA pol II, RNA polymerase II

domains of potential functional significance include an N-terminal RING-finger domain, a negatively charged region in the C terminus, and C-terminal sequences now known as BRCT domains that are partially homologous to yeast RAD9 and a cloned p53-binding protein.[186,231] The presence of these motifs is consistent with the ability of BRCA1 to activate gene transcription *in vitro*.[232,233] Evidence that BRCA1 is a component of the RNA polymerase II transcription complex,[234] and interacts physically with RNA helicase[235] and the transcriptional activators p53,[236,237] CtIP[238] and c-myc,[239] further supports its possible function in transcriptional regulation. Thus, mutational inactivation of *BRCA1* might be expected to affect the expression of other genes, involved presumably in the regulation of growth or differentiation in breast and ovarian epithelium. The expression pattern of Brca1 during mouse development,[240,241] and the cell cycle-[242,243] and hormone-regulated[244,245] expression of BRCA1, suggest a relationship with differentiation and cell proliferation. Finally, functional studies demonstrating the ability of BRCA1 expression to inhibit growth, suppress tumorigenesis and induce apoptosis support the classification of BRCA1 as a tumor suppressor protein.[246–248]

Evidence has also accumulated to suggest a role for BRCA1 in the cellular response to DNA damage. This function was originally inferred from data showing BRCA1 co-localization *in vivo* and physical association *in vitro* with the RAD51 protein, known to function in the repair of double-strand DNA breaks, implying a role for BRCA1 in the control of recombination and genomic integrity.[249] The ability of several distinct DNA-damaging agents to cause changes in the subnuclear localization and phosphorylation state of BRCA1 supports this hypothesis.[250–252] Additionally, data derived from the study of embryonic tissues and cells from mice rendered nullizygous for *Brca1* provide strong evidence for the role of Brca1 in the response to DNA damage. An embryonic lethal phenotype is observed in mice with a homozygous null mutation in *Brca1*, suggesting its requirement for embryonic cellular proliferation prior to gastrulation.[253] Partial rescue of this developmental lethality is achieved by simultaneous knock-out of either the *p53* or the *p21* genes,[254,255] one interpretation of these data is that the accumulation of DNA damage in *Brca1* knock-outs leads to the arrest of cell division mediated by p53 and p21, and that their concomitant knock-out allows additional cell division to take place before the eventual lethality. Finally, embryonic stem cells from *Brca1*-nullizygous mice are defective in transcription-coupled repair of oxidative DNA damage, and hypersensitive to ionizing radiation and hydrogen peroxide,[256] and mouse embryo fibroblasts with a partial loss of function phenotype display chromosomal abnormalities associated with centrosome amplification.[257,258] Most recently, the inducible expression of BRCA1 in human cells was found to result in the p53-independent induction of GADD45 expression, and the JNK/SAPK-dependent activation of programmed cell death.[259]

The *BRCA2* gene product exhibits many functional similarities to that of *BRCA1*. Like BRCA1, the *BRCA2* gene product is a 460-kDa nuclear phosphoprotein[260,262] that interacts physically with p53 and RAD51.[262] Expression of both genes appears to be co-ordinately regulated during cell cycle progression[260,263] and in response to estrogen[245] in human cells, and during proliferation and development in embryonic and adult mouse tissues.[264–266] Preliminary findings suggest that BRCA2 may also function as a transcription factor, as a small portion of the protein shares homology with the known transcription factor c-Jun, and is capable of activating transcription *in vitro*.[267]

Substantial evidence also exists for the function of BRCA2 in DNA repair. Studies of the role of Brca2 in mouse development indicate that loss of the protein confers radiation hypersensitivity, consistent with its interaction with the Rad51 protein involved in repair of double-strand DNA breaks.[268] Remarkably, BRCA1 and BRCA2 proteins may thus be involved in the same biochemical pathway, mediated by RAD51, regulating genomic integrity.[249,268,269] Additional studies with the *Brca2*-nullizygous mouse model provide further data implicating Brca2 in the response to DNA damage. The *Brca2* knock-out mouse also displays an embryonic lethal phenotype, indicating a critical role for Brca2 in cellular proliferation during embryogenesis.[255,268,270] This phenotype is partially rescued in *Brca2/p53* nullizygotes,[255] again implying a role for Brca2 in the cellular response to DNA damage. In *Brca2* nullizygotes carrying mutations that allow survival to adulthood, the animals display inefficient DNA repair following X-irradiation and an increased susceptibility to tumors.[271,272] Using embryonic fibroblasts from these mice, an increased sensitivity to mutagens and the accumulation of chromosomal abnormalities were also observed.[273] Using the human pancreatic carcinoma cell line Capan-1, which is homozygous for the *BRCA2* 6174delT mutation, BRCA2 was demonstrated to mediate sensitivity to radiation and drugs that induce double-strand DNA breaks,[261,274] and, furthermore, it was shown that the BRC repeat motif of BRCA2 mediates the physical interaction with RAD51 necessary for the normal response to DNA damage.[261]

THE CELL CYCLE

Although cancer cells possess many abnormal properties, deregulation of the normal constraints on cell proliferation lies at the heart of malignant transformation. A tumor may increase in size through any one of three mechanisms involving alterations pertaining to the cell cycle: shortening of the time of transit of cells through the cycle, a decrease in the rate of cell death or the re-entry of quiescent cells into the cycle. In most human cancers, all three mechanisms appear to be important in regulating tumor growth rate, a critical parameter in determining the biological aggressiveness of a tumor.[275] The classical cell cycle model, consisting of a DNA synthesis (S) phase, a mitosis (M) phase and two gap (G_1 and G_2) phases, has now been elucidated in molecular detail (Figure 3). Critical components of the cycle include the cyclins, cyclin-dependent kinases, inhibitors of cyclin-dependent kinases, and the Rb, p53 and E2F proteins. Many of the protein products of oncogenes and tumor suppressor genes are directly linked to biochemical pathways involving growth factor signalling and control of progression through the cell cycle.

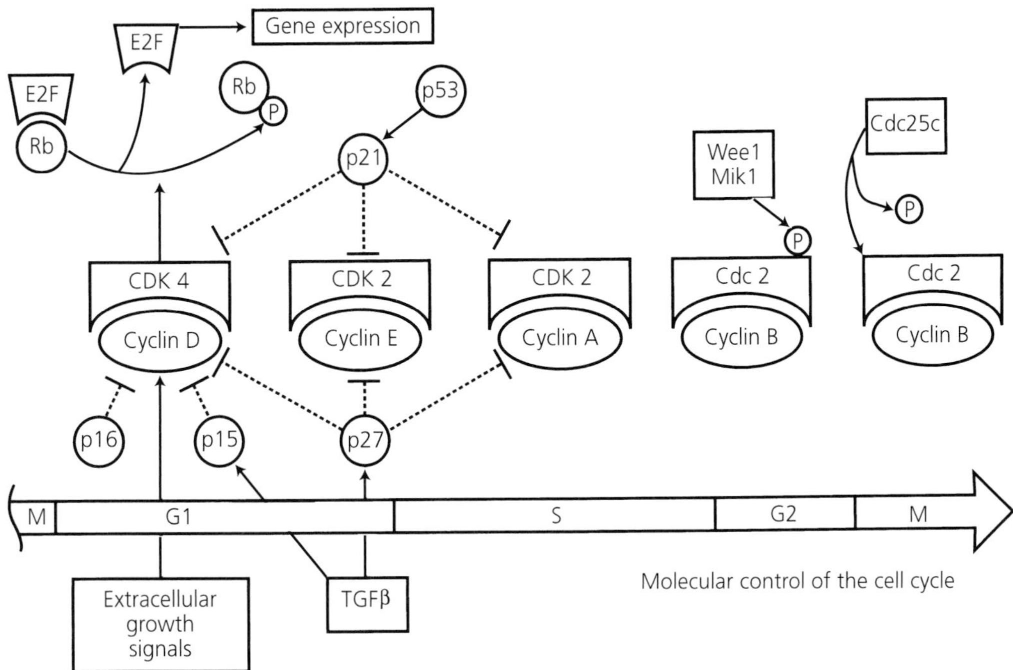

Figure 3 Molecular control of cell cycle progression. A linear version of various stages of the cell cycle is shown with the various cyclin/cyclin-dependent kinase (CDK) complexes corresponding to the stages that they control. Rb, retinoblastoma gene product; Cdc,; TGF-β, transforming growth factor-β; P, phosphorylation

The current model of cell cycle control holds that the transitions between different cell cycle states are regulated at check-points.[276] A first crucial step in the cell cycle occurs late in the G_1 phase at the so-called restriction point, when a cell commits to completing the cycle.[277] In order to pass through this point and enter the S phase, growth-promoting signals transduced from the cell surface to the nucleus cause a rapid and transient elevation in the levels of D-type cyclins (in early G_1) and cyclin E (in late G_1). There are three forms of cyclin D, which are in part cell-type specific; most cells express D3 and either D1 or D2.[278] These cyclins combine with and activate enzymes known as cyclin-dependent kinases (CDKs), primarily CDK4 with the D-type cyclins and CDK2 with cyclin E. CDK4 transfers phosphate groups from adenosine triphosphate (ATP) to the Rb tumor suppressor protein. Rb is hypophosphorylated throughout the G_1 phase, is phosphorylated just before the S phase and remains hyperphosphorylated until late M phase.[279] The Rb protein binds to and sequesters transcription factors critical for the G_1 to S transition, notably E2F, and their release following Rb phosphorylation leads to the expression of genes responsible for further cell cycle progression.[280] It is postulated that the cyclin D–CDK4 complex regulates progression through G_1, while the cyclin E–CDK2 complex regulates the G1 to S transition.[281]

Various CDK inhibitor (CDKI) proteins also play a crucial role in the process of G_1 progression.[281,282] Among these are proteins known as p15 (INK4B or MTS2), p16 (INK4 or MTS1), p21 (SDI1 or WAF1) and p27 (Kip1). The CDKIs act through the formation of stable complexes with cyclin–CDK dimers, disrupting the catalytic function of CDKs. All four of these CDKIs bind to the cyclin D–CDK4/6 dimers, while p21 and p27 also associate with the cyclin E–CDK2 dimer. Several factors are known to regulate expression of the CDKI proteins. Transforming growth factor-β (TGF-β) causes a rapid increase in levels of p15 and p27 mRNA and protein, indicating that these CDKIs are responsible for arresting cells in G_1 in response to this antimitogenic cytokine. The E2F transcription factor up-regulates p16 expression, suggesting the presence of a feedback loop, wherein repression of p16 expression by Rb hypophosphorylation leads to increased activity of CDK4 and phosphorylation of Rb. Transcriptional regulation of p21 is accomplished primarily by the p53 tumor suppressor protein, but may be affected by various activated growth factor receptors as well.

The cyclin E–CDK2 complex mediates progression out of G_1, and cyclin A expression increases dramatically with onset of the S phase. Cyclin A–CDK2 function then appears to be required for ongoing DNA replication, and again for the G_2 to M transition. Available evidence suggests that the cyclin A–CDK2 complex participates in the assembly, activation or regulation of DNA replication structures.[283] An additional function of cyclin A may be in programmed cell death.[282] The activity of cyclin A-dependent protein kinases is increased in cells undergoing apoptosis, and overexpression of cyclin A induces apoptosis in cells cultured in low-serum conditions.

Passage through G_2 and traversal of the G_2 to M check-point are mediated by cyclins B1 and B2 in complexes with the Cdc2 kinase.[284] Cyclin B–Cdc2 complexes accumulate in an inactive state during the S and G_2 phases. The Cdc2 kinase component is kept inactive through phosphorylation by Wee1/Mik1-related protein kinases. At the end of G_2, a phosphatase known as Cdc25C dephosphorylates and activates Cdc2, allowing the transition into mitosis. In normal cells, DNA damaged by radiation or alkylating agents prevents dephosphorylation of Cdc2, resulting in G_2 arrest. Several ubiquitin-dependent proteolysis events, including the destruction of B-type cyclins, allow the cell to progress completely through mitosis and complete the cell cycle.

Clearly, there are many points in this cell cycle where mutational activation or hyperactivity of cyclins and their associated kinases, or mutational inactivation or hypoactivity of CDKIs, would be expected to exert an oncogenic stimulus. Both cyclins and CDKIs are represented on the list of genes mutated in human cancers, and aberrant expression or activity of these proteins is common in tumors of many types. The gene encoding cyclin D1, *CCND1*, is located on chromosome 11q13 in a region that is amplified in several cancers. Amplification and overexpression are observed most commonly in carcinomas of the breast, lung, stomach and esophagus, while overexpression alone is seen in a much larger number of cancers.[281,285] This oncogene has also been designated *PRAD1* because of its overexpression resulting from a translocation in benign parathyroid adenomas, and as *BCL1* because of a different translocation leading to its overexpression in certain B cell lymphomas.

The gene encoding p16, *CDKN2A*, is located on chromosome 9p21 in a region that is deleted in many solid tumor types. This tumor suppressor gene is most often disrupted by large homozygous deletions, but may also be inactivated through point mutations.[281,286] Germline mutations of *CDKN2A* are also responsible for the majority of the familial melanoma kindreds that show genetic linkage to chromosome 9p21.[287,288] In humans, the genes encoding p16 and p15 lie in tandem on chromosome 9p, and homozygous deletions that include both genes have been observed in several tumor types.[281]

Two prototypical human tumor suppressor genes, *Rb1* and *Tp53*, encode proteins that play pivotal roles in G_1 to S cell cycle progression. Both molecules participate in biochemical pathways that eventually converge on regulation of the E2F transcription factor. Inactivating mutations of *Tp53* are the most common molecular genetic alterations known in human cancers.[289] Loss of p53 function leads to reduced levels of p21 and hyperactivity of both cyclin D–CDK and cyclin E–CDK complexes, hyperphosphorylation of Rb and elevated levels of E2F. Mutational inactivation of the *Rb1* gene itself, which is seen in several tumor types, would have the same end result; in addition to retinoblastomas and osteosarcomas that are seen in patients with inherited *Rb1* mutations, sporadic retinoblastomas and sarcomas, most if not all small-cell lung carcinomas, and a portion of non-small-cell lung, bladder and breast carcinomas exhibit somatic *Rb1* mutations.[10] The molecular targets of E2F and its related transcription factors are becoming known in increasingly greater detail,[290,291] providing a coherent view of the pathway through which Rb and p53 converge in negative control of the cell cycle progression. Binding sites for E2F are present in genes implicated in the induction of S phase, including those encoding thymidine kinase, the proto-oncogenes *myc* and *myb*, dihydrofolate reductase and DNA polymerase-α.[292]

REFERENCES

1. Watson JD. In Angier N, ed. *Natural Obsessions: The Search for the Oncogene*. Boston: Houghton Mifflin Co., 1988: 12.
2. Bishop JM. Cancer: the rise of the genetic paradigm. *Genes Dev* 1995 9: 1309–1315.
3. Weinberg RA. Oncogenes, antioncogenes, and the molecular basis of multistep carcinogenesis. *Cancer Res* 1989 49: 3713–3721.
4. Boyd J, Barrett JC. Genetic and cellular basis of multistep carcinogenesis. *Pharmacol Ther* 1990 46: 469–486.
5. Bishop JM. Molecular themes in oncogenesis. *Cell* 1991 64: 235–248.
6. Vogelstein B, Kinzler KW. The multistep nature of cancer. *Trends Genet* 1993 9: 138–141.
7. Armitage P, Doll R. The age distribution of cancer and a multi-stage theory of carcinogenesis. *Br J Cancer* 1954 8: 1–12.
8. Renan MJ. How many mutations are required for tumorigenesis? Implications from human cancer data. *Mol Carcinog* 1993 7: 139–146.
9. Hunter T. Cooperation between oncogenes. *Cell* 1991 64: 249–270.
10. Weinberg RA. Tumor suppressor genes. *Science* 1991 254: 1138–1146.
11. Hofstra RMW, Landsvater RM, Ceccherini I *et al.* A mutation in the *RET* proto-oncogene associated with multiple endocrine neoplasia type 2B and sporadic medullary thyroid carcinoma. *Nature* 1994 367: 375–378.
12. Schmidt L, Duh F-M, Chen F *et al.* Germline and somatic mutations in the tyrosine kinase domain of the *MET* proto-oncogene in papillary renal carcinomas. *Nat Genet* 1997 16: 68–73.
13. Bishop JM. Cellular oncogenes and retroviruses. *Annu Rev Biochem* 1983 52: 301–354.
14. Knudson AG. Hereditary cancer, oncogenes, and antioncogenes. *Cancer Res* 1985 45:1437–1443.
15. Cavenee WK, Dryja TP, Phillips RA *et al.* Expression of recessive alleles by chromosomal mechanisms in retinoblastoma. *Nature* 1983 305: 779–784.
16. Nowell P. The clonal evolution of tumor cell populations. *Science* 1976 194: 23–28.

17. Loeb LA. A mutator phenotype in cancer. *Cancer Res* 2001 61: 3230–3239.
18. Lengauer C, Kinzler KW, Vogelstein B. Genetic instability in colorectal cancers. *Nature* 1997 386: 623–627.
19. Fearon ER, Vogelstein B. A genetic model for colorectal tumorigenesis. *Cell* 1990 61: 759–767.
20. Parazzini F, La Vecchia C, Bocciolone L, Franceschi S. The epidemiology of endometrial cancer. *Gynecol Oncol* 1991 41: 1–16.
21. Deligdisch L, Holinka CF. Endometrial carcinoma: two diseases? *Cancer Detect Prev* 1987 10: 237–246.
22. Kurman RJ, Zaino RJ, Norris HJ. Endometrial carinoma. In Kurman RJ, ed. *Blaustein's Pathology of the Female Genital Tract* 3rd edn. New York: Springer-Verlag, 1994: 439–486.
23. Boyd J. Estrogen as a carcinogen: the genetics and molecular biology of human endometrial carcinoma. In Huff JE, Boyd J, Barrett JC, eds. *Cellular and Molecular Mechanisms of Hormonal Carcinogenesis: Environmental Influences.* New York: John Wiley and Sons, 1996: 151–173.
24. Mizuuchi H, Nasim S, Kudo R *et al.* Clinical implications of K-*ras* mutations in malignant epithelial tumors of the endometrium. *Cancer Res* 1992 52: 2777–2781.
25. Sasaki H, Nishii H, Takahashi H *et al.* Mutation of the Ki-*ras* protooncogene in human endometrial hyperplasia and carcinoma. *Cancer Res* 1993 53: 1906–1910.
26. Enomoto T, Fujita M, Inoue M *et al.* Alterations of the *p53* tumor suppressor gene and its association with activation of the c-K-*ras*-2 protooncogene in premalignant and malignant lesions of the human uterine endometrium. *Cancer Res* 1993 53: 1883–1888.
27. Duggan BD, Felix JC, Muderspach LI, Tsao J-L, Shibata DK. Early mutational activation of the c-Ki-*ras* oncogene in endometrial carcinoma. *Cancer Res* 1994 54: 1604–1607.
28. Ito K, Watanabe K, Nasim S *et al.* K-*ras* point mutations in endometrial carcinoma: effect on outcome is dependent on age of patient. *Gynecol Oncol* 1996 63: 238–246.
29. Berchuck A, Rodriguez G, Kinney RB *et al.* Overexpression of HER-2/*neu* in endometrial cancer is associated with advanced disease stage. *Am J Obstet Gynecol* 1991 164: 15–21.
30. Hetzel DJ, Wilson TO, Keeney GL *et al.* HER-2/*neu* expression: a major prognostic factor in endometrial carcinoma. *Gynecol Oncol* 1992 47: 179–185.
31. Czerwenka K, Lu Y, Heuss F. Amplification and expression of the c-*erbB*-2 oncogene in normal, hyperplastic, and malignant endometria. *Int J Gynecol Pathol* 1995 14: 98–106.
32. Esteller M, Garcia A, Martinez i Palones JM, Cabero A, Reventos J. Detection of c-*erbB*-2/*neu* and fibroblast growth factor-3/INT-2 but not epidermal growth factor gene amplification in endometrial cancer by differential polymerase chain reaction. *Cancer* 1995 75: 2139–2146.
33. Saffari B, Jones LA, el-Naggar A *et al.* Amplification and overexpression of the HER-2/*neu* (c-*erbB*2) in endometrial cancers: correlation with overall survival. *Cancer Res* 1995 55: 5693–5698.
34. Okamoto A, Sameshima Y, Yamada Y *et al.* Allelic loss on chromosome 17p and *p53* mutations in human endometrial carcinoma of the uterus. *Cancer Res* 1991 51: 5632–5636.
35. Kohler MF, Berchuck A, Davidoff AM *et al.* Overexpression and mutation of *p53* in endometrial carcinoma. *Cancer Res* 1992 52: 1622–1627.
36. Risinger JI, Dent GA, Ignar-Trowbridge D *et al.* *p53* gene mutations in human endometrial carcinoma. *Mol Carcinog* 1992 5: 250–253.
37. Honda T, Kato H, Imamura T *et al.* Involvement of *p53* gene mutations in human endometrial carcinomas. *Int J Cancer* 1993 53: 963–7.
38. Ito K, Watanabe K, Nasim S *et al.* Prognostic significance of *p53* overexpression in endometrial cancer. *Cancer Res* 1994 54: 4667–4670.
39. Inoue M, Okayama A, Fujita M *et al.* Clincopathological characteristics of *p53* overexpression in endometrial cancers. *Int J Cancer* 1994 58: 14–19.
40. Soong R, Knowles S, Williams KE *et al.* Overexpression of *p53* protein is an independent prognostic indicator in human endometrial carcinoma. *Br J Cancer* 1996 74: 562–567.
41. Khalifa MA, Mannel RS, Haraway SD, Walker J, Min KW. Expression of EGFR, HER-2/neu, p53, and PCNA in endometrioid, serous papillary, and clear cell endometrial adenocarcinomas. *Gynecol Oncol* 1994 53: 84–92.
42. Prat J, Oliva E, Lerma E, Vaquero M, Matias-Guiu X. Uterine papillary serous adenocarcinoma: a

10-case study of p53 and c-*erbB*-2 expression and DNA content. *Cancer* 1994 74: 1778–1783.
43. King SA, Adas AA, LiVolsi VA *et al.* Expression and mutation analysis of the *p53* gene in uterine papillary serous carcinoma. *Cancer* 1995 75: 2700–2705.
44. Zheng W, Cao P, Zheng M, Kramer EE, Godwin TA. *p53* overexpression and bcl-2 persistence in endometrial carcinoma: comparison of papillary serous and endometrioid subtypes. *Gynecol Oncol* 1996 61: 164–174.
45. Tashiro H, Isacson C, Levine R *et al.* *p53* gene mutations are common in uterine serous carcinoma and occur early in their pathogenesis. *Am J Pathol* 1997 150: 177–185.
46. Kohler MF, Nishii H, Humphrey PA *et al.* Mutation of the *p53* tumor-suppressor gene is not a feature of endometrial hyperplasias. *Am J Obstet Gynecol* 1993 169: 690–694.
47. Risinger JI, Berchuck A, Kohler MF *et al.* Genetic instability of microsatellites in endometrial carcinoma. *Cancer Res* 1993 53: 5100–5103.
48. Burks RT, Kessis TD, Cho KR, Hedrick L. Microsatellite instability in endometrial carcinoma. *Oncogene* 1994 9: 1163–1166.
49. Duggan BD, Felix JC, Muderspach LI *et al.* Microsatellite instability in sporadic endometrial carcinoma. *J Natl Cancer Inst* 1994 86: 1216–1221.
50. Peiffer SL, Herzog TJ, Tribune DJ *et al.* Allelic loss of sequences from the long arm of chromosome 10 and replication errors in endometrial cancers. *Cancer Res* 1995 55: 1922–1926.
51. Katabuchi H, van Rees B, Lambers AR *et al.* Mutations in DNA mismatch repair genes are not responsible for microsatellite instability in most sporadic endometrial carcinomas. *Cancer Res* 1995 55: 5556–5560.
52. Kobayashi K, Matsushima M, Koi S *et al.* Mutational analysis of mismatch repair genes, *hMLH1* and *hMSH2*, in sporadic endometrial carcinomas with microsatellite instability. *Jpn J Cancer Res* 1996 87: 141–145.
53. Gurin CC, Federici MG, Kang L, Boyd J. Causes and consequences of microsatellite instability in endometrial carcinoma. *Cancer Res* 1999 59: 462–466.
54. Kowalski LD, Mutch DG, Herzog TJ, Rader JS, Goodfellow PJ. Mutational analysis of *MLH1* and *MSH2* in 25 prospectively acquired RER+ endometrial cancers. *Genes Chromosomes Cancer* 1997 18: 219–227.
55. Simpkins SB, Bocker T, Swisher EM *et al.* *MLH1* promoter methylation and gene silencing is the primary cause of microsatellite instability in sporadic endometrial cancers. *Hum Mol Genet* 1999 8: 661–666.
56. Maxwell GL, Risinger JI, Alvarez AA, Barrett JC, Berchuck A. Favorable survival associated with microsatellite instability in endometrioid endometrial cancers. *Obstet Gynecol* 2001 97: 417–422.
57. Jovanovic AS, Boynton KA, Mutter GL. Uteri of women with endometrial carcinoma contain a histopathological spectrum of monoclonal putative precancers, some with microsatellite instability. *Cancer Res* 1996 56: 1917–1921.
58. Levine RL, Cargile CB, Blazes MS *et al.* *PTEN* mutations and microsatellite instability in complex atypical hyperplasia, a precursor lesion to uterine endometrioid carcinoma. *Cancer Res* 1998 58: 3254–3258.
59. Tashiro H, Lax SF, Gaudin PB *et al.* Microsatellite instability is uncommon in uterine serous carcinoma. *Am J Pathol* 1997 150: 75–79.
60. Lax SF, Kendall B, Tashiro H, Slebos RJ, Hedrick L. The frequency of p53, K-*ras* mutations, and microsatellite instability differs in uterine endometrioid and serous carcinoma: evidence of distinct molecular genetic pathways. *Cancer* 2000 88: 814–824.
61. Li J, Yen C, Liaw D *et al.* *PTEN*, a putative protein tyrosine phosphatase gene mutated in human brain, breast, and prostate cancer. *Science* 1997 275: 1943–1947.
62. Steck PA, Pershouse MA, Jasser SA *et al.* Identification of a candidate tumour suppressor gene, *MMAC1*, at chromosome 10q23.3 that is mutated in multiple advanced cancers. *Nat Genet* 1997 15: 356–362.
63. Liaw D, Marsh DJ, Li J *et al.* Germline mutations of the *PTEN* gene in Cowden disease, an inherited breast and thyroid cancer syndrome. *Nat Genet* 1997 16: 64–67.
64. Nagase S, Yamakawa H, Sato S, Yajima A, Horii A. Identification of a 790-kilobase region of common allelic loss in chromosome 10q25–q26 in human endometrial cancer. *Cancer Res* 1997 57: 1630–1633.

65. Tashiro H, Blazes MS, Wu R *et al.* Mutations in *PTEN* are frequent in endometrial carcinoma but rare in other common gynecological malignancies. *Cancer Res* 1997 57: 3935–3940.
66. Kong D, Suzuki A, Zou T-T *et al.* *PTEN* is frequently mutated in primary endometrial carcinomas. *Nat Genet* 1997 17: 143–144.
67. Risinger JI, Hayes AK, Berchuck A, Barrett JC. *PTEN/MMAC1* mutations in endometrial cancers. *Cancer Res* 1997 57: 4736–4738.
68. Imamura T, Arima T, Kato H *et al.* Chromosomal deletions and K-*ras* gene mutations in human endometrial carcinomas. *Int J Cancer* 1992 51: 47–52.
69. Fujino T, Risinger JI, Collins NK *et al.* Allelotype of endometrial carcinoma. *Cancer Res* 1994 54: 4294–4298.
70. Jones MH, Koi S, Fujimoto I *et al.* Allelotype of uterine cancer by analysis of RFLP and microsatellite polymorphisms: frequent loss of heterozygosity on chromosome arms 3p, 9q, 10q, and 17p. *Genes Chromosomes Cancer* 1994 9: 119–123.
71. Tsao SW, Mok SCH, Knapp RC *et al.* Molecular genetic evidence for a unifocal origin for human serous ovarian carcinomas. *Gynecol Oncol* 1993 48: 5–10.
72. Muto MG, Welch WR, Mok SC-H *et al.* Evidence for a multifocal origin of papillary serous carcinoma of the peritoneum. *Cancer Res* 1995 55: 490–492.
73. Mok SC, Bell DA, Knapp RC *et al.* Mutation of K-*ras* protooncogene in human ovarian epithelial tumors of borderline malignancy. *Cancer Res* 1993 53: 1489–1492.
74. Teneriello MG, Ebina M, Linnoila RI *et al.* p53 and Ki-*ras* gene mutations in epithelial ovarian neoplasms. *Cancer Res* 1993 53: 3103–3108.
75. Ichikawa Y, Nishida M, Suzuki H *et al.* Mutation of K-*ras* protooncogene is associated with histological subtypes in human mucinous ovarian tumors. *Cancer Res* 1994 54: 33–35.
76. Slamon DJ, Godolphin W, Jones LA *et al.* Studies of the HER-2/*neu* proto-oncogene in human breast and ovarian cancer. *Science* 1989 244: 707–712.
77. Berchuck A, Kamel A, Whitaker R *et al.* Overexpression of HER-2/*neu* is associated with poor survival in advanced epithelial ovarian cancer. *Cancer Res* 1990 50: 4087–4091.
78. Rubin SC, Finstad CL, Wong GY *et al.* Prognostic significance of HER-2/*neu* expression in advanced epithelial ovarian cancer: a multivariate analysis. *Am J Obstet Gynecol* 1993 168: 162–169.
79. Rubin SC, Finstad CL, Federici MG *et al.* Prevalence and significance of HER-2/*neu* expression in early epithelial ovarian cancer. *Cancer* 1994 73: 1456–1459.
80. Felip E, Del Campo JM, Rubio D *et al.* Overexpression of c-*erbB*-2 in epithelial ovarian cancer: prognostic value and relationship with response to therapy. *Cancer* 1995 75: 2147–2152.
81. Okamoto A, Sameshima Y, Yokoyama S *et al.* Frequent allelic losses and mutations of the *p53* gene in human ovarian cancer. *Cancer Res* 1991 51: 5171–5176.
82. Marks JR, Davidoff AM, Kerns BJ *et al.* Overexpression and mutation of p53 in epithelial ovarian cancer. *Cancer Res* 1991 51: 2979–2984.
83. Jacobs IJ, Kohler MF, Wiseman RW *et al.* Clonal origin of epithelial ovarian carcinoma: analysis by loss of heterozygosity, *p53* mutation, and X-chromosome inactivation. *J Natl Cancer Inst* 1992 84: 1793–1798.
84. Kohler MF, Marks JR, Wiseman RW *et al.* Spectrum of mutation and frequency of allelic deletion of the *p53* gene in ovarian cancer. *J Natl Cancer Inst* 1993 85: 1513–1519.
85. Wertheim I, Muto MG, Welch WR *et al.* *p53* gene mutation in human borderline epithelial ovarian tumors. *J Natl Cancer Inst* 1994 86: 1549–1551.
86. Berchuck A, Kohler MF, Hopkins MP *et al.* Overexpression of p53 is not a feature of benign and early-stage borderline epithelial ovarian tumors. *Gynecol Oncol* 1994 52: 232–236.
87. Henriksen R, Strang P, Wilander E *et al.* p53 expression in epithelial ovarian neoplasms: relationship to clinical and pathological parameters, Ki-67 expression and flow cytometry. *Gynecol Oncol* 1994 53: 301–306.
88. Levesque MA, Katsaros D, Yu H *et al.* Mutant p53 protein overexpression is associated with poor outcome in patients with well or moderately differentiated ovarian carcinoma. *Cancer* 1995 75: 1327–1338.
89. Herod JJO, Eliopoulos AG, Warwick J *et al.* The prognostic significance of Bcl-2 and p53 expression in ovarian carcinoma. *Cancer Res* 1996 56: 2178–2184.

90. Osborne RJ, Leech V. Polymerase chain reaction allelotyping of human ovarian cancer. *Br J Cancer* 1994 69: 429–438.
91. Iwabuchi H, Sakamoto M, Sakunaga H *et al.* Genetic analysis of benign, low-grade, and high-grade ovarian tumors. *Cancer Res* 1995 55: 6172–6180.
92. Phillips NJ, Ziegler MR, Radford DM *et al.* Allelic deletion on chromosome 17p13.3 in early ovarian cancer. *Cancer Res* 1996 56: 606–611.
93. Obata K, Morland SJ, Watson RH *et al.* Frequent *PTEN/MMAC* mutations in endometrioid but not serous or mucinous epithelial ovarian tumors. *Cancer Res* 1998 58: 2095–2097.
94. Maxwell GL, Risinger JI, Tong B *et al.* Mutation of the *PTEN* tumor suppressor gene is not a feature of ovarian cancers. *Gynecol Oncol* 1998 70: 13–16.
95. Palacios J, Gamallo C. Mutations in the β-catenin gene (*CTNNB1*) in endometrioid ovarian carcinomas. *Cancer Res* 1998 58: 1344–1347.
96. Sato T, Saito H, Morita R *et al.* Allelotype of human ovarian cancer. *Cancer Res* 1991 51: 5118–5122.
97. Cliby W, Ritland S, Hartmann L *et al.* Human epithelial ovarian cancer allelotype. *Cancer Res* 1993 53: 2393–2398.
98. Yang-Feng TL, Han H, Chen KC *et al.* Allelic loss in ovarian cancer. *Int J Cancer* 1993 54: 546–551.
99. Takahashi H, Behbakht K, McGovern PE *et al.* Mutation analysis of the *BRCA1* gene in ovarian cancers. *Cancer Res* 1995 55: 2998–3002.
100. Jacobs IJ, Smith SA, Wiseman RW *et al.* A deletion unit on chromosome 17q in epithelial ovarian tumors distal to the familial breast/ovarian cancer locus. *Cancer Res* 1993 53: 1218–1221.
101. Godwin AK, Vanderveer L, Schultz DC *et al.* A common region of deletion on chromosome 17q in both sporadic and familial epithelial ovarian tumors distal to BRCA1. *Am J Hum Genet* 1994 55: 666–677.
102. Lee JH, Kavanagh JJ, Wildrick DM, Wharton JT, Blick M. Frequent loss of heterozygosity on chromosomes 6q, 11, and 17 in human ovarian carcinomas. *Cancer Res* 1990 50: 2724–2728.
103. Ehlen T, Dubeau L. Loss of heterozygosity on chromosomal segments 3p, 6q and 11p in human ovarian carcinomas. *Oncogene* 1990 5: 219–223.
104. Vandamme B, Lissens W, Amfo K *et al.* Deletion of chromosome 11p13–11p15.5 sequences in invasive human ovarian cancer is a subclonal progression factor. *Cancer Res* 1992 52: 6646–6652.
105. Lu KH, Weitzel JN, Kodali S *et al.* A novel 4-cM minimally deleted region on chromosome 11p15.1 associated with high grade nonmucinous epithelial ovarian carcinomas. *Cancer Res* 1997 57: 387–390.
106. Kim TM, Benedict WF, Xu H-J *et al.* Loss of heterozygosity on chromosome 13 is common only in the biologically more aggressive subtypes of ovarian epithelial tumors and is associated with normal retinoblastoma gene expression. *Cancer Res* 1994 54: 605–609.
107. Dodson MK, Cliby WA, Xu H-J *et al.* Evidence of functional RB protein in epithelial ovarian carcinomas despite loss of heterozygosity at the *RB* locus. *Cancer Res* 1994 54: 610–613.
108. Takahashi H, Chiu H-C, Bandera CA *et al.* Mutations of the *BRCA2* gene in ovarian carcinomas. *Cancer Res* 1996 56: 2738–2741.
109. Bandera CA, Takahashi H, Behbakht K *et al.* Deletion mapping of two potential chromosome 14 tumor suppressor gene loci in ovarian carcinoma. *Cancer Res* 1997 57: 513–515.
110. Chenevix-Trench G, Leary J, Kerr J *et al.* Frequent loss of heterozygosity on chromosome 18 in ovarian adenocarcinoma which does not always include the DCC locus. *Oncogene* 1992 7: 1059–1065.
111. Bryan EJ, Watson RH, Davis M *et al.* Localization of an ovarian cancer tumor suppressor gene to a 0.5-cM region between *D22S284* and *CYP2D*, on chromosome 22q. *Cancer Res* 1996 56: 719–721.
112. Cheng JQ, Godwin AK, Bellacosa A *et al.* *AKT2*, a putative oncogene encoding a member of a subfamily of protein-serine/threonine kinases, is amplified in human ovarian carcinomas. *Proc Natl Acad Sci USA* 1992 89: 9267–9271.
113. Anzick SL, Kononen J, Walker RL *et al.* AIB1, a steroid receptor coactivator amplified in breast and ovarian cancer. *Science* 1997 277: 965–968.
114. Shayesteh L, Lu Y, Kuo W-L *et al.* *PIK3CA* is implicated as an oncogene in ovarian cancer. *Nat Genet* 1999 21: 99–102.
115. Wright TC, Ferenczy A, Kurman RJ. Carcinoma and other tumors of the cervix. In Kurman RJ, ed. *Blaustein's Pathology of the Female Genital Tract*, 4th edn. New York: Springer-Verlag, 1994: 279–326.

116. Howley PM. Principles of carcinogenesis: viral. In DeVita VT, Hellman S, Rosenberg SA, eds. *Cancer: Principles and Practice of Oncology*, 4th edn. Philadelphia: JB Lippincott Co., 1993: 182–199.
117. Werness BA, Levine AJ, Howley PM. Association of human papillomavirus types 16 and 18 E6 proteins with p53. *Science* 1990 248: 76–79.
118. Scheffner M, Werness BA, Huibregtse JM, Levine AJ, Howley PM. The E6 oncoprotein encoded by human papillomavirus types 16 and 18 promotes the degradation of p53. *Cell* 1990 63: 1129–1136.
119. Dyson N, Howley P, Munger K, Harlow E. The human papillomavirus-16 E7 oncoprotein is able to bind to the retinoblastoma gene product. *Science* 1989 243: 934–937.
120. Munger K, Werness BA, Dyson N *et al.* Complex formation of human papillomavirus E7 proteins with the retinoblastoma tumor suppressor gene product. *EMBO J* 1989 8: 4099–4105.
121. Riou G, Barrois M, Sheng Z-M, Duvillard P, Lhomme C. Somatic deletions and mutations of c-Ha-*ras* gene in human cervical cancers. *Oncogene* 1988 3: 329–333.
122. Koulos JP, Wright TC, Mitchell MF *et al.* Relationships between c-Ki-*ras* mutations, HPV types, and prognostic indicators in invasive endocervical adenocarcinomas. *Gynecol Oncol* 1993 48: 364–369.
123. Kihana T, Tsuda H, Teshima S *et al.* Prognostic significance of the overexpression of c-*erbB*-2 protein in adenocarcinoma of the uterine cervix. *Cancer* 1994 73: 148–153.
124. Mitra AB, Murty VVVS, Pratap M, Sodhani P, Chaganti RSK. *ERBB2* (*HER2/neu*) oncogene is frequently amplified in squamous cell carcinoma of the uterine cervix. *Cancer Res* 1994 54: 637–639.
125. Mandai M, Konishi I, Koshiyama M *et al.* Altered expression of *nm23-H1* and c-*erbB*-2 proteins have prognostic significance in adenocarcinoma but not in squamous cell carcinoma of the uterine cervix. *Cancer* 1995 75: 2523–2529.
126. Fujita M, Inoue M, Tanizawa O, Iwamoto S, Enomoto T. Alterations of the *p53* gene in human primary cervical carcinoma with and without human papillomavirus infection. *Cancer Res* 1992 52: 5323–5328.
127. Paquette RL, Lee YY, Wilczynski SP *et al.* Mutations of *p53* and human papillomavirus infection in cervical carcinoma. *Cancer* 1993 72: 1272–1280.
128. Park DJ, Wilczynski SP, Paquette RL, Miller CW, Koeffler HP. *p53* mutations in HPV-negative cervical carcinoma. *Oncogene* 1994 9: 205–210.
129. Kurvinen K, Tervahaut A, Syrjanen S, Chang F, Syrjanen K. The state of the *p53* gene in human papillomavirus (HPV)-positive and HPV-negative genital precursor lesions and carcinomas as determined by single-strand conformation polymorphism analysis and sequencing. *Anticancer Res* 1994 14: 177–181.
130. Busby-Earle RM, Steel CM, Williams AR, Cohen B, Bird CC. *p53* mutations in cervical carcinogenesis-low frequency and lack of correlation with human papillomavirus status. *Br J Cancer* 1994 69: 732–737.
131. Benjamin I, Saigo P, Finstad C *et al.* Expression and mutational analysis of *P53* in stage IB and IIA cervical cancers. *Am J Obstet Gynecol* 1996 175: 1266–1271.
132. Ocadiz R, Sauceda R, Cruz M, Graef AM, Gariglio P. High correlation between molecular alterations of the c-*myc* oncogene and carcinoma of the uterine cervix. *Cancer Res* 1987 47: 4173–4177.
133. Baker VV, Hatch KD, Shingleton HM. Amplification of the c-*myc* proto-oncogene in cervical carcinoma. *J Surg Oncol* 1988 39: 225–228.
134. Larson AA, Kern S, Sommers RL *et al.* Analysis of replication error (RER^+) phenotypes in cervical carcinoma. *Cancer Res* 1996 56: 1426–1431.
135. Misra BC, Srivatsan ES. Localization of HeLa cell tumor-suppressor gene to the long arm of chromosome 11. *Am J Hum Genet* 1989 45: 565–577.
136. Srivatsan ES, Misra BC, Venugopalan M, Wilczynski SP. Loss of heterozygosity for alleles on chromosome 11 in cervical carcinoma. *Am J Hum Genet* 1991 49: 868–877.
137. Kohno T, Takayama H, Hamaguchi M *et al.* Deletion mapping of chromosome 3p in human uterine cervical cancer. *Oncogene* 1993 8: 1825–1832.
138. Mitra AB, Murty VVVS, Li RG *et al.* Allelotype analysis of cervical carcinoma. *Cancer Res* 1994 54: 4481–4487.
139. Mullokandov MR, Kholodilov NG, Atkin NB *et al.* Genomic alterations in cervical carcinoma: losses of chromosome heterozygosity and human

papilloma virus tumor status. *Cancer Res* 1996 56: 197–205.

140. Wistuba II, Montellano FD, Milchgrub S *et al.* Deletions of chromosome 3p are frequent and early events in the pathogenesis of uterine cervical carcinoma. *Cancer Res* 1997 57: 3154–3158.
141. Boltenberg A, Furgyik S, Kullander S. Familial cancer aggregation in cases of adenocarcinoma corporis uteri. *Acta Obstet Gynecol Scand* 1990 69: 249–258.
142. Sandles LG, Shulman LP, Elias S *et al.* Endometrial adenocarcinoma: genetic analysis suggesting heritable site-specific uterine cancer. *Gynecol Oncol* 1992 47:167–171.
143. Lynch HT, Lynch J, Conway T, Watson P, Coleman RL. Familial aggregation of carcinoma of the endometrium. *Am J Obstet Gynecol* 1994 171: 24–27.
144. Wijnen J, de leeuw W, Vasen H *et al.* Familial endometrial cancer in female carriers of *MSH6* germline mutations. *Nat Genet* 1999 23: 142–144.
145. Lynch HT, Conway T, Lynch J. Hereditary ovarian cancer: pedigree studies, part II. *Cancer Genet Cytogenet* 1991 52: 161-183.
146. Narod SA, Madlensky L, Bradley L *et al.* Hereditary and familial ovarian cancer in Southern Ontario. *Cancer* 1994 74: 2341–2346.
147. Steichen-Gersdorf E, Gallion HH, Ford D *et al.* Familial site-specific ovarian cancer is linked to *BRCA1* on 17q12–21. *Am J Hum Genet* 1994 55: 870–875.
148. Vasen HFA, Mecklin J-P, Meera Khan P, Lynch HT. The international collaborative group on hereditary non-polyposis colorectal cancer (ICG-HNPCC). *Dis Colon Rectum* 1991 34: 424–425.
149. Watson P, Lynch HT. Extracolonic cancer in hereditary nonpolyposis colorectal cancer. *Cancer* 1993 71: 677–685.
150. Watson P, Vasen HF, Mecklin JP, Jarvinen H, Lynch HT. The risk of endometrial cancer in hereditary nonpolyposis colorectal cancer. *Am J Med* 1994 96: 516–520.
151. Bewtra C, Watson P, Conway T, Read-Hippee C, Lynch HT. Hereditary ovarian cancer: a clinicopathological study. *Int J Gynecol Pathol* 1992 11: 180–187.
152. Peltomäki P, Aaltonen LA, Sistonen P *et al.* Genetic mapping of a locus predisposing to human colorectal cancer. *Science* 1993 260: 810–812.
153. Lindblom A, Tannergård P, Werelius B, Nordenskjöld M. Genetic mapping of a second locus predisposing to hereditary non-polyposis colon cancer. *Nat Genet* 1993 5: 279–282.
154. Nystrom-Lahti M, Parsons R, Sistonen P *et al.* Mismatch repair genes on chromosomes 2p and 3p account for a major share of hereditary nonpolyposis colorectal cancer families evaluable by linkage. *Am J Hum Genet* 1994 55: 659–665.
155. Nicolaides NC, Papadopoulos N, Liu B *et al.* Mutations of two *PMS* homologues in hereditary nonpolyposis colon cancer. *Nature* 1994 371: 75–80.
156. Fishel R, Kolodner R. Identification of mismatch repair genes and their role in the development of cancer. *Curr Opin Genet Dev* 1995 5: 382–395.
157. Marra G, Boland CR. Hereditary nonpolyposis colorectal cancer: the syndrome, the genes, and historical perspectives. *J Natl Cancer Inst* 1995 87: 1114–1125.
158. Miyaki M, Konishi M, Tanaka K *et al.* Germline mutation of *MSH6* as the cause of hereditary nonpolyposis colorectal cancer. *Nat Genet* 1997 17: 271–272.
159. Peltomaki P, Vasen HFA, Cancer ICGoHNC. Mutations predisposing to hereditary nonpolyposis colorectal cancer: database and results of a collaborative study. *Gastroenterology* 1997 113: 1146–1158.
160. Kolodner RD. Mismatch repair: mechanisms and relationship to cancer susceptibility. *Trends Biochem Sci* 1995 20: 397–401.
161. Leach FS, Nicolaides NC, Papadopoulos N *et al.* Mutations of a *mutS* homolog in hereditary nonpolyposis colorectal cancer. *Cell* 1993 75: 1215–1225.
162. Hemminki A, Peltomaki P, Mecklin J-P *et al.* Loss of the wild-type *MLH1* gene is a feature of hereditary nonpolyposis colorectal cancer. *Nat Genet* 1994 8: 405–410.
163. Liu B, Nicolaides NC, Markowitz S *et al.* Mismatch repair gene defects in sporadic colorectal cancers with microsatellite instability. *Nat Genet* 1995 9: 48–55.
164. Loeb LA. Microsatellite instability: marker of a mutator phenotype in cancer. *Cancer Res* 1994 54: 5059–5063.
165. Modrich P. Mismatch repair, genetic stability, and cancer. *Science* 1994 266: 1959–1960.

166. Kane MF, Loda M, Gaida GM *et al.* Methylation of the *hMLH1* promoter correlates with lack of expression of hMLH1 in sporadic colon tumors and mismatch repair-defective human tumor cell lines. *Cancer Res* 1997 57: 808–811.
167. Herman JG, Umar A, Polyak K *et al.* Incidence and functional consequences of *hMLH1* promoter hypermethylation in colorectal carcinoma. *Proc Natl Acad Sci USA* 1998 95: 6870–6875.
168. Leung SY, Yuen ST, Chung LP *et al. hMLH1* promoter methylation and lack of *hMLH1* expression in sporadic gastric carcinomas with high-frequency microsatellite instability. *Cancer Res* 1999 59: 159–164.
169. Fleisher AS, Esteller M, Wang S *et al.* Hypermethylation of the *hMLH1* gene promoter in human gastric cancers with microsatellite instability. *Cancer Res* 1999 59: 1090–1095.
170. Markowitz S, Wang J, Myeroff L *et al.* Inactivation of the type II TGF-β receptor in colon cancer cells with microsatellite instability. *Science* 1995 268: 1336–1338.
171. Parsons R, Myeroff L, Liu B *et al.* Microsatellite instability and mutations of the transforming growth factor β type II receptor gene in colorectal cancer. *Cancer Res* 1995 55: 5548–5550.
172. Rampino N, Yamamoto H, Ionov Y *et al.* Somatic frameshift mutations in the *BAX* gene in colon cancers of the microsatellite mutator phenotype. *Science* 1997 275: 967–969.
173. Ouyang H, Shiwaku HO, Hagiwara H *et al.* The *insulin-like growth factor II receptor* gene is mutated in genetically unstable cancers of the endometrium, stomach, and colorectum. *Cancer Res* 1997 57: 1851–1854.
174. Scwartz S, Yamamoto H, Navarro M *et al.* Frameshift mutations at mononucleotide repeats in *caspase-5* and other target genes in endometrial and gastrointestinal cancer of the microsatellite mutator phenotype. *Cancer Res* 1999 59: 2995–3002.
175. Lynch HT, Harris RE, Guirgis HA *et al.* Familial association of breast/ovarian carcinoma. *Cancer* 1978 41: 1543–1548.
176. Go RCP, King M-C, Bailey-Wilson J, Elston RC, Lynch HT. Genetic epidemiology of breast cancer and associated cancers in high risk families. I. Segregation analysis. *J Natl Cancer Inst* 1983 71: 455–461.
177. Schildkraut JM, Risch N, Thompson WD. Evaluating genetic association among ovarian, breast, and endometrial cancer: evidence for a breast/ovarian cancer relationship. *Am J Hum Genet* 1989 45: 521–529.
178. Easton DF, Bishop DT, Ford D, Crockford GP. Genetic linkage analysis in familial breast and ovarian cancer: results from 214 families. The Breast Cancer Linkage Consortium. *Am J Hum Genet* 1993 52: 678–701.
179. Narod SA, Ford D, Devilee P *et al.* An evaluation of genetic heterogeneity in 145 breast–ovarian cancer families. *Am J Hum Genet* 1995 56: 254–264.
180. Hall JM, Lee MK, Newman B *et al.* Linkage of early-onset familial breast cancer to chromosome 17q21. *Science* 1990 250: 1684–1689.
181. Narod SA, Feunteun J, Lynch HT *et al.* Familial breast–ovarian cancer locus on chromosome 17q12–q23. *Lancet* 1991 338: 82–83.
182. Wooster R, Neuhausen SL, Mangion J *et al.* Localization of a breast cancer susceptibility gene, *BRCA2*, to chromosome 13q12–13. *Science* 1994 265: 2088–2090.
183. Narod S, Ford D, Devilee P *et al.* Genetic heterogeneity of breast–ovarian cancer revisited. *Am J Hum Genet* 1995 57: 957–958.
184. Moslehi R, Chu W, Karlan B *et al. BRCA1* and *BRCA2* mutation analysis of 208 Ashkenazi Jewish women with ovarian cancer. *Am J Hum Genet* 2000 66: 1259–1272.
185. Risch HA, McLaughlin JR, Cole DEC *et al.* Prevalence and penetrance of germline BRCA1 and BRCA2 mutations in a population series of 649 women with ovarian cancer. *Am J Hum Genet* 2001 68: 700–710.
186. Miki Y, Swensen J, Shattuck-Edens D *et al.* A strong candidate for the breast and ovarian cancer susceptibility gene *BRCA1*. *Science* 1994 266: 66–71.
187. Castilla LH, Couch FJ, Erdos MR *et al.* Mutations in the *BRCA1* gene in families with early-onset breast and ovarian cancer. *Nat Genet* 1994 8: 387–391.
188. Simard J, Tonin P, Durocher F *et al.* Common origins of *BRCA1* mutations in Canadian breast and ovarian cancer families. *Nat Genet* 1994 8: 392–398.
189. Friedman LS, Ostermyer EA, Szabo CI *et al.* Confirmation of *BRCA1* by analysis of germline

mutations linked to breast and ovarian cancer in ten families. *Nat Genet* 1994 8: 399–404.

190. Matsushima M, Kobayashi K, Emi M *et al.* Mutation analysis of the *BRCA1* gene in 76 Japanese ovarian cancer patients: four germline mutations, but no evidence of somatic mutation. *Hum Mol Genet* 1995 4: 1953–1956.

191. Stratton JF, Gayther SA, Russell P *et al.* Contribution of *BRCA1* mutations to ovarian cancer. *N Engl J Med* 1997 336: 1125–1130.

192. Berchuck A, Heron KA, Carney ME *et al.* Frequency of germline and somatic *BRCA1* mutations in ovarian cancer. *Clin Cancer Res* 1998 4: 2433–2437.

193. Breast Cancer Information Core. http://wwwn-hgrinihgov/Intramural_research/Lab_transfer/Bic, 2001.

194. Smith SA, Easton DF, Evans DG, Ponder BA. Allele losses in the region 17q12–21 in familial breast and ovarian cancer involve the wild-type chromosome. *Nat Genet* 1992 2: 128–131.

195. Merajver SD, Frank TS, Xu J *et al.* Germline *BRCA1* mutations and loss of the wild-type allele in tumors from families with early-onset breast and ovarian cancer. *Clin Cancer Res* 1995 1: 539–544.

196. Roa BB, Boyd AA, Volcik K, Richards CS. Ashkenazi Jewish population frequencies for common mutations in *BRCA1* and *BRCA2*. *Nat Genet* 1996 14: 185–187.

197. Struewing JP, Abeliovich D, Peretz T *et al.* The carrier frequency of the *BRCA1* 185delAG mutation is approximately 1% in Ashkenazi Jewish individuals. *Nat Genet* 1995 11: 198–200.

198. Peelen T, van Vliet M, Petrij-Bosch A *et al.* A high proportion of novel mutations in *BRCA1* with strong founder effects among Dutch and Belgian hereditary breast and ovarian cancer families. *Am J Hum Genet* 1997 60: 1041–1049.

199. Petrij-Bosch A, Peelen T, van Vliet M *et al.* *BRCA1* genomic deletions are major founder mutations in Dutch breast cancer patients. *Nat Genet* 1997 17: 341–345.

200. Huusko P, Paakkonen K, Launonen V *et al.* Evidence of founder mutations in Finnish *BRCA1* and *BRCA2* families. *Am J Hum Genet* 1998 62: 1544–1548.

201. Tonin PN, Mes-Masson A-M, Futreal PA *et al.* Founder *BRCA1* and *BRCA2* mutations in French Canadian breast and ovarian cancer families. *Am J Hum Genet* 1998 63: 1341–1351.

202. Russell SEH, Hickey GI, Lowry WS, Atkinson RJ. Allele loss from chromosome 17 in ovarian cancer. *Oncogene* 1990 5: 1581–1583.

203. Foulkes W, Black D, Solomon E, Trowsdale J. Allele loss on chromosome 17q in sporadic ovarian cancer. *Lancet* 1991 338: 444–445.

204. Futreal PA, Liu Q, Shattuck-Eidens D *et al.* *BRCA1* mutations in primary breast and ovarian cancers. *Science* 1994 266: 120–122.

205. Merajver SD, Pham TM, Caduff RF *et al.* Somatic mutations in the *BRCA1* gene in sporadic ovarian tumours. *Nat Genet* 1995 9: 439–443.

206. Saito H, Inazawa J, Saito S *et al.* Detailed deletion mapping of chromosome 17q in ovarian and breast cancers: 2-cM region on 17q21.3 often and commonly deleted in tumors. *Cancer Res* 1993 53: 3382–3385.

207. Wooster R, Bignell G, Lancaster J *et al.* Identification of the breast cancer susceptibility gene *BRCA2*. *Nature* 1995 378: 789–792.

208. Tavtigian SV, Simard J, Rommens J *et al.* The complete *BRCA2* gene and mutations in chromosome 13q-linked kindreds. *Nat Genet* 1996 12: 333–337.

209. Phelan CM, Lancaster JM, Tonin P *et al.* Mutation analysis of the *BRCA2* gene in 49 site-specific breast cancer families. *Nat Genet* 1996 13: 120–122.

210. Couch FJ, Farid LM, DeShano ML *et al.* *BRCA2* germline mutations in male breast cancer cases and breast cancer families. *Nat Genet* 1996 13: 123–125.

211. Thorlacius S, Olafsdottir G, Tryggvadottir L *et al.* A single *BRCA2* mutation in male and female breast cancer families from Iceland with varied cancer phenotypes. *Nat Genet* 1996 13: 117–119.

212. Easton DF, Steele L, Fields P *et al.* Cancer risks in two large breast cancer families linked to *BRCA2* on chromosome 13q12–13. *Am J Hum Genet* 1997 61: 120–128.

213. Struewing JP, Hartge P, Wacholder S *et al.* The risk of cancer associated with specific mutations of *BRCA1* and *BRCA2* among Ashkenazi Jews. *N Engl J Med* 1997 336: 1401–1408.

214. Ford D, Easton DF, Stratton M *et al.* Genetic heterogeneity and penetrance analysis of the *BRCA1* and *BRCA2* genes in breast cancer families. *Am J Hum Genet* 1998 62: 676–689.

215. Foster KA, Harrington P, Kerr J *et al.* Somatic and germline mutations of the *BRCA2* gene in sporadic ovarian cancer. *Cancer Res* 1996 56: 3622–3625.
216. Abeliovich D, Kaduri L, Lerer I *et al.* The founder mutations 185delAG and 5382insC in *BRCA1* and 6174delT in *BRCA2* appear in 60% of ovarian cancer and 30% of early-onset breast cancer patients among Ashkenazi women. *Am J Hum Genet* 1997 60: 505–514.
217. Collins N, McManus R, Wooster R *et al.* Consistent loss of the wild-type allele in breast cancers from a family linked to the *BRCA2* gene on chromosome 13q12–13. *Oncogene* 1995 10: 1673–1675.
218. Gudmundsson J, Johannesdottir G, Bergthorsson JT *et al.* Different tumor types from *BRCA2* mutation carriers show wild-type chromosome deletions on 13q12–q13. *Cancer Res* 1995 55: 4830–4832.
219. Neuhausen S, Gilewski T, Norton L *et al.* Recurrent *BRCA2* 6174delT mutations in Ashkenazi Jewish women affected by breast cancer. *Nat Genet* 1996 13: 126–128.
220. Oddoux C, Struewing JP, Clayton CM *et al.* The carrier frequency of the *BRCA2* 6174delT mutation among Ashkenazi Jewish individuals is approximately 1%. *Nat Genet* 1996 14: 188–190.
221. Boyd J, Sonoda Y, Federici MG *et al.* Clinicopathologic features of *BRCA*-linked and sporadic ovarian cancer. *J Am Med Assoc* 2000 283: 2260–2265.
222. Thorlacius S, Sigurdsson S, Bjarnadottir H *et al.* Study of a single *BRCA2* mutation with high carrier frequency in a small population. *Am J Hum Genet* 1997 60: 1079–1084.
223. Lerer I, Wang T, Peretz T *et al.* The 8765delAG mutation in *BRCA2* is common among Jews of Yemenite extraction. *Am J Hum Genet* 1998 63: 272–274.
224. Gallion HH, Powell DE, Morrow JK *et al.* Molecular genetic changes in human epithelial ovarian malignancies. *Gynecol Oncol* 1992 47: 137–142.
225. Scully R, Livingston DM. In search of the tumour-suppressor functions of BRCA1 and BRCA2. *Nature* 2000 408: 429–432.
226. Welsch PL, Owens KN, King M-C. Insights into the functions of BRCA1 and BRCA2. *Trends Genet* 2000 16: 69–74.
227. Scully R, Ganesan S, Brown M *et al.* Location of BRCA1 in human breast and ovarian cancer cells. *Science* 1996 272: 123–126.
228. Chen Y, Farmer AA, Chen C-F *et al.* BRCA1 is a 220-kDa nuclear phosphoprotein that is expressed and phosphorylated in a cell cycle-dependent manner. *Cancer Res* 1996 56: 3168–3172.
229. Coene E, Van Oostveldt P, Willems K, van Emmelo J, De Potter CR. BRCA1 is localized in cytoplasmic tube-like invaginations in the nucleus. *Nat Genet* 1997 16: 122–124.
230. Wilson CA, Ramos L, Villasenor MR *et al.* Localization of human BRCA1 and its loss in high-grade, non-inherited breast carcinomas. *Nat Genet* 1999 21: 236–240.
231. Koonin EV, Altschul S, Bork P. BRCA1 protein products: functional motifs. *Nat Genet* 1996 13: 266–268.
232. Chapman MS, Verma IM. Transcriptional activation by BRCA1. *Nature* 1996 382: 678–679.
233. Monteiro ANA, August A, Hanafusa H. Evidence for a transcriptional activation function for *BRCA1* C-terminal region. *Proc Natl Acad Sci USA* 1996 93: 13595–13599.
234. Scully R, Anderson SF, Chao DM *et al.* BRCA1 is a component of the RNA polymerase II holoenzyme. *Proc Natl Acad Sci USA* 1997 94: 5605–5610.
235. Anderson S, Schlegel B, Nakajima T, Wolpin E, Parvin J. BRCA1 protein is linked to the RNA polymerase II holoenzyme complex via helicase A. *Nat Genet* 1998 19: 1–3.
236. Ouchi T, Monteiro ANA, August A, Aaronson SA, Hanafusa H. BRCA1 regulates p53-dependent gene expression. *Proc Natl Acad Sci USA* 1998 95: 2302–2306.
237. Zhang H, Somasundaram K, Peng Y *et al.* BRCA1 physically associates with p53 and stimulates its transcriptional activity. *Oncogene* 1998 16: 1713–1721.
238. Yu X, Wu LC, Bowcock AM, Aronheim A, Baer R. The C-terminal (BCRT) domains of *BRCA1* interact *in vivo* with CtIP, a protein implicated in the CtBP pathway of transcriptional repression. *J Biol Chem* 1998 273: 25388–25392.
239. Wang Q, Zhang H, Kajino K, Greene MI. BRCA1 binds c-*myc* and inhibits its transcriptional and transforming activity in cells. *Oncogene* 1998 17: 1939–1948.

240. Marquis ST, Rajan JV, Wynshaw-Boris A *et al.* The developmental pattern of *BRCA1* expression implies a role in differentiation of breast and other tissues. *Nat Genet* 1995 11: 17–26.
241. Lane TF, Deng C, Elson A *et al.* Expression of BRCA1 is associated with terminal differentiation of ectodermally and mesodermally derived tissues in mice. *Genes Dev* 1995 9: 2712–2722.
242. Vaughn JP, Davis PL, Jarboe MD *et al.* BRCA1 expression is induced before DNA synthesis in both normal and tumor-derived breast cells. *Cell Growth Differ* 1996 7: 711–715.
243. Gudas JM, Li T, Nguyen H *et al.* Cell cycle regulation of BRCA1 messenger RNA in human breast epithelial cells. *Cell Growth Differ* 1996 7: 717–723.
244. Gudas JM, Nguyen H, Li T, Cowan KH. Hormone-dependent regulation of BRCA1 in human breast cancer cells. *Cancer Res* 1995 55: 4561–4565.
245. Spillman MA, Bowcock AM. BRCA1 and BRCA2 mRNA levels are coordinately elevated in human breast cancer cells in response to estrogen. *Oncogene* 1996 13: 1639–1645.
246. Rao VN, Shao N, Ahmad M, Reddy ES. Antisense RNA to the putative tumor suppressor gene *BRCA1* transforms mouse fibroblasts. *Oncogene* 1996 12: 523–528.
247. Holt JT, Thompson ME, Szabo C *et al.* Growth retardation and tumour inhibition by *BRCA1*. *Nat Genet* 1996 12: 298–302.
248. Shao N, Chai YL, Shyam E, Reddy P, Rao VN. Induction of apoptosis by the tumor suppressor protein BRCA1. *Oncogene* 1996 13: 1–7.
249. Scully R, Chen J, Plug A *et al.* Association of BRCA1 with Rad51 in mitotic and meiotic cells. *Cell* 1997 88: 265–275.
250. Scully R, Chen J, Ochs RL *et al.* Dynamic changes of BRCA1 subnuclear location and phosphorylation state are initiated by DNA damage. *Cell* 1997 90: 425–435.
251. Ruffner H, Verma IM. BRCA1 is a cell cycle-regulated nuclear phosphoprotein. *Proc Natl Acad Sci USA* 1997 94: 7138–7143.
252. Thomas JE, Smith M, Tonkinson JL, Rubinfeld B, Polakis P. Induction of phosphorylation of BRCA1 during the cell cycle and after DNA damage. *Cell Growth Differ* 1997 8: 801–809.
253. Hakem R, de la Pompa JL, Sirard C *et al.* The tumor suppressor gene *Brca1* is required for embryonic cellular proliferation in the mouse. *Cell* 1996 85: 1009–1023.
254. Hakem R, de la Pompa JL, Elia A, Potter J, Mak TW. Partial rescue of $Brca1^{5-6}$ early embryonic lethality by *p53* or *p21* null mutation. *Nat Genet* 1997 16: 298–302.
255. Ludwig T, Chapman DL, Papaioannou VE, Efstratiadis A. Targeted mutations of breast cancer susceptibility gene homologs in mice: lethal phenotypes of *Brca1*, *Brca2*, *Brca1/Brca2*, *Brca1/p53*, and *Brca2/p53* nullizygous embryos. *Genes Dev* 1997 11: 1226–1241.
256. Gowen LC, Avrutskaya AV, Latour AM, Koller BH, Leadon SA. BRCA1 required for transcription-coupled repair of oxidative DNA damage. *Science* 1998 281: 1009–1012.
257. Hsu L-C, White RL. BRCA1 is associated with the centrosome during mitosis. *Proc Natl Acad Sci USA* 1998 95: 12983–12988.
258. Xu X, Weaver Z, Linke SP *et al.* Centrosome amplification and a defective G_2–M cell cycle checkpoint induce genetic instability in *BRCA1* exon 11 isoform-deficient cells. *Mol Cell* 1999 3: 389–395.
259. Harkin DP, Bean JM, Miklos D *et al.* Induction of *GADD45* and JNK/SAPK-dependent apoptosis following inducible expression of *BRCA1*. *Cell* 1999 97: 575–586.
260. Bertwistle D, Swift S, Marston NJ *et al.* Nuclear location and cell cycle regulation of the BRCA2 protein. *Cancer Res* 1997 57: 5485–5488.
261. Chen P-L, Chen C-F, Chen Y *et al.* The BRC repeats in BRCA2 are critical for RAD51 binding and resistance to methyl methanesulfonate treatment. *Proc Natl Acad Sci USA* 1998 95: 5287–5292.
262. Marmorstein LY, Ouchi T, Aaronson SA. The BRCA2 gene product functionally interacts with p53 and RAD51. *Proc Natl Acad Sci USA* 1998 95: 13869–13874.
263. Wang SC, Lin SH, Su LK, Hung MC. Changes in BRCA2 expression during progression of the cell cycle. *Biochem Biophys Res Commun* 1997 234: 247–251.
264. Rajan JV, Wang M, Marquis ST, Chodosh LA. *Brca2* is coordinately regulated with *Brca1* during proliferation and differentiation in mammary epithelial cells. *Proc Natl Acad Sci USA* 1996 93: 13078–13083.
265. Connor F, Smith A, Wooster R *et al.* Cloning, chromosomal mapping and expression pattern of

the mouse *Brca2* gene. *Hum Mol Genet* 1997 6: 291–300.

266. Rajan JV, Marquis ST, Gardner HP, Chodosh LA. Developmental expression of *Brca2* colocalizes with *Brca1* and is associated with proliferation and differentiation in multiple tissues. *Dev Biol* 1997 184: 385–401.
267. Milner J, Ponder B, Hughes-Davies L, Seltmann M, Kouzarides T. Transcriptional activation functions in BRCA2. *Nature* 1997 386: 772–773.
268. Sharan SK, Morimatsu M, Albrecht U *et al.* Embryonic lethality and radiation hypersensitivity mediated by Rad51 in mice lacking *Brca2*. *Nature* 1997 386: 804–810.
269. Chen J, Silver DP, Walpita D *et al.* Stable interaction between the products of the *BRCA1* and *BRCA2* tumor suppressor genes in mitotic and meiotic cells. *Mol Cell* 1998 2: 317–328.
270. Suzuki A, de la Pompa JL, Hakem R *et al. Brca2* is required for embryonic cellular proliferation in the mouse. *Genes Dev* 1997 11: 1242–1252.
271. Connor F, Bertwistle D, Mee PJ *et al.* Tumorigenesis and a DNA repair defect in mice with a truncating *Brca2* mutation. *Nat Genet* 1997 17: 423–430.
272. Friedman LS, Thistlethwaite FC, Patel KJ *et al.* Thymic lymphomas in mice with a truncating mutation in *Brca2*. *Cancer Res* 1998 58: 1338–1343.
273. Patel KJ, Yu VPCC, Lee H *et al.* Involvement of *Brca2* in DNA repair. *Mol Cell* 1998 1: 347–357.
274. Abbott DW, Freeman ML, Holt JT. Double-strand break repair deficiency and radiation sensitivity in *BRCA2* mutant cancer cells. *J Natl Cancer Inst* 1998 90: 978–985.
275. Baserga R. *The Biology of Cell Reproduction*. Cambridge: Harvard University Press, 1985.
276. Nurse P. Ordering S phase and M phase in the cell cycle. *Cell* 1994 79: 547–550.
277. Sherr CJ. G_1 phase progression: cycling on cue. *Cell* 1994 79: 551–555.
278. Sherr CJ. Mammalian G_1 cyclins. *Cell* 1993 73: 1059–1065.
279. Hinds PW, Weinberg RA. Tumor suppressor genes. *Curr Opin Genet Dev* 1994 4: 135–141.
280. Nevins JR. E2F: a link between the Rb tumor suppressor protein and viral oncoproteins. *Science* 1992 258: 424–429.
281. Cordon-Cardo C. Mutation of cell cycle regulators: biological and clinical implications for human neoplasia. *Am J Pathol* 1995 147: 545–560.
282. Hunter T, Pines J. Cyclins and cancer II: cyclin D and CDK inhibitors come of age. *Cell* 1994 79: 573–582.
283. Heichman KA, Roberts JM. Rules to replicate by. *Cell* 1994 79: 557–562.
284. Dunphy WG. The decision to enter mitosis. *Trends Cell Biol* 1994 4: 202–207.
285. Arnold A. The cyclin D1/*PRAD1* oncogene in human neoplasia. *J Invest Med* 1995 43: 543–549.
286. Pollock PM, Pearson JV, Hayward NK. Compilation of somatic mutations of the *CDKN2* gene in human cancers: non-random distribution of base substitutions. *Genes Chromosomes Cancer* 1996 15: 77–88.
287. Hussussian CJ, Struewing JP, Goldstein AM *et al.* Germline p16 mutations in familial melanoma. *Nat Genet* 1994 8: 15–21.
288. Kamb A, Shattuck-Eidens D, Eeles R *et al.* Analysis of the p16 gene (*CDKN2*) as a candidate for the chromosome 9p melanoma susceptibility locus. *Nat Genet* 1994 8: 22–26.
289. Greenblatt MS, Bennett WP, Hollstein M, Harris CC. Mutations in the *p53* tumor suppressor gene: clues to cancer etiology and molecular pathogenesis. *Cancer Res* 1994 54: 4855–4878.
290. La Thangue NB. E2F and the molecular mechanisms of early cell-cycle control. *Biochem Soc Trans* 1996 24: 54–59.
291. Sanchez I, Dynlacht BD. Transcriptional control of the cell cycle. *Curr Opin Cell Biol* 1996 8: 318–324.
292. Johnson DJ, Schwarz JK, Cress WD, Nevins JR. Expression of transcription factor E2F1 induces quiescent cells to enter S phase. *Nature* 1993 365: 349–352.

Chapter 7

Angiogenesis and reproduction

Stephen Smith

INTRODUCTION

This book seeks to identify new developments that increase our understanding of the processes of reproduction. Many of these advances were derived from improvements in techniques such as the growth of molecular biology. However, one development that has occurred over the past 10 years has arisen from new insights into a basic process that underlies all aspects of reproduction. This is the study of how blood vessels form and what it is that controls their growth, development and subsequent demise. The application of these insights into vascular biology will have profound effects for our understanding of reproductive biology and may lead to new treatments in the future.

BLOOD VESSELS

Development

The formation of blood vessels in the embryonic and extraembryonic tissues is called vasculogenesis.[1] In this process, the blood vessels form *de novo*. The initial vasculature arises in the yolk sac, embryo and trophoblast. This process must be seen to be different from that of angiogenesis, in which new blood vessels arise from pre-existing vessels. Both vasculogenesis and angiogenesis play critical roles in reproductive biology.

Vasculogenesis

Mesenchymal cells aggregate and form blood islands.[1–3] These blood islands consist of the progenitors of both the hemopoietic and vascular system.[4] It is possible that they both share the same progenitor cells, called hemangioblasts.[5] The cell lineages begin to divide into hemopoietic stem cells from which the hemopoietic system is derived, and angioblasts from which endothelial cells are derived.[6] Transgenic experiments, in which individual genes have been deleted from the embryo, have begun to identify the genes involved in these early stages of cell commitment. One of the earliest genes that is induced, and which defines the endothelial cell precursors, is kinase domain receptor (KDR) or the vascular endothelial cell growth factor (VEGF) receptor-2 (VEGFR-2). This gene has seven immunoglobulin repeats in its extracellular domain, a transmembrane region and a tyrosine kinase domain in the intracellular sequence. When activated by its ligand VEGF-A, it promotes endothelial proliferation and prevents apoptosis. Induction of this gene is the first known step in defining the endothelial cell.[7] Recent studies suggest that it achieves this goal by promoting the migration of mesenchymal cells into the blood islands, but does not seem to be needed to promote the differentiation of the endothelial cell.[8] The differentiation of the endothelial cell involves the sequential and orderly induction of other genes, some of which are restricted to expression in endothelial cells (e.g. tie-1, tie-2, VE-cadherin, platelet endothelial cell adhesion molecule) and some of which are found on other cell types (e.g. ephrins, intercellular adhesion molecule).[5,9]

Blood vessels may be broadly divided into those that consist mostly of endothelial cells, for example capillaries, and others that establish contact with mesenchymal cells to provide support for the vessel (pericytes).[2] In some cases, this may simply include individual pericytes, but, in others, the pericytes may have differentiated further into vascular smooth muscle cells.[10] Recent studies suggest that not only can the blood island cells differentiate into hemopoietic or endothelial cells, but that they may also be able to differentiate into pericytes.

Vasculogenesis is responsible for establishment of the vascular tree,[11] but both vasculogenesis and angiogenesis are then responsible for the growth of the vascular tree in the developing organs of the fetus.

Angiogenesis

Angiogenesis is the formation of blood vessels from pre-existing vessels.[12,13] This process is rare in adults, except in the case of the female reproductive tract.[14,15] Blood vessels are remarkably stable in the adult, and endothelial cells may live for several years. This is altered if the vessel or organ becomes damaged. In this case, angiogenesis becomes a key part in the process of tissue repair. In women, this necessity arises every month in the reproductive years with ovulation[16] and menstruation.[15] Implantation similarly requires profound tissue remodelling as the maternal decidua prepares itself for association with the fetal vasculature in the formation of the placenta (Smith, unpublished data). Indeed the principal function of the placenta is to permit the maternal and fetal vasculature to exchange nutrients and waste products.

The formation of new blood vessels from existing ones occurs in three ways.[17] The earliest to be described was sprouting angiogenesis. In this process, endothelial cells at the site of new vessel growth undergo proliferation and migration away from the parent vessel. The surrounding extracellular matrix is dissolved and the new cells begin to form tubes. Alternatively, proliferation of the endothelial cells may arise within the vessel (intussusception), and, when they form a contact with the opposite side of the vessel lumen, they essentially form two lumens within the original single lumen. This process seems to be the main system for the growth of vessels in endometrium.[18,19] Finally, endothelial cells may roll along the vessel towards the end of the vessel where is it growing.

It was initially thought that angiogenesis required proliferation of endothelial cells in order to provide the new cells needed in the new blood vessel. More recently, it has been suggested that bone marrow may contain angioblasts that migrate to the site of angiogenesis and there differentiate into mature endothelial cells. Around 20% of the cells in the new vessel may be derived from these 'adult' progenitor cells.[20]

A key factor in the regulation of the blood vessels is the relationship between the endothelial cells and the pericytes. Capillaries with limited pericytic associations are less stable than vessels that have such an association.[21] These 'immature' vessels are sensitive to changes in growth factor levels that may result in apoptosis and regression of the vessel, which does not occur when pericytes are present.

REGULATION OF VASCULOGENESIS AND ANGIOGENESIS

Angiogenic growth factors

A critical group of growth factors has been identified that regulate the growth, differentiation and maintenance of blood vessels. These growth factors mediate the dialog that occurs between the endothelial cells and the pericytes. Four families of genes, the vascular endothelial growth and fibroblast growth factor families, the angiopoietins and the ephrins, are the most important (Figure 1). These factors are all expressed in reproductive tissues.

Vascular endothelial growth factor

The VEGF family consists of six members, VEGF-A, -B, -C, -D and -E and placental growth factor (PlGF).[23] The first described molecule was VEGF-A, which has five splice variants.[24] The molecule is transcribed from eight exons, and the first four exons are expressed in all of the splice variants. VEGF 189 consists of all the carboxy-terminal exons, 5, 6, 7 and 8. VEGF 165 contains exons 5, 6 and 8, VEGF 145 has exons 5, 7 and 8 and VEGF 121 has both exons 6 and 7 spliced out. Human endometrium expresses all splice variants of VEGF-A.[25] The site of expression of VEGF-A changes during the menstrual cycle.[26] In the proliferative phase, *in situ* hybridization and immunohistochemistry demonstrate VEGF-A in both glandular epithelial and stromal cells.[25] After ovulation, the expression of VEGF-A in the stromal cells declines dramatically, and is only found in surface epithelial cells. Furthermore, the secretion of VEGF-A is directed towards the luminal surface of the epithelium.[27] VEGF-A is stimulated by estradiol, and the promoter region of VEGF-A contains estrogen response elements. However, oxygen tension is more important in regulating VEGF-A expression.[28] *In situ* hybridization for VEGF-A is greatest in the glandular cells from tissue removed at menstruation, when the tissue has been exposed to significant hypoxia. Similarly, incubation of both glandular and stromal cells from endometrium

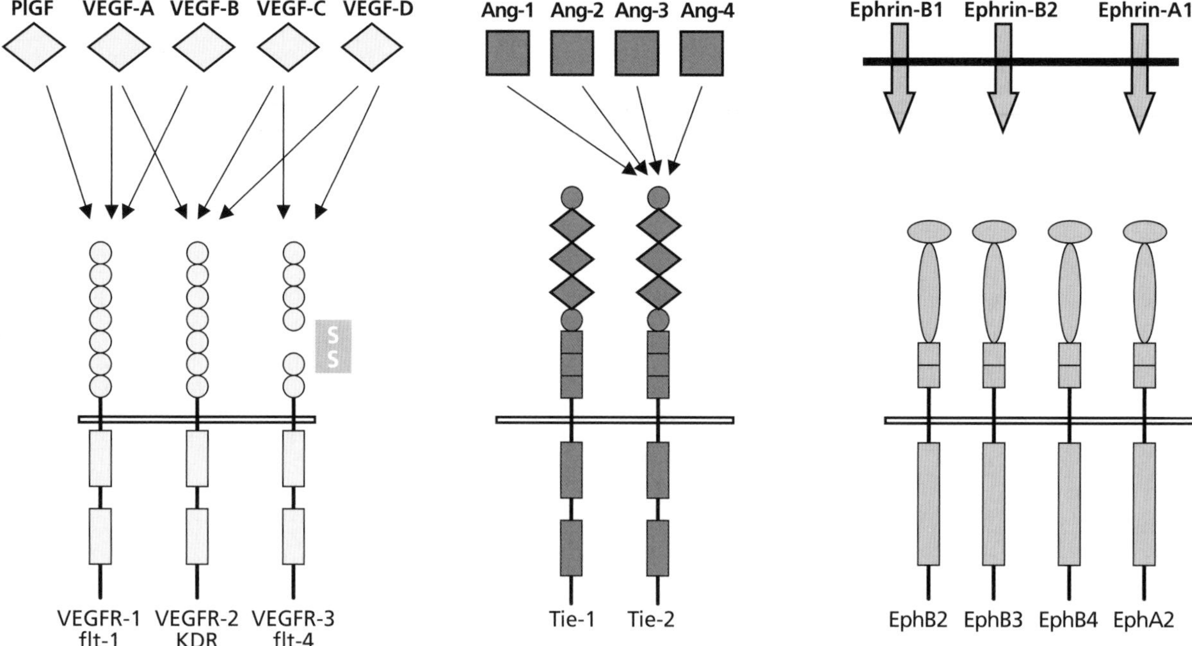

Figure 1 Regulation of blood vessel growth and structure. Diagram depicting the vascular endothelial cell growth factor (VEGF), angiopoietin and ephrin families of genes and their receptors. PlGF, placental growth factor; VEGFR, VEGF receptor; KDR, kinase domain receptor; Ang, angiopoietin. Reproduced from reference 22, with permission

results in a significant increase in mRNA levels encoding VEGF-A. This effect is mediated in part by up-regulation of VEGF-A expression. This is probably mediated by hypoxia inducible factor-1α (HIF-1α), which binds the arylhydrocarbon nuclear transporter molecule, forming a complex with the arylhydrocarbon receptor.[29] This promotes transcription of the VEGF gene, after binding the HIF-1α response element situated 500 bp upstream of the VEGF initiation codon. In addition, hypoxia results in the induction of nuclear proteins that bind the 3' non-coding region of the gene to promote stability of the mRNA.

VEGF-A induces proliferation of endothelial cells after phosphorylation of VEGFR-2. In addition, VEGF-A prevents apoptosis by several mechanisms that include increased expression of bcl-2[30] and akt-1.[31,32] A series of critical knock-out experiments have defined the role that the VEGF family plays in the establishment of the cardiovascular system.

Recently, other members of this family have been described. They include two members, VEGF-C[33,34] and VEGF-D,[35] that exert their effects predominantly on endothelial cells of the lymphatic system.[36] They do this not only by being ligands for the receptor VEGFR-2, but also by using another VEGF receptor, the fms-like tyrosine kinase 4 or VEGFR-3 receptor. The site of expression of these growth factors is different from that of VEGF-A in the endometrium. The site of VEGF-C expression is restricted to the uterine natural killer cells. These are the cells that migrate into and proliferate in the endometrium in the mid- to late secretory phase of the menstrual cycle.[37] The cells are situated around the spiral arterioles and beneath the surface epithelial cells. It is in this region that a rich plexus of capillary vessels is formed, and into which the early embryo implants. VEGFR-3 is not expressed in most of the endothelial cells of the endometrium, although VEGFR-1 and -2 are. Thus, the VEGF-C expressed by the natural killer cells is unlikely to be involved in the regulation of blood vessel growth in the endometrium. VEGF-D is expressed at low levels by stromal cells, although its expression does not alter during the cycle. While placental growth factor is expressed in the placenta, it is also surprisingly present in the uterine natural killer cells.

Fibroblast growth factor family

The fibroblast growth factor (FGF) family consists of many members,[38–40] three of which are expressed in the endometrium, FGF1, -2 and -4.[41,42] Immunohistochemical staining for FGF1 and -2 is most intense in the epithelial cells, but low levels of expression may be found in stromal fibroblasts.[41] Part of the action of FGFs in inducing angiogenesis is mediated by up-regulation of the VEGFR-2 receptor,[43–45] but FGF4 increases the expression of VEGF-A itself.[40] Conversely, VEGF promotes angiogenesis in a synergistic manner with FGF by releasing FGF from the extracellular matrix. This action is mediated through exon 6 of the VEGF molecule. This synergy is reversed as VEGFs promote the release of FGF from the extracellular matrix, an action dependent on the use of exon 6 of the splice variants of VEGF (see below).[46]

Angiopoietins

Recently, the critical importance of the dialog between the pericytes and the endothelial cells has become apparent. The identification of the angiopoietin family, containing at least four members, has further enhanced the complexity of blood vessel development and maintenance.[47–49] Angiopoietin-1 (Ang-1) is expressed by vascular smooth muscle cells, and binds to its cognate receptor, tie-2, on the endothelium. This results in the blood vessel becoming more stable. The vessel becomes resistant to changes in angiogenic growth factor levels at the site of the vessel. However, the physiological antagonist of Ang-1, Ang-2, also binds the tie-2 receptor. In these circumstances, the blood vessel undergoes atrophy, but, when the levels of VEGF-A are present, the vessel will undergo angiogenesis.[50]

Ang-1 is expressed at low levels in stromal fibroblasts,[51] but expression declines throughout both phases of the menstrual cycle. Ang-2 is not expressed in proliferative-phase endometrium. However, expression increases in the secretory phase of the cycle, but is restricted in the cells of the endometrium to the uterine natural killer cells. In addition, endothelial cells also demonstrate immunohistochemical staining.[52] Approximately 70% of these cells express Ang-2, particularly those situated beneath the luminal epithelium and around the vascular smooth muscle cells. These are the cells that also express VEGF-C and PlGF.

Ephrins

Ephrins are molecules that direct neuronal guidance.[53] They are cell surface receptors that mediate signals by cell-to-cell contact. The development of the somite is associated with both neuronal and vascular growth.[54] Surprisingly, endothelial cells express ephrins.[9] Furthermore, the origin of the endothelial cell dictates the types of ephrins expressed. Thus, endothelial cells characteristic of the arteriolar circulation express a different set of ephrins from those from the venular circulation. In this way, the capillary circulation is able to discriminate, and provides the means for arteriolar to venular flow. Ephrins are expressed in human endometrium but not just in the endothelial cells. Thus, it is possible that stromal expression of ephrins may in some way mediate vascular growth in the endometrium.

ANGIOGENESIS IN THE REPRODUCTIVE TRACT

Endometrium

Menstruation in women and primates is the process whereby the functional, upper two-thirds of the endometrium are shed, on the withdrawal of progesterone from an estrogen-primed endometrium.[55] It is preceded by 4–24 h of vasoconstriction of the spiral arterioles.[56] The spiral arterioles arise from the arcuate arteries in the myometrium, and pass into the basal part of the endometrium[57] (Figure 2). At this level, straight basal arterioles pass horizontally to supply the basal third of the endometrium. The vessels then pass towards the uterine lumen. They become thinner as they pass superficially. As the endometrium regresses in the latter stage of the secretory phase of the cycle, the vessels undergo striking coiling, and are termed spiral arterioles.

These vessels undergo intense vasoconstriction during the first 24 h of menstruation. The basal aspects of the spiral arterioles stain for several markers of the cytoskeleton, including α and γ smooth muscle actin and smooth muscle myosin.[58] This staining is retained for α smooth muscle actin in the more peripheral vessels, although the proportion of endothelial cells not associated with pericytes increases in the superficial zone. It is assumed that, in the basal layers, these cells are vascular smooth muscle cells, but that, in the superficialis, they are pericytic cells. Many of the stromal cells in the basalis also stain for α smooth muscle actin. Platelet-derived growth factor secreted from the endothelial cells may be the agent that promotes the migration of these stromal cells towards the vessel.[59] γ Smooth muscle actin and smooth muscle myosin are not found in the superficial aspect of the vessels. The vessels continue to lose their pericytic associations such that, in the subepithelial plexus, capillaries provide the main type of vessel. These capillaries then lead to venules that take the blood back towards the arcuate veins. These venules may form lake-like structures, and are more fragile than the arteriolar vessels.

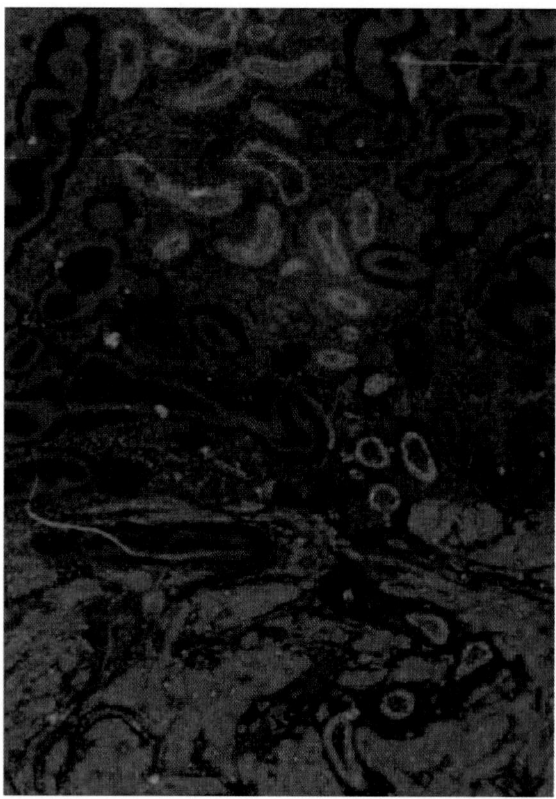

Figure 2 Immunofluorescent staining of a spiral arteriole in the basal part of the endometrium. The red staining identifies α smooth muscle actin and is present in the lower section in the myometrium. It can be seen to invest the spiral blood vessel as it passes superficially into the endometrium. The green staining is for CD31 that identifies endothelial cells

Initial studies suggested little change in the vessels in the endometrium throughout the menstrual cycle. Microvascular density did change during the cycle, and yet proliferative activity of endothelial cells, as judged by immunohistochemical staining, reached a peak in the mid- to late proliferative phase of the cycle.[18] This is partly explained by the increasing volume of the endometrium itself. More recently, clear differences have been identified between the activities of endothelial cells in the superficial functionalis layer and the basalis.[18] Proliferative activity does not alter in the basalis during the cycle, but does change in the functionalis.[60] The basal endometrium is not shed and so new blood vessels are not required. However, blood vessels are needed in the growing superficial layer. Initial studies failed to identify sprouting angiogenesis, but, using staining for integrin $\alpha_5\beta_3$ (that identifies proliferating endothelial cells), it was seen that sites of angiogenesis were characterized by intussusceptive angiogenesis.[18]

As with the endothelial cells, the proliferative activity of the vascular smooth muscle cells changes during the menstrual cycle, reaching a peak at the mid- to late proliferative phase.[60] Pericytes in the basalis have the characteristics of vascular smooth muscle cells, staining for α smooth muscle actin, γ smooth muscle actin and myosin smooth muscle actin. The superficial pericyte cells stain only for α smooth muscle actin. Fibroblasts in the basalis layer also stain for α smooth muscle actin, suggesting that the pericyte cells are derived from these basal fibroblasts.

Ovary

The most dramatic example of angiogenesis in the reproductive tract occurs in the ovary at the time of ovulation.[16,61] During follicular development, small capillaries grow into the thecal compartment of the follicle and are responsible for the transport of estrogen away from the follicle. Granulosa cells secrete large amounts of VEGF-A. Follicular rupture is followed by reorganization of blood vessels in the corpus luteum. Theca and granulosa cells combine to form the large and small luteal cells. At the same time, new blood vessels grow into the wound, organize and begin to attract pericytic cells. Elevated levels of VEGF-A are found in the corpus luteum,[62] and in mice[63] and monkeys the function of the corpus luteum can be impaired by antiangiogenic agents. At the end of the cycle, levels of Ang-2 increase dramatically, and those of VEGF-A decline. This is associated with atrophy of the corpus luteum blood vessels, and is most likely the cause of corpus luteal regression.[48]

Placenta

Vasculogenesis and angiogenesis occur in the placenta. Fetal mesenchymal cells migrate into the trabecular core of the early placenta, where they develop into angioblasts.[11] These vessels grow into the peripheral sections of the placenta, and growth of the floating villi arises because of the development of the capillaries that lie just beneath the trophoblast.

Distinct from the development of the floating villi, anchoring villi not only serve to secure the attachment of the placenta to the decidua, but are responsible for the remodelling of the maternal spiral arterioles. Extravillous trophoblast migrates into and around the spiral arteriole. The cells migrate into the vessel. The original endothelial cells of the spiral arterioles undergo apoptosis, and are replaced by the trophectoderm cells that express an endothelial phenotype.[64] At the same time, the vascular smooth muscle cells of the spiral arterioles also undergo apoptosis, resulting in a significant reduction in the muscle layer. This is the molecular basis of re-modelling of the spiral arterioles that changes them from a high-flow, high-capacitance system to a low-flow, low-capacity vascular network.

Surprisingly, the single most abundant product of the human placenta is an antiangiogenic protein. This is a truncated version of VEGFR-1, which binds VEGF but blocks its action.[65] The role of this molecule is yet to be determined. Mice that express VEGFR-1, in which the intracytoplasmic tyrosine kinase has been removed, are normal. Thus, the presence of the full-length VEGFR-1 is not needed for normal development. This has led to the suggestion that VEGFR-1 acts as a sump, regulating angiogenesis by limiting the biological availability of VEGF-A. If extra VEGF is given to mice during pregnancy, it causes the placenta to be abnormal, with disruption of the vessels in the labyrinthine layer. The vessels are larger than normal, and deposits of fibrin are found around the vessels.[66] In the human placenta, most of the VEGF-A is expressed by macrophages in the maternal and fetal sides of the circulation[67].

ANGIOGENESIS AND DISEASES OF REPRODUCTION

Benign disease

Abnormal uterine bleeding

Observational studies suggest that disturbances in angiogenesis might underlie heavy bleeding at the time of menstruation (menorrhagia). The endometrium is a rich source of VEGF-A (Figure 3). Most women presenting for medical advice with abnormal periods complain of heavy flow in the presence of a regular menstrual cycle. In these cases, bleeding arises from a desquamating endometrium. The shedding of the endometrium is induced by the expression of matrix metalloproteinases, released from stromal fibroblasts by the up-regulation of transforming growth factor-β (TGF-β) on withdrawal of progesterone.[68,69] The blood vessels in these circumstances lose their supporting cellular matrix. Control of the bleeding is assumed to arise by two mechanisms.

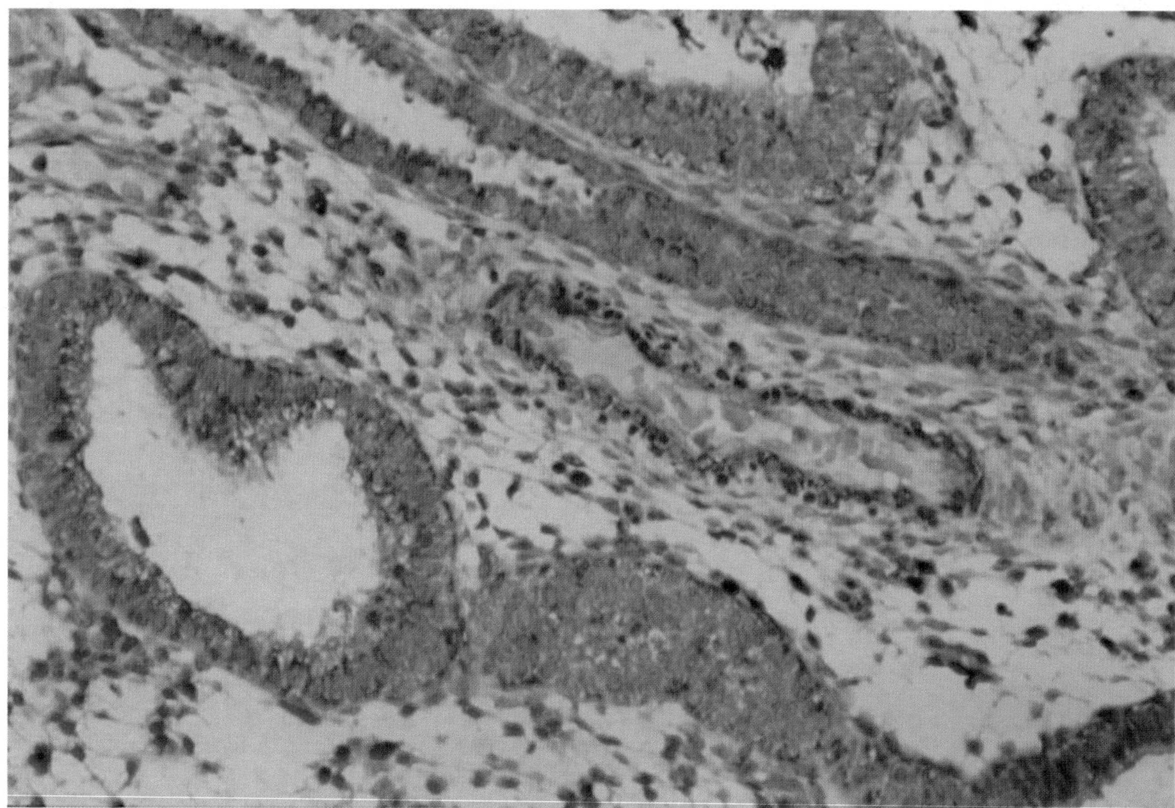

Figure 3 Immunohistochemical identification of VEGF-A in endometrium removed in the secretory phase of the menstrual cycle. Protein is present in the glandular epithelium and in the vascular smooth muscle cells of endometrial venules, as shown in the center of the figure

First, the spiral arterioles constrict, thus preventing loss of blood from their distal ends. Several agents have been implicated in this process, including prostaglandins and endothelins.[70,71] Second, platelet fibrin clots form at the end of the vessels, plugging them. Antifibrinolytic agents such as tranexamic acid reduce menstrual blood loss by 50%.[72] However, in addition, the vascular smooth muscle coat of the straight arterioles is reduced, in vessels identified in uteri removed from women with proven heavy bleeding.[58] The proliferative activity of the vascular smooth muscle cells in women with menorrhagia is reduced, compared with women with normal menstrual flow,[60] and the proliferative index of endothelial cells is increased in women with measured heavy periods.[18]

In addition to menorrhagia, bleeding can occur during the menstrual cycle, and especially in women taking oral contraceptives or hormone replacement therapy. In these cases, the superficial blood vessels and particularly the venules are abnormal, being dilated and more fragile than in women without bleeding. Exogenous steroids alter VEGF-A expression, and this may be the cause of the abnormal structure of the blood vessels.

Endometriosis

Endometriosis arises from the deposition of endometrium in the peritoneal cavity. The lesions are characterized by extensive angiogenesis around the lesion. Peritoneal fluid of women with the disease has enhanced angiogenic activity.[73,74] However, there is no significant release of VEGF-A from the endometrium of women with the disease. Another feature of the disease is the presence of increased numbers of activated macrophages, both in the peritoneal fluid and infiltrating into the lesions.[75] These activated macrophages secrete increased amounts of VEGF-A (Figure 4). The cause of activation of the macrophages is unknown, but is likely to reflect activation by T lymphocytes, although the trigger for this activation remains a mystery.

Early red lesions of endometriosis have a greater microvascular density than the aging blue and white lesions.[76] This suggests active angiogenesis at the start of the lesion that declines as the natural progression of the lesion declines.

Malignant disease

All of the major malignancies of the female reproductive tract have altered angiogenesis around, and as part of, the lesion that relates to prognosis. The very early stages of cervical intraepithelial neoplasia are associated with increased angiogenesis.[77] The classical observation at colposcopy arises from the increased prominence of the blood vessels, and VEGF-A levels are increased in the cervix of such patients. The E7 component of herpes simplex virus stimulates VEGF-A expression. In cervical malignancy that has become invasive, prognosis is related to microvascular density.[77] One suggestion is that the predilection for tumors to spread either by the hematogenous route or by the lymphatic route is determined by the type of VEGF that they express. Tumors that express VEGF-A induce hematogenous angiogenesis, while those that express VEGF-C and VEGF-D are more likely to spread by the lymphatic endothelial route. The role of angiogenesis in cancer is complex. First, the epithelial cells must undergo a form of malignant transformation in which cell cycle regulation is altered. They must then pass through the basement membrane to become invasive,[13] but the crucial step is the recruitment of host blood vessels to the lesion[78] and the subsequent shedding of malignant cells into these blood vessels that forms the cellular basis of the malignant dissemination of the disease. This is clearly seen in cervical carcinoma. Prognosis is linked to the levels of growth factor expression and the subsequent growth of microvascular blood vessels. These vessels have individual characteristics. Blood vessels within the tumor are classically 'immature' in that they are numerous, do not have pericytic associations and are 'leaky'. Conversely, the surrounding host vessels are numerous, but retain the pericytic associations not found in the tumor blood vessels. This is important, because it means that tumor blood vessels may be more susceptible to antiangiogenic agents than the host vessels.

In a similar way to cervical cancer, microvascular density is increased in the endometrium of women with endometrioid carcinoma, the cancer that constitutes about 10% of endometrial malignancies.[80] VEGF-A levels are increased in carcinoma cells, but none of the other VEGFs are increased. Surprisingly, VEGF-A is not increased in cases of complex atypical hyperplasia, and demonstrates a molecular difference between a premalignant and a malignant condition. VEGF-B, however, is increased in these cases, and microvascular density is also increased in the endometrium of these women. VEGF-B binds the VEGFR-1 receptor. Activation of VEGFR-1 does not induce endothelial cell proliferation, and the role of VEGFR-1 may be to act as a sink for VEGF. Thus, binding of increased amounts of VEGF-B, which does not bind VEGFR-2, may release more VEGF-A to bind to this receptor.[81] Thus, increased expression of VEGF-B may be angiogenic but by an indirect route, increasing the availability of VEGF-A to the tumor blood vessels.[82]

Ovarian cancer is also associated with a poorer prognosis when microvascular density is increased. Serous cystadenocarcinoma has increased levels of VEGF-A and VEGF-B expression, but not of VEGF-C or VEGF-D[83,84] (Figure 5). The mechanism of increased VEGF-A expression is varied. In some cases, polymorphisms of tumor suppressor genes result in increased levels of VEGF-A. Alternatively, tumor-associated macrophages invade the lesion and secrete VEGF-A.[85] Tumors also recruit host blood vessels by overexpressing Ang-2.

Pregnancy

Pre-eclampsia is characterized by activation of the endothelium,[86] resulting in edema and intravascular coagulation. VEGF levels are increased in the serum of women suffering from pre-eclampsia.[87,88]

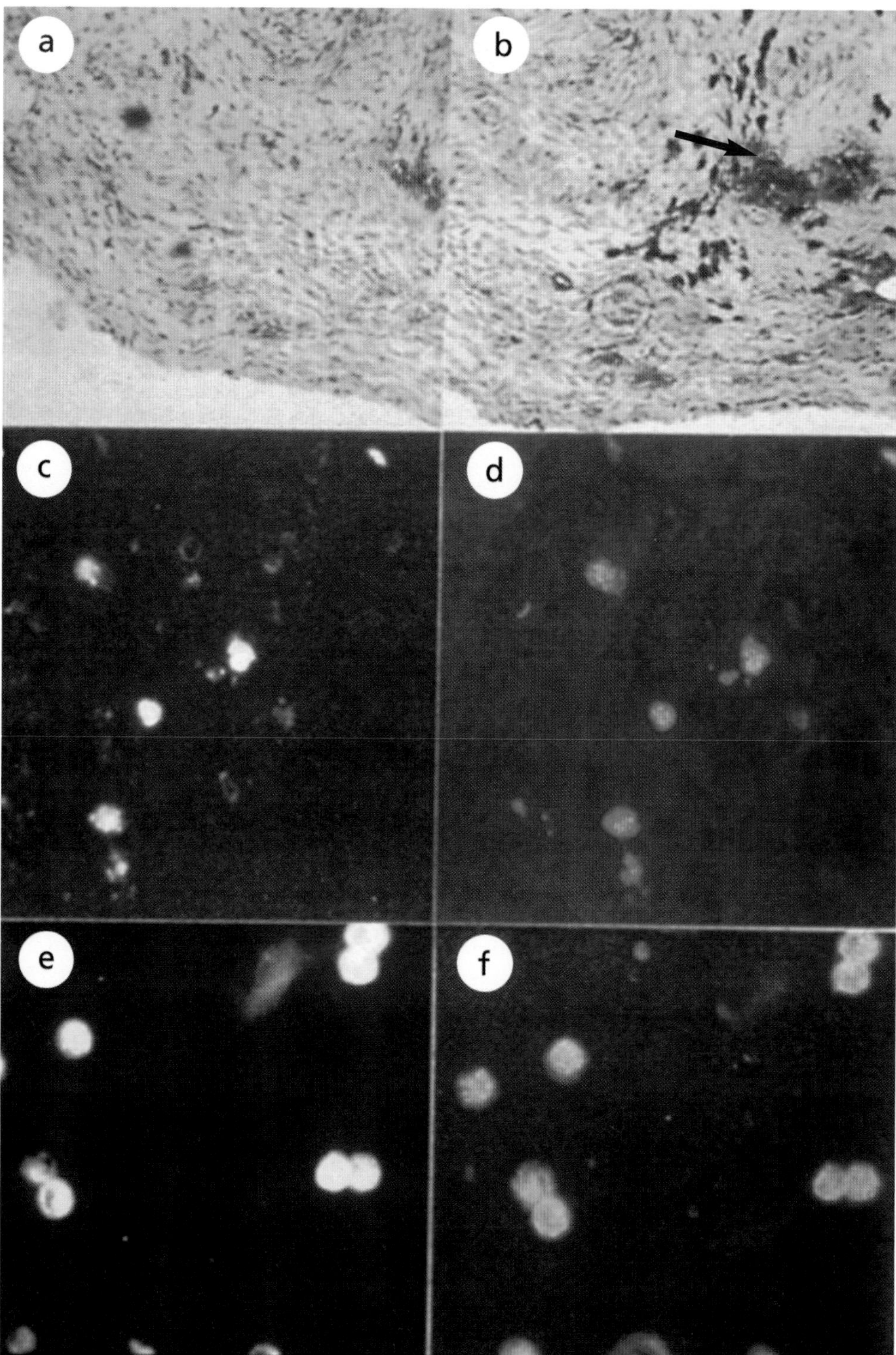

Figure 4 Immunohistochemical staining (arrowed, brown) for VEGF in an endometriotic explant (b), and negative control (a). Immunofluorescent staining for CD14 (to identify macrophages) in the stroma of an endometriotic explant (c) co-stained for VEGF-A (d). Immunofluorescent staining for CD14 (e) and VEGF-A (f) in isolated macrophages obtained from the peritoneal cavity. Reproduced from reference 75, with permission

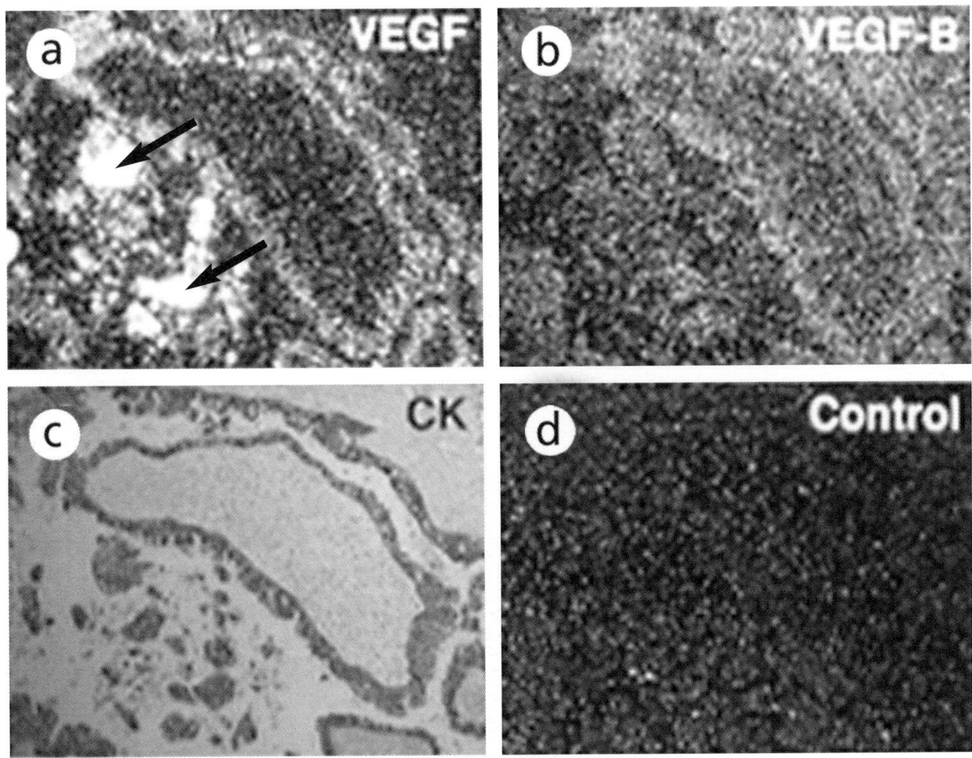

Figure 5 *In situ* hybridization for VEGF-A (a) and VEGF-B (b) in human ovarian malignancy. The high levels of expression for VEGF-A are found focally (arrowed), presumably reflecting local areas of hypoxia. The overexpression for VEGF-B is diffuse, reflecting a regulation that is not mediated by local conditions. (c) Cytokeratin, (d) control

Furthermore, the effects of plasma removed from women with pre-eclampsia on vascular tone can be mimicked by VEGF.[89] However, another member of the VEGF family, placental growth factor, is suppressed in the serum of women who develop pre-eclampsia.[90] This striking observation is assumed to arise from a reduced production of PlGF because of failed trophoblast invasion. This might further exacerbate a change in the soluble VEGFR-blocking protein/VEGF ratio, as PlGF is less active than VEGF. Thus, pre-eclampsia may arise because of increased activity of VEGF-A, exacerbated by a significant reduction of PlGF. Measurement of PlGF in the mid-trimester is an effective predictor of the onset of pre-eclampsia, before the disease becomes manifest.

Other disorders of pregnancy may be associated with altered VEGF-A expression. Placentas of women who deliver at altitude are larger than placentas obtained from women who deliver at sea-level. These placentas are characterized by increased vessel area in the floating villi that is associated with increased VEGF-A expression. Speculation remains as to whether disorders of fetal growth associated with large or small placentas may also arise from abnormal angiogenesis. However, what is clear is that oxygen tension and the release of free oxygen radicals mediates VEGF-A expression, which further regulates vascular structure in the placenta.

Other growth factors that mediate angiogenesis

Much attention is focused on agents that induce angiogenesis, but the usual state is for tight regulation of angiogenesis. The endometrium and placenta contain several proteins that inhibit angiogenesis, including thrombospondins 1 and

4,[91] endostatin,[92] angiostatin, platelet factor-4 and TGF-β.[93] Thrombospondin-1, a potent inhibitor of angiogenesis, is present in the stromal fibroblasts of the endometrium, and is induced by progesterone.[94] The eventual control of angiogenesis reflects the balance between inhibitory and stimulatory growth factors.[95]

Microarray analysis of multiple gene expression in benign and malignant ovaries shows such a complex pattern of gene expression. While angiogenic growth factors such as VEGF-A are increased in serous papillary adenocarcinoma, antiangiogenic agents such as the thrombospondins are also increased.

The overall effect on the blood vessel will be the interplay at the molecular and cellular levels between the factors promoting growth and those constraining it. Gene expression profiles will define the function of endothelial cells, pericytes and surrounding host stromal cells. It remains to be seen whether these patterns can be identified and subsequently related to vascular structure.

CLINICAL OPPORTUNITIES FOR THE FUTURE

The increased understanding of the roles that vasculogenesis, angiogenesis and arteriogenesis play in female reproduction raises the opportunity to develop new treatments for many of the common problems that reduce the quality of life for women:

1) To disrupt the ovarian or menstrual cycle by targeting the blood vessels as a means of developing new contraceptive or postcoital agents;

2) To identify new agents for treating benign diseases of women such as menorrhagia, endometriosis and fibroids;

3) To define the angiogenic factors that control the development, growth and therefore the function of the normal placenta and to determine how this is disrupted in pregnancies complicated by pre-eclampsia and intrauterine growth restriction.

Fertility regulation

Several opportunities arise to use antiangiogenic agents as contraceptives. Antiangiogenic agents inhibit the development of the corpus luteum when administered just after ovulation. In addition, these agents may inhibit vascular edema and vascular growth in the endometrium, necessary for successful implantation. Pregnancy is characterized by the release of very high levels of the anti-VEGF, soluble receptor.[65] This provides reassurance that serious systemic side-effects are most unlikely. Even if the embryo has implanted, it may be possible to disrupt the early stages of implantation. While this may not be acceptable to some on the basis of belief, it would for many other women provide an effective 'morning after' approach. The main concern here is that antiangiogenic agents such as thalidomide have serious teratogenic effects, but this does not arise because of inhibition of the VEGF system.

Benign gynecological disease

Endometriosis arises from the deposition of desquamated endometrium into the peritoneal cavity at the time of menstruation. The explant achieves implantation and is maintained by developing a rich blood supply.[15] It may be possible to block this mechanism. In a similar manner, it may be possible to inhibit blood vessel growth to the explants. Such a therapy may be effective in its own right, or in conjunction with ablative surgery.

The medical treatment of fibroids may be enhanced by the use of antiangiogenic agents.[96] Fibroids that are large and producing pressure symptoms may still need to be removed surgically, but most women present with smaller fibroids of below 10 weeks' growth.[97] In these cases, it may be possible to inject antiangiogenic agents into the fibroid that would block the vasculature and result in regression of the mass. This is currently possible in a slightly different way by uterine artery ablation. However, it would be possible to inject submucosal fibroids via the hysteroscope, removing the need for dangerous procedures to core out or laser the intramural fibroid.

Complications of pregnancy

An important element of antenatal care is the detection and management of pre-eclampsia. Measurement of PlGF in the serum of such women, along with other markers such as activins and inhibins, provides a highly specific warning blood test of the subsequent onset of pre-eclampsia.[90] Larger studies are needed to determine the efficacy of these tests in larger populations, but, if found to be successful, they would transform the care of the pregnant woman in both the developed and the developing world.

Not only may these developments in the understanding of vascular biology lead to new tests, but it may also be possible to use them for therapy. It is becoming clear that a key feature in the regulation of the action of VEGF-A is the co-expression of agents that limit the bioactivity of the molecule. Soluble VEGFR-1 is expressed in high amounts by the trophoblast, and this agent binds free VEGF-A and blocks its action. When excess VEGF-A is administered to mice, the blood vessels in the labyrinthine layer are dilated and become more leaky, resulting in the deposition of fibrin.[66] It may be possible to use this or other agents to reduce the increased levels of VEGF-A found in the blood of women with pre-eclampsia[87] and activation of the coagulation system.

Cancer

Several trials are under way to investigate the use of antiangiogenic agents as adjunct therapy for cancer.[14] These studies include intravenous, intraperitoneal and oral administration of compounds that block angiogenesis. Several phase-1 and -2 clinical trials are under way. Few of these studies have been published, but early reports suggest a clinical benefit. Side-effects are related to neurological complications. This is an interesting observation because there is increasing evidence of a link between the regulation of blood vessel growth and neuronal development. Axonal-guidance molecules, such as semaphorin III and IV, share the neuropilin-1 and -2 receptors with members of the VEGF family.[98–100] Second, cell surface ligands and activators, collectively termed ephrins, are potent mediators of neuronal development and endothelial cell positioning,[9] and the PlGF knock-out mouse shows changes in neuronal development. It is unlikely that this approach will be successful in its own right, but, administered in conjunction with other strategies such as surgery, radiotherapy or chemotherapy, it is likely to provide the fourth major arm to oncology therapy.

CONCLUSION

The study of reproductive biology has a long, distinguished and productive history. It has led to the most profound changes in individual, social and economic life. It has permitted women and men to control their fertility and has led to the creation of life itself. It has come as some surprise to find, at such a late stage, that vascular biology plays such a crucial part in all of the processes of reproduction, and that it can be manipulated to provide possible new treatments in the future. The next 30 years will determine whether this dream can become reality.

REFERENCES

1. Risau W, Flamme I. Vasculogenesis. *Annu Rev Cell Dev Biol* 1995 11: 73–91.
2. Beck L Jr, D'Amore PA. Vascular development: cellular and molecular regulation. *FASEB J* 1997 11: 365–373.
3. Drake CJ, Fleming PA. Vasculogenesis in the day 6.5 to 9.5 mouse embryo. *Blood* 2000 95: 1671–1679.
4. Shalaby F, Rossant J, Yamaguchi TP *et al*. Failure of blood-island formation and vasculogenesis in Flk-1 deficient mice. *Nature* 1995 376: 62–66.
5. Cox CM, Poole TJ. Angioblast differentiation is influenced by the local environment: FGF-2 induces angioblasts and patterns vessel formation in the quail embryo. *Dev Dyn* 2000 218: 371–82.
6. Carmeliet P, Moons L, Dewerchin M *et al*. Insights in vessel development and vascular disorders using targeted inactivation and transfer of vascular endothelial growth factor, the tissue factor receptor, and the plasminogen system. *Ann NY Acad Sci* 1997 811: 191–206.
7. Kroll J, Waltenberger J. The vascular endothelial growth factor receptor KDR activates multiple

signal transduction pathways in porcine aortic endothelial cells. *J Biol Chem* 1997 272: 32521–32527.
8. Poole TJ, Finkelstein EB, Cox CM. The role of FGF and VEGF in angioblast induction and migration during vascular development. *Dev Dyn* 2001 220: 1–17.
9. Yancopoulos GD, Klagsbrun M, Folkman J. Vasculogenesis, angiogenesis, and growth factors: ephrins enter the fray at the border. *Cell* 1998 93: 661–664.
10. Hirschi KK, D'Amore PA. Pericytes in the microvasculature. *Cardiovasc Res* 1996 32: 687–698.
11. Asan E, Kaymaz FF, Cakar AN, Dagdeviren A, Beksac MS. Vasculogenesis in early human placental villi: an ultrastructural study. *Ann Anat* 1999 181: 549–554.
12. Folkman J, Shing Y. Angiogenesis. *J Biol Chem* 1992 267: 10931–10934.
13. Hanahan D. Signaling vascular morphogenesis and maintenance. *Science* 1997 277: 48–50.
14. Smith SK. Angiogenesis and ovarian cancer. *Contemp Rev Obstet Gynaecol* 1997 9: 53–59.
15. Smith SK. Angiogenesis, vascular endothelial growth factor and the endometrium. *Hum Reprod Update* 1998 4: 509–519.
16. Goede V, Schmidt T, Kimmina S, Kozian D, Augustin HG. Analysis of blood vessel maturation processes during cyclic ovarian angiogenesis. *Lab Invest* 1998 78: 1385–1394.
17. Flamme I, Risau W. Mechanism of blood vessel formation. *Ann Anat* 1995 177: 493–502.
18. Rogers PA, Lederman F, Taylor N. Endometrial microvascular growth in normal and dysfunctional states. *Hum Reprod Update* 1998 4: 503–508.
19. Rogers PA, Plunkett D, Affandi B. Perivascular smooth muscle α-actin is reduced in the endometrium of women with progestin-only contraceptive breakthrough bleeding. *Hum Reprod* 2000 15 (Suppl 3): 78–84.
20. Carmeliet P. Developmental biology. One cell, two fates. *Nature* 2000 408: 43–45.
21. Benjamin LE, Hemo I, Keshet E. A plasticity window for blood vessel remodelling is defined by pericyte coverage of the preformed endothelial network and is regulated by PDGF-B and VEGF. *Development* 1998 125: 1591–1598.
22. Yancopoulis GD, Davis S, Gale NW, Rudge JS, Wiegand SJ, Holash J. Vascular-specific growth factors and blood vessel formation. *Nature* 2000 407: 242–248.
23. Amoroso A, Del Porto F, Di Monaco C, Manfredini P, Afeltra A. Vascular endothelial growth factor: a key mediator of neoangiogenesis. A review. *Eur Rev Med Pharmacol Sci* 1997 1: 17–25.
24. Tischer E, Mitchell R, Hartman T *et al.* The human gene for vascular endothelial growth factor: multiple protein forms are encoded through alternative exon splicing. *J Biol Chem* 1991 266: 11947–11954.
25. Charnock-Jones DS, Sharkey AM, Rajput-Williams J *et al.* Identification and localization of alternately spliced mRNAs for vascular endothelial growth factor in human uterus and estrogen regulation in endometrial carcinoma cell lines. *Biol Reprod* 1993 48: 1120–1128.
26. Shifren JL, Tseng JF, Zaloudek CJ *et al.* Ovarian steroid regulation of vascular endothelial growth factor in the human endometrium: implications for angiogenesis during the menstrual cycle and in the pathogenesis of endometriosis. *J Clin Endocrinol Metab* 1996 81: 3112–3118.
27. Hornung D, Lebovic DI, Shifren JL, Vigne JL, Taylor RN. Vectorial secretion of vascular endothelial growth factor by polarized human endometrial epithelial cells. *Fertil Steril* 1998 69: 909–915.
28. Sharkey AM, Day K, McPherson A *et al.* Vascular endothelial growth factor expression in human endometrium is regulated by hypoxia. *J Clin Endocrinol Metab* 2000 85: 402–409.
29. Maltepe E, Keith B, Arsham AM, Brorson JR, Simon MC. The role of ARNT2 in tumor angiogenesis and the neural response to hypoxia. *Biochem Biophys Res Commun* 2000 273: 231–238.
30. Katoh O, Takahashi T, Oguri T *et al.* Vascular endothelial growth factor inhibits apoptotic death in hematopoietic cells after exposure to chemotherapeutic drugs by inducing MCL1 acting as an antiapoptotic factor. *Cancer Res* 1998 58: 5565–5569.
31. Gerber HP, McMurtrey A, Kowalski J *et al.* Vascular endothelial growth factor regulates endothelial cell survival through the phosphatidylinositol 3'-kinase/Akt signal transduction pathway: requirement for Flk-1/KDR activation. *J Biol Chem* 1998 273: 30336–30343.

32. Dimmeler S, Zeiher AM. Akt takes center stage in angiogenesis signaling. *Circ Res* 2000 86: 4–5.
33. Joukov V, Pajusola K, Kaipainen A *et al.* A novel vascular endothelial growth factor, VEGF-C, is a ligand for the Flt4 (VEGFR-3) and KDR (VEGFR-2) receptor tyrosine kinases. *EMBO J* 1996 15: 290–298.
34. Jeltsch M, Kaipainen A, Joukov V *et al.* Hyperplasia of lymphatic vessels in VEGF-C transgenic mice. *Science* 1997 276: 1423–1425.
35. Achen MG, Jeltsch M, Kukk E *et al.* Vascular endothelial growth factor D (VEGF-D) is a ligand for the tyrosine kinases VEGF receptor 2 (Flk1) and VEGF receptor 3 (Flt4). *Proc Natl Acad Sci USA* 1998 95: 548–553.
36. Enholm B, Jussila L, Karkkainen M, Alitalo K. Vascular endothelial growth factor-C: a growth factor for lymphatic and blood vascular endothelial cells. *Trends Cardiovasc Med* 1998 8: 292–297.
37. King A, Burrows T, Verma S, Hiby S, Loke YW. Human uterine lymphocytes. *Hum Reprod Update* 1998 4: 480–485.
38. Basilico C, Moscatelli D. The FGF family of growth factors and oncogenes. *Adv Cancer Res* 1992 59: 115–165.
39. Chen C, Spencer TE, Bazer FW. Fibroblast growth factor-10: a stromal mediator of epithelial function in the ovine uterus. *Biol Reprod* 2000 63: 959–966.
40. Deroanne CF, Hajitou A, Calberg-Bacq CM, Nusgens BV, Lapiere CM. Angiogenesis by fibroblast growth factor 4 is mediated through an autocrine up-regulation of vascular endothelial growth factor. *Cancer Res* 1997 57: 5590–5597.
41. Ferriani RA, Charnock-Jones DS, Prentice A, Thomas EJ, Smith SK. Immunohistochemical localization of acidic and basic fibroblast growth factors in normal human endometrium and endometriosis and the detection of their mRNA by polymerase chain reaction. *Hum Reprod* 1993 8: 11–16.
42. Fujimoto J, Hori M, Ichigo S, Tamaya T. Expression of basic fibroblast growth factor and its mRNA in uterine endometrial cancers. *Invasion Metast* 1995 15: 203–210.
43. Pepper MS, Ferrara N, Orci L, Montesano R. Potent synergism between vascular endothelial growth factor and basic fibroblast growth factor in the induction of angiogenesis *in vitro*. *Biochem Biophys Res Commun* 1992 189: 824–831.
44. Schneeberger SA, Hjelmeland LM, Tucker RP, Morse LS. Vascular endothelial growth factor and fibroblast growth factor 5 are colocalized in vascular and avascular epiretinal membranes. *Am J Ophthalmol* 1997 124: 447–454.
45. Pepper MS, Mandriota SJ. Regulation of vascular endothelial growth factor receptor-2 (Flk-1) expression in vascular endothelial cells. *Exp Cell Res* 1998 241: 414–425.
46. Poltorak Z, Cohen T, Sivan R *et al.* VEGF145, a secreted vascular endothelial growth factor isoform that binds to extracellular matrix. *J Biol Chem* 1997 272: 7151–7158.
47. Davis S, Aldrich TH, Jones PF *et al.* Isolation of angiopoietin-1, a ligand for the TIE2 receptor, by secretion-trap expression cloning. *Cell* 1996 87: 1161–1169.
48. Maisonpierre PC, Suri C, Jones PF *et al.* Angiopoietin-2, a natural antagonist for Tie2 that disrupts *in vivo* angiogenesis. *Science* 1997 277: 55–60.
49. Valenzuela DM, Griffiths JA, Rojas J *et al.* Angiopoietins 3 and 4: diverging gene counterparts in mice and humans. *Proc Natl Acad Sci USA* 1999 96: 1904–1909.
50. Asahara T, Chen D, Takahashi T *et al.* Tie2 receptor ligands, angiopoietin-1 and angiopoietin-2, modulate VEGF-induced postnatal neovascularization. *Circ Res* 1998 83: 233–240.
51. Li XF, Charnock-Jones DS, Zhang E *et al.* Angiogenic growth factor messenger ribonucleic acids in uterine natural killer cells. *J Clin Endocrinol Metab* 2001 86: 1823–1834.
52. Krikun G, Schatz F, Finlay T *et al.* Expression of angiopoietin-2 by human endometrial endothelial cells: regulation by hypoxia and inflammation. *Biochem Biophys Res Commun* 2000 275: 159–163.
53. Flanagan JG, Vanderhaeghen P. The ephrins and Eph receptors in neural development. *Annu Rev Neurosci* 1998 21: 309–345.
54. Holland SJ, Peles E, Pawson T, Schlessinger J. Cell-contact-dependent signalling in axon growth and guidance: Eph receptor tyrosine kinases and receptor protein tyrosine phosphatase β. *Curr Opin Neurobiol* 1998 8: 117–127.
55. Smith SK. Antiprogestogens and endometrial function. In d'Arcangues C, Newton JR, Frazer IS, Odlind V, eds. *Contraception and Mechanisms in*

Endometrial Bleeding. Cambridge: Cambridge University Press. 1990: 337–345.

56. Markee JE. Morphological basis for menstrual bleeding. *Bull NY Acad Med* 1948 18: 159.
57. Ludwig H, Metzger H, Frauli M. Endometrium: tissue remodelling and regeneration. In d'Arcangues C, Fraser IS, Newton JR, Odlind V, eds. *Contraception and Mechanisms of Endometrial Bleeding*. Cambridge: Cambridge University Press. 1990: 441–466.
58. Abberton KM, Healy DL, Rogers PA. Smooth muscle α actin and myosin heavy chain expression in the vascular smooth muscle cells surrounding human endometrial arterioles. *Hum Reprod* 1999 14: 3095–3100.
59. Thommen R, Humar R, Misevic G *et al.* PDGF-BB increases endothelial migration and cord movements during angiogenesis *in vitro*. *J Cell Biochem* 1997 64: 403–413.
60. Abberton KM, Taylor NH, Healy DL, Rogers PA. Vascular smooth muscle cell proliferation in arterioles of the human endometrium. *Hum Reprod* 1999 14: 1072–1079.
61. Reynolds LP, Grazul-Bilska AT, Redmer DA. Angiogenesis in the corpus luteum. *Endocrine* 2000 12: 1–9.
62. Charnock-Jones DS, Zhang EG, Licence D, Malik S, Chan CLK, Smith SK. Localisation and regulation of expression of angiopoietin 1 and angiopoietin 2 in human endometrium throughout the menstrual cycle. *J Soc Gynecol Invest* 1999 6 (Suppl): 73A.
63. Ferrara N, Chen H, Davis-Smyth T *et al.* Vascular endothelial growth factor is essential for corpus luteum angiogenesis. *Nat Med* 1998 4: 336–340.
64. Zhou Y, Fisher SJ, Janatpour M *et al.* Human cytotrophoblasts adopt a vascular phenotype as they differentiate: a strategy for successful endovascular invasion? *J Clin Invest* 1997 99: 2139–2151.
65. Clark DE, Smith SK, He Y *et al.* A vascular endothelial growth factor antagonist is produced by the human placenta and released into the maternal circulation. *Biol Reprod* 1998 59: 1540–1548.
66. He Y, Smith SK, Day KA, Clark DE, Licence DR, Charnock-Jones DS. Alternative splicing of vascular endothelial growth factor (VEGF)-R1 (FLT-1) pre-mRNA is important for the regulation of VEGF activity. *Mol Endocrinol* 1999 13: 537–545.
67. Charnock-Jones DS, Sharkey AM, Boocock CA *et al.* Vascular endothelial growth factor receptor localization and activation in human trophoblast and choriocarcinoma cells. *Biol Reprod* 1994 51: 524–530.
68. Salamonsen LA. Matrix metalloproteinases and endometrial remodelling. *Cell Biol Int* 1995 18: 1139–1144.
69. Marbaix E, Kokorine I, Moulin P, Donnez J, Eeckhout Y, Courtoy PJ. Menstrual breakdown of human endometrium can be mimicked *in vitro* and is selectively and reversibly blocked by inhibitors of matrix metalloproteinases. *Proc Natl Acad Sci USA* 1996 93: 9120–9125.
70. Smith SK. Prostaglandins and menstrual dysfunction. *J Obstet Gynaecol* 1988 8 (Suppl 1): S20–S22.
71. Cameron IT, Bacon CR, Collett GP, Davenport AP. Endothelin expression in the uterus. *J Steroid Biochem Mol Biol* 1995 53: 209–214.
72. Preston JT, Cameron IT, Adams EJ, Smith SK. Comparative study of tranexamic acid and norethisterone in the treatment of ovulatory menorrhagia. *Br J Obstet Gynaecol* 1995 102: 401–406.
73. Oosterlynck DJ, Meuleman C, Sobis H, Vandeputte M, Koninckx PR. Angiogenic activity of peritoneal fluid from women with endometriosis. *Fertil Steril* 1993 59: 778–782.
74. McLaren J, Prentice A, Charnock-Jones DS, Smith SK. Vascular endothelial growth factor (VEGF) concentrations are elevated in peritoneal fluid of women with endometriosis. *Hum Reprod* 1996 11: 220–223.
75. McLaren J, Prentice A, Charnock-Jones DS *et al.* Vascular endothelial growth factor is produced by peritoneal fluid macrophages in endometriosis and is regulated by ovarian steroids. *J Clin Invest* 1996 98: 482–489.
76. Nisolle M, Casanas-Roux F, Anaf V, Mine JM, Donnez J. Morphometric study of the stromal vascularization in peritoneal endometriosis. *Fertil Steril* 1993 59: 681–684.
77. Guidi AJ, Abu-Jawdeh G, Berse B *et al.* Vascular permeability factor (vascular endothelial growth factor) expression and angiogenesis in cervical neoplasia. *J Natl Cancer Inst* 1995 87: 1237–1245.
78. Tjalma W, Van Marck E, Weyler J *et al.* Quantification and prognostic relevance of angiogenic parameters in invasive cervical cancer. *Br J Cancer* 1998 78: 170–174.

79. Holash J, Maisonpierre PC, Compton D *et al.* Vessel cooption, regression, and growth in tumors mediated by angiopoietins and VEGF. *Science* 1999 284: 1994–1998.
80. Kaku T, Kamura T, Kinukawa N *et al.* Angiogenesis in endometrial carcinoma. *Cancer* 1997 80: 741–747.
81. Carmeliet P, Collen D. Role of vascular endothelial growth factor and vascular endothelial growth factor receptors in vascular development. *Curr Top Microbiol Immunol* 1999 237: 133–158.
82. Yokoyama Y, Sato S, Futagami M *et al.* Prognostic significance of vascular endothelial growth factor and its receptors in endometrial carcinoma. *Gynecol Oncol* 2000 77: 413–418.
83. Boocock CA, Charnock-Jones DS, Sharkey AM *et al.* Expression of vascular endothelial growth factor and its receptors flt and KDR in ovarian carcinoma. *J Natl Cancer Inst* 1995 87: 506–516.
84. Sowter HM, Corps AN, Evans AL, Clark DE, Charnock-Jones DS, Smith SK. Expression and localization of the vascular endothelial growth factor family in ovarian epithelial tumors. *Lab Invest* 1997 77: 607–614.
85. Salvesen HB, Akslen LA. Significance of tumour-associated macrophages, vascular endothelial growth factor and thrombospondin-1 expression for tumour angiogenesis and prognosis in endometrial carcinomas. *Int J Cancer* 1999 84: 538–543.
86. Roberts JM. Endothelial dysfunction in preeclampsia. *Semin Reprod Endocrinol* 1998 16: 5–15.
87. Baker PN, Krasnow J, Roberts JM, Yeo KT. Elevated serum levels of vascular endothelial growth factor in patients with preeclampsia. *Obstet Gynecol* 1995 86: 815–821.
88. Sharkey AM, Cooper JC, Balmforth JR *et al.* Maternal plasma levels of vascular endothelial growth factor in normotensive pregnancies and in pregnancies complicated by pre-eclampsia. *Eur J Clin Invest* 1996 26: 1182–1185.
89. Brockelsby J, Hayman R, Ahmed A, Warren A, Johnson I, Baker P. VEGF via VEGF receptor-1 (Flt-1) mimics preeclamptic plasma in inhibiting uterine blood vessel relaxation in pregnancy: implications in the pathogenesis of preeclampsia. *Lab Invest* 1999 79: 1101–1111.
90. Torry DS, Wang HS, Wang TH, Caudle MR, Torry RJ. Preeclampsia is associated with reduced serum levels of placenta growth factor. *Am J Obstet Gynecol* 1999 179: 1539–1544.
91. Sheibani N, Newman PJ, Frazier WA. Thrombospondin-1, a natural inhibitor of angiogenesis, regulates platelet-endothelial cell adhesion molecule-1 expression and endothelial cell morphogenesis. *Mol Biol Cell* 1997 8: 1329–1341.
92. O'Reilly MS, Boehm T, Shing Y *et al.* Endostatin: an endogenous inhibitor of angiogenesis and tumor growth. *Cell* 1997 88: 277–285.
93. Mandriota SJ, Pepper MS. Vascular endothelial growth factor-induced *in vitro* angiogenesis and plasminogen activator expression are dependent on endogenous basic fibroblast growth factor. *J Cell Sci* 1997 110: 2293–2302.
94. Iruela-Arispe ML, Porter P, Bornstein P, Sage EH. Thrombospondin-1, an inhibitor of angiogenesis, is regulated by progesterone in the human endometrium. *J Clin Invest* 1996 97: 403–412.
95. Iruela-Arispe ML, Dvorak HF. Angiogenesis: a dynamic balance of stimulators and inhibitors. *Thromb Haemost* 1997 78: 672–677.
96. Casey R, Rogers PA, Vollenhoven BJ. An immunohistochemical analysis of fibroid vasculature. *Hum Reprod* 2000 15: 1469–1475.
97. Smith SK. Fibroids. In Rainsbury PA, Viniker DA, eds. *Practical Guide to Reproductive Medicine*. Carnforth, UK: Parthenon Publishing Group, 1997: 395–408.
98. Soker S, Takashima S, Hua Quan M, Neufeld G, Klagsbrun M. Neuropilin-1 is expressed by endothelial and tumor cells as an isoform-specific receptor for vascular endothelial growth factor. *Cell* 1998 92: 735–745.
99. Migdal M, Huppertz B, Tessler S *et al.* Neuropilin-1 is a placenta growth factor-2 receptor. *J Biol Chem* 1998 273: 22272–22278.
100. Giger RJ, Urquhart ER, Gillespie SKH, Levengood DV, Ginty DD, Kolodkin AL. Neuropilin-2 is a receptor for semaphorin IV: insight into the structural basis of receptor function and specificity. *Neuron* 1999 21: 1079–1092.

Chapter 8

Hormones and cancer

Sandra Timm Pearce and V. Craig Jordan

INTRODUCTION

Breast, prostate, endometrial, testicular, ovarian, thyroid and osteosarcoma cancers are considered to be hormone-related cancers.[1] The growth of these cancers is stimulated by hormones during at least part of their progression.

Many mechanisms have been proposed to account for hormonal carcinogenesis. One mechanism involves endogenous and exogenous hormones that stimulate cell proliferation and differential gene expression through interactions with a hormone receptor. By increasing the number of cell divisions, the chance of random mutations that contribute to the malignant phenotype increases. Another mechanism involves the hormone itself acting as a carcinogen, as is the case with estrogen and breast cancer. Estrogen can be metabolized to the catecholestrogens 2-hydroxyestradiol and 4-hydroxyestradiol, which are capable of redox cycling.[2] This redox cycling produces oxygen radicals known to damage DNA by causing single-strand breaks and 8-hydroxylation of guanine bases.[2] These estrogen metabolites are produced in tissues susceptible to estrogen-associated cancer, leading to DNA damage and mutations. Other modified estrogens can bind to DNA to form estrogen–DNA adducts. Estrogen can also induce changes in chromosome number, structural chromosomal aberrations, microsatellite instability and deletions.[3] These two mechanisms, receptor-mediated cell proliferation and estrogen-induced tumor initiation, may not be sufficient to cause cancer, but may contribute to neoplastic transformation by contributing to an accumulation of DNA damage.

Because the involvement of estrogen and the estrogen receptor (ER) in breast cancer has been well studied, this system is used as a detailed example of the involvement of hormones in cancer. Subsequent discussion involves endometrial cancer and prostate cancer.

BREAST CANCER

Breast cancer statistics

Each year, the American Cancer Society compiles cancer statistics predictions for Americans in the coming year.[4] In 2001, it is estimated that there will be 1 268 000 new cancer cases in males and females, of which 192 200 will be females with invasive breast cancer. In females, breast cancer will account for the largest number of new cancer cases (31%), followed by lung and bronchus cancer (13%) and colon and rectum cancer (11%). One in eight women has the probability of developing breast cancer during her lifetime. In terms of cancer deaths among all age groups, the leading cause of death in women is lung and bronchus cancer (67 300 women or 25%), and the second leading cause of death is breast cancer (40 200 women or 15%). However, breast cancer is the leading cause of death among women aged 20–59 years. Although these statistics appear disheartening, the number of reported cases of breast cancer has been declining and survival rates have been increasing. Tamoxifen and early effective therapy are credited for enhanced survival.[5] In summary, breast cancer is a prevalent cancer among women that accounts for a large number of new cases and deaths each year.

Non-genetic and inherited risk factors

Many risk factors have been associated with the development of breast cancer (Table 1). The risk factors comprise both non-genetic or environmental factors as well as inherited genetic factors. The theme in all of the risk factors is that the longer a woman is exposed to endogenous ovarian hormones, the higher is her risk of developing breast cancer.

Non-genetic factors include topics ranging from reproductive issues to the environment.[6,7] The first reproductive factor is age at menarche. An early age at menarche (less than 12 years of age) is associated with an increased breast cancer risk because the exposure to endogenous hormones is greater. A shorter menstrual cycle is also associated with an increased risk because more cycles are possible throughout life. Pregnancy in general, as well as a younger age at first full-term pregnancy, also predicts a reduced risk of breast cancer, although an initial increase in risk is observed for the first 10–15 years following pregnancy. In addition, a greater number of births is related to a reduced

Table 1 Breast cancer risk factors

Factors that increase breast cancer risk
Early age at menarche
Short menstrual cycle length
Nulliparity
Spontaneous abortion
Late age at menopause
Alcohol consumption
Ionizing radiation
High socioeconomic status
Postmenopausal obesity
Hormone replacement therapy
Family history
Current use of oral contraceptives
BRCA gene mutation
Factors that decrease breast cancer risk
Younger age at first full-term pregnancy
Larger number of births
Breast-feeding (with a longer duration)
Physical activity

risk. Women who breast feed, especially for longer durations, show a decreased risk. Spontaneous abortion could also increase breast cancer risk, because the differentiation of breast cells is interrupted. Late menopause (greater than 55 years of age) is another risk factor for the development of breast cancer and, on average, the risk of breast cancer increases by 3% per year of menopausal delay.

Because of the link between endogenous levels of hormones and breast cancer risk, much discussion has surrounded the use of postmenopausal hormone replacement therapy (HRT). The age at menopause is retrospectively defined after 1 year of no menses, and the median age of menopause is 51 years. The menopause is accompanied by symptoms such as hot flushes, depression, headaches, insomnia, fluid retention, lack of energy, stiff joints and an upset stomach, among others.[8] Normal changes that occur after the menopause include an increase in cardiovascular disease and a decrease in bone density. These symptoms and changes are due to a reduction in the production of estrogen and progesterone in the ovary. HRT consists of prescribed progestins and estrogens intended to replace this decrease in ovarian hormones and alleviate menopausal symptoms. In the early years of HRT (up until the late 1970s), estrogens were usually administered alone, but current formulations consist of both estrogens and progestins. The presumed beneficial aspects of HRT are a reduction in the incidence of cardiovascular disease, increases in bone density, decreases in total cholesterol and low-density lipoprotein (LDL), and increases in high-density lipoprotein (HDL). However, recent research raises concerns about the detrimental effect of HRT for women who are at risk for coronary heart disease.[9]

Well-documented risks are also associated with HRT. In a large compilation of studies involving 52 705 women with breast cancer and 108 411 women without breast cancer, the Collaborative Group on Hormonal Factors in Breast Cancer reported that the risk of breast cancer is increased in women using HRT, and increases with longer use.[10] However, after HRT use is discontinued, the increased risk disappears after about 5 years.

Approximately 77% of women of reproductive age in the US have taken oral contraceptives.[11] The combination formulations (COCs: combined oral contraceptives), consisting of an estrogen and a progestin, are widely used and the most effective. The estrogen component can cause side-effects such as nausea, breast tenderness and fluid retention, whereas the progestin component can cause weight gain, acne and nervousness.[11] A series of studies combined by the Collaborative Group on Hormonal Factors in Breast Cancer showed a small increase in breast cancer risk while women are taking combined oral contraceptives.[12] However, 10 years after the cessation of use, there is no increased risk.

Diet has also been shown to influence breast cancer risk. For example, alcohol increases endogenous estrogen levels, thereby increasing breast cancer risk. Phytoestrogens found in soy products may reduce breast cancer risk. They bind the ER and are hypothesized to block the actions of endogenous estrogens. Physical activity has been shown to reduce breast cancer risk, most likely by reducing circulating levels of sex hormones. Physical activity is also important because weight gain is associated with an increase in estrogen production and an increase in postmenopausal breast cancer.

Environmental factors may also contribute to an increased breast cancer risk. Ionizing radiation, as evidenced by Hiroshima and Nagasaki atomic bomb survivors and radiation therapy patients, increases breast cancer risk. Higher risk is associated with a higher dose and exposure before 10 years of age.[6] Environmental pollutants, electromagnetic fields and smoking have been hypothesized to increase breast cancer risk, but conclusive studies have not been conducted. The lifetime risk of breast cancer is decreased with lower socioeconomic status, because these women are likely to have more children at a younger age. The risk of breast cancer is lower among Asian, Hispanic and American Indian women than in American white women. Interestingly, the risk of breast cancer for foreign women increases upon migration to the USA, suggesting that environmental and life-style factors influence risk.

Genetic factors also contribute to breast cancer risk. It is estimated that 5–10% of all breast cancers are attributed to inherited genetic factors.[13] One of the most studied breast cancer susceptibility genes is BRCA1 (see Chapter 6). The BRCA1 gene produces a large nuclear protein of 1863 amino acids, which contains a large number of *Alu* repeats. Most of the mutations in BRCA1 are found in the germ line, and are therefore not somatic mutations. Normal BRCA1 function is likely to involve the repair of oxidative DNA damage as well as the cellular processes of transcription, apoptosis, the cell cycle, and DNA repair and development. In support of these functions, BRCA1 has been shown to interact with a series of proteins, including RAD 51 (a protein implicated in DNA recombination and repair), p53, c-Myc, BAP-1, BARD1, retinoblastoma susceptibility gene RB1, RNA polymerase holoenzyme and the transcription factor CREB (cyclic adenosine monophosphate (cAMP) response element binding protein).[13,14] In addition, BRCA1 tumors are often ER- and progesterone receptor (PR)-negative. Surprisingly, early oophorectomy reduces breast cancer incidence in BRCA1 carriers.[15] BRCA2 is another tumor suppressor gene that is approximately twice as large as BRCA1. There are many similarities between BRCA1 and BRCA2, and the two proteins have been found to associate with each other. Further studies will determine the precise relationship between these two genes and their relationship to breast cancer.

Estrogens and breast cancer

One of the earliest reports of a relationship between breast cancer and ovarian hormones appeared in an 1896 issue of *The Lancet*.[16] A young woman with advanced breast cancer had a cancerous growth on her thorax. Eight months after removal of the ovaries, her tumor had regressed. Subsequent studies[17] showed, however, that only one in three women would respond to oophorectomy. The reason for this was to remain unknown until the discovery of the ER[18] and the subsequent development of the ER assay to predict the hormonal responsiveness of breast cancer.[19,20]

Breast cancers can be divided into two groups: those whose growth is hormone-dependent and those that are not responsive to hormones. In general, the hormone-responsive tumors are estrogen receptor positive, and these ER+ tumors represent 60–75% of all breast cancers.[21] Of the ER+ tumors, 60% respond to antiestrogen treatment, whereas 40% do not respond.[22] With time, many breast cancers progress from a hormone-responsive to a hormone-resistant state. However, because a large proportion of breast cancers are ER+, much study has focused on the estrogen receptor.

The normal roles of estrogen and the estrogen receptor

The nuclear receptor superfamily includes members such as steroid receptors (glucocorticoid receptor, androgen receptor, progesterone receptor, mineralocorticoid receptor, estrogen receptor), thyroid receptor, vitamin D receptor and retinoic acid receptor.[23,24] These receptors are transcription factors that stimulate the expression of target genes.

As a member of the nuclear receptor superfamily, the ER is involved in many functions in the body. These include the maintenance of bone density, protection against heart disease, the production of female secondary sex characteristics, menstrual cycle control, pregnancy, breast development and sexual behavior. This makes the ER an integral part of many normal physiological processes.

The first ER cloned was ERα from MCF-7 human breast cancer cells.[25] This facilitated a subsequent detailed structural and functional analysis of ERα. Using degenerate polymerase chain reaction (PCR) primers directed to the more conserved regions of the nuclear receptors (DNA- and ligand-binding domains), a second isoform, ERβ, was cloned from a rat prostate cDNA library.[26] Mouse[27] and human[28] isoforms have also been cloned. Various other isoforms of ERβ have also been identified,[29,30] and the functions of each are under investigation. Notably, the human ERα gene is located on chromosome 6, whereas the ERβ gene is localized on chromosome 14.[31] This is the first example of a steroid hormone receptor consisting of two isoforms that are encoded by separate genes.[32]

With the cloning of these genes came the ability to express the receptors in mammalian cell culture systems and analyze their functional domains, to dissect the similarities and differences between ERα and ERβ.

Comparison of ERα and ERβ structures

The ER comprises six different functional domains, A–F, which were delineated by amino acid homology and functional similarity (Figure 1). The A/B domain contains AF-1, a constitutively active ligand-independent activation function. Little homology is observed in the AF-1 domains of ERα and ERβ, suggesting that ERα and ERβ may have differing patterns of gene activation. The C

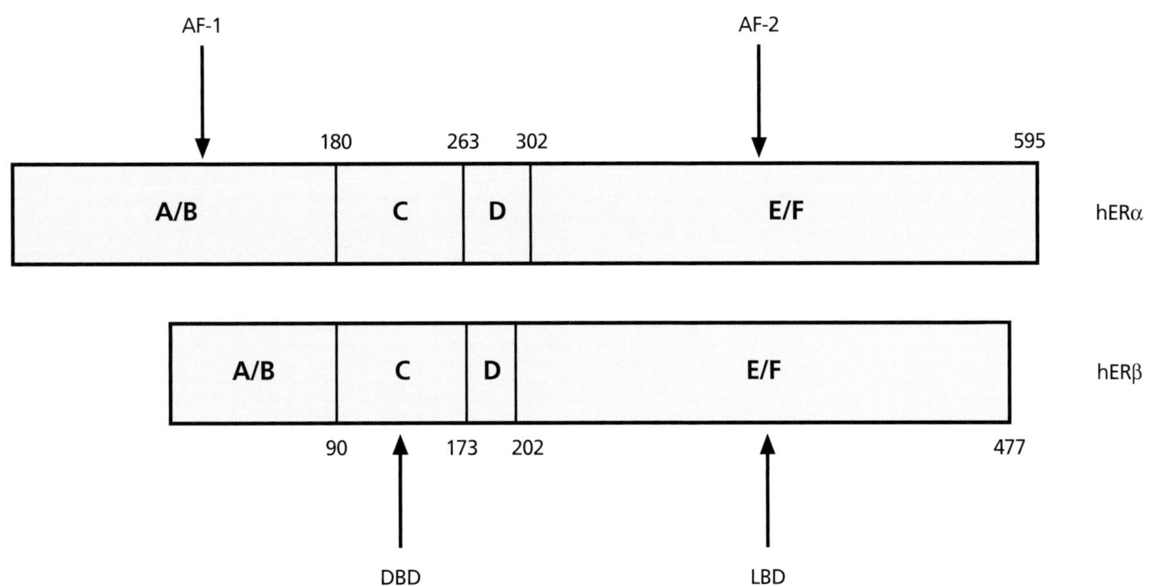

Figure 1 Structural comparison between estrogen receptors (ER) α and β. The structure of the 595-amino-acid human ERα gene (hERα) is compared with the 477-amino-acid human ERβ gene (hERβ). The positions of the two activation functions, AF-1 and AF-2, are shown, as well as the positions of the DNA-binding domain (DBD) and the ligand-binding domain (LBD)

domain is most highly conserved and constitutes the DNA-binding region. The nuclear localization signal of the ER is located in the D domain. Finally, the E/F region contains AF-2, a second activation function that is ligand-dependent. The E/F domain is also responsible for dimerization and the recruitment of co-regulators, and serves as the ligand-binding domain. Comparison between ERα and ERβ reveals a high degree of conservation in the DNA-binding domains (96%) and ligand-binding domains (58%), whereas little conservation is observed in the A/B domain.

Mechanism of ERα and ERβ actions

Both classical and non-classical pathways for gene activation have been described for the ER. In the classical model (Figure 2a), the inactive ER is part of a complex with a heat shock protein. The estrogen ligand is readily able to penetrate the plasma membrane so, upon binding of estrogen to the ER, the complex dissociates and the ER becomes active. ERα and ERβ can homodimerize or heterodimerize[33,34] and bind to a specific region of DNA termed the estrogen response element (ERE), located in the 5' untranslated region of estrogen-responsive genes. The estrogen-bound receptor then interacts with transcription factors, co-regulator proteins and components of the general transcription machinery to activate target gene expression. Genes whose expression is enhanced by estrogen and contain an ERE are vitellogenin, progesterone receptor, prolactin, pS2, c-*fos*, oxytocin, cathepsin D,[23] vascular endothelial growth factor (VEGF)[35] and lactoferrin.[36]

The second pathway of ER gene activation involves targeting genes through activator protein-1 (AP1) (Figure 2b). AP1 sites are regions of DNA that bind the transcription factors Fos and Jun.[37,38] The ER does not bind this DNA region, but interacts indirectly with Fos and Jun to activate target gene transcription. It has been shown that estrogen stimulates ERα-mediated gene transcription through AP1 elements, whereas estrogen inhibits ERβ-mediated transcription through AP1 elements.[37] Selective estrogen receptor modulators (SERMs) such as tamoxifen and raloxifene (Figure 3), or pure antiestrogens such as ICI 182 780, increase ERβ-mediated gene transcription through AP1 elements.[37]

A final pathway involves ER–Sp1 protein interactions at GC-rich promoter elements.[39] As with the AP1 pathway, the ER does not bind the GC-rich DNA region directly, but interacts with the Sp1 protein that does bind the DNA directly (Figure 2c). Other genes influenced by estrogen that do not contain an ERE include epidermal growth factor (EGF), EGF receptor (EGFR), cyclin D1 and BRCA1.[23] Regulation of these genes could occur through the AP1 or Sp1 pathway.

Therefore, many factors contribute to the molecular pharmacology of the ER, including the gene promoter, receptor subtype, receptor ligand, cell context and co-regulator proteins.[40]

Tissue distribution of ERα and ERβ

The expression patterns of ERα and ERβ differ depending on the tissue analyzed. Total RNA was isolated from 6–8-week-old male rats, whereas uterus and ovary RNA were isolated from 8-week-old female rats for reverse transcriptase (RT)-PCR analysis.[41] The highest ERα levels were detected in the epididymis, testis, uterus, ovary, kidney, adrenal and pituitary. Lower levels were observed in the prostate, bladder, liver, thymus and heart. In contrast, high ERβ levels were present in the ovary and prostate, with moderate levels in the testis, uterus, bladder and lung. Low expression was observed in the pituitary, epididymis, thymus, various brain sections and spinal cord. RT-PCR carried out on mouse mammary glands indicates moderate expression of ERα but not ERβ.[42] However, it should be noted that expression patterns among different species vary.[42] These different expression patterns of the ERs suggest tissue-specific ER functions.

Subcellular localization of ERα and ERβ

Green fluorescent protein (GFP) is a protein that contains a fluorescent chromophore,[43] so, when GFP is fused to a protein of interest, the real-time localization, independent of antibodies, can be elucidated. In addition to its use as a tag for cellular localization in living cells,[44] GFP can also be

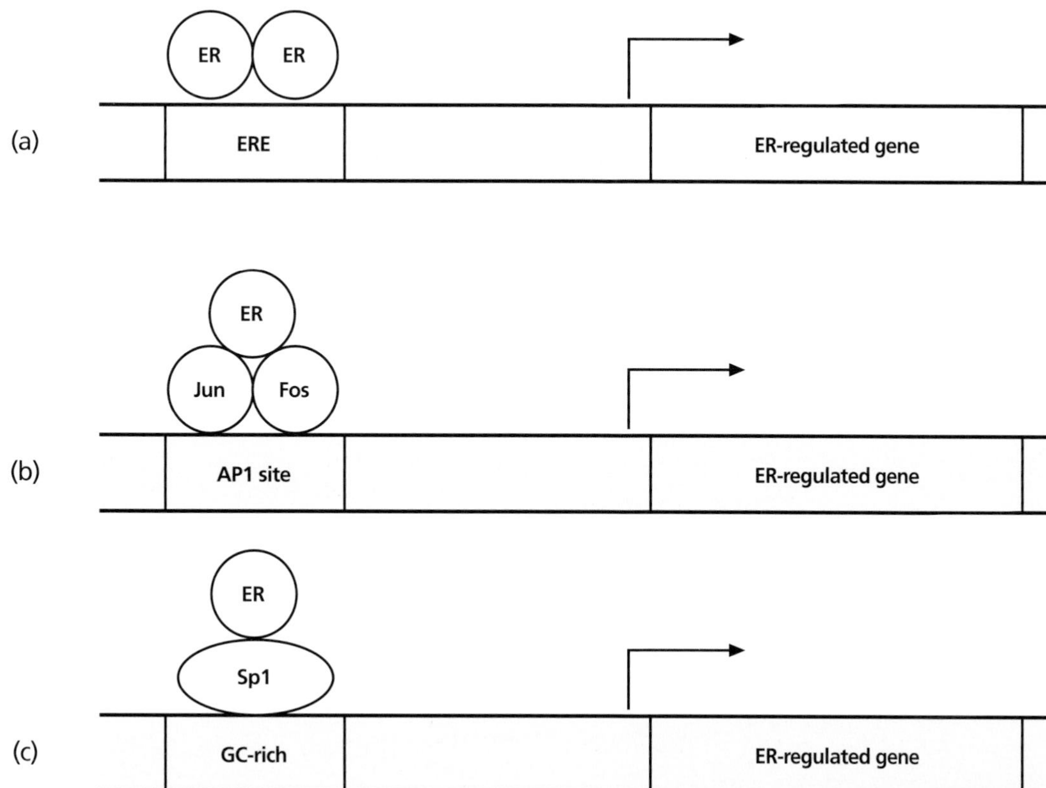

Figure 2 Pathways of estrogen receptor (ER) target gene activation. (a) The ER dimerizes and binds to the estrogen response element (ERE) to activate target gene expression in the classical pathway. (b) In a non-classical pathway, the ER interacts with the transcription factors Jun and Fos to activate gene expression thorough the activator protein-1 (AP1) DNA site. (c) In another non-classical pathway, the ER binds to the Sp1 protein to activate gene expression

used as a gene expression reporter and a cell lineage marker.

The discovery of GFP facilitated many experiments in the field of cell biology, which were also applied to the ER. Human ERα was tagged at the N-terminus with the S65T variant of GFP. Compared to wild-type GFP, this mutant has a six-fold brighter fluorescence, resists photobleaching and matures faster.[44,45] The localization was observed in a variety of human breast cancer cell lines, including the ER-positive cell lines MCF-7 and T47D, and the ER-negative cell lines MDA-MB-435A and MDA-MB-231.[46] At steady state in the absence of ligand, GFP–ERα localizes to the nucleus. Upon treatment of the cells with 17β-estradiol, the receptor redistributes within the nucleus to a more punctate localization.

Other studies provide additional insight into the subcellular localization of ERα. Upon treatment of mouse ERα-transfected African green monkey kidney cells with the pure antiestrogen ICI 182 780 (Figure 3), cytoplasmic staining was visible.[47] Using two approaches, the authors showed that the ER shuttles between the nucleus and the cytoplasm, and ICI 182 780 inhibits the nuclear uptake of the receptor.

Estrogen receptor variants and mutations

One possible mechanism of cancer development and progression could involve ER mutations; mRNA transcript analysis has shown entire exon deletions, smaller deletions, point mutations, truncations and insertions.[48] Many of these mutated

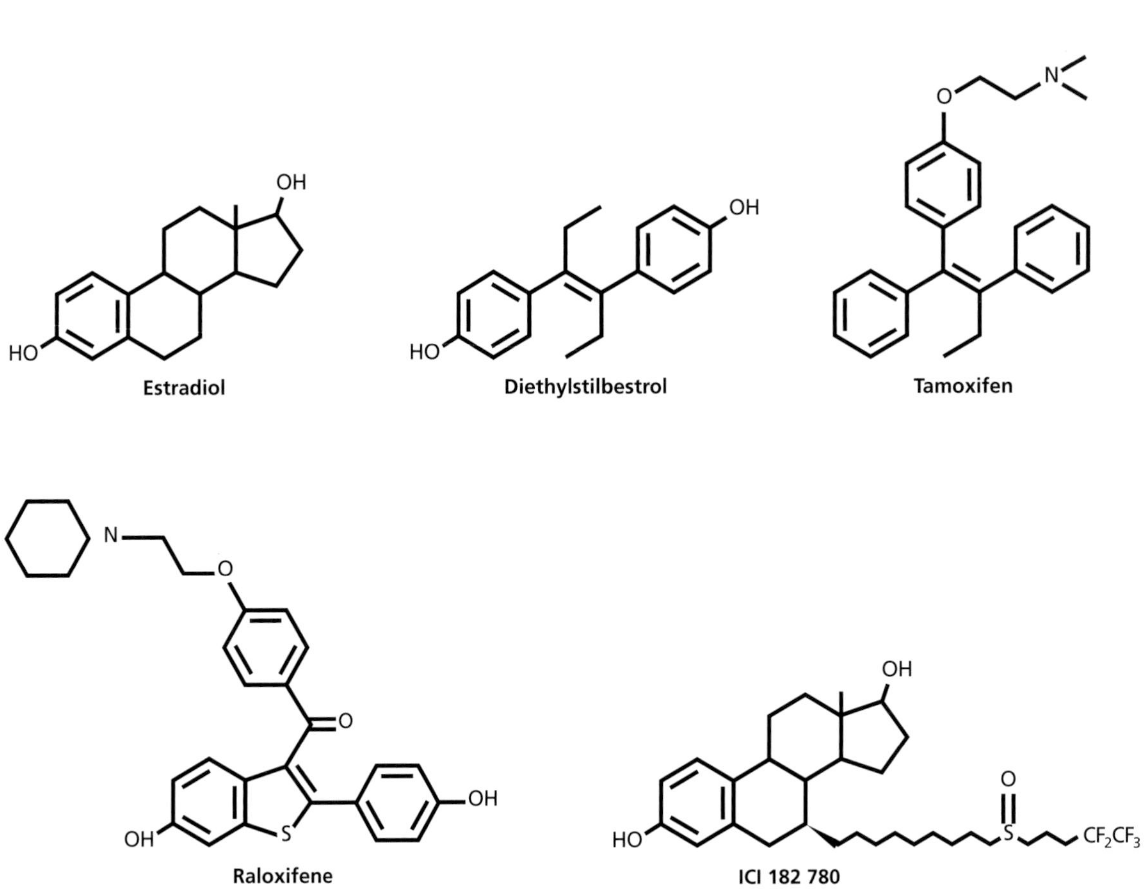

Estradiol

Diethylstilbestrol

Tamoxifen

Raloxifene

ICI 182 780

Figure 3 Structure of compounds acting at the estrogen receptor. Estradiol and diethylstilbestrol are estrogenic compounds, whereas tamoxifen and raloxifene are selective estrogen receptor modulators (SERMs). ICI 182 780 is a pure antiestrogen

transcripts are translated and expressed in normal breast tissue and in breast cancer in conjunction with the wild-type ER.[48] These variants could have a dominant negative activity or altered functions that contribute to cancer.

One example of an ERα mutation is the K303R somatic mutation found in typical breast hyperplasias.[49] This residue is located between the hinge domain and the ligand-binding domain of the ER. This mutation confers a hypersensitivity to estrogen, because stably transfected MCF-7 cells will grow faster with lower exposure to estrogen when compared with untransfected cells. The K303R mutation also enhances binding of the ER to the co-regulator transcription intermediary factor-2 (TIF-2) in the presence of lower amounts of estrogen.

Only one example of an ER mutation in humans has been described to date.[50] A man was shown to have a cytosine–thymine substitution at codon 157 in ERα, resulting in a premature stop codon. The translated protein would therefore be truncated and lack the DNA and hormone-binding regions. The patient presented with osteoporosis, unfused epiphyses and elevated serum estrogen, among other phenotypes. These results provide evidence for the critical role of estrogen in bone development and mineralization, as well as epiphyseal maturation in males.

Insight into estrogen receptor functions from knock-out mouse studies

Much of what is known about the *in vivo* functions of estrogen and the ER is a result of transgenic mouse studies. Knock-out mouse studies, where genes of interest have been deleted, have been performed to analyze the roles of ERα and ERβ in the general development and physiology of the mouse,[42,51–56] and to analyze the role of the ER in breast cancer. It is interesting to note that a loss of ERα and/or ERβ is not lethal, and the mice survive to adulthood.

The mammary glands of the adult ERα knock-out mouse (αERKO) appear essentially like those of a newborn mouse, indicating that rudimentary mammary glands can develop independent of estrogen and ERα but a fully differentiated gland requires ERα.[53] This information, combined with the high expression levels of ERα in the mammary gland, points towards ERα being the mediator of estrogen in the mammary gland. To assess the role of ERα in mammary gland carcinogenesis, αERKO mice were crossed with mouse mammary tumor virus (MMTV)-*Wnt-1* transgenic mice (*Wnt-1* TG). The *Wnt-1* transgene encodes secretory glycoproteins that stimulate cell proliferation and differentiation, resulting in mammary carcinomas.[42] Despite the lack of ERα, the αERKO/*Wnt-1* transgenic mice produced mammary tumors, although with a delayed onset.[57] This indicates that *Wnt-1* mammary tumor induction can occur in ERα-negative mammary tissue.

Treatment and prevention of breast cancer

With advances in the understanding of basic ER biology as described above, breast cancer treatments have also progressed. When used as a first-line therapy, pharmacological doses of estrogen induce remission in 30–35% of patients with advanced breast cancer.[21] Although the exact mechanism is not known, estrogen is likely to cause down-regulation of the ER and induce apoptotic mechanisms. Although the structure of diethylstilbestrol (DES) resembles that of 17β-estradiol (Figure 3), the compounds have different biological effects. DES was initially used in the 1940s to prevent adverse pregnancy outcomes such as abortion, premature delivery, hypertension and diabetes.[58] The most common side-effect of estrogen therapy is nausea, but vomiting, dizziness, headache, breast discomfort and weight gain may also be present.[59] Prenatal exposure to DES is associated with abnormalities in the vagina and cervix among women, whereas males exhibit testicular cancer and epididymal cysts.[59] High-dose DES was used effectively during the 1950s–1970s to treat certain cases of breast cancer and prostate cancer, to alleviate pain and symptoms. However, tamoxifen has almost completely replaced DES for the treatment of breast cancer, and antiandrogens are used to treat prostate cancer.

Because many of the effects of estrogen are mediated through the ER, the idea behind endocrine therapy is to reduce the synthesis of estrogen or interfere with the action of estrogen at the ER to produce antitumor effects. To reduce the amount of estrogen in the body, an ovariectomy can be performed in premenopausal women. In postmenopausal women, an adrenalectomy removes the major source of estrogen precursors. Other therapeutic approaches currently used are summarized and explained in Figure 4. Sustained-release luteinizing hormone releasing hormone (LHRH) agonists are used to suppress ovarian estrogen production in premenopausal women, and aromatase inhibitors are used to prevent the conversion of androgens (testosterone) to estrogen in postmenopausal women. Receptor antagonists, or antiestrogens, block the binding of estrogen to the ER and are therefore antagonists of estrogen action. Antiestrogens include SERMs such as tamoxifen and toremifene, and pure antiestrogens such as ICI 182 780 (Fulvestrant). SERMs block estrogen–ER interactions in specific tissues, whereas ICI 182 780 causes premature destruction of the ER. Although the use of superagonists and aromatase inhibitors is attracting much current attention in treating advanced breast cancer in pre- and postmenopausal women, there are 30 years of experience with tamoxifen.

Tamoxifen (Figure 3) was approved in 1975 by the Committee for the Safety of Medicines in the UK, and in 1977 by the US Food and Drug Administration, as the first useful antiestrogen for the treatment of advanced breast cancer in postmenopausal women.[60] The tamoxifen–ER complex exhibits target-site specificity, so it is classified as a SERM (see Chapter 19). Tamoxifen is interpreted as an estrogenic signal in tissues such as bone and liver, but as an antiestrogen in other tissues such as breast.[61] Because tamoxifen acts by

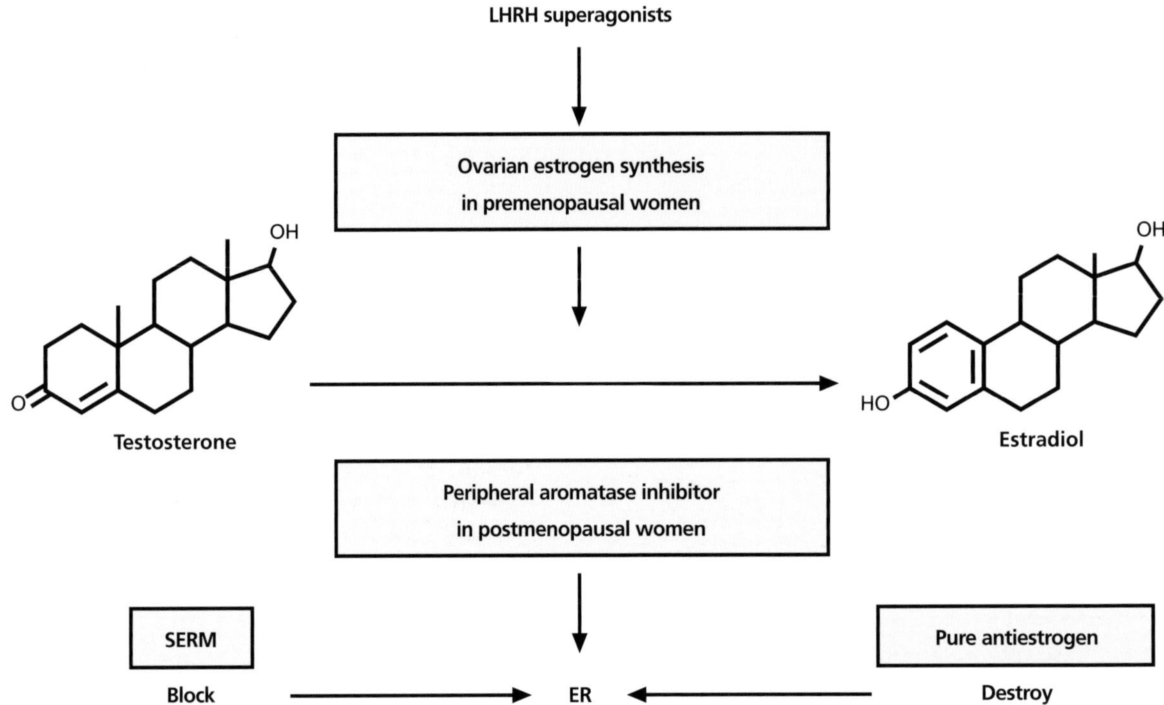

Figure 4 Multiple mechanisms of cancer therapy targeting the estrogen receptor. Luteinizing hormone releasing hormone (LHRH) superagonists interfere with ovarian estrogen synthesis in premenopausal women, whereas aromatase inhibitors block the conversion of estrogen to testosterone in postmenopausal women. This pathway interferes with the interaction of estrogen with the estrogen receptor (ER). Selective estrogen receptor modulators (SERMs) act by locking the action of estrogen at the ER, whereas pure antiestrogens act by destroying the ER

blocking the binding of estrogen to the ER, patients with ER-positive breast cancers are more likely to respond to tamoxifen treatment.

In 1998, the Early Breast Cancer Trialists' Collaborative Group compiled data from 55 individual trials involving 37 000 women.[62] This information encompassed 87% of the available data on the adjunct use of tamoxifen to treat breast cancer after surgery. After using tamoxifen for 1 year, 2 years or 5 years, there was an 18%, 25% or 42% reduction in breast cancer recurrence rate and a 10%, 15% or 22% reduction in mortality, respectively. Among women with ER-positive tumors, the reduction in recurrence rate with 1 year, 2 years or 5 years of tamoxifen was 21%, 28% or 50%, respectively, with mortality reductions of 14%, 18% or 28%. In summary, for each tamoxifen duration, there is a reduction in breast cancer recurrence rate and mortality, with an increased benefit with longer tamoxifen administration and for those women with ER-positive tumors. Approximately half of the tumors in women under age 50 years and three-quarters of the tumors in women over age 50 years are ER-positive. Therefore, in general, it is assumed that two-thirds of breast cancer tumors are ER-positive. The presence of node-positive and node-negative disease did not affect the above results.

The study also analyzed the incidence of contralateral breast cancer. The reduction in incidence rate of contralateral breast cancer for women taking tamoxifen for 1 year was 13%, for 2 years was 26% and for 5 years was 47%. Therefore, 5 years of tamoxifen therapy decreases the risk by half and is preferred for treatment.

In 1998, a study by the National Surgical Adjuvant Breast and Bowel Project (NSABP) reported an evaluation of tamoxifen as a breast cancer preventive for high-risk women.[63] A total of 13 388 pre- and postmenopausal women were randomly assigned to receive 20 mg per day of tamoxifen or placebo for 5 years. Among the tamoxifen-treated group, 124 cases of invasive and non-invasive breast cancers occurred, whereas 244 occurred in the placebo group. This represents a highly significant reduction in breast cancer risk of 50% with tamoxifen treatment. Along with the benefits of tamoxifen administration are various side-effects, the most notable being an increased risk of endometrial cancer. Participants receiving tamoxifen had a 2.53 times overall greater risk of developing an invasive endometrial cancer than did the placebo group, with the greatest risk occurring in women over age 50 years. (Tamoxifen does not increase the risk of endometrial cancer in premenopausal women.) However, there was no evidence of an increased incidence of other types of cancer with tamoxifen treatment. An increased risk of pulmonary emboli, deep vein thrombosis and cataracts was seen in women taking tamoxifen. Positive benefits of tamoxifen treatment included a decrease in cholesterol levels and fewer osteoporotic fracture events, because tamoxifen increases bone density. The conclusion of the study was that a significant benefit was afforded to high-risk women taking tamoxifen for 5 years for the prevention of breast cancer. Therefore, in 1998, tamoxifen became the first drug to be approved for the prevention of breast cancer.[64]

Raloxifene (Figure 3) is another SERM currently used for the prevention of osteoporosis. The Study of Tamoxifen and Raloxifene (STAR) trial is comparing the ability of raloxifene and tamoxifen to prevent breast cancer in high-risk postmenopausal women.[60] Preliminary results of the study are anticipated in 2006.

In summary, the antiestrogen tamoxifen interferes with the binding of estrogen to the ER in the breast. This molecular event translates into a decrease in breast cancer recurrence and mortality, as well as breast cancer prevention.

ENDOMETRIAL CANCER

In 2001, it is estimated that 38 300 new cases of endometrial cancer will be diagnosed, which represents 6% of all new cancer cases in females.[4] Endometrial cancer ranks fourth among newly diagnosed cancers, behind breast, lung and bronchus, and colon and rectum cancers. Throughout her lifetime, a woman has a 1 in 37 chance of developing endometrial cancer. It is also estimated that 6600 women (2% of the women with cancer) will die from endometrial cancer in 2001.

In general, two types of endometrial cancer have been described. Type I is associated with hyperestrogenic states, and expresses estrogen and progesterone receptors.[65] Type II is not associated with hyperestrogenic states, and functional ER is rarely expressed. ERα is expressed at higher levels than ERβ in normal and malignant endometrium. As a tumor progresses, ERα levels decrease, whereas ERβ levels remain constant.[65] This suggests that the ratio of ERα/ERβ may be associated with endometrial cancer progression.

Much like hormone-dependent breast cancer, the theory behind hormone-dependent endometrial cancer is that estrogen stimulates the mitotic activity of the endometrium, whereas progestins induce differentiation. Increased cell division increases the probability of random mutations, leading to cancer. Therefore, estrogen probably acts as a tumor promoter, rather than a carcinogen, in the endometrium.

The established risk factors in endometrial cancer are related to the fact that exposure to estrogens, unopposed by progestins, promotes endometrial cancer development.[1] In the body, this occurs during the first half of the menstrual cycle. Factors that increase the risk of endometrial cancer include late menopause, obesity, estrogen replacement therapy, early menarche, diabetes and nulliparity.[7] Protective factors include pregnancy and COCs.

In previous years, it was common to prescribe estrogens alone for postmenopausal HRT. However, after 10 years of unopposed estrogen intake, the risk of endometrial cancer increases by ten-fold, compared with women not taking estrogens.[7,66] Subsequently, estrogens were combined with progestins to prevent this increased risk. In addition, COCs have been shown to reduce the risk of ovarian and endometrial cancers. A 40% reduction in risk occurs after a few years of COC use, and the decrease in risk persists for many years, even after the use of contraceptives stops.[7]

Endometrial cancer treatment includes surgery, radiation and systemic treatments such as endocrine therapy. Although tamoxifen increases the risk of endometrial cancer in postmenopausal women, the risk is more evident in those women with a high body mass index and who have had prior estrogen replacement therapy.[67]

PROSTATE CANCER

Prostate cancer is predicted to account for 198 100 new cases in 2001, representing 31% of the newly diagnosed cancers in men.[4] This makes prostate cancer the most common cancer in men. These statistics are comparable to those for breast cancer in women. Men have a 1 in 6 chance of developing prostate cancer throughout their lifetime. Some 31 500 men are expected to die from prostate cancer in 2001, ranking prostate cancer as the second leading cause of cancer death among men, with lung and bronchus cancer being the leading cause.

Much like estrogens in breast cancer, androgens have a role in promoting the growth of the prostate gland and a role in prostate cancer.[68] Testosterone is synthesized and secreted by the testes in response to luteinizing hormone and is metabolized to dihydrotestosterone (DHT), the active form of the hormone. DHT stimulates cell proliferation and inhibits programmed cell death, which can lead to tumor growth. In general, prostate cancers are initially hormone-dependent and progress to a hormone-independent state.

The approaches to treating prostate cancer are similar to those for breast cancer, in that therapy aims to reduce the amount of androgen or interfere with the action of androgen at the androgen receptor (AR). Orchiectomy is an effective means of eliminating testicular androgen secretion. One early treatment strategy for prostate cancer was the use of DES (Figure 3). DES inhibits the release of LHRH from the hypothalamus, but it has negative cardiovascular side-effects at moderate to high doses.[68] Current methods of treatment include LHRH analogs, which are designed to stop LHRH secretion and androgen production. The steroidal antiandrogen cyproterone acetate (CPA) and the non-steroidal antiandrogens flutamide, nilutamide and biclutamide are used to block the interaction of androgen with the AR. Total androgen blockade methods are designed to block both testicular and adrenal sources of androgen, and include a combination of the above treatments.

Because androgens interact with the AR, current research studies are focused on the molecular actions of the AR complex. The frequency of mutations in primary prostate cancers is low, whereas many mutations are found in metastatic prostate cancers.[69] Many of these mutations are point mutations in the DNA-binding domain and ligand binding domain of the AR.[70] Interestingly, certain mutations in the AR render prostate cancer cells sensitive to estrogen, androgen and antiandrogens. For example, the LNCaP prostate cancer cell line contains a threnine–alanine mutation at amino acid 868.[71] HeLa cells transfected with wild-type AR showed a low response to estrogen and CPA at an androgen-regulated reporter gene, whereas an increased response was seen with the mutated AR in the presence of estrogen, CPA and hydroxyflutamide.[71–73] In addition, hydroxyflutamide, CPA and nilutamide stimulated the growth of LNCaP cells at higher doses.[73] This mutation has been reported in patients with prostate cancer,[74] and is important because these antiandrogens could stimulate the growth of prostate tumors containing specific mutations.[74]

Although androgens are largely implicated in prostate cancer, estrogens may also play a role. ERβ was first cloned from a rat prostate cDNA library,[26] and is expressed at high levels in the prostate.[41] Specifically, ERβ localizes to epithelial cells in the human prostate, whereas ERα localizes to the stromal compartment.[75] ERβ knock-out male mice (βERKO) showed epithelial hyperplasia in the collecting duct of the prostate,[52] so ERβ may protect against abnormal growth. The normal role of ERβ in the prostate may also involve the induction of genes in the antioxidant pathway that detoxify substrates.[30] A potential role for ERα was also elucidated using a rodent neonatal estrogenization model.[76] In this model, the administration of estrogens to developing neonatal mice predisposed the mice to prostate tumor formation later in life. ERα and ERβ knock-out mice showed that ERα is involved in mediating the estrogen imprint on the prostate.[76] If ER-mediated events contribute to prostate cancer, antiestrogen therapy may have clinical benefits.

In addition to dissecting the role of the ER in prostate cancer and elucidating the signal transduction mechanisms involved, future directions in prostate cancer research include the development of therapies targeting the 20–30% of patients[75] who do not respond to endocrine treatment.

SUMMARY

The link between hormones and cancer has been confirmed throughout the 20th century. Knowledge of the similar molecular mechanisms of hormone action in breast cancer, endometrial cancer and prostate cancer has not only provided an insight into the regulation of normal physiology, but also provided clues to design therapies for the treatment and prevention of cancer. For example, the use of tamoxifen as the endocrine therapy of choice presents the oncologist with a safe and effective option to treat breast cancer. Nevertheless, the small risk of endometrial cancer in postmenopausal women, and the increased risk of thromboembolic events, have resulted in the development of aromatase inhibitors to block estrogen synthesis and pure antiestrogens that have no estrogen-like effects (Figures 3 and 4). Clinical testing and the broad application of these new agents will provide the oncologist with improved treatments with few side-effects.

REFERENCES

1. Henderson BE, Feigelson HS. Hormonal carcinogenesis. *Carcinogenesis* 2000 21: 427–433.
2. Liehr JG. Role of DNA adducts in hormonal carcinogenesis. *Regul Toxicol Pharmacol* 2000 32: 276–282.
3. Liehr JG. Is estradiol a genotoxic mutagenic carcinogen? *Endocr Rev* 2000 21: 40–54.
4. Greenlee RT, Hill-Harmon MB, Murray T, Thun M. Cancer Statistics, 2001. *CA* 2001 51: 15–36.
5. Peto R, Boreham J, Clarke M, Davies C, Beral V. UK and USA breast cancer deaths down 25% in year 2000 at ages 20–69 years. *Lancet* 2000 355: 1822.
6. Willett WC, Rockhill B, Hankinson SE, Hunter DJ, Colditz GA. Epidemiology and nongenetic causes of breast cancer. In Harris JR, Lippman ME, Morrow MD, Osborne CK, eds. *Diseases of the Breast*. Philadelphia: Lippincott Williams & Wilkins, 2000: 175–220.
7. Persson I. Estrogens in the causation of breast, endometrial and ovarian cancers – evidence and hypotheses from epidemiological findings. *J Steroid Biochem Mol Biol* 2000 74: 357–364.
8. Odell WD, Burger HG. Menopause and hormone replacement. In DeGroot LJ, Jameson JL, eds. *Endocrinology*. Philadelphia: WB Saunders Company, 2001: 2153–2162.
9. Hulley S, Grady D, Bush T *et al.* Randomized trial of estrogen plus progestin for secondary prevention of coronary heart disease in postmenopausal women. Heart and Estrogen/progestin Replacement Study (HERS) Research Group. *J Am Med Assoc* 1998 280: 605–613.
10. Collaborative Group on Hormonal Factors in Breast Cancer. Breast cancer and hormone replacement therapy: collaborative reanalysis of data from 51 epidemiological studies of 52 705 women with breast cancer and 108 411 women without breast cancer. *Lancet* 1997 350: 1047–1059.
11. Mishell DR. Contraception. In DeGroot LJ, Jameson JL, eds. *Endocrinology*. Philadelphia: WB Saunders Company, 2001: 2163–2171.
12. Collaborative Group on Hormonal Factors in Breast Cancer. Breast cancer and hormonal contraceptives: collaborative reanalysis of individual data on 53 297 women with breast cancer and 100 239 women without breast cancer from 54 epidemiological studies. *Lancet* 1996 347: 1713–1727.
13. DeMichele A, Weber BL. Inherited genetic factors. In Harris JR, Lippman ME, Morrow MD, Osborne CK, eds. *Diseases of the Breast*. Philadelphia: Lippincott Williams & Wilkins, 2000: 221–236.
14. Hilakivi-Clarke L. Estrogens, BRCA1, and breast cancer. *Cancer Res* 2000 60: 4993–5001.

15. Rebbeck TR, Levin AM, Eisen A *et al.* Breast cancer risk after bilateral prophylactic oophorectomy in BRCA1 mutation carriers. *J Natl Cancer Inst* 1999 91: 1475–1479.
16. Beatson G. On the treatment of inoperable cases of the carcinoma of the mamma: suggestions for a new method of treatment, with illustrative cases. *Lancet* 1896 2: 104–107.
17. Boyd S. On oophorectomy in cancer of the breast. *Br Med J* 1900 2: 1161–1167.
18. Jensen EV, Jacobson HI. Basic guides to the mechanism of estrogen action. *Rec Prog Horm Res* 1962 18: 387–414.
19. Jensen EV, Block GE, Smith S, Kyser K, DeSombre ER. Estrogen receptors and breast cancer response to adrenalectomy. *Natl Cancer Inst Monogr* 1971 34: 55–70.
20. McGuire WL, Carbone PP, Vollmer EP. Estrogen receptors in human breast cancer: an overview. In McGuire WL, Carbone PP, Sears ME, Escher GC, eds. *Estrogen Receptors in Human Breast Cancer*. New York: Raven Press, 1975: 1–7.
21. Miller WR. *Estrogen and Breast Cancer*. Austin TX: RG Landes Company, 1996.
22. Pardee AB, Ford HL, Biswas DK, Martin KJ, Sager R. Expression genetics of hormone dependent human tumors. In Li JL, Li SA, Daling JR, eds. *Hormonal Carcinogenesis III*. New York: Springer-Verlag, 2001: 37–43.
23. Muramatsu M, Inoue S. Estrogen receptors: how do they control reproductive and nonreproductive functions? *Biochem Biophys Res Commun* 2000 270: 1–10.
24. Whitfield GK, Jurutka PW, Haussler CA, Haussler MR. Steroid hormone receptors: evolution, ligands, and molecular basis of biologic function. *J Cell Biochem* 1999 32/33(Suppl): 110–122.
25. Greene GL, Gilna P, Waterfield M *et al.* Sequence and expression of human estrogen receptor complementary DNA. *Science* 1986 231: 1150–1154.
26. Kuiper GG, Enmark E, Pelto-Huikko M, Nilsson S, Gustafsson JA. Cloning of a novel receptor expressed in rat prostate and ovary. *Proc Natl Acad Sci USA* 1996 93: 5925–5930.
27. Tremblay GB, Tremblay A, Copeland NG *et al.* Cloning, chromosomal localization, and functional analysis of the murine estrogen receptor β. *Mol Endocrinol* 1997 11: 353–365.
28. Mosselman S, Polman J, Dijkema R. ERβ: identification and characterization of a novel human estrogen receptor. *FEBS Lett* 1996 392: 49–53.
29. Ogawa S, Inoue S, Watanabe T *et al.* The complete primary structure of human estrogen receptor β (hERβ) and its heterodimerization with ERα *in vivo* and *in vitro*. *Biochem Biophys Res Commun* 1998 243: 122–126.
30. Chang WY, Prins GS. Estrogen receptor-β: implications for the prostate gland. *Prostate* 1999 40: 115–124.
31. Enmark E, Pelto-Huikko M, Grandien K *et al.* Human estrogen receptor β-gene structure, chromosomal localization, and expression pattern. *J Clin Endocrinol Metab* 1997 82: 4258–4265.
32. Gustafsson JA. Estrogen receptor β – a new dimension in estrogen mechanism of action. *J Endocrinol* 1999 163: 379–383.
33. Cowley SM, Hoare S, Mosselman S, Parker MG. Estrogen receptors α and β form heterodimers on DNA. *J Biol Chem* 1997 272: 19858–19862.
34. Pace P, Taylor J, Suntharalingam S, Coombes RC, Ali S. Human estrogen receptor β binds DNA in a manner similar to and dimerizes with estrogen receptor α. *J Biol Chem* 1997 272: 25832–25838.
35. Hyder SM, Nawaz Z, Chiappetta C, Stancel GM. Identification of functional estrogen response elements in the gene coding for the potent angiogenic factor vascular endothelial growth factor. *Cancer Res* 2000 60: 3183–3190.
36. Liu Y, Teng CT. Estrogen response module of the mouse lactoferrin gene contains overlapping chicken ovalbumin upstream promoter transcription factor and estrogen receptor-binding elements. *Mol Endocrinol* 1992 6: 355–364.
37. Paech K, Webb P, Kuiper GG *et al.* Differential ligand activation of estrogen receptors ERα and ERβ at AP1 sites. *Science* 1997 277: 1508–1510.
38. Kushner PJ, Agard DA, Greene GL *et al.* Estrogen receptor pathways to AP-1. *J Steroid Biochem Mol Biol* 2000 74: 311–317.
39. Saville B, Wormke M, Wang F *et al.* Ligand-, cell-, and estrogen receptor subtype (α/β)-dependent activation at GC-rich (Sp1) promoter elements. *J Biol Chem* 2000 275: 5379–5387.
40. Katzenellenbogen BS, Katzenellenbogen JA. Estrogen receptor α and estrogen receptor β: regulation by selective estrogen receptor modulators and importance in breast cancer. *Breast Cancer Res* 2000 2: 335–344.

41. Kuiper GG, Carlsson B, Grandien K *et al.* Comparison of the ligand binding specificity and transcript tissue distribution of estrogen receptors α and β. *Endocrinology* 1997 138: 863–870.
42. Couse JF, Korach KS. Estrogen receptor null mice: what have we learned and where will they lead us? *Endocr Rev* 1999 20: 358–417.
43. Prasher DC. Using GFP to see the light. *Trends Genet* 1995 11: 320–323.
44. Gerdes HH, Kaether C. Green fluorescent protein: applications in cell biology. *FEBS Lett* 1996 389: 44–47.
45. Cubitt AB, Heim R, Adams SR *et al.* Understanding, improving and using green fluorescent proteins. *Trends Biochem Sci* 1995 20: 448–455.
46. Htun H, Holth LT, Walker D, Davie JR, Hager GL. Direct visualization of the human estrogen receptor α reveals a role for ligand in the nuclear distribution of the receptor. *Mol Biol Cell* 1999 10: 471–486.
47. Dauvois S, White R, Parker MG. The antiestrogen ICI 182 780 disrupts estrogen receptor nucleocytoplasmic shuttling. *J Cell Sci* 1993 106: 1377–1388.
48. Murphy LC, Dotzlaw H, Leygue E *et al.* Estrogen receptor variants and mutations. *J Steroid Biochem Mol Biol* 1997 62: 363–372.
49. Fuqua SA, Wiltschke C, Zhang QX *et al.* A hypersensitive estrogen receptor-α mutation in premalignant breast lesions. *Cancer Res* 2000 60: 4026–4029.
50. Smith EP, Boyd J, Frank GR *et al.* Estrogen resistance caused by a mutation in the estrogen-receptor gene in a man. *N Engl J Med* 1994 331: 1056–1061.
51. Couse JF, Hewitt SC, Bunch DO *et al.* Postnatal sex reversal of the ovaries in mice lacking estrogen receptors α and β. *Science* 1999 286: 2328–2331.
52. Krege JH, Hodgin JB, Couse JF *et al.* Generation and reproductive phenotypes of mice lacking estrogen receptor β. *Proc Natl Acad Sci USA* 1998 95: 15677–15682.
53. Lubahn DB, Moyer JS, Golding TS *et al.* Alteration of reproductive function but not prenatal sexual development after insertional disruption of the mouse estrogen receptor gene. *Proc Natl Acad Sci USA* 1993 90: 11162–11166.
54. Ogawa S, Chan J, Chester AE *et al.* Survival of reproductive behaviors in estrogen receptor β gene-deficient (βERKO) male and female mice. *Proc Natl Acad Sci USA* 1999 96: 12887–12892.
55. Ogawa S, Chester AE, Hewitt SC *et al.* From the cover: abolition of male sexual behaviors in mice lacking estrogen receptors α and β (α β ERKO). *Proc Natl Acad Sci USA* 2000 97: 14737–14741.
56. Ogawa S, Lubahn DB, Korach KS, Pfaff DW. Behavioral effects of estrogen receptor gene disruption in male mice. *Proc Natl Acad Sci USA* 1997 94: 1476–1481.
57. Bocchinfuso WP, Hively WP, Couse JF, Varmus HE, Korach KS. A mouse mammary tumor virus-Wnt-1 transgene induces mammary gland hyperplasia and tumorigenesis in mice lacking estrogen receptor-α. *Cancer Res* 1999 59: 1869–1876.
58. Herbst AL. Behavior of estrogen-associated female genital tract cancer and its relation to neoplasia following intrauterine exposure to diethylstilbestrol (DES). *Gynecol Oncol* 2000 76: 147–156.
59. Marselos M, Tomatis L. Diethylstilboestrol: I. Pharmacology, toxicology and carcinogenicity in humans. *Eur J Cancer* 1992 28A: 1182–1189.
60. Jordan VC, Morrow M. Tamoxifen, raloxifene, and the prevention of breast cancer. *Endocr Rev* 1999 20: 253–278.
61. Jordan V. Tamoxifen: a personal retrospective. *Lancet Oncol* 2000 1: 43–49.
62. Early Breast Cancer Trialists' Collaborative Group. Tamoxifen for early breast cancer: an overview of the randomised trials. *Lancet* 1998 351: 1451–1467.
63. Fisher B, Costantino JP, Wickerham DL *et al.* Tamoxifen for prevention of breast cancer: report of the National Surgical Adjuvant Breast and Bowel Project P-1 Study. *J Natl Cancer Inst* 1998 90: 1371–1388.
64. Jordan VC. Progress in the prevention of breast cancer: concept to reality. *J Steroid Biochem Mol Biol* 2000 74: 269–277.
65. Emons G, Fleckenstein G, Hinney B, Huschmand A, Heyl W. Hormonal interactions in endometrial cancer. *Endocr Relat Cancer* 2000 7: 227–242.
66. Greenwald P, Caputo TA, Wolfgang PE. Endometrial cancer after menopausal use of estrogens. *Obstet Gynecol* 1977 50: 239–243.
67. Bernstein L, Deapen D, Cerhan JR *et al.* Tamoxifen therapy for breast cancer and

endometrial cancer risk. *J Natl Cancer Inst* 1999 91: 1654–1662.

68. Garnick MB. Hormonal therapy in the management of prostate cancer: from Huggins to the present. *Urology* 1997 49: 5–15.
69. Avila DM, Zoppi S, McPhaul MJ. The androgen receptor (AR) in syndromes of androgen insensitivity and in prostate cancer. *J Steroid Biochem Mol Biol* 2001 76: 135–142.
70. Buchanan G, Tilley WD. Androgen receptor structure and function in prostate cancer. In Li JL, Li SA, Daling JR, eds. *Hormonal Carcinogenesis III*. New York: Springer-Verlag, 2001: 333–341.
71. Veldscholte J, Ris-Stalpers C, Kuiper GG *et al.* A mutation in the ligand binding domain of the androgen receptor of human LNCaP cells affects steroid binding characteristics and response to anti-androgens. *Biochem Biophys Res Commun* 1990 173: 534–540.
72. Veldscholte J, Berrevoets CA, Brinkmann AO, Grootegoed JA, Mulder E. Anti-androgens and the mutated androgen receptor of LNCaP cells: differential effects on binding affinity, heat-shock protein interaction, and transcription activation. *Biochemistry* 1992 31: 2393–2399.
73. Veldscholte J, Berrevoets CA, Ris-Stalpers C *et al.* The androgen receptor in LNCaP cells contains a mutation in the ligand binding domain which affects steroid binding characteristics and response to antiandrogens. *J Steroid Biochem Mol Biol* 1992 41: 665–669.
74. Fenton MA, Shuster TD, Fertig AM *et al.* Functional characterization of mutant androgen receptors from androgen-independent prostate cancer. *Clin Cancer Res* 1997 3: 1383–1388.
75. Denis LJ, Griffiths K. Endocrine treatment in prostate cancer. *Semin Surg Oncol* 2000 18: 52–74.
76. Prins GS, Birch L, Couse JF *et al.* Estrogen imprinting of the developing prostate gland is mediated through stromal estrogen receptor α: studies with αERKO and βERKO mice. *Cancer Res* 2001 61: 6089–6097.

Chapter 9

Immunogenetics: genetic regulation of immunity in pregnancy

Joan S. Hunt

INTRODUCTION

The immune system is designed to maintain the integrity of the individual by preventing the incursion of foreign DNA/RNA. In mammalian pregnancy, the fetus is internalized in the uterus, resulting in an intimate juxtapositioning of maternal (decidua) and fetal (placenta, chorion membrane) tissues during pregnancy (Figure 1). The mother's immune system must be silenced or bypassed for pregnancy to go forward, because molecules on fetal cells derived from the father's genes would be perceived by the mother as foreign; the mother's reaction would then be to mount an immune graft rejection response. Fortunately, evolution has designed effective mechanisms for preventing this major disaster.

Mothers and babies co-operate to create a safe environment (Figure 2). In the uterus, leukocyte populations are re-assorted such that only certain types of leukocytes are allowed access during pregnancy. Furthermore, these cells produce a new spectrum of soluble molecules at the maternal–fetal interface that benefits rather than endangers semi-allogeneic pregnancy. This revision of the uterine climate is due to environmental programming by pregnancy hormones and the cytokines they stimulate.

In the embryo/fetus, trophoblast cells derived from the trophectoderm layer of the blastocyst form the placenta and chorion membrane. These extraembryonic tissues encase the developing embryo entirely, and therefore bear the major responsibility for protecting the fetal semi-allograft from the maternal immune response. In trophoblast cells, the expression of cell surface molecules, which control the activity of immune cells and immunological molecules, is tightly regulated. Underlying their unique patterns of expression are two powerful genetic mechanisms, i.e. developmental programming and gene selection.

This chapter first outlines the elements of the immune system and the changes that occur during

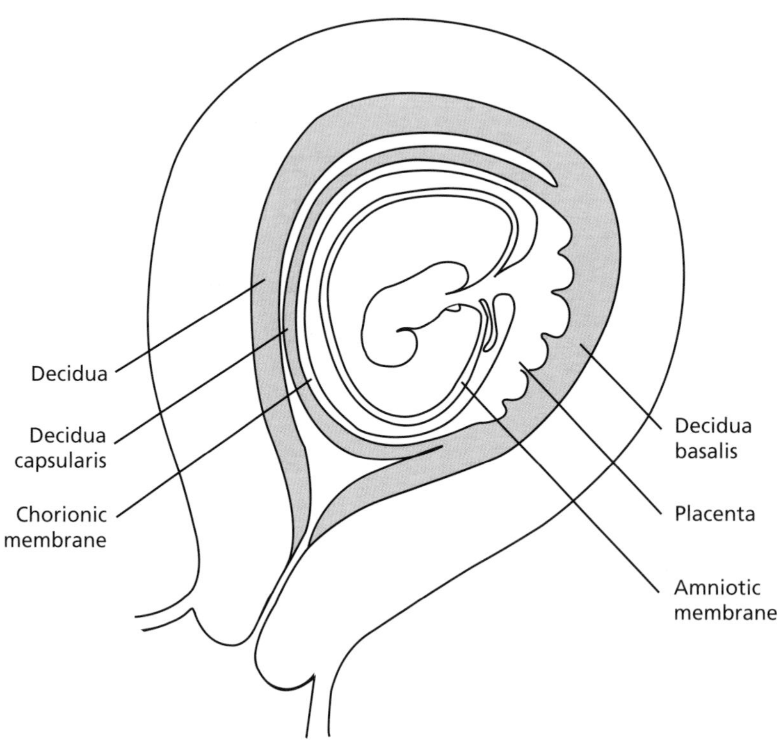

Figure 1 Schematic drawing of the pregnant uterus showing positioning of the maternal tissues, decidua basalis and decidua capsularis (shaded) in relation to the fetal tissues, placenta and chorionic membrane

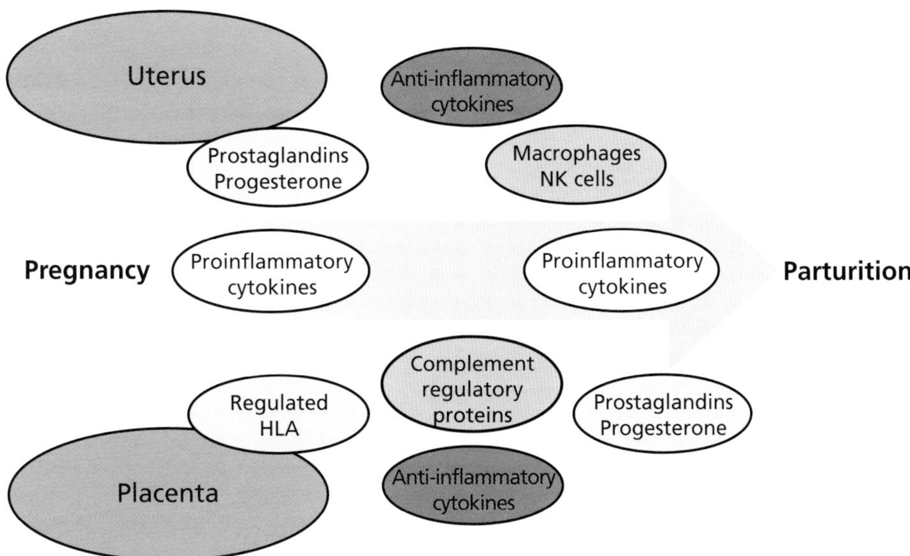

Figure 2 Representation of the postulate that both the mother (uterus) and the fetally derived extraembryonic membranes (placenta) contribute to immunomodulation at the maternal–fetal interface. HLA, human leukocyte antigen; NK, natural killer

pregnancy, then discusses the human leukocyte antigen (HLA) system, where there is strong experimental evidence for genetic programming that influences the immunobiology of pregnancy. The phrase 'genetic programming' is meant to convey the idea that a predictable, cell- and/or tissue-specific pattern of gene expression prevails throughout embryonic and fetal development into the adult animal. Final comments are made regarding other genetic programs at the maternal–fetal interface, and the likelihood that these might constitute risks to a successful pregnancy.

BACKGROUND

Two basic types of immunity: innate immunity and acquired immunity

The two fundamentally different types of immunity, innate and acquired,[1,2] are relevant to studies of the maternal–fetal interface. Innate immunity is a feature of pregnancy endometrium (decidua), which is invaded by migrating trophoblast cells. Acquired immunity may be encountered by trophoblast cells in the floating villi, which are continuously bathed by maternal blood. By studying the features of immunity, possibilities for learning about modulations of gene expression to overcome the genetic barrier of semi-allogeneic pregnancy may be identified.

Cellular elements of innate immunity and relationships with pregnancy

When implantation occurs, inflammation ensues. Resident leukocytes become activated, and more phagocytic cells are recruited to the site by chemical mediators termed chemokines.[3] Most of these cells, which are mainly polymorphonuclear leukocytes (neutrophils, eosinophils), are short-lived and die within 1–3 days. Studies in mice have shown that this phase of pregnancy is characterized by proinflammatory cytokines such as tumor necrosis factor-α (TNF-α), interleukin-1 (IL-1) and interferon-γ (IFN-γ). Inflammation dies down and the inflammation-associated cytokines diminish within a very few days.

Innate immunity characterizes the pregnant uterus throughout most of gestation. Although the endometrium of the cycling uterus contains all of the usual elements of mucosal acquired immunity (T and B lymphocytes, macrophages, precursor natural killer (NK) cells, mast cells), the T and B lymphocytes are generally excluded from the decidua. Instead, the decidua is populated by comparatively long-lived leukocytes of the innate immune system, primarily macrophages and NK-like cells.[4] In women, the macrophages stay in place throughout pregnancy, whereas the NK cells disappear during the second trimester. These NK cells are a selected subset; they lack the usual CD16 marker of blood NK cells and, instead, most often express high levels of neural cell adhesion molecules (NCAMs) (CD56).[5] Other types of lymphocytes have been reported in the human pregnant endometrium, including T lymphocytes with a mucosal type of antigen receptor called the γ/δ T cell receptor (TcR) and NKT lymphocytes that are intermediate to NK and T cells.[6,7]

It is the macrophages and NK cells whose recognition of the invading trophoblast is most important in the decidua. Both have well-established mechanisms for killing unwanted cells. As described below, it appears that several uterine and trophoblast products, including HLA class Ib molecules, may prevent these cells from killing trophoblasts within and adjacent to the decidua.[8,9]

Acquired, antigen-specific immunity and relationships with pregnancy

It is critical to understand the fairly complex components of the acquired immune system, because fetal trophoblast cells in floating villi of the placenta are continuously bathed in maternal blood containing both naive and antigen-specific immune cells, which might attack. Furthermore, some trophoblast cells detach from the placenta and circulate in mothers or lodge in maternal tissues, where they might serve as inflammatory or immunogenic stimuli.[10]

T lymphocytes Acquired immunity is a specific property of lymphocytes.[1,2] The ability to respond in a highly specific manner to antigens is conferred by cell surface receptors, which are different for the two main types of lymphocytes, T lymphocytes and B lymphocytes. T lymphocytes carry antigen-specific TcR, and B lymphocytes bear membrane-bound antigen-specific receptors that are antibody molecules. Both TcR and antibodies are extremely diverse, and confer the ability to recognize at least 10^6 different antigens.

T cells are subdivided into two major categories, T helper lymphocytes (Th, CD4+) and cytotoxic T lymphocytes (CTLs, also known as T cytotoxic/suppressor cells or Tc/s, CD8+). The Th cells come in two varieties: Th1 cells facilitate the activation of antigen-presenting cells (APCs) and development of CTLs; Th2 cells facilitate the development of B cells. These two functionally distinct subsets of T cells produce different profiles of cytokines. Th1 cells typically produce interleukin-2 (IL-2), lymphotoxin-α (LT-α, also known as tumor necrosis factor-β) and IFN-γ, whereas Th2 cells typically produce IL-3, IL-4 and IL-10. There is considerable cross-over of cytokines in the human Th subset, and there are other distinct subsets such as T_0 and NKT, with even less well-defined cytokine production patterns.

Figure 3 is a simplified diagram showing that Th cells are central to the development of both arms of the immune response. The two functional subsets of Th cells, Th1 and Th2, specifically drive one or another arm. Th1-type cytokines drive cell-mediated immunity, and Th2-type cytokines drive humoral immunity. These cytokines have reciprocal negative feedback pathways; for example, the Th2-type cytokine, IL-10, inhibits production of the Th1-type cytokine, TNF-α and vice versa.

T helper cells do not kill; cytotoxicity is an exclusive property of CTLs. CTLs utilize LT-α and other toxic members of the TNF supergene family,[11,12] as they recognize foreign entities on cell surfaces and direct killing in a cell : cell manner through death domain-containing receptors.

Pregnancy is thought to be a Th2-type phenomenon, at least in mice,[13] with down-regulation of the CTL response and up-regulation of the antibody response. In support of this idea, IL-10 and other immunosuppressive substances such as transforming growth factor-β (TGF-β) and prostaglandin E_2 have been identified in the blood of both pregnant women and pregnant rodents.

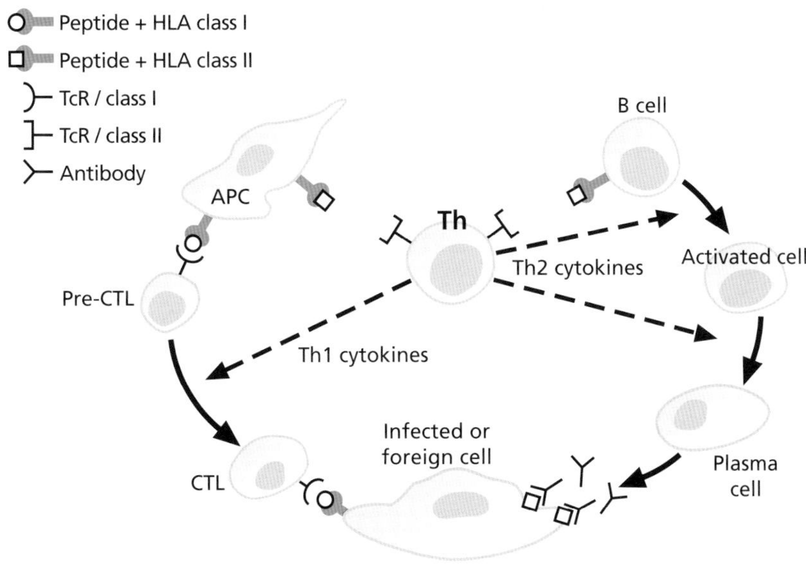

Figure 3 Schematic illustration of pathways leading to the production of killer T lymphocytes and antibody molecules. HLA, human leukocyte antigen; TcR, γ/δ T cell receptor; APC, antigen-presenting cell; CTL, cytotoxic T lymphocyte; Th, T helper lymphocyte. Reproduced from reference 9, with permission

Th1-type cytokines are not absent; they are present at low levels throughout pregnancy, and are likely to have important roles in placental development and function, since many regulate other genes.[14]

CTLs directed towards paternally derived antigens are difficult to detect in pregnant women,[15] suggesting that diversion into a Th2-type profile occurs in women. Of equal, perhaps greater importance, trophoblast cells at the maternal–fetal interface protect themselves. These unique cells down-regulate or repress synthesis of the antigens that might serve as stimulators of graft rejection, i.e. paternal HLA.[16–18] This is a clear case of genetic modification in which differential expression is achieved by permitting only certain genes to be transcribed/translated, and is widely held to be a key feature of successful pregnancy.

B lymphocytes Binding of antigen within the context of HLA class II antigens, together with cytokines provided by Th2 cells, drives a progression of events that culminates in an end-stage plasma cell producing antibodies of the same specificity as the membrane-bound receptor antibody (Figure 3). Antibodies are conveyed to the cell surface in secretory vesicles and are promptly secreted. Antibodies can neutralize bacterial exotoxins, activate complement, attract neutrophils (chemotaxis) and coat foreign cells and particles (opsonization). Opsonization promotes phagocytosis (ingestion) of the foreign material by the phagocytes.

It is the complement activation function of the antibody molecule that causes cell damage. Intriguingly, it has been known for many decades that mothers routinely synthesize antibodies to paternal HLA, but none damage the placenta. It is nearly certain that this, again, is a consequence of genetic modifications in the trophoblast cells. These specialized cells express exceptionally high levels of the complement regulatory proteins that interfere with the complement cascade, CD46 (membrane cofactor protein, MCP) and decay accelerating factor (DAF).[19,20] Recent studies in genetically deficient mice have shown clearly that, in the absence of one of the complement regulatory genes, *Crry*, the embryo is destroyed by maternal complement-fixing antibodies.[21]

Antigen-presenting cells The acquired immune response is initiated when an APC displays a peptide to a T cell within the context of class I or class II HLA (Figures 3 and 4). Several types of cells present antigens. These include the classical APCs, which are monocytes/macrophages and dendritic cells, as well as B lymphocytes. Most APCs internalize and process antigens. The processing of antigen by APCs includes endocytosis of protein, partial digestion of protein in endolysosomes into small peptides, loading of the peptides into the clefts of HLA class I and class II molecules, and transport to the cell surface. At the cell surface, the HLA–peptide complexes are recognized by either Th cells, which require peptide and HLA class II antigens, or precursor CTLs, which require peptide and HLA class I antigens (Figure 4). T lymphocytes and APCs are in close physical proximity as a consequence of the interconnections formed by the peptide-presenting complex–TcR, and also the interconnections formed by cell adhesion molecules (Figure 4). A molecule called CD80 (B7) on the APC sends an essential co-stimulatory signal to the T cell through a cell surface molecule called CD28 (Figure 4). In the absence of this signal, tolerance (no response) can develop to the peptide antigen. Other APC–lymphocyte interactions involve cell adhesion molecules and the CD4 (on Th cells) and CD8 (on CTL) structures.

The soluble inter- and intracellular communication molecules called cytokines are essential to maximum development of the immune response. IL-1 from APCs stimulates Th cells to produce an autocrine (self) growth factor, IL-2, and also to increase their expression of IL-2 receptors, thus amplifying the response. APCs are also the sources of IL-12, a potent lymphocyte stimulator, as well as IL-10, which down-regulates the immune response.

Antigen presentation in the decidua does not appear to be interrupted by genetic modifications that repress cell surface HLA in decidual APCs[22,23] or their expression of co-stimulatory molecules (Petroff and colleagues, unpublished data). However, these cells may be inhibited by soluble molecules in the uterine environment, such as progesterone, prostaglandins and perhaps also HLA-G.[8,9]

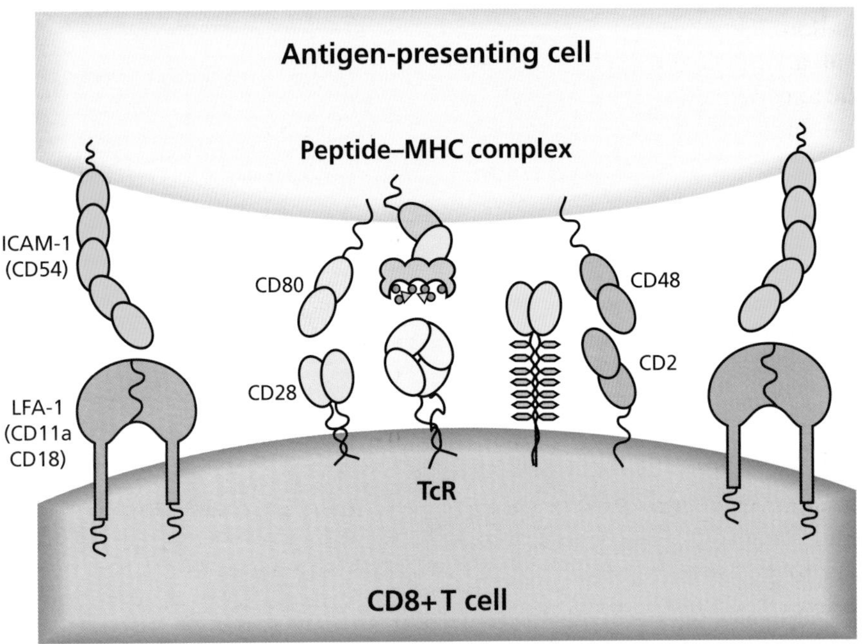

Figure 4 Antigen presentation is achieved when antigen-presenting cells and CD8+ T cells interact through peptides presented by major histocompatibility complex (MHC) antigens and the T cell receptor (TcR). Other molecules contribute to stability and specificity of the interaction. ICAM-1, intercellular adhesion molecule-1; LFA-1, lymphocyte function-associated antigen-1

HLA class I antigens

The HLA antigens are functionally unique, having both beneficial and potentially detrimental effects. Although these cell surface structures are required for host protection against infectious agents, when they are foreign to the host, as may be the case with paternal antigens in semi-allogeneic pregnancy, they are fully capable of stimulating graft rejection.

HLA class I antigens are cell surface molecules encoded within the major histocompatibility complex (MHC) located on the short arm of chromosome 6.[1,2,16] The proteins are characterized by approximately 100 amino acid 'domains' and disulfide bonds, which loop the domains. The class I antigens usually have one transmembrane heavy chain (~45 kDa) associated non-covalently with a light chain (β2-microglobulin, ~12 kDa). HLA class I glycoproteins carrying foreign peptides in the light chain–heavy chain cleft are recognized by the TcR on CD8+ precursor CTL (Figure 4). Differences between class I antigens create a major barrier associated with transplantation of donor organs, because immune cells recognize non-self alleles as foreign.

Class I molecules come in two varieties, class Ia (HLA-A, -B, -C) and class Ib (HLA-E, -F, -G). Both are expressed co-dominantly on the cell surface. That is, maternal and paternal chromosomes 6 contribute equally to cell surface expression. Class Ia heavy chains are present on essentially all nucleated cells, and are highly polymorphic. Polymorphism means that multiple alleles are present at each locus, such that the protein structures of the molecules are slightly different from allele to allele. By contrast, class Ib molecules have restricted tissue distribution and low polymorphism.

HLA class II (HLA-D) antigens

The class II antigens have two transmembrane heavy chains, are also highly polymorphic and also constitute barriers to transplantation.[1,2] They are expressed primarily on APCs and B lymphocytes, but may be expressed on other cells during certain diseases. The class II antigens form homo- rather than heterodimers. HLA class II (HLA-D) glycoproteins carrying foreign peptides in the homodimeric heavy chain cleft are recognized by the TcR on CD4+ T helper lymphocytes.

MOLECULAR REGULATION OF IMMUNITY IN PREGNANCY: HLA ANTIGENS IN TROPHOBLAST

Many multigene families undergo programming and selection for specific members during gestation. The hemoglobin genes, which have fetal and adult forms to permit appropriate oxygen binding, illustrate programming.[24] Selection is illustrated in the imprinting process, which allows preferential expression of certain paternal genes in the extraembryonic membranes.[25]

Even the puzzling immunological paradox of pregnancy may be resolved by developmental programming and gene selection. Control over the HLA antigens expressed by trophoblast cells, which is not regulated by imprinting,[26] is an excellent example, as described in detail below. These cell surface structures are the main mediators of graft rejection, but HLA is so firmly regulated that there are no known instances of inappropriate expression in fetal tissues as a cause of immune-mediated pregnancy failure.

Trophoblast subpopulations

Trophoblast cells are derived from the trophectoderm layer of the blastocyst.[27] These cells are not uniform; they differentiate in a stage-specific manner. Figure 5 shows that the cytotrophoblast is the precursor cell, and is driven into one of two pathways of differentiation. Some merge to form the multinucleated syncytium. The syncytium is an uninterrupted cell layer that, at term, would cover 9 m^2 if spread. The syncytium is responsible for bidirectional transport of nutrients and wastes, for hormone production and for protecting inner cell mass-derived cells from blood-borne maternal immune cells.

Other cytotrophoblasts proliferate and burst from the floating placental villi to form columns that contact the decidua (Figures 5 and 6). These extravillous trophoblastic cells invade into the modified endometrial stroma known as the

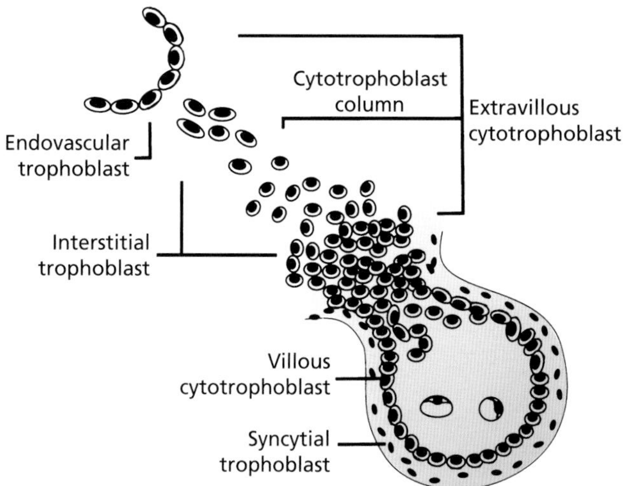

Figure 5 Schematic drawing illustrating the various trophoblast subpopulations found in the early-gestation placenta

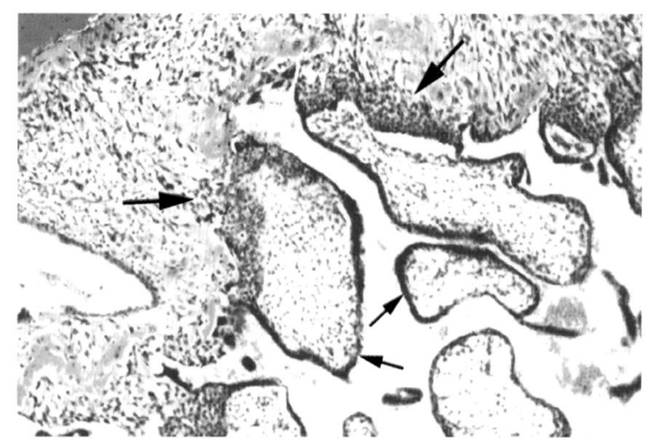

Figure 6 First-trimester placental attachment to the decidua. The photomicrograph shows both cross-sections of floating villi (small arrows) and areas where trophoblastic columns contact and infiltrate the decidua (large arrows). Hematoxylin- and eosin-stained section, original magnification ×250. From the Boyd Collection, courtesy of G. Burton, University of Cambridge. Reproduced from reference 28, with permission

decidua, migrate through the tissue, reach the spiral arteries and replace the endothelial cells in the vessels. Ultimately, the extravillous cells cease migrating and regress to form the chorion membrane as pregnancy progresses to term. Development and differentiation appear to be largely pre-programmed, with differential expression of genes being characteristic of temporal stages, but are also influenced by environmental conditions. For example, transcription factors in trophoblast are regulated by the availability of oxygen,[29] which is dramatically elevated around the placenta with development of the maternal blood lacunae during the middle to late stages of the first trimester.

Developmental programming

HLA class I antigens are weakly expressed in placental villous mesenchymal cells as early as the first trimester.[30] Their density increases towards adult levels as pregnancy progresses to term. HLA class II antigens are slower to develop, just becoming evident in the second trimester on placental macrophages. Thus, developmental programming in inner cell mass-derived cells constituting the placental stroma may help to evade the maternal response when their protective layers of trophoblast are denuded, a normal event in placentas.

In contrast to inner cell mass-derived cells, trophoblast never expresses adult levels or patterns of HLA. As described below, specific HLA class I genes are selected for expression from among the many members of this multigene family, whereas

HLA class II genes are essentially unexpressed in trophoblast. The absence of class II may be due to repressor elements on promoter and other elements in the class II gene sequences.[31] More studies are necessary to elucidate the nature of the restriction, which is effective even in the presence of class II inducers such as IFN-γ.[32]

HLA class I in trophoblast

The discovery of how the HLA genes and the glycoproteins derived from these genes are expressed in trophoblast cells is an often told story.[16–18] In summary, expression in this unique cell lineage is a function of stage of gestation, state of differentiation and anatomical location (Figure 7). Trophectoderm-derived cells that are routinely exposed to maternal blood as well as precursor cells are HLA class I-negative, whereas trophoblast cells exposed to maternal tissues are HLA class I-positive.

In early-gestation syncytiotrophoblast and villous cytotrophoblast cells, the class I message may be present, but membrane-bound proteins are invariably absent.[33] Later, neither subpopulation contains class I mRNA or protein.[34] Cytotrophoblast cells in early placentas that are proliferating in preparation for column formation are negative for both message and protein.[35] By contrast, the cytotrophoblast cells distal to the villus in the cytotrophoblastic shell are HLA class I-positive. Class I positivity is retained in the chorion membrane cytotrophoblast cells, which are derived from the invading extravillous cytotrophoblasts.

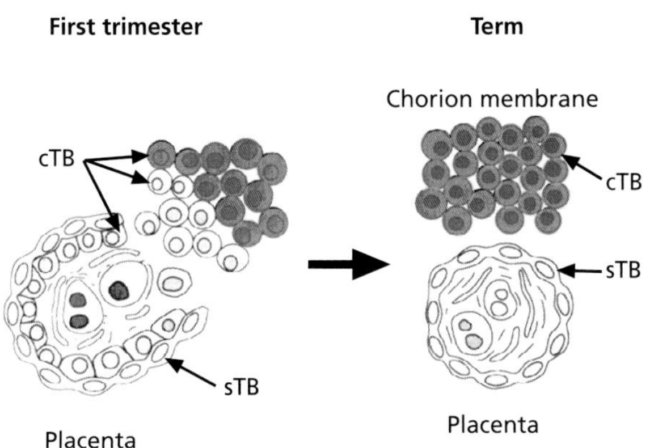

Figure 7 Schematic illustration of human leukocyte antigen (HLA) class I antigens (dark shade) in subpopulations of trophoblast cells at early and late stages of gestation. cTB, cytotrophoblast; sTB, syncytiotrophoblast. Reproduced from reference 28, with permission

HLA class I gene selection

Trophectoderm-derived cells are unique in their ability to select specific members of the HLA class I multigene family for expression. Only three members of the HLA class I gene family are present in trophoblast; one is a class Ia gene, *HLA-C*, and two are class Ib genes, *HLA-E* and *HLA-G*.[36] *HLA-A* and *-B*, which are highly polymorphic and are commonly present on eukaryotic cells, are absent.

HLA-C remains an enigma. Although the gene is polymorphic, and it is well known that proteins derived from this gene are present on trophoblast cells, whether it has any influence on immune cells or the immunobiology of pregnancy is not known. This may be because expression is transient and at a comparatively low level. HLA-E is an interesting molecule; it uses the leader sequences of other class I proteins to hitch a ride to the cell surface.[37] HLA-E appears to interact with NK cell inhibitory receptors, and to drive these cells into an anergic condition.[38] Since, in the early to middle stages of pregnancy, the decidua is packed with NK cells that would attack HLA-negative cells (Figure 8),[1,2] expression of a class I molecule is very important. In contrast, HLA-G, which was the first to be discovered on trophoblast cells,[39] has been intensively studied. Much has been uncovered about this interesting gene, as described in detail below, although the major functions of the spectrum of proteins remain unclear.

Importantly, nothing is known of how these three genes are selected for expression in trophoblast from among the members of the HLA class I multigene family. They appear to be expressed in synchrony, so whatever mechanism constitutes the 'on' signal appears to be operative for all three genes.

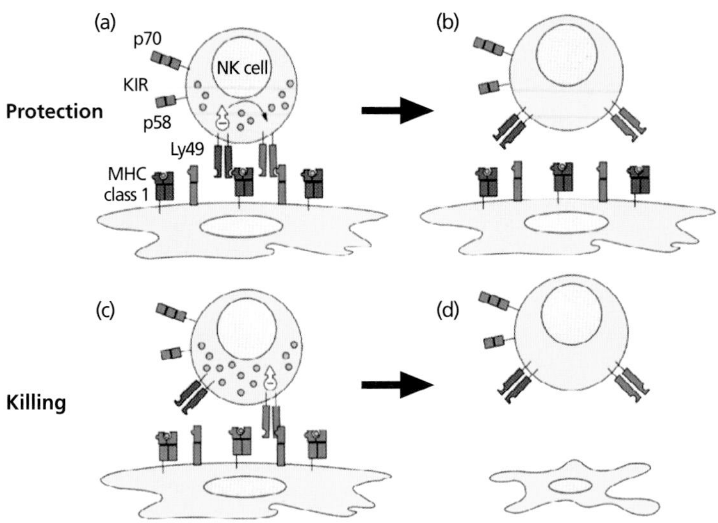

Figure 8 Recognition and killing of human leukocyte antigen (HLA) class I-negative cells by natural killer (NK) cells. The presence of HLA class I on the target cell (a) provides a negative regulatory signal and (b) prevents killing. In the absence of HLA class I, NK cells recognize (c) and kill (d) target cells. KIR, killer inhibitory receptor; MHC, major histocompatibility complex

Immunogenetics of *HLA-G*

The *HLA-G* gene has a number of unique features.[16–18,36] It has a shortened cytoplasmic tail, and may or may not be capable of transducing signals. In contrast to other class I HLA, its promoter region contains a large deletion in the enhancer A/IFN consensus region, which cripples IFN-induced transcription, and the 'gamma activated site' or GAS sequence has a single nucleotide substitution that prevents activation.[40] As a consequence, and unlike other HLA class I genes, *HLA-G* is only weakly enhanced by exposure to interferons.[41]

Unlike other HLA class I genes, alternative splicing is a feature of the HLA-G message. Figure 9 shows some of the major transcripts that have been identified and the proteins that result. Most notably, some transcripts encode cell surface proteins and others encode soluble proteins. This is because of a stop codon in intron 4 that precludes translation into the transmembrane region. All of the transcripts are readily identified in placentas and trophoblast cell lines. Because the antibodies to HLA-G are poorly characterized regarding binding site and specificity,[42] expression in adult cells and tissues remains uncertain.

At present, mechanisms underlying the production of multiple transcripts from this gene are unknown.

Functions of HLA-G

The range of functions of HLA-G is also unclear. The very fact that the maternal–fetal interface features this gene, which has low polymorphism, suggests that its substitution for *HLA-A*, *-B* is critical to the maintenance of semi-allogeneic pregnancy. Yet it is not yet known whether mothers recognize the products of the different alleles that have been identified.[43]

HLA-G appears to interact with macrophage immunoglobulin-like transcript (ILT)-4 receptors,[44] and could convey inhibitory signals. This might prevent activation of the macrophages in the decidua and the subsequent killing of migrating trophoblastic cells. As noted above, other suppressive substances such as progesterone and prostaglandins could have similar effects.

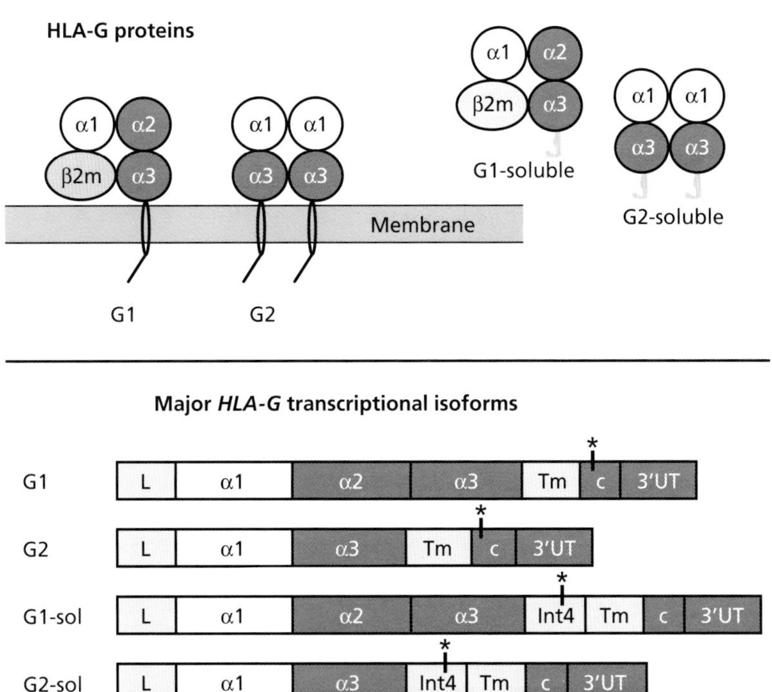

Figure 9 Alternative splicing of the *HLA-G* gene and generation of multiple isoforms of HLA-G. *, stop codon. Note that the G1 isoform binds the light chain, β2-microglobulin (β2m), thus forming heterodimers, whereas the G2 isoform forms homodimers. HLA-G, human leukocyte antigen-G; L, leader sequence; Tm, transmembrane region; c, cytoplasmic tail; 3'UT, 3' untranslated region; Int4, intron 4; sol, soluble

Presumably, these interactions would be cell : cell involvements and would rely on membrane-bound HLA-G on trophoblasts, but this does not preclude utilization of soluble HLA-G, which is receiving considerable attention.

Two groups have identified soluble HLA-G in the sera of pregnant women (Figure 10),[45,46] and it has recently been shown that recombinant soluble HLA-G1 and -G2 have dramatic effects on cytokines and other genes expressed in leukocytes (Hunt and colleagues, unpublished data). Earlier experiments employing soluble HLA-G1 partially purified from trophoblastic tumor cells documented apoptosis of T lymphocytes through the Fas/FasL pathway,[47] which has been well demonstrated as a function of other soluble HLA class I molecules. Soluble HLA-G seems also to be involved in transendothelial migration of leukocytes.[48] An immunoregulatory role for soluble

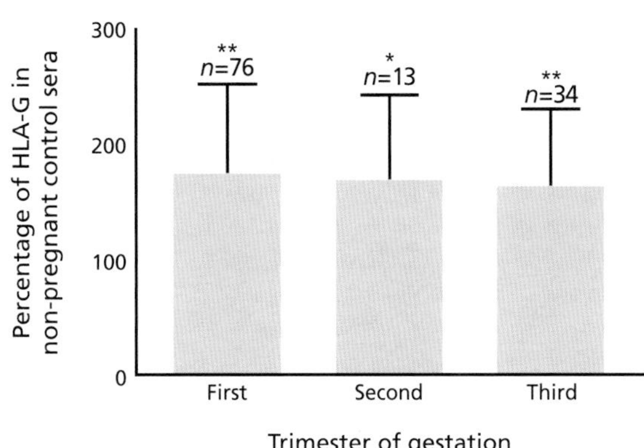

Figure 10 Capture enzyme-linked immunosorbent assay (ELISA) values showing that higher levels of soluble HLA-G are present in the sera of women who are pregnant than in women who are not. *$p < 0.004$, **$p < 0.001$, significantly different from non-pregnant values. Reproduced from reference 46, with permission

HLA-G is indicated in the finding that levels are higher in mothers who have successful pregnancies following assisted reproductive technology, as well as in heart transplant recipients who have less rejection.[45,49] HLA-G transfected into pig islets increases graft acceptance.[50]

Although it is possible that each of the isoforms derived from alternatively spliced messages has a different function, it seems more likely that there is at least some redundancy. Support for this idea has been provided by Ober and colleagues, who found that pregnancy goes forward in the absence of HLA-G1. The placentas of these women contain very little HLA-G1 protein (Figure 11).[51]

OTHER INSTANCES OF IMMUNOGENETIC CONTROL

Immunogenetic control of fertility through the HLA system is not limited to evasion via regulated expression in trophoblast. In studies of an inbred population, the Hutterites, Ober and colleagues have shown that disparities between maternal and paternal HLA class I alleles are important; when identical, delays in establishing pregnancies occur.[52] This suggests a major role for a limited degree of allorecognition by the mother.

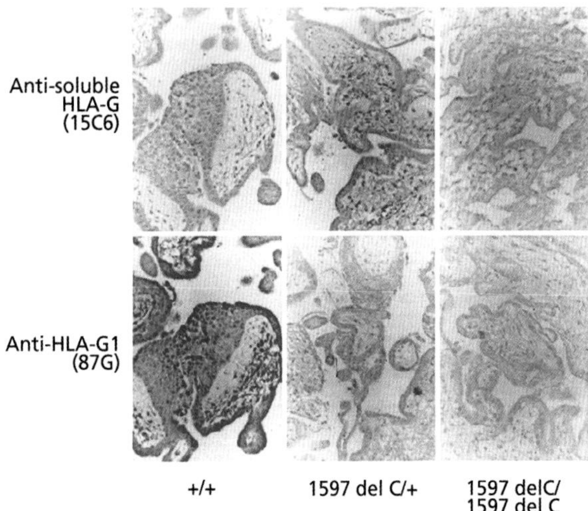

Figure 11 Immunohistochemical experiment demonstrating reduced staining for HLA-G in placentas from fetuses carrying homozygous or heterozygous deletions that result in failure to synthesize HLA-G1. Reproduced from reference 51, with permission

Not surprisingly, there are instances of genetic polymorphisms in cytokine genes that result in pregnancy problems. However, the results are unpredictable. Pre-eclampsia, which is characterized by high levels of TNF-α mRNA,[53] is not associated with the TNF2 allele, which produces higher levels of transcripts. Instead, pre-eclampsia is associated with the TNF1 allele.[53–55] 'The jury is still out' on relationships between TNF-α alleles and increased risk of preterm premature rupture of the fetal membranes, with Roberts and co-workers[56] reporting that women carrying the TNF2 allele more often rupture, and Dizon-Townson and associates[57] failing to associate TNF2 relationships with preterm delivery. It has been reported that the IL-1 receptor antagonist allele 2 may have a role in idiopathic recurrent miscarriage.[58]

Other genes may also be involved. Although the functionally important IL-10 promoter polymorphism appears not to be a major genetic regulator in recurrent spontaneous abortions, the *HLA-G* genotype is thought to be associated with this disorder of pregnancy.[59,60] Both the *HLA-G*01013* and the **0105N* dysfunctional null allele were considered to be risk factors.

Although it remains to be determined whether genetic testing could assist in predicting the risk of problems in pregnancy, or whether remedial measures might be effective, considerable effort is being expended in identifying potential relationships.

FUTURE PERSPECTIVES

Despite the abundance of information that has been gathered on the genetic aspects of pregnancy, infertility problems have not been overcome and pregnancy losses continue, with little improvement. Clearly, more emphasis must be placed on applying the principles of reproductive immunogenetics to practical problems.

Yet more must be learned, and the information already at hand must be better understood, before modern methods of molecular medicine can be used effectively to intervene when fertility is low or pregnancy is threatened. Studies on transgenic mice have been informative, as have analyses of human population genetics and the evaluation of genetic polymorphisms. Where to go from here and what methods to use are the central questions. Scientists are currently enthralled with DNA microarray technology and proteomics. These approaches are immensely powerful, and it is easy to envision how experiments using the technologies might be designed to identify immunological features of pregnancy. For example, uterine and/or placental transcripts or peptides in homozygous pregnancies might be compared with transcripts/peptides in heterozygous pregnancies. The power of bioinformatics will undoubtedly be crucial when analyzing the superabundance of data forthcoming from these types of experiments.

Despite the many obstacles still to be surmounted, it is heartening to consider that molecular approaches might ultimately yield new information relative to treatments for suboptimal fertility, a condition that may be caused in some instances by inappropriate immunological relationships between the mother and the fetus during pregnancy.

ACKNOWLEDGEMENTS

This work was supported in part by grants from the National Institutes of Health (HD24212, HD26429, HD29156, HD35859, AI38502) to J.S.H. and by core facilities of the University of Kansas Mental Retardation Research Center (HD02528) and the P30/U54 Center for Reproductive Sciences (HD33994). The author gratefully acknowledges the contributions of the many colleagues, post-doctoral fellows, graduate students and technical assistants who have participated in these studies, and thanks M.J. Soares and J.L. Nelson for close reading of the manuscript.

REFERENCES

1. Alberts B, Bray D, Lewis J *et al.* The immune system. In Alberts B, Bray D, Lewis J *et al.*, eds. *Molecular Biology of the Cell*, 3rd edn. New York: Garland Publishing 1994: 1195–1254.
2. Janeway CA, Travers P, Walport M, Capra JD. *Immunobiology*, 4th edn. New York: Elsevier Ltd/Garland Publishing 1999.
3. Murdoch C, Finn A. Chemokine receptors and their role in inflammation and infectious diseases. *Blood* 2000 95: 3032–3043.
4. Bulmer JN. Immune cells in decidua. In Kurpisz M, Fernandez N, eds. *Immunology of Human Reproduction*. Oxford: BIOS Scientific Publishers 1995; 313–334.
5. King A, Burrows T, Hiby VS, Loke YW. Human uterine lymphocytes. *Hum Reprod Update* 1998 4: 480–485.
6. Dang Y, Heyborne KD. Cutting edge: regulation of uterine NKT cells by a fetal class I molecule other than CD1. *J Immunol* 2001 166: 3641–3644.
7. Mincheva-Nilsson L, Kling M *et al.* γ/δ T cells of human early pregnancy decidua: evidence for local proliferation, phenotypic heterogeneity, and extrathymic differentiation. *J Immunol* 1997 159: 3266–3277.
8. Hunt JS, Hutter H. Current theories on protection of the fetal semiallograft. In Hunt JS, ed. *HLA and the Maternal–Fetal Relationship*. Austin, TX: Landes Publishing, 1996: 27–50.
9. Hunt JS, Johnson PM. Immunology of reproduction. In Knobil E, Neill JD, eds. *Encyclopedia of Reproduction*. New York: Academic Press, 1997 2: 798–806.
10. Redman CWG, Sacks GP, Largent IL. Preeclampsia: an excessive maternal inflammatory response to pregnancy. *Am J Obstet Gynecol* 1999 180: 499–506.
11. Ashkenazi A, Dixit VM. Apoptosis control by death and decoy receptors. *Curr Opin Cell Biol* 1999 11: 255-260.
12. Phillips TA, Ni J, Hunt JS. Death-inducing tumor necrosis factor (TNF) superfamily ligands and receptors are transcribed in human placentas, cytotrophoblasts, placental macrophages and placental cell lines. *Placenta* 2001 22: 663–672.
13. Wegmann TG, Lin H, Guilbert L, Mosmann TR. Bidirectional cytokine interactions in the maternal–fetal relationship: is successful pregnancy a TH2 phenomenon? *Immunol Today* 1993 14: 353–356.
14. Hunt JS, Chen HL, Miller L. Tumor necrosis factors: pivotal factors in pregnancy? *Biol Reprod* 1996 54: 554–562.

15. Sargent IL, Redman CW. Maternal cell-mediated immunity to the fetus in human pregnancy. *J Reprod Immunol* 1985 7: 95–104.
16. Hunt JS, Orr HT. HLA and maternal–fetal recognition. *FASEB J* 1992 6: 2344-2348.
17. Le Bouteiller P, Mallet V. HLA-G and pregnancy. *Rev Reprod* 1997 2: 7–13.
18. Ober C. HLA and pregnancy: the paradox of the fetal allograft. *Am J Hum Genet* 1998 62: 1–5.
19. Hsi BL, Hunt JS, Atkinson JP. Detection of complement regulatory proteins on subpopulations of human trophoblast. *J Reprod Immunol* 1991 19: 209–223.
20. Holmes CH, Simpson KL, Wainwright SD. Preferential expression of the complement regulatory protein decay accelerating factor at the fetomaternal interface during human pregnancy. *J Immunol* 1990 00: 3099–3105.
21. Xu C, Mao D, Holers VM *et al.* A critical role for murine complement regulator *Crry* in fetomaternal tolerance. *Science* 2000 287: 498–501.
22. Bulmer JN, Morrison L, Smith JC. Expression of class II MHC gene products by macrophages in human uteroplacental tissue. *Immunology* 1988 63: 707–714.
23. Lessin DL, Hunt JS, King CR, Wood GW. Antigen expression by cells near the maternal–fetal interface. *Am J Reprod Immunol Microbiol* 1988 16: 1–7.
24. Wong SC, Ali MA, Benzie R. The intrauterine diagnosis of hemoglobin disorders. *Clin Perinatol* 1984 11: 283–308.
25. Barton SC, Surani MA, Norris ML. Role of paternal and maternal genomes in mouse development. *Nature* 1984 311: 374–376.
26. Lenfant F, Fort M, Rodriguez AM *et al.* Absence of imprinting of HLA class Ia genes leads to coexpression of biparental alleles on term human trophoblast cells upon IFN-γ induction. *Immunogenetics* 1998 47: 297–304.
27. Benirschke K, Kaufmann P. *Pathology of the Placenta*, 3rd edn. New York: Springer-Verlag, 1995.
28. Hunt JS, Tung KKS. Immunological aspects of fertility and infertility. In Austen KF, Frank MM, Atkinson JP, Cantor H, eds. *Samter's Immunologic Diseases*, 6th edn. Baltimore: Lippincott, Williams & Wilkins, 2001: 771–785.
29. Caniggia I, Hostachfi H, Winter J. Hypoxia-inducible factor-1 mediates the biological effects of oxygen on human trophoblast differentiation through TGFβ3. *J Clin Invest* 2000 105: 577–587.
30. Lessin DL, Hunt JS, King CR. Antigen expression by cells near the maternal–fetal interface. *Am J Reprod Immunol Microbiol* 1988 16: 1–7.
31. Murphy SP, Tomasi TB. Absence of MHC class II antigen expression in trophoblast cells results from a lack of class II transactivator (CIITA) gene expression. *Mol Reprod Dev* 1998 51: 1–12.
32. Hunt JS, Andrews GK, Wood GW. Normal trophoblasts resist induction of class I HLA. *J Immunol* 1987 138: 2481–2487.
33. Hunt JS, Fishback JL, Chumbley G, Loke YW. Identification of class I MHC mRNA in human first trimester trophoblast cells by *in situ* hybridization. *J Immunol* 1990 144: 4420–4425.
34. Hunt JS, Fishback JL, Andrews GK, Wood GW. Expression of class I HLA genes by trophoblast cells: analysis by *in situ* hybridization. *J Immunol* 1988 140: 1293–1299.
35. Hunt JS, Hsi BL, King CR, Fishback JL. Detection of class I MHC mRNA in subpopulations of first trimester cytotrophoblast cells by *in situ* hybridization. *J Reprod Immunol* 1991 19: 315–323.
36. Le Bouteiller P, Solier C, Proll J, Aguerre-Girr M, Fournel S, Lenfant F. Placental HLA-G protein expression *in vivo*: where and what for? *Hum Reprod Update* 1999 5: 223–233.
37. Lee N, Goodlett DR, Ishitani A, Marquardt H, Geraghty DE. HLA-E surface expression depends on binding of TAP-dependent peptides derived from certain HLA class I signal sequences. *J Immunol* 1998 160: 4951–4960.
38. Lee N, Llano M, Carretero M *et al.* HLA-E is a major ligand for the natural killer inhibitory receptor CD94/NKG2A. *Proc Natl Acad Sci USA* 1998 95: 5199–5204.
39. Ellis SA, Sargent IL, Redman CW, McMichael AJ. Evidence for a novel HLA antigen found on human extravillous trophoblast and a choriocarcinoma cell line. *Immunology* 1986 59: 595–601.
40. Chu W, Gao J, Murphy WJ, Hunt JS. A candidate interferon-γ activated site (GAS element) in the HLA-G promoter does not bind nuclear proteins. *Hum Immunol* 1999 60: 1113–1118.
41. Yang Y, Geraghty DE, Hunt JS. Cytokine regulation of HLA-G expression in human trophoblast cells. *J Reprod Immunol* 1995 29: 179–195.

42. Blaschitz A, Hutter H, Leitner V *et al.* Reaction patterns of monoclonal antibodies to HLA-G in human tissues and on cell lines: a comparative study. *Hum Immunol* 2000 61: 1074–1085.
43. Ober C, Aldrich CL. HLA-G polymorphisms: neutral evolution or novel function? *J Reprod Immunol* 1997 36: 1–21.
44. Colonna M, Samaridis J, Cella M *et al.* Human myelomonocytic cells express an inhibitory receptor for classical and nonclassical MHC class I molecules. *J Immunol* 1998 160: 3096–3100.
45. Pfeiffer KA, Rebmann V, Passler M *et al.* Soluble HLA levels in early pregnancy after *in vitro* fertilization. *Hum Immunol* 2000 61: 559–564.
46. Hunt JS, Jedhav L, Chu W, Geraghty DE, Hunt JS. Soluble HLA-G circulates in mothers during pregnancy. *Am J Obstet Gynecol* 2000 183: 682–688.
47. Fournel S, Aguerre-Girr M, Huc X *et al.* Cutting edge: soluble HLA-G1 triggers CD95/CD95 ligand-mediated apoptosis. *J Immunol* 2000 164: 6100–6104.
48. Dorling A, Monk NJ, Lechler RI. HLA-G inhibits the transendothelial migration of human NK cells. *Eur J Immunol* 2000 30: 586–593.
49. Lila N, Carpentier A, Amrein C *et al.* Implication of HLA-G molecule in heart-graft acceptance. *Lancet* 2000 355: 2138.
50. Sasaki H, Xu XC, Mohanakumar T. HLA-E and HLA-G expression on porcine endothelial cells inhibits xenoreactive human NK cells through CD94/NKG2-dependent and -independent pathways. *J Immunol* 1999 163: 6301–6305.
51. Ober C, Aldrich B, Rosinsky A *et al.* HLA-G1 protein expression is not essential for fetal survival. *Placenta* 1998 19: 127–132.
52. Ober CL, Martin AO, Simpson JL *et al.* Shared HLA antigens and reproductive performance among Hutterites. *Am J Hum Genet* 1983 35: 994–1004.
53. Chen G, Wilson R, Wang SH *et al.* Tumour necrosis factor-α (TNF-α) gene polymorphism and expression in pre-eclampsia. *Clin Exp Immunol* 1996 104: 154–159.
54. Dizon-Townson DS, Major H, Ward K. A promoter mutation in the tumor necrosis factor α gene is not associated with preeclampsia. *J Reprod Immunol* 1998 38: 55–61.
55. Livingston JC, Park V, Barton JR *et al.* Lack of association of severe preeclampsia with maternal and fetal mutant alleles for tumor necrosis factor alpha and lymphotoxin α genes and plasma tumor necrosis factor α levels. *Am J Obstet Gynecol* 2001 184: 1273–1277.
56. Roberts AK, Monzon-Bordonaba F, Van Deerlin PG *et al.* Association of polymorphism within the promoter of the tumor necrosis factor α gene with increased risk of preterm premature rupture of the fetal membranes. *Am J Obstet Gynecol* 1999 180: 1297–1302.
57. Dizon-Townson DS, Major H, Varner M *et al.* A promoter mutation that increases transcription of the tumor necrosis factor-α gene is not associated with preterm delivery. *Am J Obstet Gynecol* 1997 177: 810–813.
58. Unfried G, Tempfer C, Schneeberger C *et al.* Interleukin-1 receptor antagonist polymorphism in women with idiopathic recurrent miscarriage. *Fertil Steril* 2001 75: 683–687.
59. Karhukorpi J, Laitinen T, Karttunen R, Tiilikainen AS. The functionally important IL-10 promoter polymorphism (–1082G→A) is not a major genetic regulator in recurrent spontaneous abortions. *Mol Hum Reprod* 2001 7: 201–203.
60. Pfeiffer KA, Fimmers R, Engels G *et al.* The HLA-G genotype is potentially associated with idiopathic recurrent spontaneous abortion. *Mol Hum Reprod* 2001 7: 373–378.

Chapter 10

Transgenic technology, cloning and germ cell transplantation

Kathleen H. Burns, Francesco J. DeMayo and Martin M. Matzuk

INTRODUCTION

Our ability to manipulate the mammalian genome is revolutionizing the ways in which we study gene function. Transgenic technology allows for designed DNA sequences to be incorporated into a genome and inherited over successive generations. In addition, early cloning and germ cell transplantation experiments promise to challenge heretofore limitations in the proliferative capacities of cells and organisms.

This chapter provides a review of these new tools, their caveats and their early – although already profound – contributions to the field of reproductive biology.

TRANSGENIC TECHNOLOGY AND INTRODUCTION OF DNA INTO THE MOUSE GENOME

A transgenic experiment introduces manipulated DNA into the genome of an organism, with the goal that it be faithfully inherited by the recipient's progeny. It is imperative that the exogenous DNA be introduced for integration early in an organism's development to allow for transmission to its germ line.

Gordon and colleagues were the first to report a procedure for producing transgenic mice.[1,2] Still widely employed, their approach involves microinjecting foreign DNA into male pronuclei of one-cell zygotes (Figure 1). Typically, transgene concatemers integrate during the one-cell stage at a single, random site within the mouse genome.[3] Embryos with DNA thus introduced are transferred to pseudopregnant females for development, and become members of the founder (F_0) generation. Each is expected to be hemizygous for a unique transgene integration that will be passed on to 50% of its F_1 progeny. Occasionally, delayed integration events (i.e. after the one-cell stage) or integration of transgene concatamers at multiple sites complicate the establishment and characterization of the transgenic strains. A number of reported transgenic mouse models exhibit reproductive findings (Table 1).

Because transgene integration takes place randomly within the recipient genome, there is a chance that the transgene will disrupt important endogenous sequences.[43] Such random insertional mutations can be very serendipitous for researchers, as they identify and mark loci with important functions. Indeed, the goal of 'gene-trapping' experiments is to generate insertional mutations at loci with an interesting pattern of expression; the strategy introduces a β-galactosidase reporter construct for random genome integration.[44–46] Insertional mutations having reproductive phenotypes are listed (Table 2).

Microinjection may be used to introduce DNA of linearized plasmids or larger bacterial artificial chromosome (BAC) and yeast artificial chromosome (YAC) DNAs. It is advisable to use constructs of genomic DNA as these are expressed more reliably than those of cDNA sequences.[58,59] The expression of plasmid transgenes depends largely on their site of genome insertion, a phenomenon known as the position effect.[60,61] Flanking sequences may silence a transgene or induce its ectopic expression. Incorporating insulator elements into a construct can reduce this effect and favor a more predictable expression pattern. Larger BAC and YAC transgenes often include *cis*-acting elements for an autonomous control of gene expression.[62,63] Therefore, the BAC/YAC copy number determines expression levels, and the position effect is minimized.

Bitransgenic regulatory systems are variations on this technology that allow researchers to direct the timing of transgene expression. These systems require two transgene constructs that function in sequence. The first transgene is the regulator construct, which encodes a transcription factor expressed under the control of a constitutive or tissue-specific promoter. The transcription factor protein product is activated or inactivated by a pharmacological agent administered by the researcher. The second transgene is the responder operon, which encodes the gene-of-interest that is only expressed in the presence of the active transcription factor. Thus, by giving or withholding a drug, biologists can toggle the expression of a gene-of-interest. The most commonly used bitransgenic systems are based on the provision of tetracycline or tetracycline derivatives (doxycycline or anhydrotetracycline); these include the *tet-off*

Day 1: PMSG injection Transgene construct prepared

Day 3: hCG injection and mating

pseudopregnant foster mothers mated with vasectomized males

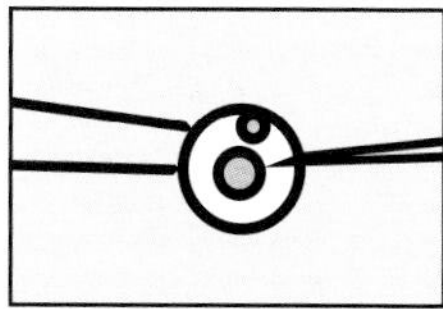

Day 4: embryo collection and DNA microinjection

embryo transfer to foster mothers

Day 23: F_0 generation is born

Day 37: tail DNA taken for Southern blot or PCR analysis

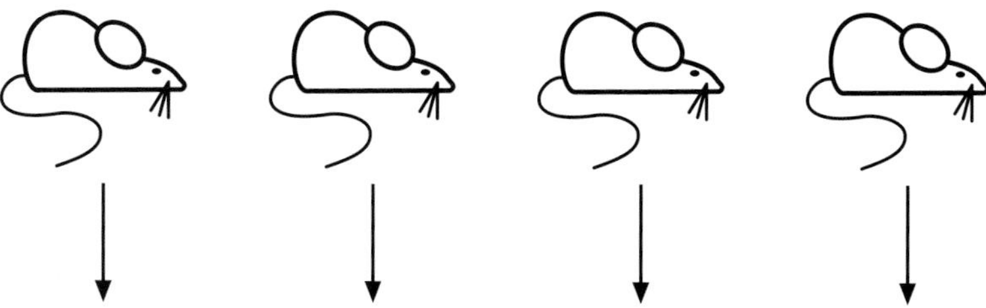

Each founder is crossed with wild-type mice to begin a unique transgenic line:

50% of the F_1 generation is expected to inherit the transgene

Figure 1 Generation of transgenic mice by DNA microinjection. Immature, adolescent female mice are superovulated using a pregnant mare serum gonadotropin–human chorionic gonadotropin (PMSG–hCG) regime and are then mated. Fertilized eggs are recovered from these females, and the transgene construct DNA is directly injected into the male pronuclei. The microinjection depends on the use of contrast optics and micromanipulators. Many factors must be considered in the choice of mouse strain, including the desire for genetic uniformity, the responsiveness of female mice to gonadotropins, the reproductive capacities of the strain, and strain-specific characteristics of the oocyte that may facilitate microinjection. The oocytes are transferred to pseudopregnant recipient females that have been mated with vasectomized males. The founder generation (F_0) transgenic mice each have a unique transgene integration, and become founders for a line of offspring that are expected to inherit the transgene with Mendelian frequency. Intercrossing members of a transgenic line allows for the establishment of homozygosity at the transgene locus. Homozygotes may exhibit a phenotype that is distinct from heterozygotes, reflecting the effects of the increased dosage of the introduced DNA or the effects of insertional mutations. PCR, polymerase chain reaction

and *tet-on* regulators (Figure 2). Variations on these regulators have been designed to enhance target gene expression[66] and increase the number of genes that can be simultaneously controlled.[67,68] In addition, bitransgenic systems have been developed which use other exogenous compounds, such as mifepristone (RU486)[69] or ecdysone (*Drosophila melanogaster* steroid hormone)[70] to control gene expression.

Several alternatives to the traditional microinjection technique have been successfully demonstrated as means for transgene delivery. These include the use of viral vectors for DNA delivery[71,72] and a variation on intracytoplasmic sperm injection (ICSI) in which transgene DNA and membrane-disrupted sperm heads are co-injected into unfertilized metaphase II oocytes.[73]

EMBRYONIC STEM CELLS

Embryonic stem (ES) cell technology is most commonly used today for modifying defined endogenous loci, for which the technique is indispensable.

Table 1 Transgenic mouse models with reproductive phenotypes: overexpresser/misexpresser models

Construct	*Reproductive phenotype*	*Reference*
mMT promoter and activin/inhibin βA sequence	males with high expression in testes are infertile; seminiferous tubules show patchy lesions and vacuoles	4
Mifepristone-inducible transgene overexpressing inhibin A from liver	females have a block in folliculogenesis at the early antral stage; males have decreased testis size	5
Rat androgen-binding protein (*Abpa*) promoter and coding sequence	increased Sertoli cell number and compromised male fertility	6
Inhibin α promoter and *Bcl2* coding sequence	increased folliculogenesis, decreased follicle apoptosis, and germ cell tumors	7
EF-1α promoter and *Bcl2* coding sequence	compromised testicular germ cell apoptosis/differentiation	8
hMT promoter and rat c-*myc* coding sequence	early arrest in spermatogenesis and apoptosis	9
Dax1 overexpresser (*Dax* regulatory region)	XY mice with a weak *Sry* allele develop as females	10
Bovine glycoprotein hormone-α promoter and diphtheria toxin coding sequence	pituitary gonadotrophs lost; mice are hypogonadal	11
Rat histone H1t promoter and diphtheria toxin coding sequence	male infertility; small testes with loss of germ cells; females fertile	12
Hydroxymethyl glutaryl CoA reductase promoter and E2F-1 transcription factor sequence	males are infertile; testicular atrophy and apoptosis in the germinal epithelium	13
β-Actin promoter and epidermal growth factor (EGF) sequence	males are sterile; few post-meiosis II gametes	14
mMT-1 promoter and follistatin (*Fst*) coding sequence	progressive infertility in both sexes; males have somatic cell and germ cell defects; females have folliculogenesis defects	15
mMT-1 promoter and β1,4-galactosyltransferase sequence	altered sperm–egg binding	16
Growth hormone overexpressers (multiple constructs studied)	compromised reproduction in both sexes	17
β-actin promoter and active human heat shock transcription factor (*Hsf1*) coding sequence	males are infertile; arrest during meiosis and germ cell apoptosis	18
mMT promoter and interferon-α_1 coding sequence	degeneration of spermatogenic cells and seminiferous tubule atrophy	19
mMT promoter and interferon-β coding sequence	male sterility; testes involuted with degeneration of spermatocytes and spermatids	20
mMT promoter and human interleukin-2 sequence	males exhibit atrophic testes and defects in spermatogenesis	21
Liver promoter and leptin coding sequence (skinny mice)	females have accelerated puberty and late-onset hypothalamic hypogonadism	22
Bovine glycoprotein hormone α promoter and bLHβ–hCG fusion	female infertility, polycystic ovaries, granulosa cell tumors	23
mMTV promoter and matrilysin (MAT) (*Mmp7*) coding sequence	progressive male infertility associated with degeneration in testes and loss of interstitial cells	24
mMT promoter and human Müllerian inhibiting substance	both sexes have defects in development of reproductive structures	25

continued

Table 1 *continued*

Construct	*Reproductive phenotype*	*Reference*
mMTV promoter and N-*ras* oncogene sequence	male infertility; sperm motility defects	26
mMTV promoter and *neu* oncogene sequence	male infertility; epithelial cell hyperplasia within the epididymis	27
Human ornithine decarboxylase overexpresser (human ODC regulatory region)	alterations in spermatogenic DNA synthesis and compromised male fertility	28
P450 aromatase overexpression	subfertility and Leydig tumor development in males	29
mMT promoter and p53 coding sequence	males exhibit variable subfertility; increased spermatid apoptosis and teratozoospermia	30
Rat proenkephalin cDNA; human promoter and flanking sequence	male subfertility or infertility; abnormal testicular morphology	31
Mouse protamine-1 promoter and avian protamine coding sequence	disrupted sperm chromatin condensation	32
Protamine 1 Δ3'UTR transgene	males are infertile; premature protamine accumulation and arrested spermatid differentiation	33
mMTV promoter and dominant negative RARα coding sequence	males are infertile or subfertile; spermatogenesis intact; squamous metaplasia of epididymis and vas deferens	34
Spermidine/spermine N1-acetyltransferase (*Sat*) promoter and coding sequence	female infertility; ovarian hypofunction and hypoplastic uteri	35
Sry promoter and coding sequence	transgenic females develop as males	36
Human GnRH promoter and SV40 T-antigen sequence	GnRH neurons do not migrate appropriately; mice are hypogonadal	37
Bovine FSHβ promoter and HSV-tk coding sequence	thymidine kinase expression in the pituitary and testis	38
Group 1 Mup promoter and HSV-tk coding sequence	male infertility	39
mMT promoter and hTGF-α coding sequence	delayed implantation and parturition	40
mMTV LTR promoter and vascular endothelial growth factor (*Vegf*) sequence	males are infertile; defects in spermatogenesis and aberrant blood vessel formation in the testes and epididymis	41
Phosphoglycerate kinase-2 promoter and v-*Mos* coding sequence	male infertility; germ cell development arrest at metaphase I of meiosis	42

mMT, mouse metallothionein; EF-1α, elongation factor-1α; hMT, human metallothionein; CoA, coenzyme A; mMTV, mouse mammary tumor virus; bLHβ, bovine luteinizing hormone β; hCG, human chorionic gonadotropin; ODC, ornithine decarboxylase; UTR, untranslated region; RARα, retinoic acid receptor α; GnRH, gonadotropin releasing hormone; FSHβ, follicle stimulating hormone β; HSV-tk, herpes simplex virus-thymidine kinase; Mup, major urinary protein; hTGF-α, human transforming growth factor-α; LTR, long terminal repeat

Use of ES cells as vectors for random transgene integration and *in vivo* expression predates this widespread application. ES cell lines have been isolated from mouse blastocysts, and these pluripotent cells can be maintained for long periods *in vitro*, mutagenized or provided with DNA for integration. Ultimately, ES cells may be reintroduced into blastocysts to give rise to somatic and germ cell lineages in the resulting chimeras (Figure 3). The technology was first described in articles by Evans and colleagues in 1986–87.[74,75]

Table 2 Transgenic mouse models with reproductive phenotypes: random insertional mutations

Gene	*Reproductive phenotype*	*Reference*
ELKL motif kinase (*Emk*)	β-gal gene trap insertion creates a null allele; homozygotes intercrossed are not fertile	47
gcd (germ-cell deficient)	infertility in homozygotes of both sexes; germ cell migration/proliferation failure	48
Histone 3.3A gene	β-gal gene trap insertion creates a hypomorphic allele; homozygotes have multiple defects including male subfertility	49
ho (hotfoot)	male homozygotes are infertile; sperm cannot penetrate the zona pellucida	50
Kisimo mouse (*Theg*)	infertility in homozygous males; asthenospermia	51
Lvs (lacking vigorous sperm)	infertility in hemizygous males; abnormal nuclear condensation during spermatogenesis	52
Morc (microrchidia)	infertility in homozygous males; early arrest in meiosis and germ cell apoptosis	53
Ods (*Odsex*, ocular degeneration with sex reversal)	XX develop as infertile males; dominant mutation associated with 150-kb deletion and increased *Sox9* expression	54
pcd (Purkinje cell degeneration)	male infertility; abnormal spermatozoa	55
sys (symplastic spermatids)	infertility in homozygous males; Sertoli cell and spermatid abnormalities	56
2 : 12 translocation associated with transgene insertion	male infertility; defects in chromosome synapsis and spermatogenesis	57

β-gal, β-galactosyltransferase

TARGETING TECHNOLOGY AND MODIFYING ENDOGENOUS LOCI

Targeting technology allows one to edit a particular endogenous DNA sequence within cells. This was first accomplished in mammalian cells by Smithies and colleagues in 1985.[76] The group used an exogenous plasmid targeting vector to insert sequences into the β-globin locus in human cancer cells maintained *in vitro*. Advances in ES cell technology in intervening years set the stage for targeted manipulations of the mouse genome. Two seminal reports of this accomplishment were published in 1987 and 1988. The first described the targeted disruption of the hypoxanthine phosphoribosyltransferase (HPRT) gene in which the cultured ES cells were selected based on the loss of HPRT activity.[77] The second provided a general targeting strategy – widely used today – whereby positive and negative selectable markers included in the targeting vector allowed for identification of ES cells with the desired recombination[78] (Figure 4).

Targeting techniques depend on regions of homology where targeting vector sequences are identical to the endogenous sequences where the exchange (cross-over) will occur. These homologous stretches serve as substrates for the cell's homologous recombination machinery. Targeted exchanges are usually rare events in comparison with random recombinations. Selection schemes allow one to isolate ES lines that have transgene integration (positive selection), while excluding those ES cells with random vector incorporation (negative selection). The targeted locus should be distinguishable from the endogenous locus by polymerase chain reaction (PCR) or Southern blot analysis, and the predicted structure confirmed. The rate of homologous recombination is influenced by length[79] and degree[80] of targeting construct homology, as well as features unique to each locus.

The goal of many targeting experiments is to disrupt a given gene in ES cells and eventually study the resultant phenotype in mice bred to homozygosity for the alteration. Northern blot,

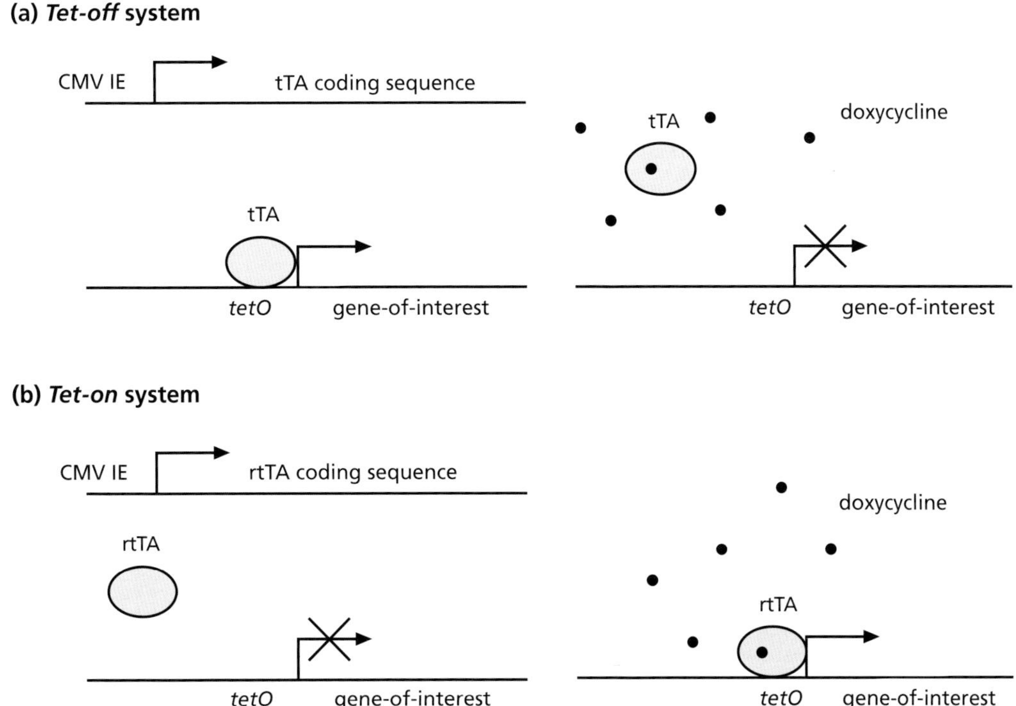

Figure 2 (a) The *tet-off* system as designed by Gossen and Bujard.[64] The cytomegalovirus (CMV) immediate early (IE) promoter/enhancer directs the constitutive expression of the tetracycline (tet) transactivator protein (tTA). The tTA is a fusion of the *Escherichia coli* tet repressor and the transcriptional activation domain of the herpes virus VP16 protein. The tTA protein binds to the Tn10 tetracycline resistance operator (*tetO*) sequence in the absence of tetracycline and recruits transcriptional machinery. Seven tandem *tetO* sequences and a minimal promoter element precede the coding sequence of a gene-of-interest, and expression takes place in the absence of doxycycline (the commonly used tetracycline-like drug). Administration of doxycycline results in drug binding and conformational changes to the tTA that abrogate expression of the responder construct – hence the nomenclature '*tet-off*'. (b) The *tet-on* system.[65] The *tet-on* bitransgenic system depends on a reverse tetracycline transactivator (rtTA) which has affinity for *tetO* only when tetracycline derivatives are provided. The responder construct gene-of-interest is expressed with the administration of these drugs – hence '*tet-on*'

reverse transcriptase (RT)-PCR or Western blot analysis should be used to show definitively that the genetic change has the intended effect. When one completely abrogates the expression of a gene, these null or 'knock-out' mice are expected to display the direct, indirect and/or compensatory *in vivo* effects of the loss-of-function. Knock-out models are leading to major conceptual breakthroughs in many fields of interest; the notable examples relevant to reproductive biology are provided here (Table 3).

One should be mindful of the possibility that unforeseeable genetic factors can influence a phenotype in a knock-out mouse model. The intended disruption may affect the expression of neighboring genes[219,220] or the operation of a long-range *cis*-acting locus control region.[221] These effects are undesirable and difficult to recognize, but may be minimized by introducing subtle mutations as described a decade ago.[222] Also, as with any *in vivo* model, genetic factors involved in a relevant pathway may affect the penetrance or expressivity of a phenotype. Different features in knock-out mice of divergent genetic backgrounds can lead to the identification of specific modifiers of a phenotype of interest.[223]

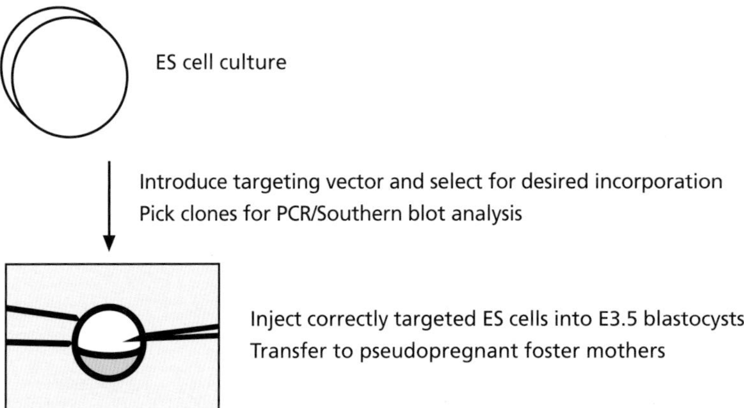
ES cell culture
Introduce targeting vector and select for desired incorporation
Pick clones for PCR/Southern blot analysis
Inject correctly targeted ES cells into E3.5 blastocysts
Transfer to pseudopregnant foster mothers

Chimeras are born and identifiable by their mosaic coat color; these are bred with wild-type mice. Germline transmission will allow for generation of heterozygotes which are intercrossed to yield homozygote knock-out mice

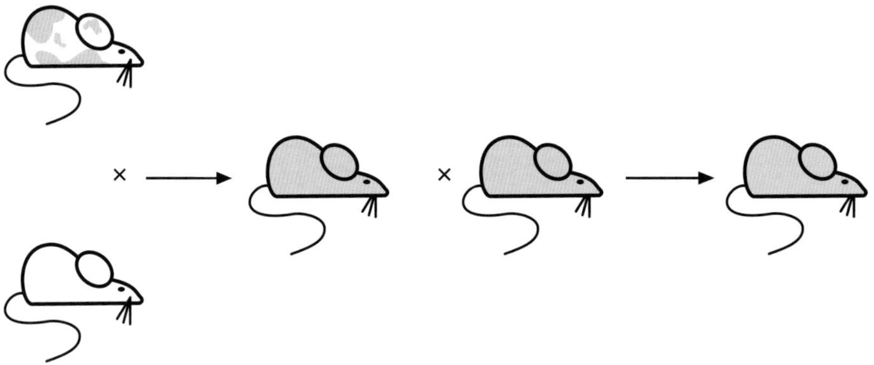

Figure 3 Generation of knock-out mice by embryonic stem (ES) cell injection. Under optimal culture conditions with fibroblast-derived feeder cells, ES cells retain pluripotency *in vitro* for multiple passages, allowing selection and characterization of a targeted disruption. ES cells, having undergone the desired recombination, are injected into day 3.5 (E3.5) blastocysts. A coat color difference between the ES cell donor strain and the blastocyst donor strain allows for the ready assessment of chimerism. Chimeras with a germline component can generate uniform heterozygote offspring when mated with wild-type mice. Finally, unless there are effects on development, crosses between heterozygotes are expected to yield knock-out mice at the Mendelian frequency of 1 : 3

FUNCTIONAL STUDIES AND KNOCK-OUT RESCUE EXPERIMENTS

One of the earliest transgenic rescue experiments was performed by Mason and colleagues.[224,225] They demonstrated that a transgene consisting of gonadotropin releasing hormone (GnRH) – GnRH-associated protein (GAP) genomic sequences could complement the functional deficiency in hypogonadal (*hpg*) mice that bear a naturally occurring truncating mutation of the GnRH sequence.

Knock-out models are used to study not only the *in vivo* function of a gene but also the potential for other related sequences to compensate for its absence. Underscoring the relevance of mouse models for understanding human biology, human follicle stimulating hormone subunit (FSHβ), zona pellucida protein 3 (ZP3) and deleted in azoospermia (*DAZ*) transgenes have been shown to rescue or rescue partially fertility defects in their respective mouse knock-out models.[226–228] Targeting in ES cells may also be employed to replace directly one

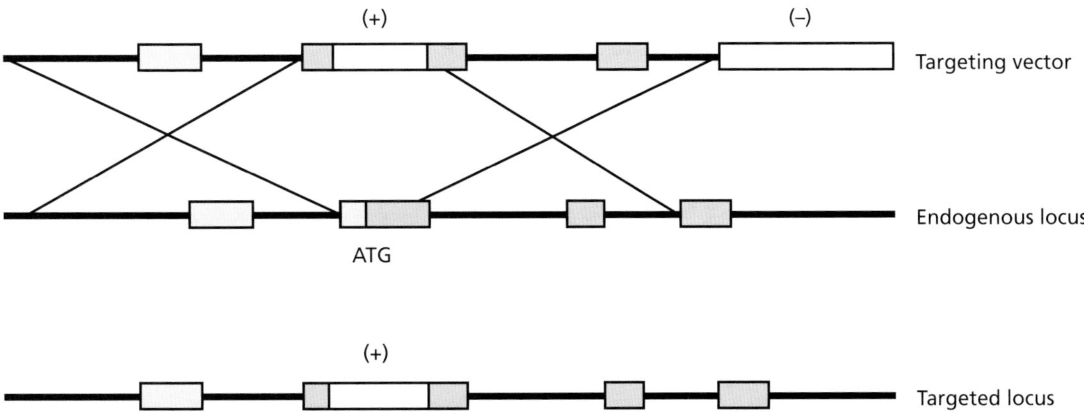

Figure 4 Replacement (Ω) targeting vector. The two regions that represent sites for homologous recombination are indicated by thick lines. The positive selection cassette (+) replaces a portion of the coding sequence and confers antibiotic resistance, allowing for the selection of ES cell clones that have incorporated the transgene. Neo^R and Puro^R are commonly used. Here the (+) cassette is also positioned to replace the endogenous initiation codon in an attempt to ensure a null allele is left at the targeted locus. The (–) cassette, if retained because of random integration, will render cells susceptible to an administered compound. Herpes simplex virus-thymidine kinase (HSV-tk) is commonly used and makes cells sensitive to gancyclovir or fialuridine [1-(2-deoxy-2-fluoro-β-D-arabinofurano-syl)-5-iodouracil] (FIAU) treatment. Together, this (+)/(–) strategy allows for the selective survival of ES cells that have undergone incorporation at the desired locus, which have the (+) cassette and have excluded the (–) cassette. ATG, start codon

Table 3 Knock-out mouse models with reproductive findings (in homozygotes unless specified)

Gene	*Reproductive phenotype*	*Reference(s)*
A-myb (*Myb11*)	male infertility with germ cell arrest; females have mammary gland defects	81
Acrosin	males exhibit delayed fertility, although sperm are capable of binding and penetrating the zona pellucida	82
Activin receptor-type IIA (*Acvr2*)	infertility in females and delayed fertility in males; small gonads	83
Activin/inhibin βB subunit (*Inhbb*)	females are subfertile and litters do not survive postnatally	84
Ahch (Dax1 or DSS-AHC region on the human X)	males are infertile with progressive degeneration of the germinal epithelium	85
Angiotensin-converting enzyme (*Ace*)	male subfertility; compromised ability of sperm to fertilize ova	86
Apaf-1 (apoptotic protease activating factor-1)	few homozygotes survive to reproduce; males that do are infertile with spermatogonial degeneration	87
Apolipoprotein B (*Apob*)	heterozygote males infertile; decreased sperm count, motility, survival time and ability to fertilize ova	88
Aryl-hydrocarbon receptor (*Ahr*)	female subfertility; early development of primordial follicles; decreased numbers of antral follicles	89, 90
Ataxia telangiectasia (*Atm*)	germ cells of both sexes degenerate; disruptions evident early in meiosis I	91
ATP-binding cassette transporter-1 (*Abca1*)	placental malformations leading to impaired embryo growth, embryo loss and neonatal death	92

continued

Table 3 *continued*

Gene	*Reproductive phenotype*	*Reference(s)*
Basigin (*Bsg*)	males are infertile with a block in spermatogenesis; females are infertile with defects in fertility and implantation	93, 94
Bax	males are infertile with a spermatogenesis block; females have an extended reproductive life span and increased primordial follicle reserve	95, 96
Bcl2	females have fewer oocytes/primordial follicles in the postnatal ovary	97
Bcl6	compromised male fertility; apoptosis in metaphase I spermatocytes	98
BCLW (*Bcl212*)	male infertility; spermatogenesis block with eventual loss of germ cells and Sertoli cells	99
Bone morphogenetic protein 8A (*Bmp8a*)	males exhibit progressive infertility; degeneration of germ cells and epididymis	100
Bone morphogenetic protein 8B (*Bmp8b*)	males infertile; germ cell proliferation/depletion defects	101
Calmegin (*Clgn*)	males are infertile; defect in sperm–zona pellucida binding	102
Camk4	males are infertile; impaired chromatin packaging during spermiogensis	103
Casein kinase IIα	males are infertile; exhibit globozoospermia (no acrosomal cap)	104
Caspase-2 (*Casp2*)	decreased apoptosis of female germ cells	105
c-mos	female subfertility; parthogenetic activation, cysts and teratomas	106, 107
c-ros	male infertility due to sperm motility defects	108, 109
Centromere protein B	males are hypogonadal and have low sperm counts; females have strain-dependent uterine epithelium defects	110, 111
C/EPBβ (CCAAT/enhancer-binding protein β)	female infertility; reduced ovulation and block in CL differentiation	112
Colony stimulating factor (*Csf1*)	subfertility in both sexes; males have reduced testosterone; females have implantation and lactation defects	113
Connexin 37 (*Gja4*)	female infertility; defects in late folliculogenesis and oocyte meiosis	114
Connexin 43 (*Gja1*)	neonatal lethality. Small ovaries and testes; decreased numbers of germ cells from E11.5	115
Crem (cAMP-responsive element modulator)	males are infertile due to defective spermatogenesis	116, 117
Cyclin A1 (*Ccna1*)	male infertility; block in spermatogenesis before the first meiotic division	118
Cyclin D2 (*Ccnd2*)	female infertility due to a failure of granulosa cell proliferation; males fertile with decreased testis size	119
Cyclin-dependent kinase 4 (*Cdk4*)	female infertility; defects in the hypothalamic–pituitary–gonadal axis	120
Cyclooxygenase 2 (*Ptgs2*)	females are mostly infertile; defects in ovulation and implantation	121, 122
Dazl (deleted in azoospermia-like autosomal)	differentiation failure and degeneration of both male and female germ cells	123
Desert hedgehog (*Dhh*)	males are infertile; likely roles for *Dhh* in early and late stages of spermatogenesis	124
Dmc1 (disrupted meiotic cDNA1 homolog)	defects in chromosome synapsis in meiosis; both males and females infertile	125, 126

continued

Table 3 *continued*

Gene	*Reproductive phenotype*	*Reference(s)*
Emx2 (empty spiracles homolog 2)	defective development of gonads and genital tracts	127
EP_2 prostaglandin E_2 receptor (*Ptger2*)	female subfertility; decreased fertilization and defects in cumulus expansion	128–130
Estrogen receptor α (ERα) (*Esr1*)	females are infertile with hemorrhagic ovarian cysts and uterine defects; males develop disruptions of the seminiferous epithelium	131, 132
Estrogen receptor β (ERβ) (*Esr2*)	females are subfertile; males are fertile, but develop prostate hyperplasia	133
Fanconi anemia complementation group C (*Fancc*)	compromised gametogenesis in males and females; impaired fertility	134, 135
Figla or FIGα (factor in the germline α)	female infertility; no primordial follicles develop at birth and oocytes die	136
Fragile-X (*Fmr1*)	males exhibit macro-orchidism	137
FSH hormone β-subunit (*Fshb*)	female infertility; pre-antral block in folliculogenesis; males fertile with decreased testis size	138
FSH receptor (*Fshr*)	female infertility; block in folliculogenesis prior to antral formation	139
β1,4-Galactosyltransferase	male infertility; defects in sperm–egg interaction; females exhibit dystocia and agalactosis	140, 141
γ-Glutamyl transpeptidase (*Ggtp*)	both males and females are hypogonadal and infertile; phenotype corrected by feeding mice *N*-acetylcysteine	142
Glycoprotein hormone α-subunit (*cga*)	males and females are infertile; hypogonadal due to FSH and LH deficiency	143
Growth differentiation factor-9 (*Gdf9*)	female infertility; folliculogenesis arrest at the one-layer follicle stage	144
Growth hormone receptor (*Ghr*)	females show delayed puberty and prolonged pregnancy	145
Heat shock protein 70-2 (*Hsp70-2*)	male infertility; meiosis defects and germ cell apoptosis	146
Heatshock transcription factor 1 (*Hsf1*)	female infertility; pre- and post-implantation defects	147
Hoxa10	variable infertility; males have cryptorchidism and females have frequent embryo loss prior to implantation	148
Hoxa11	infertility in both sexes; females have uterine defects; males have malformed vas deferens and undescended testes	149
Inhibin α (*Inha*)	female infertility; male secondary infertility; granulosa/Sertoli tumors	150
Insulin-like growth factor-I (*Igf1*)	females are hypogonadal and infertile; impaired antral follicle formation	151
Insulin receptor substrate-2 (*Irs2*)	females infertile; small, anovulatory ovaries with reduced numbers of follicles	152
Interleukin 11 (*Il11*)	female infertility; compromised implantation and decidualization	153
JunD (*Jund1*)	males are infertile; anomalous hormone levels and sperm structure defects	154
Leukemia inhibitory factor (*Lif*)	females infertile; failed implantation	155
Mismatch repair gene homolog (*Pms2*)	males are infertile; abnormal chromosome synapsis in meiosis	156
Mlh1 (MutL homolog 1)	infertility in both sexes; defects in meiosis and genome instability	157, 158

continued

Table 3 *continued*

Gene	*Reproductive phenotype*	*Reference(s)*
Msh5 (MutS homolog 5)	infertility in both sexes; prophase I meiotic defects with aberrant chromosome synapsis and apoptosis	159, 160
Müllerian inhibiting substance (*Amh*)	uterus development in males causes obstruction and secondary infertility; females exhibit early depletion of primordial follicles	161, 162
Müllerian inhibiting substance receptor	male subfertility; Müllerian duct causes physical blockage	163
$Na^{(+)}$–$K^{(+)}$–$2Cl^{(-)}$ co-transporter	males are infertile; low spermatid counts and compromised sperm transport	164
NGFI-A transcription factor (*Egr1*)	female infertility; LH deficiency	165, 166
Neuronal helix-loop-helix-2	males are infertile and hypogonadal; females are fertile when reared with males	167
NIRKO mice (neuronal insulin receptor knock-out)	mice exhibit hypothalamic hypogonadism; impaired spermatogenesis and follicle maturation	168
Osp-11/Claudin-11	males are infertile; no tight junctions between Sertoli cells	169
Ovo	males have reduced fertility and hypogenitalism	170
Oxytocin (*Oxt*)	females unable to nurse offspring	171
P450 aromatase (*Cyp19*)	females are infertile; ovaries do not form CL; males develop progressive infertility with defects in spermatogenesis	172, 173
P2X1 receptor (*P2rx1*)	male infertility; oligospermia and defective vas deferens contraction	174
$p19^{Ink4d}$ (*Cdkn2d*)	males are fertile despite testicular atrophy and germ cell apoptosis	175
$p27^{Kip1}$ (*Cdkn1b*)	female infertility with CL differentiation failure and granulosa cell hyperplasia; males fertile with testicular hyperplasia	176, 177
p53 (*Trp53*)	males are compromised in recovering spermatogenesis after irradiation	178
PC4 (testicular germ cell protease)	males are infertile; sperm have impaired fertilization ability	179
Phosphatidylinositol 3'-kinase	males are infertile; defects in proliferation and increased apoptosis of spermatogonia	180
Phosphatidylinositol glycan, class A (*Piga*)	chimeric males have abnormal testes, epididymis and seminal vesicles	181
Progesterone receptor (*Pgr*)	female infertility; ovulation failure, uterine implantation defects and mammary defects	182
Prolactin (*Prl*)	females are infertile with irregular estrous cycles	183
Prolatin receptor (*Prlr*)	males infertile; females infertile related to compromised ovulation, fertilization and preimplantation development	184
Prostaglandin F receptor (*Ptgfr*)	females do not undergo parturition; failed luteolysis	185
Protein phosphatase-1 catalytic subunit γ (*Pp1cc*)	male infertility; defects in spermiogenesis	186
Retinoic acid receptor α (*Rara*)	male infertility with seminiferous tubule degeneration	187
Retinoic acid receptor γ (*Rarg*)	male infertility; squamous metaplasia of the seminal vesicles and prostate	188
Retinoid X receptors (*Rxrb*)	male infertility; germ cell maturation defects and tubular degeneration	189
Rho GDIα (GDP dissociation inhibitor)	both sexes infertile; impaired spermatogenesis in males; post-implantation pregnancy defects in females	190

continued

Table 3 *continued*

Gene	*Reproductive phenotype*	*Reference(s)*
Scavenger receptor, class B1 (*Srb1*)	female infertility; defects in oocyte maturation and early embryo development	191, 192
SCP3 (synaptonemal complex protein 3) (*Sycp3*)	males are infertile; defects in chromosome synapsis during meiosis; germ cell apoptosis	193
Sp4 transcription factor (*Sp4*)	males are infertile; defects in reproductive behavior	194
Sperm-1	males are subfertile; defect in haploid sperm function	195
SPO11 homolog (*Spo11*)	infertility in both sexes; defects in meiosis; spermatocytes undergo apoptosis in early prophase; oocytes are lost soon after birth	196, 197
Steroid 5α-reductase type 1 (*Srd5a1*)	female infertility; defects in parturition	198, 199
Steroidogenic acute regulatory protein (*Star*)	males have female external genitalia; both sexes die of adrenocortical insufficiency	200
Steroidogenic factor-1 (SF-1) (*Nr5a1*)	gonadal agenesis in both sexes	201
Superoxide dismutase 1 (*Sod1*) (copper–zinc superoxide dismutase)	females are subfertile; folliculogenesis defect; failure to maintain pregnancy	202, 203
Telomerase	progressive infertility in both sexes; males show germ cell apoptosis; females have few oocytes and uterine abnormalities	204
TIAR RNA-binding protein	PGCs are lost by embryonic day 13.5; no spermatogonia or oogonia develop	205
TLS (translocated in liposarcoma) (*Fus1*)	males are infertile; defects in spermatocyte chromosome pairing	206
Tnp1 (transition protein 1)	male subfertility; abnormal chromosome condensation and reduced sperm motility	207
Tyro-3 family receptors (*Tyro3*, *Axl* and *Mer*)	triple knock-out males are infertile; loss of differentiating germ cells due to Sertoli cell dysfunction; triple knock-out females are fertile but exhibit granulosa cell degeneration	208
Ubiquitin-conjugating DNA repair enzyme (HR6B)	males are sterile with alterations in chromatin structure	209
VASA (*Ddx4*)	infertile males; defective proliferation/differentiation of primordial germ cells	210
Wilms tumor homolog (*Wt1*)	embryonic lethality with gonadal agenesis	211
Wnt4	female infertility; ovaries are depleted of oocytes and contain Leydig cells; Müllerian ducts do not form	212
Wnt7a	females show abnormal development of oviducts and uterus; males do not have Müllerian duct regression	213
Zfx (zinc finger protein, X-linked)	reduced germ cell numbers in both sexes; defects in mitotic proliferation	214
Zona pellucida protein (*Zp1*)	reduced female fertility; defects post-fertilization	215
Zona pellucida proetin (*Zp2*)	female infertility; fragile oocytes	216
Zona pellucida protein (*Zp3*)	female infertility; fragile oocytes	217, 218

DSS-AHC, dosage-sensitive sex reversal–adrenal hypoplasia congenita; ATP, adenosine triphosphate; CL, corpus luteum; cAMP, cyclic adenosine monophosphate; FSH, follicle stimulating hormone; LH, luteinizing hormone; GDP, guanosine diphosphate

coding sequence with another, creating a 'knock-in' model. This is a powerful technique for examining differences in the potential functions of related genes. For example, mice null for activin/inhibin βA (*Inhba–/–*) die neonatally owing to craniofacial defects that prevent suckling. This phenotype can be rescued by replacing the activin/inhibin βA coding sequence with that of the activin/inhibin βB gene, conferring the activin/inhibin βA expression pattern on this related sequence (63% amino acid identity). Interestingly, the homozygous knock-in mice demonstrate enlarged external genitalia, hypogonadism and diminished female fertility, indicating unique and previously unrecognized functions of the activin/inhibin βA protein product in reproduction.[229] Combinations of knock-out and bitransgenic models are also being used to investigate whether a specific spatiotemporal expression pattern is key to a gene function. For example, a mifeprisone-inducible system regulating ectopic expression of inhibin A subunits can avert testicular tumorigenesis in inhibin α-null mice as long as the transgene expression is maintained.[5]

SITE-SPECIFIC RECOMBINATION TECHNOLOGY

Recombinases from P1 bacteriophage (Cre) and yeast (FLP) have been shown to mediate double-stranded exchanges between defined DNA sequences when these targets are introduced into mammalian cells.[230,231] Recombinases and their target sites are used in transgenic systems in the design of tissue-specific knock-outs.[232] An endogenous gene to be deleted is first flanked by tandem *loxP* sites, 34-bp target sequences for Cre recombinase. Expression of the Cre protein in a tissue-specific and/or inducible manner will mediate deletion of the sequence between the *loxP* sites, whereas, in tissues without Cre activity, the intervening sequence remains intact. The introduction of subtle mutations or replacement alleles at a targeted locus represent similar applications of the Cre/*loxP* system (Figure 5). This approach was used recently to introduce a missense mutation within the Kit receptor (Kit^{Y719F}), thereby revealing roles for a specific aspect of Kit signalling in spermatogenesis and oogenesis.[233]

The *loxP* sites can be recognized by Cre recombinase over large distances within the genome, allowing the recombination system to be harnessed to engineer large chromosomal rearrangements.[234,235] Long-range recombination can be used to generate large chromosome deletions, duplications and inversions. The final result of the recombination depends on the orientations of the *loxP* sites with respect to one another and with respect to the centromere, as well as whether recombination occurs between sites on one chromatid (*cis*) or two sister chromatids (*trans*).

GERM CELL TRANSPLANTATION TECHNOLOGY

Spermatogonial stem cell transplantation is an emerging technology developed by Brinster and colleagues. The first reports were issued in 1994 when they demonstrated that a heterogeneous mouse testicular cell suspension could be injected into the seminiferous tubules of a sterile mouse recipient to give rise to active spermatogenesis and, ultimately, sperm capable of fertilization.[236,237] Moreover, donor germ cells can be cryopreserved[238] or cultured *in vitro* for long periods[239] before injection. Interspecies spermatogonial transplantation and germ cell exchanges between different mouse lines with defects in spermatogenesis[240] are under way to study germ cell–somatic cell interaction in the testis. Amazingly, rat germ cells introduced into the tubules of sterile, immunocompromised mice successfully complete spermatogenesis.[241] Germ cells of the hamster,[242] dog,[243] rabbit[243] and large domesticated animals[244] injected into recipient mice can survive, but do not successfully develop.

An exchange between germ cell and somatic cell components of the gonad is also possible in the case of the female. Eppig and colleagues[245] separated germ and somatic cells from trypsinized ovaries of newborn mice by using *in vitro* conditions in which somatic cells adhere to a plate and oocytes

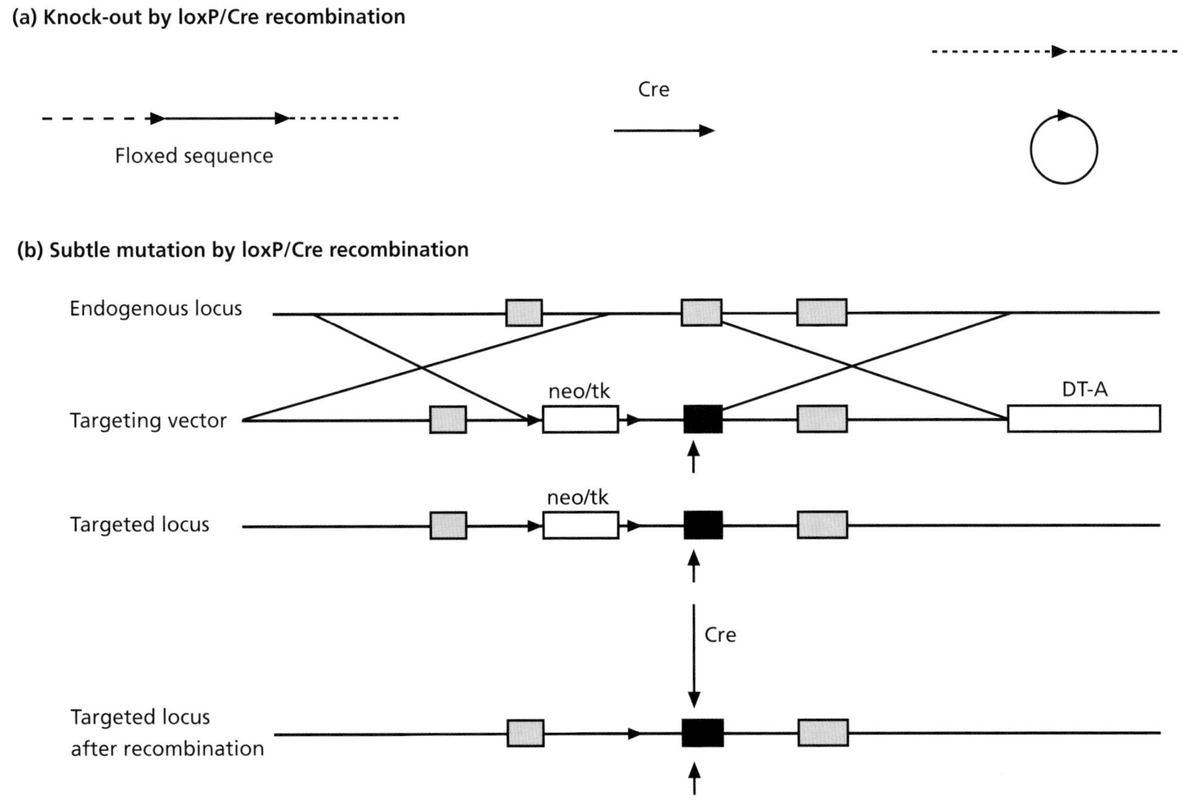

(a) Knock-out by loxP/Cre recombination

Cre

Floxed sequence

(b) Subtle mutation by loxP/Cre recombination

Endogenous locus

neo/tk

DT-A

Targeting vector

neo/tk

Targeted locus

Cre

Targeted locus after recombination

Figure 5 Cre/*loxP*-mediated recombination. (a) Tissue-specific knock-out. Two *l*ocus of crossover *x* in *P1* sites (*loxP* sites) have been introduced in tandem orientations flanking a sequence to be deleted. This is the 'floxed' allele. A promoter driving the Cre recombinase transgene determines the temporal and tissue specificity of the deletion. When the Cre transgene is expressed, active recombinase enzyme will promote a recombination between the *loxP* sequences. The result is an excision and circularization of the intervening sequence. In the absence of Cre the functional 'floxed' allele remains intact. (b) Introduction of a subtle mutation using Cre/*loxP*. A targeting vector replaces a portion of coding sequence with the desired allele (here the second exon; arrow) and introduces a 'floxed' positive/negative dual selection cassette (neomycin/tk). Two-part selection for the incorporation of the dual cassette and for the loss of a second cassette (here, the diphtheria toxin A (DT-A) gene fragment) ensures a homologous and not random introduction. Finally, Cre recombinase electroporation and selection for loss of the positive/negative cassette promotes the excision of the markers, and the locus is left with only the subtle change of codon replacement

do not. Germ cells and somatic cells from different individuals can reaggregate into 'chimeric ovaries', which can be implanted into immunocompromised mice to study the progression of folliculogenesis. The technique has been used to investigate a pathology in the LTXBO strain of mice wherein spontaneous parthenogenetic activation of oocytes leads to the development of ovarian teratomas. It was found that LTXBO oocytes exhibit the activation even when surrounded by somatic cells of other mouse strains, while LTXBO somatic components do not induce the activation of oocytes from other strains. Thus, the defect is localized to the LTXBO oocytes. Xenogenic exchanges have also been tried. Oocytes from mice maintain species-specific characteristics during development within rat follicles; these mouse gametes in the context of rat somatic cells can be recovered, *in vitro* fertilized and implanted into pseudopregnant recipients for successful development.[246] These germ cell transplantation technologies offer a new tool to manipulate the haploid genome and study the complexities of gamete production in multiple species.

CLONING WHOLE MAMMALIAN GENOMES BY NUCLEAR TRANSPLANTATION

The genetic tools described in this chapter are expanding our ability to manipulate complex genomes, and perhaps no technology symbolizes this to the public more than experiments that asexually reproduce genetically identical animals. Mammalian cloning was first accomplished in mice by replacing the pronuclei of a zygote with a nucleus from an early-stage embryo.[247] Cloning receives renewed interest today, as Wilmut and colleagues[248] have demonstrated in sheep that karyoplasts of some fully differentiated somatic cells retain pluripotency and can be injected into enucleated oocytes to prompt successful development. Transgenic sheep have been produced using this method by introducing transgenes into the somatic cells *in vitro*, selecting for the transgene integration and producing viable animals by nuclear transfer.[249] Similar experiments using mice show that the genetic material of fully differentiated adult cells can be used to derive cloned animals.[250,251] Although many technical difficulties and ethical concerns remain to be addressed, the development of cloning technology will continue in the future, and these studies will address intriguing questions pertaining to cell fate determination, chromatin remodelling and the field of epigenetics.

GENETICS, DEVELOPMENTAL BIOLOGY AND REPRODUCTIVE MEDICINE

It is clear that the development of transgenic technology, germ cell transplantation and nuclear transfer techniques has been contingent upon early discoveries in developmental and reproductive biology. We are now students of the conciliance of classical genetics and these physiological sciences. Although the technologies described herein are revolutionizing diverse fields of study, those like ourselves with an interest in reproductive medicine are uniquely positioned to appreciate and contribute to their ongoing invention and future promise.

ACKNOWLEDGEMENTS

We thank Dr Julia A. Elvin and Dr T. Rajendra Kumar for their assistance in preparing the tables presented. We also thank Ms Shirley Baker for her assistance with manuscript preparation. Transgenic mouse research in the Matzuk laboratory has been supported by Wyeth Ayerst Research, National Institutes of Health (NIH) Grants CA60651, HD32067 and HD33438, and the Specialized Cooperative Centers Program in Reproduction Research (HD07495). One of the authors (K.H.B.) is a student in the Medical Scientist Training Program at Baylor College of Medicine, and is supported in part by NIH grant T32GM07330 and The National Eye Institute grant T32EY07102.

REFERENCES

1. Gordon JW, Scangos GA, Plotkin DJ, Barbosa JA, Ruddle FH. Genetic transformation of mouse embryos by microinjection of purified DNA. *Proc Natl Acad Sci USA* 1980 77: 7380–7384.
2. Gordon JW, Ruddle FH. Integration and stable germ line transmission of genes injected into mouse pronuclei. *Science* 1981 214: 1244–1246.
3. Brinster RL, Chen HY, Trumbauer M, Senear AW, Warren R, Palmiter RD. Somatic expression of herpes thymidine kinase in mice following injection of a fusion gene into eggs. *Cell* 1981 27: 223–231.
4. Tanimoto Y, Tanimoto K, Sugiyama F *et al.* Male sterility in transgenic mice expressing activin βA subunit gene in testis. *Biochem Biophys Res Commun* 1999 259: 699–705.
5. Pierson MP, DeMayo FJ, Matzuk MM, Tsai SY, O'Malley BWO. Regulable expression of inhibin A in wild-type and inhibin a null mice. *Mol Endocrinol* 2000 14: 1075–1085.
6. Larriba S, Esteban C, Toran N *et al.* Androgen binding protein is tissue-specifically expressed and biologically active in transgenic mice. *J Steroid Biochem Mol Biol* 1995 53: 573–578.
7. Hsu SY, Lai RJ, Finegold M, Hsueh AJ. Targeted overexpression of *Bcl-2* in ovaries of transgenic mice leads to decreased follicle apoptosis, enhanced folliculogenesis, and increased germ cell tumorigenesis. *Endocrinology* 1996 137: 4837–4843.

8. Furuchi T, Masuko K, Nishimune Y, Obinata M, Matsui Y. Inhibition of testicular germ cell apoptosis and differentiation in mice misexpressing *Bcl-2* in spermatogonia. *Development* 1996 122: 1703–1709.
9. Suzuki M, Abe K, Yoshinaga K, Obinata M, Furusawa M. Specific arrest of spermatogenesis caused by apoptotic cell death in transgenic mice. *Genes Cells* 1996 1: 1077–1086.
10. Swain A, Narvaez V, Burgoyne P, Camerino G, Lovell-Badge R. *Dax1* antagonizes *Sry* action in mammalian sex determination. *Nature* 1998 391: 761–767.
11. Kendall SK, Saunders TL, Jin L *et al.* Targeted ablation of pituitary gonadotropes in transgenic mice. *Mol Endocrinol* 1991 5: 2025–2036.
12. Bartell JG, Fantz DA, Davis T, Dewey MJ, Kistler MK, Kistler WS. Elimination of male germ cells in transgenic mice by the diphtheria toxin A chain gene directed by the histone H1t promoter. *Biol Reprod* 2000 63: 409–416.
13. Holmberg C, Helin K, Sehested M, Karlstrom O. E2F-1-induced p53-independent apoptosis in transgenic mice. *Oncogene* 1998 17: 143–155.
14. Wong RW, Kwan RW, Mak PH, Mak KK, Sham MH, Chan SY. Overexpression of epidermal growth factor induced hypospermatogenesis in transgenic mice. *J Biol Chem* 2000 275: 18297–18301.
15. Guo Q, Kumar TR, Woodruff T, Hadsell LA, DeMayo FJ, Matzuk MM. Overexpression of mouse follistatin causes reproductive defects in transgenic mice. *Mol Endocrinol* 1998 12: 96–106.
16. Youakim A, Hathaway HJ, Miller DJ, Gong X, Shur BD. Overexpressing sperm surface β 1,4-galactosyltransferase in transgenic mice affects multiple aspects of sperm–egg interactions. *J Cell Biol* 1994 126: 1573–1583.
17. Bartke A, Cecim M, Tang K, Steger RW, Chandrashekar V, Turyn D. Neuroendocrine and reproductive consequences of overexpression of growth hormone in transgenic mice. *Proc Soc Exp Biol Med* 1994 206: 345–359.
18. Nakai A, Suzuki M, Tanabe M. Arrest of spermatogenesis in mice expressing an active heat shock transcription factor 1. *EMBO J* 2000 19: 1545–1554.
19. Hekman AC, Trapman J, Mulder AH, van Gaalen JL, Zwarthoff EC. Interferon expression in the testes of transgenic mice leads to sterility. *J Biol Chem* 1988 263: 12151–12155.
20. Iwakura Y, Asano M, Nishimune Y, Kawade Y. Male sterility of transgenic mice carrying exogenous mouse interferon-β gene under the control of the metallothionein enhancer–promoter. *EMBO J* 1988 7: 3757–3762.
21. Ohta M, Mitomi T, Kimura M, Habu S, Katsuki M. Anomalies in transgenic mice carrying the human interleukin-2 gene. *Tokai J Exp Clin Med* 1990 15: 307–315.
22. Yura S, Ogawa Y, Sagawa N *et al.* Accelerated puberty and late-onset hypothalamic hypogonadism in female transgenic skinny mice overexpressing leptin. *J Clin Invest* 2000 105: 749–755.
23. Risma KA, Clay CM, Nett TM, Wagner T, Yun J, Nilson JH. Targeted overexpression of luteinizing hormone in transgenic mice leads to infertility, polycystic ovaries, and ovarian tumors. *Proc Natl Acad Sci USA* 1995 92: 1322–1326.
24. Rudolph-Owen LA, Cannon P, Matrisian LM. Overexpression of the matrix metalloproteinase matrilysin results in premature mammary gland differentiation and male infertility. *Mol Biol Cell* 1998 9: 421–435.
25. Behringer RR, Cate RL, Froelick GJ, Palmiter RD, Brinster RL. Abnormal sexual development in transgenic mice chronically expressing mullerian inhibiting substance. *Nature* 1990 345: 167–170.
26. Mangues R, Seidman I, Pellicer A, Gordon JW. Tumorigenesis and male sterility in transgenic mice expressing a MMTV/N-*ras* oncogene. *Oncogene* 1990 5: 1491–1497.
27. Guy CT, Cardiff RD, Muller WJ. Activated *neu* induces rapid tumor progression. *J Biol Chem* 1996 271: 7673–7678.
28. Hakovirta H, Keiski A, Toppari J *et al.* Polyamines and regulation of spermatogenesis: selective stimulation of late spermatogonia in transgenic mice overexpressing the human ornithine decarboxylase gene. *Mol Endocrinol* 1993 7: 1430–1436.
29. Fowler KA, Gill K, Kirma N, Dillehay DL, Tekmal RR. Overexpression of aromatase leads to development of testicular Leydig cell tumors: an *in vivo* model for hormone-mediated testicular cancer. *Am J Pathol* 2000 156: 347–353.
30. Allemand I, Anglo A, Jeantet AY, Cerutti I, May E. Testicular wild-type p53 expression in transgenic mice induces spermiogenesis alterations ranging

from differentiation defects to apoptosis. *Oncogene* 1999 18: 6521–6530.

31. O'Hara BF, Donovan DM, Lindberg I *et al.* Proenkephalin transgenic mice: a short promoter confers high testis expression and reduced fertility. *Mol Reprod Dev* 1994 38: 275–284.
32. Rhim JA, Connor W, Dixon GH, *et al.* Expression of an avian protamine in transgenic mice disrupts chromatin structure in spermatozoa. *Biol Reprod* 1995 52: 20–32.
33. Lee K, Haugen HS, Clegg CH, Braun RE. Premature translation of protamine 1 mRNA causes precocious nuclear condensation and arrests spermatid differentiation in mice. *Proc Natl Acad Sci USA* 1995 92: 12451–12455.
34. Costa SL, Boekelheide K, Vanderhyden BC, Seth R, McBurney MW. Male infertility caused by epididymal dysfunction in transgenic mice expressing a dominant negative mutation of retinoic acid receptor α1. *Biol Reprod* 1997 56: 985–990.
35. Pietila M, Alhonen L, Halmekyto M, Kanter P, Janne J, Porter CW. Activation of polyamine catabolism profoundly alters tissue polyamine pools and affects hair growth and female fertility in transgenic mice overexpressing spermidine/spermine N_1-acetyltransferase. *J Biol Chem* 1997 272: 18746–18751.
36. Koopman P, Gubbay J, Vivian N, Goodfellow P, Lovell-Badge R. Male development of chromosomally female mice transgenic for *Sry*. *Nature* 1991 351: 117–121.
37 Radovick S, Wray S, Lee E *et al.* Migratory arrest of gonadotropin-releasing hormone neurons in transgenic mice. *Proc Natl Acad Sci USA* 1991 88: 3402–3406.
38. Markkula MA, Hamalainen TM, Zhang F, Kim KE, Maurer RA, Huhtaniemi IT. The FSH β-subunit promoter directs the expression of herpes simplex virus type 1 thymidine kinase to the testis of transgenic mice. *Mol Cell Endocrinol* 1993 96: 25–36.
39. Al-Shawi R, Burke J, Jones CT, Simons JP, Bishop JO. A Mup promoter–thymidine kinase reporter gene shows relaxed tissue-specific expression and confers male sterility upon transgenic mice. *Mol Cell Biol* 1988 8: 4821–4828.
40. Das SK, Lim H, Wang J, Paria BC, BazDresch M, Dey SK. Inappropriate expression of human transforming growth factor (TGF)-α in the uterus of transgenic mouse causes downregulation of TGF-β receptors and delays the blastocyst-attachment reaction. *J Mol Endocrinol* 1997 18: 243–257.
41. Korpelainen EI, Karkkainen MJ, Tenhunen A *et al.* Overexpression of VEGF in testis and epididymis causes infertility in transgenic mice: evidence for nonendothelial targets for VEGF. *J Cell Biol* 1998 143: 1705–1712.
42. Rosenberg MP, Aversa CR, Wallace R, Propst F. Expression of the v-*Mos* oncogene in male meiotic germ cells of transgenic mice results in metaphase arrest. *Cell Growth Differ* 1995 6: 325–336.
43. Costantini F, Radice G, Lee JL, Chada KK, Perry W, Son HJ. Insertional mutations in transgenic mice. *Prog Nucleic Acid Res Mol Biol* 1989 36: 159–169.
44. Wurst W, Rossant J, Prideaux V *et al.* A large-scale gene-trap screen for insertional mutations in developmentally regulated genes in mice. *Genetics* 1995 139: 889–899.
45. Gossler A, Joyner AL, Rossant J, Skarnes WC. Mouse embryonic stem cells and reporter constructs to detect developmentally regulated genes. *Science* 1989 244: 463–465.
46. Friedrich G, Soriano P. Promoter traps in embryonic stem cells: a genetic screen to identify and mutate developmental genes in mice. *Genes Dev* 1991 5: 1513–1523.
47. Bessone S, Vidal F, Le Bouc Y, Epelbaum J, Bluet-Pajot MT, Darmon M. EMK protein kinase-null mice: dwarfism and hypofertility associated with alterations in the somatotrope and prolactin pathways. *Dev Biol* 1999 214: 87–101.
48. Pellas TC, Ramachandran B, Duncan M, Pan SS, Marone M, Chada K. Germ-cell deficient (*gcd*), an insertional mutation manifested as infertility in transgenic mice. *Proc Natl Acad Sci USA* 1991 88: 8787–8791.
49. Couldrey C, Carlton MB, Nolan PM, Colledge WH, Evans MJ. A retroviral gene trap insertion into the histone 3.3A gene causes partial neonatal lethality, stunted growth, neuromuscular deficits and male sub-fertility in transgenic mice. *Hum Mol Genet* 1999 8: 2489–2495.
50. Gordon JW, Uehlinger J, Dayani N *et al.* Analysis of the hotfoot (*ho*) locus by creation of an insertional mutation in a transgenic mouse. *Dev Biol* 1990 137: 349–358.

51. Yanaka N, Kobayashi K, Wakimoto K *et al.* Insertional mutation of the murine kisimo locus caused a defect in spermatogenesis. *J Biol Chem* 2000 275: 14791–14794.
52. Magram J, Bishop JM. Dominant male sterility in mice caused by insertion of a transgene. *Proc Natl Acad Sci USA* 1991 88: 10327–10331.
53. Watson ML, Zinn AR, Inoue N *et al.* Identification of *morc* (microrchidia), a mutation that results in arrest of spermatogenesis at an early meiotic stage in the mouse. *Proc Natl Acad Sci USA* 1998 95: 14361–14366.
54. Bishop CE, Whitworth DJ, Qin Y *et al.* A transgenic insertion upstream of *sox9* is associated with dominant XX sex reversal in the mouse. *Nat Genet* 2000 26: 490–494.
55. Krulewski TF, Neumann PE, Gordon JW. Insertional mutation in a transgenic mouse allelic with Purkinje cell degeneration. *Proc Natl Acad Sci USA* 1989 86: 3709–3712.
56. MacGregor GR, Russell LD, Van Beek ME *et al.* Symplastic spermatids (*sys*): a recessive insertional mutation in mice causing a defect in spermatogenesis. *Proc Natl Acad Sci USA* 1990 87: 5016–5020.
57. Gordon JW, Pravtcheva D, Poorman PA, Moses MJ, Brock WA, Ruddle FH. Association of foreign DNA sequence with male sterility and translocation in a line of transgenic mice. *Somat Cell Mol Genet* 1989 15: 569–578.
58. Palmiter RD, Sandgren EP, Avarbock MR, Allen DD, Brinster RL. Heterologous introns can enhance expression of transgenes in mice. *Proc Natl Acad Sci USA* 1991 88: 478–482.
59. Brinster RL, Allen JM, Behringer RR, Gelinas RE, Palmiter RD. Introns increase transcriptional efficiency in transgenic mice. *Proc Natl Acad Sci USA* 1988 85: 836–840.
60. Palmiter RD, Brinster RL. Germ-line transformation of mice. *Annu Rev Genet* 1986 20: 465–499.
61. Al-Shawi R, Kinnaird J, Burke J, Bishop JO. Expression of a foreign gene in a line of transgenic mice is modulated by a chromosomal position effect. *Mol Cell Biol* 1990 10: 1192–1198.
62. Yang XW, Model P, Heintz N. Homologous recombination based modification in *Escherichia coli* and germline transmission in transgenic mice of a bacterial artificial chromosome. *Nat Biotechnol* 1997 15: 859–865.
63. Peterson KR, Clegg CH, Huxley C *et al.* Transgenic mice containing a 248-kb yeast artificial chromosome carrying the human β-globin locus display proper developmental control of human globin genes. *Proc Natl Acad Sci USA* 1993 90: 7593–7597.
64. Gossen M, Bujard H. Tight control of gene expression in mammalian cells by tetracycline-responsive promoters. *Proc Natl Acad Sci USA* 1992 89: 5547–5551.
65. Gossen M, Freundlieb S, Bender G, Muller G, Hillen W, Bujard H. Transcriptional activation by tetracyclines in mammalian cells. *Science* 1995 268: 1766–1769.
66. Shockett P, Difilippantonio M, Hellman N, Schatz DG. A modified tetracycline-regulated system provides autoregulatory, inducible gene expression in cultured cells and transgenic mice. *Proc Natl Acad Sci USA* 1995 92: 6522–6526.
67. Baron U, Freundlieb S, Gossen M, Bujard H. Co-regulation of two gene activities by tetracycline via a bidirectional promoter. *Nucleic Acids Res* 1995 23: 3605–3606.
68. Hofmann A, Nolan GP, Blau HM. Rapid retroviral delivery of tetracycline-inducible genes in a single autoregulatory cassette. *Proc Natl Acad Sci USA* 1996 93: 5185–5190.
69. Wang Y, O'Malley J, Tsai SY, O'Malley BW. A regulatory system for use in gene transfer. *Proc Natl Acad Sci USA* 1994 91: 8180–8184.
70. No D, Yao T-P, Evans RM. Ecdysone-inducible gene expression in mammalian cells and transgenic mice. *Proc Natl Acad Sci USA* 1996 93: 3346–3351.
71. Jahner D, Haase K, Mulligan R, Jaenisch R. Insertion of the bacterial *gpt* gene into the germ line of mice by retroviral infection. *Proc Natl Acad Sci USA* 1985 82: 6927–6931.
72. Kanegae Y, Lee G, Sato Y *et al.* Efficient gene activation in mammalian cells by using recombinant adenovirus expressing site-specific Cre recombinase. *Nucleic Acids Res* 1995 23: 3816–3821.
73. Perry AC, Wakayama T, Kishikawa H *et al.* Mammalian transgenesis by intracytoplasmic sperm injection. *Science* 1999 284: 1180–1183.
74. Robertson E, Bradley A, Kuehn M, Evans M. Germ-line transmission of genes introduced into cultured pluripotential cells by retroviral vector. *Nature* 1986 323: 445–448.

75. Kuehn MR, Bradley A, Robertson EJ, Evans MJ. A potential animal model for Lesch–Nyhan syndrome through introduction of HPRT mutations into mice. *Nature* 1987 326: 295–298.
76. Smithies O, Gregg RG, Boggs SS, Koralewski MA, Kucherlapati RS. Insertion of DNA sequences into the human chromosomal β-globin locus by homologous recombination. *Nature* 1985 317: 230–234.
77. Thomas KR, Capecchi MR. Site-directed mutagenesis by gene targeting in mouse embryo-derived stem cells. *Cell* 1987 51: 503–512.
78. Mansour SL, Thomas KR, Capecchi MR. Disruption of the proto-oncogene *int-2* in mouse embryo-derived stem cells: a general strategy for targeting mutations to non-selectable genes. *Nature* 1988 336: 348–352.
79. Hasty P, Rivera-Perez J, Bradley A. The length of homology required for gene targeting in embryonic stem cells. *Mol Cell Biol* 1991 11: 5586–5591.
80. te Riele H, Maandag ER, Berns A. Highly efficient gene targeting in embryonic stem cells through homologous recombination with isogenic DNA constructs. *Proc Natl Acad Sci USA* 1992 89: 5128–5132.
81. Toscani A, Mettus RV, Coupland R *et al.* Arrest of spermatogenesis and defective breast development in mice lacking A-*myb*. *Nature* 1997 386: 713–717.
82. Adham IM, Nayernia K, Engel W. Spermatozoa lacking acrosin protein show delayed fertilization. *Mol Reprod Dev* 1997 46: 370–376.
83. Matzuk MM, Kumar TR, Bradley A. Different phenotypes for mice deficient in either activins or activin receptor type II. *Nature* 1995 374: 356–360.
84. Vassalli A, Matzuk MM, Gardner HAR, Lee K-F, Jaenisch R. Activin/inhibin βB subunit gene disruption leads to defects in eyelid development and female reproduction. *Genes Dev* 1994 8: 414–427.
85. Yu RN, Ito M, Saunders TL, Camper SA, Jameson J. Role of *Ahch* in gonadal development and gametogenesis. *Nat Genet* 1998 20: 353–357.
86. Krege JH, John SW, Langenbach LL *et al.* Male-female differences in fertility and blood pressure in ACE-deficient mice. *Nature* 1995 375: 146–148.
87. Honarpour N, Du C, Richardson JA, Hammer RE, Wang X, Herz J. Adult *Apaf-1*-deficient mice exhibit male infertility. *Dev Biol* 2000 218: 248–258.
88. Huang LS, Voyiaziakis E, Chen HL, Rubin EM, Gordon JW. A novel functional role for apolipoprotein B in male infertility in heterozygous apolipoprotein B knockout mice. *Proc Natl Acad Sci USA* 1996 93: 10903–10907.
89. Benedict JC, Lin TM, Loeffler IK, Peterson RE, Flaws JA. Physiological role of the aryl hydrocarbon receptor in mouse ovary development. *Toxicol Sci* 2000 56: 382–388.
90. Abbott BD, Schmid JE, Pitt JA *et al.* Adverse reproductive outcomes in the transgenic *Ah* receptor-deficient mouse. *Toxicol Appl Pharmacol* 1999 155: 62–70.
91. Barlow C, Liyanage M, Moens PB *et al. Atm* deficiency results in severe meiotic disruption as early as leptonema of prophase I. *Development* 1998 125: 4007–4017.
92. Christiansen-Weber TA, Voland JR, Wu Y *et al.* Functional loss of *ABCA1* in mice causes severe placental malformation, aberrant lipid distribution, and kidney glomerulonephritis as well as high-density lipoprotein cholesterol deficiency. *Am J Pathol* 2000 157: 1017–1029.
93. Kuno N, Kadomatsu K, Fan Q-W *et al.* Female sterility in mice lacking the *basigin* gene, which encodes a transmembrane glycoprotein belonging to the immunoglobulin superfamily. *Fed Eur Biochem Sci* 1998 425: 191–194.
94. Igakura T, Kadomatsu K, Kaname T *et al.* A null mutation in basigin, an immunoglobulin superfamily member, indicates its important roles in peri-implantation development and spermatogenesis. *Dev Biol* 1998 194: 152–165.
95. Knudson CM, Tung KSK, Tourtellotte WG, Brown GAJ, Korsmeyer AJ. *Bax*-deficient mice with lymphoid hyperplasia and male germ cell death. *Science* 1995 270: 96–99.
96. Perez GI, Robles R, Knudson CM, Flaws JA, Korsmeyer SJ, Tilly JL. Prolongation of ovarian lifespan into advanced chronological age by *Bax*-deficiency. *Nat Genet* 1999 21: 200–203.
97. Ratts VS, Flaws JA, Kolp R, Sorenson CM, Tilly JL. Ablation of *bcl-2* gene expression decreases the numbers of oocytes and primordial follicles established in the postnatal female mouse gonad. *Endocrinology* 1995 136: 3665–3668.

98. Kojima S, Hatano M, Okada S *et al.* Testicular germ cell apoptosis in *Bcl6*-deficient mice. *Development* 2001 128: 57–65.
99. Ross AJ, Waymire KG, Moss JE *et al.* Testicular degeneration in *Bclw*-deficient mice. *Nat Genet* 1998 18: 251–256.
100. Zhao G-Q, Liaw L, Hogan BLM. Bone morphogenetic protein 8A plays a role in the maintenance of spermatogenesis and the integrity of the epididymis. *Development* 1998 125: 1103–1112.
101. Zhao G-Q, Deng K, Labosky PA, Liaw L, Hogan BLM. The gene encoding bone morphogeneetic protein 8B is required for the initiation and maintenance of spermatogenesis in the mouse. *Genes Dev* 1996 10: 1657–1669.
102. Ikawa M, Wada I, Kominami K *et al.* The putative chaperone calmegin is required for sperm fertility. *Nature* 1997 387: 607–611.
103. Wu JY, Ribar TJ, Cummings DE, Burton KA, McKnight GS, Means AR. Spermiogenesis and exchange of basic nuclear proteins are impaired in male germ cells lacking *Camk4*. *Nat Genet* 2000 25: 448–452.
104. Xu X, Toselli PA, Russell LD, Seldin DC. Globozoospermia in mice lacking the casein kinase II α' catalytic subunit. *Nat Genet* 1999 23: 118–121.
105. Bergeron L, Perez GI, Macdonald G *et al.* Defects in regulation of apoptosis in caspase-2-deficient mice. *Genes Dev* 1998 12: 1304–1314.
106. Colledge WH, Carlton MB, Udy GB, Evans MJ. Disruption of *c-mos* causes parthenogenetic development of unfertilized mouse eggs. *Nature* 1994 370: 65–68.
107. Hashimoto N, Watanabe N, Furuta Y *et al.* Parthenogenetic activation of oocytes in c-*mos*-deficient mice. *Nature* 1994 370: 68–71.
108. Yeung CH, Sonnenberg-Riethmacher E, Cooper TG. Infertile spermatozoa of c-*ros* tyrosine kinase receptor knockout mice show flagellar angulation and maturational defects in cell volume regulatory mechanisms. *Biol Reprod* 1999 61: 1062–1069.
109. Yeung CH, Wagenfeld A, Nieschlag E, Cooper TG. The cause of infertility of male c-*ros* tyrosine kinase receptor knockout mice. *Biol Reprod* 2000 63: 612–618.
110. Hudson DF, Fowler KJ, Earle E *et al.* Centromere protein B null mice are mitotically and meiotically normal but have lower body and testis weights. *J Cell Biol* 1998 141: 309–319.
111. Fowler KJ, Hudson DF, Salamonsen LA *et al.* Uterine dysfunction and genetic modifiers in centromere protein B-deficient mice. *Genome Res* 2000 10: 30–41.
112. Sterneck E, Tessarollo L, Johnson PF. An essential role for C/EBPβ in female reproduction. *Genes Dev* 1997 11: 2153–2162.
113. Cohen PE, Zhu L, Pollard JW. Absence of colony stimulating factor-1 in osteopetrotic ($csfm^{op}/csfm^{op}$) mice disrupts estrous cycles and ovulation. *Biol Reprod* 1997 56: 110–118.
114. Simon AM, Goodenough DA, Li E, Paul DL. Female infertility in mice lacking connexin 37. *Nature* 1997 385: 525–529.
115. Juneja SC, Barr KJ, Enders GC, Kidder GM. Defects in the germ line and gonads of mice lacking connexin 43. *Biol Reprod* 1999 60: 1263–1270.
116. Blendy JA, Kaestner KH, Weinbauer GF, Nieschlag E, Schutz G. Severe impairment of spermatogenesis in mice lacking the CREM gene. *Nature* 1996 380: 162–165.
117. Nantel F, Monaco L, Foulkes NS *et al.* Spermiogenesis deficiency and germ-cell apoptosis in CREM-mutant mice. *Nature* 1996 380: 159–162.
118. Liu D, Matzuk MM, Sung WK, Guo Q, Wang P, Wolgemuth DJ. Cyclin A1 is required for meiosis in the male mouse. *Nat Genet* 1998 20: 377–388.
119. Sicinski P, Donaher JL, Gene Y *et al.* Cyclin D2 is an FSH-responsive gene involved in gonadal cell proliferation and oncogenesis. *Nature* 1996 384: 470–474.
120. Rane SG, Dubus P, Mettus RV *et al.* Loss of *Cdk4* expression causes insulin-deficient diabetes and *Cdk4* activation results in β-islet cell hyperplasia. *Nat Genet* 1999 22: 44–52.
121. Dinchuk JE, Car BD, Focht RJ *et al.* Renal abnormalities and an altered inflammatory response in mice lacking cyclooxygenase II. *Nature* 1995 378: 406–409.
122. Lim H, Paria BC, Das SK *et al.* Multiple female reproductive failures in cyclooxygenase 2-deficient mice. *Cell* 1997 91: 197–208.
123. Ruggiu M, Speed R, Taggart M *et al.* The mouse Dazla gene encodes a cytoplasmic protein essential for gametogenesis. *Nature* 1997 389: 73–77.

124. Bitgood MJ, Shen L, McMahon AP. Sertoli cell signaling by Desert hedgehog regulates the male germline. *Curr Biol* 1996 6: 298–304.

125. Pittman DL, Cobb J, Schimenti KJ *et al.* Meiotic prophase arrest with failure of chromosome synapsis in mice deficient for *Dmc1*, a germline-specific *RecA* homolog. *Mol Cell* 1998 1: 697–705.

126. Yoshida K, Kondoh G, Matsuda Y, Habu T, Nishimune Y, Morita T. The mouse *RecA*-like gene *Dmc1* is required for homologous chromosome synapsis during meiosis. *Mol Cell* 1998 1: 707–718.

127. Miyamoto N, Yoshida M, Kuratani S, Matsuo I, Aizawa S. Defects of urogenital development in mice lacking *Emx2*. *Development* 1997 124: 1653–1664.

128. Kennedy CRJ, Zhang Y, Brandon S *et al.* Salt-sensitive hypertension and reduced fertility in mice lacking the prostaglandin EP_2 receptor. *Nat Med* 1999 5:217–220.

129. Tilley SL, Audoly LP, Hicks EH *et al.* Reproductive failure and reduced blood pressure in mice lacking the EP_2 prostaglandin E_2 receptor. *J Clin Invest* 1999 103: 1539–1545.

130. Hizaki H, Segi E, Sugimoto Y *et al.* Abortive expansion of the cumulus and impaired fertility in mice lacking the prostaglandin E receptor subtype EP(2). *Proc Natl Acad Sci USA* 1999 96: 10501–10506.

131. Lubahn DB, Moyer JS, Golding TS, Couse JF, Korach KS, Smithies O. Alteration of reproductive function but not prenatal sexual development after insertional disruption of the mouse estrogen receptor gene. *Proc Natl Acad Sci USA* 1993 90: 11162–11166.

132. Hess RA, Bunick D, Lee KH *et al.* A role for oestrogens in the male reproductive system. *Nature* 1997 390: 509–512.

133. Krege JH, Hodgin JB, Couse JF *et al.* Generation and reproductive phenotypes of mice lacking estrogen receptor β. *Proc Natl Acad Sci USA* 1998 95: 15677–15682.

134. Chen M, Tomkins DJ, Auerbach W *et al.* Inactivation of *Fac* in mice produces inducible chromosomal instability and reduced fertility reminiscent of Fanconi anaemia. *Nat Genet* 1996 12: 448–451.

135. Whitney MA, Royle G, Low MJ *et al.* Germ cell defects and hematopoietic hypersensitivity to γ-interferon in mice with a targeted disruption of the Fanconi anemia C gene. *Blood* 1996 88: 49–58.

136. Soyal SM, Amleh A, Dean J. FIG(α), a germ cell-specific transcription factor required for ovarian follicle formation. *Development* 2000 127: 4645–4654.

137. Kooy RF, D'Hooge R, Reyniers E *et al.* Transgenic mouse model for the fragile X syndrome. *Am J Med Genet* 1996 64: 241–245.

138. Kumar TR, Wang Y, Lu N, Matzuk MM. Follicle stimulating hormone is required for ovarian follicle maturation but not male fertility. *Nat Genet* 1997 15: 201–204.

139. Dierich A, Sairam MR, Monaco L *et al.* Impairing follicle-stimulating hormone (FSH) signaling *in vivo*: targeted disruption of the FSH receptor leads to aberrant gametogenesis and hormonal imbalance. *Proc Natl Acad Sci USA* 1998 95:13612–13617.

140. Lu Q, Shur BD. Sperm from β1,4-galactosyltransferase-null mice are refractory to ZP3-induced acrosome reactions and penetrate the zona pellucida poorly. *Development* 1997 124: 4121–4131.

141. Lu Q, Hasty P, Shur BD. Targeted mutation in β1,4-galactosyltransferase leads to pituitary insufficiency and neonatal lethality. *Dev Biol* 1997 181: 257–267.

142. Kumar TR, Wiseman AL, Kala G, Kala SV, Matzuk MM, Lieberman MW. Reproductive defects in γ-glutamyl transpeptidase-deficient mice. *Endocrinology* 2000 141: 4270–4277.

143. Kendall SK, Samuelson LC, Saunders TL, Wood RI, Camper SA. Targeted disruption of the pituitary glycoprotein hormone α-subunit produces hypogonadal and hypothyroid mice. *Genes Dev* 1995 9: 2007–2019.

144. Dong J, Albertini DF, Nishimori K, Kumar TR, Lu N, Matzuk MM. Growth differentiation factor-9 is required during early ovarian folliculogenesis. *Nature* 1996 383: 531–535.

145. Zhou Y, Xu BC, Maheshwari HG *et al.* A mammalian model for Laron syndrome produced by targeted disruption of the mouse growth hormone receptor/binding protein gene (the Laron mouse). *Proc Natl Acad Sci USA* 1997 94: 13215–13220.

146. Dix DJ, Allen JW, Collins BW *et al.* Targeted gene disruption of *Hsp70-2* results in failed meiosis,

germ cell apoptosis, and male infertility. *Proc Natl Acad Sci USA* 1996 93: 3264–3268.

147. Xiao X, Zuo X, Davis AA *et al.* HSF1 is required for extra-embryonic development, postnatal growth and protection during inflammatory responses in mice. *EMBO J* 1999 18: 5943–5952.

148. Satokata I, Benson G, Maas R. Sexually dimorphic sterility phenotypes in *Hoxa10*-deficient mice. *Nature* 1995 374: 460–463.

149. Hsieh-Li HM, Witte DP, Weinstein M *et al.* *Hoxa 11* structure, extensive antisense transcription, and function in male and female fertility. *Development* 1995 121: 1373–1385.

150. Matzuk MM, Finegold MJ, Su J-GJ, Hsueh AJW, Bradley A. α-Inhibin is a tumor-suppressor gene with gonadal specificity in mice. *Nature* 1992 360: 313–319.

151. Baker J, Hardy MP, Zhou J *et al.* Effects of an *Igf1* gene null mutation on mouse reproduction. *Mol Endocrinol* 1996 10: 903–918.

152. Burks DJ, de Mora JF, Schubert M *et al.* IRS-2 pathways integrate female reproduction and energy homeostasis. *Nature* 2000 407: 377–382.

153. Robb L, Li R, Hartley L, Nandurkar HH, Koentgen F, Begley CG. Infertility in female mice lacking the receptor for interleukin II is due to a defective uterine response to implantation. *Nat Med* 1998 4: 303–308.

154. Thepot D, Weitzman JB, Barra J *et al.* Targeted disruption of the murine *junD* gene results in multiple defects in male reproductive function. *Development* 2000 127: 143–153.

155. Stewart CL, Kaspar P, Brunet LJ *et al.* Blastocyst implantation depends on maternal expression of leukemia inhibitory-function. *Nature* 1992 359: 76–79.

156. Baker SM, Bronner CE, Zhang L *et al.* Male mice defective in the DNA mismatch repair gene *PMS2* exhibit abnormal chromosome synapsis in meiosis. *Cell* 1995 82: 309–319.

157. Edelmann W, Cohen PE, Kane M *et al.* Meitoic pachytene arrest in *MLH1*-deficient mice. *Cell* 1996 85: 1125–1134.

158. Baker SM, Plug AW, Prolla TA *et al.* Involvement of mouse *Mlh1* in DNA mismatch repair and meiotic crossing over. *Nat Genet* 1996 13: 336–341.

159. de Vries SS, Baart EB, Dekker M *et al.* Mouse MutS-like protein Msh5 is required for proper chromosome synapsis in male and female meiosis. *Genes Dev* 1999 13: 523–531.

160. Edelmann W, Cohen PE, Kneitz B *et al.* Mammalian MutS homologue 5 is required for chromosome pairing in meiosis. *Nat Genet* 1999 21: 123–127.

161. Behringer RR, Finegold MJ, Cate RL. Mullerian-inhibiting substance function during mammalian sexual development. *Cell* 1994 79: 415–425.

162. Durlinger AL, Kramer P, Karels B *et al.* Control of primordial follicle recruitment by anti-Mullerian hormone in the mouse ovary. *Endocrinology* 1999 140: 5789–5796.

163. Mishina Y, Rey R, Finegold MJ *et al.* Genetic analysis of the Mullerian-inhibiting substance signal transduction pathway in mammalian sexual differentiation. *Genes Dev* 1996 10: 2577–2587.

164. Pace AJ, Lee E, Athirakui K, Coffman TM, O'Brien DA, Koller BH. Failure of spermatogenesis in mouse lines deficient in the Na(+)-K(+)-2Cl(–) cotransporter. *J Clin Invest* 2000 105: 441–450.

165. Lee SL, Sadovsky Y, Swirnoff AH *et al.* Luteinizing hormone deficiency and female infertility in mice lacking the transcription factor NGFI-A (*Egr-1*). *Science* 1996 273: 1219–1221.

166. Topilko P, Schneider-Maunoury S, Levi G *et al.* Multiple pituitary and ovarian defects in Krox-24 (NGFI-A, *Egr-1*)- targeted mice. *Mol Endocrinol* 1998 12: 107–122.

167. Good DJ, Porter FD, Mahon KA, Parlow AF, Westphal H, Kirsch IR. Hypogonadism and obesity in mice with a targeted deletion of the *Nhlh2* gene. *Nat Genet* 1997 15: 397–401.

168. Bruning JC, Gautam D, Burks DJ *et al.* Role of brain insulin receptor in control of body weight and reproduction. *Science* 2000 289: 2122–2125.

169. Gow A, Southwood CM, Li JS *et al.* CNS myelin and sertoli cell tight junction strands are absent in *Osp/Claudin-11* null mice. *Cell* 1999 99: 649–659.

170. Dai X, Schonbaum C, Degenstein L, Bai W, Mahowald A, Fuchs E. The *ovo* gene required for cuticle formation and oogenesis in flies is involved in hair formation and spermatogenesis in mice. *Genes Dev* 1998 12: 3452–3463.

171. Nishimori K, Young LJ, Guo Q, Wang Z, Insel TR, Matzuk MM. Oxytocin is required for nursing but is not essential for parturition or reproductive

behavior. *Proc Natl Acad Sci USA* 1996 93: 11699–11704.

172. Robertson KM, O'Donnell L, Jones ME *et al.* Impairment of spermatogenesis in mice lacking a functional aromatase (*cyp 19*) gene. *Proc Natl Acad Sci USA* 1999 96: 7986–7991.
173. Fisher CR, Graves KH, Parlow AF, Simpson ER. Characterization of mice deficient in aromatase (ArKO) because of targeted disruption of the *cyp19* gene. *Proc Natl Acad Sci USA* 1998 95: 6965–6970.
174. Mulryan K, Gitterman DP, Lewis CJ *et al.* Reduced vas deferens contraction and male infertility in mice lacking P2X1 receptors. *Nature* 2000 403: 86–89.
175. Zindy F, van Deursen J, Grosveld G, Sherr CJ, Roussel MF. *INK4d*-deficient mice are fertile despite testicular atrophy. *Mol Cell Biol* 2000 20: 372–378.
176. Nakayama K, Ishida N, Shirane M *et al.* Mice lacking p27(*Kip1*) display increased body size, multiple organ hyperplasia, retinal dysplasia, and pituitary tumors. *Cell* 1996 85: 707–720.
177. Fero ML, Rivkin M, Tasch M *et al.* A syndrome of multiorgan hyperplasia with features of gigantism, tumorigenesis, and female sterility in $p27^{Kip1}$-deficient mice. *Cell* 1996 85: 733–744.
178. Hendry JH, Adeeko A, Potten CS, Morris ID. P53 deficiency produces fewer regenerating spermatogenic tubules after irradiation. *Int J Radiat Biol* 1996 70: 677–682.
179. Mbikay M, Tadros H, Ishida N *et al.* Impaired fertility in mice deficient for the testicular germ-cell protease PC4. *Proc Natl Acad Sci USA* 1997 94: 6842–6846.
180. Blume-Jensen P, Jiang G, Hyman R, Lee KF, O'Gorman S, Hunter T. Kit/stem cell factor receptor-induced activation of phosphatidylinositol 3'-kinase is essential for male fertility. *Nat Genet* 2000 24: 157–162.
181. Lin SR, Yu IS, Huang PH, Tsai CW, Lin SW. Chimaeric mice with disruption of the gene coding for phosphatidylinositol glycan class A (*Pig-a*) were defective in embryogenesis and spermatogenesis. *Br J Haematol* 2000 110: 682–693.
182. Lydon JP, DeMayo FJ, Funk CR *et al.* Mice lacking progesterone receptor exhibit pleiotropic reproductive abnormalities. *Genes Dev* 1995 9: 2266–2278.
183. Horseman ND, Zhao W, Montecino-Rodriguez E *et al.* Defective mammopoiesis, but normal hematopoiesis, in mice with a targeted disruption of the prolactin gene. *EMBO J* 1997 16: 6926–6935.
184. Ormandy CJ, Camus A, Barra J *et al.* Null mutation of the prolactin receptor gene produces multiple reproductive defects in the mouse. *Genes Dev* 1997 11: 167–178.
185. Sugimoto Y, Yamasaki A, Segi E *et al.* Failure of parturition in mice lacking the prostaglandin F receptor. *Science* 1997 277: 681–683.
186. Varmuza S, Jurisicova A, Okano K, Hudson J, Boekelheide K, Shipp EB. Spermiogenesis is impaired in mice bearing a targeted mutation in the protein phosphatase 1cγ gene. *Dev Biol* 1999 205: 98–110.
187. Lufkin T, Lohnes D, Mark M *et al.* High postnatal lethality and testis degeneration in retinoic acid receptor α mutant mice. *Proc Natl Acad Sci USA* 1993 90: 7225–7229.
188. Lohnes D, Kastner P, Dierich A, Mark M, LeMeur M, Chambon P. Function of retinoic acid receptor γ in the mouse. *Cell* 1993 73: 643–658.
189. Kastner P, Mark M, Leid M *et al.* Abnormal spermatogenesis in RXRβ mutant mice. *Genes Dev* 1996 10: 80–92.
190. Togawa A, Miyoshi J, Ishizaki K *et al.* Progressive impairment of kidneys and reproductive organs in mice lacking Rho GDIα. *Oncogene* 1999 18: 5373–5380.
191. Rigotti A, Trigatti BL, Penman M, Rayburn H, Herz J, Krieger M. A targeted mutation in the murine gene encoding the high density lipoprotein (HDL) receptor scavenger receptor class B type I reveals its key role in HDL metabolism. *Proc Natl Acad Sci USA* 1997 94: 12610–12615.
192. Trigatti B, Rayburn H, Vinals M *et al.* Influence of the high density lipoprotein receptor SR-BI on reproductive and cardiovascular pathophysiology. *Proc Natl Acad Sci USA* 1999 96: 9322–9327.
193. Yuan L, Liu JG, Zhao J, Brundell E, Daneholt B, Hoog C. The murine SCP3 gene is required for synaptonemal complex assembly, chromosome synapsis, and male fertility. *Mol Cell* 2000 5: 73–83.
194. Supp DM, Witte DP, Branford WW, Smith EP, Potter SS. Sp4, a member of the Sp1-family of zinc finger transcription factors, is required for

normal murine growth, viability, and male fertility. *Dev Biol* 1996 176: 284–299.

195. Pearse RV, Drolet DW, Kalla KA, Hooshmand F, Bermingham JR, Rosenfeld MG. Reduced fertility in mice deficient for the POU protein sperm-1. *Proc Natl Acad Sci USA* 1997 94: 7555–7560.

196. Baudat F, Manova K, Yuen JP, Jasin M, Keeney S. Chromosome synapsis defects and sexually dimorphic meiotic progression in mice lacking *spo11*. *Mol Cell* 2000 6: 989–998.

197. Romanienko PJ, Camerini-Otero RD. The mouse *spo11* gene is required for meiotic chromosome synapsis. *Mol Cell* 2000 6: 975–987.

198. Mahendroo MS, Cala KM, Russell DW. 5α-reduced androgens play a key role in murine parturition. *Mol Endocrinol* 1996 10: 380–392.

199. Mahendroo MS, Cala KM, Landrum CP, Russell DW. Fetal death in mice lacking 5α-reductase type I caused by estrogen excess. *Mol Endocrinol* 1997 11: 1–11.

200. Caron KM, Soo SC, Wetsel WC, Stocco DM, Clark BJ, Parker KL. Targeted disruption of the mouse gene encoding steroidogenic acute regulatory protein provides insights into congenital lipoid adrenal hyperplasia. *Proc Natl Acad Sci USA* 1997 94: 11540–11545.

201. Luo X, Ikeda Y, Parker KL. A cell-specific nuclear receptor is essential for adrenal and gonadal development and sexual differentiation. *Cell* 1994 77: 481–490.

202. Matzuk MM, Dionne L, Guo Q, Kumar TR, Lebovitz RM. Ovarian function in superoxide dismutase 1 and 2 knockout mice. *Endocrinology* 1998 139: 4008–4011.

203. Ho YS, Gargano M, Cao J, Bronson RT, Heimler I, Hutz RJ. Reduced fertility in female mice lacking copper–zinc superoxide dismutase. *J Biol Chem* 1998 273: 7765–7769.

204. Lee H-W, Blasco MA, Gottlieb GJ, Horner JW, Greider CW, DePinho RA. Essential role of mouse telomerase in highly proliferative organs. *Nature* 1998 392: 569–577.

205. Beck AR, Miller IJ, Anderson P, Streuli M. RNA-binding protein TIAR is essential for primordial germ cell development. *Proc Natl Acad Sci USA* 1998 95: 2331–2336.

206. Kuroda M, Sok J, Webb L *et al.* Male sterility and enhanced radiation sensitivity in TSL (–/–) mice. *EMBO J* 2000 19: 453–462.

207. Yu YE, Zhang Y, Unni E *et al.* Abnormal spermatogenesis and reduced fertility in transition nuclear protein 1-deficient mice. *Proc Natl Acad Sci USA* 2000 97: 4683–4688.

208. Lu Q, Gore M, Zhang Q *et al.* Tyro-3 family receptors are essential regulators of mammalian spermatogenesis. *Nature* 1999 398: 723–728.

209. Roest HP, van Klaveren J, de Wit J *et al.* Inactivation of the HR6B ubiquitin-conjugating DNA repair enzyme in mice causes male sterility associated with chromatin modification. *Cell* 1996 86: 799–810.

210. Tanaka SS, Toyooka Y, Akasu R *et al.* The mouse homolog of *Drosophila* VASA is required for the development of male germ cells. *Genes Dev* 2000 14: 841–853.

211. Kreidberg JA, Sariola H, Loring JM *et al.* WT-1 is required for early kidney development. *Cell* 1993 74: 679–691.

212. Vainio S, Heikkila M, Kispert A, Chin N, McMahon AP. Female development in mammals is regulated by *Wnt-4* signalling. *Nature* 1999 397: 405–409.

213. Parr BA, McMahon AP. Sexually dimorphic development of the mammalian reproductive tract requires *Wnt-7a*. *Nature* 1998 395: 707–710.

214. Luoh S-W, Bain PA, Polakiewicz RD *et al.* *Zfx* mutation results in small animal size and reduced germ cell number in male and female mice. *Development* 1997 124: 2275–2284.

215. Rankin T, Talbot P, Lee E, Dean J. Abnormal zonae pellucidae in mice lacking *ZP1* result in early embryonic loss. *Development* 1999 126: 3847–3855.

216. Soyal S, Rankin T, Dean J. *Molecular Genetics of the Mammalian Zona Pellucida: Targeted Mutagenesis and Fertility*. Bethesda MD: NIDDKD, National Institutes of Health, June 1999: 10–11q.

217. Rankin T, Familari M, Lee E, *et al.* Mice homozygous for an insertional mutation in the *Zp3* gene lack a zona pellucida and are infertile. *Development* 1996 122: 2903–2910.

218. Liu C, Litscher ES, Mortillo S *et al.* Targeted disruption of the *mZP3* gene results in production of eggs lacking a zona pellucida and infertility in female mice. *Proc Natl Acad Sci USA* 1996 93: 5431–5436.

219. Olson EN, Arnold HH, Rigby PW, Wold BJ. Know your neighbors: three phenotypes in null

mutants of the myogenic bHLH gene *MRF4*. *Cell* 1996 85: 1–4.
220. Ohno H, Goto S, Taki S *et al.* Targeted disruption of the CD3 eta locus causes high lethality in mice: modulation of *Oct-1* transcription on the opposite strand. *EMBO J* 1994 13: 1157–1165.
221. Pham CT, MacIvor DM, Hug BA, Heusel JW, Ley TJ. Long-range disruption of gene expression by a selectable marker cassette. *Proc Natl Acad Sci USA* 1996 93: 13090–13095.
222. Hasty P, Ramirez-Solis R, Krumlauf R, Bradley A. Introduction of a subtle mutation into the *Hox-2.6* locus in embryonic stem cells. *Nature* 1991 350: 243–246.
223. Dietrich WF, Lander ES, Smith JS *et al.* Genetic identification of *Mom-1*, a major modifier locus affecting Min- induced intestinal neoplasia in the mouse. *Cell* 1993 75: 631–639.
224. Mason AJ, Pitts SL, Nikolics K *et al.* The hypogonadal mouse: reproductive functions restored by gene therapy. *Science* 1986 234: 1372–1378.
225. Mason AJ, Hayflick JS, Zoeller RT *et al.* A deletion truncating the gonadotropin-releasing hormone gene is responsible for hypogonadism in the *hpg* mouse. *Science* 1986 234: 1366–1371.
226. Kumar TR, Low MJ, Matzuk MM. Genetic rescue of follicle-stimulating hormone β-deficient mice. *Endocrinology* 1998 139: 3289–3295.
227. Rankin TL, Tong ZB, Castle PE *et al.* Human *ZP3* restores fertility in *Zp3* null mice without affecting order-specific sperm binding. *Development* 1998 125: 2415–2424.
228. Slee R, Grimes B, Speed RM *et al.* A human *DAZ* transgene confers partial rescue of the mouse *Dazl* null phenotype. *Proc Natl Acad Sci USA* 1999 96: 8040–8045.
229. Brown CW, Houston-Hawkins DE, Woodruff TK, Matzuk MM. Insertion of *Inhbb* into the *Inhba* locus rescues the *Inhba*-null phenotype and reveals new activin functions. *Nat Genet* 2000 25: 453–457.
230. Lakso M, Sauer B, Mosinger B *et al.* Targeted oncogene activation by site-specific recombination in transgenic mice. *Proc Natl Acad Sci USA* 1992 89: 6232–6236.
231. O'Gorman S, Fox DT, Wahl GM. Recombinase-mediated gene activation and site-specific integration in mammalian cells. *Science* 1991 251: 1351–1355.
232. Gu H, Marth JD, Orban PC, Mossmann H, Rajewsky K. Deletion of a DNA polymerase β gene segment in T cells using cell type-specific gene targeting. *Science* 1994 265: 103–106.
233. Kissel H, Timokhina I, Hardy MP *et al.* Point mutation in kit receptor tyrosine kinase reveals essential roles for kit signaling in spermatogenesis and oogenesis without affecting other kit responses. *EMBO J* 2000 19: 1312–1326.
234. Ramirez-Solis R, Liu P, Bradley A. Chromosome engineering in mice. *Nature* 1995 378: 720–724.
235. Zheng B, Mills AA, Bradley A. A system for rapid generation of coat color-tagged knockouts and defined chromosomal rearrangements in mice. *Nucleic Acids Res* 1999 27: 2354–2360.
236. Brinster RL, Avarbock MR. Germline transmission of donor haplotype following spermatogonial transplantation. *Proc Natl Acad Sci USA* 1994 91: 11303–11307.
237. Brinster RL, Zimmermann JW. Spermatogenesis following male germ-cell transplantation. *Proc Natl Acad Sci USA* 1994 91: 11298–112302.
238. Avarbock MR, Brinster CJ, Brinster RL. Reconstitution of spermatogenesis from frozen spermatogonial stem cells. *Nat Med* 1996 2: 693–696.
239. Nagano M, Avarbock MR, Leonida EB, Brinster CJ, Brinster RL. Culture of mouse spermatogonial stem cells. *Tissue Cell* 1998 30: 389–397.
240. Ogawa T, Dobrinski I, Avarbock MR, Brinster RL. Transplantation of male germ line stem cells restores fertility in infertile mice. *Nat Med* 2000 6: 29–34.
241. Clouthier DE, Avarbock MR, Maika SD, Hammer RE, Brinster RL. Rat spermatogenesis in mouse testis. *Nature* 1996 381: 418–421.
242. Ogawa T, Dobrinski I, Avarbock MR, Brinster RL. Xenogeneic spermatogenesis following transplantation of hamster germ cells to mouse testes. *Biol Reprod* 1999 60: 515–521.
243. Dobrinski I, Avarbock MR, Brinster RL. Transplantation of germ cells from rabbits and dogs into mouse testes. *Biol Reprod* 1999 61: 1331–1339.
244. Dobrinski I, Avarbock MR, Brinster RL. Germ cell transplantation from large domestic animals into mouse testes. *Mol Reprod Dev* 2000 57: 270–279.
245. Eppig JJ, Wigglesworth K, Hirao Y. Metaphase I arrest and spontaneous parthenogenetic activa-

tion of strain LTXBO oocytes: chimeric reaggregated ovaries establish primary lesion in oocytes. *Dev Biol* 2000 224: 60–68.

246. Eppig JJ, Wigglesworth K. Development of mouse and rat oocytes in chimeric reaggregated ovaries after interspecific exchange of somatic and germ cell components. *Biol Reprod* 2000 63: 1014–1023.

247. McGrath J, Solter D. Nuclear transplantation in the mouse embryo by microsurgery and cell fusion. *Science* 1983 220: 1300–1302.

248. Wilmut I, Schnieke AE, McWhir J, Kind AJ, Campbell KH. Viable offspring derived from fetal and adult mammalian cells. *Nature* 1997 385: 810–813.

249. Schnieke AE, Kind AJ, Ritchie WA *et al.* Human factor IX transgenic sheep produced by transfer of nuclei from transfected fetal fibroblasts. *Science* 1997 278: 2130–2133.

250. Wakayama T, Perry AC, Zuccotti M, Johnson KR, Yanagimachi R. Full-term development of mice from enucleated oocytes injected with cumulus cell nuclei. *Nature* 1998 394: 369–374.

251. Wakayama T, Yanagimachi R. Cloning of male mice from adult tail-tip cells. *Nat Genet* 1999 22: 127–128.

Hormone synthesis, action and signal transduction

Philippe Bouchard

This section, written by leading experts in the field of endocrinology, deals with the components of the hypothalamic–pituitary–gonadal axis, as well as their signalling. The biology of steroid synthesis and their mechanisms of action are also comprehensively discussed.

Progress in the area of hormonal control of reproduction has been meteoric over the last 10 years. New diseases have been recognized, such as gonadotropin-releasing factor (GnRH) receptor gene mutations (2–8% of congenital hypogonadotropic hypogonadisms) as well as gonadotropin gene and receptor gene mutations. Progress in these areas has allowed the design of new tools such as GnRH analogs and, very recently, non-peptidic ligands of the GnRH receptor gene have been developed as well as chimeric or long-acting gonadotropins. Orally active ligands of luteinizing hormone (LH) and follicle-stimulating hormone (FSH) receptors are likely candidates for future development. The mutations in LH/FSH genes or in their receptors have served as models to understand their respective physiology, establishing, for example, the role of FSH as the most important switch in the cyclic recruitment of follicular selection. The role of FSH in spermatogenesis has also been clarified and appears to be of importance.

In parallel, steroid synthesis and the multiple genes involved have been elucidated, thus allowing a better knowledge of congenital defects such as the steroid acute regulatory protein (StAR) defect but also in the understanding of the local production of steroids, the so-called intracrinology. This is particularly important for androgen synthesis in the prostate or for estrogen synthesis in the example of breast aromatase, where the local blockade of this estrogen source has allowed considerable progress in the treatment of breast cancers. In addition, patients with mutations in the aromatase gene, who are unable to make estrogens, have significant but reversible bone loss and cardiovascular abnormalities which have allowed a better understanding of the pleiotropic effects of estrogens. Steroid action and, in particular, the key role of coactivators and corepressors as modulators of tissue selectivity, are fundamental to the effects of all steroid-specific receptor modulators. Finally, the cytokine/growth factor system plays a crucial role at all levels of reproductive physiology. Their mechanisms of action, as well as the pathology involved, are being unravelled.

The expertise presented in this section is essential, if not mandatory, for all reproductive endocrinologists working in the field today and in the years to come.

Chapter 11

Gonadotropin-releasing hormones and their receptors

Robert P. Millar

INTRODUCTION AND BACKGROUND

Gonadotropin-releasing hormone (GnRH), which is the central initiator of the reproductive hormonal cascade, was first isolated from mammalian hypothalami as the decapeptide (pGlu-His-Trp-Ser-Tyr-Gly-Leu-Arg-Pro-Gly.NH_2).[1–3] GnRH is processed in specialized neurons of the hypothalamus from a precursor polypeptide by enzymic processing, and packaged in storage granules that are transported down axons to the external zone of the median eminence.[4,5] GnRH is released as synchronized pulses from the nerve endings of about 1000 neurons into the hypophyseal portal system every 30–120 min, to stimulate the biosynthesis and secretion of luteinizing hormone (LH) and follicle stimulating hormone (FSH) from pituitary gonadotropes.[4] Each GnRH pulse stimulates a pulse of LH, but FSH pulses are less clear. The frequency of pulses is highest at the ovulatory LH surge and lowest during the luteal phase of the ovarian cycle. The asynchronous patterns of LH and FSH release result from changes in GnRH pulse frequency, modulating effects of gonadal steroid and peptide hormones on FSH and LH responses to GnRH, and differences in the half-lives of the two hormones. Moreover, while LH is stored and largely dependent on GnRH for secretion, FSH tends to be constitutively secreted and more dependent on biosynthesis for secretion.

CURRENT CONCEPTS AND CLINICAL RELEVANCE

Low doses of GnRH (pg/ml) delivered in a pulsatile fashion equivalent to that found in the portal vessels restore fertility in hypogonadal men and women, and are also effective in the treatment of delayed puberty.[6–12] However, high doses of GnRH or agonist analogs desensitize the gonadotrope, with a resultant decrease in LH and FSH and a decline in ovarian and testicular function.[6–13] This desensitization phenomenon is extensively applied in clinical medicine for the treatment of a wide range of diseases[6–13] (Table 1). GnRH peptide antagonists also inhibit the reproductive system through competition with endogenous GnRH, but the doses required are higher than the desensitizing agonist doses, and antagonists are currently less extensively employed. The development of non-peptide, orally active GnRH antagonists[14] will probably lead to the replacement of agonists, as they avoid the undesirable stimulation that precedes desensitization. In addition to the therapeutic applications, GnRH analogs hold promise as new-generation male and female contraceptives, in conjunction with steroid hormone replacement.[15,16]

The extensive clinical applications of GnRH analogs have attracted detailed studies of the physiology, cell biology and molecular function of the hormone, to enhance our understanding of the entire system and enable the optimal application of analog therapies.

This chapter describes the structure of various forms of GnRH and their evolution, their tissue distribution and putative functions, the design and actions of GnRH analogs, their interaction with the cognate receptors and the intracellular events mediating GnRH actions on the gonadotrope.

Table 1 Clinical applications of gonadotropin-releasing hormone (GnRH) and GnRH analogs

Pulsatile GnRH (stimulation)
Infertility: stimulates gamete and hormone production
Cryptorchidism: descent of testes
Delayed puberty: advances puberty

GnRH agonists and antagonists (inhibition)
Contraception: inhibition of ovulation and spermatogenesis with add-back sex steroid hormones
Hormone-dependent diseases
- prostate cancer
- benign prostatic hypertrophy
- breast cancer
- endometriosis
- uterine fibroids
- premenstrual syndrome
- polycystic ovarian syndrome
- hirsutism
- acne vulgaris
- precocious puberty
- acute intermittent porphyria

Infertility: inhibition of endogenous gonadotropin together with controlled administration of exogenous gonadotropin, especially in induction of ovulation in assisted reproduction techniques

STRUCTURAL VARIANTS OF GONADOTROPIN-RELEASING HORMONE

Although the isolated mammalian hypothalamic GnRH was thought to be a unique structure with a primary role in regulating LH and FSH, it became apparent that diverse forms exist in vertebrates.[17,18] This led to the structural identification of 15 different forms[19–30] (Figure 1). It also became apparent that GnRHs are distributed in a wide range of tissues in vertebrates, where they apparently have a diversity of functions including neuroendocrine (e.g. growth hormone release in certain fish species), paracrine (e.g. in placenta and gonads), autocrine (e.g. GnRH neurons, immune cells, breast and prostatic cancer cells) and neurotransmitter/neuromodulatory roles in the central and peripheral nervous system (e.g. sympathetic ganglion, mid-brain).[6,13,19–27,31–33] Although the privacy of cellular communication is maintained in all these functional modalities, and a single molecular entity is theoretically capable of serving these roles, it is evident that at least two, and usually three, forms of GnRH are present in the majority of the vertebrate species studied.[19–27,33,34] The most ubiquitous is chicken GnRH II, which was first isolated from chicken brain.[35] Since the chicken GnRH II structure is totally conserved from bony fish to man, this is probably the earliest evolved form and has critical functions. We have designated this form type II GnRH, while the hypophysiotropic form is designated type I.[33,34] In many vertebrate species a third form of GnRH occurs (localized to the forebrain in fish), and is

	1	2	3	4	5	6	7	8	9	10	
Mammal	pGlu	His	Trp	Ser	Tyr	Gly	Leu	Arg	Pro	Gly	NH_2
Guinea pig	pGlu	His	Tyr	Ser	Tyr	Gly	Val	Arg	Pro	Gly	NH_2
Chicken	pGlu	His	Trp	Ser	Tyr	Gly	Leu	Gln	Pro	Gly	NH_2
Rana	pGlu	His	Trp	Ser	Tyr	Gly	Leu	Trp	Pro	Gly	NH_2
Seambream	pGlu	His	Trp	Ser	Tyr	Gly	Leu	Ser	Pro	Gly	NH_2
Salmon	pGlu	His	Trp	Ser	Tyr	Gly	Trp	Leu	Pro	Gly	NH_2
Medaka	pGlu	His	Trp	Ser	Phe	Gly	Leu	Ser	Pro	Gly	NH_2
Catfish	pGlu	His	Trp	Ser	His	Gly	Leu	Asn	Pro	Gly	NH_2
Herring	pGlu	His	Trp	Ser	His	Gly	Leu	Ser	Pro	Gly	NH_2
Dogfish	pGlu	His	Trp	Ser	His	Gly	Trp	Leu	Pro	Gly	NH_2
Chicken II	pGlu	His	Trp	Ser	His	Gly	Trp	Tyr	Pro	Gly	NH_2
Lamprey III	pGlu	His	Trp	Ser	His	Asp	Trp	Lys	Pro	Gly	NH_2
Lamprey I	pGlu	His	Tyr	Ser	Leu	Glu	Trp	Lys	Pro	Gly	NH_2
Tunicate I	pGlu	His	Trp	Ser	Asp	Tyr	Phe	Lys	Pro	Gly	NH_2
Tunicate II	pGlu	His	Trp	Ser	Leu	Cys	His	Ala	Pro	Gly	NH_2

Figure 1 Primary amino acid sequences of naturally occurring gonadotropin-releasing hormone (GnRH) structural variants spanning approximately 600 million years of evolution. The boxed regions show the conserved NH_2- and COOH-terminal residues, which play important functional roles. Non-conserved residues are either unimportant or convey ligand-selectivity for a particular GnRH receptor. Note that the GnRHs are named according to the species in which they were first discovered and they may be represented in more than one species. For example, mammalian GnRH is widely present in amphibians and primitive bony fish, and chicken GnRH II is present in most vertebrate species, including man

designated type III GnRH.[36] Analysis of the genes encoding the GnRHs supports this general classification.[36]

STRUCTURE OF GnRH AND ANALOGS

The amino acid sequences of the 15 GnRHs from vertebrate and protochordate species reveal features that have been conserved for more than 500 million years of evolution (Figure 1). There has been conservation in peptide length (ten amino acids) and in the NH_2 terminus (pGlu-His-Trp-Ser) and COOH terminus (Pro-Gly.NH_2) (Figure 2), indicating that these features are critically important for receptor binding and activation. This is borne out by structure–activity data from thousands of analogs that were developed largely on an empirical basis. Indeed, cognisance of evolutionary constraints on the GnRH structure helps to identify the functionally important residues, and would have obviated a considerable degree of the endeavor to produce agonists and antagonists.

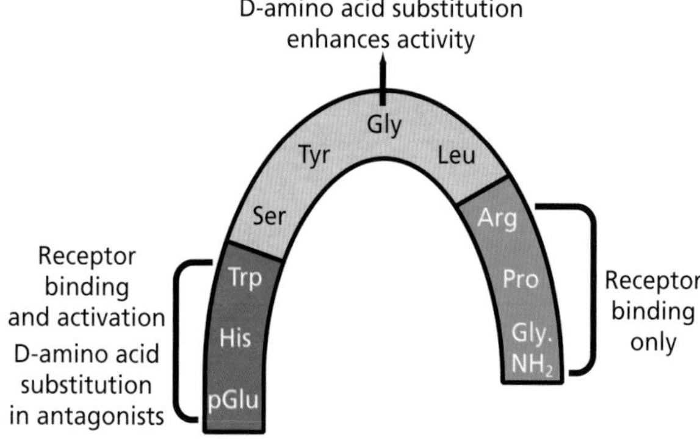

Figure 2 Schematic representation of mammalian gonadotropin-releasing hormone (GnRH) in the folded conformation in which it is bound to the GnRH pituitary receptor. The molecule is bent around the achiral glycine in position six. Substitution with D-amino acids in this position stabilizes the folded conformation, and increases binding affinity and decreases metabolic clearance. This feature is incorporated in all agonist and antagonist analogs. The NH_2 (red) and COOH (green) termini are involved in receptor binding. The NH_2 terminus alone is involved in receptor activation, and substitutions in this region produce antagonists

Position eight is the most variable amino acid (Arg, Gln, Ser, Asn, Leu, Tyr, Lys, Ala, Trp), followed by positions six (Gly, Asp, Glu, Tyr, Cys), five (Tyr, Phe, His, Leu, Asp) and seven (Leu, Trp, Phe, His, Val). The considerable variation in position eight suggests that virtually any residue is tolerated in this position. However, for some time this has been known not to be the case for the mammalian pituitary type I receptor,[33,37] which requires Arg in position eight, and recent work on cloned non-mammalian receptors also indicates certain specificities for the amino acid in this position.[24,38,39] Thus, this residue seems to play an important role in ligand-selectivity of the different GnRH receptors.

The mammalian pituitary GnRH receptor appears to be more stringent in its requirements of ligand conformation.[19,22,24,33,39] Studies have indicated that the conserved NH_2- and COOH-terminal domains of GnRH are closely apposed when mammalian GnRH binds its receptor, resulting from a β-II-type turn involving residues 5–8[33,37] (Figure 2). This is partly due to intramolecular interactions with the side-chain of Arg8, as various studies, including Trp fluorescence,[40] computer simulations using the technique of conformational memories[41] and nuclear magnetic resonance (NMR),[42] have shown that substitution of Arg8 (for example with Gln8 as in chicken GnRH I) results in a more extended structure, with a loss of predominance of the folded conformers and a low biological activity. Yet these extended forms (e.g. Gln8 GnRH) have high activity in many non-mammalian GnRH receptors,[33,38,39,43] in spite of their low activity at the mammalian receptor. The β-II-type turn conformation of GnRH also appears to be induced in part by the interaction of Arg8 with an acidic residue (D302) in extracellular loop 3 (EC3) of the mammalian receptor[33,44] (Figure 3). Substitution of a D-amino acid for Gly6 enhances the β-II-type conformation, and increases the activity of Arg8 GnRH about ten-fold in mammals. The D-amino acid substitution overcomes the deleterious effects of Arg8 substitution (for example with Gln8), such that binding affinity for the mammalian receptor is increased almost 1000-fold.[39,44]

The amino-terminal residues of GnRH are involved in receptor activation, and modification of

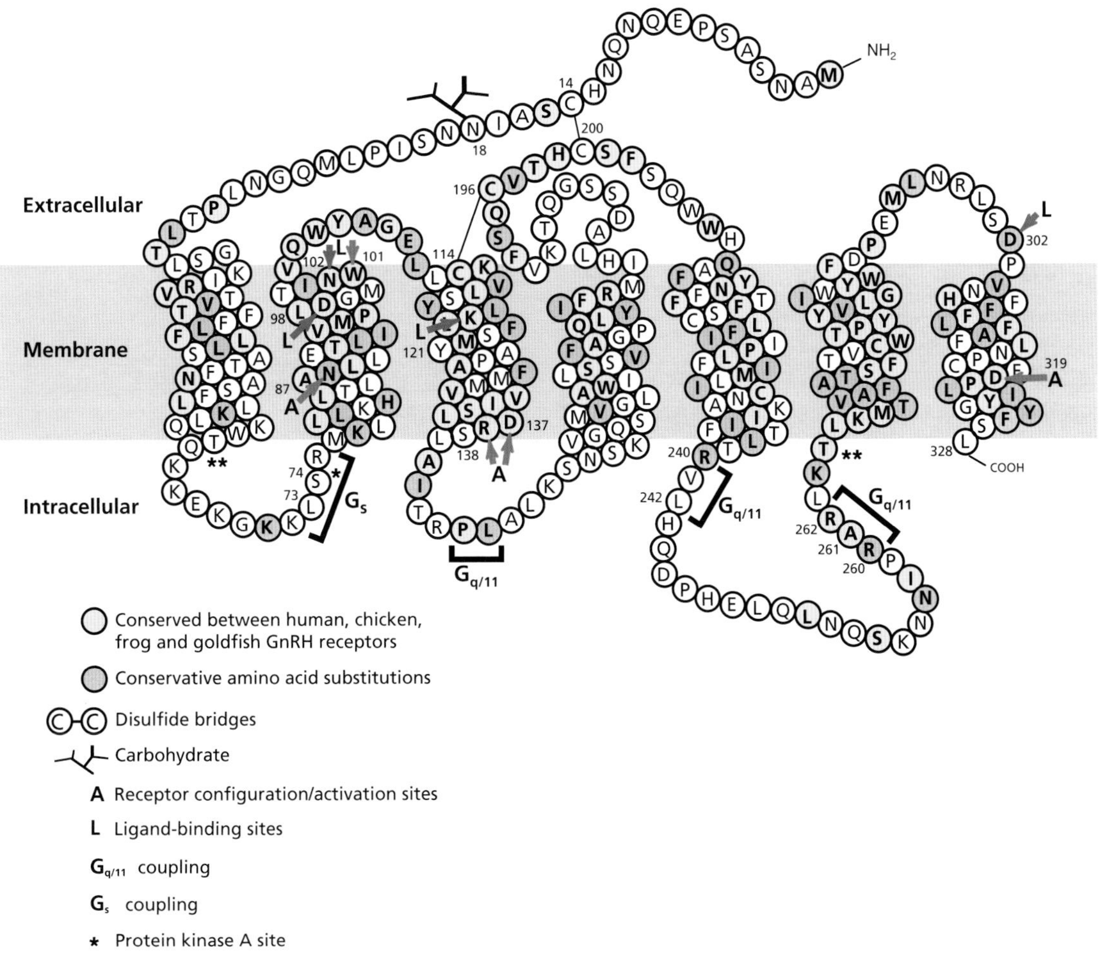

Figure 3 Two-dimensional representation of the human gonadotropin-releasing hormone (GnRH) receptor showing amino acids conserved between cloned vertebrate GnRH receptors (yellow) and conservative substitutions (blue). Putative ligand-binding sites and residues important in receptor configuration, activation and G-protein coupling are indicated. Note that the conservation of residues on single faces of transmembrane domains suggests those whose side-chains probably face into the hydrophilic interior of the pocket formed by the helices (see Figure 6). Glycosylation, phosphorylation and disulfide bridge sites are also shown

these residues in GnRH produces analogs with antagonistic properties[33,37] (Figure 2). As in agonists, substitution of Gly6 with a D-amino acid enhances the activity of the antagonists. Since the antagonists have high binding affinity, the loss of amino-terminal contacts, which activate the receptor, are presumably compensated by new contacts made by the substituted amino acids.

GONADOTROPIN-RELEASING HORMONES HAVE BEEN RECRUITED FOR DIVERSE FUNCTIONS

GnRH was thought to function exclusively as a stimulator of gonadotropin release from the vertebrate pituitary, but it soon emerged that the peptide has diverse functions in vertebrates and protochordates. This notion arose from revelations that:

(1) Multiple forms of GnRH are present in tissues of single species;

(2) GnRH is expressed in extrahypothalamic regions of the nervous system (e.g. mid-brain, spinal cord, sympathetic ganglia) and in non-neural tissue (e.g. gonads, placenta, breast);

(3) GnRH receptors are present in extrapituitary tissues and GnRH affects cell activity in these tissues;

(4) GnRH stimulates the release of other pituitary hormones (for example growth hormone in fish);

(5) GnRH is present in organisms that do not have a pituitary.

It is interesting to speculate on the possible earliest function of GnRH. The isolation of two forms of GnRH in neural tissue of a tunicate and their activation of the gonads[45] suggest that direct regulation of the gonads was an early evolved function, and that the neuroendocrine role in regulating the pituitary was a later evolutionary development. The presence of GnRH and GnRH receptors in the gonads of various vertebrate species may reflect this early function. Neurons are probably some of the earliest cells in evolution to elaborate GnRH peptides. However, the stimulation of sexual reproduction in yeast by the α-mating factor, which appears to be structurally related to GnRH, may point to an even more ancient example of the direct effects of GnRH on reproduction.[19] A receptor recently isolated from *Drosophila melanogaster* is more closely related to the GnRH receptor than other G-protein-coupled receptors (GPCRs),[46] further supporting the notion that GnRH and its receptor have an early evolutionary origin. In *D. melanogaster*, it is even more intriguing that receptor orthologs of LH and FSH receptors are present.[47] Two GnRH receptor orthologs with 20–24% amino acid identity are also present in the primitive nematode, *Caenorhabditis elegans*, along with gonadotropin receptor orthologs (Millar and colleagues, unpublished data).

It is plausible, therefore, that GnRH peptides were originally involved in cellular communication in sexual reproduction of simple unicellular and multicellular organisms. Later, they were recruited for expression in nerve cells, to translate external and internal signals into activation of reproduction, initially by acting directly on germ cells, and subsequently via pituitary gonadotrope activation. While the peptide has been co-opted as a regulator at a number of levels in the reproductive system (hypothalamus, gonad, breast, placenta), there is apparently considerable plasticity in also recruiting it as a regulator in non-reproductive tissues (e.g. adrenal, extrahypothalamic brain, the immune system, retina, pancreas).[19–22]

EVOLUTIONARILY CONSERVED TYPE II GnRH

The second form of GnRH identified from chicken brain (chicken GnRH II, type II GnRH) (Figure 1) is ubiquitous in vertebrates from primitive bony fish to man.[19–27,36,48] This complete conservation of structure over 500 million years suggests that the type II GnRH has an important function, and a discriminating receptor (or receptors) that has selected against any structural change in the ligand. This points to essential functions yet to be definitively identified. The wide distribution of type II GnRH in the central and peripheral nervous systems suggests a neurotransmitter/neuromodulatory role. This has been thoroughly demonstrated by the inhibition of M-currents in the bullfrog sympathetic ganglion, which sensitizes neurons to depolarization.[49] Type II GnRH is present in amphibian sympathetic ganglia, and the receptors are highly selective for the type II peptide.[50]

Since GnRH had been shown to have direct effects on sexual arousal in rodents,[51–53] and type II GnRH is localized in brain areas associated with reproductive behavior, it was suggested that this may be a role for the peptide.[19–21] Type II GnRH and an analog were found to be potent stimulators of reproductive behavior in ring doves[22,23] and song sparrows.[54] Recently, the cognate receptor for type II GnRH was cloned from the marmoset, and found to be distributed in those areas of primate brain associated with reproductive behaviors.[55]

In addition to its apparent role as a neuromodulator in the nervous system, type II GnRH[36] and its receptor[55] are present in non-neural reproductive tissues such as the prostate. GnRH binding sites and antiproliferative effects of GnRH analogs have been described in reproductive tissue tumors and their cell lines.[6,8,11,13] Interestingly, the GnRH binding sites, signalling and pharmacological effects of analogs were not completely in accordance with these being via type I GnRH receptors, but corresponded to type II GnRH receptors.[13,55]

PRIMARY STRUCTURES OF GnRH RECEPTORS

The amino acid sequence of the GnRH receptor was first deduced for the mouse receptor cloned from the pituitary αT3 gonadotrope cell line.[56] This sequence was confirmed,[57] and provided the basis for the cloning of GnRH pituitary receptors from the rat,[58–60] human,[61,62] (Figure 3), sheep,[63,64] cow,[65] and pig,[66] which share over 80% amino acid identity. Homologs of the mammalian GnRH receptors have also been cloned from a marsupial (possum),[67] catfish,[43] goldfish (two forms),[38] *Rana*,[68] *Xenopus*,[69] chicken,[70] Medaka,[71] striped bass,[72] trout,[73] cichlid,[74] Japanese eel,[75] amberjack (CAB 65407) and rubber eel (AD 49750). The non-mammalian receptors with greatest homology to the mammalian pituitary receptors have 42–47% amino acid identity with the mammalian receptors, but 58–67% identity amongst each other. These are all designated type I GnRH receptors (Figures 4 and 5). It is not yet altogether clear from homology comparisons that this classification is correct, but similarities in microdomains (e.g. EC3) support this. Since the evolutionary time separating amphibians and mammals is similar to that separating amphibians and bony fish, the poor conservation of sequence of the mammalian type I GnRH receptor compared with the non-mammalian receptors implies a sudden acceleration in evolutionary change in mammals. This appears to have been driven by the loss of the carboxy-terminal tail. In the goldfish, there are two isoforms (type Ia and type Ib), which have 70% amino acid identity but differ in that type Ia has SH3 domains in the carboxy-terminal tail and potential for coupling to mitogen-activated protein (MAP) kinases (Maudsley, unpublished data). The presence of three GnRH forms in most vertebrate species suggested the existence of three cognate GnRH receptor subtypes. As the EC3 domain is a major determinant of receptor selectivity for the GnRH structural variants, we used degenerate oligonucleotides to the conserved boundary transmembrane domains to amplify this domain from genomic DNA from various vertebrates.[34] These sequences were then used to identify a human putative type II GnRH receptor[76–78] and then clone *Rana*,[68] *Xenopus* (Troskie and colleagues, unpublished data), marmoset,[55] macaque and green monkey[79] receptors (Figure 4). The approach also allowed the cloning of type III GnRH receptors from *Rana*.[68] The findings suggest an early evolution of the three GnRH receptor subtypes in vertebrates, which parallels that of the GnRH ligands (Figure 4). GnRH receptor orthologs have been identified in *D. melanogaster*[46] and *C. elegans* (Swanson and colleagues, unpublished data), indicating a very early evolutionary origin. The genealogy of representative GnRH receptors is shown in Figure 5.

GnRH receptors have the characteristic features of GPCRs (Figures 3 and 4). The NH_2-terminal domain is followed by seven putative α-helical transmembrane (TM) domains connected by three extracellular loop (EC) domains and three intracellular loop (IC) domains (Figures 3 and 4). The extracellular domains and superficial regions of the TMs are usually involved in the binding of peptide hormones such as GnRH; the TMs are believed to be involved in conformational change associated with signal propagation (receptor activation), while the intracellular domains are involved in interacting with G-proteins and other proteins for intracellular signal transduction.

A unique feature of the mammalian GnRH receptor is the absence of a carboxy-terminal tail present in all other GPCRs and in all of the non-mammalian GnRH receptors. This is, therefore, a recently evolved feature, which presumably serves an important role in the functioning of the mammalian GnRH receptor (see below).

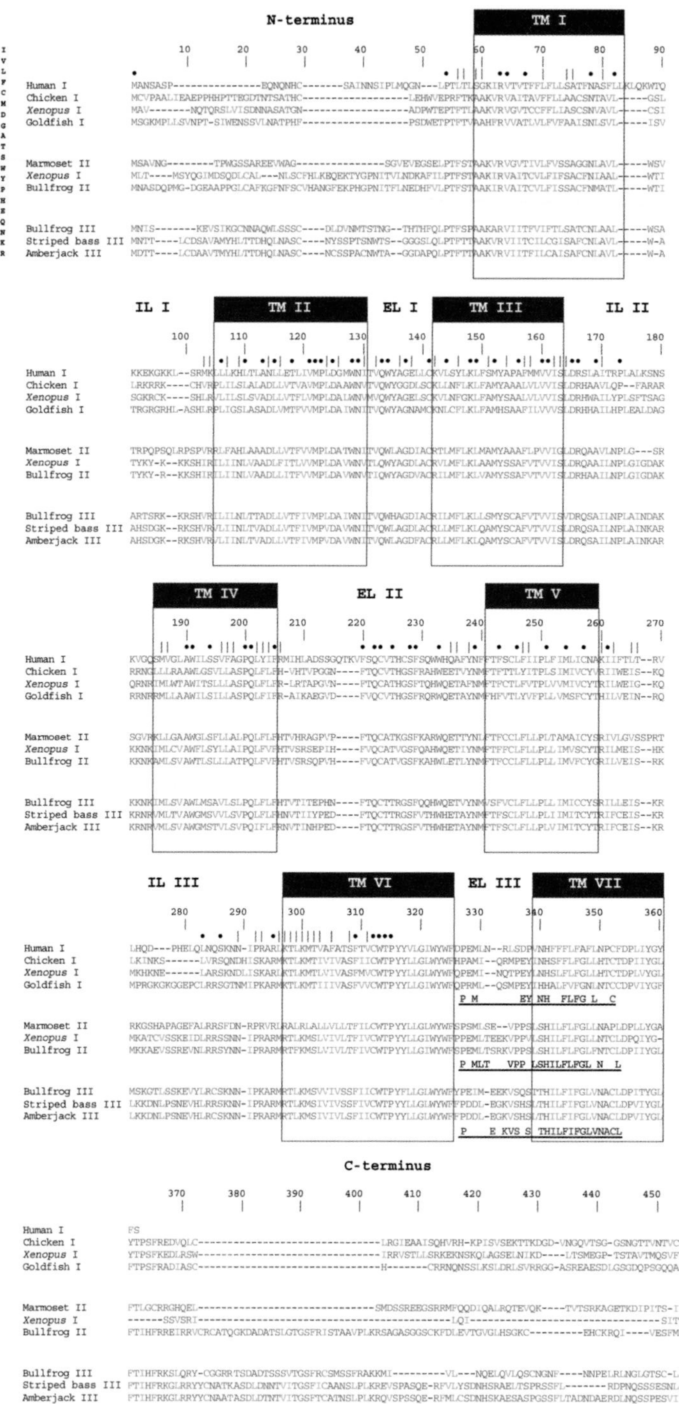

Figure 4 Alignment of representative gonadotropin-releasing hormone (GnRH) types I, II and III receptors from selected species. The transmembrane-domains are boxed and the intracellular loops (IL) and extracellular loops (EL) indicated. The overall identity (filled circle) or homology (vertical line) is indicated above. The consensus for the most characteristic domain (EL III going into TM VII) of the three receptor types is shown. This domain was used to clone the three receptor types. The colors of the amino acids are graded from red (most hydrophobic) through to blue (most hydrophilic). Note that TM domains are predominantly hydrophobic and loop domains hydrophilic

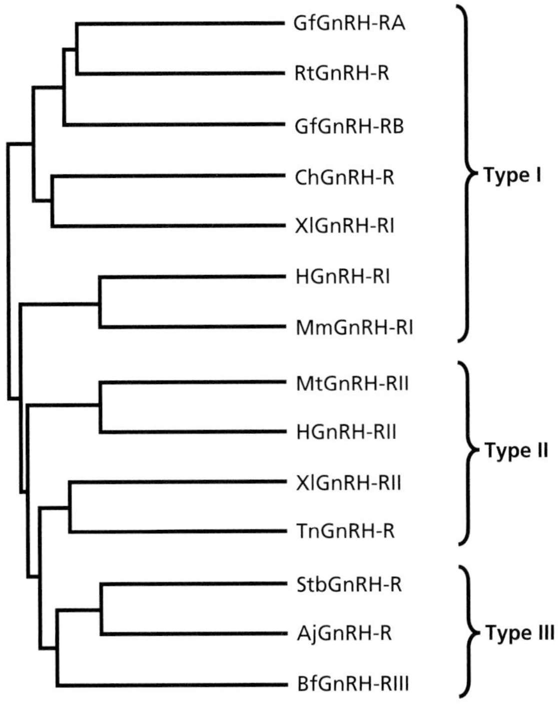

Figure 5 Phylogenetic tree of some representative gonadotropin-releasing hormone (GnRH) receptors. The receptors cluster into four groups: mammalian type I, non-mammalian type I, type II and type III. Although the mammalian type I is very different from the non-mammalian type I, there are greater similarities in microdomains compared with the type II and type III receptors. There have been rapid evolutionary changes in the mammalian type I receptor, which appear to have resulted from the loss of the carboxy-terminal tail driving co-ordinated structural changes. Gf, goldfish; Rt, rainbow trout; Ch, chicken; Xl, *Xenopus laevis*; H, human; Mm, mouse; Mt, marmoset; Tn, *Typhlonectes natans* (rubber eel); Stb, striped bass; Aj, amberjack; Bf, bullfrog

The conservation of amino acids during evolution from bony fish to mammals is likely to identify those residues that are crucial for GnRH receptor function.[24,33] These are shown in Figure 3 and include those involved in receptor activation and W164, P223, P282 and P320, which are conserved in all of the rhodopsin family of GPCRs.

GnRH RECEPTOR GENES

The structures of the genes encoding the mouse,[79] human,[80–82] chicken,[70] *Xenopus*,[69] medaka[71] and eel[75] GnRH receptors have been elucidated. In all genes, there are introns in the coding regions of TM4 and IC3. There are numerous splice variants[56,63,83–85] whose function may be regulatory, as some truncated receptor variants affect the expression of the functional receptor.[84,85] The chicken, *Xenopus* and one of the medaka receptors have an additional intron in the amino-terminal domain on the 5' untranslated region (UTR). The promoters of the rat and human receptors have been analyzed, and the genes are highly regulated by GnRH and its signalling molecules, and by gonadal, steroid and peptide hormones.[86–90]

TERTIARY STRUCTURE OF THE MAMMALIAN TYPE I GnRH RECEPTOR

A knowledge of the three-dimensional structure of the GnRH receptor is essential for a complete understanding of its molecular functioning. To date, it has been possible to crystallize only the rhodopsin GPCR and obtain X-ray structural information.[91] Other structural information on GPCRs is derived from low-resolution electron microscopy of bacteriorhodopsin and rhodopsin.[33,87,92,93] Structural information for other GPCRs relies on molecular models,[33,94–98] which are based on rhodopsin structural information.

The development of a GnRH receptor molecular model is based on initial positioning of the TM helices, as in rhodopsin, and then refinement of the angles, kinking and side-chain orientation of the TMs based on the specific amino acids constituting the GnRH receptor TMs.[33] The validity of the model and proposed interactions of the TM side-chains can be tested by site-directed mutagenesis. An example is our observation that two residues that are highly conserved in GPCRs, Asp in TM2 and Asn in TM7, appear to have undergone reciprocal mutation to Asn87 and Asp318 in the mouse GnRH receptor (Asp319 in the human) (Figure 3). This suggests that the two residues interact with each other. Mutation of Asn87 in

TM2 to Asp abolished receptor function, but a second mutation in TM7, recreating the arrangement found in other GPCRs (Asp87, Asn318), regenerated ligand binding.[99] This restoration of binding by reciprocal mutation demonstrates that the side-chains of two residues in helices TM2 and -7 have complementary roles in maintaining the structure of the receptor, and occupy the same microenvironment within the receptor. Although the structural integrity of the receptor is restored by these reciprocal mutations, receptor activation machinery is impaired (see below).

The seven TM helical domains are known, from physical structure studies in the rhodopsins, to be arranged in a tight bundle enclosing a hydrophilic pocket and surrounded by the hydrophobic membrane environment[33,91–98] (Figure 6). The evolutionary conservation of residues along a distinct face of the TM domains in the various GnRH receptors is clearly evident in Figure 3. This suggests that the conserved, more hydrophilic faces are oriented towards the hydrophilic pocket formed by the helical bundle. This is supported by studies on the TM2/TM7 interaction of Asn and Asp, since Asn87 is clearly part of the conserved hydrophilic face of TM2[99] (Figure 3).

The relative positioning of TM3 and TM4 could be deduced by the demonstration that Cys114 in EC1 and Cys196 in EC2 form a disulfide bridge (Figure 3). This was determined by a combination of photoaffinity labelling with photoactive GnRH, followed by protease digestion, reduction of S–S bonds and separation of the receptor fragments by gel electrophoresis.[100] The study also indicated that Cys14 in the NH_2-terminal domain and Cys200 in EC2 form a second disulfide bridge, thus further defining the position of the NH_2 terminus and EC2 loop structures.

Glycosylation sites have been shown at Asn4 and Asn18 in the mouse and Asn18 in the human receptors.[101,102] Glycosylation does not influence the receptor binding affinity of GnRH, but increases the number of receptors on the cell membrane. Introduction of the additional mouse receptor glycosylation site in the human receptor increased receptor number.[102]

Although significant progress has been made in establishing a molecular model of the transmembrane helix bundle of the GnRH receptor, our knowledge of the structure of the extracellular and intracellular loops is scant. Considerable effort has been put into establishing programs to define loop structures (for example, based on sequences for loop structures established from X-ray crystallography). However, these are not applicable to large loop sequences. Moreover, the known structures of the rhodopsin loops may be quite different from those of other GPCRs. Thus, a future challenge is determination of the structure of the loops in the GnRH receptors. Some progress has been made in establishing the structure of EC3, by NMR structural analysis of a synthetic peptide of EC3 anchored by cross-links similar to the distance between TM6 and TM7.[103]

BINDING OF GnRH TO THE MAMMALIAN TYPE I GnRH RECEPTOR

In the course of the targeted mutation of almost one-third of all the amino acid residues of the GnRH receptor,[24] considerable advances in identifying putative ligand contact sites in the mammalian GnRH receptor have been made (Figures 3 and 6).

Aspartate302 (D302)

Arg8 of mammalian GnRH is essential for high-affinity binding and selectivity of the mammalian GnRH receptor,[33,38,39,104] but not in non-mammalian vertebrates.[33,38,104] Flanagan and colleagues[44] sought to identify amino acids in the mammalian receptor responsible for this selectivity, and postulated that acidic amino acids would be candidates through ionic interaction with the positive Arg side-chain. Mutation of all extracellular acidic amino acids to their isosteric amides revealed that only mutation of Glu301 to Gln301 in EC3 of the mouse receptor resulted in an appropriate decrease in affinity of 100-fold. This mutation resulted in a loss of selectivity between Arg8 and Gln8 GnRH and improved binding of Glu8 GnRH, as expected.[44] This is entirely consistent with the interaction of Arg8 in native GnRH with Glu301 in EC3. Owing to the presence of an additional amino acid in EC2 of the human receptor,

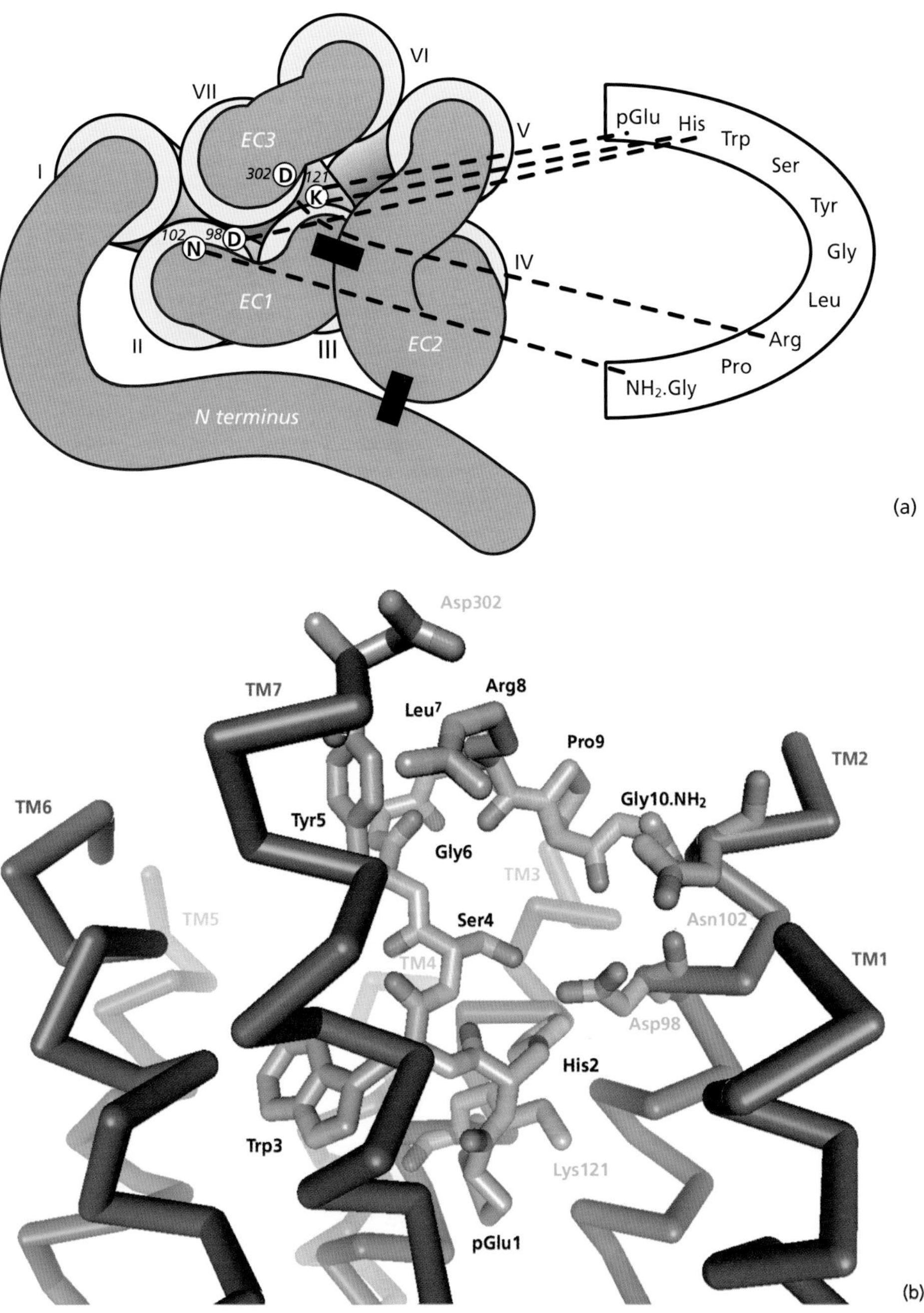

Figure 6 Gonadotropin-releasing hormone (GnRH) binding to its receptor. (a) A schematic representation of GnRH interaction with the human GnRH receptor. The receptor is viewed from above, and shows the transmembrane (TM) helices as a cluster of cylinders (yellow; going into the page) that encompass the hydrophilic pocket and are surrounded by the light hydrophobic membrane environment. The TM helices are connected by the extracellular loops (EC) (red). The dark bands represent the disulfide bridges stabilizing extracellular domains. GnRH is shown in its folded conformation interacting via pGlu1, His2, Arg8 and Gly10.NH_2 with cognate sites D98, K121, D302 and N102 in the receptor. (b) A molecular model of the human GnRH receptor in which only the essential elements are shown, to reveal the interactions of GnRH in a folded conformation with the receptor residues described above. Gf, goldfish; Rt, rainbow trout; Ch, chicken; Xl, *Xenopus laevis*; H, human; Mm, mouse; Mt, marmoset; Tn, *Typhlonectes natans* (rubber eel); Stb, striped bass; Aj, amberjack; Bf, bull-frog

the equivalent residue is Asp302 (Figures 3 and 4), which has the same property as Glu301.[105,106] These data, therefore, support the concept that Glu301 (Asp302 in the human) GnRH receptor determines selectively for Arg8. These residues are able to interact when GnRH is docked to the GnRH receptor model (Figure 6). Since analogs that are constrained in the β-II conformation (see structure of GnRH) bind with high affinity, regardless of the amino acid in position eight and the Glu301 in the mouse receptor, it appears that the role of Arg8 interaction with the acidic residue in the receptor is to induce or select the β-II conformation of the ligand,[105] which allows its binding to the other sites listed below.

Lysine121 (K121)

We considered that the highly conserved Asp in TM3 of the biogenic amine receptors, which interacts with the positively charged amine head group, may play a similar role in the GnRH receptor. The equivalent residue is Lys121 in the GnRH receptor, and mutation to Asp, Ala or Leu leads to a total loss in agonist binding, while mutation to Gln results in a 1000-fold reduction in agonist binding affinity, without affecting antagonist binding.[107] We proposed that the Lys121 interacts with His2 or pGlu1 of GnRH (Figure 6) by a charge-strengthened hydrogen bond. This was confirmed,[106] and shown to be feasible (particularly for pGlu1) in molecular models.[33,106]

Asparagine102 (N102)

A putative interaction of Asn102, located near the extracellular surface of TM2, in forming a hydrogen bond with the C=O moiety of Gly10.NH_2 in GnRH is postulated (Figures 3 and 6). Mutation to Ala results in a 100–1000-fold loss of potency in Gly10NH_2 GnRH analogs in stimulating phosphoinositol hydrolysis, while having much less effect on the potency of NH–CH_2–CH_2(10) analogs,[108] which appear to make an alternative contact. Another study confirmed these findings,[106] and showed that mutation of the adjacent Trp101 also resulted in a much greater decline in binding affinity of Gly10.NH_2 analogs than for the NH–CH_2–CH_2(10) analogs. However, this group concluded that the effects of the Trp101 mutation were indirect in distorting the binding pocket. Interestingly, antagonist binding was affected only slightly by the mutations, indicating that the antagonist binding site differs.[106]

Aspartate98 (D98)

In mutating all extracellular acidic residues as putative interacting sites for Arg8 of GnRH, it was noted that mutation of Asp98 to Asn resulted in a large decrease in inositol phosphate production.[44] Further study has revealed that mutation of Asp98 to Asn has little effect on the ED_{50} of Trp2 GnRH while causing a large increase in the ED_{50} of the native ligand, which has His2.[109] It appears that the NH of Trp can substitute for the His NH in the wild-type receptor and is still accessible for interaction with Asn in the mutant, while the His NH cannot interact owing to the longer side-chain. These findings suggest that Asp98 may also interact with GnRH, possibly through His2.

In summary, four putative ligand interaction sites in the GnRH receptor (namely Asp302(Arg8), Lys121(pGlu1,His2), Asn102 (Gly10.NH2) and Asp98 (pGlu1,His2)) have been identified. All of these sites have been conserved in vertebrate type I GnRH receptors cloned from bony fish, amphibian and bird species (Figures 3 and 4). It is interesting that the acidic residue in EC3 is conserved in nonmammals, as these receptors are not selective for Arg8. We have shown that the insertion of Pro before the acidic residue, as opposed to after it (Figure 4), is responsible for this (Fromme and colleagues, unpublished data). Although a number of other natural[110–112] and experimental mutations[113] affect GnRH binding, it is uncertain whether they represent ligand contact sites or are affecting configuration, expression or stability of the receptor.

RECEPTOR ACTIVATION

The molecular mechanisms of ligand-mediated receptor activation are best understood for rhodopsin, but are only partially elucidated for other GPCRs.

The propagation of the hormone message by the receptor to the signal-transduction pathway within the cell involves a change in receptor conformation.[114] For GPCRs, the active conformation is related to a ternary complex consisting of hormone, receptor and G-protein. This model includes an initial binding step common to both agonists and antagonists, followed by a transition step, exclusive to agonists, which leads to the formation of the ternary complex. The model also allows for the spontaneous formation of a receptor–G-protein complex, which has a higher affinity for agonist ligands and is stabilized by the binding of agonists. When guanosine triphosphate (GTP) binds to the G-protein, the receptor returns to the low-affinity conformation and the complex dissociates.[115] A revised model proposes that receptors fluctuate between an inactive R conformation and an active R* conformation.[116] Agonist binding shifts the equilibrium towards R*. The R* conformation has a high affinity for agonists, and is the only form that can bind G-proteins. The models are essentially the same in that they both require conformational change in the receptor: one ligand-induced (conformation induction) and the other ligand-stabilized (conformation selection). Studies of the GnRH receptor have provided insight into the mechanism of activation of this receptor.

Interaction of Asn/Asp in TM II/VII

Mutation of Asn87 and Asp318 in the mouse GnRH receptor (Asn87 and Asp319 in the human) (Figure 3) revealed that the Asp318 is involved in receptor activation (signal propagation), as the mutants Asn87,Asn318 and Asp87,Asn318 both retained good ligand binding but poor stimulation of inositol phosphate production.[99] These findings indicate that the unusual arrangement of Asp in TM7 in the GnRH receptor is an essential component of ligand-mediated receptor activation, normally subserved by the conserved Asp in TM2 of other GPCRs and the non-mammalian receptors,[117] which have Asp in both TM2 and TM7 (Figure 4). This seems to be an intermediate arrangement as the *Drosophila* GnRH receptor homolog has the usual GPCR Asn in TM7 combined with Asp in TM2.[46]

Interestingly, the presence of an Asp in TM7 of the GnRH receptor facilitates coupling to phospholipase C (PLC) via Gq/11, but prevents coupling to phospholipase D (PLD) by the small monomeric G-protein.[118] Mutation to Asn, as in the majority of GPCRs, recreates this coupling to phospholipase D. Thus the reverse arrangement of Asn and Asp in TM2 and -7 in the GnRH receptor appears to have been selected to allow PLC coupling and prevent PLD coupling.[118] Further exploration by mutation of TM2 Asn and TM7 Asp to various amino acids has confirmed that the TM2 Asn is essential for configuring and expression of the receptor, while the TM7 Asp is essential only for receptor activation.[119]

Asp/Arg interaction in TM III

The highly conserved motif DRXXXI/V at the intracellular end of TM3 has also been implicated in receptor activation in the GnRH receptor.[120,121] In the molecular model, Asp137 and Arg138 (DR) (Figure 3) appear to be capable of a charge interaction,[120] and this has been confirmed in the crystal structure of rhodopsin in the inactive conformation.[91] Disruption of this by mutating Asp to an uncharged residue conveys constitutive activity and increased coupling efficiency, presumably through the release of Arg to interact with other residues.[120] The Ile-located one turn of an α-helix below the Arg appears to play a role in 'caging' the Arg side-chain for this interaction, by sterically limiting its movement. Mutation to small residues (e.g. Ala) results in some uncoupling.[120] The Arg is crucial for coupling, as mutation to Gln leads to very poor coupling efficiency.[120] It is proposed that the Arg side-chain is involved in a triad interaction with the TM2 Asn and TM7 Asp in stabilizing the active conformation of the receptor[120] (Figure 7). Thus, an essential element of activation of the GnRH receptor and other GPCRs is the protonation of Asp132 to release Arg138 for interaction with the Asp/Asn residues in TM2 and -7. Since the conserved Asn in TM1 of rhodopsin interacts with Asp in TM2 in the inactive state,[91] this residue may also play a role in the transmembrane domain network involved in receptor activation.

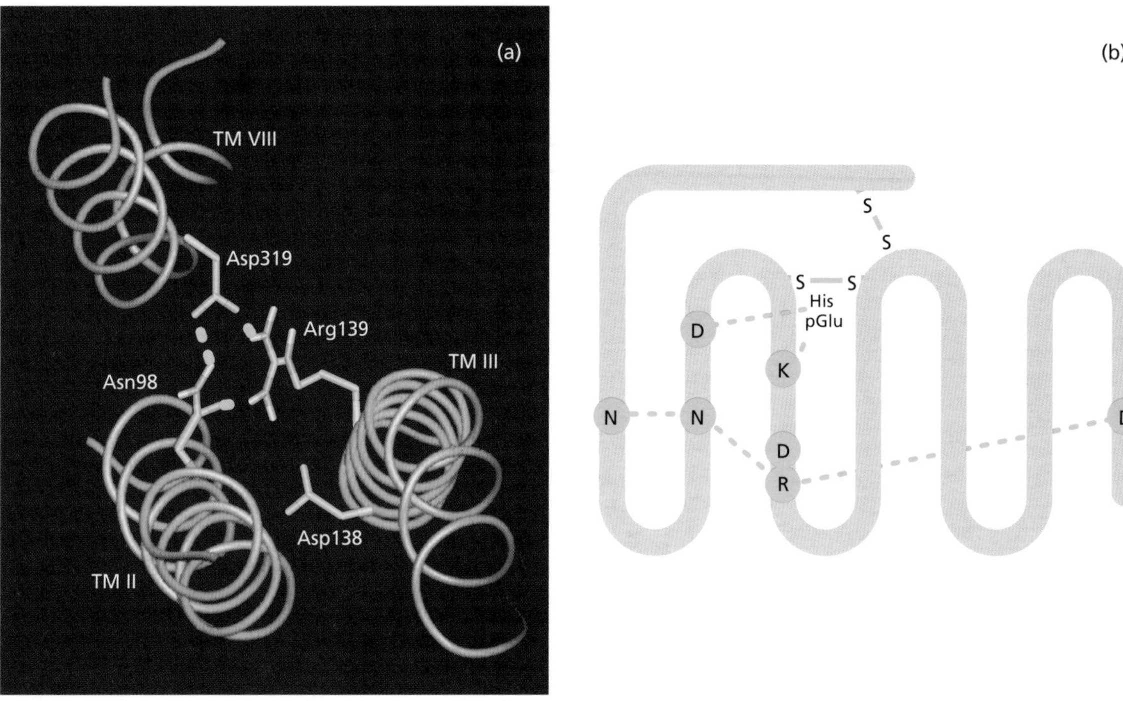

Figure 7 Activation of the gonadotropin-releasing hormone (GnRH) receptor. (a) A three-dimensional model of the transmembrane domains TM II, TM III and TM VII shows how the protonation of Asp138 breaks the ionic bond with Arg139 to allow it to form hydrogen bonds with Asp98 and Asp319 in the active conformation of the human GnRH receptor. (b) A schematic model showing the relationship between GnRH NH_2-terminal contact sites with the receptor and the residues involved in receptor activation. The interaction of the two GnRH residues known to be involved with receptor activation (pGlu,His) with D98 and K121 are postulated to stabilize the active state of the receptor, in which the ionic bond between D138 and R139 is broken to allow R139 interaction with N98 and D310. These interactions are thought to give rise to changes in the orientations of the transmembrane domains, which translate into conformational changes in the intracellular loops to allow coupling to signalling proteins (e.g. Gq/11). The disulfide bridges appear to play a role in connecting the conformational changes between TMs

RECEPTOR SIGNAL TRANSDUCTION

Intracellular coupling

The intracellular loops and carboxy-terminal tail have been implicated in the coupling of GPCRs to their cognate G-proteins, but their specific involvement varies amongst different receptors. The conservation of the carboxy-terminal sequence of IC3 in vertebrate GnRH receptors[24] (Figure 3) suggested that this region may be crucial for coupling to the primary mediator, the Gq/11 heterotrimeric G protein. A series of cassette substitutions covering the whole of IC3 confirmed this hypothesis (Wakefield and colleagues, unpublished data). Within this region, Ala261 was identified as an important residue (equivalent to the Ala in IC3 of biogenic amine receptors), which, upon mutation to large residues, induces constitutive activity of the receptor.[122] However, mutation of Ala261 to bulky amino acids resulted in an opposite effect in the GnRH receptor, namely uncoupling of the receptor and failure to generate inositol phosphate.[123] Mutation of the evolutionarily conserved adjacent basic amino acid (Arg262) to Ala was also shown to result in uncoupling (Wakefield and colleagues, unpublished data), and natural mutations of Arg262 have been shown to cause uncoupling in receptors of families with hypogonadotropic hypogonadism.[110–112]

The GnRH receptors have the motif DRXXXI/VXXPL at the start of IC2, which is conserved in GPCRs and plays a role in coupling. The importance of the DRXXXI element in receptor activation (see above) appears to extend to Pro146,Leu147. Mutation of the preceding Arg145 to Pro causes uncoupling,[124] presumably because this mutation introduces a Pro–Pro motif known to disrupt secondary structure. GnRH receptors demonstrate the capacity to activate other heterotrimeric G-proteins (e.g. G_s), resulting in the production of cyclic adenosine monophosphate (cAMP).[125] These studies have established that a sequence in IC2 (KKLSR) is a G_s-recognition motif (BBXXB, where B is a basic amino acid), and that mutation of certain of these residues leads to uncoupling of cAMP production but not of inositol phosphate production.[125] Several other studies demonstrated GnRH receptor coupling to G_s in GGH3 cells[126] and insect cells.[127]

Recent studies have shown that GPCRs can signal intracellularly through adaptor proteins other than G-proteins (e.g. β-arrestin).[128] It is possible, therefore, that the GnRH receptors with carboxy-terminal tails may signal through β-arrestin and through src homology domains, which have been identified.

Desensitization

Exposure of the gonadotrope to continuous high doses of GnRH agonists results in the phenomenon of desensitization. Gonadotrope desensitization comprises many contributing phenomena, including GnRH receptor down-regulation, receptor uncoupling from cognate G-proteins, additional downstream uncoupling (e.g. protein kinase C, Ca^{2+} stores, inositol triphosphate receptors), inhibition of gonadotropin synthesis, and alterations in the glycosylation of gonadotropins rendering them biologically inactive or even antagonistic. At the level of the GnRH receptor, agonist activation leads to two potential mechanisms of desensitization. For many GPCRs, agonist stimulation results in the activation of protein kinases, which phosphorylate IC domains, resulting in uncoupling from G-proteins and internalization of the receptor.[129–132] GnRH receptors have a number of putative phosphorylation sites in the IC domains which may serve this function (Figure 3). However, the mammalian GnRH receptors lack the carboxy-terminal IC domain, which is the prime target for phosphorylation by G-protein-coupled receptor kinases and protein kinases. Phosphorylation in this region facilitates the docking of arrestins, thus inducing G-protein uncoupling and receptor internalization. Hence, the absence of the carboxy-terminal tail in the mammalian GnRH receptors is associated with a lack of rapid desensitization in inositol phosphate production[133–135] and a relatively slow rate of receptor internalization, seven-fold less than that of the chicken GnRH receptor which has a carboxy-terminal tail.[136] Removal of the carboxy-terminal tail from the chicken receptor greatly reduces the rate of internalization to that found in the human receptor.[136] Addition of the carboxy-terminal tail of the thyrotropin-releasing hormone receptor to the rat GnRH receptor conveys both rapid desensitization of inositol phosphate production and a more rapid internalization.[137]

It appears that physiological requirements have selected for removal of the carboxy-terminal tail during evolution of the mammalian GnRH receptor, to prevent rapid desensitization and internalization. A possible driving force is the physiological need of a prolonged LH surge for ovulation in mammals.[136] Thus, although pharmacological doses of GnRH do produce desensitization of the gonadotrope, and this phenomenon has extensive clinical application, the GnRH receptor is less susceptible to ligand-mediated desensitization than other GPCRs.

Signalling pathways

The coupling of the GnRH receptor to intracellular-signalling pathways has been intensely investigated and comprehensively reviewed (see references 86, 89, 90, 94, 138–143). Although a wide range of intracellular signalling pathways have been implicated in GnRH action, there is a general consensus that the primary pathway is the activation of Ca^{2+}-dependent phospholipase Cβ (PLCβ) through the Gq/11 guanine nucleotide-binding protein (Figure 8). PLCβ hydrolyzes phosphatidylinositol-4-5-biphosphate (PIP_2) to inositol-1-4-5-triphosphate (IP_3) and diacylglycerol (DAG).

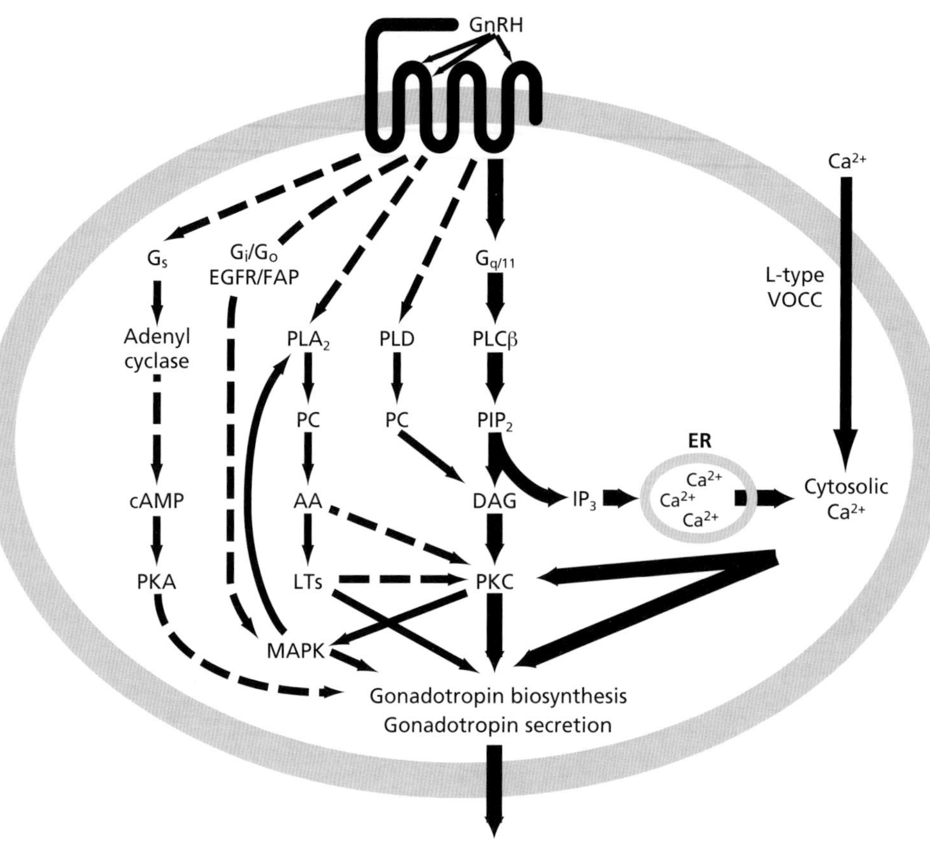

Figure 8 Signalling pathways of mammalian gonadotropin-releasing hormone (GnRH) type I receptor. The primary signalling pathway shown by heavy arrows involves the activation of phospholipase C (PLC) via the Gq/11 G-protein and the hydrolysis of phosphatidylinositol-1-4-biphosphate (PIP_2) to diacyl glycerol (DAG), which activates protein kinase C (PKC), and to inositol-1-4-5-triphosphate (IP_3), which stimulates Ca^{2+} release from the endoplasmic reticulum (ER). A second major effect is the entry of extracellular Ca^{2+} through voltage-operated Ca^{2+} channels (VOCCs). The GnRH receptor also activates phospholipase D (PLD), which hydrolyzes phosphatidyl choline (PC) to DAG, and phospholipase A_2 (PLA_2), which generates arachidonic acid (AA) and leukotrienes (LTs) from PC. PLA_2 activation may be mediated via a primary stimulation of mitogen-activated protein kinase (MAPK). DAG activates protein kinase C (PKC), which in turn can activate MAPK. MAPK isoforms may also be stimulated via Gi/o G-proteins, and may involve focal adhesion proteins (FAPs) and epidermal growth factor receptor (EGFR) and monomeric G-proteins such as ras. Broken lines show pathways that are less certain, and include the activation of adenyl cyclase through G_s resulting in the production of cyclic adenosine monophosphate (cAMP) and activation of protein kinase A (PKA), which has been demonstrated only in pituitaries from specific physiological states and cells transfected with the GnRH receptor. The products of the diverse pathways affect gonadotropin biosynthesis and secretion. Note that many of the signalling events occur in the plasma membrane, where enzymes (e.g. PLCβ, PLD, PLA_2, adenyl cyclase) and substrates (e.g. PIP_2) are anchored. For simplicity, this is not shown

IP_3 binds to receptors in the endoplasmic reticulum to release Ca^{2+} transiently from these intracellular stores, and this elicits a rapid spike of LH release. DAG activates protein kinase C (PKC), which phosphorylates proteins involved in the more sustained release of LH and in gonadotropin biosynthesis. A second major effect of GnRH binding to its receptor is the activation of L-type voltage-operated Ca^{2+} channels (VOCCs), which results in the influx of extracellular Ca^{2+} required for recharging intracellular stores and the prolonged (second phase) of LH release (Figure 8).

An important observation was that GnRH stimulates DAG production biphasically in pituitary cells, and at a relatively high molar ratio relative to IP_3.[144] This suggests that GnRH stimulates DAG production from a source independent of PLC. Phospholipase D (PLD) was found to be activated in a gonadotrope cell line a few minutes after PLC activation.[94,143,144] Since its substrate phosphatidylcholine is present at almost 100 times higher concentrations in the plasma membrane than that of PIP_2, GnRH stimulation of PLD may be the main source of sustained DAG production. PLD is known to be activated by the small G-protein ARF (adenosine diphosphate (ADP)-ribosylation factor). However, the GnRH receptor does not activate PLD through ARF owing to the substitution of the conserved Asn in TM7 with Asp (see above).[118] Intriguingly, mutation of Asp to Asn in TM7 and Asn to Asp in TM2 in the GnRH receptor, which restores the conventional GPCR arrangement, conveys ARF/PLD coupling. Thus, the GnRH receptor has been evolved so as not to couple via ARF, and the activation of PLD must be via other means.

Another potential signalling pathway of GnRH action is through phospholipase A_2 (PLA_2). GnRH stimulates arachidonic acid and its 5-lipoxygenase products, the leukotrienes, in pituitary cells some minutes after the rise in IP_3.[86,89,90,94] Although PLA_2 might be activated directly by the GnRH receptor, it is likely that activation is by mitogen-activated protein kinase (MAPK), which is known to phosphorylate and activate PLA_2. A number of studies have shown that GnRH indirectly activates MAPK in pituitary cells and gonadotrope cell lines via PKC phosphorylation of Raf-1.[94] MAPK may also be activated by Gi/o G-proteins, and may involve focal adhesion proteins (e.g. integrins), the epidermal growth factor receptor and monomeric G-proteins such as ras.

All of the signalling pathways outlined above result potentially in the activation of PKC isoforms (Figure 8). However, there are distinct differences in the timing and the PKC isoform activated, such that a sequential, co-ordinated and sustained activation of PKCs occurs. The early DAG and Ca^{2+} generated by PLC action might activate Ca^{2+}-dependent PKCs, while late DAG generated by PLD might activate Ca^{2+}-independent PKCs. Similarly, the more slowly generated products of PLA_2 such as arachidonic acid and leukotrienes are known to activate specific PKC isoforms (for a comprehensive review, see reference 94).

Finally, GnRH has been shown to stimulate adenyl cyclase, with the resultant production of cAMP and activation of protein kinase A (PKA). However, the response of pituitary cells is variable, and influenced by sex and other factors,[145,146] so the importance of this pathway remains uncertain. Recently, the rat GnRH receptor has been shown to couple to adenyl cyclase via G_s in COS cells[125] and in GGH3 cells,[147] and insect cells[127] transfected with the GnRH receptor.

GnRH regulation of gonadotropin gene expression

The distal targets of the signalling pathways, PKC isoforms, Ca^{2+} calmodulin, MAPK, leukotrienes and PKA, are all variously involved in gonadotropin gene expression, biosynthesis[86,94,148] and exocytosis.[142,149]

The mobilization of Ca^{2+} and activation of PKC resulting from the primary GnRH signalling pathway have both been reported to increase α-, LHβ- and FSHβ-subunit mRNA levels, and PKC response elements have been identified in the promoters of these subunits.[86,94,148] Although both Ca^{2+} and PKC increase LHβ and FSHβ independently, together they apparently do not do so.[94] The more protracted activation of PKC resulting from the products of the PLD and PLA_2 pathways presumably contributes to the increased gonadotropin subunit mRNA levels. GnRH activation of MAPK also stimulates the expression of subunit mRNA,[86,94,148] and, since PKC can activate MAPK, part or all of the PKC effects may be via MAPK. These effects may be mediated via distinct GnRH-response elements in the gene promoters.[150,151] Several studies have demonstrated an increase in mRNA of all the subunits on cAMP stimulation.[86,148] Although all of these signalling pathways have been shown to stimulate gonadotropin subunit mRNA, it is only the PLC/PKC/Ca^{2+} pathway that has been consistently shown to be activated by GnRH in gonadotropes from different species, sexes and physiological states. A potential role for the other pathways is therefore dependent on their expression in the gonadotrope being studied.

The relative stimulation of gonadotropin subunits in the rat pituitary is influenced by GnRH pulse frequency. High pulse frequencies stimulate LHβ mRNA levels more than FSHβ, while the converse is true at low pulse frequencies. Thus, the combination of alterations in pulse frequency together with different phasing and duration of the various signalling pathways provides the potential for fine regulation of gonadotropin secretion.[147] Feedback by gonadal steroid and peptide hormones[26,148] at the gonadotrope provides additional modulation.

CONCLUSIONS AND FUTURE PERSPECTIVES

The fundamental role of hypothalamic GnRH in the reproductive system through stimulating pituitary gonadotropin secretion has made it a prime drug target for the treatment of infertility and sex hormone-dependent diseases and for novel contraception. It is now clear that GnRHs have been coopted during evolution for other functions, in addition to regulating gonadotropins. The identification of structural variants of GnRH in extrahypothalamic tissues, and the discovery of their cognate GnRH receptor types, are providing considerable insight into novel physiological and pathophysiological roles of GnRHs in diverse processes. Moreover, a detailed molecular delineation of the interaction of these GnRHs with their cognate receptors is contributing to the development of novel GnRH therapeutics, including nonpeptide antagonists with potential oral application.

REFERENCES

1. Schally AV, Arimura A, Baba Y *et al.* Isolation and properties of the FSH and LH-releasing hormone. *Biochem Biophys Res Commun* 1971 43: 393–399.
2. Matsuo H, Baba Y, Nair RM *et al.* Structure of the porcine LH- and FSH-releasing hormone. I. The proposed amino acid sequence. *Biochem Biophys Res Commun* 1971 43: 1334–1339.
3. Baba Y, Matsuo H, Schally AV. Structure of the porcine LH- and FSH-releasing hormone. II. Confirmation of the proposed structure by conventional sequential analyses. *Biochem Biophys Res Commun* 1971 44: 459–463.
4. Fink G. Gonadotropin secretion and its control. In Knobil E, Neill J, eds. *The Physiology of Reproduction*. New York: Raven Press, 1988: 1349–1377.
5. Seeburg PH, Mason AJ, Stewart TA *et al.* The mammalian GnRH gene and its pivotal role in reproduction. *Recent Prog Horm Res* 1987 43: 69–98.
6. Millar RP, King JA, Davidson JS *et al.* Gonadotrophin-releasing hormone – diversity of functions and clinical applications. *South Afr Med J* 1987 72: 748–755.
7. Casper RF. Clinical uses of gonadotropin-releasing hormone analogues. *Can Med Assoc J* 1991 144: 153–158.
8. Conn PM, Crowley WF Jr. Gonadotropin-releasing hormone and its analogues. *N Engl J Med* 1991 324: 93–103.
9. Barbieri RL. Clinical applications of GnRH and its analogues. *Trends Endocrinol Metab* 1992 3: 30–34.
10. Moghissi KS. Clinical applications of gonadotropin-releasing hormones in reproductive disorders. *Endocrinol Metab Clin North Am* 1992 21: 125–140.
11. Conn PM, Crowley WF Jr. Gonadotropin-releasing hormone and its analogs. *Annu Rev Med* 1994 45: 391–405.
12. Filicori M. Gonadotrophin-releasing hormone agonists. A guide to use and selection [Published erratum appears in *Drugs* 1994 48: 326]. *Drugs* 1994 48: 41–58.
13. Emons G, Schally AV. The use of luteinizing hormone releasing hormone agonists and antagonists in gynaecological cancers. *Hum Reprod* 1994 9: 1364–1379.
14. Millar RP, Zhu Y-F, Chen C, Struthers RS. Progress towards the development of non-peptide orally-active gonadotropin-releasing hormone (GnRH) antagonists: therapeutic implications. *Br Med Bull* 2000 56: 761–772.
15. Nieschlag E, Behre HM, Weinbauer GF. Hormonal male contraception: a real chance? In

Nieschlag E, Habenicht U-F, eds. *Spermatogenesis–Fertilization–Contraception. Molecular, Cellular and Endocrine Events in Male Reproduction*. Berlin: Springer-Verlag, 1992: 477–501.

16. Fraser HM. GnRH analogues for contraception. *Br Med Bull* 1993 49: 62–72.
17. King JA, Millar RP. Heterogeneity of vertebrate luteinizing hormone-releasing hormone. *Science* 1979 206: 67–69.
18. King JA, Millar RP. Comparative aspects of luteinizing hormone-releasing hormone structure and function in vertebrate phylogeny. *Endocrinology* 1980 106: 707–717.
19. Millar RP, King JA. Structural and functional evolution of gonadotropin-releasing hormone. *Int Rev Cytol* 1987 106: 149–182.
20. Millar RP, King JA. Evolution of gonadotropin-releasing hormone: multiple usage of a peptide. *News Physiol Sci* 1988 3: 49–53.
21. King JA, Millar RP. Evolution of gonadotropin-releasing hormone. *Trends Endocrinol Metab* 1992 3: 339–346.
22. King JA, Millar RP. Evolutionary aspects of gonadotropin-releasing hormone and its receptor. *Cell Mol Neurobiol* 1995 15: 5–23.
23. King JA, Millar RP. Coordinated evolution of GnRHs and their receptors. In Parhar IS, Sakuma Y, eds. *GnRH Neurones: Gene to Behavior*. Toyko: Brain Shuppan, 1997: 51–77.
24. Millar RP, Troskie B, Sun YM *et al.* Plasticity in the structural and functional evolution of GnRH: a peptide for all seasons. Presented at the *Thirteenth International Congress of Comparative Endocrinology*, Yokohama, Japan, 1997: 15–27.
25. Sherwood N. The GnRH family of peptides. *Trends Neurosci* 1987 10: 129–132.
26. Sherwood NM, Lovejoy DA. The origin of the mammalian form of GnRH in primitive fishes. *Fish Physiol Biochem* 1989 7: 85–93.
27. Sherwood NM, Lovejoy DA, Coe IR. Origin of mammalian gonadotropin-releasing hormones. *Endocr Rev* 1993 14: 241–254.
28. Jimenz-Linan M, Rubin BS, King JC. Examination of guinea pig luteinizing-hormone-releasing-hormone gene reveals a unique decapeptide and existence of two transcripts in the brain. *Endocrinology* 1997 138: 423–430.
29. Yoo MS, Kang HM, Choi HS *et al.* Molecular cloning, distribution and pharmacological characterisation of a novel gonadotropin-releasing hormone ([Trp8]GnRH) in frog brain. *Mol Cell Endocrinol* 2000 164: 197–204.
30. Okubo K, Amano M, Yoshiura Y, Suetake H, Aida K. A novel form of gonadotropin-releasing hormone in the medaka, *Oryzias latipes*. *Biochem Biophys Res Commun* 2000 16: 298–303.
31. Hsueh AJ, Schaeffer JM. Gonadotropin-releasing hormone as a paracrine hormone and neurotransmitter in extra-pituitary sites. *J Steroid Biochem* 1985 23: 757–764.
32. Jennes L, Conn PM. Gonadotropin-releasing hormone and its receptors in rat brain. *Front Neuroendocrinol* 1994 15: 51–77.
33. Sealfon SC, Weinstein H, Millar RP. Molecular mechanisms of ligand interaction with the gonadotropin-releasing hormone receptor. *Endocr Rev* 1997 18: 180–205.
34. Troskie B, Illing N, Rumbak E *et al.* Identification of three putative GnRH receptor subtypes in vertebrates. *Gen Comp Endocrinol* 1998 112: 296–302.
35 Miyamoto K, Hasegawa Y, Nomura M. Identification of the second gonadotropin-releasing hormone in chicken hypothalamus: evidence that gonadotropin secretion is probably controlled by two distinct gonadotropin-releasing hormones in avian species. *Proc Natl Acad Sci USA* 1984 81: 3874–3878.
36. White RB, Eisen JA, Kasten TL *et al.* Second gene for gonadotropin-releasing hormone in humans. *Proc Natl Acad Sci USA* 1997 95: 305–309.
37. Karten MJ, Rivier JE. Gonadotropin-releasing hormone analog design. Structure–function studies toward the development of agonists and antagonists: rationale and perspective. *Endocr Rev* 1986 7: 44–66.
38. Illing N, Troskie BE, Nahorniak CS *et al.* Two gonadotropin-releasing hormone receptor subtypes with distinct ligand selectivity and differential distribution in brain and pituitary in the goldfish (*Carassius autatus*). *Proc Natl Acad Sci USA* 1999 96: 2526–2531.
39. Millar RP, Flanagan CA, Milton RC *et al.* Chimeric analogues of vertebrate gonadotropin-releasing hormones comprising substitutions of the variant amino acids in positions 5, 7, and 8. Characterization of requirements for receptor binding and gonadotropin release in mammalian

and avian pituitary gonadotropes. *J Biol Chem* 1989 264: 21007–21013.

40. Milton RC, King JA, Badminton MN *et al.* Comparative structure activity studies of mammalian [Arg8]LHRH and chicken [Gln8]LHRH by fluorimetric titration. *Biochem Biophys Res Commun* 1983 111: 1082–1088.
41. Guarnieri F, Weinstein H. Conformational memories and the exploration of biologically relevant peptide conformations: an illustration for the gonadotropin-releasing hormone. *J Am Chem Soc* 1996 118: 5580–5589.
42. Maliekal J, Jackson GE, Flanagan CA *et al.* Solution conformations of gonadotropin releasing hormone (GnRH) and [Gln8] GnRH. *South Afr J Chem* 1997 50: 217–219.
43. Tensen C, Okuzawa K, Blomenrohr M *et al.* Distinct efficacies for two endogenous ligands on a single cognate gonadoliberin receptor. *Eur J Biochem* 1997 243: 134–140.
44. Flanagan CA, Becker II, Davidson JS *et al.* Glutamate 301 of the mouse gonadotropin-releasing hormone receptor confers specificity for arginine 8 of mammalian gonadotropin-releasing hormone. *J Biol Chem* 1994 269: 22636–22641.
45. Powell JFF, Reska-Skinner SM, Prakash MO *et al.* Two new forms of gonadotropin-releasing hormone in a protochordate and the evolutionary implications. *Neurobiology* 1996 93: 10461–10464.
46. Hauser F, Sondergaard L, Grimmelikhuijzen CJ. Molecular cloning, genomic organization and developmental regulation of a novel receptor from *Drosophila melanogaster* structurally related to gonadotropin-releasing hormone receptors for vertebrates. *Biochem Biophys Res Commun* 1998 249: 822–828.
47. Hauser F, Nothacker HP, Grimmelikhuijzen CJ. Molecular cloning, genomic organization, and developmental regulation of a novel receptor from *Drosophila melanogaster* structurally related to members of the thyroid-stimulating hormone, follicle-stimulating hormone, luteinizing hormone/choriogonadotropin receptor family from mammals. *J Biol Chem* 1997 272: 1002–1010.
48. Lescheid DW, Terasawa E, Abler LA *et al.* A second form of gonadotropin-releasing hormone (GnRH) with characteristics of chicken GnRH-II is present in the primate brain. *Endocrinology* 1997 138: 5618–5629.
49. Jones SW. Chicken II luteinizing hormone-releasing hormone inhibits the M-current of bullfrog sympathetic neurons. *Neurosci Lett* 1987 80: 180–184.
50. Troskie B, King JA, Millar RP *et al.* Chicken GnRH II-like peptides and a GnRH receptor selective for chicken GnRH II in amphibian sympathetic ganglia. *Neuroendocrinology* 1997 65: 396–402.
51. Moss RL. Actions of hypothalamic–hypophysiotropic hormones on the brain. *Annu Rev Physiol* 1979 41: 617–631.
52. Muske LE. Evolution of gonadotropin-releasing hormone (GnRH) neuronal systems. *Brain Behav Evol* 1993 42: 215–230.
53. Rissman EF, Li X, King JA *et al.* Behavioral regulation of gonadotropin-releasing hormone production. *Brain Res Bull* 1997 44: 459–464.
54. Maney DL, Richardson RD, Wingfield JC. Central administration of chicken gonadotropin-releasing hormone-II enhances courtship behavior in a female sparrow. *Horm Behav* 1997 32: 11–18.
55. Millar RM, Lowe S, Conklin D *et al.* A novel mammalian receptor for the evolutionarily conserved Type II gonadotropin-releasing hormone. *Proc Natl Acad Sci* 2001 98: 9636–9641.
56. Tsutsumi M, Zhou W, Millar RP *et al.* Cloning and functional expression of a mouse gonadotropin-releasing hormone receptor. *Mol Endocrinol* 1992 6: 1163–1169.
57. Reinhart J, Mertz LM, Catt KJ. Molecular cloning and expression of cDNA encoding the murine gonadotropin-releasing hormone receptor. *J Biol Chem* 1992 267: 21281–21284.
58. Kaiser UB, Zhao D, Cardona GR *et al.* Isolation and characterization of cDNAs encoding the rat pituitary gonadotropin-releasing hormone receptor. *Biochem Biophys Res Commun* 1992 189: 1645–1652.
59. Perrin MH, Bilezikjian LM, Hoeger C *et al.* Molecular and functional characterization of GnRH receptors cloned from rat pituitary and a mouse pituitary tumor cell line. *Biochem Biophys Res Commun* 1993 191: 1139–1144.
60. Eidne KA, Sellar RE, Couper G *et al.* Molecular cloning and characterisation of the rat pituitary

gonadotropin-releasing hormone (GnRH) receptor. *Mol Cell Endocrinol* 1992 90: R5–R9.

61. Chi L, Zhou W, Prikhozhan A *et al.* Cloning and characterization of the human GnRH receptor. *Mol Cell Endocrinol* 1993 91: R1–R6.
62. Kakar SS, Musgrove LC, Devor DC *et al.* Cloning, sequencing, and expression of human gonadotropin releasing hormone (GnRH) receptor. *Biochem Biophys Res Commun* 1992 189: 289–295.
63. Illing N, Jacobs GF, Becker II *et al.* Comparative sequence analysis and functional characterization of the cloned sheep gonadotropin-releasing hormone receptor reveal differences in primary structure and ligand specificity among mammalian receptors. *Biochem Biophys Res Commun* 1993 196: 745–751.
64. Brooks J, Taylor PL, Saunders PT *et al.* Cloning and sequencing of the sheep pituitary gonadotropin-releasing hormone receptor and changes in expression of its mRNA during the estrous cycle. *Mol Cell Endocrinol* 1993 94: R23–R27.
65. Kakar SS, Rahe CH, Neill JD. Molecular cloning, sequencing, and characterizing the bovine receptor for gonadotropin releasing hormone (GnRH). *Domest Anim Endocrinol* 1993 10: 335–342.
66. Weesner GD, Matteri RL. Rapid communication: nucleotide sequence of luteinizing hormone-releasing hormone (LHRH) receptor cDNA in the pig pituitary. *J Anim Sci* 1994 72: 1911.
67. King JA, Fidler A, Lawrence S *et al.* Cloning and expression, pharmacological characterization and internalization kinetics of the pituitary GnRH receptor in a metatherian species of mammal. *Gen Comp Endocrinol* 2000 117: 439–448
68. Wang L, Bogerd J, Choi HS *et al.* Three distinct types of GnRH receptor characterized in the bullfrog. *Proc Natl Acad Sci USA* 2001 98: 361–366.
69. Troskie BE, Hapgood JP, Millar RP, Illing N. cDNA cloning, gene expression and ligand selectivity of a novel GnRH receptor expressed in the pituitary and midbrain of *Xenopus laevis*. *Endocrinology* 2000 141: 1764–1771.
70. Sun YM, Flanagan CA, Illing N *et al.* A chicken gonadotropin-releasing hormone receptor that confers agonist activity to mammalian antagonists: identification of D-Lys^6 in the ligand and extracellular loop two of the receptor as determinants. *J Biol Chem* 2000 276: 7754–7761
71. Okubo K, Nagata S, Ko R *et al.* Identification and characterization of two distinct GnRH receptor subtypes in a teleost, the Medaka *Oryzias latipes*. *Endocrinology* 2001 142: 4729–4739
72. Alok D, Hassin S, Kumar RS *et al.* Characterization of a pituitary GnRH-receptor from a perciform fish, *Morone saxatilis*: functional expression in a fish cell line. *Mol Cell Endocrinol* 2000 168: 65–75.
73. Madigou T, Mañanos-Sanchez E, Hulshof S *et al.* Cloning, tissue distribution, and central expression of the gonadotropin-releasing hormone receptor in the rainbow trout (*Oncorhynchus mykiss*). *Biol Reprod* 2000 63: 1857–1866.
74. Robison RR, White RB, Illing N. Gonadotropin-releasing hormone receptor in the teleost *Haplochromis burtoni*: structure, location, and function. *Endocrinology* 2001 142: 1737–1743.
75. Okubo K, Suetake H, Usami T, Aida K. Molecular cloning and tissue specific expression of a gonadotropin-releasing hormone receptor in the Japanese eel. *Gen Comp Endocrinol* 2000 119: 181–192.
76. Millar R, Conklin D, Lofton-Day C *et al.* A novel human GnRH receptor homolog gene: abundant and wide tissue distribution of the antisense transcript. *J Endocrinol* 1999 162: 117–126.
77. Conklin DC, Rixon MW, Kuestner RE *et al.* Cloning and gene expression of a novel human ribonucleoprotein. *Biochim Biophys Acta* 2000 1492: 465–469.
78. Faurholm B, Millar RP, Katz AA. The genes encoding for the Type II gonadotropin-releasing hormone receptor and the ribonucleoprotein RBM8A overlap in two genomic loci. *Genomics* 2001 78: 15–18.
79. Neill JD, Duck LW, Sellers JC, Musgrove LC. A gonadotropin-releasing hormone (GnRH) receptor specific for GnRH II in primates. *Biochem Biophys Res Commun* 2001 282: 1012–1018.
80. Fan NC, Jeung EB, Peng C *et al.* The human gonadotropin-releasing hormone (GnRH) receptor gene: cloning, genomic organization and chromosomal assignment. *Mol Cell Endocrinol* 1994 103: R1–R6.
81. Fan NC, Peng C, Krisinger J *et al.* The human gonadotropin-releasing hormone receptor gene:

complete structure including multiple promoters, transcription initiation sites, and polyadenylation signals. *Mol Cell Endocrinol* 1995 107: R1–R8.

82. Kakar SS. Molecular structure of the human gonadotropin-releasing hormone receptor gene. *Eur J Endocrinol* 1997 137: 183–192.
83. Zhou W, Sealfon SC. Structure of the mouse gonadotropin-releasing hormone receptor gene: variant transcripts generated by alternative processing. *DNA Cell Biol* 1994 13: 605–14.
84. Grosse R, Schöneberg T, Schultz G *et al.* Inhibition of gonadotropin-releasing hormone receptor signaling by expression of a splice variant of the human receptor. *Mol Endocrinol* 1997 11: 1305–1318.
85. Wang, L, Oh DY, Bogerd J *et al.* Inhibitory activity of alternative splice variants of the bullfrog GnRH receptor-3 on wild-type receptor signalling. *Endocrinology* 2001 142: 4015–4025.
86. Kaiser UB, Conn PM, Chin WW. Studies of gonadotropin-releasing hormone (GnRH) action using GnRH receptor-expressing pituitary cell lines. *Endocr Rev* 1997 18: 46–70.
87. Sealfon SC, Millar RP. The gonadotropin-releasing hormone receptor: structural determinants and regulatory control. In Charlton HM, eds. *Oxford Reviews of Reproductive Biology.* New York: Oxford University Press, 1994 17: 255–283.
88. Turgeon JL, Kimura Y, Waring DW *et al.* Steroid and pulsatile gonadotropin-releasing hormone (GnRH) regulation of luteinizing hormone and GnRH receptor in a novel gonadotrope cell line. *Mol Endocrinol* 1996 10: 439–450.
89. Stojilkovic SS, Reinhart J, Catt KJ. Gonadotropin-releasing hormone receptors: structure and signal transduction pathways. *Endocr Rev* 1994 15: 462–499.
90. Stojilkovic SS, Catt KJ. Expression and signal transduction pathways of gonadotropin-releasing hormone receptors. *Recent Prog Horm Res* 1995 50: 161–205.
91. Palczewski K, Kumasaka T, Hori T. Crystal structure of rhodopsin: a G protein-coupled receptor. *Science* 2000 289: 739–745.
92. Flanagan CA, Millar RP, Illing N. Advances in understanding gonadotrophin-releasing hormone receptor structure and ligand interactions. *Rev Reprod* 1997 2: 113–120.
93. Schertler GF, Villa C, Henderson R. Projection structure of rhodopsin. *Nature* 1993 362: 770–772.
94. Naor Z, Harris D, Shacham S. Mechanism of GnRH receptor signaling: combinatorial cross-talk of Ca^{2+} and protein kinase C. *Front Neuroendocrinol* 1998 19: 1–19.
95. Baldwin JM. The probable arrangement of the helices in G protein-coupled receptors. *EMBO J* 1993 12: 1693–1703.
96. Lesk AM, Boswell DR. Homology modelling: inferences from tables of aligned sequences. *Curr Opin Struct Biol* 1992 2: 242–247.
97. Donnelly D, Johnson MS, Blundell TL *et al.* An analysis of the periodicity of conserved residues in sequence alignments of G-protein coupled receptors. Implications for the three-dimensional structure. *FEBS Lett* 1989 251: 109–116.
98. Ballesteros JA, Weinstein H. Analysis and refinement of criteria for predicting the structure and relative orientations of transmembranal helical domains. *Biophys J* 1992 62: 107–109.
99. Zhou W, Flanagan C, Ballesteros JA *et al.* A reciprocal mutation supports helix 2 and helix 7 proximity in the gonadotropin-releasing hormone receptor. *Mol Pharmacol* 1994 45: 165–170.
100. Davidson JS, Assefa D, Pawson A *et al.* Irreversible activation of the gonadotropin-releasing hormone receptor by photoaffinity cross-linking: localization of attachment site to Cys residue in N-terminal segment. *Biochemistry* 1997 36: 12881–12889.
101. Davidson JS, Flanagan CA, Zhou W *et al.* Identification of N-glycosylation sites in the gonadotropin-releasing hormone receptor: role in receptor expression but not ligand binding. *Mol Cell Endocrinol* 1995 107: 241–245.
102. Davidson JS, Flanagan CA, Davies PD *et al.* Incorporation of an additional glycosylation site enhances expression of functional human gonadotropin-releasing hormone receptor. *Endocrine* 1996 4: 207–212.
103. Petry R, Craik D, Haaima G *et al.* Secondary structure of the third extracellular loop responsible for ligand selectivity of a mammalian gonadotropin-releasing hormone receptor. Submitted for publication
104. Millar RP, King JA. Synthesis, luteinizing hormone-releasing activity, and receptor binding

of chicken hypothalamic luteinizing hormone-releasing hormone. *Endocrinology* 1983 113: 1364–1369.

105. Fromme BJ, Katz AA, Roeske RW, Millar RP, Flanagan CA. Role of aspartate 7.32 (302) of the human gonadotropin-releasing hormone receptor in stabilizing a high-affinity ligand conformation. *Mol Pharmacol* 2001 60: 1280–1287.
106. Hoffmann SH, Laak TT, Kühne R, Reiländer H, Beckers T. Residues within transmembrane helices 2 and 5 of the human gonadotropin-releasing hormone receptor contribute to agonist and antagonist binding. *Mol Endocrinol* 2000 14: 1099–1115.
107. Zhou W, Rodic V, Kitanovic S *et al.* A locus of the gonadotropin-releasing hormone receptor that differentiates agonist and antagonist binding sites. *J Biol Chem* 1995 270: 18853–18857.
108. Davidson JS, McArdle CA, Davies P *et al.* Asn102 of the gonadotropin-releasing hormone receptor is a critical determinant of potency for agonists containing C-terminal glycinamide. *J Biol Chem* 1996 271: 15510–15514.
109. Rodic V, Flanagan C, Millar R *et al.* Role of Asp2.61(98) in agonist complexing with the human gonadotropin-releasing hormone receptor. *Soc Neurosci Abstr* 1996 22: 1302.
110. de Roux N, Young J, Misrahi M *et al.* A family with hypogonadotropic hypogonadism and mutations in the gonadotropin-releasing hormone receptor. *N Engl J Med* 1997 337: 1597–1602.
111. de Roux N, Young J, Brailly-Tabard S *et al.* The same molecular defects of the gonadotropin-releasing hormone-receptor determine a variable degree of hypogonadism in affected kindred. *J Clin Endocrinol Metab* 1999 84: 567–572.
112. Bertherat J. Gonadotropin-releasing hormone receptor gene mutation: a new cause of hereditary hypogonadism and another mutated G-protein-coupled receptor. *Eur J Endocrinol* 1998 138: 621–622.
113. Chauvin S, Bérault A, Lerrant Y, Hibert M, Counis R. Functional importance of transmembrane helix 6 Trp^{279} and exoloop 3 Val^{299} of rat gonadotropin-releasing hormone receptor. *Mol Pharmacol* 2000 57: 625–633
114. Kenakin T. *Pharmacologic Analysis of Drug–Receptor Interaction*, 2nd edn. New York: Raven Press, 1993.
115. De Lean A, Stadel JM, Lefkowitz RJ. A ternary complex model explains the agonist-specific binding properties of the adenylate cyclase-coupled β-adrenergic receptor. *J Biol Chem* 1980 255: 7108–7117.
116. Samama P, Cotecchia S, Costa T *et al.* A mutation-induced activated state of the β2-adrenergic receptor. Extending the ternary complex model. *J Biol Chem* 1993 268: 4625–4636.
117. Blomenrohr M, Bogerd J, Leurs R *et al.* Differences in structure–function relations between nonmammalian and mammalian gonadotropin-releasing hormone receptors. *Biochem Biophys Res Commun* 1997 238: 517–522.
118. Mitchell R, McCulloch D, Lutz E *et al.* Rhodopsin-family receptors associate with small G proteins to activate phospholipase D. *Nature* 1998 392: 411–414.
119. Flanagan CA, Zhou W, Chi L *et al.* The functional microdomain in transmembrane helices 2 and 7 regulates expression, activation and coupling pathways of the gonadotrophin-releasing hormone receptor. *J Biol Chem* 1999 274: 28880–28886.
120. Ballesteros J, Kitanovic S, Guarnieri F *et al.* Functional microdomains in G-protein-coupled receptors. The conserved arginine-cage motif in the gonadotropin-releasing hormone receptor. *J Biol Chem* 1998 273: 10445–10453.
121. Arora KK, Cheng Z, Catt KJ. Mutations of the conserved DRS motif in the second intracellular loop of the gonadotropin-releasing hormone receptor affect expression, activation, and internalization. *Mol Endocrinol* 1997 11: 1203–1212.
122. Kjelsberg MA, Cotecchia S, Ostrowski J *et al.* Constitutive activation of the α1B-adrenergic receptor by all amino acid substitutions at a single site. Evidence for a region which constrains receptor activation. *J Biol Chem* 1992 267: 1430–1433.
123. Myburgh DB, Millar RP, Hapgood JP. Alanine-261 in intracellular loop III of the human gonadotropin-releasing hormone receptor is crucial for G-protein coupling and receptor internalization. *Biochem J* 1998 331: 893-896.
124. Chi L, Davidson JS, Zhou W *et al.* Mutations of the second intracellular loop domain of the GnRH receptor. Presented at the *76th Annual Meeting of the Endocrine Society*, Anaheim, CA, 1994: abstr 159.

125. Arora KK, Krsmanovic LZ, Mores N *et al.* Mediation of cyclic AMP signaling by the first intracellular loop of the gonadotropin-releasing hormone receptor. *J Biol Chem* 1998 273: 25581–25586.
126. Ulloa-Aguirre A, Stanislaus D, Arora V *et al.* The third intracellular loop of the rat gonadotropin-releasing hormone receptor couples the receptor to Gs- and G(q/11)-mediated signal transduction pathways: evidence from loop fragment transfection in GGH3 cells. *Endocrinology* 1998 139: 2472–2478.
127. Delahaye R, Manna PR, Berault A *et al.* Rat gonadotropin-releasing hormone receptor expressed in insect cells induces activation of adenylyl cyclase. *Mol Cell Endocrinol* 1997 135: 119–127.
128. Luttrell LM, Ferguson SS, Daaka Y *et al.* β-Arrestin-dependent formation of β2 adrenergic receptor–src protein kinase complexes. *Science* 1999; 283: 655–661.
129. Sibley DR, Benovic JL, Caron MG *et al.* Regulation of transmembrane signaling by receptor phosphorylation. *Cell* 1987 48: 913–922.
130. Leeb-Lundberg LMF, Cotecchia S, DeBlasi A *et al.* Regulation of adrenergic receptor function by phosphorylation. I. Agonist-promoted desensitization and phosphorylation of α1-adrenergic receptors coupled to inositol phospholipid metabolism in DDT1 MR-2 smooth muscle cells. *J Biol Chem* 1987 262: 3098–3105.
131. Nussenzveig DR, Heinflink M, Gershengorn MC. Agonist-stimulated internalization of the thyrotropin-releasing hormone receptor is dependent on two domains in the receptor carboxyl terminus. *J Biol Chem* 1993 268: 2389–2392.
132. Benya RV, Fathi Z, Battey JF *et al.* Serines and threonines in the gastrin-releasing peptide receptor carboxyl terminus mediate internalization. *J Biol Chem* 1993 268: 20285–20290.
133. Davidson JS, Wakefield IK, Millar RP. Absence of rapid desensitization of the mouse gonadotropin-releasing hormone receptor. *Biochem J* 1994 300: 299–302.
134. McArdle CA, Forrest-Owen W, Willars G *et al.* Desensitization of gonadotropin-releasing hormone action in the gonadotrope-derived α T3-1 cell line. *Endocrinology* 1995 136: 4864–4871.
135. Anderson L, McGregor A, Cook JV *et al.* Rapid desensitization of GnRH-stimulated intracellular signalling events in α T3-1 and HEK-293 cells expressing the GnRH receptor. *Endocrinology* 1995 136: 5228–5231.
136. Pawson AJ, Katz A, Sun Y-M *et al.* Contrasting internalization kinetics of human and chicken gonadotropin-releasing hormone receptors mediated by C-terminal tail. *J Endocrinol* 1997 156: R009–R012.
137. Heding A, Vrecl M, Bogerd J *et al.* Gonadotropin-releasing hormone receptors with intracellular carboxy-terminal tails undergo acute desensitization of total inositol phosphate production and exhibit accelerated internalization kinetics. *J Biol Chem* 1998 273: 11472–11477.
138. Hille BL, Tse A, Tse FW *et al.* Signalling mechanisms during the response of pituitary gonadotropes to GnRH. *Recent Prog Horm Res* 1995 50: 75–95.
139. Clayton RN. Regulation of gonadotrophin subunit gene expression. *Hum Reprod* 1993 8 (Suppl 2): 29–36.
140. Kiesel L. Molecular mechanisms of gonadotrophin releasing hormone-stimulated gonadotrophin secretion. *Hum Reprod* 1993 8 (Suppl 2): 23–28.
141. Conn PM, Janovick JA, Stanislaus D *et al.* Molecular and cellular bases of gonadotropin-releasing hormone action in the pituitary and central nervous system. *Vitam Horm* 1995 50: 151–214.
142. Anderson L. Intracellular mechanisms triggering gonadotrophin secretion. *Rev Reprod* 1996 1: 193–202.
143. Naor Z. Signal transduction mechanisms of Ca^{2+} mobilizing ligands: the case of GnRH. *Endocr Rev* 1990 11: 326–353.
144. Zheng L, Stojilkovic SS, Hunyady L *et al.* Sequential activation of phospholipase-C and -D in agonist-stimulated gonadotrophs. *Endocrinology* 1994 134: 1446–1454.
145. Borgeat P, Chavancy G, Dupont A *et al.* Stimulation of adenosine 3':5'-cyclic monophosphate accumulation in anterior pituitary gland *in vitro* by synthetic luteinizing hormone-releasing hormone. *Proc Natl Acad Sci USA* 1972 69: 2677–2681.
146. Naor Z, Fawcett CP, McCann SM. Differential effects of castration and testosterone replacement

on basal and LHRH-stimulated cAMP and cGMP accumulation and on gonadotropin release from the pituitary of the male rat. *Mol Cell Endocrinol* 1979 14: 191–198.

147. Stanislaus D, Pinter JH, Janovick JA, Conn PM. Mechanisms mediating multiple physiological responses to gonadotropin-releasing hormone. *Mol Cell Endocrinol* 1998 144: 1–10.

148. Brown P, McNeilly AS. Transcriptional regulation of pituitary gonadotrophin subunit genes. *Rev Reprod* 1999 4: 117–124.

149. Davidson JS, Wakefield I, van der Merwe PA *et al.* Mechanisms of luteinizing hormone secretion: new insights from studies with permeabilized cells. *Mol Cell Endocrinol* 1991 76: C33–C38.

150. Robertson MS, Misra-Press A, Laurance ME, Stork PJS, Maurer RA. A role for mitogen-activated protein kinase in mediating activation of the glycoprotein hormone α subunit promoter by gonadotrophin-releasing hormone. *Mol Cell Biol* 1995 17: 3531–3539.

151. Pernasetti F, Vasilyev V, Rosenberg SB *et al.* Cell specific transcriptional regulation of follicle-stimulating hormone-β by activin and gonadotropin-releasing hormone in the LβT2 pituitary gonadotrope cell model. *Endocrinology* 2001 142: 2284–2295.

Chapter 12

Gonadotropins and gonadotropin receptors

Manuel Tena-Sempere and Ilpo T. Huhtaniemi

BACKGROUND

The gonadotropins, i.e. luteinizing hormone (LH), follicle stimulating hormone (FSH) and chorionic gonadotropin (CG), are structurally related glycoproteins composed of two non-covalently linked subunits: a common α-subunit and a hormone-specific β-subunit. They are produced in gonadotroph cells of the anterior pituitary gland (LH and FSH) and in the placenta (CG, only in primates and a few Equidae), and they act co-ordinately to regulate gonadal differentiation, growth and endocrine function in both sexes.[1,2] The actions of gonadotropins are mediated through binding to specific cell surface receptors that belong to the large family of G protein-coupled, seven transmembrane helix receptors. The transduction of the hormonal message and biological actions of LH/CG (LH receptor) and FSH (FSH receptor) in specific target cells occurs using cyclic adenosine monophosphate (cAMP) as the main, although not the only, intracellular second messenger.[3,4]

Gonadotropin receptors are primarily expressed in gonads in a strict pattern of cell-specific distribution. In addition, there is recent evidence for ubiquitous expression of LH receptors in extragonadal tissues, but this phenomenon remains with undefined physiological significance.[5] In the testis, LH receptors are present in Leydig cells, and mediate the stimulatory action of LH on testosterone production, which in turn is essential for the attainment of extragonadal (genital differentiation, secondary sex characteristics and anabolic effects) and intragonadal (spermatogenesis) functions of androgens in the male.[6,7] FSH receptors are expressed in Sertoli cells and, before puberty, FSH stimulates the proliferation of this cell type,[8] thereby indirectly determining the finite size of the adult testis, as one Sertoli cell can support a fixed number of germ cells. In addition, FSH has specific stimulatory effects on the first stages of spermatogonial maturation, thus increasing the quantity of sperm production.[7] In contrast, recent data on men and mice with inactivating FSH ligand and receptor mutations do not support the previous contention that FSH action is absolutely required for the pubertal initiation of spermatogenesis (see below).[9,10]

In the ovary, LH receptors are expressed in theca, granulosa and luteal cells. In theca cells, LH elicits the production of aromatizable androgens,[11] whereas, in the mature preovulatory follicle, LH triggers ovulation by inducing the rupture of the follicular wall. After ovulation, LH stimulates luteinization and progesterone production of the corpus luteum.[12] FSH receptors are expressed solely in granulosa cells, where they convey the pleiotropic actions of FSH on the ovary, which include promotion of follicular growth, selection of preovulatory follicles, stimulation of estrogen production by increasing the aromatization of theca-derived androgens, and induction of LH receptors upon granulosa cell luteinization.[12] In addition, if pregnancy begins, CG, which can be considered a superagonist of LH, rescues the corpus luteum from regression, and supports progesterone production of the corpus luteum of pregnancy.[2] The other known role of CG is to stimulate fetal testicular steroidogenesis during the critical period of genital masculinization between weeks 9 and 14 of pregnancy.[13]

In recent years, the application of molecular biological techniques has vastly expanded our knowledge of basic aspects of the structure–function relationships of gonadotropins and their receptors, as well as the clinical implications of their dysfunction. This chapter summarizes the current concepts and reviews some potential future perspectives in this field of research.

GENERAL STRUCTURE OF GONADOTROPINS AND GONADOTROPIN RECEPTORS

Of the four glycoprotein hormones, LH, FSH and thyroid stimulating hormone (TSH) are produced in the pituitary gland, while the LH homolog CG originates from placental trophoblast. Structurally, they are all relatively large proteins (molecular mass 30–40 kDa), consisting of a common α-subunit and a hormone-specific β-subunit that are associated through non-covalent interactions[1,2] (Figure 1). The mature α-subunit consists of 92 amino acid residues and is encoded by a single gene, comprising four exons, which in the human is localized on chromosome 6q12.21. The

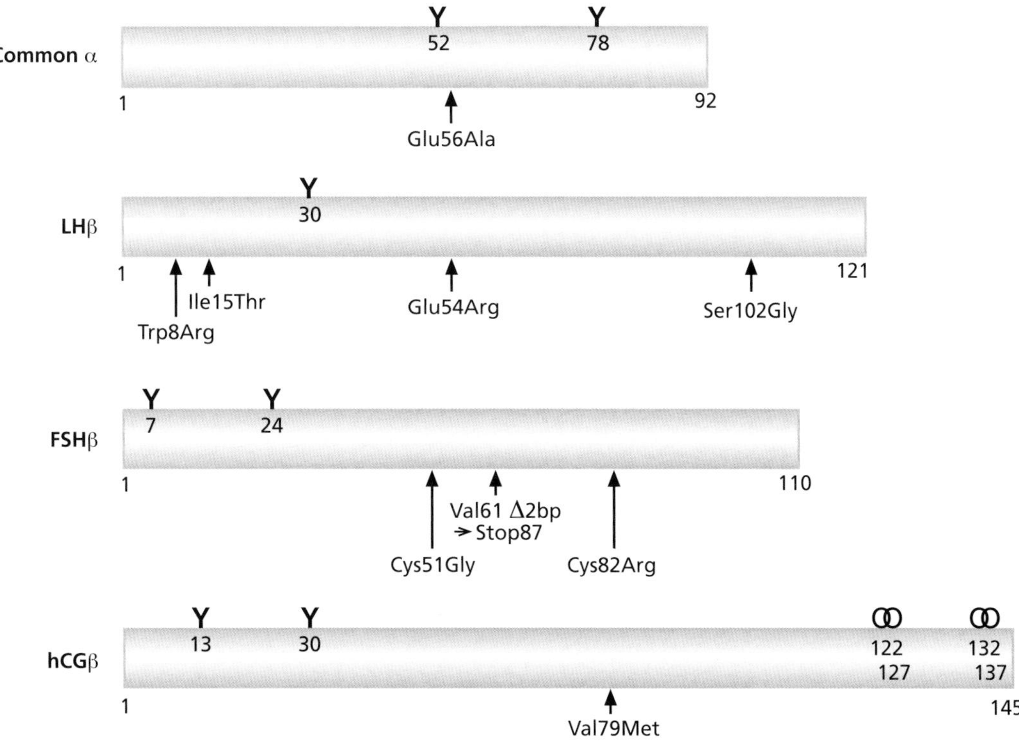

Figure 1 Schematic presentation of structures of the human gonadotropin subunits. The length of each mature protein product is indicated, taking the first amino acid of the mature protein as 1. For each subunit, the relative positions of the N-linked (Y) and O-linked (O) glycosylation sites are depicted. In addition, the mutations reported to date in the primary structure of gonadotropin subunits are indicated. LH, luteinizing hormone; FSH, follicle stimulating hormone; hCG, human chorionic gonadotropin. Data from reference 14

β-subunits are responsible for the functional specificity of the hormones. However, they show considerable amino acid homology, ranging in humans from 32% for the LH–TSH pair to 83% for the LH–CG pair (excluding the non-homologous C-terminal extension of CG). The β-subunit genes are composed of three exons and they are located on different chromosomes: the LH/CGβ gene cluster on chromosome 19q13.32, FSHβ on chromosome 11p13 and TSHβ on chromosome 1p13.[15] The LH/CGβ gene cluster consists of one LHβ gene and six CGβ genes and pseudogenes, although most of the steady-state (human) hCGβ mRNA appears to be transcribed from genes 3, 5 and 8.[16] In fact, the hCGβ gene is thought to have evolved recently from the LHβ gene through a frameshift mutation in its last exon, causing extension of the reading frame. The mature β-subunit proteins range in length from 110 to 145 amino acid residues. The most divergent subunit is the hCGβ protein, which is larger than the LHβ, owing to a C-terminal extension of 24 amino acids. The latter is considered to be responsible for the longer circulating half-life and higher biopotency of hCG over LH (see below).

Analysis of the crystal structure of chemically deglycosylated CG has revealed that, despite their different amino acid sequences, the overall structures of the α- and β-subunits are rather similar, and both contain a so-called cysteine knot structure.[2,15] This structure, similar to that found in some remotely related signalling molecules such as transforming growth factor-β (TGF-β), nerve growth factor (NGF) and platelet-derived growth factor (PDGF), is formed by four polypeptide chains and three disulfide bonds. The dimerization of the two subunits by non-covalent interactions is stabilized by a segment of the β-subunit that extends like a 'seat-belt' around the α-subunit and is 'locked' by a disulfide bridge.[2]

As predicted by the similarity between their ligands, the amino acid sequences and structural organization of the gonadotropin receptors are highly homologous. They belong to the large family of G protein-coupled receptors, all having a transmembrane domain that consists of seven plasma membrane-traversing α-helices connected by three extracellular and three intracellular loops, responsible for interaction with G proteins and signal transduction[3,4] (Figure 2). A distinctive feature of the LH, FSH and TSH receptors (i.e. the subfamily of glycoprotein hormone receptors) is that they possess an unusually large extracellular domain at the N-terminus, responsible for specific binding of the hormone ligand. This is in contrast to most other members of the G protein-coupled receptors, in which the extracellular domain is fairly short and the ligand-binding site is cleft in the transmembrane region. Worthy of note, however, is that, despite selective and specific high-affinity binding to their respective ligands, LH and hCG both bind to the same LH receptor. The LH and FSH receptor genes are composed of 11 and ten exons, respectively, where the first ten or nine exons encode the extracellular domain, while the seven transmembrane segments and the G protein-coupling domain are encoded by the last exon.[3,4] In humans, LH and FSH receptor genes are located on chromosome 2p21–16.

In both receptors, the 5'-terminal part of the open reading frame of exon 1 encodes the signal peptide that directs the protein to the luminal side of the endoplasmic reticulum, and eventually to the extracellular side of the plasma membrane. A distinctive feature of the extracellular domain of gonadotropin receptors is the presence of a number of imperfect leucine-rich repeats, which are encoded by exons 2–9.[3,4] These repeats have an approximate length of 20–25 amino acids and show an amphiphilic β-sheet/α-helix structure.

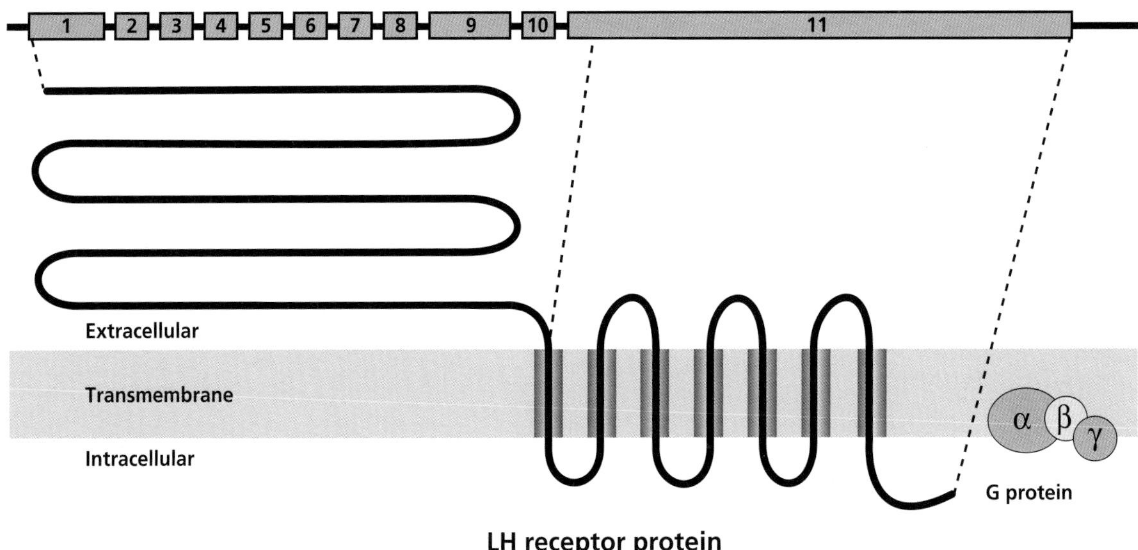

Figure 2 Schematic representation of the human luteinizing hormone (LH) receptor gene and protein. In the top of the figure, LH receptor gene structure is indicated: exons 1–11 are drawn to scale, approximately, and depicted by numbered rectangles. With respect to the protein, the structural domains of the LH receptor, namely extracellular, transmembrane and intracellular, are depicted. The long broken lines depict the part of the receptor protein encoded by exon 11. A similar structural organization applies to the follicle stimulating hormone (FSH) receptor, except that its gene contains ten exons; exons 1–9 encode the extracellular region and exon 10 the transmembrane and intracellular domains

Interestingly, the same structure has been identified in a number of proteins related to protein–protein interactions.[2–4] Indeed, the current model describes the binding domain of the gonadotropin receptors as a horseshoe-like structure, in which the leucine-rich repeats direct their β-sheets to the inner face and their α-helices to the outer face of the U-shaped region.[2] The tertiary conformation of this binding domain depends, at least partly, on the presence of disulfide bonds between several cysteine residues in the extracellular region essential for ligand binding and membrane insertion of the receptors.

The extracellular ligand-binding domain of the gonadotropin receptors is connected to the transmembrane signalling domain by a hinge region. Whether this structure has functions other than serving as a connecting peptide remains to be clarified. In the marmoset monkey, exon 10 of the LH receptor gene, which encodes the N-terminal part of such a hinge region, is completely spliced out from the mature mRNA; yet the marmoset LH receptor appears to function normally.[17] However, if exon 10 is deleted from the human LH receptor, the transit of the mutated receptor to the cell membrane is hampered, thus casting doubts on the possibility that such a region may act merely as a spacer, allowing correct location of the extracellular domain.[18] Moreover, the part of the hinge region closest to the first transmembrane segment is well conserved between the glycoprotein hormone receptors, thus suggesting a specific role. Interestingly, deletion of exon 10 was recently detected in a hypogonadal male, and his testicular testosterone production was able to respond to CG, but not to LH stimulation.[19]

This hinge region is followed by the transmembrane domain with its seven membrane-spanning α-helices, connected by three extracellular and three intracellular loops. This region is responsible for interaction with G proteins and signal transduction.[3,4] In addition, there is some evidence for low-affinity binding of LH to this region of the cognate receptor. Overall, a molecular model for LH receptor activation proposes that the binding of LH to its receptor is a multistep process, in which binding to a high-affinity site in the extracellular domain might be followed by interaction with a low-affinity site in the C-terminal half of the receptor that triggers signal transduction. Finally, the C-terminal half of the gonadotropin receptors ends in a short cytoplasmic tail that contains several potential substrates for receptor phosphorylation, a phenomenon involved in the homologous and heterologous modulation of receptor function. This involves the cessation of hormonal stimulation, by mediating the uncoupling of the receptor from its effector systems as well as down-regulation of the number of cell surface receptors (see below).

STRUCTURE–FUNCTION RELATIONSHIPS OF GONADOTROPINS AND GONADOTROPIN RECEPTORS

In recent years, detailed analyses of the structure–function relationships of gonadotropins and gonadotropin receptors have been carried out. Such analyses have taken advantage of the possibility to modify selectively the primary amino acid sequence of the receptors and their ligands through conventional molecular biological techniques, such as site-directed mutagenesis and construction of truncated and chimeric molecules. In addition, techniques to modify the pattern of glycosylation of the nascent proteins, through either chemical or genetic manipulations, have been used. In this way, the function(s) of any given residue/portion of the functional proteins could be dissected. Overall, these studies have considerably expanded our knowledge of the functional anatomy of the gonadotropins and their receptors. In this section, some of the most relevant functional roles of specific structures of these molecules are reviewed.

Concerning gonadotropins, there are ten cysteine residues in the human α-subunit protein and 12 in the mature β-subunits. Most of these residues are involved in the formation of disulfide bonds essential for maintaining the subunits in active conformation. Thus, the disulfide linkages are critical for formation of the knots that are required for proper subunit conformation.[2] In addition, disulfide binding is essential for function of the 'seat-belt' loop responsible for the α–β heterodimer assembly and stabilization. Moreover, although disulfide bonds not involved in the cysteine-knot formation can be eliminated without destroying hormone function, these alterations cause minor changes in the rate of subunit assembly and hormone efficiency.

In addition, there are two N-linked glycosylation sites in the common α-subunit and FSHβ subunits and one in LHβ. N-linked oligosaccharides are attached to the protein core of gonadotropin subunits at the asparagine of a specific glycosylation signal sequence (Asn–X–Ser/Thr). These carbohydrate moieties appear to have a role in subunit assembly during hormone synthesis, clearance of gonadotropins from serum and efficiency of signal transduction. Concerning the latter feature, deglycosylated gonadotropins are able to bind their cognate receptors but not able to trigger the hormonal signal. In addition to N-linked glycosylation, the C-terminal extension of the CGβ subunit contains four additional O-linked glycosylation sites not present in LHβ, which contribute to the longer circulating half-life and higher biopotency of hCG over LH. Accordingly, the circulating half-life of CG is over 24 h, whereas that of LH is 30–60 min. Another structural feature of LH contributing to its faster rate of elimination is the terminal sulfation of its carbohydrate moieties, as there is a liver receptor eliminating sulfated glycoproteins.[20] Because the carbohydrate termini of FSH are mainly sialylated, this gonadotropin stays longer in the circulation, with a half-life of 4–5 h. There is a certain degree of microheterogeneity in carbohydrate moieties of the secreted gonadotropins, and this feature is highly variable depending on the functional state of the hypothalamic–pituitary–gonadal axis.[21] Indeed, the circulating gonadotropins are a mixture of hormone molecules with varying half-lives and bioactivities. This may constitute a mechanism for the plasticity of the biological actions of gonadotropins,[2] although the clinical significance of this phenomenon still remains obscure.

Regarding the gonadotropin receptors, their extracellular domain presents a number of potential sites for N-linked glycosylation.[3,4] There are six sites in the LH receptor: Asn99, 174, 195, 291, 299 and 313; and four sites in the FSH receptor: Asn191, 199, 293 and 318, although the last site is not conserved among species (the amino acids in the receptor proteins are numbered by taking the first methionine of the signal peptide as 1). Although the gonadotropin receptors are actually glycosylated, the extent and the importance of this event in receptor function are not completely elucidated, and some apparently contradictory results have been presented.[3,4]

Also relevant for receptor function is the presence of several cysteine residues in the distal and middle region of the extracellular domain of the LH receptor. These give rise to disulfide bonds that are essential for maintaining proper conformation of the binding domain. Other cysteines are located close to, and within, the transmembrane region, and they may play a role in insertion of the receptor into the cell membrane.[22] In addition, detailed molecular analyses have elucidated the roles of specific amino acid residues or discrete domains within the primary sequence of gonadotropin receptors in the processing/trafficking of the nascent receptor molecule, as well as in its ligand binding and signal transduction. For example, a number of amino acid substitutions in the LH and FSH receptors cause intracellular trapping of the mutant proteins, pointing out their role in the proper folding and/or transport of the receptors to the cell surface.[23] Concerning ligand binding, studies using deleted and chimeric receptors have demonstrated that the binding region of gonadotropin receptors presents both facilitatory and inhibitory binding determinants.[24] The former are responsible for high-affinity binding to the proper ligand (in the case of LH receptor, leucine-rich repeats 1–8; in the case of FSH receptor, repeats 1–11), and the latter block the binding of inappropriate ligands (in the case of LH receptor, exons 2–4 and 7–9; in the case of FSH receptor, exons 5–6).[2] Site-directed mutagenesis studies have demonstrated that, as well as those in the N-terminal extracellular region, specific amino acid residues outside the ligand-binding domain participate in the ligand–receptor interaction.

Finally, the functional role of specific amino acids within the primary structure of the gonadotropin receptors in signal transduction has been assessed. In this sense, the LH and FSH receptors are mainly coupled to G_s, the G protein that activates the various adenylyl cyclases, with resulting elevation of intracellular cAMP levels. However, coupling to G_i proteins has also been demonstrated. In this context, several amino acid residues in the transmembrane region, but also in the extracellular domain, have been identified as essential to elicit cAMP responses after receptor activation.[25,26] Overall, there are specific areas within the receptor structure that are critical for activation of the cAMP pathway, and probably distinct domains of the gonadotropin receptors are responsible for ligand binding and signal transduction. In addition, both gonadotropin receptors are also able to activate other signal transduction pathways, involving increased phosphatidylinositide turnover and inositol triphosphate (IP_3) production, elevated intracellular Ca^{2+} and activation of mitogen-activated protein kinases.[14] Thus, coupling to $G_q/11$ and G13 proteins has been demonstrated for the porcine LH receptor,[27] and to G_i2 for the murine receptor.[28] Because activation of these alternative intracellular pathways usually takes place at higher hormone concentrations than that of the cAMP pathway, their physiological relevance may be restricted to situations of elevated gonadotropin levels, such as the preovulatory surge and pregnancy.[26] Interestingly, the ability of gonadotropins to activate different intracellular signals has been related to the plasticity of their biological actions. It is tempting to speculate that the activation of different signals after receptor activation is carried out through different loci within the primary structure of the receptors. This seems to be the case with the LH receptor, where divergence of the cAMP- and IP_3-signalling pathways was detected at or near to Lys583, located in exoloop 3. Mutations in this residue were incompatible with cAMP accumulation after receptor activation, but permitted, at least in some cases, IP_3 formation.[29] It remains to be shown whether LH can act at different sites of the binding domain of its receptor in order to activate cAMP and/or IP_3 signals selectively. It would be intriguing if the different gonadotropin isoforms triggered a different array of second messenger responses, explaining in this way the pleomorphism of their actions in different physiological states.

NORMAL AND ABERRANT REGULATION OF GONADOTROPIN PRODUCTION AND ACTION

It is very clear that gonadotropins are the main 'driving force' of gonadal function, since patients with failure of gonadotropin secretion or action (i.e. receptor defect) are always hypogonadal. Therefore, when we consider the regulation of gonadotropin action, we have to take into account factors maintaining gonadotropin secretion and the expression of their receptors in gonadal target cells. The context in which gonadotropin functions are taking place is the hypothalamic–pituitary–gonadal axis, which is a classical example of an endocrine regulatory circuit, with feedforward and feedback regulatory events (Figure 3). Notably, in addition to the endocrine regulation of the hypothalamic–pituitary–gonadal axis, a complex network of paracrine and autocrine, intrapituitary and intragonadal, regulatory mechanisms participate in the regulation of gonadotropin secretion and action.[30–32] However, such local actions have mainly been demonstrated in various *in vitro* experimental settings, and the extent of their physiological role remains largely open.

The maintenance of gonadotropin secretion is critically dependent on pulsatile secretion, at 1–2-h intervals, of the hypothalamic gonadotropin-releasing hormone (GnRH), a decapeptide that has its receptors in gonadotroph cells of the anterior pituitary gland. GnRH secretion is under positive and negative actions of gonadal hormones (see below), but also many central nervous system neurotransmitters affect GnRH secretion: for example, opiates are inhibitory and catecholamines are stimulatory. The same GnRH peptide stimulates the release of both FSH and LH, although a separate FSH-releasing peptide has also been proposed, yet conclusive evidence for its existence has not been provided. The GnRH effects are mediated through a G protein-associated receptor, and its signal transduction entails activation of phospholipase C, generation of inositol triphosphate and diacylglycerol second messengers, increase of cytoplasmic free calcium and activation of protein kinase C.[33] The other endocrine factors contributing to gonadotropin secretion are

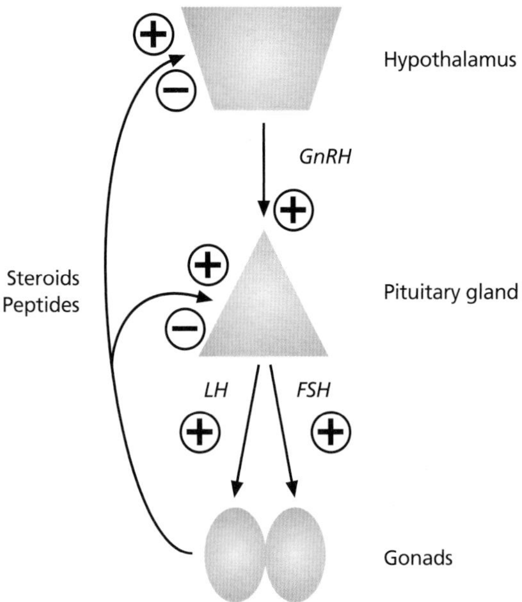

Figure 3 Functional organization of the hypothalamic–pituitary–gonadal axis. Feedforward and feedback loops are depicted. Hypothalamic gonadotropin-releasing hormone (GnRH) stimulates luteinizing hormone (LH) and follicle stimulating hormone (FSH) synthesis and secretion by gonadotroph cells from the anterior pituitary. The gonadotropins, through specific receptors, act co-ordinately to regulate gonadal differentiation, growth and endocrine function. In turn, gonadal sex steroids and peptides exert their negative and/or positive feedback actions both on GnRH secretion at the hypothalamic level and directly on gonadotropin gene expression at the pituitary

gonadal steroid and peptide hormones. The steroid hormones, androgens, estrogen and progesterone, exert their negative feedback actions both on GnRH secretion at the hypothalamic level and directly on gonadotropin gene expression in pituitary gonadotrophs. In addition, estradiol has a positive feedback effect upon the hypothalamic and pituitary levels, resulting in the preovulatory increase of gonadotropin levels, which culminates in the ovulatory secretion peak of LH. Of the ovarian peptide hormones, inhibin is a physiologically significant negative regulator of FSH secretion at the pituitary level.[34] The other peptide of this family, activin, and its binding protein, follistatin, exert regulatory effects on gonadotropin secretion and action mainly in para/autocrine fashion within the anterior pituitary and gonads.[34] There are also multiple interactions between the different pituitary cell types in the modulation of each other's secretory function, as extensively reviewed recently.[32]

The molecular pathogenesis of several diseases affecting the regulation of gonadotropin secretion is known today. A good example is Kallmann syndrome, presenting with a phenotype of hypogonadotropic hypogonadism and anosmia.[35] The syndrome is due to impaired migration of the GnRH neurons in embryonic life from the olfactory placode to their final location in the hypothalamus. This migration is disturbed in the most common X-linked form of the syndrome because of mutation in an extracellular matrix protein, anosmin, resulting in altered development of the olfactory bulbs and tracts. No mutations of the GnRH gene are known in humans to date, but recently several mutations of the GnRH receptor have been documented.[36] These cases form the most common group of hypogonadotropic hypogonadism without anosmia.

The pivotal position of gonadotropin receptors in the reproductive axis makes them a suitable target for tuning gonadal function. The regulation of gonadotropin receptor function can be divided into two types: homologous, when the circulating gonadotropin levels influence the levels of their cognate receptor; and heterologous, when other endocrine or para/autocrine factors exert their regulatory inputs.[3,30–32] These modulatory effects can be both stimulatory (sensitization) and inhibitory (desensitization), and they may directly affect the number of cell surface receptors and the expression of their genes (up- and down-regulation) and/or the post-receptor signalling systems (uncoupling).[3,4] Physiological gonadotropin levels are able to maintain or up-regulate their cognate receptor levels, but supraphysiological, and especially prolonged, elevation results in receptor down-regulation, which is accompanied by uncoupling of the signal transduction system. It has recently been shown that the cytoplasmic tail and the binding of β-arrestin to the third cytoplasmic loop play a role in desensitization of the LH receptor function.[37] At least two mechanisms are involved in gonadotropin receptor down-regulation, i.e. increase in the rate of receptor internalization and decrease in the level of receptor gene expression, that apparently take place in this chronological order.

A plethora of heterologous regulators of gonadotropin receptor expression and function have been reported.[3,4] For example, epidermal growth factor (EGF) and phorbol ester can mimic the down-regulation effect of LH on its receptor. This heterologous regulation has many sexually dimorphic features. In the rat testis, FSH receptor mRNA expression and function fluctuate according to the seminiferous epithelial cycle, pointing to a complex auto/paracrine regulatory network between germ and Sertoli cells. In the ovary, functional cyclicity is associated with the level of gonadotropin receptor expression, and a number of heterologous signals are implicated in the regulation. For instance, the induction of LH receptor expression in luteinizing granulosa cells occurs through synergistic action of FSH and estradiol.[12] Similarly, prolactin up-regulates, together with LH, the LH receptor expression in rodent corpora lutea and Leydig cells. Physiologically, testicular LH receptors appear to be constitutively expressed, and most of the regulatory effects identified are inhibitory. Other heterologous factors playing roles in the regulation of gonadotropin receptor expression are insulin-like growth factor-I (IGF-I), TGF-β_1 and -β_2, activin, inhibin, follistatin, basic fibroblast growth factor, GnRH and TGF-α.[3,14,30] The physiological role of this complicated regulatory network remains to be explored, but the possibility remains that it just reflects largely biological redundancy in the multitude of regulatory signals.

The above regulatory mechanisms have mainly been elucidated through classical physiological studies, by treating animals with putative regulatory factors or by challenging cultured gonadal cells with them. The recent unravelling of the promoter sequences of gonadotropin receptor genes has made it possible to address the regulation of their expression in a more specific way at the level of gene function. A variety of transcription factors have been shown to affect LH and FSH receptor expression. These studies have, for instance, demonstrated direct effects of thyroid hormones, progesterone, prolactin, IGF-I, FSH, nuclear orphan receptors and transcription factors Sp1, Sp2 and Ap2 on LH receptor promoter function, and of transcription factors SF-1 and upstream stimulatory factor (USF) on the FSH receptor promoter (for examples, see references 3, 38 and 39). Notably, the mechanisms for the regulation of gonadotropin receptor gene expression do not only involve modulation of the transcriptional activity, as part of these regulatory events is carried out via modulation of mRNA stability. Finally, the gonadotropin receptor messages appear in a complex pattern of alternative splicing, but its functional significance still remains unresolved.[3,4]

MUTATIONS IN GONADOTROPIN AND GONADOTROPIN RECEPTOR GENES

Mutations in gonadotropin and gonadotropin receptor genes are understandably rare, owing to their deleterious effects on fertility. However, in recent years it has become evident that such a group of mutations can constitute the molecular basis for some rare reproductive disorders. The mutations stay in the genetic pool, because fertility of the heterozygous carriers is not compromised. Although the genetic basis of the pathogenesis of male and female infertility is discussed in other chapters, it is worthy of note that the identification of mutations in gonadotropin and gonadotropin receptor genes has not only elucidated the molecular pathogenesis of some forms of infertility. Indeed, by displaying distinct phenotypes, these conditions have turned out to be very illustrative regarding some controversies about the role of gonadotropins in the regulation of reproduction. Moreover, detailed analyses of the functional impact of naturally occurring mutations in gonadotropins and their receptors have helped to elucidate further several facets in the structure–function relationships of the native proteins. These findings, corroborated by genetically manipulated animal models, are summarized briefly below.

The mutations identified to date in the genes encoding gonadotropins and gonadotropin receptors have been reviewed in detail recently.[4,14] Overall, the hormone ligand mutations found to date usually represent inactivating or loss-of-function mutations, whereas the receptor mutations can also be of the gain-of-function type. In addition, besides the clear-cut disease-causing mutations, several polymorphisms have been identified in gonadotropin and gonadotropin receptor genes. These structural changes may or may not cause alterations in gene function or structure of the encoded protein, and, consequently, they often have mild phenotypic expression.

The currently known mutations and polymorphisms in gonadotropins are summarized in Figure 1 and Table 1. Concerning the common α-subunit, the only genetic alteration so far reported is a single Glu56→Ala amino acid substitution in an α-subunit secreted ectopically by a human carcinoma; the mutated protein was unable to associate with the β-subunit.[40] In fact, although yet unproven, the lack of germline mutations in the α-subunit gene has been taken as an indication that such changes are lethal, as in humans they would impair, in addition to pituitary gonadotropins, the formation of CG and TSH. Concerning the LHβ subunit, only one true mutation causing total functional inactivation of the protein has been described. It was a homozygous A-to-G missense mutation in codon 54, causing a Glu→Arg substitution. Co-expression of the mutated LHβ gene with the normal α-subunit resulted in the formation of immunoreactive LH α–β heterodimers devoid of biological activity, because of the inability to bind to the LH receptor.[41] In the case of the FSHβ subunit, a total of five subjects (three women and two men) with different inactivating mutations of the FSHβ gene have so far been described in the literature (Table 1). Overall, alterations induced by these mutations resulted in truncated or structurally defective FSHβ proteins, with the inability to associate with the common α-subunit to form bioactive α–β dimers.[14] For the CGβ subunit, several polymorphisms have been detected in the hCGβ/LHβ gene complex, but their relevance remains to be elucidated. Indeed, in gene 5 of the CGβ gene cluster, the one most highly expressed, a total of six polymorphisms were detected, yet they were, with the exception of one, either silent or located in introns. The one located in exon 3 of CGβ gene 5

Table 1 Currently known mutations and polymorphisms, altering protein structure, that have been detected in gonadotropin subunit genes. Data from reference 14

Gene	*Location*	*Type*	*Base change*	*Amino acid change*	*Effect at protein level*
Common α	exon 3	missense	$GA^{239}G \rightarrow GCG$*	$Glu^{56}Ala$	no association with β-subunit
LHβ	exon 3	missense	$CA^{221}G \rightarrow CGG$	$Glu^{54}Arg$	absent bioactivity; normal immunoreactivity
	exon 2	two missense mutations in same allele	$T^{82}GG \rightarrow CGG$	$Trp^{8}Arg$	poorly detected by α/β specific antibodies
			$AT^{104}C \rightarrow ACC$	$Ile^{15}Thr$	increased *in vitro* bioactivity; decreased circulatory $T_{1/2}$
	exon 3	missense	$G^{1502}GT \rightarrow AGT$	$Gly^{102} \rightarrow Ser$	slightly elevated *in vitro* bioactivity
CGβ	exon 3	missense	$G^{295}TG \rightarrow ATG$	$Val^{79}Met$	inefficient assembly with common α
FSHβ	exon 3	2-bp deletion/premature stop codon	$GTG \rightarrow GX^{236,237}$	$stop^{87}$	no bioactivity or immunoreactivity
	exon 3	missense	$TG^{206}T \rightarrow GGT$	$Cys^{51}Gly$	no bioactivity or immunoreactivity
	exon 3	missense	$T^{298}GT \rightarrow CGT$	$Cys^{82}Arg$	no bioactivity or immunoreactivity

* Apparently a somatic mutation in tumor; LH, luteinizing hormone; CG, chorionic gonadotropin; FSH, follicle stimulating hormone; $T_{1/2}$, half-life

resulted in the amino acid substitution Val79→Met, which reduced its efficiency to assemble with the common α-subunit; yet this mutation was found in heterozygous form in 4.2% of randomly chosen healthy subjects.[14] The possibility exists that other mutations in the CGβ gene cluster may be lethal, owing to defective production of biologically active CG.

Concerning polymorphic variants of the LHβ subunit, Pettersson and colleagues[42] described a healthy woman with two children, whose LH was undetectable using a monoclonal antibody directed against an antigenic epitope present only in the intact LH α–β dimer; yet the LH bioactivity was normal in accordance with her fertility.[42] Upon sequencing, the LHβ gene of the subject was found to represent a genetic variant (V) allele of the LHβ gene with two missense mutations: Trp8→Arg (TGG→CGG) and Ile15→Thr (ATC→ACC).[14] In all samples analyzed to date from various populations, a complete linkage of the two mutations has been demonstrated. The world-wide frequency and the functional relevance of such a polymorphic variant of LHβ have been extensively analyzed.[33] Although the V-LHβ gene has been detected in a wide range of populations of different ethnic and geographical origins, its frequency appears to be highest in the northern European populations and aboriginal Australians, whereas lower frequencies are found in Asian populations and American Indians. Concerning the functional features of the V-LHβ protein, the Ile15→Thr mutation adds a second oligosaccharide side-chain to Asn13 of the V-LHβ subunit. Detailed analyses demonstrated that V-LH and its recombinant form are more active than wild-type hormone in *in vitro* bioassay, with lower ED_{50} (effective dose in 50% of the population) and about 20% higher maximum effect.[14] In contrast, V-LH shows a clearly shorter half-life in circulation than the native LH (26 vs. 48 min).[43] Overall, evidence from *in vivo* and clinical studies indicates that, despite higher bioactivity *in vitro*, the V-LH represents a functionally weaker form of the hormone because of its shorter half-life. Interestingly, there are eight point mutations in the first 650 nucleotides of the 5'-flanking region of the V-LHβ promoter that segregate with the two point mutations in the coding sequence.[44] These mutations result in approximately 50% enhancement of the promoter activity, thus tending to compensate for a faster rate of elimination of V-LH. Interestingly, the particular phenotype(s) of individuals carrying this polymorphic variant of LH appear to be dependent on the genetic background and, hence, there are differences among different populations. For instance, V-LH has been associated with subfertility in Japan, but less clearly in Caucasian populations.[14]

Mutations in gonadotropin receptor genes, both activating and inactivating, have been identified with vastly different phenotypic effects. The currently known mutations in LH and FSH receptors are depicted in Figures 4 and 5, respectively. Regarding the LH receptor, a relatively large number of activating mutations have been reported to date.[14] Initially, these activating mutations were mostly located in a hot-spot comprising transmembrane segment 6 and cytoloop 3. Further genetic analyses, however, revealed new gain-of-function mutations located in other transmembrane segments, with the exception to date of transmembrane domains 4 and 7. Activating mutations of the LH receptor gene are involved in the pathogenesis of the LH-independent, familiar form of male-limited precocious puberty (FMPP), also termed testotoxicosis. This disease, inherited in autosomal dominant pattern, is characterized by precocious and gonadotropin-independent activation of testosterone production, and drastically advanced onset of puberty. A total of 16 different activating mutations within exon 11 of the LH receptor gene have so far been described in FMPP patients.[14,45] In addition, up to 15 different inactivating mutations of the LH receptor have been reported to date.[14] Interestingly, their pattern of distribution within the primary structure of the LH receptor appears to be much more scattered than that of the activating mutations. Whereas the constitutive activation only takes place when an amino acid is mutated in a location participating in signal transduction, it seems that loss-of-function mutation may derive from a greater variety of missense substitutions, frameshift mutations, small deletions or insertions. These could impair the binding domain, signal transduction and/or expression of mature receptors at the cell surface.

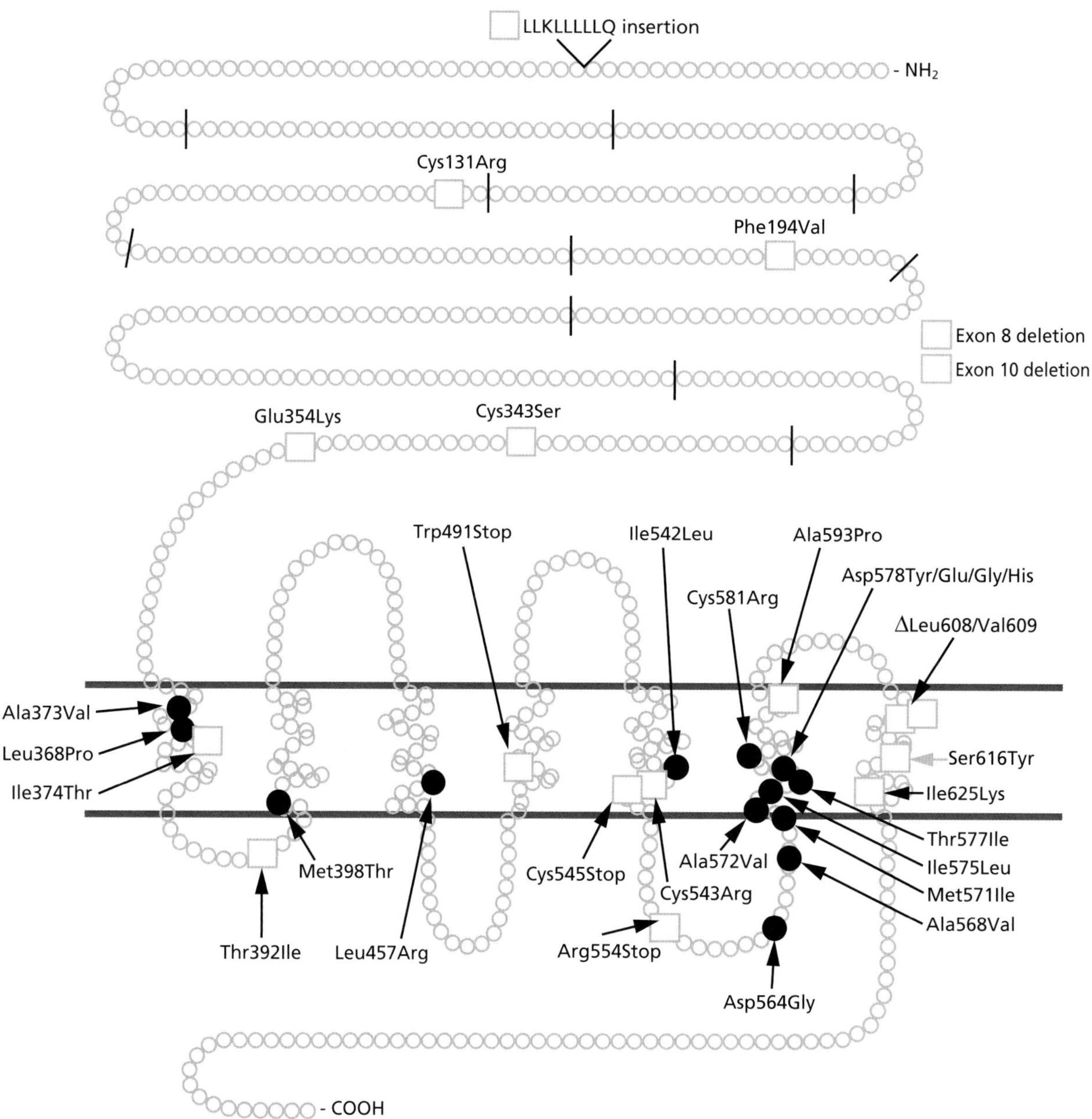

Figure 4 Diagrammatic presentation of the activating and inactivating mutations detected to date in the human luteinizing hormone (LH) receptor. The blue circles depict activating and the blue rectangles inactivating mutations. The short lines across the amino acid chain separate the 11 exons. Data from references 14 and 45

From the clinical standpoint, inactivating mutations of the LH receptor are the basis of a rare form of male pseudohermaphroditism associated with Leydig cell hypoplasia (LCH). Notably, the phenotypic presentation of LCH ranges from mild forms of undervirilization with micropenis and hypospadias to complete sex reversal due to the extremely low testosterone levels. In fact, the relative severity of the phenotype appears to be related to the degree of receptor inactivation, and thus the

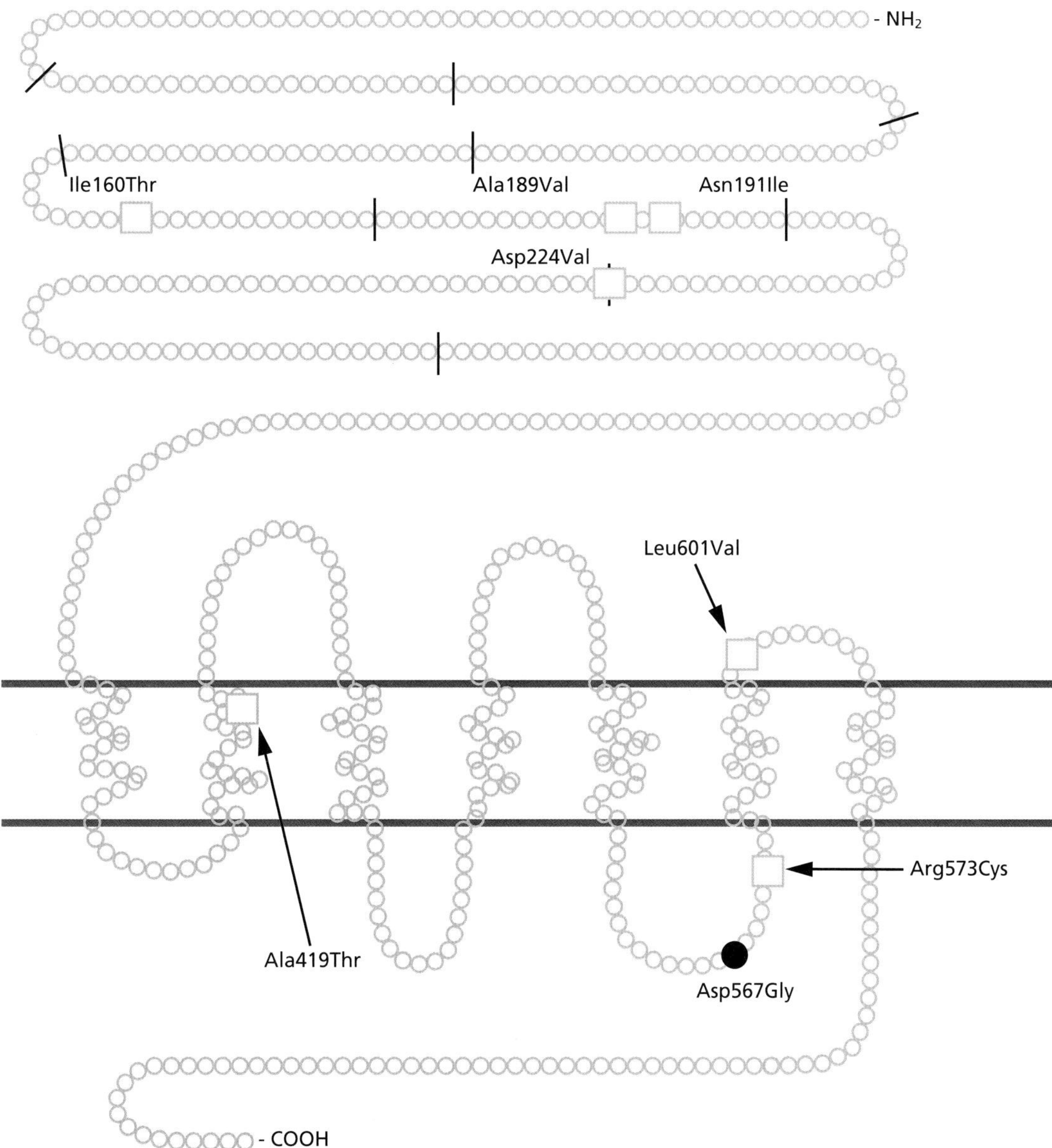

Figure 5 Diagrammatic presentation of the activating and inactivating mutations detected to date in the human follicle stimulating hormone (FSH) receptor. The blue circles represent activating and the blue rectangles inactivating mutations. The short lines across the amino acid chain separate the ten exons. Data from reference 14

loss of responsiveness to LH/CG induced by the mutation. The female phenotype of inactivating LH receptor mutations is milder, presenting with anovulatory infertility.

Concerning the FSH receptor, no convincing reports are yet available on gain-of-function mutations. However, a recent report of a genetically modified FSH receptor with constitutive activation *in vitro* indicates that these types of mutations are possible.[46] In this scenario, identification of novel clear-cut activating FSH receptor mutations in humans is possibly a matter of time. A few inactivating mutations have been identified in the human FSH receptor. The first was detected in the context of hereditary hypergonadotropic ovarian dysgenesis (ODG).[47] It consisted of a C566→T transition of exon 7 of the FSH receptor, predicting an Ala189→Val substitution. Functional expression of the mutated receptor revealed a severely reduced number of plasma membrane FSH receptors and impairment of cAMP and IP_3 responses after FSH stimulation. Interestingly, however, the equilibrium association constant of the mutant receptor was similar to that of the wild-type. It was subsequently shown that the mutated receptor protein was sequestered inside the cells, with apparent improper folding and trafficking to the cell surface (Huhtaniemi and colleagues, unpublished observation). It is worth noting that the Ala189→Val mutation is located in a stretch of five amino acids (Ala189PheAsnGlyThr) that is perfectly conserved among species and is also present in human LH and TSH receptors, and which contains a functional N-linked glycosylation site. The importance of this area for proper receptor function is emphasized by the observation that introduction of a similar Ala to Val substitution disrupted the functionality of human LH receptor and mouse FSH receptor.[48] A second inactivating mutation (heterozygous Asn191→Ile) was discovered in the same area of the FSH receptor.[4] In addition, other partially inactivating loss-of-function mutations in the human FSH receptor gene have been described.[14] As was the case with inactivating LH receptor mutations, their pattern of distribution is scattered, as single amino acid changes in exon 6, exon 7, exon 9, transmembrane segments 2 and 6 and cytoloop 3 have been reported.

Finally, several polymorphisms have been found in the LH receptor gene; one inserts a Leu–Gln sequence in the signal peptide, two are in exon 10 (one abolishes a glycosylation site) and the fourth neutral one is in exon 11.[49] No significant functional consequences of these polymorphisms have been reported. At least two allelic variants of the human FSH receptor gene have been identified.[4] One polymorphic site is located at position 307 of the extracellular domain, which can be occupied either by Ala or Thr, whereas position 680 in the intracellular domain can be occupied either by Asn or Ser; the two gene alleles are Thr307/Asn640 and Ala307/Ser680. A recent study demonstrated, in women undergoing controlled ovarian stimulation for *in vitro* fertilization (IVF), that those homozygous for the latter allele needed significantly higher doses of FSH than homozygotes for the other variant allele.[50] Hence, the ovarian response to FSH stimulation may depend on the genotype of the cognate receptor.

As stated above, mutational analyses have significantly contributed to our current knowledge of the structure–function relationships of gonadotropin receptors. For example, the fact that most of the activating mutations found to date in the LH receptor are located in a hot-spot comprising transmembrane segment 6 and cytoloop 3 is suggestive that this area is highly relevant in G protein coupling. In fact, this contention has recently been substantiated by the use of a synthetic peptide approach. A peptide designed to mimic the lower portion of transmembrane segment 6 of the LH receptor was able to activate adenylyl cyclase by interaction with G_s proteins,[51] whereas a peptide corresponding to an area in cytoloop 3 of the FSH receptor was able to interact with G protein and modulate signal transduction.[52] Moreover, from a general standpoint, it is apparent that, in basal conditions, the gonadotropin receptors are restrained in an inactive state, and that those amino acids substituted in the constitutively active mutants are likely to be involved in stabilization of the resting receptor into inactive conformation. Upon ligand–receptor interaction, the receptor is subjected to conformational changes that lead to an active form of the receptor, capable of stimulating G proteins.

In addition, evaluation of phenotypes of individuals carrying gonadotropin receptor mutations has helped to confirm/revisit the physiological roles of LH and FSH in reproductive function. For example, inactivating mutations in the LH receptor gene of genetically male individuals lead to Leydig cell hypoplasia, with a wide spectrum of phenotypic presentations ranging from pseudohermaphroditism with female external genitalia to hypogonadism with partial male differentiation, depending on the degree of impairment of receptor function. However, the XX counterparts present with normal pubertal development with primary amenorrhea, and absence of preovulatory follicles and luteinization.[53] Because of the more severe male phenotype, it is apparent that the impact of the absence of LH action during development is more deleterious for the male than for the female, although the end-point, i.e. infertility, is similar in both sexes. This is in agreement with previous observations on refractoriness of the fetal and neonatal ovary, but not the testis, to gonadotropic regulation, and with the major role of fetal testosterone production in masculinization.[2] Mice with targeted disruption of the LH receptor gene were recently generated, and, interestingly, they were found to present with a close phenocopy of their human counterparts carrying inactivating LH receptor mutations, except for the lack of pseudohermaphroditism.[54] The latter indicates that in the mouse, but not in the human, intrauterine sex differentiation is independent of LH receptor activation. Inactivating mutations in the FSH receptor gene induce primary ovarian failure with complete infertility in female patients.[47] Similarly, women carrying missense mutations in their FSHβ gene show primary amenorrhea and infertility.[26] In contrast, only partial suppression of the spermatogenic function, but not complete azoospermia or infertility, was observed in men homozygous for an inactivating mutation in the FSH receptor gene.[10] These phenotypes are similar to those presented by mice with targeted disruption of the FSHβ subunit gene.[9] In this model, female individuals are infertile owing to ovarian dysgenesis, but males, despite reduced testicular weight and spermatogenesis, maintain their fertility. Similarly, FSH receptor-null mutant mice were recently generated, and they showed similar phenotypic presentation to humans carrying inactivating mutations in the FSH receptor gene: females were infertile due to arrested follicular maturation, whereas males were fertile, but had small testes and incomplete suppression of spermatogenesis.[55] These observations indicate that FSH is necessarily essential for female reproduction but not absolutely required for qualitatively preserved male fertility. The only apparent discrepancy between human and mouse models is the azoospermia reported in two men with inactivating FSHβ subunit mutation.[14] However, the small number of human cases, as well as the possibility of another dysfunction causing azoospermia, may explain this difference.

FUTURE PERSPECTIVES

In recent years, detailed evaluation of structure–function relationships of gonadotropins and gonadotropin receptors by means of current molecular biology techniques has vastly expanded our knowledge of the functional anatomy of receptors and their cognate ligands. This has paved the way for better understanding of the mechanisms of ligand–receptor interaction and transduction of hormonal signals. In addition, from a clinical standpoint, the most revealing, intriguing and interesting findings concerning molecular aspects of the hypothalamic–pituitary–gonadal function have been related to the recent discovery of a number of mutations in gonadotropin and gonadotropin receptor genes. Most of the possible permutations of these are known today. In this sense, the largest number of cases has been detected in connection with activating and inactivating mutations of the LH receptor gene. In contrast, only one male with an inactivating LHβ mutation has been reported to date, and it would be intriguing to identify females with an inactivating LHβ mutation. The recent generation of a knock-out model for the LH receptor gene in mice (LuRKO) has roughly confirmed the phenotypes of the individual with a single LHβ inactivation, and both females and males with complete forms of inactivating LH receptor mutations. The only clear difference is the normal masculinization of the LuRKO mice, which is in line with the known gonadotropin-independence of mouse, but not of human, fetal Leydig cells. A mouse model of activating LH receptor mutation would be needed to resolve the conundrum of the apparent absence of a female human phenotype with this condition.

Less is known about the functional consequences of aberrant FSH and FSH receptor function. The inactivating mutations of the human FSH receptor as well as the mouse models for FSHβ or FSH receptor gene knock-out are in perfect harmony, showing infertility due to arrested follicular maturation in females and suppressed fertility in males. Likewise, the three reported females with FSHβ inactivation are in accordance with the previous phenotypes. In contrast, the two men with FSHβ mutation are azoospermic, which differs from the other human and mouse models of disturbed FSH function. It is apparent that additional cases of men with FSHβ mutation are needed to clarify the situation. This may be difficult if they usually present with only mild suppression of fertility, as is the case with men with FSH receptor mutation.

Both gonadotropin and gonadotropin receptor genes contain polymorphisms. Mild phenotypic effects have been described in connection with them. One polymorphism may have too mild an effect on gonadotropin function to cause a clear phenotype, but it is feasible to hypothesize that combinations of the polymorphisms, for example one in the hormone and another in its cognate receptor, could bring about effects that differ clearly from a situation in which both genes are of the wild type. Exploration of functional impact of the different alleles of gonadotropin and gonadotropin receptor genes will provide an interesting challenge for future research.

The finding of activating TSH receptor mutations in thyroid adenomas suggests that similar mutations might occur in LH and FSH receptors in gonadal tumors. Constitutive activation of the cAMP pathway by activating mutations in $G_s\alpha$ protein has been found to be oncogenic in human ovarian stromal and testicular Leydig cell tumors,[14] but such a connection has not been reported with activating gonadotropin receptor mutations constitutively increasing cAMP production. Recently, another type of activating LH receptor mutation was found in connection with Leydig cell adenomas: the second messenger constitutively elevated is IP_3 instead of cAMP.[56] Whether the cell-proliferation and, in general, growth-promoting effects of gonadotropins, and thereby their possible tumorigenic effects, are mediated by signalling pathways other than cAMP remains an interesting question for future research. Although fairly extensive studies have been carried out, no FSH receptor mutations have been identified in granulosa cell or Sertoli cell tumors. However, the tumor-promoting potential of gonadotropins has been observed in a number of genetically modified mouse models, including those with increased LH secretion, inhibin-α promoter-driven simian virus-40 (SV40) T-antigen expression and knock-out of the inhibin-α gene.[57,58] Hence, the tumorigenicity of gonadotropins and their molecular mechanisms need to be more thoroughly studied, especially in view of the theory of gonadotropin action as a promoter of human ovarian carcinogenesis. The chronically elevated gonadotropin levels in postmenopausal women may, after all, have some effects.

The physiology and the pathophysiology of gonadotropin function have been previously characterized by classical physiological and biochemical methods. The novel information brought to light by recent molecular approaches has elucidated totally new aspects of these functions. It is likely that new facets of gonadotropin function will be unravelled in the future. On the one hand, research into the importance of genetic variability of gonadotropin action, and, on the other hand, novel genetically modified animal models, will probably be the focus of future research.

ACKNOWLEDGEMENTS

The original work of the authors on the topic of this review was supported by grants from the Academy of Finland, the Sigrid Jusélius Foundation and the Ahokas Foundation. One of the authors (M.T.-S.) was partially supported by a postdoctoral grant from DGICYT (Ministerio de Educación y Cultura, Spain).

REFERENCES

1. Gharib SD, Wierman ME, Shupnik MA, Chin WW. Molecular biology of pituitary gonadotropins. *Endocr Rev* 1990 11: 177–199.
2. Moyle WR, Campbell RK. Gonadotropins. In Adashi EY, Rock JA, Rosenwaks Z, eds. *Reproductive Endocrinology, Surgery, and Technology*. Philadelphia: Lippincott-Raven, 1996: 683–724.
3. Segaloff DL, Ascoli M. The lutropin/choriogonadotropin receptor. . . 4 years later. *Endocr Rev* 1993 14: 324–342.

4. Simoni M, Gromoll J, Nieschlag E. The follicle-stimulating hormone receptor: biochemistry, molecular biology, physiology, and pathophysiology. *Endocr Rev* 1997 18: 739–773.
5. Rao CV. The beginning of a new era in reproductive biology and medicine: expression of low levels of functional luteinizing hormone/human chorionic gonadotropin receptors in non-gonadal tissues. *J Physiol Pharmacol* 1996 47 (Suppl 1): 41–5.
6. Josso N. Sexual determination. In Adashi EY, Rock JA, Rosenwaks Z, eds. *Reproductive Endocrinology, Surgery, and Technology.* Philadelphia: Lippincott-Raven, 1996: 59–74.
7. Sharpe RM. Regulation of spermatogenesis. In Knobil E, Neill JD, eds. *The Physiology of Reproduction.* New York: Raven Press, 1994: 1363–1435.
8. Orth JM. The role of follicle-stimulating hormone in controlling Sertoli cell proliferation in testes of fetal rats. *Endocrinology* 1984: 115 1284–1255.
9. Kumar TR, Wang Y, Lu N, Matzuk MM. Follicle-stimulating hormone is required for ovarian follicular maturation but not for male fertility. *Nat Genet* 1997 15: 201–204.
10. Tapanainen JS, Aittomäki K, Min J, Vaskivuo T, Huhtaniemi I. Men homozygous for an inactivating mutation of the follicle-stimulating hormone (FSH) receptor gene present variable suppression of spermatogenesis and fertility. *Nat Genet* 1997 15: 205–206.
11. Hillier SG, Whitelaw PF, Smyth CD. Follicular oestrogen synthesis: the 'two-cell, two-gonadotrophin' model revisited. *Mol Cell Endocrinol* 1994 100: 51–54.
12. Richards JS, Hedin L. Molecular aspects of hormone action in ovarian follicular development, ovulation and luteinization. *Annu Rev Physiol* 1988 50: 144–463.
13. Huhtaniemi I, Warren DW. Ontogeny of pituitary–gonadal interactions. Current advances and controversies. *Trends Endocrinol Metab* 1990 1: 356–362.
14. Themmen APN, Huhtaniemi IT. Mutations of gonadotropins and gonadotropin receptors: elucidating the physiology and pathophysiology of pituitary–gonadal function. *Endocr Rev* 2000 21: 551–583.
15. Bousfield GR, Perry WM, Ward DN. Gonadotropins. Chemistry and biosynthesis. In Knobil E, Neill JD, eds. *The Physiology of Reproduction.* New York: Raven Press, 1994: 1749–1792.
16. Bo M, Boime I. Identification of the transcriptionally active genes of the chorionic gonadotropin β gene cluster *in vivo*. *J Biol Chem* 1992 267: 3179–3184.
17. Zhang FP, Rannikko AS, Manna PR, Fraser HM, Huhtaniemi IT. Cloning and functional expression of the luteinizing hormone receptor complementary deoxyribonucleic acid from the marmoset monkey testis: absence of sequences encoding exon 10 in other species. *Endocrinology* 1997 138: 2481–2490.
18. Zhang FP, Kero J, Huhtaniemi I. The unique exon 10 of the human luteinizing hormone receptor is necessary for expression of the receptor protein at the plasma membrane in the human luteinizing hormone receptor, but deleterious when inserted into the human follicle-stimulating hormone receptor. *Mol Cell Endocrinol* 1998 142: 165–174.
19. Gromoll J, Eiholzer U, Nieschlag E, Simoni M. Male hypogonadism caused by homozygous deletion of exon 10 of the luteinizing hormone receptor: differential action of the luteinizing hormone (LH) and human chorionic gonadotropin (hCG). *J Clin Endocrinol Metab* 2000 85: 2281–2286.
20. Roseman DS, Baenziger JU. Molecular basis of lutropin recognition by the mannose/GalNAc-4-SO_4 receptor. *Proc Natl Acad Sci USA* 2000 97: 9949–9954.
21. Ulloa-Aguirre A, Midgley AR Jr, Beitins IZ, Padmanabhan V. Follicle-stimulating isohormones: characterization and physiological relevance. *Endocr Rev* 1995 16: 765–787.
22. Zhang R, Buczko E, Dufau ML. Requirement of cysteine residues in exons 1 to 6 of the extracellular domain of the luteinizing hormone receptor for gonadotropin binding. *J Biol Chem* 1996 271: 5755–5760.
23. Rozell TG, Wang H, Liu X, Segaloff DL. Intracellular retention of mutant gonadotropin receptors results in loss of binding activity of the follitropin receptor but not the lutropin/choriogonadotropin receptor. *Mol Endocrinol* 1995 9 1727–1736.

24. Ji I, Ji TH. Differential roles of exoloop 1 in the human follicle-stimulating hormone receptor in hormone binding and receptor activation. *J Biol Chem* 1995 270: 15970–15973.
25. Huang J, Puett D. Identification of two amino acid residues on the extracellular domain of the luteinizing hormone/choriogonadotropin receptor important for signaling. *J Biol Chem* 1995 270: 30023–30028.
26. Fernandez LM, Puett D. Lys^{583} in the third extracellular loop of the lutropin/choriogonadotropin receptor is critical for signaling. *J Biol Chem* 1996 271: 925–930.
27. Rajagopalan-Gupta RM, Mukherjee S, Zhu X *et al.* Roles of G_i and $G_q/11$ in mediating desensitization of the luteinizing hormone/choriogonadotropin receptor in porcine ovarian follicular membranes. *Endocrinology* 1999 140: 1612–1621.
28. Kuehn B, Gudermann T. The luteinizing hormone receptor activates phospholipase C via preferential coupling to G_i2. *Biochemistry* 1999 38: 12490–12498.
29. Gilchrist RL, Ryu K-S, Ji I, Ji TH. The luteinizing hormone/chorionic gonadotropin receptor has distinct transmembrane conductors for cAMP and inositol phosphate signals. *J Biol Chem* 1996 271: 19283–19287.
30. Saez JM. Leydig cells: endocrine, paracrine, and autocrine regulation. *Endocr Rev* 1994 15: 574–626.
31. Leung PC, Steele GL. Intracellular signaling in the gonads. *Endocr Rev* 1992 13: 476–498.
32. Schwartz J. Intercellular communication in the anterior pituitary. *Endocr Rev* 2000 21: 488–513.
33. Naor Z, Harris D, Shacham S. Mechanism of GnRH receptor signaling: combinatorial cross-talk of Ca^{2+} and protein kinase C. *Front Neuroendocrinol* 1998 19: 1–19.
34. de Kretser DM, Phillips DJ. Mechanisms of protein feedback on gonadotropin secretion. *J Reprod Immunol* 1998 39: 1–12.
35. Rugarli EI. Kallmann syndrome and the link between olfactory and reproductive development. *Am J Hum Genet* 1999 65: 943–948.
36. de Roux N, Young J, Misrahi M, Schaison G, Milgrom E. Loss of function mutations of the GnRH receptor: a new cause of hypogonadotropic hypogonadism. *J Pediatr Endocrinol Metab* 1999 12: 267–275.
37. Mukherjee S, Palczewski K, Gurevich V, Benovic JL, Banga JP, Hunzicker-Dunn M. A direct role for arrestins in desensitization of the luteinizing hormone/choriogonadotropin receptor in porcine ovarian follicular membranes. *Proc Natl Acad Sci USA* 1999 96: 493–498.
38. Heckert LL, Sawadogo M, Daggett MA, Chen JK. The USF proteins regulate transcription of the follicle-stimulating hormone receptor but are insufficient for cell-specific expression. *Mol Endocrinol* 2000 14: 1836–1848.
39. Levallet J, Koskimies P, Rahman N, Huhtaniemi I. The promoter of murine follicle-stimulating hormone receptor: functional characterization and regulation by transcription factor steroidogenic factor 1. *Mol Endocrinol* 2001 15: 80–92.
40. Nishimura R, Shin J, Ji I *et al.* A single amino acid substitution in an ectopic α subunit of a human carcinoma choriogonadotropin. *J Biol Chem* 1986 261: 10475–10477.
41. Weiss J, Axelrod L, Whitcomb RW, Harris PE, Crowley WF, Jameson JL. Hypogonadism caused by a single amino acid substitution in the β subunit of luteinizing hormone. *N Engl J Med* 1992 326: 179–183.
42. Pettersson K, Ding YQ, Huhtaniemi I. An immunologically anomalous luteinizing hormone variant in a healthy woman. *J Clin Endocrinol Metab* 1992 74: 164–171.
43. Haavisto AM, Pettersson K, Bergendahl M, Virkamäki A, Huhtaniemi I. Occurrence and biological properties of a common genetic variant of luteinizing hormone. *J Clin Endocrinol Metab* 1995 80: 1257–1263.
44. Jiang M, Pakarinen P, Zhang FP *et al.* A common polymorphic allele of the human luteinizing hormone β-subunit gene: additional mutations and differential function of the promoter sequence. *Hum Mol Genet* 1999 8: 2037–2046.
45. Latronico AC, Shinozaki H, Guerra G Jr *et al.* Gonadotropin-independent precocious puberty due to luteinizing hormone receptor mutation in Brazilian boy: a novel constitutively activating mutation in the first transmembrane helix. *J Clin Endocrinol Metab* 2000 85: 4799–4805.
46. TaoYX, Abell AN, Liu X, Nakamura K, Segaloff DL. Constitutive activation of G protein-coupled receptors as a result of selective substitution of a conserved leucine residue in transmembrane helix III. *Mol Endocrinol* 2000 14: 1272–1282.

47. Aittomäki K, Dieguez Lucena JL, Pakarinen P *et al.* Mutation in the follicle-stimulating hormone receptor gene causes hereditary hypergonadotropic ovarian failure. *Cell* 1995 82: 959–968.
48. Tena-Sempere M, Manna PR, Huhtaniemi I. Molecular cloning of mouse follicle-stimulating hormone receptor complementary deoxyribonucleic acid: functional expression of alternatively spliced variants and receptor inactivation by a C566T transition in exon 7 of the coding sequence. *Biol Reprod* 1999 60: 1515–1527.
49. Wu SM, Hallermeier K, Rennert OM, Chan WY. Polymorphisms in the coding exons of the human luteinizing hormone receptor gene. Mutations in brief no. 124 [Online]. *Hum Mutat* 1998 11: 333–334.
50. Perez Mayorga M, Gromoll J, Behre HM, Gassner C, Nieschlag E, Simoni M. Ovarian response to follicle-stimulating hormone (FSH) stimulation depends on the FSH receptor gene. *J Clin Endocrinol Metab* 2000 85: 3365–3369.
51. Abell AN, Segaloff DL. Evidence for the direct involvement of transmembrane region 6 of the lutropin/choriogonadotropin receptor in activating G_s. *J Biol Chem* 1997 272: 14586–14591.
52. Grasso P, Dexiel MR, Riechert LE Jr. Synthetic peptides corresponding to residues 551 to 555 and 650 to 653 of the rat testicular follicle-stimulating hormone (FSH) receptor are sufficient for post-receptor modulation of Sertoli cell responsiveness to FSH stimulation. *Regul Pept* 1995 60: 177–183.
53. Themmen APN, Brunner HG. Luteinizing hormone receptor mutations and sex differentiation. *Eur J Endocrinol* 1996 134: 533–540.
54. Zhang FP, Poutanen M, Wilbertz J, Huhtaniemi I. Normal prenatal but arrested postnatal development of luteinizing hormone receptor knockout (LuRKO) mice. *Mol Endocrinol* 2001 15: 172–183.
55. Dierich A, Sairam MR, Monaco L *et al.* Impairing follicle-stimulating hormone (FSH) signaling *in vivo*: targeted disruption of the FSH receptor leads to aberrant gametogenesis and hormonal imbalance. *Proc Natl Acad Sci USA* 1998 95: 13612–13617.
56. Liu G, Duranteau L, Carel JC, Monroe J, Doyle DA, Shenker A. Leydig-cell tumors caused by an activating mutation of the gene encoding the luteinizing hormone receptor. *N Engl J Med* 1999 341: 1731–1736.
57. Nishimori K, Matzuk MM. Transgenic mice in the analysis of reproductive development and function. *Rev Reprod* 1996 1: 203–212.
58. Kero J, Poutanen M, Zhang F-P *et al.* Elevated luteinizing hormone (LH) in transgenic mice induces functional LH receptor expression and steroidogenesis in adrenal cortex. *J Clin Invest* 2000 105: 633–641.

Chapter 13

Steroidogenesis

Walter L. Miller and Synthia H. Mellon

INTRODUCTION

The biosynthesis of steroid hormones is crucial for human reproduction. In addition to the classic steroidogenic organs – the gonads, adrenal gland and placenta – many other tissues also synthesize and/or modify steroid hormones into a broad array of final products that act through both classical nuclear receptors and through non-classical cell surface receptors. This chapter first describes the various steroidogenic enzymes and cofactors, highlighting their activities and deficiency states, then describes the essential features of the steroidogenic pathways in each principal steroidogenic tissue and cell type.

STEROID HORMONE SYNTHESIS

Early steps: cholesterol uptake, storage and transport

Steroidogenic cells can synthesize cholesterol *de novo* from acetate, but most of their supply of cholesterol comes from plasma low density lipoproteins (LDLs) derived from dietary cholesterol.[1] Adequate concentrations of LDL will suppress 3-hydroxy-3-methylglutaryl coenzyme A (HMG CoA) reductase, the rate-limiting enzyme in cholesterol synthesis. Adrenocorticotropic hormone (ACTH) and luteinizing hormone (LH), which stimulate adrenal and gonadal steroidogenesis, also stimulate the activity of HMG CoA reductase (the rate-limiting enzyme in cholesterol biosynthesis), the production of LDL receptors and the uptake of LDL cholesterol. LDL cholesterol esters are taken up by receptor-mediated endocytosis, then are stored directly or converted to free cholesterol and used for steroid hormone synthesis.[2] Storage of cholesterol esters in lipid droplets is controlled by the action of two opposing enzymes, cholesterol esterase (cholesterol ester hydrolase) and cholesterol ester synthetase. ACTH and LH, acting through cyclic adenosine monophosphate (cAMP), stimulate the esterase and inhibit the synthetase, thus increasing the availability of free cholesterol for steroid hormone synthesis.[2] Mutations in the cholesterol esterase (lysosomal acid lipase) cause Wolman disease (cholesterol ester storage disease).[3]

Cytochrome P450

Most steroidogenic enzymes are members of the cytochrome P450 group of oxidases.[4] 'Cytochrome P450' is a generic term for a large number of oxidative enzymes, all of which have about 500 amino acids and contain a single heme group; they are termed P450 (pigment 450) because all absorb light at 450 nm in their reduced states. Most cytochrome P450 enzymes are found in the endoplasmic reticulum of the liver, where they metabolize endogenous and exogenous toxins, drugs, xenobiotics and environmental pollutants. Despite this huge variety of substrates, the preliminary human genome sequence revealed only 62 genes encoding P450 enzymes.[5] Thus, most, if not all, P450 enzymes can metabolize multiple substrates, catalyzing a broad array of oxidations. This theme recurs with each steroidogenic P450 enzyme.

Six distinct P450 enzymes are involved in steroidogenesis (Figure 1). P450scc, found in mitochondria, is the cholesterol side-chain cleavage enzyme catalyzing the series of reactions formerly termed '20,22 desmolase'. Two distinct isozymes of P450c11, P450c11β and P450c11AS, also found in mitochondria, catalyze adrenal 11β-hydroxylase, 18-hydroxylase and 18-methyl oxidase activities. P450c17, found in the endoplasmic reticulum of adrenals, gonads and embryonic brain, catalyzes both 17α-hydroxylase and 17,20-lyase activities. Adrenal P450c21, found in the endoplasmic reticulum, catalyzes the 21-hydroxylation of both glucocorticoids and mineralocorticoids. In the ovaries (and elsewhere), P450aro in the endoplasmic reticulum catalyzes aromatization of androgens to estrogens. In addition, the biosynthesis of $1,25(OH)_2$ vitamin D, a sterol hormone, requires a series of mitochondrial cytochrome P450 enzymes.[6]

Hydroxysteroid dehydrogenases

The hydroxysteroid dehydrogenases (HSDs) have molecular masses of about 35–45 kDa, do not have heme groups, and require reduced nicotinamide–adenine dinucleotide (NADH) or its

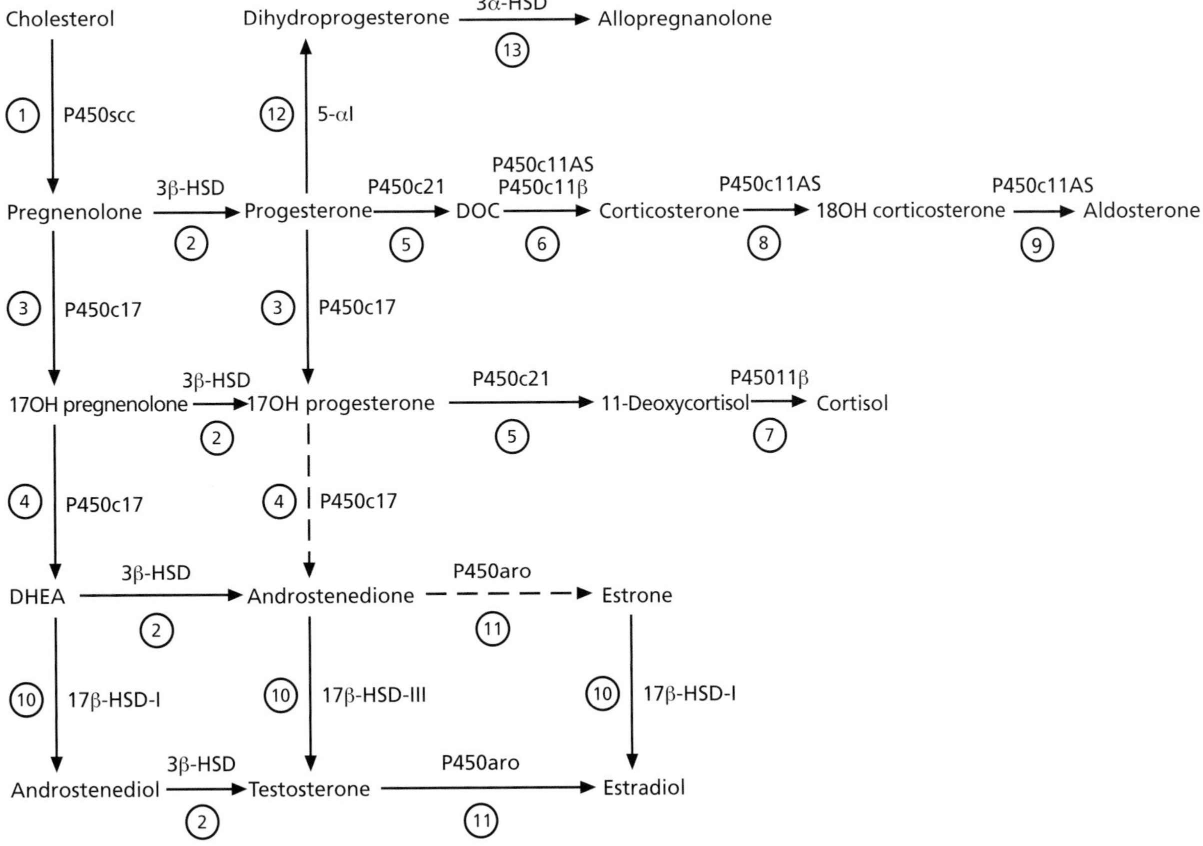

Figure 1 Principal pathways of human adrenal steroid hormone synthesis. Other quantitatively and physiologically minor steroids are also produced. The names of the enzymes are shown by each reaction, and the traditional names of the enzymatic activities given below correspond to the circled numbers. Reaction 1: mitochondrial cytochrome P450scc mediates 20α-hydroxylation, 22-hydroxylation and scission of the C20,22 carbon bond. Reaction 2: 3βHSD mediates 3βhydroxysteroid dehydrogenase and isomerase activities, converting Δ^5 steroids to Δ^4 steroids. Reaction 3: P450c17 catalyzes the 17α-hydroxylation of pregnenolone to 17-hydroxypregnenolone (17OH pregnenolone) and of progesterone to 17OH progesterone. Reaction 4: the 17,20-lyase activity of P450c17 converts 17OH pregnenolone to dehydroepiandrosterone (DHEA); only insignificant amounts of 17OH progesterone are converted to Δ^4-androstenedione by human P450c17, although this reaction occurs in other species. Reaction 5: P450c21 catalyzes the 21-hydroxylation of progesterone to deoxycorticosterone (DOC) and of 17OH progesterone to 11-deoxycortisol. Reaction 6: DOC is converted to corticosterone by the 11-hydroxylase activity of P450c11AS in the zona glomerulosa and by P450c11β in the zona fasciculata. Reaction 7: 11-deoxycortisol undergoes 11β-hydroxylation by P450c11β to produce cortisol in the zona fasciculata. Reactions 8 and 9: the 18-hydroxylase and 18-oxidase activities of P450c11AS convert corticosterone to 18OH corticosterone and aldosterone, respectively, in the zona glomerulosa. Reactions 10 and 11 are found principally in the testes and ovaries. Reaction 10: 17β-HSD-I converts DHEA to androstenediol and estrone to estradiol, while 17β-HSD-III converts androstenedione to testosterone. Reaction 11: testosterone may be converted to estradiol and androstenedione may be converted to estrone by P450aro. Reactions 12 and 13 are found principally in the nervous system. Reaction 12: type 1 5α-reductase (5α-I) converts progesterone to 5α-dihydroprogesterone. Reaction 13: one or more isozymes of 3α-HSD convert dihydroprogesterone to the anesthetic steroid allopregnanolone

phosphate (NADPH) as cofactors.[7] Whereas most steroidogenic reactions catalyzed by P450 enzymes are due to the action of a single form of P450, each of the reactions catalyzed by HSDs can be catalyzed by two or more isozymes, which are often very different. There are two families of HSDs: the short chain dehydrogenases include the 3β-hydroxysteroid dehydrogenases, the two 11β-hydroxysteroid dehydrogenases and a series of 17β-hydroxysteroid dehydrogenases; the aldo–keto reductase family includes the 3α- and 20α-hydroxysteroid dehydrogenases and type 5 17β-HSD.[7]

P450scc

Conversion of cholesterol to pregnenolone in mitochondria is the first, rate-limiting and hormonally regulated step in the synthesis of all steroid hormones.[4] This involves three distinct chemical reactions: 20α-hydroxylation, 22-hydroxylation and scission of the cholesterol side-chain to yield pregnenolone and isocaproic acid. Early studies showed that 20-hydroxycholesterol, 22-hydroxycholesterol and 20,22-hydroxycholesterol could all be isolated from bovine adrenals in significant quantities, suggesting that three separate and distinct enzymes were involved. However, protein purification studies and *in vitro* reconstitution of enzymatic activity show that a single protein, termed P450scc (where 'scc' refers to the side-chain cleavage of cholesterol), encoded by a single gene on chromosome 15, catalyzes all the steps between cholesterol and pregnenolone.[4,8] These three reactions occur on a single active site[4] that is in contact with the hydrophobic bilayer membrane.[4] Homozygous deletion of the gene for P450scc in the rabbit eliminates all steroidogenesis,[9] confirming that all steroidogenesis is initiated by this one enzyme. Homozygous inactivation of P450scc is probably incompatible with human fetal development, and haploinsufficiency of P450scc due to *de novo* heterozygous mutation causes a late-onset form of congenital lipoid adrenal hyperplasia.[10]

Transport of electrons to P450scc: adrenodoxin reductase and adrenodoxin

P450scc functions as the terminal oxidase in a mitochondrial electron transport system. Electrons from NADPH are accepted by a flavoprotein, termed adrenodoxin reductase, that is loosely associated with the inner mitochondrial membrane. Adrenodoxin reductase transfers the electrons to an iron–sulfur protein termed adrenodoxin, which is found in the mitochondrial matrix or loosely adherent to the inner mitochondrial membrane.[4,11] Adrenodoxin then transfers the electrons to P450scc (Figure 2). Adrenodoxin reductase and adrenodoxin serve as generic electron transport proteins for all mitochondrial P450s (for example the vitamin D 1α-, 24- and 25-hydroxylases), and not just those involved in steroidogenesis; hence, these proteins are also termed ferredoxin oxidoreductase and ferredoxin. Adrenodoxin forms a 1 : 1 complex with adrenodoxin reductase, then dissociates, then subsequently reforms an analogous 1 : 1 complex with P450scc or P450c11, thus functioning as an indiscriminate diffusable electron shuttle mechanism.[4,12] The single human adrenodoxin reductase gene[13] and the single, functional adrenodoxin gene[14] are expressed in all human tissues,[15,16] indicating that there are generic mitochondrial electron-transfer proteins whose roles are not limited to steroidogenesis.

Steroidogenic acute regulatory protein

The chronic regulation of steroidogenesis by ACTH is at the level of gene transcription, but the acute regulation, such as in response to the LH surge or to infusion of ACTH, is at the level of cholesterol access to P450scc.[17] When either steroidogenic cells or intact rats are treated with inhibitors of protein synthesis such as cycloheximide, the acute steroidogenic response is eliminated, indicating that a short-lived, cycloheximide-sensitive protein acts at the level of the mitochondrion as a specific trigger to the acute steroidogenic response.[17] The identity of this trigger was determined with the cloning of the steroidogenic acute regulatory protein (StAR). StAR was first identified as a family of short-lived 30- and 37-kDa phosphoproteins that were rapidly synthesized when steroidogenic cells were stimulated with tropic hormone.[17] Mouse StAR was then cloned from Leydig MA-10 cells and could induce steroidogenesis when transfected back into these cells.[18] The central role of StAR was proven by showing that it promoted steroidogenesis in non-steroidogenic COS-1 cells cotransfected with StAR

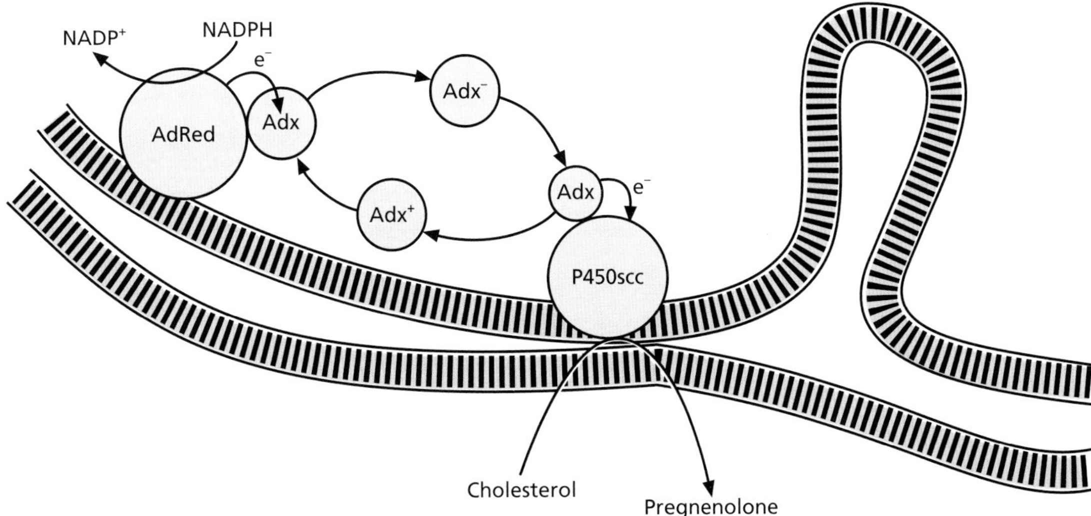

Figure 2 Electron transport to mitochondrial forms of cytochrome P450. Adrenodoxin reductase (AdRed), a flavoprotein loosely bound to the inner mitochondrial membrane, accepts electrons (e^-) from reduced nicotinamide–adenine dinucleotide phosphate (NADPH), converting it to $NADP^+$. These electrons are passed to adrenodoxin (Adx), an iron–sulfur protein in solution in the mitochondrial matrix that functions as a freely diffusable electron shuttle mechanism. Electrons from charged adrenodoxin (Adx^-) are accepted by any available cytochrome P450 such as P450c11, or P450scc shown here. The uncharged adrenodoxin (Adx^+) may then again be bound to adrenodoxin reductase to receive another pair of electrons. For P450scc, three pairs of electrons must be transported to the P450 to convert cholesterol to pregnenolone. The flow of cholesterol into the mitochondria is facilitated by steroidogenic acute regulatory protein (StAR), which is not shown in this figure

and the cholesterol side-chain cleavage enzyme system,[19] and by finding that mutations of StAR caused congenital lipoid adrenal hyperplasia.[19] Thus, StAR is the acute trigger required for the rapid flux of cholesterol from the outer to the inner mitochondrial membrane that is needed for the acute response of aldosterone to angiotensin II, of cortisol to ACTH and of sex steroids to an LH pulse. A mitochondrial surface protein, termed the peripheral benzodiazepine receptor, may also play a permissive role in choleserol flux into mitochondria.[20]

Some steroidogenesis is independent of StAR. When non-steroidogenic cells are transfected with StAR and the P450scc system, they convert cholesterol to pregnenolone at about 14% of the StAR induced rate;[19,21] furthermore, the placenta[8] and the brain[22] utilize mitochondrial P450scc to initiate steroidogenesis but the placenta does not express StAR.[23] The mechanism of StAR-independent steroidogenesis is unknown. It is possible that it occurs spontaneously, without any triggering protein, or some other protein may exert StAR-like activity to promote cholesterol flux. The mechanism of StAR's action is unknown, but it is clear that StAR acts on the outer mitochondrial membrane and does not need to enter the mitochondria to be active, as deletion of up to 62 N-terminal residues confines StAR to the cytoplasm without reducing activity.[24] StAR appears to undergo structural changes while interacting with the outer mitochondrial membrane.[25] The carboyl half of the recently described protein MLN64 (N-218 MLN64) is structurally related to StAR, exhibits StAR activity *in vitro* and is cleaved from full-length MLN64 in the placenta,[26,27] and hence may play a role in placental steroidogenesis.

3β-Hydroxysteroid dehydrogenase/$\Delta^5 \rightarrow \Delta^4$ isomerase

Once pregnenolone is produced from cholesterol, it may undergo 17α-hydroxylation by P450c17 to yield 17-hydroxypregnenolone, or it may be converted to progesterone, the first biologically important steroid in the pathway. A single 42-kDa microsomal enzyme, 3β-hydroxysteroid dehydrogenase (3β-HSD) catalyzes both conversion of the hydroxyl group to a keto group on carbon 3 and the isomerization of the double bond from the B ring (Δ^5 steroids) to the A ring (Δ^4 steroids)[28]. Thus, a single enzyme converts pregnenolone to progesterone, 17α-hydroxypregnenolone to 17α-hydroxyprogesterone, dehydroepiandrosterone (DHEA) to androstenedione, and androstenediol to testosterone.[29] Each of these reactions occurs with the same enzyme kinetics.[30] As is typical of hydroxysteroid dehydrogenases, there are two isozymes of 3β-HSD, encoded by separate genes. The 3β-HSD-I and -II genes are duplicated copies with 93% nucleotide sequence identity, but with different patterns of expression. 3β-HSD-II is expressed almost exclusively in the adrenals and gonads while 3β-HSD-I is expressed in placenta, breast and 'extraglandular' tissues.[31] Consistent with this, all known mutations deleting 3β-HSD activity are in the 3β-HSD-II gene, and cause adrenal insufficiency and pseudohermaphroditism; mutations in 3β-HSD-I would presumably interfere with placental production of progesterone, and hence would cause spontaneous abortion.

P450c17

Both pregnenolone and progesterone may undergo 17α-hydroxylation to 17α-hydroxypregnenolone and 17α-hydroxyprogesterone (17OHP), respectively. 17α-Hydroxyprogesterone may also undergo scission of the C17,20 carbon bond to yield DHEA; however, very little 17OHP is converted to androstenedione because the human P450c17 enzyme catalyzes this reaction at only 3% of the rate for conversion of 17α-hydroxypregnenolone to DHEA.[32] These reactions are all mediated by a single enzyme, P450c17. This P450 is bound to the smooth endoplasmic reticulum, where it accepts electrons from a P450 oxidoreductase. As P450c17 has both 17α-hydroxylase activity and 17,20-lyase activity, it is the key branch point in steroid hormone synthesis. If neither activity of P450c17 is present, pregnenolone is converted to progesterone and mineralocorticoids; if 17α-hydroxylase activity is present but 17,20-lyase activity is not, pregnenolone is converted to the glucocorticoid cortisol; if both activities are present, pregnenolone is converted to DHEA, which is the obligate precursor of sex steroids (Figure 1).

17α-Hydroxylase and 17,20-lyase were once thought to be separate enzymes. The adrenals of prepubertal children synthesize ample cortisol but virtually no sex steroids (i.e. have 17α-hydroxylase activity but not 17,20-lyase activity), until adrenarche initiates the production of adrenal androgens (i.e. turns on 17,20-lyase activity). Furthermore, patients have been described as lacking 17,20-lyase activity but retaining normal 17α-hydroxylase activity. However, purification of pig testicular microsomal P450c17 to homogeneity and *in vitro* reconstitution of enzymatic activity show that both 17α-hydroxylase and 17,20-lyase activities reside in a single protein,[33] and cells transformed with a vector expressing P450c17 cDNA acquire both 17α-hydroxylase and 17,20-lyase activities.[34] P450c17 is encoded by a single gene on chromosome 10q24.3[35] that is structurally related to the genes for P450c21 (21-hydroxylase).[36]

Thus, the distinction between 17α-hydroxylase and 17,20-lyase is functional and not genetic or structural. The 17α-hydroxylase reaction is equally efficient with both Δ^5-pregnenolone and Δ^4-progesterone, but human P450c17 strongly prefers Δ^5-17α-hydroxypregnenolone for 17,20-bond scission, consistent with the large amounts of DHEA secreted by both the fetal and adult adrenal.[32] Catalysis of the Δ^5-17,20-lyase reaction also requires a high abundance of P450 oxidoreductase and cytochrome b_5.

Electron transport to P450c17: P450 oxidoreductase and cytochrome b_5

All microsomal cytochrome P450 enzymes, including P450c17, P450c21 and P450aro, receive electrons from a membrane-bound flavoprotein, termed P450 oxidoreductase. P450 oxidoreductase receives two electrons from NADPH and transfers them to the P450.[37] Electron transfer for the lyase reaction is promoted by the action of cytochrome b_5 as an allosteric factor rather than as an alternative electron donor.[32] 17,20-Lyase activity also requires the phosphorylation of serine residues on P450c17 by a cAMP-dependent protein kinase[38] (Figure 3). The availability of electrons appears to determine whether P450c17 performs only 17α-hydroxylation, or also performs 17,20 bond scission, as increasing the molar abundance of P450 oxidoreductase and cytochrome b_5 to P450c17 *in vitro* or *in vivo* increases the ratio of 17,20-lyase activity to 17α-hydroxylase activity (the testis contains three to four times more P450 oxidoreductase activity than does the adrenal). Thus, the regulation of 17,20-lyase activity, and consequently of DHEA production, depends on factors that facilitate the flow of electrons to P450c17: high concentrations of P450 oxidoreductase, the presence of cytochrome b_5 and serine phosphorylation of P450c17.

P450c21

In the adrenal cortex, progesterone and 17-hydroxyprogesterone may be 21-hydroxylated by P450c21 to yield deoxycorticosterone and 11-deoxycortisol, respectively (Figure 1). Disorders of P450c21 cause about 95% all cases of congenital adrenal hyperplasia.[39,40] The clinical symptoms associated with this common genetic disease are complex and devastating. Decreased cortisol and aldosterone synthesis often leads to sodium loss, potassium retention and hypotension, which in turn will lead to cardiovascular collapse and death

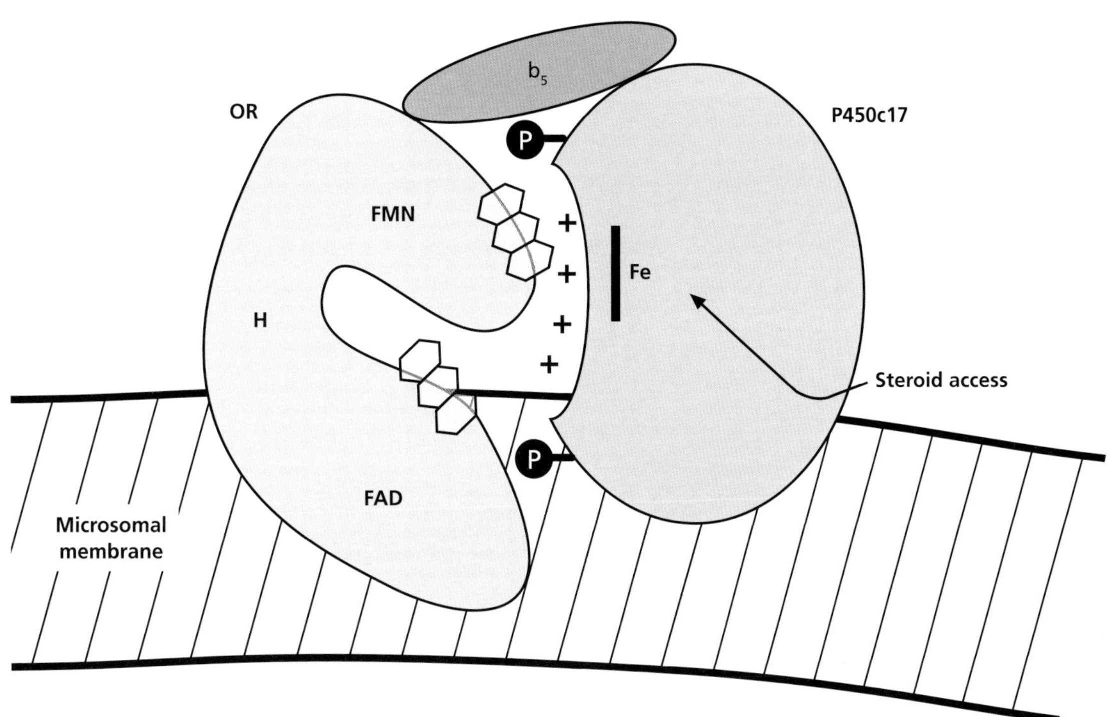

Figure 3 Electron transport to microsomal forms of cytochrome P450. This figure shows the interactions of P450c17 with its redox partners; the interactions of other P450 proteins may be simpler, as most microsomal P450 proteins are not phosphorylated and do not interact with cytochrome b_5. The flavin–adeninedinucleotide (FAD) moiety of P450 oxidoreductase (OR) picks up electrons from reduced nicotinamide–adenine dinucleotide phosphate (NADPH) (not shown) and transfers them to its flavin mononucleotide (FMN) moiety. The FAD and FMN groups are in distinct protein domains connected by a flexible protein hinge (H). The FMN domain approaches the redox partner binding site of the P450, shown as a concave region. The active site containing the steroid lies on the side of the plane of the heme ring (Fe) opposite the redox partner binding site. Cytochrome b_5 allosterically facilitates the interaction between OR and P450c17 to favor 17,20-lyase activity. Phosphoserine residues (P) also promote 17,20-lyase activity

within the month after birth if not treated appropriately. Decreased synthesis of cortisol *in utero* leads to overproduction of ACTH and consequent overstimulation of adrenal steroid synthesis; as the 21-hydroxylase step is impaired, 17OHP accumulates because P450c17 converts only miniscule amounts of 17OHP to androstenedione. However, 17-hydroxypregnenolone also accumulates and is converted to DHEA, and subsequently to androstenedione and testosterone, resulting in severe prenatal virilization of female fetuses.[39] Characterization of the P450c21 protein and gene cloning show that there is only one 21-hydroxylase encoded by a single functional gene on chromosome 6p21.[41,42] This gene lies in the middle of the major histocompatibility locus; hence, disorders of adrenal 21-hydroxylation are closely linked to specific human leukocyte antigen (HLA) types. P450c21 employs the same P450 oxidoreductase used by P450c17 to transport electrons from NADPH. 21-Hydroxylase activity has also been described in many adult and fetal extra-adrenal tissues, but this extra-adrenal 21-hydroxylation is not mediated by the P450c21 enzyme found in the adrenal;[43] the nature of the enzyme(s) responsible for extra-adrenal 21-hydroxylation remains unknown. Thus, patients lacking adrenal 21-hydroxylase activity may still have appreciable concentrations of 21-hydroxylated steroids in their plasma.

P450c11β and P450c11AS

Two closely related enzymes, P450c11β and P450c11AS, catalyze the final steps in the synthesis of both glucocorticoids and mineralocorticoids.[44,45] These two isozymes have 93% amino acid sequence identity and are encoded by tandemly duplicated genes on chromosome 8q21–22. Like P450scc, the two forms of P450c11 are found on the inner mitochondrial membrane, and use adrenodoxin and adrenodoxin reductase to receive electrons from NADPH. P450c11β converts 11-deoxycortisol to cortisol and 11-deoxycorticosterone to corticosterone; this is the classical 11β-hydroxylase, which is abundantly expressed in the adrenal zona fasciculata. P450c11AS is expressed at low levels in the zona glomerulosa, where it has 11β-hydroxylase, 18-hydroxylase and 18-methyl oxidase (aldosterone synthase) activities; thus, P450c11AS is able to catalyze all the reactions needed to convert deoxycorticosterone to aldosterone.

P450c11β is encoded by a gene (CYP11B1) that is primarily induced by ACTH via cAMP, and is suppressed by glucocorticoids such as dexamethasone. P450c11AS is encoded by a gene (CYP11B2) that is primarily regulated by angiotensin II and the potassium ion. The existence of two distinct functional genes is confirmed by the identification of mutations in each that cause distinct genetic disorders of steroidogenesis. Patients with mutations in P450c11β have classical 11β-hydroxylase deficiency but can still produce aldosterone; patients with disorders in P450c11AS have rare forms of aldosterone deficiency (commonly termed the corticosterone methyl oxidase deficiencies) while retaining the ability to produce cortisol.

17β-Hydroxysteroid dehydrogenases

Androstenedione is converted to testosterone, DHEA is converted to androstenediol and estrone is converted to estradiol by a series of short-chain dehydrogenases called 17β-hydroxysteroid dehydrogenases (17β-HSDs), sometimes also termed 17-oxidoreductase or 17-ketosteroid reductase, depending on the direction of the reaction being described.[46,47] This is the most complex and confusing step in steroidogenesis because: (1) there are several different 17β-HSDs; (2) some are preferential oxidases while others are preferential reductases; (3) they differ in their substrate preference and sites of expression; (4) there is inconsistent nomenclature, especially with the rodent enzymes; (5) some proteins termed 17β-HSD actually have very little 17β-HSD activity, and are principally involved in other reactions.[7] Human type I 17β-HSD (17β-HSD-I), also known as estrogenic 17β-HSD, is a cytosolic protein first isolated and cloned from the placenta, where it produces estriol, and is expressed in ovarian granulosa cells where it produces estradiol.[48,49] 17β-HSD-I is also expressed in the endometrium, testis, fat, liver and prostate. 17β-HSD-I catalyzes its reductase activity using NADPH as its cofactor. The three-dimensional structure of human 17β-HSD-I has been determined by X-ray crystallography.[50,51] No genetic deficiency syndrome for 17β-HSD-I has been described.

17β-HSD-II is a microsomal oxidase that uses oxidized nicotinamide–adenine dinucleotide (NAD^+) to inactivate estradiol to estrone and testosterone to Δ^4-androstenedione. 17β-HSD-II is found in the placenta, secretory endometrium, ovary, liver, small intestine and prostate. In contrast to 17β-HSD-I, which is found in placental syncytiotrophoblast cells, 17β-HSD-II is expressed in endothelial cells of placental intravillous vessels, consistent with its apparent role in defending the fetal circulation from transplacental passage of maternal estradiol or testosterone.[52] No deficiency state for 17β-HSD-II has been reported.

17β-HSD-III, the androgenic form of 17β-HSD, is a microsomal enzyme that is apparently expressed only in the testis.[53] This is the enzyme that is disordered in the classical syndrome of male pseudohermaphroditism that is often termed 17-ketosteroid reductase deficiency.[53]

An enzyme termed 17β-HSD-IV was initially identified as an NAD^+-dependent oxidase with activities similar to 17β-HSD-II, but this peroxisomal protein is primarily an enoyl-CoA hydratase and 3-hydroxyacyl-CoA dehydrogenase.[54] 17β-HSD-V, originally cloned as a 3α-hydroxysteroid dehydrogenase, catalyzes the reduction of Δ^4-androstenedione to testosterone,[55] but its precise role is unclear, and no deficiency state has been described.

Steroid sulfotransferase and sulfatase

Steroid sulfates may be synthesized directly from cholesterol sulfate or may be formed by sulfation of steroids by cytosolic sulfotransferases. Steroid sulfates may thus be hydrolyzed back to the native steroid by steroid sulfatase. Deletions in the steroid sulfatase gene on chromosome Xp22.3 cause X-linked ichthyosis.[56] In the fetal adrenal and placenta, diminished or absent sulfatase deficiency reduces the pool of free DHEA available for placental conversion to estrogen, resulting in low concentrations of estriol in the maternal blood and urine. The accumulation of steroid sulfates in the stratum corneum of the skin causes the ichthyosis. Steroid sulfatase is also expressed in the fetal rodent brain, possibly converting peripheral dehydroepiandrosterone sulfate (DHEAS) to neuroactive DHEA.[57,58]

Aromatase: P450aro

Estrogens are produced by the aromatization of androgens by the single microsomal aromatase, P450aro.[59,60] The gene for P450aro, located on chromosome 15q21.1, is unusually large and uses several different promoter sequences, transcriptional start sites and alternatively chosen first exons to encode aromatase mRNA in different tissues under different hormonal regulation. Aromatase expression in extraglandular tissues, especially fat, can covert adrenal androgens to estrogens. Aromatase in the epiphyses of growing bone converts testosterone to estradiol, accelerating epiphyseal maturation and terminating growth.[60] Although it has traditionally been thought that aromatase activity is needed for embryonic and fetal development, infants and adults with genetic disorders in this enzyme have been described, showing that fetoplacental estrogen is not needed for normal fetal development.[61]

5α-Reductase

Testosterone can be converted to a more potent androgen, dihydrotestosterone (DHT), by 5α-reductase, found in testosterone's target tissues. There are two distinct forms of 5α-reductase. The type 1 enzyme, found in the scalp, brain and other peripheral tissues, is encoded by a gene on chromosome 5; the type 2 enzyme, the predominant form found in male reproductive tissues, is encoded by a structurally related gene on chromosome 2p23.[62] The syndrome of 5α-reductase deficiency, a disorder of male sexual differentiation, is due to mutations in the gene encoding the type 2 enzyme.[63] The type 1 and 2 genes show an unusual pattern of developmental regulation of expression. The type 1 gene is not expressed in the fetus, then is expressed briefly in the skin of the newborn, and then remains unexpressed until its activity and protein are again found after puberty. The type 2 gene is expressed in fetal genital skin, in the normal prostate, and in prostatic hyperplasia and adenocarcinoma. Thus, the type 1 enzyme may be responsible for the pubertal virilization seen in patients with classic 5α-reductase deficiency, and may be involved in male pattern baldness.[62]

3α-Hydroxysteroid dehydrogenases

The 3α-HSDs are monomeric 37-kDa proteins that are NAD(P)(H) dependent, and have similar three-dimensional structures.[64–66] The reactions catalyzed by 3α-HSD are stereospecific, and involve the interconversion of a carbonyl with a hydroxyl group. To date, at least four distinct types of human cytosolic 3α-HSDs have been identified, and have been called AKR1C1–AKR1C4. These forms are known as type I 3α-HSD (AKR1C4), type 2 3α-HSD (AKR1C3), type 3 3α-HSD (AKR1C2) and 20α(3α)-HSD (AKR1C1).[7] All the human 3α-HSDs prefer NADPH as cofactor,[67] and hence are predominantly reductases. The relative tissue distribution of these human forms indicates that the liver is the only tissue that contains all four isoforms at similar levels, while other tissues vary in their concentration of the different 3α-HSD isoforms. In addition, 3α-HSD type 1 is found virtually exclusively in the liver.[67]

All 3α-HSD cDNAs share high sequence identity (over 80% at both amino acid and nucleic acid levels). The genes encoding the human types 1 and 2 enzymes span approximately 20 and 16 kb, respectively, and both contain nine exons of the same size and share equivalent intron/exon boundaries.[68] Human type 1 3α-HSD has a lower Michaelis constant (K_m) value for 5α-DHT than does type 2 3α-HSD. Thus, a major role for 3α-HSD type 1 in the liver is protection against circulating androgen excess.

The 3α-HSDs have multiple activities and substrate specificities. For example, the type 2_{brain} uses 5α-reduced androgens as substrates, rather than 5α-reduced progesterone, while the type 3 enzyme uses both androgens and progestins as substrates.[69] The type 2_{brain} and type 3 3α-HSDs have 20α-HSD and 17β-HSD activities as well.

The catalytic activity of each 3α-HSD isoform, together with the tissue-specific expression of each isoform, may indicate the function that a particular enzyme has in each tissue. In the prostate, 3α-HSD is involved in inactivating DHT, by converting it to the weak androgen 3α-androstenediol, while, in the nervous system, 3α-HSD is involved in activating 5α-reduced steroids, such as 5α-dihydroprogesterone to the potent neurosteroid 3α,5α-tetrahydroprogesterone (allopregnanolone). Thus, this enzyme is a key regulator of both steroid hormone receptor and ion gated receptor occupancy and action. Hepatic 3α-HSD plays a crucial step in the synthesis of bile acids, and is responsible for the production of 5β-cholestane-3α,7α-diol, a committed precursor of bile acids.[70]

11β-Hydroxysteroid dehydrogenases

Although certain steroids are categorized as glucorticoids or mineralocorticoids, cloning and expression of the 'mineralocorticoid' (glucocorticoid type II) receptor showed that it has equal affinity for both aldosterone and cortisol. However, cortisol does not act as a mineralocorticoid *in vivo*, even though cortisol concentrations can exceed aldosterone concentrations by 100–1000-fold. In mineralocorticoid-responsive tissues such as the kidney, cortisol is enzymatically inactivated by conversion to cortisone, which cannot activate the mineralocorticoid receptor.[71] The interconversion of cortisol and cortisone is mediated by two isozymes of 11β-hydroxysteroid dehydrogenase (11β-HSD), each of which has both oxidase and reductase activity, depending on the cofactor available.[72] The type I enzyme (11β-HSD-I) is expressed mainly in glucocorticoid-responsive tissues such as the liver, testis, lung and proximal convoluted tubule. 11β-HSD-I can catalyze both the oxidation of cortisol to cortisone using oxidized nicotinamide–adenine dinucleotide phosphate ($NADP^+$) as its cofactor (K_m 1.6 μmol/l), and the reduction of cortisone to cortisol using NADPH as its cofactor (K_m 0.14 μmol/l); the reaction catalyzed depends on which cofactor is available, but the enzyme can only function with high (micromolar) concentrations of steroid.[73] 11β-HSD-II catalyzes only the oxidation of cortisol to cortisone using NADH, and can function with low (nanomolar) concentrations of steroid (K_m 10–100 nmol/l).[74] 11β-HSD-II is expressed in mineralocorticoid-responsive tissues such as the kidney, and thus serves to 'defend' the mineralocorticoid receptor by inactivating cortisol to cortisone. Traditional mineralocorticoids, such as aldosterone or deoxycorticosterone, are not substrates for 11β-HSD-II, and hence can exert a mineralocorticoid effect. Thus, 11β-HSD-II prevents cortisol from overwhelming renal mineralocorticoid receptors.[71] In the placenta and other fetal tissues, 11β-HSD-II[75,76] also inactivates cortisol, thus defending the fetus from high maternal concentrations of glucocorticoids. The placenta also has abundant $NADP^+$ favoring the oxidative action of 11β-HSD-I, so in the placenta both enzymes protect the fetus from high maternal concentrations of cortisol.[72]

TISSUE-SPECIFIC PATTERNS OF STEROIDOGENESIS

Adrenal gland

Fetal adrenal steroidogenesis

Fetal adrenocortical steroidogenesis probably begins at about week 6 of gestation. Fetuses affected with genetic lesions in adrenal steroidogenesis can produce sufficient adrenal androgen to virilize a female fetus to a nearly-male appearance, and this masculinization of the genitalia is complete by the 12th week of gestation. Thus, there is substantial fetal adrenal gland steroidogenesis between weeks 6 and 12. The definitive zone of the fetal adrenal gland produces small amounts of steroid hormones according to the pathways shown in Figure 1. In contrast, the large fetal zone of the adrenal is relatively deficient in 3β-HSD-II activity because it contains very little mRNA for this enzyme.[77] The fetal adrenal also has relatively abundant 17,20-lyase activity of P450c17; low 3β-HSD and high 17,20-lyase activity account for the huge amount of DHEA produced by the fetal adrenal. The fetal adrenal also has considerable sulfotransferase activity but little steroid sulfatase activity, favoring conversion of DHEA to DHEAS. The resulting DHEAS cannot be a substrate for adrenal 3β-HSD-II; instead, it is secreted, 16α-hydroxylated in the fetal liver, and then acted on by placental 3β-HSD-I, 17β-HSD-I and P450aro to produce estriol, or the substrates can bypass the liver to yield estrone and estradiol. Placental estrogens inhibit adrenal 3β-HSD activity, providing a feedback system to promote production of DHEAS.[78] Fetal adrenal steroids account for 50% of the estrone and estradiol and 90% of the estriol in the maternal circulation.

Although the fetoplacental unit produces huge amounts of DHEA, DHEAS and estriol, as well as other steroids, they do not appear to play an essential role. Successful pregnancy is wholly dependent on placental synthesis of progesterone, which suppresses uterine contractility and prevents spontaneous abortion; however, fetuses with genetic disorders of adrenal and gonadal steroidogenesis develop normally, reach term gestation and undergo normal delivery.[79] Mineralocorticoid production is required postnatally, estrogens are not required and androgens are only needed for male sexual differentiation.[79] It is not clear whether human fetal development requires any glucocorticoids, but, if so, the small amount of maternal cortisol that escapes placental inactivation suffices.[79,80]

Adrenal zonation

The adult adrenal cortex is divided anatomically into an outermost zona glomerulosa (ZG) which lies immediately below the capsule, followed by the zona fasciculata (ZF) and zona reticularis (ZR), then the adrenal medulla where catecholamines are synthesized. Many investigators also identify an intermediate or transitional zone a few cell layers in thickness between the ZG and ZF. These zones were first characterized morphologically, but are now generally defined according to the presence of zone-specific enzymes or functions. As there is no unambiguously agreed-upon marker that differentiates the ZF and ZR, there is some ambiguity in this distinction. Although the three zones are usually described as distinct layers of the cortex, numerous immunohistochemical studies show substantial mixing among the zones. Thus, 'fingers' of ZG cell (as defined by the presence of the ZG-specific enzyme P450c11AS) extend down through the ZF to the ZR,[81] and islands of medullary tissue, defined by staining for catecholamine biosynthetic enzymes, are found scattered throughout the ZR and ZF. Therefore, although adrenal zonation has an anatomical basis, it is best considered in functional terms.

Zona glomerulosa

The distinction between the ZG and the other zones is clear. The ZG is the sole site of expression of P450c11AS, and hence is the sole site of aldosterone biosynthesis. The ZG is also characterized by lack of expression of P450c17 and by expression of receptors for angiotensin II. Receptors for ACTH are also found in the ZG, and ACTH appears to play a permissive role in mineralocorticoid biosynthesis. Angiotensin II acts through cell surface receptors to activate the protein kinase A (PKA) and Ca^{2+}/calmodulin pathways. These intracellular second messengers stimulate the transcription of the gene for P450scc by activating different P450scc promoter elements from those used by the ACTH/cAMP pathway. Angiotensin II and ACTH also stimulate the transcription of the genes for 3β-HSD-II and for P450c21, but these have not yet been studied in detail. Most importantly, angiotensin II stimulates transcription of the gene for P450c11AS but not the transcription of the closely related gene for P450c11β. Thus, angiotensin II stimulates the pathway cholesterol → pregnenolone → progesterone → 11-deoxycorticosterone → corticosterone → 18-hydroxycorticosterone → aldosterone, which is accomplished by the expression of only four enzymes: P450scc, 3β-HSD-II, P450c21 and P450c11AS.[45] Severe mutations in any of these steps would be predicted to cause aldosterone deficiency and a consequent salt-losing syndrome, but, in the case of mutations in P450c11AS, the clinical findings may be mitigated by synthesis of deoxycorticosterone in the ZF, as deoxycorticosterone has substantial mineralocorticoid activity.

Zona fasciculata

The ZF is the principal site of cortisol synthesis, and is characterized by the presence of P450c11β and P450c17, the absence of P450c11AS and the apparent absence of receptors for angiotensin II. Recent evidence also suggests that the ZF is relatively deficient in cytochrome b_5, possibly explaining why 17-hydroxypregnenolone in the ZF is almost entirely converted to 17-hydroxyprogesterone, with relatively little conversion to DHEA.[82,83] The principal pathway of steroidogenesis in the ZF is cholesterol → pregnenolone → 17-hydroxypregnenolone → 17-hydroxyprogesterone → 11 deoxycortisol → cortisol. However, some pregnenolone is also converted to progesterone and thence to deoxycorticosterone and corticosterone; in rodents, which do not express P450c17 in their adrenal gland,[84] this is the principal pathway of steroidogenesis. Similarly, humans who have P450c17 deficiency do not show signs of severe glucocorticoid deficiency because they produce sufficient corticosterone to meet their glucocorticoid requirements. As corticosterone is a weaker glucocorticoid than cortisol, these individuals produce large quantities of corticosterone. As a consequence of increased corticosterone production, they secrete more of the precursor deoxycorticosterone, and often present with deoxycorticosterone-induced mineralocorticoid hypertension, with low concentrations of aldosterone. The low aldosterone level in mineralocorticoid hypertension may appear paradoxical, but it is the logical result predicted by adrenal physiology. Thus, P450c17 deficiency leads to cortisol deficiency and increased ACTH, which drives ZF production of deoxycorticosterone and corticosterone. The deoxycorticosterone causes salt retention and increased intravascular volume, thus suppressing the production of renin and consequently suppressing angiotensin II, so that the ZG then produces very little aldosterone. A similar form of deoxycorticosterone-induced mineralocorticoid hypertension with suppressed aldosterone is seen in P450c11β deficiency.

However, the most common genetic disorder of the ZF is 21-hydroxylase (P450c21) deficiency. As there is a single functional gene for P450c21 expressed in all three zones of the adrenal cortex, mutations that destroy all 21-hydroxylase activity (e.g. P450c21 gene deletions, frameshifts, premature stop codons, etc.) will eliminate both mineralocorticoid and glucocorticoid synthesis. The decreased cortisol results in increased ACTH secretion, stimulating pregnenolone synthesis, which is then diverted to DHEA, androstenedione and testosterone, which will virilize affected 46,XX genetic females prenatally. Certain amino acid replacement (missense) mutations in P450c21 will result in an enzyme that retains a small amount of activity. This causes the 'simple virilizing' form of 21-hydroxylase deficiency in which salt loss is not a clinical feature. In these patients, the low level of enzymatic activity catalyzed by the mutant enzyme is sufficient to produce nearly normal amounts of aldosterone (albeit with elevated plasma renin activity), but minimal cortisol. Thus, the cortisol-deficient ACTH-based drive to produce virilizing adrenal androgens remains, but sufficient aldosterone is produced to prevent a salt-wasting crisis.

Zona reticularis and adrenarche

The ZR does not appear until about 3 years of age, and is not fully developed until adolescence. Immunocytochemical studies show that the ZR lacks P450c11AS, and is relatively (but not completely) deficient in 3β-HSD-II, P450c21 and steroid sulfatase, and is relatively enriched in steroid sulfotransferase and cytochrome b_5, thus fostering the 17,20-lyase activity of P450c17. The morphological development of the ZR precedes the secretion of its principal products DHEA and DHEAS. Based on the logic of symmetry and parsimony, it has been suggested that a polypeptide, variously termed CASH (cortical androgen-stimulating hormone) or AASH (adrenal androgen-stimulating hormone), ought to stimulate the production of DHEA in the ZR, but no convincing evidence for such a factor has been offered. Thus, the rise in synthesis of adrenal C19 steroids ('adrenarche'), which precedes puberty and is independent of the gonads or gonadotropins,[84] appears to require a series of developmentally programmed intracellular events. The current view of adrenarche may be summarized as follows. First, a developmentally timed trigger, possibly insulin-like growth factor-I (IGF-I), stimulates proliferation of the ZR and the synthesis of the enzymes needed for androgen synthesis. Second, 3β-HSD in the ZR remains low, favoring the production of Δ^5 steroids, but the 17,20-lyase activity of P450c17 is turned on, favoring conversion of 17-hydroxypregnenolone to DHEA. This increased lyase activity probably requires increased synthesis of cytochrome b_5 and increased serine phosphorylation of P450c17. Third, the ZR (but not the ZF) produces a steroid sulfotransferase but little or no steroid sulfatase, favoring the conversion of DHEA to DHEAS. This system is so efficient that, by age 20–30 years, the adrenal secretes more DHEAS than cortisol. After age 30, adrenal secretion of DHEA and DHEAS slowly falls, so that, by age 70–80 years, DHEAS concentrations are similar to those in children 10–12 years old (adrenopause) (Figure 4).

Ovarian steroidogenesis

Ovarian biosynthesis of steroids involves two principal cell types: the granulosa cell and the theca cell. Granulosa cells contain P450scc and hence convert cholesterol to pregnenolone, principally under the stimulation of follicle stimulating hormone (FSH); granulosa cells also contain 3β-HSD-II, and thus can secrete progesterone, principaly in response to the LH surge. Luteinized human granulosa cells in culture respond to LH in a dose-responsive fashion, with increased synthesis of P450scc and increased secretion of progesterone.[84] Granulosa cells also express 17β-HSD-I and aromatase, and thus produce

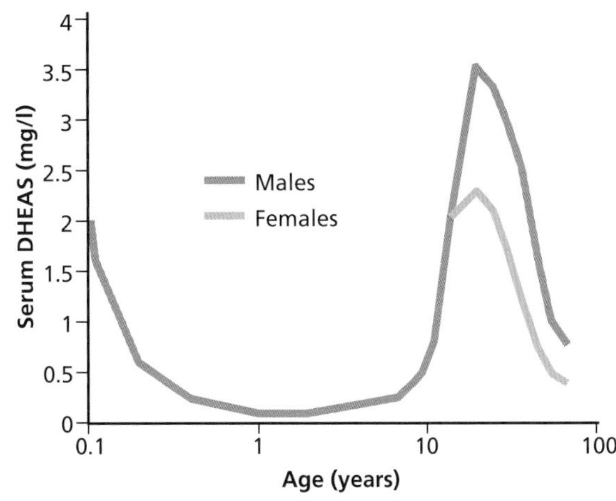

Figure 4 Concentrations of dehydroepiandrosterone sulfate (DHEAS) as a function of age. Note that the *x*-axis uses a log scale

estradiol,[49] but they do not express P450c17, so pregnenolone cannot be converted to C19 precursors of estrogens.[84] Thus, all ovarian conversion of C21 steroids to C19 steroids is catalyzed by P450c17 in theca cells. In addition to P450c17, theca cells express P450scc, and 3β-HSD-II, so they can express the steroidogenic pathway cholesterol → DHEA → androstenedione, but theca cells do not express P450aro, so they cannot convert C19 steroids to estrogens.

Unlike other steroidogenic tissues, the fetal ovary is steroidogenically quiescent. Although explanted fetal ovarian tissue can be induced to express steroidogenic enzymes and catalyze steroidogenic conversions *in vitro*, little if any steroidogenesis occurs in the fetal ovary *in vivo*. Thus, the mRNAs for P450scc and P450c17 were minimal or undetectable in 15–20-week human fetal ovarian tissues.[86] Furthermore, there is no identified syndrome of deficient fetal ovarian steroidogenesis. Hence, if the fetal ovary makes any steroid, it is quantitatively minimal and physiologically inconsequential.

Testicular steroidogenesis

Steroidogenesis in both the fetal and the adult testis occurs in the Leydig cells, following the same pathway. Under the stimulation of LH/human chorionic gonadotropin (hCG), these cells express P450scc, P450c17, 3β-HSD-II and 17β-HSD-III, and hence their principal pathway of steoidogenesis is cholesterol → pregnenolone → 17-hydroxypregnenolone → DHEA → androstenedione → testosterone. The Leydig cells secrete rather little progesterone or 17OHP, which has traditionally been interpreted to mean that 17OHP is converted to androstenedione by these cells, and indeed this conversion can be documented in rodent and bovine Leydig cells. However, it is also clear that bovine and rodent P450c17 can catalyze 17,20-lyase activity with Δ^4 substrates, thus converting 17OHP to androstenedione, while it is equally clear that human P450c17 has minimal Δ^4-17,20-lyase activity. This may suggest suppression of 3β-HSD-II activity with pregnenolone and 17-hydroxypregnenolone as substrates, but no such mechanism has been described, and, at least *in vitro*, 3β-HSD-II has the same K_m and maximum velocity (V_{max}) for all three Δ^5 substrates.[30] Thus, the mechanism by which the Leydig cell suppresses production of progesterone and 17OHP is unclear.

Placental steroidogenesis

Placental steroidogenesis occurs in trophoblast cells. Like ovarian granulosa cells, these cells are steroidogenically incomplete: they express P450scc and 3β-HSD-I, so they can convert cholesterol to progesterone. This placentally produced progesterone is required to maintain pregnancy after the middle of the first trimester. Before that time, the maternal corpus luteum of pregnancy produces progesterone, which is required to suppress uterine contractility (probably acting through cell surface receptors rather than via the conventional nuclear progesterone receptor). As the corpus luteum involutes, progesterone synthesis switches to the placenta – the 'luteoplacental shift'. Thus, maternal ovariectomy during early pregnancy causes spontaneous abortion and ovariectomy after the first trimester does not, but the administration of antiprogestins can cause abortion at any time. Consistent with the indispensable role of placental progesterone, homozygous defects have never been found in any of the enzymes required for placental progesterone synthesis: P450scc, adrenodoxin, adrenodoxin reductase or 3β-HSD-I (all cases of 3β-HSD deficiency involve 3β-HSD-II in the adrenal and gonad). Similarly, placental steroidogenesis does not require StAR, and pregnancy proceeds normally when the fetus (and placenta) carry homozygous StAR mutations.[21] The function of StAR in the placenta may be replaced by that of MLN64, which has StAR-like activity and is cleaved to a StAR-like peptide by the placenta.[27]

The placenta also produces substantial amounts of estrogen, principally estriol. Because the placenta lacks P450c17, it must receive C19 steroid precursors from the fetal adrenal gland. The principal secretory product of the fetal adrenal is DHEAS, most of which is 16α-hydroxylated by the fetal liver, to produce 16α-hydroxy-DHEAS. In the placenta, the sequential actions of steroid sulfatase, 3β-HSD-I, 17β-HSD-I and P450aro produce estriol. Small amounts of fetal adrenal DHEAS (or

DHEA) may escape hepatic 16α-hydroxylation and are acted upon by placental 3β-HSD-I, 17β-HSD-I and P450aro to become estradiol, or simply by 3β-HSD-I and P450aro without the action of 17β-HSD-I to become estrone. Irrespective of the biosynthetic pathway, most of these placental estrogens enter the maternal circulation. Thus, the placenta excretes unwanted fetal adrenal DHEA(S) into the maternal circulation as estrogens. Normal cord blood contains a ratio of estrone : estradiol : estriol of about 2400 : 900 : 14 000 (ng/dl), but, when the fetal adrenal cannot produce DHEA, the values fall to 100 : 80 : 10 (ng/dl);[87] thus, the maternal circulation provides very little estrogen to the fetus. The fetal adrenal cannot produce DHEA when the fetus has defects in StAR or P450c17, and the placenta cannot produce estrogens when the fetus lacks P450aro. All of these conditions are well described, and all are compatible with normal fetal development, term pregnancy and normal parturition. Hence, it is clear that the fetoplacental production of estrogen plays no essential role in pregnancy.[79]

Neurosteroidogenesis

The brain is a steroidogenic organ. Observations in the 1980s showed that pregnenolone, DHEA and their sulfate and lipoidal esters were present in higher concentrations in tissue from the nervous system (brain and peripheral nerve) than in plasma, and that these steroids remained in the nervous system long after gonadectomy or adrenalectomy.[88,89] These results suggested that steroids might either be synthesized *de novo* in the central and peripheral nervous systems, or might accumulate in those structures. Such steroids were named 'neurosteroids', to refer to their unusual origin and to differentiate them from steroids derived from more classical steroidogenic organs such as gonads, adrenals and placenta. The finding of steroids in the brain was striking. The view that steroids are actually made in the brain received powerful support when several laboratories established unequivocally that the enzymes found in classical steroidogenic tissues are found in the nervous system.[90] However, the nervous system contains additional steroid-modifying enzymes as well. In particular, the presence of 5α-reductase and 3α-HSD may ensure that progesterone is converted to the neuroactive steroid 3α,5α-tetrahydroprogesterone, otherwise known as allopregnanolone.

Neurosteroids exert several biological actions in the brain during embryogenesis, as well as in adults. These have been reported in several species including the human. One of the earliest observations on the neural action of neurosteroids was the report of the remarkable anesthetic property of progesterone.[91] Mechanisms by which progesterone and its derivatives act in the nervous system include both genomic actions, mediated by nuclear steroid receptors, and non-genomic actions, mediated by neurotransmitter receptors (for example γ-aminobutyric acid (A) ($GABA_A$)).[92–94] Neurosteroids derived from P450c17 activity, DHEA and DHEAS, are not known to act on nuclear steroid receptors but do modulate activity of $GABA_A$, *N*-methyl-D-aspartate (NMDA) and σ receptors.[95–99] The non-genomic actions of neurosteroids in the nervous system have been covered by several reviews.[90,93,100–103]

Neurosteroids can mediate their effects through neurotransmitter receptors, through classic steroid hormone receptors or through effects on membranes, which alter the function of receptors associated with or localized to the plasma membrane.[104] In addition to its other functions, progesterone plays a role in the functional maturation of the nervous system. Progesterone produced *in situ* in the nervous system is involved in the repair of injured peripheral nerves, and this effect is mediated by the nuclear progesterone receptor. In the injured rodent sciatic nerve, progesterone promotes myelinization[105] by enhancing proteins involved in this process: the peripheral myelin protein 22 and the protein zero.[106] The initiation of myelinization may also be correlated with increased expression of P450scc and 3β-HSD,[107] which are required for progesterone synthesis, but other studies suggest that reduced progesterone derivatives, rather than progesterone, modulate expression of these proteins.[108]

Allopregnanolone is a potent allosteric modulator of $GABA_A$ receptors. 5α-Derivatives of progesterone, like allopregnanolone and 5α-dihydroprogesterone, as well as pregnenolone and pregnenolone sulfate, bind to $GABA_A$ receptors at sites distinct from those for GABA, benzodiazepines and barbiturates,[109–111] where they act as allosteric agonists to induce chloride currents.[93] Different neurosteroids may act as agonists or antagonists at $GABA_A$ receptors, and may also display different binding characteristics. Steroids that are agonists at $GABA_A$ receptors act as sedative-hypnotics, anticonvulsants and anxiolytics in animal models.

The assembly of different $GABA_A$ receptor subunits into a functional $GABA_A$ receptor may be differentially sensitive to neurosteroids. Furthermore, neurosteroids may modulate the expression of different $GABA_A$ receptor subunits, thereby autoregulating their function at these receptors. Changes have been noted between plasma concentrations of allopregnanolone and the concentrations of $GABA_A$ receptor subunit in the brain, before, during and after pregnancy.[112,113] Similarly, pregnancy, expression of neurosteroidogenic enzymes and uterine $GABA_A$ function have also been correlated.[114]

Animal models of progesterone withdrawal, which are aimed at modelling human menstrual cycles, also suggest that neurosteroids modulate $GABA_A$ receptor subunit composition and function.[115] In rats, allopregnanolone withdrawal but not progesterone withdrawal resulted in increased expression of the α_4 subunit of the $GABA_{A+}$ receptor. Incorporation of this subunit into $GABA_A$ receptors altered its sensitivity to allopregnanolone, resulting in a receptor that was less sensitive to allopregnanolone. This may relate to changes in mood and behavior seen in some women during the late luteal phase of the menstrual cycle, when allopregnanolone levels drop. It may also relate to catamenial epilepsies and increased seizure susceptibility occurring during the late luteal phase of the menstrual cycle, and postpartum.

Neurosteroids such as pregnenolone sulfate, DHEA and DHEAS, modulate NMDA receptor activity[58,116–119]. Allopregnanolone sulfate, but not allopregnanolone, acts as a negative allosteric modulator of NMDA receptors,[118] while DHEA, pregnenolone and their sulfate esters are considered to be positive allosteric modulators of NMDA receptors. Unlike $GABA_A$ receptor interactions, the interaction of neurosteroids with the NMDA receptor is not well documented, and no specific interaction sites have been described.

Neurosteroids may be involved in the maturation of the nervous system. In primary cultures of pure embryonic neocortical neurons, DHEA, but not DHEAS, caused growth of Tau-1 immunopositive axons, whereas DHEAS, but not DHEA, caused growth of MAP-2-immunopositive dendrites.[58] These effects occurred at very low steroid concentrations (10^{-12} to 10^{-10} mol/l), and could not be mimicked by other steroids upstream or downstream of DHEA in the steroidogenic pathway, indicating that DHEA and DHEAS mediate these effects. Furthermore, stimulation of these neurons with DHEA resulted in an increase in intracellular calcium in the absence of KCl or NMDA. These effects could be abolished by both MK801 and D-AP5, inhibitors of NMDA receptor function, suggesting a direct interaction of DHEA at the NMDA receptor. The mechanism by which DHEAS mediates its effects are unknown.

DHEA can also act to protect neurons from ischemic insults,[120,121] which may be relevant to the decline of DHEA(S) during aging. In addition, allopregnanolone is protective against picrotoxin treatment.[122] The mechanism(s) by which DHEA, DHEAS and allopregnanolone exert their neuroprotective effects are unknown.

REFERENCES

1. Gwynne JT, Strauss JF III. The role of lipoproteins in steroidogenesis and cholesterol metabolism in steroidogenic glands. *Endocr Rev* 1982 3: 299–329.
2. Strauss JF, III, Miller WL. Molecular basis of ovarian steroid synthesis. In Hillier SG, ed. *Ovarian Endocrinology*. Blackwell Scientific: Oxford, UK, 1991: 25–72
3. Anderson RA, Byrum RS, Coates PM, Sando GN. Mutations at the lysosomal acid cholesteryl ester hydrolase gene locus in Wolman disease. *Proc Natl Acad Sci USA* 1994 91: 2718–2722.
4. Miller WL. Molecular biology of steroid hormone synthesis. *Endocr Rev* 1988 9: 295–318.

5. Venter JC, Adams MD, Myers EW *et al.* The sequence of the human genome. *Science* 2001 291: 1304–1351.
6. Miller WL, Portale AA. Genetic disorders of vitamin D biosynthesis. *Endocrinol Metab Clin North Am* 1999 28: 825–840.
7. Penning TM. Molecular endocrinology of hydroxysteroid dehydrogenases. *Endocr Rev* 1997 18: 281–305.
8. Chung B, Matteson KJ, Voutilainen R, Mohandas TK, Miller WL. Human cholesterol side-chain cleavage enzyme, P450scc: cDNA cloning, assignment of the gene to chromosome 15, and expression in the placenta. *Proc Natl Acad Sci USA* 1986 83: 8962–8966.
9. Yang X, Iwamoto K, Wang M *et al.* Inherited congenital adrenal hyperplasia in the rabbit is caused by a deletion in the gene encoding cytochrome P450 cholesterol side-chain cleavage enzyme. *Endocrinology* 1993 132: 1977–1982.
10. Tajima T, Fujieda K, Kouda N, Nakae J, Miller WL. Heterozygous mutation in the cholesterol side chain cleavage enzyme (P450scc) gene in a patient with 46,XY sex reversal and adrenal insufficency. *J Clin Endocrinol Metab* 2001 86: 3820–3825.
11. Hanukoglu I, Suh BS, Himmelhoch S, Amsterdam A. Induction and mitochondrial localization of cytochrome P450scc system enzymes in normal and transformed ovarian granulosa cells. *J Cell Biol* 1990 111: 1373–1381.
12. Coghlan VM, Vickery LE. Expression of human ferredoxin and assembly of the [2Fe–2S] center in *Escherichia coli*. *Proc Natl Acad Sci USA* 1989 86: 835–839.
13. Lin D, Shi Y, Miller WL. Cloning and sequence of the human adrenodoxin reductase gene. *Proc Natl Acad Sci USA* 1990 87: 8516–8520.
14. Chang C-Y, Wu D-A, Lai C-C, Miller WL, Chung B. Cloning and structure of the human adrenodoxin gene. *DNA* 1988 7: 609–615.
15. Picado-Leonard J, Voutilainen R, Kao L *et al.* Human adrenodoxin: cloning of three cDNAs and cycloheximide enhancement in JEG-3 cells. *J Biol Chem* 1988 263: 3240–3244.
16. Brentano ST, Black SM, Lin D, Miller WL. cAMP post-transcriptionally diminishes the abundance of adrenodoxin reductase mRNA. *Proc Natl Acad Sci USA* 1992 89: 4099–4103.
17. Stocco DM, Clark BJ. Regulation of the acute production of steroids in steroidogenic cells. *Endocr Rev* 1996 17: 221–244.
18. Clark BJ, Wells J, King SR, Stocco DM. The purification, cloning and expression of a novel luteinizing hormone-induced mitochondrial protein in MA-10 mouse Leydig tumor cells. Characterization of the steroidogenic acute regulatory protein (StAR). *J Biol Chem* 1994 269: 28314–28322.
19. Lin D, Sugawara T, Strauss JF III *et al.* Role of steroidogenic acute regulatory protein in adrenal and gonadal steroidogenesis. *Science* 1995 267: 1828–1831.
20. Papadopoulos V. Peripheral-type benzodiazepine/diazepam binding inhibitor receptor: biological role in steroidogenic cell function. *Endocr Rev* 1993 14: 222–240.
21. Bose HS, Sugawara T, Strauss JF III, Miller WL. The pathophysiology and genetics of congenital lipoid adrenal hyperplasia. *N Engl J Med* 1996 335: 1870–1878.
22. Mellon SH, Deschepper CF. Neurosteroid biosynthesis: genes for adrenal steroidogenic enzymes are expressed in the brain. *Brain Res* 1993 629: 283–292.
23. Sugawara T, Holt JA, Driscoll D *et al.* Human steroidogenic acute regulatory protein (StAR): functional activity in COS-1 cells, tissue-specific expression, and mapping of the structural gene to 8p11.2 and an expressed pseudogene to chromosome 13. *Proc Natl Acad Sci USA* 1995 92: 4778–4782.
24. Arakane F, Sugawara T, Nishino H *et al.* Steroidogenic acute regulatory protein (StAR) retains activity in the absence of its mitochondrial targeting sequence: implications for the mechanism of StAR action. *Proc Natl Acad Sci USA* 1996 93: 13731–13736.
25. Bose HS, Whittal RM, Baldwin MA, Miller WL. The active form of the steroidogenic acute regulatory protein, StAR, appears to be a molten globule. *Proc Natl Acad Sci USA* 1999 96: 7250–7255.
26. Moog-Lutz C, Tomasetto C, Régnier CH *et al.* MLN64 exhibits homology with the steroidogenic acute regulatory protein (StAR) and is overexpressed in human breast carcinomas. *Int J Cancer* 1997 71: 183–191.

27. Bose HS, Whittal RM, Huang MC, Baldwin MA, Miller WL. N-218 MLN64, a protein with StAR-like steroidogenic activity is folded and cleaved similarly to StAR. *Biochemistry* 2000 39: 11722–11731.
28. Thomas JL, Myers RP, Strickler RC. Human placental 3β-hydroxy-5-ene-steroid dehydrogenase and steroid 5 → 4-ene-isomerase: purification from mitochondria and kinetic profiles, biophysical characterization of the purified mitochondrial and microsomal enzymes. *Steroid Biochem* 1989 33: 209–217.
29. Lorence MC, Murry BA, Trant JM, Mason JI. Human 3β-hydroxysteroid dehydrogenase/$\Delta^5 \rightarrow \Delta^4$ isomerase from placenta: expression in nonsteroidogenic cells of a protein that catalyzes the dehydrogenation/isomerization of C21 and C19 steroids. *Endocrinology* 1990 126: 2493–2498.
30. Lee TC, Miller WL, Auchus RJ. Medroxyprogesterone acetate and dexamethasone are competitive inhibitors of different human steroidogenic enzymes. *J Clin Endocrinol Metab* 1999 84: 2104–2110.
31. Morel Y, Mébarke F, Rhéaume E *et al.* Structure–function relationships of 3β-hydroxysteroid dehydrogenase: contribution made by the molecular genetics of 3β-hydroxysteroid dehydrogenase deficiency. *Steroids* 1997 62: 176–184.
32. Auchus RJ, Lee TC, Miller WL. Cytochrome b_5 augments the 17,20 lyase activity of human P450c17 without direct electron transfer. *J Biol Chem* 1998 273: 3158–3165.
33. Nakajin S, Shinoda M, Haniu M, Shively JE, Hall PF. C_{21} steroid side-chain cleavage enzyme from porcine adrenal microsomes. Purification and characterization of the 17α-hydroxylase/$C_{17,20}$ lyase cytochrome P450. *J Biol Chem* 1984 259: 3971–3976.
34. Zuber MX, Simpson ER, Waterman MR. Expression of bovine 17α-hydroxylase cytochrome P450 cDNA in non-steroidogenic (COS-1) cells. *Science* 1986 234: 1258–1261.
35. Fan YS, Sasi R, Lee C *et al.* Localization of the human CYP17 gene (cytochrome P450 17α) to 10q24.3 by fluorescence *in situ* hybridization and simultaneous chromosome banding. *Genomics* 1992 14: 1110–1111.
36. Picado-Leonard J, Miller WL. Cloning and sequence of the human gene encoding P450c17 (steroid 17α-hydroxylase/17,20 lyase): similarity to the gene for P450c21. *DNA* 1987 6: 439–448.
37. Wang M, Roberts DL, Paschke R *et al.* Three-dimensional structure of NADPH-cytochrome P450 reductase: prototype for FMN- and FAD-containing enzymes. *Proc Natl Acad Sci USA* 1997 94: 8411–8416.
38. Zhang L, Rodriguez H, Ohno S, Miller WL. Serine phosphorylation of human P450c17 increases 17,20 lyase activity: implications for adrenarche and for the polycystic ovary syndrome. *Proc Natl Acad Sci USA* 1995 92: 10619–10623.
39. Morel Y, Miller WL. Clinical and molecular genetics of congenital adrenal hyperplasia due to 21-hydroxylase deficiency. *Adv Hum Genet* 1991 20: 1–68.
40. White PC, Speiser PW. Congenital adrenal hyperplasia due to 21-hydroxylase deficiency. *Endocr Rev* 2000 21: 245–291.
41. Higashi Y, Yoshioka H, Yamane M, Gotoh O, Fujii-Kuriyama Y. Complete nucleotide sequence of two steroid 21-hydroxylase genes tandemly arranged in human chromosome: a pseudogene and genuine gene. *Proc Natl Acad Sci USA* 1986 83: 2841–2845.
42. White PC, New MI, Dupont B. Structure of the human steroid 21-hydroxylase genes. *Proc Natl Acad Sci USA* 1986 83: 5111–5115.
43. Mellon SH, Miller WL. Extra-adrenal steroid 21-hydroxylation is not mediated by P450c21. *J Clin Invest* 1989 84: 1497–1502.
44. White PC, Curnow KM, Pascoe L. Disorders of steroid 11β-hydroxylase isozymes. *Endocr Rev* 1994 15: 421–438.
45. Fardella CE, Miller WL. Molecular biology of mineralocorticoid metabolism. *Annu Rev Nutr* 1996 16: 443–470.
46. Labrie F, Luu-The V, Lin SX *et al.* The key role of 17β-hydroxysteroid dehydrogenases in sex steroid biology. *Steroids* 1997 62: 148–158.
47. Moghrabi N, Andersson S. 17β-Hydroxysteroid dehydrogenases: physiological roles in health and disease. *Trends Endocrinol Metab* 1998 9: 265–270.
48. Peltoketo H, Isomaa V, Mäenlavsta O, Vihko R. Complete amino acid sequence of human placental 17β-hydroxysteroid dehydrogenase deduced from cDNA. *FEBS Lett* 1988 239: 73–77.
49. Tremblay Y, Ringler GE, Morel Y *et al.* Regulation of the gene for estrogenic 17-ketosteroid reduc-

tase lying on chromosome 17cen→q25. *J Biol Chem* 1989 264: 20458–20462.

50. Ghosh D, Pleuteu VZ, Zhu DW *et al.* Structure of human estrogenic 17β-hydroxysteroid dehydrogenase at 2.2 Å resolution. *Structure* 1995 3: 503–513.
51. Sawicki MW, Erman M, Puranen T, Vihko P, Ghosh D. Structure of the ternary complex of human 17β-hydroxysteroid dehydrogenase type 1 with 3-hydroxyestra-1,3,5,7-tetraen-17-one (equilin) and NADP+. *Proc Natl Acad Sci USA* 1999 96: 840–845.
52. Takeyama J, Sasano H, Suzuki T *et al.* 17β-Hydroxysteroid dehydrogenase types 1 and 2 in human placenta: an immunohistochemical study with correlation to placental development. *J Clin Endocrinol Metab* 1998 83: 3710–3715.
53. Geissler WM, Davis DL, Wu L *et al.* Male pseudohermaphroditism caused by mutations of testicular 17β-hydroxysteroid dehydrogenase 3. *Nat Genet* 1994 7: 34–39.
54. Leenders F, Tesdorpf JG, Markus M *et al.* Porcine 80-kDa protein reveals intrinsic 17β-hydroxysteroid dehydrogenase, fatty acyl-CoA-hydratase/dehydrogenase, and sterol transfer activities. *J Biol Chem* 1996 271: 5438–5442.
55. Dufort I, Rheault P, Huang XF, Soucy P, Luu-The V. Characteristics of a highly labile human type 5 17β-hydroxysteroid dehydrogenase. *Endocrinology* 1999 140: 568–574.
56. Ballabio A, Shapiro LJ. Steroid sulfatase deficiency and X-linked icthiosis. In Schriver CR, Beaduet AL, Sly WS, Valle D, eds. *The Metabolic and Molecular Basis of Inherited Disease*. New York: Mc-Graw-Hill, 1995: 2999–3022
57. Compagnone NA, Salido E, Shapiro LJ, Mellon SH. Expression of steroid sulfatase during embryogenesis. *Endocrinology* 1997 138: 4768–4773.
58. Compagnone NA, Mellon SH. Dehydroepiandrosterone: a potential signalling molecule for neocortical organization during development. *Proc Natl Acad Sci USA* 1998 95: 4678–4683.
59. Simpson ER, Mahendroo MS, Means GD *et al.* Aromatase cytochrome P450, the enzyme responsible for estrogen biosynthesis. *Endocr Rev* 1994 15: 342–355.
60. Grumbach MM, Auchus RJ. Estrogen: consequences and implications of human mutations in synthesis and action. *J Clin Endocrinol Metab* 1999 84: 4677–4694.
61. Conte FA, Grumbach MM, Ito Y, Fisher CR, Simpson ER. A syndrome of female pseudohermaphroditism, hypergonadotropic hypogonadism, and multicystic ovaries associated with missense mutations in the gene encoding aromatase (P450arom). *J Clin Endocrinol Metab* 1994 78: 1287–1292.
62. Thigpen AE, Silver RI, Guileyardo JM *et al.* Tissue distribution and ontogeny of steroid 5α-reductase isozyme expression. *J Clin Invest* 1993 92: 903–910.
63. Wilson JD. The role of androgens in male gender role behavior. *Endocr Rev* 1999 20 726–737.
64. Jez JM, Bennett MJ, Schlegel BP, Lewis M, Penning TM. Comparative anatomy of the aldo–keto reductase superfamily. *Biochem J* 1997 326: 625–636.
65. Jez JM, Flynn TG, Penning TM. A new nomenclature for the aldo–keto reductase superfamily. *Biochem Pharmacol* 1997 54: 639–647.
66. Penning TM, Bennett MJ, Smith-Hoog S *et al.* Structure and function of 3α-hydroxysteroid dehydrogenase. *Steroids* 1997 62: 101–111.
67. Penning TM, Burczynski ME, Jez JM *et al.* Human 3α-hydroxysteroid dehydrogenase isoforms (AKR1C1–AKR1C4) of the aldo–keto reductase superfamily: functional plasticity and tissue distribution reveals roles in the inactivation and formation of male and female sex hormones. *Biochem J* 2000 351: 67–77.
68. Khanna M, Qin KN, Wang RW, Cheng KC. Substrate specificity, gene structure, and tissue-specific distribution of multiple human 3α-hydroxysteroid dehydrogenases. *J Biol Chem* 1995 270: 20162–20168.
69. Griffin LD, Mellon SH. Selective serotonin reuptake inhibitors directly alter activity of neurosteroidogenic enzymes. *Proc Natl Acad Sci USA* 1999 96: 13512–13517.
70. Danielsson H, Sjovall J. Bile acid metabolism. *Annu Rev Biochem* 1975 44 233–253.
71. Funder JW, Pearce PT, Smith R, Smith I. Mineralocorticoid action: target tissue specificity is enzyme, not receptor, mediated. *Science* 1988 242: 583–585.
72. White PC, Mune T, Agarwal AK. 11β-Hydroxysteroid dehydrogenase and the

syndrome of apparent mineralocorticoid excess. *Endocr Rev* 1997 18: 135–156.

73. Moore CCD, Mellon SH, Murai J, Siiteri PK, Miller WL. Structure and function of the hepatic form of 11β-hydroxysteroid dehydrogenase in the squirrel monkey, an animal model of glucocorticoid resistance. *Endocrinology* 1993 133: 368–375.
74. Rusvai E, Náray-Fejes-Tóth A. A new isoform of 11β-hydroxysteroid dehydrogenase in aldosterone target cells. *J Biol Chem* 1993 268: 10717–10720.
75. Krozowski Z, MaGuire JA, Stein-Oakley AN *et al.* Immunohistochemical localization of the 11β-hydroxysteroid dehydrogenase type II enzyme in human kidney and placenta. *J Clin Endocrinol Metab* 1995 80: 2203–2209.
76. Hirasawa G, Sasono H, Suzuki T *et al.* 11β-Hydroxysteroid dehydrogenase type 2 and mineralocorticoid receptor in human fetal development. *J Clin Endocrinol Metab* 1999 84: 1453–1458.
77. Voutilainen R, Ilvesmaki V, Miettinen PJ. Low expression of 3β-hydroxy-5-ene steroid dehydrogenase gene in human fetal adrenals *in vivo*; adrenocorticotropin and protein kinase C-dependent regulation in adrenocortical cultures. *J Clin Endocrinol Metab* 1991 72: 761–767.
78. Fujieda K, Faiman C, Feyes FI, Winter JSD. The control of steroidogenesis by human fetal adrenal cells in tissue culture: IV. The effects of exposure to placental steroids. *J Clin Endocrinol Metab* 1982 54: 89–94.
79. Miller WL. Steroid hormone biosynthesis and actions in the materno-feto-placental unit. *Clin Perinatol* 1998 25: 799–817.
80. Kari MA, Raivio KO, Stenman U-H, Voutilainen R. Serum cortisol, dehydroepiandrosterone sulfate, and sterol-binding globulins in preterm neonates: effects of gestational age and dexamethasone therapy. *Pediatr Res* 1996 40: 319–324.
81. Oshima T, Suzuki H, Hata JI, Mitani F, Ishmura Y. Zone-specific expression of aldosterone synthase cytochrome P450 and cytochrome P45011β in rat adrenal cortex: histochemical basis for the functional zonation. *Endocrinology* 1992 130: 2971.
82. Yanase T, Sasano H, Yubisui T *et al.* Immunohistochemical study of cytochrome b_5 in human adrenal gland and in adrenocortical adenomas from patients with Cushing's syndrome. *Endocr J* 1998 45: 89–95.
83. Mapes S, Corbin C, Tarantal A, Conley A. The primate adrenal zona reticularis is defined by expression of cytochrome b_5, 17α-hydroxylase/17,20-lyase cytochrome P450 (P450c17) and NADPH-cytochrome P450 reductase (reductase) but not 3β-hydroxysteroid dehydrogenase/Δ^{5-4} isomerase (3β-HSD). *J Clin Endocrinol Metab* 1999 84: 3382–3385.
84. Voutilainen R, Tapanainen J, Chung B, Matteson KJ, Miller WL. Hormonal regulation of P450scc (20,22-desmolase) and P450c17 (17α-hydroxylase/17,20-lyase) in cultured human granulosa cells. *J Clin Endocrinol Metab* 1986 63: 202–207.
85. Sklar CA, Kaplan SL, Grumbach MM. Evidence for dissociation between adrenarche and gonadarche: Studies in patients with idiopathic precocious puberty, gonadal dysgenesis, isolated gonadotropin deficiency, and constitutionally delayed growth and adolescence. *J Clin Endocrinol Metab* 1980 51: 548–556.
86. Voutilainen R, Miller WL. Developmental expression of genes for the steroidogenic enzymes P450scc (20,22 desmolase), P450c17 (17α-hydroxylase/17,20 lyase) and P450c21 (21-hydroxylase) in the human fetus. *J Clin Endocrinol Metab* 1986 63: 1145–1150.
87. Saenger P, Klonari Z, Black SM *et al.* Prenatal diagnosis of congenital lipoid adrenal hyperplasia. *J Clin Endocrinol Metab* 1995 80: 200–205.
88. Corpéchot C, Robel P, Axelson M, Sjovall J, Baulieu EE. Characterization and measurement of dehydroepiandrosterone sulfate in rat brain. *Proc Natl Acad Sci USA* 1981 78: 4704–4707.
89. Corpéchot C, Synguelakis M, Talha S *et al.* Pregnenolone and its sulfate ester in the rat brain. *Brain Res* 1983 270: 119–125.
90. Compagnone NA, Mellon SH. Neurosteroids: biosynthesis and function of these novel neuromodulators. *Front Neuroendocrinol* 2000 21: 1–58.
91. Seyle H. The anesthetic effects of steroid hormones. *Proc Soc Exp Biol Med* 1941 46: 106–112.
92. Majewska MD, Harrison NL, Schwartz RD, Barker JL, Paul SM. Steroid hormone metabolites are barbiturate-like modulators of the GABA receptor. *Science* 1986 232: 1004–1007.

93. Lambert JJ, Belelli D, Hill-Venning C, Peter JA. Neurosteroids and $GABA_A$ receptor function. *Trends Pharmacol Sci* 1995 16: 295–303.
94. Harrison NL, Simmonds MA. Modulation of the GABA receptor complex by a steroid anaesthetic. *Brain Res* 1984 323: 287–292.
95. Majewska MD, Demirgoren S, Spivak CE, London ED. The neurosteroid dehydroepiandrosterone sulfate is an allosteric antagonist of the $GABA_A$ receptor. *Brain Res* 1990 526: 143–146.
96. ffrench-Mullen JM, Spence KT. Neurosteroids block Ca^{2+} channel current in freshly isolated hippocampal CA1 neurons. *Eur J Pharmacol* 1991 202: 269–272.
97. Urani A, Privat A, Maurice T. The modulation by neurosteroids of the scopolamine-induced learning impairment in mice involves an interaction with σ1 (σ1) receptors. *Brain Res* 1998 799: 64–77.
98. Spivak CE. Desensitization and noncompetitive blockade of $GABA_A$ receptors in ventral midbrain neurons by a neurosteroid dehydroepiandrosterone sulfate. *Synapse* 1994 16: 113–122.
99. Monnet FP, Mahe V, Robel P, Baulieu EE. Neurosteroids, via σ receptors, modulate the [3H]norepinephrine release evoked by *N*-methyl-D-aspartate in the rat hippocampus. *Proc Natl Acad Sci USA* 1995 92: 3774–3778.
100. Baulieu EE, Robel P. Neurosteroids: a new brain function? *J Steroid Biochem Mol Biol* 1990 37: 395–403.
101. Majewska M. Neurosteroids: $GABA_A$-agonistic and $GABA_A$-agonistic modulators of the $GABA_A$ receptor. In Costa E, Paul SM, eds. *Neurosteroids and Brain Function*. New York: Thieme, 1991 8: 109–117
102. Mellon SH. Neurosteroids: biochemistry, modes of action, and clinical relevance. *J Clin Endocrinol Metab* 1994 78: 1003–1008.
103. Lambert JJ, Belelli D, Hill-Venning C, Callachan H, Peters JA. Neurosteroid modulation of native and recombinant $GABA_A$ receptors. *Cell Mol Neurobiol* 1996 16: 155–174.
104. Rupprecht R, Holsboer R. Neuroactive steroids: mechanisms of action and neuropsychopharmocological perspectives. *Trends Neurosci* 1999 22: 410–416.
105. Koenig HL, Schumacher M, Ferzaz B *et al.* Progesterone synthesis and myelin formation by Schwann cells. *Science* 1995 268: 1500–1503.
106. Desarnaud F, Do Thi AN, Brown AM *et al.* Progesterone stimulates the activity of the promoters of peripheral myelin protein-22 and protein zero genes in Schwann cells. *J Neurochem* 1998 71: 1765–1768.
107. Chan JR, Phillips LJN, Glaser M. Glucocorticoids and progestins signal the initiation and enhance the rate of myelin formation. *Proc Natl Acad Sci USA* 1998 95: 10459–10464.
108. Melcangi RC, Magnaghi V, Cavarretta I, Martini L, Piva F. Age-induced decrease of glycoprotein P_0 and myelin basic protein gene expression in the rat sciatic nerve: repair by steroid derivatives. *Neuroscience* 1998 85: 569–578.
109. Gee KW. Steroid modulation of the GABA/benzodiazepine receptor-linked chloride ionophore. *Mol Neurobiol* 1988 2: 291–317.
110. Turner DM, Ransom RW, Yang JS, Olsen RW. Steroid anesthetics and naturally occurring analogs modulate the γ-aminobutyric acid receptor complex at a site distinct from barbituates. *J Pharmacol Exp Ther* 1989 248: 960–966.
111. Lan NC, Chen JS, Belelli D, Pritchett DB *et al.* A steroid recognition site is functionally coupled to an expressed GABA(A)-benzodiazepine receptor. *Eur J Phamacol* 1990 188: 403–406.
112. Concas A, Pierobon P, Mostallino MC *et al.* Modulation of γ-aminobutyric acid (GABA) receptors and the feeding response by neurosteroids in *Hydra vulgaris*. *Neuroscience* 1998 85: 979–988.
113. Concas A, Mostallino MC, Porcu P *et al.* Role of brain allopregnanolone in the plasticity of γ-aminobutyric acid type A receptor in rat brain during pregnancy and after delivery. *Proc Natl Acad Sci USA* 1998 95: 13284–13289.
114. Fujii E, Mellon SH. Regulation of uterine γ-aminobutyric acid(A) receptor subunit expression throughout pregnancy. *Endocrinology* 2001 142: 1770–1777.
115. Smith SS, Gong QH, Hsu F-C, *et al.* $GABA_A$ receptor α_4 subunit suppression prevents withdrawal properties of an endogenous steroid. *Nature* 1998 392: 926–929.
116. Bowlby MR. Pregnenolone sulfate potentiation of *N*-methyl-D-aspartate receptor channels in hippocampal neurons. *Mol Pharmacol* 1993 43: 813–819.

117. Wu FS, Gibbs TT, Farb DH. Pregnenolone sulfate: a positive allosteric modulator at the *N*-methyl-D-aspartate receptor. *Mol Pharmacol* 1991 40: 333–336.
118. Park-Chung M, Wu FS, Farb DH. 3α-Hydroxy-5β-pregnan-20-one sulfate: a negative modulator of the NMDA-induced current in cultured neurons. *Mol Pharmacol* 1994 46: 146–150.
119. Fahey JM, Lindquist DG, Pritchard G, Miller LG. Pregnenolone sulfate potentiation of NMDA-mediated increase in intracellular calcium in cultured chick cortical neurons. *Brain Res* 1995: 669: 183–188.
120. Kimonides VG, Khatibi NH, Svendsen CN, Sofroniew MV, Herbert J. Dehydroepiandrosterone (DHEA) and DHEA-sulfate (DHEAS) protect hippocampal neurons against excitatory amino acid-induced neurotoxicity. *Proc Natl Acad Sci USA* 1998 95: 1852–1857.
121. Guarneri P, Russo D, Cascio C *et al.* Pregnenolone sulfate modulates NMDA receptors, inducing and potentiating acute excitotoxicity in isolated retina. *J Neurosci Res* 1998 54: 787–797.
122. Brinton RD. The neurosteroid 3α-hydroxy-5α-pregnan-20-one induces cytoarchitectural regression in cultured fetal hippocampal neurons. *J Neurosci* 1994 14: 2763–2774.

Chapter 14

Aromatase: glandular and extraglandular activity

Linda R. Nelson and Serdar E. Bulun

INTRODUCTION

In reproductive-aged women, estrogen, a C_{18} steroid, has many functions in its traditional role as an endocrine signal from the granulosa cells of the ovary. The physiological functions of estrogen in women include development of secondary sexual characteristics, regulation of gonadotropin secretion for ovulation, preparation of tissues for progesterone response, maintenance of bone mass, regulation of lipoprotein synthesis, prevention of urogenital atrophy, regulation of insulin responsiveness and maintenance of cognitive function. The critical functions of estrogen in men, however, were largely unanticipated until recently, when this steroid was found to be essential for fusion of epiphyses and maintenance of bone mass in young adult men.[1–3] An enzyme termed aromatase in the endoplasmic reticulum of cells catalyzes the biosynthesis of estrogen from androgens. The principal sites of aromatase expression in women are the ovarian granulosa cells in the premenopausal woman, the placental syncytiotrophoblast in the pregnant woman and the adipose and skin fibroblasts in the postmenopausal woman.[4] However, in both postmenopausal women and men, aromatization of C_{19} steroids in peripheral tissues (adipose and skin) is the primary mechanism for estrogen formation.[5,6] In men, it has been estimated that testicular steroidogenesis accounts for 15% of the circulating level of estrogen.[6] Aromatase expression in the ovary, placenta, adipose tissue and skin gives rise to the formation of sufficient amounts of estrogen to allow measurement in the peripheral blood. However, the local tissue concentrations of estrogen may be significantly higher owing to local aromatization, and, thus, there may be intracrine or paracrine effects in addition to the classical endocrine effects. In all of these sites of estrogen production, the substrate for aromatase (i.e. the steroid precursor) is synthesized in a different cell type from that expressing aromatase. This precursor steroid-producing cell may be distal or adjacent; for example, in the case of adipose tissue and skin, the precursor is distal, since it is primarily circulating androstenedione derived from the adrenal cortex. In the case of the placenta, the principal precursor is 16α-hydroxyandrostenedione, derived from 16α-hydroxydehydroepiandrosterone sulfate produced by the concerted actions of the fetal adrenal and fetal liver. On the other hand, in ovarian granulosa cells, the precursor is derived primarily from the theca cells that surround the preovulatory follicle.[4] In the brain, various cell types including neurons and astrocytes express aromatase, and testosterone was shown to stimulate aromatase activity in brain tissue.[7–9] In Japanese quails and mice, brain aromatase expression was correlated with increased sexual activity, which was more apparent in males compared with females.[7,10,11] The role of brain aromatase in human sexual behavior is currently not known. There has been a recent surge of interest in the possible role that estrogen plays in the cognitive functioning of the human brain. The short-term effects of sudden estrogen loss on cognitive tests and the preliminary evidence that there may be a positive effect of postmenopausal estrogen replacement on the prevention of dementia are intriguing. If the role of estrogen is confirmed with further research, then brain aromatase may play a role in the cognitive function.[12] Finally, aromatase expression in human bone was demonstrated in osteoblasts, chondrocytes and adipose fibroblasts.[13,14] The physiological role of local aromatase expression in the bone in the prevention of osteoporosis is currently not known. Note that the terms ‘extraglandular’ and ‘peripheral’ are used interchangeably for some of the body sites of estrogen formation in this chapter, and denote tissues other than the ovary and placenta. It is implied that extraglandular (or peripheral) tissues do not contain a full complement of steroidogenic enzymes, and thus are dependent on circulating C_{19} steroids, such as androstenedione, as substrate for aromatase activity on circulating C_{19} steroids, such as androstenedione. (Based on this definition, the placenta may also be considered extraglandular. To avoid confusion, however, we continue to use the words extraglandular or peripheral in reference to tissues other than the ovary and placenta, as originally described.) Features of aromatase activity in different tissues are discussed, and the relevance to selected disorders in both women and men is outlined.

AROMATASE EXPRESSION AND REGULATION

In the human, usage of alternative and partially tissue-specific promoters in the placenta (promoter I.1), adipose tissue (promoters I.4, I.3 and II) and ovary (promoter II) regulates aromatase expression. Activation of these promoters, and thus aromatase expression, in these tissues is controlled by various hormones.[4] In ovarian granulosa cells, follicle stimulating hormone (FSH) stimulates the activation of promoter II via a cyclic adenosine monophosphate (cAMP)-dependent signalling pathway. In adipose fibroblasts, glucocorticoids together with members of the interleukin (IL)-6 cytokine family give rise to activation of promoter I.4, whereas treatment with cAMP analogs or prostaglandin E_2 (PGE_2) switches the promoter use to I.3 and II in these cells. In the brain, a tissue-specific promoter (1f) is used along with fibroblast (I.4) and ovarian (II) type promoters.[15] Cellular and molecular mechanisms responsible for the regulation of brain aromatase expression are not well delineated. The fibroblast type promoter (I.4) is primarily used for aromatase expression in bone.[13,14] As in adipose fibroblasts, promoter I.4 in osteoblastic cells is activated by cytokines in the presence of glucocorticoids.[14] Thus, aromatase expression in the human is regulated by extremely complex molecular mechanisms in a cell-specific manner. However, it should be noted that, regardless of the promoter used, the encoded aromatase protein is identical in all these cells. Consequently, alternative promoter use does not affect the structure of aromatase enzyme that catalyzes estrogen formation.

AROMATASE EXPRESSION IN ADIPOSE TISSUE

MacDonald and co-workers first addressed the significance of human adipose tissue as a major source of estrogen production and demonstrated that, in both women and men, with advancing age, there is a progressive increase in the efficiency with which circulating androstenedione is converted to estrone.[16] We have subsequently shown that this age-related elevation of peripheral estrogen formation is associated with increases in both aromatase activity and mRNA levels in the adipose tissue.[17,18] This is congruent with the observation that aromatase activity in adipose stromal cells is regulated primarily by changes in the level of mRNA encoding aromatase.[19] In addition, the peripheral conversion rate of androstenedione to estrone and adipose tissue aromatase activity and mRNA levels all increased by a similar factor (two- to four-fold) when women in their 20s were compared with those in their 60s. This correlation clearly suggests that the primary site of peripheral estrogen formation in older women is the adipose tissue. Aromatase mRNA levels were found to increase significantly with advancing age in adipose tissue samples from all body sites tested, including buttocks, thighs and abdomen. The highest adipose tissue aromatase mRNA levels were found in the hip, followed by the thigh, and the lowest in the abdomen. Adipose tissue aromatase expression in these three sites was primarily under the control of promoter I.4, which is regulated by cytokines and glucocorticoids.[20] In a separate group of 12 women who ranged in age from 17 to 67 years, breast adipose tissue aromatase mRNA levels showed a tendency to increase with advancing age ($R = 0.569$); however, the p value (0.053) for this correlation remained at the limit of statistical significance.

Obesity *per se* was not an independent factor affecting aromatase mRNA levels in the adipose tissue. Thus, the increase in extraglandular estrogen formation as a function of obesity is probably due to an increased bulk of adipose tissue.[17,18] Both aromatase activity and mRNA in adipose tissue occur primarily in adipose fibroblasts, but not in mature adipocytes.[21,22] The fibroblast-to-adipocyte ratios in the hip, abdomen and breast, however, showed neither age-related nor region-specific differences.[23] Therefore, the observed increases in aromatase expression with advancing age or by region were not found to result from changes in fibroblast-to-adipose ratios. Specific aromatase expression per fibroblast may increase with age. One possible mediator may be serum levels of one of the most potent inducers of aromatase activity in adipose fibroblasts, namely IL-6, since these were noted to increase with advancing age.[24–26]

AROMATASE EXPRESSION IN SKIN

Aromatase expression in the skin occurs primarily in hair follicles and sebaceous glands.[27] Upon placement of pieces of skin in a culture dish, fibroblast-like cells grow as a molecular culture to confluence over a few weeks, and express aromatase. The exact origin of these cells in skin tissue, however, is not known. Glucocorticoids, cAMP analogs, growth factors and cytokines[28] regulate aromatase expression in these cells. Glucocorticoids in the presence of serum (or possibly cytokines) induce aromatase expression via promoter I.4.[29] Finally, the expression of 17β-hydroxysteroid dehydrogenase type 1, the enzyme that catalyzes the conversion of estrone to the potent estrogen estradiol, was demonstrated in the skin.

The relationship of aromatase expression in the skin to advancing age has not been studied extensively. However, aromatase expression in skin fibroblasts possibly accounts for a significant proportion of peripheral estrogen formation in postmenopausal women. The relationship between peripheral aromatization and body weight is highly suggestive of a significant role of adipose tissue in this process, but an increase in weight also results in an increase of body surface area; thus, it is very difficult accurately to estimate the relative contributions of these two tissues to extraglandular estrogen formation.

AROMATASE EXPRESSION IN BONE

Aromatase expression has been demonstrated in osteoblasts and chondrocytes from both fetal and adult bone tissues.[13,14] However, another group of investigators failed to demonstrate aromatase mRNA in normal bone tissue from the femoral neck.[30] Aromatase expression in cultured osteoblastic cells seems to be regulated by glucocorticoids and cytokines primarily via promoter I.4.[15,31] Activities of both aromatase and a reductive type 17β-hydroxysteroid dehydrogenase, which permit local estradiol formation, were demonstrated in osteoblastic cells.[32] In summary, the presence or absence of sufficient levels of aromatase activity in bone tissue *in vivo* is still controversial. It is, however, tempting to speculate that postmenopausal bone loss may be modified by levels of local aromatase activity in this tissue.

AROMATASE EXPRESSION IN BRAIN

Aromatase is expressed in many regions of the brain.[7,9,33] Aromatase expression in brain tissues of birds and rodents regulates the activation of sexual behavior.[7,10,11] Aromatase is expressed in many sites and cell types in the human brain, although its exact role is still under investigation.[9,34] Aromatase expression in this tissue seems to be primarily regulated by a brain-specific promoter termed 1f.[15] The role of aromatase expression in the brain in relation to the cognitive health of reproductive-aged and postmenopausal women is an exciting concept, and further research is needed in this field. Several recent studies were highly suggestive of a beneficial role of postmenopausal estrogen replacement in improved cognition, reduced risk for dementia and improvement in the severity of dementia.[12]

AROMATASE EXPRESSION IN ENDOMETRIOSIS

The delivery of estrogen to endometriotic implants has been assumed to occur via the peripheral circulation in an endocrine fashion. We and others, however, have recently demonstrated markedly high levels of aromatase P450 mRNA and activity in pelvic endometriotic implants.[35,36] Moreover, PGE_2, which is produced in very high levels in endometriotic tissues, was found to be the most potent inducer of aromatase activity in endometriosis-derived stromal cells.[37] The production of PGE_2 in eutopic endometrial stromal cells, in turn, was demonstrated to be greatly stimulated by cytokines and 17β-estradiol via enhancement of cyclo-oxygenase-2 (COX-2) expression.[38–40] Additionally, endometriotic tissue expresses 17β-hydroxysteroid dehydrogenase type 1 that is capable of converting the product of the aromatase reaction, estrone, to the most potent estrogen, estradiol. Collectively, these data are supportive of the model that alterations in the expression of aromatase and COX-2 in endometriosis may lead to increased local concentrations of estradiol. In fact, higher concentrations of estradiol have been detected in the peritoneal fluid of women with endometriosis, compared with normal controls.[38] This would represent a mechanism whereby the endometriotic implant stimulates its own local estrogen production in a paracrine fashion.

Endometriosis in postmenopausal women represents an extremely aggressive form of this disease, characterized by complete progesterone resistance and extraordinarily high levels of aromatase expression.[41,42] We evaluated a 57-year-old postmenopausal woman who presented with recurrent severe endometriosis after hysterectomy and bilateral salpingo-oophorectomy. Two additional laparotomies were performed, owing to severe pelvic pain and bilateral ureteral obstruction giving rise to left renal atrophy and right hydronephrosis. Recurrent pelvic endometriosis evident by a 30-mm vaginal lesion visible on speculum examination did not respond to an oral megestrol acetate treatment of 4 months. Anastrozole (an aromatase inhibitor) was used to treat endometriosis in this woman. Anastrozole (1 mg daily) was administered orally for 9 months together with calcium and alendronate (a non-estrogenic inhibitor of bone resorption). Circulating levels of estradiol were reduced to approximately 50% of the baseline value after treatment with anastrozole. Pain rapidly decreased and completely disappeared after the second month of treatment, and the vaginal lesion regressed by the end of 9 months of treatment. Interestingly, the high pretreatment levels of aromatase mRNA noted in the endometriotic tissue biopsy became undetectable in a rebiopsy specimen after 6 months of treatment. The main side-effect was a decrease in the lumbar spine bone density by 6.2% after 9 months of treatment. No other side-effects were noted. The use of an aromatase inhibitor in the treatment of endometriosis in this patient was extraordinarily successful in elimination of pain, and near-complete eradication of implants associated with severe endometriosis not responsive to other therapy. The occurrence of significant bone loss, despite the addition of alendronate to the treatment regimen in this particular case, should be studied further in large clinical trials. In addition to the expected inhibition of aromatase enzyme activity by anastrozole, the reason for the disappearance of aromatase mRNA expression in the lesion may be that absence of estrogen at the level of the lesion should decrease the local biosynthesis of PGE_2, which, in turn, would decrease aromatase expression. We conclude that the recently developed potent aromatase inhibitors are candidate drugs in the treatment of endometriosis that is resistant to standard regimens.

AROMATASE EXPRESSION IN OTHER ENDOMETRIAL DISEASES

Müllerian-derived tissues are targets of estrogen action. Because aromatase is expressed in extraglandular tissues, we have investigated the regulation of expression of the aromatase gene in estrogen-dependent neoplasia or disorders that involve Müllerian-derived tissues. First, we were unable to detect any aromatase activity or mRNA in disease-free endometrium.[43] On the other hand, aromatase expression was demonstrable in the disease states of this tissue, namely endometriosis and endometrial cancer.[35,44] Endometrial cancer occurs mostly in postmenopausal women, and grows in response to estrogen. MacDonald, Siiteri and co-workers in the 1970s demonstrated that the fractional conversion of androstenedione to estrone in humans increases as a function of obesity and aging,[5,16,45] and, thus, extraglandular tissues (adipose and skin) are a source of estrogens in postmenopausal women. This parameter was shown to correlate positively with excess body weight in both pre- and postmenopausal women, and may be increased as much as ten-fold in morbidly obese postmenopausal women. Furthermore, the fractional conversion of androstenedione to estrone is also increased with aging. This increase with both obesity and aging bears a striking relationship to the incidence of endometrial cancer, which is more commonly observed in elderly obese women.[46] It is now generally accepted that the continuous production of estrogen by the adipose tissue in such women is one of the causative factors of this condition. Thus, understanding the tissue sites of estrogen formation in postmenopausal women with endometrial cancer is extremely important. Aromatase expression in peripheral tissues (adipose and skin) may give rise to sufficient blood levels of estradiol to stimulate postmenopausal endometrial growth and uterine bleeding, which is an endocrine effect.[5,18,46] One may envision that estrogen of peripheral origin may also cause endometrial hyperplasia and cancer. In addition, local aromatase activity in endometrial cancer may give rise to extremely high local concentrations of estrogen to stimulate growth further (paracrine or intracrine effects). The relative contributions of estrogen formed peripherally and locally are not known at the moment. These data, however, are strongly suggestive that aromatase inhibitors may have a significant role in the future treatment of endometrial cancer.

AROMATASE EXPRESSION IN BREAST DISEASE

Evidence is also suggestive of a role of estrogens produced by adipose tissue in the pathogenesis of breast cancer. For example, it was shown that aromatase activity of breast adipose tissue obtained from sites proximal to a tumor was higher than aromatase activity of breast adipose tissue removed from sites distal to the tumor.[47] We confirmed these findings by demonstrating colocalization of highest levels of aromatase mRNA to quadrants bearing tumors, compared with other quadrants, in 70% of mastectomy specimens.[48] The usefulness of estrogen antagonists as well as inhibitors of aromatase in the management of breast cancer has long been recognized. The implication of adipose tissue estrogen biosynthesis in the development of breast cancer is apparent from the palliative effects of adrenalectomy. By decreasing the substrate, circulating androstenedione, adrenalectomy can suppress adipose tissue estrogen biosynthesis. The product of aromatase activity in adipose tissue, namely estrone, is only weakly estrogenic, and must therefore be converted locally to estradiol to attain full estrogenic activity. Evidence from several laboratories is indicative that 17β-hydroxysteroid dehydrogenase type 1 present in breast adipose and tumor tissues is capable of this conversion.[49,50]

Epidemiological studies are indicative of strong environmental factors in the incidence of breast cancer. Notably indicated are Western-style diets with high fat content, diets that predispose to obesity.[51] Fat distribution may be a contributing factor in the etiology of breast cancer. Increased waist-to-hip ratio in women was implicated to be a risk factor for breast cancer.[52] However, conflicting results also exist.[53] Our report is in agreement with those of a previous study in that the adipose tissue of the buttocks contains much higher levels of aromatase mRNA than that of the abdomen or thighs.[18,54] Therefore, even if postmenopausal women with central obesity are at higher risk for breast cancer, there may be additional factors from the central adipose tissue that interact to potentiate the effects of estrogens on the breast.

The breast stroma is largely composed of adipose tissue, which is a blend of mature adipocytes and undifferentiated fibroblasts. Vascular and neural tissues constitute very small proportions of the breast stroma. Aromatase expression in the breast almost exclusively occurs in the fibroblast component of the adipose tissue.[21,22] The fibroblast-to-adipocyte ratio in adipose tissue displays large variations from one region to another within the same breast, or from one individual to another.[55] These variations in the distribution of fibroblasts determine local estrogen biosynthesis in the breast.[48,55] Local estrogen biosynthesis in breast adipose tissue may influence the growth of breast tumors. O'Neill and colleagues demonstrated that the breast quadrant displaying the highest level of aromatase activity is most frequently the one containing the tumor.[56] This suggested that breast cancers develop preferentially in anatomical sites in which there is the highest aromatase activity. We confirmed this result by showing a significant positive correlation between adipose tissue aromatase mRNA levels and the presence of a malignancy in breast quadrants.[48]

We also quantified by computerized morphometry the histological components of adipose tissue samples from each quadrant in these mastectomy specimens. The distribution of fibroblast-to-adipocyte ratios significantly correlated with the distribution of aromatase mRNA levels, in those quadrants containing highest proportions of fibroblasts matched to highest mRNA levels.[48] Thus, breast quadrants bearing malignant tumors contain the highest numbers of fibroblasts. Moreover, in breast adipose tissues of cancer-free women, the highest fibroblast-to-adipocyte ratios and aromatase mRNA transcript levels were found in the outer regions.[55] This distribution pattern correlates positively with the most common site of carcinoma in the breast in large series, which is the outer region. Since breast cancer occurs in regions of the breast with the highest levels of aromatase expression, the presence of high fibroblast content and increased aromatase expression in the outer region of the disease-free breast may predispose this site to the development of breast cancer.

As indicated above, tissue-specific expression of the aromatase gene is determined, in part, by the use of tissue-specific promoters. Adipose tissue employs the cytokine plus glucocorticoid-induced promoter I.4 or the cAMP-induced promoters I.3 and II for aromatase expression. There is evidence that promoter switching is a mechanism contributing to the increased aromatase expression in adipose tissue surrounding a breast tumor. Aromatase gene transcription in breast adipose tissue of cancer-free individuals utilizes preferably promoter I.4, implicating a role of glucocorticoids and members of the IL-6 cytokine family. Breast adipose tissue containing a carcinoma preferentially utilizes promoters II and I.3 for aromatase expression. These are regulated by factors that lead to increased cAMP formation. These factors are presumably secreted by the malignant epithelial cells.[57–59] One candidate factor is PGE_2, since this prostaglandin is a very potent stimulator of aromatase expression in cultured adipose fibroblasts via promoter II.[60] The *in vivo* role of PGE_2 in aromatase expression in adipose fibroblasts proximal to a breast cancer has not been fully elucidated. The search for cAMP-dependent mechanisms responsible for aromatase excess in breast cancer, therefore, continues. However, there is no doubt that there is a role for local estrogen production, and, clinically, aromatase inhibitors have been widely used to treat breast cancer.[61]

AROMATASE DEFICIENCY STATES

Until the 1990s, aromatase deficiency had been considered incompatible with life, and this belief may have inhibited the investigation of this diagnosis in suspected cases. Following the first description in 1991 of a Japanese newborn girl with an aromatase P450 (P450arom) gene defect,[61,62] there have been several recent reports in the world literature describing aromatase deficiency.[1,2,61,63,64] Estrogen biosynthesis in all these patients was virtually absent, giving rise to a number of anticipated as well as unexpected symptoms. As a result, we now know that aromatase deficiency is an autosomal recessive condition manifest in 46,XX fetuses in the form of female pseudohermaphroditism, and, in the case of adult men, tall stature with eunuchoid proportions due to unfused epiphyses. In fact, the essential role of estrogen as a determinant of height and bone mass was understood only after the description of an estrogen-resistant[3] and two aromatase-deficient men.[1,2] The male phenotype resulting from aromatase deficiency is similar in many aspects to the phenotype described in a case of estrogen receptor deficiency. Estrogens also appear to play a role in glucose metabolism. Smith and colleagues[3] described an extremely tall man with estrogen resistance who had incomplete epiphyseal closure, with a history of continued linear growth into adulthood, osteoporosis and glucose intolerance. This was caused by a homozygous point mutation in the estrogen receptor gene.

The two men with aromatase deficiency thus far described went through an unremarkable pubertal development.[1,2] Although their heights were regarded as normal at adolescence, the linear growth continued into adulthood. When first evaluated for aromatase deficiency, they were 24 and 38 years old and 204 and 190 cm tall, and both had severely retarded bone ages and eunuchoid body proportions. Diffuse bone demineralization and lack of epiphyseal fusion were detected by X-radiography in the knees and pelvis. Epiphyseal fusion and maintenance of bone mass appear to be two consequences of estrogen action in men. Estrogens have been shown to exert their effects on osteoblasts and osteoclasts through regulation of cytokine expression in bone.[65,66] Thus, it is conceivable that the low plasma concentrations of estrogen found ordinarily in normal men may sufficiently suppress the expression of osteoporotic cytokines in bone. In addition, the physiological role of local production of estrogen from the large circulating pool of androgens within bone cells is not known as yet. A synergistic role, at the level of the bone-forming units, for testosterone at amounts found in adult men in this antiresorptive action of estrogen is also very likely.

There were also gonadal abnormalities in the above two men: macro-orchidism in one[2] and micro-orchidism in the other.[1] The aromatase-deficient man with micro-orchidism also suffered from infertility due to severe oligoasthenospermia, although his two brothers without P450arom gene defects were also found to have similar sperm problems. It is likely, therefore, that the severe oligoasthenospermia in this aromatase-deficient man resulted from another familial disorder in addition to the proband aromatase deficiency. Luteinizing hormone (LH), FSH and testosterone levels in both aromatase-deficient patients were markedly elevated.[1,2]

Another crucial role of estrogens in men has recently been demonstrated to be the regulation of lipoprotein synthesis.[67] Aromatase inhibitor treatment of normal men was observed to decrease serum high-density lipoprotein (HDL) cholesterol levels.[67] An abnormally low HDL/low-density lipoprotein (LDL) cholesterol ratio in an aromatase-deficient man[2] emphasizes this critical role of estrogen action in men, which is the prevention of cardiovascular disease. In this case, fasting insulin and total LDL cholesterol levels were reported to be increased, whereas HDL cholesterol was low. Despite evidence of insulin resistance, glucose levels were normal and estrogen administration lowered insulin levels.[2] Although the mechanism of insulin resistance in aromatase deficiency is not known, it is not likely to be caused by elevated testosterone, because the estrogen-resistant man[3] had a much more severe degree of insulin resistance despite a normal serum testosterone level.

The role of estrogen in the sex steroid–gonadotropin feedback mechanism is clarified by observations in patients with aromatase deficiency. For example, the hypergonadotropic state during the first 2 years of life of an aromatase-deficient girl was markedly exaggerated, emphasizing the negative feedback role of ovarian estrogen during this period. In aromatase-deficient men, elevated concentrations of serum FSH and LH, despite the strikingly high testosterone level, support an important negative feedback role of estrogen, either of peripheral origin or synthesized locally in the brain from C_{19} steroid precursors or from both sources, in the regulation of FSH and LH.[68,69] In fact, levels of serum gonadotropins and testosterone decreased after low-dose estrogen replacement in an aromatase-deficient man.[1]

The possible impact of lack of estrogen action in the brain on the psychosexual development of both women and men with aromatase deficiency was considered in two reports.[2,63] Thus far, no evidence is available to suggest gender-identity problems in these patients. At this point, it appears that the libido of these patients has not been adequately evaluated.

Finally, one can envision mutations in the *CYP19* (P450arom) gene causing milder degrees of aromatase deficiency analogous to other steroidogenic P450 defects, for example late-onset 21-hydroxylase deficiency. These may give rise to a whole host of symptoms with varying degrees of severity, such as polycystic ovary disease and incomplete pubertal development in women, and infertility, insulin resistance, tallness with eunuchoid proportions and tendency to develop cardiovascular disease in both men and women. In addition, since tissue-specific expression of the *CYP19* gene is regulated in part by use of tissue-specific promoters,[4] mutations in any of these promoter regions or splice-junctions may result in loss of estrogen formation in one organ only, such as the placenta, ovary, brain or adipose tissue. Analysis of such cases would provide insight into the relative roles of estrogens produced in each of these sites.

SUMMARY

Aromatase is a key enzyme for estrogen formation in human tissues. In women during their reproductive years, the ovarian granulosa cells are an important site of estrogen formation for local as well as endocrine signalling to a host of tissues, including the uterus, skin appendages, brain, bone, vascular system and breast. There is, however, an increasing awareness of the importance of local production of estrogens from precursor androgens within these target tissues themselves. This process occurs in both men and women. In men, aging is accompanied by a much slower decrement in androgen and estrogen levels in circulating and extraglandular compartments. In women, when ovarian aromatase expression is halted after the menopause, age-related increases in aromatase expression in peripheral tissues (adipose and skin) modify the severity of estrogen deficiency by elevating plasma estrogen levels. In fact, in obese women, elevated circulating estradiol may persist at sufficient levels to cause postmenopausal uterine bleeding, endometrial hyperplasia and even cancer. Local expression of aromatase in other tissues (brain and bone) may also have physiologically significant consequences, such as maintenance of cognitive function and bone mass. In disease states that are exacerbated by estrogen stimulation, the standard therapy has been to suppress the endocrine source of the hormone, for example using a gonadotropin-releasing hormone analog to suppress estradiol production by the ovary in patients with estrogen-dependent diseases such as endometriosis. However, this therapy does not impact upon the local production of estradiol within the tissue itself, and thus may not adequately address the problem. In the future, it is likely that there will be therapies designed to address both the circulating and the local sources of estrogen, to maximize the response. In addition, perhaps the combination of inhibitors of aromatase activity and the selective agonist/antagonist receptor modulators will allow us to regulate estrogen in selected tissues more precisely. These therapies of the future hold promise for maintaining health and treating disease in both men and women.

ACKNOWLEDGEMENT

The authors thank Margarita Guerrero for expert editorial assistance.

REFERENCES

1. Caranci C, Qin K, Simoni M *et al.* Effect of testosterone and estradiol in a man with aromatase deficiency. *N Engl J Med* 1997 337: 91–95
2. Morishima A, Grumbach MM, Simpson ER, Fisher C, Qin K. Aromatase deficiency in male and female siblings caused by a novel mutation and the physiological role of estrogens. *J Clin Endocrinol Metab* 1995 80: 3689–3698.
3. Smith EP, Boyd J, Frank GR *et al.* Estrogen resistance caused by a mutation in the estrogen-receptor gene in a man. *N Engl J Med* 1994 331: 1056–1061.
4. Simpson ER, Mahendroo MS, Means GD *et al.* Aromatase cytochrome P450, the enzyme responsible for estrogen biosynthesis. *Endocr Rev* 1994 15: 342–355.
5. Grodin JM, Siiteri PK, MacDonald PC. Source of estrogen production in postmenopausal women. *J Clin Endocrinol Metab* 1973 36: 207-214.
6. MacDonald PC, Madden JD, Brenner PF, Wilson JD, Siiteri PK. Origin of estrogen in normal men and in women with testicular feminization. *J Clin Endocrinol Metab* 1979 49: 905–916.
7. Balthazart J, Foidart A, Hendrick JC. The induction by testosterone of aromatase activity in the preoptic area and activation of copulatory behavior. *Physiol Behav* 1990 47: 83–94.
8. Balthazart J, Absil P. Identification of catecholaminergic inputs to and outputs from aromatase-containing brain areas of the Japanese quail by tract tracing combined with tyrosine hydroxylase immunocytochemistry. European Graduate School for Neurosciences, University of Liège, Laboratory of Biochemistry, Belgium. *J Comp Neurol* 1997 382: 402–428.
9. Garcia-Segura LM, Wozniak A, Azcoitia I, Rodriguez JR, Hutchison RE, Hutchison JB. Aromatase expression by astrocytes after brain

injury: implications for local estrogen formation in brain repair [in Process citation]. *Neuroscience* 1999 89: 567–578.

10. Fisher CR, Graves KH, Parlow AF, Simpson ER. Characterization of mice deficient in aromatase (ArKO) because of targeted disruption of the *CYP19* gene. *Proc Natl Acad Sci USA* 1998 95: 6965–6970.
11. Honda S, Harada N, Ito S, Takagi Y, Maeda S. Disruption of sexual behavior in male aromatase-deficient mice lacking exons 1 and 2 of the *CYP19* gene. *Biochem Biophys Res Commun* 1998 252: 445–449.
12. Yaffe K, Dawya G, Lieberburg I, Grady D. Estrogen in postmenopausal women: effects on cognitive function and dementia. *J Am Med Assoc* 1998 279: 688–695.
13. Sasano H, Uzuki M, Sawai T *et al*. Aromatase in human bone tissue. *J Bone Miner Res* 1997 12: 1416–1423
14. Shozu M, Simpson ER. Aromatase expression of human osteoblast-like cells. *Mol Cell Endocrinol* 1998 139: 117–129.
15. Honda S, Harada N, Takagi Y. Novel exon 1 of the aromatase gene specific for aromatase transcripts in human brain. *Biochem Biophys Res Commun* 1994 198: 1153–1160.
16. Hemsell DL, Grodin JM, Brenner PF, Siiteri PK, MacDonald PC. Plasma precursors of estrogen. II. Correlation of the extent of conversion of plasma androstenedione to estrone with age. *J Clin Endocrinol Metab* 1974 38: 476–479.
17. Cleland WH, Mendelson CR, Simpson ER. Effects of aging and obesity on aromatase activity of human adipose cells. *J Clin Endocrinol Metab* 1985 60: 174–177.
18. Bulun SE, Simpson ER. Competitive RT-PCR analysis indicates levels of aromatase cytochrome P450 transcripts in adipose tissue of buttocks, thighs, and abdomen of women increase with advancing age. *J Clin Endocrinol Metab* 1988 7: 1379–1385.
19. Evans CT, Corbin CJ, Saunders CT, Merrill JC, Simpson ER, Mendelson CR. Regulation of estrogen biosynthesis in human adipose stromal cells: effects of dibutyryl cyclic AMP, epidermal growth factor, and phorbol esters on the synthesis of aromatase cytochrome P-450. *J Biol Chem* 1987 262: 6914–6920.
20. Agarwal VR, Ashanullah CI, Simpson ER, Bulun SE. Alternatively spliced transcripts of the aromatase cytochrome P450 (*CYP19*) gene in adipose tissue of women. *J Clin Endocrinol Metab* 1996 82: 70–74.
21. Ackerman GE, Smith ME, Mendelson CR, MacDonald PC, Simpson ER. Aromatization of androstenedione by human adipose tissue stromal cells in monolayer culture. *J Clin Endocrinol Metab* 1981 53: 412–417.
22. Price T, Aitken J, Head J, Mahendroo MS, Means GD, Simpson ER. Determination of aromatase cytochrome P450 messenger RNA in human breast tissues by competitive polymerase chain reaction (PCR) amplification. *J Clin Endocrinol Metab* 1992 74: 1247–1252.
23. Rink JD, Simpson ER, Barnard JJ, Bulun SE. Cellular characterization of adipose tissue from various body sites of women. *J Clin Endocrinol Metab* 1996 81: 2443–2447.
24. Zhao Y, Nichols JE, Bulun SE, Mendelson CR, Simpson ER. Aromatase P450 gene expression in human adipose tissue: role of a Jak/STAT pathway in regulation of the adipose-specific promoter. *J Biol Chem* 1995 270: 16449–16457.
25. Daynes RA, Araneo BA, Ershler WB, Maloney C, Li GZ, Ryu SY. Altered regulation of IL-6 production with normal aging. *J Immunol* 1993 150: 5219–5230.
26. Wei J, Xu H, Davies JL, Hemmings GP. Increase of plasma IL-6 concentration with age in healthy subjects. *Life Sci* 1992 51: 1953–1956.
27. Sawaya ME, Penneys NS. Immunohistochemical distribution of aromatase and 3β-hydroxysteroid dehydrogenase in human hair follicle and sebaceous gland. *J Cutan Pathol* 1992 19: 309–314.
28. Emoto N, Ling N, Baird A. Growth factor-mediated regulation of aromatase activity in human skin fibroblasts. *Proc Soc Exp Biol Med* 1991 196: 351–358.
29. Harada N. A unique aromatase (P-450arom) mRNA formed by alternative use of tissue-specific exons 1 in human skin fibroblasts. *Biochem Biophys Res Commun* 1992 189: 1001–1007.
30. Lea CK, Ebrahim S, Tennant S, Flanagan AM. Aromatase cytochrome P450 transcripts are detected in fractured human bone but not in normal skeletal tissue. *Bone* 1997 21: 433–440.
31. Shozu M, Zhao Y, Bulun SE, Simpson ER. Multiple splicing events involved in regulation of

human aromatase expression by a novel promoter I.6. *Endocrinology* 1998 139: 1610–1617.

32. Eyre LJ, Bland R, Bujalska IJ, Sheppard MC, Stewart PM, Hewison M. Characterization of aromatase and 17β-hydroxysteroid dehydrogenase expression in rat osteoblastic cells. *J Bone Miner Res* 1998 13: 996–1004.
33. Naftolin F, Horvath TL, Jakab RL, Leranth C, Harada N, Balthazart J. Aromatase immunoreactivity in axon terminals of the vertebrate brain. An immunocytochemical study on quail, rat, monkey and human tissue. *Neuroendocrinology* 1996 63: 149–155.
34. Sasano H, Takashashi K, Satoh F, Nagura H, Harada N. Aromatase in the human central nervous system. *Clin Endocrinol (Oxf)* 1998 48: 325–329.
35. Noble LS, Simpson ER, Johns A, Bulun SE. Aromatase expression in endometriosis. *J Clin Endocrinol Metab* 1996 81: 174–179.
36. Kitawaki J, Noguchi T, Amatsu T *et al.* Expression of aromatase cytochrome P450 protein and messenger ribonucleic acid in human endometriotic and adenomyotic tissues but not in normal endometrium. *Biol Reprod* 1997 57: 514–519.
37. Noble LS, Takayama K, Putman JM *et al.* Prostaglandin E_2 stimulates aromatase expression in endometriosis-derived stromal cells. *J Clin Endocrinol Metab* 1997 82: 600–606.
38. DeLeon FD, Vijayakumar R, Brown M, Rao CV, Yussman MA, Schultz GL. Peritoneal fluid volume, estrogen, progesterone, prostaglandin, epidermal growth factor concentrations in patients with and without endometriosis. *Obstet Gynecol* 1986 68: 189–194.
39. Ishihara O, Matsuoka K, Kinoshita K, Sullivan M, Elder M. Interleukin-1β-stimulated PGE_2 production from early first trimester human decidual cells is inhibited by dexamethasone and progesterone. *Prostaglandins* 1995 49: 15–26.
40. Huang JC, Dawood MY, Wu KK. Regulation of cyclooxygenase-2 gene in cultured endometrial stromal cells by sex steroids. *Am Soc Reprod Med* 1996 5-5 (abstr)
41. Lessey BA, Metzger DA, Haney AF, McCarty KS Jr. Immunohistochemical analysis of estrogen and progesterone receptors in endometriosis: comparison with normal endometrium during the menstrual cycle and the effect of medical therapy. *Fertil Steril* 1989 51: 409–415.
42. Takayama K, Zeitoun K, Gunby RT, Sasano H, Carr BR, Bulun SE. Treatment of severe postmenopausal endometriosis with an aromatase inhibitor. *Fertil Steril* 1998 69: 709–713.
43. Bulun SE, Mahendroo MS, Simpson ER. Polymerase chain reaction amplification fails to detect aromatase cytochrome P450 transcripts in normal human endometrium or decidua. *J Clin Endocrinol Metab* 1993 76: 1458–1463.
44. Bulun SE, Economos K, Miller D, Simpson ER. *CYP19* (aromatase cytochrome P450) gene expression in human malignant endometrial tumors. *J Clin Endocrinol Metab* 1994 79: 1831–1834.
45. Edman CD, MacDonald PC. Effect of obesity on conversion of plasma androstenedione to estrone in ovulatory and anovulatory young women. *Am J Obstet Gynecol* 1978 130: 456–461.
46. MacDonald PC, Edman CD, Hemsell DL, Porter JC, Siiteri PK. Effect of obesity on conversion of plasma androstenedione to estrone in postmenopausal women with and without endometrial cancer. *Am J Obstet Gynecol* 1978 130: 448–455.
47. Miller WR. Aromatase activity in breast tissue. *J Steroid Biochem Mol Biol* 1991 39: 783–790.
48. Bulun SE, Price TM, Mahendroo MS, Aitken J, Simpson ER. A link between breast cancer and local estrogen biosynthesis suggested by quantification of breast adipose tissue aromatase cytochrome P450 transcripts using competitive polymerase chain reaction after reverse transcription. *J Clin Endocrinol Metab* 1993 77: 1622–1628
49. James VHT, Reed MJ, Lai LC *et al.* Regulation of estrogen concentrations in human breast tissues. *Ann NY Acad Sci* 1990 595: 373–378
50. Malet C, Vacca A, Kuttnen F, Mauvais-Jarvis PJ. 17β-Estradiol dehydrogenase (E_2DH) activity in T47D cells. *J Steroid Biochem Mol Biol* 1991 39: 769–775
51. Barbosa JC, Shults TD, Filley SJ, Nieman DC. The relationship among adiposity, diet and hormone concentrations in vegetarian and nonvegetarian postmenopausal women. *Am J Clin Nutr* 1990 51: 798–803.
52. Ballard-Barbash R, Schatzkin A, Carter CL *et al.* Body fat distribution and breast cancer in the Framingham Study. *J Natl Cancer Inst* 1990 82: 286–290.

53. Sellers TA, Kushi LH, Potter JD *et al.* Effect of family history, body-fat distribution, and reproductive factors on the risk of postmenopausal breast cancer. *N Engl J Med* 1992 326: 1323–1329.
54. DeRidder CM, Bruning PF, Zonderland ML *et al.* Body fat mass, body fat distribution, and plasma hormones in early puberty in females. *J Clin Endocrinol Metab* 1990 70: 888–893.
55. Bulun SE, Sharda G, Rink J, Sharma S, Simpson ER. Distribution of aromatase P450 transcripts and adipose fibroblasts in the human breast. *J Clin Endocrinol Metab* 1996 81: 1273–1277.
56. O'Neill JS, Elton RA, Miller WR. Aromatase activity in adipose tissue from breast quadrants: a link with tumor site. *Br Med J* 1988 296: 741–743.
57. Zhou C, Zhou D, Esteban J *et al.* Aromatase gene expression and its exon I usage in human breast tumors. Detection of aromatase messenger RNA by reverse transcription-polymerase chain reaction. *J Steroid Biochem Mol Biol* 1996 59: 163–171
58. Harada N, Honda S. Molecular analysis of aberrant expression of aromatase in breast cancer tissues. *Breast Cancer Res Treat* 1998 49(Suppl 1): S15–S21, discussion S33–S37.
59. Agarwal VR, Bulun SE, Leitch M, Rohrich R, Simpson ER. Use of alternative promoters to express the aromatase cytochrome P450 (*CYP19*) gene in breast adipose tissues of cancer-free and breast cancer patients. *J Clin Endocrinol Metab* 1996 81: 3843–3849.
60. Zhao Y, Agarwal VR, Mendelson CR, Simpson ER. Estrogen biosynthesis proximal to a breast tumor is stimulated by PGE_2 via cyclic AMP, leading to activation of promoter II of the *CYP19* (aromatase gene). *Endocrinology* 1996 150: 51S-57S.
61. Shozu M, Akasofu K, Harada T, Kubota Y. A new cause of female pseudohermaphroditism: placental aromatase deficiency. *J Clin Endocrinol Metab* 1991 72: 560–566.
62. Harada N, Ogawa H, Shozu M *et al.* Biochemical and molecular genetic analyses on placental aromatase (P450arom) deficiency. *J Biol Chem* 1992 267: 4781–4785.
63. Conte FA, Grumbach MM, Ito Y, Fisher CR, Simpson ER. A syndrome of female pseudohermaphroditism, hypergonadotropic hypogonadism, and multicystic ovaries associated with missense mutations in the gene encoding aromatase (P450arom). *J Clin Endocrinol Metab* 1994 78: 1287–1292.
64. Portrat-Doyen S, Forest MG, Nicolino M *et al.* Aromatase (P450arom) deficiency associated with a novel mutation (R457X) in the CYP19 gene in a third case of female pseudohermaphroditism (FPH). Presented at the *10th International Congress on Endocrinology*, San Francisco, CA, June 1996: 586.
65. Horowitz MC. Cytokines and estrogen in bone: anti-osteoporotic effects. *Science* 1993 260: 626–627.
66. Jilka RL, Hangoc G, Girasole G *et al.* Increased osteoclast development after estrogen loss: mediation by interleukin-6. *Science* 1992 257: 88–91.
67. Bagatell CJ, Knopp RH, Rivier JE, Bremner WJ. Physiological levels of estradiol stimulate plasma high density $lipoprotein_2$ cholesterol levels in normal men. *J Clin Endocrinol Metab* 1994 78: 855–861.
68. Finkelstein JS, O'Dea LS, Whitcomb RW, Crowley WF. Sex steroid control of gonadotropin secretion in the human male. II. Effects of estradiol administration in normal and GnRH deficient men. *J Clin Endocrinol Metab* 1991 74: 621–628.
69. Bagatell CJ, Dahl KD, Bremner WJ. The direct pituitary effect of testosterone to inhibit gonadotropin secretion in men is partially mediated by aromatization to estradiol. *J Androl* 1994 15: 15–21.

Chapter 15

Steroid hormone receptors

Albert O. Brinkmann

INTRODUCTION

It has become well established in the past three decades that steroid hormones act via regulation of gene expression. Steroid hormones (i.e. androgens, estrogens, glucocorticoids, mineralocorticoids and progestogens) are relatively small, hydrophobic molecules, and enter their target cells by a simple diffusion process. Inside the target cells, their action is mediated by specific nuclear receptor proteins, which belong to a superfamily of ligand-modulated transcription factors that regulate homeostasis, reproduction, development and differentiation. This family includes receptors for steroid hormones, thyroid hormone, all-*trans* and 9-*cis* retinoic acid, 1,25-dihydroxy-vitamin D, ecdysone and peroxisome proliferator-activated receptors.[1] In addition, an increasing number of nuclear proteins have been identified with a protein structure homologous with that of nuclear receptors, but without a known ligand. These so-called 'orphan' receptors form an important subfamily of transcription factors, acting either unliganded or with yet unknown endogenous ligands.[2] Steroid receptor proteins have equilibrium dissociation constants (K_d) in the order of 10^{-9}–10^{-10} mol/l, and a typical steroid target cell contains about 10 000 receptor molecules.

The original cytoplasmic localization of steroid receptors in a target cell was based on the observation that, in the absence of hormone in target tissues from either adrenalectomized, ovariectomized, castrated or immature animals, the receptor could be isolated from the 'cytosol' fraction. This subcellular localization has resulted in a two-step model for steroid hormone action: first, binding of the hormone to the cytoplasmic receptor, and second, transformation of the hormone–receptor complex to the DNA-binding state and simultaneous translocation of the complex to the nucleus.[3] With the availability of specific antibodies for steroid hormone receptors and more sophisticated subcellular fractionation techniques, it became clear that the two-step model had to be re-evaluated, because most steroid receptors reside in the nucleus in the absence of hormone and undergo a conformational change to a tight nuclear-binding state upon binding of the hormone.[4]

The current model for steroid hormone action, moreover, involves a multistep mechanism as depicted in Figure 1. Upon entry of the steroid into the target cell, binding to the cognate receptor takes place, followed by dissociation of heat shock proteins in the cytoplasm (see section relating to heat shock proteins below), accompanied simultaneously by a conformational change of the receptor protein resulting in a transformation and a translocation to the nucleus. This mechanism reflects the situation for the glucocorticoid receptor. The estradiol and vitamin D receptors reside predominantly in the nucleus in the absence of hormone. These receptors become activated in the nucleus upon binding of their ligands. For the androgen and progesterone receptors, an intermediary mechanism has been proposed. Upon binding in the nucleus to specific DNA sequences, the receptor dimerizes with a second molecule, and the homodimer entity recruits further additional proteins (coactivators, general transcription factors, RNA–polymerase II), resulting in specific activation of transcription at discrete sites on the chromatin.

BACKGROUND

Functional domain structure of steroid receptors

Since the first cloning in 1985 of a full-length cDNA of a steroid hormone receptor (the glucocorticoid receptor), considerably more information has become available with respect to the molecular structure of steroid hormone receptors.[5–13] Comparative structural and functional analysis of nuclear hormone receptors has revealed a common structural organization in four different functional domains: an NH_2-terminal (N-terminal) domain, a DNA-binding domain, a hinge region and a ligand-binding domain (Figure 2).

The non-conserved N-terminal domain is involved in cell type-specific transcriptional regulation and is very variable in size (Figure 2). This domain also appears to be highly immunogenic.[14] The DNA-binding domain is the best conserved among the members of the receptor superfamily. It is characterized by a high content of basic amino

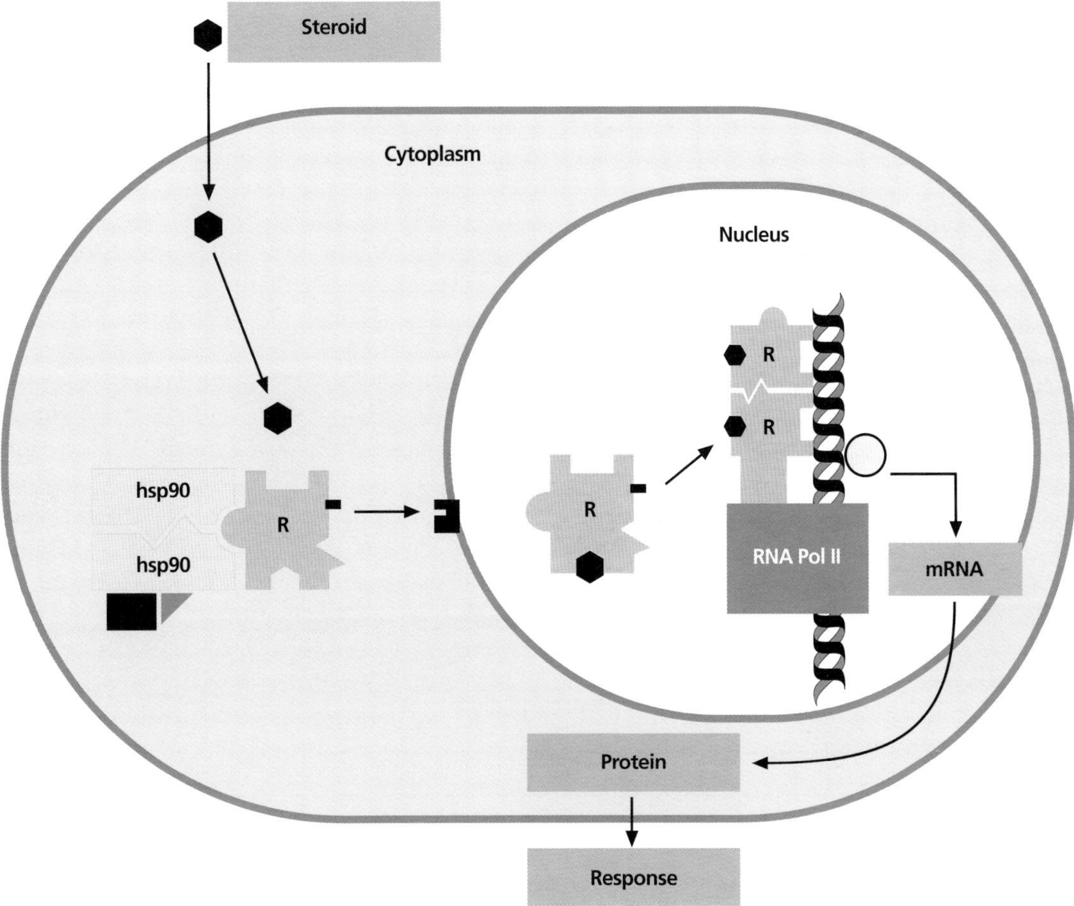

Figure 1 Simplified model of steroid hormone action. The key protein is the steroid hormone receptor (R, in dark blue), which binds heat shock proteins (hsp90, in pale blue). The receptor enters the nucleus via an intrinsic nuclear localization signal. Upon steroid hormone binding, which may occur either in the cytoplasm or in the nucleus, the heteromeric receptor–hsp90 complex dissociates, and subsequently the receptor binds as a dimer to specific DNA sequences. The binding to DNA at specific sites triggers mRNA synthesis and consequently protein synthesis, which finally results in a response. RNA Pol II, RNA–polymerase II

acids and by nine conserved cysteine residues. Detailed structural information has been published on the crystal structure of the DNA-binding domain of the glucocorticoid receptor complexed with DNA.[15] This structural information might also be representative for the other members of the steroid hormone receptor family. Briefly, the DNA-binding domain has a compact, globular structure in which two substructures can be distinguished. Both substructures contain centrally one zinc atom, which interacts via co-ordination bonds with four cysteine residues (Figure 3). The two zinc co-ordination centers are both C-terminally flanked by an α-helix.[15] The two zinc clusters are structurally and functionally different and are encoded by two different exons. The α-helix of the most N-terminal-located zinc cluster interacts directly with nucleotides of the hormone response element in the major groove of the DNA. Three amino acid residues at the N-terminus of this α-helix are responsible for the specific recognition of the DNA sequence of the responsive element[16] (Figure 3). These three amino acid residues (Gly, Ser, Val) are identical in the androgen, progesterone, glucocorticoid and mineralocorticoid receptors, and differ from the residues at the homologous positions in

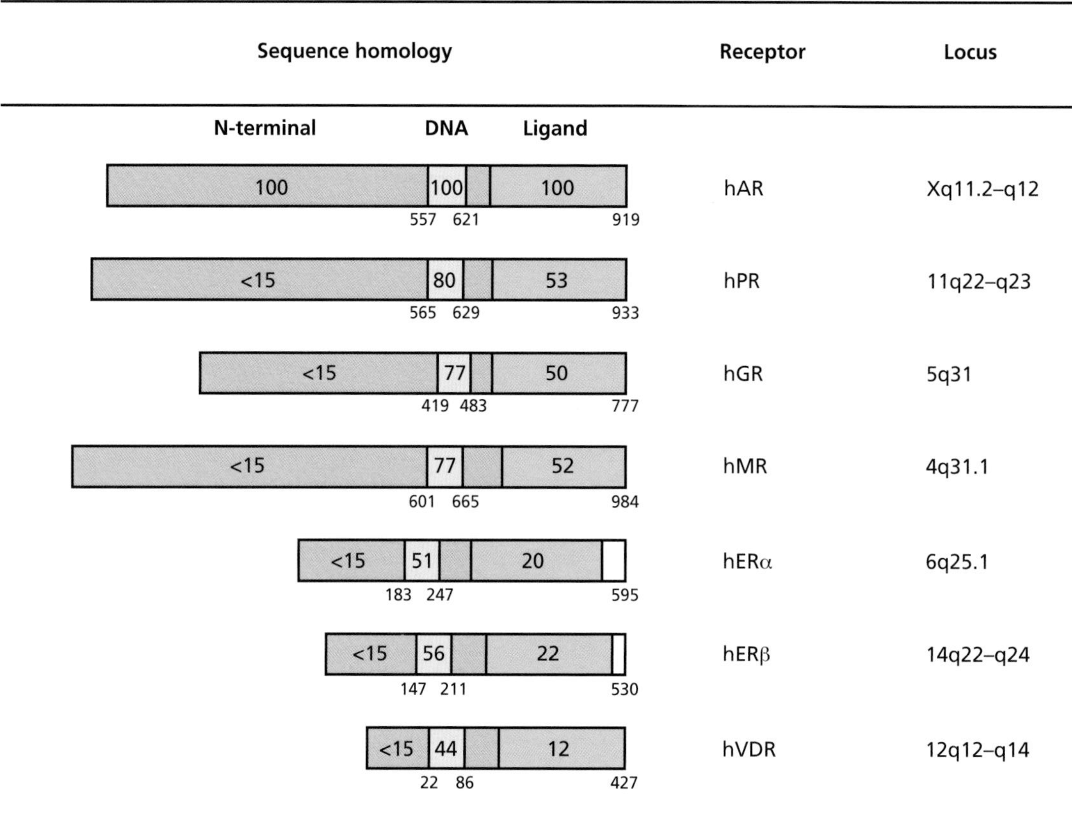

Figure 2 Sequence homology between the androgen receptor (hAR), progesterone receptor (hPR), glucocorticoid receptor (hGR), mineralocorticoid receptor (hMR), estrogen receptor α (hERα), estrogen receptor β (hERβ) and vitamin D receptor (hVDR). The hinge region (open box) is located between the DNA-binding domain (pale blue box) and the ligand-binding domain (green box). The sequence and size of the hinge region are not conserved among steroid hormone receptors. The N-terminal domain (gray box) is highly variable in size and composition. The chromosomal localization of the genes encoding these human steroid receptors is indicated. For hERβ, the longest open reading frame of the cDNA, encoding 530 amino acid residues, is represented. From references 5–13

the estradiol receptor. It is not surprising, therefore, that the androgen, progesterone, glucocorticoid and mineralocorticoid receptors can recognize the same response element. For the hormone- and tissue-specific responses of the different receptors, additional determinants are needed. Important in this respect are DNA sequences flanking the hormone response element, receptor interactions with other proteins and receptor concentrations. The second zinc cluster motif is supposed to be involved in protein–protein interactions such as receptor dimerization.[15]

Between the DNA-binding domain and the ligand-binding domain, a non-conserved hinge region is located, which is also variable in size in the different steroid receptors (Figure 2). The hinge region can be considered as a flexible link between the ligand-binding domain and the rest of the receptor molecule. It also contains the nuclear localization signal.

Finally, the second-most conserved region is the hormone-binding domain. This domain is encoded by approximately 250 amino acid residues in the C-terminal end of the molecule (Figure 2). Crystallographic studies of the ligand-binding domains of several nuclear receptors, including those from estradiol receptors α and β, progesterone receptor, androgen receptor and vitamin D receptor, have provided much more insight into the three-dimensional structure of this domain.[17–21] The entire ligand-binding structure is

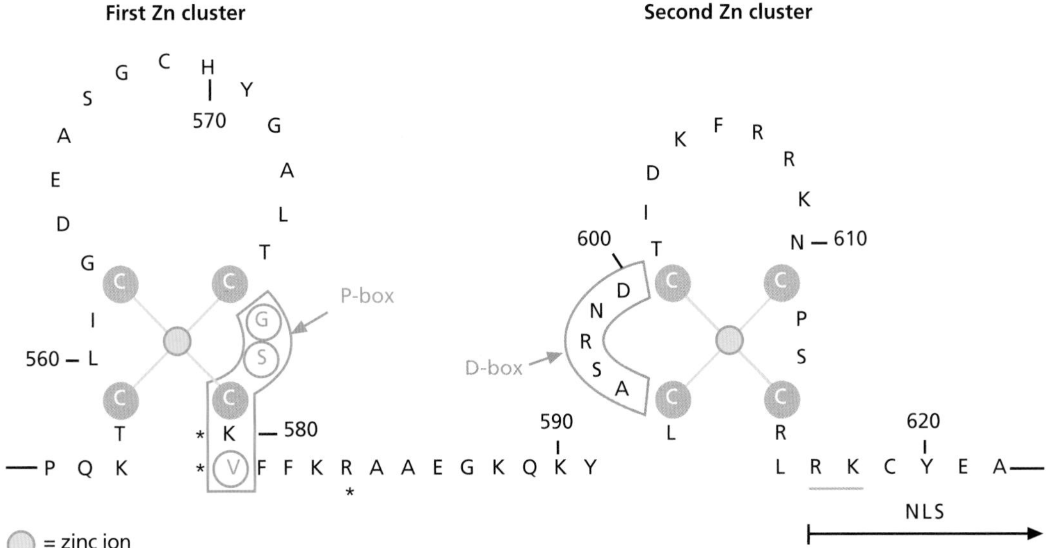

Figure 3 Sequence and functional motifs of the androgen receptor DNA-binding domain (AR-DBD). The amino acid sequence of the AR-DBD is shown in single-letter code. The domain consists of two zinc cluster modules. The first (N-terminal) zinc cluster contains the P-box (proximal box, in red), of which three residues (red circled) determine hormone response element (HRE) recognition specificity. The second zinc cluster contains the D-box (distal box, in green), in which amino acids are located that are involved in protein–protein interactions with a second receptor molecule in the homodimer complex. The second zinc cluster also contains the first part of the nuclear localization signal (NLS, blue underlined). Asterisks indicate the residues that most likely make base-pair contacts in the HRE half-site

composed of 11 α-helices (H_1, H_3–H_{12}) and two short antiparallel β-strands, folded into an antiparallel α-helical sandwich. Interesting and informative are structural differences of the ligand-binding domain in the presence and absence of ligand. The liganded receptor has a more compact structure owing to the folding back of helix 12 upon ligand binding. Wurtz and colleagues proposed a general mechanism for nuclear receptor activation, in which hormone-induced conformational changes within the ligand-binding domain result in a close contact of helix 12 and helix 4, thereby creating an interaction surface that allows binding of coactivators to the ligand-dependent activation domain.[22] The ligand-binding cavity in the estrogen receptor α consists of several helices, and forms a hydrophobic pocket which is completely partitioned from the external environment. Interestingly, upon binding of the antiestrogen raloxifene, the folding of the ligand-binding domain is different from that upon binding an estrogen agonist (Figure 4).[17] In particular, helix 12 is displaced, and protrudes now from the binding pocket (Figure 4b), causing a different three-dimensional structure, with consequences for the interaction surface of the ligand-binding domain and for the recruitment of coactivators or co-repressors. Similar structural observations were made for the ligand-binding domain of the estrogen receptor β in the presence of a full antagonist.[21]

Deletions in the ligand-binding domain abolish hormone binding partly or completely, depending on the type of steroid receptor.[23,24] Deletions in the N-terminal domain and DNA-binding domain do not affect hormone binding. Deletion of the hormone-binding domain leads to a constitutively active receptor protein with *trans*-activation capacity depending on the receptor type, promoter context and cell type.[23,24] Thus, it appears that the hormone-binding domain acts as a repressor of the *trans*-activation function in the absence of hormone. The C-terminal domain also participates in other receptor functions, such as receptor dimerization, and in interaction with heat shock proteins.[25,26]

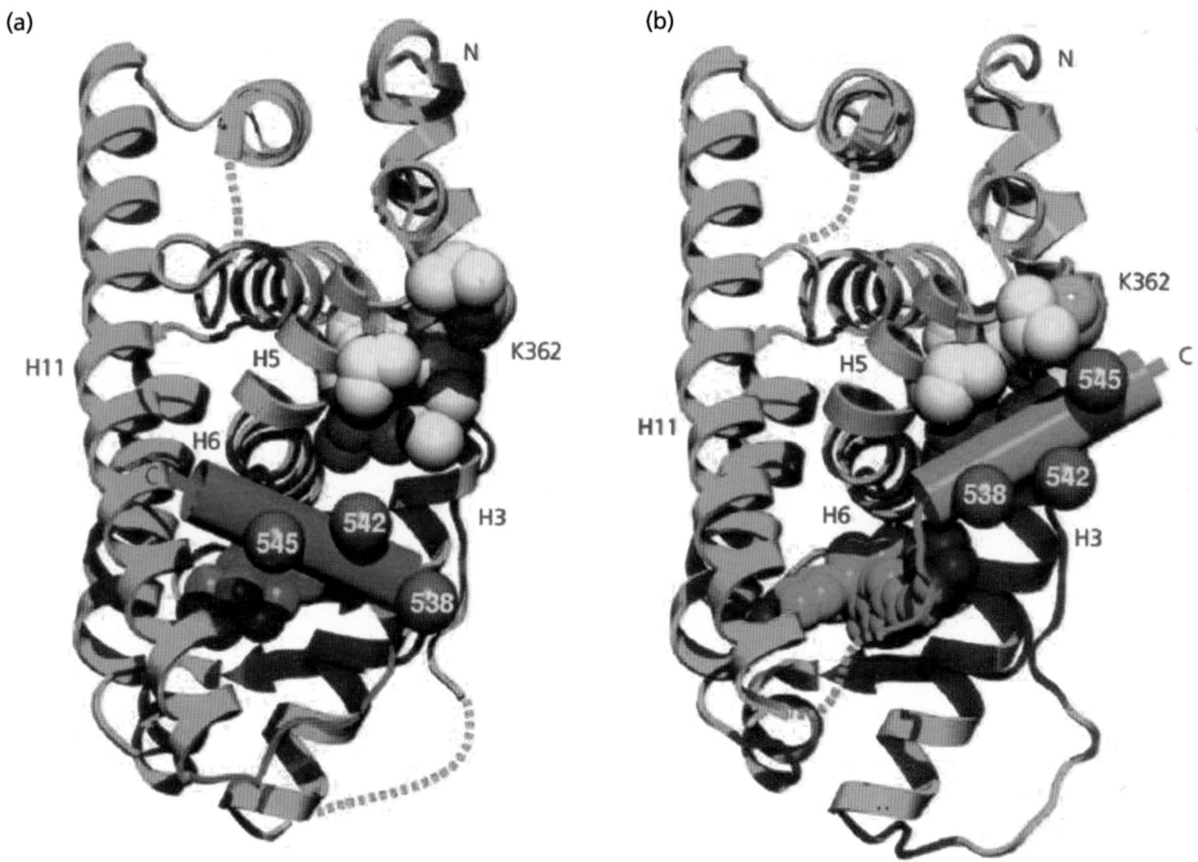

Figure 4 Three-dimensional structure of the ligand-binding domain of the estrogen receptor α. In (a) the liganded receptor with estradiol and in (b) the liganded receptor with the antiestrogen raloxifene are represented. The majority of the structural elements are shown in red. Helix 12 is drawn as a cylinder, colored blue in the estradiol complex (a) or green in the raloxifene complex (b). The displacement of helix 12 in the case of a bound antiestrogen is clearly depicted in this model (b). Reprinted with permission from *Nature* 1997 389: 753–758, copyright (1997) MacMillan Magazines Ltd.

Steroid hormone receptor genes

The organization of the genes encoding the different receptors is highly conserved. The protein-encoding region of the receptor genes is divided over eight exons.[27–29] Also, for the human estrogen receptor α and the glucocorticoid receptor, 5'-non-encoding exons have been identified.[30,31] The sequence encoding the N-terminal domain is present in one large exon. The DNA-binding zinc clusters are encoded by two small exons; the information for the ligand-binding domain is distributed over five different exons. Although intron sizes vary considerably between the different genes, all encoding exon/intron boundaries are at exactly the same positions in the various steroid hormone receptor genes.[27–29] The chromosal localization of the various steroid hormone receptor genes is indicated in Figure 2.

Heat shock proteins and steroid hormone receptors

It has been known for more than two decades that steroid hormone receptors are associated with other proteins upon isolation from target tissues under low-salt conditions. In the isolated large complexes (molecular masses of approximately

300 kDa and with sedimentation constants of 8–9 S on sucrose gradients), it appears that several different heat shock proteins (hsp90, hsp70 and hsp56), together with other proteins, are complexed with steroid receptors.[26,32,33] In intact target cells in the presence of hormone and at physiological temperatures, these complexes dissociate rapidly.[33] Several functions have been attributed to the association of heat shock proteins with steroid hormone receptors. Folding of the correct hormone-binding pocket upon synthesis of the receptor molecule at the ribosome has been suggested for the glucocorticoid receptor, in studies with yeast mutants lacking hsp90 or with diminished levels of hsp90.[34] Another possible function could be prevention of the interaction of the receptor molecule with DNA in the absence of hormone. No association with heat shock proteins has been found for the vitamin D receptor. This receptor appears to be more tightly bound in the nucleus even in the absence of hormone.

Steroid receptor phosphorylation

Steroid hormone receptors are phosphoproteins in the absence of ligand, and become hyperphosphorylated in the presence of hormone.[35,36] The hormone-induced extra phosphorylation, which is two- to seven-fold, is a rapid process.[37–40] All steroid receptors are phosphorylated at more than one site. Most phosphorylation sites are located in the N-terminal domain, and phosphorylation occurs mainly on serine residues.[37,38] In only a few cases does phosphorylation occur on threonine residues. Tyrosine residue phosphorylation has been found only for the estrogen receptor.[41,42] Phosphorylation of this tyrosine residue in the ligand-binding domain is required to maintain the estrogen receptor in a transcriptionally inactive state in the absence of ligand.[42] To date, six different kinases (estrogen receptor kinase, protein kinase A (PKA), protein kinase C (PKC), casein kinase II, DNA-dependent kinase, (Ser-Pro)-directed kinases) have been reported to phosphorylate steroid receptors. Several studies have been performed to uncover a physiological role for receptor phosphorylation in steroid hormone action. The following receptor functions or activities linked to phosphorylation have been suggested: receptor association with heat shock proteins, activation of hormone binding, nuclear import, subnuclear localization, nucleocytoplasmic shuttling, modulation of binding to hormone response elements, receptor dimerization, interactions with other transcription factors, and receptor half-life (receptor turnover and recycling).[43–45]

DNA binding of steroid hormone receptors

Upon hormone binding, the hormone–receptor complex undergoes a conformational change resulting in a higher affinity for DNA-enhancer sequences, termed hormone response elements, which are located within or adjacent to the promoter of their target genes. Hormone response elements consist of two hexameric 'half sites'.[46] The hexamers are arranged as an imperfect palindrome separated by three nucleotides, for the steroid hormone response elements, and as direct repeats separated by a varying number of nucleotides (0–6), for the vitamin D receptor.[47] The spacing between the two half-sites suggests that the two interaction sites for the receptor are on the same face of the DNA, separated by about one turn of the DNA helix. This supports the concept that the receptor binds to the response element as a homodimer or as a heterodimer, with each receptor monomer binding to one half-site in the response element. Significant variation can occur in a number of nucleotides in the response elements, depending upon the specific promoter in which it is found. The core sequence (5'-TGTTCT-3') for the glucocorticoid, mineralocorticoid, progesterone and androgen response elements is the same, but slight modifications in this sequence and additional flanking nucleotides modify the affinity of the core sequence to render response elements more selective. The steroid hormone–receptor complex is supposed to transduce the steroidogenic signal via protein–DNA interactions and by protein–protein interactions with other transcription factors.[48–50] This results in the formation of a stable pre-initiation complex near the transcription start site of the target gene, which allows efficient transcription initiation by RNA–polymerase II. Steroid hormone receptors might achieve this in the presence of hormones by stimulating the assembly of the pre-initiation complex, or by stabilizing the complex. This interaction can be direct or indirect, involving coactivators that mediate synergism between different transcription factors.

CURRENT CONCEPTS

Steroid receptor regulators

It has been known for some time that interaction of nuclear receptors with basal transcription factors is necessary for the control of hormone-dependent transcription activation. However, more recently it became clear that not only components of the basal transcription machinery are necessary, but also additional protein factors different from the basal factors are involved. The necessity of these factors in the response was demonstrated in so-called 'auto-squelching' experiments, in which overexpression of a nuclear receptor in a particular cell line can result in attenuation of transcription instead of a further increase. These additional factors belong to a family of three distinct, but related, so-called 'p160 coactivators'.[51] A fourth group of interacting proteins is a recently identified category of co-integrator proteins (cAMP response element binding protein (CREB) binding protein (CBP)/p300 and p300/CBP-associated factor (pCAF)).[52–54] These proteins are supposed to form the bridge between nuclear receptors and coactivators in the transcription initiation complex and act synergistically. These large, multi-interacting proteins can therefore perform a function different from that of coactivators, because these co-integrator proteins can also interact with components of other signal transduction pathways. The current model of the transcription activating complex therefore consists of: activator proteins (nuclear receptors), basal transcription factors (TFIID = complex of TATA-box binding protein (TBP) + several transcription activating factors (TAFs)), co-activators (steroid receptor coactivator-1 (SRC-1), transcription intermediary factor 2 (TIF2), glucocorticoid receptor interacting protein-1 (GRIP-1), androgen receptor associated protein-70 (ARA-70), receptor interacting protein 140 (RIP-140), receptor associated coactivator (RAC3)) and co-integrators (CBP/p300, pCAF).

The recent finding that several coactivators for nuclear receptors, including those for steroid hormone receptors, harbor intrinsic histone acetyltransferase (HAT) activity, has provided new insight into the molecular mechanisms by which nuclear receptors can control transcription regulation.[55] The targeted acetylation of chromatin can be considered an essential step in the activating mechanism of steroid receptors by coactivators. Upon hormone binding, steroid receptors bind to a hormone response element in chromatin-repressed DNA, and subsequently recruit a set of coactivators with intrinsic histone acetylase activity (SRC-1 (p160), peroxisome proliferator activating receptor-γ binding protein (PBP), p300) as is illustrated for the estrogen receptor α (ERα)-mediated transcription activation in a cyclic model[56] (Figure 5). The consequence of histone acetylation is relief of the transcriptionally repressed chromatin, by allowing transcription factors and RNA–polymerase II access to recognition elements. Consequently, the transcription pre-initiation complex is recruited at the correct site and transcription is initiated (Figure 5). Upon transcription initiation, the C-terminal domain of the RNA–polymerase II becomes phosphorylated, allowing elongation, while, in the complex, p300 is replaced by CBP, bringing in pCAF. Through acetylation of p160 by CBP, the complex dissociates, and both ERα and p160 are released. Subsequently, CBP and pCAF disassemble and the cycle will start again. Also, a dynamic model has been proposed for glucocorticoid receptor-mediated transcription activation, based on photobleaching experiments with a green fluorescent protein-tagged receptor molecule.[57] In this model, the receptor undergoes a continuous exchange between chromatin regulatory elements and the nucleoplasmic compartment, whereby the ligand is constantly available.

Tissue-specific responsiveness

An intriguing new aspect to the field of steroid hormone action is the beginning of the unravelling of selective tissue specificity of steroid responses.

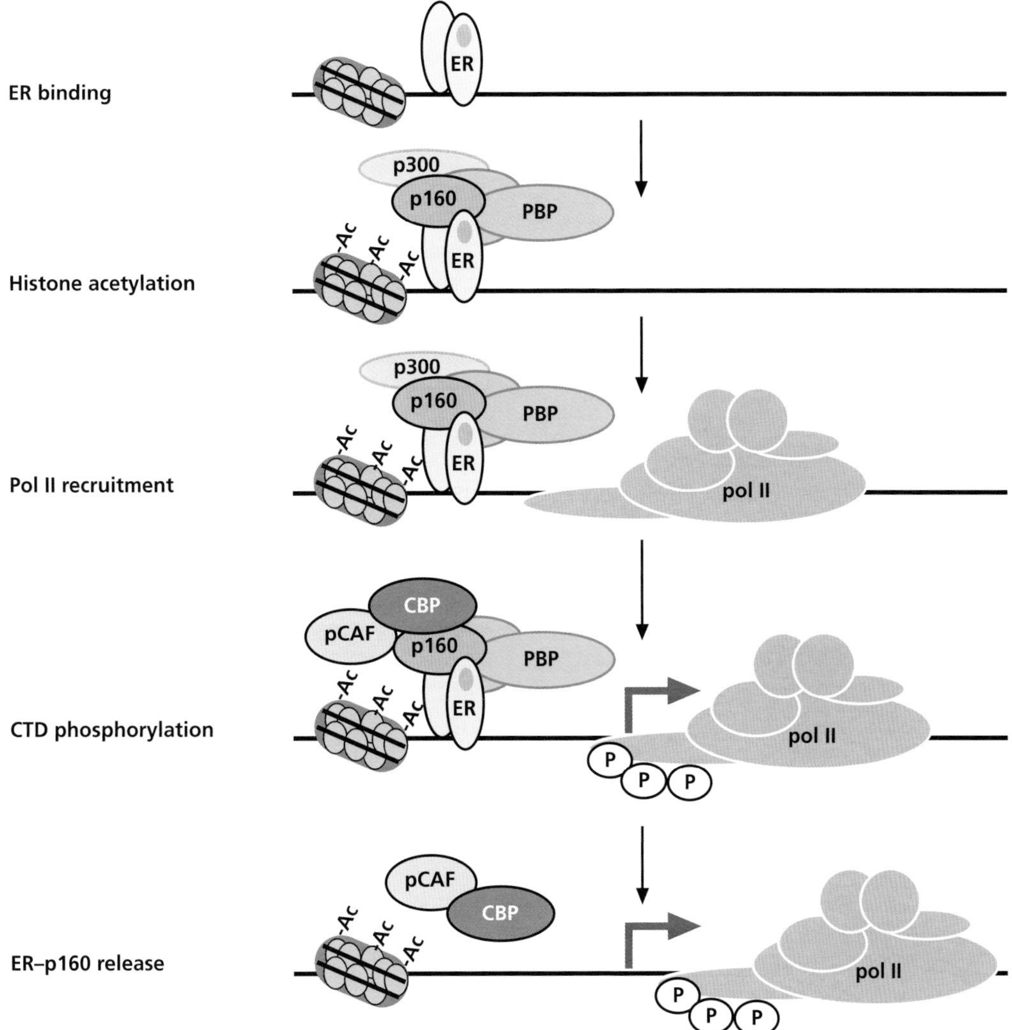

Figure 5 Model for transcriptional activation by estradiol via the estradiol receptor (ER) and the role of coactivators. Upon binding of estradiol to the receptor, the complex binds specifically as a homodimer to specific hormone response elements on the chromatin-repressed DNA (ER binding). Simultaneously, several protein factors, so-called coactivators (peroxisome proliferator activating receptor-γ binding protein (PBP), p300, p160) are recruited, some (p300, p160) with intrinsic histone acetyltransferase activities. Coactivation can occur by disruption of DNA repression via histone acetylation (Ac) (Histone acetylation). Subsequently, the RNA–polymerase II (pol II) complex is recruited at the correct site and transcription is initiated (pol II recruitment). Transcription elongation can occur upon phosphorylation (P) of the C-terminal domain (CTD) of the RNA–polymerase II, and, in the receptor initiation complex, p300 is replaced by cAMP response element binding protein (CREB) binding protein (CBP) and p-coactivating factor (pCAF), two other co-regulators with histone acetyltransferase activity (CTD phosphorylation). Finally, the association between the ER dimer and p160 is disrupted owing to extra acetylation of p160, resulting in release of ER and p160, and transcription elongation can proceed (ER–p160 release). Subsequently, a new cycle can start by binding of ER. Adapted from reference 56

In this respect, the discovery of a second estrogen receptor (ERβ) has increased our insight into the complex phenomenon of the different actions of estrogens in several estrogen target tissues.[58] The ERα is predominantly expressed in the uterus, ovary, pituitary, kidney, adrenal gland, testis and epididymis, while ERβ is expressed in the prostate, bladder, bone, brain, uterus, ovary (granulosa cells) and testis (developing spermatids).[59,60] This suggests that expression patterns of both receptor types are under tissue-specific regulation. Furthermore, in those tissues expressing both types of receptor, heterodimers can be expected, while, in other tissues expressing either one or the other type, homodimers exist. Consequently, heterodimers of ERα and ERβ constitute a novel estrogen-dependent mechanism for gene regulation. Another new aspect is the ligand specificity of ERα and ERβ. Although a variety of synthetic and naturally occurring estrogenic compounds have similar relative affinities for both receptor types, clear differences were also observed for certain ligands.[61,62] In the case of the progesterone receptor (PR), two receptor isoforms (A and B) have distinctly different reproductive functions, as was established after selective ablation of the PRA isoform.[63] Also, the ratio of PRA/PRB, modulated in transgenic mice models, can influence the phenotypic expression of progestogen action in the mammary gland.[64]

POTENTIAL CLINICAL RELEVANCE

If, in the cascade between hormone binding and gene expression, an essential step is lacking, an aberrant hormone response can be expected. This kind of steroid hormone resistance can be the cause of severe pathological situations. Defects in steroid hormone synthesis have been extensively studied, and have been characterized at the molecular level in enzymes involved in steroid biosynthesis. Another important cause can be situated at the steroid receptor level. Post-receptor defects, such as defective hormone response element or a defective transcription factor essential for function, are very rare.[65,66] Interestingly, individuals with a heterozygous mutation in the gene encoding the co-integrator protein CBP exhibit severe developmental defects in the so-called Rubinstein–Taybi syndrome, indicating that this protein is physiologically maintained at a limited concentration.[67]

Since the cloning of steroid receptor cDNAs and elucidation of the structure of the corresponding genes, extensive analysis has been performed of these receptor genes in patients with steroid hormone resistance syndromes. At least three pathological situations are associated with abnormal androgen receptor structure and function: androgen insensitivity syndrome (AIS), spinal and bulbar muscular atrophy (SBMA) and prostate cancer. In the X-linked androgen insensitivity syndrome, defects in the androgen receptor gene have prevented the normal development of both internal and external male structures in 46,XY individuals. Complete or gross deletions of the androgen receptor gene have not been found frequently in persons with the complete androgen insensitivity syndrome. Point mutations at several different sites in exons 2–8 encoding the DNA- and androgen-binding domains have been reported for partial and complete forms of androgen insensitivity[68,69] (Figure 6). A relatively high number of mutations were reported in two different clusters in exon 5 and in exon 7. The number of mutations in exon 1 is extremely low, and no mutations have been reported in the hinge region, located between the DNA-binding domain and the ligand-binding domain (Figure 6). The X-linked spinal and bulbar muscular atrophy (SBMA; Kennedy's disease) is associated with an expanded length (more than 40 residues) of one of the polyglutamine stretches in the N-terminal domain of the androgen receptor.[70] In prostate cancer patients, the highest percentage of mutations in the androgen receptor gene can be detected in late-stage, hormone-refractory prostate cancer.[71] For a limited number of mutations, it has been shown that the ligand responsiveness to progestogens, estrogens, glucocorticoids and even antiandrogens is increased.[72–74] In a relatively small number of reports, the syndrome of familial cortisol resistance has been correlated with mutations in the gene encoding the human glucocorticoid receptor.[75–78] All reported mutations affect ligand binding and are either heterozygous or homozygous.

It has been thought for a long time that mutations in the estrogen receptor gene would be lethal. However, recently, the molecular cause of severe estrogen resistance has been reported in a 46,XY patient with osteoporosis, unfused epiphyses and continuing growth in adulthood.[79] In this man, a homozygous mutation in the estradiol receptor gene was found, resulting in a premature stop

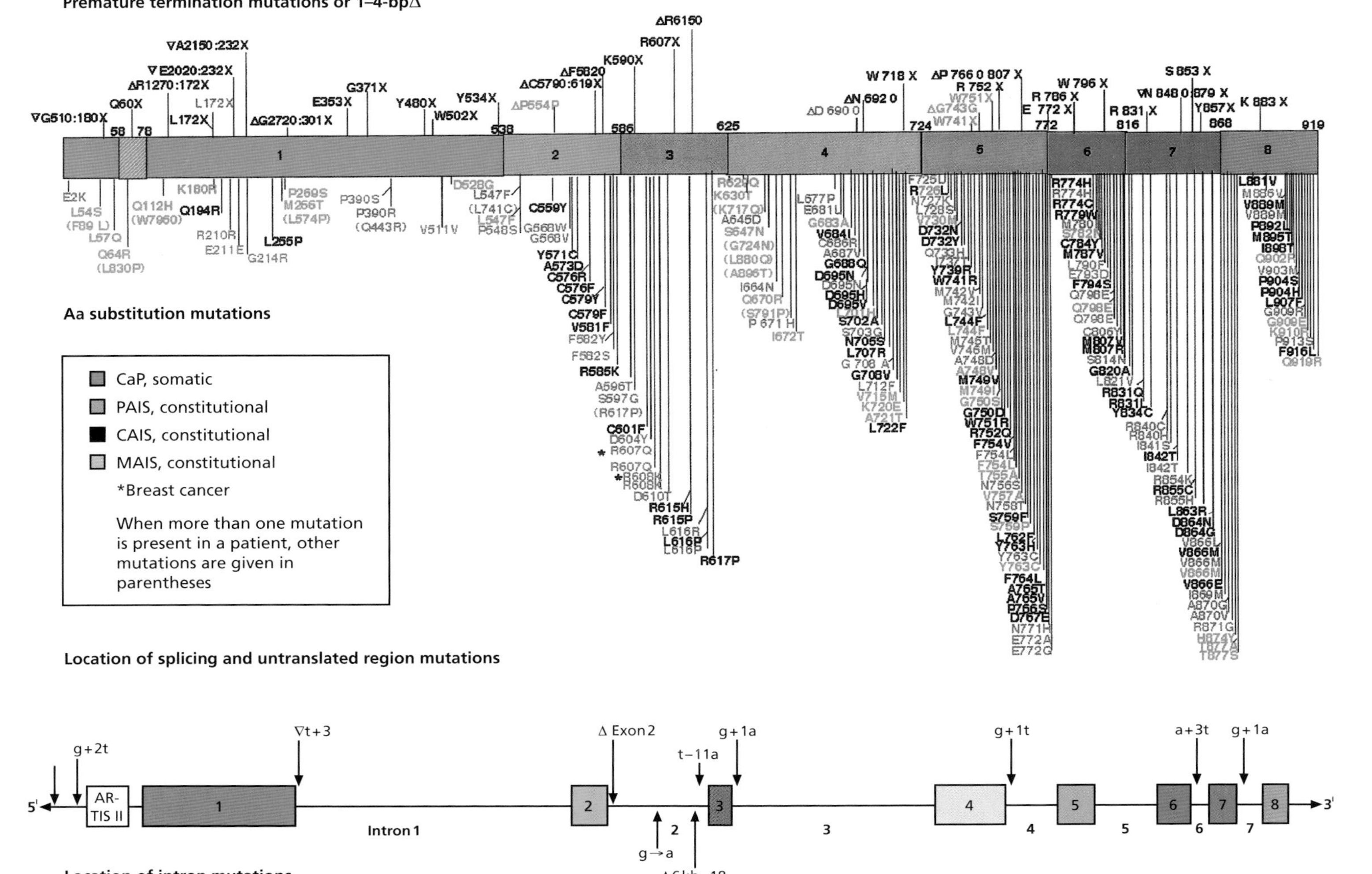

Figure 6 Overview of mutations in the androgen receptor (AR) gene, as deduced from the AR gene mutations database (http://www.mcgill.ca/androgendb/). Indicated are the positions of amino acid (Aa) substitutions, small deletions Δ and insertions ∇ identified in subjects with complete (CAIS, black), partial (PAIS, green) or mild (MAIS, blue) androgen insensitivity syndrome (AIS) as well as in prostate cancer (CaP, red). The different exons of the androgen receptor gene are represented by different colors and numbered 1–8. Amino acid residues are represented by single-letter code. The lower part of the figure represents mutations found in splice donor and splice acceptor sites in the androgen receptor gene of individuals with CAIS. Nucleotides (t, a, g) are represented by lower-case letters, while the sign (+ or –) before the number indicates the number of nucleotides downstream or upstream of the splice site. AR-TIS II, AR-transcription initiation site II (Reprinted with the permission of Dr Bruce Gottlieb, McGill University, Montreal, Canada)

codon in the DNA-binding domain of the receptor protein. This finding illustrates that estrogen receptor mutations need not be lethal, and that estrogens are important in the male for normal skeletal growth and development.

Heterozygous loss-of-function mutations in the mineralocorticoid receptor cause pseudohypoaldosteronism type I, a disease featuring salt wasting and hypotension.[80] Also, a gain-of-function mutation in the mineralocorticoid receptor has been found in a family with affected individuals having early onset of severe hypertension.[81]

No complete loss-of-function mutation in the human progesterone receptor gene has been described. This suggests strongly that complete loss of the progesterone receptor might lead to embryonic lethality. However, mice with an inactivating mutation in the progesterone receptor gene, introduced by homologous recombination in embryonic stem cells, develop normally, although female homozygous mice are infertile due to abnormalities in the reproductive system.[82] Furthermore, in addition to pregnancy, ovulation, luteinization and mammary gland development are also impaired.[82]

FUTURE PERSPECTIVES

In the near future, a number of breakthroughs with respect to gene regulation by nuclear hormone receptors can be expected. Since the three-dimensional structure of the DNA-binding domain of the glucocorticoid receptor was established in 1991, and the three-dimensional structure of the ligand-binding domain of the estradiol receptor in 1997, it can be expected that the next challenge will be the resolution of the three-dimensional structure of the N-terminal domain. It can also be expected that this might be accomplished first for the retinoic acid, vitamin D, thyroid hormone and estrogen receptors. This structural knowledge will reveal new information about possible structural and functional subdomains in the N-terminus, and the influence of the ligand on their conformation.

In the forthcoming years, more steroid receptor-specific coactivators and transcription factors will be identified and further characterized, in particular those involved in tissue-specific hormone responses. The interaction of these transcription factors with wild-type and mutant nuclear receptors will give a better understanding of the pleiotropic actions of steroid hormones in health and disease. For this purpose, the development of different, specific, steroid target cell lines is a prerequisite.

ACKNOWLEDGEMENTS

The author gratefully acknowledges MacMillan Magazines Ltd and Dr Roderick Hubbard and Dr Ashley Pike (York University, UK), who granted permission to include Figure 4. Dr Bruce Gottlieb (McGill University, Montreal, Canada) is gratefully acknowledged for his permission to include Figure 6.

REFERENCES

1. Evans RM. The steroid and thyroid hormone receptor superfamily. *Science* 1988 240: 889–895.
2. Enmark E, Gustafsson JA. Orphan nuclear receptors – the first eight years. *Mol Endocrinol* 1996 10: 1293–1307.
3. Jensen EV, Suzuki T, Kawashima T *et al.* A two-step mechanism for the interaction of estradiol with rat uterus. *Proc Natl Acad Sci USA* 1968 59: 632–638.
4. King WJ, Greene GL. Monoclonal antibodies localize oestrogen receptor in the nuclei of target cells. *Nature* 1984 307: 745–747.
5. Hollenberg SM, Weinberger C, Ong ES *et al.* Primary structure and expression of a functional human glucocorticoid receptor cDNA. *Nature* 1985 318: 635–641.
6. Green S, Walter P, Kumar V *et al.* Human oestrogen receptor cDNA: sequence expression and homology to v-*erb*-A. *Nature* 1986 320: 134–139.
7. Arriza JL, Weinberger C, Cerelli G *et al.* Cloning of human mineralocorticoid receptor complementary DNA: structural and functional kinship with the glucocorticoid receptor. *Science* 1987 237: 268–275.
8. Misrahi M, Atger M, d'Auriol L *et al.* Complete amino acid sequence of the human progesterone receptor deduced from cloned cDNA. *Biochem Biophys Res Commun* 1987 143: 740–748.
9. Trapman J, Klaassen P, Kuiper GGJM *et al.* Cloning, structure and expression of a cDNA

encoding the human androgen receptor. *Biochem Biophys Res Commun* 1988 153: 241–248.

10. Baker AR, McDonnell DP, Hughes M *et al.* Cloning and expression of full-length cDNA encoding human vitamin D receptor. *Proc Natl Acad Sci USA* 1988 85: 3294–3298.
11. Kuiper GG, Enmark E, Pelto-Huikko M, Nilsson S, Gustafsson JA. Cloning of a novel receptor expressed in rat prostate and ovary. *Proc Natl Acad Sci USA* 1996 93: 5925–5930.
12. Mosselman S, Polman J, Dijkema R. ERβ: identification and characterization of a novel human estrogen receptor. *FEBS Lett* 1996 392: 49–53.
13. Ogawa S, Inoue S, Watanabe T *et al.* The complete primary structure of human estrogen receptor β (hERβ) and its heterodimerization with ERα *in vivo* and *in vitro*. *Biochem Biophys Res Commun* 1998 243: 122–126.
14. Van Laar JH, Voorhorst-Ogink MM, Zegers ND *et al.* Characterization of polyclonal antibodies against the N-terminal domain of the human androgen receptor. *Mol Cell Endocrinol* 1989 67: 29–38.
15. Luisi BF, Xu WF, Otwinowski Z *et al.* Crystallographic analysis of the interaction of the glucocorticoid receptor with DNA. *Nature* 1991 352: 497–505.
16. Green S, Kumar V, Theulaz I, Wahli W, Chambon P. The N-terminal DNA-binding zinc finger of the oestrogen and glucocorticoid receptors determines target gene specificity. *EMBO J* 1988 7: 3037–3044.
17. Brzozowski AM, Pike ACW, Dauter Z *et al.* Molecular basis of agonism and antagonism in the oestrogen receptor. *Nature* 1997 389: 753–758.
18. Williams SP, Sigler PB. Atomic structure of progesterone complexed with its receptor. *Nature* 1998 393: 392–396.
19. Matias PM, Donner P, Coelho R *et al.* Structural evidence for ligand specificity in the binding domain of the human androgen receptor; implications for pathogenic gene mutations. *J Biol Chem* 2000 275: 26164–26171.
20. Rochel N, Wurtz JM, Mitschler A, Klaholz B, Moras D. The crystal structure of the nuclear receptor for vitamin D bound to its natural ligand. *Mol Cell* 2000 5: 173–179.
21. Pike ACW, Brzozowski AM, Hubbard RE *et al.* Structure of the ligand-binding domain of oestrogen receptor β in the presence of a partial agonist and a full antagonist. *EMBO J* 1999 18: 4608–4618.
22. Wurtz JM, Bourguet W, Renaud JP *et al.* A canonical structure for the ligand-binding domain of nuclear receptors. *Nat Struct Biol* 1996 3: 87–94.
23. Jenster G, van der Korput JAGM, van Vroonhoven C *et al.* Domains of the human androgen receptor involved in steroid binding, transcriptional activation, and subcellular localization. *Mol Endocrinol* 1991 5: 1396–1404.
24. Godowski PJ, Rusconi S, Miesfeld R, Yamamoto KR. Glucocorticoid receptor mutants that are constitutive activators of transcriptional enhancement. *Nature* 1987 325: 365–368.
25. Fawell SE, Lees JA, White R, Parker MG. Characterization and colocalization of steroid binding and dimerization activities in the mouse estrogen receptor. *Cell* 1990 60: 953-962.
26. Pratt WB, Toft DO. Steroid receptor interactions with heat shock protein and immunophilin chaperones. *Endocr Rev* 1997 18: 361–377.
27. Ponglikitmongkol M, Green S, Chambon P. Genomic organization of the human oestrogen receptor gene. *EMBO J* 1988 7: 3385–3388.
28. Kuiper GGJM, Faber PW, van Rooij HCJ *et al.* Structural organization of the human androgen receptor gene. *J Mol Endocrinol* 1989 2: R1–R4.
29. Encio IJ, Detera-Wadleigh SD. The genomic structure of the human glucocorticoid receptor. *J Biol Chem* 1991 266: 7182–7188.
30. Keaveney M, Klug J, Dawson MT *et al.* Evidence for a previously unidentified upstream exon in the human oestrogen receptor gene. *J Mol Endocrinol* 1991 6: 111–115.
31. Zong J, Ashraf J, Thompson EB. The promoter and first, untranslated exon of the human glucocorticoid receptor gene are GC rich but lack consensus glucorticoid receptor element sites. *Mol Cell Biol* 1990 10; 5580–5585.
32. Bresnick EH, Dalman FC, Pratt WB. Direct stoichiometric evidence that the untransformed M_r 300 000, 9S, glucocorticoid receptor is a core unit derived from a larger heteromeric complex. *Biochemistry* 1990 29: 520–527.
33. Veldscholte J, Berrevoets CA, Brinkmann AO, Grootegoed JA, Mulder E. Anti-androgens and the mutated androgen receptor of LNCaP cells: differential effects on binding affinity, heat-shock protein interaction, and transcription activation. *Biochemistry* 1992 31: 2393–2399.
34. Picard D, Khursheed B, Garabedian MJ *et al.* Reduced levels of hsp90 compromise steroid

receptor activation *in vivo*. *Nature* 1990 348: 166–168.

35. Orti E, Bodwell JE, Munck A. Phosphorylation of steroid hormone receptors. *Endocr Rev* 1992 13: 105–128.

36. Van Laar JH, Berrevoets CA, Trapman J, Zegers ND, Brinkmann AO. Hormone-dependent androgen receptor phosphorylation is accompanied by receptor transformation in human lymph node carcinoma of the prostate cells. *J Biol Chem* 1991 266: 3734–3738.

37. Kuiper GGJM, De Ruiter PE, Trapman J *et al.* Localization and hormonal stimulation of phosphorylation sites in the LNCaP-cell androgen receptor. *Biochem J* 1993 291: 95–101.

38. Sheridan PL, Krett NL, Gordon JA, Horwitz KB. Human progesterone receptor transformation and nuclear down-regulation are independent of phosphorylation. *Mol Endocrinol* 1988 2: 1329–1342.

39. Beck CA, Weigel NL, Edwards DP. Effects of hormone and cellular modulators of protein phosphorylation on transcriptional activity, DNA-binding, and phosphorylation of human progesterone receptors. *Mol Endocrinol* 1992 6: 607–620.

40. Mendel DB, Bodwell JE, Munck A. Molybdate-stabilized nonactivated glucocorticoid-receptor complexes contain a 90 kDa steroid binding phosphoprotein that is lost on activation. *J Biol Chem* 1987 263: 6695–6702.

41. Auricchio F. Phosphorylation of steroid receptors. *J Steroid Biochem* 1989 32: 613–622.

42. White R, Sjoberg M, Kalkhoven E, Parker MG. Ligand independent activation of the oestrogen receptor by mutation of a conserved tyrosine. *EMBO J* 1997 16: 1427–1435.

43. Lange CA, Shen T, Horwitz KB. Phosphorylation of human progesterone receptors at serine-294 by mitogen-activated protein kinase signals their degradation by the 26S proteasome. *Proc Natl Acad Sci USA* 2000 97: 1032–1037.

44. Tremblay A, Tremblay GB, Labrie F, Giguere V. Ligand-independent recruitment of SRC-1 to estrogen receptor β through phosphorylation of activation function AF-1. *Mol Cell* 1999 3: 513–519.

45. Rogatsky I, Trowbridge JM, Garabedian MJ. Potentiation of human estrogen receptor β transcriptional activation through phosphorylation of serines 104 and 106 by the cyclin A–CDK2 complex. *J Biol Chem* 1999 274: 22296–22302.

46. Naar AM, Boutin JM, Lipkin SM *et al.* The orientation and spacing of core DNA-binding motifs dictate selective transcriptional responses to three nuclear receptors. *Cell* 1991 65: 1267–1279.

47. Umesone K, Murakami KK, Thompson CC, Evans RM. Direct repeats as selective response elements for the thyroid hormone, retinoic acid, and vitamin D3 receptors. *Cell* 1991 65: 1255–1266.

48. Freedman LP. Increasing the complexity of coactivation in nuclear receptor signaling. *Cell* 1999 97: 5–8.

49. Glass CK, Rosenfeld MG. The coregulator exchange in transcriptional functions of nuclear receptors. *Genes Dev* 2000 14: 121–141.

50. Reichardt HM, Kaestner KH, Tuckermann J *et al.* DNA binding of the glucocorticoid receptor is not essential for survival. *Cell* 1998 93: 531–541.

51. McKenna NJ, Lanz RB, O'Malley BW. Nuclear receptor coregulators: cellular and molecular biology. *Endocr Rev* 1999 20: 321–344.

52. Kamei Y, Xu L, Heinzel T *et al.* A CBP integrator complex mediates transcriptional activation and AP-1 inhibition by nuclear receptors. *Cell* 1996 85: 403–414.

53. Chakravarti D, LaMorte VJ, Nelson MC *et al.* Role of CBP/P300 in nuclear receptor signalling. *Nature* 1996 383: 99–103.

54. Blanco JC, Minucci S, Lu J *et al.* The histone acetylase PCAF is a nuclear receptor coactivator. *Genes Dev* 1998 12: 1638–1651.

55. Jenster G, Spencer TE, Burcin MM *et al.* Steroid receptor induction of gene transcription: a two-step model. *Proc Natl Acad Sci USA* 1997 94: 7879–7884.

56. Shang Y, Hu X, DiRenzo J, Lazar MA, Brown M. Cofactor dynamics and sufficiency in estrogen receptor-regulated transcription. *Cell* 2000 103: 843–852.

57. McNally JG, Müller WG, Walker D, Wolford R, Hager GL. The glucocorticoid receptor: rapid exchange with regulatory sites in living cells. *Science* 2000 287: 1262–1265.

58. Paige LA, Christensen DJ, Grøn H *et al.* Estrogen receptor (ER) modulators each induce distinct conformational changes in ERα and ERβ. *Proc Natl Acad Sci USA* 1999 96: 3999–4004.

59. Krege JH, Hodgin JB, Couse JF *et al.* Generation and reproductive phenotypes of mice lacking estrogen receptor β. *Proc Natl Acad Sci USA* 1998 95: 15677–156842.

60. Weihua Z, Saji S, Mäkinen S, Cheng G *et al.* Estrogen receptor (ER) β, a modulator of ERα in the uterus. *Proc Natl Acad Sci USA* 2000 97: 5936–5941.
61. Kuiper GGJM, Carlsson B, Grandien K *et al.* Comparison of the ligand binding specificity and transcript tissue distribution of estrogen receptors α and β. *Endocrinology* 1997 138: 863–870.
62. Kuiper GGJM, Lemmen JG, Carlsson B *et al.* Interaction of estrogenic chemicals and phytoestrogens with estrogen receptor β. *Endocrinology* 1998 139: 4252–4263.
63. Mulac-Jericevic B, Mullinax RA, DeMayo FJ, Lydon JP, Conneely OM. Subgroup of reproductive functions of progesterone mediated by progesterone receptor-B isoform. *Science* 2000 289: 1751–1754.
64. Shyamala G, Yang X, Silbertsein G, Barcellos-Hoff MH, Dale E. Transgenic mice carrying an imbalance in the native ratio of A to B forms of progesterone receptor exihibit developmental abnormalities in mammary glands. *Proc Natl Acad Sci USA* 1998 95: 696–701.
65. New MI, Nimkarn S, Brandon DD *et al.* Resistance to several steroids in two sisters. *J Clin Endocrinol Metab* 1999 84: 4454–4464.
66. Adachi M, Takayanagi R, Tomura A *et al.* Androgen-insensitivity syndrome as a possible coactivator disease. *N Engl J Med* 2000 343: 856–862.
67. Petrij F, Giles RH, Dauwerse HG *et al.* Rubinstein–Taybi syndrome caused by mutations in the transcriptional co-activator CBP. *Nature* 1995 376: 348–351.
68. Quigley CA, De Bellis A, Marschke KB *et al.* Androgen receptor defects: historical, clinical and molecular perspectives. *Endocr Rev* 1995 16: 271–321.
69. Gottlieb B, Beitel LK, Lumbroso R, Pinsky L, Trifiro M. Update of the androgen receptor gene mutations database. *Hum Mutat* 1999 14: 103–114.
70. La Spada AR, Wilson E, Lubahn D, Harding A, Fischbeck KH. Androgen receptor gene mutations in X-linked spinal and bulbar muscular atrophy. *Nature* 1991 352: 77–79.
71. Jenster G, Trapman J, Brinkmann AO. The androgen receptor. In Burris T, McCabe E, eds. *Nuclear Receptors and Genetic Disease*. London: Academic Press, 2000: 137–177.
72. Veldscholte J, Ris-Stalpers C, Kuiper GGJM *et al.* A mutation in the ligand binding domain of the androgen receptor of human LNCaP cells affects steroid binding characteristics and response to antiandrogens. *Biochem Biophys Res Commun* 1990 173: 534–540.
73. Zhao X-Y, Malloy PJ, Krishnan AV *et al.* Glucocorticoids can promote androgen-independent growth of prostate cancer cells through a mutated androgen receptor. *Nat Med* 2000 6: 703–706.
74. Brinkmann AO, Trapman J. Prostate cancer schemes for androgen escape. *Nat Med* 2000 6: 628–629.
75. Hurley DM, Accili D, Stratakis CA *et al.* Point mutation causing a single amino acid substitution in the hormone binding domain of the glucocorticoid receptor in familial glucocorticoid resistance. *J Clin Invest* 1991 87: 680–686.
76. Malchov DM, Brufsky A, Reardon G *et al.* A mutation of the glucocorticoid receptor in primary cortisol resistance. *J Clin Invest* 1993 91: 1918–1925.
77. Karl M, Lamberts SWJ, Koper JW *et al.* Cushing's disease preceded by generalized glucocorticoid resistance: clinical consequences of a novel, dominant-negative glucocorticoid receptor mutation. *Proc Assoc Am Physicians* 1996 108: 296–307.
78. De Lange P, Koper JW, Huizenga NATM *et al.* Differential hormone-dependent transcriptional activation and repression by naturally occurring human glucocorticoid receptor variants. *Mol Endocrinol* 1997 11: 1156–1164.
79. Smith EP, Boyd J, Frank GR *et al.* Estrogen resistance caused by a mutation in the estrogen receptor gene in a man. *N Engl J Med* 1994 331: 1056–1061.
80. Geller DS, Rodriguez-Soriano J, Boado AV *et al.* Mutations in the mineralocorticoid receptor gene cause autosomal dominant pseudohypoaldosteronism type I. *Nat Genet* 1998 19: 279–281.
81. Geller DS, Farhi A, Pinkerton N *et al.* Activating mineralocorticoid receptor mutation in hypertension exacerbated by pregnancy. *Science* 2000 289: 119–123.
82. Lydon JP, DeMayo FJ, Funk CR *et al.* Mice lacking progesterone receptor exhibit pleiotropic reproductive abnormalities. *Genes Dev* 1995 9: 2266–2278.

Chapter 16

Cytokines, growth factors and their receptors

Stacey C. Chapman and Teresa K. Woodruff

INTRODUCTION

Hormones, growth factors and cytokines communicate information between cells in multicellular organisms to control growth and development, defend against infection and injury, and regulate cell differentiation. Growth factors and cytokines make up a large proportion of all signalling molecules in the body. Signalling by these and other protein factors relies on the interaction of the protein factor, or ligand, with specific cell-surface receptors to initiate a variety of complex signalling pathways that convert the 'analog' signal into the 'digital' read-out reflected in intracellular changes.

As with all physiological systems, the reproductive axis depends on a diverse array of hormones, growth factors and cytokines to control every aspect of normal reproductive function. This dependency is illustrated in the fertility-related defects that arise in humans with gene mutation or protein malfunction of cellular signalling pathway components.[1] Other chapters in this volume detail the specific functions of the hormones as they relate to reproductive function. The purpose of this chapter is to describe the fundamental components of cytokine and growth factor signal transduction pathways, and to highlight representative ligands and their signalling mechanisms.

BACKGROUND

Three major growth factor and cytokine receptor classes have been defined: the receptor tyrosine kinases (RTKs), the receptor-associated (or non-receptor) protein tyrosine kinases (PTKs) and the receptor serine kinases (RSKs) (Figure 1).[2–4] Although the specific signalling mechanisms initiated by these receptor families are distinct, all involve ligand binding to cell surface receptors,

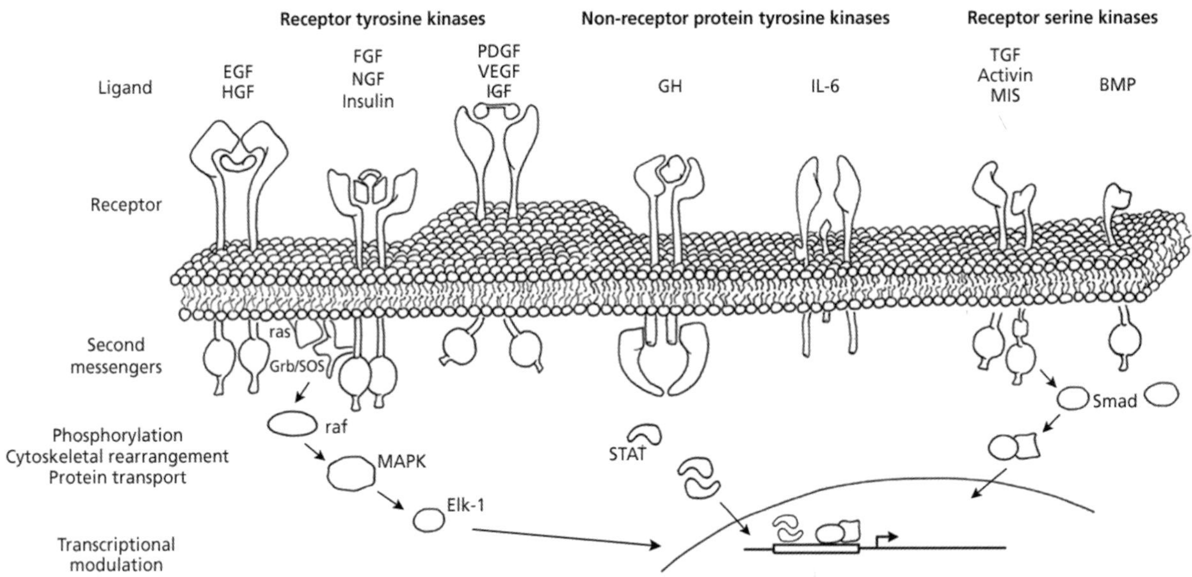

Figure 1 The three major growth factor and cytokine receptor classes and common themes in signal transduction. The receptor tyrosine kinases (RTKs), the non-receptor protein tyrosine kinases (PTKs) and the receptor serine kinases (RSKs) bind their respective ligands at the cell surface and undergo conformational changes, usually dimerization, to transduce a signal inside the cell. A variety of second messengers are activated and propagate signals to effect a number of outcomes, including phosphorylation of other signalling or effector proteins, changes in the cytoskeleton, regulation of protein transport and modulation of transcription in the nucleus. EGF, epidermal growth factor; HGF, hepatic growth factor; FGF, fibroblast growth factor; NGF, nerve growth factor; PDGF, platelet-derived growth factor; VEGF, vascular endothelial growth factor; IGF, insulin-like growth factor; GH, growth hormone; IL-6, interleukin-6; TFG, transforming growth factor; MIS, Müllerian inhibiting substance; BMP, bone morphogenic protein; Grb/SOS, growth factor receptor-binding protein/son of sevenless; MAPK, mitogen-activated protein kinase; STAT, signal transducers and activators of transcription

which results in the activation (phosphorylation) of co-receptors and cytoplasmic signal-transducing proteins, the 'second messengers'.[5,6] Second messengers are functionally diverse, and include kinases, tyrosine phosphatases (PTPases), lipases, guanosine triphosphatases (GTPases) or adaptor/docking proteins that carry the signal from the 'first messenger', the receptor, to the downstream signalling and effector proteins that will execute the proper cellular response. It is now recognized that intracellular signalling molecules are modular in nature, that is, they consist of several domains necessary for protein–protein recognition and cellular localization (Figure 2a).[2] Signalling proteins contain one or more of these recognition domains, which allow for proper localization and protein interaction between components within a signalling pathway. Some common domains include the src homology 2 (SH2) domain and the phosphotyrosine binding (PTB) domain, which recognize and interact with phosphotyrosine residues. Another src homology domain, SH3, and the tryptophan repeat domain, WW, interact with distinct proline-rich sequences. At least two domains, the plextrin homology (PH) domain and the FYVE domain (named after four representative proteins: Fab1p, YOTB, Vac1p and EEA1) recognize membrane phosphoinositides and can localize signalling molecules to the cell membrane.

Some signalling proteins contain several recognition domains, and can act as signalling scaffolds by binding multiple signalling proteins and localizing them to membrane-bound receptors via membrane component recognition motifs, such as the FYVE domain (Figure 2b). For example, in mammals, the JUN N-terminal kinase (JNK) interacting protein (JIP-1) colocalizes several intracellular signal-transducing protein kinases, and this may promote efficient sequential activation and enhance cellular signalling.[7] Thus, scaffolds increase the efficiency of signal transduction by

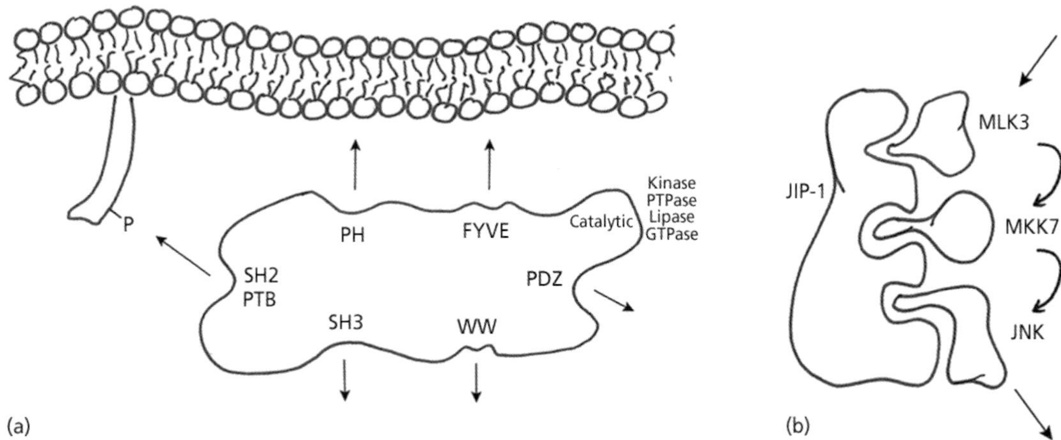

Figure 2 (a) Intracellular signalling molecules are modular. The src homology (SH) domain containing protein SH2 and the phosphotyrosine binding domain (PTB) recognize and interact with phosphotyrosine residues (P) on activated receptor tyrosine kinases. SH3 domains interact specifically with proline-rich sequences (PXXP) on target proteins. The WW domain of conserved tryptophan residues also recognizes proline-rich regions PPXY or prolines near phosphoserine or phosphothreonine residues. The PSD-95/Dlg/ZO-1 (PDZ) domain recognizes C-terminal hydrophobic residues. At least two domains recognize phosphoinositides and can localize intracellular signalling proteins to the membrane: the plextrin homology (PH) domain, which interacts with phosphoinositol (PtdIns)-(4,5)-P2 and PtdIns-(3,4,5)-P3, and the FYVE domain, named after four representative proteins: Fab1p, YOTB, Vac1p and EEA1, which interacts with PtdIns-3-P. (b) Scaffold proteins can contain multiple recognition domains to colocalize several components of a signalling pathway. The only known mammalian signalling scaffold, JUN N-terminal kinase interacting protein (JIP-1), colocalizes the sequential components of a mitogen-activated signalling cascade: a mitogen-activated kinase kinase kinase, mixed-lineage kinase 3 (MLK3), phosphorylates and activates mitogen-activated kinase kinase 7 (MKK7). MKK7 in turn phosphorylates and activates JUN N-terminal kinase (JNK). PTPase, tyrosine phosphatase; GTPase, guanosine triphosphatase

co-localizing the components of a signalling cascade within the cell.

Receptor-activated second messengers initiate signalling cascades, in which each step of a signalling pathway amplifies the initial signal of a single ligand–receptor interaction through the activation of several cytoplasmic signal-transducing proteins (Figure 3a). The end-points of these signalling cascades are diverse, and include alterations in cell metabolism, rearrangement of the cytoskeleton and modulation of cell proliferation, protein transport and gene expression.[8]

One of the notable features of cytokine and growth factor signal-transduction pathways is the conservation of the signalling proteins throughout all eukaryotes. Orthologous molecules involved in the RTK, cytokine receptor and RSK signalling pathways are found from nematodes to the vertebrates. In fact, the completed genome sequences of the worm, *Caenorhabditis elegans*, and yeast, *Saccharomyces cerevisiae*, revealed a surprising number of signalling kinases that are shared with humans.[9] In fact, based on amino acid sequences, a direct lineage can be traced from the *C. elegans* receptor encoded by *Daf4*, to the human type-II activin receptors.[10]

Cells exist in a constantly changing milieu of growth factors, cytokines and hormones, which

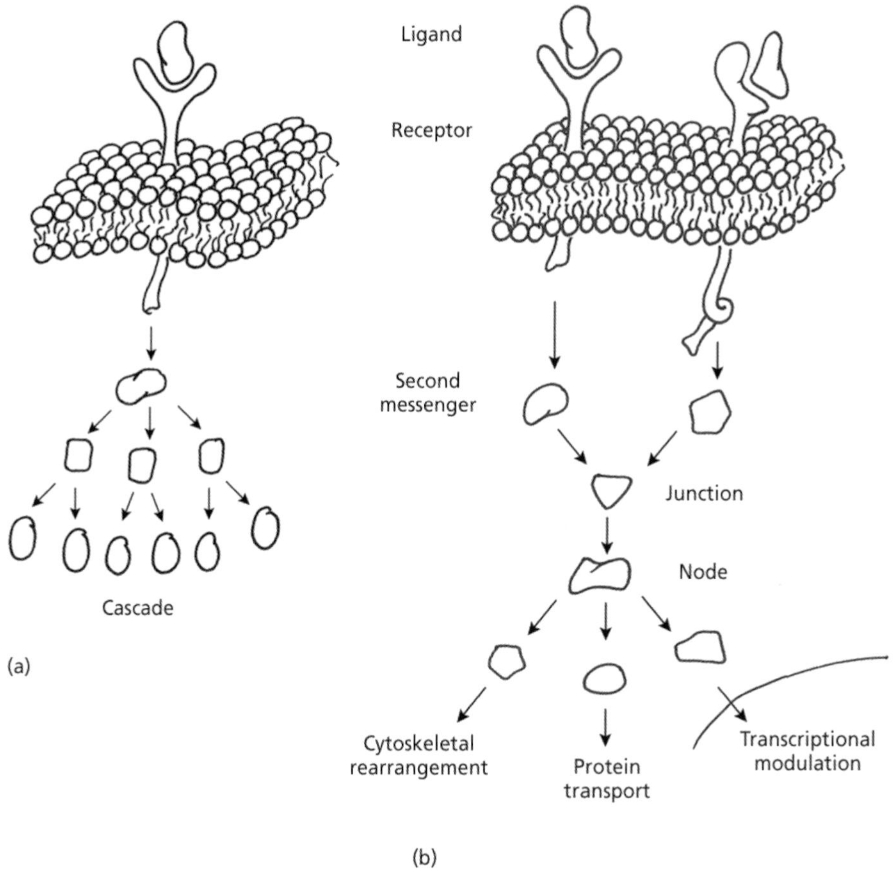

Figure 3 (a) Signalling cascades of receptor-activated second messengers amplify the initial signal of a single ligand–receptor interaction. (b) Junctions and nodes in intracellular signalling networks. Cells that must integrate simultaneous signals can utilize signalling junctions, or proteins that are downstream targets of multiple signals. Junctions can lead to the synergistic amplification of two positive signals, or the balancing of two opposing signals. One signal can initiate several cellular outcomes by passing through nodes. Nodes can amplify a signal by activating several downstream signalling components at a time, or can split the signal and lead to the activation of several distinct signalling pathways by one ligand

must be recognized, sorted and integrated, to respond appropriately to the environment. It is not surprising, then, that cellular signal-transduction systems are not simple linear pathways, but consist of extensive and complex networks of several interacting pathways. Multiple distinct signals can be integrated by a cell through 'junctions', components of a signalling network that are able to receive signals from several upstream sources and activate a common downstream target (Figure 3b).[11] Alternatively, junctions may be sites where two opposing signals are balanced against one another to achieve a higher degree of control over signalling events. On the other hand, one signal can be amplified through 'nodes', points in a signalling cascade where the signal is split to effect several outcomes. The epidermal growth factor receptor (EGFR) serves as a node for EGF signalling by stimulating a number of intracellular signal transducers and pathways to effect a mitogenic, or cellular growth, response and to modulate intracellular trafficking events.[12]

The 'cross-talk' initiated at junctions and nodes of signalling networks allows for efficient and tight control of opposing simultaneous signals. Bone morphogenic protein (BMP) is important in bone development and remodelling, and acts through the Smad1 intracellular signalling protein.[13] The mitogen-activated protein kinase (MAPK) signalling pathway is initiated by a number of growth factors, including EGF. MAPK negatively regulates Smad1 activity; thus, a tight regulation of bone growth and morphogenesis is possible in a cell in which the opposing BMP and EGF signals are integrated and balanced.[14]

An increasing number of human diseases have been characterized that involve mutations and inappropriate regulation of the agonists, receptors and coactivators of the cytokine and growth factor family.[1] However, a redundancy of signal-transduction pathways prevents the total malfunction of complex organisms when only one of the pathway components is faulty. Parallel 'back-up' systems generally permit continued function of cells, unless a second debilitating mutation renders the system incapable of normal activity. For example, insulin-like growth factor (IGF) signals are transduced by a receptor tyrosine kinase that activates one of two second messengers: the insulin receptor substrate (IRS) or the shc adaptor protein.[15,16] Cell death is prevented by activation of the alternative shc-mediated signalling pathway in cells which lack the IRS protein.[17] The presence of redundant pathways is especially prominent in the reproductive axis, and many of these systems have been identified as a consequence of single-gene deletion studies.

An additional concept important to the understanding of cytokine and growth factor signal transduction is the principle of signal thresholds. During development, local ligand concentration is used to specify tissue differentiation. For instance, *Xenopus goosecoid* (*Xgsc*) and *Xenopus brachyury* (*Xbra*) are both expressed in the dorsal marginal zone of early *Xenopus* embryos, and specific spatial expression of these transcription factors is necessary for proper dorsal–ventral patterning.[18] A high concentration of the peptide hormone activin in the developing *Xenopus* embryo leads to the stimulation of *Xgsc* gene expression, while a low concentration activates the transcription of *Xbra*. Thus, local differences in activin concentration play a vital role in the development of the dorsal–ventral axis in *Xenopus*.

The degree of cellular response to a ligand is also specified by the availability of receptor and signalling pathway component proteins. Of course, the cellular expression of particular signalling components determines the cell's responsiveness to a ligand, and serves as the basis for tissue specificity. In addition, signalling by ligands that require the formation of large receptor complexes can be modulated by the availability and assembly of receptor components. For example, interleukin-6 (IL-6) receptor subunits exist in both membrane-tethered and soluble forms, and the assembly of the multimeric cell surface IL-6 receptor depends on the concentration balance between the two subunit forms.[19] Thus, receptor availability and assembly are just as important as local ligand concentration when determining the downstream cellular response.

Finally, signal down-regulation is an integral part of homeostatic regulation of cell signalling by growth factors and cytokines. Inactivation of a signal can be accomplished in several ways: the activation of phosphatases, the active removal of receptors from the cell surface or the initiation of opposing, inhibitory signalling pathways. Phosphatases abrogate signalling by reversing the activating phosphorylation of receptors and intracellular second messengers.[20,21] The active removal of cytokine and growth factor receptors from the cell surface by internalization in clathrin-coated pits also down-regulates signalling.[22,23] Negative modulation of signalling by the mammalian growth hormone (GH) receptor has also been shown to occur through the ubiquitin/proteosome mechanism.[24,25] Downstream of the receptor, inhibitory molecules of opposing signalling pathways can also down-regulate signalling. Activin and transforming growth factor-β (TGF-β) are able to regulate their own signal transduction negatively through the activation of the Smad6 and -7 proteins. These Smads interfere directly with the signalling capabilities of the activin/TGF-β second messengers, Smad2 and -3, thereby abrogating activin/TGF-β signalling.[26–28]

The remainder of this chapter describes the signal transduction mechanisms of the three major growth factor and cytokine receptor kinase families, and some examples of their roles in normal reproductive physiology and disease states. Although the RTKs, receptor-associated tyrosine kinases and RSKs are distinct, they share the primary features of using signalling networks and parallel, redundant signalling pathways, as well as exhibiting sensitivity to signalling thresholds and the various mechanisms of signal down-regulation.

INTRINSIC RECEPTOR TYROSINE KINASE SIGNALLING

Family action

The ligands that interact with transmembrane tyrosine kinase proteins include the EGFs, insulin and IGF-I, hepatic growth factor (HGF), vascular endothelial growth factors (VEGFs), the Eph family, fibroblast growth factors (FGFs), the nerve growth factor family (NGFs) and the platelet-derived growth factor family (PDGFs).[29] The RTK ligands are dimers of identical subunits (e.g. VEGF, HGF, insulin) or heterodimers of highly related subunits (for example, PDGF AA is a homodimer of two A subunits, PDGF BB is a homodimer of two B subunits, and PDGF AB is a heterodimer of the A and B subunits). In general, and within the reproductive tract, the RTK ligands function as autocrine (effect changes in the same cell in which ligand is produced) or paracrine (effect changes in nearby cells) regulators of mitosis, growth, angiogenesis and differentiation in specific target cells. The elimination or abnormal expression of genes encoding RTK ligands results in gross developmental abnormalities, tumor formation or neonatal lethality.[2,30] In the development of the mammary gland, lack of functional IGF or its receptor, IGFR, leads to loss of mammary tissue differentiation, while overexpression results in abnormal growth and, in some cases, tumor formation.[31]

The generic RTK receptor contains a large extracellular domain, a single transmembrane spanning region and an intracellular tyrosine kinase domain (Figure 1).[32] The extracellular domains of RTK family members are diverse, and lack an overall consensus motif. Excluding the insulin receptor and the HGF receptor, RTKs are monomeric. The insulin receptor is dimeric, and is cotranslationally cleaved into an α-chain and a β-chain, which assemble into α–β complexes stabilized by covalent modification (Figure 5b). The HGF receptor is post-translationally processed, resulting in the assembly of a short N-terminal protein domain covalently associated with the N-terminus of the membrane-bound receptor. Upon ligand binding, RTKs dimerize and undergo autophosphorylation of tyrosines on the opposite receptor subunit.[29] Downstream of the receptor, second messengers recognize and associate with the phosphorylated tyrosine residues on activated receptors through their modular SH2 and PTB domains (Figure 2).[33] In turn, receptor-activated second messengers propagate the extracellular signal by initiating one or more intracellular response pathways.

Epidermal growth factor

Epidermal growth factor is involved in epithelial cell differentiation and mesenchymal–epithelial cell interaction in the developing and adult body axis, and some members of this family are specifically involved in the development of central and peripheral nerve tracts.[34,35] Gene deletion of EGF results in a wide array of abnormalities resulting from abnormal or delayed epithelial cell development. EGF and other members of this protein family (TGF-α, neuregulin, cripto) are cleaved from large precursor proteins (for example, that for EGF contains 1217 amino acids) that are tethered to the membrane by a single transmembrane domain followed by a short cytoplasmic tail (Figure 4).[36,37] Mature EGF is 6 kDa in size and binds to the 170-kDa plasma membrane EGF receptor (ErbB1). Four Erb RTKs have been identified in mammals, and are able to form various homo- and heterodimers that bind specific members of the large EGF protein family, thereby permitting signal discrimination in a wide array of tissues.[38,39] Both membrane-bound and mature EGF regulate cell responses through ErbB1 homodimers. The activated ErbB1 RTK transduces EGF signals through members of at least three families of protein kinases, the stress-activated protein kinases (SAPKs), the JNKs or the extracellular signal-

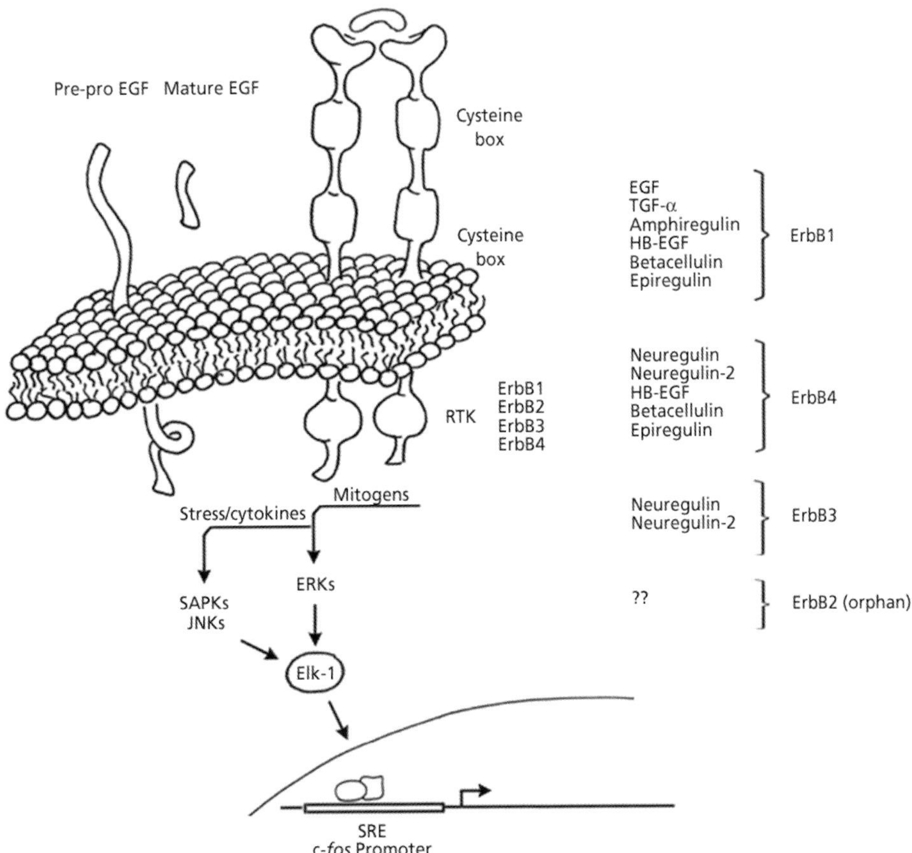

Figure 4 The epidermal growth factor (EGF) family of ligands and tyrosine kinase receptors. EGF ligands are cleaved from large, membrane-bound precursor proteins (pre-pro EGF). Soluble, mature EGF ligands bind to one of four EGF receptors (ErbB1–ErbB4). EGF receptors can form homo- or heterodimeric complexes upon ligand binding, providing some level of signalling specificity within this family of ligands. Ligand-bound EGF receptors dimerize and can activate members of at least three families of protein kinases: the stress-activated protein kinases (SAPKs), the Janus kinases (JNKs) or the extracellular signal-regulated kinases (ERKs). These kinases phosphorylate and activate the transcriptional cofactor, Elk-1, which translocates to the nucleus and binds responsive elements in the promoters of EGF-responsive genes. Shown here is the serum response element (SRE) in the promoter of the immediate early gene, c-*fos*. RTK, receptor tyrosine kinase; TGF-α, transforming growth factor-α; HB-EGF, heparin-binding EGF-like growth factor

regulated kinases (ERKs). Activation of these kinases leads to the induction of cellular growth and proliferation through the activation of transcriptional cofactors such as Elk-1. Similar to many cell surface receptors, EGF signalling can be down-regulated by internalization of the ligand-bound EGF receptor via clathrin-coated pits,[40] or by ubiquitination and targeting to the proteosome.[41]

EGF is a product of the ovary, and acts as an autocrine and paracrine regulator of a wide variety of gonadal functions, including steroidogenesis, induction of gonadotropin receptors, maturation of the oocyte and the stimulation of other growth factors such as IGF-I.[42–44] EGF receptors are found on all cell types in the ovary, and are regulated in ovarian tissue by gonadotropins during the normal reproductive cycle.[45,46]

Insulin and insulin-like growth factor-I

Insulin is a product of the pancreatic β-cell, and is essential to the management of glucose metabolism.[47] Insulin is synthesized as a single-chain pro-hormone, and, within the β-cell secretory granule, the pro-hormone is processed into a dimeric mature ligand and the bioinactive C-peptide (Figure 5a).[48] The molecular basis for insulin action through the insulin RTK is now well characterized, although subtle nuances are found in specific target tissues. The insulin receptor is a tetramer composed of two extracellular α-subunits and two extracellular β-subunits linked by disulfide bonds[49] (Figure 5b). The cytoplasmic domain of the insulin receptor exhibits intrinsic tyrosine kinase activity. Upon insulin binding, the insulin RTK undergoes conformational changes that result in a tight association between receptor subunits.[50] This is in contrast to the ligand-induced dimerization of monomeric RTKs such as Erb1. Autophosphorylation of tyrosine residues on the cytoplasmic face of the insulin receptor activates the RTK and leads to the phosphorylation and activation of the membrane-localized IRS proteins.[5] IRS proteins are modular, and contain SH2 domains that recognize and interact with phosphotyrosines of the insulin RTK (Figure 5b). Activated IRS proteins act as an insulin-signalling node by initiating several divergent signalling pathways, all of which work to achieve the same end: the efficient uptake, storage and metabolism of glucose.[51,52] Activation of phosphatidylinositol 3-kinase (PI-/3-K) by IRS leads to the activation of protein kinase C (PKC) and the stimulation of glucose transport through the GLUT4 transporter in insulin target tissues. Alternatively, IRS proteins can activate protein kinase B (PKB) and glycogen synthase kinase-3 (GSK-3) to increase glycogen synthesis. In addition, interaction of activated IRS with the growth factor receptor-binding protein-2/son of sevenless (Grb2/SOS) adaptor proteins leads to changes in gene transcription via the MAPK pathway. This pathway involves the immediate interaction of a small GTPase, ras, with Grb/SOS at the cell membrane, which activates a series of intracellular dual-specificity protein kinases: raf (mitogen activated protein kinase (MAPK)/ERK kinase (MEK) kinase or MAPK kinase kinase), the MEKs (MAPK/ERK kinases or MAPK kinases) and the MAPKs. Finally, insulin signalling can be negatively modulated through the activation of the SH2 domain-containing protein tyrosine phosphatase-2 (SHP2) by IRS, resulting in the dephosphorylation of IRS and the attenuation of Grb2/SOS and PI-3-kinase signalling.[53] Insulin signalling is also down-regulated by the internalization of insulin–insulin RTK complexes from the cell surface.[54]

The dominant stimulus for insulin secretion is increased postprandial blood glucose levels. Insulin-dependent (type I) diabetes mellitus results from inadequate insulin production in the face of a glucose challenge, while resistance to insulin action in target tissues can lead to the development of non-insulin-dependent (type II) diabetes mellitus.[55] In insulin-resistant patients, the levels of insulin become greatly elevated as the body attempts to compensate for the lack of insulin target tissue response. Insulin resistance is a common symptom of polycystic ovarian syndrome (PCOS).[56] Approximately 10% of women of reproductive age succumb to this disease, the symptoms of which also include chronic anovulation, hyperandrogenism, obesity and dyslipidemia. Whether insulin participates in the onset or progression of the disease is controversial, but it is clear that the ovaries of women with PCOS manifest an under-

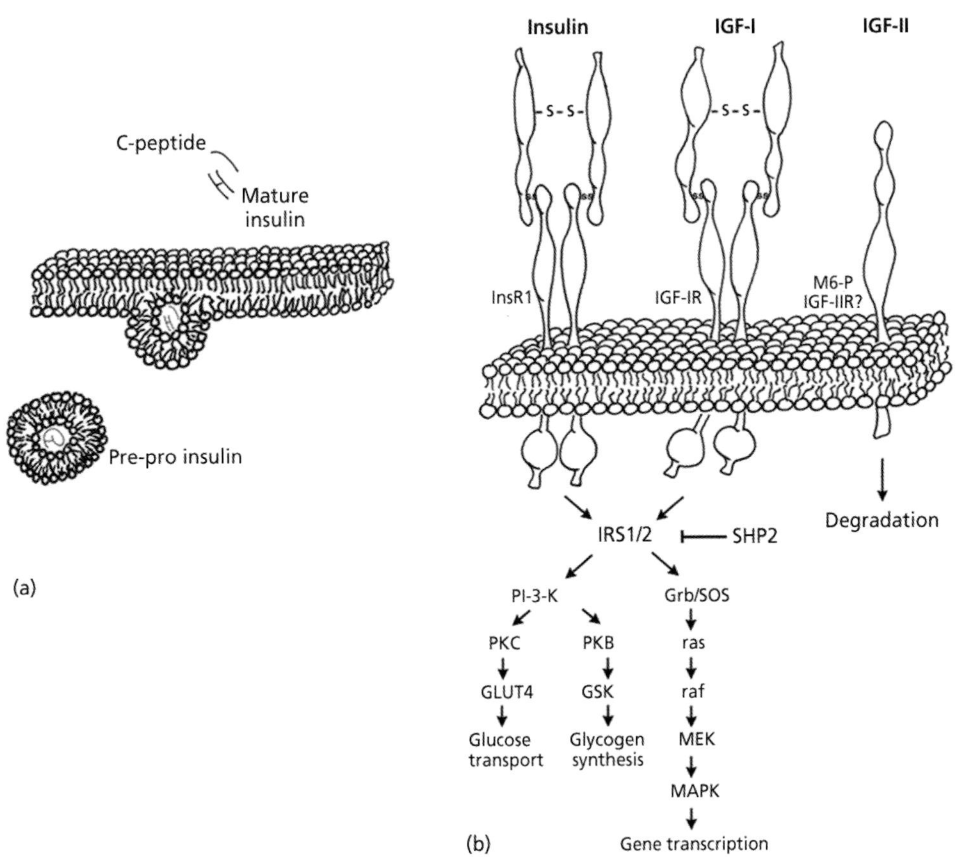

Figure 5 Insulin, insulin-like growth factors (IGFs) and signalling through the insulin receptor tyrosine kinases. (a) Insulin is synthesized as single-chain pre-pro insulin and is processed into dimeric mature ligand and the bioinactive C-peptide. (b) The insulin receptor (InsR1) is a tetramer that undergoes conformational changes upon ligand binding. The activated insulin receptor tyrosine kinase (RTK) phosphorylates and activates the insulin receptor substrate (IRS) proteins that in turn activate several downstream pathways. These include the activation of phospatidylinositol-3-kinase (PI-3-K), which leads to an increase in glucose transport and glycogen synthesis, and the activation of the mitogen-activated protein kinase (MAPK) cascade via the growth factor receptor-binding protein/son of sevenless (Grb/SOS) adaptor protein and the modulation of gene transcription. IGF-I signals through the IGF-I receptor (IGFIR) and can also activate the IRS proteins. Although no known IGF-II specific receptor has been identified, IGF-II can initiate its own degradation by binding to the mannose-6-phosphate (M6-P) receptor. PKC, protein kinase C; GLUT4, glucose transporter protein; PKB, protein kinase B; GSK, glycogen synthase kinase; SHP2, SH-containing protein tyrosine phosphatase-2; MEK, MAPK kinase

lying insulin resistance that compounds the ovulatory dysfunction.

Two closely related ligands, IGF-I and IGF-II, are structurally similar to insulin and are synthesized and secreted by most cell types (Figure 5b).[57] The IGF-I receptor is similar in structure to the insulin receptor and has been shown to interact with IRS1, thus raising questions regarding specificity of insulin and IGF signalling.[16,58,59] In fact, insulin and IGF-I are used interchangeably as granulosa cell survival factors and mitogens.[60] IGF-I is a product of the ovarian granulosa cell, and is regulated by both pituitary and gonadal hormones as well as by locally produced IGF binding proteins (IGFBPs).[61] Although an IGF-II-specific receptor has not been identified, IGF-II can bind to the mannose-6-phosphate (M6-P) receptor and initiate its own degradation.[58]

IGF-I signalling can be negatively modulated through the action of six distinct IGFBPs, which bind and neutralize active IGF-I.[62] The regulation of IGFBPs in the ovary is dynamic throughout the rat estrous and human menstrual cycles, indicating a temporal regulation of IGF-I action in the ovary.[63] In addition to its mitogenic effects on the ovarian granulosa cell, IGF-I acts in a synergistic manner with luteinizing hormone (LH) to induce androgen production in cultured theca and Leydig cells.[61]

THE CYTOKINE RECEPTORS AND RECEPTOR-ASSOCIATED TYROSINE KINASE SIGNALLING

Family action

The cytokine receptor family is involved in the regulation of growth and basal metabolism (growth hormone (GH), prolactin (Prl)), hematopoiesis (interleukins (Ils), erythropoietin (Epo)), cell development and differentiation (leukemia inhibitory factor (LIF)) and viral surveillance and defense (interferons (IFNs)).[64] The cytokine receptors are grouped into two classes: the class I hematopoietin receptors and the class II interferon receptors.[3,65] Both classes of cytokine receptors have large extracellular domains dominated by fibronectin type III domain repeats. Cytokine receptor complexes assemble from several different membrane-spanning proteins, making ligand binding extremely complicated.[66] Similar to RTKs, ligand binding to cytokine receptors induces receptor dimerization.[67] However, cytokine receptors lack intrinsic signalling activity, and instead rely on intracellular non-receptor tyrosine kinases to transduce a signal.[68] The receptor-associated kinases involved in cytokine signalling are members of the Janus kinase family of non-receptor tyrosine kinases: JAK1, JAK2, JAK3 and Tyk2. Some degree of cytokine signalling specificity is achieved through the selective activation of JAK proteins by various ligand-bound receptor complexes.[69] Cytokine signals are transduced through a non-covalent interaction between the dimerized cytokine receptor and two JAK proteins. Receptor-associated JAK kinases undergo transphosphorylation, and also phosphorylate residues on the cytokine receptors. The phosphorylated residues serve as docking sites for a variety of SH2-containing second messengers, including the signal tranducers and activators of transcription (STAT) proteins.[70,71] JAK phosphorylation of STATs leads to the formation of STAT dimers that translocate to the nucleus, where they bind to target genes and regulate transcription. STAT binding motifs are prevalent throughout the genome, giving rise to the notion that the assembly of multiple STATs or the association of STAT with other transcriptional coactivators provides some specificity for regulation of cytokine target gene activity.

Growth hormone

Growth hormone is produced by the somatotropes of the pituitary, and is the endocrine hormone primarily responsible for cellular growth.[72,73] GH acts directly on target cells, or indirectly through the stimulation of IGF-I synthesis by the liver. The growth hormone receptor (GHR) is a class I cytokine receptor, and has a large extracellular domain composed of fibronectin-like repeats (Figure 6a).[74,75] Surprisingly, only one growth hormone molecule binds two molecules of GHR, indicating that each receptor monomer recognizes a different binding motif on the single GH ligand. The necessity for ligand-induced GHR dimerization for signalling is confirmed by naturally occurring mutations in the receptor subunits that prevent dimerization and produce a GH-resistant phenotype, indicative of Laron-type dwarfism.[76]

The prolactin receptor, erythropoietin receptor and IL-2 receptor transduce signals in a manner analogous to GHR.[77] Assembly of GH receptor dimers engages downstream signal transduction through JAK2 non-receptor tyrosine kinase. The close proximity of ligand-bound GHR dimer creates a cytoplasmic receptor face, with motifs that are recognized and bound by inactive JAK2. Subsequently, JAK2 undergoes autophosphorylation and also transphosphorylates tyrosines within the cytoplasmic tails of the GHR dimer. Once phosphorylated, the GHR–JAK complex binds and phosphorylates the STAT proteins, primarily

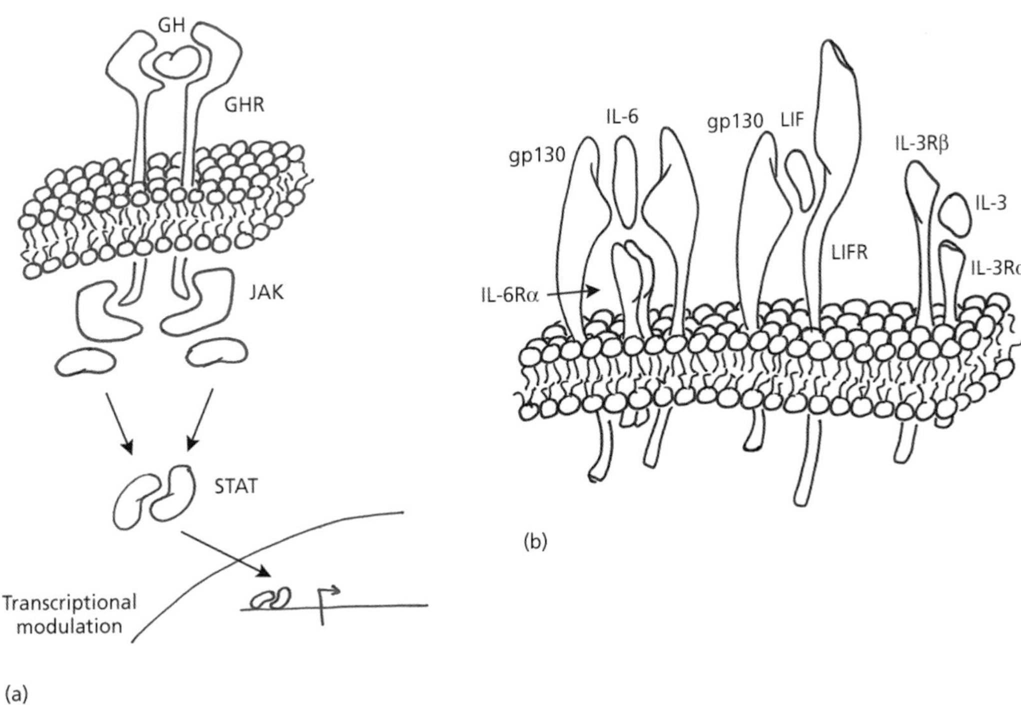

Figure 6 (a) Mechanism of growth hormone signal transduction. Growth hormone (GH) binds to its receptor, GHR, and induces receptor dimerization. Activated GHR dimers are recognized by the non-receptor protein tyrosine kinase, Janus kinase (JAK), which binds the intracellular tails of GHR. JAK in turn activates the downstream signalling proteins, the signal transducers and activators of transcription (STAT) proteins. STATs translocate to the nucleus and modulate GH-responsive target gene transcription. (b) Variations on signalling by class I cytokine receptors. Interleukin-6 (IL-6) binds a heteromeric receptor complex comprising IL-6Rα and gp130 proteins. Leukocyte inhibitory factor (LIF) signals are transduced through a receptor LIFR and gp130 assembly. IL-3 binds with high specificity to the IL-3α receptor and recruits a common β-chain (IL-3Rβ)

STAT5, which translocate to the nucleus to modulate gene expression.

Although not traditionally characterized as a reproductive hormone, GH receptors have been localized to both granulosa and theca cells within the ovary.[78] The effects of GH on the reproductive axis vary among species, and it is still unclear in *in vivo* studies whether effects of GH administration are indicative of a direct GH effect or an indirect effect of GH-stimulated IGF-I production in the liver. However, *in vitro* experiments on cultured ovarian follicles have demonstrated that GH treatment augments the effects of follicle stimulating hormone (FSH) and LH on follicular growth and steroidogenesis.

Interleukin-3, interleukin-6 and leukocyte inhibitory factor

A variation on the cytokine signalling theme is exemplified by interleukin-6 (IL-6) and LIF interaction with their respective class I receptors, the IL-6 receptor (IL-6Rα) and the LIF receptor (LIFR) (Figure 6b).[79] These receptors provide high-affinity binding sites and tissue specificity to their respective ligands only in the presence of a co-receptor, gp130. LIF-stimulated LIFR–gp130 assembly stimulates a JAK1/2–STAT signal transduction pathway.

LIF may be an important factor in the establishment of pregnancy and the viability of the early embryo.[80] LIF mRNA is detected in the uterine tissue of mice, sheep and humans, and its expression peaks at or near the time of blastocyst implantation. Furthermore, LIF knock-out female mice produce normal blastocysts which fail to implant; this may be due to a loss in the effects of LIF on the proliferation of the trophoblast layer that is necessary for the initiation of embryonic interaction with the uterine wall. In humans, LIF may play a later role in maintenance of the placenta and pregnancy, since treatment of cultured trophoblast with LIF stimulates expression of fibronectin (involved in attachment of the placenta to the uterus) and human chorionic gonadotropin, necessary for maternal recognition of pregnancy.

IL-6 also signals through a heteromeric receptor complex comprising IL-6Rα and gp130. However, the prevailing model describes a stoichiometry of two IL-6 ligands to two IL-6Rα and two gp130 receptors to form a hexameric complex.[81] Tetrameric complexes of homodimerized gp130 and two molecules of IL-6 have also been proposed.

Another class I ligand, IL-3, also utilizes multiple receptor subunits to activate the JAK–STAT signalling pathway.[70] IL-3 binds to its ligand-specific α-chain (IL-3α), which dimerizes with a common β-chain it shares with granulocyte-macrophage colony-stimulating factor (GM-CSF) and IL-5. JAK3 activity is initiated once the receptor complex is assembled.

RECEPTOR SERINE KINASE SIGNALLING

Family action

The TGF-β superfamily of proteins utilizes the third major class of kinase receptors, the serine kinase receptors (Figure 7). This superfamily comprises a large group of peptide growth factors including activin, inhibin, the bone morphogenic proteins (BMPs), anti-Müllerian substance (AMH, or Müllerian inhibiting substance (MIS)) and gonadal differentiation factor-9 (GDF-9), all of which regulate a variety of cellular functions including cell proliferation, differentiation and death.[82] Activin and TGF-β play broad roles in tissue proliferation in a number of target tissues, while the actions of other TGF-β ligands can be extremely tissue-restricted. For example, the gonadal peptide hormone inhibin acts on the anterior pituitary to inhibit FSH release, and MIS acts in a negative manner on the developing Müllerian duct.[83,84] The TGF-β superfamily ligands are homodimers or heterodimers of subunits that contain a highly conserved array of cysteine amino-acid residues.[85] The association between ligand subunits is either covalent (TGF-β, activin, inhibin, BMP) or non-covalent (GDF-9, BMP-15). The ligands of this family bind a discrete number of receptor molecules, and this interaction is modified by a growing number of accessory molecules or co-receptors that participate in cell-type specificity.[86]

Two families of RSKs have been identified (type I and type II), and conserved orthologs are found in *C. elegans*, *Drosophila melanogaster* and mammals.[10,87] Each ortholog can function identically in heterologous systems with complete fidelity, underscoring the evolutionary conservation of the signal-transduction mechanism for this superfamily. The ligand-binding type II receptors are composed of a small extracellular domain, a single transmembrane spanning region and an intracellular serine–threonine kinase domain (Figure 7). Ligand binding causes oligomerization and autophosphorylation of at least two type II receptors, and subsequent recruitment of two type I receptors into a large receptor complex. The juxtamembrane glycine- and serine-rich domain (GS-domain) of the type I receptor family is the immediate substrate for activated type II receptor kinase activity. To date, six type I receptors have been identified, each with ligand selectivity. In contrast to other family members, BMPs bind with low affinity to their type I receptors (BMPRIA and BMPRIB); however, full signal-transducing activity still depends on the presence of both type I and type II receptors in a ligand-regulated heteromeric receptor complex.[88]

An emerging theme in receptor serine kinase signalling concerns the actions of a variety of soluble and membrane-tethered accessory proteins or co-receptors.[86] These proteins are able to control the access of the available pool of TGF-β

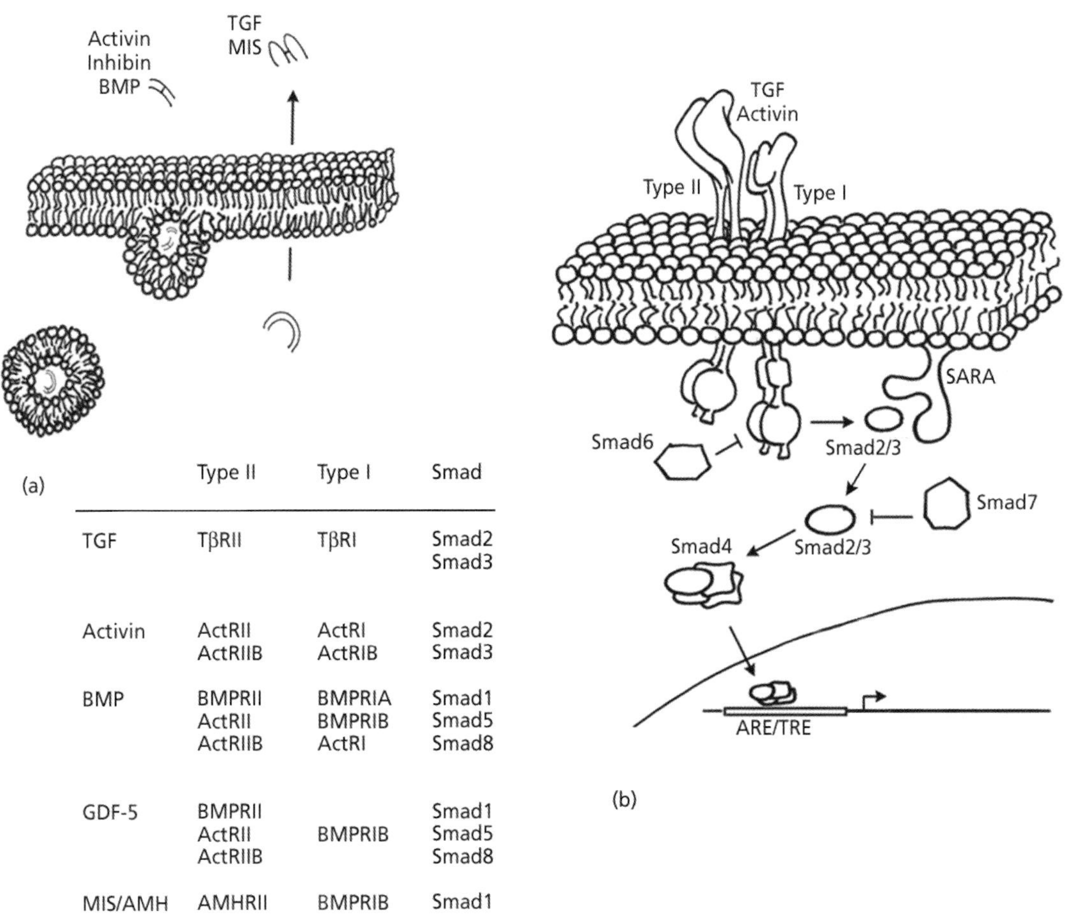

	Type II	Type I	Smad
TGF	TβRII	TβRI	Smad2 Smad3
Activin	ActRII ActRIIB	ActRI ActRIB	Smad2 Smad3
BMP	BMPRII ActRII ActRIIB	BMPRIA BMPRIB ActRI	Smad1 Smad5 Smad8
GDF-5	BMPRII ActRII ActRIIB	BMPRIB	Smad1 Smad5 Smad8
MIS/AMH	AMHRII	BMPRIB	Smad1

Figure 7 Signal transduction by the transforming growth factor-β (TGF-β) superfamily of ligands and their receptor serine kinases. (a) Ligands in the TGF-β superfamily are homodimers or heterodimers, which associate in a covalent (TGF-β, activin, inhibin, bone morphogenic protein (BMP)) or non-covalent (gonadal differentiation factor-5 (GDF-5), BMP-15) manner. (b) Signal transduction by these ligands occurs through two types of serine receptor kinases, the type II ligand-binding receptor and the type I signalling receptor. Signals are propagated by the intracellular Smad proteins, which can act as transcriptional cofactors. The table lists the known receptors and Smad proteins utilized by each ligand in this family. MIS, Müllerian inhibiting substance; SARA, Smad anchor for receptor activation; ARE/TRE, activin/TGF-β response element; AMH, anti-Müllerian hormone

superfamily ligands and signalling proteins to individual receptor components or to the type II/type I receptor complexes. The importance of these accessory proteins, which can be soluble, bioneutralizing or accessory proteins, or membrane-associated co-receptors, lies in their ability to restrict hormone action by rapidly neutralizing specific ligands, confining the action of ligands or colocalizing signalling components within the cell. Consistent with this theme, the actions of activin, nodal and TGF-β are specified and restricted by a variety of accessory proteins and co-receptors.[86] For example, the Smad anchor for receptor activation (SARA) is localized to the plasma membrane by a FYVE domain (Figure 2a) and interacts with the TGF-β second messenger, Smad2 (Figure 7). SARA enhances TGF-β signalling by localizing Smad2 proximal to the activated TGF-β receptor complex, where it can be efficiently phosphorylated.

Downstream of activated RSKs are the Smad proteins (Figure 7).[89] The term 'Smad' is a combination and contraction of the *C. elegans* TGF-β signalling protein, Sma, and its *D. melanogaster ortholog*, Mad. Smad proteins are phosphorylated by activated ligand–receptor complexes on a conserved C-terminal motif. The Smad proteins have type I receptor selectivity, which permits the large variety of type I/type II receptor complexes specifically to regulate a diverse array of activities in many tissues. BMP signals are propagated by the receptor-specific Smad1, -5 and -8, while Smad2 and -3 are phosphorylated as a consequence of receptor activation by TGF-β or activin. Smad4 is a 'common Smad' that associates with receptor-specific Smads to facilitate translocation of the Smad complex to the nucleus, where it acts as a transcriptional cofactor. Smad4 is able to bind DNA in conjunction with a diverse array of transcriptional cofactors, thus providing additional specificity of TGF-β superfamily action at the level of the gene. As mentioned above, Smad6 and -7 are inhibitory Smads that inhibit the TGF-β signalling pathway. Smad6 blocks phosphorylation of receptor Smads directly at the level of the receptor, while Smad7 blocks assembly of phosphorylated receptor-specific Smads into a complex with Smad4. The Smad proteins are also the target of an E3 ubiquitin ligase, Smurf1, that may contribute to down-regulation of BMP signalling by targeting Smad1 and Smad5 to the ubiquitin/proteosome mechanism.

Activin and transforming growth factor-β

Activin is a dimer of two β subunits, and binds to one of two activin type II receptors (ActRIIA and ActRIIB), which activate the activin type-IB receptor (ActRIB or Alk4).[90] Activin was originally identified as a stimulator of pituitary FSH release, but was later shown to act as a neuronal survival factor, an inductive factor in mesoderm development and a potent stimulator of hemoglobin production.[91] The action of activin can be blocked by a binding protein known as follistatin.[92] Two forms of follistatin have been identified in different cell types: a soluble, bioneutralizing form and a membrane-tethered form. The two forms have different affinities for activin, and are capable of either slowing or blocking activin's effects on a target cell. Interestingly, follistatin is highly regulated in the pituitary gland. As a result, many of activin's effects on pituitary FSH may be governed not by local production of activin, but rather through the availability of the ligand as determined by relative follistatin levels.[93]

Some of activin's actions can be blocked by its functional antagonist, inhibin.[94] Inhibin is composed of a β subunit, which it shares with activin, and a dissimilar α subunit. In physiological cases in which the α subunit is produced in excess relative to the β subunit, the formation of αβ inhibin dimers is favored over the production of the ββ activin dimer. Furthermore, the β subunit of the inhibin dimer permits inhibin to bind the type II activin receptor; however, the inhibin α subunit blocks further assembly of a higher-ordered type I/type II activin receptor complex. In addition, two inhibin-binding proteins, InhBP and betaglycan, have been shown to modulate the assembly and signalling capacity of the activin heteromeric receptor complex.[95,96] Thus, inhibin is able to modulate activin signal transduction through regulation of the activin dimer assembly, restriction of activin's access to its receptor complex and, via inhibin-binding co-receptors, antagonism of activin receptor assembly in certain signalling systems.

TGF-β is produced as an inactive hormone, and is controlled by a protease that cleaves the precursor protein from the bioactive C-terminal tail.[97] This 'on-board' protease ensures that the ligand is not active until needed by recipient cells. Several isoforms of TGF-β are expressed in mammalian cells, and, while highly homologous, have different binding affinities for the TGF-β type II (TβRII) and type I (TβRI or Alk5) receptors.[98] In fact, TGF-β2 has a low affinity for the TGF-β type II receptor except in the presence of an additional co-receptor, the type III TGF-β receptor, also called betaglycan.[99]

Although the signalling pathway of TGF-β and activin through their specific RSKs is well characterized, an ongoing dilemma involves the specificity of signalling through the Smad2 and -3 proteins. Both activin and TGF-β utilize Smad2 and -3 as intracellular second messengers, so how are signals by these two ligands perceived differently by a cell? It is possible that different motifs are recognized in the Smad proteins by the TGF-β and activin type I receptors, although this remains to be experimentally proven. Alternatively, other unidentified protein signalling components may be involved in ligand-specific responsiveness by Smad2 and -3.

SUMMARY

Although growth factor and cytokine receptors have been classified into three major groups, there are basic mechanisms underlying all growth factor and cytokine signalling. Ligand binding leads to necessary receptor conformational changes, intracellular signals are amplified by cascades of signalling proteins and cross-talk exists between distinct signalling pathways. These concepts provide a foundation on which we can expand our understanding of the growing number of signalling mechanisms employed by cells, both in general and in reproductive tissues. Many questions remain regarding signalling specificity and the nature of cross-talk occurring between signalling pathways. Transgenic technologies, coupled with the emerging field of functional genomics, will help to untangle the complex intricacies of cellular signal-transduction pathways.

REFERENCES

1. Bernard DJ, Woodruff TK. Genetic approaches to the study of pituitary follicle-stimulating hormone regulation. In Matzuk M, Brown CW, Kumar TR, eds. *Transgenics in Endocrinology: Contemporary Endocrinology*. Totowa, NJ: Humana Press, 2001: 297–316.
2. Avruch J. Receptor tyrosine kinases. In DeGroot LJ, Jameson JL, Burger HG *et al.*, eds. *Endocrinology*. Philadelphia: WB Saunders, 2001: 25–47.
3. Mathews LS, Gaddy-Kurten D. Hormone signalling via cytokine and receptor serine kinases. In Groot LJ, Jameson JL, Burger HG, *et al.*, eds. *Endocrinology*. Philadelphia: WB Saunders, 2001: 49–57.
4. Hunter T. A thousand and one protein kinases. *Cell* 1987 50: 823–829.
5. Granner DK. Hormones of the pancreas and gastrointestinal tract. In Murray RK, Granner DK, Mayes PA, Rodwell VW, eds. *Harper's Biochemistry*. Connecticut: Appleton and Lange, 1996: 581–598.
6. Sibley DR, Benovic JL, Caron MG, Lefkowitz RJ. Regulation of transmembrane signaling by receptor phosphorylation. *Cell* 1987 48: 913–922.
7. Whitmarsh AJ, Cavanagh J, Tournier C, Yasuda J, Davis RJ. A mammalian scaffold complex that selectively mediates MAP kinase activation. *Science* 1998 281: 1671–1674.
8. Karin M. Signal transduction from cell surface to nucleus in development and disease. *FASEB J* 1992 6: 2581–2590.
9. Plowman GD, Sudarsanam S, Bingham J, Whyte D, Hunter T. The protein kinases of *Caenorhabditis elegans*: a model for signal transduction in multicellular organisms. *Proc Natl Acad Sci USA* 1999 96:13603–13610.
10. Newfeld SJ, Wisotzkey RG, Kumar S. Molecular evolution of a developmental pathway: phylogenetic analyses of transforming growth factor-β family ligand, receptors, and Smad signal transducers. *Genetics* 1999 152: 783–795.
11. Jordan JD, Landau EM, Iyengar R. Signaling networks: the origins of cellular multitasking. *Cell* 2000 103: 193–200.
12. Carpenter G. The EGF receptor: a nexus for trafficking and signaling. *Bioessays* 2000 22: 687–707.
13. Miyazono K. Signal transduction by bone morphogenic protein receptors: functional roles of Smad proteins. *Bone* 1999 25: 91–93.
14. Kretzschmar M, Doody J, Massague J. Opposing BMP and EGF signaling pathways converge on the TGF-β family mediator Smad1. *Nature* 1997 389: 618–622.
15. LeRoith D. Regulation of proliferation and apoptosis by the insulin-like growth factor I receptor. *Growth Horm IGF Res* 2000 10 (Suppl A): S12–S13.
16. LeRoith D. Insulin-like growth factor I receptor signaling – overlapping or redundant pathways? *Endocrinology* 2000 141: 1287–1288.
17. Dews M, Prisco M, Peruzzi M *et al.* Domains of the insulin-like growth factor I receptor required for the activation of extracellular signal-regulated kinases. *Endocrinology* 2000 141: 1289–1300.
18. Dale L. Development: morphogen gradients and mesodermal patterning. *Curr Biol* 1997 7: R698–R700.
19. Falus A. Cytokine receptor architecture, structure and genetic assembly. *Immunol Lett* 1995 44: 221–223.
20. Byon JC, Kenner KA, Kusari AB, Kusari J. Regulation of growth factor-induced signaling by

protein-tyrosine-phosphatases. *Proc Soc Exp Biol Med* 1997 216: 1–20.

21. Chernoff J. Protein tyrosine phosphatases as negative regulators of mitogenic signaling. *J Cell Physiol* 1999 180: 173–181.
22. Ceresa BP, Schmid SL. Regulation of signal transduction by endocytosis. *Curr Opin Cell Biol* 2000 12: 204–210.
23. Sorkin A, Waters CM. Endocytosis of growth factor receptors. *Bioessays* 1993 15: 375–382.
24. Hicke L. Gettin' down with ubiquitin: turning off cell-surface receptors, transporters and channels. *Trends Cell Biol* 1999 9: 107–112.
25. van Kerkhof P, Sachse M, Klumperman J, Strous GJ. Growth hormone receptor ubiquitination coincides with recruitment to clathrin-coated membrane domains. *J Biol Chem* 2001 276: 3778–3784.
26. Afrakhte M, Moren A, Jossan S *et al.* Induction of inhibitory Smad6 and Smad7 mRNA by TGF-β family members. *Biochem Biophys Res Commun* 1998 249: 505–511.
27. Nakao A, Afrakhte M, Moren A *et al.* Identification of Smad7, a TGFβ-inducible antagonist of TGF-β signaling. *Nature* 1997 389: 631–635.
28. Imamura T, Takase M, Nishihara A, *et al.* Smad6 inhibits signaling by the TGF-β superfamily. *Nature* 1997 389: 622–626.
29. Schlessinger J. Cell signaling by receptor tyrosine kinases. *Cell* 2000 103: 211–225.
30. Mendelsohn J, Lippman ME. Principles of molecular cell biology of cancer: growth factors. In DeVita VT Jr, Hellman S, Rosenberg SA, eds. *Cancer: Principles and Practices of Oncology*. Philadelphia: Lippincott, Williams, and Wilkins, 2001: 114–133.
31. Hadsell DL, Bonnette SG. IGF and insulin action in the mammary gland: lessons from transgenic and knockout models. *J Mammary Gland Biol Neoplasia* 2000 5: 19–30.
32. Hubbard SR, Till JH. Protein tyrosine kinase structure and function. *Annu Rev Biochem* 2000 69: 373–398.
33. Sudol M. From Src homology domains to other signaling modules: proposal of the 'protein recognition code'. *Oncogene* 1998 17: 1469–1474.
34. Skeath JB. The *Drosophila* EGF receptor controls the formation and specification of neuroblasts along the dorsal–ventral axis of the Drosophila embryo. *Development* 1998 125: 3301–3312.
35. Sapir A, Schweitzer R, Shilo BZ. Sequential activation of the EGF receptor pathway during *Drosophila* oogenesis establishes the dorsoventral axis. *Development* 1998 125: 191–200.
36. Breyer JA, Cohen S. The epidermal growth factor precursor isolated from murine kidney membranes. Chemical characterization and biological properties. *J Biol Chem* 1990 265: 16564–16570.
37. Luetteke NC, Lee DC. Transforming growth factor α: expression, regulation and biological action of its integral membrane precursor. *Semin Cancer Biol* 1990 1: 265–275.
38. Sweeney C, Carraway KL. Ligand discrimination by ErbB receptors: differential signaling through differential phosphorylation site usage. *Oncogene* 2000 19: 5568–5573.
39. Riese DJ, Stern DF. Specificity within the EGF family/ErbB receptor family signaling network. *Bioessays* 1998 20: 41–48.
40. Vieira AV, Lamaze C, Schmid SL. Control of EGF receptor signaling by clathrin-mediated endocytosis. *Science* 1996 274: 2086–2089.
41. Stang E, Johannessen LE, Knardal SL, Madshus IH. Polyubiquitination of the epidermal growth factor receptor occurs at the plasma membrane upon ligand-induced activation. *J Biol Chem* 2000 275: 13940–13947.
42. Qu J, Nisolle M, Donnez J. Expression of transforming growth factor-α, epidermal growth factor, and epidermal growth factor receptor in follicles of human ovarian tissue before and after cryopreservation. *Fertil Steril* 2000 74: 113–121.
43. Hillier SG. A role for epidermal growth factor in the follicular paracrine system? *Clin Endocrinol (Oxf)* 1990 33: 427–428.
44. Leung PC, Steele GL. Intracellular signaling in the gonads. *Endocr Rev* 1992 13: 476–498.
45. Moreno-Cuevas J, Khan-Dawood FS. Epidermal growth factor receptors in rat ovarian tissue. *Tissue Cell* 1997 29: 55–62.
46. Feng P, Knecht M, Catt K. Hormonal control of epidermal growth factor receptors by gonadotropins during granulosa cell differentiation. *Endocrinology* 1987 120: 1121–1126.
47. Cheatham B, Kahn CR. Insulin action and the insulin signalling network. *Endocr Rev* 1995 16: 117–142.

48. Halban PA. Proinsulin processing in the regulated and the constitutive secretory pathway. *Diabetologia* 1994 37: S65–S72.
49. Czech MP. The nature and regulation of the insulin receptor: structure and function. *Annu Rev Physiol* 1985 47: 357–381.
50. Ottensmeyer FP, Beniac DR, Luo RZ, Yip CC. Mechanism of transmembrane signaling: insulin binding and the insulin receptor. *Biochemistry* 2000 39: 12103–12112.
51. Withers DJ, White M. Perspective: the insulin signaling system – a common link in the pathogenesis of type 2 diabetes. *Endocrinology* 2000 141: 1917–1921.
52. Pessin JE, Saltiel AR. Signaling pathways in insulin action: molecular targets of insulin resistance. *J Clin Invest* 2000 106: 165–169.
53. Elchebly M, Cheng A, Tremblay ML Modulation of insulin signaling by protein tyrosine phosphatases. *J Mol Med* 2000 78: 473–482.
54. Di Guglielmo GM, Drake PG, Baass PC *et al.* Insulin receptor internalization and signalling. *Mol Cell Biochem* 1988 182: 59–63.
55. Kahn CR. Insulin action, diabetogenesis, and the cause of type II diabetes. *Diabetes* 1994 43: 1066–1084.
56. Dunaif A. Insulin resistance and the polycystic ovary syndrome: mechanism and implications for pathogenesis. *Endocr Rev* 1997 18: 774–800.
57. LeRoith D, Roberts CT Jr. Insulin-like growth factors and their receptors in normal physiology and pathological states. *J Pediatr Endocrinol* 1993 6: 251–255.
58. LeRoith D, Werner H, Beitner-Johnson D, Roberts CT Jr. Molecular and cellular aspects of the insulin-like growth factor I receptor. *Endocr Rev* 1995 16: 143–163.
59. Dey BR, Frick K, Lopaczynski W, Nissley SP, Furlanetto RW. Evidence for the direct interaction of the insulin-like growth factor receptor with IRS-1, Shc, and Grb10. *Mol Endocrinol* 1996 10: 631–641.
60. Gong JG, McBride D, Bramley TA, Webb R. Effects of recombinant bovine somatotrophin, insulin-like growth factor-I and insulin on the proliferation of bovine granulosa cells *in vitro*. *J Endocrinol* 1993 139: 67–75.
61. Wang H-S, Chard T. IGFs and IGF-binding proteins in the regulation of ovarian and endometrial function. *J Endocrinol* 1999 161: 1–13.
62. Baxter RC. Insulin-like growth factor (IGF)-binding proteins: interactions with IGFs and intrinsic bioactivities. *Am J Physiol Endocrinol Metab* 2000 278: E967–E976.
63. Monget P, Besnard N, Huet C, Pisselet C, Monniaux D. Insulin-like growth factor-binding proteins and ovarian folliculogenesis. *Horm Res* 1996 45: 211–217.
64. Touw IP, De Koning JP, Ward AC, Hermans MH. Signalling mechanisms of cytokine receptors and their perturbances in disease. *Mol Cell Endocrinol* 2000 160: 1–9.
65. Neet K, Hunter T. Vertebrate non-receptor protein-tyrosine kinase families. *Genes Cells* 1996 1:147–169.
66. Stahl N, Yancopoulos GD. The alphas, betas, and kinases of cytokine receptor complexes. *Cell* 1993 74: 587–590.
67. Heldin CH. Dimerization of cell surface receptors in signal transduction. *Cell* 1995 80: 213–223.
68. Kishimoto T, Taga T, Akira S. Cytokine signal transduction. *Cell* 1994 76: 253–262
69. Silva CM, Lu H, Weber MJ, Thorner MO. Differential tyrosine phosphorylation of JAK1, JAK2, and STAT1 by growth hormone and interferon-γ in IM-9 cells. *J Biol Chem* 1994 269: 27532–27539.
70. Reddy EP, Korapati A, Chaturvedi P, Rane S. IL-3 signalling and the role of Src kinases, JAKs and STATs: a covert liaison unveiled. *Oncogene* 2000 19: 2532–2547.
71. Ihle JN. STATs: signal transducers and activators of transcription. *Cell* 1996 84: 331–334.
72. Strobl JS, Thomas MJ. Human growth hormone. *Pharmacol Rev* 1994 46: 1-34.
73. Argente J, Pozo J, Chowen JA. The growth hormone axis: control and effects. *Horm Res* 1996 45 (Suppl): 9–11.
74. Kopchick JJ, Andry JM. Growth hormone (GH), GH receptor, and signal transduction. *Mol Genet Metab* 2000 71: 293–314.
75. Thomas MJ. The molecular basis of growth hormone action. *Growth Horm IGF Res* 1998 8: 3–11.
76. Duquesnoy P, Sobrier ML, Duriez B *et al.* A single amino acid substitution in the exoplasmic domain of the human growth hormone (GH) receptor confers familial GH resistance (Laron syndrome) with positive GH-binding activity by abolishing

receptor homodimerization. *EMBO J* 1994 13: 1386–1395.

77. Carter-Su C, Smit LS. Signalling via JAK tyrosine kinases: growth hormone receptor as a model system. *Recent Prog Horm Res* 1998 53: 61–82.
78. Hull KL, Harvey S. Growth hormone: roles in female reproduction. *J Endocrinol* 2001 168: 1–23.
79. Auernhammer CJ, Melmed S. Leukemia-inhibitory factor – neuroimmune modulator of endocrine function. *Endocr Rev* 2000 21: 313–345.
80. Vogiagis D, Salamonsen LA. The role of leukaemia inhibitory factor in the establishment of pregnancy. *J Endocrinol* 1999 160: 181–190.
81. Grotzinger J, Kernebeck T, Kallen KJ, Rose-John S. IL-6 cytokine receptor complexes: hexamer, tetramer or both? *Biol Chem* 1999 380: 803–813.
82. Woodruff TK. Regulation of cellular and system function by activin. *Biochem Pharmacol* 1997 55: 953–963.
83. de Jong FH, Grootenhuis AJ, Klaij IA, Van Beurden WM. Inhibin and related proteins: localization, regulation, and effects. *Adv Exp Med Biol* 1990 274: 271–293.
84. Ingraham HA, Hirokawa Y, Roberts LM, Mellon SH, McGee E. Autocrine and paracrine Müllerian inhibiting substance hormone signaling in reproduction. *Recent Prog Horm Res* 2000 55: 53–67.
85. McDonald NQ, Hendrickson WA. A structural superfamily of growth factors containing a cysteine knot motif. *Cell* 1993 73: 421–424.
86. Bernard DJ, Chapman SC, Woodruff TK. An emerging role for co-receptors in inhibin signal transduction. *Mol Cell Endocrinol* 2001 180: 55–62.
87. Piek E, Heldin CH, Ten Dijke P. Specificity, diversity, and regulation in TGF-β superfamily signaling. *FASEB J* 1999 13: 2105–2124.
88. Liu F, Ventura F, Doody J, Massague J. Human type II receptor for bone morphogenic proteins (BMPs): extension of the two-kinase receptor model to the BMPs. *Mol Cell Biol* 1995 15: 3479–3486.
89. Wrana JL, Attisano L. The Smad pathway. *Cyt Growth Factor Rev* 2000 11: 5–13.
90. Mathews LS. Activin receptors and cellular signaling by the receptor serine kinase family. *Endocr Rev* 1994 15: 310–325.
91. Pangas SA, Woodruff TK. Activin signal transduction pathways. *Trends Endocrinol Metab* 2000 11: 309–314.
92. Hashimoto O, Kawasaki N, Tsuchida K *et al.* Difference between follistatin isoforms in the inhibition of activin signalling: activin neutralizing activity of follistatin isoforms is dependent on their affinity for activin. *Cell Signal* 2000 12: 565–571.
93. Besecke LM, Guendner MJ, Sluss PA *et al.* Pituitary follistatin regulates activin-mediated production of follicle-stimulating hormone during the rat estrous cycle. *Endocrinology* 1997 138: 2841–2848.
94. Bernard DJ, Chapman SC, Woodruff TK. Mechanisms of inhibin signal transduction. *Recent Prog Horm Res* 2001 56: 417–450.
95. Lewis KA, Gray PC, Blount AL *et al.* Betaglycan binds inhibin and can mediate functional antagonism of activin signalling. *Nature* 2000 404: 411–414.
96. Chapman SC, Woodruff TK. Modulation of activin signal transduction by inhibin B and inhibin-binding protein (InhBP). *Mol Endocrinol* 2001 15: 668–679.
97. Saharinen J, Hyytiainen M, Taipale J, Keski-Oja J. Latent transforming growth factor-β binding proteins (LTBPs) – structural extracellular matrix proteins for targeting TGFβ action. *Cyt Growth Factor Rev* 1999 10: 99–117.
98. Massague J. The transforming growth factor-β family. *Annu Rev Cell Biol* 1990 6: 597–641.
99. Lopez-Casillas F, Wrana JL, Massague J. Betaglycan presents ligand to the TGF β signaling receptor. *Cell* 1993 73: 1435–1444.

Molecular pharmacology

Anthony J. Rutherford

Nowhere is the impact of molecular biology more evident in clinical practice than in applied pharmacology. Tools that allow an understanding of the molecular structure of naturally occurring hormones provide a mechanism to closely examine their component parts, to help unravel their essential structure–function relationships. Modifications to these base molecules produce dramatic changes in function. This section comprises three chapters looking at each level of the hypothalamic–pituitary–gonadal axis, starting with gonadotropin-releasing hormone (GnRH) analogs, which modify pituitary gonadotropin release, the manufacture of pituitary and placental gonadotropins, and their mimics, which affect ovarian and testicular function, and finally, selective estrogen receptor modulators (SERMs), which alter end-organ response.

The structure of native GnRH, a relatively short decapeptide, has been known since 1971. Substitution of the glycine in position 6 by an unnatural D amino acid changes the natural degradation process, producing a compound with a longer half-life and a much greater binding affinity for the GnRH receptor. These GnRH agonists bind, activate, then continue to occupy the GnRH receptor, causing desensitization. This results in an initial rise in gonadotropin secretion, followed by a fall of both luteinizing hormone (LH) and follicle stimulating hormone (FSH) levels. Further modifications to GnRH, introducing multiple amino acid substitutions, produce a completely different class of compounds – the GnRH antagonists. These all lack histidine in position 2, which is essential for biological activity. They bind tightly to the receptor but do not effect a response, acting as competitive inhibitors of native GnRH, blocking the release of LH and FSH. Thus, GnRH analogs are employed where FSH and LH suppression may be beneficial. Originally, the agonists were confined to use in prostate carcinoma, but their role has expanded to benign disease in the female, and as a means of suppressing endogenous LH release in superovulation protocols. Long-acting agonists are available, which make them particularly useful for prolonged therapy. GnRH antagonist use is limited by a relatively short half-life, although they have proved very effective and patient-friendly as part of a simplified superovulation strategy.

Gonadotropins are much more complex molecules, consisting of a common α and a non-covalently bonded specific β subunit, each coded by different genes. All β subunits are glycosylated and contain quite diverse N-linked oligosaccharides, essential for their respective hormone-specific activity. In addition, chorionic gonadotropin contains O-linked oligosaccharides. Most human pituitary gonadotropins have oligosaccharides, which contain terminal sulfates, apart from FSH which contains sialic acid residues. The heterogeneity of the circulating glycoproteins is partly determined by the patterns of sulfation or sialylation. The ability to manufacture recombinant gonadotropins has opened exciting avenues of research, allowing an understanding of the complex structure–function relationships of the two subunits. By introducing specific modifications in the coding sequence of the gonadotropin genes, the residues essential for subunit assembly, secretion, biological activity and half-life have been identified. This has allowed manipulation of the two subunits to produce various novel gonadotropins and gonadotropin mimics. By combining the carboxyl terminal peptide from human chorionic gonadotropin-β (hCGβ) with the FSHβ subunit, a potent long-acting FSH has been created. In further genetic experiments the C-terminus of the α and the N-terminus of the β subunits were fused to form a single-chain molecule, which appears to have similar levels of biological activity to the native hormone. This important finding has introduced the possibility of producing mini gonadotropins, which potentially could be administered intranasally. Another model, where two β chains (FSH and CG), were attached to one α chain, produced a chimera that could bind and activate both receptors. This gives the possibility of producing analogs with dual functions, with variable ratios of intrinsic activity.

The final chapter covers the SERMs, non-steroidal compounds that affect the estrogen receptors (ERα and ERβ) in opposite ways in differing tissues, having a positive estrogenic effect at some sites and an antiestrogenic effect at others. The best-known SERMs are tamoxifen and raloxifene. Although the precise mechanism of action is still unknown, evaluation of the effect of the SERMs on ERα has shown that the ER surface amino acid D351 plays a crucial role. In addition, differing proportions of the two ER receptor types could influence the effect of the SERM at specific target sites. Their principal application is in postmenopausal patients where some estrogenic activity is desired to prevent osteoporosis and lower the risks of ischemic heart disease, but where estrogenic activity on the breast would be detrimental by increasing the risk of breast carcinoma. The first drug demonstrated to have this differential effect was tamoxifen. This decreases the incidence of breast carcinoma in high-risk women yet protects against osteoporosis. Raloxifene prevents osteoporosis and significantly reduces the risk of newly diagnosed breast carcinoma, with the additional benefit of lowering levels of circulating cholesterol. At present, a number of new multifunctional compounds are in the process of development.

These chapters demonstrate that molecular biology has already had a significant impact in clinical pharmacology. The pace of change is truly remarkable, and over the next decade the medications we employ in reproductive medicine are likely to change beyond recognition, with significant benefits for both men and women.

Chapter 17

Gonadotropin-releasing hormone analogs

Nathalie Chabbert-Buffet, Bart C.J.M. Fauser and Philippe Bouchard

BACKGROUND

The pivotal role of the neurohormone, gonadotropin-releasing hormone (GnRH I), in the control of reproduction has triggered the development of new therapeutics, such as the administration of pulsatile GnRH with minipumps to treat patients with hypothalamic hypogonadism, and the suppression of luteinizing hormone (LH) and follicle stimulating hormone (FSH) and gonadal function through inhibition of GnRH action with GnRH analogs.

GnRH analogs were synthesized in the months following elucidation of the structure of GnRH in 1971 by Schally and his co-workers, Baba, Arimura and Matsuo.[1] Several thousands of molecules were produced, with the initial hope that these analogs would improve fertility. By 1979, it was realized that the repeated administration of GnRH superanalogs (GnRH agonists) was in fact producing a decrease in gonadal function, associated with a significant drop in sex steroid levels following an initial increase.[2] It was several years before this phenomenon could be explained, when the measurement of LH improved through the use of immunoradiometric and immunofluorimetric assays. These were able to show that, in parallel to the decrease of testosterone or estradiol following agonist treatment, LH levels were decreased, and not maintained as was initially wrongly assumed based upon radioimmunossay results. Finally, it was also shown that the decrease in LH and the corresponding sex steroid was due to desensitization of the GnRH receptor at the level of the pituitary gonadotrope cell membranes, exactly as if GnRH was administered in a continuous fashion and not in a pulsatile manner as physiology requires.[3,4]

CURRENT CONCEPTS

Treatment with a GnRH agonistic analog results in an initial rise in gonadotropin secretion because of its binding to GnRH receptors and their subsequent activation. Secondary to the continuous occupation of the receptors, LH levels decrease, while FSH levels also decrease but to a lesser degree, owing to the desensitization process.[5] If the agonists are administered for several months, LH levels remain suppressed, while FSH returns to normal and eventually may rise to supraphysiological levels.[6] The mechanisms underlying the desensitization and the changes in FSH levels, or even in the α subunit, are poorly understood. However, it is known that the desensitization, as mentioned above, is not specific to agonistic analogs and can also be observed with continuous administration of GnRH itself. Also, it does not require a decrease in receptor numbers or in calcium-mediated events.[7] Until recently, the use of GnRH analogs has been restricted to GnRH agonists. Their indications are now well established and concern the suppression of LH-dependent estradiol/testosterone-dependent disorders. The world market for these drugs represents more than a billion US dollars annually. Their use comprises mainly the palliative treatment of advanced prostate cancers (over 50% of the market), and the adjunctive therapy of premenopausal receptor-positive breast cancers. In gynecology, GnRH agonists are used in the preoperative treatment of uterine leiomyomas and in the treatment of severe endometriosis. A significant indication is now represented by assisted reproductive technologies (25% of the market). Other indications are marginal and include: gonadotropin-dependent precocious puberty, delay of onset of normal puberty, hyperandrogenism, menometrorrhagia, catamenial disorders and endometrial hyperplasia.

Many agonists are commercially available, and they all derive from the substitution of glycine 6 by an unnatural D amino acid. The glycine 6 is a site of enzymatic proteolysis. This substitution increases the half-life of the compound to a duration much longer than that of the native hormone (3 min), enabling it to desensitize GnRH cellular mechanisms of action through a non-calcium, non-receptor-mediated mechanism. In addition, GnRH agonists have a binding affinity for GnRH receptors that is 100–200 times higher than that of native GnRH. All products have similar mechanisms of action and potencies. Differences in efficacy are mainly due to the route of administration. For example, intranasal formulations require multiple daily administrations and are less efficient than subcutaneous or depot preparations. Recent progress in the field is linked to the development of new, long-acting formulations of agonists such as 3-month depot preparations or even longer-term experimental formulations.

GnRH antagonists were discovered as early as 1972. Their mechanism of action is completely different from that of agonists since, although they bind to GnRH receptors with high affinity, they are unable to activate the transducing capacity of the receptors because of their inactive nature. Thus, they behave as competitive inhibitors.[8–10] The first antagonistic analogs are now available (cetrorelix, ganirelix), and their use is thus far restricted to suppression of the LH surge in superovulation protocols in assisted reproduction techniques such as *in vitro* fertilization (IVF) and intracytoplasmic sperm injection (ICSI).[11,12] GnRH antagonists result from multiple amino-acid substitutions in the native GnRH structure. As mentioned above, the antagonists are inactive but able to bind to GnRH receptors with high affinity, hence behaving as competitive inhibitors of GnRH binding.[8] The site of binding of GnRH antagonists seems to differ from that of the agonists.[13] In addition, receptor dimerization that occurs following GnRH agonist binding does not seem to occur following antagonist binding.[7] Subcutaneous administration of GnRH antagonists results in the rapid decrease of bioactive as well as immunoreactive LH, followed by a secondary decrease in FSH levels. The latter effect is significant following several injections of antagonists, probably because of the long half-life of FSH.[14] This confirms that there is no differential control for LH and FSH, and justifies the terminology of GnRH rather than luteinizing hormone-releasing hormone (LHRH). It is still debated whether GnRH receptor levels decrease following antagonist administration. Schally and colleagues have reported a significant decrease in GnRH receptors, 3–6 h following cetrorelix administration in the rat. This decrease was even higher than that observed following agonist treatment. Receptor gene transcription was also significantly decreased.[15] The clinical significance of this phenomenon is still unclear, as other authors have reported that, regardless of duration of treatment with antagonists, a response to exogenous GnRH was maintained.[16]

GnRH antagonists (Table 1) are produced by the substitution of between four and six amino acids by unnatural D amino acids. Since 1972, when the first antagonist was synthesized by the removal of histidine in position 2 (an amino acid essential for biological action), many antagonists have been produced.[8,17] Their development was limited, however, because of side-effects related to their ability to release histamine from the skin at the site of injection, or eventually to provoke systemic reactions.[18] This is especially true for the DArg6 class of antagonists. New generations of GnRH antagonists devoid of significant histamine-releasing properties are now available for clinical use.[19] The more advanced compounds include cetrorelix and

Table 1 Structures of gonadotropin-releasing hormone (GnRH) antagonists

Abarelix†	Ac-DNal-DCpa-DPal-Ser-N-Me-Tyr-DAsn-Leu-ILys-Pro-DAla-NH_2
Acyline††	Ac-DNal-DCpa-DPal-Ser-Aph-DAph(Ac)-Leu-ILys-Pro-DAla-NH_2
Antarelix‡	Ac-DNal-DCpa-DPal-Ser-Tyr-DHci-Leu-ILys-Pro-DAla-NH_2
Iturelix*	Ac-DNal-DCpa-DPal-Ser-Lys(Nic)-DLys(Nic)-Leu-ILys-Pro-DAla-NH_2
Azaline B	Ac-DNal-DCpa-DPal-Ser-Aph(Atz)-DAph(Atz)-Leu-ILys-Pro-DAla-NH_2
Cetrorelix**	Ac-DNal-DCpa-DPal-Ser-Tyr-DCit-Leu-Arg-Pro-DAla-NH_2
Ganirelix***	Ac-DNal-DCpa-DPal-Ser-Tyr-DHArg(Et_2)-Leu-Harg(Et_2)-Pro-DAla-NH_2
Nal-Glu	Ac-DNal-DCpa-DPal-Ser-Arg-DGlu(AA)-Leu-Arg-Pro-DAla-NH_2
FE 200486‡‡	Ac-DNal-DCpa-DPal-Ser-Aph(Hor)-DAph(Cba)-Leu-ILys-Pro-DAla-NH_2
GnRH	pGlu-His-Trp-Ser-Tyr-Gly-Leu-Arg-Pro-Gly-NH_2

Products already on the market: *Antide®, Lab. Ares Serono; **Cetrotide®, Lab. Ares Serono; ***Antagon®, Orgalutran®, Lab. Organon: half-life 12.8 h single dose, 16.2 h multiple dose

Products in development: †Praecis Pharmaceuticals: peptide polymer complex for sustained delivery, ††patent assigned to Salk Institute, developed by National Institutes of Health; ‡Teverelix®, Lab. Europeptides; ‡‡Lab. Ferring ganirelix,[20,21] and other antagonists under development are abarelix, antarelix (teverelix), FE 200486 and acyline.[22] The antagonists suppress sex steroids to castration levels at milligram doses. This difference from agonists, which require microgram doses for their action, is probably due to the fact that competitive inhibitors need to occupy all GnRH receptors in a continuous manner. The pharmacokinetic properties of GnRH antagonists remain somewhat of a mystery, since the plasma half-life can vary from a few hours to more than 30 h or longer, depending on the dose administered and is also related to gel formation, probably at the site of injection. Gel formation is under the influence of plasma NaCl concentrations. This intrinsic 'depot' property has been used in the manufacture of long-acting forms of antagonists. Although many companies claim to have produced long-acting preparations, no long-term reproducible data have been reported to date. It remains established, however, that the antagonist effect and duration of action are proportional to its plasma level.

Finally, GnRH receptors have been described on gonadotrope cells (GnRHr I),[23,24] while the presence of extrapituitary receptors in humans, including the GnRHr II recently cloned, remains controversial.[24,26] Although many studies have produced data which support a role of GnRH and its analogs *in vitro*, no convincing *in vivo* extrapituitary effect of GnRH analogs has been demonstrated.[27–31]

GnRH agonists in clinical practice

Consistent with their mechanism of action, GnRH agonists are used in the treatment of gonadotropin- and/or sex steroid-dependent pathological conditions. This represents three main therapeutic areas: assisted reproductive technologies (ART), sex steroid-dependent tumors and central, gonadotropin-dependent precocious puberty. Among sex steroid-dependent tumors, metastatic prostate cancer is the main indication, but metastatic breast cancer, fibroids and endometriosis are also indications for GnRH agonist therapy. Except in the case of ART where the GnRH agonist treatment is of short duration, the two main limitations to such treatment are the consequences of sex steroid deprivation. The immediate consequence is hot flushes, while the longer-term complication is bone loss. In benign diseases, these complications can be prevented by the addition of sex steroids to the GnRH agonist (add-back therapy) without compromise of therapeutic efficacy. In malignancy, hot flushes are unavoidable and unpreventable, while osteoporosis can be treated by bisphosphonates (limiting bone resorption).

In vitro fertilization

During the physiological menstrual cycle, terminal ovarian follicular growth is modulated by gonadotropins (FSH in the early follicular phase and LH in the late follicular phase), allowing follicular rescue from atresia. Gonadotropin secretion is under the control of the pulsatile hypothalamic secretion of GnRH. The role of GnRH during the mid-cycle LH surge in women is still unclear. In particular, the existence of a GnRH surge has not been demonstrated in women. However, GnRH seems to be necessary to initiate and maintain the LH surge, although the amount of GnRH necessary is still unknown. The role of a direct effect of estradiol at the level of the hypothalamus has been documented in ewes.[32] GnRH agonist treatment as discussed above induces a short-term pituitary stimulation ('flare-up') followed by pituitary desensitization, which inhibits the synthesis and release of gonadotropins.

During controlled ovarian hyperstimulation (COH), one major cause of cycle cancellation is the occurrence of a premature LH peak, owing to the very high estradiol levels obtained during multifollicular growth (8–20% of women).[33] It has thus been postulated that a GnRH agonist could be used to block the endogenous LH surge by suppressing endogenous gonadotropin synthesis and release. This would avoid premature granulosa cell luteinization and thus allow properly scheduled triggering of ovulation by exogenous human chorionic gonadotropin (hCG)/LH. Furthermore, a need for standardization and scheduling has arisen with the rapidly increasing number of patients. GnRH agonists were first studied in patients with a history of poor response to gonadotropins alone. The protocols were then soon applied to all women undergoing IVF. Different protocols are used, according to the duration of GnRH agonist administration. The three main types are long, short and ultrashort (or flare-up) regimens (Figure 1). The long regimen was the first to be used, and is currently the most common. The administration of GnRH beginning in the luteal phase preceding COH achieves very low LH levels during the follicular phase of the stimulated cycle, and prevents the endogenous LH surge. This regimen implies a longer duration of GnRH agonist administration and higher doses of gonadotropins. It is thus more expensive than other protocols. However, it allows a better follicular yield as well as good patient scheduling. The short regimen was based on the idea that endogenous gonadotropins could be used in association with exogenous gonadotropins, allowing cheaper treatment protocols. In these regimens, GnRH agonists are administered from day 1 of ovarian stimulation (Figure 1).

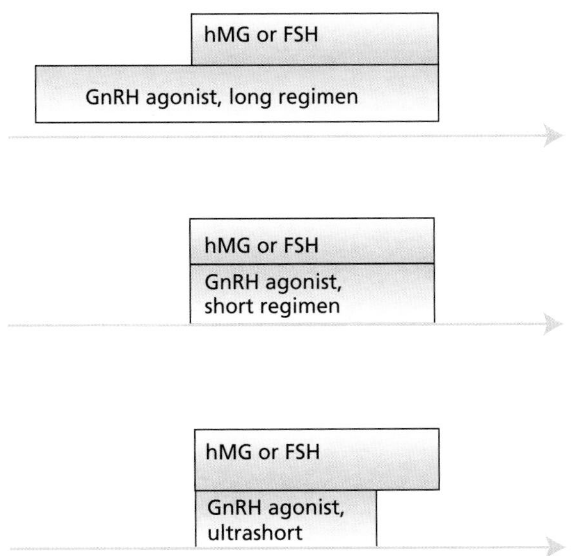

Figure 1 Schematic representation of the most widely used protocols in assisted reproductive technologies. In the long protocol, the gonadotropin-releasing hormone (GnRH) agonist is started before stimulation, in the mid-luteal phase of the pretreatment cycle, while gonadotropins are started on day 7 of the subsequent cycle. In short protocols, GnRH agonist and gonadotropins are started simultaneously on day 7 of the stimulated cycle. hMG, human menopausal gonadotropin; FSH, follicle stimulating hormone. Adapted from reference 34

The different yields of these protocols for ovarian stimulation were compared with those of gonadotropins alone in a meta-analysis of 15 trials.[35] Briefly, cycle cancellation was reduced after GnRH agonist compared with regimens using clomiphene, human menopausal gonadotropin (hMG) or FSH (odds ratio (OR) 0.33 for short protocols, 95% confidence interval (CI) 0.25–0.44; OR 1.13 for long protocols, 95% CI 0.706–1.827), mainly due to prevention of the endogenous LH surge, although data were heterogeneous. The number of ampules was significantly increased when gonadotropins were given with a GnRH agonist (31 vs. 19 for short protocols). More oocytes were retrieved (6.76 vs. 5.25 for short protocols). However, comparison of long and short regimens showed a marginal improvement in the number of oocytes retrieved in long-regimen cycles. Overall, an 80–127% clinical pregnancy rate improvement was estimated. The common OR for multiple pregnancy was 2.56 after GnRH agonist use. Finally, the incidence of spontaneous abortion was comparable in all groups.

Data are still controversial regarding the issue of ovarian hyperstimulation syndrome (OHSS) occurrence. The risk of OHSS varies between 2 and 6% of cycles depending on patient and protocol types. OHSS may be a very severe life-threatening complication of IVF.[36] The use of hCG to trigger ovulation is one of the possible causes of OHSS. The use of a GnRH agonist instead of hCG to trigger ovulation allows a significant reduction of OHSS, but this protocol cannot be used in a patient already receiving a GnRH agonist for pituitary desensitization and prevention of a premature LH surge. The association of a GnRH antagonist for prevention of the premature LH surge and a GnRH agonist to trigger ovulation may be an efficient solution for the prevention of OHSS (see section on GnRH antagonists below). Among the compounds commercially available (Table 2), no major difference has been shown in terms of efficacy or side-effects. The development of sustained-release formulations has improved patient compliance and routine organization of ART,

Table 2 Commercially available gonadotropin-releasing hormone (GnRH) agonists: molecules, dosages and administration forms

Agonist	*Brand*	*Route of administration*	*Dosage*	*Frequency*
Histrelin	Supprelin®	sc	5–100 μg	daily
Buserelin	Suprefact®	sc or nasal	1200 μg	3 times/day
	Bigonist®	sc implant	6.3 mg	every 8 weeks
Goserelin	Zoladex®	sc implant	3.6 or 10.8 mg	every 4 or 12 weeks
Nafarelin	Synarel®	nasal	200 μg	1–3 times/day
Triptorelin	Decapeptyl®	sc, depot	0.1, 3 or 11.25 mg	0.1 mg/day, 3 mg/28 days or 11.25 mg/12 weeks
Leuprorelin	Lupron®	sc, im	1 mg	daily
	Lupron® depot	sc, depot	7.5 mg	7.5 mg/28 days
	Lupron depot-3®	sc, depot	22.5 mg	22.5 mg/3 months
	Lupron depot-4®	sc, depot	30 mg	30 mg/4 months
	Viadur™	implant		1 implant/12 months
	Lucrin	sc	5 mg/ml	1 mg/day
	EnanthoneLP 3.75®	sc, im	3.75 mg	3.75 mg/28 days
	EnanthoneLP 11.25®	sc, im	11.25 mg	11.25 mg/3 months

sc, subcutaneous; im, intramuscular

although many groups still prefer to use long protocols including daily subcutaneous administration of the analog. The time necessary to reach pituitary desensitization has been shown to be 10–12 days, no matter which formulation is used.[37] On the other hand, pituitary activity was suppressed for a much longer time period with depot forms of GnRH (2 months after the last administration), compared with daily administration (7 days after the last injection). The effects of depot forms may thus persist into the luteal phase of the stimulated cycle and in early pregnancy. However, no difference was observed in terms of cycle cancellation, pregnancy rate or obstetric and neonatal outcome.[38]

Optimized regimens of COH still do not result in clinical pregnancy rates of over 27% per transfer in IVF cycles, and 26.8% per transfer in ICSI cycles, at least in centers where the number of embryos transferred is limited to three.[39] GnRH analogs have been suggested to exert direct ovarian effects, depending on the type of analog, the type of regimen applied and ovarian status at the time of exposure.[27] However, no direct proof of the clinical relevance of this putative effect has been provided to date.

Elaborating lower cost regimens is another of today's clinical research goals. Reduced doses of triptorelin (100 μg daily subcutaneously, reduced to 50 μg daily from day 1 of FSH stimulation) allow appropriate pituitary desensitization, but do not improve IVF cycle outcome when compared with the depot formulation (one injection of 3 mg depot triptorelin).[40]

Prostate cancer

Prostate cancer is androgen-dependent, as shown by the very low incidence of this form of cancer in eunuchs and in men with inherited deficiency of type 2 5α-reductase, the enzyme that converts testosterone to dihydrotestosterone (DHT), the most active androgen in the prostate. Hormone suppression in prostate cancer is mainly achieved by orchiectomy (since more than 90% of circulating androgens originate from the testis in men). Administration of high doses of estrogens to suppress gonadotropin secretion (and, secondarily, testicular androgen production) has been shown to be efficient in reducing symptoms and tumor size.[41] However, orchiectomy is usually poorly accepted by patients, and estrogen administration is complicated by a very high risk of venous thromboembolism. Additionally, it has been demonstrated that very low levels of plasma androgens could lead to DHT production in the prostate (with levels in castrated men remaining at 40% of those observed in untreated men), and that very low amounts of DHT could stimulate prostate protein synthesis. The potential role of adrenal androgens as a source of prostatic DHT was thus highlighted.[42,43] In fact, under the influence of adrenocorticotropic hormone (ACTH), the adrenals produce androstenedione, dehydroepiandrosterone (DHEA) and its sulfate (DHEAS), in addition to cortisol. In turn, DHEA can be locally transformed into DHT. The development of GnRH agonists and non-steroidal antiandrogens has thus renewed interest in the old concept of combined androgen blockade.

Prostate cancer is not considered to be surgically curable once it has invaded the gland capsule, a situation estimated to have occurred in 7–12% of patients at the time of tumor discovery. Androgen deprivation (medical and surgical) alleviates symptoms in 80–90% of patients and may prolong survival.[44] However, nearly all patients with advanced disease who respond to androgen deprivation relapse after 18–24 months. Randomized trials comparing leuprolide or goserelin with treatment with diethylstilbestrol or surgical castration have all shown equivalent efficacy.[44–49] A meta-analysis has further reported a comparable efficacy of all GnRH agonists.[45] The effect of long-term combined androgen blockade (CAB) remains controversial. Two randomized trials reported a benefit in using GnRH-based CAB. The Southwest Oncology Group (SWOG) study, comparing leuprolide plus placebo with leuprolide plus flutamide (CAB group), reported improved overall survival (35.5 vs. 28.3 months) and event-free survival (16.5 vs. 13.9 months) in the CAB group.[50] The European Organization for Radiation Therapy in Cancer (EORTC) study compared goserelin plus flutamide (CAB group) with orchiectomy, and showed a 25-week increase in time to progression and a 7-month improvement of overall survival in the CAB group.[51–53] Other well-designed large trials have not confirmed these data. The Danish Prostate Cancer Group, in a study comparable in design to the EORTC study, did not show any overall survival improvement.[54] Pooled analysis of data from the two trials did not show an improvement in survival duration. Three trials compared goserelin with goserelin plus flutamide and did not show improvements in progression rate or survival.[55–57] Two meta-analyses, including trials comparing CAB with androgen suppression alone, showed no substantial improvement with CAB (2–3% improvement of survival rate after 5 years of therapy).[51–58] Recommendations for treatment of patients with metastatic prostate cancer include an initial CAB to prevent the 'flare-up' consequences, followed by monotherapy with GnRH agonist, showing equivalent efficacy to surgical castration as first-line therapy. The evaluation of quality of life should be the main criterion in decision-making.

GnRH agonists are proposed in association with neoadjuvant hormone therapy in patients before definitive therapy for early prostate cancer. Goserelin in association with radiotherapy was compared with radiotherapy alone in a 3-year randomized study in 415 men with T1–4 N0 stage disease.[59] A significant 5-year overall survival (79 vs. 62%) and relapse-free survival were demonstrated in the combined therapy group. Several prospective studies have evaluated the efficiency of GnRH agonist use prior to radical prostatectomy. They suggest a positive effect of the treatment on risk of progression, time to progressive disease and disease-free survival. Long-term data are still missing. The Fourth International Conference on Neoadjuvant Hormonal Therapy of Prostate Cancer[60] concluded that neoadjuvant therapy may be considered investigational, and without proven benefit for patients undergoing radical prostatectomy.

The twice-daily intranasal administration and the daily subcutaneous administration methods were rapidly abandoned owing to poor acceptability and compliance. Depot intramuscular formulations are usually used for these long-term treatments, and allow LH and testosterone suppression for 5 weeks to 4 months, depending on the agonist and the formulation. The newer leuprolide formulation includes a micropump, Viadur™ system in a cylinder, and allows minidose delivery for 1 year (Table 2).

In patients with severe urethral obstruction, or severe, painful vertebral metastases, GnRH agonist must be associated with an antiandrogen treatment (flutamide, nilutamide, bicalutamide) started 7 days before the agonist treatment. This is needed at the initiation of therapy, since the initial pituitary release of gonadotropins under the influence of the agonist can be responsible for a severe worsening of symptoms (flare-up effect). Cyproterone acetate and ketoconazole have also been used for this purpose in association with GnRH agonist treatment; however, the former still has weak androgenic properties and the latter is a blocker of CYP 17, the enzyme needed to form C19 steroids, but is not potent enough to eradicate intratissular DHT production. Impotence occurs in most men treated with GnRH agonists. Hot flushes can be profound, and more severe than those experienced after surgical castration; they occur in 5–30% of cases. Various therapeutic trials have been conducted using megestrol acetate, serotonin reuptake inhibitors and cyproterone acetate, and all are reported to have some efficiency in suppressing hot flushes.

Long-term (more than 6 months) GnRH treatment results in osteoporosis. In cases of prostate cancer, testosterone (which would prevent osteoporosis after aromatization to estradiol) add-back therapy is obviously contraindicated. The prevention and treatment of osteoporosis in these cases can rely on bisphosphonates and calcium.

Breast cancer

In premenopausal women with metastatic breast cancer, oophorectomy results in objective responses in approximately one-third of unselected patients.[61] GnRH agonist treatment can alternatively be proposed to women who do not wish to undergo surgery. Therapy with goserelin has been shown to be as effective as ovariectomy in estrogen receptor- and progesterone receptor-positive untreated metastatic breast cancer in terms of response rate, failure-free survival and overall survival.[62] Although no initial estradiol rise has been detected in women with advanced breast cancer treated with GnRH agonists, reactions may occur in some patients that are possibly related to gonadotropin release.[62,63] The incidence of tumor flare was five times more common (16 vs. 3%) in the goserelin group.[62]

The combination of GnRH agonist with tamoxifen, a non-steroidal selective estrogen receptor modulator, has been studied as first-line treatment in premenopausal metastatic breast cancer. Tamoxifen significantly reduces mortality and reduces the risk of contralateral disease when administered in early breast cancer.[64] Responses to tamoxifen in advanced breast cancer range between 16 and 56%, mostly in estrogen receptor-positive tumors.[65] A meta-analysis of four trials suggested that combined therapy with tamoxifen and GnRH agonist was more effective than GnRH agonist alone, with a higher response rate (39 vs. 30%) and a significant 30% reduction in the risk of progression or death.[66] Combined therapy can thus be offered to premenopausal women with advanced breast cancer as a first-line therapy. Recently, the addition of aromatase inhibitors to antiestrogens has provided very interesting results, since this treatment appears to be useful in suppressing also the local production of estrogens in the vicinity of or within the tumor itself.[67]

Fibroids

The treatment of fibroids is limited to symptomatic (i.e. painful, hemorrhagic or infertility-related) disease. The main therapy is surgical removal (hysterectomy or myomectomy). GnRH agonists are proposed as preoperative medical therapy to minimize intra- and postoperative morbidity, and to decrease the overall cost of treatment. In fact, GnRH agonist therapy leads to a mean decrease in uterine volume of 35–50% after 12 weeks of treatment.[68–71] This renders vaginal hysterectomy more frequently feasible; the procedure is clearly associated with shorter hospital stay and recovery times. The benefit of preoperative GnRH agonist therapy for women undergoing abdominal myomectomy is less clear.[70] Another benefit of GnRH agonist use in fibroids is menstrual suppression, which, associated with iron supplementation, allows repletion of iron stores and normalization of hemoglobin levels.[72] Subsequent autologous blood transfusion may then be performed in these patients. Finally, longer-term treatment may be offered to patients close to the menopause who are unwilling to undergo hysterectomy. In this case, prevention of osteoporosis must also be taken into account.

Endometriosis

Endometriosis, the development of ectopic endometrial implants, may be complicated by pain and/or infertility. GnRH agonists have been evaluated in several randomized trials, and have been shown to be efficient in relieving pain,[73] although other hormonal therapeutic agents (such as progestins, oral contraceptives or danatrol) have shown comparable efficacy. The efficacy of GnRH agonists associated with add-back therapy is comparable to that of GnRH agonist-alone treatment. Evaluation of the treatment of infertility in women with endometriosis has demonstrated that none of the currently used drugs (danatrol, GnRH agonists, oral contraceptives) result in greater fertility than placebo. Furthermore, these drugs may delay fertility, since women cannot become pregnant while receiving treatment. IVF is of value in women with advanced endometriosis, but has not shown significant efficacy compared with no treatment, in women with less severe disease.

Central precocious puberty

Precocious puberty occurs in girls before 8 years of age and in boys before 9, and can be due to premature activation of the hypothalamic–pituitary–gonadal axis (true central precocious puberty), or to a primary gonadal spontaneous activation. The main sequela is a reduced final height. In the large majority of cases where no curable cause of central, gonadotropin-dependent precocious puberty is found, GnRH analogs are very effective in suspending the progression of pubertal development, thereby minimizing the psychological consequences of premature puberty. They rapidly suppress gonadotropins and sex steroid levels.[74] Their effect is reversible, and the resumption of normal menstrual cycles and fertility has been reported after cessation of GnRH agonist treatment.[75,76] However, GnRH agonists are not effective in restoring a normal adult height. In children with precocious puberty, the pubertal growth spurt occurs before sufficient childhood growth has occurred. Growth hormone levels are similar to those observed in children with normally timed puberty. Bone age is advanced, and both height prediction and final height are reduced. GnRH agonists are ineffective in restoring height potential,[77] although they allow an improvement in final height, compared with predicted height at the onset of treatment.[78] The level of improvement is significantly, positively correlated with treatment duration, mid-parental height, predicted height at the onset of treatment and growth velocity during the last year of treatment. Improvement in final height is inversely correlated with delay in the onset of treatment, age at the beginning of treatment, breast stage at the onset of treatment, bone age at both initiation and end of treatment, and difference between bone age and chronological age at the initiation of treatment. GnRH agonist treatment may thus be effective in children with rapidly progressing central precocious puberty, accelerated bone maturation and compromised adult height, regardless of bone age or chronological age at the start of treatment. Once treatment is considered appropriate, it should be started quickly, and continued for as long as possible.

The association of growth hormone treatment with GnRH agonist in non-growth hormone deficient children is also ineffective in restoring final height,[79] and irreversible polycystic ovarian appearance has been observed in girls after combined treatment with GnRH agonists and growth hormone. In girls with advanced puberty (between ages 8 and 10 with bone age greater than 10.9 years), GnRH agonists are ineffective in restoring final height.[80] Peak bone mass alterations have been described in girls with central precocious puberty treated with GnRH agonists, and calcium supplementation at the beginning of GnRH agonist treatment may help to prevent this side-effect.[81]

Miscellaneous

The potential direct physiological role of GnRH in endometrial and ovarian cancer is currently the focus of clinical trials. No significant beneficial effects have so far been reported.

Hirsutism and dysfunctional uterine bleeding are two conditions for which GnRH agonist therapy has been evaluated. However, in these situations, therapy should be based on better tolerated and more cost-effective alternative treatments.

Following the discovery of the putative role of GnRH as a growth factor in various tumors, Schally and colleagues[82] have proposed the use of cytotoxic analogs of GnRH consisting of doxorubicin or 2-pyrrolinodoxorubicin linked with GnRH agonist [DLys6]LHRH, which functions as a carrier. The above authors thus propose to use these cytotoxic analogs in tumors where high-affinity GnRH receptors can act as targets for these compounds, in breast, prostate, ovarian and endometrial cancers. These treatments require assessment of their effect.

Safety issues

The main drawbacks of the treatment remain bone loss and vasomotor symptoms, in addition to the usual consequences of withdrawal of estradiol or testosterone in women or men, respectively. The induced bone loss is greater than that observed after a natural menopause, averaging 1% per month during a 6-month treatment period. After 6 months of GnRH agonist treatment, there is a significant loss of bone mineral density (BMD) of 1.5–11.8% in the lumbar spine, regardless of the measurement technique used. Loss of femoral neck or distal radius BMD is variable, reaching 0.9–4% and 4–4.6%, respectively. The total recovery of BMD is not usually observed within 6 months following discontinuation of treatment.[83,84] In treated women, the addition of estrogen–progestin preparations to agonist treatment gives the advantages of estrogen treatment, provided that the symptoms of the disease which indicated the treatment do not recur. Other effective treatments are possible, such as with bisphosphonates or, eventually, parathyroid hormone. However, the cost of the procedure then becomes difficult to accept. This is the basis of the use of these compounds for no more than 3 months, for benign lesions, without add-back treatment.

Concern has been expressed that exposure to fertility drugs may increase the risk of ovarian as well as breast and uterus-body cancers. The 5-year follow-up of a very important cohort of women (n = 29 700), treated (n = 20 656) or not (n = 9044) for infertility with IVF, in ten Australian centers[85] has shown that treated women seem to demonstrate a transient increase in the risk of breast (standardized incidence ratio 2.64) or uterine (standardized incidence ratio 4.59) cancer being diagnosed in the first year after treatment. However, the responsibility of IVF in the cause of these cancers has not been demonstrated. These results may also be the consequence of closer follow-up of these women. Furthermore, the impact of the use of GnRH agonists has not been evaluated. Also, unexplained infertility was associated with an increased risk of having ovarian or uterine cancer diagnosed, independently of the treatment received, which may have introduced bias in previous studies.

It has been debated many times in the past whether agonists could have a direct impact on tissues other than the desensitization of GnRH receptors at the pituitary level. There is now an agreement suggesting that this is not the case.

GnRH antagonists in clinical practice

In clinical practice, GnRH antagonists do not produce significant desensitization of GnRH receptors, although this is the case *in vitro*.[86,87] As discussed above, these antagonists are able to suppress gonadotropin secretion in a rapid and dramatic manner. It is also clear that GnRH antagonists are more potent in suppressing LH and FSH than are GnRH agonists.[14] The suppression of gonadotropin production is observed within hours, and lasts 10–100 h depending on the dose administered, since antagonists have an intrinsic property of increased half-life in relation to dose administered. This phenomenon is generally considered to be due to precipitation of the product and a subsequent depot effect. Obviously, there is no flare-up following antagonist administration, and the gonadotropin suppression can be overridden by exogenous administration of native GnRH or GnRH agonist. In normal women,

GnRH antagonist administration is able to postpone the LH surge if administered at the end of the follicular phase before occurrence of the surge, or even at the time of the surge.[5,88,89] Although it is impossible in humans to proceed to direct measurement of GnRH levels in the portal circulation, this effect of antagonist treatment is clearly related to the key role of GnRH during the surge, as has been demonstrated in the rat, the ewe and, more recently, in non-human primates (all species in which the secretion of GnRH is accessible to direct determination).[32,89–94] The only outstanding controversy concerns the amount of GnRH secreted at the time of the surge.[95] Recent evidence would support that GnRH has a permissive action, and that increasing levels are not required to trigger the gonadotropin surge.

On the basis of this information, it seems likely that GnRH antagonists are of interest essentially in circumstances where gonadotropin suppression needs to be profound and immediate. Long-term applications must wait until depot preparations are available on the market, provided that the final cost remains acceptable. Clinical applications include, therefore, controlled ovarian stimulation for IVF, in which one or two injections of GnRH antagonist or longer protocols using repeated daily administration prevent LH surges.[12,96,97] The value of the first type of regimen, using a single or eventually two injections, is in suppressing the LH surge without a long-lasting depletion of endogenous gonadotropins, while the multiple-dose regimen (although using low doses) provides a longer-term and eventually deeper suppression of LH, which can produce lower plasma estradiol levels and, as a consequence, eventually reduced endometrial quality when gonadotropin regimens devoid of LH are used.[98] The single-dose regimen requires more attention in terms of monitoring follicular development and estradiol production, but allows the completion of multifollicular growth without the use of high doses of hMG/FSH. Indeed, in all studies, the number of ampules of hMG/FSH was reduced by about 40%, compared with agonist treatment. In most cases, one injection of antagonist administered on day 8 is enough. However, in slow responders, a repeat injection may be needed. With the use of the antagonist cetrorelix, it has been shown that each injection allows 3–4 days of treatment with gonadotropins without an LH surge. The dose that seems least effective in this regimen is 3 mg cetrorelix subcutaneously. Other researchers, in particular the group of Diedrich, have proposed a prolonged treatment with the antagonist, starting on day 7 of stimulation, until the triggering of oocyte release with hCG.[99,100] Yet other groups have used a similar protocol with different antagonists.[101,102] The results obtained with the long protocol are also excellent in terms of pregnancy rate. Obviously, with this regimen, the dose of hMG/FSH required is more than in the single- or dual-administration protocol. Interestingly, in this latter, 'long' antagonist protocol, the dose of antagonist required may be dramatically reduced to doses well below 1 mg (0.25 mg) per day.

Further advantages of antagonists in IVF include the possibility of triggering ovulation with native GnRH or GnRH agonists,[103–105] and the possible avoidance of luteal supplementation owing to the short half-life of the compound, the rapid recovery following cessation of the antagonist and the possibility of using the antagonist in minimal-stimulation or in natural (unstimulated) cycles.[106] Such a procedure certainly seems very promising, since women do not receive high doses of gonadotropins, and since ICSI can probably improve the efficacy. In a natural cycle, monofollicular development is associated with a high risk of cancellation caused by premature LH surges. As proposed by Rongieres-Bertrand and colleagues, this can be prevented by the co-administration of an antagonist and a low dose of hMG/FSH on day 8 or 9 of the follicular phase.[107] The gonadotropin administration would compensate for the decrease in endogenous gonadotropin following antagonist administration. The 'minimal-stimulation' approach concerns low-dose hMG/FSH administration restricted to the mid- to late follicular phase.[108–112]

The use of antagonist analogs in IVF needs additional study, in particular to select new protocols of administration and eventually to select patients who can benefit most from this treatment. Although the antagonists allow flexible stimulation procedures, results to date, in some centers at least, may appear to be less successful than those observed with agonists. This seems mainly due to the necessity for the physician to learn a new procedure and to adapt gonadotropin doses. Also, concerning the potent suppression of endogenous LH, it is clear that the potentially poorer results obtained with the antagonists may be related to the use of recombinant FSH, which may impair follicular growth or the estradiol increase as well as its subsequent impact on the endometrium. Careful titration of the antagonist may be necessary. Finally, comparison with agonist analogs remains difficult, because the reasons behind the efficacy (increase in pregnancy rate by 70%, compared with protocols without agonists) of these drugs in addition to suppression of the LH surge, remain unclear.

As discussed below, antagonists have been compared with agonists in IVF in several studies, all showing non-significant differences.[113] For instance, it is still unknown at this stage whether agonists modify the kinetics of follicular growth significantly.[114] It is well accepted that agonists, especially in the so-called long protocols, the present gold standard, allow easy organization of the procedure. However, comparisons between short, ultrashort and long protocols were unable convincingly to show significant differences, other than the long protocol is preferred for practical reasons. Also, a long period of gonadotropin suppression has not been proven to improve follicular recruitment and subsequent growth.

Because antagonists allow a significant reduction in duration of the treatment cycle, there is no doubt that they are more 'user-friendly', compared with the agonists. They also reduce the risk of ovarian hyperstimulation. In a recent meta-analysis of five trials comparing a fixed GnRH antagonist protocol with a long GnRH agonist protocol, Al-Inany and Aboulghar[115] reported excellent suppression of the LH surges in both groups, but lower pregnancy rates when antagonists were used (OR 0.78, 95% CI 0.62–0.97). The absolute treatment effect was 5%, and the required treatment number was $n = 20$. However, in this meta-analysis, the risk of severe OHSS was dramatically reduced with antagonist use (OR 0.36). These figures clearly suggest that antagonists are still very promising, and should be studied further using flexible regimens. A standardized procedure does not exist, and patients should be treated according to their individual characteristics. On this basis, GnRH antagonists represent a very valuable contribution to IVF.

Outstanding concerns include the potential direct effects of GnRH agonists and antagonists on ovarian function and the endometrium. Although this issue is as old as the use of GnRH, no data supporting a significant deleterious effect of these drugs have thus far been produced. Although the absence of proof is clearly not the proof of absence of effect, it must be kept in mind that there is no reason to believe that antagonists would behave differently from agonists. Interestingly, recent data suggest that frozen embryos obtained from antagonist-treated cycles are normal and yield a number of ongoing pregnancies, which is not different from agonist-treated cycles.[116]

Finally, the issue of fetal risk has also been raised. Although no clear answer has been given to this question, the risk of malformation appears not to differ from that of other ART procedures, and babies born after the use of antagonists also appear to be as healthy.[117,118]

In addition to the successful use of antagonists in IVF, there may be favorable effects in gonadotropin ovarian stimulation protocols for intrauterine insemination. This has not been studied as yet. It is also unknown whether chronic administration of antagonists will produce better results, compared with agonists, in hormone-dependent diseases such as polycystic ovarian syndrome, breast cancer, prostate cancer, endometriosis and uterine myomas. However, preliminary observations suggest that antagonists are more potent suppressors of bioactive LH and FSH. This potent suppression may provide better results in some indications such as precocious puberty or hormone-dependent cancers. Male contraception has also been obtained by combining the daily administration of an antagonist with androgen supplementation.[119,120] However, this procedure remains too expensive and cumbersome to be widely used. All applications besides ovulation control are still uncertain, and results need to be confirmed. Finally, for these chronic disorders, provided that GnRH antagonists allow better clinical results (which is uncertain with the exception of male contraception, where antagonists are clearly superior to agonists), a depot preparation of antagonist is absolutely mandatory. Such a long-acting form is still under investigation, and requires further craftsmanship. Alternatively, new therapeutic regimens such as repeated treatments by antagonist, separated by treatment-free intervals, must be assessed carefully. Intermittent treatments, although promising because of their comfort, still have not been proven to be efficient and therefore safe in patients with hormone-dependent cancers. Other indications might include preparation for hysteroscopy, or eventually to reduce breast density for mammography. Many authors have reported the idea that antagonists could be useful in women with polycystic ovaries and in poor responders. There is no reason to believe that antagonists have additional benefits besides the flexibility of their use.

In conclusion, GnRH antagonists are powerful suppressors of LH and FSH. A new generation of antagonists such as cetrorelix and ganirelix is already on the market, and other preparations are under development. Antagonists appear to be very useful in ovarian stimulation protocols, in which they suppress the LH surge. Although their use is already known to produce good although slightly lower pregnancy rates than with the agonists, more research is needed to achieve better results. The main qualities are absence of flare-up, reduction of ovarian hyperstimulation, triggering of the LH surge by GnRH, potential absence of luteal support (in single-dose protocols) and, above all, flexibility of use, allowing protocols with reduced doses of gonadotropins and at lower cost.

SUMMARY AND FUTURE PERSPECTIVES

GnRH agonists all derive from the substitution of glycine 6, a site of enzymatic proteolysis, by an unnatural D amino acid, increasing the half-life of the compound. GnRH agonist administration results in an initial stimulation with a short-term increase in gonadotropins, followed by a long-term desensitization responsible for a fall in steroid secretion. Their clinical applications are rather well defined, and include mainly the palliative treatment of advanced prostate cancers (over 50% of the market) and the adjunctive therapy of premenopausal receptor-positive breast cancers. In gynecology, GnRH agonists are used in the preoperative treatment of uterine leiomyomas and in the treatment of severe endometriosis. A significant indication is now represented by assisted reproductive technologies (ART) (25% of the market). Other indications are marginal, and include gonadotropin-dependent precocious puberty, delay of onset of normal puberty, hyperandrogenism, menometrorrhagia, catamenial disorders and endometrial hyperplasia.

In prostate cancer, the results are equivalent between surgical castration and agonist treatment in stage C and D prostate cancers: the median survival was 1141 days in the agonist-treated group versus 1003 days in the surgery group. The interesting finding that GnRH agonist treatment is also useful in locally advanced prostate cancer, in association with radiotherapy, is encouraging. In premenopausal patients with advanced or recurrent breast cancer, clinical complete and partial responses were observed in 40–70%, compared with 0–22% in postmenopausal women. In ART, GnRH agonists have allowed easier control of ovarian hyperstimulation and scheduling of patients. In central precocious puberty, treatment with GnRH agonists resulted in a gain of 4.2–14.1 cm in height, depending on the onset of puberty.

Future strategies include the development of slow-release preparations for the treatment of sex steroid-dependent cancers, as well as the use of cytotoxic GnRH agonists in the same area. In ART, the development of new, low-dose ultrashort protocols could allow us to reduce the costs of ovarian stimulation, therefore enabling ART to be used also in developing countries.

GnRH antagonists result from multiple amino-acid substitutions in the native GnRH structure, and, as they are inactive but able to bind to GnRH receptors with high affinity, the antagonists behave as competitive inhibitors of GnRH binding to its receptors and allow immediate suppression of gonadotropins and sex steroids. Their clinical applications are still being studied, particularly in ART where they are more 'user-friendly' and lead to a dramatic improvement in the risk of ovarian hyperstimulation syndrome. Although antagonists are orally active at 1000 times the subcutaneous or intravenous active dose, orally active antagonists remain to be studied. However, the search has begun for a non-peptidic ligand of the GnRH receptor, using screening programs for small molecules that bind to the receptor. To date, no product has reached clinical application, owing to major difficulties ranging from instability to absorption and elimination. Several hits have been obtained on a total of 5000 molecules tested. The present compounds are arylquinolone or thienopyridine derivatives, which are effective at several mg/kg doses.[121] Two compounds are currently in phase I in Japan and the USA. Other researchers are aiming to identify the bioactive conformation of GnRH antagonists which can be used to design non-peptide, orally active molecules.[122]

REFERENCES

1. Baba Y, Arimura A, Matsuo H, Schally AV. Structure of the porcine LH and FSH releasing hormone. II Confirmation of the proposed structure by conventional sequential analysis. *Biochem Biophys Res Commun* 1971 44: 459–463.
2. Cusan L, Auclair C, Belanger A *et al.* Inhibitory effects of long term treatment with a LHRH agonist on the pituitary–gonadal axis in male and female rats. *Endocrinology* 1979 104: 1369–1376.
3. Knobil E. The neuroendocrine control of the menstrual cycle. *Recent Prog Horm Res* 1980 36: 53–88.
4. Casper RF. Clinical uses of gonadotropin-releasing hormone analogues. *C M A J* 1991 144: 153–158.
5. Leroy I, d'Acremont M, Brailly-Tabard S, Frydman R, de Mouzon J, Bouchard P. A single injection of a gonadotropin-releasing hormone (GnRH) antagonist (cetrorelix) postpones the luteinizing hormone (LH) surge: further evidence for the role of GnRH during the LH surge. *Fertil Steril* 1994 62: 461–467.
6. Santen RJ, Manni A, Harvey H. Gonadotropin releasing hormone (GnRH) analogs for the treatment of breast and prostatic carcinoma. *Breast Cancer Res Treat* 1986 7: 129–145.
7. Conn PM, Crowley WF Jr. Gonadotropin-releasing hormone and its analogs. *Annu Rev Med* 1994 45: 391–405.
8. Karten MJ, Rivier JE. Gonadotropin-releasing hormone analog design. Structure–function studies toward the development of agonists and antagonists: rationale and perspective. *Endocr Rev* 1986 7: 44–66.
9. Bouchard P, Garcia E. Comparison of the mechanisms of action of LHRH analogs and steroids in the treatment of endometriosis. *Contrib Gynecol Obstet* 1987 16: 260–265.
10. Bouchard P, Wolf JP, Hajri S. Inhibition of ovulation: comparison between the mechanism of action of steroids and GnRH analogues. *Hum Reprod* 1988 3: 503–506.
11. Bouchard P. [GnRH antagonists. Therapeutic and practical aspects]. *J Gynecol Obstet Biol Reprod* 1990 19: 607–608.
12. Bouchard P, Fauser BC. Gonadotropin-releasing hormone antagonist: new tools vs. old habits. *Fertil Steril* 2000 73: 18–20.
13. Zhou W, Rodic V, Kitanovic S *et al.* A locus of the gonadotropin-releasing hormone receptor that differentiates agonist and antagonist binding sites. *J Biol Chem* 1995 270: 18853–18857.
14. Pavlou SN, Brewer K, Farley MG *et al.* Combined administration of a gonadotropin-releasing hormone antagonist and testosterone in men induces reversible azoospermia without loss of libido. *J Clin Endocrinol Metab* 1991 73: 1360–1369.
15. Kovacs M, Schally AV, Csernus B, Rekasi Z. Luteinizing hormone-releasing hormone (LH–RH) antagonist cetrorelix down-regulates the mRNA expression of pituitary receptors for LH-RH by counteracting the stimulatory effect of endogenous LH-RH. *Proc Natl Acad Sci USA* 2001 98: 1829–1834.
16. Gordon K, Hodgen GD. Evolving role of GnRH antagonists. *Trends Endocrinol Metab* 1992 3: 259–263.
17. Felberbaum RE, Ludwig M, Diedrich K. Clinical application of GnRH-antagonists. *Mol Cell Endocrinol* 2000 166: 9–14.
18. Sundaram K, Didolkar A, Thau R, Chaudhuri M, Schmidt F. Antagonists of luteinizing hormone releasing hormone bind to rat mast cells and

induce histamine release. *Agents Actions* 1988 25: 307–313.

19. Felberbaum R, Ludwig M, Diedrich K. *Agonists and Antagonists, Formulation and Indication in GnRH Analogues.* Philadelphia: WB Saunders Company, 2001.
20. Reissmann T, Schally AV, Bouchard P, Riethmuller H, Engel J. The LHRH antagonist cetrorelix: a review. *Hum Reprod Update* 2000 6: 322–331.
21. Gillies PS, Faulds D, Balfour JA, Perry CM. Ganirelix. *Drugs* 2000 59: 107–111.
22. Blithe DL. Applications for GnRH antagonists. *Trends Endocrinol Metab* 2001 12: 238–240.
23. Chi L, Zhou W, Prikhozhan A *et al.* Cloning and characterization of the human GnRH receptor. *Mol Cell Endocrinol* 1993 91: R1–6.
24. Illing N, Jacobs GF, Becker OV II *et al.* Comparative sequence analysis and functional characterization of the cloned sheep gonadotropin-releasing hormone receptor reveal differences in primary structure and ligand specificity among mammalian receptors. *Biochem Biophys Res Commun* 1993 196: 745–751.
25. Millar R, Lowe S, Conklin D *et al.* A novel mammalian receptor for the evolutionarily conserved type II GnRH. *Proc Natl Acad Sci USA* 2001 98: 9636–9641.
26. Neill J, Duck LW, Sellers JC, Mugrove LC. A GnRH receptor specific for GnRH II in primates. *Biochem Biophys Res Commun* 2001 282: 1012–1018.
27. Hanssens RM, Brus L, Cahill DJ, Huirne JA, Schoemaker J, Lambalk CB. Direct ovarian effects and safety aspects of GnRH agonists and antagonists. *Hum Reprod Update* 2000 6: 505–518.
28. Mannaerts B, Gordon K. Embryo implantation and GnRH antagonists: GnRH antagonists do not activate the GnRH receptor. *Hum Reprod* 2000 15: 1882–1883.
29. Ortmann O, Diedrich K. Pituitary and extrapituitary actions of gonadotrophin-releasing hormone and its analogues. *Hum Reprod* 1999 14(Suppl 1): 194–206.
30. Weiss JM, Oltmanns K, Gurke EM *et al.* Actions of gonadotropin-releasing hormone antagonists on steroidogenesis in human granulosa lutein cells. *Eur J Endocrinol* 2001 144: 677–685.
31. Kang SK, Tai CJ, Nathwani PS, Choi KC, Leung PC. Stimulation of mitogen activated protein kinase by GnRH human granulosa-luteal cells. *Endocrinology* 2001 142: 671–679.
32. Moenter SM, Caraty A, Karsch FJ. The estradiol-induced surge of gonadotropin-releasing hormone in the ewe. *Endocrinology* 1990 127: 1375–1384.
33. Jones GS, Acosta AA, Garcia JE, Bernardus RE, Rosenwaks Z. The effect of follicle-stimulating hormone without additional luteinizing hormone on follicular stimulation and oocyte development in normal ovulatory women. *Fertil Steril* 1985 43: 696–702.
34. Filicori M CG, Arnone R, Falbo A *et al.* Endocrine and clinical characteristics of different GnRH agonist regimens used for gonadotropin ovulation induction. In Filicori M, Flamigni C, eds. *Treatments with GnRH Analogs: Controversies and Perspectives.* Carnforth, UK: Parthenon Publishing, 1996: 183–187
35. Hughes EG. Meta-analysis and the critical appraisal of infertility literature. *Fertil Steril* 1992 57: 275–277.
36. Casper RF. Ovarian hyperstimulation: effects of GnRH analogues. Does triggering ovulation with gonadotrophin-releasing hormone analogue prevent severe ovarian hyperstimulation syndrome? *Hum Reprod* 1996 11: 1144–1146.
37. Porcu E, Dal Prato L, Seracchioli R, Fabbri R, Longhi M, Flamigni C. Comparison between depot and standard release triptoreline in *in vitro* fertilization: pituitary sensitivity, luteal function, pregnancy outcome, and perinatal results. *Fertil Steril* 1994 62: 126–132.
38. Porcu E, Filicori M, Dal Prato L *et al.* Comparison between depot leuprorelin and daily buserelin in IVF. *J Assist Reprod Genet* 1995 12: 15–19.
39. Nygren KG, Andersen AN. Assisted reproductive technology in Europe, 1998. Results generated from European registers by ESHRE. *Hum Reprod* 2001 16: 2459–2471.
40. Dal Prato L, Borini A, Trevisi MR, Bonu MA, Sereni E, Flamigni C. Effect of reduced dose of triptorelin at the start of ovarian stimulation on the outcome of IVF: a randomized study. *Hum Reprod* 2001 16: 1409–1414.
41. Huggins C. Endocrine-induced regression of cancers. *Science* 1967 156: 1050–1054.
42. Labrie F, Dupont A, Belanger A *et al.* New hormonal therapy in prostatic carcinoma: combined treatment with an LHRH agonist and

an antiandrogen. *Clin Invest Med* 1982 5: 267–275.
43. Labrie F, Dupont A, Simard J, Luu-The V, Belanger A. Intracrinology: the basis for the rational design of endocrine therapy at all stages of prostate cancer. *Eur Urol* 1993 24(Suppl 2): 94–105.
44. Robson DN. How is androgen-dependent metastatic prostate cancer best treated? *Hematol Oncol Clin North Am* 1996 10: 727–747.
45. Seidenfeld J, Samson DJ, Hasselblad V *et al.* Single-therapy androgen suppression in men with advanced prostate cancer: a systematic review and meta-analysis. *Ann Intern Med* 2000 132: 566–577.
46. The Leuprolide Study Group. Leuprolide versus diethylstilbestrol for metastatic prostate cancer. *N Engl J Med* 1984 1311: 1281–1286.
47. Turkes AO, Peeling WB, Griffiths K. Treatment of patients with advanced cancer of the prostate: phase III trial, zoladex against castration; a study of the British Prostate Group. *J Steroid Biochem* 1987 27: 543–549.
48. Vogelzang NJ, Chodak GW, Soloway MS *et al.* Goserelin versus orchiectomy in the treatment of advanced prostate cancer: final results of a randomized trial. Zoladex Prostate Study Group. *Urology* 1995 46: 220–226.
49. Kaisary AV, Tyrrell CJ, Peeling WB, Griffiths K. Comparison of LHRH analogue (Zoladex) with orchiectomy in patients with metastatic prostatic carcinoma. *Br J Urol* 1991 67: 502–508.
50. Crawford ED, Eisenberger MA, McLeod DG *et al.* A controlled trial of leuprolide with and without flutamide in prostatic carcinoma. *N Engl J Med* 1989 321: 419–424.
51. Denis L, Murphy GP. Overview of phase III trials on combined androgen treatment in patients with metastatic prostate cancer. *Cancer* 1993 72(Suppl 12): 3888–3895.
52. Denis LJ, Carnelro de Moura JL, Bono A *et al.* Goserelin acetate and flutamide versus bilateral orchiectomy: a phase III EORTC trial (30853). EORTC GU Group and EORTC Data Center. *Urology* 1993 42: 119–129.
53. Denis LJ, Keuppens F, Smith PH *et al.* Maximal androgen blockade: final analysis of EORTC phase III trial 30853. EORTC Genito-Urinary Tract Cancer Cooperative Group and the EORTC Data Center. *Eur Urol* 1998 33: 144–151.
54. Iversen P, Suciu S, Sylvester R, Christensen I, Denis L. Zoladex and flutamide versus orchiectomy in the treatment of advanced prostatic cancer. A combined analysis of two European studies, EORTC 30853 and DAPROCA 86. *Cancer* 1990 66(Suppl 5): 1067–1073.
55. Fourcade RO, Cariou G, Coloby P *et al.* Total androgen blockade with Zoladex plus flutamide vs. Zoladex alone in advanced prostatic carcinoma: interim report of a multicenter, double-blind, placebo-controlled study. *Eur Urol* 1990 18(Suppl 3): 45–47.
56. Tyrrell CJ, Altwein JE, Klippel F *et al.* A multicenter randomized trial comparing the luteinizing hormone-releasing hormone analogue goserelin acetate alone and with flutamide in the treatment of advanced prostate cancer. The International Prostate Cancer Study Group. *J Urol* 1991 146: 1321–1326.
57. Boccardo F, Rubagotti A, Barichello M *et al.* Bicalutamide monotherapy versus flutamide plus goserelin in prostate cancer patients: results of an Italian Prostate Cancer Project study. *J Clin Oncol* 1999 17: 2027–2038.
58. Prostate Cancer Trialists' Collaborative Group. Maximum androgen blockade in advanced prostate cancer: an overview of the randomised trials. *Lancet* 2000 355: 1491–1498.
59. Bolla M, Gonzalez D, Warde P *et al.* Improved survival in patients with locally advanced prostate cancer treated with radiotherapy and goserelin. *N Engl J Med* 1997 337: 295–300.
60. Garnick MB, Fair WR, Goldenberg SL *et al.* Fourth International Conference on Neoadjuvant Hormonal Therapy of Prostate Cancer: Overview Consensus Statement. *Mol Urol* 1999 3: 171–174.
61. Veronesi U, Pizzocaro G, Rossi A. Oophorectomy for advanced carcinoma of the breast. *Surg Gynecol Obstet* 1975 141: 569–570.
62. Taylor CW, Green S, Dalton WS *et al.* Multicenter randomized clinical trial of goserelin versus surgical ovariectomy in premenopausal patients with receptor-positive metastatic breast cancer: an intergroup study. *J Clin Oncol* 1998 16: 994–999.
63. Dowsett M, Jacobs S, Aherne J, Smith IE. Clinical and endocrine effects of leuprorelin acetate in pre- and postmenopausal patients with advanced breast cancer. *Clin Ther* 1992 14(Suppl A): 97–103.

64. Osborne C. Tamoxifen in the treatment of breast cancer [Review]. *N Engl J Med* 1998 339: 1609–1618.
65. Hayes DF, Henderson IC, Shapiro CL. Treatment of metastatic breast cancer: present and future prospects [Review]. *Semin Oncol* 1995 22(2 Suppl 5): 5–19.
66. Klijn JG, Blamey RW, Boccardo F, Tominaga T, Duchateau L, Sylvester R. Combined tamoxifen and luteinizing hormone-releasing hormone (LHRH) agonist versus LHRH agonist alone in premenopausal advanced breast cancer: a meta-analysis of four randomized trials. *J Clin Oncol* 2001 19: 343–353.
67. Goss PE, Strasser K. Tamoxifen resistant and refractory breast cancer: the value of aromatase inhibitors. *Drugs* 2002 62: 957–966.
68. Friedman AJ, Barbieri RL, Doubilet PM, Fine C, Schiff I. A randomized, double-blind trial of a gonadotropin releasing-hormone agonist (leuprolide) with or without medroxyprogesterone acetate in the treatment of leiomyomata uteri. *Fertil Steril* 1988 49: 404–409.
69. Friedman AJ, Hoffman DI, Comite F, Browneller RW, Miller JD. Treatment of leiomyomata uteri with leuprolide acetate depot: a double-blind, placebo-controlled, multicenter study. The Leuprolide Study Group. *Obstet Gynecol* 1991 77: 720–725.
70. Friedman AJ, Rein MS, Harrison-Atlas D, Garfield JM, Doubilet PM. A randomized, placebo-controlled, double-blind study evaluating leuprolide acetate depot treatment before myomectomy. *Fertil Steril* 1989 52: 728–733.
71. Schlaff WD, Zerhouni EA, Huth JA, Chen J, Damewood MD, Rock JA. A placebo-controlled trial of a depot gonadotropin-releasing hormone analogue (leuprolide) in the treatment of uterine leiomyomata. *Obstet Gynecol* 1989 74: 856–862.
72. Stovall TG, Muneyyirci-Delale O, Summitt RL Jr, Scialli AR. GnRH agonist and iron versus placebo and iron in the anemic patient before surgery for leiomyomas: a randomized controlled trial. Leuprolide Acetate Study Group. *Obstet Gynecol* 1995 86: 65–71.
73. Olive DL. Treatment of endometriosis [Review]. *N Engl J Med* 2001 345: 266–275.
74. Sklar CA, Rothenberg S, Blumberg D, Oberfield SE, Levine LS, David R. Suppression of the pituitary–gonadal axis in children with central precocious puberty: effects on growth, growth hormone, insulin-like growth factor-I, and prolactin secretion. *J Clin Endocrinol Metab* 1991 73: 734–738.
75. Kauli R, Kornreich L, Laron Z. Pubertal development, growth and final height in girls with sexual precocity after therapy with the GnRH analogue D-TRP-6-LHRH. A report on 15 girls, followed after cessation of gonadotrophin suppressive therapy. *Horm Res* 1990 33: 11–17.
76. Manasco PK, Pescovitz OH, Feuillan PP *et al.* Resumption of puberty after long term luteinizing hormone-releasing hormone agonist treatment of central precocious puberty. *J Clin Endocrinol Metab* 1988 67: 368–372.
77. Manasco PK, Pescovitz OH, Hill SC *et al*. Six-year results of luteinizing hormone releasing hormone (LHRH) agonist treatment in children with LHRH-dependent precocious puberty. *J Pediatr* 1989 115: 105–108.
78. Klein KO, Barnes KM, Jones JV, Feuillan PP, Cutler GB Jr. Increased final height in precocious puberty after long-term treatment with LHRH agonists: the National Institutes of Health experience. *J Clin Endocrinol Metab* 2001 86: 4711–4716.
79. Bridges NA, Cooke A, Healy MJ, Hindmarsh PC, Brook CG. Ovaries in sexual precocity. *Clin Endocrinol (Oxf)* 1995 42: 135–140.
80. Bouvattier C, Coste J, Rodrigue D *et al.* Lack of effect of GnRH agonists on final height in girls with advanced puberty: a randomized long-term pilot study. *J Clin Endocrinol Metab* 1999 84: 3575–3578.
81. Antoniazzi F, Bertoldo F, Lauriola S *et al.* Prevention of bone demineralization by calcium supplementation in precocious puberty during gonadotropin-releasing hormone agonist treatment. *J Clin Endocrinol Metab* 1999 84: 1992–1996.
82. Plonowski A, Schally AV, Nagy A. Inhibition of *in vivo* proliferation of MDA-PCa-2b human prostate cancer by a targeted cytotoxic analog of luteinizing hormone-releasing hormone AN-207. *Cancer Lett* 2002 176: 57–63.
83. Dawood MY. Impact of medical treatment of endometriosis on bone mass. *Am J Obstet Gynecol* 1993 168: 674–684.
84. Fogelman I. Gonadotropin-releasing hormone agonists and the skeleton. *Fertil Steril* 1992 57: 715–724.

85. Venn A, Watson L, Bruinsma F, Giles G, Healy D. Risk of cancer after use of fertility drugs with *in-vitro* fertilisation. *Lancet* 1999 354: 1586–1590.
86. Dubourdieu S, Charbonnel B, D'Acremont MF, Carreau S, Spitz IM, Bouchard P. Effect of administration of a gonadotropin-releasing hormone (GnRH) antagonist (Nal–Glu) during the periovulatory period: the luteinizing hormone surge requires secretion of GnRH. *J Clin Endocrinol Metab* 1994 78: 343–347.
87. Lahlou N, Delivet S, Bardin CW, Roger M, Spitz IM, Bouchard P. Changes in gonadotropin and α-subunit secretion after a single administration of gonadotropin-releasing hormone antagonist in adult males. *Fertil Steril* 1990 53: 898–905.
88. Ditkoff EC, Cassidenti DL, Paulson RJ *et al.* The gonadotropin-releasing hormone antagonist (Nal-Glu) acutely blocks the luteinizing hormone surge but allows for resumption of folliculogenesis in normal women. *Am J Obstet Gynecol* 1991 165: 1811–1817.
89. Christin-Maitre S, Olivennes F, Dubourdieu S *et al.* Effect of gonadotrophin-releasing hormone (GnRH) antagonist during the LH surge in normal women and during controlled ovarian hyperstimulation. *Clin Endocrinol (Oxf)* 2000 52: 721–726.
90. Xia L, Vugt D, Alstor EJ, Luckhaus J, Ferrin M. A surge of GnRH accompanies the estradiol-induced gonadotropin surge in the rhesus monkey. *Endocrinology* 1992 131: 2812–2820.
91. Norman L, Rivier J, Vale W, Spies HG. Inhibition of estradiol induced gonadotropin release in ovariectomized rhesus macaques by a GnRH antagonist. *Fertil Steril* 1986 45: 288–291.
92. Clarke IJ. Variable patterns of GnRH secretion during the estrogen-induced LH surge in ovariectomized ewes. *Endocrinology* 1993 133: 1624–1632.
93. Karsch FJ, Bowen JM, Caraty A, Evans NP, Moenter SM. Gonadotrophin-releasing hormone requirements for ovulation. *Biol Reprod* 1997 56: 303–309.
94. Caraty A, Antoine C, Delaleu B *et al.* Nature and bioactivity of gonadotropin-releasing hormone (GnRH) secreted during the GnRH surge. *Endocrinology* 1995 136: 3452–3460.
95. Martin KA, Welt CK, Taylor AE, Smith JA, Crowley WF Jr, Hall JE. Is GnRH reduced at the mid cycle surge in the human? Evidence from a GnRH-deficient model. *Neuroendocrinology* 1998 67: 363–369.
96. Frydman R, Cornel C, de Ziegler D, Taieb J, Spitz IM, Bouchard P. Spontaneous luteinizing hormone surges can be reliably prevented by the timely administration of a gonadotrophin releasing hormone antagonist (Nal-Glu) during the late follicular phase. *Hum Reprod* 1992 7: 930–933.
97. Diedrich K, Ludwig M, Felberbaum RE. The role of gonadotropin-releasing hormone antagonists in *in vitro* fertilization. *Semin Reprod Med* 2001 19: 213–220.
98. Olivennes F, Belaisch-Allart J, Emperaire JC *et al.* Prospective, randomized, controlled study of in vitro fertilization–embryo transfer with a single dose of a luteinizing hormone-releasing hormone (LH-RH) antagonist (cetrorelix) or a depot formula of an LH-RH agonist (triptorelin). *Fertil Steril* 2000 73: 314–320.
99. Ludwig M, Felberbaum R, Diedrich K. *The Use of LHRH Antagonists: Multiple-Dose Administration.* Philadelphia: WB Saunders Company, 2001.
100. Felberbaum RE, Albano C, Ludwig M *et al.* Ovarian stimulation for assisted reproduction with hMG and concomitant midcycle administration of the GnRH antagonist cetrorelix according to the multiple dose protocol: a prospective uncontrolled phase III study. *Hum Reprod* 2000 15: 1015–1020.
101. The Ganirelix Dose-Finding Study Group. A double-blind, randomized, dose-finding study to assess the efficacy of the gonadotrophin-releasing hormone antagonist ganirelix (Org 37462) to prevent premature luteinizing hormone surges in women undergoing ovarian stimulation with recombinant follicle stimulating hormone (Puregon). *Hum Reprod* 1998 13: 3023–3031.
102. European Orgalutran Study Group. Treatment with the GnRH antagonist ganirelix in women undergoing ovarian stimulation with recombinant FSH is effective, safe and convenient: result of a controlled, randomized, multicenter trial. *Hum Reprod* 2000 15: 1490–1498.
103. Olivennes F, Fanchin R, Bouchard P, Taieb J, Frydman R. Triggering of ovulation by a gonadotropin-releasing hormone (GnRH) agonist in patients pretreated with a GnRH antagonist. *Fertil Steril* 1996 66: 151–153.
104. de Jong D, Van Hooren EG, Macklon NS, Mannaerts BM, Fauser BC. Pregnancy and birth after GnRH-agonist treatment for induction of

oocyte maturation in a woman undergoing ovarian stimulation for ICSI, using a GnRH antagonist (Orgalutran/Antagon) to prevent a premature LH surge: a case report. *J Assist Reprod Genet* 18 2001: 30–33.
105. Fauser BC, de Jong D, Olivennes F *et al.* Endocrine profiles after triggering the final oocyte maturation with GnRH agonist after treatment with the GnRH antagonist ganirelix during ovarian stimulation for IVF: hormonal profiles. *J Clin Endocrinol Metab* 2002 87: 709–715.
106. de Jong D, Macklon NS, Fauser BC. A pilot study involving minimal ovarian stimulation for *in vitro* fertilization: extending the 'follicle-stimulating hormone window' combined with the gonadotropin-releasing hormone antagonist cetrorelix. *Fertil Steril* 2000 73: 1051–1054.
107. Rongieres-Bertrand C, Olivennes F, Righini C *et al.* Revival of the natural cycles in *in-vitro* fertilization with the use of a new gonadotrophin-releasing hormone antagonist (cetrorelix): a pilot study with minimal stimulation. *Hum Reprod* 1999 14: 683–688.
108. Fauser BC, van Heusden AM. Manipulation of human ovarian function: physiological concepts and clinical consequences. *Endocr Rev* 1997 18: 71–106.
109. Fauser BC, Devroey P, Yen SS *et al.* Minimal ovarian stimulation for IVF: appraisal of potential benefits and drawbacks. *Hum Reprod* 1999 14: 2681–2686.
110. de Jong D, Macklon NS, Fauser BC. Minimal ovarian stimulation for IVF: extending the FSH window. In Jansen R, Mortimer D, eds. *Towards Reproductive Certainty*. Carnforth, UK: Parthenon Publishing, 2000: 195–199.
111. Macklon NS, Fauser BC. Regulation of follicle development and novel approaches to ovarian stimulation for IVF. *Hum Reprod Update* 2000 6: 1–6.
112. Hohmann F, Laven JS, de Jong FH, Eijkemans MJ, Fauser BC. Low-dose exogenous FSH initiated during the early, mid or late follicular phase can induce multiple dominant follicle development. *Hum Reprod* 2001 16: 846–854.
113. The European and Middle East Orgalutran Study Group. Comparable clinical outcome using the GnRH antagonist ganirelix or a long protocol of the GnRH agonist triptorelin for the prevention of premature LH surges in women undergoing ovarian stimulation. *Hum Reprod* 2001 16: 644–651.
114. de Jong D, Macklon NS, Eijkemans MJ, Mannaerts BM, Coelingh Bennink HJ, Fauser B, for the Ganirelix Dose-Finding Study Group. Dynamics of the development of multiple follicles during ovarian stimulation for IVF using recombinant FSH and various doses of GnRH antagonist ganirelix. *Fertil Steril* 2001 75: 688–693.
115. Al-Inany H, Aboulghar M. Gonadotrophin-releasing hormone antagonists for assisted conception [Cochrane Review]. *Cochrane Database Syst Rev* 2001 4.
116. Kol S. Embryo implantation and GnRH antagonists: GnRH antagonists in ART: lower embryo implantation? *Hum Reprod* 2000 15: 1881–1882.
117. Olivennes F, Mannaerts B, Struijs M, Bonduelle M, Devroey P. Perinatal outcome of pregnancy after GnRH antagonist (ganirelix) treatment during ovarian stimulation for conventional IVF or ICSI: a preliminary report. *Hum Reprod* 2001 16: 1588–1591.
118. Ludwig M, Riethmuller-Winzen H, Felberbaum RE *et al.* Health of 227 children born after controlled ovarian stimulation for *in vitro* fertilization using the luteinizing hormone-releasing hormone antagonist cetrorelix. *Fertil Steril* 2001 75: 18–22.
119. Anderson RA. Hormonal contraception in the male. *Br Med Bull* 2000 56: 717–728.
120. Behre HM, Kliesch S, Lemcke B, von Eckardstein S, Nieschlag E. Suppression of spermatogenesis to azoospermia by combined administration of GnRH antagonist and 19-nortestosterone cannot be maintained by this non-aromatizable androgen alone. *Hum Reprod* 2001 16: 2570–2577.
121. Young JR, Huang SX, Chen I *et al.* Quinolones as gonadotropin releasing hormone (GnRH) antagonists: simultaneous optimization of the C(3)-aryl and C(6)-substituents. *Bioorg Med Chem Lett* 2000 10: 1723–1727.
122. Fromme BJ, Katz AA, Roeske RW, Millar RP, Flanagan CA. Role of aspartate7.32(302) of the human gonadotropin-releasing hormone receptor in stabilizing a high-affinity ligand conformation. *Mol Pharmacol* 2001 60: 1280–1287.

Chapter 18

Gonadotropins and gonadotropin mimics

Albina Jablonka-Shariff, Wiebe Olijve and Irving Boime

INTRODUCTION

The family of glycoprotein hormones includes pituitary follitropin or follicle stimulating hormone (FSH), lutropin or luteinizing hormone (LH) and thyrotropin or thyroid stimulating hormone (TSH), and the placental chorionic gonadotropin (CG). They are heterodimers composed of two non-covalently linked subunits, α and β, which are encoded by separate genes located on different chromosomes. Within a species, the α subunit is common to all four hormones, whereas the β subunit confers the unique biological specificity for each hormone. Formation of the heterodimer occurs in the endoplasmic reticulum from independently synthesized subunits, and is required for biological activity of these hormones.

The complexity of the gonadotropin structure has been a reservoir for elucidating a variety of structure–function questions. This chapter reviews some key structure–function determinants of the glycoprotein hormone family, and discusses the use of site-directed mutagenesis and gene transfer to create potential novel gonadotropin agonists and antagonists for diagnostic and therapeutic use.

THE BIOLOGICAL ACTIVITY OF GONADOTROPINS

The pituitary gonadotropins are critical for the regulation of gonadal function and reproduction in mammals. These hormones are released under the control of numerous factors, including hypothalamic gonadotropin releasing hormone and steroids.[1,2] FSH and LH stimulate the female and male gonads to regulate sexual maturation and reproductive function; FSH acts on the granulosa cells of the ovary and the Sertoli cells of the testis indirectly to promote germ cell development in these gonadal tissues.[1,3] LH binds to the Leydig cells of the testis and the theca cells of the ovary to regulate local and peripheral concentrations of gonadal steroid hormones. LH is essential for the resumption of meiosis and ovulation, causing rupture of the preovulatory follicle and release of the ovum. CG synthesized by the placenta syncytiotrophoblast is crucial for stimulating the progesterone production from the corpus luteum during early pregnancy.[2] TSH binds to receptors on the thyroid gland and regulates thyroid hormone synthesis.[4]

STRUCTURAL FEATURES OF THE GONADOTROPINS

The common α subunit

The α subunit is expressed from a single gene of 8–16.5 kb in size, and is composed of four exons.[5–7] This gene encodes a 92-amino acid polypeptide with two N-linked oligosaccharides attached to asparagine (Asn) residues 52 and 78 (Figure 1). The subunit contains ten cysteine residues that maintain its structural integrity by forming five disulfide bonds.[8,9]

The hormone-specific β subunit

Although the β subunits determine the biological specificity for each hormone, there is a high degree of sequence similarity between them.[2,9,10] This is most apparent for the LHβ and CGβ subunits, which share greater than 80% sequence identity in the first 114 amino acids. This similarity is responsible for the binding of LH and CG to a common gonadal receptor.[11,12] Based on their extensive sequence identity, it was suggested that CGβ evolved from an ancestral LHβ gene.[13,14] A single base-pair deletion at codon 114 led to the generation of a mRNA encoding the CGβ subunit composed of 145 amino acids, compared with the 121 amino acids of the LHβ subunit.[13,14] All β subunits contain 12 cysteine residues, which form six conserved intrachain disulfide bonds. The length of the β subunit polypeptide varies from 111 amino acids for the FSHβ subunit to 145 amino acids for the CGβ subunit (Figure 1). In many species, a single gene encodes the LHβ,[14–16] FSHβ[17,18] and TSHβ[19,20] subunits, whereas a multigene family encodes the CGβ subunit.[14–16]

Research into the structure and function of gonadotropins was greatly facilitated by the recent report of the human chorionic gonadotropin (hCG) crystal structure[21,22] (Figure 2). These data revealed that the α and β subunits both contain a cysteine knot, similar to a variety of growth factors.[23] This is a cluster of disulfide bonds that

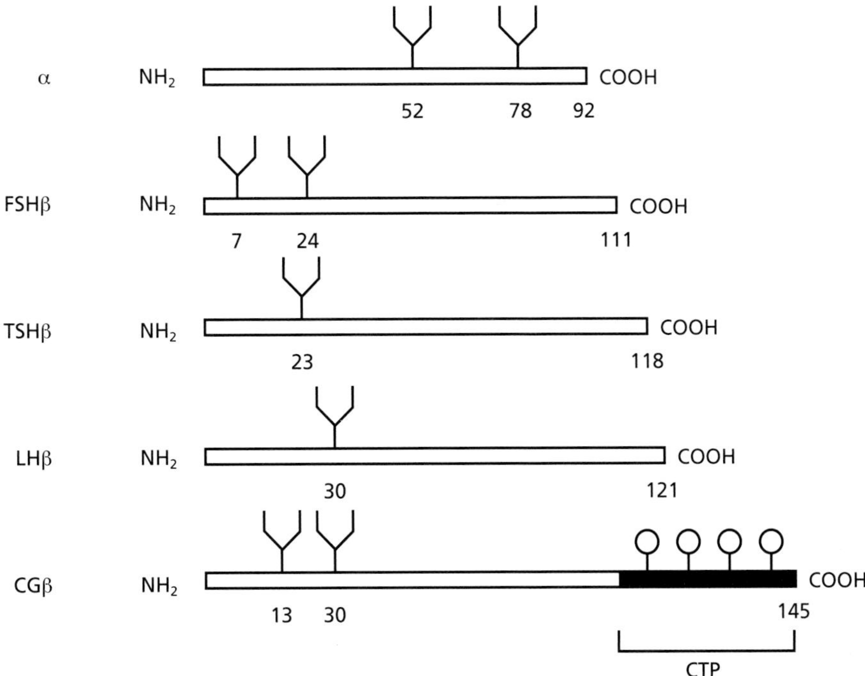

Figure 1 Glycosylation sites in the α and β subunits. The asparagine-linked oligosaccharides (N) are indicated by the dibranched symbol, and the O designation in the chorionic gonadotropin β (CGβ) subunit corresponds to the serine-linked oligosaccharides in the carboxy-terminal region. The numbers under each site indicate the amino acid residue to which the oligosaccharide is attached. FSH, follicle stimulating hormone; TSH, thyroid stimulating hormone; LH, luteinizing hormone; CTP, carboxy-terminal peptide

forms the basic scaffold of the subunit and is critical for β–α combination.[23] Figure 2 shows that the CGβ and α subunits have remarkably similar three-dimensional structures. Each subunit has an elongated shape, with two β-hairpin loops on one side of the central cysteine knot and a long loop on the other side. The non-covalent interaction between the two subunits is apparently stabilized by a stretch of amino acids at the carboxyl end of the β subunit that extends like a 'seatbelt' around the α subunit and is 'locked' by a disulfide bridge. A similar crystal structure has recently been reported for human FSH.[24] The ability to generate crystallographic data is critical for testing models of ligand–receptor interaction, and for determinants associated with the assembly steps. Deducing such structural information from hormone–receptor complexes is essential for the design of agonists and antagonists.

The β subunits are glycosylated and contain N-linked oligosaccharides. The CGβ subunit is distinguished from the other β subunits by the presence of O-linked oligosaccharides (Figure 1). There is one N-linked oligosaccharide at Asn30 of LHβ and at Asn23 of the TSHβ subunit. The CGβ and FSHβ subunits contain two N-linked oligosaccharides at Asn13 and -30, and -7 and -24, respectively (Figure 1). In contrast to LHβ, the CGβ subunit contains a carboxy-terminal peptide (CTP),[2,9,25,26] presumably due to a frameshift mutation at codon 114 in the human LHβ gene.[14] This extension contains four O-linked oligosaccharides.[27]

N- and O-linked oligosaccharides

Hormone-specific processing

Despite the sequence similarity of the polypeptide chains of the glycoprotein hormones, the N-linked oligosaccharides show extensive structural diversity.[2,28–30] The general structures are shown in Figure 3. In the case of CG, the N-linked carbohydrates terminate exclusively with sialic acid. The

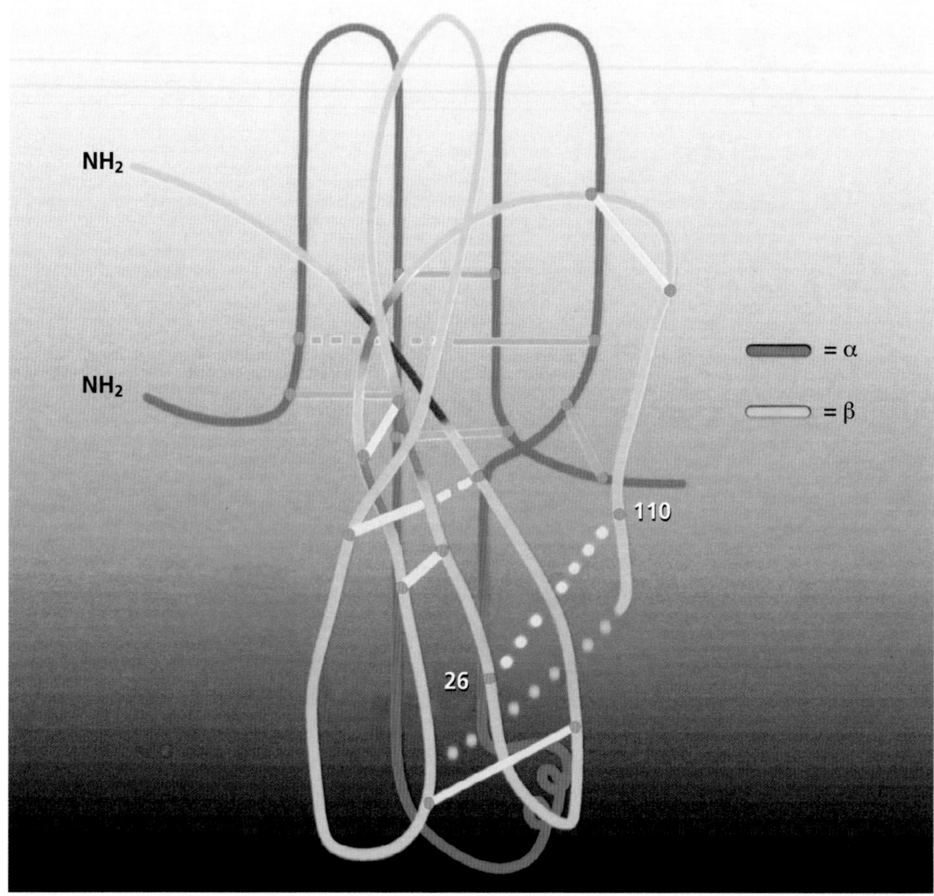

Figure 2 Diagram of the deglycosylated chorionic gonadotropin (CG) crystal structure (see references 21 and 22). Intrasubunit disulfide bonds are shown in orange and yellow for α and CBβ subunits, respectively. The sequence 26–110 corresponds to the 'seatbelt' region of the CGβ subunit

majority of pituitary glycoprotein hormones from several species contain terminal sulfates on their oligosaccharide units, except for FSH, which contains less sulfate and more sialic acid; human FSH oligosaccharides contain only terminal sialic acid[31–33] (Figure 3).

The diversity of oligosaccharide structures indicates that combination of the α subunit with a hormone-specific β subunit determines further steps in specific processing of the N-linked oligosaccharides. This is especially relevant in the case of FSH and LH: both are synthesized in the same cell,[34] and share the same α subunit, but their oligosaccharides differ in both branching and terminal modifications (Figure 4). This conclusion was supported by results from mammalian cell transfection experiments with the gonadotropin subunit genes.[35] In such a system, the only variable was the β subunit introduced with the common α subunit, providing a convenient system to study the effect of subunit assembly on oligosaccharide processing. There is, however, a tissue-specific component regarding carbohydrate maturation. For example, the placenta lacks the enzymes responsible for sulfating the glycoprotein hormones.[36,37]

Bioactivity and half-life of circulating hormones

N-linked oligosaccharides have been implicated in intracellular events such as folding, subunit assembly and secretion, and in the biological activity of

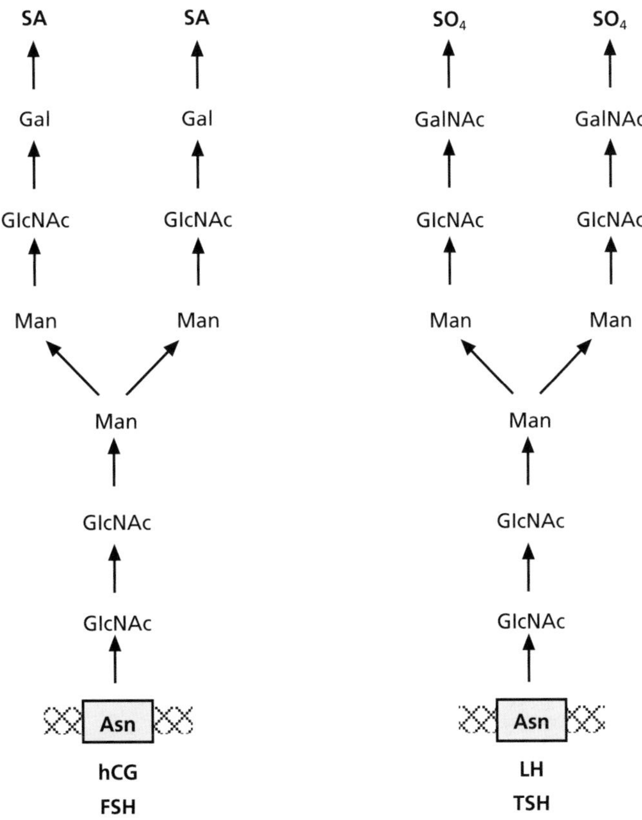

Figure 3 Distribution of SO_4 and sialic acid (SA) in the glycoprotein hormones; GlcNAc, N-acetyl glucosamine; GalNAc, N-acetyl galactosamine; hCG, human chorionic gonadotropin; FSH, follicle stimulating hormone; LH, luteinizing hormone; TSH, thyroid stimulating hormone

glycoprotein hormones.[38–40] Mutagenesis experiments showed that the oligosaccharides in the CGβ subunit were important for proper folding and disulfide bond pairing, critical for efficient assembly with the α subunit.[41] The carbohydrate units apparently do not have a direct link to the receptor, but rather maintain the proper interaction of the peptide component with the receptor-transducing domain.[41,42]

The best understood role of the oligosaccharides concerns the circulatory stability of the hormones. The half-life is primarily dependent on the type of terminal modification of their N-linked oligosaccharides. The presence of terminal sulfate leads to more rapid clearance of the hormones by hepatic cells, compared with the sialylated glycoforms.[43–49] Sulfated bovine LH has a five-fold greater clearance rate than that of the corresponding sialylated hormone.[46] A rapid clearance system may be necessary to maintain pulsatile levels of hormones, and prevent down-regulation of the receptor from a sustained level of the glycoprotein hormones.

The variations in the pattern of sialylation and sulfation contribute to the heterogeneity of the glycoprotein hormones, as observed by isoelectric focusing (Figure 5). These isoforms differ in biopotency and the circulatory half-life *in vivo*.[50–56] Some of the circulating forms of TSH[57] and human FSH[58] behave *in vitro* as antagonists, although no *in vivo* studies have been performed to confirm such activities.

O-linked oligosaccharides

The CGβ subunit contains O-linked oligosaccharides linked to the four serine residues 121, 127,

132 and 138 (Figure 6). Recombinant DNA methods demonstrated that abolishing the O-linked glycans or removing the CTP bearing the O-linked oligosaccharides did not affect receptor binding and signal transduction of hCGβ containing the truncated β subunit.[59–62] However, the heterodimer containing the truncated hCG subunit was three-fold less potent *in vivo*, compared with the native hormone, in inducing ovulation in rats.[60,63] Therefore, the carboxy-terminal extension of the CGβ subunit is critical for the circulatory half-life of hCG.

MUTATIONS IN THE GONADOTROPIN GENES

The presence of naturally occurring mutations in the glycoprotein hormone family represents another example of structure–function overlap with clin-

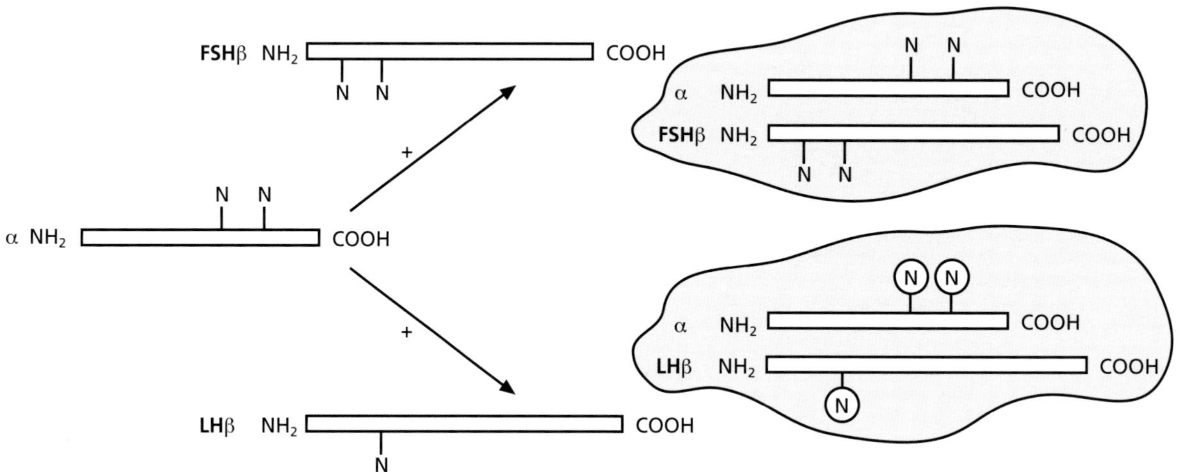

Figure 4 Schematic diagram of co-transfection experiments with the common α subunit and either follicle stimulating hormone β (FSHβ) or luteinizing hormone β (LHβ) subunit. N corresponds to the asparagine-linked oligosaccharides. The N of the FSHβ/α dimer is different from the N of the LHβ/α dimer

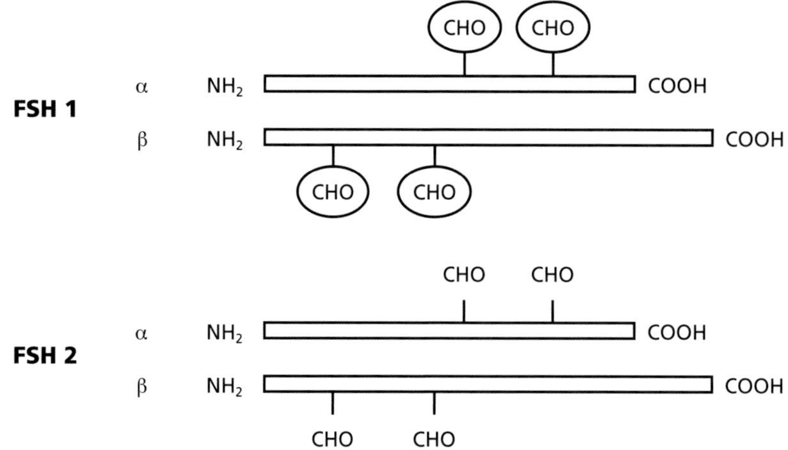

Figure 5 Variations in oligosaccharide structure and heterogeneity of a glycoprotein hormone. Variable quantities of SO_4, sialic acid and branching change the overall charge, and affect biological activity. Follicle stimulating hormone (FSH) 1 and FSH 2 denote two hypothetical species. CHO, carbohydrate oligosaccharide

ical pathophysiology. A number of diseases at the hypothalamic and pituitary levels can impair the synthesis and secretion of gonadotropins.[64–66] Aberrations in the hypothalamic regulation of gonadotropin synthesis can result in hypogonadotropic hypogonadism and anosmia. It has been reported that suppressed gonadotropin action in a patient was caused by aberrant glycosylation.[67] This carbohydrate-deficient syndrome resulted in high circulating levels of immunoreactive FSH with very low bioactivity.

There are several other examples of mutations in the genes encoding the β subunits of LH,[68–70] FSH[71,72] and TSH[73–75] that result in clinical disorders. A single amino acid substitution (Glu to Arg substitution in codon 54) in the LHβ subunit is associated with hypogonadism in homozygous males. Heterozygotes have a high incidence of infertility revealed by two point mutations in the LHβ gene, which result in heterodimers that have a decreased bioactivity, presumably due to a short half-life.[66,68–70,76,77] Moreover, patients expressing a FSHβ subunit containing a 2-bp deletion in exon three of the FSHβ gene exhibit primary amenorrhea and decreased fecundity.[66,71,72] In the amenorrhea case, a missense T to C mutation caused a Cys51 to Gly transition in the mature FSHβ protein, and cells transfected with this mutant failed to produce immunoreactive FSH dimer. A single base deletion in the TSH cysteine codon 105 was shown to cause hypothyroidism, characterized by mental and growth retardation.[66,73–75]

No mutations have yet been reported for the CGβ subunit or the common α subunit.[66] The only genetic alteration so far reported in the α subunit is a single Glu56 to Ala amino acid substitution in a subunit ectopically secreted by a human carcinoma.[78] This mutated protein failed to associate with the β subunit, and appeared to have significantly higher molecular weight than the native α subunit. It was proposed that this mutation altered the structure, resulting in the formation of homodimers and/or altered glycosylation, which could then be responsible for the subunit's increased size and its inability to assemble with LHβ. Most of the polymorphisms detected in the CGβ gene were either silent or located in intron regions.[79,80] However, one genetic variant identified in exon 3 of gene 5 (G to A transition) resulted in a CGβ subunit with impaired ability to assemble with the α subunit and fold correctly.[81]

CLINICAL RELEVANCE OF RECOMBINANT-DERIVED ANALOGS

A major benefit that arises from the structure–function data described above is that the concepts developed from such work can lead to the design of clinically useful agents. Human FSH, partially purified from urinary human menopausal gonadotropin, has been used clinically for decades to stimulate follicular maturation in women with chronic anovulatory syndrome[82] or luteal phase deficiency.[83] In males, a combination of FSH and LH has been used in a variety of conditions related to male infertility. Advances in biotechnology and gonadotropin genetic engineering have established industrial production of the therapeutic-quality FSH. The efficacy of this recombinant material is comparable to that of urinary FSH in stimulating ovarian follicular development in

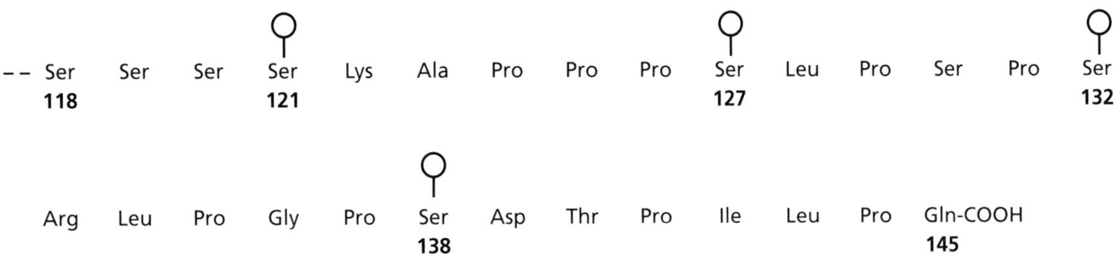

Figure 6 The carboxy-terminal peptide (CTP) of the chorionic gonadotropin β (CGβ) subunit. The positions of the serine (O)-linked oligosaccharides in the last 28 amino acids of the subunit are shown

women undergoing *in vitro* fertilization protocols.[84–86] These pure preparations avoid the risk of prion or slow-virus contamination that might result with extracts from pituitary tissue or postmenopausal urine.

Site-directed mutagenesis represents a powerful approach for designing therapeutic gonadotropin analogs. The introduction of specific modification in the coding sequences of gonadotropin genes has permitted analysis of residues essential for subunit assembly and secretion,[38,63,87,88] biological activity,[23,40,41,63,89] the circulatory half-life[45,46,60,90] and the role of carbohydrates.[38–42,46] Such information allows the testing of predictions regarding the functional properties of the glycoprotein hormones, and becomes crucial for engineering potent agonists and antagonists. For example, the CTP of the hCGβ subunit (Figures 1 and 6) is essential for the relatively long half-life of hCG in the circulation.[28,91,92] Currently, a major issue regarding the clinical use of FSH to induce follicle development is its short half-life *in vivo*, which necessitates multiple injections.[93,94] Using recombinant DNA techniques, the CTP of the hCGβ was fused to the human FSHβ subunit coding sequence, thereby generating a longer-acting FSH agonist that exhibited a prolonged circulating half-life and enhanced bioactivity *in vivo* (Figure 7).[90] Treatment with such long-acting gonadotropins would be less stressful for the patient, requiring fewer injections. Similar results were observed with a TSH dimer containing a CTP in the TSHβ subunit.[57,95]

Recently, the heterodimer bearing the FSHβ–CTP subunit was tested in a phase I multicenter study comprising of 13 hypogonadotropic male subjects.[96] In this study, the FSH–CTP administration did not lead to detectable antibody formation. Moreover, compared with the wild type recombinant FSH, the half-life of FSH–CTP was increased two to three times.[96] Thus, at this stage of development, the FSH–CTP analog represents a potentially new therapeutic agent in the development of a more convenient regimen for treating male and female infertility.

Because of the potential for receptor desensitization, which is associated with long-acting ligands, superactive analogs with enhanced receptor-binding affinity could, in certain cases, be more desirable. Such analogs were constructed for human TSH by introducing certain crucial residues seen in subunits from different animal species into the human subunit.[97]

A major issue concerning the use of recombinant FSH in assisted reproduction protocols is the potential for ovarian hyperstimulation (resulting, for example, in multiple births). Therefore, for

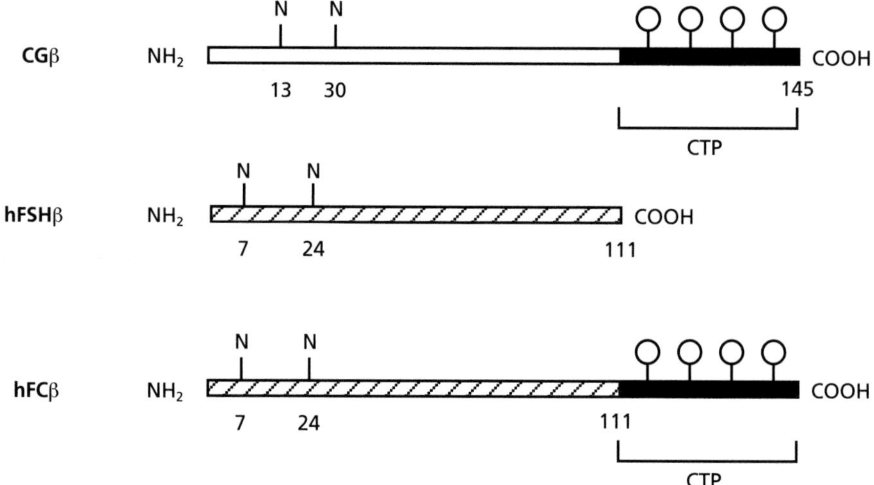

Figure 7 Design of long-acting follicle stimulating hormone (FSH) agonists. The DNA encoding the carboxy-terminal peptide (CTP) sequence was ligated to the 3' end of the human FSHβ subunit mini gene to produce the chimera human FCβ (FSHβ with the CTP). This construct was co-transfected with the gene of the common α subunit into Chinese hamster ovary cells to generate the heterodimeric analog. CG, chorionic gonadotropin

certain regimens, a short-acting FSH analog could be advantageous. A short-acting analog with a more rapid clearance than the native molecule can be created by deleting one or more of the N-linked carbohydrate groups from the α and/or β subunits. This would result in enhanced metabolic clearance, and consequently lower activity *in vivo*.[98]

DESIGN OF NOVEL GONADOTROPIN MIMICS

Single-chain analogs bearing a single α and β subunit

The structure–function relationship studies of multimeric proteins, having non-covalently associated subunits, have often been hindered owing to mutagenesis-induced defects in subunit association. It is clear that gonadotropin subunit assembly is vital to the function of the hormone. First, only dimers are biologically active. Second, the conformation of the heterodimer complex signals the addition of hormone-specific oligosaccharides that contribute to the circulatory half-life and to signal transduction.[31,32] Third, the secretion efficiency of the dimers is determined by the β subunit.[35] A model was designed to bypass the subunit assembly step, which would potentially result in expanding the structure–function studies of the hormones. To address this point, the β and α subunits were genetically linked to form single chains,[99–101] and CG was chosen as the prototype;[99] the C-terminus of CGβ subunit was genetically fused to the N-terminus of the α subunit (CGβ–α) (Figure 8). The design of such tethered forms was based on data that the free C-terminus of the α subunit and N-terminus of the β subunit are important for receptor binding.[102–108] In addition, the CTP of the CGβ subunit was exploited as a 'natural' linker to bridge the subunits. Because the CTP has a high content of proline and serine residues, it has little secondary structure, and presumably would allow the α subunit to orient appropriately with the CGβ subunit domains. Further studies also indicated that this linker facilitates secretion of the fusion protein without significant influence on the bioactivity.[109–111] Expression of the tethered CGβ–α gene in Chinese hamster ovary cells[99] and insect cells[100] revealed that the secretion of the single chain and its biological activity *in vivo* and *in vitro* were comparable to those of the native heterodimer. Based on this model, the single chains of LH, FSH and TSH were also constructed.[109–112] Using the CTP segment as a linker, the resulting LHβ–α, FSHβ–α, and TSHβ–α single-chain biological activities were comparable to those of the corresponding heterodimers. Moreover, the single chains exhibited increased heat resistance *in vitro*, compared with the stability of the heterodimer incubated under the same conditions.[109–112] Another approach to single-chain construction is to link the two subunits through non-natural intersubunit disulfide bonds. This was achieved by introducing pairs of cysteine residues in the α and CGβ subunits.[113] The rationale for the design of such mutants is the enhanced stability when additional disulfide bonds are introduced.[114] These mutants were biologically active and exhibited enhanced thermostability, compared with the native heterodimer.[113]

The tether model also offers a novel approach to glycoprotein hormone structure–function analyses that would otherwise be difficult to perform with the heterodimers. Studies using the single-chain model have shown that a tight association between α and β domains is not required for receptor recognition.[23,115] Moreover, it has been demonstrated that residues in the α subunit that are critical for heterodimeric assembly are not essential for bioactivity, since they can be replaced in the single chain without affecting the biological activity of the variant.[116] That the structural features created by the α–β association are not critical for receptor activation has important implications for the production of recombinant products, for example, the design of biologically active mini-gonadotropins. A successful example of a 'minimized' gonadotropin was CG, which was reduced by one-third of its molecular weight.[117] This truncated form exhibited biological activity similar to that of the native heterodimer. The ability to reduce highly complex heterodimeric glycoprotein hormones to smaller proteins would be a major step towards providing agents that can be administered in alternative dosage forms, such as nasal rather than parenteral administration.

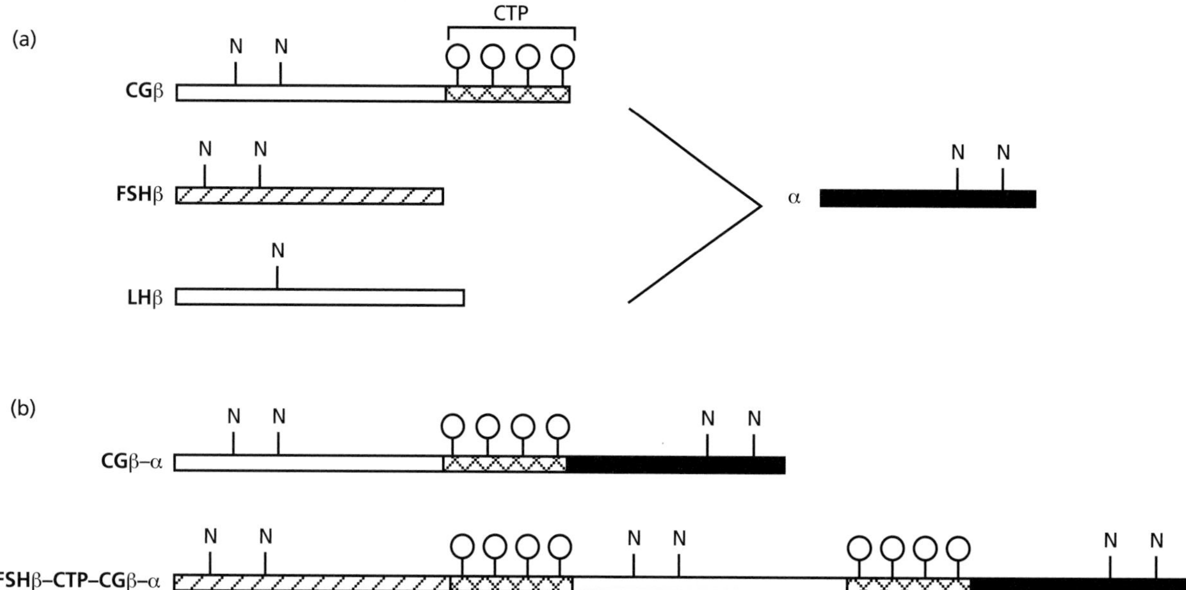

Figure 8 Schematic diagram of the single-chain gonadotropin analogs. (a) Structure of the α and β subunits; (b) structures of the single-chain models. CG, chorionic gonadotropin; FSH, follicle stimulating hormone; LH, luteinizing hormone; CTP, carboxy-terminal peptide

Modelling single-chain variants with multiple domains

It has been demonstrated that multiple contact points of the αβ heterodimer with the receptor are necessary to induce the signal. Moreover, α–β domains can exist in different conformations, suggesting that the receptor can recognize different forms of the ligand. To explore further structure–function studies of the hormones and their cognate receptors, a covalently linked hormone–receptor single chain was generated.[118] The CG tether was linked to the N-terminus of the mature LH/CG receptor sequence via the CTP sequence. Such chimeras activated cells *in vitro*, and demonstrated a stable interaction between the ligand and its receptor. This unique analog could be useful for identifying functionally relevant ligand–receptor interactions, and also for elucidating the downstream components in the signalling cascade.

Previous studies have shown that the contact sites between the α subunit and hormone-specific β subunit are not identical.[119,120] This observation implies that multi-subunit interactions can occur, i.e. one α subunit can interact intracellularly with two different β subunits, forming a transient intermediate. This issue is relevant for FSH and LH, which are both synthesized in the same cell,[34] and presumably the α subunit co-mingles with a mixed population of β subunits. To address this issue and to acquire further information on the conformational relationships between the ligand and the receptor, a novel chimera was constructed in which the α subunit was covalently linked to two tandemly arranged β subunits (Figure 8b).[121] The triple-domain single chain in the orientation FSHβ–CGβ–α displayed high-affinity binding to both FSH and LH/hCG receptors. The ability of this large structure to interact with either receptor further supported the hypothesis that sufficient flexibility exists in both the ligand and the receptor to establish a functional unit. It is unclear how the three-subunit domains orient with respect to one another to generate a biologically active molecule. One possibility is that the α subunit can interact with two β subunits simultaneously, and achieve a heterodimeric-like configuration with each hormone unit. This result would be consistent with the hypothesis that a multimeric gonadotropin could be generated in the secretory pathway. Such dual interactions would presumably not be stable, but rather transitory heterodimeric complexes. On the other hand, the molecule can be linear, and the receptor recognizes individual subunits, bringing them together in the receptor pocket. The latter implies that a single molecule possesses both FSH and CG activity.

Dually acting gonadotropins offer a potential tool for improving fertility protocols. A number of studies have reported that some LH activity is necessary for supporting FSH-induced follicular development.[122–125] As independent entities, differences in the *in vivo* response to LH and FSH are plausible, given their distinctive biochemical features and the variations in metabolic clearance by the patient population. However, as a single entity, there could be more effective control of the half-life and duration of LH and FSH activity, in addition to the beneficial effect of both activities being administered in a single injection. Furthermore, selectively altering the ratio of FSH/LH activities within the same molecule has important clinical implications. Hypogonadotropic hypogonadism patients undergoing FSH therapy require exogenous LH for an adequate follicular response. However, for a fixed dose of FSH, it has been shown that there are patient to patient variations in the dose of LH required to promote optimal follicular development.[126] Similarly, changing the FSH activity during the stimulation phase could also be clinically beneficial. Thus, the availability of a wide set of different active dual gonadotropin analogs would provide a very efficient system/method to calibrate fertility protocols adapted accurately to individual patients' requirements.

To construct bifunctional analogs with variable ratios of intrinsic activities of FSH: hCG/LH within the same molecule, a specific structural feature of the LHβ subunit was selectively modified.[127] In a preliminary series of structure–function experiments using triple-domain chimeras, truncating the carboxyl end of the LHβ subunit altered receptor binding. Based on this initial finding, FSHβ–CTP–LHβ–α variants differing in the length of the carboxy-terminal region of the LHβ subunit were constructed (Figure 9). All of the analogs were secreted by transfected carbohydrate oligosaccharide cells, and exhibited comparable wild-type FSH receptor binding affinity and signalling activity.[127] In contrast, the LH activity was variable, and correlated with the length of the carboxyl terminus of the LHβ subunit. These results demonstrate that the relative potencies of the two gonadotropin activities within the single-chain analog can be selectively modified. The data suggest that such analogs will be useful therapeutically in cases where it will be desirable to have variable LH action and a constant FSH activity.

LABORATORY TO CLINIC: RECOMBINANT FSH

We have discussed above how the use of recombinant DNA technology to study the structure–function biology of gonadotropins can be used to design potential therapeutic agents. The success of such technology is now manifest in the production of a recombinant glycoprotein hormone. Partially purified FSH has been used for decades to treat infertility and, until recently, such preparations were isolated from human postmenopausal urine, which contains LH and other contaminating proteins. Although a variety of treatment regimens have been tried and efforts made to titrate dosages for *in vitro* fertilization protocols, multiple gestations still often occur. As discussed above, the glycoprotein hormones display significant charge heterogeneity due to structural differences in the Asn-linked carbohydrates. Since these structural isoforms display different bioactivities, such species of FSH seen in commercial preparations may contribute to the complications seen with human menopausal gonadotropin administration. It would be advantageous to have some source of homogeneous FSH that could be standardized with respect to mass and bioactivity. Transfection of the FSH subunit genes into heterologous cells offers a potential source for producing large quantities of relatively homogeneous glycoprotein hormones.

Production of recombinant FSH in cell culture

There are several key issues that need to be considered for converting a recombinant DNA product in

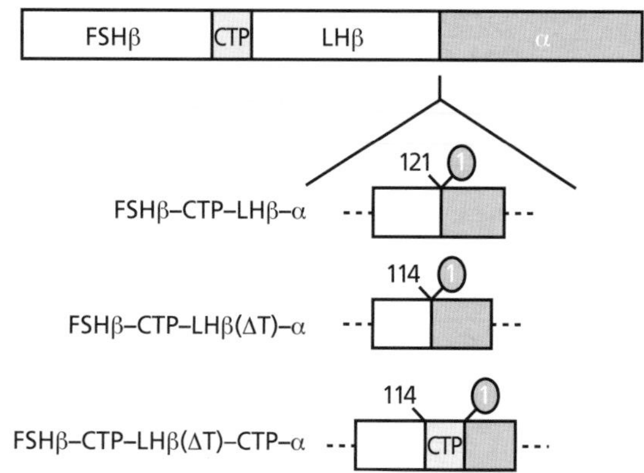

Figure 9 Schematic diagram of triple domain FSHβ–CTP–LHβ(ΔT)–α mutant and its luteinizing hormone β (LHβ) variants. FSHβ–CTP–LHβ(ΔT)–α lacks the carboxy-terminal heptapeptide of LHβ, and FSHβ–CTP–LHβ(ΔT)–CTP–α contains the human chorionic gonadotropin β (hCGβ) subunit carboxy-terminal peptide (CTP) segment, replacing the heptapeptide. FSH, follicle stimulating hormone

the laboratory to a therapeutic agent. Genetic stability is an important requirement for cell clones to be used in a production process for a therapeutic protein. A Chinese hamster ovary cell line expressing FSH was subjected to such detailed genetic analysis. Since the vectors used do not propagate independently, they must integrate into the chromosomes to be stably passed to the daughter cells during growth. Proof of integration was obtained by fluorescence *in situ* hybridization of a Chinese hamster ovary clone. Such analyses are critical to demonstrate stability of the integrated gene(s). Genetic abnormalities, e.g. translocations of the FSH expression consette to other chromosomes, can be detected. The data indicated the integration of the α and FSHβ DNA at a single position in the Chinese hamster ovary cell genome.

The culturing of mammalian cells on a production scale is technically much more difficult than microbial fermentation. This is due to the low growth rate (usually with population doubling times of 16–24 h for Chinese hamster ovary cells) and the fragility of mammalian cells. Furthermore, they usually require growth factors, such as are present in fetal calf serum. Because of the complexity of sera, the use of such a component in production media should be avoided. Therefore, either cells must be adapted to grow on serum-free media or, as an alternative, growth of cells can be achieved with serum-free medium.

The Chinese hamster ovary cell line is an anchorage-dependent cell line, and thus a proper surface must be provided for growth of the cells. To obtain a favorable surface/volume ratio, cells are grown on small collagen-coated beads. The use of microcarriers in cell culture also provides an opportunity for easy physical separation of the cells from the culture supernatant. Central to this perfusion system is the bioreactor that ensures aseptic operation and optimal growth conditions for mammalian cells, such as temperature, pH and dissolved oxygen. A bioreactor is a glass or stainless steel tank that contains a high-performance stirring system. The entire unit is computerized, enabling the growth of mammalian cells (and FSH production) for extended periods. A perfusion process has the advantages of short residence time and high cell concentration. FSH was isolated from pooled culture supernatant by a series of chromatographic steps. The overall recovery of the purification procedure was about 50%, and the final product was obtained as a lyophilized powder.

In terms of clinical efficacy, recombinant FSH was reported to be more efficacious than urinary FSH in inducing multifollicular development and achieving an ongoing pregnancy, and is now clinically available.

SUMMARY

In summary, the complexity of the gonadotropin structure has been a reservoir of fascinating structure–function concepts to investigate. The use of recombinant DNA methods to evaluate these issues has led to the generation of recombinant FSH as replacement therapy. As discussed above, a new generation of agonists (and antagonists) should be forthcoming to aid the clinician and patient in more comfortable and efficient treatment of reproductive disorders.

ACKNOWLEDGEMENT

The authors thank Dr Raj Kumar for his critical comments.

REFERENCES

1. Catt KJ, Dufau ML. Gonadotropic hormones: biosynthesis, secretion, receptors, and actions. In Yen SSC, Jaffe RB, eds. *Reproductive Endocrinology*. Philadelphia: WB Saunders, 1991: 105–155.
2. Bousfield GR, Perry WM, Ward DN. Gonadotropins: chemistry and biosynthesis. In Knobil E, Neil JD, eds. *The Physiology of Reproduction*. New York: Raven Press, 1994: 1749–1785.
3. Hsueh AJW, Bicsak TA, Jia X-C *et al*. Granulosa cells as hormone targets: the role of biologically active follicle-stimulating hormone in reproduction. *Rec Prog Horm Res* 1989 45: 209–277.
4. Burrow GN. *Reproductive Endocrinology*. Philadelphia: WB Saunders, 1991: 555–575.
5. Fiddes JC, Goodman HM. The gene encoding the common α subunit of the four human glycoprotein hormones. *J Mol Appl Genet* 1981 1: 3–18.
6. Boothby M, Ruddon RW, Anderson C, McWilliams D, Boime I. A single gonadotropin α-subunit gene in normal tissue and tumor-derived cell lines. *J Biol Chem* 1981 256: 5121–5127.
7. Stewart F, Thomson JA, Leigh SEA, Warwick JM. Nucleotide (cDNA) sequence encoding the horse gonadotropin α-subunit. *J Endocrinol* 1987 115: 341–346.
8. Mise T, Bahl OP. Assignment of disulfide bonds in the α subunit of human chorionic gonadotropin. *J Biol Chem* 1980 255: 8516–8522.
9. Pierce JG, Parsons TF. Glycoprotein hormones: structure and function. *Annu Rev Biochem* 1981 50: 465–495.
10. Strickland TW, Puett D. Contribution of subunits to the function of luteinizing hormone/human chorionic gonadotropin recombinants. *Endocrinology* 1981 109: 1933–1942.
11. Tena-Sempere M, Huhtaniemi IP. Gonadotropin receptors. In Fauser BCJM, ed. *Molecular Biology in Reproductive Medicine*. Carnforth, UK: Parthenon Publishing, 1999: 165–199.
12. Jia X-C, Oikawa M, Bo M *et al*. Expression of human luteinizing hormone (LH) receptor: interaction with LH and chorionic gonadotropin from human but not equine, rat, and ovine species. *Mol Endocrinol* 1991 5: 759–768.
13. Fiddes JC, Talmadge K. Structure, expression, and evolution of the genes for the human glycoprotein hormones. *Rec Prog Horm Res* 1984 40: 43–78.
14. Boorstein WR, Vamvakapoulos NC, Fiddes JC. Human chorionic gonadotropin β-subunit is encoded by at least eight genes arranged in tandem and inverted pairs. *Nature* 1982 300: 419–422.
15. Talmadge K, Boorstein WR, Fiddes JC. The human genome contains seven genes for the β-subunit of chorionic gonadotropin but only one gene for the β-subunit of luteinizing hormone. DNA 1983 2: 281–289.
16. Policastro PF, Daniels-McQueen S, Carle G, Boime I. A map of the hCGβ–LHβ gene cluster. *J Biol Chem* 1986 261: 5907–5916.
17. Watkins PC, Eddy R, Beck AK *et al*. DNA sequence and regional assignment of the human follicle-stimulating hormone β-subunit gene to the short arm of human chromosome 11. *DNA* 1987 6: 205–212.
18. Keene JL, Matzuk MM, Otani T *et al*. Expression of biologically active human follitropin in Chinese hamster ovary cells. *J Biol Chem* 1989 264: 4769–4775.
19. Whitfield GK, Powers RE, Gurr JA, Wolf O, Kourides IA. *Frontiers in Thyroidology*. New York: Plenum Publishing, 1986: 173–176.
20. Hayashizaki Y, Miyai K, Kata K, Matsubara K. Molecular cloning of the human thyrotropin-β subunit gene. *FEBS Lett* 1985 188: 394–400.
21. Lapthorn AJ, Harris DC, Littlejohn A *et al*. Crystal structure of human chorionic gonadotropin. *Nature* 1994 369: 455–461.
22. Wu H, Lustbader JW, Liu Y, Canfield RE, Hendrickson WA. Structure of human chorionic gonadotropin at 2.6 Å resolution from MAD analysis of the selenomethionyl protein. *Structure* 1994 2: 545–558.
23. Ben-Menahem D, Kudo M, Pixley MR *et al*. The biologic action of single-chain choriogonadotropin is not dependent on individual disulfide bonds of the β subunit. *J Biol Chem* 1997 272: 6827–6830.
24. Fox KM, Dias JA, Van Roey P. Three-dimensional structure of human follicle-stimulating hormone. *Mol Endocrinol* 2001 15: 378–389.
25. Birken S, Canfield RE. Isolation and amino acid sequence of COOH-terminal fragments from the β subunit of human choriogonadotropin. *J Biol Chem* 1977 252: 5386–5392.

26. Bousfield GR, Sugino H, Ward DN. Demonstration of a COOH-terminal extension on equine lutropin by means of a common acid-labile bond in equine lutropin and equine chorionic gonadotropin. *J Biol Chem* 1985 260: 9531–9533.
27. Kessler MJ, Mise T, Ghai RD, Bahl OP. Structure and location of the O-glycosidic carbohydrate units of human chorionic gonadotropin. *J Biol Chem* 1979 254: 7909–7914.
28. Parsons TF, Pierce JG. Oligosaccharide moieties of glycoprotein hormones: bovine lutropin resists enzymatic deglycosylation because of terminal O-sulfated N-acetylhexosamines. *Proc Natl Acad Sci USA* 1980 77: 7089–7093.
29. Hortin G, Natowicz M, Pierce JG, Baenziger J, Parsons T, Boime I. Metabolic labeling of lutropin with [^{35}S] sulfate. *Proc Natl Acad Sci USA* 1981 78: 7468–7472.
30. Anumula KR, Bahl OP. Biosynthesis of lutropin in ovine pituitary slices: incorporation of [^{35}S] sulfate in carbohydrate units. *Arch Biochem Biophys* 1983 220: 645–651.
31. Green ED, Baenziger JU. Asparagine-linked oligosaccharides on lutropin, follitropin, and thyrotropin. I. Structural elucidation of the sulfated and sialylated oligosaccharides on bovine, ovine, and human pituitary glycoprotein hormones. *J Biol Chem* 1988 263: 25–35.
32. Green ED, Baenziger JU. Asparagine-linked oligosaccharides on lutropin, follitropin, and thyrotropin. II. Distributions of sulfated and sialylated oligosaccharides on bovine, ovine, and human pituitary glycoprotein hormones. *J Biol Chem* 1988 263: 36–44.
33. Renwick AG, Mizuochi T, Kochibe N, Kobata A. The asparagine-linked sugar chains of human follicle-stimulating hormone. *J Biochem* 1987 101: 1209–1221.
34. Child GV. Studies of hormone storage and secretion in the multi potential gonadotrope. In Labrie F, Prouex L, eds. *Endocrinology*. Amsterdam: Elsevier, 1984: 499–503.
35. Corless CL, Matzuk MM, Ramabhadran TV, Krichevsky A, Boime I. Gonadotropin β subunits determine the rate of assembly and the oligosaccharide processing of hormone dimer in transfected cells. *J Cell Biol* 1987 104: 1173–1181.
36. Green ED, Gruenebaum J, Bielinska M, Baenziger JU, Boime I. Sulfation of lutropin oligosaccharides using a cell-free system. *Proc Natl Acad Sci USA* 1984 81: 5320–5324.
37. Smith PL, Baenziger JU. A pituitary N-acetylgalactosamine transferase that specifically recognizes glycoprotein hormones. *Science* 1988 242: 930–933.
38. Matzuk MM, Boime I. The role of the asparagine-linked oligosaccharides of the a subunit in the secretion and assembly of human chorionic gonadotropin. *J Cell Biol* 1988 106: 1049–1059.
39. Matzuk MM, Boime I. Site-specific mutagenesis defines the intracellular role of the asparagine-linked oligosaccharides of chorionic gonadotropin β subunit. *J Biol Chem* 1988 263: 17106–17111.
40. Kaetzel DM, Virgin JB, Clay CM, Nilson JH. Disruption of N-linked glycosylation of bovine luteinizing hormone β-subunit by site-directed mutagenesis dramatically increases its intracellular stability but does not affect biological activity of the secreted heterodimer. *Mol Endocrinol* 1989 3: 1765–1774.
41. Feng W, Matzuk MM, Mountjoy K, Bedows E, Ruddon RW, Boime I. The asparagine–linked oligosaccharides of the human chorionic gonadotropin β subunit facilitates correct disulfide bond pairing. *J Biol Chem* 1995 270: 11851–11859.
42. Matzuk MM, Keene JL, Boime I. Site-specificity of the chorionic gonadotropin N-linked oligosaccharides in signal transduction. *J Biol Chem* 1989 264: 2409–2414.
43. Van Hall EV, Vaitukaitis JL, Ross GT, Hickman JW, Ashwell G. Immunological and biological activity of hCG following progressive desialylation. *Endocrinology* 1971 88: 456–464.
44. Van Hall EV, Vaitukaitis JL, Ross GT, Hickman JW, Ashwell G. Effects of progressive desialylation on the rate of disappearance of immunoreactive hCG from plasma in rats. *Endocrinology* 1971 89: 11–15.
45. Morell AG, Gregoriadis G, Scheinberg IH, Hickman J, Ashwell G. The role of sialic acid in determining the survival of glycoproteins in the circulation. *J Biol Chem* 1971 246: 1461–1467.
46. Baenziger JU, Kumar S, Brodbeck RM, Smith, PL, Beranek MC. Circulatory half-life but not interaction with the lutropin/chorionic gonadotropin receptor is modulated by sulfation

of bovine lutropin oligosaccharides. *Proc Natl Acad Sci USA* 1992 89: 334–338.

47. Fiete D, Srivastava V, Hindsgaul O, Baenziger JU. A hepatic reticuloendothelial cell receptor specific for SO_4-4GalNAcβ1,4GlcNAcβ1,2Manα that mediates rapid clearance of lutropin. *Cell* 1991 67: 1103–1110.
48. Smith P, Bousfield GR, Kumar S, Fiete D, Baenziger JU. Equine lutropin and chorionic gonadotropin bear oligosaccharides terminating with SO_4-4-GalNAc and Siaα 2,3Gal, respectively. *J Biol Chem* 1993 268: 795–802.
49. Thotakura R, Desai RK, Bates LG, Cole ES, Pratt BM, Weintraub BD. Biological activity and metabolic clearance of a recombinant human thyrotropin produced in Chinese hamster ovary cells. *Endocrinology* 1991 128: 341–348.
50. Wakabayashi K. Heterogeneity of rat luteinizing hormone revealed by radioimunoassay and electrofocusing studies. *Endocrinol Jpn* 1977 24: 473–485.
51. Chappel SC, Ulloa-Aguirre A, Ramalay JA. Sexual maturation in female rats: time-related changes in the isoelectric focusing pattern of anterior pituitary follicle-stimulating hormone. *Biol Reprod* 1983 28: 196–205.
52. Wide L. Median charge and charge heterogeneity of human pituitary FSH, LH and TSH. I. Zone electrophoresis in agarose suspension. *Acta Endocrinol (Copenh)* 1985 109: 181–189.
53. Matteri RL, Papkoff H. Microheterogeneity of equine follicle-stimulating hormone. *Biol Reprod* 1988 38: 324–331.
54. Cole AL. Occurrence and properties of glycoprotein hormone free subunits. In Keel BA, Grotjan HE Jr, eds. *Microheterogeneity of Glycoprotein Hormones*. Boca Raton, FL: CRC Press, 1989: 149–184.
55. Wilson CA, Leigh AJ, Chapman AJ. Gonadotropin glycosylation and function. *J Endocrinol* 1990 125: 3–14.
56. Sergi I, Papandreou M-J, Medri G, Canonne C, Verrier B, Ronin C. Immunoreactive and bioactive isoforms of human thyrotropin. *Endocrinology* 1991 128: 3259–3268.
57. Joshi LR, Weintraub BD. Naturally occurring forms of thyrotropin with low bioactivity and altered carbohydrate content act as competitive antagonists to more bioactive forms. *Endocrinology* 1983 113: 2145–2154.
58. Dahl KD, Bicsak TA, Hsueh AJW. Naturally occurring antihormones: secretion of FSH antagonists by women treated with a GnRH analog. *Science* 1988 239: 72–74.
59. Matzuk MM, Krieger M, Corless CL, Boime I. Effects of preventing O-linked glycosylation on the secretion of human chorionic gonadotropin in Chinese hamster ovary cells. *Proc Natl Acad Sci USA* 1987 84: 6354–6358.
60. Matzuk MM, Hsueh AJW, LaPolt P, Tsafriri A, Keene JL, Boime I. The biological role of the carboxyl-terminal extension of human chorionic gonadotropin β-subunit. *Endocrinology* 1990 126: 376–383.
61. el Deiry S, Kaetzel D, Kennedy G, Nilson J, Puett D. Site-directed mutagenesis of the human chorionic gonadotropin β subunit: bioactivity of a heterologous hormone, bovine α-human des-(122–145) β. *Mol Endocrinol* 1989 3: 1523–1528.
62. Kalyan NK, Bahl OP. Role of carbohydrate in human chorionic gonadotropin. Effect of deglycosylation on the subunit interaction and on its *in vitro* and *in vivo* biological properties. *J Biol Chem* 1983 258: 67–74.
63. Chen F, Puett D. Delineation via site-directed mutagenesis of the carboxyl-terminal region of human choriogonadotropin β required for subunit assembly and biological activity. *J Biol Chem* 1991 266: 6904–6908.
64. Seminara SB, Hayes FJ, Crowley WF Jr. Gonadotropin-releasing hormone deficiency in the human (idiopathic hypogonadotropic hypogonadism and Kallmann's syndrome): pathophysiological and genetic considerations. *Endocr Rev* 1998 19: 521–539.
65. Thorner MO, Lee Vance M, Laws ER Jr, Horvath E, Kovacs K. The anterior pituitary. In Wilson JD, Foster DW, Kronenberg HM, Larsen PR, eds. *Williams Textbook of Endocrinology*, 9th edn. Philadelphia: WB Saunders, 1998: 249–340.
66. Themmen A, Huhtaniemi IT. Mutations of gonadotropins and gonadotropin receptors: elucidating the physiology and pathophysiology of pituitary–gonadal function. *Endocr Rev* 2000 21: 551–583.
67. Keir G, Winchester BG, Clayton P. Carbohydrate-deficient glycoprotein syndromes: inborn errors of protein glycosylation. *Ann Clin Biochem* 1999 36: 20–36.

68. Weiss J, Axelrod L, Witcomb RW, Harris PE, Crowley WF, Jameson JL. Hypogonadism caused by a single amino acid substitution in the β subunit of luteinizing hormone. *N Engl J Med* 1992 326: 179–183.

69. Suganuma N, Furui K, Kikkawa F, Tomoda Y, Furuhashi M. Effects of the mutations ($Trp^8 \rightarrow Arg$ and $Ile^{15} \rightarrow Thr$) in human luteininzing hormone (LH) β subunit LH bioactivity *in vitro* and *in vivo*. *Endocrinology* 1996 137: 831–838.

70. Haavisto A-M, Pettersson K, Bergendhal M, Virkamaki A, Huhtaniemi I. Occurrence and biological properties of a common genetic variant of luteinizing hormone. *J Clin Endocrinol Metab* 1995 80: 1257–1263.

71. Matthews CH, Borgato S, Beck-Peccoz P *et al.* Primary amenorrhoea and infertility due to a mutation in the β subunit gene of follicle-stimulating hormone. *Nat Genet* 1993 5: 83–86.

72. Phillip M, Arbelle JE, Segev Y, Parvari R. Male hypogonadism due to a mutation in the gene for the β-subunit of follicle-stimulating hormone. *N Engl J Med* 1998 338: 1729–1732.

73. Hayashizaki Y, Hiroaka Y, Endo Y, Matsubara K. Thyroid-stimulating hormone (TSH) deficiency caused by a single substitution in the CAGYC region of the β-subunit. *EMBO J* 1989 8: 2291–2296.

74. Dacou-Voutetakis C, Feltquate DM, Drakopoulou M, Kourides IA, Dracopoli NC. Familial hypothyroidism caused by a nonsense mutation in the thyroid-stimulating hormone β-subunit gene. *Am J Hum Genet* 1990 46: 988–993.

75. Medeiros-Neto G, Herodotou DT, Rajan S *et al.* A circulating biologically inactive thyrotropin caused by a mutation in the β subunit gene. *J Clin Invest* 1996 97: 1250–1256.

76. Hakola K, Pierroz DD, Aebi AC, Vuagnat BA, Aubert ML, Huhtaniemi I. Dose and time relationships of intravenously injected rat recombinant luteinizing hormone and testosterone secretion in the male rat. *Biol Reprod* 1998 59: 338–343.

77. Pierroz DD, Aebi AC, Huhtaniemi IT, Aubert ML. Many LH peaks are needed to physiologically stimulate testosterone secretion: modulation by fasting and NPY. *Am J Physiol* 1999 276: E603–E610.

78. Nishimura R, Shin J, Ji I *et al.* A single amino acid substitution in an ectopic α subunit of a human carcinoma choriogonadotropin. *J Biol Chem* 1986 261: 10475–10477.

79. Roach DJ, Layman LC, McDonough PG, Lanclos KD, Wall SW, Wilson JT. Identification of restriction-fragment length polymorphism for the human chorionic gonadotropin-β/luteinizing hormone-β gene cluster. *Fertil Steril* 1992 58: 914–918.

80. Layman LC, Edwards JL, Osborne WE *et al.* Human chorionic gonadotropin-β gene sequences in women with disorders of hCG production. *Mol Hum Reprod* 1997 3: 315–320.

81. Miller-Lindholm A, Bedows E, Bartels CF, Ramey J, Maclin V, Ruddon RW. A naturally occurring genetic variant in the human chorionic gonadotropin-β gene is assembly inefficient. *Endocrinology* 1999 140: 3496–3506.

82. Worley RJ. Ovulation induction. In Garcia CR, Mastroianni L, Anelar RD, Dubin L, eds. *Current Therapy of Infertility – 3.* Toronto: Decker, 1988 3: 106–110.

83. Lightman A, Jones EE, Boyers SP. Ovulation induction: human menopausal gonadotropins. In Decherney AH, Palan ML, Lee RD, Boyers SP, eds. *Decision Making in Infertility*. Toronto: Decker, 1988: 32–33.

84. Jones GS, Acosta A, Garcia JE, Bernardus RE, Rosenwaks Z. The effect of follicle-stimulating hormone without additional luteinizing hormone on follicular stimulation and oocyte development in normal ovulatory women. *Fertil Steril* 1985 43: 696–702.

85. Albert PJ, Schlafke J, Kaesemann H, Gille J. Pregnancy following induction of ovulation with pure FSH after suppression of endogenous gonadotropins with subcutaneous buserelin. *Arch Gynecol Obstet* 1987 241: 53–56.

86. Olijve W, De Boer W, Mulders JWM, Van Wezenbeek PMGF. Molecular biology and biochemistry of human recombinant follicle stimulating hormone (Puregon). *Mol Hum Reprod* 1996 2: 371–382.

87. Matzuk MM, Spangler MM, Camel M, Suganuma N, Boime I. Mutagenesis and chimeric genes define determinants in the β-subunits of human chorionic gonadotropin and lutropin for secretion and assembly. *J Cell Biol* 1989 109: 1429–1438.

88. Suganuma N, Matzuk MM, Boime I. Elimination of disulfide bonds affects assembly and secretion

of the human chorionic gonadotropin β-subunit. *J Biol Chem* 1989 264: 19302–19307.

89. Chen F, Wang Y, Puett D. Role of the invariant aspartic acid 99 of human chorionic gonadotropin β in receptor binding and biological activity. *J Biol Chem* 1991 266: 19357–19361.
90. Fares F, Suganuma N, Nishimori K, LaPolt PS, Hsueh AJW, Boime I. Design of a long-acting follitropin agonist by fusing the C-terminal sequence of the chorionic gonadotropin β subunit to the follitropin β subunit. *Proc Natl Acad Sci USA* 1992 89: 4304–4308.
91. Mizouchi T, Nishimura R, Taniguchi T, Utsunomiya T, Mochizuki M, Kobata A. Comparison of carbohydrate structure between human chorionic gonadotropin present in urine of patients with trophoblastic diseases and healthy individuals. *Jpn J Cancer Res (Gann)* 1985 76: 752–759.
92. Kobata A. Structures, function, and transformational changes of the sugar chains of glycohormones. *J Cell Biochem* 1988 37: 79–90.
93. Sowers JR, Pecary AE, Hershman JM, Kanter M, Distefano JJ. Metabolism of exogenous human chorionic gonadotropin in men. *J Endocrinol* 1979 80: 83–89.
94. Amin HK, Hunter WM. Human pituitary follicle-stimulating hormone: distribution, plasma clearance and urinary excretion as determined by radioimmunoassay. *J Endocrinol* 1970 48: 307–317.
95. Joshi LR, Murata Y, Wondisford FE, Szkudlinski MW, Desai R, Weintraub BD. Recombinant thyrotropin containing a β-subunit chimera with the human chorionic gonadotropin-β carboxyl terminus is biologically active, with a prolonged plasma half-life; role of carbohydrate in bioactivity and metabolic clearance. *Endocrinology* 1995 136: 3839–3848.
96. Bouloux PMG, Handelsman DJ, Jockenhövel F *et al.* First human exposure to FSH-CTP in hypogonadotrophic hypogonadal males. *Hum Reprod* 2001 16: 1592-1597.
97. Szkudlinski MW, Teh NG, Grossmann M, Tropea JE, Weintraub BD. Engineering human glycoprotein hormone superactive analogues. *Nat Biotech* 1996 14: 1257–1263.
98. Galway AB, Hsueh AJW, Keene JL, Yamato M, Fauser BCJM, Boime I. *In vitro* and *in vivo* bioactivity of recombinant human FSH mutants and partially deglycosylated variants secreted by transfected eukaryotic cell lines. *Endocrinology* 1990 127: 93–100.
99. Sugahara T, Pixley MR, Minami S, Perlas E, Ben-Menahem D, Hsueh AJW, Boime I. Biosynthesis of a biologically active single peptide chain containing the human common α and chorionic gonadotropin β subunits in tandem. *Proc Natl Acad Sci USA* 1995 92: 2041–2045.
100. Narayan P, Wu C, Puett D. Functional expression of yoked human chorionic gonadotropin in baculovirus-infected insect cells. *Mol Endocrinol* 1995 9: 1720–1726.
101. Heikoop JC, van Beuningen-de Vaan MM, van den Boogaart P, Grootenhuis PD. Evaluation of subunit truncation and the nature of the spacer for single chain human gonadotropins. *Eur J Biochem* 1997 245: 656–662.
102. Parsons TF, Pierce JG. Biologically active covalently cross-linked glycoprotein hormones and the effects of modification of the COOH-terminal region of their α subunit. *J Biol Chem* 1979 254: 6010–6015.
103. Merz WE. Studies of the specific role of the subunits of choriogonadotropin for biological, immunological and physical properties of the hormone: digestion of the α-subunit with carboxypeptidase. *Eur J Biochem* 1979 101: 541–553.
104. Gordon WL, Ward DN. Structural aspects of luteinizing hormone action. In Ascoli M, ed. *Luteinizing Hormone Action and Receptors.* Boca Raton, FL: CRC Press, 1985: 173–198.
105. Charlesworth MC, McCormick DJ, Madden BJ, Ryan RJ. Inhibition of human choriotropin binding to receptor by human α peptides: a comprehensive synthetic approach. *J Biol Chem* 1987 262: 13409–13416.
106. Bielinska M, Pixley MR, Boime I. Site-directed mutagenesis identifies two receptor-binding domains in the human chorionic gonadotropin a subunit. *J Cell Biol* 1990 111: 330a (abstr 1844).
107. Yoo J, Ji I, Ji T. Conversion of lysine 91 to methionine or glutamic acid in human choriogonadotropin α results in the loss cAMP inducibility. *J Biol Chem* 1991 266: 17741–17743.
108. Chen F, Wang Y, Puett D. The carboxy-terminal region of the glycoprotein hormone α-subunit: contributions to receptor binding and signaling in human chorionic gonadotropin. *Mol Endocrinol* 1992 6: 914–919.

109. Garcia-Campayo V, Sato A, Hirsh B *et al.* Design of stable biologically active recombinant lutropin analogs. *Nat Biotech* 1997 15: 663–667.
110. Grossman M, Wong R, Szkudlinski MW, Weintraub BD. Human thyroid-stimulating hormone (hTSH) subunit gene fusion produces hTSH with increased stability and serum half-life and compensates for mutagenesis-induced defects in subunit association. *J Biol Chem* 1997 22: 21312–21316.
111. Sugahara T, Grootenhuis PDJ, Sato A *et al.* Expression of biologically active fusion genes encoding the common α subunit and the follicle-stimulating hormone β subunit: role of linker sequence. *J Biol Chem* 1996 271: 10445–10448.
112. Fares FA, Yamabe S, Ben-Menahem D, Pixley MR, Hsueh AJW, Boime I. Conversion of thyrotropin heterodimer to a biologically active single-chain. *Endocrinology* 1998 139: 2459–2464.
113. Heikoop JC, Boogart PVD, Mulder JWM, Grootenhuis PDJ. Structure-based design and protein engineering of intersubunit disulfide bonds in gonadotropins. *Nat Biotech* 1997 15: 658–662.
114. Matsumara M, Matthews BM. Stabilization of functional proteins by introduction of multiple disulfide bonds. *Methods Enzymol* 1991 202: 336-356.
115. Sato A, Perlas E, Ben-Menahem D *et al.* Cysteine knot of the gonadotropin α subunit is critical for intracellular behavior but not for *in vitro* biological activity. *J Biol Chem* 1997 272: 18098–18103.
116. Jackson AM, Berger P, Pixley MR, Klein C, Hsueh AJW, Boime I. The biological action of choriogonadotropins is not dependent on the complete native quaternary interactions between the subunits. *Mol Endocrinol* 1999 13: 2175–2188.
117. Heikoop JC, Huisman-de Winkel B, Grootenhuis PDJ. Towards minimized gonadotropins with full bioactivity. *Eur J Biochem* 1999 261: 81–83.
118. Wu C, Narayan P, Puett D. Protein engineering of a novel constitutively active hormone receptor complex. *J Biol Chem* 1996 271: 31638–31642.
119. Furuhashi M, Suzuki S, Suganuma N. Disulfide bonds 7-31 and 59-87 of the α-subunit plays a different role in assembly of human chorionic gonadotropin and lutropin. *Endocrinology* 1996 137: 4196–4200.
120. Grossman M, Szkudlinski MW, Dias JA, Xia H, Wong R, Puett D, Weintraub BD. Site-directed mutagenesis of amino acid 33–34 of the glycoprotein hormones suggests that cyclic adenosine 3' 5'–monophosphate production and growth promotion are potentially dissociable functions of human thyrotropin. *Mol Endocrinol* 1996 10: 769–779.
121. Kanda M, Jablonka-Shariff A, Sato A *et al.* Genetic fusion of an α-subunit gene to the follicle-stimulating hormone and chorionic gonadotropin β-subunit genes: production of bifunctional protein. *Mol Endocrinol* 1999 13: 1873–1881.
122. Couzinet B, Lestrat N, Brailly S, Forest M, Schaison G. Stimulation of ovarian follicular maturation with pure follicle stimulating hormone in women with gonadotropin deficiency. *J Clin Endocrinol Metab* 1988 66: 552–556.
123. Shoham Z, Balen A, Patel A, Jacobs HS. Results of ovulation induction using human menopausal gonadotropin or purified follicle-stimulating hormone in hypogonadotropic hypogonadism patients. *Fertil Steril* 1991 56: 1048–1053.
124. Schoot DC, Harlin J, Shoham Z *et al.* Recombinant human follicle-stimulating hormone and ovarian response in gonadotropin-deficient women. *Hum Reprod* 1994 9: 1237–1242.
125. Balash J, Miro F, Burzaco T *et al.* The role of luteinizing hormone in human follicle development and oocyte fertility: evidence from *in vitro* fertilization in a woman with long-standing hypogonadotrophic hypogonadism and using recombinant human follicle stimulating hormone. *Hum Reprod* 1995 10: 1678–1683.
126. The European Recombinat Human LH Study Group. Recombinant human luteinizing hormone (LH) to support recombinant human follicle-stimulating hormone (FSH)-induced follicular development in LH- and FSH-deficient anovulatory women: a dose-finding study. *J Clin Endocrinol Metab* 1998 83: 1507–1514.
127. Garcia-Campayo V, Boime I. Independent activities of FSH and LH structurally confined in a single polypeptide: selective modification of the relative potencies of the hormones. *Endocrinology* 2001 142: 5203–5211.

Chapter 19

Selective estrogen receptor modulators

Bin Chen and V. Craig Jordan

INTRODUCTION

With an aging population and demands for an active, disease-free retirement, a conceptual shift in health care is occurring. Chemoprevention, a term first used by Sporn and co-workers in the 1970s,[1,2] holds promise for an increased, cancer-free life. However, the term can be expanded to encompass the application of chemopreventives as multifunctional drugs to prevent breast and endometrial cancer, osteoporosis and coronary heart disease (CHD). In the mid-1980s, it was recognized that the non-steroidal antiestrogens were actually antiestrogenic at some sites, but showed estrogen-like behavior at other sites. For example, tamoxifen and raloxifene (Figure 1) prevented mammary carcinogenesis in the rat,[3] but also maintained bone density in ovariectomized rats.[4] The tamoxifen–estrogen receptor (ER) complex acts selectively as an antiestrogen at some sites, but as an estrogen at others.[5,6] This chapter traces the evolution of this idea during the past decade, which has resulted in the first selective estrogen receptor modulators (SERMs), tamoxifen and raloxifene being approved as chemopreventives for breast cancer and osteoporosis, respectively, in well women.

CLINICAL BASIS FOR SELECTIVE ESTROGEN RECEPTOR MODULATOR ACTION

In 1936, Lacassagne[7] suggested that, if breast cancer was caused by a special hereditary sensitivity to estrogen, then the disease could be prevented by developing a therapeutic antagonist to estrogen action in the breast. However, there were no therapeutic antagonists of estrogen at that time, nor was there an aim to design drug molecules. Jensen and Jacobson[8] first described the ER (now known as ERα) in estrogen target tissues, and subsequently proposed the ER assay to identify those breast cancer patients who were most likely to respond to endocrine therapy.[9]

Tamoxifen is an antiestrogen in breast cancer[10] with low toxicity. These properties led to the development of tamoxifen as an adjuvant therapy,[11] and the discovery that adjuvant tamoxifen treatment improved the survival of ER-positive breast cancer patients.[12] The expanded clinical use of tamoxifen over the past 25 years also caused an intense evaluation of the pharmacology of the non-steroidal antiestrogens in the laboratory. This clinical and laboratory knowledge paved the way for the evaluation of SERMs for multiple applications in well women.[11,13,14]

The National Surgical Breast and Bowel Project (NSABP) P1 study began in 1992 in the USA and Canada, and was designed to test the worth of tamoxifen as a breast cancer preventive in a prospective randomized clinical trial.[15] High-risk women were identified based on patient age, the Gail model[16] or a history of lobular carcinoma *in situ* treated by biopsy alone. The Gail model is available as a computer program that calculates a woman's 5-year and lifetime risks of breast cancer on the basis of age, number of first-degree relatives with breast cancer, presence of atypical hyperplasia, number of breast biopsies, age at menarche, stage of menopause and parity. Patients were randomly assigned to receive placebo ($n = 6707$) or tamoxifen 20 mg/day ($n = 6681$) for 5 years. Tamoxifen reduced the risk of invasive breast cancer by 49% ($p < 0.00001$) and reduced the risk of non-invasive breast cancer by 50% ($p < 0.002$). The cumulative incidence of breast cancer over 69 months was 43.4 per 1000 women and 22 per 1000 women in the two groups, respectively, for invasive cancer, and 15.9 per 1000 women and 7.7 per 1000 women, respectively, for non-invasive cancer. Tamoxifen reduced the occurrence of ER-positive tumors by 69%, but there was no difference in the occurrence of ER-negative tumors. Overall, the results of the NSABP-P1 study demonstrated the SERM principle, i.e. an estrogen-like effect on bone in reducing hip fractures and a predicted increase in endometrial cancer in postmenopausal volunteers (from one per 1000 women to four per 1000 women). Additionally, studies in postmenopausal breast cancer patients demonstrated that tamoxifen maintained a slightly increased bone density[17] and reduced levels of circulating cholesterol.[18] Indeed, there were suggestions that tamoxifen might reduce the risk of CHD,[19] but this has not been confirmed in other studies.[20,21] Tamoxifen is

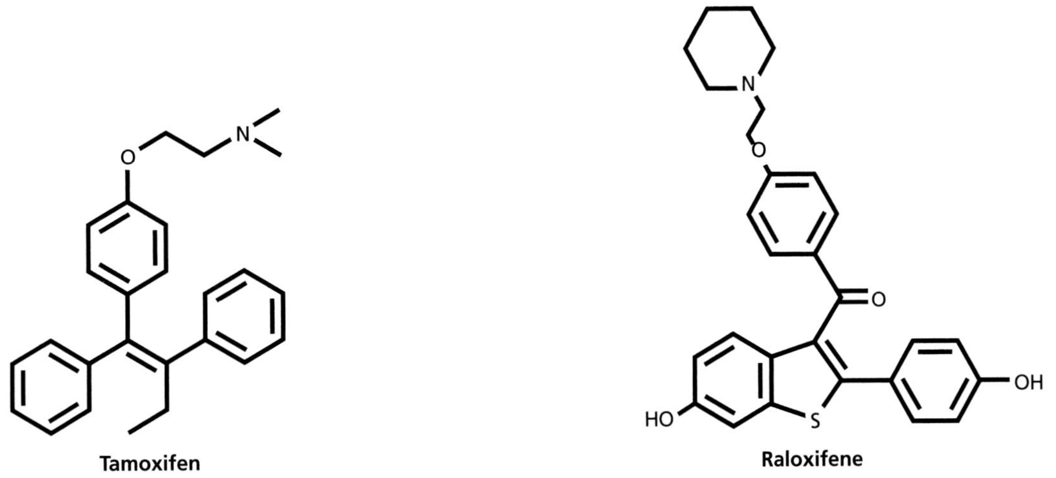

Figure 1 Structures of the clinically useful selective estrogen receptor modulators (SERMs) tamoxifen and raloxifene

available in the USA for the reduction of breast cancer incidence in high-risk women.

During the 1990s, novel agents were sought that would behave as multifunctional drugs. On the basis of laboratory studies,[3,4] it was realized that non-steroidal antiestrogens produced estrogen-like effects in bone and lowered cholesterol, but exerted antiestrogenic effects in breast. Clearly, a drug could be developed that would prevent osteoporosis or CHD in postmenopausal women, but would reduce the risk of breast cancer as a side-effect.[11] Most important, new agents were necessary to reduce the small risk of endometrial cancer observed with tamoxifen in postmenopausal women.[22] The result of this paradigm shift was raloxifene.

The pivotal study, Multiple Outcomes of Raloxifene Evaluation (MORE), was undertaken in 1994 to evaluate the effects of raloxifene on the skeleton.[23] The study randomized 7705 postmenopausal osteoporotic women in 25 countries to 60 mg/day or 120 mg/day raloxifene and placebo. The risk of vertebral fractures was decreased by 30–50% among women treated with raloxifene for 36 months.

Raloxifene treatment also reduced the incidence of breast cancer in the MORE trial.[24] At a median follow-up of 40 months, 54 women were diagnosed with breast cancer in the study population. Raloxifene reduced the risk of newly diagnosed invasive breast cancer by 76% in postmenopausal women treated with raloxifene for osteoporosis. A follow-up evaluation continues to demonstrate the beneficial effect of raloxifene in reducing breast cancer incidence in the MORE trial.[25]

Additionally, raloxifene lowers the level of circulating cholesterol[26] and appears not to stimulate increases in endometrial thickness.[24,27,28] Raloxifene is being tested in the Raloxifene Use for The Heart (RUTH) trial to determine whether a SERM can reduce the risk of CHD in high-risk women, and in the Study of Tamoxifen and Raloxifene (STAR) to establish its worth as a breast cancer preventive in high-risk postmenopausal women. A comparison of the clinical pharmacologies of tamoxifen and raloxifene in postmenopausal women is summarized in Figure 2.

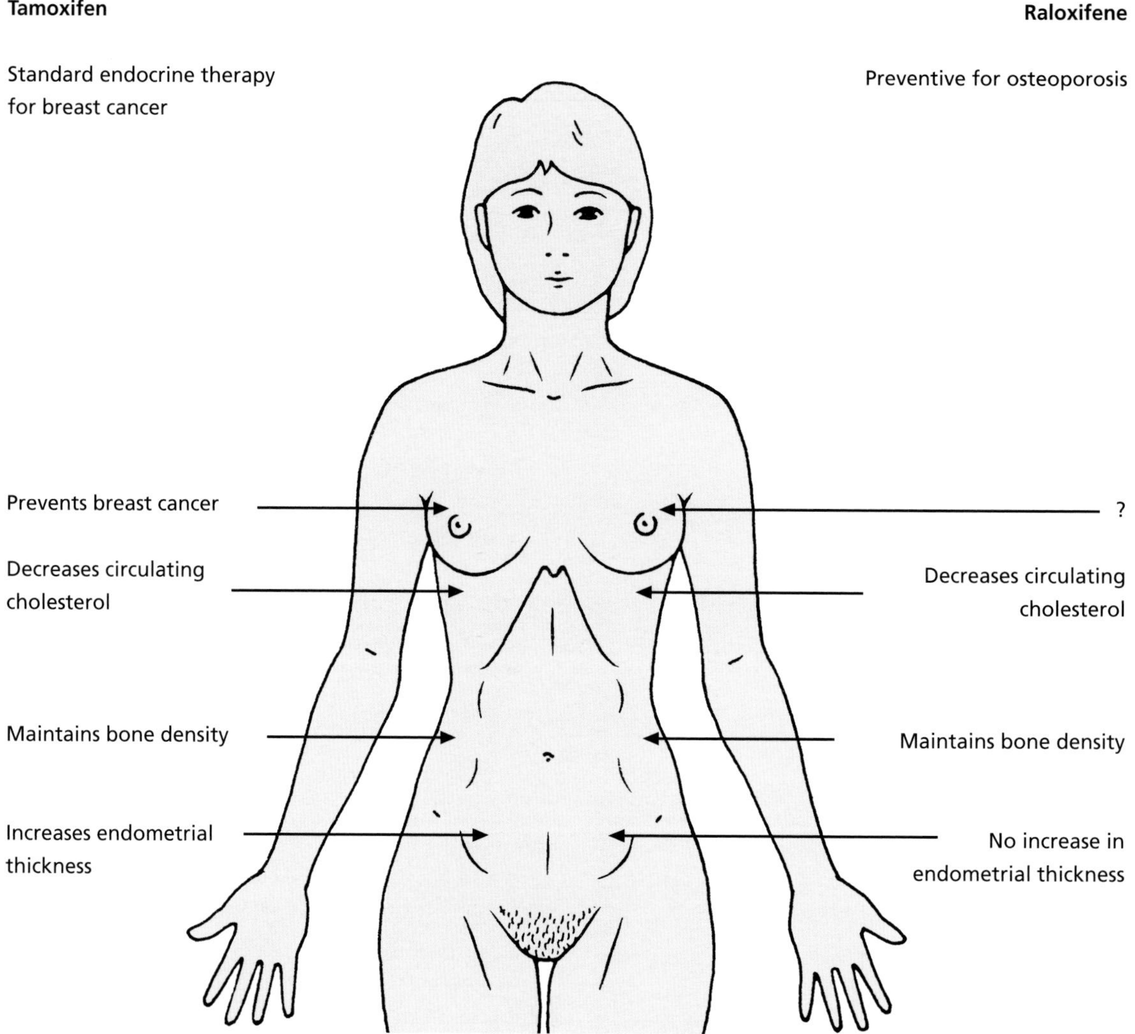

Figure 2 Comparison of the clinical pharmacologies of tamoxifen and raloxifene in the postmenopausal woman

MOLECULAR MECHANISM OF ACTION OF SELECTIVE ESTROGEN RECEPTOR MODULATORS

The clinical success of SERMs has encouraged investigation of the molecular mechanism of action of the compounds, to understand the target site-specific actions of the drugs as a prelude to developing a series of novel agents.

At this point, it is not possible to describe all the mechanisms of SERMs at a target site, because the relative importances of co-regulator proteins and pathways are not known. The proportion of ERs, co-regulator proteins and pathways must be determined for each target site so that targeted drug development can evolve rationally. Nevertheless, the use of the reverse transcriptase-polymerase chain reaction (RT-PCR)[29,30] and the current development of new monoclonal[31–33] antibodies to the ERβ protein will enhance knowledge of the target site requirements for the modulation of estrogen-like and antiestrogenic functions.

There are a number of decision points that will ultimately determine the biological response to a SERM (Figure 3). Two structurally related ERs, referred to as ERα and ERβ,[34] have some degree of

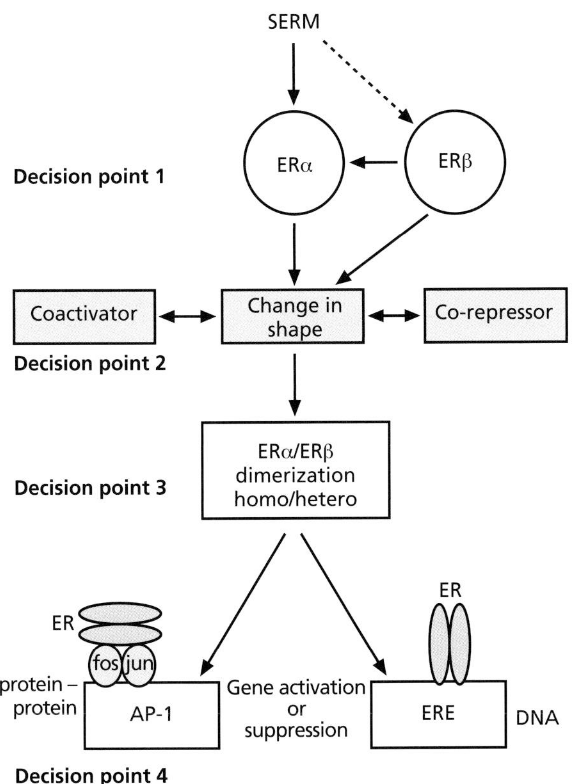

Figure 3 Molecular mechanism of selective estrogen receptor modulator (SERM) action. The SERM can choose to bind to either ERα or ERβ, which then changes shape so that the complex can bind either coactivators or co-repressors. The complexes can either hetero- or homodimerize and initiate gene activation at the target site via an activating protein-1 (AP-1) site (protein–protein interaction) or at an estrogen responsive element (ERE) directly at DNA

homology with each other. A SERM, therefore, has a choice of receptor molecules. Both ERs have a ligand-binding and a DNA-binding domain, and can directly bind to DNA to activate gene transcription. There are, however, differences in the activating functions (AFs) that can alter the SERM–ER complex, resulting in increased or decreased estrogenicity. Tamoxifen appears to be more antiestrogenic when complexed with ERβ, compared with ERα.[35] It is now clear that the ligand programs the shape of the ER complex so that coactivators or co-repressors can bind to the external surface of a SERM–ER complex.[36] Coactivators will aid signal transduction, whereas co-repressors will block transduction. At the time when a transcriptional complex is to be formed, the SERM–ERα/β complexes must decide whether to homo- or heterodimerize, before initiating gene transcription. Finally, there is evidence to suggest that SERMs can modulate gene transcription through two mechanisms, either via an activating protein (AP)-1 pathway, when a protein–protein interaction occurs with fos and jun at the AP-1 site, or the SERM–ER complex can activate or silence an estrogen response element (ERE) directly on DNA.[37]

The chance finding of the D351Y ER mutation in a tamoxifen-stimulated breast tumor[38] that enhanced the estrogen-like actions of SERMs[39–41] was an invaluable starting point in deciphering SERM modulation of the ER. The change in biology with the mutant ER has subsequently been clarified and supported by X-ray crystallography.[42–44]

Structure–function relationships of ERα have been examined by stable transfections of mutant ERs into ER-negative breast cancer cells, using transforming growth factor (TGF)-α as a target gene *in situ* to evaluate the estrogenicity of SERMs.[39,40,45] It is now known that the antiestrogenic side-chain of a SERM interacts differently with the surface amino acid D351.[42,43,46] In the case of 4-hydroxytamoxifen, it is hypothesized that the surface charge of D351 is not neutralized or shielded by the antiestrogenic side-chain, so in a coactivator-rich environment these proteins can bind to the SERM–ER complex to initiate TGF-α transcription. In contrast, the side-chain of raloxifene shields D351 and produces only an antiestrogenic complex.[47] The hypothesis that the interaction of the antiestrogen side-chain and D351 programs the ER–SERM complex for either estrogenic or antiestrogenic actions has been tested by swapping amino acid 351 or altering the antiestrogenic side-chain. It is possible to silence estrogen action of the 4-hydroxytamoxifen–ER complex by substituting a glycine for aspartate at amino acid 351.[48] It is reasoned that the 4-hydroxytamoxifen–ER complex must bind coactivators at a novel site other than the traditional AF-2 site, because this is blocked by helix 12 in the 4-hydroxytamoxifen–ER complex.[48] The new activating site for coactivator binding on the SERM–ER complex is called AF-2b,[48] distinct from AF-2 used by estrogens in the ligand-binding domain.[43] The raloxifene–ER complex can be made more estrogen-like by substituting a larger, negatively charged amino acid, such as tyrosine, at amino acid 351.[49] The amino acid now reaches over the antiestrogen side-chain of raloxifene to allow a coactivator to bind to the AF-2b site. The other approach to structure–function relationships of SERMs is to change the side-chain of tamoxifen. The compound GW 7604 has a carboxylic side-chain that repels aspartate 351, thereby disrupting the AF-2b coactivator docking site.[50] The compound completely alters the intrinsic activity of the SERM–ERα complex from estrogen-like to antiestrogenic (see below for further information about GW 7604 and its prodrug GW 5638).

Overall, these data regarding ERα in a single context (MDA-MB-231 breast cancer cells) demonstrate how the SERM–ERα complex can be modulated through aspartate 351. However, it is known that ERα can heterodimerize with a SERM–ERβ complex, and this could affect the estrogen-like actions of SERMs in a target site. It is known that the 4-hydroxytamoxifen–ERβ complex is less estrogen-like than the 4-hydroxytamoxifen–ERα complex.[35] If high concentrations of ERβ exist at a specific target site, tamoxifen could be less estrogen-like at that site. Nevertheless, the exact roles and interaction of ERα and ERβ in target sites have yet to be determined. These concepts and possibilities have recently been summarized.[47]

At present, a number of agents are being evaluated in the laboratory and the clinic for multiple applications in women's health.

NEW SELECTIVE ESTROGEN RECEPTOR MODULATORS

The compound CP 336 156 (Figure 4) is a new nonsteroidal estrogen agonist/antagonist, which binds to the human ERα with high affinity and has excellent oral bioavailability.[51] CP 336 156 decreases serum levels of cholesterol and reduces bone loss in ovariectomized rats.[52] The compound has potential as a SERM for postmenopausal women, and is being tested as a preventive for osteoporosis.

GW 5638 (Figure 4) is a tamoxifen derivative with a novel carboxylic acid side-chain, with no uterotropic activity in the rat.[53] GW 5638 has a high affinity to ER, and acts as an ER antagonist, inhibiting the agonist activity of estrogen and tamoxifen *in vitro*.[54] However, this compound is a full ER agonist in the bone and the cardiovascular system. Recent data suggest that the molecular mechanism of action of GW 7604 – the metabolite of GW 5638 – is different from that of compounds previously reported.[55] GW 7604 interacts with the critical amino acid, aspartate 351, on the surface of the ER, and disrupts the external charge so that an interaction with coactivators does not occur.[50] This observation makes the compound highly interesting, and suggests that it may have applications in the treatment of tamoxifen-resistant breast cancer.[56] GW 5638 is currently being tested in animal models prior to clinical evaluation.

LY 353381 (arzoxifene) (Figure 4) is an analog of raloxifene, and it was shown to be very potent in preventing the effects of ovariectomy on serum cholesterol level and bone, while maintaining estrogen antagonist effects on the uterus.[57] Arzoxifene is not completely cross-resistant with tamoxifen in laboratory models,[58] and the drug is currently being evaluated as a treatment for breast cancer.[59]

EM 800 is a new, non-steroidal, orally active antiestrogen that exerts pure antagonistic action. It has been shown to prevent dimethylbenz(a)anthracene (DMBA)-induced mammary tumors while reducing serum triglyceride and cholesterol levels, without an adverse effect on bone mass.[60] EM 800 also completely prevented estrogen-stimulated tumor growth in ovariectomized nude mice bearing ZR-75-1 xenografts.[61] Simultaneous administration with tamoxifen further enhanced antitumor activity, and EM 800 can completely reverse the stimulatory effects of tamoxifen. EM 800 is a prodrug, and the active metabolite EM 652 (Figure 4) is being developed clinically.[62] Originally, it was claimed that EM 652 was an orally active pure antiestrogen,[62,63] but its molecular mechanism of action at the ER is identical to that of raloxifene.[64] Like raloxifene, EM 652 maintains bone density in the rat,[65] and the compound is being pursued for applications as a SERM.

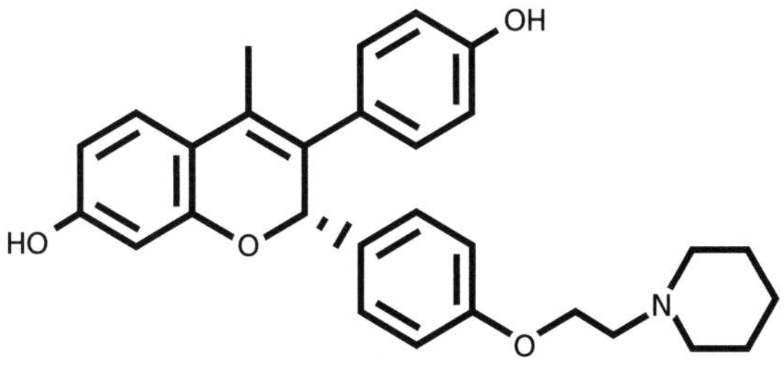

EM 652

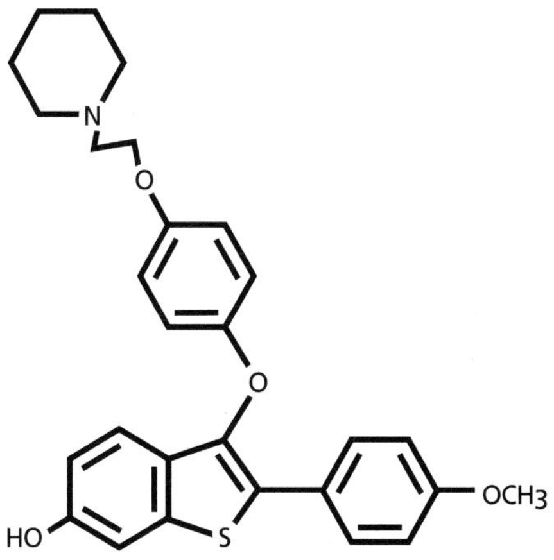

Arzoxifene

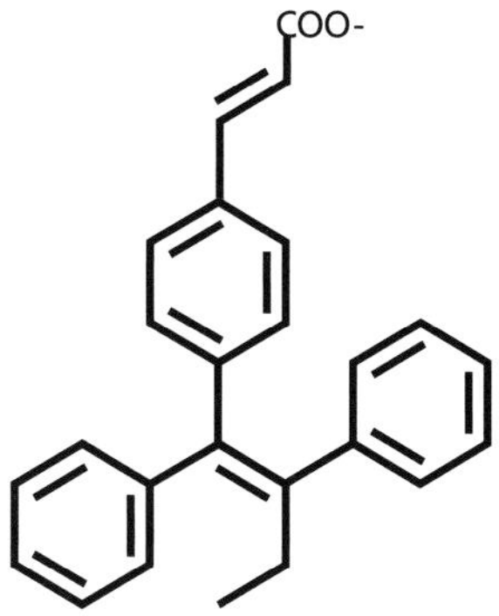

GW 5638

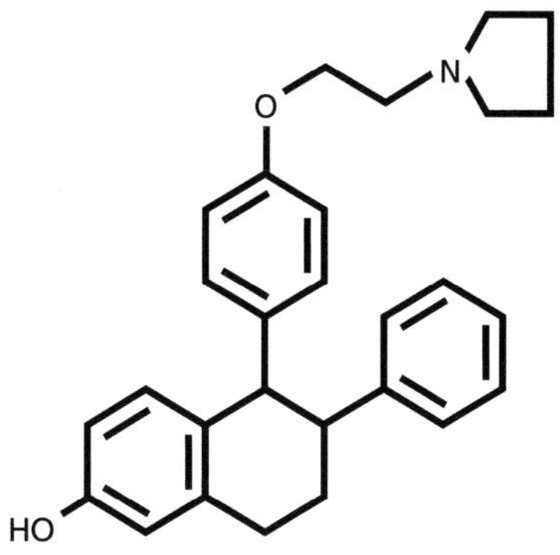

CP 336 156

Figure 4 A selection of new selective estrogen receptor modulators (SERMs) currently being evaluated in the laboratory and the clinic for applications in breast cancer treatment, in the prevention of osteoporosis or as a chemopreventive for breast cancer in high-risk women

SUMMARY

During the past decade, a new group of clinically useful drugs, the SERMs, have shown promise for the prevention of breast cancer and osteoporosis. This clinical proof of principle, and emerging new knowledge about the potential molecular pathways whereby estrogens and SERMs can modulate gene activation, hold promise for a menu of new multifunctional agents for women's health. However, with this innovation must come an awareness of physicians regarding the potential for cross-resistance in breast cancer treatment that will influence treatment choices. In other words, a woman who develops breast cancer while taking raloxifene to prevent osteoporosis must subsequently consider options other than tamoxifen for treatment. Aromatase inhibitors[66] or the new pure antiestrogen ICI 182 780[67] may provide the only appropriate intervention.

REFERENCES

1. Sporn MB. Approaches to prevention of epithelial cancer during the preneoplastic period. *Cancer Res* 1976 36: 2699–2702.
2. Sporn MB, Dunlop NM, Newton DL, Smith JM. Prevention of chemical carcinogenesis by vitamin

A and its synthetic analogs (retinoids). *Fed Proc* 1976 35: 1332–1338.
3. Gottardis MM, Jordan VC. Antitumor actions of keoxifene and tamoxifen in the *N*-nitrosomethylurea-induced rat mammary carcinoma model. *Cancer Res* 1987 47: 4020–4024.
4. Jordan VC, Phelps E, Lindgren JU. Effects of anti-estrogens on bone in castrated and intact female rats. *Breast Cancer Res Treat* 1987 10: 31–35.
5. Jordan VC, Robinson SP. Species-specific pharmacology of antiestrogens: role of metabolism. *Fed Proc* 1987 46: 1870–1874.
6. Gottardis MM, Robinson SP, Satyaswaroop PG, Jordan VC. Contrasting actions of tamoxifen on endometrial and breast tumor growth in the athymic mouse. *Cancer Res* 1988 48: 812–815.
7. Lacassagne A. Hormonal pathogenesis of adenocarcinoma of the breast. *Am J Cancer* 1936 27: 217–225.
8. Jensen EV, Jacobson HI. Basic guides to the mechanism of estrogen action. *Recent Prog Horm Res* 1962 18: 387–414.
9. Jensen EV, Block GE, Smith S, Kyser K, DeSombre ER. Estrogen receptors and breast cancer response to adrenalectomy. *Natl Cancer Inst Monogr* 1971 34: 55–70.
10. Cole MP, Jones CT, Todd ID. A new anti-oestrogenic agent in late breast cancer. An early clinical appraisal of ICI 46474. *Br J Cancer* 1971 25: 270–275.
11. Lerner LJ, Jordan VC. Development of antiestrogens and their use in breast cancer: Eighth Cain Memorial Award Lecture. *Cancer Res* 1990 50: 4177–4189.
12. Early Breast Cancer Trialists' Collaborative Group. Tamoxifen for early breast cancer: an overview of the randomised trials. *Lancet* 1998 351: 1451–1467.
13. Jordan VC. Designer estrogens. *Sci Am* 1998 279: 60–67.
14. Jordan VC, Morrow M. Raloxifene as a multifunctional medicine? *Br Med J* 1999 319: 331–332.
15. Fisher B, Costantino JP, Wickerham DL *et al.* Tamoxifen for prevention of breast cancer: report of the National Surgical Adjuvant Breast and Bowel Project P-1 Study. *J Natl Cancer Inst* 1998 90: 1371–88.
16. Gail MH, Costantino JP, Bryant J *et al.* Weighing the risks and benefits of tamoxifen treatment for preventing breast cancer [Published erratum appears in *J Natl Cancer Inst* 2000 92: 275]. *J Natl Cancer Inst* 1999 91: 1829–1846.
17. Love RR, Mazess RB, Barden HS *et al.* Effects of tamoxifen on bone mineral density in postmenopausal women with breast cancer. *N Engl J Med* 1992 326: 852–856.
18. Love RR, Wiebe DA, Newcomb PA. Effects of tamoxifen on cardiovascular risk factors in postmenopausal women. *Ann Intern Med* 1991 115: 860–864.
19. McDonald CC, Stewart HJ. Fatal myocardial infarction in the Scottish adjuvant tamoxifen trial. The Scottish Breast Cancer Committee. *Br Med J* 1991 303: 435–437
20. Costantino JP, Kuller LH, Ives DG, Fisher B, Dignam J. Coronary heart disease mortality and adjuvant tamoxifen therapy. *J Natl Cancer Inst* 1997 89: 776–782.
21. Reis SE, Constantino JP, Wickerham DL, Tan-Chiu E, Wang J, Kavanah M. Cardiovascular effects of tamoxifen in women with and without heart disease: breast cancer prevention trial. National Surgical Adjuvant Breast and Bowel Project Breast Cancer Prevention Trial Investigators. *J Natl Cancer Inst* 2001 93: 16–21.
22. Assikis VJ, Neven P, Jordan VC, Vergote I. A realistic clinical perspective of tamoxifen and endometrial carcinogenesis. *Eur J Cancer* 1996 32A: 1464–1476.
23. Ettinger B, Black DM, Mitlak BH *et al.* Reduction of vertebral fracture risk in postmenopausal women with osteoporosis treated with raloxifene: results from a 3-year randomized clinical trial. Multiple Outcomes of Raloxifene Evaluation (MORE) Investigators. *J Am Med Assoc* 1999 282: 637–645.
24. Cummings SR, Eckert S, Krueger KA *et al.* The effect of raloxifene on risk of breast cancer in postmenopausal women: results from the MORE randomized trial. Multiple Outcomes of Raloxifene Evaluation. *J Am Med Assoc* 1999 281: 2189–2197.
25. Cauley JA, Norton L, Lippman ME *et al.* Continued breast cancer risk reduction in postmenopausal women treated with raloxifene: 4-year results from the MORE trial. *Breast Cancer Res Treat* 2001 65: 125–134.

26. Walsh BW, Kuller LH, Wild RA *et al.* Effects of raloxifene on serum lipids and coagulation factors in healthy postmenopausal women. *J Am Med Assoc* 1998 279: 1445–1451.
27. Fugere P, Scheele WH, Shah A, Strack TR, Glant MD, Jolly E. Uterine effects of raloxifene in comparison with continuous-combined hormone replacement therapy in postmenopausal women. *Am J Obstet Gynecol* 2000 182: 568–574.
28. Delmas PD, Bjarnason NH, Mitlak BH *et al.* Effects of raloxifene on bone mineral density, serum cholesterol concentrations, and uterine endometrium in postmenopausal women. *N Engl J Med* 1997 337: 1641–1647.
29. Kuiper GG, Enmark E, Pelto-Huikko M, Nilsson S, Gustafsson JA. Cloning of a novel receptor expressed in rat prostate and ovary. *Proc Natl Acad Sci USA* 1996 93: 5925–5930.
30. Moore JT, McKee DD, Slentz-Kesler K *et al.* Cloning and characterization of human estrogen receptor β isoforms. *Biochem Biophys Res Commun* 1998 247: 75–78.
31. Su JL, McKee, DD, Ellis B *et al.* Production and characterization of an estrogen receptor β subtype-specific mouse monoclonal antibody. *Hybridoma* 2000 19: 481–487.
32. Fuqua SA, Schiff R, Parra I *et al.* Expression of wild-type estrogen receptor β and variant isoforms in human breast cancer. *Cancer Res* 1999 59: 5425–5428.
33. Leav I, Lau K, Adams J *et al.* Comparative studies of the estrogen receptors β and α and the androgen receptor in normal human prostate glands, dysplasia, and in primary and metastatic carcinoma. *Am J Pathol* 2001 159: 79–92.
34. Kuiper GG, Carlsson B, Grandien K *et al.* Comparison of the ligand binding specificity and transcript tissue distribution of estrogen receptors α and β. *Endocrinology* 1997 138: 863–870.
35. Hall JM, McDonnell DP. The estrogen receptor β-isoform (ERβ) of the human estrogen receptor modulates ERα transcriptional activity and is a key regulator of the cellular response to estrogens and antiestrogens. *Endocrinology* 1999 140: 5566–5578.
36. Shang Y, Hu X, DiRenzo J, Lazar MA, Brown M. Cofactor dynamics and sufficiency in estrogen receptor-regulated transcription. *Cell* 2000 103: 843–852
37. Paech K, Webb P, Kuiper GG *et al.* Differential ligand activation of estrogen receptors ERα and ERβ at AP1 sites. *Science* 1997 277: 1508–1510.
38. Wolf DM, Jordan VC. The estrogen receptor from a tamoxifen stimulated MCF-7 tumor variant contains a point mutation in the ligand binding domain. *Breast Cancer Res Treat* 1994 31: 129–138.
39. Catherino WH, Wolf DM, Jordan VC. A naturally occurring estrogen receptor mutation results in increased estrogenicity of a tamoxifen analog. *Mol Endocrinol* 1995 9: 1053–1063.
40. Levenson AS, Catherino WH, Jordan VC. Estrogenic activity is increased for an antiestrogen by a natural mutation of the estrogen receptor. *J Steroid Biochem Mol Biol* 1997 60: 261–268.
41. Levenson AS, Tonetti DA, Jordan VC. The oestrogen-like effect of 4-hydroxytamoxifen on induction of transforming growth factor α mRNA in MDA-MB-231 breast cancer cells stably expressing the oestrogen receptor. *Br J Cancer* 1998 77: 1812–1819.
42. Brzozowski AM, Pike AC, Dauter Z *et al.* Molecular basis of agonism and antagonism in the oestrogen receptor. *Nature* 1997 389: 753–758.
43. Shiau AK, Barstad D, Loria PM *et al.* The structural basis of estrogen receptor/coactivator recognition and the antagonism of this interaction by tamoxifen. *Cell* 1998 95: 927–937.
44. Pike AC, Brzozowski AM, Hubbard RE *et al.* Structure of the ligand-binding domain of oestrogen receptor β in the presence of a partial agonist and a full antagonist. *EMBO J* 1999 18: 4608–4618.
45. Jiang SY, Jordan VC. Growth regulation of estrogen receptor-negative breast cancer cells transfected with complementary DNAs for estrogen receptor. *J Natl Cancer Inst* 1992 84: 580–591.
46. Levenson AS, Jordan VC. The key to the antiestrogenic mechanism of raloxifene is amino acid 351 (aspartate) in the estrogen receptor. *Cancer Res* 1998 58: 1872–1875.
47. Jordan VC. Selective estrogen receptor modulation: a personal perspective. *Cancer Res* 2001 61: 5683–5687.
48. MacGregor Schafer J, Liu H, Bentrem DJ, Zapf JW, Jordan VC. Allosteric silencing of activating function 1 in the 4-hydroxytamoxifen–estrogen receptor complex is induced by substituting

glycine for aspartate at amino acid 351. *Cancer Res* 2000 60: 5097–5105.

49. Liu H, Lee ES, De Los Reyes A, Zapf JW, Jordan VC. Silencing and reactivation of the selective estrogen receptor modulator–estrogen receptor α complex. *Cancer Res* 2001 61: 3632–3639.
50. Bentrem, DJ, Dardes RC, Liu H, MacGregor-Schafer JI, Zapf JW, Jordan VC. Molecular mechanism of action at estrogen receptor α of a new clinically relevant antiestrogen (GW 7604) related to tamoxifen. *Endocrinology* 2001 142: 838–846.
51. Rosati RL, Da Silva Jardine P, Cameron KO *et al.* Discovery and preclinical pharmacology of a novel, potent, nonsteroidal estrogen receptor agonist/antagonist, CP-336 156, a diaryltetrahydronaphthalene. *J Med Chem* 1998 41: 2928–2931.
52. Ke HZ, Paralkar VM, Grasser WA *et al.* Effects of CP-336 156, a new, nonsteroidal estrogen agonist/antagonist, on bone, serum cholesterol, uterus and body composition in rat models. *Endocrinology* 1998 139: 2068–2076.
53. Willson TM, Henke BR, Momtahen TM *et al.* 3-[4-(1,2-Diphenylbut-1-enyl)phenyl]acrylic acid: a non-steroidal estrogen with functional selectivity for bone over uterus in rats. *J Med Chem* 1994 37: 1550–1552.
54. Willson TM, Norris JD, Wagner BL *et al.* Dissection of the molecular mechanism of action of GW 5638, a novel estrogen receptor ligand, provides insights into the role of estrogen receptor in bone. *Endocrinology* 1997 138: 3901–3911.
55. Wijayaratne AL, Nagel SC, Paige LA *et al.* Comparative analyses of mechanistic differences among antiestrogens. *Endocrinology* 1999 140: 5828–5840.
56. Connor CE, Norris JD, Broadwater G *et al.* Circumventing tamoxifen resistance in breast cancers using antiestrogens that induce unique conformational changes in the estrogen receptor. *Cancer Res* 2001 61: 2917–2922.
57. Sato M, Turner CH, Wang T, Adrian MD, Rowley E, Bryant HU. LY353 381.HCl: a novel raloxifene analog with improved SERM potency and efficacy *in vivo*. *J Pharmacol Exp Ther* 1998 287: 1–7.
58. MacGregor-Schafer J, Lee E-S, Dardes RC *et al.* Analysis of cross-resistance of the selective estrogen receptor modulators arzoxifene (LY353 381) and LY117 018 in tamoxifen-stimulated breast cancer xenografts. *Clin Cancer Res* 2001 7: 2505–2512.
59. Munster PN, Buzdar A, Dhingra K *et al.* Phase I study of a third-generation selective estrogen receptor modulator, LY353 381.HCI, in metastatic breast cancer. *J Clin Oncol* 2001 19: 2002–2009.
60. Luo S, Labrie C, Belanger A, Candas B, Labrie F. Prevention of development of dimethylbenz(a)anthracene (DMBA)-induced mammary tumors in the rat by the new nonsteroidal antiestrogen EM 800 (SCH57050). *Breast Cancer Res Treat* 1998 49: 1–11.
61. Couillard S, Gutman M, Labrie C, Belanger A, Candas B, Labrie F. Comparison of the effects of the antiestrogens EM 800 and tamoxifen on the growth of human breast ZR-75-1 cancer xenografts in nude mice. *Cancer Res* 1998 58: 60–64.
62. Labrie F, Labrie C, Belanger A *et al.* EM 652 (SCH 57068), a third generation SERM acting as pure antiestrogen in the mammary gland and endometrium. *J Steroid Biochem Mol Biol* 1999 69: 51–84.
63. Tremblay A, Tremblay GB, Labrie C *et al.* EM 800, a novel antiestrogen, acts as a pure antagonist of the transcriptional functions of estrogen receptors α and β. *Endocrinology* 1998 139: 111–118.
64. MacGregor-Schafer J, Liu H, Tonetti DA, Jordan VC. The interaction of raloxifene and the active metabolite of the antiestrogen (SC57068) with the human estrogen receptor (ER). *Cancer Res* 1999 59: 4308–4313.
65. Martel C, Picard S, Richard V, Belanger A, Labrie C, Labrie F. Prevention of bone loss by EM 800 and raloxifene in the ovariectomized rat. *J Steroid Biochem Mol Biol* 2000 74: 45-56.
66. Goss PE, Gwyn KM. Current perspectives on aromatase inhibitors in breast cancer. *J Clin Oncol* 1994 12: 2460–2470.
67. Howell A, DeFriend DJ, Robertson JF *et al.* Pharmacokinetics, pharmacological and anti-tumour effects of the specific anti-oestrogen ICI 182 780 in women with advanced breast cancer. *Br J Cancer* 1996 74: 300–308.

Gamete and embryo biology

André Van Steirteghem

This section on 'Gamete and embryo biology' describes female and male gametogenesis. The key processes in oogenesis, from the emergence of germ cells in the fetus up to the formation of a fertilizable oocyte, are described, including current information on oocyte growth and cytoplasmic and nuclear maturation. The development of the oocyte is closely linked to events occurring in granulosa cells. Knowledge has increased regarding genes and proteomes, which are required for normal oogenesis. Epigenetic changes, such as genomic imprinting, also influence future events. The molecular events controlling meiosis in the oocyte are currently only partially understood. These advances in knowledge may have a role in more efficient assisted reproductive technology and the development of new contraceptive techniques.

The molecular and cellular aspects of spermatogenesis have been best studied in experimental models. Genes and proteins involved in some of the key events of spermatogenesis and spermiogenesis have been characterized. Spermatogenic gene expression involves switches to expression of homologous genes encoding somatic proteins by making use of promoters. Switching of X-chromosomal and autosomal genes is frequent for spermatogenesis. Alternative splicing for several mRNAs can result in the production of testis-specific proteins. Protein ubiquitination is an important mechanism to control modification and breakdown of proteins.

The current understanding of the cellular and molecular basis of the different steps in the fertilization process is reviewed, including sperm activation and incorporation, activation of the oocyte, block to polyspermy, pronucleus formation, nuclear fusion, pronuclear migrations and genomic union. Because of the limited research material of human origin, fertilization is studied using oocytes from a non-human primate model, the rhesus monkey. The better understanding of the cellular and molecular events of successful fertilization by intracytoplasmic sperm injection (ICSI) may lead to improvements in the clinical use of ICSI, including its safety and efficacy.

A mouse model has been used to study molecular and cellular events during preimplantation development. The maternal-to-embryonic transition occurs in different steps: formation of the embryonic genome, the reprogramming of gene expression, the transcription processes in maternal and paternal pronuclei, the onset of zygotic gene activation, and the genetic interactions controlling blastomere survival in the early embryo. The compaction process involves the establishment of an epithelium and cell polarity. E-cadherin is responsible for the calcium-dependent events of compaction. Blastocyst development is reviewed, including the origin of the inner cell mass and trophectoderm cells, the origin and function of the blastocele and the gene expression in the blastocyst.

Current knowledge on implantation and endometrial function is reviewed. Estrogen and progesterone receptors in the endometrium are key factors in endometrial development, in view of embryo implantation. Pinopodes and cell adhesion molecules such as carbohydrate epitopes, mucins, integrins and the trophinin–tastin–lystine complex play a role in apposition and attachment of the embryo to the endometrium. Several growth factors and cytokines, i.e. epidermal growth factor, leukemia inhibitory factor, interleukin-2 and interleukin-1, mediate the effects of estradiol and progesterone on the endometrium. Some *HOX* genes and calcitonin also play a role in endometrial development. Several factors such as progesterone, cyclic AMP and prostaglandins influence the endometrial stromal decidualization. The endometrium has a role in placental invasion through enzyme inhibition, extracellular matrix proteins, insulin-like growth factors (IGFs) and IGF-binding protein-1 and transforming growth factor-β. Several factors control angiogenesis in the endometrium. Some of the many secretory-phase endometrial proteins, such as prolactin, progestogen-associated endometrial protein, placental protein 14 and glycodelin, have a role in water transport and immunomodulation of the endometrium. The role of endometrial macrophages, T cells and natural killer cells in implantation is only partially defined.

Application of genetics and molecular biology has led to some understanding of early embryonic patterning. This advance came from work in the fruit fly *Drosophila melanogaster* and numerous other organisms. This includes the discovery of the Spemann organizer and its role in induction of tissue to follow a certain organization, establishment of primordial germ layers, hierarchy of gene activation and establishment of certain patterns in development. The identity of a region of an embryo is defined by the hometic genes initially studied in the fruit fly but also in, for example, mouse and human. Other important signalling pathways play a role during development such as *Wingless*, *hedgehog*, fibroblast growth factors and transforming growth factor β.

Chapter 20

Female gametogenesis

Roger Gosden, Hugh Clarke and David Miller

INTRODUCTION

The mature oocyte is one of the largest and rarest cells in the body. Its remarkable life-history involves migration, multiplication, growth, differentiation and the reduction divisions of meiosis. It is a paradoxical cell, being at the same time totipotent, implying an ability to generate the full cellular diversity of the mature organism including the fetal membranes and chorion cells, and highly specialized, being the only cell in the female body which can undergo meiosis and fertilization and generate a new individual. These properties are genetically latent in the majority of oocytes in the ovary, and do not emerge until the cells have undergone a lengthy period of growth and cytoplasmic and nuclear maturation. During this phase, the oocyte secretes a thick extracellular membrane, the zona pellucida, increases the numbers of cytoplasmic organelles and accumulates a molecular program for post-fertilization development. Gene transcription is suspended after recommencing meiosis until the 4–8-cell stage 2–3 days later. Hence, the old maxim that embryogenesis is rooted in oogenesis still holds true. This chapter describes key processes in oogenesis, starting with the emergence of germ cells in the fetus and ending with the formation of a fertile gamete. The importance of this cell for reproductive medicine is difficult to exaggerate, because it is central and indispensible in fertility, even in an era when animals are being cloned.

BACKGROUND

Primordial germ cells (PGCs) are the stem cells which give rise to oocytes and spermatozoa. They can be visualized by alkaline phosphatase staining at as early as 4 weeks of gestation in human embryos but, curiously, PGCs first take up residence in an extraembryonic site, the yolk sac, like the primitive hematopoietic stem cells. Subsequently, they migrate by a combination of morphogenetic movements and self-propulsion along the hindgut to the genital ridge, where they stay. Some of the cells fail to reach their destination and others end up, and perish, in ectopic sites. PGCs are social cells, for they migrate in small groups and increase in number *en route*. Their survival and navigation depend on signals in the local environment, including laminin and cytokines such as *kit*-ligand and transforming growth factor-β1. However, they are not sexually dimorphic at this stage and human gonads are 'sexually indifferent' until approximately 6 weeks of age.

When PGCs have reached the gonad, they are termed oogonia, by analogy with spermatogonia which give rise to male germ cells and persist throughout life. In the ovary, from the second trimester until the end of the third, oogonia progressively enter prophase of meiosis I after reduplicating their nuclear DNA for the last time. By 20–22 weeks of gestation, the fetal ovary possesses a spectrum of stages from oogonia to diplotene oocytes, and the most advanced stages begin to associate with somatic cells in the cortical cords.[1] Meanwhile, the number of oogonia continues to dwindle until they have vanished completely. While some germ cells are differentiating, a larger number are degenerating by apoptosis, and, by birth, only 20% of the original 6–7 million cells have survived.[2]

Germ cells separate from each other as they become pinched off by pre-granulosa cells to form the developmental units of the ovary, the primordial follicles. A few follicles initiate growth before birth, but the great majority remain dormant for an indeterminate period of up to 60 years. The signal that initiates the growth of the follicle is unknown, but is apparently not follicle stimulating hormone (FSH), since primary follicle recruitment continues after hypophysectomy and the smallest follicles do not express FSH receptors. Development of the oocyte and granulosa cells is integrated throughout the growth phase until a ripe Graafian follicle emerges.[3] Cytogenetic evidence suggests that the first-formed oocytes may also be first to be ovulated, but otherwise there are no hints of a production-line, and the ovarian morphology suggests that follicles initiate growth randomly.

CURRENT CONCEPTS

Oocyte growth

Size and developmental competence

The growth of somatic cells is limited by the intervention of mitosis, but, in oocytes, cytoplasmic cleavage is suspended, and they continue to expand up to 100-fold in volume, reaching a diameter of 120 μm in humans. Such enormous growth serves to provide newly fertilized eggs with enough organelles and key molecules to support early development until the embryo is autonomous. Oocytes do not need to be as large as in amphibians, or contain abundant yolk, because the embryos develop in the protected and nutrient-rich environment of the Fallopian tube and develop a placenta. They possess very limited energy stores in the forms of glycogen and lipids, and specific yolk proteins have never been identified in mammals.

Nevertheless, size matters, and the ability to complete meiotic maturation and undergo successful post-fertilization development depends on oocyte growth. When mouse oocytes below a critical size were transferred from follicles to culture, they were unable to undergo nuclear maturation.[4] Slightly larger oocytes were able to progress to metaphase I, but then became arrested. Only when nearly fully grown were they able to complete maturation to metaphase II. Thus, meiotic competence is acquired during growth. Its molecular basis is not fully understood, although competent oocytes contain more $p34^{cdc2}$, cyclin B and cdc25 than incompetent ones.[5] The ability to reach metaphase II and undergo fertilization is not, however, necessarily an assurance of the ability to develop as an embryo.[6] Although the basis of embryonic competence is not known, it is thought to be associated with the accumulation during oocyte growth of mRNAs, and the synthesis of proteins required to drive early embryonic events. The molecular inventory of the cell, represented by its transcriptome and proteome, is likely to be much more predictive of developmental competence than either morphology or size, but we neither know the identity of all the key molecules nor yet have methods for measuring them non-invasively. Nevertheless, recent reports on the identification of genes in primordial germ cells and preimplantation embryos hold considerable promise in addressing these gaps.[7] The first indication of the mature human oocyte proteome, following serial analysis of gene expression (SAGE), is also a welcome development in this respect.[8]

Cytoplasmic organelles

There is a paucity of cytoplasmic organelles in non-growing oocytes of primordial follicles (Figure 1a), but during growth their numbers increase. By the time the oocyte is fully grown, there are approximately 10^5 mitochondria, which will not undergo fission again until the blastocyst stage. These organelles are the sole source of mitochondrial DNA in the embryo, because the mitochondria of the sperm degenerate following ubiquitination of mitochondrial sheath proteins during cleavage stages. Mitochondria also change in appearance during oogenesis, starting as elongated forms with abundant cristae and becoming more ovoid with fewer, arching cristae in mature oocytes, indicating less respiratory activity.[9]

Likewise, the Golgi apparatus undergoes morphological evolution during oocyte growth. Starting as a few flattened sacs, it increases in size and dilates, and segments migrate to the periphery for exporting the prodigious amounts of glycoproteins needed to make the zona pellucida. The endoplasmic reticulum is abundant in oocytes, although rough endoplasmic reticulum and polyribosomes are scarce. Scattered throughout the cytoplasm are multivesicular bodies, crystalloids and, more notably in rodents, fibrous lattices. These inclusions have a variety of origins and functions, and some may be derived by endocytosis of proteins and extracellular fluid. Endocytosis is important for the comparatively enormous oocytes in amphibians because yolk proteins, or vitellogenins, are not synthesized locally but are imported from the bloodstream. In mammalian oocytes, there are many coated pits and endocytotic vesicles, but the smaller cell mass demands less external support. Membrane-bound exocytotic vesicles known as cortical granules are assembled during growth and transported to the periphery of

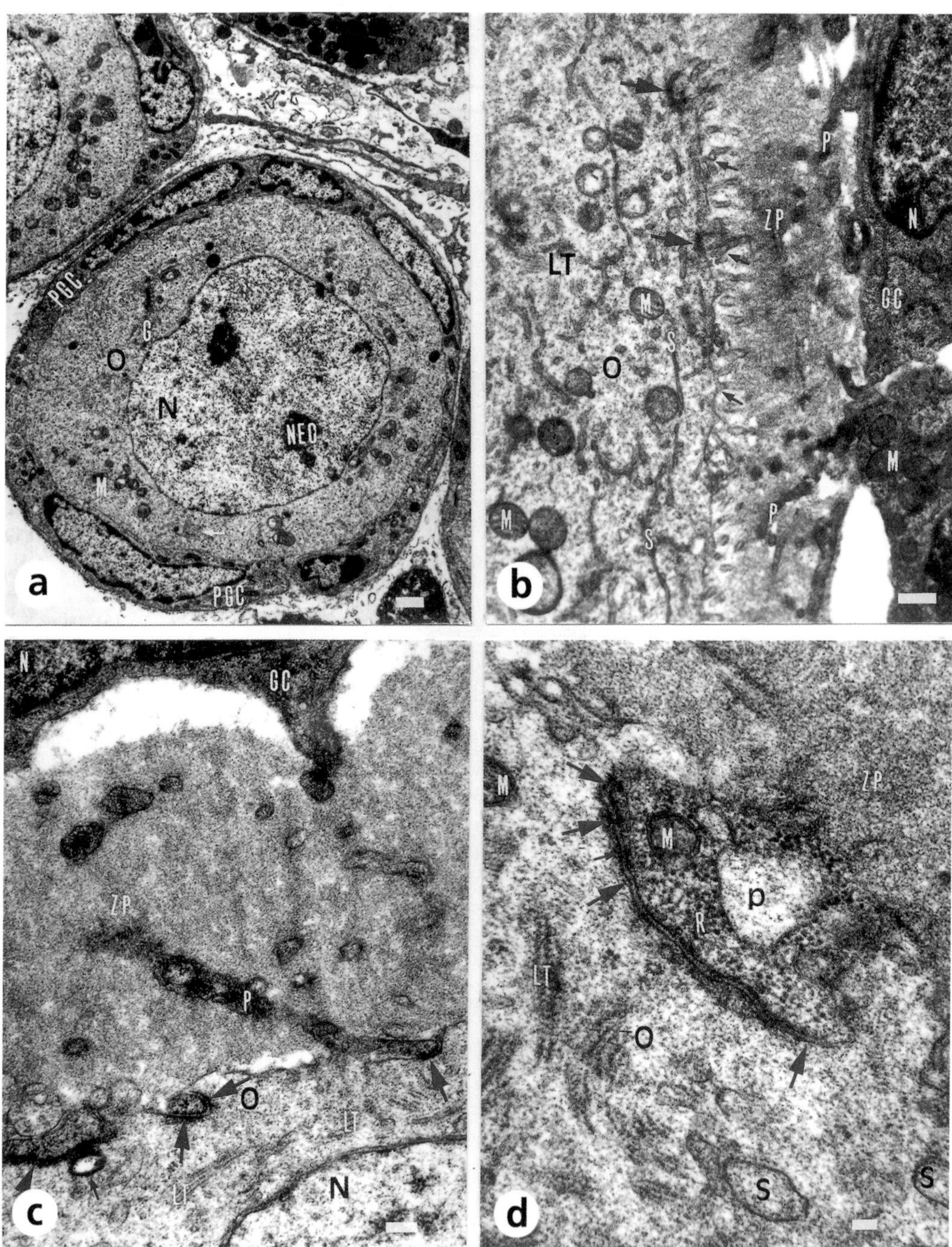

Figure 1 Ultrastructure of oocyte development (mouse). (a) In a primordial follicle, the oocyte (O) occupies the bulk and contains a large nucleus (N) with finely dispersed chromatin and a prominent nucleolus (NEO). Mitochondria (M) and smooth endoplasmic reticulum (arrows) are scattered in the cytoplasm. The pre-granulosa cells (PGC) are flattened and closely apposed to the oocyte membrane (bar = 1 μm). (b) In a multilaminar growing follicle, microvilli are evenly spaced on the oocyte's membrane, or oolemma, lying to the left of the zona pellucida (ZP) (bar = 0.5 μm). (c) Numerous granulosa cell processes (P) traverse the zona to contact the oocyte of a small antral follicle. Junctional apparatus occurs at the oocyte–granulosa cell process interface (bar = 0.25 μm). (d) At higher magnification, a foot-like process of a granulosa cell can be seen deeply indenting the oocyte's surface and contacting a large area of junctional complexes, which are mainly of the adherens type but include gap junctions (arrows). There are ribosomes (R), smooth endoplasmic reticulum (S) and fibrillar lattices (LT) as well as endoplasmic reticulum and mitochondria in the cytoplasm (bar = 0.1 μm)

the oocyte.[10] These granules are thought to play a role in blocking polyspermy after fertilization by releasing proteases to harden the zona pellucida by modifying two of its constituent proteins, ZP2 and ZP3.[11]

Centrioles are required in somatic cells to form asters for mitosis, but in oocytes (with the exception of rodents) they disappear during mid-growth stages and have to be inherited from the fertilizing sperm.[12] At metaphase, the oocyte contains an eccentrically placed spindle apparatus opposite the polar body, and mitochondria cluster around the spindle to meet local energetic needs. Otherwise, there are no morphological signs of polarity in the cytoplasm and, in humans at least, the fine layer of microvilli is spread uniformly in the cell surface (Figure 1b). Prior to ovulation, 'foot-processes' from the corona radiata cells project through the zona pellucida to contact the surface of the oocyte (Figure 1c and d). The cells are metabolically coupled via gap junctions containing connexin-37, which permit regulated traffic of small molecules such as amino acids, sugars and nucleotides (i.e. < 1 kDa). Larger molecules such as mRNAs and proteins are excluded, and the communication link breaks down completely at meiotic maturation.

Cellular polarity

The eggs of oviparous species in frogs, fish and flies are asymmetrical and their polarity specifies cell fate and the axis in the embryo.[13] This polarity is associated with mRNAs which are highly localized in the cell owing to sequences in their 3'-untranslated region (UTR) that bind specific proteins. In contrast, mammalian eggs appear to be much more uniform (marsupials less so), at least until the polar body is extruded and defines an 'animal pole'. Until recently, it had been assumed that oogenesis in mammals is organized in a fundamentally different way, with cell fate and the body axis determined by cellular interactions rather than being predetermined by molecular organization in the egg. This assumption has been questioned.[14] The answer has a practical bearing on the potential impact of embryo biopsy and apoptosis of blastomeres, as well as on the benefits and risks of cytoplasmic transfer and other invasive procedures.

A number of polarized molecules have come to light, including leptin and STAT3 close to the spindle.[15] Furthermore, there is correspondence between marker molecules in eggs and their subsequent localization in embryos, suggesting that cellular mosaicism is conserved during the early stages of development. However, experimental studies have indicated that, while an animal–vegetal axis is established in mouse zygotes, there is developmental plasticity, and removal of portions of cytoplasm or one or two blastomeres does not disrupt normal development.[16] These results are reassuring for assisted reproductive technologies (ART) since they confirm that the mammalian egg can compensate for cytoplasmic damage or disturbance.

Formation and storage of a molecular program

A fully grown mouse oocyte contains about 0.5 ng of RNA and 25 ng of protein;[9] a human oocyte probably contains five times as much of each. These molecules accumulate during oocyte growth, which is a slow process lasting several months in humans. The quantities are minute compared to the voluminous eggs of amphibians but, in relation to cell volume, they are rather equivalent. At the end of the growth phase, while the cell awaits a signal to resume meiosis, RNA synthesis almost halts, although protein synthesis continues. Since gene transcription does not resume until the four-cell stage in humans,[17] there is an interval of several days before and after fertilization when the cells depend wholly on the maternal endowment of proteins and on translation of mRNAs that were stored during oogenesis. The shutdown of transcription is associated with condensation of chromatin around the nucleolus,[18] which can be revealed by vital stains for DNA and serves as a marker of the maturity and transcriptional status of living oocytes.

Control of the disposition of RNA and protein synthesis is subtle. Accumulation and timely expression of this molecular program are fundamental for oocyte survival and developmental competence. Storage of mRNAs is a characteristic of both male and female germ cells, and they both depend on translational regulation to prepare for embryogenesis. Some mRNAs are translated early or continuously during oogenesis because they carry essential metabolic or structural functions (e.g. β-actin, ZP3). Others are recruited for translation around the time of meiotic maturation in full-sized oocytes because they are needed for cell cycle control (e.g. C-MOS). Finally, some are reserved for roles after fertilization, and the histones are an example of this type.

Histones are DNA-binding proteins and part of the basic structural unit of chromatin – the nucleosome. Protamines are bound to the DNA in sperm nuclei and, because they inactivate transcription by condensing the chromatin, must be replaced rapidly by histones after fertilization. The embryo also needs histones to assemble its newly replicated DNA into chromatin. Large amounts of mRNA encoding the histones accumulate in oocytes and, while some is translated and degraded during growth, the rest is stockpiled to meet the needs of the zygote. Histone mRNAs are unusual in having the typical long poly(A) tail replaced by a stem loop structure in the 3'-UTR. A protein known as stem-loop binding protein (SLBP) regulates histone translation by binding to the stem-loop,[19] and it appears that a burst in SLBP synthesis just before fertilization activates translation of histone mRNAs in the embryo.

Since timely expression of gene products is vital during the long and complex life-history of the oocyte, special regulatory mechanisms have evolved to control it. mRNAs that are not required immediately are stored and 'masked' in ribonucleoprotein (RNP) particles, which are probably equivalent to 'nuage' material seen in oocytes under the microscope. Masked mRNAs reserved for later use are bound to proteins, and their short poly(A) tails are regulated by sequences in the 3'-UTR.[20] Polyadenylation is closely correlated with competence for translation, and is regulated by a highly conserved signal sequence in the 3'-UTR (AAUAAA). Masked mRNAs, such as C-MOS and tissue-type plasminogen activator (tPA), also bear a distinctive sequence 5' to the polyadenylation site. The sequence motifs are uracil-rich (UUUUUAU), and are called the cytoplasmic polyadenylation element (CPE) or adenylation control element (ACE). mRNAs containing an ACE are dormant until meiotic maturation, when their translation is activated; conversely, mRNAs lacking it are translated in growing oocytes but become silenced when maturation begins. The presence or absence of an ACE sequence therefore determines the stage of oogenesis at which an mRNA is translated.

The significance of the poly(A) tail length can be illustrated by tPA, which is normally about 250 nucleotides long but is stored with approximately 40 adenine residues until the time of oocyte maturation. Translational silencing is probably due to a *trans*-acting translational repressor protein.[20,21] In addition, there may also be physical masking when mRNAs are merged in RNP particles,[22] or perhaps in association with cytoskeletal elements to produce large RNP particles combined with translation factors. Packaged RNA excludes ribosomes from access to regulatory elements, causing the synthesis and translation of RNA into protein to become uncoupled. In conclusion, more than one mechanism operates to control the storage and stability of mRNAs before oocyte maturity and their rapid degradation and replacement afterwards. Many of these regulatory processes have parallels in the testis, where postmeiotic gene expression during the later stages of spermiogenesis is wholly dependent on the translation of long-lived, stored mRNAs resulting from the earlier shutdown of transcription (see Chapter 21).

Zona pellucida

The zona pellucida is a glycoprotein coat up to 15 μm in thickness, enclosing the developing oocyte and serving to protect the oocyte and early embryo from physical harm and exclude microorganisms, to bind spermatozoa and to prevent polyspermy after fertilization. The physiological and molecular characteristics of the zona have been investigated in considerable detail.

The zona is secreted by the growing oocyte, although, in some species, there may be a contribution from granulosa cells. It amounts to 5 ng of protein, or about 17% of the total cell protein, and contains three molecules, called ZP1, ZP2 and ZP3, which have molecular weights, respectively, of 200 kDa, 120 kDa and 83 kDa in the mouse. ZP3 acts as the primary receptor for binding the sperm head and inducing the acrosome reaction. This highly specific reaction results from post-translational modifications of the ZP components, all of which have sites for N-linked glycosylation. Apparently, the O-linked (serine/threonine) class of oligosaccharides on ZP2 and -3 is important for receptor specificity.[23] ZP2 acts as a secondary receptor for acrosome-reacted sperm, and facilitates penetration of the zona matrix.

Each of the proteins is encoded by a separate gene, and all three genes become transcriptionally active from the earliest stages of oocyte growth. The corresponding mRNAs are very abundant and, by mid-growth, represent 1.5% of the total poly(A) RNA in the oocyte. By the time of ovulation, they have fallen to 5% of peak levels and soon become extinguished. Gene transcription is controlled by the transcription factor figα, which binds to a DNA sequence that is present in the promoter region of all three genes. figα is expressed only in female germ cells and, when the gene is inactivated by targeted mutation, the oocytes die before primordial follicles have been formed.[24] Thus, several genes required specifically for oocyte development are apparently regulated co-ordinately by figα. If *zp-3* expression is inactivated by mutagenesis or antisense oligonucleotides, the zona matrix fails completely to form, even if ZP1 and ZP2 proteins are present.[25] What is more, the oocytes of these mutant mice (*Zp3* –/–) are infertile, even though neither oocyte growth nor meiotic maturation are affected. Human infertility is very rarely associated with absence of the zona, but cases are presumably due to a new mutation in a member of the *zp* family.

The zona genes are by no means unique in being expressed specifically in oocytes. Powerful new technologies employing cDNA libraries, serial analysis of gene expression (SAGE) libraries, subtractive hybridization, differential RNA display and DNA database screening for homologous genes are revealing a growing list of genes that are expressed either exclusively in oocytes or in both male and female germ cells. In addition, many more genes not wholly specific to the germ line have been found by spontaneous mutation or targeted mutagenesis in mice to be essential for gametogenesis, or establishing a normal follicular population (Table 1). Knowledge gleaned from such animal models helps to predict phenotypes in humans and may eventually lead to new genetic tests for infertility and ovarian failure.

Table 1 Principal examples of gene products in female germ cells or granulosa cells that are mandatory for normal oogenesis

Gene expression	*Role(s)*
C-Kit, DAZLA, DIA, FIGα, FMR1 repeats, GATA-4, SF-1, TIAR	required for the formation, multiplication and migration of primordial germ cells
BMP-4, E-cadherin, β_1-integrin	involved in germ cell colonization of the gonad
ATR/ATM, DMCR, MLH1, MSH1	synapsis, recombination and meiotic check-points
BCL-2, BCL-X, BAX, FAS ligand, FIGα, HNF-1α	survival or apoptosis of female germ cells
Cyclin D2, estrogen receptor, Müllerian regression hormone, WT1	hormonal signalling and control
C/ERPβ, connexin-37, connexin-43, GDF-9, GDF9-B (BMP-15), FSH receptor, LH receptor, IGF-1, IGFBP-I, SOD-2	sustained follicular growth and differentiation
MOS, COX-2, ENOS, spindlin, tPA	meiotic maturation and ovulation
OCT-3	involved in maintaining totipotency and required for embryo development
FIGα, ZP-1, ZP-2, ZP-3	regulation and formation of the zona pellucida

FSH, follicle stimulating hormone; LH, luteinizing hormone; IGF-I, insulin-like growth factor-I; IGFBP-1, IGF binding protein-1; COX-2, cyclo-oxygenase-2; tPA, tissue plasminogen activator

Growth regulation

The mechanisms regulating oocyte growth are poorly understood. We are almost wholly ignorant about the process of growth initiation and why one primordial follicle begins to grow when its neighbors remain dormant. When ovarian cortical slices are cultured or transplanted to heterotopic sites in the body, many more follicles than normal begin to grow. This suggests that growth initiation is under inhibitory control, but the identity of all the molecules involved is unknown. In addition, when oocytes are grown *in vitro,* they attain the same maximum size as those developing *in vivo*, and this implies that oocyte growth is a strictly controlled process.

The initiation and completion of oocyte growth require granulosa cells to be in intimate contact, and denuded germ cells fail to grow. The granulosa cells, which in Graafian follicles should be called cumulus cells, probably promote growth through several pathways. First, as noted earlier, the cells are metabolically coupled by gap junctions. The foot projections from the cumulus granulosa cells that traverse the zona enable cell communication to be maintained until preovulatory maturation. If gap junctional communication between the oocyte and cumulus cells is prevented by deleting the gene encoding connexin-37, oocytes halt in mid-growth and display numerous ultrastructural abnormalities.[26]

Since oocytes can grow partially in the absence of functioning gap junctions, other communication pathways apparently exist. These are most likely growth factors secreted by cumulus granulosa cells, or expressed on their surfaces where they can interact with receptors on the oocyte's cell membrane to promote growth. KIT is a plasma membrane tyrosine-kinase receptor whose activation triggers specific intracellular signalling pathways according to the cell type. KIT is expressed in premeiotic germ cells and at all stages of oocyte growth, and its ligand (*kit*-ligand or KL) is expressed in granulosa cells.[27] KL is thought to be necessary for oocyte growth, and is down-regulated as the cell reaches full size.[28]

Oocytes are not mere passengers during follicular development, but actively influence their own fate by secreting molecules that affect the phenotype of surrounding somatic cells. Growth-differentiation factor-9 (GDF-9) is a member of the transforming growth factor-β (TGF-β) family, synthesized by the oocyte and essential for proliferation of the granulosa cells and the ability of cumulus cells to secrete mucopolysaccharides.[29] In the absence of GDF-9, oocytes grow in the defective follicles, but they are infertile and abnormally large,[30] possibly because they fail to down-regulate KL expression.[28] The general conclusion from the above studies is that follicular cells are mutually interdependent, and normal oocyte growth depends critically on signals and metabolites from granulosa cells.

Epigenetic changes

Growing oocytes contain 4C DNA and are diploid, with one set of homologous chromosomes inherited from the mother and the other from the father (Figure 2). In somatic cells, only one of the two X chromosomes is transcriptionally active, the other having been inactivated shortly after embryonic implantation (the 'Lyon hypothesis'). This gene dosage compensation mechanism brings transcription into balance with male cells, which only possess one X chromosome. The mechanism is controlled by X-linked genes and involves hypermethylation of cytosines in the DNA of genes on the inactive X chromosome. While there is a slight tendency for reactivation to occur in somatic cells, the oocyte is exceptional in reactivating its silent X chromosome during embryogenesis and before the onset of meiosis. Apparently, both X chromosomes must be active for optimal oogenesis in humans, and the pure 45,XO karyotype is associated with ovarian dysgenesis.[31]

Genomic imprinting is another type of epigenetic modification of the oocyte's genome involving the autosomes. Imprinting is also associated with monoallelic expression by hypermethylation of the inactive allele, the major enzyme responsible being DNA methyltransferase. An oocyte-specific form of this enzyme exists, and its absence leads to loss of imprints during embryogenesis and almost 100% fetal mortality.[32] The phenomenon of imprinting came to light during studies of zygote nuclear transplantation in mice. They revealed that exchanging either the male or the female

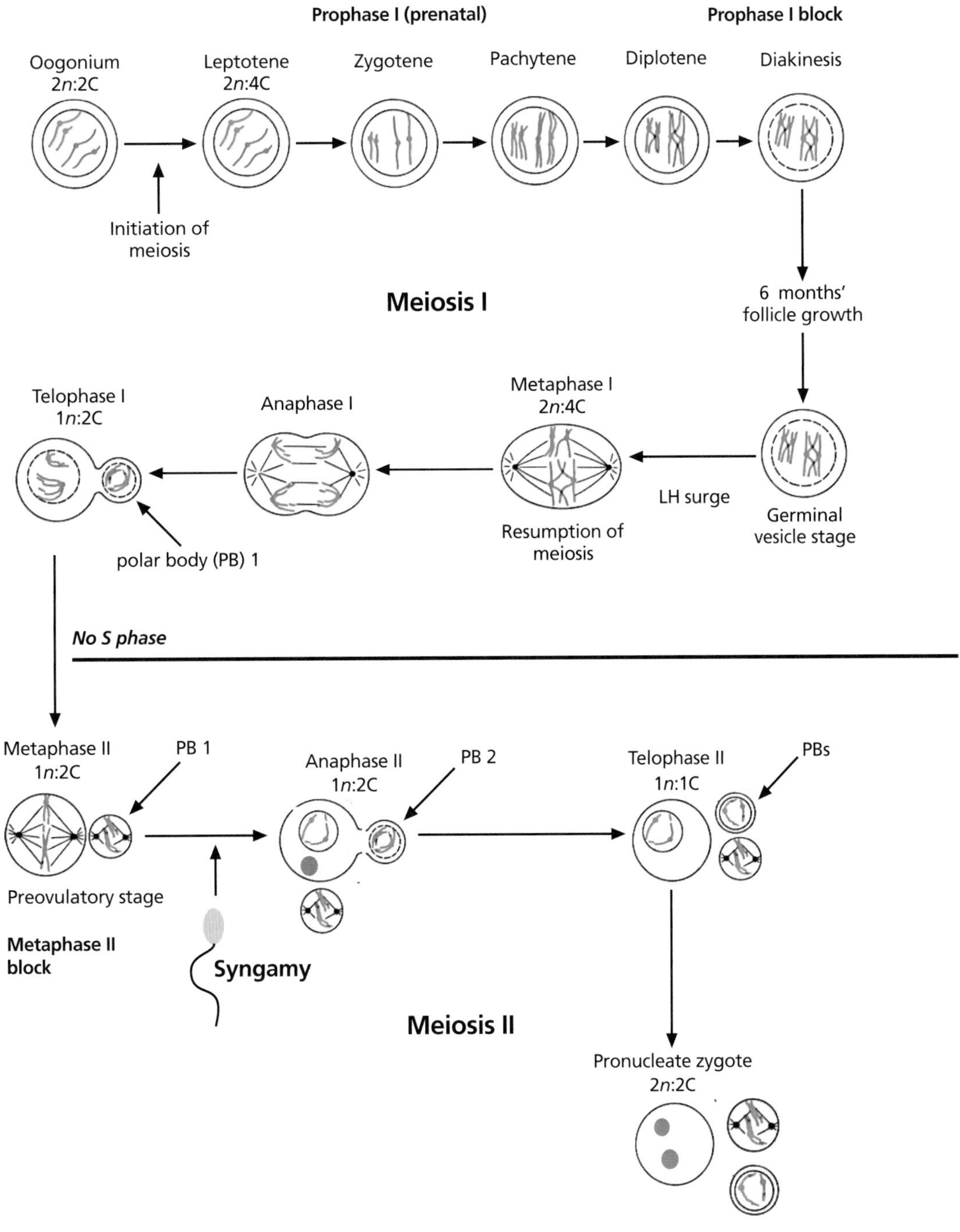

Figure 2 Stages of meiosis in oogenesis from the oogonial stage in the fetal ovary to ovulation in the adult. LH, luteinizing hormone

pronucleus with a pronucleus of the opposite type to make an identical pair always resulted in pregnancy failure. Embryonic phenotypes were different in gynogenones and androgenones, reflecting differential expression at imprinted loci. More than 40 imprinted genes are now suspected to exist, some of which are regulators of fetal growth (for example, insulin-like growth factor (IGF) II and its receptor). Mutations in these genes may be responsible for certain types of cancer, and defective expression has been linked with large-lamb and -calf syndrome after embryo manipulation in culture, and with abortion and fetal malformations after embryo cloning.

Primordial germ cells are almost globally demethylated, and specific DNA methylation patterns responsible for suppressing alleles are acquired during gametogenesis. At the *H19* locus, for instance, the paternally derived gene is methylated and silenced, whereas the maternal copy is expressed. Since epigenetic marks are formed during oocyte growth, vigilance is needed as technologies emerge for growing follicular oocytes *in vitro*. Concern that imprints may be labile in cultured cells has been heightened by evidence of biallelic expression of *H19* in mouse embryos growing in a simple medium.[32] This field is set for rapid expansion, with possible implications for spontaneous pregnancy wastage as well as ART.

Cytoplasmic and nuclear maturation

Cytological changes

Meiosis is one of the most intensively studied, biologically important and exclusive features of germ cells. In oogenesis, it involves two reduction divisions, which are interrupted twice and generate a haploid oocyte. In contrast to spermatogenesis, which distributes the sparse cytoplasm equally, reduction divisions in oocytes are not equatorial, so that the bulk of cytoplasm can be reserved for one of the products. The process of meiosis in females is also extraordinarily long, since it begins in the fetus and oocytes may not complete it until over 50 years later.

Early in the second trimester of pregnancy, the first oogonia cease dividing mitotically, undergo a final round of DNA synthesis to generate a 4C quota, and enter preleptotene of meiotic prophase (Figure 2). Subsequently, the chromosomes condense to form visible threads and thicken, with chromomeres appearing in zygotene. By pachytene, the chromatids are thicker, and form bivalents with synaptonemal complexes, which signify that DNA strands are engaging in homologous recombination. Interestingly, this is the stage when most germ cells are lost by apoptosis. At the resting stage of diplotene, the chiasmata are clearly visible, but the chromosomes are unlike the 'lampbrush' forms in amphibian oocytes, where there is massive amplification of ribosomal RNA synthesis to meet the higher demands of protein synthesis. No further nuclear maturation occurs in the germ cells unless they are stimulated in a preovulatory Graafian follicle or entering an abortive division in an atretic follicle.

Fully grown oocytes in large follicles remain developmentally arrested until they receive a signal to resume nuclear maturation. During this phase, the cell cycle advances from G_2 to M phase, and the oocyte undergoes changes preparing it for fertilization and embryonic development. These processes are called nuclear and cytoplasmic maturation.[6]

The first indication that nuclear maturation has begun is the disappearance of the oocyte nucleus, called the germinal vesicle (GV) because of its large, pale appearance. Breakdown of the GV membrane leads to dispersal of nuclear contents in the cytoplasm. The chromosomes become condensed and arranged on a spindle and the first meiotic division follows immediately, resulting in segregation of homologous chromosomes and emission of the first polar body. One set of homologs is retained in the germ cell and assembles on a second spindle at metaphase II. At this point, the cell cycle becomes arrested again until fertilization triggers completion of the second meiotic division and entry into the mitotic cell cycle.

Cytoplasmic maturation encompasses many dynamic changes in the oocyte, and is vital for successful fertilization and for embryo competence. The accumulation of mRNAs and proteins needed for post-fertilization development has been described. In most species, probably including humans, there is a need for the synthesis of new proteins and post-translational modification of existing proteins during maturation. These changes enable the oocyte to become 'activated' after sperm penetration, which triggers a series of Ca^{2+} oscillations lasting for 3–4 h.[34] Ca^{2+}-activated signalling pathways cause the destruction of cyclin B, allowing the cell cycle to resume. They also stimulate migration and exocytosis of the cortical granules at the oocyte's surface for blocking polyspermy. If sperm penetrate immature oocytes, however, the Ca^{2+} oscillations are smaller and briefer,[35,36] and the cortical reaction is much attenuated.[37] Hence, the sperm chromatin fails to form a pronucleus,[38] although this abnormality is also linked to the low glutathione activity in immature oocytes.[39] The Ca^{2+}-releasing activity is enhanced during maturation by the synthesis of inositol triphosphate and accumulation of endoplasmic reticulum near the cortex.[40] These changes are thought to enhance the Ca^{2+}-releasing activity of the oocyte, and illustrate how changes in protein synthesis and organelles work together during maturation.

Molecular control of meiosis

The physiological trigger for resuming meiosis is the hormonal surge of luteinizing hormone (LH). This hormone acts via receptors induced in granulosa cells, which, in turn, relay signals to the oocyte. Evidence of both positive and negative signals exists. Since maturation occurs spontaneously when fully grown oocytes are incubated in hormone-free culture medium, the follicular environment is thought actively to inhibit maturation. Cyclic adenosine monophosphate (cAMP) analogs prevent maturation *in vitro*, which suggests that the follicular granulosa cells inhibit maturation *in vivo* by maintaining an elevated level of cAMP in the oocyte. LH acts either by reducing cAMP in oocytes or by transmitting a signal that bypasses its inhibitory effect. A C29 sterol, called follicular fluid meiosis-activating sterol (FF-MAS), is a candidate for this role because it is synthesized by granulosa cells and can override cAMP *in vitro*.[41]

While meiosis exhibits unique characteristics, many of its components are the same as in mitosis, especially in the second division. The immature oocyte contains $p34^{cdc2}$ (known as H1 kinase and as maturation- or metaphase-promoting factor (MPF)), cyclin B (its regulatory subunit) and cdc25 (a phosphatase), which have accumulated during oocyte growth.[5] The $p34^{cdc2}$–cyclin B complex triggers the G_2–M cell cycle transition in somatic cells. In oocytes, $p34^{cdc2}$ is stored in an inactive form as a result of phosphorylation at three sites by the protein kinases Myt1 and Wee1. For maturation to begin, cdc25 must dephosphorylate $p34^{cdc2}$ on two of these sites. How cdc25 becomes activated in maturing oocytes is not yet clear, although it is known that new protein synthesis is not required. Active $p34^{cdc2}$ catalyzes, perhaps indirectly via an auto-amplification loop, the rapid accumulation of more of the same molecules, and this causes cells to enter M phase. Once maturation has reached metaphase I, a portion of the cyclin B becomes degraded, leading to a decrease in $p34^{cdc2}$ activity and permitting the first meiotic division to occur.[42] Following meiosis I, cyclin B then re-accumulates, leading to a rise in $p34^{cdc2}$ activity. This increase is necessary for the oocyte to enter metaphase II. If cyclin B fails to re-accumulate, $p34^{cdc2}$ activity remains low, and the oocyte enters interphase following completion of the first meiotic division and may begin parthenogenetic development. Figure 3 depicts the mechanism schematically.

The oocyte halts temporarily at metaphase II because cyclin B is stabilized, whereas its degradation normally permits the cell to exit from metaphase. MOS and the mitogen-activated protein (MAP) kinases, $p42^{erk2}$ and $p44^{erk1}$, are protein kinases that play key roles in cyclin B stabilization. mRNA for MOS is stored in immature oocytes, and synthesis of the protein begins at an early stage of maturation. MOS remains present during maturation but disappears after fertilization. Oocytes are the only cells that express the c-*mos* gene, and its deletion in mice causes them to pass through metaphase II without halting and begin parthenogenetic development.[43,44] Thus, MOS protein is a molecular brake which is required for oocytes while they are pausing for fertilization. The other protein kinases, Erk1 and Erk2, are inactive in immature oocytes until newly synthesized MOS protein phosphorylates and thereby activates

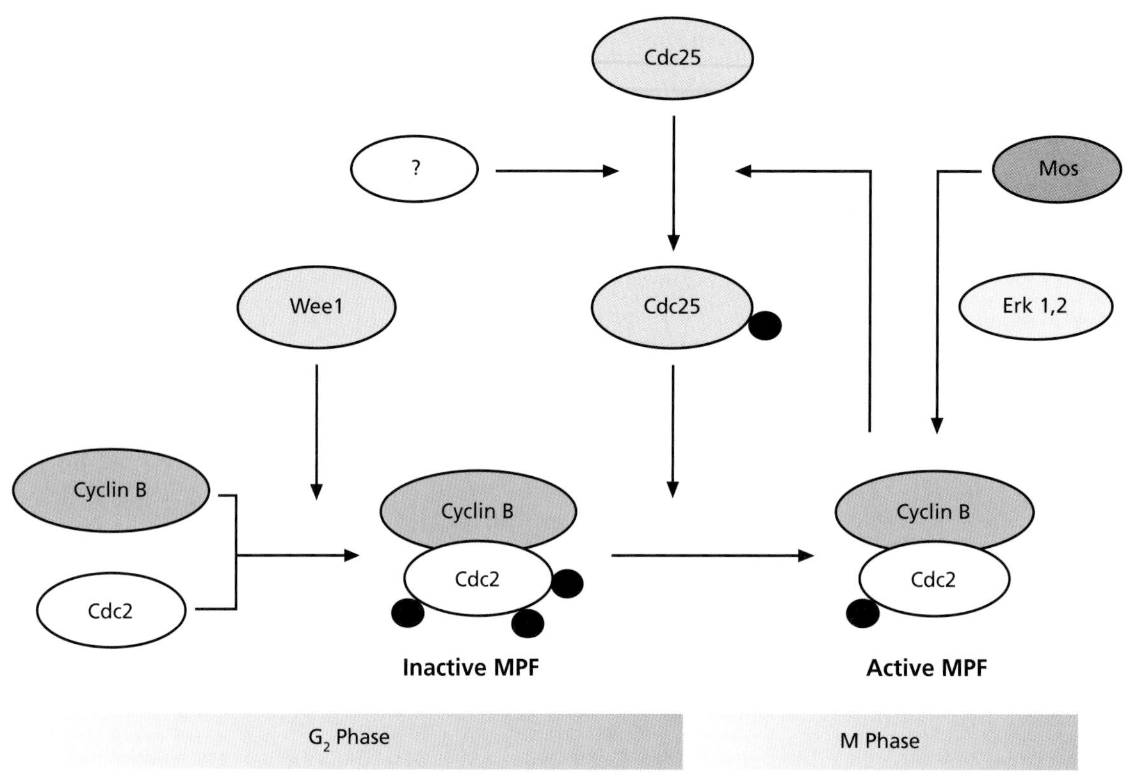

Figure 3 Molecules involved in the regulation of meiotic maturation in oocytes (see text for explanation). MPF, metaphase-promoting factor

them.[45] Erk protein kinases appear to be critical mediators because their activation is both necessary and sufficient for maintaining meiotic arrest at metaphase II.[46]

POTENTIAL CLINICAL RELEVANCE

Reproductive medicine

ARTs have revolutionized the treatment of infertility, but success rates have improved slowly over the two decades since the advent of clinical *in vitro* fertilization (IVF). The chance of a full-term pregnancy after standard IVF treatment is still little more than 20%, and is inversely related to the age of the egg. The majority of eggs and embryos from controlled stimulation cycles have limited developmental competence and, on average, 25–30 oocytes must be available to achieve a live birth in young women, and rather more in older women. Normally, more than one embryo is transferred at a time, to compensate for the mixed quality of embryos, and no infallible indicators of developmental competence have been identified. Therein lies responsibility for the dramatic rise in multiple pregnancies associated with ART. Until either the developmental competence of eggs is improved or the best-quality embryos can be reliably identified for transfer, IVF success rates are unlikely to change much. Advances in embryo culture media, although very desirable, cannot improve the quality of eggs that are intrinsically defective.

Current hopes of improving IVF success and reducing the number of embryos transferred are pinned on selection of embryos at the blastocyst stage, when viability is more assured. Trials are also under way to screen blastomeres using fluorescent *in situ* hybridization (FISH) to identify euploid embryos and to overcome deficiencies in the oocyte's cytoplasm. Where the prospects of success with IVF are poor because of advanced reproductive age (more than 40 years, for example) or idiopathic factors, patients are considered for egg donation. Cytoplasmic transfer or germinal vesicle nucleus transfer aims to 'rescue' the haploid genome by improving its environment, perhaps by transferring healthy mitochondria. However, the pathophysiological bases of oocyte aging are not well understood, and the complexity of oogenesis indicates that many factors, including those from somatic cells, could contribute. As the full inventory of genes required for oogenesis becomes known and their actions understood, diagnosing genetic causes of infertility and predicting environmental interactions will become possible. Furthermore, this progress will help to anticipate the side-effects of micromanipulation and to optimize hormone stimulation protocols and culture media. In principle, it is desirable to aim for the least invasive treatments, but it is conceivable that some causes of cytoplasmic incompetence might be treated by microinjection to overcome a molecular deficiency. However, this remedy would come too late for genes that exert their actions early in oogenesis, unless a front-end technology to culture small oocytes is developed.

A culture technology for growing and maturing oocytes can bring benefits to patients long before it is feasible and safe to transfer informational molecules to germ cells. Indeed, harvesting immature oocytes for *in vitro* maturation (IVM) is already practiced clinically in a few centers. This technology can avoid the need to stimulate patients hormonally, bringing benefits of economy, safety and simplicity. In the future, we imagine that culture technology might extend to the entire span of oogenesis, but this could be risky since epigenetic modifications to the genome occur during oocyte growth. Research into the molecular development of oocytes also warns about other advanced technologies. For example, when DDK female mice are outcrossed with males of another strain, there are nuclear–cytoplasmic incompatibilities, and there might be corresponding problems for cytoplasmic transfer between eggs of different women. Last, it is difficult to overstate the importance of oocyte aging at a time when the average age of child-bearing is continuing to rise. A better understanding of why this cell declines in numbers and quality before others might lead to improving the prospects of genetic parenthood for women of later reproductive years, and reducing the frequency of aneuploid conceptions.

Contraceptive strategies

Steroidal contraception is highly effective, and currently used by almost 100 million women worldwide. But ovulation suppression is induced indirectly, and the drugs have side-effects which, very occasionally, can be serious. A drug that specifically targets the oocyte to render it incapable of growth or maturation and fertility could theoretically be better for women's health, and may not even disturb the natural menstrual cycle. The prospects of a radical shift from conventional contraception in the near future are remote because research and development costs are high, but, as knowledge advances, new candidate molecules emerge for contraception research.

Antifertility vaccines have been generated to test the effects of immunizing against zona pellucida antigens in experimental animals. The antibodies block sperm binding and penetration with high efficiency, although variable antibody titers and T cell-mediated autoimmune destruction of small follicles present major obstacles in practice. Another strategy is suggested by the unique nature of meiosis in germ cells. Mos protein is an unsatisfactory target because depletion triggers a high frequency of parthenogenetic activation, but phosphodiesterase-3 inhibitors provide a more promising avenue. These drugs block meiotic maturation without affecting either the length of the estrous cycle or the numbers of ova shed at ovulation in rats. If a highly specific formulation free of cardiotoxicity is found, clinical development may be able to go ahead. This is far from an exhaustive list of possibilities, and other potential targets include cortical granules and sperm binding sites on the cell membrane. Also, there is the more revolutionary approach of inhibiting the recruitment of primordial follicles. While steroid replacement would be required to prevent menopausal symptoms owing to profound inhibition of follicle growth, this strategy could provide a wide margin of contraceptive safety plus the option of slowing down the rate of follicular attrition and delaying the menopause.

SUMMARY AND FUTURE PERSPECTIVES

A single chapter cannot reflect the enormous advances that have been made towards understanding the biology of oocyte development. Until recently, progress was limited by the scarcity of the cells and low sensitivity of technologies. The arrival of new technology and genomics has stimulated research, while clinical IVF has provided spare human oocytes for study. We are becoming more aware that the oocyte's life-history and the development of the early embryo are a continuum. Post-fertilization anomalies and aneuploidy must therefore be understood in terms of the molecules and organelles that have been produced during oogenesis. What is more, the controlled development of the oocyte can never be understood fully in isolation from the granulosa cells, which exert influences at every stage. In the next few years, all the genes involved in oocyte fertility will be identified, and the oocyte proteome will be explored to reveal the role and localization of protein molecules. This knowledge will evolve into functional genomics through experimental mutagenesis and cell culture models, and gene expression profiling will reveal the scheduling of transcription. Attention will have to be paid to post-translational modifications of proteins from phosphorylation, glycosylation, etc. to understand the fine control of development. The benefits of these advances for knowledge about the causes of infertility and birth defects and development of new contraceptive techniques will be enormous, and there will be spin-offs from a better understanding of how the oocyte achieves the remarkable feat of nuclear reprogramming.

ACKNOWLEDGEMENTS

We thank NV Organon and the Canadian Institutes of Health Research for supporting research into oocyte development in our laboratories. Natalie Habra assisted with the illustrations.

REFERENCES

1. Byskov AG. Differentiation of mammalian embryonic gonad. *Physiol Rev* 1986 66: 71–117.
2. Baker TG. A quantitative and cytological study of germ cells in human ovaries. *Proc R Soc London (Biol)* 1963 158: 417–433.
3. Gougeon A. Regulation of ovarian follicular development in primates – facts and hypotheses. *Endocr Rev* 1996 17: 121–155.
4. Sorenson RA, Wassarman PM. Relationship between growth and meiotic maturation of the mouse oocyte. *Dev Biol* 1976 50: 531–536.
5. Kanatsu-Shinohara M, Schultz RM, Kopf GS. Acquisition of meiotic competence in mouse oocytes: absolute amounts of p34cdc2, cyclin B1, cdc25C, and wee1 in meiotically incompetent and competent oocytes. *Biol Reprod* 2000 63: 1610–1616.
6. Eppig JJ, Schultz RM, O'Brien M, Chesnel F. Relationship between the developmental programs controlling nuclear and cytoplasmic maturation of mouse oocytes. *Dev Biol* 1994 164: 1–9.
7. Adjaye J, Daniels R, Monk, M. The construction of cDNA libraries from human single preimplantation embryos and their use in the study of gene expression during development. *J Assist Reprod Genet* 1998 15: 344–348.
8. Neilson L, Andalibi A, Kang D *et al.* Molecular phenotype of the human oocyte by PCR-SAGE. *Genomics* 2000 63: 13–24.
9. Wassarman PM. The mammalian ovum. In Knobil E, Neill J, eds. *The Physiology of Reproduction*. New York: Raven Press, 1988: 69–102.
10. Ducibella T, Duffy P, Buetow J. Quantification and localization of cortical granules during oogenesis in the mouse. *Biol Reprod* 1994 50: 467–473.
11. Wassarman PM. Profile of a mammalian sperm receptor. *Development* 1990 108: 1–17.

12. Schatten G. The centrosome and its mode of inheritance: the reduction of the centrosome during gametogenesis and its restoration during fertilization. *Dev Biol* 1994 165: 299–335.
13. Davidson EH. *Gene Activity in Early Development*, 3rd edn. Orlando, FL: Academic Press, 1986.
14. Edwards RG, Beard HK. Oocyte polarity and cell determination in early mammalian embryos. *Mol Hum Reprod* 1997 3: 863–905.
15. Antczak M, van Blerkom J. Oocyte influences on early development: the regulatory proteins leptin and STAT3 are polarized in mouse and human oocytes and differentially distributed within the cells of the preimplantation stage embryo. *Mol Hum Reprod* 1997 3: 1067–1086.
16. Ciemerych MA, Mesnard D, Zernica-Goetz M. Animal and vegetal poles of the mouse egg predict the polarity of the embryonic axis, yet are nonessential for development. *Development* 1998 127: 3467–3474.
17. Braude P, Bolton V, Moore S. Human gene expression first occurs between the four- and eight-cell stages of preimplantation development. *Nature* 1988 332: 459–461.
18. Bouniol-Baly C, Hamraoui L, Guibert J, *et al.* Differential transcriptional activity associated with chromatin configuration in fully grown mouse germinal vesicle oocytes. *Biol Reprod* 1999 60: 580–587.
19. Wang ZF, Whitfield MF, Ingledue TC, Dominski Z, Marzluff WF. The protein that binds the 3'-end of histone mRNA; a novel RNA-binding protein required for histone pre-mRNA processing. *Genes Dev* 1996 10: 3028–3040.
20. Richter JD. Cytoplasmic polyadenylation in development and beyond. *Microbiol Mol Biol Rev* 1999 63: 446–456.
21. Stutz A, Conne B, Huarte J *et al.* Masking, unmasking, and regulated polyadenylation in the translational control of a dormant mRNA in mouse oocytes. *Genes Dev* 1998 12: 2535–2548.
22. Matsumoto K, Wassarman KM, Wolffe AP. Nuclear history of a pre-mRNA determines the translational activity of cytoplasmic mRNA. *EMBO J* 1998 17: 2107–2121.
23. Miller DJ, Macek MB, Shur BD. Complementarity between sperm surface β-1,4-galactosyl-transferase and egg-coat ZP3 mediates sperm–egg binding. *Nature* 1992 357: 589–593.
24. Soyal SM, Amleh A, Dean J. FIGα, a germ cell-specific transcription factor required for ovarian follicle development. *Development* 2000 127: 4645–4654.
25. Rankin T, Familari M, Lee E *et al.* Mice homozygous for an insertional mutation in the *Zp3* gene lack a zona pellucida and are infertile. *Development* 1996 122: 2903–2910.
26. Simon AM, Goodenough DA, Li E, Paul DL. Female infertility in mice lacking connexin-37. *Nature* 1997 385: 525–529.
27. Kissel H, Timokhina T, Hardy MP *et al.* Point mutation in Kit receptor tyrosine kinase reveals essential roles for Kit signaling in spermatogenesis and oogenesis without affecting other Kit responses. *EMBO J* 2000 19: 1312–1326.
28. Joyce IM, Clark AT, Pendola FL, Eppig JJ. Comparison of recombinant growth differentiation factor-9 and oocyte regulation of KIT ligand messenger ribonucleic acid expression in mouse ovarian follicles. *Biol Reprod* 2000 63: 1669–1675.
29. Dong J, Albertini DF, Nishimori K *et al.* Growth differentiation factor-9 is required during early ovarian folliculogenesis. *Nature* 1996 383: 531–535.
30. Carabatsos MJ, Elvin J, Matzuk MM, Albertini DF. Characterization of oocyte and follicle development in growth differentiation factor-9-deficient mice. *Dev Biol* 1998 204: 373–384.
31. Singh RP, Carr DH. The anatomy and histology of XO human embryos and fetuses. *Anat Rec* 1966 155: 369–384.
32. Howell CY, Bestor TH, Ding F *et al.* Genomic imprinting disrupted by a maternal effect mutation in the Dnmt1 gene. *Cell* 2001 104: 829–838.
33. Doherty AS, Mann MR, Tremblay KD, Bartolomei MS, Schultz RM. Differential effects of culture on imprinted H19 expression in the preimplantation mouse embryo. *Biol Reprod* 2000 62: 1526–1535.
34. Miyazaki S, Shirakawa H, Nakada K, Honda Y. Essential role of the inositol 1,4,5-trisphosphate receptor/Ca^{2+} release channel in Ca^{2+} waves and Ca^{2+} oscillations at fertilization of mammalian eggs. *Dev Biol* 1993 158: 62–78.
35. Mehlmann LM, Kline D. Regulation of intracellular calcium in the mouse egg: calcium release in response to sperm or inositol trisphosphate is enhanced after meiotic maturation. *Biol Reprod* 1994 51: 1088–1098.

36. Chung A, Swann K, Carroll J. The ability to develop normal Ca(2+) transients in response to spermatozoa develops during the final stages of oocyte growth and maturation. *Hum Reprod* 2000 15: 1389–1395.
37. Ducibella T, Buetow J. Competence to undergo normal fertilization-induced cortical activation develops after metaphase I of meiosis in mouse oocytes. *Dev Biol* 1994 165: 95–104.
38. McLay DW, Clarke HJ. The capacity to remodel sperm nuclei into functional chromatin is acquired during oocyte maturation in the mouse independently of activation. *Dev Biol* 1997 186: 73–84.
39. Moor RM, Dai Y, Lee C, Fulka J Jr. Oocyte maturation and embryonic failure. *Hum Reprod Update* 1998 4: 223–236.
40. Mehlmann LM, Mikoshiba K, Kline D. Redistribution and increase in cortical inositol 1,4,5-trisphosphate receptors after meiotic maturation of the mouse oocyte. *Dev Biol* 1996 180: 489–498.
41. Grøndahl C, Hansen TH, Marky-Nielsen K, Ottesen JL, Hyttel P. Human oocyte maturation *in vitro* is stimulated by meiosis-activating sterol. *Hum Reprod* 2000 15 (Suppl 5): 3–10.
42. Hampl A, Eppig JJ. Analysis of the mechanism(s) of metaphase I arrest in maturing mouse oocytes. *Development* 1995 121: 925–933.
43. Colledge WH, Carlton MB, Udy GB, Evans MJ. Disruption of c-*mos* causes parthenogenetic development of unfertilized mouse eggs. *Nature* 1994 370: 65–68.
44. Hashimoto N, Watanabe N, Furuta Y *et al.* Parthenogenetic activation of oocytes in c-*mos*-deficient mice. *Nature* 1994 370: 68–71.
45. Verlhac MH, Kubiak JZ, Weber M *et al.* Mos is required for MAP kinase activation and is involved in microtubule organization during mouse meiosis. *Development* 1996 122: 815–822.
46. Gross SD, Schwab MS, Taieb FE *et al.* The critical role of the MAP kinase pathway in meiosis II in *Xenopus* oocytes is mediated by p90(Rsk). *Curr Biol* 2000 10: 430–438.

Chapter 21

Male gametogenesis

Willy M. Baarends and J. Anton Grootegoed

INTRODUCTION

The two testes of a fertile man can produce more than 1000 spermatozoa per second, for many years. This requires spermatogenesis to proceed according to a strictly controlled program, in a well-organized epithelium. Control and organization of spermatogenesis are based on an intricate interplay between the developing germ cells and the somatic Sertoli cells.

Spermatogenesis involves mitotic expansion of stem cells, meiotic recombination of genetic information and the generation of spermatids containing a haploid genome through the meiotic divisions. This is followed by construction of the highly specialized sperm cell, which is capable of bringing the haploid genome to the oocyte, specific binding to the zona pellucida and subsequent fusion with the oocyte.

The developmental series of events during spermatogenesis that leads to the enclosure of a haploid genome in the head of the spermatozoon requires specific control of gene expression. This control of gene expression will show different properties during the mitotic, meiotic and postmeiotic phases of spermatogenesis. However, these phases form a continuum, and spermatogonia and spermatocytes seem to have some foresight. This is most evident for the transition of meiotic spermatocytes into postmeiotic spermatids, since spermatocytes already express a number of genes that encode proteins essential for sperm function rather than for meiotic events.

This chapter focuses on molecular and cellular aspects of spermatogenesis, in particular in the mouse. In recent years, much knowledge of mammalian gametogenesis has been gained by making use of mouse models, but it should be noted that the mechanisms of spermatogenesis in the mouse and the human show quite a number of differences. Nevertheless, research into the molecular and cellular biology of spermatogenesis in the mouse provides tools and information, to improve our understanding of human male infertility phenotypes, and to identify novel targets for new methods of contraception.

APPLIED TECHNOLOGY

Germ cells and Sertoli cells

Most studies on isolated germ cells and Sertoli cells have been performed using testes from the mouse, rat and hamster. The tools and information that are being generated can be applied to evaluate human testis biopsies, using techniques such as immunohistochemistry, *in situ* hybridization and detection of apoptosis (see Chapter 5). In addition, when human homologs of mouse genes are identified, a genetic analysis can be performed, to evaluate whether the function of the respective genes is compromised in infertile human males.

Enzymes such as collagenase, trypsin and hyaluronidase are used to dissociate testis tissue into single cells and small clusters of cells. Purification of germ cell types is carried out using velocity sedimentation at unit gravity in an albumin gradient (Staput) or centrifugal elutriation in a specialized counter-flow rotor, followed by separation on the basis of differences in cell density using Percoll gradient centrifugation. The composition of the incubation medium needs adaptation. For example, isolated spermatocytes and round spermatids will require exogenous lactate and/or pyruvate, to prevent an energy crisis related to lack of glycolytic activity in these cells.[1] *In vitro* survival of spermatogenic cells is improved when the cells are in direct contact with Sertoli cells.[2] However, it remains a challenge to define culture conditions that support long-term survival and development of a sufficient number of spermatogenic cells.

Stem cell transplantation (see Chapter 10) is a novel tool which can be used, for example, to evaluate whether a primary defect in spermatogenesis in a mouse model is present either in the germ cells or in the supporting somatic cell lineages.[3]

Sertoli cells showing some mitotic activity can be obtained from very young animals. Most often, mitotically inactive Sertoli cells are isolated from testes of immature animals, when the first wave of spermatogenesis has not yet resulted in the formation of spermatids. During this phase of initiation of spermatogenesis, the Sertoli cells show marked responses to follicle stimulating hormone (FSH). When cultured on an extracellular matrix, the Sertoli cells can maintain a columnar shape.[4]

Isolation and culture of Sertoli cells from adult animals with intact spermatogenesis are much more complicated, in view of the relatively small number of Sertoli cells and their extensive interaction with all germ cell types including the elongated spermatids.[5]

Gene expression

To study the possible role of a gene in spermatogenesis, the first step usually taken is to study the tissue specificity and cellular localization of mRNA expression. For a tissue as complex as the testis, a useful approach is *in situ* hybridization, in particular when transcription of the respective gene results in a relatively high mRNA level. The expression of mRNAs can be analyzed in molecular detail using Northern blotting, RNAse protection or reverse transcriptase-polymerase chain reaction (RT-PCR), for isolated cell types or testis tissue containing different populations of spermatogenic cells, for example by taking testes from immature mice at different developmental time points during initiation of spermatogenesis. In general, mRNA expression studies yield important information, but evaluation of the cellular and subcellular localization of the encoded proteins is crucial to determine the spatial and temporal aspects of gene expression. In particular for spermatogenesis, it is of the upmost importance to look for the proteins and not only the mRNAs, in view of the marked translational control of gene expression in spermatocytes and spermatids (see 'Control of splicing and translation of mRNAs' below).

Immunohistochemical or cytochemical studies of protein expression, using improved fluorescent detection methods, yield detailed information about (co)localization of proteins and about changes in chromatin configuration during spermatogenesis. In the near future, even more exciting results are to be expected from techniques that make use of small fluorescent proteins, such as green fluorescent protein (GFP), coupled to a protein of interest. DNA constructs encoding proteins carrying a fluorescent tag can be designed to result in expression at specific steps of spermatogenesis in transgenic mice. Subsequently, the intracellular location, movement and interactions of fluorescent proteins can also be traced in living cells, using multiphoton laser scanning microscopy.[6]

Transgenic and gene knock-out approaches (see Chapter 10) have started to reveal the role of a number of different genes in gametogenesis, and lead to a better understanding of some of the mechanisms involved in control of spermatogenic gene expression.

ENDOCRINE AND CELLULAR ASPECTS

Endocrine control of spermatogenesis

Spermatogenesis can be viewed as a developmental process for which all necessary stem cells and supporting cell types have been formed during embryonic/fetal development. The actual initiation of spermatogenesis is postponed to a later point in time, namely the onset of puberty. Then, the rising circulating levels of luteinizing hormone (LH) and FSH will induce marked activation of testicular Leydig cells and Sertoli cells, resulting in the stimulation of androgen production and initiation of spermatogenesis. Some aspects of this endocrine control of spermatogenesis can be summarized as follows.

The principal hormonal regulators of spermatogenesis, testosterone and FSH, do not act on the germ cells directly. Rather, Sertoli cells are the main target for hormonal control of spermatogenesis. Sertoli cells show marked expression of the nuclear androgen receptor, which is a ligand-activated transcription factor, and of the G protein-coupled FSH receptor at the cell surface.[7–10]

Testicular androgen action is quite complex, and relatively little is known. This is somewhat surprising, in view of the predominant role of androgens in hormonal control of spermatogenesis. A major challenge is to identify genes that show direct and specific regulation of their transcription by testosterone in Sertoli cells. One candidate gene is *Pem*, which encodes a homeodomain protein.[11] Androgenic control of spermatogenesis is a long-term process, and the continuous testicular action of androgens may control an intricate network of gene expression events in Sertoli cells. In addition, androgen action on the peritubular cells may result in the release of local signalling factors, which travel across the basal membrane to act on Sertoli cells.[12] Stimulation of adenylate cyclase activity in Sertoli cells by FSH results in up-regulation of the transcription of several cyclic adenosine monophosphate (AMP) responsive genes. In particular for immature Sertoli cells, a pronounced response is observed, such as for the gene *LRPR1*, which encodes a protein with a yet unknown function.[13] Sertoli cells can respond to FSH very rapidly, within 30 min, but the overall effect of FSH involves long-term stimulation.[10] The relative importance of FSH in initiation and maintenance of spermatogenesis has been the subject of extensive debate. Observations on *FSH β-subunit* gene and *FSH receptor* gene knock-out mice[14–16] have shown that spermatogenesis can occur in the absence of FSH action. However, quantitatively, spermatogenesis is impaired. In the human male, the role of FSH seems to be somewhat more prominent, but more observations on the phenotypes resulting from mutational loss of FSH action[17] (see Chapter 21) are needed, to obtain a clear picture of the role of FSH in human testis function.

Taken together, it can be concluded that testosterone and FSH act on Sertoli cells in a synergistic manner, and that the more long-term effect of this synergistic action is dependent on the performance of a local regulatory network.[12,18] This network stimulates Sertoli cells to become the mature and fully active cells which can support spermatogenesis, and which can adequately respond to interactive signals from the developing germ cells.

The spermatogenic epithelium

In male and female embryos, a number of embryonic cells enter the germ cell lineage to become primordial germ cells, the stem cells of gametogenesis.[19] At a subsequent stage of embryonic/fetal development, the primordial germ cells reach the undifferentiated anlagen of the gonads through migration, which is stimulated by the interaction of stem cell factor (SCF) with Kit, a cell surface receptor with intrinsic tyrosine kinase activity. The germ cells encounter SCF along their migration route, and the interaction of SCF with the receptor Kit at the surface of the germ cells stimulates their proliferation and survival.[19] After arrival of the germ cells in the undifferentiated gonads and differentiation of the gonads into testes, the primordial germ cells are enclosed in testicular tubules together with the precursor Sertoli cells. In this tubular environment, entry of the male germ cells into the meiotic prophase is inhibited until postnatal initiation of spermatogenesis.[20]

In newborns, the spermatogenic epithelium is composed of immature Sertoli cells that enclose a relatively small number of descendants of the embryonic primordial germ cells, the gonocytes and the undifferentiated spermatogonia, which have the potential for long-term mitotic proliferation. In contrast, mitotic activity of Sertoli cells slows down during early postnatal development, followed by complete and permanent mitotic silencing. The number of Sertoli cells is a limiting factor for the final length of the testicular tubules and the spermatogenic output of the mature testis.[18] Upon initiation of spermatogenesis, the spermatogonia make contact with the basal membrane and are periodically mitotically active[21] (arrow 3 in Figure 1).

The role of Sertoli cells in the organization of the spermatogenic epithelium is highlighted by the formation of the Sertoli cell barrier, also called the blood–testis barrier. This barrier allows Sertoli cells to create a specific microenvironment in which meiosis and postmeiotic development take place. Shortly after entrance into the prophase of meiosis, the early primary spermatocytes are transported across the Sertoli cell barrier into the adluminal compartment, leaving space in the basal compartment for the next generation of spermatogonia to enter into meiosis. The passage across the Sertoli cell barrier is a remarkable achievement, the more so because the spermatocytes are not single cells, but form syncytia in which the cells are connected through cytoplasmic bridges.[23] These syncytia start to be formed at the early A spermatogonia stage, through incomplete cytokinesis, and more than 100 spermatocytes can be interconnected in a single syncytium. When the syncytium moves towards the adluminal compartment, the Sertoli cell barrier is opened on the adluminal side and simultaneously is closed again on the basal side, to

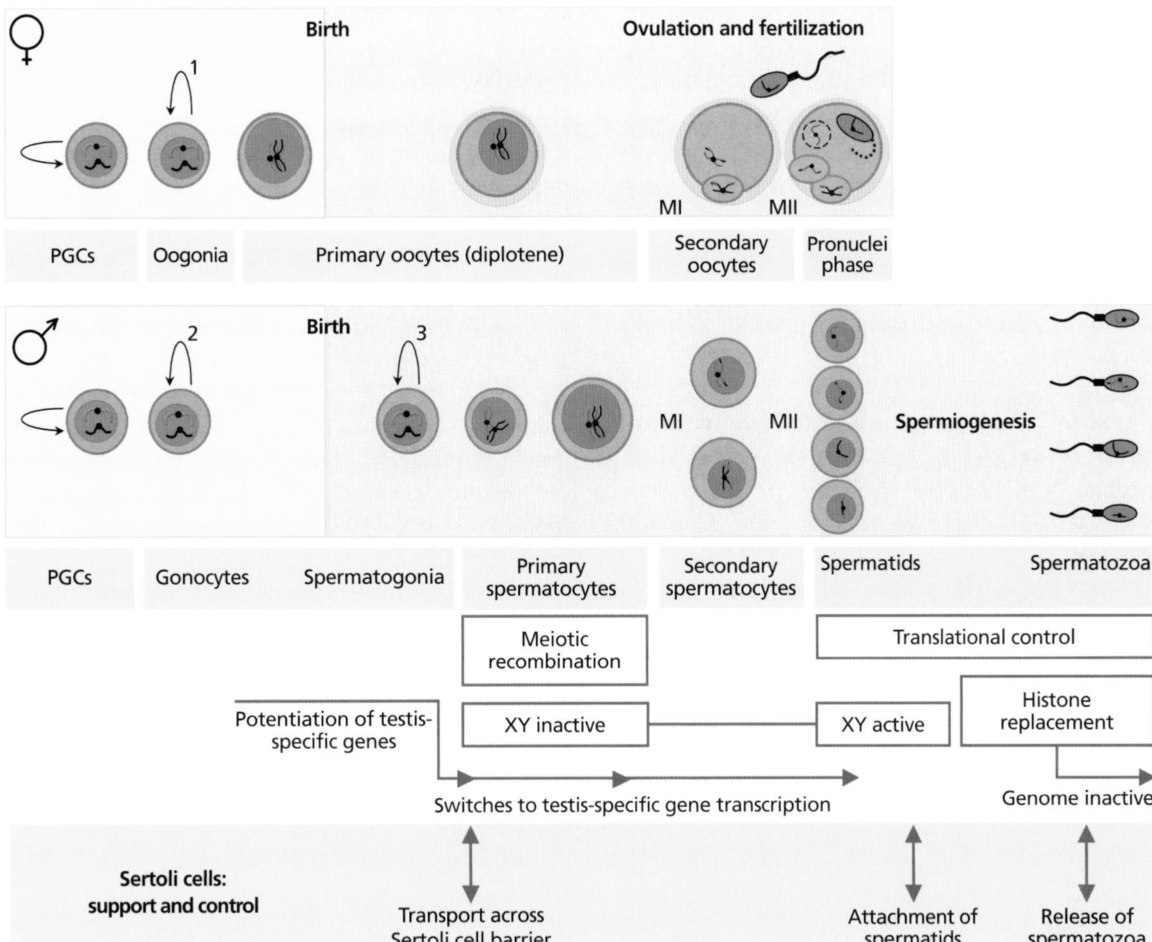

Figure 1 Overview of key events in spermatogenesis (an outline of oogenesis is also shown, for comparison). Primordial germ cells (PGCs) undergo mitotic proliferation during fetal development. In the fetal ovary, the PGCs form a population of oogonia, and the cell number increases by further mitotic proliferation (arrow 1). Subsequently, the oogonia enter the meiotic prophase, and become arrested in the diplotene stage of primary oocytes. Following the initial proliferation, there is a substantial decline of the number of germ cells, and around birth the ovaries contain a limited supply of diplotene-stage (dictyate) oocytes, in primordial follicles. During follicle development, the oocytes express genes that are not expressed during spermatogenesis, such as the genes encoding the proteins of the zona pellucida (see Chapter 20). Around ovulation, the first meiotic division is completed, and sperm entry triggers the metaphase (M II)-secondary oocytes to complete the second meiotic division (see Chapter 20).

In the fetal testis, the PGCs are enclosed in the tubules, and form a population of gonocytes with proliferative activity (arrow 2). Entry into the meiotic prophase is inhibited. At birth, the tubules contain Sertoli cells and gonocytes. Towards the start of reproductive life, the gonocytes give rise to a population of spermatogonia, which constitute a proliferating and differentiating stem cell pool throughout reproductive life (arrow 3). At the beginning of the meiotic prophase, the spermatocytes are transported across the Sertoli cell barrier. The X and Y chromosomes form the heterochromatic sex body, and all chromosomes are engaged in meiotic recombination. Changes in the structure of specific chromatin areas may potentiate expression of testis-specific genes, already at the early steps of spermatogenesis (foresight of spermatogonia and spermatocytes). Gene expression switches during the later steps of spermatogenesis are described in the text. The two meiotic divisions yield four haploid spermatids. The following differentiation process of these haploid cells, spermiogenesis, requires intensive interaction with Sertoli cells. The developing spermatids become firmly attached at the future head region to Sertoli cells (ectoplasmic junctional specializations). The X and Y chromosomes are transiently reactivated, before the whole genome is silenced owing to compaction of the DNA by the histone-to-protamine transition. Protein synthesis at later steps of spermiogenesis requires control of mRNA stability and translation. Spermiogenesis ends with the release of testicular spermatozoa into the lumen of the testicular tubules.

The haploid genomes generated through oogenesis and spermatogenesis are functionally non-equivalent during embryonic development, effectuated by genomic imprinting[22]

prevent at all times leaks in the barrier. Evidently, this requires precise control of proteolytic enzymes that dissolve the tight junctions of the barrier.[24]

Upon initiation of spermatogenesis during postnatal life, the entry of cohorts of cells into spermatogenesis, each time starting with a new generation of differentiating A1 spermatogonia originating from undifferentiated A spermatogonia, occurs at regular time intervals.[21] In addition, the different phases and steps of spermatogenesis take a precise length of time. Consequently, the cellular composition of the spermatogenic epithelium is dynamic, but also highly ordered, resulting in the spermatogenic cycle of defined associations of germ cell types (stages).[18,23,25]

MEIOSIS AND SPERMIOGENESIS

Meiosis

Throughout male reproductive life, undifferentiated A spermatogonia develop into differentiating spermatogonia. This involves a defined series of mitotic divisions. In murine species, this series leads from undifferentiated A spermatogonia to differentiated A1, A2, A3, A4, intermediate (In) and finally B spermatogonia. In the human, the mitotically active spermatogonia population is named A pale spermatogonia, as opposed to a more silent A dark spermatogonia population. In establishing a population of spermatogonia of proper composition and size, programmed cell death or apoptosis plays an important role.[18,21,26] The last spermatogenic mitotic division, that of the B spermatogonia, results in the formation of preleptotene primary spermatocytes. This mitotic step is often considered the point of entry into meiosis, but it should be noted that A1 spermatogonia have already embarked on a one-way path leading to the meiotic prophase.

The spermatogenic meiotic prophase starts with formation of the preleptotene primary spermatocytes, which then perform the last DNA replication of spermatogenesis ($2n$, 2C becomes $2n$, 4C; $2n$ is diploid, $1n$ is haploid and 1C is the amount of DNA in a haploid gamete). The meiotic prophase in spermatocytes takes a relatively long period of time, in the order of 2 weeks, with variation between species, and during the greater part of this period the spermatocytes are in the pachytene stage. During leptotene and zygotene, homology search and pairing (synapsis) of homologous chromosomes take place. Meiotic recombination occurs in pachytene spermatocytes. The resulting chiasmata (cross-overs) are rapidly resolved during the short diplotene stage, which is dramatically different from the long-lasting diplotene or dictyate stage as it occurs in oocytes. The spermatogenic diplotene stage ($2n$, 4C) leads directly to the first meiotic division, to generate the haploid secondary spermatocytes ($1n$, 2C), followed immediately by the second meiotic division, resulting in haploid spermatids ($1n$, 1C) (Figure 1).

In recent years, a large number of genes have been cloned which encode proteins involved in chromatin dynamics and DNA repair mechanisms in many cell types. It appears that a number of these proteins are also found in association with meiotic chromatin. DNA double-strand breaks in somatic cells can be repaired through a homologous recombination mechanism.[27] Proteins involved in this mechanism also function during meiotic recombination of non-sister chromatids.[28] However, meiotic recombination in addition requires the actions of several meiosis-specific proteins, such as the Spo11 protein. This protein causes induction of DNA double-strand breaks, in the initiation of meiotic recombination. Spo11 is conserved from yeast to man. Mutation of the gene encoding this protein in mice results in defective pairing of homologous chromosomes in both male and female meiosis.[29,30] In wild-type mice, Spo11 is found in association with the synaptonemal complex,[30] the specialized structure that links together the paired homologous chromosomes during the meiotic prophase (Figure 2).

At different steps in the meiotic prophase, the synaptonemal complex seems to be a kind of meeting point for DNA repair proteins, known for their functions in various DNA repair mechanisms.[28] To give an example, the DNA mismatch repair protein MLH1 localizes to the synaptonemal complex in mid-pachytene spermatocytes and oocytes.[33] The MLH1 protein forms foci, corresponding to the cross-over sites (Figure 2). This

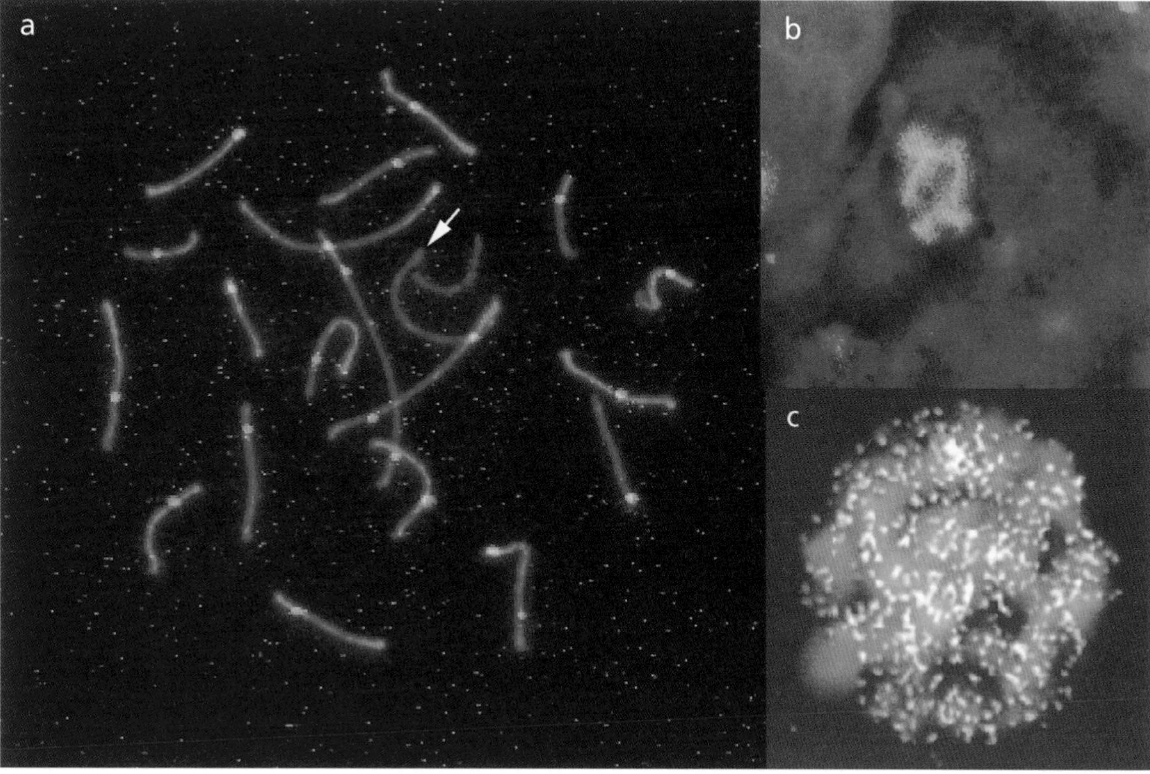

Figure 2 A few properties of primary spermatocytes, at the pachytene stage of the meiotic prophase. The chromatin of a mouse spermatocyte is spread on a glass slide, and immunostaining for SCP3 shows the synaptonemal complexes between the homologous chromosomes (a). The green foci represent the DNA mismatch repair protein MLH1, which is found at cross-over sites. The paired and unpaired regions of the X and Y chromosomes also bind SCP3 (arrowhead in (a)), and the paired pseudoautosomal regions of X and Y can have a MLH1 focus (not in this nucleus). The unpaired regions of X and Y bind the cell-cycle check-point protein Atr (green fluorescent signal in (b)), whereas this protein is not found on the paired autosomes, when the homologous pairing is complete.[31] The transcriptional activity of a spermatocyte nucleus is shown in (c). The white grains represent RNA synthesis, visualized by incubation of the cell with [^{3}H]uridine.[32] The sex body chromatin is not covered by grains, illustrating that the X and Y chromosomes are transcriptionally inactive

protein is essential for male and female fertility, in addition to its role in maintaining genome stability in somatic cells.[33,34] Interestingly, germ cell-specific members of the mismatch repair gene family have been identified, named MSH4 and MSH5. Inactivation of either one of these two genes in mice results in male and female infertility, with major impairment of progression through the meiotic prophase.[35–37] One of the structural components of the synaptonemal complex is the protein SCP3[38] (Figure 2). In *SCP3* knock-out mice, the meiotic prophase in the male animals is greatly disturbed, resulting in infertility.[39] Surprisingly, the female *SCP3* knock-out animals are fertile. This emphasizes that there are differences between the meiotic prophase in spermatocytes and oocytes.

The mechanisms for arrest of meiosis at diplotene and metaphase II, as operational in oocytes, are not found in spermatocytes. In normal spermatogenesis, the spermatocytes go through the meiotic prophase and the meiotic divisions without making a halt. However, this does not exclude the existence of cell cycle check-point mechanisms during spermatogenesis, similar to the pachytene check-point described for yeast.[40] Pachytene arrest, and loss of spermatocytes, is observed in several mouse gene knock-out models with a defect in the meiotic prophase.[40] Also, meiotic germ cell loss contributes to the relatively poor efficiency of human spermatogenesis,[41] and spermatocytic arrest is a relatively frequent finding in testis biopsies from infertile males.[42] Cloning of human homologs of genes involved in meiotic chromatin dynamics, such as the identification of the human *MSH4* and *MSH5* genes,[43,44] will generate information about the factors involved in control of the human meiotic prophase.

Spermiogenesis

The process of spermiogenesis takes about 3 weeks, with variation between species. The differentiation into the highly specialized sperm cells is one of the most remarkable cell developmental processes that can be observed in biological systems. It involves phases of acrosome development, nuclear elongation and condensation, formation of middle piece and tail, and reduction of cytoplasmic volume. Evidently, this process requires synthesis of many specific gene products, and, in addition, there is pronounced support by Sertoli cells. As soon as the developing spermatids start to elongate, a firm attachment of the future head region of the spermatids to Sertoli cells is established, through formation of so-called ectoplasmic junctional specializations.[5,23] The formation of these cell-to-cell attachments requires expression of cell adhesion molecules such as cadherins and integrins.[45,46] Shortly before spermiation, the Sertoli cells play an active role in reduction of the cytoplasmic volume of the elongated spermatids, and the remains of the Golgi apparatus that was actively engaged in building the acrosome are found in the residual body.[5,23] When the adhesive contacts between spermatids and Sertoli cells are destructed, and the cytoplasmic bridges interconnecting the spermatids are finally broken, the spermatids, or testicular spermatozoa, are released into the tubular lumen. Spermatozoa acquire fertilizing capacity upon transit through the epididymis.[47]

During spermatid development, the nucleus elongates and condenses. The fully condensed nucleus of spermatids contains tightly compacted DNA, following replacement of nucleosomal histones (somatic and testis-specific isoforms) by transition proteins and subsequently by protamines.[48] Specific aspects of the histone-to-protamine replacement are being studied in mouse models. For example, targeted deletion of the *Tnp1* gene, which encodes the major transition protein TP1, results in an abnormal pattern of chromatin condensation and reduced fertility.[49]

It seems that the histone-to-protamine replacement process in human spermatids is less complete than that observed in experimental animals such as the rat and mouse. Chromatin packaging in human spermatozoa shows marked inter-sperm and inter-individual variation, with persistence of variable amounts of histones in the protamine–DNA complex.[50–52] The presence of histones in human spermatozoa might play a functional role, for example in reactivation of the male haploid genome in the male pronucleus. However, it can also be suggested that persistence of histones in spermatozoa is a negative hallmark of sperm quality, and may result in poor chromatin packaging and enhanced genome instability.

SPERMATOGENIC GENE EXPRESSION

Introduction

The gene expression program of spermatogenesis includes testis-specific expression of a variety of genes, which encode proteins that take part in spermatogenesis and/or function in spermatozoa. Several proteins encoded by these genes are quite similar to proteins found in somatic cells, and can be considered testicular isotypes or isozymes. Clear examples are found in the energy-yielding machinery. Spermatogenic cells express testis-specific isozymes of widely expressed somatic proteins, to assemble the energy-yielding machinery of the spermatozoa.[1] To give one such example, a testis-specific isoform of glyceraldehyde 3-phosphate dehydrogenase (GAPD) has been identified, named GAPDS in the mouse and GAPD2 in the human. In both the mouse and human, this glycolytic enzyme was found to be located in the principal piece of the sperm flagellum.[53,54]

Many other testicular proteins are also related to somatic proteins, but appear to exert a more specialized function. One example is the gene encoding fertilin-β, which shows testis-specific expression.[55] Functional importance has been demonstrated; mice lacking fertilin-β show fertilization defects.[56] Fertilin-β is a member of the extensive ADAM (a disintegrin and metalloprotease) family of membrane proteins, and this member has gained a specialized function in fertilization[47,57,58] (see Chapter 22). Another example relates to the heat-shock proteins (HSPs), which act in somatic cells as molecular chaperones in a variety of cellular processes through their involvement in folding, transport, assembly and disassembly of proteins and protein complexes. The germ cell-specific HSP70-2 protein associates with the synaptonemal complex and is involved in disassembly of the synaptonemal complex at the end of the meiotic prophase.[59] The importance of this protein is demonstrated by the phenotype of the *Hsp70-2* knock-out mouse, showing male infertility owing to a block in progression through the spermatogenic meiotic prophase.[60,61]

The spermatogenic gene expression profile, leading to the highly specific proteome (expression of the genome at the protein level), is based on a variety of mechanisms, briefly discussed in the following paragraphs.

Switches to homologous genes and variant gene products

Gene switching during spermatogenesis leads to testis-specific expression of genes encoding testis-specific isotypic proteins (Figure 3a). Evolutionary gene duplication events are at the basis of this type of switching. Examples include the gene encoding the testis-specific isoenzyme GAPDS (see above) and also the gene encoding the lactate dehydrogenase subunit C (which forms the isoenzyme $LDHC_4$).[1,53,62] The homologous genes encoding the somatic proteins (GAPD, LDHA/B) become silenced during spermatogenesis. However, it should be noted that, for several other genes, the switching is incomplete, meaning that the genes encoding the somatic isoproteins are not completely silenced. This is found, for example, for genes encoding somatic and testis-specific variants of the histones H1, H2A, H2B and H3.[48,63] Starting in spermatogonia and ending with the histone-to-protamine transition in spermatids, the chromatin of spermatogenic cells contains a combination of testis-specific and somatic histones. The significance of the testis-specific isoforms is not clear. Knock-out of the gene encoding testis-specific H1 in the mouse resulted in up-regulation of the expression of somatic H1 isoforms, and did not lead to infertility.[64,65]

The human genome contains about 30 000–40 000 protein-encoding genes, which is only twice as many as for the worm or fly.[66,67] However, many of the mammalian genes are complex, with alternative splicing events generating different protein products from one gene. There are numerous reports about alternative splicing of testicular mRNAs, and often the mRNA splice variants can encode functional proteins. For several genes, alternative splicing may generate mRNAs that encode testis-specific proteins. This is found, among others, for the *CREM* gene (see 'Control of gene transcription' below) (Figure 3d). In addition, one gene can encode somatic and testis-specific proteins when a testis-specific intronic promoter is used. Among others, this is observed for calspermin, which is transcribed from an intronic promoter in the *Camk4* gene, normally encoding Ca^{2+}/calmodulin kinase IV[68] (Figure 3b).

Switches to autosomal retroposons

Meiotic pairing of the X and Y chromosomes in spermatocytes is limited to the short pseudoautosomal region of the two sex chromosomes. During the greater part of the meiotic prophase, the X and Y chromosomes are heterochromatic, forming the so-called sex body, which is transcriptionally inactive[69] (Figures 1 and 2). In marked contrast, the meiotic prophase of oocytes shows activity of two X chromosomes. The sex body may have a function to stabilize the unpaired regions of the X and Y chromosomes. In addition, or alternatively, synthesis of X-chromosomal gene products might be incompatible with progression through the meiotic prophase of spermatogenesis.[70,71] After completion of the meiotic divisions, transcriptional reactivation of the X and Y chromosomes takes place in the haploid round spermatids.[72] Transport of gene products through the intercellular cytoplasmic

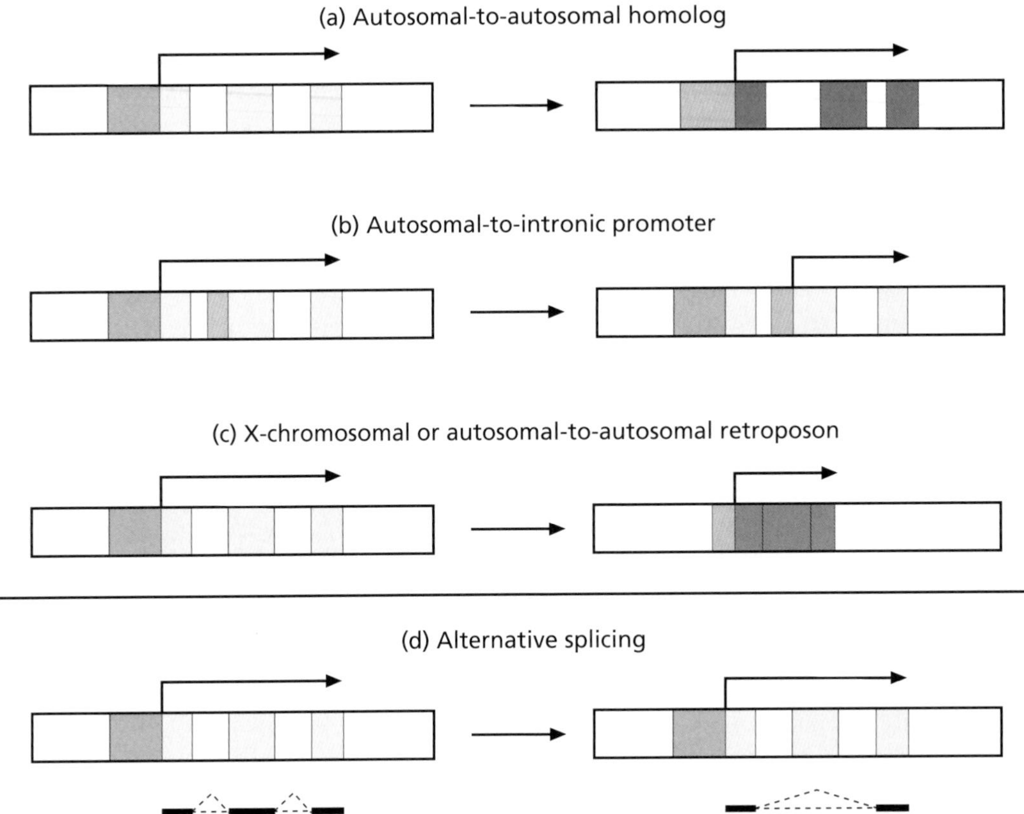

Figure 3 Switches in gene expression during spermatogenesis. Spermatogenic gene expression involves switches to expression of homologous genes encoding isotypes of somatic proteins. Such homologous genes are transcribed from a promoter, which shows specific or relatively high activity in spermatogonia and/or spermatocytes and/or spermatids (a).

By making use of an intronic promoter, the spermatogenic cells can produce a testis-specific protein from a gene which is expressed also in somatic cell types (b). Switching of X-chromosomal genes to expression of autosomal and functional retroposed genes is a quite frequent finding for mammalian spermatogenesis. Such a switching can also occur for autosomal genes (c). Finally, alternative splicing is very frequently observed in spermatogenic cells, and, for several mRNAs, this can result in the production of testis-specific proteins (d). The different switches are schematically depicted for a hypothetical gene containing three exons (light-green boxes). The approximate positions of a general (somatic) promoter (orange box) and a testis-specific (germ cell-specific) promoter (red box) are also shown

bridges ensures that the haploid spermatids are functionally diploid,[73] probably also for gene products encoded by the sex chromosomes.

Formation of the sex body in early primary spermatocytes results in the loss of expression of several X-chromosomal genes. The phosphoglycerate kinase PGK1 isoform is encoded by *Pgk1* on the X chromosome, and is required for glycolysis in somatic cells. However, following sex body formation and X chromosome inactivation in spermatocytes, the *Pgk1* gene is silenced. This lack of *Pgk1* expression during the meiotic prophase is compensated by expression of the autosomal gene *Pgk2*, which encodes the isoenzyme PGK2. The *Pgk2* gene is an intron less gene, a functional retroposon.[74,75] It may have originated from mature mRNA transcribed from *Pgk1* that has been inserted into the DNA at another chromosomal site through viral intervention, at some point in evolution. Usually such intronless retroposons become mutated and lose their function, but the autosomal *Pgk2* has become of vital importance for the provision of energy requirements in spermatogenic cells and spermatozoa, and has gained possession of a testis-specific promoter[76] (Figure 3c). The pyruvate dehydrogenase (PDH) E1α-subunit in spermatocytes and spermatids is also encoded by an autosomal retroposon, *Pdha2*, which compensates for the absence of the isotypic subunit encoded by the X chromosome.[77] Similar observations have been made for several other genes.[78–81] However, it should be noted that X chromosome inactivation is not the only factor that drives expression of functional retroposed genes in spermatogenesis, because a switch to spermatogenic expression of retroposed genes is also observed for autosomal-to-autosomal retroposition[82] (Figure 3c).

Control of gene transcription

Gene transcription is determined by the chromatin structure and the presence of specific transcription regulatory proteins, transcription factors. In several transgenic mouse models, it has been shown that a relatively small promoter region (in the order of 300 bp) can be sufficient to direct expression of a reporter gene at the correct spermatogenic stage, corresponding to the time of expression of the endogenous gene. This was found, for example, for promoter sequences of *SP10* (a gene encoding an acrosomal protein), which direct expression in round spermatids immediately following completion of the meiotic divisions.[83] Specificity and timing of gene expression in spermatogenic cells will require developmental control of the synthesis of a set of transcription factors. Spermatogenic gene expression may also require the absence, rather than the presence, of certain transcription factors, which effectuate repression of testis-specific genes in somatic cells.[84] Furthermore, it is likely that control of spermatogenic gene expression, including quantitative aspects, requires additional and more intricate mechanisms, including a role for extended gene enhancer/promoter regions, DNA methylation and specific aspects of chromatin structure. Changes in chromatin structure have been found to occur in the 5' upstream region of several testis-specific genes, prior to the expression of these genes[85] (Figure 1).

There are a number of genes encoding transcription factors that show quite exclusive expression during spermatogenesis, such as the homeobox-containing genes *Hox1.4* and *Esx1*.[86,87] An interesting spermatogenic transcription factor, for which much information is available, is the factor CREMτ (cyclic AMP response element modulator-τ). CREMτ is translated from a testis-specific transcript, which is generated through alternative splicing upon transcription of the *CREM* gene in postmeiotic cells. At earlier steps of spermatogenesis, and in somatic cell types, *CREM* gene transcription yields a repressor CREM protein. In contrast, the postmeiotic CREMτ, or activator CREM protein, functions as a transcriptional activator.[88] Activator CREM action has been implicated in the regulation of a number of testis-specific gene promoters which contain a cyclic AMP response element (CRE).[89,90] Male mice homozygous for an inactivating mutation of the *CREM* gene are infertile, owing to a failure to complete spermatogenesis beyond the round spermatid steps.[91,92] The protein ACT (activator of CREM in testis) is also expressed specifically in spermatids, and acts as a coactivator in the presence of CREM activator isoforms, by protein–protein interaction and stimulation of the transcriptional activity of activator CREM.[93] These observations emphasize that testis-specific transcription factors and coactivators are involved in control of the spermatogenic gene expression and cell differentiation program.

Control of splicing and translation of mRNAs

The histone-to-protamine transition and chromatin compaction in condensing spermatids finally make the complete sperm genome transcriptionally inactive. However, there is also a need for synthesis of several proteins after chromatin compaction in spermatids has started. By making use of mechanisms to control translation of mRNAs, protein synthesis in late spermatids can occur up to several days after silencing of gene transcription. For certain, this is highly relevant for expression of the protamine genes, because transcription of the protamine genes should occur well before protamine protein synthesis results in chromatin compaction. Untranslated mRNA sequences and RNA-binding proteins are involved in the mechanisms which ensure that specific mRNAs are not translated until several days after gene transcription.[94,97] The importance of this translational control has been demonstrated in a transgenic mouse model, in which lack of translational control of a transgenic modified protamine mRNA results in premature protamine synthesis and nuclear condensation, and thereby in arrest of spermatid development.[98] The mechanism of translational control also requires the action of specific RNA-binding proteins to initiate translational activation. The protein Prbp (encoded by the gene *Tarbp2*) is involved in this translational activation. Disruption of the *Tarbp2* gene results in failure to translate protamine mRNAs, when the protamines need to be synthesized.[99] Consequently, in these *Tarbp2*-deficient mice, there is delayed replacement of transition proteins by protamines, and failure of late spermatid development.

It should be noted that RNA-binding proteins appear to play an important role in spermatogenesis also at earlier steps in the process, in spermatogonia and spermatocytes, well before the histone-to-protamine transition results in the arrest of spermatogenic transcription. This is best demonstrated by human male infertility resulting from loss of Y-chromosomal *RBM* genes, which encode RNA-binding proteins. In spermatocytes, the RBM proteins are possibly involved in control of testis-specific pre-mRNA splicing.[100] The human Y-chromosomal *DAZ* genes also encode RNA-binding proteins, which are essential for human male fertility.[101] The mouse genome lacks Y-chromosomal *DAZ* genes, but the human and mouse genomes both contain an autosomal homolog, which in the mouse is named *Dazla*. Disruption of the *Dazla* gene in the mouse results in complete absence of gamete production, in males and also in females.[102] This emphasizes that certain aspects of the manifold roles of RNA-binding proteins (in control of pre-mRNA splicing, and mRNA transport, stability and translation) are not confined to spermiogenesis, but play a crucial role also at premeiotic steps of male and female gametogenesis.

Post-translational control of protein stability

Evidently, not only control of transcription and translation, but also post-translational control of the activity and stability of proteins determine the proteome of spermatogenic cells. This section focuses on protein ubiquitination, which is an important mechanism to control modification and breakdown of proteins.

Ubiquitin is a small protein composed of 76 amino acids. In spite of its small size, the functional importance of this protein is paramount. Ubiquitin is present in all cells, and has essential roles in a variety of basic cellular processes. These roles require covalent coupling of ubiquitin to cellular substrates, resulting in mono- or poly-ubiquitination, through a complex enzymatic pathway. Poly-ubiquitination signals proteolytic breakdown of substrate proteins by the proteasome, whereas mono-ubiquitination may serve other functions.[103]

Mutation of the widely expressed human *USP9Y* gene (also named *DFFRY*), a Y-chromosomal gene which encodes an enzyme that catalyzes a reverse ubiquitination reaction (a de-ubiquitination reaction), does not result in a somatic phenotype, but only affects spermatogenesis. *USP9Y* mutation in man results in partial arrest of spermatogenesis at the spermatocyte stage, with few postmeiotic germ cells still being present.[104] The first gene knock-out mouse model for a component of the ubiquitin system, the *HR6B* knock-out, showed no apparent phenotype, apart from male-limited infertility.[105] This *HR6B* gene is a mammalian homolog of the gene *RAD6* in the yeast *Saccharomyces cerevisiae*. In yeast, the *RAD6* gene encodes a ubiquitin-conjugating enzyme (E2 enzyme). *RAD6* null mutants show a pleiotropic phenotype, including defects in gene silencing, and impaired sporulation.[106] In mammalian species, two highly conserved homologs of this gene have been identified: the X-chromosomal *HR6A* gene and the autosomal *HR6B* gene.[107] It is important to note that the genes encoding these two proteins are expressed in virtually all cell types. When *HR6A* and *HR6B* knock-out mice were generated, no major phenotype was observed for somatic tissues, which is explained by functional redundancy of the somatic functions of the proteins encoded by these two

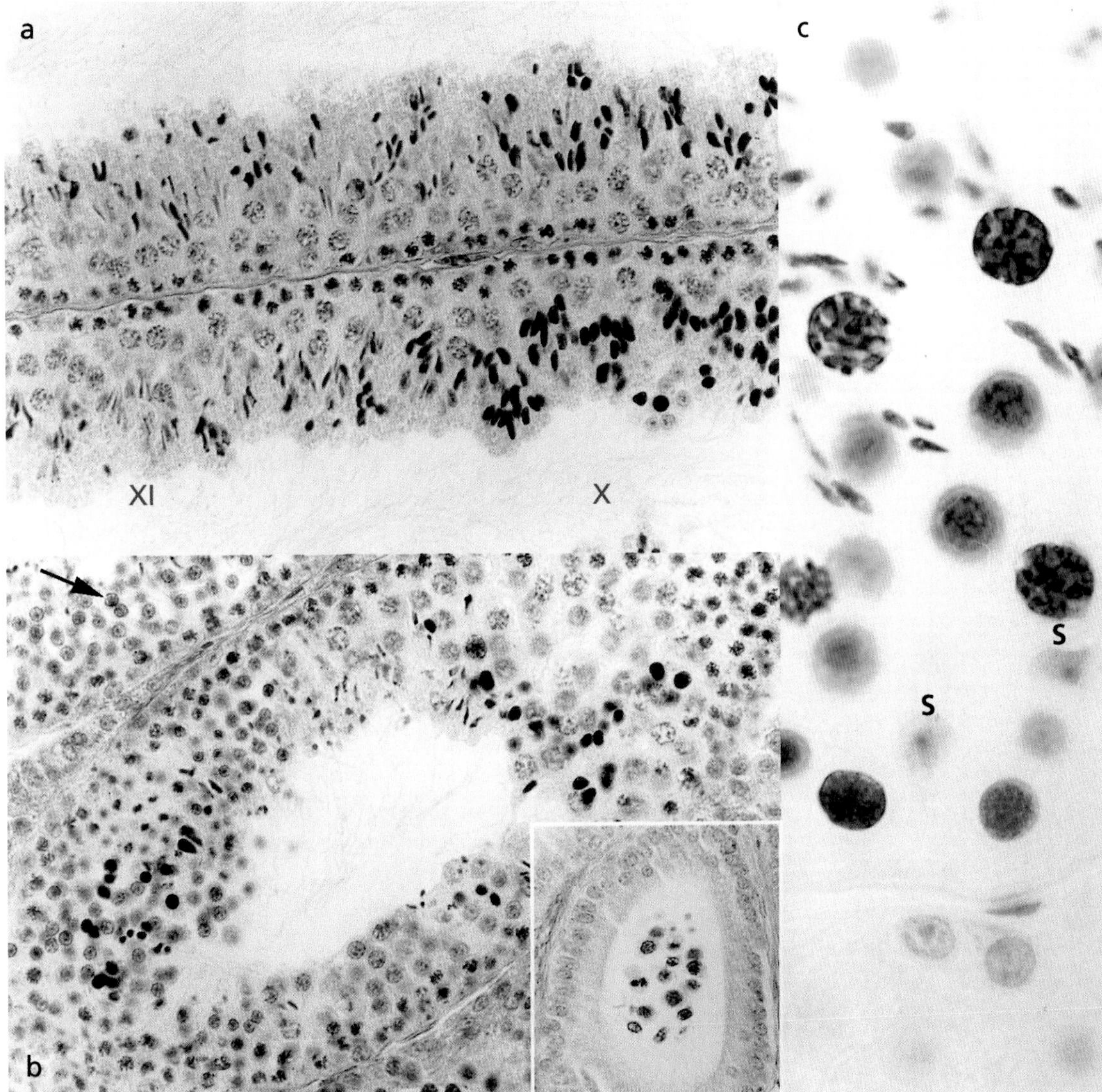

Figure 4 Immunohistochemical localization of testis-specific histone 2B. In the mouse, immunoexpression of testis-specific histone 2B (tH2B) is detected in early spermatocytes, but, in pachytene spermatocytes and round spermatids, the chromatin structure largely prevents immunodetection of tH2B.[108] An increase in immunostaining of tH2B is observed in elongating spermatids (step 10), shortly before the histones will be replaced by the transition proteins. Panel (a) shows the diminution of tH2B in condensing spermatids, when the cells develop from stage X to stage XI (from right to left) of the spermatogenic cycle, in two neighboring tubules in a wild-type mouse testis. In the *HR6B* knock-out mouse (described in the text), there is severe impairment of spermatogenesis, in particular during the chromatin reorganization in condensing spermatids. Panel (b) shows the irregular organization of the spermatogenic epithelium in these knock-out mice, and the overall impairment of spermatogenesis during the spermatid condensation steps. The chromatin structure of round spermatids is also affected, as indicated by the tH2B immunostaining of these cells (arrow in (b)). Furthermore, immature round germ cells with positive staining for tH2B are released from the epithelium, and are found in the epididymis (inset in (b)). In the *HR6B* knock-out epididymis, only a few spermatozoa are found, and most of these are abnormal.[105]

Localization of testis-specific histone 2B (tH2B) can also be observed in the human testis (c). Immunostaining of tH2B in a histological section of a testicular biopsy reveals the presence of tH2B in spermatocytes, and also at an earlier step of spermatogenesis, in spermatogonia.[52] During spermatid elongation and condensation, a characteristic staining pattern illustrates the gradual and regional disappearance of tH2B from the nucleus. The somatic Sertoli cells (nucleus 'S') and peritubular/interstitial cells are completely negative, illustrating that the protein tH2B, like many other proteins, is expressed exclusively in the developing germ cells

genes (the proteins show 96% amino acid sequence identity). The combined functions of these proteins are highly important, because *HR6A/HR6B* double knock-out animals are not obtained; this condition results in very early embryonic death. The role of the mammalian *RAD6* homologs in gametogenesis is complex and intriguing. Whereas the *HR6B* knock-out mice demonstrate impaired spermatogenesis and male-restricted infertility (Figure 4b), the HR6A knock-out mice show a maternal-effect infertility (two-cell stage block of embryonic development) and normal male fertility.[109]

Protein targets for the RAD6-dependent ubiquitination probably include histones H2A and H2B, and most likely also other chromosomal proteins.[110,111] In mammalian spermatogenesis, the ubiquitin-conjugating activity of HR6B might play some role in the histone-to-protamine transition which takes place in condensing spermatids. In addition, HR6B-dependent ubiquitination may target proteins involved in control of gene expression and chromatin dynamics during the meiotic prophase.

This, along with other observations, points to a very important role of the ubiquitin system in spermatogenesis.[112]

POTENTIAL CLINICAL RELEVANCE AND FUTURE PERSPECTIVE

Basic science projects provide information that is relevant to improve our understanding of human gametogenesis. The mouse gene knock-out technology makes it possible to determine the possible role of a gene product. When the respective gene is highly conserved from mouse to human, the potential clinical relevance of the human gene homologs can be evaluated. The importance of identifying genetic causes of male infertility has gained clinical interest, with the introduction of the intracytoplasmic sperm injection (ICSI) technique of assisted reproduction. Characterization of genetic causes of male infertility will improve the quality of diagnosis and counselling in the context of the application of ICSI. However, it is anticipated that many cases of human infertility will prove to be a multigene disorder, and will also involve gene–environment interactions.

Second, and not less important, we should take advantage of the current increase in knowledge about molecular and cellular aspects of spermatogenesis to look for new avenues in contraception. There is certainly a need to develop new methods of contraception, in particular, methods targeting the male, in addition to an approach based on hormonal control. Spermatogenesis involves many specific genes and gene products. Advances in genomics, proteomics and high-throughput screening technology should lead to identification of targets and lead compounds, enabling development of novel non-hormonal methods for the control of male fertility.

The staggering increase in information about the genomes of *Drosophila*,[113] human[66,67] and other species will be followed by the generation of detailed genomic and proteomic expression profiles of genes in spermatogenic cells. The feasibility of this is demonstrated by the description of the meiotic transcriptome in yeast,[114,115] the global profile of germline gene expression in *Caenorhabditis elegans*[116] and the computational and microarray analysis of transcription in the *Drosophila* testis.[117] It will be a major challenge to establish a firm link between the power of bioinformatical databases, gene expression profile analysis, and biological and clinical questions regarding normal and impaired spermatogenesis.

ACKNOWLEDGEMENTS

We thank our laboratory colleagues, past and present, for their valuable contributions to our research on spermatogenesis. Current research is supported by the Dutch Science Foundation (NWO) through GB-MW (Medical Sciences).

REFERENCES

1. Grootegoed JA, DenBoer PJ. Energy metabolism of spermatids: a review. In Hamilton DW, Waites GM, eds. *Cellular and Molecular Events in Spermiogenesis*. Cambridge: Cambridge University Press, 1989: 193–216.

2. Toebosch AMW, Brussee R, Verkerk A, Grootegoed JA. Quantitative evaluation of the maintenance and development of spermatocytes and round spermatids in cultured tubule fragments from immature rat testis. *Int J Androl* 1989 12: 360–374.
3. Johnston DS, Russell LD, Griswold MD. Advances in spermatogonial stem cell transplantation. *Rev Reprod* 2000 5: 183–188.
4. Dym M. Basement membrane regulation of Sertoli cells. *Endocr Rev* 1994 15: 102–115.
5. Russell LD. Role of Sertoli cells in spermiation. In Russell LD, Griswold MD, eds. *The Sertoli Cell*. Clearwater, FL: Cache River Press, 1993: 269–303.
6. Denk W, Svoboda K. Photon upmanship: why multiphoton imaging is more than a gimmick. *Neuron* 1997 18: 351–357.
7. Grootegoed JA, Peters MJ, Mulder E, Rommerts FF, Van der Molen HJ. Absence of a nuclear androgen receptor in isolated germinal cells of rat testis. *Mol Cell Endocrinol* 1977 9: 159–167.
8. Bremner WJ, Millar MR, Sharpe RM, Saunders PT. Immunohistochemical localization of androgen receptors in the rat testis: evidence for stage-dependent expression and regulation by androgens. *Endocrinology* 1994 135: 1227–1234.
9. Roijen van JH, Assen van S, Kwast van der T *et al.* Androgen receptor immunoexpression in the testes of subfertile men. *J Androl* 1995 16: 510–516.
10. Simoni M, Gromoll J, Nieschlag E. The follicle-stimulating hormone receptor: biochemistry, molecular biology, physiology, and pathophysiology. *Endocr Rev* 1997 18: 739–773.
11. Sutton KA, Maiti S, Tribley WA *et al.* Androgen regulation of the *Pem* homeodomain gene in mice and rat Sertoli and epididymal cells. *J Androl* 1998 19: 21–30.
12. Skinner MK. Cell–cell interactions in the testis. *Endocr Rev* 1991 12: 45–77.
13. Slegtenhorst-Eegdeman KE, Post M, Baarends WM, Themmen AP, Grootegoed JA. Regulation of gene expression in Sertoli cells by follicle-stimulating hormone (FSH): cloning and characterization of *LRPR1*, a primary response gene encoding a leucine-rich protein. *Mol Cell Endocrinol* 1995 108: 115–124.
14. Kumar TR, Wang Y, Naifang L, Matzuk MM. Follicle stimulating hormone is required for ovarian follicle maturation but not male fertility. *Nat Genet* 1997 15: 201–204.
15. Dierich A, Sairam MR, Monaco L *et al.* Impairing follicle-stimulating hormone (FSH) signaling *in vivo*: targeted disruption of the FSH receptor leads to aberrant gametogenesis and hormonal imbalance. *Proc Natl Acad Sci USA* 1998 95: 13612–13617.
16. Abel MH, Wootton AN, Wilkins V, Huhtaniemi I, Knight PG, Charlton HM. The effect of a null mutation in the follicle-stimulating hormone receptor gene on mouse reproduction. *Endocrinology* 2000 141: 1795–1803.
17. Themmen APN, Huhtaniemi IT. Mutations of gonadotropins and gonadotropin receptors: elucidating the physiology and pathophysiology of pituitary–gonadal function. *Endocr Rev* 2000 21: 551–583.
18. Sharpe RM. Regulation of spermatogenesis. In Knobil E, Neill JD, eds. *The Physiology of Reproduction*. New York: Raven Press, 1994: 1363–1434.
19. Wylie C. Germ cells. *Curr Opin Genet Dev* 2000 10: 410–413.
20. McLaren A, Southee D. Entry of mouse embryonic germ cells into meiosis. *Dev Biol* 1997 187: 107–113.
21. de Rooij DG, Grootegoed JA. Spermatogonial stem cells. *Curr Opin Cell Biol* 1998 10: 694–701.
22. Surani MA. Imprinting and the initiation of gene silencing in the germ line. *Cell* 1998 93: 309–312.
23. Kretser de DM, Kerr JB. The cytology of the testis. In Knobil E, Neill JD, eds. *The Physiology of Reproduction*. New York: Raven Press, 1994: 1177–1290.
24. Wright WW, Zabludoff SD, Penttila TL, Parvinen M. Germ cell–Sertoli cell interactions: regulation by germ cells of the stage-specific expression of CP-2/cathepsin L mRNA by Sertoli cells. *Dev Genet* 1995 16: 104–113.
25. Clermont Y. Kinetics of spermatogenesis in mammals: seminiferous epithelium cycle and spermatogonial renewal. *Physiol Rev* 1972 52: 198–236.
26. Meng X, Lindahl M, Hyvonen ME *et al.* Regulation of cell fate decision of undifferentiated spermatogonia by GDNF. *Science* 2000 287: 1489–1493.

27. Kanaar R, Hoeijmakers JH, van Gent DC. Molecular mechanisms of DNA double strand break repair. *Trends Cell Biol* 1998 8: 483–489.
28. Baarends WM, van der Laan R, Grootegoed JA. DNA repair mechanisms and gametogenesis. *Reproduction* 2001 121: 31–39.
29. Baudat F, Manova K, Yuen JP, Jasin M, Keeney S. Chromosome synapsis defects and sexually dimorphic meiotic progression in mice lacking *Spo11*. *Mol Cell* 2000 6: 989–998.
30. Romanienko PJ, Camerini-Otero RD. The mouse *Spo11* gene is required for meiotic chromosome synapsis. *Mol Cell* 2000 6: 975–987.
31. Keegan KS, Holtzman Da, Plug AW *et al.* The Atr and Atm protein kinases associate with different sites along meiotically pairing chromosomes. *Genes Dev* 1996 10: 2383–2388.
32. Grootegoed JA, Grolle-Hey AH, Rommerts FFG, Van der Molen HJ. Ribonucleic acid synthesis *in vitro* in primary spermatocytes isolated from rat testis. *Biochem J* 1977 168: 23–31.
33. Baker SM, Plug AW, Prolla TA *et al.* Involvement of mouse MLH1 in DNA mismatch repair and meiotic crossing over. *Nat Genet* 1996 13: 336–342.
34. Edelmann W, Cohen P, Kane M *et al.* Meiotic pachytene arrest in MLH1-deficient mice. *Cell* 1996 85: 1125–1134.
35. de Vries SS, Baart EB, Dekker M *et al.* Mouse MutS-like protein MSH5 is required for proper chromosome synapsis in male and female meiosis. *Genes Dev* 1999 13: 523–531.
36. Edelmann W, Cohen PE, Kneitz B *et al.* Mammalian MutS homologue 5 is required for chromosome pairing in meiosis. *Nat Genet* 1999 21: 123–127.
37. Kneitz B, Cohen PE, Avdievich E *et al.* MutS homolog 4 localization to meiotic chromosomes is required for chromosome pairing during meiosis in male and female mice. *Genes Dev* 2000 14: 1085–1097.
38. Heyting C. Synaptonemal complexes: structure and function. *Curr Opin Cell Biol* 1996 8: 389–396.
39. Yuan L, Liu JG, Zhao J, Brundell E, Daneholt B, Hoog C. The murine SCP3 gene is required for synaptonemal complex assembly, chromosome synapsis, and male fertility. *Mol Cell* 2000 5: 73–83.
40. Roeder GS, Bailis JM. The pachytene checkpoint. *Trends Genet* 2000 16: 395–403.
41. Johnson L. Efficiency of spermatogenesis. *Microsc Res Tech* 1995 32: 385–422.
42. Martin-du Pan RC, Campana A. Physiopathology of spermatogenic arrest. *Fertil Steril* 1993 60: 937–946.
43. Paquis-Flucklinger V, Santucci-Darmanin S, Paul R, Saunieres A, Turc-Carel C, Desnuelle C. Cloning and expression analysis of a meiosis-specific MutS homolog: the human MSH4 gene. *Genomics* 1997 44: 188–194.
44. Bocker T, Barusevicius A, Snowden T *et al.* hMSH5: a human MutS homologue that forms a novel heterodimer with hMSH4 and is expressed during spermatogenesis. *Cancer Res* 1999 59: 816–822.
45. Palombi F, Salanova M, Tarone G, Farini D, Stefanini M. Distribution of β1 integrin subunit in rat seminiferous epithelium. *Biol Reprod* 1992 47: 1173–1182.
46. Johnson KJ, Patel SR, Boekelheide K. Multiple cadherin superfamily members with unique expression profiles are produced in rat testis. *Endocrinology* 2000 141: 675–683.
47. Blobel CP. Functional processing of fertilin: evidence for a critical role of proteolysis in sperm maturation and activation. *Rev Reprod* 2000 5: 75–83.
48. Kistler WS, Henriksen K, Mali P, Parvinen M. Sequential expression of nucleoproteins during rat spermiogenesis. *Exp Cell Res* 1996 225: 374–381.
49. Yu YE, Zhang Y, Unni E *et al.* Abnormal spermatogenesis and reduced fertility in transition nuclear protein 1-deficient mice. *Proc Natl Acad Sci USA* 2000 97: 4683–4688.
50. Gatewood JM, Cook GR, Balhorn R, Bradbury EM, Schmid CW. Sequence specific packaging of DNA in human sperm chromatin. *Science* 1987 236: 962–964.
51. Gatewood JM, Cook GR, Balhorn R, Schmid CW, Bradbury EM. Isolation of four core histones from human sperm chromatin representing a minor subset of somatic histones. *J Biol Chem* 1990 265: 20662–20666.
52. Roijen van JH, Ooms MP, Spaargaren M *et al.* Immunoexpression of testis-specific histone 2B (TH2B) in human spermatozoa and testis tissue. *Hum Reprod* 1998 13: 1559–1566.

53. Bunch DO, Welch JE, Magyar PL, Eddy EM, O'Brien DA. Glyceraldehyde 3-phosphate dehydrogenase-S protein distribution during mouse spermatogenesis. *Biol Reprod* 1998 58: 834–841.
54. Welch JE, Brown PL, O'Brien DA *et al.* Human glyceraldehyde 3-phosphate dehydrogenase-2 gene is expressed specifically in spermatogenic cells. *J Androl* 2000 21: 328–338.
55. Vidaeus CM, von Kapp-Herr C, Golden WL, Eddy RL, Shows TB, Herr JC. Human fertilin beta: identification, characterization, and chromosomal mapping of an ADAM gene family member. *Mol Reprod Dev* 1997 46: 363–369.
56. Cho C, Bunch DO, Faure JE *et al.* Fertilization defects in sperm from mice lacking fertilin beta. *Science* 1998 281: 1857–1859.
57. Primakoff P, Myles DG. The ADAM gene family: surface proteins with adhesion and protease activity. *Trends Genet* 2000 16: 83–87.
58. Wassarman PM, Jovine L, Litscher ES. A profile of fertilization in mammals. *Nat Cell Biol* 2001 3: E59–E64.
59. Allen JW, Dix DJ, Collins BW *et al.* HSP70-2 is part of the synaptonemal complex in mouse and hamster spermatocytes. *Chromosoma* 1996 104: 414–421.
60. Dix DJ, Allen JW, Collins BW *et al.* HSP70-2 is required for desynapsis of synaptonemal complexes during meiotic prophase in juvenile and adult mouse spermatocytes. *Development* 1997 124: 4595–4603.
61. Eddy EM. Role of heat shock protein HSP70-2 in spermatogenesis. *Rev Reprod* 1999 4: 23–30.
62. Li S, Zhou W, Doglio L, Goldberg E. Transgenic mice demonstrate a testis-specific promoter for lactate dehydrogenase, LDHC. *J Biol Chem* 1998 273: 31191–31194.
63. Meistrich ML, Bucci LR, Trostle-Weige PK, Brock WA. Histone variants in rat spermatogonia and primary spermatocytes. *Dev Biol* 1985 112: 230–240.
64. Lin Q, Sirotkin A, Skoultchi AI. Normal spermatogenesis in mice lacking the testis-specific linker histone H1t. *Mol Cell Biol* 2000 20: 2122–2128.
65. Drabent B, Saftig P, Bode C, Doenecke D. Spermatogenesis proceeds normally in mice without linker histone H1t. *Histochem Cell Biol* 2000 113: 433–442.
66. Lander ES, Linton LM, Birren B *et al.* Initial sequencing and analysis of the human genome. *Nature* 2001 409: 860–921.
67. Venter JC, Adams MD, Myers EW *et al.* The sequence of the human genome. *Science* 2001 291: 1304–1351.
68. Sun Z, Sassone-Corsi P, Means AR. Calspermin gene transcription is regulated by two cyclic AMP response elements contained in an alternative promoter in the calmodulin kinase IV gene. *Mol Cell Biol* 1995 15: 561–571.
69. Monesi V. Differential rate of ribonucleic acid synthesis in the autosomes and sex chromosomes during male meiosis in the mouse. *Chromosoma* 1965 17: 11–21.
70. Lifschytz E, Lindsley DL. The role of X chromosome inactivation during spermatogenesis. *Proc Natl Acad Sci USA* 1972 69: 182–186.
71. McKee BD, Handel MA. Sex chromosomes, recombination, and chromatin conformation. *Chromosoma* 1993 102: 71–80.
72. Hendriksen PJM, Hoogerbrugge JW, Themmen APN *et al.* Postmeiotic transcription of X and Y chromosomal genes during spermatogenesis in the mouse. *Dev Biol* 1995 170: 730–733.
73. Braun RE, Behringer RR, Peschon JJ, Brinster RL, Palmiter RD. Genetically haploid spermatids are phenotypically diploid. *Nature* 1989 337: 373–376.
74. Boer PH, Adra CN, Lau YF, McBurney MW. The testis-specific phophoglycerate kinase gene *pgk-2* is a recruited retroposon. *Mol Cell Biol* 1987 7: 3107–3112.
75. McCarrey JR, Thomas K. Human testis-specific PGK gene lacks introns and possesses characteristics of a processed gene. *Nature* 1987 326: 501–505.
76. Robinson MO, McCarrey JR, Simon MI. Transcriptional regulatory regions of testis-specific PGK2 defined in transgenic mice. *Proc Natl Acad Sci USA* 1989 86: 8437–8441.
77. Dahl H-H, Brown RM, Hutchison WM, Maragos C, Brown GK. A testis-specific variant of the mouse pyruvate dehydrogenase E1α subunit. *Genomics* 1990 8: 225–232.
78. Eddy EM. 'Chauvinist genes' of male germ cells: gene expression during mouse spermatogenesis. *Reprod Fertil Dev* 1995 7: 695–704.
79. Hendriksen PJM, Hoogerbrugge JW, Baarends WM *et al.* Testis-specific expression of a

functional retroposon encoding glucose-6-phosphate dehydrogenase in the mouse. *Genomics* 1997 41: 350–359.

80. Hart PE, Glantz JN, Orth JD, Poynter GM, Salisbury JL. Testis-specific murine centrin, Cetn1: genomic characterization and evidence for retroposition of a gene encoding a centrosome protein. *Genomics* 1999 60: 111–120.

81. Sedlacek Z, Munstermann E, Dhorne-Pollet S *et al.* Human and mouse XAP-5 and XAP-5-like (X5L) genes: identification of an ancient functional retroposon differentially expressed in testis. *Genomics* 1999 61: 125–132.

82. Kleene KC, Mastrangelo MA. The promoter of the Poly(A) binding protein 2 (Pabp2) retroposon is derived from the 5'-untranslated region of the Pabp1 progenitor gene. *Genomics* 1999 61: 194–200.

83. Reddi PP, Flickinger CJ, Herr JC. Round spermatid-specific transcription of the mouse *SP-10* gene is mediated by a 294-base pair proximal promoter. *Biol Reprod* 1999 61: 1256–1266.

84. Xu W, Cooper GM. Identification of a candidate c-*mos* repressor that restricts transcription of germ cell-specific genes. *Mol Cell Biol* 1995 15: 5369–5375.

85. Kramer JA, McCarrey JR, Djakiew D, Krawetz SA. Differentiation: the selective potentiation of chromatin domains. *Development* 1998 125: 4749–4755.

86. Wolgemuth DJ, Viviano CM, Gizang-Ginsberg E, Frohman MA, Joyner AL, Martin GR. Differential expression of the mouse homeobox-containing gene *Hox-1.4* during male germ cell differentiation and embryonic development. *Proc Natl Acad Sci USA* 1987 84: 5813–5817.

87. Li Y, Lemaire P, Behringer R. *Esx1*, a novel X chromosome-linked homeobox gene expressed in mouse extraembryonic tissues and male germ cells. *Dev Biol* 1997 188: 85–95.

88. Sassone-Corsi P. Transcriptional checkpoints determining the fate of male germ cells. *Cell* 1997 88: 163–166.

89. Sun Z, Means A. An intron facilitates activation of the calspermin gene by the testis-specific transcription factor CREM tau. *J Biol Chem* 1995 270: 20962–20967.

90. Zhou Y, Sun Z, Means AR, Sassone-Corsi P, Bernstein KE. cAMP-response element modulator tau is a positive regulator of testis angiotensin converting enzyme transcription. *Proc Natl Acad Sci USA* 1996 93: 12262–12266.

91. Blendy JA, Kaestner KH, Weinbauer GF, Nieschlag E, Schütz G. Severe impairment of spermatogenesis in mice lacking the *CREM* gene. *Nature* 1996 380: 162–165.

92. Nantel F, Monaco L, Foulkes NS *et al.* Spermiogenesis deficiency and germ-cell apoptosis in *CREM*-mutant mice. *Nature* 1996 380: 159–165.

93. Fimia GM, De Cesare D, Sassone-Corsi P. CBP-independent activation of CREM and CREB by the LIM-only protein ACT. *Nature* 1999 398: 165–169.

94. Hecht NB. The making of a spermatozoon: a molecular perspective. *Dev Genet* 1995 16: 95–103.

95. Kleene KC. Patterns of translational regulation in the mammalian testis. *Mol Reprod Dev* 1996 43: 268–281.

96. Fajardo MA, Haugen HS, Clegg CH, Braun RE. Separate elements in the 3' untranslated region of the mouse protamine 1 mRNA regulate translational repression and activation during murine spermatogenesis. *Dev Biol* 1997 191: 42–52.

97. Wu XQ, Hecht NB. Mouse testis brain ribonucleic acid-binding protein/translin colocalizes with microtubules and is immunoprecipitated with messenger ribonucleic acids encoding myelin basic protein, alpha calmodulin kinase II, and protamines 1 and 2. *Biol Reprod* 2000 62: 720–725.

98. Lee K, Haugen HS, Clegg CH, Braun RE. Premature translation of protamine 1 mRNA causes precocious nuclear condensation and arrests spermatid differentiation in mice. *Proc Natl Acad Sci USA* 1995 92: 12451–12455.

99. Zhong J, Peters AH, Lee K, Braun RE. A double-stranded RNA binding protein required for activation of repressed messages in mammalian germ cells. *Nat Genet* 1999 22: 171–174.

100. Venables J, Eperon I. The roles of RNA-binding proteins in spermatogenesis and male infertility. *Curr Opin Genet Dev* 1999 9: 346–354.

101. Cooke HJ. Y chromosome and male infertility. *Rev Reprod* 1999 4: 5–10.

102. Ruggiu M, Speed R, Taggart M *et al.* The mouse *Dazla* gene encodes a cytoplasmic protein essential for gametogenesis. *Nature* 1997 389: 73–77.

103. Ciechanover A, Orian A, Schwartz AL. Ubiquitin-mediated proteolysis: biological regulation via destruction. *BioEssays* 2000 22: 442–451.

104. Sun C, Skaletsky H, Birren B *et al.* An azoospermic man with a *de novo* point mutation in the Y-chromosomal gene *USP9Y*. *Nat Genet* 1999 23: 429–432.

105. Roest HP, Klaveren van J, Wit de J *et al.* Inactivation of the HR6B ubiquitin-conjugating DNA repair enzyme in mice causes a defect in spermatogenesis associated with chromatin modification. *Cell* 1996 86: 799–810.

106. Lawrence C. The *RAD6* repair pathway in *Saccharomyces cerevisiae*: what does it do, and how does it do it? *BioEssays* 1994 16: 253–258.

107. Koken MH, Reynolds P, Jaspers-Dekker I *et al.* Structural and functional conservation of two human homologs of the yeast DNA repair gene *RAD6*. *Proc Natl Acad Sci USA* 1991 88: 8865–8869.

108. Unni E, Mayerhofer A, Zhang Y, Bhatnagar YM, Russell LD, Meistrich MK. Increased accessibility of the N-terminus of testis-specific histone TH2B to antibodies in elongating spermatids. *Mol Reprod Dev* 1995 42: 210–219.

109. Grootegoed JA, Baarends WM, Roest HP, Hoeijmakers JH. Knockout mouse models and gametogenic failure. *Mol Cell Endocrinol* 1998 145: 161–166.

110. Baarends WM, Hoogerbrugge JW, Roest HP *et al.* Histone ubiquitination and chromatin remodeling in mouse spermatogenesis. *Dev Biol* 1999 207: 322–333.

111. Robzyk K, Recht J, Osley MA. RAD6-dependent ubiquitination of histone H2B in yeast. *Science* 2000 287: 501–504.

112. Baarends WM, van der Laan R, Grootegoed JA. Specific aspects of the ubiquitin system in spermatogenesis. *J Endocrinol Invest* 2000 23: 597–604.

113. Adams MD, Celniker SE, Holt RA *et al.* The genome sequence of *Drosophila melanogaster*. *Science* 2000 287: 2185–2195.

114. Chu S, DeRisi J, Eisen M *et al.* The transcriptional program of sporulation in budding yeast. *Science* 1998 282: 699–705.

115. Primig M, Williams RM, Winzeler EA *et al.* The core meiotic transcriptome in budding yeasts. *Nat Genet* 2000 26: 415–423.

116. Reinke V, Smith HE, Nance J *et al.* A global profile of germline gene expression in *C. elegans*. *Mol Cell* 2000 6: 605–616.

117. Andrews J, Bouffard GG, Cheadle C, Lu J, Becker KG, Oliver B. Gene discovery using computational and microarray analysis of transcription in the *Drosophila melanogaster* testis. *Genome Res* 2000 10: 2030–2043.

Chapter 22

Fertilization

Laura Hewitson, Calvin Simerly and Gerald Schatten

INTRODUCTION

Fertilization is the process that culminates in the union of one, and only one, sperm nucleus with the egg nucleus within the activated egg cytoplasm. For it to occur successfully, several events must occur:

(1) Sperm preparation;

(2) Incorporation of the sperm into the egg cytoplasm, which includes the binding between, and fusion of, plasma membranes of the egg and the sperm;

(3) Egg activation and the initiation of the first cell cycle;

(4) Formation of the sperm and egg nuclei (male and female pronuclei, respectively);

(5) Migrations of the pronuclei that lead to genomic union;

(6) Initiation of first division and early development.

While an understanding of the molecular and cell biological basis of fertilization is central in creating a complete body of knowledge regarding reproduction, the study of fertilization itself has served as an intellectual foundation on which the entire field of cell biology has been constructed. With *in vitro* fertilization of mammals, including humans, now routine, it has been possible to solve many of the critically important problems of fertilization at the molecular and cell biological levels. However, with the recent introduction of novel forms of assisted reproduction, we are now faced with new challenges. This chapter focuses on the frontiers of knowledge regarding the molecular and cell biological basis of fertilization, and its relevance to clinical reproductive medicine.

BACKGROUND: THE CELLULAR AND MOLECULAR BASIS OF FERTILIZATION

The deliberate creation of human zygotes and embryos for scientific research is complicated by ethical, political, financial and practical issues. Nevertheless, a detailed understanding of the cellular and molecular events during human fertilization is essential for the field of developmental biology, as well as for clinical applications including infertility treatments, contraception and the avoidance of developmental abnormalities. The motility and cytoskeletal rearrangements essential for the successful union of the sperm and egg nuclei during fertilization are poorly understood in primates, including humans. The vast bulk of the knowledge in mammals rests on murine models, and it is not clear whether this information can be extrapolated to fertilization in non-human and human primates. In order to address this key step in primate fertilization and to avoid the complexities in working with fertilized human zygotes, studies are now exploring the molecular foundations of fertilization using oocytes from a non-human primate, the rhesus monkey.

APPLIED TECHNOLOGY

The fields of cell, developmental, molecular and reproductive biology and genetics emerged from studies on eggs at fertilization (reviewed in reference 1). These early studies relied on systems such as those of frogs and sea urchins, in which fertilization *in vitro* and embryo culture occurred using simple solutions like sea water or pond water at room temperature. The recent design of methods for routinely and reliably obtaining *in vitro* fertilization of many mammals now permits detailed experimentation on molecular and structural features of development in mammals. These investigations have led to many important and unexpected basic discoveries, including genomic imprinting,[2–5] gametic recognition involving unique receptors and galactosyltransferases,[6–8] atypical maternal inheritance pattern of the centrosome in mice (reviewed in reference 9) and both paternal and maternal inheritance of mitochondria,[10] and unexpected signal-transduction pathways for fertilization and cell cycle regulation.[11,12]

The justification for performing primate studies using rhesus material is quite strong. Hormonal regulation in the rhesus monkey for gamete production closely parallels that of humans. In addition, the cytoskeletal arrangements observed during meiotic maturation and fertilization in rhesus oocytes closely mirror events observed in humans.[13,14] Rhesus material also avoids the complexities of working with fertilized human zygotes. Finally, there is a growing and accurate literature (reviewed in reference 15) and, for fertilization researchers, the clarity and malleability of the rhesus oocyte is excellent. Notwithstanding these virtues, however, all mammalian oocytes and embryos provide unusual challenges for the researcher interested in detecting cytoplasmic or nuclear structural features: they are large cells (80–150 μm), some oocytes are nearly opaque (e.g. bovine, canine), and the extracellular zona pellucida and cumulus cells can cause problems in introducing the imaging probe (e.g. antibody, calcium-sensitive dye) and in its removal for later intracellular detection.

The methods used in the detection of cytoskeletal and nuclear architectural structures in primate oocytes during fertilization have been previously described (Figures 1 and 2).[13,14,16–18] Immunocytochemical investigations of microtubule organization and DNA configurations after both *in vitro* fertilization and intracytoplasmic sperm injection (ICSI) have been carried out using both conventional and laser-scanning confocal microscopy.[14,16–17] These studies demonstrated that the rhesus follows a paternal centrosome inheritance pattern, as found for all mammals except rodents (reviewed in reference 9). Microtubules assembled from the introduced sperm centrosome were found to be essential for pronuclear apposition during primate fertilization. The results from ICSI fertilization events in the rhesus further demonstrated how similar the monkey model is to the human, suggesting that this non-human primate model will be valuable for exploring sperm nuclear remodelling following injection into the cytoplasm. In the course of these rhesus investigations, new methods for inducing parthenogenetic activation in monkey oocytes were developed, and a better understanding of the mechanisms of these activation protocols, through calcium imaging of fertilization and activation, can be reliably demonstrated.[14]

In addition to cytoskeletal investigations, microinjecting probes to detect rarer structural components such as γ-tubulin in the centrosome,[19] and for the investigation of the function of particular proteins, for example motors in the kinetochore,[20] can be applied in primates. As well as investigations by static immunocytochemistry, the detection of cytoskeletal dynamics in living oocytes can be approached using fluorescence recovery after photobleaching (FRAP)[21] or fluorescence activated cytochemistry. Finally, calcium ion imaging using both conventional and confocal microscopy can be artfully used to explore the role of cations during oocyte activation.[14,22,23]

It is well known that an increase in cytosolic calcium occurs at fertilization in several species, including the sea urchin, *Xenopus*, mouse, hamster, rabbit, bovine and human, suggesting that calcium plays a central role in all of fertilization (reviewed in reference 24). Methods used to explore cytoskeletal regulation by intracellular calcium ion imaging and the role calcium plays in triggering meiotic resumption and developmental progression in the rhesus have recently been developed.[14] Research characterizing the pattern of calcium dynamics during *in vitro* insemination and parthenogenetic activation in rhesus oocytes indicates that calcium dynamics during insemination in monkeys is similar to that reported recently for the human,[23] suggesting that the rhesus may be an excellent model for understanding events during human fertilization.

CURRENT CONCEPTS

Sperm activity

The process of fertilization is initiated upon recognition of the sperm that is in the vicinity of the egg.[7] This cellular recognition event leads to the sperm acrosome reaction that results in the externalization of the contents of the acrosomal vesicle. These contents include hydrolytic enzymes that permit the sperm to penetrate the outer investments of the egg (the zona pellucida in mammals, the egg jelly in most other species), as well as proteins and glycoproteins that bind the sperm to the egg, usually in a species-specific manner (e.g.

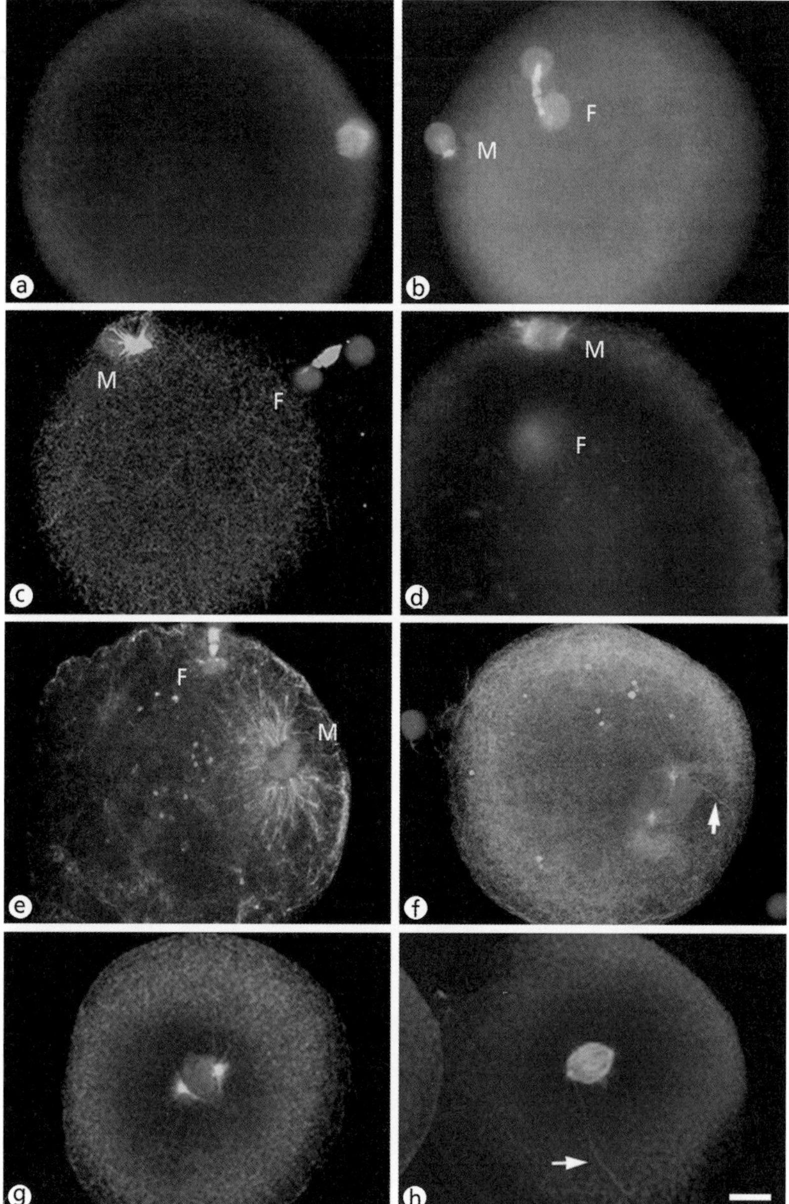

Figure 1 Microtubule and DNA organization during human fertilization. (a) The meiotic spindle in mature, unfertilized human oocytes is anastral, oriented radially to the cell surface, and asymmetric, with a focused pole abutting the cortex and a broader pole facing the cytoplasm. No other microtubules are detected in the cytoplasm of the unfertilized human oocyte. (b–d) Shortly after sperm incorporation (3–6.5 h post-insemination), sperm astral microtubules assemble around the base of the sperm head, as the inseminated oocytes complete second meiosis and extrude the second polar body. The close association of the meiotic mid-body identifies the female pronucleus. Short, sparse, disarrayed cytoplasmic microtubules can also be observed in the cytoplasm following confocal microscopic observations of these early activated oocytes (c). (e) As the male pronucleus continues to decondense in the cytoplasm, the microtubules of the sperm aster enlarge, circumscribing the male pronucleus. (f) By 15 h post-insemination, the centrosome splits and organizes a bipolar microtubule array that emanates from the tightly apposed pronuclei. The sperm tail is associated with an aster (arrow). (g) At first mitotic prophase (16.5 h post-insemination), the male and female chromosomes condense separately as a bipolar array of microtubules marks the developing first mitotic spindle poles. (h) By prometaphase, when the chromosomes begin to intermix on the metaphase equator, a barrel shaped, anastral spindle forms in the cytoplasm. The sperm axoneme (arrow) remains associated with a small aster found at one of the spindle poles. M, male pronucleus; F, female pronucleus; bar, 10 μm. Reproduced with permission from reference 13

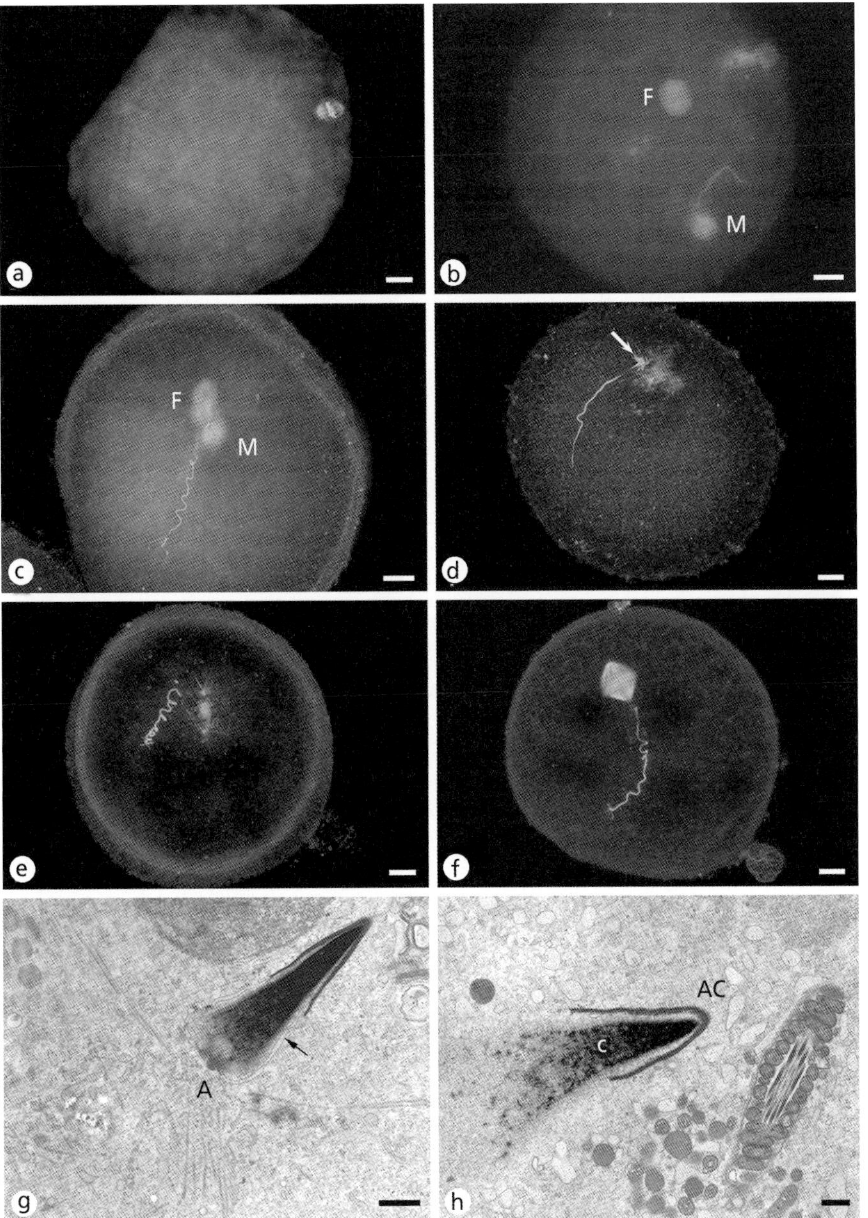

Figure 2 Conventional (a and b) and laser-scanning (c–f) fluorescence microscopy and transmission electron microscopy (g and h) of rhesus oocytes following fertilization by intracytoplasmic sperm injection (ICSI). (a) The second meiotic spindle of a sham-injected oocyte shows no apparent damage. (b) Following sperm injection, microtubules assemble close to the sperm head, which begins decondensation. (c) As the male and female pronuclei decondense, the microtubules elongate to fill the entire cytoplasm. (d) By prophase, most of the cytoplasmic microtubules disassemble so that just a small aster remains associated with the adjacent male and female pronuclei. (e) By prometaphase, microtubules form a bipolar structure opposite the duplicated and split centrosomes. (f) A fusiformed anastral spindle slightly eccentric within the cytoplasm forms during metaphase as the condensed chromosomes become aligned at the equator. (g) The residual acrosome causes irregular decondensation of sperm chromatin during the early stages of sperm disassembly. (h) A large aster of microtubules develops around the proximal centriole of the injected sperm despite the asynchronous sperm nuclear decondensation. Note that a new double membrane is being assembled around the base and equatorial region of this sperm. M, male; F, female; AC, acrosome; C, chromatin; A, aster; bars, a–f, 10 μm; bars g and h, 500 nm. Modified with permission from reference 16

bindin in sea urchins, several proteins in mammals including galactosyltransferase and fertilin). Capacitation is the event prerequisite to development of the ability to undergo the acrosome reaction in mammals.

The swimming of the sperm is also influenced by the egg. Perhaps the most extreme case is found in systems such as that of the horseshoe crab, *Limulus*, in which the sperm is immotile until it senses that it is in the vicinity of the egg. In most other systems, egg proteins and glycoproteins affect sperm motility. In mammals, the sperm swims in long curvilinear tracks, probably to span the distances within the female reproductive tract, until it nears the egg, when the faster 'homing' pattern known as hyperactivation occurs. The figure-of-eight trajectory during hyperactivation is thought to constitute part of the sperm's precise location of the oocyte.

As the sperm is perhaps the most highly differentiated cell, its components are reduced to the barest minimum. In addition to the loss of most cytoplasmic organelles, except for the mitochondria essential for swimming, spermatogenesis results in the segregation and packaging of the haploid sperm nucleus into the compact, inactive nucleus with its lean, nuclear envelope largely devoid of nuclear pore complexes and modified nuclear lamins. In addition, the sperm centrosome (the cell's microtubule organizing center) is reduced during spermatogenesis to that of a 'precentrosome', capable of recruiting from maternal stores the essential gene products necessary for mitotic spindle pole formation during fertilization and early development (reviewed in reference 9).

From the egg's perspective, the sperm is essential for contributing three critical components:

(1) The paternal haploid genome;

(2) The signal to initiate the metabolic activation of the egg;

(3) The centrosome, which directs microtubule assembly within the inseminated egg and leads to the union of the sperm and egg nuclei in the activated egg cytoplasm, as well as the formation of the mitotic spindles during development (reviewed in reference 9).

Sperm incorporation

The physical incorporation of the sperm into the egg cytoplasm occurs simultaneously with egg activation and the cortical reaction. In humans, the sperm rests tangentially against the egg surface as it enters. Microfilaments, polymers assembled from maternal actin, assemble at the oocyte's cortex overlying the decondensing sperm nucleus to form the 'incorporation cone'. Evidence that microfilaments are vital for sperm incorporation includes correlation with the new microfilament assembly specifically by the entering sperm, as well as inhibition of sperm tail incorporation by drugs that prevent egg microfilament assembly (reviewed in reference 25).

Metabolic activation of the egg

Oocytes and eggs are metabolically quiescent, and sperm–egg binding and fusion initiate a cascade of events that transforms the dormant egg into the dynamic, animated zygote. While there are still diverse opinions on the precise manner in which the sperm activates this cascade, it is clear in all fertilization systems that an elevation in intracellular calcium ion concentration is the central messenger in communicating the activating signal (reviewed in references 24 and 26). Sperm incorporation into an egg in which the calcium increase is prevented does not result in activation, whereas an artificial increase of intracellular calcium ion concentration leads to activation and parthenogenetic development.

At the beginning of the 20th century, the pioneering fertilization researcher Jacques Loeb asked the incredibly chauvinistic question: 'How does the sperm save the life of the egg?' Strong evidence is provided by two, often disparate, teams of researchers (reviewed in reference 11). On the one hand, the egg possesses receptors for sperm glycoproteins in its membrane, and compelling data indicate that the sperm might well behave as an honorary hormone or neurotransmitter, activating the egg through well-understood signal-transduction pathways. In this model, binding of the sperm to the receptor in the egg alters the receptor so that it might trigger the G-protein cascade or activate tyrosine kinase pathways that then lead to the activation of phospholipase C and the production of inositol triphosphate, which mobilizes internal calcium. In many eggs, the calcium essential for activation is internally sequestered, while, in others, the opening of plasma membrane calcium channels leads to the influx of external calcium.

Alternatively, the sperm may activate the egg not by binding and modifying receptors in the egg membrane, but by direct injection into the egg cytoplasm of a sperm-derived protein or factor referred to as 'oscillin'.[27,28] Likely candidates have been isolated from several species of sperm, and some stimulate egg activation across species lines.[29–31] This theory helps to explain the clinical success of ICSI, in which the sperm is introduced directly into the human egg cytoplasm, and it also avoids the danger of a sperm that is incapable of being incorporated activating the egg by binding to a receptor.

Still other researchers are convinced that fertilization might have evolved multiple overlapping pathways to ensure the success of egg activation, and that both the receptor pathway and the oscillin protein results are accurate and valid.

All eggs are arrested prior to sperm incorporation, but the precise stage during meiosis varies from species to species. In vertebrates, the eggs of most species arrest at metaphase of second meiosis, while others arrest at other stages of egg maturation. Sea urchins are among the rare groups in which the eggs complete meiotic maturation and arrest with a haploid female pronucleus (egg nucleus). Because most eggs are actually at arrested stages of meiosis, they are usually referred to as 'oocytes.' Cell cycle arrest in vertebrate eggs is due to the presence of a 'cytostatic factor', which is known to involve the c-*mos* gene product as well as a mitogen-activated protein (MAP) kinase cascade. The elevation in intracellular calcium at sperm incorporation destroys the cytostatic factor and leads to the destruction of the cyclin portion of the 'M-phase promoting factor'. With the biodegradation of cyclin, the metaphase-II arrested oocyte is capable of entering anaphase and the first cell cycle.

The molecular mechanism permitting fusion between the plasma membranes of the sperm and the egg is not well understood.[7,8] Sperm proteins such as fertilin[32,33] are critical for the adhesion of the sperm to the egg, usually in a species-specific fashion. While theories have been advanced that these or similar proteins might mediate fusion, this conclusion awaits further investigation.

Block(s) to polyspermy

Fertilization is defined as the fusion of one, and only one, sperm nucleus with the egg nucleus. Different eggs avoid the pathological condition of polyspermy in various ways, although most rely on the secretion of cortical granules to modify the egg's extracellular matrix to prevent the incorporation of supernumerary sperm (reviewed in reference 34). Secretion of the cortical granules (the 'cortical reaction') prevents polyspermy by proteolytically digesting the sites for sperm binding, and by also enzymatically modifying the egg's coat to toughen and harden it. In mammals, this results in modification of the zona pellucida, while, in many lower systems, the vitelline layer closely adherent to the egg's plasma membrane is elevated and laminated to form the 'fertilization coat' after the cortical reaction.

Perhaps the best understood fast block to polyspermy is the rapid, electrical block of many invertebrates and some vertebrates.[35] In this system, the successful sperm initiates the opening of ion channels (such as Na^+ channels in sea urchins), resulting in reversal of the membrane potential that prevents additional sperm fusion until the time expected for establishment of the cortical reaction blockage.

Pronucleus formation and, at times, nuclear fusion

Since the sperm enters the egg with a highly condensed, inactive nucleus, and also since most eggs are actually arrested oocytes with condensed meiotic chromosomes, fertilization also leads to the decondensation and remodelling of the parental genomes (reviewed in reference 36). In many systems, the sperm nucleus is modified during spermiogenesis so that the typical histone chromosome proteins are reduced by protamines, leading to a further tighter packaging of the DNA from the usual nucleosome pattern (reviewed in reference 37). In all systems, the paternal DNA binds maternal histones and other proteins, and is invested with a new nuclear envelope of maternal origin containing nuclear lamins and nuclear pore complexes. The completion of fertilization is signalled by genomic union. In most nonmammalian species, the decondensed male and female pronuclei migrate into close apposition, and then the two nuclear membranes circumscribing each pronucleus fuse. This results in formation of the diploid 'zygote nucleus'. While data are limited on the molecular mechanisms involved in plasma membrane fusion, they are virtually completely missing in the case of nuclear fusion. In mammals, the pronuclei do not fuse during interphase; instead, the adjacent pronuclei undergo separate nuclear breakdowns at first mitosis, and the parental chromosomes intermix as they align at the metaphase plate of the first mitotic spindle. Consequently, the events of mammalian fertilization start during second meiosis and are not complete until first mitosis.

Pronuclear migrations and genomic union

Following successful incorporation of the sperm into the egg and subsequent egg activation, and coincident with pronucleus formation, a new motility system assembles within the egg, and its organization is directed by the newly introduced sperm centrosome (reviewed in reference 9). The microtubule-based 'sperm aster' is a radially arrayed three-dimensional structure organized by the sperm centrosome, found adjacent to and affixed to the sperm nucleus. The sperm aster has three essential roles. First, its growth subjacent to the egg cortex pushes the sperm nucleus towards the egg center. Next, when the distal ends of the dynamic microtubules contact the egg nucleus, it undergoes a sudden and swift translocation to the center of the sperm aster where it meets the sperm nucleus. Finally, the continuing elongation of the sperm astral microtubules moves the now adjacent pronuclei towards the egg center.

The role of the sperm aster during fertilization cannot be overestimated, and certain forms of fertilization failure appear to be due to defects in the organization or functioning of the sperm aster.[13,18,38] It is organized by the paternally inherited sperm centrosome, the structure of which is not yet well understood. The sperm centrosome (Figure 3) attracts and binds numerous maternal proteins, especially γ-tubulin.[39] These transform the inactive precursor sperm centrosome into the microtubule-nucleating, reproducing 'zygote centrosome'. The paternal inheritance of the centrosome occurs in all systems, including humans. Rodents and parthenogenetic systems appear to be the only exception and they rely on a maternally inherited centrosome (reviewed in reference 9).

Following the migration and union of the pronuclei, the centrosome duplicates and separates to become the future two poles for the first mitotic spindle. The DNA undergoes replication during interphase, and a new cascade of cell cycle-mediating events leads to the initiation of first division and early development.

POTENTIAL CLINICAL RELEVANCE

Fertilization in humans is the process that culminates when the parental genomes unite within the activated oocyte. For this to occur successfully, a series of events must accurately unfold. Failure of these events at any stage is lethal for the zygote, and may well be the source of undiagnosed types of human infertility.[13] While sperm behavior and the initial encounters of the gametes have been studied in humans, the motility of the sperm and egg nuclei (male and female pronuclei, respectively) within the oocyte, which leads to their union, is less well understood. The role of microtubules and other structural and motility arrays within the inseminated human oocyte during union of the pronuclei and the first cell cycle is vital.

An irony with regard to ICSI is that all the currently employed sperm analysis assays could now be viewed as obsolete (reviewed in reference 40).

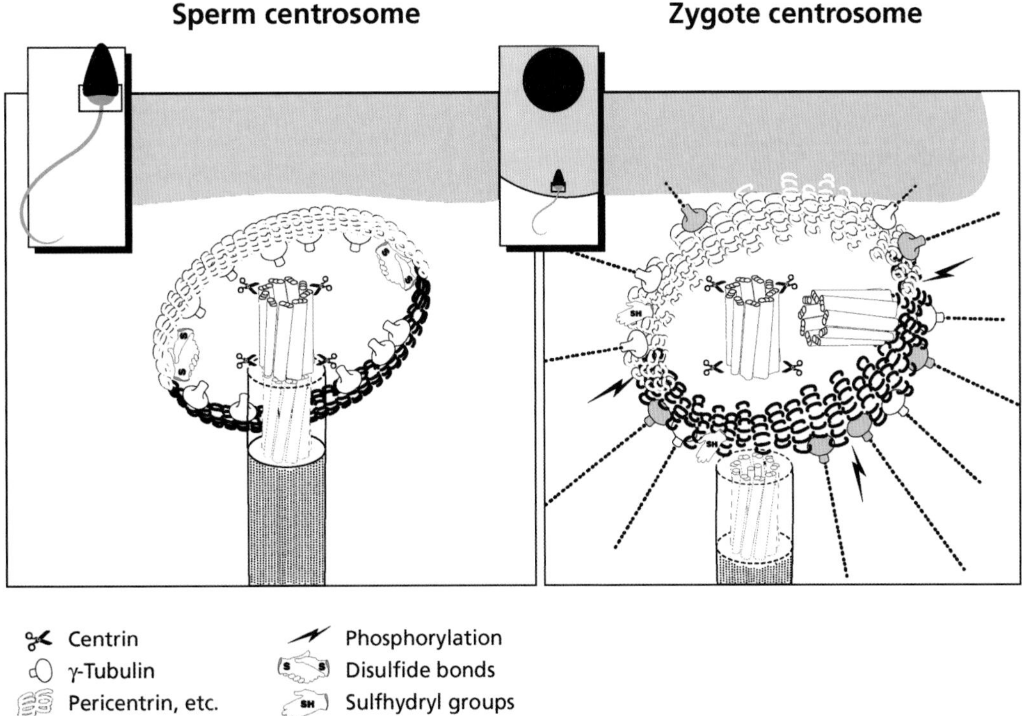

Figure 3 Molecular dissection of the human sperm centrosome and its reconstruction in the zygote. The sperm centrosome (left): the human sperm centrosome has centrin concentrated in one or two focal sites, corresponding to the centrioles. γ-Tubulin is not apparent in mature human sperm, but becomes detectable after 'centrosomal priming' of the sperm with disulfide reducing agents; this is a novel type of cytoplasmic capacitation. γ-Tubulin is also detectable on Western blots with intact or sonicated human sperm. The centrosome is not phosphorylated and the sperm tail microtubules extend from a centriole. The coiled-coil infrastructure of the centrosome probably anchors the centrosome to the sperm nucleus and regulates the exposure of, and binding sites for, γ-tubulin. The zygote centrosome (right): after permeabilization and incubation in extracts from *Xenopus* oocytes, the human sperm becomes phosphorylated and heavily immunoreactive with antibodies to γ-tubulin. The γ-tubulin found in the human sperm is probably a combination of some paternal and largely maternal protein. The binding of calcium ions, released during the transient increase during egg activation, to centrin is predicted to result later in a centrin-induced severing of the doublet sperm tail microtubules from the triplet microtubules of the centriole. Perhaps the severing of the tail microtubules from the basal body frees the basal body complex so that it can bind additional γ-tubulin and undergo transformation into a centriole. In humans, calcium-mediated centrin excision does not lead to complete sperm tail dissociation from the centriole–centrosome complex. The coiled-coil domains of the centrosome are drawn as unravelling, expanding and everting in the zygote; this exposes paternal γ-tubulin and also exposes binding sites for maternal γ-tubulin. The halo of γ-tubulin nucleates the microtubules that assemble into the sperm aster. Reproduced with permission from reference 9

This is because conventional semen analysis parameters and function tests are designed to investigate sperm populations, but the results do not necessarily reflect the properties of a single sperm selected for ICSI. The number of sperm produced no longer matters if fertilization is successful when a single sperm is aspirated and injected into the egg. Sperm motility is irrelevant, because the sperm is moved by the biologist, not by its own locomotion. Tests that measure the ability of the sperm to penetrate the egg's outer investments ('the hemizona assay' and the sperm penetration assay using hamster oocytes)[41] are now largely irrelevant.[42,43] Even the proper shape of the sperm is not necessarily a predictor of a successful outcome of fertilization.[44] A new generation of sperm assays that explore the interaction of the sperm within the egg cytoplasm have now been proposed,[9,13,45] and these may become relevant in the ICSI era.

FUTURE PERSPECTIVES

The general perception of the manner in which innovative assisted reproductive technologies (ART) are introduced is the initiation of a carefully controlled series of experiments in an animal model, such as the mouse. Only after the principle has been proven can one consider confirmatory studies in mammals closely related to humans, such as rhesus monkeys or other non-human primates. With this background of a peer-reviewed body of well-established published facts, there are now sufficient foundation and rationale to propose a clinical investigation to a responsible human subjects institutional review board (IRB). The IRB weighs the benefits and risks of new methods to human subjects, and then considers the appropriate informed consent procedures for this particular case in which the actual outcome of the experiments cannot be foreseen by the physicians and scientists. Only after a large number of clinical studies, performed at multiple sites and peer-reviewed, can the efficacy and safety of the innovative approach be clearly evaluated. At that time, the potential therapy can be responsibly offered to suitable beneficiaries.

In the case of the clinical introduction and global acceptance of ICSI and other innovative ART therapies, this sensible scenario did not occur, and, ironically, could not have occurred (reviewed in reference 40). The discovery and development of ICSI followed the precise opposite path, with the clinical treatment pioneering the approach. Notwithstanding its success, questions remain about the dangers of passing on traits responsible for male infertility,[46] sex and autosomal chromosome aberrations,[47–49] and mental, physical and reproductive normalcy[50,51] after ICSI. Fundamental research in relevant animal models still lags behind the clinical achievements, so biomedical researchers are only now beginning to understand the cellular and molecular events that permit successful fertilization by ICSI.

Why worry about ICSI and other ART approaches?

This argument could be advanced by those impressed with the many babies born following ICSI, who show nearly the same percentages of major and minor abnormalities as those found during natural fertilization or *in vitro* fertilization (IVF). These statistics, arguably, are even more impressive when considered in the light of the poor quality and quantities of sperm available for ICSI.[52–57] Perhaps the few reports of increased frequency of chromosomal anomalies[47–49,58,59] need to be confirmed, or do not raise cautionary warnings sufficient to undermine the most significant advance in the treatment of male infertility.

During the past two decades, the IVF field has contributed to the births of thousands of babies. The health of these children is within the boundaries of all the normal physiological and psychological parameters established through decades of neonatal surveillance.[60,61] Evaluation of the IVF statistics is complicated somewhat by the fact that these children are from families that tend to be socioeconomically privileged, and that the IVF offspring are only now approaching reproductive age. While, thankfully, all the available data suggest there are no apparent health risks for ICSI offspring, some problems may not become apparent for decades. For example, in some infertile men, errors in the DNA found in their testes cannot be repaired; this DNA-repair problem is also found in malignant tumor cells of some cancer patients. Since these men are unable to conceive naturally, their infertility might be nature's way of stopping the propagation of genetic defects.[62] However, with the use of ICSI for assisted fertilization, even men with the most severely defective sperm can become fathers. If the cause of the father's infertility lies in faulty DNA-repair genes, then the resulting child could not only be infertile but might be at an increased risk for cancer in later life. This unfortunately means that would-be parents, who have no alternative than to use ICSI if they want a genetically related child, will have to make their decision knowing that it could take many more years of research before there is sufficient scientific data to reassure them of their decision.

Variations that occur during ICSI and other ART approaches

Intracytoplasmic sperm injection differs from natural fertilization or IVF in several fundamental ways. These include the selection of successful sperm by the embryologist, rather than by normal competition among the hundreds of actively swimming sperm at the egg's surface. The selection and aspiration of sperm into the glass needle for microinjection also expose the sperm to non-physiological chemical and physical treatments. Chemical treatments sometimes include exposure to polyvinylpyrrolidone used to immobilize the sperm, as well as culture medium that may contain the potentially carcinogenic pH-indicator, phenol red. Physical treatments of the sperm include 'scoring', used to immobilize the sperm tail, so that it can be sucked into the microinjection needle. The light from the microscope is necessary for selection of the sperm, and the search for the best of a few sperm could involve elongated light exposure. Recent surprising results using a rhesus ICSI model for transgenesis demonstrated that exogenous DNA bound to the sperm's surface is retained and transferred into the egg during ICSI, but is lost during IVF.[63] Rhodamine-labelled DNA encoding the green fluorescence protein (GFP) gene binds avidly to sperm, and the rhodamine signal, while lost at the egg surface during IVF, remains as a brilliant marker on the microinjected sperm within the egg cytoplasm after ICSI. The transgene is expressed in preimplantation embryos produced by ICSI, but not IVF, as early as at the four-cell stage, and the percentage of expressing embryos and the number of expressing cells increase during embryogenesis to the blastocyst stage. While this technique may prove to be a powerful strategy for producing transgenic primates, the ability to transfer foreign DNA into oocytes during ICSI but not IVF raises the concern that ICSI could also transmit infectious material.

Prior to ICSI, the cumulus cells are removed from the oocyte by exposure to hyaluronidase and by mechanical stripping. The oocyte is then positioned for ICSI with the polar body at a right angle to the site of sperm injection, to avoid injecting the sperm too close to the spindle region, which is thought to reside close to the polar body. However, dynamic imaging of rhodamine-labelled spindle microtubules of donated human oocytes undergoing mechanical stripping demonstrated that the first polar body can be as much as 95° displaced from the spindle, compared with 20° for control oocytes.[64] Since the first polar body is not anchored firmly, mechanical manipulations such as those required for cumulus cell removal may result in its lateral displacement within the perivitelline space. Therefore, polar body positioning might be an unreliable indicator of the position of the meiotic spindle,[65] as has also been observed with hamster oocytes using polarization microscopy.[66]

The egg inseminated by ICSI undergoes a series of unusual events. These include physical penetration of the egg by the microinjection needle, resulting in significant deformation of the egg from a sphere into a biconcave disk. While the egg recovers from this deformation, microscopic blemishes on the egg's surface can be imaged by scanning electron microscopy for several hours post-injection.[67] Prior to injection of the sperm into the egg cytoplasm, suction is applied to the microneedle so that the membrane of the egg is breached and some egg cytoplasm is drawn into the needle. This ensures that the sperm will be injected into the egg proper, and not lost to the exterior. However, this results in the regional homogenization of cytoplasm, which raises concerns regarding the mixing of cytoplasmic contents among various compartments of the egg. For example, some proteins needed for the sperm's earliest transformation are prestored in the egg's cortex (the region just underneath the plasma membrane). This region would normally be first in direct contact with the incorporated sperm. Finally, the injected sperm undergoes an unusual nuclear remodelling (Figure 4).[16,64,68] The acrosome may remain intact; consequently, the digestive enzymes usually released to the exterior are instead internalized. While most healthy cells deal with hydrolytic enzymes in specific ways, the introduction of the concentrated acrosomal enzymes raises concerns regarding the manner in which they are immediately neutralized upon their introduction into the egg, as well as their ultimate fates.

The perinuclear theca, normally lost during sperm incorporation,[69] persists throughout pronucleus formation, constricting the apical region of the sperm head and thereby resulting in an asynchrony in DNA decondensation. Additionally, the retention of vesicle-associated membrane protein

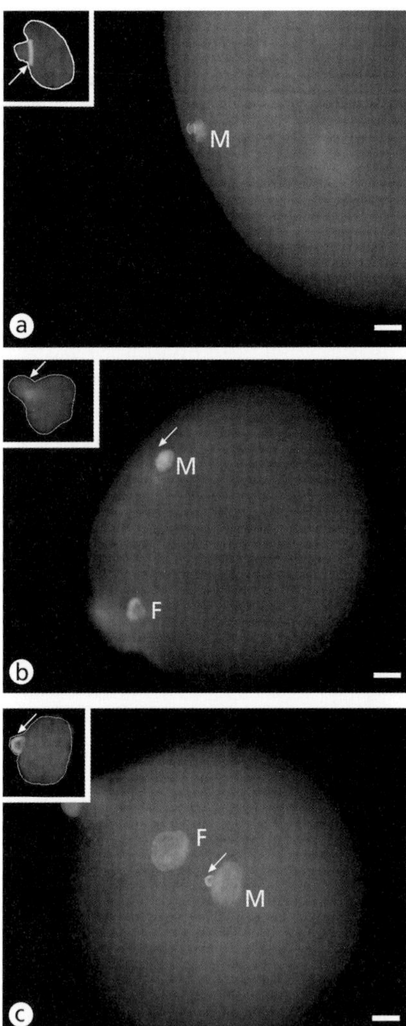

Figure 4 Unusual nuclear remodelling after intracytoplasmic sperm injection (ICSI). (a) Vesicle-associated membrane protein (VAMP) is detected as a constricting ring (inset, arrow) around the male pronucleus, separating the decondensed DNA from the still condensed apical region. (b) The nuclear mitotic apparatus protein NuMA is excluded from the condensed, apical region of the male pronucleus. A microtubule aster emanates from the developing male pronucleus. Inset: asynchronous chromatin decondensation after ICSI (arrow) as detected by Hoechst labelling. (c) The perinuclear theca (arrow) is observed up to 12 h post-ICSI, constricting the apical region of DNA (and inset). M, male pronucleus; F, female pronucleus; bar, a, 20 μm; bars b and c, 10 μm. Reproduced with permission from reference 64

(VAMP) on the injected sperm further separates the condensed and decondensing regions of DNA by forming a two-piece collar around the sperm head.[68] While these unusual modifications do not prevent pronucleus formation, the onset of DNA synthesis and pronuclear migration might be delayed after ICSI[69] with, as yet, unknown consequences. Furthermore, the finding that the X chromosome might be preferentially located in the apical region of the human sperm head[70] may provide the first connection between the ICSI technique and the higher reported rates of sex chromosome anomalies in ICSI babies.

The positioning of the sperm after ICSI, in contrast to its normal entrance at the egg surface, may expose the sperm to a different cytoplasmic environment. The use of time-lapse video microscopy to study pronuclear migration during fertilization has revealed that a sperm injected by ICSI usually migrates to a more cortical cytoplasmic position before attracting the female pronucleus, whereas a sperm entering the oocyte following IVF remains in a cortical position during sperm aster assembly (D. Takahashi, personal communication). Positioning the sperm in the center of the oocyte may delay access to critically important proteins involved in the formation of the sperm aster, the microtubule-based motility machinery that unites the sperm and egg nuclei within the inseminated egg (reviewed in references 9 and 13). Preliminary results examining the roles of the proteins during ICSI underscore the concerns about altered choreography in the development and union of the sperm and egg nuclei. For example, NuMA, a nuclear protein important during interphase and a mitotic apparatus protein important during meiosis and mitosis (reviewed in reference 71), is not found in sperm, but maternal NuMA enters the sperm nucleus as it decondenses during fertilization. After ICSI, however, because the sperm nucleus decondenses in an asynchronous manner, NuMA enters the basal, decondensing region of the sperm nucleus, but is initially excluded from the apical, condensed region.[64] Variations in the dynamics of fertilization similar to this might well lead to a diminished ability of the oocyte to express, or be exposed to, important paternal genes or gene products, thus leading to improper embryo formation.

Cell and molecular challenges during ICSI and other ART approaches

In spite of several encouraging retrospective studies during the past few years,[49,60,61,72–74] it is not yet possible to conclude absolutely that there are no long-lasting and unanticipated consequences of ICSI. Perhaps, the concerns raised in this chapter will turn out to be of minimal clinical importance, possibly because abnormal embryos due to very high aneuploidy rates in immature sperm or those with extensive oocyte damage, do not develop properly and are lost prior to implantation. This hopefully cautious view, based on the substantial gaps in our knowledge, is in contrast to the swift developments in the use of testicular sperm and immature spermatogenic cells (spermatids, spermatid nuclei, spermatocytes) for the clinical treatment of male infertility.[75–78] Prior to the advent of ICSI, nature exerted strict selection criteria at the sperm–egg interface (zona pellucida, sperm plasma membrane), barriers that we are just beginning to understand. But now, direct injection of sperm selected by the embryologist, and not by sperm–oocyte interplay and gamete self-selection, means that these biological restrictions are relaxed. The genetic make-up of the resultant offspring is of great importance from the point of view of future generations, particularly in the light of emerging literature demonstrating connections between abnormal sperm and mutations and deletions.[57,79–81] Perhaps the Y-chromosome deletions are among those best explored[80,82–84] and are particularly relevant since a son conceived by ICSI may also suffer from male infertility.[46,85,86]

Possibilities of cytoplasmic damage to both the sperm and the egg during ICSI are not yet well understood, and the mixing of components arising from the various compartments upon sperm introduction could lead to arrested fertilization. Damage to the sperm's mitochondria might release mitochondrial DNA (mtDNA) or other constituents, affecting later egg development. The ultimate fate of the paternal mitochondria is not clear, although paternal mtDNA has been shown to persist from the two-cell to the blastocyst stage in IVF embryos,[87] whereas in abnormal IVF and ICSI embryos paternal mtDNA was only found in those generated by IVF.[88] Perhaps the paternal mitochondria, otherwise marked for destruction by ubiquitination[89] following entry into the egg cortex, will lead to biparental, rather than maternal, inheritance of mitochondria. The clinical consequences of this extranuclear inheritance cannot be predicted, but, with the recent introduction of ooplasmic transfer, which uses donated cytoplasm from one oocyte injected into the recipient oocyte during ICSI,[90] they may be grounds for concern. Preliminary data suggest that this technique can result in sustained mtDNA heteroplasmy in the offspring, representing both the donor and the recipient.[91] It will be several decades before we know whether these offspring will transmit both mitochondrial genotypes to their offspring.

ICSI is in great demand among infertile couples. In men with severe male infertility, for whom ejaculated sperm are scarce or sperm motility is low, ICSI is the only effective fertility treatment at present. ICSI also provides hope for oligospermic or azoospermic men, whether the problem is due to insufficient sperm production or to an interruption of conductance of the vas deferens between the epididymis and the urethra. In men with such a blockage, the concentration of sperm that may be recovered by needle aspiration from the epididymis, or from the rete testis, is variable. In the case of diminished sperm production, sperm may be accessed directly from testicular tissue, which in most azoospermic men contains a few developed sperm. When no sperm can be found, oocyte fertilization has also been attempted by injection of elongated[76,92] or round[77,92] spermatids. However, problems in the correct identification of spermatids prior to injection has led to considerable controversy surrounding their use,[93–95] with some countries even issuing a moratorium on the use of these immature germ cells for fertilization. Owing to the low rate of success of these procedures, there are no definitive data yet available on the offspring.

Ethical considerations for ICSI and other ART approaches

Whereas the technological advances of ICSI provide an important opportunity for infertile men to father children, there are serious ethical concerns. First, severe infertility in some men is related to genetically inherited diseases, since the emerging data link congenital lack of vas deferens with cystic fibrosis, and azoospermia with deletions of the long arm of the Y chromosome. Thus, in male offspring conceived by ICSI, there is the danger that infertility and related congenital disorders will be propagated.[46,83,85,96] Arguments put forward by proponents of the safety of ICSI are based on the fact that, in approximately 2500 ICSI children (the oldest of whom are about 8 years of age) who were systematically followed, the rate of congenital malformations was about the same as in the normal population. However, the rates of transmitted chromosomal aberrations, and the risk of *de novo*, mainly sex-chromosome abnormalities are somewhat higher.[49,73] Those who are concerned see an additional risk: there are no data as yet regarding the physical and mental development of these children, nor regarding the cancer rates, or congenital malformation frequencies in their children.[62] Bowen and colleagues[51] have suggested that there may be a moderate risk of developmental problems in ICSI babies, although this may be overcome by the age of 2 years. Even if the long-term effects of ICSI are not detrimental, it would seem desirable to limit the genetic impact of ICSI fertilization at the traditional evolutionary level by injecting selected mature sperm. We believe that, in order to eliminate sperm-related risks, methods should be developed for the selection of mature sperm for ICSI. These techniques should be non-invasive and applicable to intact, viable sperm. The ultimate goal is selection of mature spermatozoa for ICSI that would have been part of the normal fertilization pool, if the diminished sperm concentration or motility had not necessitated mechanical injection, thus bypassing the natural sperm–zona selection process.

The improvement of sperm selection is important for other reasons. In addition to male infertility, ICSI is now introduced for couples with female factor infertility in which the sperm–zona interaction is diminished owing to oocyte dysfunction. Also, because ICSI overrides both sperm and oocyte defects and eliminates the need for diagnostic tests to identify whether there is a male or female factor leading to infertility, several fertility centers plan to utilize ICSI exclusively for all IVF procedures.

While ICSI has been an extremely important development for couples suffering from male infertility, and ICSI with immature germ cells may hold additional promises, many critical questions need to be answered regarding the injection of immature sperm. These include concerns about the physiological, behavioral and psychological integrity of the ICSI offspring, whether they will experience unexpected degenerative diseases in later life and the reproductive potential of the ICSI offspring, particularly sons. The proposal to use immature testicular spermatogenic cells raises another set of problems, including accurate definitions and criteria for cells selected, and whether alteration of the normal pattern of genomic imprinting might result in propagating diseases such as Angelman syndrome or Beckwith–Wiedemann syndrome. One of the first reports examining the pattern of DNA methylation in ICSI children determined that there does not appear to be a higher risk of DNA-methylation defects.[97]

In summary, while medical historians will conclude that this period has been one of 'ART before science', it is imperative to remedy the situation with relevant research aimed towards discovering the fundamental cellular and molecular aspects of ICSI. This information may well lead to improvements in the safety of ICSI, and also inform and reassure infertile patients, and their physicians, about the safety and efficacy of this powerful, but still experimental, therapeutic approach.

SUMMARY

As fertilization is the process that bridges the fields of reproductive and developmental biology, a full understanding of the cellular and molecular events taking place during fertilization is critical. In addition, the solution of many clinical problems ranging from sophisticated infertility treatments (reviewed in reference 40) to contraception hinges on a more complete knowledge underlying this crucial but, in many ways, still mysterious and miraculous process.

ACKNOWLEDGEMENTS

The authors would like to thank their many colleagues who have contributed to this research, especially Dr D. Battaglia, Dr A. Chan, Dr J. Fanton, Dr M. Luetjens, Dr J. Ramalho-Santos, Dr P. Sutovsky, E. Jacoby, C. Martinovich, C. Payne and D. Takahashi. The sponsorship of the National Institutes of Health in support of the fundamental research cited in this chapter is gratefully acknowledged.

REFERENCES

1. Boveri T. *Zellen-studieren: Ueber die Natur der Centrosomen*. Jena, Germany: Fisher, 1901: IV.
2. Rossant J. Interspecific cell markers and lineage in mammals. *Philos Trans R Soc (London) B* 1985 312: 91–100.
3. Solter D. Differential imprinting and expression of maternal and paternal genomes. *Ann Rev Genet* 1988 22: 127–146.
4. Surani MAH, Kothary R, Allen ND *et al*. Genome imprinting and development in the mouse. *Development* 1990 90 (Suppl): 89–98.
5. John RM, Surani MA. Genomic imprinting, mammalian evolution, and the mystery of egg-laying mammals. *Cell* 2000 101: 585–588.
6. Shur BD, Neely CA. Plasma membrane association, purification, and partial characterization of mouse sperm β, 1,4 galactosyltransferase. *J Biol Chem* 1988 263: 17706–17714.
7. Snell WJ, White JM. The molecules of mammalian fertilization. *Cell* 1996 85: 629–637.
8. Wassarman PM. Mammalian fertilization: molecular aspects of gamete adhesion, exocytosis, and fusion. *Cell* 1999 96: 175–183.
9. Schatten G. The centrosome and its mode of inheritance: the reduction of the centrosome during gametogenesis and its restoration during fertilization. *Dev Biol* 1994 165: 299–335.
10. Gyllensten U, Wharton D, Josefsson A, Wilson AC. Paternal inheritance of mitochondrial DNA in mice. *Nature* 1991 352: 255–257.
11. Jaffe LA. First messengers at fertilization. *J Reprod Fertil* 1990 42 (Suppl): 107–116.
12. Clark EA, Brugge JS. Integrins and signal transduction pathways: the road taken. *Science* 1995 268: 233–239.
13. Simerly C, Wu G, Zoran S *et al*. The paternal inheritance of the centrosome, the cell's microtubule-organizing center, in humans and the implications for infertility. *Nat Med* 1995 1: 47–53.
14. Wu G, Simerly C, Zoran S *et al*. Microtubule and chromatin configurations during fertilization and early development in rhesus monkeys, and regulation by intracellular calcium ions. *Biol Reprod* 1996 55: 269–270.
15. Trounson A, Bongso A. Fertilization and development in humans. *Curr Top Dev Biol* 1996 32: 59–101.
16. Hewitson LC, Simerly C, Tengowski MW *et al*. Chromatin and DNA configurations after intracytoplasmic sperm injection in the rhesus monkey: failures and successes. *Biol Reprod* 1996 55: 271–280.
17. Sutovsky P, Hewitson LC. Simerly C *et al*. Intracytoplasmic sperm injection for rhesus monkey fertilization results in unusual chromatin, cytoskeletal, and membrane events, but eventually leads to pronuclear development and sperm aster assembly. *Hum Reprod* 1996 11: 1703–1712.
18. Rawe VY, Olmedo SB, Nodar FN *et al*. Cytoskeletal organization defects and abortive activation in human oocytes after IVF and ICSI failure. *Mol Hum Reprod*. 2000 6: 510–516.
19. Palacios MJ, Joshi HC, Simerly C *et al*. Dynamic reorganization of γ-tubulin during murine fertilization. *J Cell Sci* 1993 104: 383–389.
20. Simerly C, Balczon R, Brinkley BR, Schatten G. Microinjected centromere antibodies interfere with chromosome movement in meiotic and mitotic mouse oocytes. *J Cell Biol* 1990 111: 1491–1504.
21. Gorbsky GJ, Simerly C, Schatten G, Borisy GG. Microtubules in the metaphase-arrested mouse oocyte turn over rapidly. *Proc Natl Acad Sci USA* 1990 87: 6049–6053.
22. Stricker SA, Centonze VE, Paddock SW, Schatten G. Confocal microscopy of fertilization-induced calcium dynamics in sea urchin eggs. *Dev Biol* 1992 149: 370–380.
23. Tesarik J, Sousa M. Comparison of Ca^{2+} responses in human oocytes fertilized by subzonal insemination and by intracytoplasmic sperm injection. *Fertil Steril* 1994 62: 1197–1204.
24. Whitaker M, Swann K. Lighting the fuse at fertilization. *Development* 1993 117:1–12.

25. Simerly C, Navara C, Wu G-J, Schatten G. Cytoskeletal organization and dynamics in mammalian oocytes during maturation and fertilization. In Grudzinskas JD, Yovic JL, eds. *Cambridge Reviews in Human Reproduction Gametes: The Oocyte*. London: Cambridge Press, 1995: 54–94.
26. Schultz RM, Kopf GS. Molecular basis of mammalian egg activation. *Curr Top Dev Biol* 1995 30: 21–62.
27. Parrington J, Swann K, Shevchenko VI, Sesay AK, Lai FA. Calcium oscillations in mammalian eggs triggered by a soluble sperm factor. *Nature* 1996 379: 364–368.
28. Swann K, Lai FA. A novel signalling mechanism for generating Ca^{2+} oscillations at fertilization in mammals. *BioEssays* 1997 19: 371–378.
29. Parrington J, Jones KT, Lai A, Swann K. The soluble sperm factor that causes Ca^{2+} release from sea-urchin (*Lytechinus pictus*) egg homogenates also triggers Ca^{2+} oscillations after injection into mouse eggs. *Biochem J* 1999 341: 1–4.
30. Witton CJ, Swann K, Carroll J, Moore HD. Injection of a boar sperm factor causes calcium oscillations in oocytes of the marsupial opossum, *Monodelphis domestica*. *Zygote* 1999 7: 271–727.
31. Stricker, SA, Swann K, Jones KT, Fissore RA. Injections of porcine sperm extracts trigger fertilization-like calcium oscillations in oocytes of a marine worm. *Exp Cell Res* 2000 257: 341–347.
32. Myles DG, Kimmel LH, Blobel CP, White JM, Primakoff P. Identification of a binding site in the disintegrin domain of fertilin required for sperm-egg fusion. *Proc Natl Acad Sci USA* 1994 91: 4195–4198.
33. Blobel CP. Functional processing of fertilin: evidence for a critical role of proteolysis in sperm maturation and activation. *Rev Reprod* 2000 5: 75–83.
34. Yanagimachi R. Mammalian fertilization. In Knobil E, Neill JD, eds. *The Physiology of Reproduction*. New York: Raven Press 1994: 189–317.
35. Jaffe, LA, Cross NL. Electrical regulation of sperm-egg fusion. *Annu Rev Physiol* 1986 48: 191–200.
36. Stricker S, Prather R, Simerly C, Schatten H, Schatten G. Nuclear architectural changes during fertilization and development. In Schatten H, Schatten G, eds. *The Cell Biology of Fertilization*. San Diego: Academic Press, 1989: 225–250.
37. Clarke HJ. Nuclear and chromatin composition of mammalian gametes and early embryos. *Biochem Cell Biol* 1992 70: 856–866.
38. Asch R, Simerly C, Ord T, Ord VA, Schatten G. The stages at which human fertilization arrests: microtubule and chromosome configurations in inseminated oocytes which failed to complete fertilization and development in humans. *Mol Hum Reprod* 1995 1: 1897–1906.
39. Oakley CD, Oakley BR. Identification of γ-tubulin, a new member of the tubulin superfamily encoded by mipA gene of *Aspergillus nidulans*. *Nature* 1989 338: 662–664.
40. Schatten G, Hewitson L, Simerly C, Sutovsky P, Huszar G. Cell and molecular biological challenges of ICSI: ART before science? *J Law Med Ethics* 1998 26: 29–37.
41. Yanagimachi R, Yanagimachi H, Rogers BJ. The use of zona free animal ova as a test system for the assessment of the fertilizing capacity of human spermatozoa. *Biol Reprod* 1976 15: 471–476.
42. Hewitson L, Haavisto A, Simerly C, Jones J, Schatten G. Microtubule organization and chromatin configurations in hamster oocytes during fertilization, parthenogenetic activation and after insemination with human sperm. *Biol Reprod* 1997 57: 967–975.
43. Terada Y, Luetjens CM, Sutovsky P, Schatten G. Atypical decondensation of the sperm nucleus, delayed replication of the male genome, and sex chromosome positioning following intracytoplasmic human sperm injection (ICSI) into golden hamster eggs: does ICSI itself introduce chromosomal anomalies? *Fertil Steril* 2000 74:454–460.
44. Liu J, Nagy Z, Joris H *et al.* Successful fertilization and establishment of pregnancies after intracytoplasmic sperm injection in patients with globozoospermia. *Hum Reprod* 1995 10: 626–629.
45. Simerly C, Zoran SS, Payne C *et al.* Biparental inheritance of γ-tubulin during human fertilization: molecular reconstitution of functional zygotic centrosomes in inseminated human oocytes and in cell-free extracts nucleated by human sperm. *Mol Biol Cell* 1999 10: 2955–2969.
46. Kent-First MG, Kol S, Muallem A *et al.* The incidence and possible relevance of Y-linked microdeletions in babies born after intracyto-

plasmic sperm injection and their infertile fathers. *Mol Hum Reprod* 1996 2: 943–950.
47. In't Veld P, Brandenburg H, Verhoeff A, Dhont M, Los F. Sex chromosomal abnormalities and intracytoplasmic sperm injection. *Lancet* 1995 773: 346.
48. Van Opstal D, Los FJ, Ramlakhan S *et al*. Determination of the parent of origin in nine cases of prenatally detected chromosome aberrations found after intracytoplasmic sperm injection. *Hum Reprod* 1997 12: 682–686.
49. Bonduelle, M, Camus M, De Vos A *et al*. Seven years of intracytoplasmic sperm injection and follow-up of 1987 subsequent children. *Hum Reprod* 1999 14 (Suppl 1): 243–264.
50. Bonduelle M, Joris H, Hofmans K, Liebaers I, Van Steirteghem A. Mental development of 201 ICSI children at 2 years of age. *Lancet* 1998 351: 1553.
51. Bowen JR, Gibson FL, Leslie GI, Saunders DM. Medical and developmental outcome at 1 year for children conceived by intracytoplasmic sperm injection. *Lancet* 1998 351: 1529–1534.
52. Palermo G, Joris H, Devroey P, Van Steirteghem AC. Pregnancies after intracytoplasmic sperm injection of single spermatozoon into an oocyte. *Lancet* 1992 340: 17–18.
53. Palermo G, Joris H, Derde MP *et al*. Sperm characteristics and outcome of human assisted fertilization by subzonal insemination and intracytoplasmic sperm injection. *Fertil Steril* 1993 59: 826–835.
54. Van Steirteghem AC, Nagy Z, Joris H. High fertilization and implantation rates after intracytoplasmic sperm injection. *Hum Reprod* 1993 8: 1061–1066.
55. Devroey P, Liu J, Nagy Z *et al*. Pregnancies after testicular sperm extraction and intracytoplasmic sperm injection in non-obstructive azoospermia. *Hum Reprod* 1995 10: 1457–1460.
56. Silber SJ. The use of epididymal sperm in assisted reproduction. In Tesarik J, ed. *Male Factor in Human Infertility*. Rome: Ares-Serono Symposia, 1994: 335–368.
57. Patrizio P. Intracytoplasmic sperm injection (ICSI): potential genetic concerns. *Hum Reprod* 1995 10: 2520–2523.
58. Wennerholm UB, Bergh C, Hamberger L *et al*. Incidence of congenital malformations in children born after ICSI. *Hum Reprod* 2000 15: 944–948.
59. Tarlatzis BC, Bili H. Intracytoplasmic sperm injection. Survey of world results. *Ann NY Acad Sci* 2000 900: 336–344.
60. Palermo GD, Colombero LT, Schattman GL, Davis OK, Rosenwaks Z. Evolution of pregnancies and initial follow-up of new-borns delivered after intracytoplasmic sperm injection. *J Am Med Assoc* 1996 276: 1893–1898.
61. Tournaye H, Van Steirteghem A. ICSI concerns do not outweigh its benefits. *J NIH Res* 1997 9: 35–40.
62. Nudell D, Castillo M, Turek PJ, Pera RR. Increased frequency of mutations in DNA from infertile men with meiotic arrest. *Hum Reprod* 2000 15:1289–1294.
63. Chan AWS, Luetjens CM, Dominko T *et al*. TransgenICSI: Foreign DNA transmission by intracytoplasmic sperm injection: injection of sperm bound with exogenous DNA results in embryonic GFP expression and live rhesus births. *Mol Hum Reprod* 2000 6: 26–33.
64. Hewitson, L. Dominko T, Takahashi D *et al*. Unique checkpoints during the first cell cycle of fertilisation after intracytoplasmic sperm injection in rhesus monkeys. *Nat Med* 1999 5: 431–433.
65. Hardarson T, Lundin K, Hamberger L. The position of the metaphase II spindle cannot be predicted by the location of the first polar body in the human oocyte. *Hum Reprod* 2000 15: 1372–1376.
66. Silva CP, Kommineni K, Oldenbourg R, Keefe DL. The first polar body does not predict accurately the location of the metaphase II meiotic spindle in mammalian oocytes. *Fertil Steril* 1999 71: 719–721.
67. Hewitson L, Simerly C, Tengowski M *et al*. The cell biological basis of intracytoplasmic sperm injection: microtubule, chromatin and membrane dynamics. *Mol Biol Cell* 1996 7: 639a.
68. Ramalho-Santos J, Sutovsky P, Oko R, Wessel GM *et al*. ICSI choreography: fate of sperm structures after monospermic ICSI and first cell cycle implications. *Hum Reprod* 2000 15:2610–2620.
69. Sutovsky P, Oko R, Hewitson L, Schatten G. The removal of the sperm perinuclear theca and its association with the bovine oocyte surface during fertilization. *Dev Biol* 1997 188:75–84.

70. Luetjens MC, Payne C, Schatten G. Non-random chromosome positioning on human sperm and sex chromosome anomalies following intracytoplasmic sperm injection. *Lancet* 1999 353: 1240.
71. Cleveland DW. NuMA: a protein involved in nuclear structure spindle assembly, and nuclear re-formation. *Trends Cell Biol* 1995 5: 60–64.
72. Bonduelle M, Legein J, Derde MP *et al.* Comparative follow-up study of 130 children born after ICSI and 130 children born after IVF. *Hum Reprod* 1995 10: 3327–3331.
73. Bonduelle M, Aytoz A, Van Assche E. Incidience of chromosomal aberrations in children born after assisted reproduction through intracytoplasmic sperm injection. *Hum Reprod* 1998 13: 781–782.
74. Van Golde R, Boada M, Veiga A *et al.* A retrospective follow-up study on intracytoplasmic sperm injection. *J Assist Reprod Genet* 1996 16: 227–232.
75. Schoysman R, Vanderzwalmen P, Nijs M, Segal L, Shoysman D. Pregnancy after fertilization with human testicular sperm. *Lancet* 1993 342: 1327–1330.
76. Fishel S, Green S, Bishop M *et al.* Pregnancy after intracytoplasmic injection of spermatid. *Lancet* 1995 345: 1641–1642.
77. Testart J, Tesarik J, Mendoza C. Viable embryos from the injection of round spermatids into oocytes. *N Engl J Med* 1995 333: 525.
78. Sofikitis N, Mantzavinos T, Loutradis D *et al.* Ooplasmic injections of secondary spermatocytes for non-obstructive azoospermia. *Lancet* 1998 351: 1177–1178.
79. Reijo R, Lee T, Salo P *et al.* Diverse spermatogenic defects in humans caused by Y chromosome deletions encompassing a novel RNA binding protein gene. *Nat Genet* 1995 10: 383–393.
80. Reijo R, Alagappan RK, Patrizio P, Page DC. Severe oligozoospermia resulting from deletions of azoospermia factor gene on Y chromosome. *Lancet* 1996 347: 1290–1293.
81. Menke DB, Mutter GL, Page DC. Expression of DAZ, an azoospermia factor candidate, in human spermatozoa. *Am J Hum Genet* 1997 60: 237–241.
82. Vogt P, Chandley AC, Hargreave TB. Microdeletions in interval 6 of the Y chromosome of males with idiopathic sterility point to disruption of *AZF*, a human spermatogenesis gene. *Hum Genet* 1992 89: 491–496.
83. Pryor JL, Kent-First M, Muallem A *et al.* Microdeletions in the Y-chromosome of infertile men. *N Engl J Med* 1997 336: 534–539.
84. Mulhall JP, Reijo R, Brown L *et al.* Azoospermic men with deletions of the *DAZ* cluster gene are capable of completing spermatogenesis, fertilization, normal embryonic development and pregnancy with retrieved testicular spermatozoa and ICSI. *Hum Reprod* 1997 2: 503–508.
85. Kent-First M, Muallem A, Shultz J *et al.* Defining regions of the Y-chromosome responsible for male infertility and identification of a fourth AZF region (AZFd) by Y-chromosome microdeletion detection. *Mol Reprod Dev* 1996 53: 27–41.
86. Kamischke A, Gromoll J, Simoni M, Behre HM, Nieschlag E. Transmission of a Y chromosomal deletion involving the deleted in azoospermia (*DAZ*) and chromodomain (*CDY1*) genes from father to son through intracytoplasmic sperm injection: case report. *Hum Reprod* 1999 14: 2320–2322.
87. St. John JC, Sakkas D, Dimitriadi K, Barnes AM, Barratt CLR, De Jonge CJ. Detection of paternal mitochondrial DNA (mtDNA) in normal, multicellular human embryos. *Fertil Steril* 1999 72:S21.
88. St. John J, Sakkas D, Dimitriadi K *et al.* Abnormal human embryos show a failure to eliminate paternal mitochondrial DNA. *Lancet* 2000 350: 961–962.
89. Sutovsky P, Moreno RD, Ramalho-Santos J *et al.* Ubiquitin tag for sperm mitochondria. *Nature* 1999 402: 371–372.
90. Cohen J, Scott R, Alikani M *et al.* Ooplasmic transfer in mature human oocytes. *Molec Hum Reprod* 1998 4: 269–280.
91. Brenner CA, Barritt JA, Willadsen S, Cohen J. Mitochondrial DNA heteroplasmy after human ooplasmic transplantation. *Fertil Steril* 2000 74: 573–578.
92. Fishel S, Green S, Hunter A *et al.* Human fertilization with round and elongated spermatids. *Hum Reprod* 1997 12: 336–340.
93. Vanderzwalmen P, Nijs M, Schoysman R *et al.* The problems of spermatid microinjection in the human: the need for an accurate morphological approach and selective methods for viable and normal cells. *Hum Reprod* 1998 13: 515–519.
94. Silber SJ, Johnson L. Are spermatid injections of any clinical value? ROSNI and ROSI revisited. Round spermatid nucleus injection and round

spermatid injection. *Hum Reprod* 1998 13: 509–515.

95. Silber SJ, Johnson L, Verheyen G, Van Steirteghem A. Round spermatid injection. *Fertil Steril* 2000 73: 897–900.

96. Phillipson GT, Petrucco OM, Matthews CD. Congenital bilateral absence of the vas deferens, cystic fibrosis mutation analysis and intracytoplasmic sperm injection. *Hum Reprod* 2000 15: 431–435.

97. Manning M, Lissens W, Bonduelle M *et al*. Study of DNA-methylation patterns at chromosome 15q11–q13 in children born after ICSI reveals no imprinting defects. *Mol Hum Reprod* 2000 6: 1049–1053.

Chapter 23

Preimplantation embryo development

Keith E. Latham and Richard M. Schultz

INTRODUCTION

This chapter describes molecular and cellular events that underlie preimplantation development. The discussion focuses on the preimplantation mouse embryo, as this is the species that is best characterized to date. The underlying principles that govern development of the preimplantation mouse embryo will probably apply to preimplantation embryos of other species.

MORPHOLOGICAL DESCRIPTION AND TIMING OF EVENTS THAT OCCUR DURING PREIMPLANTATION DEVELOPMENT

Preimplantation development is a unique feature of mammalian development, and has several inherent advantages when compared with the external development that characterizes most non-mammalian species. For example, the preimplantation embryo develops in an environment free from predators. The nutrition for the preimplantation embryo is derived from the maternal reproductive tract. Thus, yolk, which is synthesized and accumulated by non-mammalian eggs and serves as an energy source, is absent in mammalian eggs and preimplantation embryos. In fact, one can view the mother as the source of yolk, i.e. energy sources.

Preimplantation development is characterized by cell proliferation without significant cell growth, a hallmark of the early cell divisions in almost all species. Thus, the volume of the preimplantation embryo remains essentially constant. Although there is little change in the amount of protein, RNAs of all species increase significantly in mass during this period.[1] Following cleavage to the two-cell stage, subsequent cleavage divisions are asynchronous and result in embryos containing an odd number of cells. All phases of the cell cycle are present, i.e. G_1, S, G_2 and M, and the length of these cell cycles is 12–20 h, which is characteristic of many somatic cells. The first two cell cycles show an interesting reciprocal allocation in the time that they spend in G_1 and G_2 (Figure 1). Whereas the length of G_1 of the first cell cycle is quite pronounced and a very short G_2/M phase is detected, the two-cell embryo has a very short period of G_1 and a lengthy G_2. G_1 also tends to be shorter than G_2 in the third and fourth cell cycles.[3] These differences in cell cycle phase length probably reflect the pace of essential events that occur during those periods.

During the first two cell cycles, the embryo undergoes the critical maternal-to-embryonic transition. During this transition, the gene expression program that promoted the development of the oocyte and its subsequent passage through meiosis is terminated, and the developmental program directed by the embryonic genome and leading to embryogenesis is initated. This transition is under the control of maternally inherited mRNAs and proteins that are deposited in the developing oocyte.

Up to the eight-cell stage, the blastomeres are quite distinct. During compaction at the eight-cell stage (see below), adhesive junctions form between the individual blastomeres, so that they can no longer clearly be resolved. Along with this morphological transition come pronounced biochemical transitions through which the blastomeres acquire characteristics resembling somatic cells, reflected in such features as ion transport, metabolism, cellular architecture and gene expression pattern.

The early cleavage divisions occur in the oviduct. The embryo enters the uterus on about day 4 following fertilization around the 16-cell morula stage. About 90 h post-ovulation, the early blastocyst stage is present and is characterized by the formation of the blastocoele, which is a fluid-filled cavity. The blastocyst (see below) is composed of two cell types. The cells of the inner cell mass (ICM) give rise to the embryo proper, as well as some extraembryonic membranes. The differentiated cells of the trophectoderm comprise a fluid-transporting epithelium that is responsible for the fluid accumulation in the blastocoele. The trophectoderm gives rise exclusively to extra-embryonic tissues.

The last event of preimplantation development is the escape (hatching) of the embryo from the zona pellucida (ZP), which occurs on day 5 following fertilization; the day that fertilization occurs is

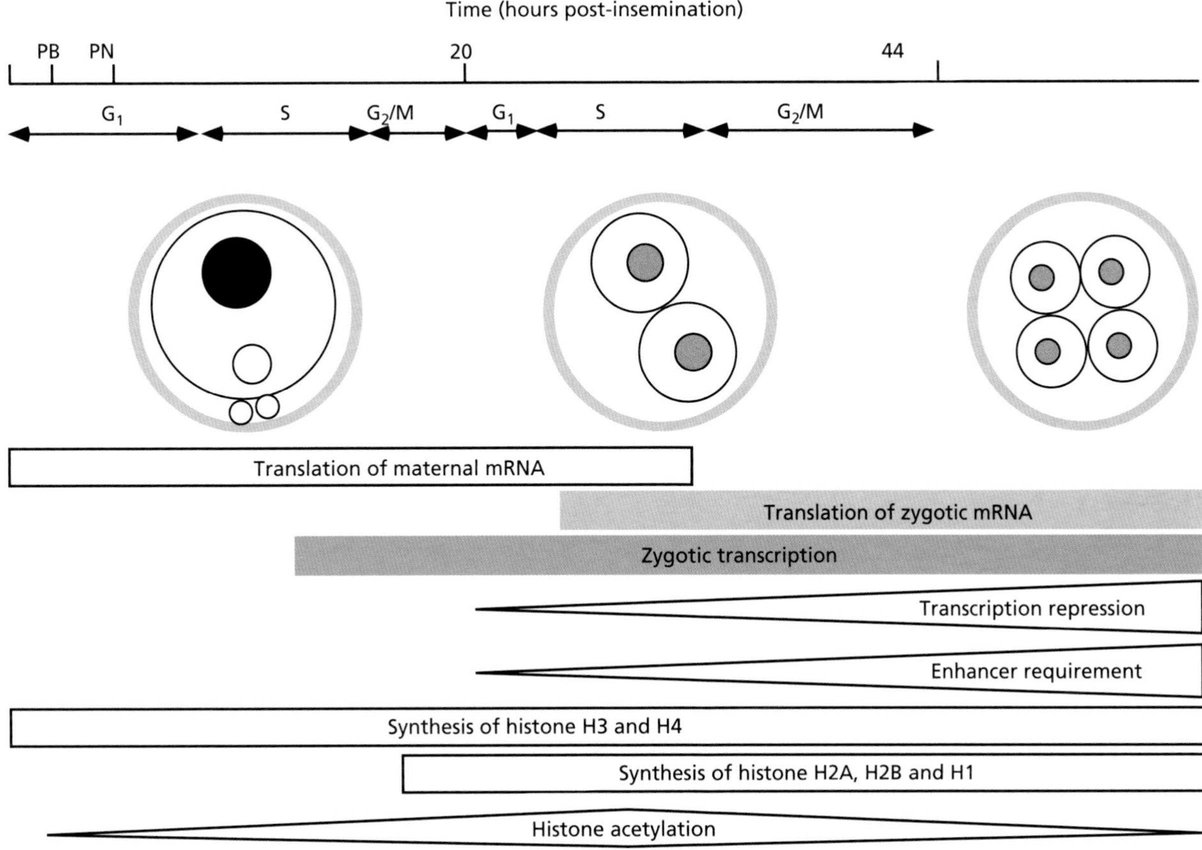

Figure 1 Schematic depiction of translational and transcriptional events that occur during the first two cell cycles. Following insemination, the second polar body (PB) is emitted within 2 h and the pronuclei form between 5 and 7 h. The paternal pronucleus (PN) (filled circle) is the larger of the two and the maternal pronucleus PN (open circle) is usually near the PB. Following cleavage both sets of parental chromosomes now reside in embryonic nuclei (gray circle). The embryo is surrounded by the zona pellucida (blue circle). Translation of maternal mRNAs drives protein synthesis during the first cell cycle. Degradation of maternal mRNAs is essentially complete by the two-cell stage and new protein synthesis is now supported by translation from zygotic transcripts. Embryonic transcription can first be detected during the S phase of the one-cell embryo. Correlated with the development of a transcriptionally repressive state is the requirement for an enhancer for efficient expression from plasmid-born reporter genes, as well as the synthesis of histone H1, which could also contribute to repression of transcription. Not shown is the decrease in histone acetylation that occurs between the two-cell and four-cell stages; this decrease could also contribute to a transcriptionally repressive state. Modified from reference 2

counted as day 1. The ZP is an extracellular coat that surrounds all mammalian eggs and mediates the initial events of fertilization, namely, sperm–zona pellucida interaction.[4] The ZP may serve other functions. Before compaction, the ZP may prevent the loosely adhering blastomeres from dissociating during embryo transport in the oviduct as well as the ectopic attachment of the developing embryo to the oviduct.[5] The ZP may also prevent leukocytes from attacking the embryo. While the ZP may serve these beneficial functions during preimplantation development, the embryo must escape from it in order to contact the endometrium. How the embryo escapes from the ZP is poorly understood. Trypsin inhibitors block hatching, and a trypsin-like protease, termed strypsin, appears to be responsible for hatching *in vitro*.[6] Other results suggest that the generation of superoxide by the blastocyst may also play a role in hatching.[7]

MATERNAL-TO-EMBRYONIC TRANSITION

Forming the embryonic genome

The first step on the path leading to embryonic control of development is the formation of the embryonic genome. Soon after fertilization, the oocyte completes meiosis and the sperm nuclear envelope breaks down. Sperm nuclear envelope breakdown is mediated by factors in the mature oocyte.[8,9] The oocyte also directs the breakdown of nuclear envelopes of microsurgically transplanted somatic cells. After breakdown of the sperm nuclear envelope, the sperm chromatin decondenses. The factors mediating sperm nucleus decondensation are expressed abundantly in the oocyte and are also expressed stage-specifically.[8] Subsequently, the maternal and paternal sets of chromosomes become segregated within two separate pronuclei. Pronuclear formation has been examined in detail in the sea urchin, in which it proceeds by the accumulation of membrane-bound vesicles around the chromatin.[9] The ability to form the paternal pronucleus is developmentally acquired, being absent from the germinal vesicle-stage oocyte, is regulated by a mitogen-activated kinase and is co-ordinated with formation of the maternal pronucleus.[8,9] Factors that promote pronucleus formation exist in limiting quantities in the oocyte.[8,9] Pronuclear formation is associated with changes in lamin composition, which in turn affect changes in nuclear envelope permeability. This increase in permeability appears to be critical for successful chromatin remodelling, DNA replication and reprogramming of gene expression.[9]

Molecular basis for the reprogramming of gene expression

The growing mammalian oocyte synthesizes and accumulates organelles, proteins, mRNAs, etc.[10] Before the maternal-to-embryonic transition, or embryonic gene activation (EGA), these maternal stores support the metabolism of the early embryo, and also control and mediate early development. EGA is critical for continued development of the embryo. Transcripts produced during EGA are of two types. Some replace maternal transcripts, the degradation of which initiates during oocyte maturation.[11] Other transcripts are expressed *de novo* following EGA, resulting in a dramatic reprogramming of the pattern of gene expression, as revealed in protein synthetic patterns observed with high-resolution two-dimensional gel electrophoresis.[12] This change in pattern of gene expression probably establishes totipotency of the blastomeres of the two-cell embryo, both of which can develop into a viable and fertile mouse if separated.

Genome activation does not comprise a single discrete event, but rather occurs in a stepwise manner.[13] In all species examined, a major genome activation event, in which the greatest number of genes becomes activated, is preceded by one or more smaller transcriptional activation events affecting a small subset of genes. In the mouse embryo, a limited amount (30–40% of that observed later) of gene expression is detectable during the G_2 portion of the one-cell stage, as revealed by expression of endogenous genes (including paternally inherited alleles), inherited transgenes and microinjected reporter constructs.[9,13] During the second cell cycle, a second minor wave of transcription occurs during the G1 period, wherein a number of genes are induced transiently.[9,13] Then, during the G_2 portion of the two-cell stage, the major genome activation event occurs, encompassing a large number of housekeeping genes and other genes.[12] In other species, such as human and cow, the major genome activation event occurs at the 4–8 cell stage, but this is also preceded by transcription as early as the one-cell stage.[13]

Some event occurring between fertilization and the first mitosis must be critical in the initial transcriptional activation and/or reprogramming of gene expression. Early one-cell-stage cytoplasm does not support transcription in exogenous, transcriptionally competent nuclei introduced microsurgically, but late one-cell-stage cytoplasm does.[13] Thus, the early one-cell embryo either lacks essential factors required for transcription or contains a transcriptional inhibitor. Protein synthesis is necessary for the one-cell embryo to become transcriptionally active.[14] This may reflect a need to synthesize essential transcription factors or proteins that regulate them. One likely candidate is a kinase that is necessary to phosphorylate the RPB1 subunit of RNA polymerase II.[13]

Chromatin structure appears to become modified during the first cell cycle to facilitate gene transcription. Nucleosomes are the functional unit of chromatin and are composed of a histone octomer (two molecules each of histones H2A, H2B, H3 and H4) and about 145 bp of DNA wrapped around this histone core. The nucleosome can inhibit the binding of transcription factors to promoter and enhancer elements. During the first round of DNA replication, with consequent nucleosome displacement, maternally derived transcription factors may gain access to their *cis*-acting DNA-binding sequences (Figure 2).[15,16] The spectrum of genes for which expression is linked to DNA replication will reflect the array of promoter and enhancer elements for each gene, and the complement of maternal transcription factors and enhancers that can bind to these elements. Consistent with this proposal, aphidicolin, which inhibits DNA replication, inhibits the maximal expression of initiation factor eIF-1A, the transcription-requiring complex (TRC) genes and many other endogenous genes.[17,18] Nevertheless, EGA is not totally dependent on the first round of DNA replication, since global transcription does initiate in aphidicolin-treated one-cell embryos.[18] The synthesis of a subset of α-amanitin-sensitive polypeptides is also inhibited by aphidicolin, but another subset is aphidicolin insensitive.[19] Thus, although some genes may require the first round of DNA replication for their maximal expression, the expression of other genes is apparently unlinked or loosely linked to DNA replication. The switching/sucrose non-fermenting (SWI/SNF) family of proteins may be critical for the expression of these aphidicolin-insensitive genes, as these proteins can remodel chromatin in the absence of DNA replication, and promote nucleosome disruption that is coupled to adenosine triphosphate (ATP) hydrolysis, resulting in the accessibility of DNA to transcription factors.[20,21]

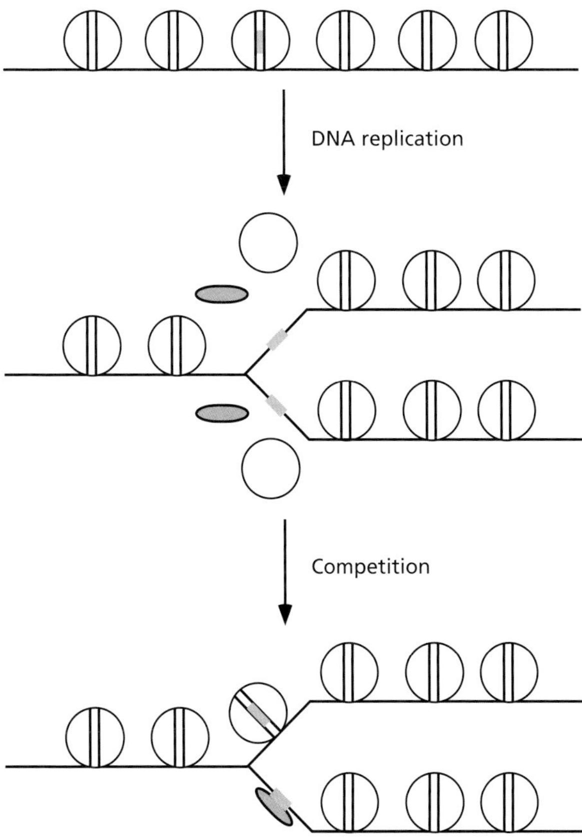

Figure 2 Schematic diagram depicting how DNA replication provides a window of opportunity for transcription factors to bind to their *cis*-acting cognate DNA-binding sequence. The chromatin template is transcriptionally inactive since transcription factors cannot gain access to the DNA-binding element (blue rectangle on third nucleosome). DNA replication disrupts the nucleosomes and hence makes the DNA-binding element accessible to both nucleosomes (open circle) and a transcription factor (gray oval). Since 145 bp of DNA wrap around the octomer core, amounts of free DNA less than this cannot support nucleosome formation and favor binding of transcription factors. If the amount of free DNA is sufficient to foster nucleosome formation, then a competition for this DNA, which harbors a DNA-binding element, will ensue. As shown in the bottom panel, in one case a nucleosome was formed, whereas in the other instance the transcription factor outcompeted nucleosome formation and hence could initiate the formation of a productive transcription complex. The diagram has been oversimplified in order to convey the central components of the model

The expression of eIF-1A, the TRC and a number of other proteins increases transiently during the two-cell stage.[12,13] The second round of DNA replication seems to be critical for the transient expression of these genes (i.e. their repression), since addition of aphidicolin to two-cell embryos prevents the decrease in eIF-4C and TRC expression.[17] The second round of DNA replication may be linked to the establishment of a transcriptionally repressive state that develops during the two-cell stage.[9,13] When aphidicolin is used to inhibit the second round of DNA replication, the total amount of bromo-uridine triphosphate (BrUTP) incorporated by these embryos is greater than that of control embryos in which this round of DNA replication occurs, a finding that also suggests that a transcriptionally repressive state develops during the two-cell stage (see below).

Development of the transcriptionally repressive state and role of enhancers in maintaining transcription

The early embryo may initially lack the ability to regulate the array of genes that are transcribed. This ability may be acquired stage-specifically through the establishment of a transcriptionally repressive state, wherein enhancers and strong promoters become essential for efficient transcription. Enhancers are *cis*-acting DNA sequences that bind transcription factors to stimulate transcription. Enhancers are located at either the 5' or the 3' end of a gene, and can function at distances many kilobases away from the transcription start site. A change in enhancer utilization accompanies EGA. The expression of a reporter gene in cleavage-arrested one-cell embryos that are chronologically at the two-cell stage is independent of the presence of an enhancer in the construct, but an enhancer is required for the efficient expression of the reporter gene following injection into a two-cell blastomere.[2] This indicates that, in the two-cell embryo, weak promoters are repressed unless they are coupled to an enhancer.

Nuclear transplantation experiments reveal that the repressive activity is present in the two-cell embryo, but not in the one-cell embryo.[22] Transplantation of an injected male pronucleus to a two-cell blastomere results in repression of the reporter gene. Transplantation of an injected two-cell nucleus into a one-cell embryo results in the maintenance of repression of the reporter gene.

A change in chromatin structure probably establishes the transcriptionally repressive state, making enhancers and/or strong promoters essential during early development. Enhancerless reporter genes injected into the nucleus of a two-cell blastomere are expressed efficiently if histone hyperacetylation is induced by treatment with agents that inhibit histone deacetylase, with a level of expression similar to that of the reporter gene that contains the enhancer.[9,13] This treatment does not increase the expression of the reporter gene that contains an enhancer. Likewise, treatment of two-cell embryos with histone deacetylase inhibitors prevents the decrease in expression of eIF-1A and the TRC that normally occurs during the two-cell stage, and increases BrUTP incorporation.[17,18] Thus, this experimentally induced change in chromatin structure relieves the repression that normally develops in the two-cell embryo.

Changes in histone H1 isotypes and histone acetylation state could underlie the change in chromatin structure.[9,13] The synthesis from maternal transcripts of histone H1, required for the condensation of chromatin into the 30-nm fiber that represses transcription, begins in the late one-cell embryo, and embryonic transcription of the somatic histone H1 gene initiates at the two-cell stage, temporally correlated with the development of the transcriptionally repressive state. Inhibiting the first round of DNA replication inhibits the expression of histone H1, and this correlates with the lack of repression of enhancerless reporter genes in cleavage-arrested, aphidicolin-treated one-cell embryos. The amount of hyperacetylated histones may also decrease between the two-cell and four-cell stages.[2] This could contribute to the conversion of transcriptionally permissive chromatin to transcriptionally repressed chromatin.

Differences in transcriptional capacity of maternal and paternal pronuclei

As indicated above, the maternal and paternal chromosomes are initially contained in separate pronuclei. This is the last time in development that the two parental sets of chromosomes are physically isolated, because, following migration of the pronuclei to the center of the fertilized egg and breakdown of the nuclear envelopes, the chromosomes congress on a metaphase plate, and mitosis results in the production of two diploid embryonic nuclei containing one copy each of both parental sets of chromosomes.

Results from reporter gene studies indicate that the paternal pronucleus (PN) supports a higher level of transcription than the maternal PN, and that efficient reporter gene expression in the paternal PN does not require an enhancer, whereas an enhancer is required for efficient expression when the maternal PN is injected with the reporter gene.[9,13] BrUTP incorporation by the paternal PN reveals that endogenous gene transcription is about 4–5 times greater than that of the female pronucleus.[18]

These differences in gene expression probably reflect differences in the composition and/or architecture of the paternal and maternal pronuclei.[13] The concentration of transcription factors (such as TATA box-binding protein (TBP), Sp1, etl-1) is greater in the paternal PN (Figure 3), and this could contribute to the enhanced rate of transcription that is supported by the male PN. Some components of the paternal PN may be obtained directly via the sperm. Sperm DNA is packaged in a highly organized manner, with DNA loop domains attached to the nuclear matrix, and specific chromosomes localized within the sperm head.[13] This architecture, which can be disrupted with detergent treatment, is necessary for successful development. The ability of exogenous DNA to incorporate into sperm DNA has also been taken as evidence for specific regions of DNA that may be uniquely packaged and available for transcription. Thus, proteins associated with the sperm nucleus may persist in the paternal pronuclei and affect early paternal gene expression. Paternal pronuclei also exhibit an initially greater amount of hyperacetylated histone H4. This difference persists throughout the G_1 phase, and may affect subsequent events that, in turn, affect transcriptional activity. Following fertilization, the sperm-bound protamines are exchanged and replaced by maternally derived histones to transform the sperm DNA into chromatin. This exchange could provide a window of opportunity for maternal transcription factors to gain access to their *cis*-DNA binding sequences that is not available to the maternal PN, in which DNA is already complexed with histones and present as nucleosomes. The paternal PN also undergoes a preferential loss of DNA methylation for the TKZ751 transgene and endogenous

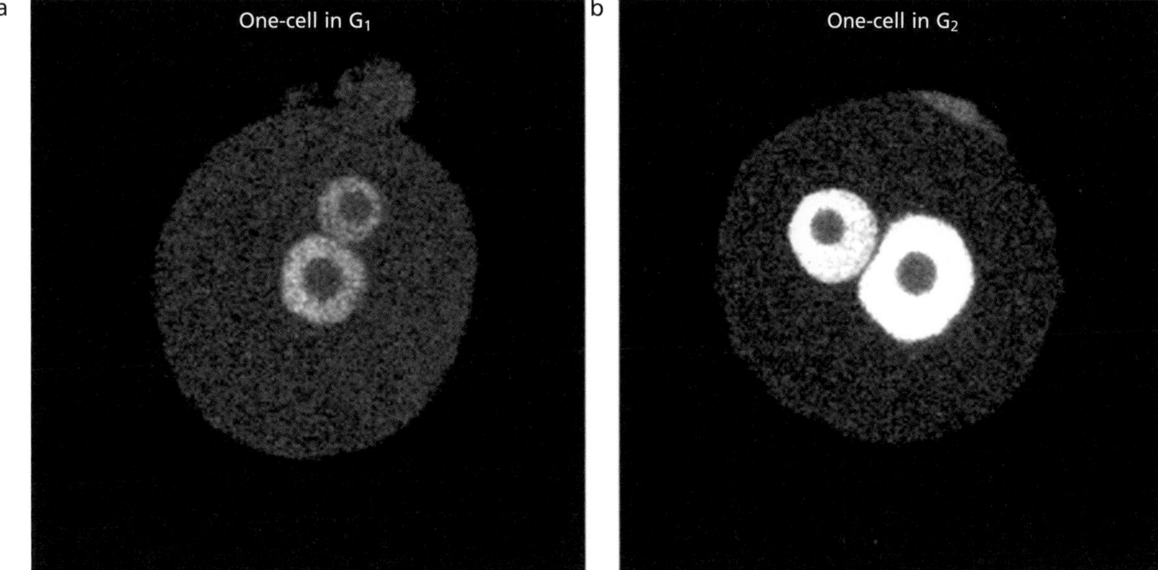

Figure 3 Laser-scanning confocal image of one-cell embryo stained for etl-1, a member of the switching/sucrose non-fermenting (SWI/SNF) family of transcription factors. Note that the staining intensity in the male pronucleus (the larger of the two pronuclei) is greater than that in the female pronucleus. The increase in staining that occurs between G_1 (a) and G_2 (b) requires DNA replication. The affinity-purified etl-1 antibody was the generous gift of Dr Achim Gossler

imprinted genes.[13] These observations collectively indicate that the transcriptional environments of the maternal and paternal pronuclei are quite different.

It should be noted that the transcriptional capacity of the maternal PN is not inherently less than that of the paternal PN. Nuclear transplantation studies indicate that the reduced rate of transcription in the maternal PN relative to the paternal PN is probably due to specific properties of the maternal PN.[22] When the maternal PN does not have to compete with the paternal PN for maternal transcription factors, the maternal PN can efficiently sequester these transcription factors to support high levels of transcription.

Molecular basis for timing of the onset of zygotic gene activation

It is critical that the embryo controls the timing of genome activation. Nuclear transplant studies reveal an ability of the one-cell cytoplasm to repress transcriptionally active genes inappropriately, with repression persisting beyond the period when these genes are normally activated.[13] Thus, delaying transcription of many genes until the two-cell stage appears to be essential for normal gene expression and may serve a protective function. Moreover, widespread transcription before the trancriptionally repressive state is established may lead to deregulated gene expression and compromise development. Two mechanisms can be envisioned through which the timing of genome activation may be controlled, one involving a DNA replication-dependent titration of inhibitory factors, and the other involving the time-dependent appearance of essential transcription factors.

Under the first scenario, the small amount of maternally derived histones may account for the early onset of transcription in mammalian eggs, in contrast to species such as *Xenopus laevis*, whose eggs harbor an extremely large pool of maternally derived histones: the *X. laevis* egg contains enough histone to support the formation of 15–20 000 nuclei.[23] In these non-mammalian species, a competition between chromatin assembly and transcription complex assembly is strongly implicated in regulating gene expression during early development.[24,25] Once the maternal histone pool is titrated out by the exponential increase in DNA following 12 rounds of DNA replication, stable basal transcription complexes can form, and transcription and the reprogramming of gene expression can initiate. Although the relevance of this competition model in regulating mammalian EGA and nuclear reprogramming is unknown, it could account for the differences in timing of EGA between species. The major EGA event in human, cow and sheep embryos occurs around the 4–8-cell stage.[13] The eggs of these species are about 3–4 times the volume of the mouse egg, and thus may contain 3–4 times the amount of histone. Titration of the maternally derived histones could thus take an additional one or two cell cycles. As a result, the major EGA event would be predicted to occur around the 4–8-cell stage, which is the case.

The second mechanism for regulating the timing of EGA relates to the time-dependent appearance/activation of transcription factors (or elimination/inactivation of transcriptional inhibitors). Translational recruitment of maternal mRNAs encoding transcription factors may occur at specific stages. Cycloheximide treatment from the late one-cell stage onwards prevents the normal onset of transcription of housekeeping genes, and reduces the expression of transcription factors such as Sp1 and TBP.[13] An analysis of polysomal mRNA populations of eggs, one-cell, and two-cell embryos revealed that the mRNAs for Sp1 and TBP did not exhibit stage-specific translational recruitment.[13] Thus, the effect of cycloheximide on the expression of Sp1 and TBP may reflect the need for continuous synthesis to maintain the necessary abundance of these factors to support transcription. In contrast to Sp1 and TBP, the transcription factor mTEAD2 is first produced at the two-cell stage entirely from maternally encoded mRNA, which appears to undergo recruitment onto polysomes coupled to rapid degradation at the two-cell stage.[13] The recruitment of mRNAs encoding such transcription factors could thus be the limiting step for transcriptional activation of genes that require those factors. The recruitment of maternal mRNAs encoding essential transcription factors could be under the control of specific mRNA-binding proteins, which in turn may be regulated post-translationally by cell cycle-dependent kinases.[13]

Genetic interactions controlling blastomere survival in the early embryo

As terminally differentiated cells, both sperm and egg are destined to die. In the absence of any other event (for example, parthenogenetic activation of the egg), neither sperm nor egg will survive indefinitely once released into the reproductive tract. The union of the two gametes, however, initiates a new life, and thus rescues the oocyte from a path leading towards death to one leading to life. A substantial incidence of apoptosis and/or cellular fragmentation occurs in oocytes.[26,27] Apoptosis and cellular fragmentation are less problematic in embryos, but nevertheless do occur, and often during the period immediately preceding the major genome activation event, indicating a role for early gene transcription in suppressing these processes.[28–30] The incidence of blastomere fragmentation in mouse two-cell stage embryos is determined by the combination of maternal and paternal genotypes, with certain combinations of parental genotypes producing an enhanced incidence of embryo fragmentation.[31] The paternal genotype effect is reduced using F_1 hybrid males, consistent with a possible role for post-fertilization transcription of the paternally inherited genome in controlling blastomere fragmentation, possibly through interactions with oocyte-derived gene products or oocyte mitochondria.

COMPACTION

Establishment of an epithelium and cell polarity

The individual blastomeres of the developing embryo are clearly discernible under the light microscope until compaction, when the blastomeres flatten on one another such that, following compaction, it is difficult to delineate clearly the boundaries between the blastomeres.[32] During compaction, a variety of junctions form between blastomeres, and include gap junctions, adherens junctions and tight junctions. Gap junctions provide channels for the direct passage of small molecules, usually < 1000 Da, between cells. These molecules tend to be metabolites (e.g. sugars, amino acids and nucleotides), but may also be second messengers such as cyclic adenosine monophosphate (cAMP). Adherens junctions foster cell–cell association via calcium-dependent interactions of the extracellular domains of members of the cadherin family. Tight junctions form a permeability seal that prevents small

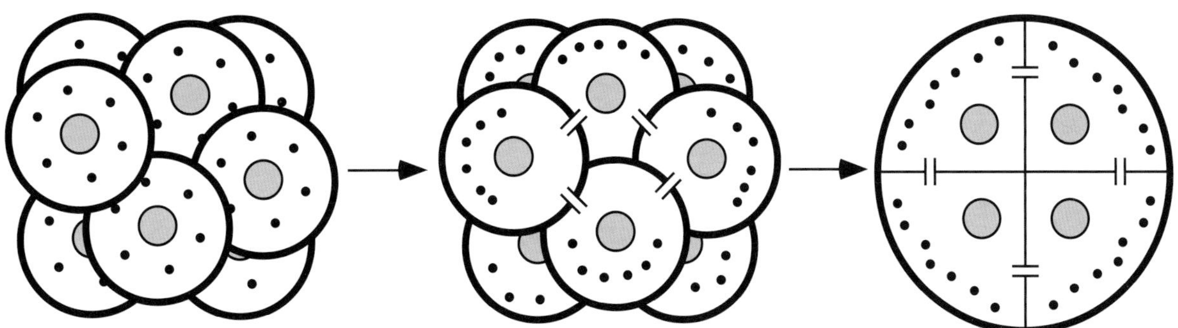

Figure 4 Schematic diagram depicting the events that occur during compaction. The thick borders of the circles represent the presence of microvilli. The small dots represent elements of cytoplasmic polarity, e.g. endocytotic vesicles and actin filaments. During the initial stages of compaction cell–cell interactions, both gap and tight junctions form (depicted by pairs of parallel lines) and cytoplasmic polarity can be observed. These early events are then followed by cell flattening and the development of surface polarity (the thick lines represent microvillar regions, whereas the thin lines represent amicrovillar regions). Also note that the nucleus becomes basolaterally localized in the fully compacted embryo. Whereas surface polarity is stable to cell division and agents that disrupt the cytoskeleton, cytoplasmic polarity is lost during mitosis and is sensitive to agents that disrupt the cytoskeleton

molecules (e.g. sodium and chloride ions) from passing between cells. The formation of these junctions transforms the loosely attached cells into a society of tightly associated, communicating, polarized epithelial cells.

The cells of the embryo become highly polarized during compaction[32] (Figure 4). The apical surface contains numerous microvilli, whereas the basolateral surface is relatively amicrovillar; this constitutes surface polarity. A cortical domain of filamentous actin concentrates under the apical surface, microtubules align parallel to the basolateral surface, and endocytotic vesicles lie between the apical region and the basolaterally located nucleus; these features constitute cytoplasmic polarity. The basolateral localization of the nucleus may account for the fact that, in the morula, the outer cells are larger than the inner cell.[33] The asymmetrical distribution of microvilli accounts for the observation that the inner cells of the morula are also relatively amicrovillar.

The cytoplasmic polarity is very sensitive to agents that disrupt the cytoskeleton, such as nocodazole and cytochalasin.[34] For example, nocodazole treatment, which does not inhibit cell flattening or junctional communication, inhibits cytoplasmic polarity. Cytochalasin D treatment, which inhibits cell flattening and junctional communication, also inhibits cytoplasmic polarity. Surface polarity, however, is much more refractory to these inhibitors. In fact, once surface polarity is established, it is stable and survives mitosis.[35] Polarity does not require extensive cell–cell contact, since it is maintained following decompaction by experimental methods (see below) and in isolated blastomeres.

Molecular basis for compaction

Compaction does not require the prior round of DNA replication, since aphidicolin-treated, cleavage-arrested four-cell embryos undergo cell flattening and polarization at about the same time as control, untreated embryos.[36] Protein synthesis during the four-cell stage is also not required for compaction.[37,38] Thus, the preimplantation embryo probably contains all of the proteins required for compaction by the two-cell stage. Interestingly, inhibition of protein synthesis results in premature compaction,[38] which suggests that embryos contain a factor that inhibits compaction; protein synthesis may be required for the expression/maintenance of this activity.

E-cadherin, also called uvomorulin, is responsible for the calcium-dependent events of compaction.[39] Antibodies to E-cadherin inhibit compaction, but do not inhibit gap junction-mediated intercellular communication or surface polarity. The surface and cytoplasmic polarity, however, are off axis, i.e. no longer directly opposite the regions of cell–cell contact. The persistence of maternal stores of E-cadherin provides a likely explanation for the observation that embryos lacking a functional E-cadherin gene nevertheless undergo compaction.[40]

E-cadherin is expressed in the egg and during all stages of preimplantation development.[41] The amount present decreases between the four-cell and eight-cell stages and then shows a dramatic increase by the blastocyst stage. E-cadherin also undergoes dramatic changes in localization.[42] In the unfertilized egg, E-cadherin is mainly localized in the cytoplasm and only trace amounts are found on the egg's surface. E-cadherin appears uniformly on the egg's surface some 6–11 h following fertilization or egg activation; E-cadherin is not associated with the emitted polar body. During compaction, E-cadherin accumulates in the region of intercellular contacts, whereas there is a general reduction in the contact-free regions. β-Catenin co-localizes with E-cadherin in the compacted eight-cell embryo.[43]

What triggers the timing of compaction is not known. A sufficient amount of a set of the components required for compaction is clearly present in the four-cell embryo, since treatment of four-cell embryos with biologically active phorbol diesters to activate protein kinase C (PKC) induces compaction within 15 min.[44] Longer exposure leads to decompaction of these treated embryos, which may be a consequence of the depolymerization of both microfilaments and microtubules.[45] Treatment of four-cell embryos with diacylglycerides, which are more natural activators of PKC, also induces compaction that is not followed, however, by decompaction. The precocious induction of compaction is inhibited by sphingosine, an inhibitor of PKC. Sphingosine also inhibits the normal compaction of eight-cell embryos. E-cadherin is involved in precocious compaction, since a monoclonal antibody to E-cadherin that blocks normal compaction also blocks phorbol diester-induced compaction of four-cell embryos. These results indicate that compaction observed in response to the PKC activation is very similar to the compaction that occurs naturally.

The aforementioned results suggest that post-translational modifications may govern the timing of compaction. Phosphorylation of E-cadherin is first clearly observed in the eight-cell embryo and thus could be involved in compaction.[41] E-cadherin phosphorylation is inhibited in embryos cultured in calcium-free medium, conditions that prevent compaction.[41] The role of PKC-induced phosphorylation in compaction, however, is not clear, since premature compaction in response to activators of PKC is not associated with an increase in E-cadherin phosphorylation. Likewise, dimethylaminopurine, which is a relatively non-specific inhibitor of protein kinases, also induces premature compaction in the absence of an increase in E-cadherin phosphorylation.[46] Reciprocally, inhibition of compaction by cytochalasin D treatment does not reduce E-cadherin phosphorylation that normally accompanies compaction. Thus, the phosphorylation of E-cadherin can be dissociated from compaction. The role of E-cadherin phosphorylation in regulating the cell adhesion properties of E-cadherin remains unclear at present, as does the molecular basis for the timing of compaction.

BLASTOCYST DEVELOPMENT

Origin of the inner cell mass and trophectoderm cells

Following compaction at the eight-cell stage, subsequent cleavage divisions allocate cells to the inside of the developing morula.[47] These inner cells are allocated between the eight-cell and 16-cell stages, and then again between the 16-cell stage and the 32-cell stage. The cleavage planes of the eight-cell-stage blastomeres are essentially random and, hence, the daughter blastomeres are allocated to the inside or remain on the outside such that about 3–4 cells occupy the interior of the 16-cell embryo. The cleavage plane of an outer 16-cell-stage blastomere, however, is not random; 80% of daughter cells remain on the outside and 20% are allocated directly to the inside. This results in 10–12 cells that constitute the ICM of the early 32-cell blastocyst. Subsequent divisions of the differentiated trophectoderm cells always result in trophectoderm descendants.

Culture of ICM cells isolated from early blastocysts can result in the formation of trophectoderm cells.[47] This potency is lost, however, by the late blastocyst stage. In contrast, differentiated trophectoderm cells never give rise to ICM cell descendants during culture. Although culture of outer cells obtained from 16-cell embryos preferentially gives rise to trophectoderm descendants, ICM progeny can also be generated. This restriction in potency is consistent with the aforementioned changes in inner cell allocation that accompanies development. These results indicate that the ICM can function as a stem cell population, i.e. give rise to trophectoderm, and lineage-marked ICM cells can be found in the trophectoderm of the blastocyst.[48] Nevertheless, results of a recent study that involved the selective replacement of some inside cells of the late morula with corresponding cells from a genetically distinct donor did not find progeny colonizing the trophectoderm.[49] Thus, while ICM cells appear to retain the ability to become trophectoderm cells following culture, it remains controversial whether the ICM cells serve *in situ* as a stem cell population for the trophectoderm.

It is also interesting to note that developmental restriction in the potency of the ICM cells correlates with developing restrictions in gap junction-mediated intercellular communication.[50] ICM cells communicate preferentially with each other and not with trophectoderm cells. Reciprocally, trophectoderm cells communicate among themselves and little with ICM cells. The formation of these separate communication compartments could be involved either in the initiation of the loss of potency of the ICM or in the maintenance of these two cell lineages.

Origin and function of the blastocoele

Between the 16-cell and 32-cell stages, the outer cells of the embryo start to transport fluid into the interior of the embryo, to produce a fluid-filled cavity called the blastocoele. Tight junctions between the trophectoderm cells are essential for blastocoele formation, because they prevent leakage of ions present in the blastocoele.

The formation of tight junctions initiates during the eight-cell stage and continues thereafter.[51] The tight junctions that form during the eight-cell stage are focal, but by the blastocyst stage the tight junctions form a continuous belt around the apicolateral region between adjacent trophectoderm cells to produce a tight permeability seal. The tight junctions also maintain the asymmetrical localization of proteins in the membranes of trophectoderm cells, for example the basolateral localization of the sodium/potassium ATPase.

Tight junctions are composed of several integral (such as occludin) and peripheral proteins (ZO-1, cingulin). While expression and assembly into junctions of the ZO-1α^- form, which is the isoform typically expressed by endothelial cells, occur from the eight-cell stage onwards, expression and assembly of the ZO-1α^+ form initiate during the late morula stage, just prior to blastocoele formation.[51] Thus, the ZO-1α^- form may be involved in maintaining membrane polarity, i.e. the asymmetrical localization of proteins in the membrane, starting in the compacted eight-cell embryo, whereas the ZO-1α^+ form may be critical to form a tight permeability seal required for blastocoele formation. E-cadherin is also required for the formation of a trophectoderm epithelium, since embryos that do not contain a functional E-cadherin gene fail to form a trophectodermal epithelium and blastocoele.[40,51]

Fluid accumulation in the blastocoele is due to an energy-dependent transepithelial sodium flux via numerous apically located channels that are permeant for sodium, for example an amiloride-sensitive sodium channel.[51,52] A basolaterally located sodium/potassium ATPase then actively pumps the intracellular sodium into the blastocoele. An osmotically driven, passive movement of water across the trophectoderm results in blastocoele formation. The movement of other ions such as chloride and bicarbonate contributes to blastocoele formation. It should be noted that, while transcripts for the α subunit of the sodium/potassium ATPase are present throughout preimplantation development, transcripts for the β subunit start to increase at the morula stage.[51] This correlates with the onset of blastocoele formation.

cAMP stimulates the rate of blastocoele expansion.[53] Growth factors such as transforming growth factor-α (TGF-α) also stimulate the rate of blastocoele expansion.[51,54] This effect may be a consequence of an increase in the activity of the sodium/potassium ATPase and/or a further tightening of the tight junctions. Calcium derived from the inositol triphosphate store may also be involved in blastocoele formation, since inhibitors of phospholipase C inhibit cavitation in a concentration-dependent and reversible manner.[55] Also in support of a role for intracellular calcium is that treatment of embryos with agents that elevate intracellular calcium (e.g. ethanol, ionomycin) accelerates the process of cavitation.[56] Whether such calcium signalling normally occurs *in vivo* and is critical for blastocoele formation, however, remains to be determined.

Blastocoele formation and expansion are essential for differentiation of the ICM and continued development. The formation of a free surface on the ICM cells facing the blastocoele may be critical for the differentiation of the outer cells of the ICM into primitive endodermal cells. The ion composition of the blastocoele fluid and proteins that are secreted by both the ICM and trophectoderm into the blastocoele fluid may be critical for ICM cell proliferation and differentiation.[57] Blastocoele expansion may also contribute to hatching, and also the initial events of implantation by enhancing the initial cell–cell contacts between the trophectoderm and endometrium.

Gene expression in the blastocyst

Transcriptional and translational events required for blastocyst formation occur only a few hours before the onset of cavitation.[51] The identities of the genes and proteins involved are unknown. Two obvious candidates, however, are the $\beta 1$ subunit of the sodium/potassium ATPase and ZO-1α^+ genes, each of which is clearly implicated in blastocyst formation, since their expression dramatically increases during this time.[51] Removal of cytoplasm from one-cell embryos results in the precocious onset of cavitation.[58] While the molecular basis for this is not known – perhaps an inhibitor of cavitation is titrated with each cell division – it would be interesting to observe whether the increased expression of the sodium/potassium ATPase and ZO-1α+ genes also occurs earlier in such experimentally manipulated embryos. Blastocyst formation does not require the previous round of DNA replication.[59]

Whereas the transcriptional events required for blastocoele formation occur near the time of cavitation, the transcriptional events that underlie hatching are completed many hours in advance, with the translational needs being completed just a few hours in advance of hatching.[37] This suggests that post-transcriptional control may be a major mechanism regulating hatching. Nevertheless, very little is known about the molecular basis for hatching.

Differences in gene expression exist between the two cell types present in the blastocyst.[60,61] These differences in gene expression are presumably intimately coupled to cell specification and the maintenance or loss of totipotency. A few ICM-specific transcripts have been identified, such as *Fgf4*.[62] ICM cell proliferation is severely impaired in embryos homozygous for the null allele of *Fgf4*, whereas trophectoderm development is relatively unaffected.[63] The retinoic acid-regulated, zinc finger gene *rex-1* is also preferentially expressed in the ICM,[64] as is the expression of *Oct4*, a placenta–ovary–uterus (POU) domain-binding transcription factor that may play a critical role in establishment of the germline (Figure 5).[65] The cytokeratin endo A, for which expression starts to increase at around the eight-cell stage,[66] becomes preferentially expressed in the trophectoderm. Similarly, the expression of the imprinted *H19* gene (maternal allele-expressed) is first detected in the blastocyst and specifically in the trophectoderm.[67]

Virtually nothing is known regarding the molecular basis for the initiation and maintenance of such differences in gene expression between ICM and trophectoderm. Although cell–cell contact is

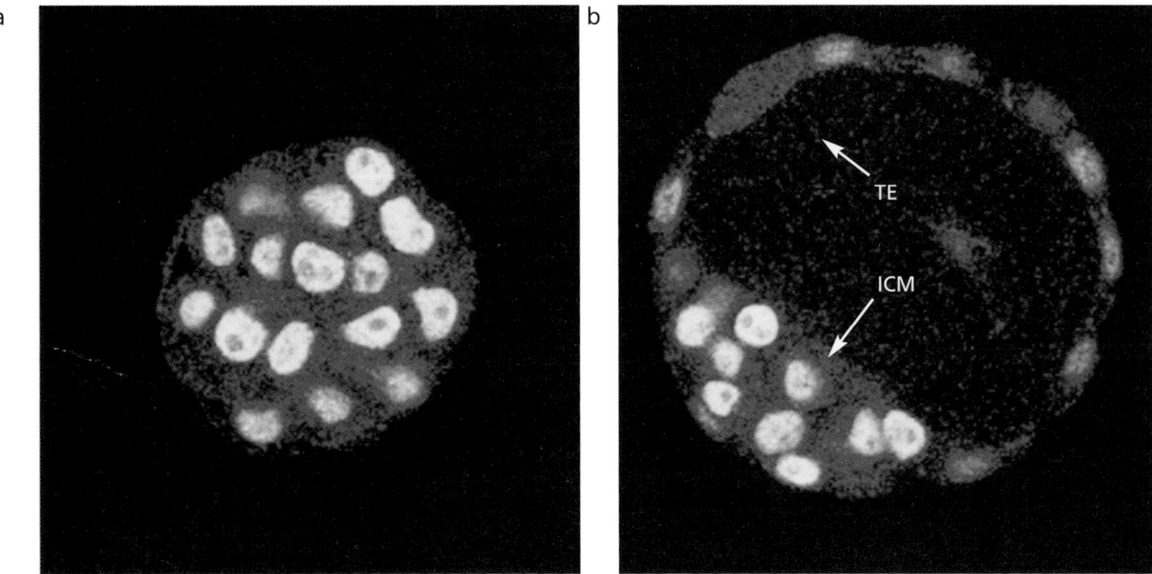

Figure 5 Laser-scanning confocal image of blastocyst stained for *Oct4*: (a) 32-cell morula, (b) blastocyst. Note that the staining of *Oct4* is largely confined to the inner cell mass (ICM). Prior to this, *Oct4* is expressed equally in virtually all cells of the morula; nuclei that are not in the full center of the confocal plane give rise to the fainter-staining nuclei present in the morula. TE, trophectoderm. The affinity-purified *Oct4* antibody was the generous gift of Dr Hans Schöler

clearly critical in cell specification, how this contact is interpreted in the context of restricted patterns of gene expression is not apparent. Catenins, which participate in the transduction of signals related to cell contacts, could provide the linkage between cell contact and changes in ICM or trophectoderm gene expression pattern.

Another possible mechanism for initiating and/or maintaining spatially restricted patterns of gene expression in the blastocyst is a change in promoter utilization. The efficient expression of a plasmid-borne reporter gene driven by the *tk* promoter and F101 enhancer does not require a TATA box in undifferentiated blastomeres.[68] In contrast, a TATA box is required for efficient expression in differentiated cells such as the oocyte. While such a change in promoter utilization could maintain high levels of expression of housekeeping genes that frequently lack a TATA box in the early embryo and result in a decrease in their expression following differentiation, such a change could also be coupled to the change in the spatial pattern of gene expression that accompanies the differentiation of the outer cells of the morula into the trophectoderm cells of the blastocyst. It is interesting to note that the ICM-specific *Oct4* gene lacks a TATA box.[65]

The expression of the transcription factor *Oct4* in the blastocyst is critical for successful development. Mouse embryos lacking a functional *Oct4* gene fail to form an ICM and differentiate exclusively along the trophectoderm lineage.[69] Interestingly, the trophectoderm cells cease proliferating, but proliferation can be made to resume by treatment with fibroblast growth factor 4 (FGF4), which is itself encoded by a target gene of *Oct4*. These observations indicate that *Oct4* is essential for formation and maintenance of the totipotent ICM. They also indicate the existence of a critical axis of communication between the ICM and trophectoderm in the form of FGF4 and other factors (see below), which are essential for the long-term proliferative potential and survival of the ICM and trophectoderm cell populations. Other studies have revealed that the precise amount of *Oct4* expressed in a given cell is a critical determinant of cell lineage. Embryo stem cells made to overexpress *Oct4* by less than two-fold will differentiate primarily along the mesoderm/primitive endoderm pathway, whereas underexpression leads to trophectoderm differentiation.[70] These results indicate that quantitative aspects of *Oct4* expression, and not just expression *per se*, are critical for *Oct4*-mediated control of lineage commitment and maintenance of developmental pluripotentiality.

X chromosome regulation in the blastocyst

The mammalian X chromosome is paternally imprinted so that the paternal X chromosome is preferentially inactivated in extraembryonic tissues. By contrast, this imprint is lost in the developing ICM of eutherian mammals. As a result, X chromosome inactivation becomes 'random' in the developing embryo proper. This process is mimicked during cloning through the use of somatic cell nuclear transfer, so that the previously inactive X chromosome of female cumulus cells remains inactive in the developing trophectoderm cells but can be expressed in the developing embryo proper.[71]

Apoptosis in the blastocyst

Apoptosis occurs in the blastocyst and is essentially restricted to the ICM.[72] The regulation of cell death in the blastocyst is likely to be critical for later development. For example, cell death may eliminate ICM cells that retain the potential to form trophectoderm. Conversely, unregulated cell death in the ICM would also compromise later development, since a critical threshold number of ICM cells is required for normal post-implantation development.[73,74] Although early blastomeres do not require survival signals, it is possible that ICM and trophectoderm cells do.

Cell division, differentiation and death in embryos are likely to be regulated by peptide growth factors. Embryos express a variety of growth factor receptors and their ligands, and exogenous growth factors can modulate many aspects of embryo development.[75] Exogenous TGF-α stimulates the rate of blastocoele expansion and protein synthesis. TGF-α also stimulates the secretion of proteins by the blastocyst and alters the pattern of gene expression.[76] Signalling mediated by the epidermal growth factor (EGF) receptor may be critical for ICM development, since one of the phenotypes observed following ablation of the EGF receptor gene is that peri-implantation lethality on day 6.5 is associated with failure of the ICM to develop.[77]

Growth factors, and TGF-α in particular, may function as cell survival factors in the preimplantation mouse embryo.[72] For example, when embryos are cultured singly in individual drops from the two-cell stage to the blastocyst stage, apoptosis in the ICM is increased by up to three-fold, when compared with control blastocysts that have developed *in vivo*. This suggests that factors present in the maternal reproductive tract may act as survival factors. These factors may also be secreted by the embryos themselves, since the incidence of apoptosis is reduced following culture of two-cell embryos to blastocyst at high density (30 embryos per drop). Moreover, exogenous TGF-α reduces the incidence of apoptosis in the ICM of blastocysts that develop from singly cultured two-cell embryos. In addition, when ICMs are removed from the blastocoele by immunosurgery and cultured singly for 24 h, they show a dramatic increase in the incidence of apoptosis that is partially reduced by adding exogenous TGF-α to the medium. TGF-α-deficient (i.e. –/–) mice are fertile and display only minor phenotypic changes that include skin architecture, hair follicle and eyelid abnormalities.[78,79] Although TGF-α –/– embryos develop into morphologically normal blastocysts *in vitro* or *in vivo* with similar size, total cell numbers, mitotic index and allocation to ICM and trophectoderm lineages, when compared with wild-type control embryos that express TGF-α, the incidence of apoptosis in the ICM for the TGF-α –/– embryos is much higher, when compared with wild-type control embryos.[72] Thus, TGF-α functions as a cell survival factor in the preimplantation mouse embryo, but other factors can clearly compensate for its absence. This observation is yet another reminder that a close examination of the phenotype can reveal subtle phenotypic changes that can, in turn, suggest functions not previously apparent or appreciated.

ACKNOWLEDGEMENTS

The research conducted in the authors' laboratories was supported by grants from the National Institutes of Health (HD 22681 and 38381).

REFERENCES

1. Schultz GA. Utilization of genetic information in the preimplantation mouse embryo. In Rossant J, Pedersen RA, eds. *Experimental Approaches to Mammalian Embryonic Development*. Cambridge University Press: Cambridge, 1986: 239–265.
2. Nothias J-Y, Majumder S, Kaneko KJ, DePamphilis ML. Regulation of gene expression at the beginning of mammalian development. *J Biol Chem* 1995 270: 22077–22080.
3. Smith RKW, Johnson MH. Analysis of the third and fourth cell cycles of mouse early development. *J Reprod Fertil* 1986 76: 393–399.
4. Wassarman PM. Zona pellucida glycoproteins. *Ann Rev Biochem* 1988 57: 414–442.
5. Modlinski JA. The role of zona pellucida in the development of mouse eggs *in vivo*. *J Embryol Exp Morphol* 1970 23: 539–547.
6. Perona RM, Wassarman PM. Mouse blastocysts hatch *in vitro* by using a trypsin-like proteinase associated with cells of the mural trophectoderm. *Dev Biol* 1986 114: 42–52.
7. Thomas M, Jain S, Kumar GP, Laloray M. A programmed oxyradical burst causes hatching of mouse blastocysts. *J Cell Sci* 1997 110: 1597–1602.
8. Yanagimachi R. Mechanisms of fertilization in mammals. In Mastroianni L, Biggers JD, eds. *Fertilization and Embryonic Development In Vitro*. New York: Plenum Press, 1981: 81–182.
9. Latham KE. Mechanisms and control of embryonic genome activation in mammalian embryos. *Int Rev Cytol* 1999 193: 71–124.
10. Schultz RM. Molecular aspects of mammalian oocyte growth and maturation. In Rossant J, Pedersen RA, eds. *Experimental Approaches to Mammalian Embryonic Development*. Cambridge University Press: Cambridge, 1986: 195–237.
11. Schultz RM. Regulation of zygotic gene activation in the mouse. *BioEssays* 1993 8: 531–538.
12. Latham KE, Garrels JI, Chang C, Solter D. Quantitative analysis of protein synthesis in mouse embryos. I. Extensive reprogramming at

the one- and two-cell stages. *Development* 1991 112: 921–932.

13. Latham KE, Schultz RM. Embryonic genome activation. *Front Biosci* 2001 6: d748–759.

14. Wang Q, Chung YG, deVries WN, Struwe M, Latham KE. Role of protein synthesis in the development of a transcriptionally permissive state in one cell stage mouse embryos. *Biol Reprod* 2001 65: 748–754.

15. Wolffe AP. Implications of DNA replication for eukaryotic gene expression. *J Cell Sci* 1991 99: 201–206.

16. Wolffe AP. The transcription of chromatin templates. *Curr Opin Genet Dev* 1994 4: 245–254.

17. Davis W Jr, De Sousa PD, Schultz RM. Transient expression of translation initiation factor eIF-4C during the 2-cell stage of the preimplantation mouse embryo: identification by mRNA differential display and the role of DNA replication. *Dev Biol* 1996 174: 190–201.

18. Aoki F, Worrad DM, Schultz RM. Regulation of endogenous transcriptional activity during the first and second cell cycles in the preimplantation mouse embryo. *Dev Biol* 1997 181: 296–307.

19. Davis W Jr, Schultz RM. Role of the first round of DNA replication in reprogramming gene expression in the preimplantation mouse embryo. *Mol Reprod Dev* 1997 47: 430–434.

20. Pazin MJ, Kadonaga JT. SWI2/SNF2 and related proteins: ATP-driven motors that disrupt protein–DNA interactions. *Cell* 1997 88: 737–740.

21. Wilson CJ, Chao DM, Imbalzano AN *et al.* RNA polymerase II holoenzyme contains SWI/SNF regulators involved in chromatin remodeling. *Cell* 1996 84: 235–244.

22 Henery CC, Miranda M, Wiekowski M, Wilmut I, DePamphilis ML. Repression of gene expression at the beginning of mouse development. *Dev Biol* 1995 169: 448–460.

23 Woodland HR, Adamson EP. The synthesis and storage of histone during oogenesis in *Xenopus laevis*. *Dev Biol* 1977 57: 118–135.

24. Newport J, Kirschner M. A major developmental transition in early *Xenopus* embryo. II. Control of the onset of transcription. *Cell* 1982 30: 687–696.

25. Prioleau M-N, Buckle RS, Méchali M. Programming of a repressed but committed chromatin structure during early development. *EMBO J* 1995 14: 5073–5084.

26. Van Blerkom, J, Davis PW. DNA strand break and phosphatidylserine redistribution in newly ovulated cultured mouse and human oocytes: occurrence and relationship to apoptosis. *Hum Reprod* 1998 13: 1317–1324.

27. Perez GI, Tao XJ, Tilly JL. Fragmentation and death (aka apoptosis) of ovulated oocytes. *Mol Hum Reprod* 1999 5: 414–420.

28. Jurisicova A, Latham KE, Casper RF, Varmuza, SL. Expression and regulation of genes associated with cell death during murine preimplantation embryo development. *Mol Reprod Dev* 1998 51: 243–253.

29. Antczak, M, Van Blerkom, J. Temporal and spatial aspects of fragmentation in early human embryos: possible effects on developmental competence and association with the differential elimination of regulatory proteins from polarized domains. *Hum Reprod* 1999 14: 429–447.

30. Hardy K. Apoptosis in the human embryo. *Rev Reprod* 1999 4: 125–134.

31. Hawes SM, Chung YG, Latham KE. Genetic and epigenetic factors affecting blastomere fragmentation in two cell stage mouse embryos. *Biol Reprod* 2001 65: 1050–1056.

32. Johnson MH, Maro B. Time and space in the mouse early embryo: a cell biological approach to cell diversification. In Rossant J, Pedersen RA, eds. *Experimental Approaches to Mammalian Embryonic Development*. Cambridge University Press: Cambridge, 1986: 35–65.

33. Reeve WJD, Kelly FP. Nuclear position in the cells of the mouse early embryo. *J Embryol Exp Morphol* 1983 75: 117–139.

34. Pratt HPM, Ziomek CA, Reeve WJD, Johnson MH. Compaction of the mouse embryo: an analysis of its components. *J Embryol Exp Morphol* 1982 70: 113–132.

35. Johnson MH, Ziomek CA. Induction of polarity in mouse 8-cell blastomeres: specificity, geometry, and stability. *J Cell Biol* 1981 91: 303–308.

36. Smith KW, Johnson MH. DNA replication and compaction in the cleaving embryo of the mouse. *J Embryol Exp Morphol* 1985 89: 133–148.

37. Kidder GM, McLachlin JR. Timing of transcription and protein synthesis underlying morphogenesis in preimplantation mouse embryos. *Dev Biol* 1985 112: 265–275.

38. Levy JB, Johnson MH, Goodall H, Maro B. The timing of compaction: control of a major

developmental transition in mouse early embryogenesis. *J Embryol Exp Morphol* 1986 95: 213–237.

39. Hyafil F, Babinet C, Jacob F. Cell–cell interactions in early embryogenesis: a molecular approach to the role of calcium. *Cell* 1981 26: 447–454.

40. Larue L, Ohsugi M, Hirchenhain J, Kemler R. E-cadherin null mutant embryos fail to form a trophectoderm epithelium. *Proc Natl Acad Sci USA* 1994 91: 8263–8267.

41. Sefton M, Johnson MH, Clayton L, McConnell JML. Experimental manipulations of compaction and their effects on the phosphorylation of uvomorulin. *Mol Reprod Dev* 1996 44: 77–87.

42. Clayton L, Stinchcombe SV, Johnson MH. Cell surface localisation and stability of uvomorulin during early mouse development. *Zygote* 1993 1: 333–344.

43. Ohsugi M, Hwang S-Y, Butz S *et al.* Expression and cell membrane localization of catenins during mouse preimplantation development. *Dev Dynamics* 1996 206: 391–402.

44. Winkel GK, Ferguson JE, Takeichi M, Nuccitelli R. Activation of protein kinase C triggers premature compaction in the four-cell stage mouse embryo. *Dev Biol* 1990 138: 1–15.

45. Bloom TL. The effects of phorbol ester on mouse blastomeres: a role for protein kinase C in compaction? *Development* 1989 106: 159–171.

46. Aghion JI, Guth-Hallonet C, Antony C *et al.* Cell adhesion and gap junction formation are induced prematurely by 6-DMAP in the absence of E-cadherin phosphorylation. *J Cell Sci* 1994 107: 1369–1379.

47. Pedersen RA. Potency, lineage, and allocation in preimplantation mouse embryos. In Rossant J, Pedersen RA, eds. *Experimental Approaches to Mammalian Embryonic Development.* Cambridge University Press: Cambridge, 1986: 3–33.

48. Winkel GK, Pedersen RA. Fate of the inner cell mass in mouse embryos as studied by microinjection lineage markers. *Dev Biol* 1988 127: 143–156.

49. Gardner RL, Nichols J, Gardner RL, Nichols J. An investigation of the fate of cells transplanted orthotopically between morulae/nascent blastocysts in the mouse. *Hum Reprod* 1991 6: 25–35.

50. Lo CW, Gilula NB. Gap junctional communication in the preimplantation mouse embryo. *Cell* 1979 18: 399–409.

51. Watson AJ, Barcroft LC. Regulation of blastocyst formation. *Front Biosci* 2001 6: d708–730.

52. Biggers JD, Bell JE, Benos DJ. Mammalian blastocyst: transport functions in a developing epithelium. *Am J Phsyiol: Cell Physiol* 1988 255: C419–C432.

53. Manejewala FM, Schultz RM. Blastocoel expansion in the preimplantation mouse embryo: stimulation of sodium uptake by cAMP and possible involvement of cAMP-dependent protein kinase. *Dev Biol* 1989 136: 560–563.

54. Dardik A, Schultz RM. Blastocoel expansion in the preimplantation mouse embryo: stimulatory effect of TGFα and EGF. *Development* 1991 113: 919–930.

55. Stachecki JJ, Armant DR. Regulation of blastocoelee formation by intracellular calcium release is mediated through a phospholipase C-dependent pathway in mice. *Biol Reprod* 1996 55: 1292–1298.

56. Stachecki JJ, Armant DR. Transient release of calcium from inositol 1,4,5-triphosphate-specific stores regulates mouse preimplantation development. *Development* 1996 122: 2485–2496.

57. Feng Y-L, Gordon JW. Removal of cytoplasm from one-celled mouse embryos induces early blastocyst formation. *J Exp Zool* 1997 277: 345–352.

58. Handyside AH, Johnson MH. Temporal and spatial patterns of synthesis of tissue-specific polypeptides in the preimplantation mouse embryo. *J Embryol Exp Morphol* 1978 44: 191–199.

59. Feldman B, Poueymirou W, Papaioannou VE, DeChiara TM, Goldfarb M. Requirement for FGF-4 for postimplantation mouse development. *Science* 1995 267: 246–249.

60. Van Blerkom J, Barton SC, Johnson MH. Molecular differentiation of the preimplantation mouse embryo. *Nature* 1976 259: 319–321.

61. Yeom YII, Fuhrmann G, Ovitt CE *et al.* Germline regulatory element of *Oct-4* specific for the totipotent cycle of embryonal cells. *Development* 1996 122: 881–894.

62. Duprey P, Morello D, Vasseur M *et al.* Expression of the cytokeratin endo A gene during early mouse embryogenesis. *Proc Natl Acad Sci USA* 1985 82: 8535–8539.

63. Doherty AS, Mann MRW, Tremblay KD, Bartolomei MS, Schultz RM. Differential effects of culture on imprinted H19 expression in the

preimplantation mouse embryo. *Biol Reprod* 2000 62: 1526–1535.

64. Majumder S, DePamphilis ML. TATA-dependent enhancer stimulation of promoter activity in mice is developmentally regulated. *Mol Cell Biol* 1994 14: 4258–4268.
65. Nichols J, Zevnik B, Anastassiadis K *et al.* Formation of pluripotent stem cells in the mammalian embryo depends on the POU transcription factor *Oct4*. *Cell* 1998 95: 379–391.
66. Niwa H, Miyazaki J, Smith AG. Quantitative expression of *Oct-3/4* defines differentiation, dedifferentiation or self-renewal of ES cells. *Nat Genet* 2000 24: 372–376.
67. Eggan K, Akutsu H, Hochedlinger K, Rideout W III, Yanagimachi R, Jaenisch R. X-chromosome inactivation in cloned mouse embryos. *Science* 2000 290: 1578–1581.
68. Brison DR, Schultz RM. Increased incidence of apoptosis in TGF-α-deficient mouse blastocysts. *Biol Reprod* 1998 59: 136–144.
69. Handyside AH, Hunter S. Cell division and death in the mouse blastocyst before implantation. *Roux's Arch Dev Biol* 1986 195: 519–526.
70. Pierce GB, Lewellyn AL, Parchment RE. Mechanism of programmed cell death in the blastocyst. *Proc Natl Acad Sci USA* 1989 86: 3654–3658.
71. Adamson ED. Activities of growth factors in preimplantation embryos. *J Cell Biochem* 1993 53: 280–287.
72. Babalola GO, Schultz RM. Modulation of gene expression in the preimplantation mouse embryo by TGF-alpha and TGF-beta. *Mol Reprod Dev* 1995 41: 133–139.
73. Threadgill DW, Dlugosz AA, Hansen LA *et al.* Targeted disruption of mouse EGF receptor: effect of genetic background on mutant phenotype. *Science* 1995 269: 230–233.
74. Luetteke NC, Qiu TH, Peiffer RL *et al.* TGF alpha deficiency results in hair follicle and eye abnormalities in targeted and waved-1 mice. *Cell* 1993 73: 263–278.
75. Mann GB, Fowler KJ, Gabriel A *et al.* Mice with a null mutation of the TGF alpha gene have abnormal skin architecture, wavy hair, and curly whiskers and often develop corneal inflammation. *Cell* 1993 73: 249–261.

Chapter 24

Implantation and endometrial function

Linda C. Giudice

INTRODUCTION

The endometrium is a dynamic tissue which develops, on a monthly basis, in response to circulating ovarian steroids that prepare it for pregnancy. Implantation involves complex interactions between the implanting embryo and the primed maternal endometrium. Histological examination of early human pregnancies reveals distinct patterns of blastocyst attachment to the endometrial surface and the underlying stroma.[1] These observations support a model (Figure 1) of implantation in humans in which the embryo apposes and attaches to the endometrial epithelium, traverses adjacent cells of the epithelial lining and invades into the endometrial stroma. The endometrium is receptive to embryonic implantation during a defined 'window' that is temporally and spatially restricted. Early studies suggest that this window resides in the mid-secretory phase, because embryos identified in secretory phase hysterectomy specimens were all free-floating before day 20 of the cycle and all were attached when the specimens were obtained after day 20.[1] Definition of the window of implantation has received substantial support from the appearance of epithelial dome-like structures ('pinopodes'; see below) that last for 1–2 days and correlate with implantation sites.[2] Additional information derives from assisted reproductive cycles in which the window of endometrial receptivity was tested by performing embryo transfers between days 16 and 24 of hormonally and histologically defined cycles.[3] For transfers within cycle days 17–19, 40.5% conceptions occurred, whereas, in cycles in which embryos were transferred on or beyond day 20, no conceptions occurred. Recently, Wilcox and colleagues[4] reported a high success rate (84%) of continuing pregnancy for embryos that implant between cycle days 22 and 24 (post-ovulation days 8–10), compared with 18% when implantation occurred 11 days or

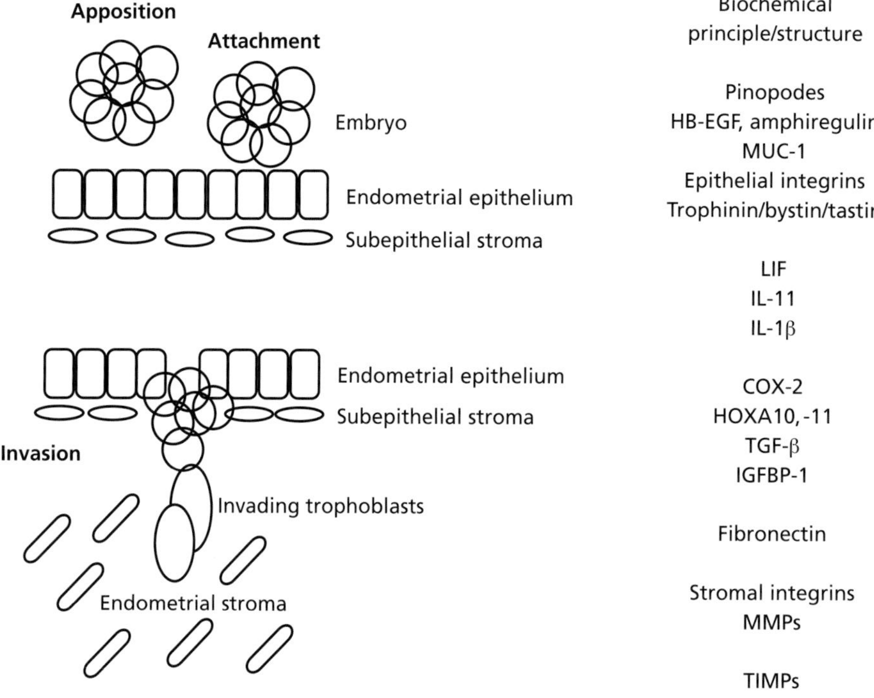

Figure 1 Schematic diagram of model of human implantation. Shown are the apposition/attachment phases and the invasive phase. Biochemical principles and structures involved in implantation are given at the right. HB-EGF, heparan-binding epidermal growth factor (EGF)-like growth factor; MUC-1, mucin-1; LIF, leukemia inhibitory factor; IL, interleukin; COX-2, cyclo-oxygenase-2; TGF, transforming growth factor; IGFBP, insulin-like growth factor (IGF) binding protein; MMP, matrix metalloproteinase; TIMP, tissue inhibitor of metalloproteinase

more after ovulation. Together, these data suggest that the window of implantation in humans spans cycle days 20–24.

The primary goals of the embryo are to ensure a secure place for itself and to invade into the maternal vasculature to access nutrients for its continued sustenance and growth. The invasive phase of implantation occurs first primarily in a non-decidualized endometrium, and it is likely that the embryo dictates changes within the maternal endometrium to facilitate its own implantation. The molecular dialog that occurs between the endometrium and the implanting conceptus involves cell–cell and cell–extracellular matrix interactions, mediated by lectins, integrins, matrix-degrading enzymes and their inhibitors, and a variety of growth factors and cytokines, their receptors and modulatory proteins. Several cell types are involved in the process of implantation, including endometrial epithelium, stroma, fibroblasts, vascular endothelium, smooth muscle and cells of the lymphoid lineage, as well as embryonic trophectoderm, placental fibroblasts and a variety of trophoblast phenotypes. Cross-talk between and among these cell types is important for endometrial receptivity, stromal decidualization and embryonic implantation. The molecular changes occurring in the endometrium during the process of human implantation are discussed herein in the context of successful implantation and implantation failure. Most of what we believe occurs during human implantation is derived from animal models, many of which have different mechanisms of implantation, compared with primates. None-theless, transgenic and gene 'knock-out' models have been useful in elucidating the importance of some effector molecules in the process of implantation, which may have relevance to humans. Additional information has derived from *in vitro* studies with human endometrial cells and explant cultures, and human trophoblasts and placental explant cultures. As a consequence, our understanding of the mechanisms underlying co-ordinated development of the embryo and the endometrium, and the establishment of successful implantation and pregnancy in humans is incomplete. Several reviews have been published recently, addressing the role of the endometrium in implantation.[5–9]

STEROID HORMONE RECEPTORS IN ENDOMETRIUM: THE BASIS OF ENDOMETRIAL DEVELOPMENT FOR IMPLANTATION

Endometrial development is dependent on steroid hormones. These hormones act via steroid hormone receptors which are transcription factors. Estrogen receptor (ER) concentrations are highest during the proliferative phase[10,11] and decrease after ovulation, reflecting the suppressive effects of progesterone on the ER. Immunostaining has revealed that the ER is present in highest amounts in the glandular epithelium during the proliferative phase.[11] Recent evidence demonstrates two forms of the estrogen receptor, ERα and ERβ. ERα is the predominant form expressed in human endometrium.[12] Early studies demonstrated peak progesterone receptor (PR) concentrations at the time of ovulation (reflecting induction of the PR by estradiol), and that the PR is most prominent in glandular epithelium in the proliferative phase, but undetectable in the mid-secretory phase.[10,12] Furthermore, stromal cells were demonstrated to have high levels of PR in the proliferative phase and throughout the secretory phase. It is now known that there are two isoforms of the PR, PRA and PRB. PRA and PRB are co-expressed in target cells in human endometrium.[12–14] In glands (Figure 2a), PRA and PRB are expressed before subnuclear vacuole formation and glycogenolysis, suggesting both isoforms in this process. In contrast, only PRB persists during the mid-secretory phase, suggesting a role for this isoform in glandular secretion. In endometrial stroma (Figure 2b), PRA predominates throughout the cycle, suggesting that it is important in progesterone action in the secretory phase. Overall, these results support the view that PRA and PRB mediate distinct pathways of progesterone action in the glandular epithelium and stroma throughout the menstrual cycle. It is the marked down-regulation of the PR that coincides with the opening of the window of implantation (see below).[11]

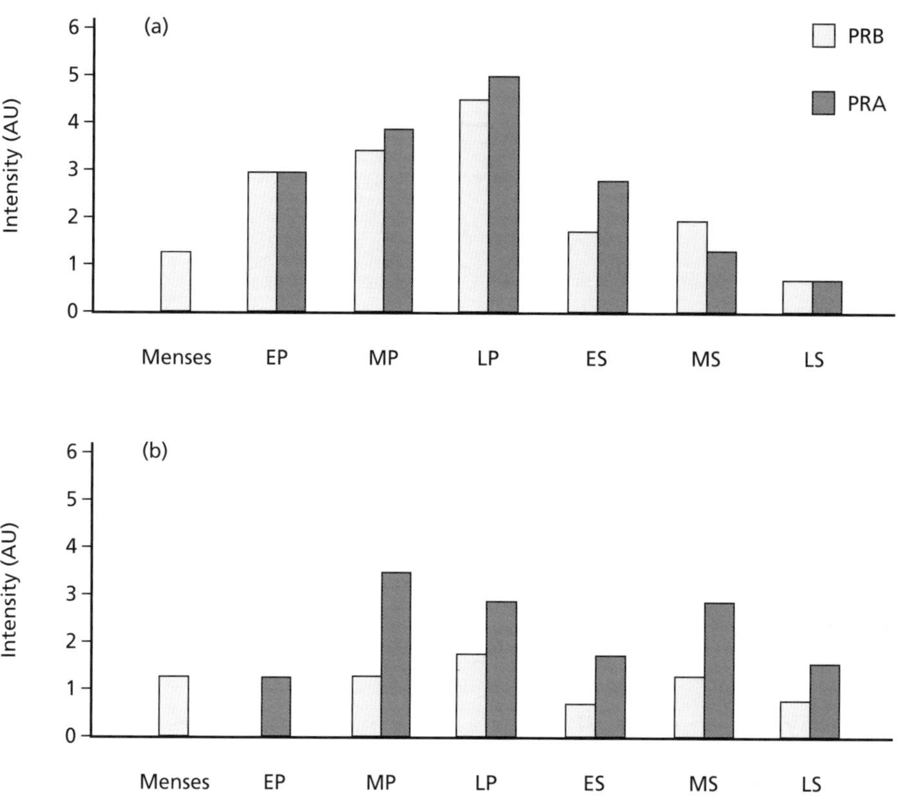

Figure 2 Relative expression of progesterone receptors A and B (PRA and PRB) in normal endometrium during the menstrual cycle: (a) glands, (b) stroma. E, early; M, mid-; L, late; P, proliferative; S, secretory. From reference 13, with permission

ROLE OF THE ENDOMETRIUM IN APPOSITION AND ATTACHMENT

Pinopodes

Dome-like structures involved in endocytosis and pinocytosis are found on the apical surface of endometrial luminal epithelium during the period of receptivity in rats and humans.[2,15] The appearance of pinopodes (from the Greek: *pino* = drink; *podia* = feet) is strictly progesterone-dependent. The function(s) of the pinopodes is unclear, but they may be involved in implantation by mechanisms that involve uptake of macromolecules, withdrawal of uterine fluid and/or facilitating adhesion of the blastocyst to the luminal epithelium. The temporal and spatial expression of pinopodes has led to the speculation that pinopode formation may define the development of uterine receptivity for blastocyst implantation.[16] Scanning electron microscopy has revealed that 78% of endometrial biopsies obtained from normally cycling women on post-ovulatory day (POD) 6 have pinopodes, compared with rare pinopodes visualized on POD 2 or 9. Pinopodes form briefly and last for 1–2 days, and their numbers correlate positively with implantation sites.[2,15] There is up to 5 days' variation among women in the days post-ovulation in natural cycles in which pinopodes form. There are abundant data demonstrating limited plasticity of the window, with advancement of the window in clomiphene- or gonadotropin-stimulated cycles (POD 5–6) and delay of the window in steroid hormone replacement cycles for donor-recipients (POD 7–8).[15,16] These observations collectively support a distinct and narrow period of endometrial specialization that coincides with the window of implantation. Furthermore, they suggest that hormonal treatment used to induce ovulation or in donor-recipient cycles can modify normal development of the prenidatory endometrium and may have a negative effect on embryonic implantation. While pinopodes are best seen by scanning electron microscopy, this technique is disadvantageous as a clinical tool because of the need for specialized equipment and the time and cost of sample processing. However, it may contribute further to our understanding of the implantation process in general, and uterine receptivity (and non-receptivity) in select groups of infertile women with, for example, unexplained infertility or repetitive miscarriage with transient implantations.

In vitro models have been developed that have given unique insight into early implantation events in humans which are difficult, if not impossible, to assess *in vivo*. *In vitro* culture of polarized human epithelial cells on a basement membrane with stromal cells underneath reveals preferential human blastocyst (trophectoderm) attachment to pinopode-presenting epithelial cells,[17] mimicking the limited *in vivo* data available.[1] Furthermore, this model provides some insight into mechanisms of trophoblast penetration of the epithelium. For example, sharing of apical junctional complexes and desmosomes between lateral plasma membranes of endometrial epithelial cells and trophoblasts as the trophoblast cells penetrate the epithelial cells to gain access to the stromal compartment has been observed.[17–19] There is recent evidence that embryonic passage through the epithelium may involve apoptosis of the latter,[20] although this is controversial.[19] Challenges for the future include determining the functions of pinopodes in the process of implantation and defining what regulates their formation and disappearance in a temporally and spatially restricted manner in the endometrium, and mechanisms underlying embryonic attachment and intrusion through the epithelium.

Cell adhesion molecules

Attachment of the trophoblast to the endometrial epithelium is unique in that this cell–cell contact occurs via respective apical cell membranes, and it has been postulated that these apical plasma membranes express unique cell adhesion molecules which mediate this initial attachment that begins the process of implantation. Carbohydrate moieties on the uterine epithelium, expressed or regulated during the window of implantation, are believed to participate in the processes of apposition and attachment of the hatching blastocyst.[21] Common features of changes in luminal epithelial cells opening the window of implantation include decreased surface negativity, increased thickness of periodic acid–Schiffe (PAS)-positive material and a more fibrillar glycocalyx structure revealed by electron microscopy. Sialic acid, which confers surface negative charge to the epithelium, is associated with non-receptivity in some species. Removal of these carbohydrate moieties or changing the surface charge to become less negative is compatible with the acquisition of uterine receptivity to implantation. In addition, there are numerous cell adhesion molecules expressed by the endometrial epithelium (and the embryo) during the window of implantation.[9,21,22] Some of the glycocalyces are receptors for extracellular matrix proteins, such as fibronectin, vitronectin, laminin and collagen. The focus of this section is on carbohydrate epitopes, mucins, integrins and a novel cell adhesion complex expressed on endometrial epithelium that is believed to be important in implantation.

Carbohydrate epitopes

The H-type-1 antigen, Fucα1-2Galβ1-3GlcNAcβ1, is expressed by human glandular and luminal epithelium, in greater amounts in the secretory than in the proliferative phase.[23] In mice, it is expressed uniformly on the apical side of the luminal epithelium, up to day 4 of pregnancy, corresponding to the period of endometrial receptivity.[24] Fully expanded, hatching mouse blastocysts bind H-type-1 antigen.[25] While it is unknown whether hatching human blastocysts bind to this structure, expression of the antigen in human endometrium indicates that, by analogy with the mouse, the H-type-1 antigen is a candidate participating in embryonic attachment during human implantation.

Another surface carbohydrate, heparan sulfate proteoglycan (HSPG), is also believed to be involved in blastocyst–endometrial epithelial interactions. In mice, HSPG binds via perlecan, a basement membrane type of HSPG that forms around the blastocyst after hatching.[26] Studies of human cell lines suggest that this interaction may also be important in human implantation. For example, binding of JAR cells (a human trophoblast cell line) to an endometrial adenocarcinoma cell line, RL-95, is heparin-dependent.[27] An HSPG-binding protein (heparin/heparan sulfate proteoglycan interacting protein (HIP)) is expressed in human endometrial epithelium, although its expression is not cycle-dependent. Nonetheless, HIP–perlecan–heparan sulfate interactions have been proposed to participate in the attachment reaction, based on high levels of perlecan immunostaining in anchoring villi, where attachment begins.[28] Also, the neural cell adhesion molecule (NCAM), an immunoglobulin superfamily cell adhesion molecule, can bind HSPG and has been detected in human embryos,[29] supporting the importance of these molecules in embryonic attachment to the epithelium. In addition to their potential participation in trophoblast attachment, HSPGs regulate growth factor morphogen signalling and may have a similar role during implantation at the maternal–fetal interface.

Mucins

Mucins are highly glycosylated, large molecules that are found on many cell surfaces, and on the apical side of endometrial epithelial cells in mice and humans. They are believed to play a role in embryo attachment.[30] MUC-1 is a highly glycosylated mucin, expressed by many epithelial cells, and in the endometrium its expression is cycle-dependent. In mice, MUC-1 disappears on cycle day 4.5, corresponding to the initial attachment phase of implantation in this species. It has been postulated that MUC-1 may be a barrier to implantation that must be removed to allow embryonic interaction with the apical surface of the luminal epithelium of the endometrium during the window of implantation. In contrast to results with the mouse, in human endometrium MUC-1 expression is up-regulated in the peri-implantation period.[31] It is possible that alterations in the carbohydrate structure of mucins may permit embryo attachment during the window of implantation, or that the upregulation of MUC-1 in human endometrium may have functional significance *per se* for embryo attachment and nidation. Some electronegative mucins may act as specific ligands for the embryo, marking the initial point of attachment, leading to later interactions mediated by integrins (see below). Another mucin, MAG (mouse ascites Golgi), appears on the luminal epithelial surface in human endometrium at the earliest phase of implantation.[32] MAG is on the surface of the endometrium only on cycle days 18–19, and, in one study, 60% of patients with unexplained infertility had abnormal expression of MAG. MAG may join a group of epithelial products that could be clinical markers for a non-receptive endometrium[32] in select groups of patients.

Integrins

Integrins are ubiquitous transmembrane glycoproteins that belong to a family of cell adhesion molecules of the immunoglobulin superfamily, cadherins, and are cell adhesion molecules with lectin-like domains. They are vital to the maintenance of cellular phenotype, and the pattern of integrin expression forms the basis for a cell's location within tissues, its shape, basal and apical polarity, and its ability to respond to steroid hormones. Integrins are the most intensely studied group of cell adhesion molecules in the endometrium. They are a family of heterodimeric, non-covalently bound α and β subunits. In addition to the transmembrane domain, there is an extracellular domain that serves as a receptor for various extracellular matrix ligands, including fibronectin, collagen and laminin. Specific recognition and binding of extracellular matrix components transmit information to the cytoskeleton, and this interaction may have major roles in promoting hormone responsiveness and genomic activation. The endometrium is unique in that it expresses both constitutive and cycle-dependent integrins.[33] There are three integrins expressed constitutively on endometrial epithelial cells: $\alpha_2\beta_1$, $\alpha_3\beta_1$ and $\alpha_6\beta_4$,[34] which is similar to the case for other tissues. These collagen/laminin receptors are believed to play a role in cell–substratum attachment. The fibronectin receptor, $\alpha_5\beta_1$, is the only integrin expressed constitutively by endometrial stromal cells. Other stromal integrins, $\alpha_1\beta_1$, $\alpha_3\beta_1$, $\alpha_6\beta_1$, $\alpha_v\beta_3$ and $\alpha_v\beta_5$, are cycle-dependent.[35] It should be noted that endometrial epithelial cells display a marked reduction in estrogen and progesterone receptors during the window of implantation.[11,36] The decline in progesterone receptors is specifically associated with initiation of $\alpha_v\beta_3$ expression. This integrin may directly participate in implantation and may serve as a marker of uterine receptivity in humans.[7,34,35] Of interest, in mice, blockage of the $\alpha_v\beta_5$ integrin adversely affects implantation.[37]

In women with histologically evident maturational delay (by endometrial biopsy) and infertility, the loss of the progesterone receptor is delayed, as is the expression of $\alpha_v\beta_3$. Medical correction of this condition restores both the timely loss of progesterone receptors and the expression of $\alpha_v\beta_3$.[35,38] These studies suggest that the presence or absence of individual integrins may serve to evaluate endometrial function during the window of implantation, and as a basis for a clinical test for uterine receptivity.[39] Endometrial expression of β_3 integrin may be associated with histological delay (i.e. classical luteal phase defect), called the 'type I defect', or with histologically normal endometrium (i.e. without histological delay), called the 'type II defect'. The latter suggests an intrinsic defect in endometrial function, and is found largely in women with minimal or mild endometriosis and in women with hydrosalpinges (reviewed in references 7 and 34). The lack of $\alpha_v\beta_3$ and the loss of $\alpha_4\beta_1$ have both been associated with unexplained infertility.[34] Since regulation of β_3 is not precisely defined, pharmacological correction of this defect is empirical. Ovarian suppression or laparoscopic ablation of pelvic endometriosis may restore $\alpha_v\beta_3$ expression in the eutopic endometrium (reviewed in reference 34).

Trophinin/tastin/bystin complex

Trophinin is an intrinsic membrane protein that is important in trophoblast cell adhesion, and tastin (trophinin assisting protein) is a cytoplasmic protein necessary for trophinin to function as an adhesion molecule, binding to it via another protein, bystin (reviewed in reference 40). Tastin associates with the cytoskeleton, which results in restricted distribution of trophinin in the plasma membrane, creating highly concentrated areas of trophinin 'patches' that may function as efficient adhesion sites.[40] In human endometrium, the trophinin/tastin/bystin complex is uniquely expressed in the epithelium and only at the expected time of implantation. In addition, trophinin has been detected in trophoblasts and endometrial epithelium in the monkey blastocyst implantation site. The spatially and temporally restricted expression of this novel cell adhesion complex suggests that it is important in the process of blastocyst attachment in primate implantation, although its precise role remains to be elucidated. Molecular approaches to elucidate the mechanism of embryo implantation involving trophinin, bystin and tastin involvement in initial attachment of blastocysts to human endometrial epithelium have recently been reviewed.[40]

GROWTH FACTORS AND CYTOKINES

Growth factors and cytokines comprise families of peptides and proteins that bind to specific cell surface receptors, resulting in cellular mitosis or differentiation, by autocrine, paracrine, juxtacrine or endocrine mechanisms. While these factors are expressed in most tissues, in endometrium many growth factors, cytokines, their receptors and their binding proteins exhibit cycle-dependence (Figure 3) (reviewed in reference 8). They are believed to mediate the effects of estradiol and progesterone on the endometrium. Many of these effector molecules are differentially expressed in endometrium during the secretory phase of the cycle (Figure 3), and, for the most part, their expression continues during early pregnancy, suggesting a major role for them in the cross-talk between the decidua and conceptus. These include members of the epidermal growth factor (EGF) family, leukemia inhibitory factor (LIF), interleukin-11 (IL-11), IL-β, transforming growth factor-β (TGF-β) and the insulin-like growth factor (IGF) family. Some of these effector molecules are believed to participate

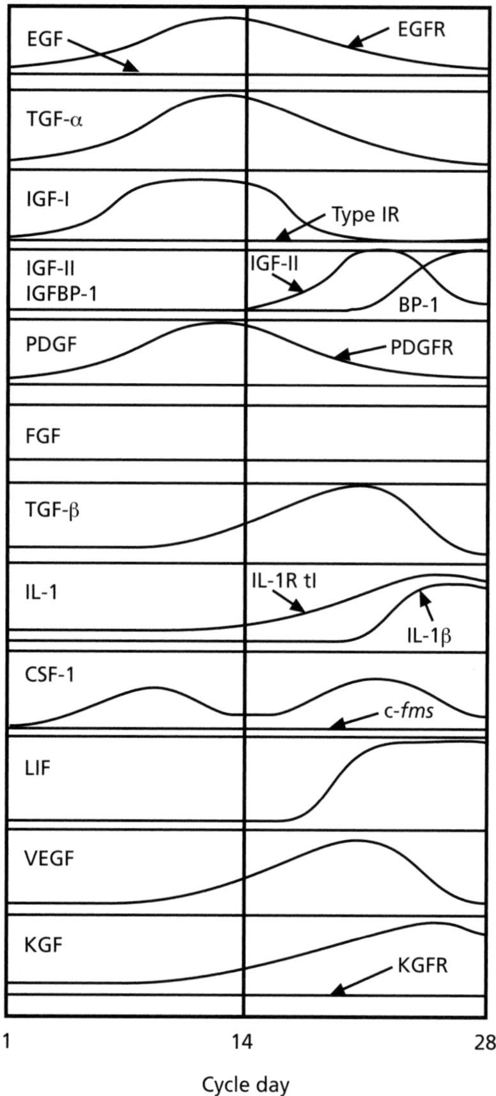

Figure 3 Cycle-dependence of growth factors and cytokines in human endometrium. For description of growth factors and cytokines, see text. EGF, epidermal growth factor; EGFR, EGF receptor; TGF-α, transforming growth factor-α; IGF-I, insulin-like growth factor-I; Type I R, type I IGF receptor; IGFBP-1, IGF binding protein-1; PGDF, platelet-derived growth factor; FGF, fibroblast growth factor; IL-1, interleukin 1; IL-1R tl, IL-1 receptor, type I; CSF, colony-stimulating factor; *c-fms*, CSF receptor; LIF, leukemia inhibitory factor; VEGF, vascular endothelial growth factor; KGF, keratinocyte growth factor; KGFR, KGF receptor

in the apposition/attachment phase of implantation, and many may participate in the process of decidualization of the stroma, thereby regulating trophoblast invasion, while others may participate in both.[10]

Epidermal growth factor

There has been an explosion of information over the past 5 years about the members of the EGF family and their receptors in the rodent, derived primarily from mouse knock-out models, rat uterine cell models and *in vitro* experiments using recombinant ligand effects on mouse embryo development.[41] There is also exciting, new information about this family in the human, relevant to potential roles in maternal–embryonic cross-talk during implantation (reviewed in reference 42). EGF family member ligands include EGF, transforming growth factor-α (TGF-α), heparin-binding EGF-like growth factor (HB-EGF), amphiregulin (AR), betacellulin (BC) and epicellulin (EP). Ligands are synthesized as transmembrane precursors that either are proteolytically cleaved to release the soluble form of the active peptide or remain membrane-bound to act via juxtacrine signalling. The peptides interact with a family of four HER/erbB membrane receptors (HER/erbB1-HERA4/erbB4) that activate downstream signalling by phosphorylation of tyrosine residues in the receptor, which serve as docking sites for recruitment of SH2 domain-containing effector molecules.[43] Actions of EGF family members that support their potential roles in implantation include them being mitogens, inducers of integrins with subsequent effects on cell motility, and chemoattractants. Figure 4 shows expression of EGF-like ligands in the peri-implantation mouse uterus. The HB-EGF gene is expressed in mouse uterine luminal epithelium surrounding the blastocyst just prior to implantation,[44,45] and HB-EGF also promotes blastocyst growth, zona hatching and trophoblast outgrowth *in vitro*.[44] HB-EGF binds to the heparan sulfate proteoglycan and the HER1 and 4 receptors on the blastocyst surface,[43] all linked to blastocyst activation within the uterus during the implantation process. In human endometrium, HB-EGF expression is in stroma during the proliferative phase, and in luminal epithelium maximally during the mid-secretory phase, coincident with the opening of the window of implantation (reviewed in reference 42). In addition, HB-EGF increases the numbers of human embryos reaching the blastocyst stage and promotes hatching,[46] suggesting that it facilitates embryo attachment to the

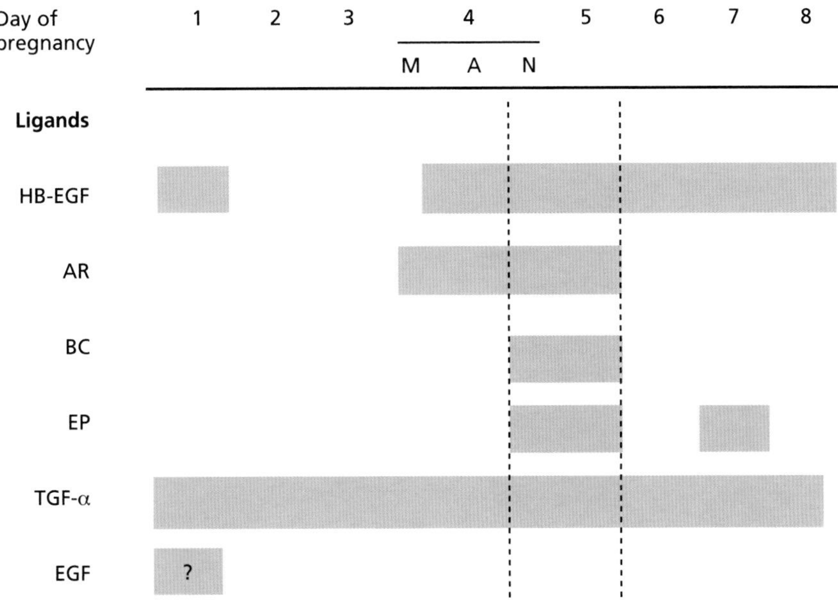

Figure 4 Expression of epidermal growth factor (EGF) ligands in the peri-implantation phase in the mouse. M, morning; A, afternoon; N, night; HB-EGF, heparin-binding epidermal growth factor (EGF)-like growth factor; AR, amphiregulin; BC, betacellulin; EP, epicellulin; TGF-α, transforming growth factor-α. Adapted from reference 45

epithelium during human implantation, as it is believed to do in the mouse. HB-EGF is also expressed in endometrial endothelial cells, suggesting that this growth factor may also be important in vascular remodelling during early pregnancy.

With regard to the other EGF ligands, amphiregulin expression overlaps with that of HB-EGF, BC and EP during days 4 and 5 of pregnancy (Figure 4). However, there are two important differences between AR and the other EGF ligands. Late on day 4 of mouse pregnancy, AR expression goes from diffuse to highly restricted to the luminal epithelium at the site of blastocyst apposition,[47] and the AR gene is up-regulated in the luminal epithelium directly in response to progesterone, independent of the presence of the blastocyst.[47] It appears that AR preferentially promotes stromal decidualization in response to progesterone by paracrine interactions between epithelium and stroma, whereas HB-EGF, for example, appears to act by paracrine mechanisms between epithelium and trophectoderm.[41,47]

Leukemia inhibitory factor

Leukemia inhibitory factor (LIF) is a pleiotropic cytokine that has both proliferative and differentiative effects on a variety of cells *in vitro*, including cells of embryonic, hematopoietic, osteoblastic and endothelial origin, and it prevents embryonic stem cells from differentiating. The actions of LIF are mediated by a high-affinity receptor which consists of two subunits, an LIF-binding subunit and a glycoprotein 130 (gp130) subunit. Glycoprotein 130 is also involved in signalling of other cytokines, including IL-6, IL-11 and ciliary neurotrophic factor (CNTF). In mouse endometrium, LIF is expressed prior to ovulation and also on the fourth day of pregnancy (where day 1 is the day of the vaginal plug).[48] LIF was originally believed to be induced in endometrial epithelium in response to the presence of an embryo. However, similar LIF expression has been observed in pseudopregnant females, suggesting that LIF expression is under maternal control.[48] The role of LIF in implantation has been shown conclusively in a mouse model lacking a functional LIF gene, achieved by gene targeting and homologous recombination. Homozygous females, when mated with normal males, produced normal blastocysts that were recoverable on day 4 from uteri. In addition, transfer of these embryos to wild-type pseudopregnant females resulted in implantation and pregnancy.[48] Although the mechanisms underlying maternal endometrial LIF expression and its role in embryonic implantation await further investigation, it is likely that LIF plays an important role in embryonic attachment to the epithelium and perhaps intrusion through the epithelium. In addition, it is important in endometrial stromal decidualization.[48]

In humans, LIF and its receptor are expressed in endometrium and decidua (reviewed in references 7 and 8). LIF mRNA is preferentially expressed in the secretory, compared with the proliferative, phase (Figure 3), and is expressed three-fold higher in glandular epithelium, compared with stroma. LIF and its receptor are expressed in human embryos,[49] suggesting a role for LIF in embryonic development in humans. Indeed, LIF acts on human cytotrophoblasts, diverting them to differentiate to the anchoring phenotype by increasing synthesis of fibronectin and decreasing human chorionic gonadotropin (hCG) production[50] and differentiating along the invasive pathway.[51] Thus, LIF may also play a role in trophoblast differentiation. There is evidence that abnormalities in LIF expression contribute to human infertility. Heterozygosity for an LIF gene mutation has been reported in a subset of infertile women.[52] Also, the levels of LIF in uterine flushings from women with unexplained infertility are significantly lower during the implantation window (day of luteinizing hormone (LH) surge plus 10), compared with fertile controls.[53] In addition, concentrations of LIF in cultured endometrial explants from women with infertility are lower than in fertile controls.[54,55] Recently, Lessey, Stewart and colleagues reported that mice treated with peritoneal fluid from infertile women with endometriosis had decreased fertility and a reduction in endometrial LIF expression.[56] Additional roles for this cytokine in human implantation are likely to be forthcoming.

Interleukin-11

Interleukin-11 is a pleiotropic cytokine that has actions in multiple cell types. Mice null for the IL-11 receptor (IL-11Rα) have a post-implantation decidualization defect in response to the implanting blastocyst.[57] Artificial decidualization, induced by oil applied to the uterine lumen in pseudopregnant mice, is also impaired in these null animals. IL-11 mRNA and receptor are expressed in decidual cells, and IL-11 mRNA expression is maximal at the time of stromal decidualization.[58] In human endometrium, IL-11 is cycle-dependent, with expression greatest in decidualizing stroma, along with constitutive expression of the IL-11Rα.[58,59] During the process of stromal decidualization, IL-11 expression precedes that of prolactin, suggesting what it is one of the early decidualization products. IL-11 is also expressed in vascular endothelium in a cycle-dependent fashion.[58] The roles of IL-11 in the vasculature and in implantation are not well defined. However, IL-11 supports production of T helper cell TH2 over TH1 cytokine production by T cells that are important in establishing and maintaining pregnancy.[60] Thus, this cytokine may be crucial to successful implantation.

Interleukin-1

The IL-1 'family' comprises IL-1α, IL-1β, IL-1 receptor antagonist (IL-1ra) and two IL-1 receptors, only one of which mediates signal transduction (IL-1R tI). IL-1α and IL-1β are homologous peptides that bind to the same receptor and mediate similar actions, whereas IL-1ra, while structurally similar to IL-1 ligands, inhibits their actions at the cellular level. The entire IL system is expressed in human endometrium. IL-1R tI mRNA is expressed in epithelial cells and is maximally expressed in the luteal phase (reviewed in references 7, 8 and 61). IL-1α and IL-1β mRNA and protein are expressed in endometrial epithelium, stroma, endothelial cells and macrophages. IL-1ra is also expressed in endometrium, primarily in epithelium. Immunoreactive IL-1α, IL-1β, IL-1ra and IL-1R tI are present in human embryos, and embryos co-cultured with endometrial epithelium (but not stroma) secrete both ligands and the antagonist. In addition to embryonic expression and apparent regulation by endometrial epithelial product(s), endometrial IL-1α and IL-1β and IL-1ra are regulated by steroid hormones.[61] Also, IL-1 in embryo culture fluid has been suggested to be a useful predictor of successful embryo implantation. In addition, human blastocysts selectively up-regulate $\alpha_v\beta_3$ integrin in human endometrial epithelial cells, compared with arrested human embryos.[62] This embryonic up-regulation of endometrial β3 integrin is mediated, in part, by the embryonic IL-1 system.[62]

Animal models have been useful in elucidating the role of IL-1β in implantation, although more questions have been raised than answered. Using a mouse model, it has been demonstrated that intraperitoneal injection of high levels of IL-1ra into pregnant mice on the third day of pregnancy results in a significant decrease in the number of implantation sites (6.7%), compared with non-injected and buffer-injected animals (59% and 74%, respectively).[63] These studies suggest that IL-1 may be necessary for implantation, and that elevated antagonist levels are detrimental to a successful pregnancy. However, implantation is normal in similar studies using different strains of mice, and also in IL-1R tI-deficient mice (reviewed in references 7 and 8). It is also normal in the absence of IL-1 converting enzyme (ICE), which converts the IL-1β precursor into its active form. These observations suggest that IL-1 may not play a key role in implantation in the rodent, or that there is redundancy in the regulation of embryonic attachment, involving IL-1 and probably other cytokines, including perhaps LIF and colony-stimulating factor-1 (CSF-1). Redundancy in cytokines proposed to participate in the implantation process is likely to be important to minimize dependence of continuation of a species on one effector molecule, and is probably not restricted just to the rodent or to IL-1β.

IL-1β decreases prolactin and IGF binding protein-1 (IGFBP-1) in human endometrial stromal cells, suggesting that this cytokine may, in part, regulate stromal decidualization.[64] However, the system appears to be quite complex with regard to stromal decidualization, including signal transduction through the NFκB signalling pathway.[65]

HOX GENES

HOX genes constitute a large gene family of transcription factors that share a common sequence, the homeobox, and that regulate embryonic morphogenesis and differentiation. HOXA10 and HOXA11 are expressed in the developing uterus. Their expression is necessary in adult endometrium for embryonic implantation, and targeted disruption of either HOXA10 or HOXA11 results in uterine factor infertility.[66,67] HOXA10 or -11 null embryos implant in wild-type controls, but neither null embryos nor wild-type embryos implant into null mice. HOXA11 null (–/–) mice have decreased endometrial glands and LIF secretion, perhaps accounting for the implantation failure.[68,69] Targeted reduction of HOXA10 gene expression, using antisense oligonucleotides, results in decreased implantation rates, whereas overexpression results in large litter sizes.[70] Taken together, these observations support the necessity of adult endometrial HOXA10 and -11 expression in endometrial receptivity in the mouse.

In human endometrium, HOXA10 and HOXA11 are steroid hormone-dependent and are expressed in endometrial glands and stroma.[71] In the mid-secretory phase, at the time of implantation, both HOX genes are up-regulated and levels remain elevated throughout the secretory phase and in early-pregnancy decidua. It is believed that HOXA10 and HOXA11 are important in embryonic development of the uterus, and also by activating endometrial downstream target genes in each menstrual cycle that are necessary for implantation. It is of interest that up-regulation of HOXA10 and -11 fails to occur in mid-secretory endometrium of women with endometriosis,[72] suggesting that implantation failure involving these transcription factors is a mechanism of infertility in some women with endometriosis.

CALCITONIN

Calcitonin has long been known as an endocrine regulator of calcium metabolism. Recently, however, it has been found to be progesterone-regulated and expressed in endometrium during the secretory phase.[73] It is expressed in epithelium and coincides with the time of implantation.[73] Studies with mice demonstrate that calcitonin may act in a paracrine fashion, communicating directly with the blastocyst and facilitating attachment and invasion.[74] Further roles for calcitonin in human endometrium, and mechanisms underlying calcitonin actions in endometrial–blastocyst interactions, await further investigation.

ENDOMETRIAL STROMAL DECIDUALIZATION

The invasive phase of implantation occurs in stroma that is decidualizing subepithelially and in the periarteriolar regions, but with many regions that are non-decidualized, and then, as invasion and pregnancy proceed, into a decidualized endometrium. Decidualization of endometrial stromal cells represents their differentiation with regard to distinct morphological appearance, accompanied by a unique biosynthetic and secretory phenotype. There is substantial evidence that progesterone and/or cyclic adenosine monophosphate (cAMP) play an important part in decidualization (see below). The decidua is a morphologically and functionally distinct tissue that persists throughout gestation and represents the maternal aspect of the maternal–fetal interface. In humans, decidualization initially begins independently of conception during the late secretory phase of normal ovulatory cycles, and is a prerequisite for successful implantation. Decidualized stromal cells synthesize and secrete 'markers' of decidualization, including prolactin and IGFBP-1, and produce extracellular matrix components typical of gestational decidual cells, including laminin, type IV collagen, fibronectin and heparan sulfate proteoglycan (reviewed in reference 75). It is important to recognize that several players participate in the process of decidualization of the stroma (Table 1), and that some of the classical 'markers' of decidualization are predominant in more advanced stages of stromal decidualization. Gene profiling using microarray technology with human endometrial stromal cells decidualized *in vitro* in response to progesterone and to cAMP demonstrates many common genes regulated by both effectors of decidualization.[59] Tables 2 and 3 demonstrate gene families in decidualized endometrial stromal cells. Unexpected findings include expression of neuromodulators in decidualized endometrial stroma, as well as up-regulation of members of the activin/inhibin family.[59] Gene profiling of human endometrial biopsy samples timed to the luteinizing hormone (LH) surge and within the window of implantation demonstrates many molecules known to be expressed, as well as multiple unexpected gene candidates.[76]

Table 1 Participants in decidualization of the endometrial stromal cell

Amphiregulin
Cyclic adenosine monophosphate (cAMP)
Cyclo-oxygenase-2
HOXA10
HOXA11
Interleukin-11
Leukemia inhibitory factor
Progesterone

Progesterone and cyclic AMP

In vitro models of decidualizing human endometrial stromal cells have given insight into some of the effector molecules important in this process (reviewed in reference 75). Markers of differentiation, including IGFBP-1 and tissue inhibitor of metalloproteinase-3 (TIMP-3), generally considered to correspond to 'full differentiation' of the stromal cell, are involved in regulating trophoblast invasion in *in vitro* models (see below). However, the functionality of stromal cells during the pathway of differentiation to a fully decidualized cell is not well understood. Also, mechanisms underlying this differentiation process are not completely understood, although it is likely that cAMP, and perhaps other modulators, are important for this process. For example, cAMP-dependent kinases in endometrium are regulated by progesterone,[77] and endometrial cAMP concentrations and adenylate cyclase activity are higher during the secretory

Table 2 Fold increase in mRNAs encoding growth factors, cytokines, chemokines, receptors and binding proteins in progesterone (P) and cyclic adenosine monophosphate (cAMP) in decidualized vs. non-decidualized human endometrial stromal cells (two patient samples)

	Patient 1		*Patient 2*	
	P	*cAMP*	*P*	*cAMP*
IL-5	30.6	4.9	3.7	1.6
IL-6	8.3	1.6	34.8	4.9
IL-7	4.9	2.7	2.6	1.7
IL-8	24.0	3.5	27.1	6.8
IL-10	4.7	1.8	3.2	14.8
IL-11	6.0	3.6	3.1	1.9
IL-13	125.6	13.7	7.4	3.5
IL-14	29.1	5.7	2.4	2.7
IGFBP-1	3.2	10.4	4.4	15.4
IGFBP-3	3.3	2.6	10.4	3.1
Insulin R	3.3	3.1	3.1	2.9
Prorelaxin H2	7.9	2.4	3.0	3.3
EGF	9.7	3.8	1.8	1.9
EGFR	9.9	3.9	3.0	1.5
Amphiregulin	1.4	4.9	1.7	5.8
GM-CSF	5.1	2.8	1.7	2.3
Endothelin-2	2.6	1.5	3.0	2.6
Endothelin R type B	1.9	3.0	1.5	8.7
VEGF	1.3	3.6	1.6	4.1
VEGFR-3 (FLT-4)	3.3	2.4	9.5	1.6
VEGFR/KDR	2.4	4.0	2.3	2.6
FLT 3 ligand	6.4	3.6	2.8	1.4
TGF-β2	8.4	9.0	2.2	2.0
TGF β R type III	6.9	3.2	1.7	1.5
MCP-1	2.1	15.2	2.1	11.5
INF-α	4.9	1.7	4.8	13.7
INF-β	8.3	5.2	8.2	4.3
TNF-α	5.5	1.2	3.2	1.4
TNF receptor-1	5.5	4.2	4.0	2.0
Neurotrophin-3	7.8	2.9	2.8	1.4
Neuromodulin	1.3	3.1	0.8	2.7
Neurotrophin-4	4.9	2.2	2.0	3.1
α-Adrenergic R 2A	2.1	1.4	3.5	4.2
Angiotensin II-1A R	3.1	1.0	2.8	1.8
Renin-binding protein	6.0	2.8	2.6	3.7
Acyl Co-A BP	3.2	2.2	5.1	1.5
Glucagon precursor	6.6	2.8	2.6	3.8
Inhibin α subunit	8.7	2.9	3.6	2.9
Inhibin/activin β_A subunit	5.0	2.8	2.8	1.6
FSHR	5.8	2.3	2.8	1.5
Placental-like GF 1 and 2	3.5	1.9	7.1	1.8
Estrogen sulfotransferase	4.1	1.1	2.4	1.9

IL, interleukin; IGFBP, insulin-like growth factor binding protein; R, receptor; EGF, epidermal growth factor; GM-CSF, granulocyte macrophage colony stimulating factor; VEGF, vascular endothelial growth factor; KDR, kinase insert domain-containing receptor; FLT, FMS-like tyrosine kinase receptor; TGF, transforming growth factor; MCP, monocyte chemotactic factor; INF, interferon; TNF, tumor necrosis factor; FSH, follicle stimulating hormone; GF, growth factor

Table 3 Fold increase in mRNAs encoding cell cycle regulators, intracellular signalling, apoptosis, DNA synthesis, repair and transcription factors in progesterone (P) and cyclic adenosine monophosphate (cAMP) decidualized vs. non-decidualized endometrial stromal cells in cells from two patients

	Patient 1		*Patient 2*	
	P	*cAMP*	*P*	*cAMP*
Oncogenes, cell cycle regulators				
c-myc	3.3	2.1	8.2	5.0
c-fms (CSF-1-R)	4.9	2.8	2.0	1.3
c-fer	3.9	3.2	1.9	3.3
c-src kinase	3.1	2.1	3.0	1.9
CDK N1C	1.5	3.0	1.9	3.4
Colorectal ca supp	3.6	7.9	1.4	3.6
Bullous pemphigoid Ag*	1.0	3.0	2.0	3.1
G_1/S-cyclin E	2.8	3.9	1.5	8.7
Cyclin 4 inhibitor 2B	4.7	1.5	3.0	2.2
Apoptosis, DNA synthesis, repair				
GADD45	3.1	3.3	3.1	2.7
RAR-ε	8.9	4.2	4.1	1.2
RAR-β	2.6	1.3	6.6	2.9
Microsomal GT	3.0	1.9	7.8	1.4
TRAIL	2.8	2.9	2.7	4.1
Intracellular signalling				
cAMP PDE4A6	4.8	1.5	2.5	1.9
SCDF 1R	3.8	5.2	4.3	2.6
Guanine nucleotide BP	3.7	1.4	3.3	2.5
CDK-activating kinase	2.5	2.7	1.0	3.2
Transcription factors, DNA BPs				
TATA-box factor	2.4	1.7	3.4	2.0
HOX2A	4.0	3.1	3.7	4.1
HOXD-3; Hox4A	3.6	2.5	2.2	2.2
HOX11	1.8	2.6	2.1	4.6
Brain-specific Hox	1.8	3.1	2.1	4.4
Homeobox hLim1	4.5	2.5	15.4	2.4
NF-ATc	3.1	2.9	3.2	4.2
Colorectal ca mut pro	2.5	9.1	5.3	7.4
Gf-inducible NP475	3.8	2.4	2.3	8.0
NF-90	1.8	3.7	4.0	1.4
Forkhead TF AFX	3.4	4.5	3.9	3.0

CSF, colony-stimulating factor; R, receptor; CDK, cyclin-dependent kinase; ca, cancer; supp, suppressor; Ag, antigen; GADD, DNA-damage-inducible gene; RAR, retinoic acid receptor; GT, glutathione transferase; TRAIL, tumor necrosis factor (TNF)-related apoptosis-inducing ligand; PDE, phosphodiesterase; SCDF, stromal cell-derived factor; BP, binding protein; NF, nuclear factor; Atc, activated T cell cytosolic component; mut pro, mutant protein; GF, growth factor; NP, nuclear protein

phase when decidualization occurs, compared with the proliferative phase.[77] Also, unique products of decidualized stromal cells, including prolactin and IGFBP-1, are induced by progesterone and/or cAMP treatment of human endometrial stromal cells in culture.[59,75,78,79] In addition, progesterone-induction of mRNAs encoding the decidual markers prolactin and IGFBP-1 is inhibited by protein kinase A inhibitors, and activated protein kinase A is required for differentiation-dependent transcription of the decidual prolactin gene in human endometrial stromal cells.[79,80] Stimulation of the cAMP pathway can affect transcription of certain genes, and involves the cAMP response element (CRE)-binding protein, CRE modulators (CREMs) and inducible cAMP early repressor (ICER), which bind specifically to the CRE region in the promoters. During the course of decidualization, human endometrial stromal cells express isoforms of CREM as well as up-regulation of ICER.[80] IGFBP-1 has a CRE in its promoter, and cAMP treatment of endometrial stromal cells *in vitro* results in induction of the IGFBP-1 gene.[59] Furthermore, stimulation of intracellular cAMP in endometrial stromal cells by prostaglandin PGE_2 and activation of endometrial stromal cAMP by relaxin,[81] with concomitant morphological changes and prolactin and IGFBP-1 production, support an important role for cAMP in endometrial stromal cell decidualization.

Prostaglandins

The physiological stimulus for cAMP in human endometrium is not known with certainty, although leading candidates are prostaglandins (PGs) and progesterone. In mouse endometrium, however, induction of stromal decidualization, which requires the presence of an embryo (or a mechanical stimulus), appears to be dependent upon PGs. Cyclo-oxygenase-2 (COX-2) is one of two isoforms of the COX enzymes that are the rate-limiting step of PG biosynthesis from arachadonic acid. Mice null for the COX-2 alleles (–/–) demonstrate multiple reproductive defects, including failure to ovulate and failed stromal decidualization.[82] Transfer of wild-type embryos into these null animals results in markedly decreased implantation rates, compared with transfer into wild-type controls. The implantation defect is partially rescued by prostacyclin (PGI_2) as well as by peroxisome proliferator activated receptor (PPAR-δ) analogs, suggesting that PGI_2 interacts with PPAR-δ for embryonic implantation and decidualization in mice.[83]

In humans, PGs enhance decidualization of endometrial stromal cells,[84] and the endometrium exhibits increased vascular permeability in response to PGs, an important precursor to decidualization of human endometrium.[85] In human endometrium, COX-1 is mainly expressed in the glandular and luminal epithelium, and COX-2 is expressed in higher amounts in luminal epithelium and perivascular cells, although not in stroma.[78,86] Interestingly, treatment with an antiprogestin significantly decreases the expression of COX-1 and -2 in glandular and luminal epithelium.[86] COX-2 expression can be stimulated in stromal cells[87] and up-regulated in glandular epithelial cells[88] by the addition of hCG. In addition, IL-1β induces COX-2 gene expression in cultured endometrial stromal cells.[89] These observations suggest complex regulation of COX-2 gene expression and PG synthesis in human endometrium, and the source of the stimuli *in vivo* has not been conclusively identified, during implantation or during decidualization, which in humans, in contrast to rodents, does not require the presence of a conceptus.

ROLE OF THE ENDOMETRIUM IN PLACENTAL INVASION

During the invasive phase of implantation, the human cytotrophoblast must be capable of adhering to the extracellular matrix and rapidly invading into the endometrial stroma. Trophoblast invasion into the maternal endometrium occurs by processes similar to those accompanying tumor invasion, requiring attachment to the extracellular matrix via cell adhesion molecules, local extracellular matrix proteolysis (by matrix metalloproteinases (MMPs)), cellular migration and inhibition of these processes.[90] In the transition to the invasive phenotype, first-trimester cytotrophoblasts undergo integrin 'switching', from $\alpha_6\beta_4$ to $\alpha_5\beta_1$ and $\alpha_1\beta_1$, which probably enable the intermediate cytotrophoblast to anchor and migrate in the maternal decidua.[91] In addition to changes in cell adhesion molecules, first-trimester cytotrophoblasts uniquely express matrix-degrading enzymes. Of the family of MMPs, the invading cytotrophoblast produces primarily 72-kDa and 92-kDa gelatinases (MMP-2 and MMP-9, respectively).[92,93] The latter is required for human cytotrophoblast invasiveness *in vitro*,[93] and the production and activation of MMP-9 in cytotrophoblasts coincide with maximal invasive behavior *in vivo*.[93,94] In fact, MMP-9 mRNA is expressed only in first-trimester invading cytotrophoblast, and is undetectable in the second and third trimesters.[92] IL-1β stimulates MMP-9 production by first-trimester cytotrophoblast *in vitro*,[95] suggesting that autocrine actions of this cytokine probably promote cytotrophoblast invasiveness in the first trimester. MMP-2 is also highly expressed in the first-trimester invading cytotrophoblast, and its expression decreases with gestational age.[92] Thus, it may also play a key role in cytotrophoblast invasiveness, although its regulation in cytotrophoblast is not well understood. Overall, the activity of these proteinases is controlled by specific inhibitors, including TIMPs, or indirectly by cytokines, and by interactions with the extracellular matrix. While the cytotrophoblast has long been considered the kingpin of the invasive process of implantation, the maternal endometrium plays a major, active role in limiting this invasion. It probably does so by several mechanisms, including producing inhibitors of trophoblast differentiation into the invasive phenotype, inhibition of autocrine or paracrine stimulators of invasion, direct inhibition of invasion and inhibition of trophoblast-derived matrix-degrading enzymes. The roles of decidual products as participants in these potential mechanisms are discussed below in the context of 'maternal restraints' on placental invasion (Figure 5).

Enzyme inhibitors

Broad-spectrum protease inhibitors (for example α_2-macroglobulin) as well as more specialized inhibitors (TIMP-1, -2 and -3) of primarily decidual (and cytotrophoblast) origin are probably important in limiting cytotrophoblast invasion.[96] Rodent deciduae express TIMPs, and these TIMPs abolish trophoblast invasion *in vitro*. However, mice with a null mutation in the TIMP-1 gene have normal fertility,[96] suggesting that TIMP-1 is not a critical inhibitor of trophoblast invasion. In humans, endometrial TIMP-1 mRNA expression is not cycle-dependent, and TIMP-2 mRNA is also relatively constant throughout the menstrual cycle (reviewed in reference 97). Current evidence supports a major role for TIMP-3 in limiting invasion. TIMP-3 mRNA in decidualized endometrial stromal cells is up-regulated by progesterone *in vitro* and *in vivo*,[59,97] and, in mice, TIMP-3 mRNA is expressed in decidua immediately adjacent to the implanting embryo.[96] Human trophoblasts with the invasive phenotype up-regulate MMP-9 and TIMP-3,[96] and *in situ* hybridization studies reveal that TIMP-3 mRNA is expressed by the decidua (as well as the invading trophoblast).[98] *In vitro*, TIMP-3 inhibits invasion of cytotrophoblasts into Matrigel.[95] Recent evidence demonstrates that two trophoblast products, IL-1β and IGF-II, inhibit TIMP-3 expression in human decidualized endometrial stromal cells, suggesting that the trophoblast promotes its own invasiveness by inhibiting maternal restraints on invasion.[99,100]

Extracellular matrix proteins

The extracellular matrix probably plays a major role in trophoblast invasiveness, because it is the substratum that supports cellular adhesion and cell–cell interactions, and because interactions with the extracellular matrix result in changes in trophoblast invasiveness. The extracellular matrix proteins, laminin and fibronectin, are important participants in the invasive process of implantation. Endometrial stromal cells, decidualized *in vitro*, secrete laminin and fibronectin, and progestin stimulates fibronectin mRNA expression and protein synthesis in these cells (reviewed in reference 75). Laminin production also increases during decidualization of rat uterine stromal cells. Laminin production by human endometrial stroma is more abundant in secretory compared with proliferative endometrium, and it increases with implantation and in early pregnancy. Laminin decreases levels of prolactin and IGFBP-1 during *in vitro* decidualization of endometrial stromal cells, suggesting that it may play a role in facilitating trophoblast invasiveness (see below).

Fibronectin, a major secretory product of decidualized endometrial stromal cells, has an Arg-Gly-

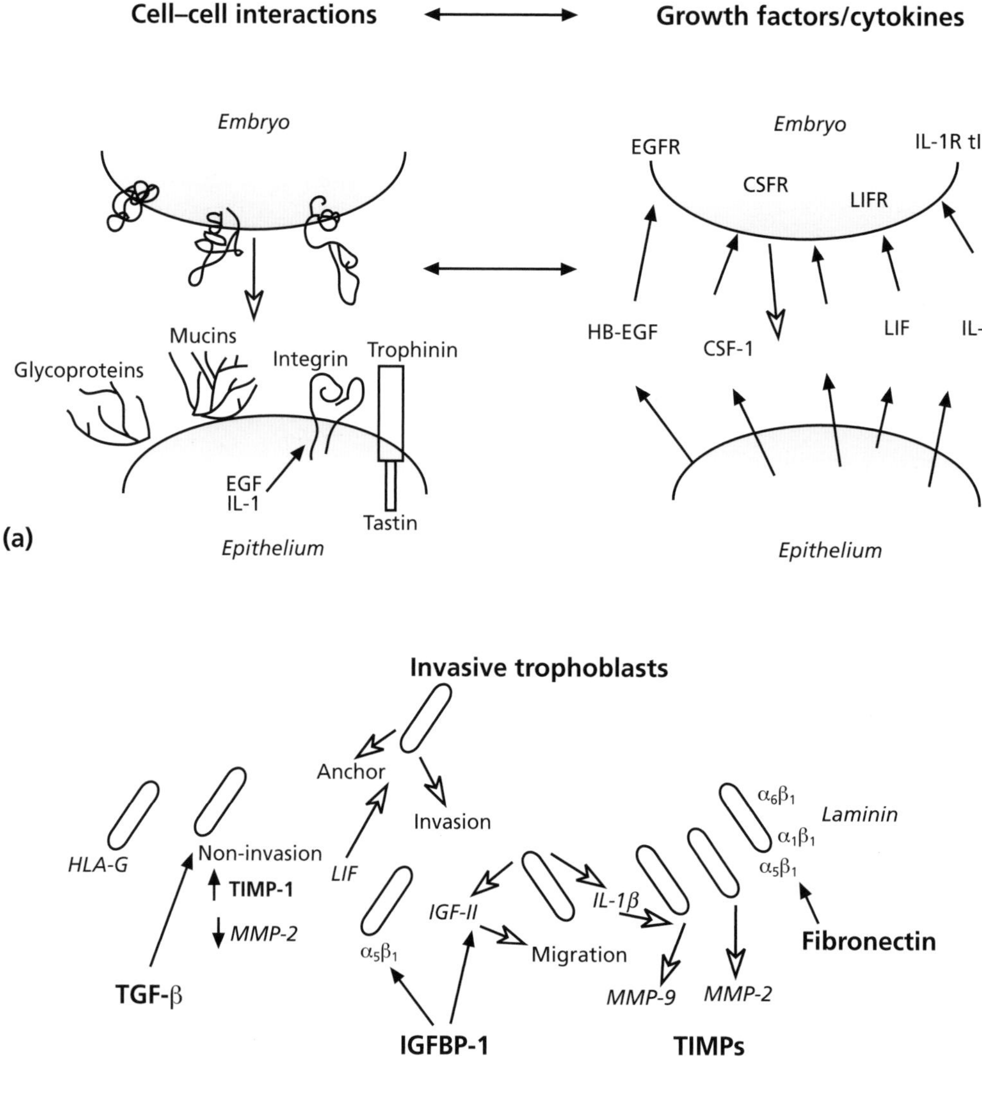

Figure 5 Schematic diagram of function of various biochemical principles within the endometrium in the process of implantation. (a) Apposition/attachment phase. Open arrows indicate possible paracrine effectors originating in the embryo. Double-headed arrows suggest interactions of growth factors and cytokines on the cell–cell interactions. (b) Invasion phase. Effector molecules shown in italics and open arrows indicate invasion facilitators. Effector molecules shown in bold indicate members of the 'maternal defense', i.e. inhibitors of trophoblast invasion. EGF, epidermal growth factor; IL-1, interleukin-1; EGFR, EGF receptor; CSFR, colony-stimulating factor (CSF) receptor; LIFR, leukemia inhibitory factor (LIF) receptor; IL-1R t1, IL-1 receptor type I; HB-EG, heparan-binding EGF-like growth factor; HLA-G, human leukocyte antigen-G; TGF, transforming growth factor; TIMP, tissue inhibitor of metalloproteinase; MMP, matrix metalloproteinase; IGF, insulin-like growth factor

Asp (RGD) sequence, and may interact with the invading trophoblast. The repertoire of adhesion receptors changes during first-trimester cytotrophoblastic differentiation along the invasive pathway *in vivo*, from $\alpha_6\beta_4$ to $\alpha_5\beta_1$ and $\alpha_1\beta_1$ fibronectin receptor (see below). *In vitro*, human first-trimester trophoblasts attach preferentially to laminin-coated surfaces. Also, addition of exogenous fibronectin inhibits cytotrophoblast invasion through a Matrigel matrix *in vitro*,[91] and addition of RGD-containing peptides inhibits cytotrophoblast attachment to fibronectin-coated culture dishes.[91,101,102] Thus, cytotrophoblast interaction with fibronectin, through the $\alpha_5\beta_1$ integrin, restrains, rather than promotes, invasion.[91] Mechanisms underlying this inhibition are not well understood, but fibronectin is likely to be one of several members of the 'maternal restraint' on trophoblast invasiveness at the maternal–fetal interface (Figure 5).

Insulin-like growth factors and IGF binding protein-1

Insulin-like growth factors (IGFs) are peptides with mitogenic, differentiative, antiapoptotic and metabolic actions. They bind to a family of high-affinity binding proteins (IGFBPs), which facilitate transport of the IGFs from their sites of synthesis to their sites of action, and which regulate IGF bioavailability to their specific receptors on target cells. In humans, IGF-I and IGF-II are mitogenic to decidualized endometrial stromal and glandular cells. The IGF-I gene is expressed primarily in late proliferative and early secretory endometrium, and is believed to mediate the mitotic actions of estradiol in this tissue. On the other hand, the IGF-II gene is expressed abundantly in mid-late secretory endometrium and early pregnancy decidua, and may be a mediator of progesterone action (reviewed in reference 8). Although this temporal expression suggests that IGF-II may play a role in early pregnancy, this has not been defined. However, IGF-II has been shown to stimulate trophoblast migration,[103] although an effect on trophoblast invasion has yet to be demonstrated.

Insulin-like growth factor binding protein-1 (IGFBP-1) is a member of a family of proteins that have high affinity for the IGF peptides, and which act primarily to inhibit IGF actions at their target cells. IGFBP-1 is a major product of secretory endometrium and decidua (reviewed in reference 75). Maternal serum levels increase throughout gestation, peaking in mid-gestation, and are believed to derive from the decidua (reviewed in reference 75). IGFBP-1 is abundantly produced in the decidual stroma, and is well situated to interact with the IGF-II-expressing invading cytotrophoblast.[104] IGFBP-1 also has IGF-independent actions, binding to cell membranes and altering cellular motility.[105] IGFBP-1 contains the Arg-Gly-Asp tripeptide motif, which binds specific integrins, including the $\alpha_5\beta_1$ integrin. The latter is uniquely expressed by the invading cytotrophoblast, of all trophoblast phenotypes at the maternal–fetal interface (see below). Thus, IGFBP-1 is also well situated in this interface to interact with the cytotrophoblast, potentially affecting the implantation process by IGF-dependent and/or -independent actions. *In vitro* studies support modulatory roles for IGFBP-1 in implantation. For example, IGFBP-1 competes with endometrial membrane receptors for binding IGF-I (reviewed in reference 75). IGFBP-1 also inhibits the binding and biological activity of IGF-I on choriocarcinoma cells, suggesting that abundant decidual IGFBP-1 may regulate trophoblast-derived IGF autocrine and paracrine actions. Recent studies demonstrate that IGFBP-1 binds to human cytotrophoblasts, binds to the $\alpha_5\beta_1$ integrin in the cytotrophoblast membrane and inhibits human cytotrophoblast invasion into decidualized human endometrial stromal cultures.[102] Cumulatively, these studies suggest that this abundant maternal decidual product interacts with the invading trophoblast and is probably one of several participants in the 'maternal restraint' effort to curb placental invasion into the maternal host (Figure 5). Whether it does this by interacting directly with the trophoblast or by inhibiting the actions of IGF-II is not known. IGFBP-1 has also been shown to increase migration of long-term passaged trophoblasts on plastic surfaces,[103,106] via mitogen-activated pathway.[107] The different actions of IGFBP-1 on migration and invasion warrant further investigation. Whether an excess of IGFBP-1 at the maternal–fetal interface is detrimental to cytotrophoblast invasion, resulting in shallow implantation or overly aggressive invasion or early pregnancy loss, is not certain at this time. Potential roles of IGF peptides and IGFBP-1 in occult endometrial defects and uterine non-receptivity await further investigation.

During human pregnancy, IGF-II is markedly down-regulated in decidualizing endometrium in the presence of intrauterine pregnancy,[104] but not with ectopic pregnancy,[108] suggesting that trophoblast products down-regulate IGF-II in the maternal endometrial stromal cells. In contrast, during pregnancy, IGF-II is exclusively and abundantly expressed by trophoblasts invading into the maternal uterine decidua,[104] a process that involves matrix degradation by MMPs.[96] IGF-II expression is highest in the invading trophoblast front.[103] Modulators of trophoblast invasion, IGFBP-1 and TIMP-3, are abundantly expressed in the maternal decidua.[98,100,104] These maternal modulators of invasion are both inhibited by IGF-II,[93,100] supporting paracrine actions of trophoblast-derived IGF-II on decidual regulators of invasion. The striking juxtaposition of IGF-II expression in the trophoblast and maternal modulators of invasion in the maternal decidua suggests that IGF-II is well placed for paracrine interactions between the placenta and decidua.

Transforming growth factor-β

Transforming growth factor-β comprises five isoforms, which share approximately 80% sequence homology. Within the endometrium, various TGF-β isoforms are expressed in specific cell types and at distinct times, suggesting that these growth factors have specific roles during the peri-implantation period.[109] In mouse uterus, TGF-β_1 is primarily expressed in endometrial luminal and glandular epithelium during the pre-implantation period, and in decidua in the post-implantation period.[108] TGF-β_2 is synthesized by epithelial, myometrial and decidual cells, and its secretion may be regulated by estradiol, whereas TGF-β_3 is primarily synthesized by smooth muscle cells of the myometrium, independent of ovarian steroids. In humans, TGF-β_1 mRNA is equally distributed in endometrial glands and stroma, and levels are lowest in the proliferative phase. They increase two-fold in the secretory phase, and about five-fold in the maternal endometrium (decidua) of pregnancy.[110] The cycle-dependence of TGF-β_1 mRNA expression suggests that it may contribute to inhibition of cellular proliferation during periods when endometrial cellular differentiation, induced by TGF-β or other growth factors, is dominant (reviewed in reference 8). In addition, the marked abundance of TGF-β mRNA expression in decidua also suggests a possible role for this growth factor in implantation (reviewed in reference 111). TGF-β_1 inhibits proliferation of cytotrophoblasts and promotes their differentiation into non-invasive syncytiotrophoblasts. The invading trophoblast secretes a variety of proteases, including plasminogen activators and MMPs. TGF-β_1 induces plasminogen activator inhibitor (PAI) mRNAs and TIMP-1 secretion, and decreases 72-kDa gelatinase secretion by cytotrophoblasts.[111] Thus, these observations suggest that TGF-β controls trophoblast invasion by regulating proteases and their inhibitors in the maternal–fetal interface. In addition, since the latent form of TGF-β_1 is activated by plasmin, it is possible that maternal decidua stores latent TGF-β in the extracellular matrix, waiting for the invading trophoblast and its production of plasmin. TGF-β could then increase PAIs and inhibit trophoblast-derived MMPs (by increasing TIMP-1). The net result would be limiting trophoblast invasion. Support of this hypothesis derives from the observation that TGF-β produced by first-trimester decidual cells in culture inhibits the *in vitro* invasiveness of first-trimester human trophoblasts, owing to the induction of TIMP-1.[110] Thus, TGF-β_1 is probably a member of the maternal restraint on trophoblast invasiveness (Figure 5).

ANGIOGENESIS

Blood vessel proliferation increases in the mid-late proliferative phase and during the secretory phase. This process includes both endothelial cells and vascular smooth muscle cells. Control of angiogenesis in the preparation of a receptive endometrium is most widely attributed to the fibroblast growth factor (FGF), vascular endothelial growth factor (VEGF) and angiopoietin (Ang) families (reviewed in reference 111). Three of the eight FGFs (FGF1, -2 and -4) are expressed in endometrium. FGF1 and -2 are abundantly expressed in endometrial epithelial cells. They up-regulate endothelial VEGFR-2 receptor. Also, VEGF promotes angiogenesis by releasing FGF from the extracellular matrix. VEGF members include VEGF-A, -B, -C, -D, -E and placental growth factor (PlGF), and several splicing variants for VEGF-A exist. These ligands bind to three receptors, VEGFR-1, -2 and -3. All members of the VEGF-A family are expressed in human endometrium (in stroma and epithelium) and are cycle-dependent.[113,114] Steroid hormones regulate VEGF expression, as do hypoxia and glucose.[115,116] Ang-1 is expressed by vascular smooth muscle cells and binds to its cognate receptor, tie-2, on the endothelium. This results in stabilization of the endothelial cell–vascular smooth muscle structure (reviewed in reference 112). Ang-1 promotes tie-2 receptor phosphorylation, which results in inhibition of apoptosis of the endothelium. Ang-2 is antagonistic to Ang-1. Ang-2 binds to the tie-2 receptor, but binding does not transmit a signal, and the endothelial cells undergo apoptosis, and, in the absence of VEGF, the vessel undergoes atrophy. Interestingly, Ang-1 expression decreases during both phases of the menstrual cycle. However, Ang-2 is up-regulated in the secretory phase, and its expression is restricted to uterine natural killer cells.[117]

While the precise mechanisms underlying angiogenesis involving FGF, VEGF and Ang family members are not well understood, there is compelling evidence that uterine natural killer cells are critical to the process of vascular remodelling in early implantation. In addition to the above discussed observations of Ang-2 expression in these cells,[116] studies with mice lacking natural killer cells demonstrate abnormalities of blood vessels at the implantation site (reviewed in reference 118). Bone marrow rescue studies with administration of natural killer cells to null animals restored the normal vascular and placental phenotype,[119,120] supporting an important role for these cells in angiogenesis and further demonstrating that peripheral natural killer cells traffic to the uterus. It has been suggested that VEGF-A promotes adhesion and migration of natural killer cells to the endometrium, and that Ang-2 expressed by the decidual natural killer cells may facilitate vascular remodelling by blocking constitutive stabilization of Ang-1, allowing the vessels to become more plastic and unstable. With regard to trophoblast remodelling of the spiral arterioles, Ang-2 is up-regulated by hypoxia and may act on endothelial tie-2 receptors. Extravillous trophoblast migrate into and around the spiral arterioles of the decidua, and they eventually replace the endothelial cells. The precise mechanisms involving Ang-2 and vascular remodelling warrant further investigation.

SECRETORY PHASE ENDOMETRIAL PROTEINS

Many proteins are secreted uniquely in secretory phase endometrium, and some whose functions are beginning to be appreciated have been discussed above. However, there are many other proteins in secretory endometrium whose functions are unknown, although their temporal expression would suggest a functional role for them in the process of implantation. Two of these, in particular, prolactin and progestin-associated endometrial protein/placental protein-14 (PEP/PP14) are discussed below, because of their potential roles in water transport and as an immune modulator, respectively.

Prolactin

Prolactin is a product of decidualized endometrium and is a progesterone-dependent protein in this tissue (reviewed in reference 75). While prolactin is known to modulate water transport across membranes, its precise function during the invasive phase of implantation, as a stromal product, is not well defined. Its temporal and spatial expression would suggest that it is not involved in the fluid transport across the epithelium during peak glandular secretory activity, since this is a stromal product and it is not expressed until peak glandular secretory activity has passed. Despite the paucity of knowledge about prolactin function in endometrium during implantation, it was hoped that it might serve as a marker of decidualization, and potentially of endometrial adequacy. Since prolactin secretion increases with progressive endometrial decidualization and histological maturation, it may be useful in identifying women with histological delay. However, at least one study did not find a significant difference in prolactin immunostaining of endometrial biopsies from patients with normal cycles and those with luteal phase defect.[121] Its potential role in unexplained infertility has not been determined.

Progestogen-associated endometrial protein/placental protein 14/glycodelin

Progestogen-associated endometrial protein (PEP), also known as placental protein-14 (PP14), pregnancy-associated endometrial α_2 globulin and glycodelin, is the major secretory protein of the epithelium of luteal phase endometrium (reviewed in reference 122). This 28-kDa dimeric glycoprotein is a secretory product that appears in the glandular lumen and in serum during the late luteal phase. Glycodelin transcription, synthesis and secretion by human epithelial cells are stimulated by progestins and antiprogestins, but not by estrogen.[123,124] Glycodelin has been examined as a potential marker for adequate corpus luteum function. Mean glycodelin levels for patients with 'in-phase' biopsies were not statistically different from those with out-of-phase biopsies, and serum levels did not adequately reflect histological changes in the endometrium. The gene encoding glycodelin is homologous to members of the β-lactoglobulin family. Other members of this family can bind retinal and retinoic acid. However, although glycodelin is homologous to members of the β-lactoglobulin family, it does not bind retinoids.[125] Glycodelin inhibits the actions of natural killer cells, which are abundant in decidualized endometrium during early pregnancy[126] (see below). Thus, there may be an immunomodulatory role for this protein during the implantation process, although it is not well understood. Besides this immunosuppressive effect, glycodelin strongly inhibits binding of sperm to the zona pellucida. Whether it has any effect on an implanting embryo is yet to be determined. The biological activities of glycodelins in reproduction and cellular differentiation have recently been reviewed.[122]

ENDOMETRIAL MACROPHAGES, T CELLS AND NATURAL KILLER CELLS

White cells, or leukocytes, constitute about 30% of the stromal cell population during early pregnancy. Unlike other immunologically active tissues, the endometrial lymphoid system is devoid of plasma cells, intraepithelial B cells, and has only rare B cells in the stroma. The predominant cell population, rather, comprises macrophages, T cells (CD8+) and granulated lymphocytes.[127] Macrophages increase in the late secretory phase and are present in the decidua throughout pregnancy. They produce cytokines, for example CSF-1 and granulocyte macrophage CSF (GM-CSF), which may regulate placental growth. Macrophages at the maternal–fetal interface may be immunosuppressants, facilitating tolerance of the new allograft. They may also have phagocytic roles, and/or be anti-inflammatory agents. T cells are present in the endometrium, although their numbers and location do not change with the menstrual cycle or in early pregnancy. The majority of endometrial T cells express CD8 suppressor/cytotoxic markers, whereas about 30% express the CD4 helper marker. T cells are a rich source of cytokines, which have the potential to influence placental growth and invasion. However, their role(s) at the maternal–fetal interface remain undefined.

The largest population of leukocytes in late secretory and early pregnancy endometrium is that of the large granular lymphocytes, which have markers consistent with an early differentiation stage of natural killer cells. They are abundant in the peri-implantation period and throughout the first trimester, after which their numbers decline markedly. While their function is not well understood, human large granular lymphocytes secrete an isoform of TGF-β which has immunosuppressive properties.[128] Trophoblasts express antigens that are recognized by immune cells. They do not express the major histocompatibility complex (MHC) class II antigens, but some express unusual MHC class I antigens, which are discriminant for self versus non-self. In particular, human trophoblasts express only a non-classical MHC class I molecule, human leukocyte antigen (HLA)-G (reviewed in reference 129). HLA-G protects against killing by natural killer cells, and, because of limited polymorphism in trophoblast HLA-G, the protein encoded by paternally derived HLA-G genes is not recognized as foreign by the maternal immune system. This provides a mechanism whereby maternal tolerance of the new allograft is achieved (reviewed in reference 129). Immunological factors in implantation and a critical assessment of proposed immunotherapies for select couples with infertility or recurrent miscarriage have recently been reviewed.[130] In addition to immune modulation, natural killer cells are believed to be important in vascular remodelling and placentation early in pregnancy (see above).

SUMMARY

The endometrium is a dynamic tissue whose cell components and molecular constituents change during the menstrual cycle and during early pregnancy. These changes reflect interactions with its environment, including fluctuating levels of circulating steroid hormones and paracrine interactions with the implanting conceptus. The apposition/attachment phase of implantation involves cell adhesion, changes in cell motility, cell surface charge and cellular differentiation. The invasive phase of implantation involves extensive paracrine signalling between the invading trophoblast and the decidual stroma, vasculature and lymphoid cells. The expression, regulation, modes of action and functions of pinopodes, glycoproteins, cytokines, growth factors and related proteins in the decidua (and in the conceptus) during the process of implantation and sustained successful pregnancy are just beginning to be understood. Challenges for the future include:

(1) A comprehensive identification of endometrial molecules that participate in endowing the epithelium with 'receptivity' and the mechanisms that govern their expression, dysregulation and misexpression;

(2) Determining whether receptivity is also a characteristic of the stroma;

(3) Investigating the players and the molecular mechanisms underlying the cross-talk that occurs between the trophoblast and the decidua;

(4) Discovering what occurs in microenvironments to facilitate and/or inhibit trophoblast invasiveness;

(5) Investigating what clinical syndromes associated with abnormalities in the endometrium and the decidua result in abnormal implantation or implantation failure.

Answers to these issues should be forthcoming in the near future, especially by means of gene profiling and proteomic approaches, which should prove valuable in facilitating the evaluation of fertility potential in women with infertility. In the wake of elucidation of the human genome as we enter the post-genomic era, genetic variations will soon be defined among individuals. This information, coupled with knowledge of the 'players' and 'regulators' should provide tools and strategies to ensure successful implantation for couples with infertility or women with implantation failure and repetitive miscarriage of non-genetic origin, as well as inhibition of implantation for contraceptive purposes.

REFERENCES

1. Hertig A, Rock JA. A description of 34 human ova within the first 17 days of development. *Am J Anat* 1956 98: 435–494.
2. Nikas G. Cell surface morphological events relevant to human implantation. *Hum Reprod* 1999 14(Suppl 2): 37–44.
3. Navot D, Scott RT, Droesch K, Veeck LL, Liu HC, Rosenwaks Z. The window of embryo transfer and the efficiency of human conception *in vitro*. *Fertil Steril* 1991 55: 114–118.
4. Wilcox AJ, Baird D, Weinberg CR. Time of implantation of the conceptus and loss of pregnancy. *N Engl J Med* 1999 340: 1796–1799.
5. Giudice LC. Genes associated with embryonic attachment and implantation and the role of progesterone. *J Reprod Med* 1999 44(Suppl): 165–171.
6. Paria BC, Lim H, Das SK, Reese J, Dey SK. Molecular signaling in uterine receptivity for implantation. *Semin Cell Dev Biol* 2000 11: 67–76.
7. Lessey BA. The role of the endometrium during embryo implantation. *Hum Reprod* 2000 15: 39–50.

8. Kao L-C, Giudice LC. Growth factors and implantation. *Infertil Reprod Med Clin North Am* 2001 12: 281–300.
9. Kimber SJ. Molecular interactions at the maternal–embryonic interface during the early phase of implantation. *Semin Reprod Med* 2001 18: 237–254.
10. Bergeron C, Ferenczy A, Toft DO, Schneider W, Shyamala G. Immunocytochemical study of progesterone receptors in the human endometrium during the menstrual cycle. *Lab Invest* 1988 59: 862–869.
11. Lessey BA, Killam AP, Metzger DA, Haney AF, Greene GL, McCarty KS. Immunohistochemical analysis of human uterine estrogen and progesterone receptors throughout the menstrual cycle. *J Clin Endocrinol Metab* 1988 67: 334–340.
12. Mangal RK, Wiehle RD, Poindexter AN III *et al.* Differential expression of uterine progesterone receptor forms A and B during the menstrual cycle. *J Steroid Biochem Mol Biol* 1997 63: 195–202.
13. Mote PA, Balleine RL, McGowan EM, Clarke CL. Colocalization of progesterone receptors A and B by dual immunofluorescent histochemistry in human endometrium during the menstrual cycle. *J Clin Endocrinol Metab* 1999 84: 2963–2971.
14. Mote PA, Balleine RL, McGowan EM, Clarke CL. Heterogeneity of progesterone receptors A and B expression in human endometrial glands and stroma. *Hum Reprod* 2000 15(Suppl 3): 48–56.
15. Nikas G, Makrigiannakis A, Hovatta O, Jones HW Jr. Surface morphology of the human endometrium: basic and clinical aspects. *Ann NY Acad Sci* 2000 900: 316–324.
16. Martel D, Frydman R, Glissant M, Maggioni C, Roche D, Psychoyos A. Scanning electron microscopy of postovulatory human endometrium in spontaneous cycles and cycles stimulated by hormone treatment. *J Endocrinol* 1987 114: 319–324.
17. Bentin-Ley U, Petersen B, Lindenberg S *et al.* Isolation and culture of human endometrial cells in a three dimensional cell culture system. *J Reprod Fertil* 1994 101: 327–332.
18. Bentin-Ley U, Sjogren A, Nilsson L *et al.* Presence of uterine pinopodes at the embryo–endometrial interface during human implantation *in vitro*. *Hum Reprod* 1999 14: 515–520.
19. Bentin-Ley U, Horn T, Sjogren A *et al.* Ultrastructure of human blastocyst–endometrial interactions *in vitro*. *J Reprod Fertil* 2000 2: 337–350.
20. Galen A, O'Conner JE, Valbuena D *et al.* The human blastocyst regulates endometrial epithelial apoptosis in embryonic implantation. *Biol Reprod* 2000 63: 430–439.
21. Kimber SJ, Spanswick C. Blastocyst implantation: the adhesion cascade. *Semin Cell Dev Biol* 2000 11: 77–92.
22. Turpeenniemi-HujanenT, Feinberg RF, Kauppila A, Puistola U. Extracellular matrix interactions in early human embryos: implications for normal implantation events. *Fertil Steril* 1995 64: 132–138.
23. Ravn V, Mandel U, Svenstrup B, Dabelsteeen E. Expression of type-2-histo-blood group carbohydrate antigens (Le(x), Le(y), and H) in normal and malignant human endometrium. *Virchow's Arch* 1994 424: 411–419.
24. Kimber SJ, Lindenberg S, Lundblad A. Distribution of some Galβ1-3(4)GlcNAc related carbohydrate antigens on the mouse uterine epithelium in relation to the peri-implantation period. *J Reprod Immunol* 1988 12: 297–313.
25. Lindenberg S, Kimber SJ, Kalin E. Carbohydrate binding properties of mouse embryos. *J Reprod Fertil* 1990;89:431-439.
26. Smith SE, French MM, Julian J, Paria BC, Dey SK, Carson DD. Expression of heparan sulfate proteoglycan (perlecan) in the mouse blastocyst is regulated during normal and delayed implantation. *Dev Biol* 1997 184: 1109–1122.
27. Rohde LH, Carson DD. Heparin-like glycosaminoglycans participate in binding of a human trophoblastic cell line (JAR) to a human uterine epithelial cell line (RL95). *J Cell Physiol* 1993 155: 185–196.
28. Rohde LH, Janpore MI, McMaster MT *et al.* Complementary expression of HIP, a cell-surface heparan sulphate binding protein, and perlecan at the human fetal–maternal interface. *Biol Reprod* 1998 58: 1075–1083.
29. Campbell S, Swann HR, Seif MW, Kimber SJ, Aplin JD. Cell adhesion molecules on the oocyte and preimplantation human embryo. *Hum Reprod* 1995 10: 1571–1578.
30. Aplin JD. The cell biology of implantation. *Placenta* 1996 17: 269–275.

31. Hey NA, Graham RA, Seif MW, Aplin JD. The polymorphic epithelial mucin MUC 1 in human endometrium is regulated with maximal expression in the implantation phase. *J Clin Endocrinol Metab* 1994 78: 337–342.

32. Feinberg RF, Kliman HJ. MAG (mouse ascites Golgi) mucin in the endometrium: a potential marker of endometrial receptivity to implantation. In Diamond MP, Osteen KG, eds. *Endometrium and Endometriosis*. Malden, MA: Blackwell Science, 1997: 131–139.

33. Clark EA, Brugge JS. Integrins and signal transduction pathways: the road taken. *Science* 1995 268: 233–239.

34. Lessey BA, Ilesami AO, Sun J *et al.* Luminal and glandular endometrial epithelium express integrins differentially throughout the menstrual cycle: implications for implantation, contraception, and infertility. *Am J Reprod Immunol* 1996 35: 195–204.

35. Lessey BA, Damjanovich L, Coutifaris C *et al.* Integrin adhesion molecules in the human endometrium: correlation with the normal and abnormal menstrual cycle. *J Clin Invest* 1992 90: 188–195.

36. Lessey BA, Yeh I, Castelbaum AJ *et al.* Endometrial progesterone receptors and markers of uterine receptivity in the window of implantation. *Fertil Steril* 1996 65: 477–83.

37. Illera MJ, Cullinan E, Gui Y *et al.* Blockage of the $\alpha_v\beta_3$ integrin adversely affects implantation in the mouse. *Biol Reprod* 2000 62: 1285–1290.

38. Creuss M, Balasch J, Ordi J *et al.* Integrin expression in normal and out-of-phase endometria. *Hum Reprod* 1998 13: 3460–3468.

39. Lessey BA, Castlebaum AJ, Harris J *et al.* Use of integrins to date the endometrium. *Fertil Steril* 2000 73: 779–787.

40. Aoki R, Fukuda MN. Recent molecular approaches to elucidate the mechanism of embryo implantation: trophinin, bystin, and tastin as molecules involved in the initial attachment of blastocysts to the uterus in humans. *Semin Reprod Endocrinol* 2000 18: 265–271.

41. Lim H, Das SK, Dey SK. ErbB genes in the mouse uterus: cell-specific signaling by epidermal growth factor (EGF) family of growth factors during implantation. *Dev Biol* 1998 204: 97–110.

42. Leach RE. EGF-like growth factor families. *Infertil Reprod Med Clin North Am* 2001 12: 359–371.

43. Schlessinger J, Ullrich A. Growth factor signaling by receptor tyrosine kinases. *Neuron* 1992 9: 383–391.

44. Das SK, Wang X, Paria BC *et al.* Heparin-binding EGF-like growth factor gene is induced in the mouse uterus temporally by the blastocyst solely at the site of its apposition: a possible ligand for interaction with blastocyst EGF-receptor in implantation. *Development* 1994 120: 1071–1083.

45. Das SK, Dan N, Wang J, Dey SK. Expression of betacellulin and epiregulin genes in the mouse uterus temporally by the blastocyst solely at the site of its apposition is coincident with the window of implantation. *Dev Biol* 1997 190: 178–190.

46. Sargent IL, Martin KL, Barlow DH. The use of recombinant growth factors to promote human embryo development in serum-free medium. *Hum Reprod* 1998 13(Suppl 4): 239–248.

47. Das SK, Chakraborty I, Paria BC *et al.* Amphiregulin is an implantation-specific and progesterone-regulated gene in the mouse uterus. *Mol Endocrinol* 1995 9: 691–705.

48. Stewart CL. Leukemia inhibitory factor and the regulation of pre-implantation development of the mammalian embryo. *Mol Reprod Dev* 1994 39: 233–238.

49. Chen HF, Shew JY, Shew JY *et al.* Expression of leukemia inhibitory factor and its receptor in preimplantation embryos. *Fertil Steril* 1999 72: 713–719.

50. Nachtigall MJ, Kliman HJ, Feinberg RF *et al.* The effect of leukemia inhibitory factor on trophoblast differentiation: a potential role in human implantation. *J Clin Endocrinol Metab* 1996 81: 801–806.

51. Bischof P, Haenggeli L, Campana A. Effect of leukemia inhibitory factor on human cytotrophoblast differentiation along the invasive pathway. *Am J Reprod Immunol* 1995 34: 225–230.

52. Giess R, Tanasescu I, Steck T, Sendtner M. Leukaemia inhibitory factor gene mutates in infertile women. *Mol Hum Reprod* 1999 5: 581–586.

53. Laird SM, Tuckerman EM, Dalton CF, Dunphy BC, Li TC, Zhang X. The production of leukaemia inhibitory factor by human endometrium: presence in uterine flushings and production by cells in culture. *Hum Reprod* 1997 12: 569–574.

54. Delage G, Moreau JF, Taupin JL *et al.* *In vivo* endometrial secretion of human interleukin for DA cells/leukaemia inhibitory factor by explant cultures from fertile and infertile women. *Hum Reprod* 1995 10: 2483–2488.
55. Hambartsoumian E. Endometrial leukemia inhibitory factor as a possible cause of unexplained infertility and multiple failures of implantation. *J Reprod Fertil* 1998 39: 137–143.
56. Illera MJ, Yuan L, Stewart CL, Lessey BD. Effect of peritoneal fluid from women and endometriosis on implantation in the mouse model. *Fertil Steril* 2000 714: 41–48.
57. Robb L, Li R, Hartley L, Nandurkar HH, Koentgen F, Begley CG. Infertility in female mice lacking the receptor for interleukin 11 is due to a defective uterine response to implantation. *Nat Med* 1998 4: 303–308.
58. Dimitriadis E, Salamonsen LA, Robb L. Expression of interleukin-11 during the human menstrual cycle coincidence with stromal cell decidualization and relationship to leukaemia inhibitory factor and prolactin. *Mol Hum Reprod* 2000 6: 907–912.
59. Popovici RM, Kao L-K, Giudice LC. Discovery of new inducible genes in *in vitro* decidualized human endometrial stromal cells using microarray technology. *Endocrinology* 2000 141: 3510–3513.
60. Schwertschlag US, Trepicchio WL, Dykstra KH, Keith JC, Turner KJ, Dorner AJ. Hematopoietic, immunomodulatory, and epithelial effects of interleukin-11. *Leukemia* 1999 13: 1307–1315.
61. Simon C, Mercader A, Frances A *et al.* Hormonal regulation of serum and endometrial Il-1α, Il-1β and Il-1ra. IL-1 endometrial microenvironment of the human embryo at the apposition phase under physiological and supraphysiological steroid level conditions. *J Reprod Immunol* 1996 31: 165–184.
62. Simon C, Gimeno MJ, Mercader A *et al.* Embryonic regulation of integrins β_3, α_4 and α_1 in human endometrial epithelial cells *in vitro*. *J Clin Endocrinol Metab* 1997 82: 2607–2616.
63. Simon C, Frances A, Piquette GN *et al.* Embryonic implantation in mice is blocked by interleukin-1 receptor antagonist. *Endocrinology* 1994 134: 521–528.
64. Frank GR, Brar AK, Jikihara H, Cedars MI, Handwerger S. Interleukin-1b and the endometrium: an inhibitor of stromal cell differentiation and possible autoregulator of decidualization in humans. *Biol Reprod* 1995 52: 184–191.
65. Strakova Z, Srusuparp S, Fazleabas AT. Interleukin-1β induces the expression of insulin-like growth factor binding protein-1 during decidualization in the primate. *Endocrinology* 2000 141: 4664–4670.
66. Hsieh-Li HM, Witte DP, Weinstein M *et al.* Hoxa11 structure, extensive antisense transcription and function in male and female infertility. *Development* 1995 121: 1373–1385.
67. Satokata I, Bhenson G, Maas R. Sexually dimorphic sterility phenotypes in Hoxa-10 deficient mice. *Nature* 1996 374: 460–463.
68. Benson GV, Lim H, Paria BC *et al.* Mechanisms of reduced fertility in Hoxa10 mutant mice: uterine homeostasis and loss of maternal Hoxa10 expression. *Development* 1996 122: 2687–2696.
69. Gendron RL, Hsieh L, Lee DW *et al.* Abnormal uterine stromal and glandular function associated with maternal reproductive defects in Hoxa-11 null mice. *Biol Reprod* 1997 56: 1123–1136.
70. Bagot C, Troy P, Taylor HS. Alteration of maternal HOXA10 expression by *in vivo* gene transfection affects implantation. *Gene Ther* 2000 7: 1278–1284.
71. Taylor HS, Igarashi P, Olive DL, Arici A. Sex steroids mediate HOXA11 expression in human peri-implantation endometrium. *J Clin Endocrinol Metab* 1999 84: 1129–1135.
72. Taylor HS, Bargot C, Kardana A *et al.* Hox gene expression is altered in the endometrium of women with endometriosis. *Hum Reprod* 1999 14: 1328–1331.
73. Zhu LJ, Cullinan-Bove K, Polihronis M, Bagchi MK, Bagchi IC. Calcitonin is a progesterone-regulated marker that forecasts the receptive state of endometrium during implantation. *Endocrinology* 1998 139: 3923–3934.
74. Wang J, Rout UK, Bagchi IC, Armant DR. Expression of calcitonin receptors in mouse preimplantation embryos and their function in the regulation of blastocyst differentiation by calcitonin. *Development* 1998 125: 4293–4302.
75. Irwin JC, Giudice LC. Decidua. In Knobil E, Neill JD, eds. *Encyclopedia of Reproduction*. San Diego: Academic Press, 1998: 823–835.
76. Kao L-C, Yang JP, Tulac SJ, Lessey BA, Giudice LC. Human endometrium gene profiles during

the window of implantation by Genechip microarray. *Endocr Soc Abstr* 2001 #P03: 154.

77. Tanaka N, Miyazaki K, Tashiro H, Mizutani H, Okamura H. Changes in adenylyl cyclase activity in human endometrium during the menstrual cycle and in human decidua during pregnancy. *Reprod Fertil* 1993 97: 33–39
78. Kim JJ, Wang J, Bambra C, Das SK, Dey SK, Fazleabas AT. Expression of cyclooxygenase-1 and -2 in the baboon endometrium during the menstrual cycle and pregnancy. *Endocrinology* 1999 40: 2672–2678.
79. Brar AK, Frank GR, Kessler CA, Cedars MI, Handwerger S. Progesterone-dependent decidualization of the human endometrium is mediated by cAMP. *Endocrine* 1997 6: 301–307.
80. Gellersen B, Kempf R, Telgmann R. Human endometrial stromal cells express novel isoforms of the transcriptional modulator CREM and upregulate ICER in the course of decidualization. *Mol Endocrinol* 1997 11: 97–113.
81. Tseng L, Gao JG, Chen R, Zhu HH, Mazella J, Powell DR. Effect of progestin, antiprogestin, and relaxin on the accumulation of prolactin and insulin-like growth factor-binding protein-1 messenger ribonucleic acid in human endometrial stromal cells. *Biol Reprod* 1992 47: 441–450.
82. Lim H, Paria BC, Das SK, Maas RM. Multiple female reproductive failures in cyclooxygenase 2-deficient mice. *Cell* 1997 91: 197–208.
83. Lim H, Gupta RA, Ma WG *et al.* Cyclooxygenase-2-derived prostacyclin mediates embryo implantation in the mouse via PPARδ. *Genes Dev* 1999 13: 1561–1574.
84. Frank GR, Brar AK, Cedars MI, Handwerger S. Prostaglandin E_2 enhances human endometrial stromal cell differentiation. *Endocrinology* 1994 134: 258–263.
85. Psychoyos A, Nikas G, Gravanis A. The role of prostaglandins in blastocyst implantation. *Hum Reprod* 1995 10(Suppl 2): 30–42.
86. Marions L, Danielsson KG. Expression of cyclo-oxygenase in human endometrium during the implantation period. *Mol Hum Reprod* 1999 5: 961–965.
87. Han SW, Lei ZM, Rao CV. Up-regulation of cyclooxygenase-2 gene expression by chorionic gonadotropin during the differentiation of human endometrial stromal cells into decidua. *Endocrinology* 1996 137: 1791–1797.
88. Zhou HG, Lei ZM, Rao CV. Treatment of human endometrial gland epithelial cells with chorionic gonadotropin/luteinizing hormone increases the expression of the cyclooxygenase-2 gene. *J Clin Endocrinol Metab* 1999 84: 3364–3377.
89. Huang JC, Liu DY, Yadollahi S *et al.* Interleukin-1 beta induces cyclooxygenase-2 gene expression in cultured endometrial stromal cells. *J Clin Endocrinol Metab* 1998 83: 538–541.
90. Fisher SJ, Damsky CH. Human cytotrophoblast invasion. *Cell Biol* 1993 4: 183–188.
91. Damsky CH, Librach C, Lim KH *et al.* Integrin switching regulates normal trophoblast invasion. *Development* 1994 120: 3657–3666.
92. Polette M, Nawrocki B, Pintiaux A *et al.* Expression of gelatinases A and B and their tissue inhibitors by cells of early and term human placenta and gestational endometrium. *Lab Invest* 1994 71: 838–846.
93. Librach CL, Werb Z, Fitzgerald ML *et al.* 92-kD type IV collagenase mediates invasion of human cytotrophoblasts. *J Cell Biol* 1991 113: 437–449.
94. Shimonovitz S, Hurwitz A, Dushnik M, Anteby E, Geva-Eldar T, Yagel S. Developmental regulation of the expression of 72 and 92 kD type IV collagenases in human trophoblasts: a possible mechanism for control of trophoblast invasion. *Am J Obstet Gynecol* 1994 171: 832–838.
95. Librach CL, Feigenbaum SL, Bass KE *et al.* Interleukin-1β regulates human cytotrophoblast metalloproteinase activity and invasion *in vitro*. *J Biol Chem* 1994 269: 17125–17131.
96. Cross JC, Werb Z, Fisher SJ. Implantation and the placenta: key pieces of the development puzzle. *Science* 1994 266: 1508–1518.
97. Salamonsen LA, Zhang J, Hampton A, Lathbury L. Regulation of matrix metalloproteinases in human endometrium. *Hum Reprod* 2000 15(Suppl 3): 112–119.
98. Higushi T, Kanzaki H, Nakayama H *et al.* Induction of tissue inhibitor of metalloproteinase 3 gene during *in vitro* decidualization of human endometrial stromal cells. *Endocrinology* 1995 136: 4973–4981.
99. Huang HY, Wen Y, Irwin JC, Kruessel JS, Soong YK, Polan ML. Cytokine mediated regulation of tissue inhibitor of metalloproteinase-1 (TIMP-1), TIMP-3, and 92 kD type IV collagenase mRNA expression in human endometrial stromal cells. *J Clin Endocrinol Metab* 1998 83: 1721–1729.

100. Irwin JC, Suen L-F, Giudice LC. Insulin-like growth factor-II regulation of endometrial stromal cell tissue inhibitor of metalloproteinase-3 and IGF binding protein-1 suggests paracrine interactions at the decidual: trophoblast interface during human implantation. *J Clin Endocrinol Metab* 2001 86: 2060–2064.

101. Kao L-C, Caltabiano S, Wu S, Strauss JF III, Kliman HJ. The human villous cytotrophoblast: interactions with extracellular matrix proteins, endocrine function, and cytoplasmic differentiation in the absence of syncytium formation. *Dev Biol* 1988 130: 693–702.

102. Irwin JC, Giudice LC. IGFBP-1 binds to the $\alpha_5\beta_1$ integrin in human cytotrophoblasts and inhibits their invasion into decidualized endometrial stromal cells *in vitro*. *GH IGF Res* 1998 8: 21–31.

103. Irving JA, Lala PK. Functional role of cell surface integrins on human trophoblast cell migration: regulation by TGF-beta, IGF-II, and IGFBP-1. *Exp Cell Res* 1995 217: 419–427.

104. Han VK, Bassett N, Walton J, Challis JRG. The expression of insulin-like growth factor (IGF) and IGF binding protein genes in the human placenta and membranes: evidence for IGF: IGFBP interactions at the feto-maternal interface. *J Clin Endocrinol Metab* 1996 81: 2680–2693.

105. Jones JI, Gockerman A, Busby WH, Clemmons DR. IGFBP-1 stimulates cell migration and binds to the $\alpha_5\beta_1$ integrin by means of its Arg-Gly-Asp sequence. *Proc Natl Acad Sci USA* 1993 90: 10553–10558.

106. Hamilton GS, Lysiak JJ, Han VKM, Lala PK. Autocrine–paracrine regulation in human trophoblast invasiveness by IGF-II and IGFBP-1. *Exp Cell Res* 1998 244: 147–156.

107. Gleeson LM, Chakraborty C, McKinnon T, Lala PK. Insulin-like growth factor-binding protein 1 stimulates human trophoblast migration by signaling through $\alpha_5\beta_1$ integrin via mitogen-activated protein kinase pathway. *J Clin Endocrinol Metab* 2001 86: 2484–2493.

108. Giudice LC, Dsupin BA, Jin IH, Vu TH, Hoffman AR. Differential expression of mRNAs encoding insulin-like growth factors and their receptors in human uterine endometrium and decidua. *J Clin Endocrinol Metab* 1993 76: 1115–1122.

109. Das SK, Flanders KC, Andrews GK, Dey SK. Expression of transforming growth factor-ß isoforms (ß2 and ß3) in the mouse uterus: analysis of the periimplantation period and effects of ovarian steroids. *Endocrinology* 1992 130: 3459–3466.

110. Chegini N, Zhao Y, Williams RS, Flanders KC. Human uterine tissue throughout the menstrual cycle expresses transforming growth factor-β 1 (TGF-β 1), TGF-β 2, TGF-β 3, and TGF-β type II receptor messenger ribonucleic acid and protein and contains [^{125}I] TGF-β 1 binding sites. *Endocrinology* 1994 235: 439–449.

111. Graham CH, Lysiak JJ, McCrae KR *et al.* Localization of transforming growth factor beta at the human fetal–maternal interface: role in trophoblast growth and differentiation. *Biol Reprod* 1992 46: 561–572.

112. Smith SK. Angiogenesis and implantation. *Infertil Reprod Med Clin North Am* 2001 12: 301–314.

113. Charnock-Jones DS, Sharkey AM, Rajput-Williams J, *et al.* Identification and localization of alternately spliced mRNAs for vascular endothelial growth factor in human uterus and estrogen regulation in endometrial carcinoma cell lines. *Biol Reprod* 1993 48: 1120–1128.

114. Shifren JL, Tseng JF, Zaloudek CJ *et al.* Ovarian steroid regulation of vascular endothelial growth factor in the human endometrium: implications for angiogenesis during the menstrual cycle and in the pathogenesis of endometriosis. *J Clin Endocrinol Metab* 1996 81: 3112–3118.

115. Popovici R, Irwin JC, Giaccia A, Giudice LC. Hypoxia and cAMP stimulate vascular endothelial growth factor (VEGF) in human endometrial stromal cells: potential relevance to menstruation and endometrial regeneration. *J Clin Endocrinol Metab* 1999 84: 2245–2248.

116. Sharkey AM, Day K, McPherson A *et al.* Vascular endothelial growth factor expression in human endometrium is regulated by hypoxia. *J Clin Endocrinol Metab* 2000 85: 402–409.

117. Li X-F, Charnock-Jones DS, Zhang E *et al.* Angiogenic growth factor mRNAs in uterine natural killer cells. *J Clin Endocrinol Metab* 2001 86: 1823–1834.

118. Croy BA, Ashkar AA, Minhas K *et al.* Can murine uterine natural killer cells give insights into the pathogenesis of preeclampsia? *J Soc Gynecol Invest* 2000 1: 12–20.

119. Croy BA, Ashkar AA, Foster RA *et al.* Histological studies of gene-ablated mice support important functional roles for natural killer cells in the

uterus during pregnancy. *J Reprod Immunol* 1997 35: 111–133.
120. Guimond MJ, Wang B, Croy BA. Engraftment of bone marrow from severe combined immunodeficient (SCID) mice reverses the reproductive deficits in natural killer cell deficient-tg epsilon 26 mice. *J Exp Med* 1998 187: 217–223.
121. Kauma S, Shapiro SS. Immunoperoxidase localization of prolactin in endometrium during normal menstrual, luteal phase defect, and correlated luteal phase defect cycles. *Fertil Steril* 1986 46: 37–42.
122. Seppala M. Glycodelins: biological activity in reproduction and differentiation. *Curr Opin Obstet Gynecol* 1999 11 :261–263.
123. Mueller MD, Vigne J-L, Vaisse C, Taylor RN. Glycodelin: a pane in the implantation window. *Semin Reprod Endocrinol* 2000 18: 289–298.
124. Taylor RN, Vigne HL, Zhang P, Hoang P, Lebovic DI, Mueller MD. Effects of progestins and relaxin on glycodelin gene expression in human endometrial cells. *Am J Obstet Gynecol* 2000 182: 841–847.
125. Koistinen H, Koistinen R, Seppala M, Burova TV, Choiset Y, Haertlae T. Glycodelin and beta-lactoglobulin, lipocalins with a high structural similarity differ in ligand binding properties. *FEBS Lett* 1999 450: 158–162.
126. Okamoto N, Uchida A, Takakura K *et al.* Suppression by human placental protein-14 of natural killer cell activity. *Am J Reprod Immunol* 1991 26: 137–142.
127. Bulmer JN, Sunderland CA. Immunohistological characterization of lymphoid cell populations in the early human placental bed. *Immunology* 1984 52: 349–357.
128. Clark DA, Vince G, Flanders KC *et al.* CD56+ lymphoid cells in human first trimester pregnancy decidua as a source of novel transforming growth factor-β_2-related immunosuppressive factors. *Human Reprod* 1994 9: 2270–2277.
129. Fisher SJ. The placenta dilemma. *Semin Reprod Endocrinol* 2000 18: 321-326.
130. Ghazeeri GS, Clark DA, Kutteh WH. Immunologic factors in implantation. *Infertil Reprod Med Clin North Am* 2001 12: 315–337.

Chapter 25

Concepts in early embryonic patterning

G. Praveen Raju, Change Tan and Peter S. Klein

INTRODUCTION

Because mammals are viviparous organisms, postimplantation development has been for the most part inaccessible to direct observation or microsurgical manipulations. Nevertheless, the field has advanced tremendously over the past several years, owing in large part to advances in genetics and molecular biology and their application to fundamental questions of development in diverse model organisms. In this review, we summarize some of the important conceptual advances as well as the technological breakthroughs that have made these advances possible. The anatomy of the mammalian embryo is described only briefly, as these details are better cataloged elsewhere.[1] Rather, our emphasis is on principles of embryonic development learned from the study of diverse organisms, including the mouse and other vertebrates, but also including invertebrates such as *Drosophila melanogaster*.

Embryology began as a purely descriptive science with Aristotle's observations of chick embryo development. In the 19th century, Karl Ernst von Baer described the similarity between many vertebrate embryos, including chick, fish, reptiles, birds and mammals shortly after gastrulation.[2] At this stage, all vertebrate embryos have gill arches, notochords, spinal cords and a very similar epidermis. As development proceeds, these tissues differentiate into structures characteristic of the given organism, with the epidermis, for example, differentiating into the scales of fish or reptiles, the feathers of birds, or the hair of mammals. Likewise, early limb development is virtually the same in all vertebrates, and only later do limbs develop into legs, arms, fins, or wings. We focus here on the parallels between different organisms, since our current understanding of early mammalian development is a montage of principles learned from these diverse organisms. The equally interesting question of what makes these organisms develop differently is not addressed here.

Embryology became an experimental science in the late 19th century and the early work in this field led to several fundamental principles that are discussed below, such as the ability of the embryo to regulate its own development and the process of embryonic induction. More recently, major advancements in the understanding of embryonic development have come from the application of genetics and molecular biology to the fundamental questions formulated and studied by classical embryologists. This advance has been led by work in the fruit fly *Drosophila melanogaster*, but important contributions have come from numerous other organisms, including the nematode *Caenorrhabditis elegans*, the mouse, the chick and the frog (*Xenopus laevis*), as well as the zebrafish, sea urchins and other model organisms.

BACKGROUND

In general, the key stages of embryonic development are similar for many different organisms (Figure 1a), although the details vary considerably, even among mammals. After fertilization, the egg undergoes a series of rapid mitotic divisions (cleavage stage). For many of the organisms discussed in this chapter, including frogs, worms, flies and chicks, the cleavage stages occur in the absence of transcriptional activity from the embryo. At the end of the cleavage stage, the multicellular embryo is termed the blastula or blastocyst (the vertebrate blastula often contains a central cavity or blastocoele).

After the midblastula transition, where the maternal control of early embryonic development transfers to embryo control, the rate of mitotic division slows and cell divisions become asynchronous. Gastrulation begins with the involution or invagination of cells into the interior of the blastula, ultimately leading to the formation of the three germ layers. The mechanics of gastrulation are complex and differ considerably between vertebrate species, but the outcome is similar. The outer germ layer, termed the ectoderm, gives rise to cells that eventually produce skin and the nervous system. The inner germ layer, the endoderm, gives rise to tissues such as the intestines, liver and pancreas. The middle germ layer, the mesoderm, will give rise to several tissues including muscle, heart, mesenchyme and blood. Thus, gastrulation culminates in the placement of the three germ layers into new positions that will allow the induction of specific organs and tissues to proceed after gastrulation.[3]

Mammals, as exemplified by the mouse, differ somewhat from the other model organisms discussed here (Figure 1b). These differences include an essentially symmetrical fertilized egg, a late first cleavage (17–20 h after fertilization) and the onset of zygotic gene transcription shortly after the first cleavage. Up to the eight-cell stage (morula), all of the cells have the same developmental potential. Interestingly, if the early embryo is split in two, two complete normal individuals can be produced, as seen with monozygotic twins. The eight-cell embryo undergoes a process termed compaction, in which the cells become more tightly packed, maximizing cell–cell contact, and cell boundaries on the surface become indistinct. The first differentiation occurs in the 16-cell embryo, so that the external cells will become trophoblast, which produces the chorion. The trophoblast cells secrete fluid into the center of the morula, leading to formation of a central cavity, the blastocoele. At this stage, the embryo has become a blastocyst, in which the entire outer cell layer, the trophectoderm, encloses the blastocoele and a cluster of cells, termed the inner cell mass (ICM). Within the ICM, cells in contact with the blastocoele give rise to the hypoblast, while cells that are not in contact with the blastocoele give rise to the epiblast. The hypoblast lines the blastocoele cavity, where it gives rise to the yolk sac endoderm, while the epiblast provides essentially all of the cells that will form the embryo proper. At gastrulation, prospective mesodermal and endodermal cells migrate through the primitive streak, an invagination that forms along the anterior–posterior axis of the primitive ectoderm. They then move anteriorly and laterally to form the definitive mesodermal and endodermal tissues of the embryo.[4,5]

In the remaining part of this chapter, the methods and techniques that have been developed and utilized to gain understanding of embryonic development at the cellular and molecular level are described. Also, several key developmental concepts that have emerged from molecular and embryological studies in various organisms are highlighted. Important signalling pathways that have been conserved in development throughout evolution are mentioned. Finally, the potential clinical relevance of these developmental studies is addressed.

APPLIED TECHNOLOGY

Characterization of mutants that alter normal embryonic development has been hindered by the difficulty in identifying the mutated genes. With the advances in molecular biology over the past three decades, it has become easier to identify and clone these genes, especially in organisms with small genomes, such as *Drosophila* and *C. elegans*, and thus to understand the role of a particular molecule in development.

In order for a gene product to play a role in the development of a particular structure or tissue, it must be expressed in the correct place and at the correct time. The technique of *in situ* hybridization has been invaluable in answering both of these questions. This technique uses a labelled nucleic acid probe that is complementary to the gene of interest.[6] Embryos are placed in fixative to immobilize mRNAs and maintain the overall structure of the embryo and then are incubated with the labelled probe, which hybridizes wherever the mRNA is expressed (Figure 2). A similar technique using antibodies allows *in situ* detection of proteins. Some of the early applications of this technique revealed the astonishing findings of repeating patterns of expression of pair-rule genes in *Drosophila* and the similar anterior–posterior organization of homeobox genes in invertebrates and vertebrates (described below).

Gain-of-function approaches

Gene transfer is an important tool for understanding gene function. There are several ways to transfer genetic information in the analysis of embryonic development. In *Xenopus*, the large size of early embryos allows for the microinjection of DNA or mRNA. Other techniques to induce overexpression of genes involve the use of viral vectors, which have certain advantages over plasmid DNA or mRNA injection. The most commonly used vectors are adenovirus, retroviruses, herpes simplex virus (HSV) and adenovirus-associated virus (AAV).[7] The advantage of these viral vectors is that they

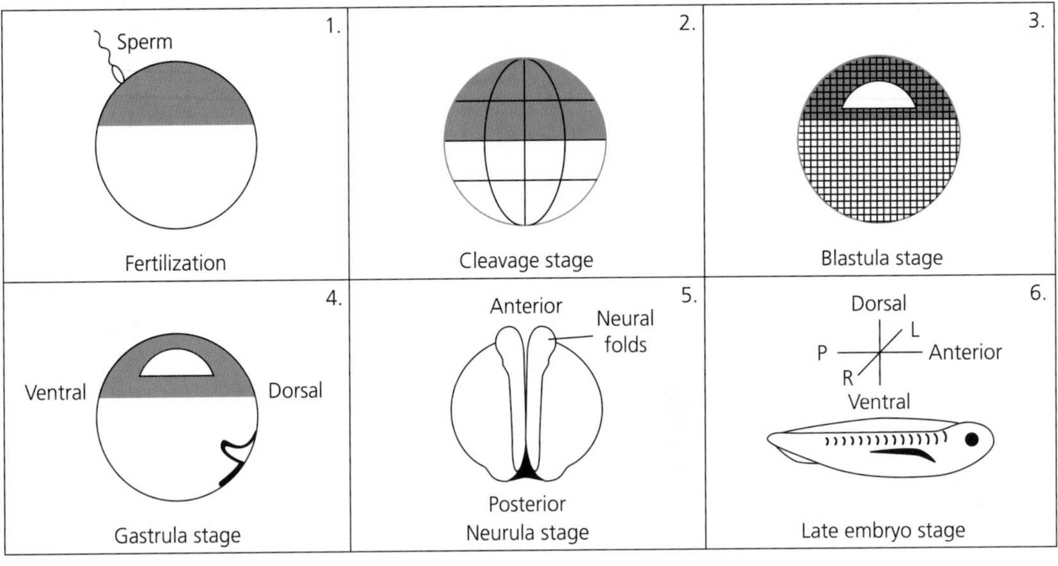

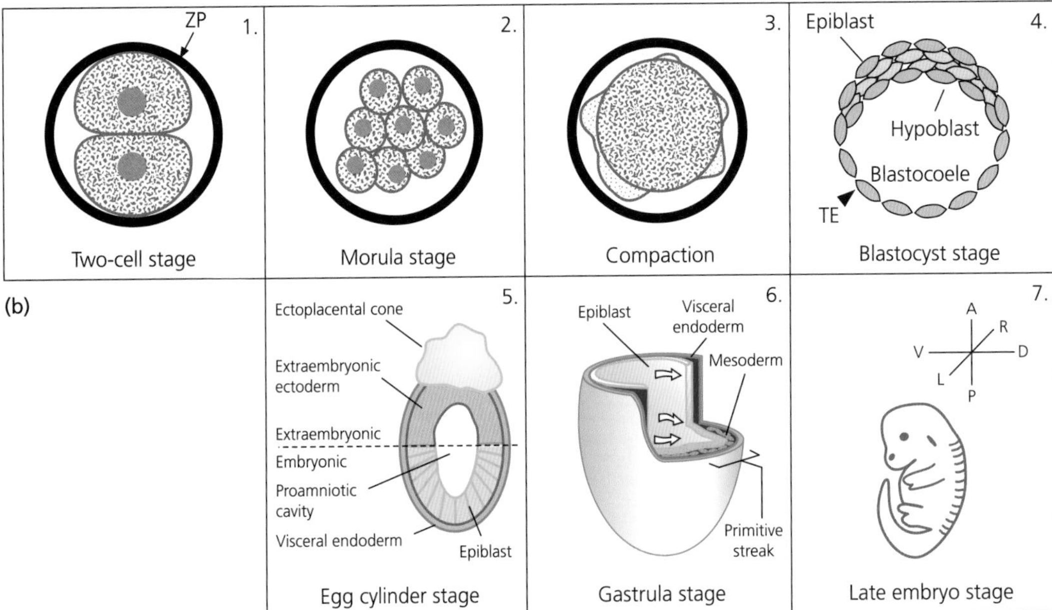

Figure 1 Overview of early development. (a) Schematic diagram of early *Xenopus* embryogenesis (from reference 3). Fertilization (1) initiates a series of rapid cleavages (2) of the embryo. Generally, zygotic gene transcription begins during the blastula stage (3) when an internal fluid-filled cavity is formed (blastocoele). Gastrulation (4) is initiated by the involution of cells on the dorsal side, termed the blastopore lip, and culminates in the correct positioning of the primary germ layers. Neurulation (5) and organogenesis are initiated as a result of proper inductive interactions between neighboring tissues, resulting in a normally patterned organism (6) (adapted from reference 118). (b) Schematic diagram of early mouse development. Zygotic transcription begins at the two-cell stage (1). At the eight-cell stage (morula) (2), cells undergo compaction (3). The blastocyst (4) is composed of an outer trophectodermal (TE) layer and an inner cell mass, comprising the hypoblast and epiblast, which forms the embryo proper. The egg cylinder stage (5) is followed by gastrulation (6), which begins with the formation of the primitive streak, and the migration of mesodermal and endodermal cells into the interior of the embryo. Organogenesis defines the major developmental events during the fetal stage (7). Adapted from reference 4

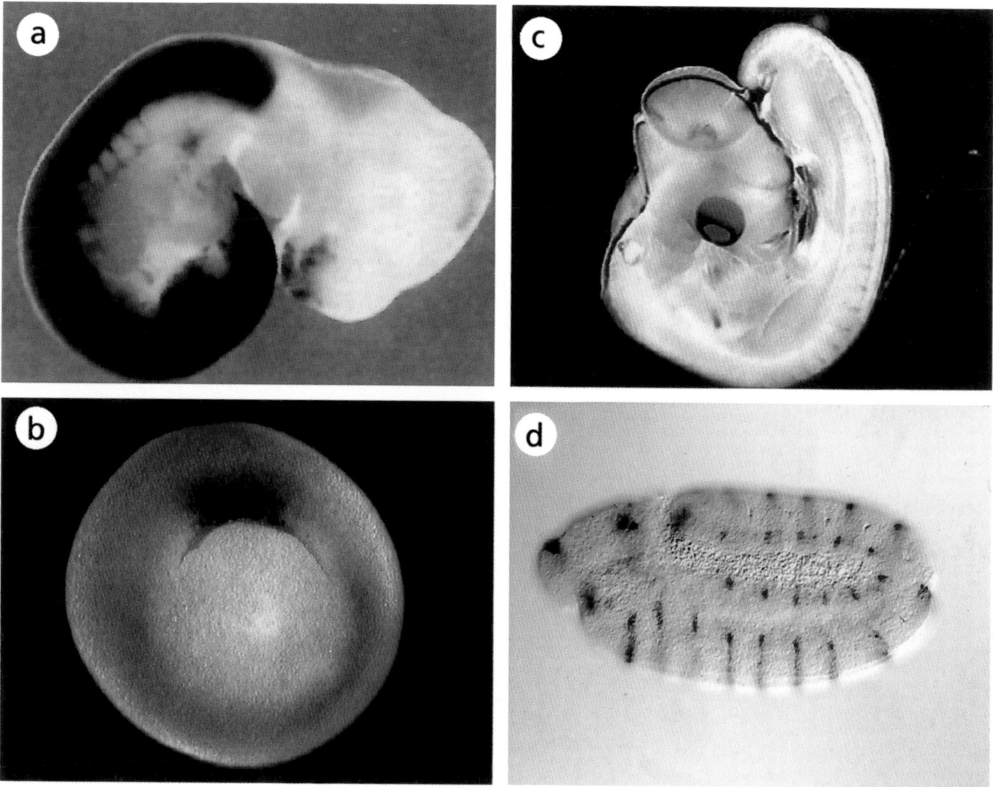

Figure 2 *In situ* hybridization. (a) *Hoxb-4* expression in mouse embryo of mid-gestational age (courtesy of Robb Krumlauf). (b) *Xwnt8* (light blue) and *Xfrizzled8* (purple) in a *Xenopus* gastrula embryo. *Xfrizzled8* expression can be detected in the dorsal blastopore lip (top). *XWnt8* expression is seen in ventral and lateral regions (courtesy of Matt Deardorff). (c) Expression of a novel *Wnt* gene in early chick embryo. Note high-level expression in the anterior neural tube (courtesy of Bob Riddle and Tony Brown). (d) Expression of *wingless* (purple) and *DWnt-4* (brown) in *Drosophila* gastrula embryo (courtesy of Betsy Wilder)

provide high-level, efficient gene expression. Retroviral vectors can be made in relatively high titer, can infect most cell types and allow for stable transgene expression. Once the RNA-containing virus enters the host cell, it makes a DNA copy of the viral genome through the enzyme reverse transcriptase, and the new double-stranded message then integrates into the host genome.

Another method to obtain overexpression of genes is through the use of transposable elements. These naturally occurring mobile elements are found in the genomes of numerous organisms, from plants to higher mammals. For example, P elements are found in the genome of *Drosophila* and can jump from one chromosomal location to another.[8] P elements can be isolated, and cloned genes can be inserted into the center of the P element. The recombinant P element is then injected into a host embryo where it integrates into the genome, creating a transgenic organism.

In mice, overexpression studies are usually performed through the use of transgenic techniques. The most common method of creating transgenic mice involves the direct injection of DNA into a pronucleus of the fertilized egg.[9,10] Alternatively, embryonic stem cells can be isolated from an early mouse embryo and cultured *in vitro*. A recombinant vector can be used to transfect or introduce the transgene into these cultured cells. Those cells that have incorporated the transgene are selected and micro-injected into a blastocyst of a host embryo. Usually, the host embryo is of another strain of mouse with different coat color to identify whether transgenic cells have been incorporated into the host. The blastocysts are reimplanted into the uterus of a pseudopregnant mouse and allowed to develop. The transferred blastocyst will develop into a chimeric mouse that can be detected by the presence of a mixed coat color. If the transgenic stem cells contributed to the germ line, the chimeric mouse can be mated with a wild-type mouse, with some of the resulting progeny being heterozygous for the transgene. By crossing two heterozygotes, a strain can be generated that is homozygous for the overexpressed transgene. The consequences of this gain-of-function approach can then be analyzed. The injected DNA can include specific regulatory elements to control expression both spatially and temporally.

Loss-of-function approaches

There are two general approaches to understanding the function of a gene through loss of function. One approach is random mutagenesis followed by screening for a specific phenotype. In this approach, the gene responsible for the phenotype must then be identified. Alternatively, a gene of interest can be specifically targeted for disruption or mutagenesis, and the effect of such manipulation on the embryo can be studied. Both of these approaches are described below.

Random mutagenesis allows disruption of genes without bias towards a particular or expected pathway. Radiation and chemical mutagenesis are two of the most common methods. While a major limitation of X-ray mutagenesis is that chromosome breaks with major deletions or rearrangements in the chromosome are often seen, these breaks can also be used to map the location of a gene of interest.[11] Chemical mutagenesis with mutagens such as ethylnitrosylurea (ENU) or ethyl methane sulfonate (EMS) can be titrated to allow for single point mutations in the genome.[12] Both approaches have proven useful in organisms from *Drosophila* to mice and zebrafish, where genes important for the establishment of the primary body axes as well as genes involved in development of specific tissues and organs have been identified.

Another method for non-targeted mutagenesis used in mice involves recombinant retroviruses or transgene insertions, both of which can integrate into the host genome to disrupt gene function. This approach is particularly useful in genetic screens since, once the desired phenotype is found, the inserted DNA can be used as a marker to clone the adjacent, disrupted gene of interest.[13,14]

Antisense-mediated disruption of gene function reduces mRNA levels by introducing DNA or RNA complementary to the mRNA of interest. The expression of the target gene is presumably reduced either through degradation of the RNA/RNA or RNA/DNA hybrids or by preventing translation of the mRNA. The antisense RNA approach has proven to be useful in *Drosophila*, *Xenopus*, chick embryos and tissue culture cells.[15–17] However, one of the limitations of the antisense approach is that the expression of the target mRNA is reduced, but the gene is not eliminated. Also, reducing mRNA levels does not always guarantee that the protein levels will be reduced, particularly if the protein has a long half-life.

In the mouse, the most reliable method to eliminate gene function has been through homologous recombination, or gene knock-outs.[18] In this technique, embryonic stem cells (ES cells), derived from the inner cell mass, are grown in culture. The coding region of the gene of interest is replaced with a neomycin resistance gene, but the 5' and 3' flanking sequences of the endogenous gene are maintained in a vector that is introduced into the ES cells. Homologous recombination to replace the endogenous gene with the disrupted gene is a rare event, but can be positively selected by growth in neomycin. A second selectable marker, the herpes virus thymidine kinase (*tk*) gene, is also included in the vector outside the regions of homology. If a non-homologous recombination event occurs, those cells can be negatively selected for by growth in gancyclovir, which is converted into a toxic intermediate by thymidine kinase. Cells containing a disrupted gene are injected into a blastocyst, which is reintroduced into the uterus of a pseudopregnant mouse to generate chimeric mice. If the recombinant ES cells contribute to the germline, the chimeras can be bred to yield mice that are null for the targeted gene. Variations on this approach allow introduction of reporter genes, such as β-galactosidase, to monitor the expression pattern of the gene, 'knock-in' strategies that introduce specific mutations and conditional knock-outs that allow disruption of gene function at specific times or in specific locations within the embryo.[19–21]

CURRENT CONCEPTS

This section surveys some of the important concepts in embryology derived from the study of diverse model organisms that are critical for the understanding of mammalian embryogenesis. This includes the discovery of the 'organizer' in vertebrates, which illustrates the fundamental idea that one tissue can instruct or induce neighboring tissue to follow a new fate and organization, the induction and establishment of the primordial germ layers, the hierarchical order of gene activation in establishing pattern and cell fate and the evidence for morphogen gradients in establishing pattern. At the end of this section, several of the highly conserved signalling pathways important for vertebrate and invertebrate embryogenesis are summarized.

Induction and the Spemann organizer

The late 19th-century experimental work of Roux, Driesch and others, including the earlier work of Spemann, helped to establish the concepts of mosaic and regulative development. Mosaic development involves the localization of determinants in the egg that are partitioned unequally as the egg divides and subsequently are responsible for the development of specific structures in the embryo. Mosaic development is more often a central feature in the development of invertebrate embryos, such as *Drosophila*, in which much of the future patterning of the embryo is established before fertilization (see below). In regulative development, individual cells have the potential to form many more structures than they are actually fated to become. At early stages of development, discrete regions of the developing embryo may contain all of the information required for the formation of a complete organism, so that removal of half of the embryo can result in a smaller, but normally patterned embryo. Mammalian embryos are primarily regulative, showing essentially complete totipotency through the first several days of development.[2]

The ability of the embryo to regulate its own development implies that the cells of the embryo (blastomeres) must communicate with each other. The ability of cells (or a group of cells) to instruct neighboring cells to follow a specific developmental pathway is a fundamental process referred to as induction, perhaps best illustrated in the organizer transplant experiments of Hans Spemann and Hilda Mangold. These experiments were carried out in amphibians, but the principles are broadly applicable, as discussed below.

In amphibians, gastrulation begins on the future dorsal side of the embryo and can be seen as a structure termed the dorsal lip of the blastopore. Spemann and Mangold showed that, when the dorsal lip of the blastopore was transplanted to the ventral side of a host embryo at the same stage of development, a secondary embryo formed, joined to the first (Figure 3a).[22,23] Furthermore, most of the tissue in the secondary axis was derived from the host cells, not the graft. Thus, host cells that had been destined to form ventral and posterior tissues were now contributing to dorsal and anterior tissues, such as brain, spinal cord, notochord and somites. This observation demonstrated that the dorsal lip instructed, or induced, ventral host cells to assume a new fate and to be organized in a new pattern. The dorsal lip was thus termed the organizer, or primary organizer, and is now referred to as the Spemann organizer.[24] The Spemann organizer is thus a signalling center that conveys two general types of inductive signals (Figure 3b). A dorsalizing signal induces neighboring mesoderm to form properly organized dorsal axial tissues. A neuralizing signal induces ectodermal cells to form neural rather than epidermal tissues. In addition, the organizer may play a role in anteriorization of the endoderm.

Signalling centers similar to the amphibian organizer are found in other vertebrates, including the mouse and the chick, where it lies at the anterior end of the primitive streak and is referred to as the node, or Hensen's node. In zebrafish, an analogous structure called the shield is localized in the dorsal marginal zone. Transplantation of the node in mouse or chick embryos induces a secondary axis similar to the activity of the Spemann organizer.[25–27] The conservation of this type of inductive

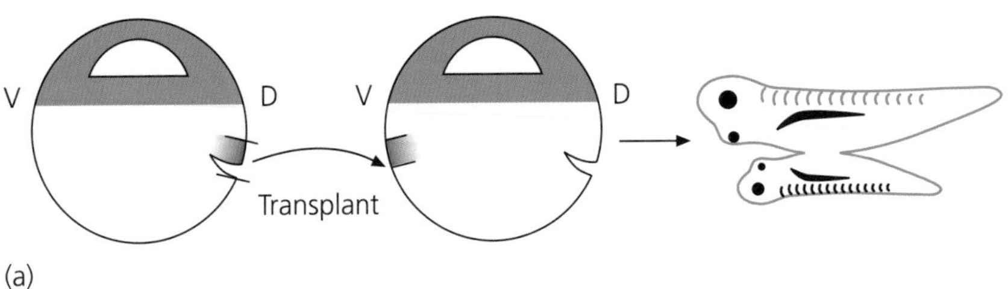

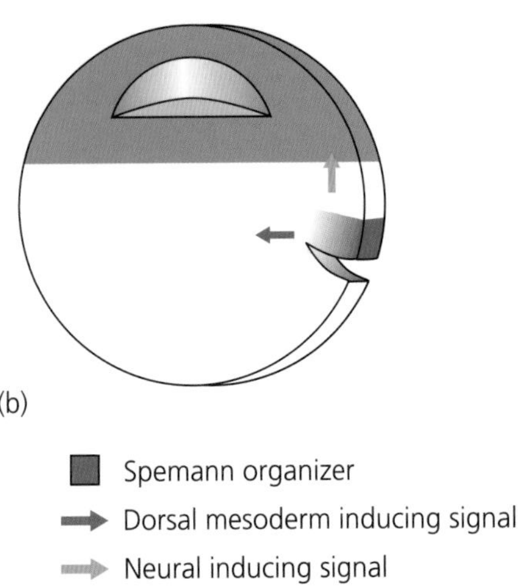

Figure 3 Spemann organizer. (a) Schematic diagram of Spemann and Mangold organizer transplant experiment. When the dorsal (D) lip of the blastopore of an amphibian embryo was grafted to the ventral (V) side of a host embryo, a second embryo formed, joined to the host. (b) Spemann organizer as a signalling center. The Spemann organizer (blue) has been proposed to provide at least two types of inducing signals. A dorsalizing signal (blue arrow) is provided to adjacent, lateral mesoderm in the equatorial region of the embryo. A neural inducing signal (red arrow) acts on prospective ectoderm in the animal pole

tissue through hundreds of millions of years of evolution underlines the fundamental importance of the organizer in animal development.

Mesoderm induction

The importance of inductive signals during early embryogenesis is also evident in the formation of the primordial germ layers. This was clearly demonstrated by the experiments of Nieuwkoop, also with amphibian embryos.[28] Prior to gastrulation, the three germ layers have already been determined. Ectoderm, which will eventually form the surface of the embryo after gastrulation, arises from cells in the upper, animal hemisphere, whereas endoderm, the innermost layer after gastrulation, arises from the vegetal hemisphere. Mesoderm, which will eventually lie between endoderm and ectoderm, is formed from cells at the equator of the blastula. Therefore, animal pole cells removed from the blastula and cultured in isolation form ectoderm only, vegetal pole explants form endoderm and equatorial cells form mesoderm. Nieuwkoop found that, when animal and vegetal pole explants were cultured together, the animal pole cells developed into mesodermal structures such as notochord, muscle, kidney and blood (Figure 4). Thus, vegetal pole cells (presumptive endoderm) provide a signal that induces animal hemisphere cells (that would otherwise become ectoderm) to become mesoderm. Similar inductive processes are responsible for the formation and patterning of specific cell types throughout development. For example, mesenchymal–epithelial interactions are critical for the induction and the patterning of the liver, pancreas, intestines, respiratory system and kidney. These inductive signals can now be attributed to secreted factors from several gene families, including transforming growth factor (TGF)-β, fibroblast growth factor, wnts and others, as is described at the end of this section.

Similar inductive signalling appears to be responsible for specifying the germ layers in other vertebrates. This has been addressed experimentally with the use of similar tissue recombinants in the chick, but the study of mesoderm induction in mice and other mammals has been limited by the inaccessibility of the embryos at these early stages. Nevertheless, it has been proposed that mesoderm in mammals is induced by signals originating in analogous tissues (Figure 5). Thus, the

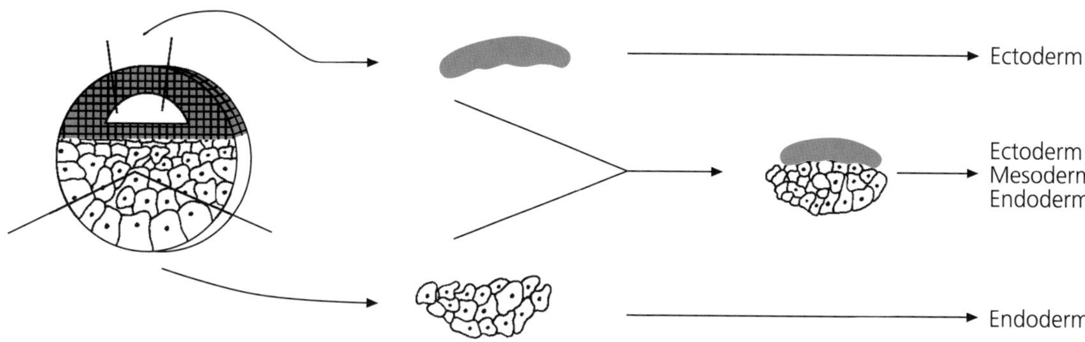

Figure 4 Nieuwkoop recombinants. Animal caps (brown) of blastula-stage amphibian embryos will form epidermis (ectodermal cell type) if cultured in isolation. When cultured in the presence of vegetal pole (white) explants, animal cap cells are induced to form mesoderm

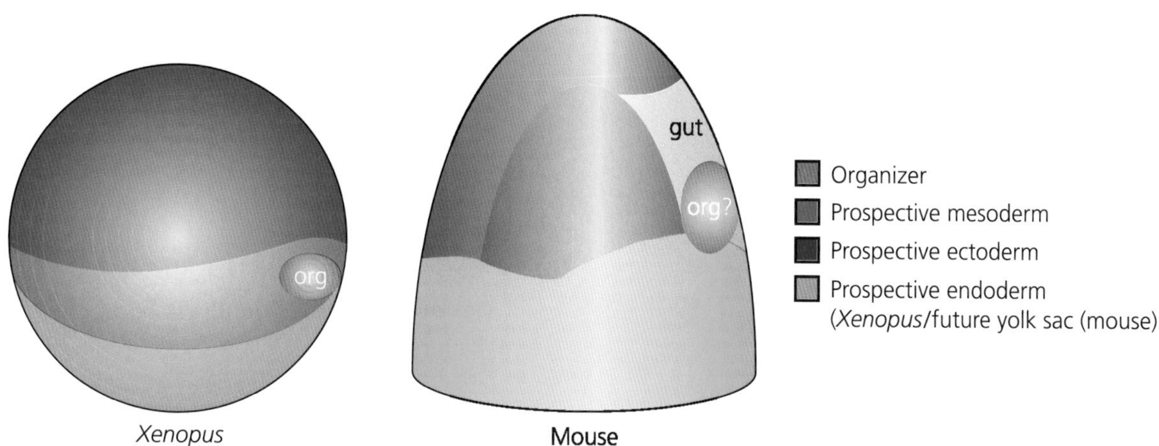

Figure 5 Comparison of fate maps of early *Xenopus* and mouse embryos. The fate maps of *Xenopus* and mouse embryos are similar in the positioning of future mesoderm between prospective ectoderm and prospective endoderm (*Xenopus*) or extraembryonic tissue (mouse). The extraembryonic tissue of mouse embryos will give rise to the yolk sac, similar to the vegetal pole of *Xenopus* embryos which are mainly composed of yolk platelets. The similarity in the positioning of prospective mesoderm suggests a common mechanism for the induction of mesoderm. Adapted from reference 119

extra-embryonic tissue of the early mouse embryo has been considered to be analogous to extra-embryonic tissues of avian embryos, the yolk mass of fish eggs and the yolk-filled presumptive endoderm of amphibian embryos; whether these extra-embryonic tissues provide inductive or patterning signals in mammals is currently an area of active research.[29,30]

A hierarchy of gene expression establishes the pattern of the embryo

Much of our current understanding of the control of embryonic pattern and development comes from the molecular and genetic analyses of the fruit fly *Drosophila melanogaster*, particularly from the pioneering work of Lewis, Wieschaus, Nusslein-Volhard and many others. To understand these fundamental contributions, a very brief summary of embryonic development of *Drosophila* is first necessary (Figure 6).

In *Drosophila* embryos, the anterior–posterior and dorsal–ventral patterns are already established in a rudimentary form by the end of oogenesis, as indicated by the characteristic shape of the egg. After fertilization, the egg undergoes a series of synchronous nuclear divisions without cell divisions, to create the syncytial blastoderm. After 13 nuclear divisions, plasma membranes grow inwards from the egg surface to enclose each nucleus, and the syncytial blastoderm becomes a cellular blastoderm. Gastrulation begins after the completion of

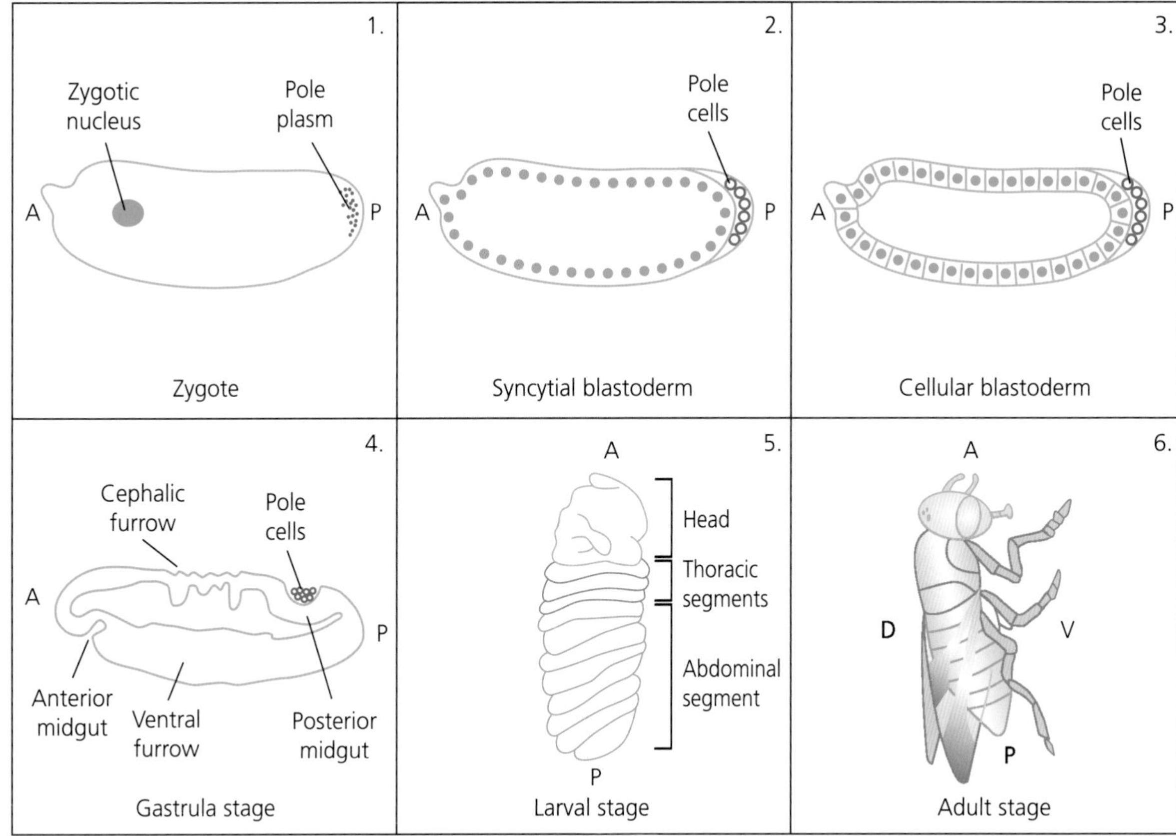

Figure 6 Overview of *Drosophila* embryogenesis. The fertilized egg (1) undergoes a series of synchronous nuclear divisions without cell divisions, to create the syncytial blastoderm (2). After 13 nuclear divisions, plasma membranes grow inward to enclose each nucleus (cellular blastoderm stage) (3). Soon after cellularization, gastrulation (4) begins as an invagination of ventral mesoderm. The embryo hatches out of the egg shell to become a larva (5), in which the segmentation of the body plan is easily observed. The larva grows and molts, then undergoes metamorphosis to produce the adult fly (6). A, anterior; P, posterior; D, dorsal; V, ventral. Adapted from reference 5

cellularization as an invagination of ventral mesoderm; during gastrulation, segmental boundaries become apparent along the anterior–posterior axis. The anteriormost segments give rise to head structures, three middle segments give rise to the thorax and the eight posterior segments give rise to the abdomen. The embryo hatches into a larva, which grows and molts before a final pupation stage, during which the larva metamorphoses into an adult fly. The surface, or cuticle, of the larva shows the regular pattern of segments, with ventral projections referred to as denticles that occupy the anterior region of each segment. Analysis of mutations that disrupt segmentation or the pattern within the segments has formed one of the cornerstones of modern developmental genetics in *Drosophila*.[31]

The patterning of the *Drosophila* embryo depends on a hierarchical system of gene activation, beginning with cues provided maternally, and continuing with zygotically expressed genes (Figure 7). The maternal genes responsible for the basic body plan can be divided into four groups: the anterior, posterior, dorsal and terminal groups.[32] We focus on the pathway used for anterior patterning as an instructive example of this hierarchy of gene regulation.

Analysis of maternal-effect mutants revealed that the gene *bicoid* is critical for anterior development. Mutants lacking *bicoid* produce embryos that lack head and thoracic structures. Such embryos can be rescued by transplanting cytoplasm from wild-type embryos, but only if it is taken from the most

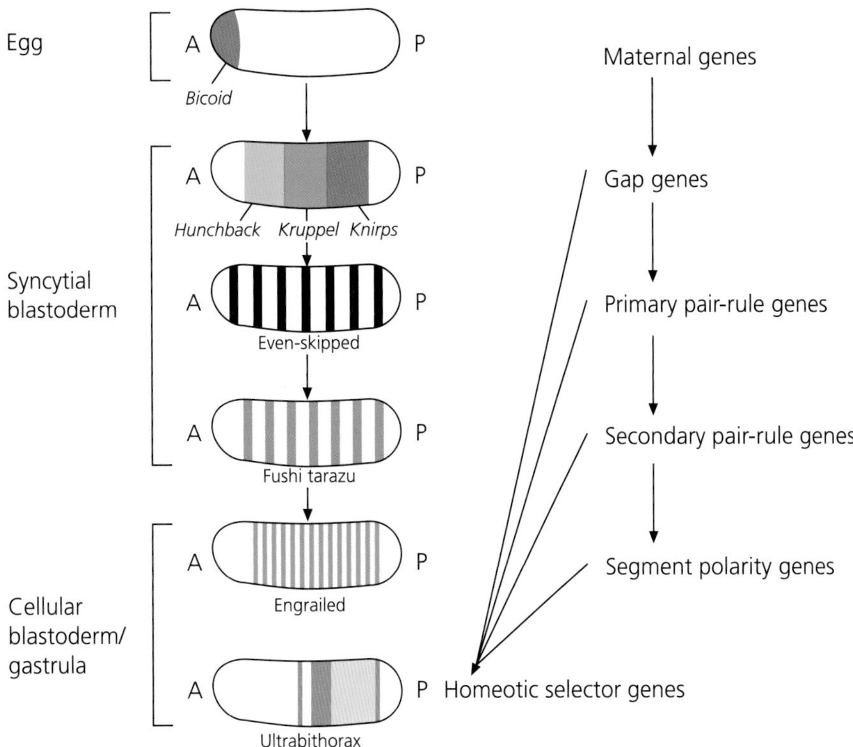

Figure 7 Hierarchical system of gene activation in *Drosophila*. The patterning of the *Drosophila* embryo is dependent on an ordered system of sequential gene expression. The patterns of representative mRNAs are shown. Maternal genes, such as *bicoid* (blue, top), turn on gap genes, which have broad expression domains. Gap genes, in turn, turn on primary pair-rule genes, which are expressed in seven evenly spaced stripes. Primary pair-rule genes activate secondary pair-rule genes, which ultimately turn on segment polarity gene expression. The precise co-ordination of these classes of genes results in the correct expression of homeotic selector genes in the proper body segments, eventually resulting in an adult fruit fly. Adapted from references 2 and 5

anterior region, suggesting that the information is localized to this region in the wild-type embryos. Microinjection of *bicoid* mRNA can also rescue *bicoid* mutants, but normal embryos form only if the RNA is injected in the anterior region. Injection of *bicoid* mRNA into other regions of the embryo causes the formation of an ectopic head and thorax. These observations show that the activity of *bicoid* must be localized to the anterior of the embryo. This was confirmed by *in situ* hybridization, which showed that *bicoid* mRNA is highly localized at the anterior tip of the early embryo (Figure 8). Bicoid protein translated from this localized mRNA diffuses from the anterior toward the posterior region of the egg to form an anterior–posterior concentration gradient. This distribution of *bicoid* is consistent with a model that predicts that a signal (a morphogen) present in a concentration gradient can induce multiple cell types from a population of cells if the cells respond differently to different concentrations of the morphogen.[33] The model was elegantly tested in *Drosophila* by the expression of varying levels of *bicoid* (by changing gene dosage, for example). In the absence of *bicoid*, anterior segments do not form. With increasing gene dosage, and consequently a higher concentration of *bicoid* at each position along the anterior–posterior axis, anterior structures form progressively further from the anterior end of the egg.[34]

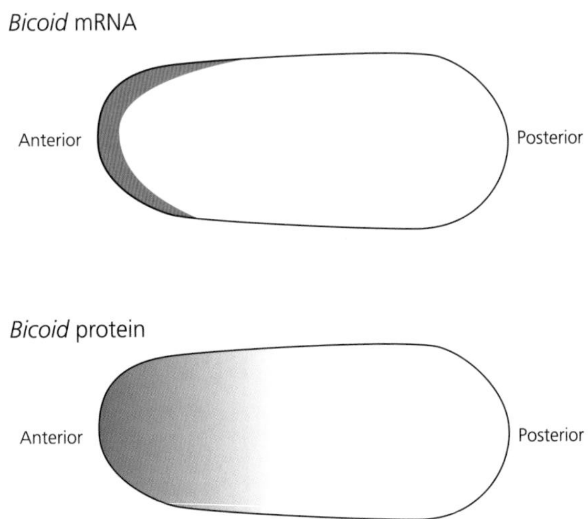

Figure 8 Gradient of *bicoid* expression. Schematic diagram of *bicoid* mRNA and protein expression in the early *Drosphila* embryo. *Bicoid* mRNA is localized to the anterior region of the early embryo. *Bicoid* protein is detected in a gradient with highest levels anteriorly and reduced expression more posteriorly

The different patterns induced by varying the *bicoid* concentration gradient are reflected in earlier changes in the pattern of expression of immediate targets of *bicoid*, zygotic transcription factors termed gap genes. Gap genes were so named because mutations in these genes resulted in the loss of large regions of the body, producing gaps in development. Representatives of this class include: *hunchback*, which is required for head and thoracic structures; *kruppel*, required for thoracic and abdominal structures; and *knirps*, required for most of the abdominal structures. The expression domains of the gap genes are consistent with their phenotypic effect, with *hunchback* in the anterior, *knirps* in the posterior and *kruppel* in between. Increasing the level of *bicoid* so that a higher concentration is present at a given position along the anterior–posterior axis results in a posterior shift in the expression domain of *hunchback*, and consequently anterior structures develop at a relatively more posterior position along the axis.

Gap genes, in turn, regulate the expression of downstream transcription factors named pair-rule genes. Pair-rule genes were identified from mutations that cause deletion of alternating segments. For example, loss-of-function mutations in *even skipped* (*eve*) lead to a loss of odd-numbered segments. These genes are expressed at the cellular blastoderm stage in seven, evenly spaced stripes along the anterior–posterior axis at positions characteristic of each pair-rule gene and corresponding to the regions deleted in a given pair-rule mutant. Pair-rule genes are divided into two groups: the primary pair-rule genes are directly regulated by the gap genes, and secondary pair-rule genes are regulated by the primary group. The secondary pair-rule genes regulate the genes in the next level of the hierarchy, the segment polarity genes.

Segment polarity genes regulate the pattern within each segment. Mutations in this class of genes cause the loss of defined pattern elements within each segment, often accompanied by inverted duplications of the remaining structures. This group of genes differs from the gap and pair-rule genes, since they are active after cellularization and therefore encode proteins involved in intercellular signalling. Also, unlike the gap and pair-rule genes, segment polarity genes, such as *engrailed* (*en*), *wingless* (*wg*) and *hedgehog* (*hh*) are highly conserved in vertebrates, playing important (although distinct) roles in mammalian development, as well as the control of proliferation (as discussed below).

Thus, anterior–posterior development of the *Drosophila* embryo begins with an ordered sequence of gene activation that subdivides the anterior–posterior axis first into anterior and posterior regions (maternal genes *bicoid* and *nanos*, for example), then into large subdivisions determined by the domains of gap gene activity, subsequently into pairs of segments dependent on pair-rule gene activity and finally into a repeating array of segments, each with an inherent pattern and each ultimately to acquire a specific identity. Loss-of-function mutations in these genes generally lead to the loss of a region of the embryo corresponding to the level in the hierarchy at which the affected gene functions. Genes at each level of this hierarchy also regulate the time and location of expression of the genes that actually define the identity of the segments, the homeotic genes, as described in the next section.

Homeotic genes

The homeotic genes of *Drosophila* were the first genes identified that define the identity of a region of the embryo.[35,36] Mutations in these genes cause the transformation of a given structure into another structure (rather than the deletion of a body segment, as seen with the genes described above). For example, mutations in *Antennapedia* result in a fly with legs in the place of the antennae (Figure 9).[37] Since the ectopic leg can be quite normal otherwise and since the rest of the fly develops normally, this observation indicates that the affected gene is required at an early step in tissue specification. Thus, while the instructions to form a leg or an antenna must be complex and distinct, the switch from one path to another can depend on the activity of a single gene.

Homeotic genes in *Drosophila* are organized into one of two gene clusters, the *Antennapedia* (ANT-C) and the *Bithorax* (BX-C) complexes.[31,38] Furthermore, the genes are organized on the chromosome in a pattern that parallels their domains of expression along the anterior–posterior axis of the embryo. All encode transcription factors that contain a highly conserved DNA-binding domain termed the homeobox (*bicoid* also contains a homeobox). The high conservation of the homeobox sequence has led to the identification of homologs in essentially all multicellular organisms from sponges to vertebrates, as well as in plants and fungi.

Figure 9 Homeotic transformation. A mutation in the *Antennapedia* gene results in ectopic expression in the head, resulting in the transformation of antennae to legs. Courtesy of Matthew Scott

In vertebrates, about 180 different homeobox genes have been cloned. They can be divided into two subfamilies, the clustered homeobox genes known as Hox genes and the non-clustered or divergent homeobox genes. The divergent homeobox genes are scattered throughout the genome and fall into a number of groups based on sequence homology such as *Pax*, *Msx*, *Emx* and *Otx* genes named after their homologs in the fly, i.e. the paired, muscle segment homeobox, empty spiracles and orthodenticle genes, respectively.

In mouse and human, Hox genes are organized into four clusters (rather than two), Hox A, B, C and D, which are localized to human chromosomes 7, 17, 12 and 2, respectively. The order of the mammalian Hox genes in each cluster is conserved, and again parallels the anterior–posterior order of gene expression in the embryo, as seen in the fly (Figure 10). The individual Hox genes within the different clusters can be aligned with each other and with the genes of the *Drosophila* homeotic complex (HOM-C) clusters. These similarities suggest that the four mammalian clusters probably arose from a single ancestral complex.

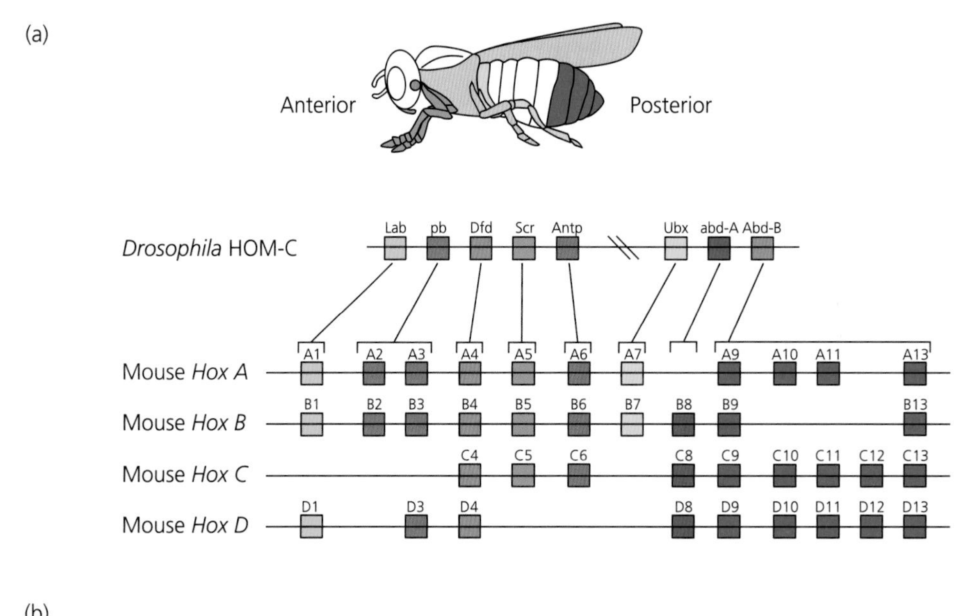

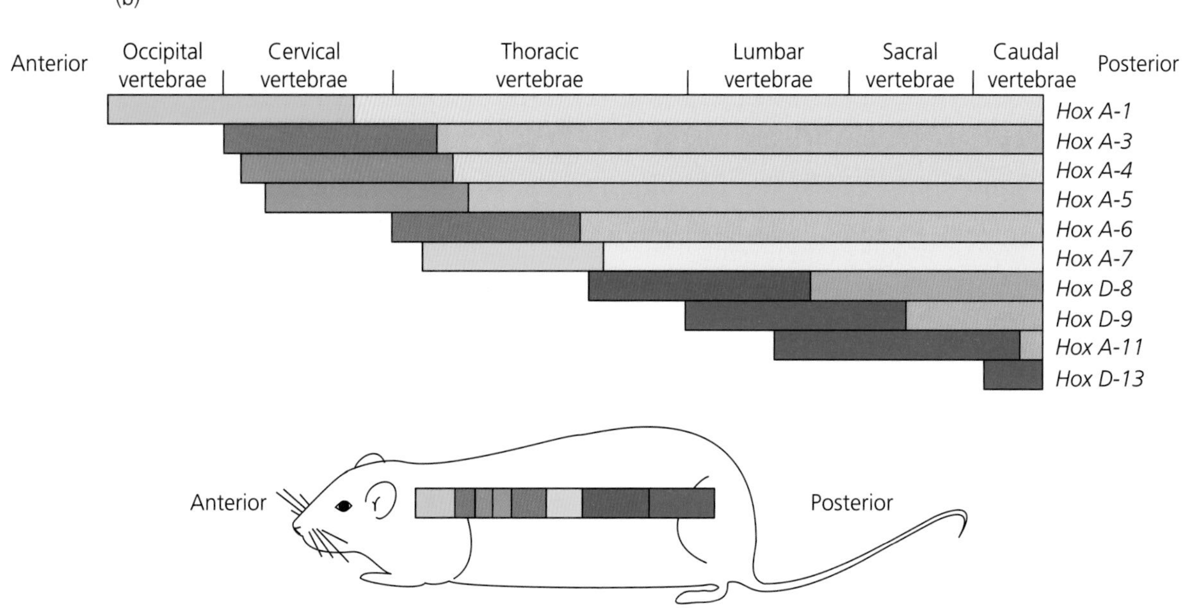

Figure 10 Conservation of homeotic genes between *Drosophila* and mouse. (a) Four *Hox* gene clusters are found in mice which are related to the homeotic complex (*HOM-C*) in *Drosophila*. Similar to the situation in *Drosophila*, the linear order of the *Hox* genes within the cluster defines the anterior–posterior expression patterns of the *Hox* genes. The mouse *Hox* genes are numbered 1–13, starting with the genes expressed most anteriorly. (b) The highest level of expression of selected *Hox* genes along the anterior–posterior axis of the mouse. Adapted from reference 38

In mammalian embryos, the earliest expression of Hox genes can be detected at gastrulation and can be seen later in all three germ layers. The available evidence suggests that mammalian Hox genes may serve similar functions in determining cell identities along the anterior–posterior axis. More than half of the known Hox genes have been inactivated by homologous recombination in the mouse. The majority of these mutations resulted in the transformation of one vertebra to a more anterior fate. Furthermore, mice with a disruption of a particular Hox gene usually show homeotic transformations in the anteriormost region where that Hox gene is expressed. This is similar to loss-of-function mutations of homeotic genes in flies where a body segment is transformed to an anterior fate. Similarly, gain-of-function mutations in homeotic genes in *Drosophila*, as well as in transgenic mice, lead to posterior homeotic transformations.[38,39] In general, more posterior-acting homeotic genes are dominant with respect to function over more anterior-acting genes. This phenomenon is called 'posterior dominance' in mice and 'phenotypic suppression' in flies. More recently, a mouse homolog of the homeotic *Drosophila Polycomb* gene, *M33*, has been disrupted and results in mutant mice with a male-to-female sex reversal phenotype.[40] Defects could be seen in early male gonadal growth with retardation of genital ridge formation.

The work in *Drosophila* so far has clearly demonstrated the power of combining genetic analysis and molecular biology to identify and characterize the genes required for the development of a complex organism. This work has established a paradigm in which the egg/embryo is divided into successively smaller domains by hierarchical activation of gene expression. Even if the genes involved were unique to *Drosophila*, the success of this approach indicates that the developmental blueprint for other organisms could be deciphered by similar approaches. However, the work in *Drosophila* has also demonstrated that several classes of genes, and indeed complete signalling pathways, are also highly conserved in higher organisms. Therefore, the lessons as well as the molecules discovered in *Drosophila* can be readily transplanted to other organisms, as seen with the homeobox genes and with the signalling pathways described below.

Important signalling pathways

As with homeotic genes, several important signals and signalling pathways have been conserved during evolution and are used at multiple stages of development within a single organism. Many of these pathways also function in the adult to regulate cell growth. *Wingless*, *hedgehog*, fibroblast growth factors and TGF-βs play decisive roles in pattern formation, as well as in neurulation and somite and limb formation. Disruption of these pathways not only perturbs normal development but can also lead to uncontrolled cell proliferation in adult tissues.

Wnt pathway

Wnts are a large family of cysteine-rich secreted proteins that control development in organisms ranging from nematodes to mammals. Wnts play essential roles in embryonic induction, the generation of cell polarity and the specification of cell fate. *Wnt* genes are defined by sequence homology to the original members of the family, murine *Wnt-1* and *Drosophila wingless*. This class of molecules are glycoproteins which are usually 350–400 amino acid residues in length, and currently are comprised of at least 16 members.[41]

The wnt signalling pathway was initially elucidated by genetic epistasis experiments in *Drosophila*, and later refined by biochemical analysis. The canonical wnt signalling pathway is shown in Figure 11. Wnts are thought to bind to a class of receptors, Frizzleds, named after the tissue polarity gene *frizzled* from *Drosophila*.[42,43] Frizzled proteins have a cysteine-rich extracellular domain, which is thought to bind wnts, seven transmembrane domains, and a cytoplasmic carboxy terminus. This topology is reminiscent of G protein-coupled receptors, raising the possibility that wnts signal through G proteins.[44]

Dishevelled was placed downstream of *wingless* by genetic epistasis experiments.[45] Members of the dishevelled family, of which there are three vertebrate members, are cytoplasmic proteins that

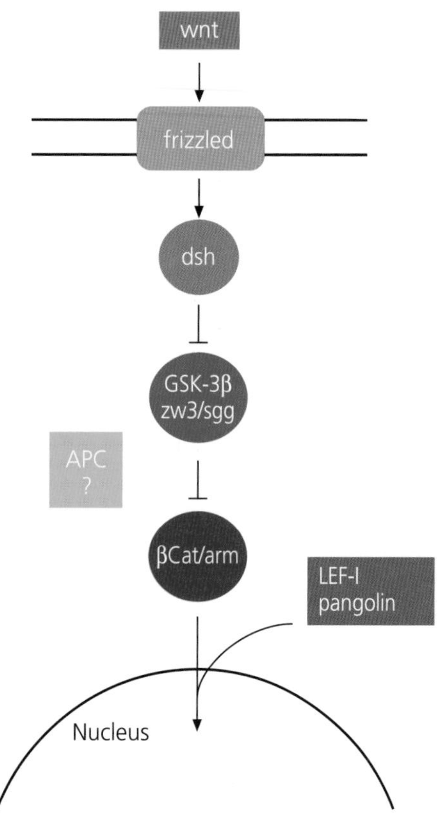

Figure 11 Wnt signalling pathway. Wnt signalling is initiated by the binding of wnt to the putative receptor, frizzled. The signal activates the cytoplasmic protein, dishevelled, which in turn inhibits the serine/threonine kinase glycogen synthase kinase-3β (GSK-3β). Inhibition of GSK-3β leads to stabilization of β-catenin protein. β-catenin then binds with DNA binding factors, such as lymphocyte enhancer binding factor-1 (LEF-1), translocates to the nucleus and turns on *wnt* response genes. The adenomatous polyposis coli (APC) protein may bind with GSK-3β and β-catenin to stimulate β-catenin degradation in the absence of wnt signals

probably function as molecular adaptors, since they contain putative protein–protein interaction domains, but no known enzymatic activity. Wnts can cause dishevelled to translocate to the cell membrane, but the mechanism by which signals are transmitted from wnt, through frizzled, to dishevelled is unclear. Dishevelled, in turn, is believed to inhibit glycogen synthase kinase-3β (GSK-3β), a cytoplasmic serine/threonine kinase which is functionally homologous to the *Drosophila* zeste-white3 gene product.[46,47] GSK-3β is a negative regulator of the wnt pathway. Wild-type GSK-3β can inhibit wnt signalling while dominant-negative GSK-3β can mimic the activity of wnts in *Xenopus* and other organisms.

Wnt signalling in multiple organisms leads to stabilization of cytoplasmic β-catenin protein (the gene is known as *armadillo* in *Drosophila*).[48] β-Catenin is well known for its role in cell adhesion, but its function in the wnt pathway is distinct. Cytoplasmic β-catenin is normally degraded rapidly in a GSK-3-dependent manner. Wnt inhibition of GSK-3β activity prevents the turnover of β-catenin, leading to its accumulation. Stabilized β-catenin can then be found in the nucleus where it binds to DNA binding factors, such as lymphocyte enhancer binding factor-1 (LEF-1), to activate transcription directly. This regulation may occur through a transcriptional activation domain that is present in β-catenin. The components of this highly conserved pathway have been defined genetically. However, direct interaction has not been demonstrated for most of the identified components, leaving open the possibility for additional intermediates to be found.

In *Xenopus*, expression of various wnt genes in early ventral blastomeres leads to induction of dorsal mesoderm and a duplicated body axis, very similar to the Spemann and Mangold organizer transplant experiments.[49] Interestingly, treatment of early *Xenopus* embryos with lithium chloride also causes dorsal axis duplication. This observation has led to the proposal that lithium activates the wnt pathway through inhibition of the antagonist GSK-3β.[50] This proposal has been confirmed *in vivo*.[51,52] In mice, ectopic expression of *CWnt8c* also leads to a duplicated axis or a severe dorsalized phenotype.[53] Despite these overexpression results, an endogenous role for a *wnt* gene in the induction of the primary axis has not been demonstrated.

Many of the known *wnts* have been disrupted in the mouse, yielding remarkable phenotypes. Mice homozygous for a null allele of *wnt-1* fail to develop midbrain and cerebellar structures, while mice lacking *wnt-3a* exhibit a defect in gastrulation which leads to axial truncation and loss of caudal somites. Disruption of *wnt-7a* results in ventralized limbs. This phenotype is consistent with the dorsal expression pattern of *wnt-7A*, and with studies in chick limb buds where ectopic expression of *wnt-7A* leads to a dorsalization of the limb. *Wnt-4* mutations result in the absence of kidneys. Ectopic overexpression studies with *wnt-4* had previously demonstrated that this gene may function in mesenchymal–epithelial transitions that occur during the formation of the kidney. Table 1 summarizes the known wnt loss-of-function phenotypes in mice. These phenotypes underscore the important roles that wnts play in early pattern formation and organogenesis.

Hedgehog pathway

Hedgehog (*hh*) was first identified as a segment polarity gene in *Drosophila*, where it is required for the maintenance of *wingless* expression and for proper patterning of the embryonic segments. More recently, vertebrate hedgehog genes have been identified and shown to play critical roles in neural induction and in somite and limb patterning.[54]

Hedgehog proteins are secreted proteins whose biologically active form is processed from a larger precursor protein. Interestingly, hedgehog precursor proteins undergo an internal autoproteolytic cleavage which results in a 19-kDa N-terminal peptide, with a novel C-terminal cholesterol ester, and a 26–28-kDa C-terminal peptide. The N-terminal peptide stays closely associated with the cell membrane, while the C-terminal peptide can freely diffuse.

Table 1 Loss-of-function mutant phenotypes of *wnt* genes in the mouse (adapted from Wnt homepage[120])

Gene	*Phenotype*	*Reference*
Wnt-1	loss of midbrain and cerebellum	McMahon 1990 Thomas 1990 Thomas 1991 Mastick 1996
Wnt-2	placental defects	Monkley 1996
Wnt-3A	somite and tailbud defects	Takada 1994 Greco 1996 Yoshikawa 1997
Wnt-4	kidney defects	Stark 1994
Wnt-7A	limb polarity defects	Parr 1995

The hedgehog signalling pathway has also been elucidated genetically in *Drosophila*. In target cells, hedgehog signalling is mediated by two transmembrane proteins, patched, which has structural similarities to ion channel and transporter proteins, and smoothened, a protein with seven transmembrane domains similar to the frizzled proteins. Patched is a negative regulator and smoothened a positive regulator of the hedgehog pathway. Hedgehog is believed to bind to patched and suppress its inhibitory activity. Since patched, in turn, is believed to inhibit smoothened, hedgehog binding indirectly activates smoothened by removing its inhibitor. This model is supported by genetic data. Double mutants lacking *hedgehog* and *patched* regain expression of hedgehog-regulated genes; since *patched* is absent, *hedgehog* is no longer needed to inhibit it. Conversely, double mutants lacking *smoothened* and *patched* do not express *hedgehog* target genes. Five putative cytoplasmic components of the hedgehog signal transduction machinery have been identified. These include the serine/threonine kinase Fused, Suppressor of Fused, Costal 2, whose molecular nature is unknown, the cAMP-dependent protein kinase A (PKA), and Cubitus interruptus, a potential transcription factor that contains several PKA consensus phosphorylation sites. Little is known regarding the biochemical mechanisms of this pathway.

Hedgehog is an inducing signal in vertebrates. In vertebrate limb buds, the zone of polarizing activity (ZPA) is an organizing center that patterns the posterior limb. Transplantation of this region to the anterior limb bud induces ectopic posterior digits. The ZPA expresses one of the three vertebrate homologs of *hedgehog*, *sonic hedgehog* (*Shh*). *Shh* appears to account for the activity of the ZPA, since anterior expression of *Shh* induces duplications of posterior digits similarly to transplantation of the ZPA.[55]

Sonic hedgehog expression begins shortly after the beginning of gastrulation in the node. In the chick, Shh expression becomes asymmetrically expressed on the left side of Hensen's node, and correlates with proper looping of the heart.[56] In the central nervous system (CNS), Shh from the notochord and floorplate has been implicated in the induction of ventral CNS fates including motor neuron differentiation. When Shh is ectopically expressed, the midbrain and hindbrain of mice and zebrafish become ventralized.[57,58] Furthermore, somites, comprising sclerotome and dermamyotome, are also patterned by Shh from the midline. The sclerotome is a compartment of the somite that gives rise to cartilage, and the dermamyotome compartment gives rise to skeletal muscle and dermis. Shh can induce the expression of sclerotomal markers such as Pax1 and Twist, at the expense of the dermamyotomal marker, Pax3.[59]

Loss-of-function studies with *Sonic hedgehog* in mice support the data from other organisms. Mice with targeted mutations in *Shh* demonstrate that this gene is required for limb patterning and motor neuron and floorplate differentiation.[60] These mice have severe cyclopia and lack ventral midline structures and motor neurons. This phenotype is very similar to that produced by the teratogen jervine, found in skunk cabbage, which causes cyclopia in sheep. It is also similar to holoprosencephaly in humans, and indeed a mutation responsible for holoprosencephaly maps to the *sonic hedgehog* gene in humans (see below).

Fibroblast growth factor pathway

Members of the fibroblast growth factor (FGF) family of molecules have been shown to play an important role in development in a variety of organisms.[61] Although they were originally identified by their ability to stimulate fibroblast proliferation in culture, FGFs and their receptors (FGFRs) have numerous important biological activities. The FGF family is composed of at least nine members in mammals. FGFs bind to receptors that belong to the receptor tyrosine kinase family, leading to autophosphorylation of the receptors and recruitment of intracellular signalling proteins. Activation of the FGFR tyrosine kinase leads to an increase in the tyrosine phosphorylation of a number of intracellular proteins such as phospholipase C-γ (PLC-γ), Shc, phosphatidyl inositol-3 kinase (PI-3 kinase), Raf-1 kinase, the mitogen-activated protein kinase (MAP kinase) and p90rsk.

FGFs are widely expressed during development. FGF appears to play a role in mesoderm induction and in early embryonic pattern formation as well as in later stages of embryogenesis. Studies in *Xenopus* showed that FGF is expressed early and can induce a subset of mesodermal cell types including mesothelium, mesenchyme and muscle.[62,63]

Further evidence for the importance of FGF signalling in early embryogenesis also comes from work in *Xenopus*. FGF signal transduction can be blocked with a dominant-negative mutant of the FGF receptor that lacks the cytoplasmic tyrosine kinase domain. Expression of this truncated FGF receptor led to the disruption of trunk and posterior structures in whole embryos.[64] Furthermore, inhibition of *ras* and *raf-1* also led to phenotypes similar to that produced by the truncated FGF receptor in *Xenopus*.[65] Moreover, activated *Raf-1*, like FGF, can induce *MyoD*, an early muscle determination gene, and muscle actin, suggesting that the mesoderm induction by FGF occurs via activation of the Ras-Raf-MAP kinase cascade.[66]

The presence of numerous FGFs and FGFRs in mammalian development has hindered the identification of the roles of specific FGF members in early development, owing to redundant and compensatory effects. However, targeted mutations in a few FGF ligands and receptors has revealed important roles for those ligands. *FGFR-1* is expressed in the primitive streak and in ectoderm and mesoderm of the anterior-distal part of the late primitive streak embryo. Mice disrupted in *FGFR-1* fail to form paraxial mesoderm, but rather have an expansion of axial mesoderm consistent with the role of FGF in mesodermal patterning.[67] Disruption of *FGF-4*, which is expressed in the distal region of the primitive streak, results in mouse embryos that implant in the uterus but do not develop further.[68] When these *FGF-4* null embryos are cultured *in vitro*, they display defects in growth and differentiation of the inner cell mass. Targeted mutations in *FGFR-3* suggest a role for this receptor in bone and limb formation, resulting in a phenotype resembling achondroplasia (see below). Double and triple knock-out mice involving several of the ligands and receptors will be useful in determining which FGF signalling components play specific roles in development and patterning in mammals.

Transforming growth factor-β superfamily

The TGF-β superfamily is comprised of a growing family of peptide growth factors, which include TGF-βs, bone morphogenetic proteins (BMPs), activins, inhibins, growth and differentiation factors (GDFs) and Müllerian inhibiting substance (MIS). TGF-β superfamily members are expressed throughout embryogenesis and have been shown to play roles in cell proliferation, differentiation, extracellular matrix production and apoptosis in numerous settings.[69,70] They consist of homo- or heterodimeric polypeptides of approximately 30 kDa which are processed from larger precursors.

TGF-β superfamily members bind to several cell surface proteins, but two major transmembrane receptors (type I and type II receptors) are important for TGF-β signal transduction. TGF-β ligands are recognized primarily by the type II receptor, which contains a constitutively active serine/threonine kinase domain. Binding of ligand promotes the formation of a heteromeric complex with the type I receptor. The type II receptor subsequently phosphorylates the type I receptor. This transphosphorylation reaction leads to the activation of the type I receptor serine/threonine kinase domain, which, in turn, phosphorylates the cytoplasmic Smad proteins. Smad proteins are members of a family of TGF-β signalling molecules first identified in *Drosophila* and *C. elegans*.[71,72] Once phosphorylated by the type I receptor, they translocate to the nucleus and appear to activate transcription directly through interaction with DNA binding proteins (Figure 12).

BMPs were first identified by their ability to induce ectopic bone formation when implanted into the soft tissue of rats.[73,74] BMPs also function

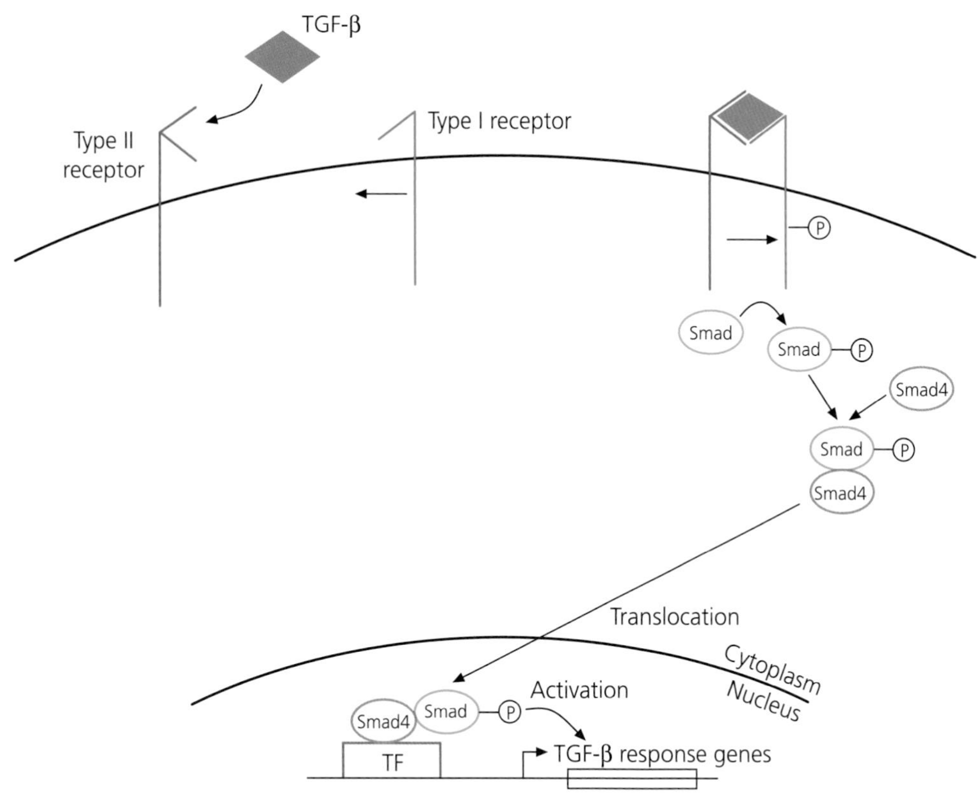

Figure 12 TGF-β signalling pathway. TGF-β signalling commences with the binding of ligand to the type II receptor. The type I receptor is recruited and subsequently phosphorylated by the type II receptor. The type I receptor then relays the signal by phosphorylating the cytoplasmic Smad proteins, resulting in their dimerization with Smad4 (DPC4). This dimeric Smad complex then translocates to the nucleus, where it is believed to cause TGF-β-dependent gene activation

in many developmental processes in invertebrates and vertebrates. The BMP-2 and BMP-4 homolog in *Drosophila*, *decapentaplegic* (*dpp*), is required for dorsal–ventral patterning in early embryogenesis, in addition to patterning of wing and leg primordia as well as the gut.[75]

In *Xenopus*, the role of BMPs in two separate processes have been elucidated. The addition of BMP-2 or BMP-4 to animal pole explants in culture leads to induction of ventral mesoderm, including blood.[76,77] Dorsal mesoderm, including muscle actin and notochord, are not induced by BMP-4. Rather, ectopic BMP-4 reduces dorsal and anterior structures, which supports a role for BMP-4 in ventral patterning. The second process in which BMP has been implicated is in neural/epidermal fate decisions. In *Xenopus*, ectodermal tissue when explanted in culture will differentiate into epidermis. When BMP signalling is inhibited, these ectodermal explants differentiate into neural tissue.[78,79] As the ectoderm is normally fated to become both epidermis and neural tissue, this suggests that the regulation of BMP signalling in the ectoderm may be important for this fate decision. In support of this hypothesis the Spemann organizer expresses extracellular antagonists of BMP-4, such as chordin, noggin and follistatin, which bind BMP protein and prevent it from binding to its receptor.[78,80] Homologous molecules have been found in other organisms.[81,82]

The functions of BMPs during development in the mouse have recently been studied. BMP-2 and BMP-4, as well as the BMP type I receptor, have been knocked out by homologous recombination. The common phenotype resulting from these studies is embryonic lethality at gestational stage 6.5–9.5. The BMP-4 ligand null homozygotes fail to undergo gastrulation.[83] Mesoderm does not appear to be formed, consistent with the studies from *Xenopus*. The homozygotes that do develop to the headfold stage have truncated or disorganized posterior structures and a reduction in extraembryonic mesoderm, including blood islands. The phenotype of the BMP type I receptor knockout indicates that no mesoderm is formed at E7.0, and the embryos were smaller than normal.[84]

GDFs are a TGF-β subfamily which are most closely related to BMPs. Mutations in several GDFs have been generated in mice. *GDF-8* is expressed in both developing and adult skeletal muscle. Disruption of the *GDF-8* gene results in mice which are significantly larger, with increased skeletal muscle mass owing to myocyte hyperplasia and hypertrophy.[85] The human homolog of GDF-5, cartilage-derived morphogenetic protein-1, has recently been associated with a recessive chondrodysplasia, which is characterized by skeletal abnormalities in the limbs.[86] Similarly, the skeletal alterations that are found in the brachypodism mouse (bp) have been shown to be a result of mutations in the *GDF-5* gene.[87] Another family member, *GDF-9*, is expressed solely in the oocyte, and disruption of the *GDF-9* gene in mice causes a block in follicular development, resulting in complete infertility.[88]

Activin was first identified as an activity from porcine follicular fluid that antagonized the activity of inhibin in assays of follicule stimulating hormone (FSH) release from pituitary cells.[89] In vertebrates, activin has been implicated in mesoderm induction and left–right patterning. In *Xenopus*, the addition of purified activin protein to animal pole explants leads to mesoderm induction in a dose-dependent manner.[90–92] At high concentrations of activin, dorsal mesoderm such as notochord and muscle actin is induced, and at lower activin concentrations, ventrolateral mesoderm such as mesothelium and mesenchyme is induced. The concentration-dependent response of apparently homogeneous ectodermal cells to activin is reminiscent of a morphogen, but neither activin nor other TGF-βs have been shown to be present in a graded fashion in vertebrates.[93,94] Similarly, Smad2, which is believed to be the intracellular signalling molecule downstream of the activin type I receptor, can also induce similar dorsal mesodermal tissues in animal pole explants.[95]

In the mouse, gene targeting of components of the activin signalling pathway has been performed. Disruption of the activin type I receptor, *ActRIB*, results in embryos that fail to gastrulate and stop growing at the egg cylinder stage.[96] Chimeric analysis of these mutant ES cells suggests that *ActRIB* is required to mediate signals in both epiblast and extraembryonic cells. When wild-type ES cells are injected into *ActRIB* null blastocysts, the extraembryonic tissue is normal but the wild-type ES cells cannot rescue the gastrulation defect in the embryo proper. The converse chimera experiment, in which *ActRIB* null ES cells are injected into wild-type ES cells, results in failure of the primitive streak to form when injected cells contribute highly to the epiblast. Embryos lacking *Smad2* display a similar early embryonic phenotype. *Smad2*-deficient mice fail to restrict the site of primitive streak formation, resulting in a defect in anterior–posterior patterning in the epiblast.[29,30] Chimeric analysis, in which wild-type ES cells are injected into *Smad2*-deficient blastocysts, does not rescue the phenotype, suggesting that a *Smad2*-dependent signal from the extraembryonic tissue is required for proper anterior–posterior patterning of the embryo.

A related TGF-β family member, *nodal*, also plays an important role in axial patterning in the mouse. *Nodal* is normally expressed in the node of the mouse embryo.[97] When *nodal* is disrupted by insertional mutagenesis, the primitive streak fails to form and there is abnormal expression of mesoderm-specific genes in the epiblast.[98]

Müllerian inhibiting substance (MIS) is an important TGF-β family member expressed in mesenchymal cells adjacent to the Müllerian duct epithelium which causes regression of the Müllerian ducts in males, an essential process for male sexual differentiation. Disruption of *MIS* or the *MIS type II receptor* results in male mice that develop as internal pseudohermaphrodites.[99] These mutant mice possess a complete male reproductive tract and also a uterus and oviducts. These disruption studies in mice stress the important roles that TGF-β family members play in early embryonic patterning, mesoderm induction and reproductive development.

POTENTIAL CLINICAL RELEVANCE

The application of molecular biology, genetics and biochemistry to the study of early embryogenesis has yielded a treasure trove of new insights into regulatory genes and signalling pathways that will have profound implications for clinical medicine in the future. However, the greatest impact has been in the field of neoplasia in adult tissues, and the relevance to clinically important diseases in early development has so far been limited (Table 2). In part this may reflect an essential role of these molecules at early steps in embryogenesis, so that mutations of these genes in humans would result in a non-viable fetus that would rarely be detected. In this section, a few developmental abnormalities linked to the above signalling pathways and observed in mammals are discussed, and then the relevance to neoplasia is briefly discussed.

Interference with sonic hedgehog signalling shows some of the most dramatic phenotypes that have been observed clinically and in the laboratory. For example, an inherited form of holoprosencephaly, characterized by variable neural and craniofacial abnormalities including cyclopia, has been mapped to the *Shh* gene in humans.[100–102] Mice

Table 2 Diseases linked to developmental genes

Gene	Phenotype
HoxA13	hand–foot–genital syndrome
HoxD13	synpolydactyly
Pax3	Waardenburg syndrome
HoxA9	acute myelogenous leukemia with t(7;11)(p15;p15)
Hox11	T-cell acute lymphocytic leukemia with t(10;14)(q24;q11)
Sonic hedgehog	holoprosencephaly
Patched	basal cell nevus syndrome
Smoothened	basal cell nevus syndrome
GLI3	*Greig cephalopolysyndactyly*
GLI3	glioblastoma
Wnt-1	mammary adenocarcinomas
β-catenin	colon carcinoma, melanoma
APC	colon carcinoma
APC	congenital hypertrophy of retinal pigment epithelium (CHRPE)
FGFR-3	achondroplasia, bone dysplasias
DPC-4	pancreatic and colon carcinoma
GDF-5 (CDMP-1)	chondrodysplasia

with a targeted disruption of *Shh* develop similar craniofacial and neural developmental defects, as well as loss of motor neurons, limb and skeletal patterning defects and other abnormalites.[60] The craniofacial abnormalites arise because the anterior neuroectoderm of the early embryo develops initially as a single field specified to form anterior structures, such as the eye. Shh, secreted by the underlying mesoderm, inhibits eye development in the midline, thereby dividing the eye field into two distinct morphogenetic fields.[103] Thus, lack of *Shh* results in the development of a single eye.

The fibroblast growth factor pathway has also been implicated in developmental abnormalities. In particular, mutations in the coding sequence of the *FGFR-3* gene cause achondroplasia, the most common form of human dwarfism.[104] Achondroplastic patients characteristically exhibit exaggerated lumbar lordosis, relative macrocephaly and minimal proliferation of the growth plate cartilage of long bones. Other mutations in the *FGFR-3* gene cause distinct bone defects, including neonatal skeletal dysplasia (thanatophoric dysplasia) and hypochondroplasia. All are dominant missense mutations that result in bone dysplasia.

Other inherited developmental anomalies linked to known genes include synpolydactyly, which has been mapped to *HoxD13*.[105] Hand–foot–genital syndrome, an autosomal dominant condition affecting bones of the hands and feet and also associated with genitourinary malformations, including bicornuate uterus, malposition of ureteral orifices and hypospadias, has been associated with mutations in *HoxA13*.[106]

The roles of many of the molecules discussed above have been more thoroughly investigated in the study of neoplasia. Members of the Wnt, Hh and FGF families have been shown to stimulate cell proliferation and to induce transformation, at least in cell culture. Conversely, TGF-β family members have been strongly implicated as negative regulators of cell growth, in addition to their other, rather variable effects on cells in culture.[107]

The hedgehog pathway also appears to play a role in control of cell proliferation in adult cells. The basal cell nevus syndrome, an autosomal dominant disease in humans characterized by multiple developmental defects and a predisposition to basal cell carcinoma, has been linked to loss of function mutations in patched, the proposed receptor for Shh, as well as activating mutations in smoothened (see Figure 12).[108,109] In addition, transgenic mice over-expressing Shh have increased basal cell carcinomas.[110] Furthermore, mutations in the downstream Shh signalling molecule GLI3, a mammalian homolog of *Cubitus interruptus*, are associated with glioblastomas and with Greig syndrome, a disorder affecting limb and craniofacial development.[111]

Wnt genes were first discovered as proto-oncogenes, since activation of the mouse *wnt-1* gene led to transformation of mammary epithelial cells.[112] Similarly, transgenic mice overexpressing wnts have an increased frequency of mammary tumors.[113] In addition, the syndrome of familial adenomatous polyposis is caused primarily by mutations in the tumor suppressor gene adenomatous polyposis coli (*APC*).[114] *APC* has been shown to inhibit β-catenin-dependent transcription, a target of wnt signalling, as shown in Figure 11. It is proposed that loss of *APC* leads to increased levels of β-catenin protein, increased transcription of *β-catenin/LEF-1* target genes, and thus increased cell proliferation. This hypothesis is strongly supported by the identification of mutations in β-catenin that prevent APC-dependent degradation of catenin. These mutations have been identified in colorectal carcinomas as well as in spontaneous melanomas.[115,116]

TGF-βs, based on their ability to cause growth arrest in the G_1 phase of the cell cycle, have been implicated as tumor suppressors. The *Smad4* gene, originally identified as Deleted in Pancreatic Carcinoma, or *DPC4*, has been implicated in both pancreatic cancer and colon carcinoma.[117] About 90% of human pancreatic carcinomas show allelic loss at chromosome 18q, which includes the *DPC4* gene. Furthermore, pancreatic carcinomas that did not have homozygous deletions at 18q21.1 did possess inactivating mutations within the *DPC4* gene, thus implicating this gene as a tumor suppressor. Smad4 has a central role in all known TGF-β signalling. Furthermore, expression of several TGF-β ligands are reduced in animal models of cancer, suggesting that loss of growth inhibition by this growth factor family may be a key step in tumorigenesis.[107] Indeed, the TGF-β-related ligand activin is able to reduce proliferation and increase erythrocytic differentiation in mouse erythroleukemia cells, in addition to its possible roles in mesoderm induction and in the regulation of FSH release from the anterior pituitary.

SUMMARY AND FUTURE PERSPECTIVES

A practical application of understanding the molecular basis of development may be in the field of tissue engineering. During embryogenesis, cell types such as neurons and cardiac myocytes are active in proliferation and upon differentiation exit the cell cycle. How these cell types leave the cell cycle is poorly understood, but if this could be regulated therapeutically, it could have enormously important implications for the treatment of ischemic and degenerative diseases, such as ischemic heart and cerebrovascular disease, as well as neurodegenerative diseases such as Alzheimer's and Parkinson's diseases. Understanding the mechanisms by which these cell types exit the cell cycle might produce the ability to grow or regenerate these postmitotic cells selectively.

Another practical benefit from the field of molecular embryology will be the identification of more genes that are mutated in inherited disorders of development. In addition to furthering our basic understanding of human development, they may also offer prenatal identification of potential developmental anomalies, such as holoprosencephaly, which occurs in 1 in 16 000 live births and is found in 1 in 250 aborted fetuses.

The integration of molecular biology, genetics and classical embryology has demonstrated that the molecules required to co-ordinate the early development of a complex organism can be identified and characterized. This field has also established fundamental paradigms, particularly from the work in *Drosophila*, that can be applied to the analysis of early development in other organisms. However, important unresolved questions remain. For example, little is known about how the many and diverse signalling pathways active in the same tissues interact with each other. In addition, few embryologists have begun to address the more complicated questions of morphogenesis, that is, what directs whole tissues to move and to change shape during development, as seen most dramatically during gastrulation and neurulation, but also on a smaller scale with the genesis of each of the internal organs. Furthermore, it remains to be seen how similar in detail mammalian development really is to that of lower vertebrates and invertebrates. These are a few of the questions that promise to occupy mammalian embryologists for the next several years.

ACKNOWLEDGEMENTS

The authors would like to thank Matt Scott, Betsy Wilder, Robb Krumlauf, Bob Riddle, Tony Brown and Matt Deardorff for providing photographs. We would also like to thank Chris Phiel, Mahmoud Najafi and Matt Deardorff for critical review of the manuscript and for helpful suggestions; and finally, Laurie Zimmerman for assistance in preparation of the manuscript.

REFERENCES

1. Moore K. *The Developing Human: Clinically Oriented Embryology*, 6th edn. Philadelphia: Saunders, 1998.
2. Gilbert SF. *Developmental Biology*, 2nd edn. Sunderland, MA: Sinauer Associates, 1998.
3. Nieuwkoop PD, Faber J. *Normal Table of Xenopus laevis (Daudin)*, 2nd edn. Amsterdam: North Holland Publishing Company, 1967.
4. Hogan B, Beddington R, Constantini F, Lacy E. *Manipulating the Mouse Embryo: A Laboratory Manual*, 2nd edn. New York: Cold Spring Harbor Laboratory Press, 1994.
5. Slack JMW. *From Egg to Embryo: Regional Specification in Early Development*. Cambridge: Cambridge University Press, 1991.
6. Pardue ML, Gall JG. Chromosomal localization of mouse satellite DNA. *Science* 1970 168: 1356–1358.
7. Robbins PD, Tahara H, Ghivizzani SC. Viral vectors for gene therapy. *Trends Biotechnol* 1998 16: 35–40.
8. Engels WR. P elements in *Drosophila*. *Curr Topics Microbiol Immunol* 1996 204: 103–123.

9. Brinster RL, Chen HY, Trumbauer M *et al.* Somatic expression of herpes thymidine kinase in mice following injection of a fusion gene into eggs. *Cell* 1981 27: 223–231.
10. Nishimori K, Matzuk MM. Transgenic mice in the analysis of reproductive development and function. *Rev Reprod* 1996 1: 203–212.
11. Thomas JW, LaMantia C, Magnuson T. X-ray-induced mutations in mouse embryonic stem cells. *Proc Natl Acad Sci USA* 1998 95: 1114–1119.
12. Nusslein-Volhard C, Wieschaus E. Mutations affecting segment number and polarity in *Drosophila*. *Nature* 1980 287: 795–801.
13. Zeng L, Fagotto F, Zhang T *et al.* The mouse Fused locus encodes Axin, an inhibitor of the Wnt signaling pathway that regulates embryonic axis formation. *Cell* 1997 90: 181–192.
14. Robertson E, Bradley A, Kuehn M, Evans M. Germ-line transmission of genes introduced into cultured pluripotential cells by retroviral vector. *Nature* 1986 323: 445–448.
15. Isaac A, Sargent MG, Cooke J. Control of vertebrate left–right asymmetry by a snail-related zinc finger gene. *Science* 1997 275: 1301–1304.
16. Heasman J, Crawford A, Goldstone K *et al.* Overexpression of cadherins and underexpression of beta-catenin inhibit dorsal mesoderm induction in early *Xenopus* embryos. *Cell* 1994 79: 791–803.
17. Rosenberg UB, Preiss A, Seifert E, Jackle H, Knipple DC. Production of phenocopies by Kruppel antisense RNA injection into *Drosophila* embryos. *Nature* 1985 313: 703–706.
18. Moreadith RW, Radford NB. Gene targeting in embryonic stem cells: the new physiology and metabolism. *J Mol Med* 1997 75: 208–216.
19. Varlet I, Collignon J, Robertson EJ. Nodal expression in the primitive endoderm is required for specification of the anterior axis during mouse gastrulation. *Development* 1997 124: 1033–1044.
20. Hanks M, Wurst W, Anson-Cartwright L, Auerbach AB, Joyner AL. Rescue of the En-1 mutant phenotype by replacement of En-1 with En-2. *Science* 1995 269: 679–682.
21. Gu H, Marth JD, Orban PC, Mossmann H, Rajewsky K. Deletion of a DNA polymerase beta gene segment in T cells using cell type-specific gene targeting. *Science* 1994 265: 103–106.
22. Hamburger V. *The Heritage of Experimental Embryology: Hans Spemann and the Organizer.* New York: Oxford University Press, 1988.
23. Spemann H. *Embryonic Development and Induction.* New York: Yale University Press, 1938.
24. Harland R, Gerhart J. Formation and function of Spemann's organizer. *Annu Rev Cell Dev Biol* 1997 13: 611–667.
25. Beddington RSP. Induction of a second neural axis by the mouse node. *Development* 1994 120: 613–620.
26. Dias MS, Schoenwolf GC. Formation of ectopic neurepithelium in chick blastoderms: age-related capacities for induction and self-differentiation following transplantation of quail Hensen's nodes. *Anat Rec* 1990 228: 437–448.
27. Storey KG, Crossley JM, DeRobertis EM, Norris WE, Stern CD. Neural induction and regionalization of the chick embryo. *Development* 1992 114: 729–741.
28. Nieuwkoop PD. The formation of mesoderm in urodelean amphibians. I. Induction by the endoderm. *Wilhelm Roux Arch EntwMech Org* 1969 162: 341–373.
29. Waldrip WR, Bikoff EK, Hoodless PA, Wrana JL, Robertson EJ. Smad2 signaling in extraembryonic tissues determines anterior–posterior polarity of the early mouse embryo. *Cell* 1998 92: 797–808.
30. Nomura M, Li E. Smad2 role in mesoderm formation, left–right patterning, and craniofacial development. *Nature* 1998 393: 786–790.
31. Lawrence P. *The Making of a Fly: the Genetics of Animal Design.* Oxford: Blackwell Scientific Publications, 1992.
32. St Johnston D, Nusslein-Volhard C. The origin of pattern and polarity in the *Drosophila* embryo. *Cell* 1992 68: 201–219.
33. Wolpert L. Positional information and the spatial pattern of cellular differentiation. *J Theor Biol* 1969 25: 1–47.
34. Driever W, Nusslein-Volhard C. The bicoid protein determines position in the *Drosophila* embryo in a concentration-dependent manner. *Cell* 1988 54: 95–104.
35. Graba Y, Aragnol D, Pradel J. *Drosophila Hox* complex downstream targets and the function of homeotic genes. *BioEssays* 1997 19: 379–388.
36. Lewis EB. A gene complex controlling segmentation in *Drosophila*. *Nature* 1978 276: 565–570.

37. Scott MP, O'Farrell PH. Spatial programming of gene expression in early *Drosophila* embryogenesis. *Annu Rev Cell Biol* 1986 2: 49–80.
38. Mark M, Rijli FM, Chambon P. Homeobox genes in embryogenesis and pathogenesis. *Pediatr Res* 1997 42: 421–429.
39. Kessel M, Balling R, Gruss P. Variations of cervical vertebrae after expression of a *Hox 1.1* transgene in mice. *Cell* 1990 61: 301–308.
40. Katoh-Fukui Y, Tsuchiya R, Shiroishi T *et al.* Male-to-female sex reversal in M33 mutant mice. *Nature* 1998 393: 688–692.
41. Cadigan KM, Nusse R. Wnt signaling: a common theme in animal development. *Genes Dev* 1997 11: 3286–3305.
42. Bhanot P, Brink M, Samos CH *et al.* A new member of the frizzled family from *Drosophila* functions as a Wingless receptor. *Nature* 1996 382: 225–30.
43. Vinson CR, Adler PN. Directional non-cell autonomy and the transmission of polarity information by the frizzled gene of *Drosophila*. *Nature* 1987 329: 549–551.
44. Slusarski DC, Yang SJ, Busa WB, Moon RT. Modulation of embryonic intracellular Ca^{2+} signaling by Wnt-5A. *Dev Biol* 1997 182: 114–120.
45. Miller JR, Moon RT. Signal transduction through b-catenin and specification of cell fate during embryogenesis. *Genes Dev* 1996 10: 2527–2539.
46. Siegfried E, Wilder EL, Perrimon N. Components of wingless signalling in *Drosophila*. *Nature* 1994 367: 76–80.
47. Bourouis M, Moore P, Ruel L *et al.* An early embryonic product of the gene shaggy encodes a serine/threonine protein kinase related to the CDC28/cdc2+ subfamily. *EMBO J* 1990 9: 2877–2884.
48. Riggleman B, Schedl P, Weischaus E. Spatial expression of the *Drosophila* segment polarity gene *armadillo* is post-transcriptionally regulated by wingless. *Cell* 1990 63: 549–560.
49. McMahon AP, Moon RT. Ectopic expression of the proto-oncogene int-1 in *Xenopus* embryos leads to duplication of the embryonic axis. *Cell* 1989 58: 1075–1084.
50. Klein PS, Melton DA. A molecular mechanism for the effect of lithium on development. *Proc Natl Acad Sci USA* 1996 93: 8455–8459.
51. Hedgepeth C, Conrad L, Zhang Z, Huang H, Lee V, Klein P. Activation of the Wnt signaling pathway: a molecular mechanism for lithium action. *Dev Biol* 1997 185: 82–91.
52. Stambolic V, Ruel L, Woodgett J. Lithium inhibits glycogen synthase kinase-3 activity and mimics wingless signalling in intact cells. *Curr Biol* 1996 6: 1664–1668.
53. Popperl H, Schmidt C, Wilson V *et al.* Misexpression of Cwnt8C in the mouse induces an ectopic embryonic axis and causes a truncation of the anterior neuroectoderm. *Development* 1997 124: 2997–3005.
54. Hammerschmidt M, Brook A, McMahon AP. The world according to hedgehog. *Trends Genet* 1997 13: 14–21.
55. Riddle RD, Johnson RL, Laufer E, Tabin C. Sonic hedgehog mediates the polarizing activity of the ZPA. *Cell* 1993 75: 1401–1416.
56. Levin M, Johnson RL, Stern CD, Kuehn M, Tabin C. A molecular pathway determining left–right asymmetry in chick embryogenesis. *Cell* 1995 82: 803–814.
57. Roelink H, Augsburger A, Heemskerk J *et al.* Floor plate and motor neuron induction by vhh-1, a vertebrate homolog of hedgehog expressed by the notochord. *Cell* 1994 76: 761–775.
58. Marti E, Bumcrot DA, Takada R, McMahon AP. Requirement of 19K form of Sonic hedgehog for induction of distinct ventral cell types in CNS explants. *Nature* 1995 375: 322–325.
59. Johnson RL, Laufer E, Riddle RD, Tabin C. Ectopic expression of Sonic hedgehog alters dorsal–ventral patterning of somites. *Cell* 1994 79: 1165–1173.
60. Chiang C, Litingtung Y, Lee E *et al.* Cyclopia and defective axial patterning in mice lacking Sonic hedgehog gene function. *Nature* 1996 383: 407–413.
61. Yamaguchi TP, Rossant J. Fibroblast growth factors in mammalian development. *Curr Opin Genet Dev* 1995 5: 485–491.
62. Kimelman D, Abraham JA, Haaparanta T, Palisi TM, Kirschner MW. The presence of fibroblast growth factor in the frog egg: its role as a natural mesoderm inducer. *Science* 1988 242: 1053–1056.
63. Slack JMW, Darlington BG, Heath JK, Godsave SF. Mesoderm induction in early *Xenopus* embryos by heparin-binding growth factors. *Nature* 1987 326: 197–200.
64. Amaya E, Musci TJ, Kirschner MW. Expression of a dominant negative mutant of the FGF receptor

disrupts mesoderm formation in *Xenopus* embryos. *Cell* 1991 66: 257–270.

65. Whitman M, Melton D. Involvement of p21ras in *Xenopus* mesoderm induction. *Nature* 1992 357: 252–254.

66. MacNicol AM, Muslin AJ, Williams LT. Raf-1 kinase is essential for early *Xenopus* development and mediates the induction of mesoderm by FGF. *Cell* 1993 73: 571–583.

67. Yamaguchi TP, Harpal K, Henkemeyer M, Rossant J. fgfr-1 is required for embryonic growth and mesodermal patterning during mouse gastrulation. *Genes Dev* 1994 8: 3032–3044.

68. Feldman B, Poueymirou W, Papaioannou VE, DeChiara TM, Goldfarb M. Requirement of FGF-4 for postimplantation mouse development. *Science* 1995 267: 246–249.

69. Massague J, Cheifetz S, Laiho M *et al.* Transforming growth factor-beta. *Cancer Surv* 1992 12: 81–103.

70. Massague J, Weis-Garcia F. Serine/threonine kinase receptors: mediators of transforming growth factor beta family signals. *Cancer Surv* 1996 27: 41–64.

71. Derynck R, Zhang Y. Intracellular signalling: the mad way to do it. *Curr Biol* 1996 6: 1226–1229.

72. Kretzschmar M, Massague J. SMADs: mediators and regulators of TGF-beta signaling. *Curr Opin Genet Dev* 1998 8: 103–111.

73. Urist MR, DeLange RJ, Finnerman GAM. Bone cell differentiation and growth factors. *Science* 1983 220: 680–686.

74. Wozney JM, Rosen V, Celeste AV *et al.* Novel regulators of bone formation: molecular clones and activities. *Science* 1988 242: 1528–1534.

75. Kingsley DM. The TGF-β superfamily: new members, new receptors, and new genetic tests of function in different organisms. *Genes Dev* 1994 8: 133–146.

76. Dale L, Howes G, Price BMJ, Smith JC. Bone Morphogenetic Protein 4: a ventralizing factor in *Xenopus* development. *Development* 1992 115: 573–585.

77. Jones CM, Lyons KM, Lapan PM, Wright CVE, Hogan BJM. DVR-4 (Bone Morphogenetic Protein-4) as a postero-ventralizing factor in *Xenopus* mesoderm induction. *Development* 1992 115: 639–647.

78. Sasai Y, Lu B, Steinbeisser H, Derobertis EM. Regulation of neural induction by the Chd and Bmp-4 antagonistic patterning signals In *Xenopus*. *Nature* 1995 376: 333–336.

79. Wilson PA, Hemmati-Brivanlou A. Induction of epidermis and inhibition of neural fate by Bmp-4. *Nature* 1995 376: 331–333.

80. Zimmerman LB, De Jesus-Escobar JM, Harland RM. The Spemann organizer signal noggin binds and inactivates bone morphogenetic protein 4. *Cell* 1996 86: 599–606.

81. Streit A, Lee KJ, Woo I *et al.* Chordin regulates primitive streak development and the stability of induced neural cells, but is not sufficient for neural induction in the chick embryo. *Development* 1998 125: 507–519.

82. McMahon JA, Takada S, Zimmerman LB *et al.* Noggin-mediated antagonism of BMP signaling is required for growth and patterning of the neural tube and somite. *Genes Dev* 1998 12: 1438–1452.

83. Winnier G, Blessing M, Labosky PA, Hogan BL. Bone morphogenetic protein-4 is required for mesoderm formation and patterning in the mouse. *Genes Dev* 1995 9: 2105–2116.

84. Mishina Y, Suzuki A, Ueno N, Behringer RR. Bmpr encodes a type 1 bone morphogenetic protein receptor that is essential for gastrulation during mouse embryogenesis. *Genes Dev* 1995 9: 3027–3037.

85. McPherron AC, Lawler AM, Lee SJ. Regulation of skeletal muscle mass in mice by a new TGF-beta superfamily member. *Nature* 1997 387: 83–90.

86. Thomas JT, Lin K, Nandedkar M *et al.* A human chondrodysplasia due to a mutation in a TGF-beta superfamily member. *Nature Genet* 1996 12: 315–317.

87. Storm EE, Huynh TV, Copeland NG *et al.* Limb alterations in brachypodism mice due to mutations in a new member of the TGF betasuperfamily. *Nature* 1994 368: 639–643.

88. Dong J, Albertini DF, Nishimori K *et al.* Growth differentiation factor-9 is required during early ovarian folliculogenesis. *Nature* 1996 383: 531–535.

89. Vale W, Hsueh A, Rivier C, Yu J. The inhibin/activin family of hormones and growth factors. In Sporn MB, Roberts AB, eds. *Peptide Growth Factors and Their Receptors I.* Berlin: Springer Verlag, 1990.

90. Asashima M, Nakano H, Shimada K *et al.* Mesodermal induction in early amphibian embryos by activin A (erythroid differentiation factor). *Roux's Arch Dev Biol* 1990 198: 330–335.
91. Smith JC, Yaqoob M, Symes K. Purification, partial characterization and biological effects of the XTC mesoderm-inducing factor. *Development* 1988 103: 591–600.
92. Thomsen G, Woolf T, Whitman M *et al.* Activins are expressed early in *Xenopus* embryogenesis and can induce axial mesoderm and anterior structures. *Cell* 1990 63: 485–493.
93. Green JBA, Smith JC. Graded changes in dose of a *Xenopus* activin A homologue elicit stepwise transitions in embryonic cell fate. *Nature* 1990 347: 391–394.
94. Green JBA, New HV, Smith JC. Responses of embryonic *Xenopus* cells to activin and FGF are separated by multiple dose thresholds and correspond to distinct axes of the mesoderm. *Cell* 1992 71: 731–739.
95. Graff JM, Bansal A, Melton DA. *Xenopus* Mad proteins transduce distinct subsets of signals for the TGF beta superfamily. *Cell* 1996 85: 479–487.
96. Gu Z, Nomura M, Simpson BB *et al.* The type I activin receptor ActRIB is required for egg cylinder organization and gastrulation in the mouse. *Genes Dev* 1998 12: 844–857.
97. Zhou X, Sasaki H, Lowe L, Hogan BL, Kuehn MR. Nodal is a novel TGF-beta-like gene expressed in the mouse node during gastrulation. *Nature* 1993 361: 543–547.
98. Conlon FL, Lyons KM, Takaesu N *et al.* A primary requirement for *nodal* in the formation and maintenance of the primitive streak in the mouse. *Development* 1994 120: 1919–1928.
99. Mishina Y, Rey R, Finegold MJ *et al.* Genetic analysis of the Mullerian-inhibiting substance signal transduction pathway in mammalian sexual differentiation. *Genes Dev* 1996 10: 2577–2587.
100. Belloni E, Muenke M, Roessler E *et al.* Identification of Sonic hedgehog as a candidate gene responsible for holoprosencephaly. *Nature Genet* 1996 14: 353–356.
101. Roessler E, Belloni E, Gaudenz K *et al.* Mutations in the human Sonic Hedgehog gene cause holoprosencephaly. *Nature Genet* 1996 14: 357–360.
102. Ming JE, Muenke M. Holoprosencephaly: from Homer to Hedgehog. *Clin Genet* 1998 53: 155–163.
103. Li H, Tierney C, Wen L, Wu JY, Rao Y. A single morphogenetic field gives rise to two retina primordia under the influence of the prechordal plate. *Development* 1997 124: 603–615.
104. Horton WA. Fibroblast growth factor receptor 3 and the human chondrodysplasias. *Curr Opin Pediatr* 1997 9: 437–442.
105. Muragaki Y, Mundlos S, Upton J, Olsen BR. Altered growth and branching patterns in synpolydactyly caused by mutations in HOXD13. *Science* 1996 272: 548–551.
106. Mortlock DP, Innis JW. Mutation of HOXA13 in hand–foot–genital syndrome. *Nature Genet* 1997 15: 179–180.
107. Massague J, Weinberg RA. Negative regulators of growth. *Curr Opin Genet Dev* 1992 2: 28–32.
108. Johnson RL, Rothman AL, Xie J *et al.* Human homolog of patched, a candidate gene for the basal cell nevus syndrome. *Science* 1996 272: 1668–1671.
109. Xie J, Murone M, Luoh SM *et al.* Activating Smoothened mutations in sporadic basal-cell carcinoma. *Nature* 1998 391: 90–92.
110. Oro AE, Higgins KM, Hu Z *et al.* Basal cell carcinomas in mice overexpressing sonic hedgehog. *Science* 1997 276: 817–821.
111. Vortkamp A, Gessler M, Grzeschik KH. GLI3 zinc-finger gene interrupted by translocations in Greig syndrome families. *Nature* 1991 352: 539–540.
112. Nusse R, Varmus HE. Many tumors induced by the mouse mammary tumor virus contain a provirus integrated in the same region of the host genome. *Cell* 1982 31: 99–109.
113. Lane TF, Leder P. Wnt-10b directs hypermorphic development and transformation in mammary glands of male and female mice. *Oncogene* 1997 15: 2133–2144.
114. Powell SM, Petersen GM, Krush AJ *et al.* Molecular diagnosis of familial adenomatous polyposis. *N Engl J Med* 1993 329: 1982–1987.
115. Morin PJ, Sparks AB, Korinek V *et al.* Activation of beta-catenin-Tcf signaling in colon cancer by mutations in beta-catenin or APC. *Science* 1997 275: 1787–1790.
116. Rubinfeld B, Robbins P, El GM *et al.* Stabilization of beta-catenin by genetic defects in melanoma cell lines. *Science* 1997 275: 1790–1792.

117. Hahn SA, Schutte M, Hoque AT *et al.* DPC4, a candidate tumor suppressor gene at human chromosome 18q21.1. *Science* 1996 271: 350–353.
118. Klein PS, Melton DA. Hormonal regulation of embryogenesis: the formation of mesoderm in *Xenopus laevis*. *Endocr Rev* 1994 15: 326–341.
119. Tam PP, Behringer RR. Mouse gastrulation: the formation of a mammalian body plan. *Mech Dev* 1997 68: 3–25.
120. Wnt homepage: www.stanford.edu/~rnusse/wnt window.html

SECTION VI

Principles of medical genetics

Joe Leigh Simpson

Two decades ago, there was a strong tendency for reproductive endocrinologists to consider their field distinct from that of clinical geneticists. Of course, investigative technologies overlapped, but the dichotomy existed to a great extent in the clinical domain. Geneticists tended to emphasize etiology and differential diagnoses of unusual syndromes found predominantly in the pediatric age group. Although endocrinologists were involved in the treatment of these disorders, they paid less attention to the etiology and to rare conditions. However, the merging of these two fields is becoming increasingly evident. This places a responsibility on the community of reproductive endocrinologists to delve into areas once considered the sole domain of geneticists.

In this section, we explore topics in genetic counselling and prenatal genetic diagnosis relevant to the current demands of reproductive endocrinology. In Chapter 26, Liebaers and Simpson discuss the principles of genetic counselling. The differing genetic etiologies are outlined, including chromosomal, single-gene, multigenic or multifactorial defects and also mitochondrial disorders and imprinting. The clinical approach traditionally used by geneticists in counselling is delineated. It is crucial that the reproductive medicine specialist appreciates the psychological impact on a couple whose child is born (or miscarried) with a genetic disorder. The manner in which geneticists elicit family history, use physical examinations and select laboratory tests is considered. In addition, specific examples of genetic counselling are given.

Traditionally, prenatal genetic diagnosis has encompassed amniocentesis and chorionic villus sampling. Although these methods have provided a reliable diagnosis of the presence of genetic abnormalities in the first or second trimester of pregnancy, they are not suitable for couples not wishing to jeopardize a pregnancy achieved at great effort or who do not want clinical termination. To this end, preimplantation genetic diagnosis has come to the fore in the last decade. In Chapter 27, Sermon and Liebaers discuss the methods available to obtain one or two cells (blastomere or polar body), and the general diagnostic approach. Of great interest is the increasing use of laboratory-based screening for the detection of chromosomal abnormalities. We are only at the start of this burgeoning field.

Finally, in Chapter 28, Kennedy discusses the genetic elucidation of complex genetic traits. Several of these are widely known to reproductive specialists, and constitute much of our work: endometriosis, polycystic ovarian disease, and leiomyoma come readily to mind. Recurrence risks point toward multifactorial/polygenic inheritance (2–7% for first-degree relatives). All these disorders are thus heritable, but not in a straightforward manner. They are the result of the cumulative effect of several different genes, probably located throughout the genome. Locating the several genes involved could allow subtypes of the disorder to be identified. In turn, recognition of heterogeneity could allow the use of targeted therapeutic regimes that would improve outcome. For example, surgery might be a suitable treatment for some disorders, whereas others might be best treated medically. However, extant approaches to find the relevant genes are proving frustrating. Kennedy outlines the strategy and pitfalls of genome-wide linkage analysis. The approach increasingly taken for endometriosis and for polycystic ovarian disease is lucidly and succinctly discussed. Emphasis is placed on sib-pair analysis, the most widely used approach, and although this section provides only an outline of the general principles of genetic counselling and delineation, many disorders are specifically discussed throughout the volume. In particular, the genetic basis of sex differentiation disorders is discussed in the next section.

Chapter 26

Genetic counselling in reproductive disorders

Inge Liebaers and Joe Leigh Simpson

INTRODUCTION

Causes of reproductive disorders are heterogeneous in women as well as in men. They can be acquired prenatally or postnatally. They can be genetic, *de novo* or inherited. Nowadays, when a couple cannot conceive spontaneously, treatment is often based on the available, often technical means to help the conception and birth of a (hopefully) healthy child. It is important, however, not to overlook the cause of the reproductive failure, to enable efficient treatment to be given to the couple, as well as information about possible genetic risks to the offspring, risks that may or may not be prevented.

BACKGROUND

Genetic counselling is a process of communication between the counsellors and counsellees. The aim of the counsellors is to inform the counsellees about the cause and course of a disorder, the treatment possibilities, the recurrence risks to the offspring, and the possible ways in which to prevent the transmission of a disorder. To be able to do this within the framework of reproductive failure, one needs to refine as much as possible the cause of a reproductive disorder. Moreover, attention should be paid to other possible genetic risks, which may not interfere with reproduction but which should be discussed before initiating any treatment, if not an attempt at reproduction. With the correct information given to the counsellees, they should be able to make their own decision concerning their future treatment.[1]

An increasing number of genes as well as genetic defects are being identified, and, therefore, more and more genetic causes of reproductive disorders are also being or will be reported.[2] However, in many cases the cause of the reproductive failure remains unknown. New diagnostic procedures to pinpoint genetic defects are being developed and should be used whenever indicated. The counselling process should continuously be adapted to remain 'state of the art'.

Genetic diseases are the consequence of an abnormality in the genome. The genome is composed of nuclear double-stranded DNA, containing around 30 000 genes, and is organized in chromosomes. As important for normal function as nuclear DNA is mitochondrial double-stranded DNA, containing 37 genes and being present in multiple copies in mitochondria in the cytoplasm of all cells.[3] In diploid somatic cells, 46 chromosomes (22 pairs of autosomes plus one pair of sex chromosomes) are present in each nucleus, while the numbers of mitochondria and of mitochondrial DNA molecules in the cytoplasm will vary from tissue to tissue. In haploid germ cells leading to gametes, only 23 chromosomes (one chromosome of each pair) are present in the nucleus. The oocyte will always contain an X chromosome in the nucleus and many mitochondrial DNA molecules in the cytoplasm; these will populate the embryonic, fetal and adult cells later on. The spermatozoon contains either an X chromosome or a Y chromosome and some mitochondrial DNA in the midpiece, which probably does not contribute to the mitochondrial DNA pool in embryonic cells. However, spermatozoa determine the sex of the embryo. Nuclear genes are positioned on chromosomes in a rather conserved way; through DNA replication during mitosis in somatic cells, and during meiosis in germ cells, they are transmitted from one cell to another and from one generation to another. Within one individual, the gene pool remains unchanged throughout life unless a mutation occurs. Within a somatic cell, such a mutation can lead to cancer or other acquired diseases, but not to a constitutional genetic disease. When a mutation occurs in a germ cell, this mutation will be transmitted to the next generation, and may, if dominant, cause a genetic disease in the next generation.[4,5]

There are several categories of genetic diseases. Chromosomal disorders can be the result of numerical aberrations; one such disorder is Klinefelter syndrome, which is the result of a meiotic non-disjunction in a female or male germ cell. The same is true for Down syndrome, although, most often, non-disjunction has occurred in the female gamete. Chromosomal disorders can also be the result of structural aberrations, usually due to interchromosomal rearrangements, for example translocations (Figure 1a and b). Intrachromosomal rearrangements can be the result of a paracentric or pericentric inversion, a duplication or a

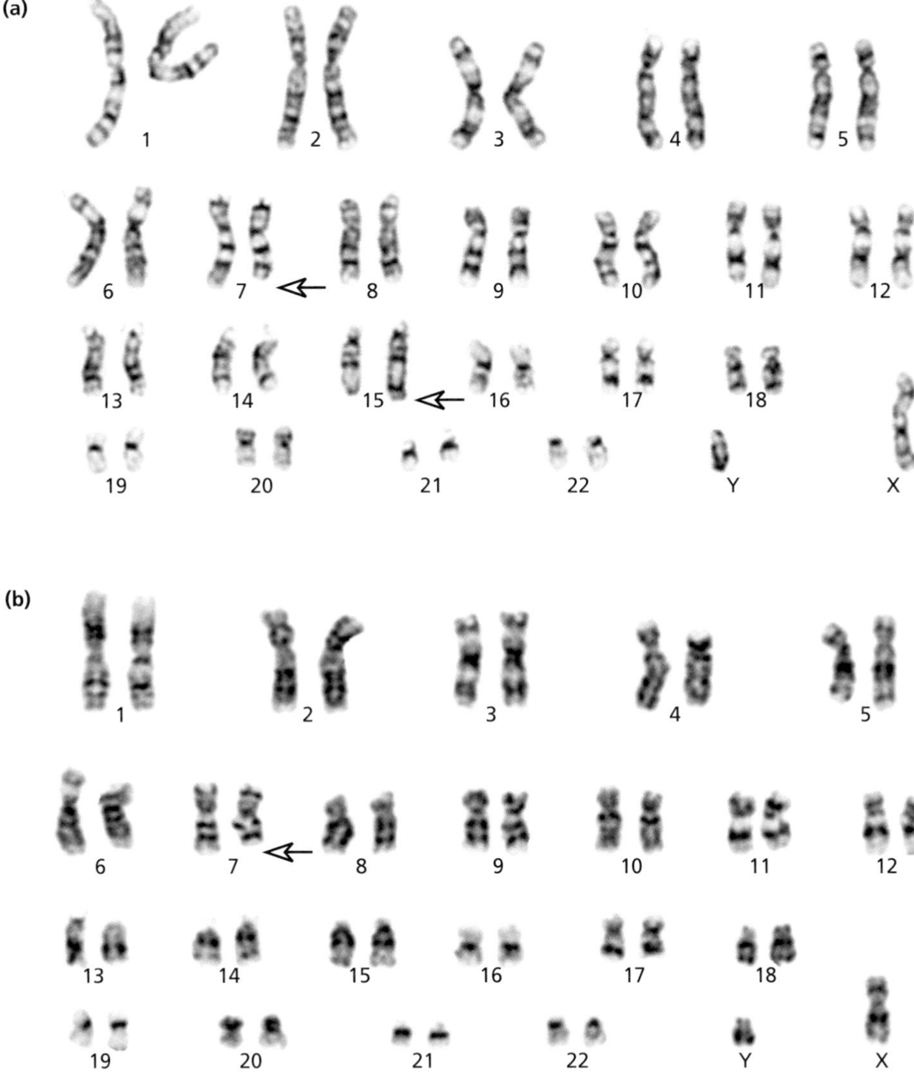

Figure 1 (a) Balanced reciprocal translocation: 46,XY,t(7;15)(q32;q26). This is the karyotype of a male with severe oligoastheno-teratospermia (OAT). He is a carrier of a balanced translocation 46,XY,t(7;15)(q32,q26). A reciprocal exchange has occurred between the terminal part of the long arm of chromosome 7 and the long arm of chromosome 15; the breakpoint in chromosome 7 is q32; the breakpoint in chromosome 15 is q26. (b) Unbalanced reciprocal translocation: 46,XY,der 7,t(7;15)(q32,q26). This is the karyotype of a male child of the man with the balanced reciprocal translocation seen in (a). This karyotype is imbalanced, and represents a partial monosomy of chromosome 7 because the child inherited the deleted chromosome 7 and the normal chromosome 15 from his father. The child presented with multiple congenital anomalies and mental retardation

deletion, which, when very small, is called a microdeletion. Approximately six out of 1000 people are born with an overt or hidden chromosomal aberration.[4,5]

Monogenic diseases are caused by gene deletions or, more often, gene mutations. Gene mutations are DNA sequence errors, such as nucleotide substitutions or nucleotide(s) deletions or duplications. These mutations lead to absent or non-functional proteins. Monogenic diseases can be autosomal or sex-linked. Autosomal disease can be recessive if both alleles at a given locus are defective, or dominant if only one allele of the gene pair is defective in such a way that the 'normal' gene cannot function properly. Sex-linked diseases are usually the result of mutated recessive genes on the X chromosome, leading to affected males and unaffected but transmitting carrier females. Approximately ten out of 1000 people are born with a monogenic disease, and altogether around 6000 monogenic diseases are known today. Therefore, each of these diseases are rare: hemochromatosis 1/1000; cystic fibrosis 1/2500; and Kennedy disease 1/50 000.[4–6]

Multifactorial genetic diseases are the most frequent, but also the less well understood, because most probably they are caused by a genetic predisposition defined by more than one gene and influenced by environmental factors. The term 'polygenic' is applied if environmental factors do not exist, and only when more than one gene is responsible. In practice, it is difficult if not impossible to distinguish between polygenic and multifactorial etiology. Many reproductive disorders are in this category.[4,5]

Mitochondrial disorders can be the result of nuclear defective genes or mitochondrial defective genes. If the defective genes are nuclear, the disease can be recessive or dominant, autosomal or sex-linked. If the defective gene is mitochondrial, transmission will always follow the female line through the oocytes.[3–7]

Finally, imprinting disorders constitute a new category of genetic diseases. These disorders are concerned with differential gene expression resulting from differential methylation, reflecting transmission through a female or a male germ cell. A typical example of an imprinted gene disorder is the Prader–Willi syndrome, which in males causes not only infertility but also other symptoms such as obesity and mental retardation.[8–10]

Genetic disorders related to reproduction can occur in the male or in the female. Known male and female genetic disorders leading to infertility or subfertility are chromosomal or monogenic in origin. The most common of these defects are listed in Tables 1 and 2.

Chromosomal disorders

Chromosomal abnormalities are numerical or structural (Table 1). Numerical aberration may involve autosomes or sex chromosomes. Autosomal trisomy usually leads to lethality, or to such severe anomalies that pregnancies are rare events. However, female Down syndrome patients with 47,XX+21 chromosomes may become pregnant, and less than half of their offspring will have a trisomy 21.[34] This means that, during meiosis of a trisomic cell, either an unequal number of

Table 1 Chromosomal aberrations in reproductive disorders

Sex chromosomes
Male
47,XXY
47,XYY
46,X/46,XY
other mosaics
46,XX
46,X,der(Y)
t(Y;autosome), etc.
Yq11 microdeletions
Female
45,X
47,XXX
45,X/46,XX
other mosaics
46,XY
46,X,der(X)
t(X;autosome), etc.
Autosomes
Robertsonian translocations: t(acrocentric chromosomes)
Reciprocal translocations: t(all chromosomes)
Inversions: pericentric/paracentric (all chromosomes)

der, derivative; t, translocation

Table 2 Monogenic defects in reproductive disorders

Condition	Gene	Locus	Transmission	References
Male				
Congenital bilateral absence of vas deferens ± cystic fibrosis	*CFTR*	7q31.2	AR	11, 12, 13–16
Kallmann syndrome	*KAL 1*	Xp22.3	XL	17, 18
Kennedy disease	*AR*	Xq11–12	XL	19
Myotonic dystrophy	*DMPK*	19p13.3	AD	20, 21
Primary ciliary dyskinesia	NN	9p21	AR	22, 23
Noonan syndromes	NN	12q24	AD	24
Hemochromatosis	*HFE*	6p21.3	AR	25, 26
Female				
Hand–foot–genital syndrome	*HOXA13*	Xp14.2–15	AD	27
Blepharophimosis–ptosis–epicanthus inversus syndrome	*FOXL-2*	3q23	AD	28
Galactosemia	*GALT*	9p13	AR	29
Fragile-X	*FMR-1*	Xq27.3	XL	30
Nijmegen breakage syndrome	*NBS1*	8q21	AR	31
Androgen insensitivity	*AR*	Xq11.12	XL	32
Congenital adrenal hyperplasia due to 21β-hydroxylase deficiency	*CYP21*	6q21.3	AR	33

AR, autosomal recessive; XL, X-linked recessive; AD, autosomal dominant; NN, not known

gametes with $n = 23$ chromosomes and $(n + 1) = 24$ chromosomes has been formed, possibly through elimination of the additional chromosome in certain gametes, or an equal number of n and $(n + 1)$ gametes was formed but the viability of the $(n + 1)$ gamete or the $(2n + 1)$ embryo was reduced. Rather frequent and well-known numerical aberrations of sex chromosomes are 47,XXY in Klinefelter syndrome, resulting in azoospermia, or 45,X in Turner syndrome resulting in oocyte depletion. In both syndromes, the karyotype can be mosaic 46,XY/47,XXY or 45,X/46,XX, or other variants; individuals with these karyotypes may sometimes produce a low number of gametes.[4,5] Theoretically, these gametes may be n, $(n + 1)$ or $(n - 1)$. Mosaicism may be limited to the gonads, and lead to recurrent aneuploid offspring. Exceptionally, the karyotype in a male can be 46,XX, owing to translocations of the sex-determining region of the Y (*SRY*) gene on one of the X chromosomes; the phenotype of these males is Klinefelter-like with azoospermia. Females with a 46,XY karyotype due to an *SRY* deletion also exist.[35,36] Other sex-chromosomal numerical aberrations seen more often in patients with reproductive failure than in fertile individuals are the 47,XYY karyotype in males and the 47,XXX karyotype in females; usually, gametes are present, although some 47,XXX women may present with premature menopause. Females with a 47,XXX genotype should theoretically yield 46,XX, 46,XY, 47,XXX and 47,XXY zygotes in equal frequencies. Some reports have claimed increased spontaneous abortions in 47,XXX women, but these claims probably reflect selection biases: cytogenetic studies were often initiated only after repetitive abortions had occurred;[37,38] the number of 47,XXX women not experiencing abortions would never be ascertained. Despite these reassuring comments, antenatal cytogenetic studies should be discussed with pregnant 47,XXX patients. 47,XYY men should theoretically produce 46,XX, 46,XY, 47,XYY and 47,XXY offspring. Few chromosomally abnormal offspring have been recognized. This observation is consistent with data obtained in sperm from an XYY male.[39] Offspring of 47,XYY males may thus be normal, because aneuploid gametes seem not to be formed.

Structural aberrations may involve translocations between the sex chromosomes and autosomes or may involve Y- or X-chromosome derivatives. Depending on the break-points involved or the extent of the deletions, reproduction may be impaired. Robertsonian translocations, reciprocal translocations and other structural aberrations of the autosomes are all seen more frequently in infertile couples, or couples with recurrent miscarriages.[4,5,40,41] If Klinefelter or Turner syndrome patients desire children, most often their only possibility is to make use of donor gametes.[42] Exceptionally, Klinefelter patients can be helped by intracytoplasmic sperm injection (ICSI) of oocytes of their partner by sperm retrieved from a testicular biopsy.[43,44] In most of the conditions described above, patience can help the couple establish an ongoing pregnancy. Counselling in the case of a chromosomal translocation remains complicated. Theoretically and very simplistically, only two out of six gametes will be balanced by containing either the two normal chromosomes involved or the two chromosomes involved in the translocation. In the other four gametes, one will observe partial trisomies or monosomies of the chromosomes involved. Thus, theoretically, 33% of the resulting embryos can develop into a 'healthy' child, but the other 67% of embryos will either not implant or lead to a miscarriage or the birth of a child with multiple congenital anomalies. In general, it is accepted that the smaller the translocated chromosomal region, the higher the likelihood that the pregnancy will go to term.[4,5,45] Finally, existing structural chromosomal aberrations may predispose to aneuploidy for chromosomes other than those involved in the translocation or other structural aberrations. Irrespective of the above, prenatal diagnosis should be offered. If extreme oligoasthenoteratospermia (OAT) exists, or if the number of miscarriages keeps increasing, *in vitro* fertilization and preimplantation diagnosis (PGD) may help the couples to conceive.[46–49]

Yq11 microdeletions (AZFa, AZFb, AZFc) are deletions of the long arm of the Y chromosome that are not visible on the conventional karyotype because they are too small. Molecular techniques must be used to detect the deleted DNA stretches containing a number of genes under study in male infertility. These deletions, when transmitted from fathers to sons, will most probably cause infertility in the next generation (see Chapter 34).[50–53]

Monogenic diseases

Numerous monogenic conditions exist that impair sex differentiation as well as reproduction, or interfere only with reproduction. A few are listed in Table 2. In the male, these include congenital bilateral absence of the vas deferens (CBAVD),[11,12] with or without cystic fibrosis (CF), Kallmann syndrome,[17,18] Kennedy disease,[19] myotonic dystrophy,[20,21] primary ciliary dyskinesia or immotile cilia syndrome,[22,23] Noonan syndrome,[24] and hemochromatosis.[25,26] These and more conditions are discussed elsewhere (see Chapter 34). Because of its relatively high incidence, CF is discussed in more detail here. Cystic fibrosis is a common autosomal recessive disorder of widely variable phenotype. The disorder is usually caused by mutations in the cystic fibrosis transmembrane regulator (*CFTR*) gene, located on chromosome 7q. Some affected individuals have severe pulmonary and pancreatic disease, and survive only a few decades, or less. Others are less severely affected. Of greatest relevance to this book is that, in the absence of renal anomalies, congenital bilateral absence of the vas deferens (CBAVD) is usually caused by mutations in the *CFTR* gene. One or more of the two dysfunctional alleles required to produce CBAVD are different from those alleles that cause CF with pulmonary and pancreatic disease.

Population screening to detect couples at risk for (severe) cystic fibrosis is practiced world-wide in countries in which disease incidence is high. In the United States, the disease incidence is 1 per 2500, and there are over 30 000 affected individuals and 8 000 000 heterozygotes. In persons of Northern European extraction, the CF heterozygote frequency is 1 per 29. Using the panel of alleles in Table 3, population screening will detect 80% of CF heterozygotes.[13] In Ashkenazi Jews, the heterozygote frequency and detection rate are 1/29 and 97%, respectively; in Hispanics, 1/46 and 57%; and in African–Americans, 1/65 and 69%. Since 2001 it has been standard in the United States to offer CF screening to all pregnant women who are white (especially of Northern European extraction) or Askhenazi Jewish. In other ethnic groups, screening should be 'made available'.[14]

The most common mutations in both CF with pulmonary and pancreatic disease and CBAVD are

ΔF508 and *W1281X*. In CF characterized by pancreatic and pulmonary disease, homozygosity often (50%) exists for *ΔF508*. In other cases, compound heterozygosity exists for two alleles conferring equal severity. In CBAVD, in contrast, one or more dysfunctional alleles confer a less deleterious phenotype.[15] *R117H* is one such mutation. Not infrequently, only one or not even a single *CFTR* mutation in CBAVD is detectable using panels encompassing up to 70 of the nearly 1000 reported CF mutations. In some cases of CBAVD, a DNA polymorphism called *5-T* (five thymines) may be present *trans* to an unequivocal *CFTR* mutation such as *ΔF508*. The *5-T* polymorphism results in decreased stability and aberrant processing of the *CFTR* mRNA transcript, resulting in levels of translated protein having 10% of expected activity. The basis is aberrant exon–intron splicing results. This low level of gene activity is sufficient to prevent pulmonary and pancreatic abnormalities but not sufficient to avoid CBAVD. Thus, *5-T* screening is generally recommended in CBAVD, but not necessarily in the general population in which the purpose of screening is to detect severe disorders. When *5-T* is *cis* (located on same chromosome) to *R117H*, however, a more severe phenotype can result; thus, in general population screening, *5-T* analysis is recommended as a reflex (automatic) test when *R117H* is detected. In the absence of *R117H* or other deleterious alleles, *5-T* analysis is not recommended because the fetus would not usually be expected to be at risk for severe CF (pancreatic and pulmonary disease).

Offspring of ICSI pregnancies sired by a CBAVD father are obviously at risk for CF, not only CBAVD like the father but more serious phenotypes. Therefore, males with CBAVD should undergo CF screening, certainly for at least the 25 most frequent mutations as shown in Table 3. The female partner must be screened for at least the same panel of alleles. The likelihood of detecting one of these more common mutant *CFTR* alleles depends on ethnic background. Even so, finding none of the common mutations confers no certainty of a baby devoid of CF. Mutations may be missed because the CF gene is large (27 exons), and intragenic perturbations can pass undetected. Mutations may also exist in the promoter gene region or be involved in translational activities (downstream effect). Table 4 gives the residual risk (likelihood) incurred by a family consisting of one known heterozygote and one partner negative for the mutant alleles in Table 3.

If the male with CBAVD proves homozygous for two dysfunctional *CFTR* alleles and his partner is heterozygous, the likelihood of any given offspring having two mutant alleles is 50% (Figure 2). The same holds if *5-T* is identified in one or the other partner. The phenotype associated with most CBAVD-derived compound heterozygosity is likely to be less severe than *ΔF508* homozygosity. This reflects the genetic principle that the phenotype in

Table 3 Recommended core mutation panel for general population cystic fibrosis (CF) carrier screening: all the mutations listed have an incidence of 0.1% or greater in the general US population

Standard mutation panel					
ΔF508	*ΔI507*	*G542X*	*G551D*	*W1282X*	*N1303K*
R553X	*621*+1G→T	*R117H*	*1717*-1G→A	*A455E*	*R560T*
R1162X	*G85E*	*R334W*	*R347P*	*711*+1G→T	*1898*+1G→A
*2184*delA	*1078*delT	*3849*+10kbC→T	*2789*+5G→A	*3659*delC	*I148T*
3120+1G→A					
Reflex tests					
I506V, I507V, F508C*					
5-T/7-T/9-T†					

*Benign variants. This test distinguishes between a CF mutation and these benign variants; *I506V*, *I507V* and *F508C* are performed only as reflex tests for unexpected homozygosity for *ΔF508* and/or *ΔI507*

†*5-T* (five thymines) in *cis* can modify *R117H* phenotype, or alone can contribute to congenital bilateral absence of vas deferens (CBAVD); *5-T* analysis is performed only as a reflex test for *R117H*-positives. A male with *5-T* homozygosity could have a child also homozygous for *5-T* (if the partner has this allele); however, the fetus would not be expected to be at risk for pulmonary and pancreatic diseases[13]

autosomal recessive disorders showing compound heterozygosity reflects the less severe of the two alleles. The reason is that even limited gene products (for example 10–15% of normal) can suffice to avoid pulmonary and pancreatic lesions. Overall, it is statistically unlikely that the offspring of ICSI parents will be homozygous *ΔF508*, but this possibility must be excluded using the strategy outlined in Table 4.

In the female, fewer such monogenic defects are known or well characterized. Some of these are mentioned below. The hand–foot–genital syndrome is characterized by genital tract duplication owing to incomplete Müllerian fusion, urinary incontinence, small feet, short first toes and fingers, and clinodactyly, as well as X-ray anomalies of hands and feet; the syndrome is autosomal dominant with variable expression.[27] The blepharophymosis–ptosis–epicanthus inversus syndrome (BPES), which is autosomal dominant, presents with a typical facial appearance and premature menopause in women.[28] Also, in galactosemia, a rare autosomal recessive metabolic disease, premature menopause is one of the symptoms.[29] More important is the knowledge that carriers of the X-linked fragile X syndrome with a CGG trinucleotide expansion in the *FMR-1* gene seem to be prone to premature menopause.[30] Screening for such a mutation in patients with premature menopause is therefore indicated since, if present, an increased risk for children with mental retardation exists. The Nijmegen breakage syndrome is another rare autosomal recessive disorder characterized by microcephaly, progressive growth retardation and intellectual impairment, immunodeficiency and amenorrhea due to hypergonadotropic hypogonadism. Reproduction is also hampered because of cancer predisposition.[31] The androgen insensivity syndrome is due to a mutation in the X-linked androgen receptor gene.[32] Counselling in these families should focus on healthy sisters of affected individuals, because they may be carriers of the mutation and have affected sons. The likelihood that a sporadic case will represent a fresh mutation is only 1 in 3; other cases will have resulted from transmission from a heterozygous mother. Women with congenital adrenal hyperplasia due to 21β-hydroxylase deficiency, an autosomal recessive trait, also have reduced fertility.[56] They present with menstrual disturbances, which usually but not always can be overcome by adequate adrenal suppression therapy.[33] For monogenic diseases, the recurrence risk figures will depend on the mode of transmission only, or on the mode of transmission in combination with the disease frequency in the population. For autosomal dominant diseases, the recurrence risk to offspring, in general, is 50%. Variable expression, *de novo* mutations and anticipation of trinucleotide expansions should be taken into account when counselling. For X-linked recessive diseases in which the mother is known to be heterozygous, the recurrence risk to the offspring of carrier females is 25%, or 1 in 2 boys. For autosomal recessive diseases, the recurrence risk to parents of an affected child is 25%; for other family members, the recurrence risk will depend on the a priori risk of the family member being a carrier (2/3 for a sib of

Table 4 Residual risk incurred by a family consisting of one known heterozygote and one partner negative for mutant alleles. Data from reference 14

Ethnic group	*Heterozygote detection rate**	*Estimated carrier risk*	
		before test	*after negative test*
Ashkenazi Jewish	97%	1/29	~1/930
European Caucasian	80%	1/29	~1/140
African–American	69%	1/65	~1/207
Hispanic–American	57%	1/46	~1/105
Asian–American	NA	1/90	NA

*Heterozygote detection rate is based on panel in Table 3, and was expressly calculated for the US population; NA, not available

an affected child; 1/2 for a sib of a parent of an affected child), and the carrier frequency in the population, which, in turn, depends on the disease frequency. (The heterozygote frequency, $2pq$, is twice the square root of the disease frequency q^2.) These risks can be further modulated or precisely calculated if DNA tests allow one to determine carrier status within a family, and, eventually, within the population (see below and Figure 2).[13,14] In many cases, prenatal diagnosis or preimplantation genetic diagnosis can be offered (see Chapter 27).[1,4,5,57]

Multifactorial diseases

Multifactorial inheritance is presumed to apply to many of the frequent disorders interfering with normal fertility in females, such as endometriosis, polycystic ovary syndrome and premature ovarian failure. However, for most of these disorders, further studies are needed to determine clearly the cause of these heterogeneous conditions.[58–62]

Mitochondrial disorders

Mitochondrial aberrations as a cause of infertility in men have been reported, but data are still limited.[63–66]

Imprinting disorders

Imprinting disorders have not been reported as causes of infertility, but they have been mentioned as a possible consequence of artificial reproductive technology using immature sperm cells.[8,9,67]

GENETIC COUNSELLING APPROACH

History intake

In medicine, the physician will, in most instances, take care of an individual presenting with a complaint. In reproductive medicine, the physician deals, most of the time, with a couple. In medical genetics, questions can be asked or problems can be posed by an individual or a couple, but in most cases a whole family is involved or may be involved. This means that, when a history is taken, this history should be extended to the family. The only way to collect correct and sufficient information is to draw a family tree (Figures 2–5), and ask specific questions about first-degree (sibs, parents, offspring) and second-degree (grandparents, grandchildren, uncles, aunts, nieces, nephews) degree relatives. Sometimes it may be necessary to see and question other relatives, or request medical reports or charts. One limitation is that patients do not always tell everything they know, or they simply do not know. In any case, and even in a case of reproductive failure, the familial history should be taken as broadly as possible, enquiring about infertility, miscarriages, malformation syndromes, diseases or mental retardation, which may be hereditary and therefore of importance when trying to reproduce and especially when asking for help to reproduce.

Physical examination

Performing a general physical examination of a couple with reproductive failure is indicated, as it is for any couple intending to have children, because an examination may point to a diagnosis related to infertility (such as Turner or Klinefelter syndrome).[4,5] Also detectable are heretofore unrecognised but common genetic diseases such as myotonic dystrophy,[20,21] which is associated with infertility in males and autosomal dominant, or neurofibromatosis, also an autosomal dominant trait, not associated with infertility but characterized by variable expression with café-au-lait spots being the landmarks.[68]

Complementary tests

In medical genetics, a whole array of complementary tests may be necessary to establish a correct diagnosis, before counselling becomes possible. In this section, specific genetic tests related to reproduction in general and to reproductive disorders in particular are discussed. Selected complementary genetic tests related to reproductive failure in the male will be based on information obtained through history, physical examination and results

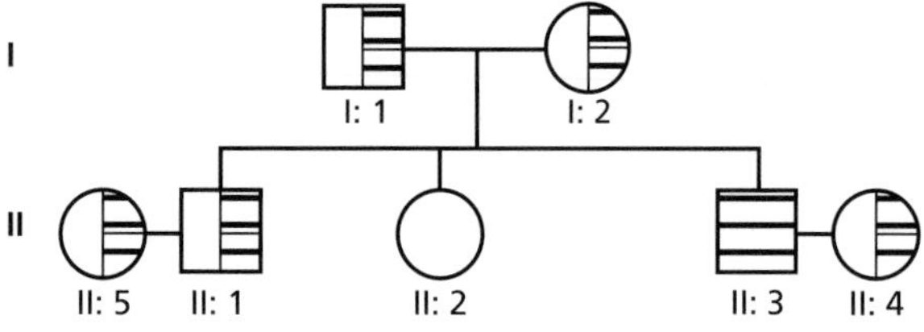

Figure 2 Cystic fibrosis family. The proband (II: 3) has cystic fibrosis (CF) as well as congenital bilateral absence of the vas deferens (CBAVD). He is homozygous for the mutation *ΔF508*; he is in reasonably good health at the age of 25 years and is married and wants children. His wife turns out to be a carrier of a CF mutation, which means that they have a 1/2 risk of having a CF child. His brother (II: 1) carries a *ΔF508* mutation and his wife is also a carrier of a CF mutation, which means that they have a risk of 1/4 of having a CF child. The CF patient needs *in vitro* fertilization/intracytoplasmic sperm injection if he wants children; preimplantation diagnosis should be offered to him. The CF carrier couple may choose between prenatal diagnosis or preimplantation diagnosis if they want to avoid the risk of having a CF child.[54,55] The healthy sister (II: 2) has a risk of 2/3 of being a carrier. Her partner to be, if a European, has a risk of 1/29 to be a carrier. Their risk of having a CF child is 2/3 x 1/29 x 1/4 = 1/174. If she turns out to be a carrier, their risk is 1/116; if he is a carrier as well, their risk is 1/4; if he is not a carrier, their risk is 1/560; if she is not a carrier, their risk is virtually 0

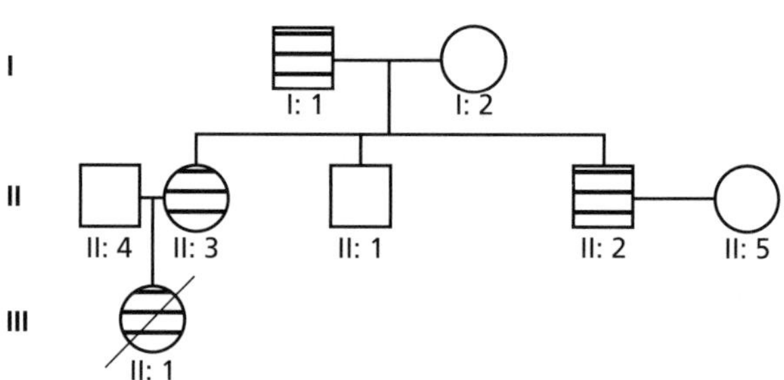

Figure 4 Myotonic dystrophy. In generation II, the proband (II: 3) had mild signs of myotonic dystrophy; she lost her first child, who had the congenital lethal form of the disease. Her brother (II: 2) had oligoasthenoteratospermia (OAT) and upon close inspection also displayed mild signs of the disease; he was also a carrier of a CTG trinucleotide expansion in the *DMPK* gene.[20,21] Preimplantation genetic diagnosis could be offered to him.[57] The father of both individuals, presenting with only cataract as a symptom, was a carrier of a smaller CTG expansion in the gene causing this autosomal dominant disease. One of his sons did not inherit the mutation

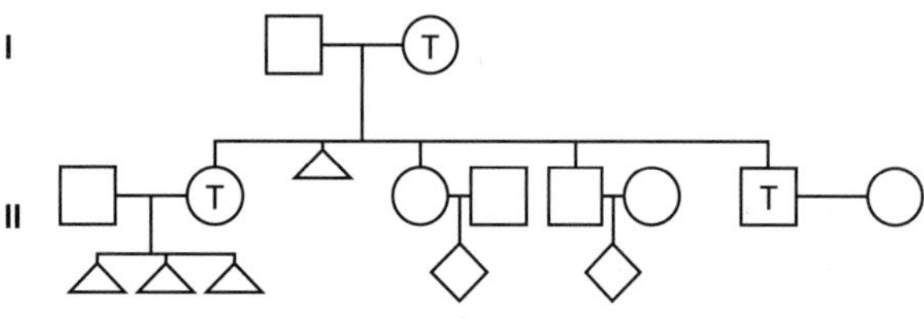

Figure 3 Familial reciprocal chromosomal translocation. The grandparents in generation I had four children and one miscarriage) triangle); two of their children in generation II were fertile and had several children spontaneously (diamond); their karyotypes were 46,XX and 46,XY normal; their daughter (square with 'T') in generation II had three miscarriages and was the carrier of a balanced reciprocal translocation 45,XX,t(7;15)(q32,q26); their son (circle with 'T') in generation II was infertile and displayed oligoasthenoteratospermia (OAT); his karyotype was also 45,XY,t(7;15)(q32,q26). It was subsequently shown that the grandmother carried the same balanced translocation as two of her children. Both translocation carrier couples may benefit from preimplantation genetic diagnosis[42–45]

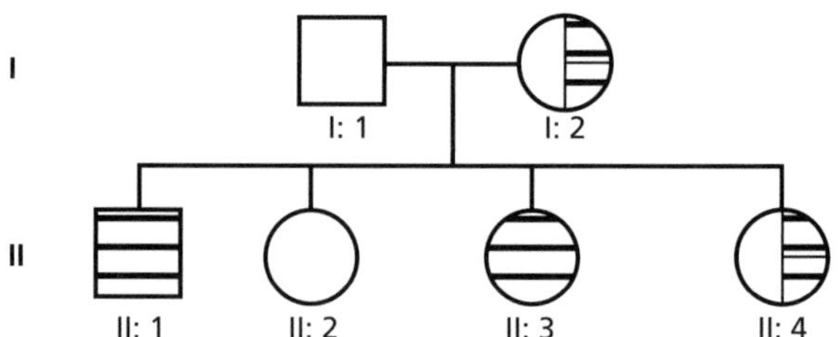

Figure 5 Fragile-X syndrome. The mother (I: 2) in generation I is a carrier of a premutation (< 200 CGG repeats). Her son (II: 1) has the fragile-X syndrome and is mentally retarded. One daughter (II: 2) is healthy and is not a carrier. One daughter (II: 3) has the full mutation and is mentally retarded. Another daughter (II: 4) is a carrier of the premutation and therefore has a risk of having an affected son. This can be avoided by offering prenatal diagnosis or preimplantation diagnosis.[57] Because this daughter is at risk for premature menopause, she should not postpone pregnancy for too long[30]

of other tests such as those to measure semen parameters and hormonal concentrations. Infertile but otherwise healthy males can present with non-obstructive or obstructive azoospermia, or oligospermia or oligoasthenoteratospermia (OAT). In males with non-obstructive azoospermia or OAT ($< 20 \times 10^6$/ml), a karyotype should be determined. Also, a search by appropriate molecular techniques for microdeletions of the AZFa,b,c region on Yq11 should be offered, certainly if the semen sample contains $< 5 \times 10^6$/ml spermatozoa.[52] In males with CBAVD without other anomalies of the urogenital tract, a search for mutations or for the *5-T* polymorphism in the *CFTR* gene should be undertaken, as discussed above.[11,12] Targeted molecular tests can be performed if other diseases such as Kallman syndrome,[17,18] Kennedy disease,[19] Steinert myotonic dystrophy,[20,21] or a mitochondrial disease[3,7] are suspected (see also Chapter 34).

Other genetic tests not related to infertility may be indicated, according to the family history or the physical examination. The aim of these investigations is to obtain a diagnosis and counsel accordingly. Finally, more sophisticated genetic tests studying male meiosis in testicular biopsies or spermatozoa, or studying Y-linked genes, belong still to the research area, and do not have to be part of routine clinical practice (see Chapter 34).[69,70] The aim of these tests is to understand more about male infertility, knowledge that may improve counselling.

Complementary molecular genetic tests related to reproductive failure in the female will also be based on information obtained through the history, the physical examination and the results of hormone tests, as well as ultrasound scanning of the ovaries. In infertile but otherwise healthy females, it may be indicated to determine a karyotype, although no convincing data exist to show that an increased incidence of aberrations occurs in this patient population. In infertile females presenting with associated symptoms such as premature ovarian failure, streak ovaries or other physical or developmental problems, specific genetic tests might include more elaborate karyotyping, analysis of the *FMR-1* gene[30] or, exceptionally, the *FOXL-2* gene (suspected BPES)[28] or the *GALT* gene (galactosemia).[29] The same is true for congenital adrenal hyperplasia[56] and androgen insensitivity, although the latter test is still available only on a research basis.[32] A mandatory test done in couples presenting with recurrent miscarriages is a karyotyping of both partners, searching for structural aberrations such as translocations or inversions (Table 1). Genetic tests based on history and physical examination but not related to infertility may also be indicated. Finally, genetic screening tests may be indicated for relatively frequent and severe diseases according to the ethnic origin of the patient, such as cystic fibrosis, Tay–Sachs disease, β-thalassemia or sickle cell disease. For CF carrier screening, precise guidelines in the United States have recently been codified, and could be useful in any venue:[16]

(1) CF carrier screening should be offered to all couples with a positive family history of CF. This extends to all Caucasian couples of European or Ashkenazi Jewish descent planning a pregnancy or seeking prenatal care. Ideally, screening is performed prior to conception or during the first or early second trimester. Information about CF screening should be provided to patients in other ethnic and racial groups. Counselling and screening should be made available to individuals in these lower-risk groups (Blacks, Hispanics and Asians) upon their request.

(2) Counselling and standardized written educational material or other formats (for example videos or interactive computer programs) can be used to inform the woman and, whenever possible, her partner. Should her partner not accompany the woman to a prenatal or preconception visit, written educational material should be provided for her to give to her partner. Women and their providers may elect to use either a simultaneous or sequential carrier screening strategy for CF. Simultaneous testing may be particularly important when there are time constraints for a decision regarding prenatal genetic diagnosis or termination of an affected pregnancy. At a minimum, the panel of 25 mutations cited in Table 3 should be included in CF screening. Provider educa-

tional materials, patient pamphlets and model informed consent forms are available.

3) Counselling for couples with positive/negative, positive/untested and positive/positive screening may require special knowledge of CF concerning range of severity, prognosis, treatment options and calculation of genetic risk. Geneticists may be needed in these scenarios, and will almost always be needed if positive/positive screening results exist. One must derive the appropriate figures and provide detailed counselling concerning prognosis and reproductive options (Figure 2). A vexing situation may involve decisions concerning disclosure of results to relatives, who could also be heterozygotes.

Thus, in general, the availability of molecular screening tests should be discussed and offered before starting any assisted reproduction treatment, to detect couples in whom both partners are carriers and consequently have a risk of 1/4 of having an affected child, and for whom prenatal diagnosis or preimplantation diagnosis can be offered.[16,71]

Diagnosis and counselling

Once the necessary information has been collected, a conclusion can be drawn. In many cases a clear-cut diagnosis will be established. Still, on too many occasions, reproductive failure will remain unexplained. In this case, genetic counselling will be limited to telling the couple that, although no cause of their reproductive failure could be uncovered, the cause can still be genetic and therefore transmittable to their offspring if their treatment is successful.[72] And, of course, in that case the offspring can most probably benefit from a similar treatment. If the treatment is *in vitro* fertilization (IVF) or IVF with intracytoplasmic sperm injection (ICSI), information should be given concerning possible problems in offspring in cases of twins and especially in cases of high-order multiple pregnancies; however, this is actually reproductive counselling rather than genetic counselling. Concerning the outcome of IVF with ICSI, the existing data indicate that the technique is, in general, a safe technique.[73] The major malformation rate (defined as malformations causing functional impairment or requiring surgical correction) of 2.3% in 1987 ICSI children is similar to that found in most assisted reproduction surveys as well as in the general population registries. However, the frequency of *de novo* chromosomal aberrations observed in fetuses through prenatal diagnosis is higher (1.66%) after ICSI than expected, compared with a control newborn population. Some autosomal trisomies may be associated with advanced maternal age, but certainly not all aberrations are related to maternal age (mean 32.5 years). Sex chromosomal aberrations in ICSI occur at a rate in the order of 0.83%, four-fold higher than the 0.2% of sex chromosomal aberrations found in a newborn population. The frequency of structural chromosomal aberrations, mainly balanced translocations, of 0.36% is also increased, compared with the incidence of 0.07% in the newborn population.[73] Existing data, not yet published, show that especially children born from men with $< 20 \times 10^6$ spermatozoa/ml display *de novo* chromosomal aberrations. Based on these and other data, the most attractive hypothesis is that this increase in chromosomal anomalies is related to sperm quality, rather than to the ICSI technique itself. For couples with reproductive disorders in which a genetic defect was identified, specific counselling is possible and mandatory. Moreover, the options for treatment should be discussed in terms of its efficiency and the possible risks to the offspring.

Psychological defenses

Psychological defenses permeate genetic counselling. If not appreciated, these defenses can impede the entire counselling process. As long as the anxiety level remains low, comprehension of information is usually not impaired. Anxiety is low in couples counselled for advanced maternal age or for an abnormality in a distant relative. Couples who have experienced a stillborn infant, an anomalous child or multiple repetitive abortions are more anxious. Their ability to retain information may be hindered.

For example, couples experiencing abnormal pregnancy outcomes manifest with the same grief reactions that occur after the death of a loved one: denial, anger, guilt, bargaining, resolution. One should recall this sequence and not attempt definitive counselling immediately after the birth of an abnormal neonate. Parents must be supported at that time, and the obstetrician should avoid discussing specific recurrence risks for fear of adding to the immediate burden. After 4–6 weeks, the couple will have begun to cope and be more receptive to counselling.

An additional psychological consideration is that of parental guilt. One naturally searches for exogenous factors that might have caused an abnormal outcome. In the process of such a search, guilt may arise. Conversely, a tendency to blame the spouse may arise. Usually guilt or blame is not warranted, but occasionally 'blame' is realistic (for example, in autosomal dominant traits). Fortunately, most couples can be assured that nothing could have prevented an abnormal pregnancy.

The psychological defenses described above help to explain the failure of ostensibly intelligent and well-counselled couples to comprehend genetic information readily.

SUMMARY AND CLINICAL RELEVANCE

Couples who want to reproduce wish to have healthy children. In the case of reproduction failure, they need medical intervention to reach this target. Good medical practice must follow rules. In assisted reproduction, it becomes even more important than with natural conception to pinpoint eventual genetic risks. Those genetic risks should be discussed before treatment is begun, to allow the couple to make informed choices and decisions.

The more common situations can be managed in the following ways:

(1) Turner patients with a 45,X karyotype or a variant karyotype may benefit from oocyte donation.[42] However, before treating, possible associated problems occurring in Turner syndrome should be evaluated.

(2) Klinefelter patients with a 47,XXY karyotype or a variant karyotype may be helped by sperm retrieval from testicular tissue. If spermatozoa are found, they can be used to fertilize oocytes of their partner using ICSI.[43] Before transferring embryos into the uterus, it is preferable to evaluate the sex chromosomes through preimplantation genetic diagnosis.[44] The couple should know that the success rate of this approach remains low, and depends on the number of spermatozoa found. Availability of donor sperm as a back-up should be discussed beforehand.

(3) Male patients with a 47,XYY karyotype or female patients with a 47,XXX karyotype may present with decreased fertility. In some cases, they may need IVF. Preimplantation genetic diagnosis may also be offered to evaluate the sex chromosomes of the embryos before transfer.

(4) Structural chromosomal aberration may be present in women with reproductive failure. In men, these aberrations are often associated with OAT. Otherwise, the couple in whom one or the other partner is a carrier may suffer from recurrent miscarriages. Nowadays, preimplantation genetic diagnosis should be offered to these couples to increase the success rate of the IVF/ICSI treatment on the one hand, and to avoid the birth of a child with an unbalanced karyotype on the other[46–49] (Figures 1 and 3).

(5) Men with microdeletions of Yq11 will obligatorily transmit these deletions to their sons, who will presumably also be infertile.[53] This information must be discussed with the couple, who might choose to have preimplantation genetic diagnosis to select only female embryos.[57]

(6) Men with CBAVD with or without CF must be told that the *CFTR* gene of their partner should be screened to detect for mutations, and to know before treatment what their risk is of having a child with CF. This risk can be as high as 1/2 or 1/4, and in these situations

preimplantation genetic diagnosis can be offered (Figure 2).[53,54,57] In situations in which the man has two *CFTR* mutations and the risk of his partner being a carrier has been reduced to 1/140 because screening found no mutation, their residual risk will still be in the order of 1/280. The choice to continue with the treatment should then be left to the couple. In other situations, the risk may be different.[74]

(7) Examples of other possible counselling situations are discussed in Figures 4 and 5.

To couples who need assisted reproduction but who have no increased genetic risk or for whom no increased risk could be determined, available information on the outcome of these novel treatments should be given. In particular, the slight increase in chromosomal aberrations seen in children after ICSI should be discussed, to allow them to decide whether or not to opt for prenatal diagnosis.[73]

FUTURE PERSPECTIVES

Taking into account the ever-increasing knowledge, as well as the ever-increasing technical possibilities, it will become possible to identify more and more genetic defects, allowing more adequate counselling. Also, more short- and long-term data on the outcome of assisted reproductive technologies (ART) will hopefully allow one to reassure couples with reproductive disorders, or pinpoint the areas of problems such as possible imprinting defects with the use of immature sperm cells.[8,9,67]

Concerning preimplantation genetic diagnosis for aneuploidy screening, it is premature to conclude that all couples undergoing assisted reproductive techniques should benefit from this procedure. Data obtained through ongoing studies should be further collected and analyzed before one can conclude that implantation rates increase, miscarriage rates decrease and the birth of children with trisomies is avoided.[75–78] Limitations of this current approach are that embryos selected for transfer have been analyzed for only 7–9 chromosome pairs, usually at least 13, 16, 18, 21, 22, X and Y. No information is available on the other chromosomes, although the use of new approaches (such as comparative genomic hybridization) could allow analysis of all chromosomes at the single-cell level.[79–81]

REFERENCES

1. Harper P. *Practical Genetic Counselling*, 5th edn. Oxford: Butterworth-Heinemann, 1999.
2. International Human Genome Sequencing Consortium. Initial sequencing and analysis of the human genome. *Nature* 2001 409: 860–921.
3. Leonard JV, Schapira AHV. Mitochondrial DNA defects. *Lancet* 2000 355: 299–304.
4. Rimoin DL, Connor JM, Peyritz RE. In Emery, Rimoin. *Principles and Practice in Medical Genetics*, 3rd edn. Edinburgh: Churchill Livingstone 1997.
5. Geletirter TD, Collins FS, Ginsburg D. *Principles of Medical Genetics*, 2nd edn. Baltimore, MD: Williams & Wilkins, 1998.
6. McKusick V. Mendelian inheritance in man. www.ncbi.nlm.nik.gov/OMIM/searchomim.html.
7. Leonard JV, Sapira AHV. Mitochondrial respiratory chain disorders. II: Neurodegenerative disorders and nuclear gene defects. *Lancet* 2000 355: 389–394.
8. Feil R, Khosla S. Genomic imprinting in mammals. *Trends Genet* 1999 15: 431–435.
9. Pfeifer K. Mechanisms of genomic imprinting. *Am J Hum Genet* 2000 67: 777–787.
10. Cassidy SB. Prader–Willi syndrome. *J Med Genet* 1997 34: 917–923.
11. Lissens W, Mercier B, Tournaye H *et al.* Cystic fibrosis and infertility caused by congenital bilateral absence of the vas deferens and related clinical entities. *Hum Reprod* 1996 11: 55–80.
12. Cuppens H, Lin W, Jaspers M *et al.* Polyvariant mutant cystic fibrosis conductance regulator genes. The polymorphic (TG)m locus explains the partial penetrance of the T5 polymorphism as a disease mutation. *J Clin Invest* 1998 101: 487–496.
13. Grody WW, Cutting GR, Katherine W *et al.* Laboratory standards and guidelines for population-based cystic fibrosis carrier screening. *Genet Med* 2001 3: 149–154.
14. Grody WW, Desnick RJ. Cystic fibrosis population carrier screening: here at last – are we ready? *Genet Med* 2001 3: 87–90.
15. Chillon M, Casals T, Mercier B *et al.* Mutations in the cystic fibrosis gene in patients with congenital

absence of the vas deferens. *N Engl J Med* 1995 332: 1475–1480.

16. ACOG/ACMG/NIH. *Preconception and Prenatal Carrier Screening for Cystic Fibrosis*. Clinical Laboratory Guidelines. Washington, DC: American College of Obstetricians and Gynecologists, 2001.
17. Rugarli El, Ballabio A. Kallmann syndrome. From genetics to neurobiology. *J Am Med Assoc* 1993 270: 2713–2716.
18. Butcher D, Behre HM, Kliesh S *et al.* Pulsatile GnRH or human chorionic gonadotropin/human menopausal gonadotropin as effective treatment for men with hypogonadotropic hypogonadism: a review of 42 cases. *Eur J Endocrinol* 1998 139: 298–303.
19. Igarashi S, Tanno U, Onodera O *et al.* Strong correlation between the number of CAG repeats in androgen receptor genes and the clinical onset features of spinal and bulbar atrophy. *Neurology* 1992 42: 2300–2302.
20. Mahadevan M, Tsilfidis C, Sabourin L *et al.* Myotonic dystrophy mutation: an unstable CTG repeat in the 3' untranslated region of the gene. *Science* 1992 255: 1253–1255.
21. Hunter A, Tsilfidis C, Mettler G *et al.* The correlation of age of onset with CTG trinucleotide repeat amplification in myotonic dystrophy. *J Med Genet* 1992 29: 774–779.
22. Afzelius BA. Immotile cilia syndrome: past, present, and prospects for the future. *Thorax* 1998 53: 894–897.
23. Pennarun G, Escudier E, Chapelin C *et al.* Loss-of-function mutations in a human gene related to *Chlamydomonas reinhardtii* dynein IC78 result in primary ciliary dyskinesia. *Am J Hum Genet* 1999 65: 1508–1519.
24. Elsawi MM, Pryor JP, Klufio G, Barnes C, Patton MA. Genital tract function in men with Noonan syndrome. *J Med Genet* 1994 31: 468–470.
25. Cazzola M, Ascari E, Barosi G *et al.* Juvenile poliopatic haemochromatosis. A life-threatening disorder presenting as hypogonadotropic hypogonadism. *Hum Genet* 1983 65: 149–154.
26. Lamon JM, Marnick SP, Roseblatt R *et al.* Idiopathic hemochromatosis in young female: a case study and review of the syndrome in young people. *Gastroenterology* 1979 76: 178–183.
27. Mortlock D, Innis J. Mutation of HOXA13 in hand–foot–genital syndrome. *Nat Genet* 1997 15: 179–181.
28. Crisponi L, Dejana M, Lui A *et al.* The putative forkhead transcription factor FOXL2 is mutated in blepharophimosis/ptosis/epicanthus inversus syndrome. *Nat Genet* 2001 27: 159–166.
29. Kaufman F, Kogut MD, Donnel GW *et al.* Ovarian failure in galactosemia. *Lancet* 1979 2: 737–738.
30. Allingham-Hawkins DJ, Babul Hirji R, Chitayati D *et al.* Fragile X premutation is a significant risk factor for premature ovarian failure: the international collaboration POF in fragile X study preliminary data. *Am J Med Genet* 1999 83: 322–325.
31. Matsuura S, Tauchi H, Nakamura A *et al.* Positional cloning of the gene for Nijmegen breakage syndrome. *Nat Genet* 1998 19: 179–181.
32. Griffin JE. Androgen resistance – the clinical and molecular spectrum. *N Engl J Med* 1992 326: 611–618.
33. Holmes-Walter D, Conway G, Honour J *et al.* Menstrual disturbance and hypersecretion of progesterone in women with congenital adrenal hyperplasia due to 21-hydroxylase deficiency. *Clin Endocrinol* 1995 43: 291–296.
34. Van de Velde E, Staquet M, Breynaert R *et al.* La descendance des mères trisomiques 21. *Am J Hum Genet* 1973 21: 187–206.
35. Weil D, Wang I, Dietrich A *et al.* Highly homologous loci on the X and Y chromosomes are hotspots for ectopic recombination resulting in XX maleness. *Nat Genet* 1994 7: 414–419.
36. Ferguson-Smith MA, Goodfellow PN. SRY and primary sex-reversal syndromes. In Scriver CR, Beaudet AL, Sly WS, Valle D, eds. *The Metabolic and Molecular Bases of Inherited Disease*. New York: McGraw-Hill, 1995: 739–748.
37. Simpson JL. Pregnancies in women with chromosomal abnormalities. In Schulman JD, Simpson JL, eds. *Genetic Diseases in Pregnancy: Material Effects and Fetal Outcome*. New York: Academic Press, 1981: 439–471.
38. Dewhurst J. Fertility in 47,XXX and 45,X patients. *J Med Genet* 1978 15: 132–135.
39. Martini E, Geraedts JPM, Liebaers I *et al.* Constitution of semen samples from XYY and XXY males as analysed by *in-situ* hybridization. *Hum Reprod* 1996 11: 1638–1643.

40. Van Assche E, Bonduelle M, Tournaye H *et al.* Cytogenetics of infertile men. *Hum Reprod* 1996 11: 1–26.
41. Gabriel-Robez O, Rumpler Y. The meiotic pairing behaviour in human spermatocytes in carriers of chromosome anomalies and their repercussions on reproductive fitness. II. Robertsonian and reciprocal translocations. A European collaborative study. *Ann Génét* 1996 39: 17–25.
42. Pados G, Camus M, Van Steirteghem A *et al.* The evolution and outcome of pregnancies from oocyte donation. *Hum Reprod* 1994 9: 538–542.
43. Tournaye H, Staessen C, Liebaers I *et al.* Testicular sperm recovery in nine 47,XXY Klinefelter patients. *Hum Reprod* 1996 11: 1644–1649.
44. Staessen C, Coonen E, Van Assche E *et al.* Preimplantation diagnosis for X and Y normality in embryos from three Klinefelter patients. *Hum Reprod* 1996 11: 1650–1653.
45. Stengel-Rutkowski S, Sten J, Gallano P. Risk estimates in balanced parental reciprocal translocations. Analysis of 120 pedigrees. *Monogr Annal Génét.* Paris: Expansion Scientifique Française, 1988.
46. Van Assche E, Staessen C, Vegetti W *et al.* Preimplantation genetic diagnosis and sperm analysis by fluorescence *in situ* hybridization for the most common reciprocal translocation t(11;22). *Mol Hum Reprod* 1999 5: 682–690.
47. Escudero T, Lee M, Carrel D *et al.* Analysis of chromosome abnormalities in sperm and embryos from two t(13;14)(q10;q10) carriers. *Prenat Diagn* 2000 20: 599–602.
48. Munné S, Sandalinas M, Escudero T *et al.* Outcome of preimplantation genetic diagnosis of translocations. *Fertil Steril* 2000 73: 1209–1218.
49. Scriven PN, O'Mahony F, Bickerstaff H *et al.* Clinical pregnancy following blastomere biopsy and PGD for a reciprocal translocation carrier: analysis of meiotic outcomes and embryo quality in IVF cycles. *Prenat Diagn* 2000 20: 587–592.
50. Vogt PH. Human chromosome deletions in Yq11, AZF candidate genes and male infertility: history and update. *Mol Hum Reprod* 1998 4: 739–744.
51. Van Landuyt L, Lissens W, Stouffs K *et al.* Validation of a simple Yq deletion screening programme in an ICSI candidate population. *Mol Hum Reprod* 2000 6: 291–297.
52. Simoni M, Bakker E, Eurling MC *et al.* Laboratory guidelines for molecular diagnosis of Y-chromosomal microdeletions. *Int J Androl* 1999 22: 292–299.
53. Page DC, Silber S, Brown LG *et al.* Men with infertility caused by AZFc deletion can produce sons by intracytoplasmic sperm injection but are likely to transmit the deletion and infertility. *Hum Reprod* 1999 14: 1722–1726.
54. Goossens V, Sermon K, Lissens W *et al.* Clinical application of preimplantation genetic diagnosis for cystic fibrosis. *Prenat Diagn* 2000 20: 571–581.
55. Dreesen JC, Jabos LJ, Bras M *et al.* Multiplex PCR of polymorphic markers flanking the CFTR-gene; a general approach for preimplantation genetic diagnosis of cystic fibrosis. *Mol Hum Reprod* 2000 6: 391–396.
56. Mulaikal RM, Migeon CJ, Rock JA. Fertility rates in female patients with congenital adrenal hyperplasia due to 21-hydroxylase deficiency. *N Engl J Med* 1987 316: 178–182.
57. Vandervorst M, Staessen C, Sermon K *et al.* The Brussels experience of more than 5 years of clinical preimplantation genetic diagnosis. *Hum Reprod Update* 2000 6: 364–373.
58. Simpson JL. Genes and chromosomes that cause female infertility. *Fertil Steril* 1985 44: 725–739.
59. Bischoff F, Simpson JL. Heritability and molecular genetic studies of endometriosis. *Hum Reprod Update* 2000 6: 37–44.
60. Campbell I, Thomas E. Endometriosis candidate genes. *Hum Reprod Update* 2001 7: 15–20.
61. Legro R, Spielman R, Urbanek M *et al.* Phenotype and genotype in polycystic ovary syndrome. *Rec Prog Horm Res* 1998 53: 217–256.
62. de Bruin J, Bovenhuis H, van Noord P *et al.* The role of genetic factors in age at natural menopause. *Hum Reprod* 2001 16: 2014–2018.
63. Folgero T, Bertheussen K, Lindal S *et al.* Mitochondrial disease and reduced sperm motility. *Hum Reprod* 1993 8: 1863–1868.
64. Lestienne P, Reynier P, Chretien ME *et al.* Oligoasthenospermia associated with multiple mitochondrial DNA rearrangements. *Mol Hum Reprod* 1997 3: 811–814.
65. Kao SH, Chao HT, Wei YH. Multiple deletions of mitochondrial DNA are associated with the decline of motility and fertility of human spermatozoa. *Mol Hum Reprod* 1998 4: 657–666.

66. St John JC, Cooke ID, Barratt CL. Mitochondrial mutations and male infertility. *Nat Med* 1997 3: 124–125.
67. Manning M, Lissens W, Bonduelle M *et al.* Study of DNA-methylation at chromosome 15q11-q13 in children born after ICSI reveals no imprinting defects. *Mol Hum Reprod* 2000 11: 1049–1053.
68. Dugoff L, Sujansky E. Neurofibromatosis type I and pregnancy. *Am J Med Genet* 1996 66: 7–10.
69. Martin R. Genetics of human sperm. *J Assist Reprod Genet* 1998 15: 240–245.
70. Barlow AL, Hulten MA. Combined immunocytogenetic and molecular cytogenetic analysis of meiosis in human spermatocytes. *Chromosome Res* 1996 4: 562–573.
71. Paw BH, Tieu PT, Kaback MM, Lim J, Neufeld EF. Frequency of three *Hex A* mutant alleles among Jewish and nonJewish carriers identified in a Tay–Sachs screening program. *Am J Hum Genet* 1990 47: 689–705.
72. Hargreave TB. Genetic basis of male fertility. *Br Med Bull* 2000 56: 650–671.
73. Bonduelle M, Camus M, De Vos A *et al.* Seven years of intracytoplasmic sperm injection and follow-up of 1987 subsequent children. *Hum Reprod* 1999 14 (Suppl 1): 243–264.
74. Liebaers I, Van Steirteghem A, Lissens W. Severe male factor: genetic consequency and recommendations for genetic testing. In Gardner D, Weissman A, Howels C, Shoham Z, eds. *Textbook of Assisted Reproductive Techniques*. London: Martin Dunitz, 2001 285–296.
75. Munné S, Morrison L, Fung J *et al.* Spontaneous abortions are reduced after preconception diagnosis of translocations. *J Assist Reprod Genet* 1998 15: 290–296.
76. Verlinsky Y, Cieslak J, Ivakhnenko V *et al.* Preimplantation diagnosis of common aneuploidies by first- and second-polar body FISH analysis. *J Assist Reprod Genet* 1998 15: 285–296.
77. Gianaroli L, Magli MC, Ferraretti AP *et al.* Preimplantation genetic diagnosis increases the implantation rate in human *in vitro* fertilization by avoiding the transfer of chromosomally abnormal embryos. *Fertil Steril* 1997 68: 1128–1131.
78. Gianaroli L, Magli MC, Ferraretti AP *et al.* Preimplantation diagnosis for aneuploidies in patients undergoing *in vitro* fertilization with a poor prognosis: identification of the categories for what should be proposed. *Fertil Steril* 1999 72: 837–844.
79. Verlinsky Y, Evsikov S. A simplified and efficient method for obtaining metaphase chromosomes from individual human blastomeres. *Fertil Steril* 1999 72: 1127–1133.
80. Willadsen S, Levron J, Munné S *et al.* Rapid visualization of metaphase chromosomes in single human blastomeres after fusion with *in-vitro* matured bovine eggs. *Hum Reprod* 1999 14: 470–475.
81. Wilton L, Williamson R, McBain J, Voullaire L. Use of comparative genomic hybridization to karyotype embryos prior to implantation: the achievement of a successful pregnancy. *Hum Reprod* 2001 16: 33.

Chapter 27

Preimplantation genetic diagnosis and screening

Karen Sermon and Inge Liebaers

INTRODUCTION

The major drawback of prenatal diagnosis (amniocentesis and chorionic villus sampling) is that, in the case of an affected fetus, the parents are faced with a decision regarding possible termination of pregnancy. Since couples at risk were requesting selection of healthy gametes, and in the wake of the development of techniques such as *in vitro* fertilization (IVF), fluorescent *in situ* hybridization (FISH) and polymerase chain reaction (PCR), researchers began to envisage the possibility of analyzing the genetic constitution of the embryo *in vitro*. Preimplantation genetic diagnosis (PGD) is a very early form of prenatal diagnosis: oocytes or embryos are biopsied during culture *in vitro*, a genetic diagnosis is carried out on the biopsied cells (a polar body in the case of oocytes and blastomeres in the case of embryos) and embryos shown to be free of the genetic disease under investigation are transferred to the mother. The first clinical PGD was reported by Handyside and colleagues, who described the sexing of preimplantation embryos at risk for a sex-linked disease followed by the births of several female, healthy babies.[1] A few years later, the diagnostic arsenal was completed by FISH, a method whereby fluorescently labelled, chromosome-specific probes are hybridized to metaphase or interphase chromosomes, thus allowing the sexing of embryos as well as aneuploidy screening[2,3] and PGD for inherited translocations.[4] From then on, the constant refining of methods for single cell analysis led to an exponential increase in the number of clinical PGDs.

APPLIED TECHNOLOGY

Biopsy methods

First and second polar body biopsy

The first and second polar bodies contain the genetic complement of the oocyte, and can thus be biopsied for detection of the genotype of the oocyte.[5] The first polar body allows for the detection of a possible cross-over event, while the second polar body shows which of the two chromatids was extruded after fertilization. The advantages of polar body biopsy are that more time is available for analysis, and that no manipulation of the embryo proper has to be carried out. In contrast, only the maternal contribution to the embryonic genome can be assessed. Thus, this method cannot be used when the male partner is affected with an autosomal dominant disease or carries a translocation, or for sexing in the case of a sex-linked disease. In the case of an autosomal recessive disease, half of the oocytes will be lost because they carry the affected gene, while, after embryo biopsy, only one out of four embryos will be discarded because it is affected. Technically, polar body biopsy is quite straightforward. The oocyte is held with a holding pipette with the polar body or bodies at the 12 o'clock position. Using a sharp needle, a slit is made in the zona pellucida, tangential to the polar bodies. With a thin pipette, the polar bodies are removed from under the zona, and transferred to a PCR tube or a glass slide.[6]

Cleavage-stage biopsy

Cleavage-stage biopsy is the most widely used technique. The disadvantages are possible moral, religious or legal objections to embryo manipulation and the assumed risk of damage to the embryo. On the basis of current knowledge, and after the birth of a limited but increasing number of children after PGD,[7,8] embryo biopsy seems to be safe. The advantage is, theoretically, that the genetic constitution of the embryo is completely formed, and thus comparable to genetic material obtained after prenatal diagnosis. However, concern remains about the frequent chromosomal mosaicism encountered in preimplantation embryos, and the repercussions this might have on the accuracy of PGD. Mosaicism in preimplantation embryos has already been described after karyotyping (between 23 and 40%),[9,10] and later detected by FISH in embryos biopsied during PGD and reanalyzed afterwards if they were not transferred.[11,12]

At our center, cleavage-stage biopsy is performed as described by Sermon and colleagues.[13,14] Embryos are evaluated in the morning of day 3 of culture *in vitro*. It is our policy to biopsy two cells from embryos with seven cells or more, because of a misdiagnosis that occurred after diagnosis using one cell, although other centers, especially those performing PGD for aneuploidy screening, limit the biopsy to one cell.[3,15] In the first step, a hole is made in the zona pellucida by applying a strong current of acidic Tyrode's solution or by using a diode laser. A larger, blunted pipette is inserted through the hole and one blastomere is aspirated partly and gently into the pipette, removed from the embryo and gently released into the medium. Possible compaction can be reversed by incubation of the embryos in Ca^{2+}- and Mg^{2+}-free medium, a procedure which does not seem to have an adverse effect on the developmental and implantation potential of the embryos.[13,16]

Blastocyst biopsy

Since the beginning of the development of PGD, the popularity of blastocyst biopsy has known many ups and downs. There are a number of advantages to blastocyst biopsy which explain why it has been considered:

(1) A larger amount of tissue can be obtained (typically about 10–30 cells) than at the cleavage stage, making diagnosis more reliable.[17]

(2) At the blastocyst stage, the embryo is already differentiated into the inner cell mass (the embryo proper) and the trophoblast, which is the tissue biopsied. This approach would be safer and also ethically more acceptable.

(3) Especially with the advent of sequential media, the rate of embryos that develop to the blastocyst stage can be expected to be higher, thus making blastocyst biopsy a reasonable alternative to cleavage-stage biopsy.[18]

An alternative to *in vitro* culture of blastocysts is uterine lavage after natural intercourse. This method was proposed for couples with normal fertility to avoid cumbersome and expensive assisted reproductive technology, but results have been disappointing because only a few viable blastocysts could be obtained.[19]

Blastocyst biopsy is performed at days 5–8 after fertilization. Before hatching, a slit is made, using a sharp needle, in the zona, to allow the embryo to hatch through. When a large enough portion of the blastocyst is hatched, usually after 18–24 h in culture, the extruding tissue is cut again with a sharp needle and transferred to a tube or glass slide.[17] Veiga and colleagues[18] are now carrying out blastocyst hatching and biopsy using a 1.48-μm diode laser. In this case also, the authors report no detrimental effect on the embryo: after collapse of the blastocoele, it is reformed within a few hours.

Diagnostic methods

Gender determination and diagnosis of chromosome aneuploidies

Two methods are currently in use to fix blastomeres on glass slides: the first is derived from the Tarkowski method of karyotyping embryos, using a hypotonic solution and acetic acid/methanol to fix the single blastomeres,[20] and the second uses HCl and Tween 20 to disrupt the cytoplasm of the cell.[21]

Two types of hybridization have been described: the direct method and the indirect method. In PGD, the direct method is preferred because it takes less time than the indirect method. In general, three types of probes can be used: repetitive sequences, which are most widely used in PGD,[21] painting probes used mainly on metaphase spreads and, finally, single-copy sequences inserted in cosmid and yeast artificial chromosome (YAC) vectors used especially to delineate translocation breakpoints in metaphase as well as interphase nuclei.[22] Probes can be made in-house[22] or are commercially available.[2]

In PGD for sexing and aneuploidy screening, up to five different colors are used for five different chromosomes.[15] The fluorescent dots visible in the specimen can be scored using an image analysis system, as is now performed in most centers, which allows for the detection of a larger number of colors.

Two FISH-related methods are primed *in situ* labelling (PRINS),[23] allowing for the detection of several chromosomes, and comparative genomic hybridization (CGH), whereby the whole chromosome complement can be analyzed.[24]

Diagnosis of single-gene defects

For this type of diagnosis, the PCR is used almost exclusively. PCR allows the amplification of well-defined DNA sequences enzymatically in an exponential way. PCR was the method used for the first clinical PGD: embryos were sexed by amplifying repetitive sequences on the Y chromosome.[1] Later, Holding and Monk showed that the specificity and efficiency of the PCR could be greatly enhanced by using nested PCR.[25] The amplification efficiencies given in most publications reporting this strategy surpass 90%. This strategy has been used by most authors describing single-cell PCR or PGD to date. For an extensive overview, refer to that by Lissens and Sermon.[26] More recently, use of the more sensitive fluorescent PCR was introduced by Findlay and colleagues[27] and ourselves.[13,14,28,29]

Contamination is an important problem when performing single-cell PCR: when the sample contains only two copies of the DNA under investigation, one copy of extraneous DNA can lead to misdiagnosis. Two sources of contamination can be distinguished: cellular sources, containing whole genomic DNA, and carry-over contamination, whereby product from a former PCR reaction finds its way into the reaction tube. A good overview of possible measures to be taken against contamination is given by Kwok and Higuchi.[30]

Navidi and Arnheim first described allele drop-out (ADO, i.e. the non-amplification of one allele in a diploid cell) on a completely theoretical basis.[31] Ray and co-workers were the first to describe ADO in blastomeres obtained after PGD for the *ΔF508* mutation in cystic fibrosis,[32] and ADO has since become a phenomenon recognized to occur in nearly every PCR tested.

A number of explanations for ADO have been put forward: the presence of a haploid cell in a diploid embryo, insufficient accessibility of the blastomere DNA after lysis,[33,34] incomplete denaturation of the DNA in the first part of the PCR cycle, and strand breaks or DNA damage.[35] Strategies other than raising the denaturation temperature can be used for reducing or at least detecting ADO. The first and, to date, most efficient method is fluorescent PCR.[13,14,27,28] A second way of detecting ADO is to assay two DNA sequences at a time, such as the mutation and a polymorphic marker linked to the gene, or two flanking polymorphic markers.[36] The two methods can be combined, adding significantly to the efficiency of the method.[37]

CLINICAL RELEVANCE

Gender selection using fluorescent *in situ* hybridization

Gender selection is the most widely used type of preimplantation genetic diagnosis, since the technology used is relatively simple and efficient, and accurate probes are available commercially. Two methods are in use: the FISH method, the most reliable and most widely used, and the PCR method, which has already led to two misdiagnoses.[7,38]

Aneuploidy screening using fluorescent *in situ* hybridization

This group of PGDs has become the most important, when counting the number of cycles. Two types of aneuploidy screening can be distinguished: by performing FISH on polar bodies, as practiced by the group of Verlinsky in Chicago,[39] and by performing FISH on blastomeres, as practiced, among other groups, by the group of Munné.[3,15] The rationale behind aneuploidy screening is that, since it is known that embryos from older women more frequently show chromosomal abnormalities, and that this partly explains the lower implantation rate in these patients, screening and selection for chromosomally normal embryos would improve implantation and, hence, pregnancy rate.

Preimplantation genetic diagnosis for translocations

Two approaches to PGD for translocations have been used until now: Munné and co-workers have used specific probes spanning the break-points, which is a cumbersome approach since new probes have to be designed for every new break-point,[22] and, more recently, the same authors have described the use of telomere probes.[40] Since these probes are commercially available for all chromosomes, preparatory work for the PGD of specific cases is significantly reduced.

Preimplantation genetic diagnosis for monogenic disease

Because of the greater technical complexity of single-cell PCR for the detection of monogenic diseases, fewer centers have consistently developed this type of diagnosis. This would also explain the observation that misdiagnoses have been reported exclusively after PCR.[7,41] Nevertheless, PGD for monogenic disease is a true alternative for prenatal diagnosis for couples at high risk of affected offspring. This is why several of the larger groups are constantly refining the methods used for diagnosis, and can now present a good track record of safety and efficiency.[8,36] In our own group, at the moment of writing, we can offer PGD for 17 different indications of monogenic diseases.[8] Usually, the decision to develop a test for a patient couple is taken on the basis of demand. This is why PGD for monogenic diseases is time-consuming and labor-intensive, and requires a well-developed molecular biology laboratory in conjunction with the IVF laboratory.

FUTURE PERSPECTIVES

Although it is stated at the beginning of this chapter that PGD is an early form of prenatal diagnosis, already it is clear that there are differences between PGD and prenatal diagnosis, especially in relation to ethical matters. One example of this is the work presented by the Chicago group at the 3rd International Symposium on Preimplantation Genetics, in which PGD was performed for Fanconi's anemia in conjunction with human leukocyte antigen (HLA)-typing in embryos so that the child born after treatment would be a healthy, compatible donor for its affected sibling.

Today, the reasons why patients opt for PGD, rather than prenatal diagnosis, are:

(1) When they have a concurrent fertility problem necessitating treatment by IVF;

(2) When they have experienced prenatal diagnosis followed by termination of pregnancy;

(3) When they have moral, emotional or religious objections to abortion, and see PGD as the only way to have unaffected children.[42]

As long as it is necessary to undergo a long, complex and costly treatment to obtain preimplantation embryos *in vitro* and to perform PGD, the vast majority of patients at risk for genetic disease will probably try prenatal diagnosis first. Technically, a number of novelties have been introduced, or are in the pipeline. CGH has now been described, using single blastomeres, by two different groups and seems to be the way forward for diagnosis of translocations and aneuploidy.[24] On the PCR front, duplex and multiplex PCR at the single-cell level will lead to a tremendous increase in the accuracy of diagnosis, obviating the need to biopsy two cells, which may compromise the viability of the embryo.[36,37,43]

Another widely awaited evolution, which the present authors strongly advocate, is the follow-up of babies born after PGD. Larger centers, such as our own and that of the Chicago group, have published data on a significant number of children.[8,44] More importantly, the European Society of Human Reproduction and Embryology (ESHRE) PGD Consortium provides practitioners in PGD with a forum in which to share their data, and in return they receive substantial information.[7,45] In the latest report of the Consortium, 26 centers reported on 162 babies, and no particular differences were found when children born after PGD were compared with children born after intracytoplasmic sperm injection.[7] Clearly, this follow-up is a *conditio sine qua non* to convince both policy makers and the public alike that PGD is safe and accurate, and can play an important role in the reproductive decisions of parents at risk for genetic disease. Already, the two publications of the Consortium are a valuable source of information for all those involved, for example, in counselling patients and interacting with governmental bodies.

REFERENCES

1. Handyside AH, Kontogianni EH, Hardy K, Winston RML. Pregnancies from biopsied human preimplantation embryos sexed by Y-specific DNA amplification. *Nature* 1990 344: 768–770.
2. Staessen C, Van Assche E, Joris H *et al.* Clinical experience of sex determination by fluorescent in-situ hybridization for preimplantation genetic diagnosis. *Mol Hum Reprod* 1999 5: 382–389.
3. Gianaroli L, Magli C, Ferraretti A, Munné S. Preimplantation diagnosis for anueploidies in patients undergoing *in vitro* fertilization with a poor prognosis: identification of the categories for which it should be proposed. *Fertil Steril* 1999 72: 837–844.
4. Van Assche E, Staessen C, Vegetti W *et al.* Preimplantation genetic diagnosis and sperm analysis by fluorescence *in-situ* hybridization for the most common reciprocal translocation t(11;22). *Mol Hum Reprod* 1999 5: 682–690.
5. Strom S, Rechitsky S, Cieslak J *et al.* Preimplantation diagnosis of single gene disorders by two-step oocyte genetic analysis. *J Assist Reprod Genet* 1997 14: 469.
6. Verlinsky Y, Kuliev A, eds. *Preimplantation Diagnosis of Genetic Disease: a New Technique for Assisted Reproduction*. New York: Wiley Liss, 1994.
7. European Society of Human Reproduction and Embryology PGD Consortium Steering Committee. ESHRE Preimplantation Genetic Diagnosis (PGD) Consortium: data collection II (May 2000). *Hum Reprod* 2000: 2673–2683.
8. Vandervorst M, Staessen C, Sermon K *et al.* The Brussels' experience of more than 5 years of clinical preimplantation genetic diagnosis. *Hum Reprod Update* 2000 6: 364–373.
9. Angell RR. Chromosome abnormalities in human preimplantation embryos. In *Development of Preimplantation Embryos and their Environment.* New York: Alan R. Liss, 1989: 181–187.
10. Zenzes MT, Casper RF. Cytogenetics of human oocytes, zygotes, and embryos after *in vitro* fertilization. *Hum Genet* 1992 88: 367–375.
11. Delhanty JDA, Griffin DK, Handyside AH *et al.* Detection of aneuploidy and chromosomal mosaicism in human embryos during preimplantation sex determination by fluorescent *in situ* hybridisation (FISH). *Hum Mol Genet* 1993 2: 1183–1185.
12. Munné S, Lee A, Rosenwaks Z, Grifo J, Cohen J. Diagnosis of major chromosome aneuploidies in human preimplantation embryos. *Hum Reprod* 1993 8: 2185–2191.
13. Sermon K, Goossens V, Seneca S *et al.* Preimplantation diagnosis for Huntington's disease (HD): clinical application and analysis of the HD expansion in affected embryos. *Prenat Diagn* 1998 18: 1427–1436.
14. Sermon K, Seneca S, Vanderfaeillie A *et al.* Preimplantation diagnosis for fragile X syndrome based on the detection of the non-expanded paternal and maternal CGG. *Prenat Diagn* 1999 19: 1223–1230.
15. Munné S, Magli C, Cohen J *et al.* Positive outcome after preimplantation diagnosis of aneuploidy in human embryos. *Hum Reprod* 1999 14: 2191–2199.
16. Dumoulin JC, Bras M, Coonen E *et al.* Effect of Ca^{2+}/Mg^{2+}-free medium on the biopsy procedure for preimplantation genetic diagnosis and further development of human embryos. *Hum Reprod* 1998 13: 2880–2883.
17. Varawalla NY, Dokras A, Old JM, Sargent IL, Barlow DH. An approach to preimplantation diagnosis of β-thalassaemia. *Prenat Diagn* 1991 11: 775–785.
18. Veiga A, Sandalinas M, Benkhalifa M *et al.* Laser blastocyst biopsy for preimplantation genetic diagnosis in the human. *J Assist Reprod Genet* 1997 14: 476–477.
19. Formigli L, Roccio C, Belotti G *et al.* Non-surgical flushing of the uterus for pre-embryo recovery: possible clinical applications. *Hum Reprod* 1990 5: 329–335.
20. Munné S, Grifo J, Cohen J, Weier H-UG. Chromosome abnormalities in human arrested preimplantation embryos: a multiple-probe FISH study. *Am J Hum Genet* 1994 55: 150–159.
21. Coonen E, Dumoulin JCM, Ramaekers FCS, Hopman AHN. Optimal preparation of preimplantation embryo interphase nuclei for analysis by fluorescence *in situ* hybridization. *Hum Reprod* 1994 9: 533–537.
22. Munné S, Fung J, Cassel MJ, Marquez C, Weier HUG. Preimplantation genetic analysis of translocations: case-specific probes for interphase cell analysis. *Hum Genet* 1998 102: 663–674.
23. Pellestor F, Girardet A, Andréo B, Lefort G, Charlieu JP. The PRINS technique: potential use

for rapid preimplantation embryo chromosome screening. *Mol Hum Reprod* 1996 2: 135–138.

24. Wells D, Delhanty J. Comprehensive chromosomal analysis of human preimplantation embryos using whole genome amplification and single cell comparative genomic hybridization. *Mol Hum Reprod* 2000 6: 1055–1062.
25. Holding C, Monk M. Diagnosis of β-thalassaemia by DNA amplification in single blastomeres from mouse preimplantation embryos. *Lancet* 1989 2: 532–535.
26. Lissens W, Sermon K. Preimplantation genetic diagnosis: current status and new developments. *Hum Reprod* 1997 12: 1756–1761.
27. Findlay I, Urquhart A, Quirke P *et al.* Simultaneous DNA 'fingerprinting', diagnosis of sex and single-gene defect status from single cells. *Hum Reprod* 1995 10: 1005–1013.
28. Sermon K, De Vos A, Van de Velde H *et al.* Fluorescent PCR and automated fragment analysis for the clinical application of preimplantation genetic diagnosis of myotonic dystrophy (Steinert's disease). *Mol Hum Reprod* 1998 4: 791–795.
29. De Vos A, Sermon K, Van de Velde H *et al.* Pregnancy after preimplantation genetic diagnosis for Charcot–Marie–Tooth disease type 1A. *Mol Hum Reprod* 1998 4: 978–984.
30. Kwok S, Higuchi R. Avoiding false positives with PCR (product review). *Nature* 1989 339: 237–238.
31. Navidi W, Arnheim N. Using PCR in preimplantation genetic disease diagnosis. *Hum Reprod* 1991 6: 836–849.
32. Ray P, Winston RML, Handyside AH. Single cell analysis for diagnosis of cystic fibrosis and Lesch–Nyhan syndrome in human embryos before implantation. *Miami Bio/Technology Short Reports: Proceedings of the 1994 Miami Bio/Technology European Symposium, Advances in Gene Technology: Molecular Biology and Human Genetic Disease*. 1994 5: 46.
33. Sermon K, Lissens W, Nagy ZP *et al.* Simultaneous amplification of the two most frequent mutations of infantile Tay–Sachs disease in single blastomeres. *Hum Reprod* 1995 10: 2214–2217.
34. Wu R, Cuppens H, Buyse I *et al.* Co-amplification of the cystic fibrosis ΔF508 mutation with the HLA DQA1 sequence in single cell PCR: implications for improved assessment of polar bodes and blastomeres in preimplantation diagnosis. *Prenat Diagn* 1993 1111–1112.
35. Ray PF, Handyside AH. Increasing the denaturation temperature during the first cycles of amplification reduces allele dropout from single cells for preimplantation diagnosis. *Mol Hum Reprod* 1996 2: 213–218.
36. Rechitsky S, Strom C, Verlinsky O *et al.* Accuracy of preimplantation diagnosis of single-gene disorders by polar body analysis of oocytes. *J Assist Reprod Genet* 1999 16: 192–198.
37. Dreesen J, Jacobs L, Bras M *et al.* Multiplex PCR of polymorphic markers flanking the CFTR gene; a general approach for preimplantation genetic diagnosis of cystic fibrosis. *Mol Hum Reprod* 2000 6: 391–396.
38. Hardy K, Handyside AH. Biopsy of cleavage stage human embryos and diagnosis of single gene defects by DNA amplification. *Arch Pathol Lab Med* 1992 116: 388–392.
39. Verlinsky Y, Cieslak J, Freidine M *et al.* Polar body diagnosis of common aneuploidies by FISH. *J Assist Reprod Genet* 1996 13: 157–162.
40. Munné S, Sandalinas M, Escudero T *et al.* Outcome of preimplantation genetic diagnosis of translocations. *Fertil Steril* 2000 73: 1209–1218.
41. Grifo JA, Tang YX, Munné S *et al.* Healthy deliveries from biopsied human embryos. *Hum Reprod* 1994 9: 912–916.
42. Liebaers I, Sermon K, Staessen C *et al.* Clinical experience with preimplantation genetic diagnosis and intracytoplasmic sperm injection. *Hum Reprod* 1998 13(Suppl 1): 186–191.
43. Lewis CM, Pinel T, Whittaker JC, Handyside AH. Controlling misdiagnosis errors in preimplantation genetic diagnosis: a comprehensive model encompassing extrinsic and intrinsic sources of error. *Hum Reprod* 2001 16: 43–50.
44. Strom CM, Levin R, Strom S, Masciangelo C, Kuliev A, Verlinsky Y. Neonatal outcome of preimplantation genetic diagnosis by polar body removal: the first 109 infants. *Pediatrics* 2000 106: 650–653.
45. European Society of Human Reproduction and Embryology PGD Consortium Steering Committee. ESHRE Preimplantation Genetic Diagnosis (PGD) Consortium: preliminary assessment of data from January 1997 to September 1998. *Hum Reprod* 1999 14: 3138–3148.

Genetic analysis of complex (multifactorial) diseases

Stephen Kennedy

INTRODUCTION

This chapter is intended to provide the reader with a brief introduction to the methods that are currently being used to identify susceptibility loci in complex genetic traits. Where possible, examples have been provided from the field of reproductive medicine, for example endometriosis and polycystic ovary syndrome (PCOS), to illustrate how these methods are being applied. Linkage and association studies are the complementary methods mainly used. In linkage studies, one attempts to identify genetic markers that are linked to the disease amongst affected members of families. Association studies involve demonstrating a statistically significant association between susceptibility to a disease and a specific allele, either within affected families or in whole populations.

BACKGROUND

What are complex genetic traits?

A condition can have a genetic basis owing to the inheritance of a single gene with varying degrees of penetrance and expressivity of the phenotype (Mendelian inheritance), a limited number of genes (oligogenic inheritance), or multiple genes each with a small influence (polygenic inheritance). The effect of multiple genes in the case of polygenic inheritance is usually not causative; rather, the genes predispose individuals, or make them more susceptible, to develop the condition. Examples of polygenic conditions include diabetes, asthma and hypertension. They are also termed 'complex' or 'multifactorial' traits because the phenotype is expressed as a result of interactions between allelic variations in multiple genes and each other, and between those genes and environmental factors. Unlike conditions inherited in a Mendelian fashion, complex traits are usually common in the general population. The phenotypes are typically described as quantitative (e.g. fasting insulin levels) or dichotomous (e.g. diabetes present or absent based upon a predefined threshold).

Understanding the relative contributions made by susceptibility genes and environmental risk factors in producing disease phenotypes is the challenge facing molecular geneticists and epidemiologists. It is an enormous challenge, given that the number of genes responsible for predisposing an individual to a disease, and the relative contributions of those genes and environmental risk factors, may vary considerably in different populations. The other major problem is that a consistent relationship between genotype and phenotype does not always exist. For instance, the same genotype can result in different phenotypes, depending upon the influence of chance or environmental factors. For example, an individual who inherits asthma-predisposing genes may not develop the disease unless exposed to environmental pollutants (incomplete penetrance). However, an individual may also develop a condition such as asthma as a consequence of some environmental exposure, without inheriting any predisposing genes (phenocopy).

The age of onset of a condition is another factor contributing to the variable expression of susceptibility genes; in other words, a phenotype such as Huntington's disease may not manifest itself in a genetically predisposed individual until that person reaches a certain age. Attempts at classifying such a person as affected or unaffected at too early an age will inevitably result in misclassification in the absence of a diagnostic test with sufficient predictive power. Expression of the phenotype may also be limited to a particular sex, which clearly represents a problem in reproductive medicine; it has led many researchers to seek possible male phenotypes related to conditions that affect only women, for example early balding in the male relatives of women with PCOS.

Sometimes, a phenotype can result from mutations in different genes (genetic heterogeneity) or different mutations at a single locus, such as breast/ovarian cancer at the *BRCA1* locus (allelic heterogeneity). Allelic heterogeneity does not interfere with gene mapping, because all families will show linkage to the same chromosomal region. However, genetic heterogeneity can cause problems in identifying susceptibility genes, because a chromosomal region may be linked to the phenotype in some families or ethnic populations and not in others. Finally, a condition that appears to be a complex trait may have subtypes with a Mendelian pattern of inheritance; for example, approximately

5% of Alzheimer's disease is inherited in large families as an early-onset form due to mutations in single genes.

Defining the phenotype

Clearly, it may be difficult to correlate genotype and phenotype precisely. Therefore, in setting out to determine the genetic basis of any disease, it is vital to have a reliable definition of phenotype, and this applies particularly in complex traits. Without a clear and consistent definition, it may be impossible to determine the relative contributions of genetic and environmental influences on disease etiology and progression. Defining phenotype precisely has been difficult in some conditions, particularly psychiatric disorders, because diagnoses are often subjective and they have a tendency to vary over time. This has also proved a problem in reproductive disorders such as endometriosis, given that different classification systems have been used over the past 30 years; new manifestations of the disease have been defined (e.g. subtle peritoneal lesions), and some researchers have argued that minimal endometriosis does not represent a disease state. Similarly, the definition of PCOS has varied considerably, ranging from amenorrhea/ oligomenorrhea plus hyperandrogenemia without ultrasound evidence of polycystic ovaries,[1] to polycystic ovaries identified using ultrasound plus one or more clinical or biochemical parameters suggestive of the syndrome (see reference 2 for review).

In many respects, the phenotype is easier to define in a surgically diagnosed condition such as endometriosis, because it is usually obvious whether disease is present or not at surgery, and histological confirmation of disease is often obtained as well. The only uncertainties in endometriosis concern whether all disease severities and types should be included under one phenotypic classification, and how to classify women who clearly have evidence of disease on gross inspection of the pelvis but in whom the histology is non-conclusive.

One approach to the problem of such uncertainty is to restrict the phenotypic definition to severe examples, which enables the disease to be defined as unequivocally as possible. Examples of restricted phenotypes in reproductive disorders are stage III–IV disease in endometriosis using the revised American Fertility Society (rAFS) classification system,[3] or the obese, hirsute woman with oligomenorrhea and polycystic ovaries who has raised androgen levels in PCOS. The approach presupposes that such traits are homogeneous; it is equally or more likely, however, that qualitative and not quantitative differences exist between such severe manifestations and the lesser forms of what is assumed to be the same condition. For example, women with polycystic ovaries who have normal body mass index and ovulatory cycles may have a completely different phenotype from that of women with classical Stein–Leventhal syndrome. Another approach is to analyze the data in multiple ways using different diagnostic schemes, but this increases the chance of false-positive (type 1) errors. Finally, it is possible to define all disease severities and types as a single phenotype and then sub-categorize, if necessary, at a later date on the basis of genotype findings. In this way, the results of genetic analyses lead to a new way of defining phenotype that may or may not justify the hypotheses made regarding disease classification at the outset of a study.

Familial clustering

Complex traits may not show a clear pattern of Mendelian inheritance, but they do have a tendency to cluster in families, which suggests a genetic basis. Examples in reproductive medicine include PCOS;[4] uterine fibroids[5] and endometriosis in both humans[6,7] and rhesus monkeys[8]. However, other factors can explain familial clustering, such as exposure to environmental or infectious agents. Ascertainment bias is another explanation for such findings if the diagnosis in a proband (defined as the person in the family through whom the remaining family members are ascertained) makes it more likely that other family members will be diagnosed. Bias can also occur if probands are not representative of affected individuals in the general population in terms of their symptoms and/or disease type, which may arise if they are ascertained through specialized routes such as tertiary referral clinics. In clinical practice, such patients typically have more extreme manifestations of a disease, which may make a genetic basis more likely. In ideal circumstances, therefore, families should be ascertained via community- or population-based registers, to reduce the possibility of introducing any forms of bias.

The extent to which a condition clusters within families is defined by its relative risk (λ_R). This is the risk of an affected first-degree (sib), second-degree (uncle/aunt, grandparent) or third-degree (cousin) relative having the disease, compared with the risk in the general population.

The term λ_S is used in the case of sibs, and is defined in the same way by the formula:

$$\lambda_S = \frac{\text{Frequency of condition in sibs of affected}}{\text{Frequency of condition in general population}}$$

Example for type 1 diabetes
Frequency of condition in sibs of affecteds = 6%
Frequency of condition in general population = 0.4%
$\lambda_S = 6/0.4 = 15$

Other examples of λ_S values for common conditions are those for Crohn's disease (~20), type 2 diabetes (~3) and rheumatoid arthritis (~6). In a condition such as endometriosis, λ_S is difficult to estimate, given that the prevalence in the general population is uncertain because the diagnosis is only reliably made at surgery. Nevertheless, it is reasonable to suggest that λ_S is approximately 6–9 for all disease severities, based upon the three studies in the literature that have reported relative risks in the sisters of affected women.[9–11] In women with the most severe forms of endometriosis, λ_S may be as high as 15, based upon the findings of magnetic resonance imaging in the sisters of women with rAFS stage III–IV disease[12] (Table 1). A higher value for λ_S in the sisters of women with more severe disease is consistent with the general finding that severe phenotypic types in most complex traits usually display the highest λ_S ratios. In PCOS, there is evidence from a recently published study to suggest that λ_S may be approximately 10.[13]

In complex traits, the extent to which individual genetic loci contribute to the overall value of λ_S may vary. Moreover, the contribution may be additive or multiplicative. Thus, the total value for λ_S is either the sum (additive model) or the product (multiplicative model) of the λ_S values for each individual locus. Multiplicative interactions represent an example of epistasis, a term that describes the phenomenon of the combined effect of two or more loci having an effect that could not be predicted by knowing their individual effects. Thus, a condition with an overall λ_S of 12 may have six contributing loci, each with an approximate λ_S of 1.5 (multiplicative), or three loci with λ_S values of 6, 4 and 2 (additive). The probability of detecting linkage is directly related to the size of λ_S, i.e. the higher the value of λ_S the easier it is to detect linkage. Clearly, therefore, if multiple loci are contributing to the phenotype, it becomes more difficult to detect linkage, especially if some loci have only a small effect and, as is invariably the case, the number of loci contributing to the phenotype and the type of interaction between them are unknown at the outset.

Table 1 Disease prevalence in the first-degree relatives of women with endometriosis

Authors	*Controls*	*Sisters*	*Mothers*	*Sisters or mothers*
Simpson *et al.*[11] (1980)	first-degree relatives of patients' husbands	9/153 (5.8%) vs. 1/104 (1%)	10/123 (8.1%) vs. 1/107 (0.9%)	—
Coxhead and Thomas[9] (1993)	first-degree relatives of women with a normal pelvis	—	—	6/64 (9.4%) vs. 2/128 (1.6%)
Moen and Magnus[10] (1993)*	first-degree relatives of women with a normal pelvis	25/523 (4.8%) vs. 1/169 (0.6%)	20/515 (3.9%) vs. 1/149 (0.7%)	—
Kennedy *et al.*[12] (1998)†	no controls used	—	—	5/35 (14.3%)

*Disease defined as endometriosis and/or adenomyosis; †disease defined as radiological evidence of lesions > 1 cm in diameter, excluding adenomyosis

POSITIONAL CLONING APPROACH

Principles of linkage analysis

The 'reverse genetics' or 'positional cloning' approach to unravel the genetic basis of a disease with unknown biochemical defects broadly involves:

(1) Chromosomal localization of the disease locus;

(2) Identification of the gene(s) and mutations;

(3) Determination of abnormal function associated with the mutations.

The localization of genes in monogenic disorders has mainly been achieved by studying the segregation of DNA markers with disease in selected pedigrees (linkage analysis). Thus, genes are identified by first mapping their approximate chromosomal localization using genetic markers without any knowledge of disease mechanisms (Figure 1 shows the major steps in the process). The next stages (which are not the subject of this review) involve searching public databases to identify a candidate 'positional' gene in that region – the so-called 'positional candidate' approach – often based upon some theoretical or actual knowledge of disease mechanisms, and showing that mutations in that gene occur more frequently in affected individuals than in controls.

Positional cloning relies upon the fact that reciprocal exchange of DNA sequences occurs by recombination (cross-over) when homologous chromosomes pair at meiosis. The probability that an exchange has occurred is defined by the distance between any two genetic loci. Thus, two loci (such as a disease gene and a genetic marker) that are very close together on a chromosome will rarely be separated by recombination. They will therefore tend to co-segregate within families, i.e. there will be a tendency, depending upon the disease's mode of inheritance, for affected family members to inherit from their parents the same chromosomal region in which the mutated gene and the genetic marker are located. Conversely, the further apart two loci are on a chromosome, the more likely it is that a cross-over will separate them. In other words, a disease gene and an unlinked genetic marker will segregate randomly, and their inheritance will not deviate from simple Mendelian rules.

The genetic distance between loci is measured by the recombination fraction (θ): two loci that show 1% recombination are defined as being 1 centimorgan (cM) apart on a genetic map. One cM is approximately equivalent to 1 megabase (Mb), i.e. 1 million base pairs, of DNA, although this figure varies between chromosomes and between the sexes as recombination is more frequent in female meiosis. The likelihood that two loci are linked is typically measured by the log ratio of odds (LOD) score. The LOD score is defined as the logarithm (base 10) of the odds ratio (z) between the likelihood that the loci are linked with a recombination fraction equal to θ and the likelihood that the loci are unlinked with a recombination fraction equal to 0.5:

$z(x) = \log_{10}$ [(likelihood of observing configuration of two loci within pedigree(s) if $\theta = x$)/(likelihood of observing configuration of two loci within pedigree(s) if $\theta = 0.5$)]

Log_{10} of the ratio of these likelihoods is calculated for each value of θ between 0.0 and 0.5, and each of the results obtained represents $z(x)$ (the LOD score), where x is a value of θ within the range of recombination fractions. In these circumstances, $z(x)$ is termed a two point LOD score, because linkage between two-loci (a disease locus and a marker locus) has been calculated. However, in practice, multiple markers can be studied at the same time, producing a multipoint LOD score. LINKAGE is a commonly used computer program that performs both two-point and multipoint analyses.[14]

The statistical approach requires the parameters – dominant or recessive inheritance, allele frequencies and penetrance (often age-related) – that are classically obtained from complex segregation analysis, to be specified within a model: hence the term parametric analysis. The analysis involves altering each parameter until the model with the maximum LOD score (MLS) is obtained. For Mendelian diseases, a LOD score greater than +3 is highly suggestive of linkage, and one less than –2 is evidence against linkage. A LOD score of 3 corresponds to a significance level of $\alpha = 0.05$ at a

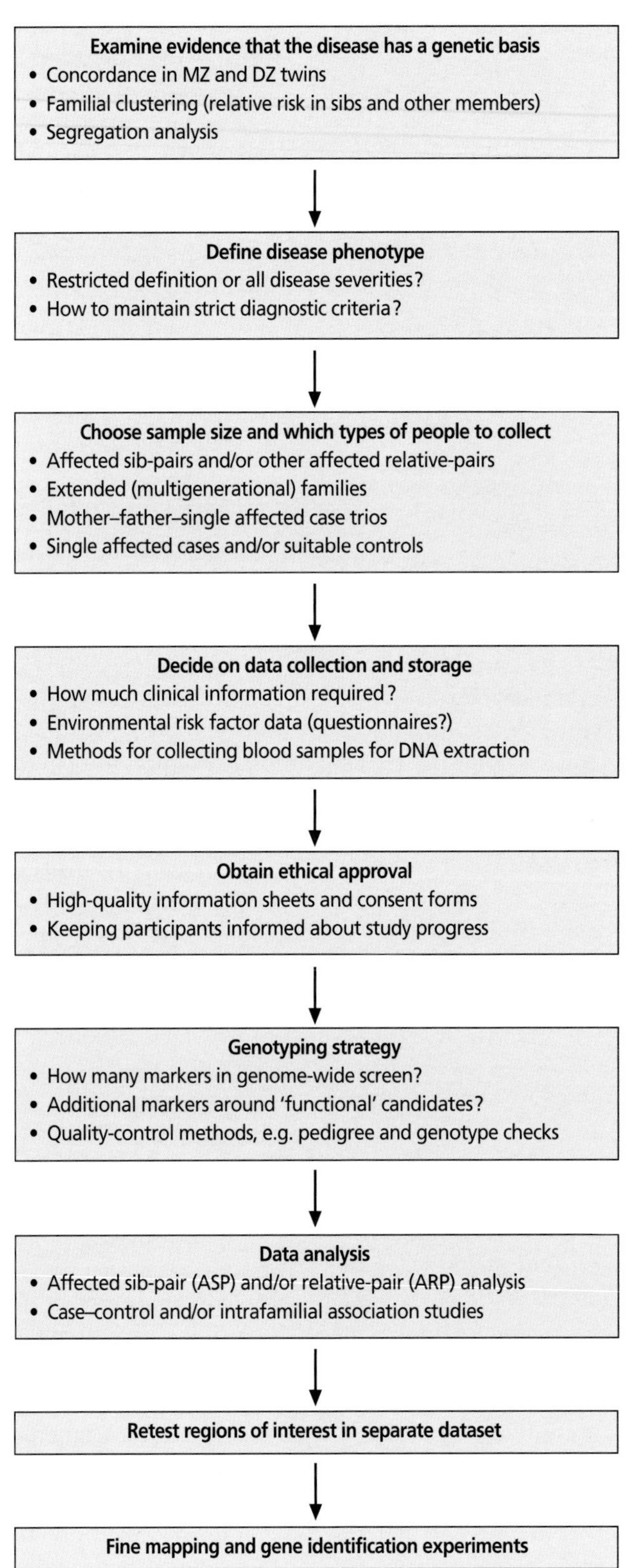

Figure 1 Outline of the major steps involved in identifying susceptibility genes in complex traits. MZ, monozygotic; DZ, dizygotic

single-point, i.e. a 0.05 chance of a false-positive result, a type I error quantified as α. The probability of false-positive and false-negative (type 2) errors inevitably rises if any of the parameters have been misspecified, particularly if the specified mode of inheritance is incorrect.

In principle, any genetic marker that is inherited in a Mendelian fashion can be used, provided that its chromosomal location is known and it is sufficiently informative or polymorphic (i.e. multiple alleles exist at the marker locus, thereby increasing the probability that any individual will be heterozygous). Restriction fragment length DNA polymorphisms (RFLPs)[15] and sequence polymorphisms defined by variable numbers of tandem repeats[16] provided a set of genetic markers that were, for the first time, sufficiently informative and evenly spaced throughout the genome, to make linkage analysis and gene mapping on a large scale feasible. Southern blotting was originally used to study such polymorphisms; since then, the use of the polymerase chain reaction (PCR) to amplify DNA from the region of interest before fragment analysis has made the process of genotyping much easier and more rapid.

For single-gene disorders, localization was a relatively straightforward process. The approach was successful in diseases such Duchenne muscular dystrophy[17] and cystic fibrosis.[18] However, it proved more difficult to apply the same statistical methods to the identification of susceptibility loci in complex traits, given that multiple genetic loci are involved. Moreover, the available markers were not sufficiently informative, and the techniques were too laborious to permit large numbers of individuals to be genotyped. The discovery of microsatellites – highly polymorphic dinucleotide, mostly (CA)*n*, tandem repeats – considerably increased the number of markers available,[19] and the development of automated DNA fragment analysis with fluorescently labelled PCR primers made high-throughput genotyping a reality[20] (Figure 2). The number of microsatellites available has since dramatically increased: in 1992, Weissenbach and colleagues reported the development of 600 microsatellite markers covering a linkage distance spanning approximately 90% of the genome.[21] Then, in 1996, a map consisting of 5264 microsatellites was published, spanning a sex-averaged genetic distance of 3699 cM with an average interval size of 1.6 cM.[22] The latest generation of markers are biallelic, single nucleotide polymorphisms (SNPs). SNPs can be studied on solid-state arrays, which allows for very high-throughput genotyping; however, their use at present is largely confined to fine-mapping experiments, not genome-wide screens.

Non-parametric linkage analysis

In contrast to the parametric methods described above, there are non-parametric methods of analysis available that do not require the mode of inheritance to be known. Data are analyzed solely on the basis of the inheritance patterns of genetic markers. The most commonly used approach to identify susceptibility loci in complex genetic traits is the affected sib-pair (ASP) method.[23] It aims to demonstrate whether identical chromosomal regions are inherited by affected sibs more commonly than would be expected by chance alone. Affected sibs and their parents are genotyped to determine how often marker alleles in the affected sibs are shared ‘identical by descent’ (IBD), i.e. inherited from a common parent. Allele sharing at a chromosomal region that is occurring more frequently than expected by chance, i.e. that is not consistent with random Mendelian segregation, is indicative of linkage. Figure 3 illustrates the basic principles of the ASP method. Assuming that four alleles (A–D) exist at a hypothetical marker locus and that both parents are heterozygous (mother A/B and father C/D), it follows that two affected sisters share the same two marker alleles IBD if they have both inherited the genotype A/C. The proportion of sharing of 0, 1 and 2 alleles from both parents, according to simple Mendelian rules of inheritance, is expected to be 25%, 50% and 25% in the absence of linkage, respectively. Similarly, there will be 50% IBD sharing from each parent of 0 or 1 allele in the absence of linkage. A number of statistical methods are available to test for linkage, which rely ideally upon unambiguous genotyping to determine the degree of allele sharing. Typically, the χ^2 goodness of fit test is used to estimate the extent to which the sharing of alleles IBD

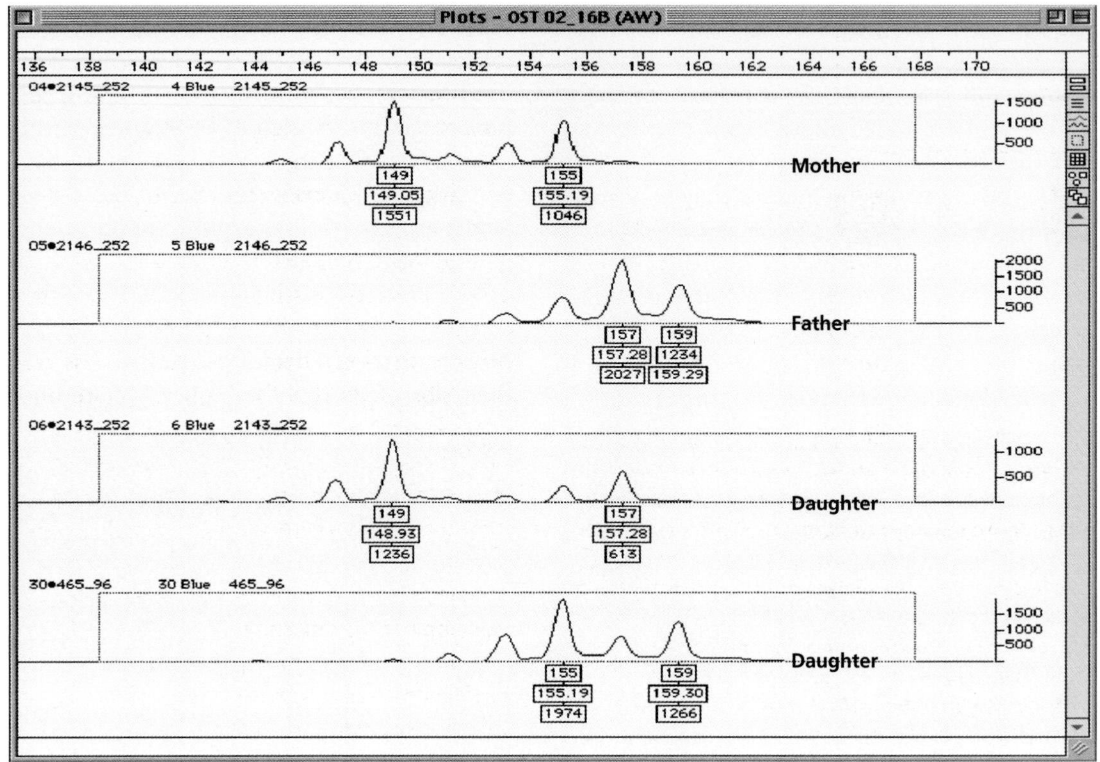

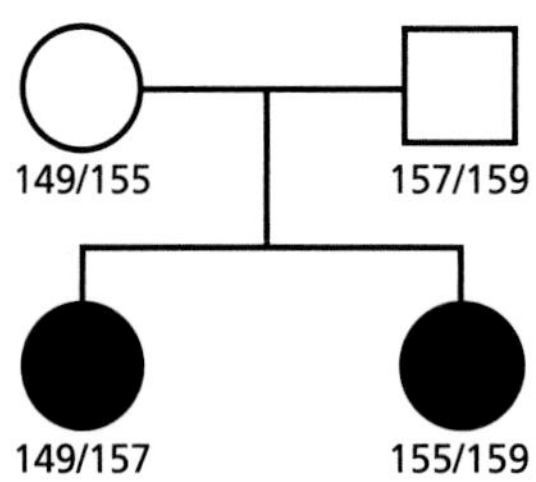

Figure 2 Output from an automated genotyping system showing four different allele sizes (149, 155, 157 and 159) at a marker locus represented by the peak fluorescent intensities for samples from a mother, father and two daughters. Each person in this affected sib-pair family is heterozygous, as illustrated in the pedigree above. Reprinted with permission from Urbanck M, Legro RS, Driscoll DA *et al.* Thirty-seven candidate genes for polycystic ovary syndrome: strongest evidence for linkage is with follistatin. *Proc Natl Acad Sci USA* 1999 86: 8573–8578. Copyright (1999) National Academy of Sciences, USA

is in excess of what would be predicted in the absence of linkage. Thus, a comparison is made between the observed percentages of sib-pairs sharing 0, 1 and 2 alleles IBD with the expected percentages 25%, 50% and 25% in the absence of linkage.

Some of the other analytical methods in use include:

(1) The percentage of sibs that share two alleles IBD can be compared with the expected percentage of 25%, or the mean value of IBD sharing can be compared with the expected value of 50%.

(2) The maximum LOD score (MLS) method utilizes genotype data from additional relatives,

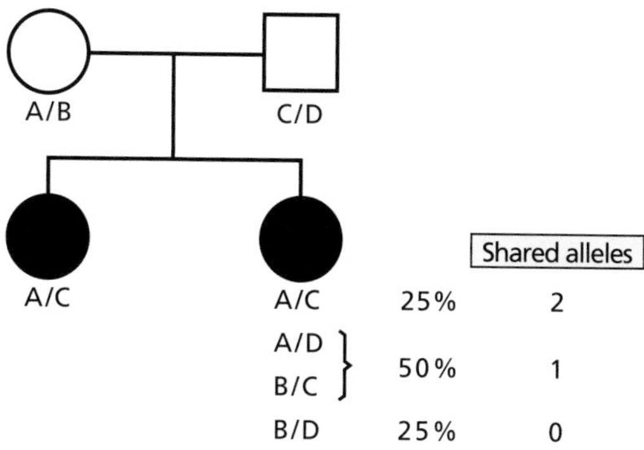

Figure 3 Allele sharing in affected sib-pairs. Two affected sisters can share 2, 1 or 0 alleles at any marker locus with 25%, 50% and 25% likelihoods in the absence of linkage, i.e. with random segregation. In the simplified example shown, there are four alleles (A, B, C and D) at a hypothetical marker locus. Both parents are heterozygous (mother's genotype A/B and father's genotype C/D). Assuming that the older affected sister has inherited the A/C genotype, then there is a 25% chance in the absence of linkage that the younger sister has the A/C genotype; a 50% chance of the A/D or B/C genotypes; and a 25% chance of the B/D genotype. Evidence of linkage of the disease gene to the marker locus would exist if the affected sisters shared the two alleles A and C, identical-by-descent (IBD) from their parents, more frequently than expected by chance alone

if parental DNA is not available in all cases, to determine the optimal IBD proportions that fit the observed data.[24,25]

(3) Non-parametric linkage (NPL) analysis allows alleles of all affected members in a pedigree to be compared simultaneously. This method is particularly useful because it allows many relatives other than sibs to be analyzed.[26]

If there are more than two affected sibs in a family, then multiple pairs arise (3 sibs = 3 pairs, 4 sibs = 6 pairs, etc.), and different formulae have been devised to take into account the fact that such pairs are not as independent as those arising in families with only two affected sibs. Typically, this involves assigning different weights to the sib-pairs depending upon the total number of affected sibs in the family, i.e. the weighting decreases as the number of affected sibs in the family rises. There are also non-parametric methods available to utilize known unaffected sibs, especially if they are extreme discordants, i.e. one sib has a normal pelvis and the other has severe endometriosis. Clearly, such methods are only useful if other sibs are prepared to undergo the diagnostic test, and if the likelihood of the unaffected sib developing the condition after undergoing the diagnostic test is low.

The ASP method can be adapted to include other affected relative-pairs (ARPs) such as first cousins. Using such relatives can be powerful: cousins sharing alleles must occur less frequently than sibs given that first cousins share one-eighth of their genetic material by chance, as opposed to sibs who share half of their genetic material. There are certain advantages in incorporating unaffected sibs in the ASP method, particularly if one or both parents are dead, in which case the unaffected sib's genotype can sometimes be used to help estimate the parental genotype(s). Methods also exist for comparing allele sharing between affected and unaffected sibs in quantitative traits such as hypertension: sibs who are extremely discordant (i.e. one sib is markedly hypertensive and the other has low blood pressure) should share very few genes IBD.

The advantages of the ASP method include:

(1) Only affected individuals need to be recruited, which is important in a disease such as endometriosis because there is no need to identify at laparoscopy those women in a family who are unaffected.

(2) Families with only two affected sisters are much easier to find than large pedigrees.

(3) The parents of women with reproductive disorders are usually alive, which is not the case in conditions such as hypertension and type 2 diabetes.

It may not always be possible to determine the amount of IBD sharing exactly if one or both parents are dead, one or both parents are homozygotes, or both parents have the same genotypes, especially if DNA from other relatives is unavailable for genotyping. In such cases, it may only be possible to determine the extent to which two sibs share the same alleles 'identical by state' (IBS) at a marker locus. Different types of analysis can then be performed to test for linkage, such as the affected pedigree member (APM) method, which only requires the affected sibs in a family to be genotyped.[27] The power of the APM method depends upon knowing the frequencies of the marker alleles in the general population, to allow the proportion of allele sharing that would be expected in the absence of linkage to be estimated. Figure 4 illustrates the basic principles. In Figure 4a, assuming that three alleles (A–C) exist at a hypothetical marker locus and that both parents are heterozygous (mother A/B and father A/C), it follows that two affected sisters share the allele A IBS if they have inherited the genotypes A/B and A/C, respectively. Only if the parental genotypes are known is it clear that no alleles are inherited IBD, because, in this case, allele A has been inherited by each sister from a different parent. In Figure 4b, the affected sisters share two alleles (A and B) IBS. However, even knowing the parental genotypes, it is not possible to determine whether 2 or 0 alleles have been inherited IBD because both parents have the same genotype (A/B).

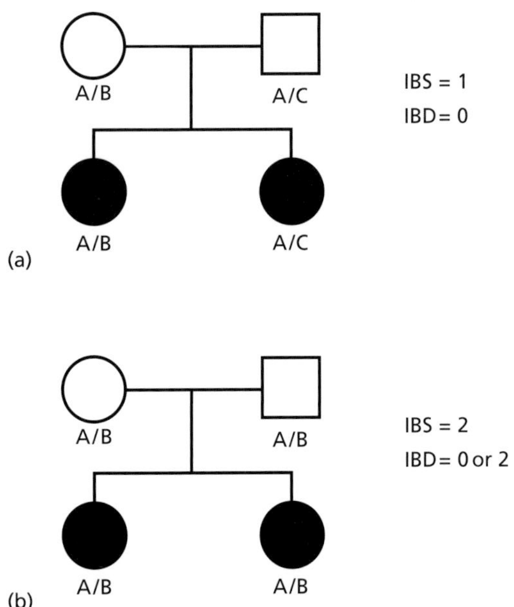

Figure 4 Examples of allele sharing in affected sibs illustrating the differences between IBS and IBD. The term 'identical by state' refers to the extent to which two individuals share the same alleles at a marker locus. In (a), there are three alleles (A, B and C) at a hypothetical marker locus; both parents are heterozygous (mother's genotype A/B and father's genotype A/C). Without knowing the parental genotypes, it is only possible to state that the affected sisters share an inherited A allele, i.e. the number of alleles inherited 'identical by state' (IBS) is 1. It is not possible to determine how many alleles have been inherited 'identical by descent' (IBD). Knowing the parental genotypes, however, allows the observation that no alleles are inherited IBD, because allele A has been inherited from the woman's father in the case of the older sister and from the woman's mother in the case of the younger sister. In (b), both parents are heterozygous at the same hypothetical marker locus (mother's and father's genotype A/B). The affected sisters share two inherited alleles 'identical by state'. However, even knowing the parental genotypes, it is not possible to determine whether 2 or 0 alleles have been inherited 'identical by descent', because both parents have the same genotype. The same situation arises when one or both parents are homozygous

Genome-wide scan

If multiple markers are used in a scan of the whole genome (so-called genome-wide screen), the chance of a spurious positive result is greater than if only a few markers are used in the region of a candidate gene. It follows that the definition of statistical significance needs to be more robust in complex genetic traits than in Mendelian diseases, and that the results of a genome-wide screen need to be replicated in separate data sets before a linkage can be considered confirmed.

Therefore, a key issue in designing a study is the size of the sample, because this will influence the power of the method to detect linkage (see Table 2

Table 2 Factors influencing the power of a genome-wide screen to detect susceptibility loci

Sample size (i.e. the number of families genotyped)
Marker allele and disease allele frequencies
Size of relative risk (λ_S)
Number of susceptibility loci and their relative contributions to the value of λ_S
Relationships between the affected family members (i.e. ASPs or ARPs)
How many markers are used in the genome-wide screen (i.e. intermarker genetic distance) and how polymorphic they are (i.e. heterozygosity)

ASP, affected sib-pair; ARP, affected relative-pair

for other important factors). To give an example, 200 sib-pair families would need to be genotyped to detect an LOD score of > 3.0 with 90% power, at a locus with a λ_S value of 3.0, using markers 10 cM apart and 75% heterozygosity. Clearly, these are complicated issues beyond the remit of this chapter, and the reader is advised to consult reviews of the subject.[28,29] Given that family collection and genotyping are both expensive, it has been suggested that the most cost-efficient method is to perform a two-stage genome-wide screen: utilizing a limited number of markers in the first stage followed by genotyping of another set of families with additional markers around regions that have reached a predetermined significance level.

Predisposing loci in multifactorial diseases such as type 1 diabetes,[30] asthma[31] and inflammatory bowel disease[32] have been mapped using this strategy. The reproductive disorder that is being most extensively investigated in this way is endometriosis.[33] The Oxford Endometriosis Gene (Oxegene) Study has identified over 1750 probands, principally from the UK and USA, from which the first 85 families containing 121 ASPs have been genotyped (~400 markers at ~10 cM), generating preliminary evidence of linkage in these families to five chromosomal regions, with non-parametric linkage (NPL) scores of > 2.[34] A similar strategy has been adopted by an Australian group: marker data have been generated (~400 markers at ~10 cM) and analyzed for a total of 289 families containing 374 sister-pairs plus other affected relatives. Suggestive linkage has been identified for one locus, and possible linkage for five other loci.[35] These two groups have recently announced that they are combining their resources to identify susceptibility genes in endometriosis.

An alternative but highly speculative approach to the genome-wide screen is to investigate 'functional' candidate genes using the ASP method. The term 'functional' candidate refers to any gene that might be implicated etiologically, based upon knowledge of its putative or actual role in mechanisms involved in disease development. Thus, Hadfield and colleagues investigated *GSTM1* – a gene previously implicated in endometriosis on the basis of the results of association studies[36,37] – and found no evidence of linkage to the chromosome 1p13 region, to which *GSTM1* has been mapped, in 52 sister-pairs with stage III–IV disease using three highly polymorphic microsatellite markers, *D1S2844*, *D1S2635* and *D1S2762*.[38] Geirsson and colleagues investigated the *GALT* gene, previously implicated in endometriosis in an association study:[39] 203 women with endometriosis (rAFS stages I–IV) and their relatives from 59 Icelandic families were genotyped using 30 microsatellite markers on chromosome 9, to which *GALT* has been mapped, but no evidence for linkage was found.[40] Urbanek and associates investigated 37 'functional' candidate genes in 150 families with PCOS/hyperandrogenemia. The strongest evidence for linkage was found with the follistatin gene, for which affected sisters showed significantly increased IBD allele sharing even after adjusting for multiple testing.[1]

Allelic association studies represent another non-parametric approach to the identification of susceptibility genes in complex traits, which is complementary to the linkage methods described above. The two principal uses of allelic association studies are fine mapping and candidate gene studies. The approach invariably involves comparing the frequency of alleles in affected individuals and suitable controls, either in the population (case–control studies) or within families (intrafamilial association studies). Allelic association may arise because the variant allele is causally implicated in the disease (i.e. it alters the biological function of the gene in such a way that it predisposes the individual to develop the disease). Examples of such association reported in complex traits include apolipoprotein E in Alzheimer's disease[41] and the insulin gene in type 1 diabetes.[42,43]

The investigation of candidate genes in endometriosis using association studies has recently been reviewed in detail.[44] In addition to the studies mentioned above implicating the *GALT N314D* polymorphism[39] and the *GSTM1* null mutation,[36,37] positive association has also been described for polymorphisms in *NAT2*[45] and the estrogen receptor gene.[46] It is unclear at present whether such results provide strong enough evidence of the functional involvement of these genes in disease etiology and/or progression, especially as false-positive results can arise in association studies simply because the sample size is too small and/or inappropriate control groups have been used.

Another possible explanation for finding allelic association is that the allele is in linkage disequilibrium with a disease-susceptibility allele. This occurs when a marker allele is situated so close to a disease-susceptibility allele that the two alleles tend to be inherited together amongst affected individuals in the population as a whole. In effect, this means that the two loci are tightly linked over small recombination distances (usually < 1 cM). Linkage disequilibrium can occur in a population when a mutation, which causes disease, arises in a common ancestor. Alleles that are closely linked to the mutation will tend to be co-inherited, but the disequilibrium will diminish over time as a result of recombination. As a general rule, therefore, the closer an allele is to the mutation, the slower the linkage disequilibrium will be lost. Thus, it has been estimated that it takes 70 generations for linkage disequilibrium between two alleles 1 Mb apart to decay by 50%.

Clearly, linkage disequilibrium is influenced by the structure of the population: the effect is most strongly seen in small, homogenous populations in which the likelihood of a founder effect is greater. It follows, therefore, that spurious allelic associations can also arise by chance, because of ethnic variations in allele frequencies between the groups studied. There are methods to study allelic association that eliminate the problem of population differences between cases and controls by using the parents of affected individuals as controls, which involves collecting mother–father and single affected case-trios (intrafamilial association studies). The tests compare the frequencies with which alleles are transmitted and not transmitted from the parents to the affected person. Examples include the haplotype relative risk (HRR) test,[47] and the transmission disequilibrium test (TDT)[48] (Figure 5). As an example in PCOS, Urbanek and colleagues tested 37 candidate gene regions at 45 loci using TDT analysis (Figure 6).[1] Evidence for association was found in the region of the *INSR* (insulin receptor) gene with allele 5 of the marker *D19S884* ($\chi^2 = 8.53$); however, the result was not statistically significant after correcting for multiple testing.

	Not transmitted	
Transmitted	*Allele 1*	*Allele 2*
Allele 1	*a*	*b*
Allele 2	*c*	*d*

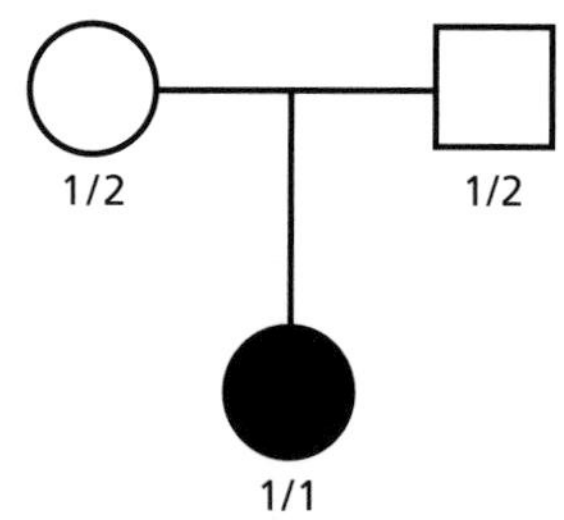

Figure 5 Transmission disequilibrium test (TDT). Mother–father–affected person trios are genotyped and the transmission of alleles within multiple families is tabulated as shown. Statistical significance is tested by the χ^2 test using the formula $\chi^2 = (b-c)^2/b+c$, where *b* is the frequency with which a heterozygous parent transmits allele 1 to the affected person, and *c* is the frequency with which a heterozygous parent transmits allele 2 to the affected person. In the example illustrated, there are two alleles (1 and 2) at a hypothetical marker locus. Both parents are heterozygous and the affected daughter has inherited an allele 1 from both her mother and her father. Thus, two allele 1s were transmitted and two allele 2s were not transmitted. In reference 1 investigating candidate genes in polycystic ovary syndrome (PCOS), the strongest evidence for association reported was seen with allele 5 of the marker *D19S884* in the region of the *INSR* (insulin receptor) gene, although this was not significant after correction for multiple testing. Ten allele 5s were transmitted and 28 were not transmitted ($\chi^2 = 8.53$; $p = 0.004$)

SUMMARY

Once the genes involved in the etiology of a complex trait have been determined, it is likely that genetic tests will be developed at a general population level to enable individuals at risk to be detected. It has been predicted that this advance will facilitate the identification of specific environmental risk factors for genetically predisposed individuals and those who do not carry susceptibility genes (phenocopies). In many cases, the risk factors are likely to be similar, although the magnitude of the effect may differ. Thus, the long-term use of hormone replacement therapy may increase the

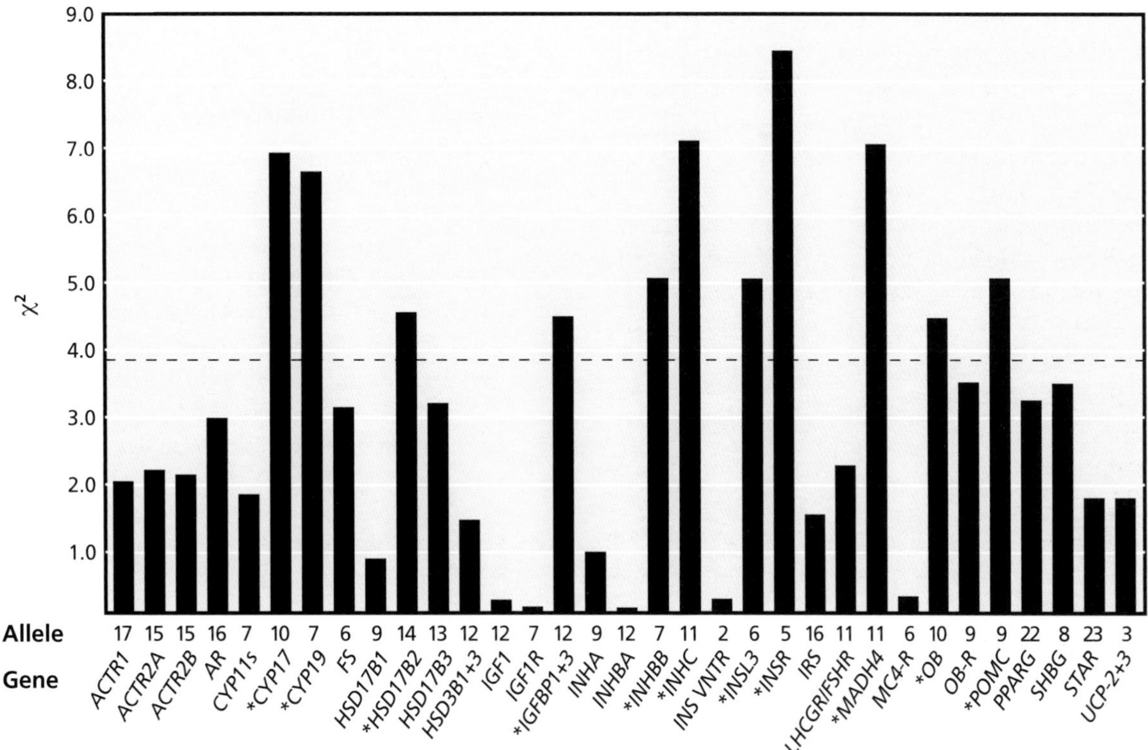

Figure 6 Summary of transmission disequilibrium test (TDT) analysis. The dashed line indicates a χ^2 value of 3.84 ($p = 0.05$). The candidate gene regions and the allele with the highest χ^2 value for each region are listed on the *x* axis. Each allele with a nominally elevated χ^2 (> 3.84) is indicated with an asterisk, and the number of transmissions tested is shown above the bar. A total of 349 alleles at 45 loci were tested. Reproduced with permission from reference 1

risk of women developing breast cancer, and the combined oral contraceptive may protect against the development of ovarian cancer, but the risks may be quantitatively different in women carrying a mutation in the *BRCA1* gene. Having determined which environmental factors make the expression of the disease phenotype in a genetically susceptible person more likely, individuals who are positive for a genetic susceptibility test will be advised to avoid certain environmental risk factors. In theory, it should be possible to identify genotype-specific risk factors. However, it is also possible that the effect of the major environmental risk factors, which have been identified by studying whole populations with traditional epidemiological methods, will far outweigh the small effect of multiple genes in complex traits. Thus, it would not be surprising if individuals who are genetically predisposed to develop lung cancer were advised not to smoke.

Such discoveries may translate into clinical reality, but the most exciting prospect must be the possibility of understanding better the genetic basis of common diseases that are highly heterogeneous, i.e. they result from disturbances in a myriad of cellular and molecular mechanisms. Therefore, such conditions can only be understood by identifying the genetic determinants contributing to the expression of the individual phenotypes. It follows that drugs will be developed that are specifically aimed at treating disease subsets who have been classified on the basis of genotyping. Thus, conditions such as PCOS that are clearly heterogeneous will, in the future, be treated with drugs targeted against the specific genes that are aberrantly regulated in the diagnostic subset to which an individual belongs. It is predicted that new classification systems based upon genetic information about individual patients will lead to the development of targeted drugs, which, in turn, will produce more effective therapy and fewer side-effects.

REFERENCES

1. Urbanek M, Legro RS, Driscoll DA *et al.* Thirty-seven candidate genes for polycystic ovary syndrome: strongest evidence for linkage is with follistatin. *Proc Natl Acad Sci USA* 1999 96: 8573–8578.
2. Franks S, Gharani N, McCarthy M. Genetic abnormalities in polycystic ovary syndrome. *Ann Endocrinol (Paris)* 1999 60: 131–133.
3. American Fertility Society. Revised American Fertility Society classification of endometriosis: 1985. *Fertil Steril* 1985 43: 351–352.
4. Franks S, Gharani N, McCarthy M. Candidate genes in polycystic ovary syndrome. *Hum Reprod Update* 2001 7: 405–410.
5. Vikhlyaeva EM, Khodzhaeva ZS, Fantschenko ND. Familial predisposition to uterine leiomyomas. *Int J Gynaecol Obstet* 1995 51: 127–131.
6. Kennedy SH, Mardon HJ, Barlow DH. Familial endometriosis. *J Assist Reprod Genet* 1995 12: 32–34.
7. Stefansson H, Geirsson RT, Jonsson H *et al.* Familial clustering of endometriosis evident from a population based study. Presented at the *World Endometriosis Congress*, London, May 2000.
8. Hadfield RM, Yudkin PL, Coe CL *et al.* Risk factors for endometriosis in the rhesus monkey (*Macaca mulatta*): a case–control study. *Hum Reprod Update* 1997 3: 109–115.
9. Coxhead D, Thomas EJ. Familial inheritance of endometriosis in a British population: a case control study. *J Obstet Gynaecol* 1993 13: 42–44.
10. Moen MH, Magnus P. The familial risk of endometriosis. *Acta Obstet Gynecol Scand* 1993 72: 560–564.
11. Simpson JL, Elias S, Malinak LR, Buttram VCJ Jr. Heritable aspects of endometriosis. I. Genetic studies. *Am J Obstet Gynecol* 1980 137: 327–331.
12. Kennedy S, Hadfield R, Westbrook C, Weeks DE, Barlow D, Golding S. Magnetic resonance imaging to assess familial risk in relatives of women with endometriosis. *Lancet* 1998 352: 1440–1441.
13. Kahsar-Miller MD, Nixon C, Boots LR, Go RC, Azziz R. Prevalence of polycystic ovary syndrome (PCOS) in first-degree relatives of patients with PCOS. *Fertil Steril* 2001 75: 53–58.
14. Lathrop GM, Lalouel JM, Julier C, Ott J. Strategies for multilocus linkage analysis in humans. *Proc Natl Acad Sci USA* 1984 81: 3443–3446.
15. Botstein D, White RL, Skolnick M, Davis RW. Construction of a genetic linkage map in man using restriction fragment length polymorphisms. *Am J Hum Genet* 1980 32: 314–331.
16. Nakamura Y, Leppert M, O'Connell P *et al.* Variable number of tandem repeat (VNTR) markers for human gene mapping. *Science* 1987 235: 1616–1622.
17. Koenig M, Hoffman EP, Bertelson CJ, Monaco AP, Feener C, Kunkel L. Complete cloning of the Duchenne muscular dystrophy (DMD) cDNA and preliminary genomic organization of the DMD gene in normal and affected individuals. *Cell* 1987 50: 509–517.
18. Kerem B, Rommens JM, Buchanan JA *et al.* Identification of the cystic fibrosis gene: genetic analysis. *Science* 1989 245: 1073–1080.
19. Weber JL, May PE. Abundant class of human DNA polymorphisms which can be typed using the polymerase chain reaction. *Am J Hum Genet* 1989 44: 388–396.
20. Reed PW, Davies JL, Copeman JB *et al.* Chromosome-specific microsatellite sets for fluorescence-based, semi-automated genome mapping. *Nat Genet* 1994 7: 390–395.
21. Weissenbach J, Gyapay G, Dib C *et al.* A second-generation linkage map of the human genome. *Nature* 1992 359: 794–801.
22. Dib C, Faure S, Fizames C *et al.* A comprehensive genetic map of the human genome based on 5264 microsatellites. *Nature* 1996 380: 152–154.
23. Suarez BK, Rice J, Reich T. The generalized sib pair IBD distribution: its use in the detection of linkage. *Ann Hum Genet* 1978 42: 87–94.
24. Risch N. Linkage strategies for genetically complex traits. III. The effect of marker polymorphism on analysis of affected relative pairs. *Am J Hum Genet* 1990 46: 242–253.
25. Risch N. Linkage strategies for genetically complex traits. II. The power of affected relative pairs. *Am J Hum Genet* 1990 46: 229–241.
26. Kruglyak L, Daly MJ, Reeve DM, Lander ES. Parametric and nonparametric linkage analysis: a

unified multipoint approach. *Am J Hum Genet* 1996 58: 1347–1363.

27. Weeks DE, Lange K. The affected-pedigree-member method of linkage analysis. *Am J Hum Genet* 1988 42: 315–326.
28. Lander ES, Schork NJ. Genetic dissection of complex traits. *Science* 1994 265: 2037–2048.
29. Risch NJ. Searching for genetic determinants in the new millenium. *Nature* 2000 405: 847–856.
30. Davies JL, Kawaguchi Y, Bennett ST *et al.* A genome-wide search for human type 1 diabetes susceptibility genes. *Nature* 1994 371: 130–136.
31. Sandford AJ, Shirakawa T, Moffatt MF *et al.* Localisation of atopy and β subunit of high-affinity IgE receptor (Fc ε RI) on chromosome 11q. *Lancet* 1993 341: 332–334.
32. Satsangi J, Parkes M, Louis E *et al.* Two stage genome-wide search in inflammatory bowel disease provides evidence for susceptibility loci on chromosomes 3, 7 and 12. *Nat Genet* 1996 14: 199–202.
33. Kennedy SH, Bennett S, Weeks DE. Affected sib-pair analysis in endometriosis. *Hum Reprod Update* 2001 7: 411–418.
34. Oxford Endometriosis Group, Oxegene Collaborative Group and Oxagen. Preliminary results from the Oxegene study: genome-wide screen for endometriosis susceptibility loci in over 100 affected sister-pairs. Presented at the *World Endometriosis Congress*, London, May 2000.
35. Treloar SA, Bahlo M, Ewen K *et al.* Suggestive linkage for endometriosis found in genome-wide scan. *Am J Hum Genet* 2000 67 (Suppl 2): 318 (abstr).
36. Baranov VS, Ivaschenko T, Bakay B *et al.* Proportion of the *GSTM1* 0/0 genotype in some Slavic populations and its correlation with cystic fibrosis and some multifactorial diseases. *Hum Genet* 1996 97: 516–520.
37. Baranova H, Bothorishvilli R, Canis M *et al.* Glutathione S-transferase M1 gene polymorphism and susceptibility to endometriosis in a French population. *Mol Hum Reprod* 1997 3: 775–780.
38. Hadfield RM, Manek S, Weeks DE *et al.* Linkage and association studies of the relationship between endometriosis and genes encoding the detoxification enzymes GSTM1, GSTT1 and CYP1A1. *Mol Hum Reprod* 2001 7:1073–1078.
39. Cramer DW, Hornstein MD, Ng WG, Barbieri RL. Endometriosis associated with the *N314D* mutation of galactose-1-phosphate uridyl transferase (*GALT*). *Mol Hum Reprod* 1996 2: 149–152.
40. Stefansson H, Einarsdottir A, Geirsson RT *et al.* Endometriosis is not associated with or linked to the GALT gene. *Fertil Steril* 2001 76: 1019–1022.
41. Pericak VM, Haines JL. Genetic susceptibility to Alzheimer disease. *Trends Genet* 1995 11: 504–508.
42. Bennett ST, Lucassen AM, Gough SC *et al.* Susceptibility to human type 1 diabetes at *IDDM2* is determined by tandem repeat variation at the insulin gene minisatellite locus. *Nat Genet* 1995 9: 284–292.
43. Bennett ST, Todd JA. Human type 1 diabetes and the insulin gene: principles of mapping polygenes. *Annu Rev Genet* 1996 30:343–370
44. Zondervan KT, Cardon LR, Kennedy SH. The genetic basis of endometriosis. *Curr Opin Obstet Gynecol* 2001 13: 309–314.
45. Baranova H, Canis M, Ivaschenko T *et al.* Possible involvement of arylamine *N*-acetyltransferase 2, glutathione S-transferases M1 and T1 genes in the development of endometriosis. *Mol Hum Reprod* 1999 5: 636–641.
46. Georgiou I, Syrrou M, Bouba I *et al.* Association of estrogen receptor gene polymorphisms with endometriosis. *Fertil Steril* 1999 72: 164–166.
47. Falk CT, Rubinstein P. Haplotype relative risks: an easy reliable way to construct a proper control sample for risk calculations. *Ann Hum Genet* 1987 51: 227–233.
48. Spielman RS, McGinnis RE, Ewens WJ. Transmission test for linkage disequilibrium: the insulin gene region and insulin-dependent diabetes mellitus (IDDM). *Am J Hum Genet* 1993 52: 506–516.

Reproductive dysfunction

Bart C.J.M. Fauser

The true revolution in molecular and genetic studies in clinical medicine can clearly be seen from the current section. The first edition of this book – released just 3 years ago – contained only two chapters in the clinical section. As predicted, the introduction of new genetic tools and molecular technologies into clinical practice in reproductive medicine has rapidly advanced the field, justifying the extension of the current section to eight chapters. However, it seems that what we witness today is only the beginning of a new era of molecular medicine, with major consequences for the prevention, diagnosis and treatment of reproductive disorders and for the further development of assisted reproductive technologies. It is becoming increasingly clear that these new developments will impact on areas of medicine going way beyond reproduction. These forefront issues such as cloning, embryonic gene therapy and human stem cells, where the potential for clinical medicine remains uncertain at this stage, are not discussed as yet.

The many topics now covered separately in this section include (a) disturbed sexual differentiation, (b) 21-hydroxylase deficiency, polycystic ovary syndrome, and premature ovarian failure in the female, and (c) genetic and endocrine factors involved in male reproductive dysfunction.

Recent advances in the unravelling of complex mechanisms underlying abnormal sexual differentiation, sex reversal due to alterations in the sex-determining gene located on the Y chromosome, and mutations in anti-Müllerian hormone are discussed. Moreover, mutations in the many enzyme systems involved in steroid biosynthesis, as well as defective steroid reception, are also covered. It is amazing to see that, in some rare disease states, up to 200 mutations have been identified and associations between genotype and phenotype established.

For a deeper understanding of the difficulties involved in genetic studies of polycystic ovary syndrome, see Chapter 28 in the previous section, dealing with the genetic analysis of complex multifactorial diseases. The continued discussion regarding the phenotype of the syndrome, the likelihood that multiple genes are involved, different environmental influences and the lack of appropriate animal models are all complicating factors hampering the tremendous efforts to unravel this disease. Genetic conditions affecting ovarian development or failure of the normally developed ovary involved in premature ovarian failure are also extensively outlined.

Advances in assisted reproductive technologies have created new opportunities to generate offspring. It should also be acknowledged that these technologies have discouraged further studies regarding the genetic/molecular background of male infertility. Moreover, concerns over the chances of transmitting infertility and related genetic abnormalities to children should be taken very seriously. The debate with regard to a decrease in sperm quality and associated environmental factors continues.

It is widely acknowledged that assisted reproductive technologies have contributed immensely in allowing couples to reproduce who would otherwise have remained childless. However, it should also be realized that these advances have elicited many moral dilemmas with regard to the responsibility of the prospective parents, treating physicians, scientists and society as a whole. These complex issues will be discussed in the concluding chapter on ethics.

Chapter 29

Male pseudohermaphroditism, true hermaphroditism and XY sex reversal

Joe Leigh Simpson

INTRODUCTION

Principles underlying reproductive embryology have been known for half a century.[1] For a quarter of a century, clinical delineation of the disorders of sex differentiation has been generally accepted.[2] Incremental progress and insight into many clinical disorders have since been ongoing, including identification in the early 1990s of the testes-determining gene (*SRY*).[3]

With the ability to apply molecular technology, additional biological and, hence, clinical advances can be expected. This chapter begins with a general discussion of reproductive embryology and the genetic control of sex determination. Molecular aspects of male pseudohermaphroditism true hermaphroditism, and 'XY females' are then discussed.

BACKGROUND

The general principles of reproductive embryology have long been recognized. Primordial germ cells originate in the endoderm of the yolk sac and migrate to the genital ridge to form the indifferent gonad. Initially, 46,XY and 46,XX gonads are indistinguishable. Indifferent gonads develop into testes if the embryo, or more specifically the gonadal stroma, is 46,XY (Figure 1). This process begins about 43 days after conception. Testes become morphologically identifiable 7–8 weeks after conception (9–10 gestational or menstrual weeks).

The first cells to become recognizable in testicular differentiation are the Sertoli cells, which organize the surrounding cells into tubules. Leydig cells and Sertoli cells secrete hormones that direct

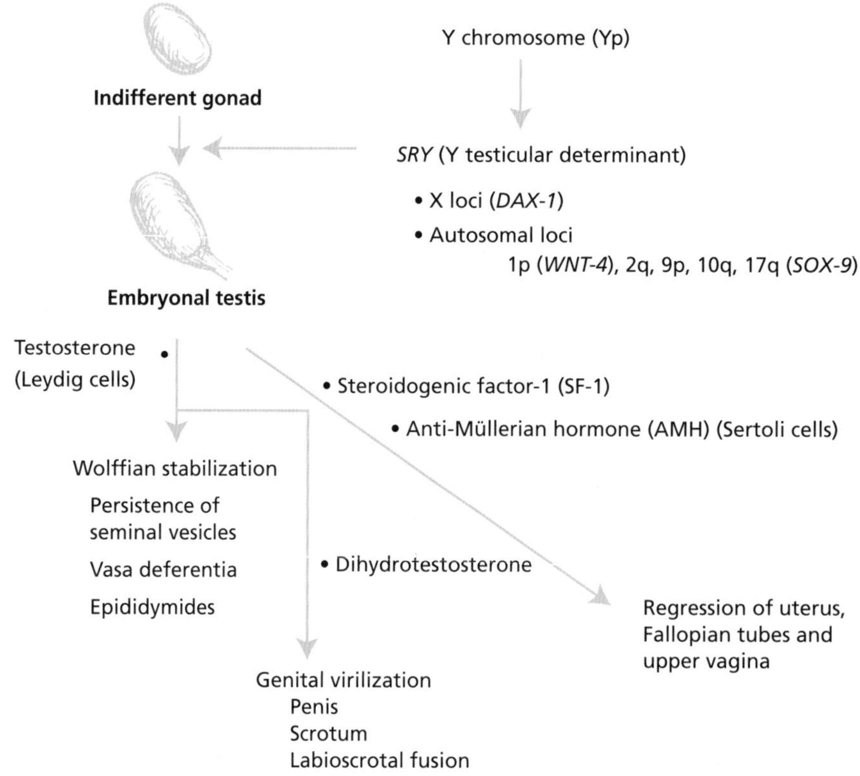

Figure 1 Schematic diagram showing major events in the genetic control of reproductive embryology. Adapted from reference 4

subsequent male differentiation (Figure 1). Fetal Leydig cells secrete an androgen – testosterone – that stabilizes Wolffian ducts and permits differentiation of vasa deferentia, epididymides and seminal vesicles. After conversion by 5α-reductase to dihydrotestosterone (DHT), external genitalia are virilized. Fetal Sertoli cells produce anti-Müllerian hormone (AMH), a glycoprotein that diffuses locally to cause regression of Müllerian derivatives (uterus and Fallopian tubes).

In the absence of a Y chromosome, the indifferent gonad develops into an ovary. Transformation into fetal ovaries begins 50–55 days after conception. Germ cells are present in 45,X embryos,[5] only to undergo atresia at a rate more rapid than that occurring in normal 46,XX embryos. Pathogenesis involves not failure of germ cell formation but increased atresia.

Internal ductal and external genital development is secondary to, but independent of, gonadal differentiation. In the absence of testosterone and AMH, external genitalia develop in female fashion. Müllerian ducts form the uterus and Fallopian tubes, and Wolffian ducts regress.

APPLIED TECHNOLOGY

With the recognition nearly 50 years ago that the pivotal step in human sex differentiation was the Y chromosome and the presence of a testes-determining factor, it became feasible to delineate clinically the various disorders of sexual differentiation. Considerable progress was soon made on the basis of deducing embryologically those steps that had become perturbed, and applying the basic principles of human genetics. For example, usually, autosomal recessive genes cause metabolic errors; autosomal dominant genes show varied expressivity and are more likely to represent a fresh mutation if a condition is so serious that reproduction is adversely affected (decreased reproductive fitness). Different modes of inheritance for two clinically similar disorders confirm that the disorders are distinct.

In the delineation of any group of birth defects, sexual differentiation included, the key underlying clinical question is always whether a given disorder constitutes a single entity or is heterogeneous. Should one lump together or split? In delineating disorders of sexual differentiation, splitting has generally proved to be preferable. For example, once, the term 'pure gonadal dysgenesis' was used for XY females, but it is now known that individuals can have the same phenotype (normal stature, hypergonadotropic hypogonadism, no somatic anomalies) with any of a dozen different etiological explanations. Using these traditional genetic approaches, the key disorders of sexual differentiation have been delineated clinically for well over a quarter of a century.[2] Some disorders proved to be cytogenetic, others Mendelian and yet others polygenic/multifactorial. Except for X-linked conditions, it was not possible to localize a gene to a given chromosome.

With the advent of the molecular era (*c.* 1985–1990), molecular technology could be applied to localize genes and determine the perturbations responsible for a particular disorder. With an existing nosology already in place, molecular advances were rapidly made. Pleasingly, only rarely have molecular investigations revealed flaws in the clinical delineation scheme which has already proved to be serviceable. Molecular progress has resulted from the application of three strategies:

(1) Localization and subsequent analysis of DNA probes based on sequences from animal genes known to have a human homolog;

(2) Genome-wide linkage analysis;

(3) Fortuitous recognition of chromosomal rearrangements that identify relevant chromosomal regions.

Progress in elucidating the genes governing disorders of adrenal and testicular biosynthesis was especially rapid, because mouse and rat genes aided the localization of homolog genes in humans. For rare disorders of sexual differentiation lacking a known metabolic basis, and thus defined only by clinical criteria (for example, a form of XY gonadal dysgenesis associated with characteristic somatic anomalies), a different approach was necessary. One option was the second approach – genome-wide scanning – a laborious but validated technique in which one hopes that a dyad of two affected individuals (affected–affected), but not other dyads in the same kindred (affected–unaffected or unaffected–unaffected), shares common polymorphic markers as a result of identity by descent. This approach can localize the relevant chromosomal region and, hence, the gene. Once the region has been found, gene banks can be scanned to identify potentially relevant sequences. If a gene seems important, mutations can be sought, and, if found, cause and effect established. This second approach proved successful for Finnish XX gonadal dysgenesis due to follicle stimulating hormone receptor (*FSHR*) mutations and for blepharophimosis–ptosis–epicanthus due to *FOXL2* mutations.

The third approach depends upon fortuitously finding a chromosomal rearrangement in a family in which the mutant gene in question is segregating. The principle is that chromosomal rearrangement perturbs the gene, which is rendered dysfunctional. An individual with a balanced rearrangement may thus show a Mendelian disorder, but otherwise be normal because gross chromosomal imbalance does not exist. In the rarer disorders of sexual differentiation, this may be the only approach possible. Once a region is identified, the approach illustrated above for genome-wide linkage analysis becomes applicable.

CURRENT CONCEPTS

The major testicular determinant is *SRY* (sex-determining region Y chromosome), located on the Y short arm near the distal Xp/Yp pseudoautosomal region (PAR).[6–8] *SRY* consists of two open reading frames, 99 and 273 amino acids in length. The pivotal sequence involves an HMG (high-mobility group) box encompassing codons 13–82 (Figure 2). This HMG box has characteristics in common with other DNA-binding sequences. When XY females (XY gonadal dysgenesis) result from point mutations involving *SRY*, the perturbation almost always occurs within the HMG box.[9–11] The exact function of *SRY* remains unclear,[11,12] but in all species *SRY* is known to bind DNA at a consensus sequence (AATAAC). *SRY* could play a role in DNA bending,[13] presumably juxtaposing more than one testicular-determining gene and, hence, facilitating transcription or interaction between gene products.

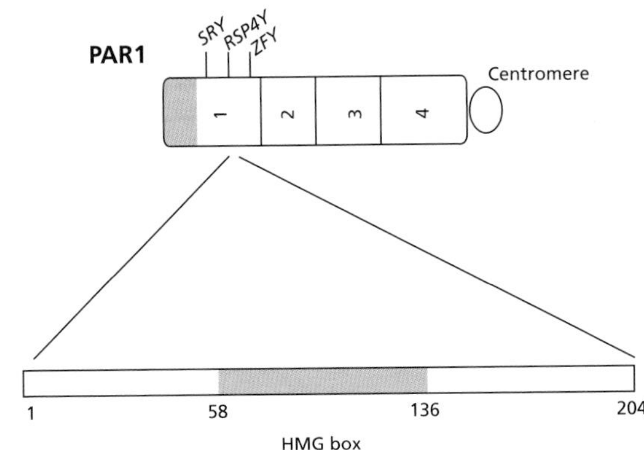

Figure 2 Relationship of the Y short arm, pseudoautosomal region (PAR) and *SRY* (sex-determining region Y chromosome). Mutations responsible for 10–15% of XY gonadal dysgenesis are usually found in the high-mobility group (HMG) box. Adapted from reference 9

Interrelationships exist between *SRY* and various autosomal and X loci, which, in aggregate, govern testicular differentiation. In one scenario, *SRY* directly represses the X-linked locus *DAX-1*, which otherwise might direct ovarian development. The basis for this hypothesis is that, in XY humans, duplication of Xp and, hence, the *DAX-1* locus at Xp21.1 is associated with ovaries. In the presence of *DAX-1* duplication, perhaps *SRY* cannot fully exert its normal action to induce testicular differentiation. There is evidence that *WNT-4*, located on 1p, regulates *DAX-1*; duplication of 1p and *WNT-4* produced XY sex reversal by up-regulation of DAX-1.[14]

In a second scenario, the pivotal autosomal gene is *SOX-9* and the campomelic dysplasia locus, or a nearby region. *SOX-9* has two activation domains, suggesting the ability to up-regulate other genes.[15] Deletion of *SOX-9* causes campomelic dysplasia and XY gonadal dysgenesis (sex reversal). Conversely, duplication of 17q23.1→q24.3, the region in which *SOX-9* is localized, resulted in a 46,XX individual having bilateral scrotal gonads (presumptive testes), a small male phallus and a perineal urethral orifice (i.e. incompletely sex-reversed XX male).[16] Insertional mutagenesis deleting a 150-kb region 5' to *SOX-9* results in XX mice having testes.[17] This suggests that *SRY* derepresses a locus to permit testicular differentiation. Mutation involving the locus could result in unscheduled derepression in XX mammals, despite lack of *SRY*, leading to unexpected testicular development in XX mammals.

Deletions of other autosomal regions preclude testicular development, and result in XY females (XY gonadal dysgenesis). These regions include 2q33 and 9p24.3 (which encodes *DMRT* (Doublesex and mab3 related transcription factor-1)), 10q26 (which encompasses steroidogenic factor-1 (SF-1))[18,19] and 11p (which encodes WT-1)[20] (see below for further comments on these regions). As mentioned above, duplication of the region of 1p encoding *WNT-4* is also associated with XY sex reversal.[14,21]

Whether primary female (ovarian) differentiation is a default (constitutive) pathway or directed by a specific gene product remains uncertain. Traditionally, the default hypothesis has been favored,[22,23] but a primary directive role in ovarian differentiation has also been proposed. However, one previously touted candidate gene, *DAX-1* (or at least its mouse homolog *Ahch*), now seems unlikely to play a role as a primary ovarian determinant. Murine null mutations for *Ahch* unexpectedly showed normal ovarian development, and also abnormal testicular germ cell defects.[24]

MALE PSEUDOHERMAPHRODITISM AND XY GONADAL DYSGENESIS

In male pseudohermaphroditism, the external genitalia of individuals with a Y chromosome fail to develop as expected for normal males. Typically, external genitalia are sufficiently ambiguous to confuse the sex of rearing, but in some disorders the phenotype extends to female external genitalia (46,XY females). Table 1 delineates the forms of male pseudohermaphroditism. Only those disorders whose molecular basis has been elucidated are emphasized. Discussed elsewhere are cytogenetic forms (45,X/46,XY) and other clinical disorders.[4,26]

Table 1 Clinical delineation of male pseudohermaphroditism. All these disorders are associated (or postulated to be associated) with genital ambiguity, but external genitalia may also be female (sex reversed) in some disorders

A. Teratogenic
Cyproterone acetate (no reported cases)
Finisteride (no reported cases)
Flutamide (no reported cases)
Progestins (unproved) (see reference 25)
B. Cytogenetic (45,X/46,XY) mosaicism
Female external genitalia
Ambiguous external genitalia
C. Testosterone biosynthetic defects (adrenal or testis)
Congenital adrenal lipoid hyperplasia (StAR deficiency) (*StAR*)
3β-ol-Dehydrogenase/3β-hydroxysteroid dehydrogenase deficiency (3β-HSD)
17α-Hydroxylase/17,20-desmolase (-lyase) deficiency (*CYP17*)
17β-Hydroxysteroid dehydrogenase/17-ketosteroid reductase (17β-HSD-3)
D. Steroidogenic factor–1 (SF-1) deficiency (FTZ1)
E. Androgen insensitivity syndrome (AIS) (ARG)
Complete (CAI)
Partial (PAI)
Mild (MAI)
F. 5α-Reductase deficiency (5RD5A2)
G. Agonadia (testicular regression)
H. Leydig cell hypoplasia (LH receptor mutation)
I. Multiple malformation syndromes
Smith–Lemli–Opitz (SLO)
Genitopalatocardiac
Meckel–Gruber
Other

StAR, steroidogenic acute regulatory protein; LH, luteinizing hormone

Testosterone biosynthetic defects

Considerable progress has been made in the molecular elucidation of the forms of male pseudohermaphroditism caused by deficiencies of the adrenal or testicular enzymes required for testosterone biosynthesis (Figure 3). These enzymes are 17α-hydroxylase/17,20 desmolase, 3β-ol-dehydrogenase, 17-ketosteroid reductase, steroid acute regulatory protein (StAR) and the other enzymes necessary to convert cholesterol to pregnenolone (congenital adrenal lipoid hyperplasia). The

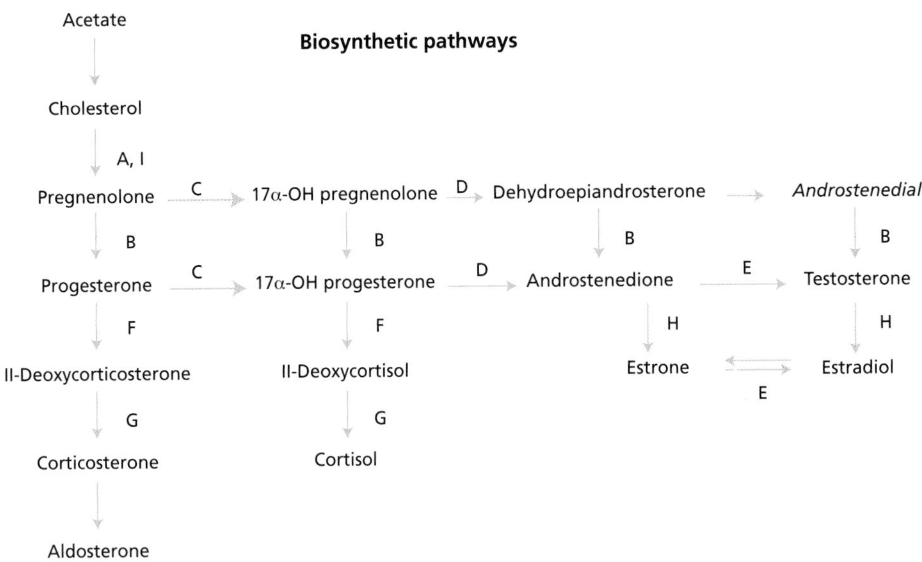

Figure 3 Important adrenal and gonadal biosynthetic pathways. Letters designate enzymes or activities required for the appropriate conversions. A, 20α-hydroxylase, and 20,22-desmolase; B, 3β-ol-dehydrogenase; C, 17α-hydroxylase; D, 17,20-desmolase; E, 17-keto-steroid reductase; F, 21-hydroxylase; G, 11β-hydroxylase; H, aromatase; I, StAR (steroid acute regulatory protein). Both 17α-hydroxylase and 17,20-desmolase activities are governed by a single gene. Adapted from reference 4

common pathogenesis involves testosterone levels insufficient to virilize external genitalia. Deficiencies of 21- or 11β-hydroxylase, the most common causes of female pseudohermaphroditism (see Chaper 30), do not cause male pseudohermaphroditism. In fact, 46,XY individuals with the above deficiencies show precocious masculinization.

Congenital adrenal lipoid hyperplasia (StAR deficiency)

Male pseudohermaphrodites with congenital adrenal lipoid hyperplasia show ambiguous or female-like external genitalia, severe sodium wasting and adrenals characterized by foamy-appearing cells filled with cholesterol.[27,28] Hyperpigmentation and respiratory distress occur in 25% of affected neonates.

Accumulation of cholesterol has long been taken as an indication that cholesterol cannot be converted to pregnenolone (Figure 3). Inheritance is presumed to be autosomal recessive on the basis of increased parental consanguinity, affected sibs of both sexes and the metabolic nature of the disorder.

The cytochrome P450 enzyme responsible for converting cholesterol to pregnenolone is P450scc (side-chain cleavage); the gene is designated *CYP11A*. P450scc converts cholesterol to pregnenolone via 20α-hydroxylase, 22α-hydroxylase and 20,22-desmolase (lyase). Located on chromosome 15, *CYP11A* is 20 kb in length and consists of nine exons. In rabbits, homozygous *CYP11A1* gene deletion was shown a decade ago to cause adrenal lipoid hyperplasia.[29] Surprisingly, perturbations of *CYP11A1* have never been observed in human congenital adrenal lipoid hyperplasia. Rather, the disorder in humans proved to be the result of perturbation of the gene encoding StAR. The StAR protein delivers precursors for cholesterol side-chain cleavage. The term 'acute' reflects the necessity of responding rapidly ('acutely') to corticotropin stimulation, specifically by producing a 30-kb mitochondrial protein in adrenal cells. Mapped to 8p11.2, the *StAR* gene spans 8 kb and consists of seven exons.[30]

The nearly 100 reported cases show ethnic predilection in Japanese and Koreans. A variety of different mutations have been reported (Figure 4). Bose and colleagues[30] sequenced 14 patients, finding 15 different mutations. Most were in exons 5, 6 and 7; all rendered StAR protein inactive *in vitro*. In Koreans and Japanese, the mutant allele is usually (80%) Gln258(stop). In Arabs, the mutation is usually (78%) Arg182Leu. Different mutations have been found in other ethnic groups, presumably reflecting a founder effect.

3β-Hydroxysteroid dehydrogenase (3β-ol-dehydrogenase) deficiency

46,XX cases have been reported, and, as expected, also show adrenal lipoid hyperplasia.[30] Two Japanese cases reported by Fujieda and colleagues[31] exhibited menarche and polycystic ovarian disease. One case was homozygous for a stop codon in exon 7 (Gln258X), whereas the other showed compound heterozygosity (Gln258X/deletion at codon that caused a frameshift, resulting in a non-functional gene product).

In this enzyme deficiency, levels of both androgens and estrogens are decreased. The major androgen synthesized is dehydroepiandrosterone (DHEA), a relatively weaker androgen than testosterone. DHEA alone is not capable of adequately virilizing the male fetus; thus, genital ambiguity occurs. Testes and Wolffian ducts differentiate normally. In addition to genital abnormalities, 3β-ol-dehydrogenase deficiency is associated with severe sodium wasting, predictably, given that both aldosterone and cortisol levels are decreased.

Affected females (46,XX) also show genital ambiguity; thus, 3β-hydroxysteroid dehydrogenase (3β-HSD) is the only enzyme that, when deficient, produces male pseudohermaphroditism in males and female pseudohermaphroditism in females.

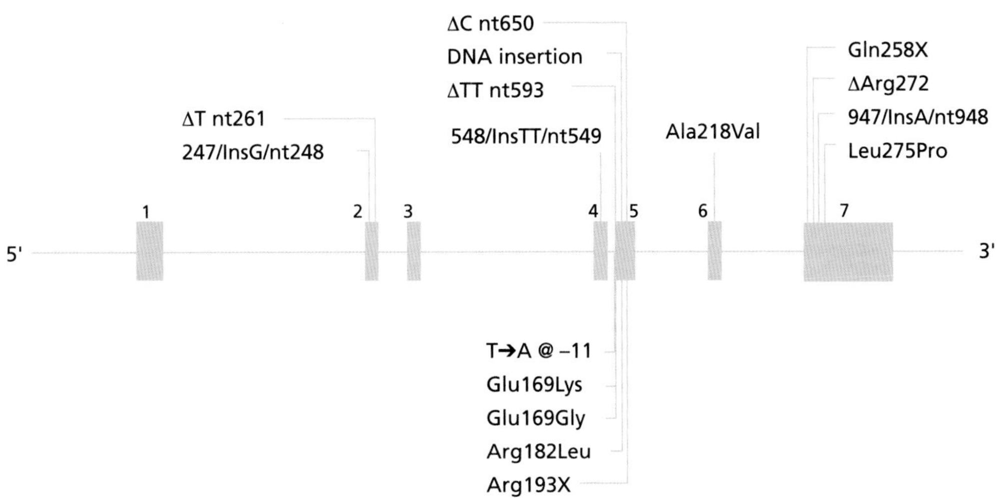

Figure 4 Diagrammatic representation of the mutations identified in the steroidogenic acute regulatory protein (*StAR*) gene associated with congenital lipoid adrenal hyperplasia. Nucleotide (nt) and amino acid numbers are given according to the cDNA sequence. Missense mutations causing amino acid substitutions are indicated by the three-letter abbreviation for the wild-type amino acid, followed by the amino acid number in the protein and the three-letter abbreviation for the substituted amino acid. X indicates a non-sense (stop) mutation. 247/InsG/nt248 is an insertion of a guanine causing a frameshift between nucleotides 247 and 248. ΔT nt261 is a deletion of a thymidine at nucleotide 261. 548/InsTT/nt549 is an insertion of two thymidines in exon 4 between nucleotides 548 and 549, causing a frameshift. ΔTT nt593 is a deletion of two thymidines at nucleotide 593, causing a frameshift. ΔC nt650 is a deletion of a cytosine at nucleotide 650, causing a frameshift. T→A @ –11 is a thymidine-to-adenine transition minus 11 nucleotides (5') from the intron 4/exon 5 junction. 947/InsA/nt948 is the insertion of an adenine between nucleotides 947 and 948, which results in a frameshift. Adapted from reference 30

3β-HSD is a microsomal enzyme. Of the five 3β-HSD genes, only type II is expressed in the adrenal cortex and gonads. Male pseudohermaphroditism thus results only from type II mutations. This gene is located on chromosome 1p13, and consists of four exons. Mutations are scattered among the four exons,[32] and have been observed in both salt-wasting and non-salt-wasting forms.

17α-Hydroxylase and 17,20-desmolase (-lyase) deficiency

17α-Hydroxylase/17,20-desmolase activity is governed by a bidirectional cytochrome P450 enzyme, encoded on 10q24→25. 46,XY cases deficient in 17α-hydroxylase/17,20-desmolase (-lyase) usually show ambiguous external genitalia, but some show female-like external genitalia.[33] Wolffian duct development and testicular development are normal. Affected 46,XX females often show hypertension, but males deficient in 17α-hydroxylase more often show normal blood pressure. Inheritance is autosomal recessive.

The *CYP17* gene is located on chromosome 10q24→25 and consists of eight exons. *CYP17* is structurally similar to *CYP21*, but no pseudogene exists. A variety of different missense mutations have been observed,[34–36] usually abolishing P450c17 function. An exception to the typical molecular heterogeneity in this disorder occurs in Mennonites, an ethnic group in which a four-base duplication beginning in codon 480 accounts for most cases.[37] The same perturbation is found in Friesland (The Netherlands), presumably from where the original mutation arose.

That a single enzyme fulfils both 17α-hydroxylase and 17,20-desmolase functions has generated considerable genetic and nosological confusion.[38] Mutations have been observed in which only 17,20-lyase function is affected, partially elucidating the confusion,[39,40] and perhaps helping to explain the family reported by Zachmann and colleagues[41] in which only 17,20-desmolase seemed to be deficient. In the rat, targeted mutagenesis and correction have identified regions responsible for specific enzymatic activities.[42]

Deficiency of 17β-hydroxysteroid dehydrogenase, type 3 (17-ketosteroid reductase)

The inability to convert dehydroepiandrosterone to testosterone is caused by a deficiency of the microsomal enzyme 17β-hydroxysteroid dehydrogenase (17β-ol-dehydrogenase), also referred to as 17-ketosteroid reductase. The enzymatic reaction is reversible; thus, designations connoting either oxidative or reductive function would be appropriate. The 17β-HSD-3 gene is located on chromosome 9.

Plasma testosterone is usually decreased; androstenedione and dehydroepiandrosterone are increased. Affected males show ambiguous or female-like external genitalia, Wolffian derivatives, bilateral testes and no Müllerian derivatives. Breast development may or may not be present, presumably reflecting the estrogen/testosterone ratio.

The 17β-HSD isozyme of relevance is type 3, consisting of 11 exons (Figure 5). Both 3β-HSD and 17β-HSD-3 are microsomal, and not cytochrome P450 enzymes. This is consistent with their action being primarily on gonads rather than adrenals. Male pseudohermaphroditism results from this enzyme deficiency. The largest aggregation of cases is in Gaza Arabs,[43,44] an ethnic group in which the most frequent mutation is a missense mutation in exon 3 (Arg80Gln). This reduces enzyme activity to 15–20% of normal.[45] 46,XX homozygous females having this mutation are asymptomatic, again consistent with 17β-HSD-3 expression limited to the testes.

Different mutations have been reported in other populations.[45] These mutations are dispersed throughout the 11 exons, with aggregation towards the 3' end (exons 9 and 10, especially). Reported mutations have included frameshifts, splice junction alterations (especially involving exon 3) and, most frequently, missense mutations. Nonsense (stop) mutations usually abolish 17β-HSD-3 activity and, hence, are associated with female external genitalia.

Deficiency of SF-1

An orphan nuclear receptor related structurally to steroid receptors, SF-1 has no known ligand (hence

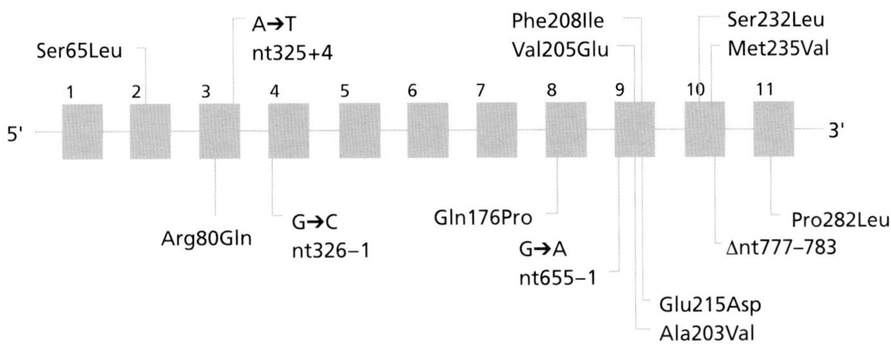

Figure 5 Diagrammatic representation of the gene encoding 17β-hydroxysteroid dehydrogenase (17β-HSD) type 3 with the mutations reported to cause 17β-HSD deficiency. The exons are represented by the numbered boxes. Missense mutations causing amino acid substitutions in the enzyme are indicated by the three-letter abbreviation for the wild-type amino acid, followed by the amino acid number in the enzyme and the three-letter abbreviation for the substituted amino acid. nt325 + 4 is a splice junction mutation, namely a transition of adenine (A) to thymidine (T) four nucleotides downstream (3') from the boundary between exon 3 and intron 3; nucleotide (nt) 325 is the closest base pair in the exon to the mutation. nt326 – 1 is a splice junction mutation, a transition of guanine (G) to cytosine (C) one nucleotide upstream (5') from the splice junction between intron 8 and exon 9. Δnt777–783 is a deletion of nucleotides 777–783. Adapted from reference 43

its designation as an 'orphan' receptor). The SF-1 gene product is encoded by the gene *FTZ1*, located on 9q33 and translated into a zinc finger protein.

Prior to recognized human cases, disruption of *FTZ1* in mice ('knock-out') was known to perturb adrenal and gonadal development, as well as hypothalamic and pituitary gonadotropes.[46,47] It was thus predicted that mutations in human *FTZ1* would perturb SF-1 receptors and cause significant abnormalities of sexual development.

The first human case of SF-1 deficiency due to mutant *FTZ1* was reported by Achermann and colleagues.[48] This unique 46,XY individual showed primary adrenal failure, female external genitalia, streak gonads and normal Müllerian derivatives responsive to hormones. Diagnosis of congenital adrenal lipoid hyperplasia was originally entertained, but the proband proved heterozygous for a 2-bp substitution in *FTZ1* at codon 35. Compound heterozygosity was presumed to exist, even though the second mutation could not be immediately detected; *SRY*, *StAR* and *DAX-1* were normal. That the mutated glycine is the last amino acid of the first zinc finger of *FTZ1* suggests disturbance of a DNA-binding site. In mice, the same (G35E) mutation fails to exert a dominant negative effect; thus, two dysfunctional alleles would be expected to be necessary to produce an adverse phenotype.

Of significance is the presence of a uterus, despite genital and gonadal abnormalities. This supports the hypothesis that SF-1 regulates repression of AMH.

5α-Reductase deficiency

Some genetic males show ambiguous external genitalia at birth, but at puberty virilize like normal males. The phenotype is characterized by phallic enlargement, increased facial hair, muscular hypertrophy and voice deepening, but no breast development. The external genitalia consist of a phallus that resembles a clitoris more than a penis, and a perineal urethral orifice. Initially called pseudovaginal perineoscrotal hypospadias, this trait was shown in 1971 to be inherited in autosomal recessive fashion.[49,50]

This disorder is the result of a deficiency of 5α-reductase,[51–53] the enzyme that converts testosterone to DHT. That intracellular 5α-reductase deficiency results in the phenotype described above is predictable, given that virilization of the external genitalia during embryogenesis requires DHT;

Wolffian differentiation requires only testosterone. However, pubertal virilization can be accomplished by testosterone alone.

Of the two 5α-reductase (*SRD5*) genes, only type II is expressed in gonads; thus, only type II (*SRD5A2*) mutations cause male pseudohermaphroditism. Located on chromosome 2p23 and consisting of five exons,[54] perturbations of *SRD5A2* typically show point mutations (Figure 6).[56] Most mutations are missense, but cases from Papua, New Guinea show deletions. Different ethnic groups show different recurrently observed mutations, scattered among the five exons. Within a given ethnic group, the mutation is usually molecularly homozygous, presumably reflecting parental consanguinity with identity by descent (founder effect).

Women (46,XX) homozygous for 5α-reductase deficiencies are fertile, and show normal ovarian function.[55] Breast development is normal.[57] Limb and pubic hair may be reduced, and menarche delayed.

Androgen insensitivity

In complete androgen insensitivity, 46,XY individuals show bilateral testes, female external genitalia, a blindly ending vagina and no Müllerian derivatives. Incidence is 2–5 per 100 000.[58,59] The phenotype is entirely predictable, given that the underlying pathogenesis involves cellular inability to respond to testosterone. AMH is synthesized normally by Sertoli cells, and the response to AMH is normal; thus, no uterus is present, as expected in XY individuals. Also predicted, given that testes are able to synthesize estrogens in an unimpeded fashion, is that affected individuals manifest breast development and pubertal feminization. Feminization occurs because circulating androgens fail to antagonize cellular action of estrogens. Luteinizing hormone (LH) is disproportionally increased, reflecting hypothalamic unresponsiveness to testosterone.

Despite pubertal feminization, some individuals with androgen insensitivity show clitoral enlargement and labioscrotal fusion; the appellation 'incomplete or partial androgen insensitivity' (PAI) (incomplete testicular feminization) is applied. A milder end of the androgen insensitivity spectrum consists of males manifesting only gynecomastia or oligospermia/azoospermia (mild androgen insensitivity, MAI).[60–63] In MAI, gynecomastia, impotence and inadequate virilization are consistently observed, but often a few sperm are recoverable.

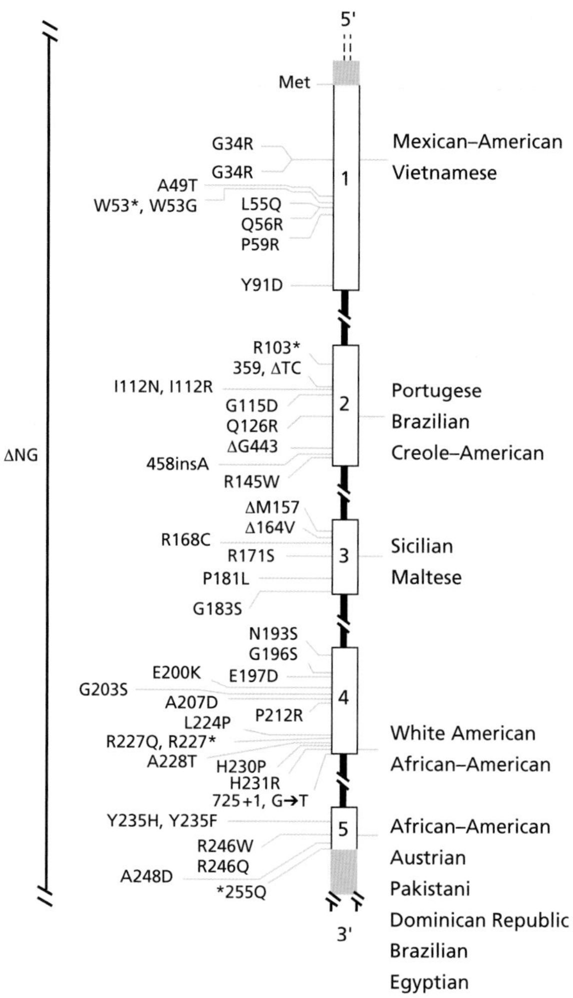

Figure 6 Mutations in the steroid 5α-reductase 2 gene and protein. A schematic diagram of the 5α-reductase 2 gene is shown in the middle. On the left are the locations of 45 different mutations. Countries of origin of recurring mutations are indicated on the right. Point mutations are represented by the amino acid symbols and the codon numbers. ΔNG represents the deletion of the gene in the New Guinea cohort. The initiating methionine is designated Met in exon 1, and the normal termination codon is designated with an asterisk in exon 5. Adapted from reference 55

Complete (CAI), partial (PAI) and mild (MAI) androgen insensitivities are inherited in X-linked recessive fashion. All result from mutations of the same androgen receptor gene, located on Xq11. Approximately 90 000 bp in length and consisting of eight exons, this gene encodes fewer than 2000 amino acids (Figure 7). Exon 1 appears to confer regulatory function. The DNA-binding domain encompasses amino acids 552–616, encoded by exon 2 and part of exon 3. The latter part of exon 3 and part of exon 4 encodes a bipartite nuclear localization signal (amino acids 617–636). The C-terminal region of 250 amino acids extending over exons 4–8 is the androgen-binding domain.[64] The androgen receptor (*AR*) gene contains two homopolymeric amino acid 'repeats' of varying length: one polyglutamine (9–36 amino acids) and the other polyglycine (10–31 amino acids).

Over 200 different *AR* gene mutations have been reported.[59,65] A registry of these mutations is maintained at McGill University (e-mail: mc33@musica.mcgill.ca). Table 2 categorizes reported mutations by genotypic class. No single perturbation has proved to be paramount (molecular heterogeneity). Large deletions and mutations causing premature message termination (stop codon) lead to non-functional androgen receptors and predictably cause complete androgen insensitivity. Single nucleotide mutations (single amino-acid substitutions) constitute the vast majority of reported mutations,[59,65] but show no invariate correlation between phenotype and molecular genotype. Exon 1 encompasses half the *AR* gene but only 10% of recognized mutations. Mutations in exons 2 and 3 (the DNA-binding domain) are more common, producing either CAI or PAI. The most frequent sites for mutations are exons 5–8, the androgen-binding region. In this region, most mutations are missense, in no seemingly predictable fashion causing PAI, CAI or even MAI. An especially vexing dilemma is that identical mutations may be associated with different phenotypes (PAI or MAI).[59,65] Presumably, multiple interactions exist, and certain amino acids are more pivotal for androgen binding than others; however, not until gene products can be studied more thoroughly (proteomics) is this confusion likely to be elucidated fully. If no mutation is found, the defect presumably involves regulation or downstream processing of the *AR* gene product.

Molecular diagnosis and heterozygote detection

Searching for the presence or absence of a specific molecular defect in the *AR* gene would be desirable clinically, but this may be possible only if the mutation has already been identified in another family member. Facile ability to sequence the entire gene for diagnostic purposes is the eventual goal, but, at present, testing is usually available only on a research basis.[65]

Heterozygote detection is invaluable for genetic counselling and prenatal genetic diagnosis. In families with only one affected male, it may be difficult to ascertain whether the case is a new (*de novo*) mutation or the result of transmission from a carrier (mother). In one-third of all CAI cases, no other family member is affected, as expected for genetically lethal X-linked recessive conditions (i.e. affected males unable to reproduce). In two-thirds of cases, the phenotypically normal mother should

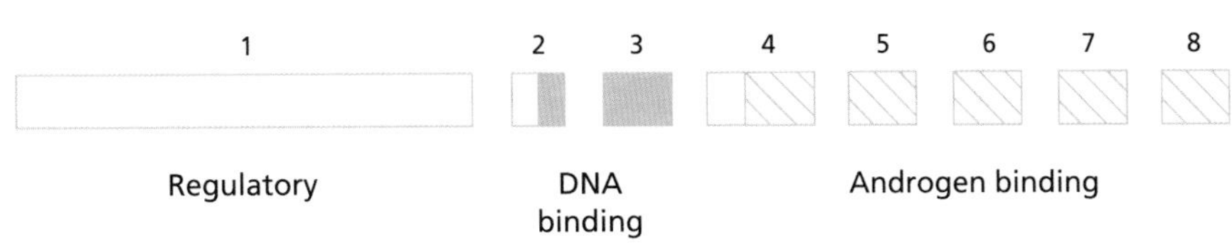

Figure 7 A linear representation of the major structure–function domains and putative subdomains of the androgen receptor gene, which consists of eight exons. Each box represents one exon

Table 2 Classes of mutations in the androgen receptor gene database maintained at McGill University, correlated with androgen insensitivity syndrome (AIS) phenotype. Modified from reference 59

Type	*AIS phenotype*	*Number of mutations*
Single-base mutations		
Amino acid substitution	complete	134
	partial	104
	mild	10
Multiple amino acid substitutions	complete	2
Splice junction substitution	complete	5
	partial	1
Splice junction insertion	complete	1
Intron substitution	complete	1
Premature termination codons	complete	22
	partial	1
Structural defects		
Complete gene deletions	complete	3
Partial gene deletions (6–30 bp, 1–7 exons)	complete	8
	mild	1
1- to 4-bp Deletions	complete	6
Intron deletion	partial	1
Splice junction deletion	complete	1
Insertions	complete	4
Base pair duplications	complete	1
Total		306

be heterozygous. Sometimes heterozygous women show sparse pubic hair and delayed puberty, but, lacking these features, the genotype of the mother of an affected case would be unknown.

In families in which a mutant *AR* gene is known to be segregating (for example, more than one affected case exists), the probability that any given female relative is a heterozygote can be estimated through linkage analysis, even if no mutations have been identified. Intragenic or flanking polymorphic markers near the *AR* gene may enable a woman's carrier status to be determined; DNA from affected and unaffected male relatives is needed, and the family must further be informative for some linked markers. If only a single affected individual is available, this approach is not feasible.

Gonadal dysgenesis and 46,XY 'females'

The prototypic XY female with gonadal dysgenesis shows normal female external genitalia, vagina, uterus and Fallopian tubes. At puberty, secondary sexual development fails to occur. Height is normal and somatic anomalies are usually not present. External appearance is identical clinically with the forms of 46,XX gonadal dysgenesis not associated with somatic anomalies: follicle stimulating hormone (FSH) and LH are elevated, estrogens are decreased. Approximately 20–30% of XY gonadal dysgenesis patients develop a dysgerminoma or gonadoblastoma, sometimes arising in the first or second decade.[66]

Failure of the testes to develop or function before 7–8 weeks' embryogenesis would produce the above phenotype, as originally described by Jost[1] in rabbits. In at least some cases, the gonads of human XY females are embryologically ovaries,[67] an observation supporting constitutive differentiation of the indifferent gonad along ovarian lines, unless directed otherwise.

XY gonadal dysgenesis without somatic anomalies can result from several distinct etiologies (Table 3), all of which may produce clinically indistinguishable individuals. In none is the sex of rearing in doubt, all being unequivocally female. Already alluded to is the mutation or deletion of *SRY*, involving the *SRY* HMG box.[11,68,69] Approximately 10–15% of sporadic XY gonadal dysgenesis cases show identifiable perturbations of *SRY*.[10,70] A second form of XY gonadal dysgenesis results from a gene segregating in the fashion expected of an X-linked recessive.[49,71–76] A third cause involves perturbation of the X-linked gene dose-sensitive sex-reversal region, which contains *DAX-1* as already alluded to and involves either duplication of Xp21.2→p22.1,[77] or up-regulation due to *WNT-4* overexpression.[14] However, only one of 27 '46,XY sex-reversal females' showed duplication of Xp21.[77] Fourth, deletions of several autosomal regions may be associated with the XY gonadal dysgenesis phenotype: 9p, 2q33, 10q26. Fifth, duplication of 1p may produce the same phenotype, acting through overexpression of the *WNT-1* gene encoded in this region.[14]

In several other conditions, molecular perturbation produces a spectrum of phenotypic expression with respect to sexual differentiation. External genitalia range from phenotypic female to genital ambiguity to nearly normal male; somatic anomalies coexist.

XY gonadal dysgenesis and WT-1

WT-1 (Wilms' tumor oncogene) is a complex gene located on chromosome 11p. Ten exons in length, at least 32 isoforms exist.[9] Peculiarities of *WT-1* include alternate translation initiation sites, alternate differential splicing in exon 5 and alternate donor acceptor sites at the exon 9 boundary. The last results in differential splicing involve the amino acids lysine (K), threonine (T) and serine (S).[78,79]

Table 3 Clinical delineation of 46,XY sex reversal (XY females)

A. Teratogenic
Cyproterone acetate (no reported cases)
Finisteride (no reported cases)
Flutamide (no reported cases)
B. Cytogenetic
45,X/46,XY (female external genitalia)
C. XY gonadal dysgenesis without somatic anomalies
Perturbations of *SRY* (HMG-box)
Duplication Xp (*DAX-1*)
X-linked recessive
Unknown etiology (no detectable molecular perturbation or heritability)
D. Autosomal deletions
Del(2p)
Del(9p) (*DMRT*)
Del(10q)
E. Autosomal duplications
Dup(1p) (*WNT-4*)
Dup(17p)
F. XY gonadal dysgenesis and Wilms' tumor oncogene (WT-1)
Denys–Drash syndrome
Frasier syndrome
G. XY gonadal dysgenesis and campomelic dysplasia (SOX-9)
H. XY gonadal dysgenesis/α-thalassemia X-chromosome (ATX)
I. Other malformation syndromes
Ectodermal anomalies (Brosnon)
Genitopalatocardiac (Gardner–Silengo–Wachtel)
Spastic paraplegia–optic atrophy–microcephaly (Teebi)
Brachioskeletal–genital
Other
J. Germ cell failure in both sexes (46,XY cases)
K. Agonadia (46,XY cases)
L. Leydig cell hypoplasia (LH receptor mutation)
M. Steroidogenic factor-1 (SF-1) deficiency (FTZ-1)
N. Testosterone biosynthetic defects
17α-Hydroxylase/17,20 desmolase deficiency
3β-ol-Dehydrogenase/3α-hydroxysteroid dehydrogenase deficiency
O. Steroidogenic factor-1 (SF-1) deficiency

HMG, high-mobility group; *DMRT*, Doublesex and mab3 related transcription factor-1; LH, luteinizing hormone

The association of nephropathy, genital ambiguity and Wilms' tumor was appreciated long before the *WT-1* gene and its perturbations were identified. The constellation of anomalies became known as the Denys–Drash syndrome.[80] Mental retardation, aniridia and Wilms' tumor next proved to be associated with deletions of 11p13, following which Gessler and colleagues[81] showed that the region contained *WT-1*. Most cases of 46,XY Denys–Drash syndrome are males with genital ambiguity and dysgenetic testes, but some are female (46,XX) with streak gonads.[82]

A related condition is Frasier syndrome, characterized by renal parenchymal disease and XY sex reversal.[83,84] Frasier syndrome and Denys–Drash syndrome are caused by different perturbations of *WT-1*.[78] In Frasier syndrome, the ratio of KTS (+) and non-KTS (–) transcripts differs between the testis and ovary.[85]

In mice, the *WT-1*-null mutation results in the failure of either sex to develop gonads (and kidneys).[86] Both *SRY* and *WT-1* are present in the testes, with the latter expressed earlier. Thus, *WT-1* may be required upstream of *SRY*, i.e. before *SRY* is expressed.[9]

XY gonadal dysgenesis and campomelic dysplasia

Campomelic dysplasia is characterized by long-bone bowing, abnormal facies and other skeletal anomalies. The locus is on 17q24.3→q25.1, and its perturbation involves the DNA-binding protein *SOX-9*.[87,88] Mutations in *SOX-9* may result in XY sex-reversal, but not obligatorily. *SOX-9* mutations may produce 46,XY gonadal dysgenesis in the absence of skeletal anomalies.[89]

Molecular etiology appears to involve nearby modifying genes or mutations in the *SOX-9* promoter region.[90] The 1-mb length of this region makes mutations difficult to detect. Olney and colleagues[91] reported interstitial deletion of 17q (q23.3→q24.3) associated with campomelic dysplasia and complete absence of *SOX-9*, suggesting haploinsufficiency. Huang and associates[16] reported duplication of 17p associated with presumptive bilateral testes and incomplete male-like external genitalia. In mice, insertional mutagenesis resulting in a 150-kb deletion of a gene (*oddsex, ods*) upstream of *SOX-9* results in unexpected testicular development in XX mice, despite the absence of a Y chromosome on *SRY*.[17] Thus, *SRY* was postulated normally to derepress this region, allowing testicular development in XY embryos. A mutation resulting in unscheduled derepression would result in the same phenotype in XX embryos.

XY gonadal dysgenesis and X-linked α-thalassemia

The X-linked gene (Xq12→q21.31) X-linked α-thalassemia (*ATX*) causes mental retardation, α-thalassemia, abnormal facies (upturned nose, carp-shaped mouth) and male pseudohermaphroditism. The phenotype in 46,XY individuals encompasses genital ambiguity as well as complete sex reversal (46,XY with female external genitalia).[92–95] *ATX* is a member of the DNA helicase family. Approximately 25 different mutations are reported.[96,97]

XY gonadal dysgenesis and other multiple malformation syndromes

Genital ambiguity (male pseudohermaphroditism) is not uncommon in other multiple malformation syndromes. These include Meckel–Gruber syndrome, Smith–Lemli–Opitz type I syndrome, brachioskeletal–genital syndrome,[98] esophageal–facial–genital syndrome[99] and additional disorders tabulated elsewhere.[100] These disorders are usually inherited in either autosomal recessive or, rarely, X-linked recessive fashion. Many other syndromes are associated with simple cryptorchidism or simple hypospadias, and sex of rearing is not in doubt. In many of these disorders, causative genes are identified, but most are not of immediate relevance here.

Of special interest is Smith–Lemli–Opitz (SLO) syndrome. In this relatively common (1 : 10 000) autosomal recessive syndrome, 46,XY individuals show genital abnormalities ranging from hypospadias to male pseudohermaphroditism.[101] In type I SLO, hypospadias is the most common genital abnormality, whereas, in type II, external genitalia are ambiguous or even female (sex reversal).[102] Both types I and II are caused by mutation of the gene that encodes the enzyme responsible for converting 7-hydroxycholesterol to cholesterol.[103,104] The most common molecular perturbation is an exon–intronic splicing defect. During pregnancy, maternal serum estriol is characterized by low to non-detectable dehydro-estriol and dehydropregnanetriol, detectable in both maternal serum and urine, whereas these compounds are non-detectable in normal pregnancies.[105] These findings make feasible the detection of an SLO pregnancy by maternal urine or serum analyte screening.

Agonadia (testicular regression)

In agonadia, the gonads are, by definition, completely absent. The phenotype thus differs from disorders in which gonads are non-functional but anatomically represented by fibrous streaks (streak gonads). Most cases of agonadia are 46,XY, but a few are 46,XX. External genitalia in 46,XY agonadia are abnormal but female-like; only rudimentary Müllerian or Wolffian derivatives present. External genitalia usually consist of a phallus about the size of a clitoris, underdeveloped labia majora and nearly complete fusion of labioscrotal folds; sometimes, the external genitalia resemble those of normal females. Somatic anomalies (craniofacial anomalies, vertebral anomalies and mental retardation) are present in about one-half of cases.

Explanations for 46,XY agonadia have usually focused on loss of testes early in embryogenesis (hence the term testicular regression preferred by some). However, any explanation must also take into account abnormal external genitalia and lack of internal genital ducts. Fetal testes could have existed transiently, sufficiently long to initiate male differentiation and suppress Müllerian differentiation but not sufficiently long to complete male differentiation. Another possibility is defective connective tissue, reflecting concomitant somatic anomalies (developmental field defect).

Heritable tendencies exist.[106] The molecular basis is uncertain, but *SRY* has not been shown to be perturbed.[10]

Leydig cell hypoplasia (luteinizing hormone receptor gene mutation)

With bilateral testes devoid of Leydig cells, 46,XY individuals have female external genitalia and no uterus. Epididymides and vasa deferentia are present, and serum LH is elevated. Affected siblings have been reported and parental consanguinity observed. Thus, autosomal recessive inheritance has long been accepted.

The molecular basis involves mutation in the luteinizing hormone receptor (*LHR*) gene, located on chromosome 2.[107] Leydig cells presumably fail to develop because LH cannot exert its normal effect during embryogenesis, leading to inadequate virilization and gonadal differentiation. Embryonic testes presumably secrete AMH, explaining the absence of a uterus as expected in 46,XY individuals. The *LHR* gene consists of 11 exons and 699 amino acids. These exons comprise intracellular domains, intracellular and extracellular loops, transmembrane domains and extracellular domains. Complete resistance to LH produces XY females, whereas partial resistance leads to males with a small penis or hypospadias. Approximately one dozen different *LHR* mutations have been found in 46,XY 'females' with Leydig cell hypoplasia, and sometimes their 46,XX sibs[107] (46,XX cases present with primary amenorrhea). Kremer and colleagues[108] reported two siblings of consanguineous parents; homozygosity for a missense mutation (–C593R) existed. In other cases, deletions, various point mutations and stop codons have been recognized.[109–111]

TRUE HERMAPHRODITISM

True hermaphrodites have both ovarian and testicular tissue. A separate ovary and a separate testis may exist, or, more often, one or more ovotestes. Most true hermaphrodites (60%) have a 46,XX chromosomal complement; however, 46,XX/46,XY, 46/XY, 46,XX/47,XXY and other complements are observed.[112] Phenotype probably varies by chromosomal constitution, but it suffices to discuss the overall phenotype of true hermaphrodites in general.

Phenotype

Without medical intervention, two-thirds of true hermaphrodites were traditionally raised as males.[2] In contrast, the external genitalia are usually ambiguous or predominantly female. Breast development typically occurs at puberty, even with predominantly male external genitalia. Gonadal tissue may be located in the ovarian, inguinal or labioscrotal regions. Interestingly, a testis or an ovotestis is more likely to be present on the right than on the left side, paralleling right-sided testicular dominance in some lower animals. The greater is the proportion of testicular tissue in an ovotestis, the greater is the likelihood of gonadal descent. In 80% of ovotestes, testicular and ovarian components are juxtaposed end to end.[112,113] Testicular tissue is softer and darker than ovarian tissue. An ovotestis may thus be detectable by inspection, palpation, ultrasound scan or magnetic resonance imaging (MRI). Spermatozoa are rarely identified; however, ostensibly normal oocytes are often present, even in ovotestes.

A uterus is usually present, albeit sometimes in bicornuate or unicornuate form. The fimbriated end of the Fallopian tube is not infrequently occluded ipsilateral to an ovotestis, and squamous metaplasia of the endocervix may occur.[113] Menstruation is not uncommon, and may be manifested as cyclic hematuria. The presence of a uterus is helpful diagnostically because, among individuals with genital ambiguity who have a Y chromosome, only 46,XY true hermaphroditism, SF-1 deficiency and 45,X/46,XY mosaicism are associated with a uterus. Overall, only 10–20% of XX true hermaphrodites lack a uterus. A uterus is more likely to be lacking in familial XX true hermaphrodites and in families in which XX males and XX true hermaphrodites coexist.[25,112] The absence of only a uterine horn usually indicates ipsilateral testis or ovotestis.

Etiology

46,XX/46,XY true hermaphroditism is presumably caused by chimerism, the presence in a single individual of two or more cell lines. By definition, each is derived from a different zygote. A mutation in the *SRY* gene has been a rare cause of some 46,XY cases.[114] However, most 46,XY cases are probably unrecognized chimeras.[112] Chimerism seems unlikely to explain more than a rare case of 46,XX true hermaphroditism. Translocation of *SRY* from the Yp to an Xp during paternal meiosis is not the explanation either, unlike the situation in 46,XX males in which 80% of cases show *SRY* on this basis. Only in exceptional cases do 46,XX true hermaphrodites show *SRY*.[114–118]

The following hypotheses could explain the testicular differentiation observed in 46,XY true hermaphrodites:

(1) Stochastic, non-random, inactivation of an X chromosome containing *SRY*, which arose through Xp/Yp interchange;[119]

(2) Interchange involving not Xp and Yp but Xq and Yp. Recall that there exist three regions of X–Y homology: Xp/Yp, Xq/Yp, Xp/Yq.[120] Xq/Yp recombination would place *SRY* on Xq near the X-inactivation gene (*XIST*), located at Xq13. Partial inactivation of *SRY* could result, mitigating against complete sex reversal and leading instead to partial testicular expression and, hence, XX true hermaphroditism. Such a case has been reported;[121]

(3) Somatic mutations of *SRY* restricted to the gonads;[122]

(4) Mendelian factors.

Mendelian factors seem highly plausible, and are probably the principal explanation. Sibships of XX true hermaphrodites are reported.[4,112,123–125] 46,XX males and 46,XX true hermaphrodites have been observed in the same kindred. In this case, 46,XX males usually show genital ambiguity, unlike sporadic 46,XX males, and 46,XX true hermaphrodites in these families atypically lack a uterus. It could be postulated that perturbation (derepression) of a normally dormant autosomal gene induces inappropriate (testicular) gonadal development in 46,XX individuals. Offered over a decade ago, this hypothesis now seems highly plausible, given the many autosomal regions known to influence testicular differentiation. The prime candidate is human 17q, which contains *SOX-9*.[16,17]

SUMMARY AND FUTURE PERSPECTIVES

Applying molecular technologies to elucidate the disorders of sex differentiation has greatly refined our knowledge, but not changed the fundamental nosology of clinical delineation and management. The pivotal step in the genetic control of sex differentiation remains the testes-determining factor on the Y chromosome, which has proved to be *SRY*. It has become clear that genes on the X chromosome and various autosomes also play key roles. The most pivotal appear to be *DAX-1* on the X and *SOX-9* on chromosome 17. Testicular differentiation seems governed not only by *SRY*, but by an autosomal gene(s) that must be derepressed by *SRY* in order to exert its expected action. The clinical significance of these genes is that physicians can expect some disorders of sexual differentiation to result from autosomal genes and autosomal chromosomal rearrangements.

It remains unclear whether primary ovarian differentiation is controlled by specific gene(s) or is constitutive, that is, a default embryonic pathway from the indifferent gonad. A once-touted candidate gene (*DAX-1*) appears not to be pivotal in humans. Irrespective of the above, the prevention of ovarian attrition (i.e. ovarian maintenance) is clearly under genetic control. Many syndromes of gonadal failure (XY and XX) are caused by autosomal genes, but in few is the gene currently known. In the foreseeable future, we can expect the identification of many more, facilitated by a host of candidate genes identified through animal knock-out technology and human genome sequencing. Some of these genes are predictable causes of gonadal failure, like those relating to the FSH receptor, LH receptor or primordial germ cell differentiation. Others have been a surprise, because superficially they seem to have little to do with reproduction. Examples include bone morphogenetic protein, DNA repair genes, heat shock proteins, cell cycle genes and oncogenes. The unifying disturbance probably involves defects in meiosis or mitosis.

Considerable confusion persists in the etiology of true hermaphroditism. Little progress has been made during the molecular era. Most 46,XX sex-reversed males are the result of *SRY* interchanged from the paternal Y to the paternal X during paternal meiosis, but no such explanation exists for XX true hermaphroditism. Only a few exceptional cases have a molecular explanation. Familial aggregates exist, suggesting that autosomal genes are pivotal. Once identified, these genes will be important for clarifying reproductive embryology.

Some principles have been gleaned through molecular analysis of the disorders of male pseudohermaphroditism and XY sex reversal. One is that molecular heterogeneity is typical. This has permitted limited correlation between molecular genotype and phenotype. Mutations producing deletions, stop codons and frameshifts result in the most severe phenotypes. Point mutations at the 3' end may be, but are not always, associated with the retention of some gene function and, hence, a less severe phenotype. The converse holds true for point mutations at the 5' end. Given that heterogeneity exists within the causative gene, it is difficult to provide a diagnosis on a molecular basis. In the future, more facile sequencing of entire genes can be expected. Now, testing is available often only on a research basis, except for a limited number of common mutant alleles. Thus, a major disappointment has been that, despite considerable progress in the elucidation and identification of genes, diagnosis is still being made in much the same way as in the early 1990s. If the molecular basis is identified, however, prenatal genetic diagnosis and heterozygote detection can be readily offered. Molecular diagnosis is likely to be most useful in disorders in which a specific mutation exists in a given ethnic group, for example 5α-reductase in certain populations.

In true hermaphroditism and the disorders of male pseudohermaphroditism not associated with a known enzymatic defect, progress has been slower. As noted above in the section on 'applied technology', progress may depend on fortuitous observations, to localize the chromosomal region containing the gene. Knowledge of human genome sequences will facilitate efforts once the region is found, but the location of the region often remains elusive. Genes in animals shown through knock-out technology to affect gonadal and genital differentiation serve as a rich repository of candidate genes. Scanning the entire human genome for informative sequences (genes) is probably fruitless, given that gonadal failure is a pedestrian clinical feature, adversely affected by hundreds of different genes acting through disparate mechanisms.

Overall, we can expect continued progress in identifying the genes responsible for sexual differentiation and its perturbation. New syndromes will doubtless be identified, but most will probably be rare. As new genes are identified and their modes of action clarified, gene product(s) can be synthesized and applied therapeutically.

REFERENCES

1. Jost A. Problems of fetal endocrinology. The gonadal and hypophyseal hormones. *Recent Prog Horm Res* 1953 8: 379–418.
2. Simpson JL. *Disorders of Sexual Differentiation: Etiology and Clinical Delineation*. New York: Academic Press, 1976: 1–466.
3. Kooperman P, Munsterberg A, Capel B *et al.* Expression of a candidate sex-determining gene during mouse testis determination. *Nature* 1990 348: 450–452.
4. Simpson JL, Elias S. *Genetics in Obstetrics and Gynecology*, 3rd edn. Philadelphia, PA: WB Saunders, 2002: in press.
5. Jirásek J. Principles of reproductive embryology. In Simpson JL, ed. *Disorders of Sexual Differentiation*. San Diego: Academic Press, 1976: 51–111.
6. Gubbay J, Collignon J, Koopman P *et al.* A gene mapping to the sex-determining region of the mouse Y chromosome is a member of a novel family of embryonically expressed genes. *Nature* 1990 346: 245–250.
7. Sinclair AH, Berta P, Palmer MS *et al.* A gene from the human sex-determining region encodes a protein with homology to a conserved DNA-binding motif. *Nature* 1990 345: 240–244.
8. Koopman P, Munsterberg A, Capel B, Vivian N, Lovell-Badge R. Expression of a candidate sex-determining gene during mouse testis differentiation. *Nature* 1990 348: 450–452.
9. McElreavey K, Fellous M. Sex determination and the Y chromosome. Seminars in Medical Genetics. *Am J Med Genet* 1999 89: 176–185.
10. Pivnick EK, Wachtel S, Woods D, Simpson JL, Bishop CE. Mutations in the conserved domain of *SRY* are uncommon in XY gonadal dysgenesis. *Hum Genet* 1992 90: 308–310.
11. Cameron F, Sinclair AH. Mutations in *SRY* and *SOX9*: testis-determining genes. *Hum Mutat* 1997 9: 388–395.
12. Ostrer H. Sexual differentiation. *Semin Reprod Med* 2000 18: 41–49.
13. Giese K, Pagel J, Grosschedl R. Distinct DNA-binding properties of the high mobility group domain of murine and human *SRY* sex-determining factors. *Proc Natl Acad Sci USA* 1994 91: 3368–3372.
14. Jordan BK, Mohammed M, Ching ST *et al.* Up-regulation of *wnt-4* signaling and dosage-sensitive sex reversal in humans. *Am J Hum Genet* 2001 68: 1102–1109.
15. McDowall S, Argentaro A, Ranganathan S *et al.* Functional and structural studies of wild type *SOX9* and mutations causing campomelic dysplasia. *J Biol Chem* 1999 274: 24023–24030.
16. Huang B, Wang S, Ning Y, Lamb AN, Bartley J. Autosomal XX sex reversal caused by duplication of *SOX9*. *Am J Med Genet* 1999 87: 349–353.
17. Bishop CE, Whitworth DJ, Qin Y *et al.* A transgenic insertion upstream of *SOX9* is associated with dominant XX sex reversal in the mouse. *Nat Genet* 2000 26: 490–494.
18. Waggoner DJ, Chow CK, Dowton SB, Watson MS. Partial monosomy of distal 10q: three new cases and a review. *Am J Med Genet* 1999 86: 1–5.
19. Veitia R, Nunes M, Rapport R *et al.* Sywers syndrome and 46,XY partial gonadal dysgenesis associated with 9p deletions and the absence of monosomy 9p syndrome. *Am J Hum Genet* 1998 63: 901–905.
20. Slavotinek A, Schwarz C, Getty JF *et al.* Two cases with interstitial deletions of chromosome 2 and sex reversal in one. *Am J Med Genet* 1999 86: 75–81.
21. Wieacker P, Missbach D, Jakubiczka S, Borgmann S, Albers N. Sex reversal in a child with the karyotype 46,XY,dup(1)(p22.3→p32.3). *Clin Genet* 1996 49: 271–273.
22. Simpson JL. Phenotypic–karyotypic correlations of gonadal determinants: current status and relationship to molecular studies. In Sperling K, Vogel F, eds. *Proceedings of 7th International*

Congress, Berlin, 1986. Heidelberg: Springer-Verlag, 1987: 224–232.

23. German J. Gonadal dimorphism explained as a dosage effect of a locus on the sex chromosomes, the gonad-differentiation locus (GDL). *Am J Hum Genet* 1988 42: 414–421.
24. Yu RN, Ito M, Saunders TL, Camper SA, Jameson JL. Role of *Ahch* in gonadal development and gametogenesis. *Nat Genet* 1998 20: 353–357.
25. Simpson JL, Kaufman R. Fetal effects of progestogens and diethylstilbestrol. In Fraser IS, Jansen RPS, Lobo RA, Whitehead MI, eds. London: Churchill Livingstone, 1998: 533–553.
26. Simpson JL. Disorders of the gonads, genital tract, and genitalia. In Emery AEH, Rimoin DL, eds. *Principles and Practice of Medical Genetics,* 4th edn. London: Harcourt Brace, 2002: in press.
27. Frydman M, Kauschansky A, Zamir R, Bonne-Tamir B. Familial lipoid adrenal hyperplasia: genetic marker data and an approach to prenatal diagnosis. *Am J Med Genet* 1986 25: 319–332.
28. Chung BC, Matteson KJ, Voutilainen R, Mohandas TK, Miller WL. Human cholesterol side-chain cleavage enzyme P450scc: cDNA cloning assignment of the gene to chromosome 15 and expression in the placenta. *Proc Natl Acad Sci USA* 1986 83: 8962–8966.
29. Yang X, Iwamoto K, Wang M *et al.* Inherited congenital adrenal hyperplasia in the rabbit is caused by a deletion in the gene encoding cytochrome P450 cholesterol side-chain cleavage enzyme. *Endocrinology* 1993 132: 1977–1982.
30. Bose HS, Sugawara T, Strauss JF III, Miller WL. The pathophysiology and genetics of congenital lipoid adrenal hyperplasia. International Congenital Lipoid Adrenal Hyperplasia Consortium. *N Engl J Med* 1996 335: 1870–1878.
31. Fujieda K, Tajima T, Nakae J *et al.* Spontaneous puberty in 46,XX subjects with congenital lipoid adrenal hyperplasia. Ovarian steroidogenesis is spared to some extent despite inactivating mutations in the steroidogenic acute regulatory protein (StAR) gene. *J Clin Invest* 1997 99: 1265–1271.
32. Simard J, Rheaume E, Mebarki F *et al.* Molecular basis of human 3β-hydroxysteroid dehydrogenase deficiency. *J Steroid Biochem Mol Biol* 1995 53: 127–138.
33. Heremans GF, Moolenaar AJ, Van Gelderen HM. Female phenotype in a male child due to 17α-hydroxylase deficiency. *Arch Dis Child* 1976 51: 721–723.
34. Yanase T. 17αhydroxylase/17,20-lyase defects. *J Steroid Biochem Mol Biol* 1995 53: 153–157.
35. Kagimoto K, Waterman MR, Kagimoto M *et al.* Identification of a common molecular basis for combined 17α-hydroxylase/17,20-lyase deficiency in two Mennonite families. *Hum Genet* 1989 82: 285–286.
36. Rumsby G, Skinner C, Lee HA, Honour JW. Combined 17α-hydroxylase/17,20-lyase deficiency caused by heterozygous stop codons in the cytochrome P450 17α-hydroxylase gene. *Clin Endocrinol (Oxf)* 1993 39: 483–485.
37. Imai T, Yanase T, Waterman MR, Simpson ER, Pratt JJ. Canadian Mennonites and individuals residing in the Friesland region of The Netherlands share the same molecular basis of 17α-hydroxylase deficiency. *Hum Genet* 1992 89: 95–96.
38. Nebert DW, Nelson DR, Adesnik M *et al.* The P450 superfamily: updated listing of all genes and recommended nomenclature for the chromosomal loci. *DNA* 1989 8: 1–13.
39. Biason-Lauber A, Leiberman E, Zachmann M. A single amino acid substitution in the putative redox partner-binding site of P450c17 as cause of isolated 17,20-lyase deficiency. *J Clin Endocrinol Metab* 1997 82: 3807–3812.
40. Geller DH, Auchus RJ, Mendonca BB, Miller WL. The genetic and functional basis of isolated 17,20-lyase deficiency. *Nat Genet* 1997 17: 201–205.
41. Zachmann M, Vollmin JA, Hamilton W, Prader A. Steroid 17,20-desmolase deficiency: a new cause of male pseudohermaphroditism. *Clin Endocrinol* 1972 1: 369–385.
42. Kitamura M, Buczko E, Dufau ML. Dissociation of hydroxylase and lyase activities by site-directed mutagenesis of the rat P45017α. *Mol Endocrinol* 1991 5: 1373–1380.
43. Andersson S, Geissler WM, Wu L *et al.* Molecular genetics and pathophysiology of β-hydroxysteroid dehydrogenase 3 deficiency. *J Clin Endocrolin Metab* 1996 81: 130–136.
44. Rösler A, Kohn G. Male pseudohermaphroditism due to 17β-hydroxysteroid dehydrogenase deficiency: studies on the natural history of the defect and effect of androgens on gender role. *J Steroid Biochem* 1983 19: 663–674.

45. Rösler A. Steroid 17β-hydroxysteroid dehydrogenase deficiency in man: an inherited form of male pseudohermaphroditism. *J Steroid Biochem Mol Biol* 1992 43: 989–1002.

46. Parker KL, Schimmer BP. Steroidogenic factor 1: a key determinant of endocrine development and function. *Endocr Rev* 1997 18: 361–377.

47. Luo X, Ikeda Y, Parkee KL. A cell-specific nuclear receptor is essential for adrenal and gonadal development and sexual differentiation. *Cell* 1994 77: 481–490.

48. Achermann JC, Ito M, Ito M, Hindmarsh PC, Jameson J. A mutation in the gene encoding steroidogenic factor-1 causes XY sex reversal and adrenal failure in humans. *Nat Genet* 1999 22: 125–126.

49. Simpson JL, New M, Peterson RE, German J. Pseudovaginal perineoscrotal hypospadias (PPSH) in sibs. *Birth Defects* 1971 7(6): 140–144.

50. Opitz JM, Simpson JL, Sarto GE *et al.* Pseudovaginal perineoscrotal hypospadias. *Clin Genet* 1972 3: 1–36.

51. Imperato-McGinley J, Guerrero L, Gauiter T, Peterson RE. Steroid 5α-reductase deficiency: an inherited form of male pseudohermaphroditism. *Science* 1974 186: 1213–1215.

52. Walsh C, Madden JD, Harrod MJ *et al.* Familial incomplete male pseudohermaphroditism, type 2. Decreased dihydrotestosterone formation in pseudovaginal perineoscrotal hypospadias. *N Engl J Med* 1974 291: 944–949.

53. Peterson RE, Imperato-McGinley J, Gautier T, Sturla E. Male pseudohermaphroditism due to steroid 5α-reductase deficiency. *Am J Med* 1977 62: 170–191.

54. Labrie F, Sugimoto Y, Luu-The V *et al.* Structure of human type II 5α-reductase gene. *Endocrinology* 1992 131: 1571–1573.

55. Wilson JD, Griffin JE, Russell DW. Steroid 5α-reductase 2 deficiency. *Endocr Rev* 1993 14: 577–593.

56. Thigpen AE, Davis DL, Milatovich A *et al.* Molecular genetics of steroid 5α-reductase deficiency. *J Clin Invest* 1992 90: 799–809.

57. Katz MD, Cai LQ, Zhu YS *et al.* The biochemical and phenotypic characterization of females homozygous for 5α-reductase-2 deficiency. *J Clin Endocrinol Metab* 1995 80: 3160–3167.

58. Grumbach MM, Conte FA. Disorders of sex differentiation. In Wilson JD, Foster DW, Kronenberg HM, Larsen PR, eds. *Williams Textbook of Endocrinology,* 9th edn. Philadelphia, PA: WB Saunders, 1998: 1303–1425.

59. Gottlieb B, Pinsky L, Beitel LK, Trifiro M. Androgen insensitivity. *Am J Med Genet* 1999 89: 210–217.

60. Cundy TF, Rees M, Evans BA *et al.* Mild androgen insensitivity presenting with sexual dysfunction. *Fertil Steril* 1986 46: 721–723.

61. Pinsky L, Kaufmam M, Killinger DW. Impaired spermatogenesis is not an obligate expression of receptor-defective androgen resistance. *Am J Med Genet* 1989 32: 100–104.

62. Tsukada T, Inoue M, Tachibana S, Nakai Y, Takebe H. An androgen receptor mutation causing androgen resistance in undervirilized male syndrome. *J Clin Endocrinol Metab* 1994 79: 1202–1207.

63. Grino PB, Isidro-Gutierrez F, Griffin JE, Wilson JD. Androgen resistance associated with a qualitative abnormality of the androgen receptor and responsive to high dose androgen therapy. *J Clin Endocrinol Metab* 1989 68: 587–584.

64. Gottlieb B, Trifiro M, Lumbroso R, Vasiliou DM, Pinsky L. The androgen receptor gene mutations database. *Nucleic Acids Res* 1996 24: 151–154.

65. Gottlieb B, Beitel LK, Lumbroso R, Pinsky L, Trifiro M. Update of the androgen receptor gene mutations database. *Hum Mutat* 1999 14: 103–114.

66. Simpson JL, Photopulos G. The relationship of neoplasia to disorders of abnormal sexual differentiation. *Birth Defects* 1976 12(1): 15–50.

67. Cussen LJ, MacMahon RA. Germ cells and ova in dysgenetic gonads of a 46,XY female dizygote twin. *Arch Dis Child* 1979 133: 373–375.

68. Hawkins JR, Taylor A, Berta P *et al.* Mutational analysis of *SRY*; nonsense and missense mutations in XY sex reversal. *Hum Genet* 1992 88: 471–474.

69. Jager RJ, Anvret M, Hall K, Scherer G. A human XY female with a frame shift mutation in the candidate testis-determining gene *SRY*. *Nature* 1990 348: 452–454.

70. McElreavey K, Vilain E, Barbaux S *et al.* Loss of sequences 3' to the testis-determining gene, *SRY* including the Y pseudoautosomal boundary associated with partial testicular determination. *Proc Natl Acad Sci USA* 1996 93: 8590–8594.

120. McElreavey sex differen 133–139.

121. Margarit E, transferred in a Y-positi Genet 2000

122. Braun A, hermaphtoc by a postzy male gonad cular geneti radic case.

71. Sternberg WH, Barclay DL, Kloepfer HW. Familial XY gonadal dysgenesis. *N Engl J Med* 1968 278: 695–700.

72. Espiner EA, Veale AMO, Sands VE, Fitzgerald P. Familial syndrome of streak gonads and normal male karyotype in five phenotypic females. *N Engl J Med* 1970 238: 6–11.

73. Simpson JL. Gonadal dysgenesis and sex chromosome abnormalities. Phenotypic/karyotypic correlations. In Vallet HL, Porter IH, eds. *Genetic Mechanisms of Sexual Development*. New York: Academic Press, 1979: 365–405.

74. Simpson JL. Pregnancies in women with chromosomal abnormalities. In Schulman JD, Simpson JL, eds. *Genetic Diseases in Pregnancy*. New York: Academic Press, 1981: 439–471.

75. German J, Simpson JL, Chaganti RSK *et al.* Genetically determined sex-reversal in 46,XY humans. *Science* 1978 202: 53–56.

76. Mann JR, Corkery JJ, Fisher HJW *et al.* The X-linked recessive form of XY gonadal dysgenesis with high incidence of gonadal cell tumours: clinical and genetic studies. *J Med Genet* 1983 20: 264–270.

77. Bardoni B, Xanaria E, Guioli S *et al.* A dose sensitive locus at chromosome Xp21 is involved in male to female sex reversal. *Nat Genet* 1994 7: 497–501.

78. Barbaux S, Niaudet P, Gubler M-C *et al.* Donor splice-site mutations in *WT1* are responsible for Frasier syndrome. *Nat Genet* 1997 17: 467–470.

79. Scharnhorst V, Dekker P, van der Eb AJ, Jochemsen A. Internal translation initiation generates novel WT1 protein isoforms with distinct biological properties. *J Biol Chem* 1999 27: 23456–23462.

80. Drash A, Sherman F, Hartmann WH, Blizzard RM. A syndrome of pseudohermaphroditism, Wilms' tumor, hypertension, and degenerative renal disease. *J Pediatr* 1970 76: 585–593.

81. Gessler M, Thomas GH, Couillin P *et al.* A deletion map of the *WAGR* region on chromosome 11. *Am J Hum Genet* 1989 44: 486–495.

82. Edidin DV. Pseudohermaphroditism, glomerulopathy, and Wilms' tumor (Drash syndrome). *J Pediatr* 1985 107: 988.

83. Frasier JE, Andres GA, Cooney DR *et al.* A syndrome of pure gonadal dysgenesis: gonadoblastoma, Wilms' tumour and nephron disease. *Lab Invest* 1964 48: 4P.

84. Simpson JL, Chaganti RSK, Mouradian J, German J. Chronic renal disease, myotonic dystrophy, and gonadoblastoma in an individual with XY gonadal dysgenesis. *J Med Genet* 1982 19: 73–76.

85. Nachtigal MW, Hirokawa Y, Enyeart-Van-houten DL *et al.* Wilms' tumor 1 and *Dax-1* modulate the orphan nuclear receptor SF-1 in sex-specific gene expression. *Cell* 1998 93: 445–454.

86. Kreidberg JA, Sariola H, Loring JM *et al.* *WT-1* is required for early kidney development. *Cell* 1993 74: 679–691.

87. Sudbeck P, Schmitz ML, Baeuerle PA, Scherer G. Sex reversal by loss of the C-terminal transactivation domain of human *SOX9*. *Nat Genet* 1996 13: 230–232.

88. Kwok C, Weller PA, Guioli S *et al.* Mutations in *SOX9*, the gene responsible for campomelic dysplasia and autosomal sex reversal. *Am J Hum Genet* 1995 57: 1028–1036.

89. Kwok C, Tyler-Smith C, Mendonca BB *et al.* Mutation analysis of the 2 kb 5' to *SRY* in XX females and Y intersex subjects. *J Med Genet* 1996 33: 465–468.

90. Pfeifer D, Kist R, Dewar K *et al.* Campomelic dysplasia translocation breakpoints are scattered over 1 mb proximal to *SOX9*: evidence for an extended control region. *Am J Hum Genet* 1999 55: 111–124.

91. Olney PN, Kean LS, Graham D, Elsas LJ, May KM. Campomelic syndrome and deletion of *SOX9*. *Am J Med Genet* 1999 84: 20–24.

92. Gibbons RJ, Suthers GK, Wilkie AOM, Buckle VJ, Higgs DR. X-linked α-thalassemia/mental retardation (ART-X) syndrome: localization to Xq12–q21.31 by X inactivation and linkage analysis. *Am J Hum Genet* 1992 51: 1136–1149.

93. Gibbons RJ, Picketts DJ, Villard L, Higgs DR. Mutations in a putative global transcriptional regulator cause X-linked mental retardation with α-thalassemia (ATR-X syndrome). *Cell* 1995 80: 837–845.

94. Reardon W, Gibbons RJ, Winter RM, Baraitser M. Male pseudohermaphroditism in sibs with the α-thalassemia/mental retardation (ATR-X) syndrome. *Am J Med Genet* 1995 55: 285–287.

95. McPherson EW, Clemens MM, Gibbons RJ, Higgs DR. X-linked α-thalassemia/mental retardation (ART-X) syndrome: a new kindred with severe

genital
sion. *A*

96. Gibbons
and her
lassemia
Am J Me

97. Gibbons
in trans
function
Genet 19

98. El-shay
genital
542–550

99. Opitz J
(dysence
drome).

100. Pinsky
Disorders
Oxford

101. Ryan
Smith–L
and bioc
558–565

102. Curry
Smith–L
congenit
phroditis
Genet 19

103. Tint GS,
lesterol
Smith–L
1994 33

104. Tint G
Smith–L
birth de
matic st
Biochem

105. Shacklet
Dehydro
candidat
Smith–L
21: 207–

106. de Grouc
Embryon
severe m
28: 154–

107. Sultan C
Kempers
eds. *F*
Amsterd

Chapter 30

Female pseudohermaphroditism and 21-hydroxylase deficiency

Maria I. New

INTRODUCTION

Congenital adrenal hyperplasia (CAH) owing to 21-hydroxylase deficiency is the most common cause of ambiguous genitalia in the newborn, accounting for over 90% of CAH cases. According to the largest newborn screening programs, 21-hydroxylase deficiency is estimated to have a world-wide frequency of approximately 1 in 15 000 people;[1] thus, CAH is also one of the most common human inborn errors of metabolism. Absence of 21-hydroxylase activity impairs cortisol synthesis by the adrenal cortex, and in its classic, or most severe form, causes ambiguous genitalia in genetic females. The decreased production of cortisol results in increased pituitary secretion of adrenocorticotropic hormone (ACTH), which stimulates both the accumulation of precursor steroids in the impeded pathways and excessive steroid synthesis in other adrenal biosynthetic pathways unaffected by the enzyme deficiency. This resultant production of excess adrenal androgens causes virilization of female genitalia *in utero*, and precocious development postnatally in both males and females. A less severe, non-classic form of 21-hydroxylase deficiency characterized by postnatal androgen excess is relatively common, occurring in approximately 1 in 100 people, and does not cause ambiguous genitalia in the newborn.[2] Glucocorticoid (and often mineralocorticoid) replacement therapy can arrest or reverse symptoms to some extent. The gene encoding the 21-hydroxylase enzyme, *CYP21*, has been isolated, and specific mutations causing most cases of CAH have been identified. Prenatal treatment for 21-hydroxylase deficiency has been used successfully for over 15 years. Advances in molecular genetics currently allow for accurate prenatal diagnosis and treatment as well as newborn screening, and the possibility for gene therapy in the near future.

BACKGROUND

The adrenal gland synthesizes cortisol in a diurnal variation under the control of ACTH, forming a negative feedback loop in which high serum cortisol inhibits, and low serum cortisol stimulates, release of ACTH from the anterior pituitary. The central nervous system determines the hypothalamic set-point for the expected plasma cortisol level, and secretion of corticotropin-releasing factor (CRH) stimulates the ACTH production. This pathway defines the hypothalamic–pituitary–adrenal axis.

A deficiency of any of the enzymes necessary for the production of cortisol leads to chronic elevations of ACTH, with overstimulation and consequent hyperplasia of the adrenal cortex. The 21-hydroxylase enzyme is produced in the endoplasmic reticulum of the zona fasciculata and zona reticularis of the adrenal cortex, and converts 17-hydroxyprogesterone (17-OHP) to 11-deoxycortisol, which is then converted to cortisol by 11β-hydroxylase (Figure 1). Complete or near-complete blocks in the activity of 21-hydroxylase represent the classic form of CAH. Decreased cortisol levels increase ACTH production and result in the accumulation of steroid precursors, especially 17-OHP, and adrenal androgens. Continued oversecretion of adrenal androgens in untreated CAH causes continued penile or clitoral enlargement, advanced bone age and tall stature in early childhood, with adult short stature caused by premature epiphyseal closure. Precocious appearance of facial, axillary and pubic hair and acne are characteristic of both classic and non-classic CAH.

Approximately three-quarters of classic CAH cases are of the salt-wasting type, in which renal salt wasting occurs from insufficient aldosterone synthesis. The other form of classic CAH is termed simple-virilizing, because normal aldosterone synthesis occurs in these patients. The pathogenic difference between salt-wasting and simple-virilizing 21-hydroxylase deficiency may result from the difference in enzyme activity caused by specific mutations in the gene for 21-hydroxylase. Expression studies have shown that as little as 1% of the normal activity of 21-hydroxylase allows adequate aldosterone biosynthesis to prevent salt wasting.[4] Signs of salt wasting in the newborn period are hyponatremia, hyperkalemia, high urinary sodium, and low serum and urinary aldosterone with elevated plasma renin activity (PRA). Affected newborn boys and girls are subject to life-threatening salt-wasting crises within the first few weeks of life.

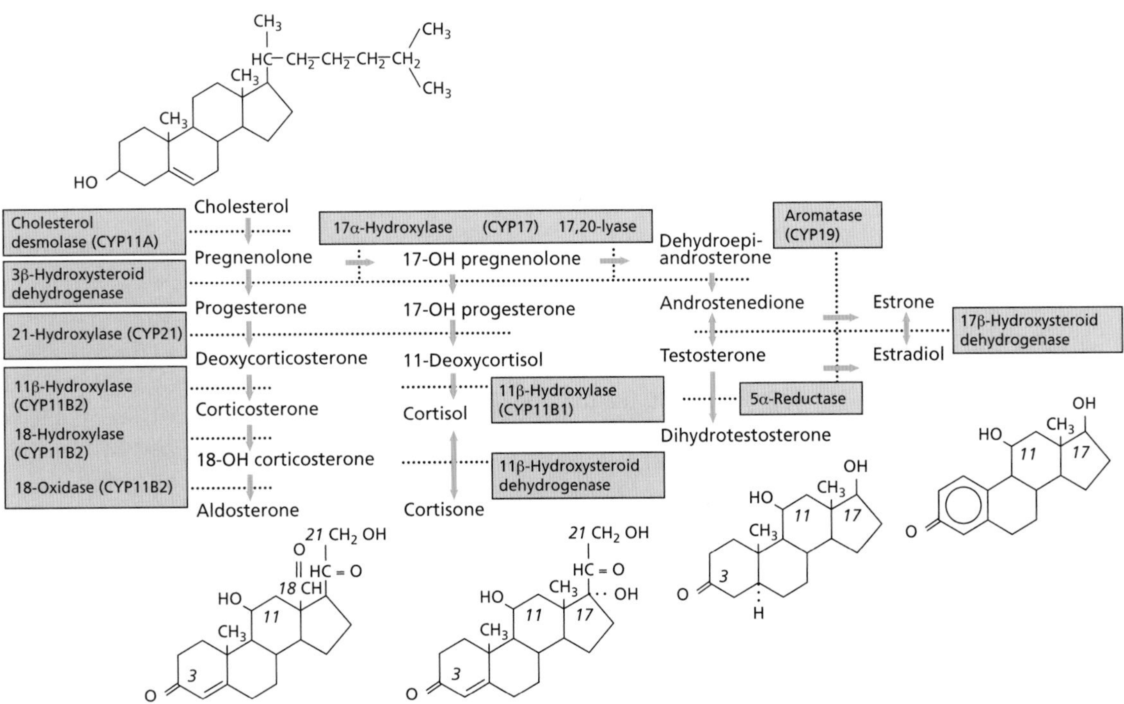

Figure 1 Schema of adrenal steroidogenesis. From reference 3 with permission

An early-morning serum concentration of 17-OHP, at the peak of ACTH secretion, may be useful as a screening test for classic 21-hydroxylase deficiency. It is important to note that the ACTH test to measure 17-OHP levels in suspected CAH cases should not be performed during the initial 24 h of life, because samples from this period are typically elevated in all infants and may yield false-positive results. The best hormonal diagnostic test for 21-hydroxylase deficiency is the ACTH stimulation test, measuring the serum concentration of 17-OHP. A logarithmic nomogram was developed to provide hormonal standards for assignment of the 21-hydroxylase deficiency type, by relating baseline to ACTH-stimulated serum concentrations of 17-OHP (Figure 2).[5] The nomogram can clearly distinguish patients with classic CAH from those with non-classic CAH as well as identify classic and non-classic heterozygotes. Patients whose hormonal values fall on the regression line within a defined group may be assigned as salt-wasting, simple-virilizing or heterozygous carriers for CAH. Radioimmunoassays are a simple and accurate method to determine directly serum hormone levels of 17-OHP, Δ^4-androstenedione, dehydroepiandrosterone, and testosterone in suspected 21-hydroxylase deficiency cases.[6]

Female pseudohermaphroditism

Congenital adrenal hyperplasia is the most common cause of ambiguous genitalia in the newborn. Females with classic CAH owing to 21-hydroxylase deficiency may present with varying degrees of virilization, represented in a five-stage classification by Prader[7] (Figure 3). The extent of virilization may range from mild clitoral enlargement to severe virilization resulting in a penile urethra, and varying degrees of fusion of the labioscrotal folds (posterior to anterior).

If medical treatment is begun early in life, a large and prominent clitoris in a newborn may shrink slightly and, as the surrounding structures

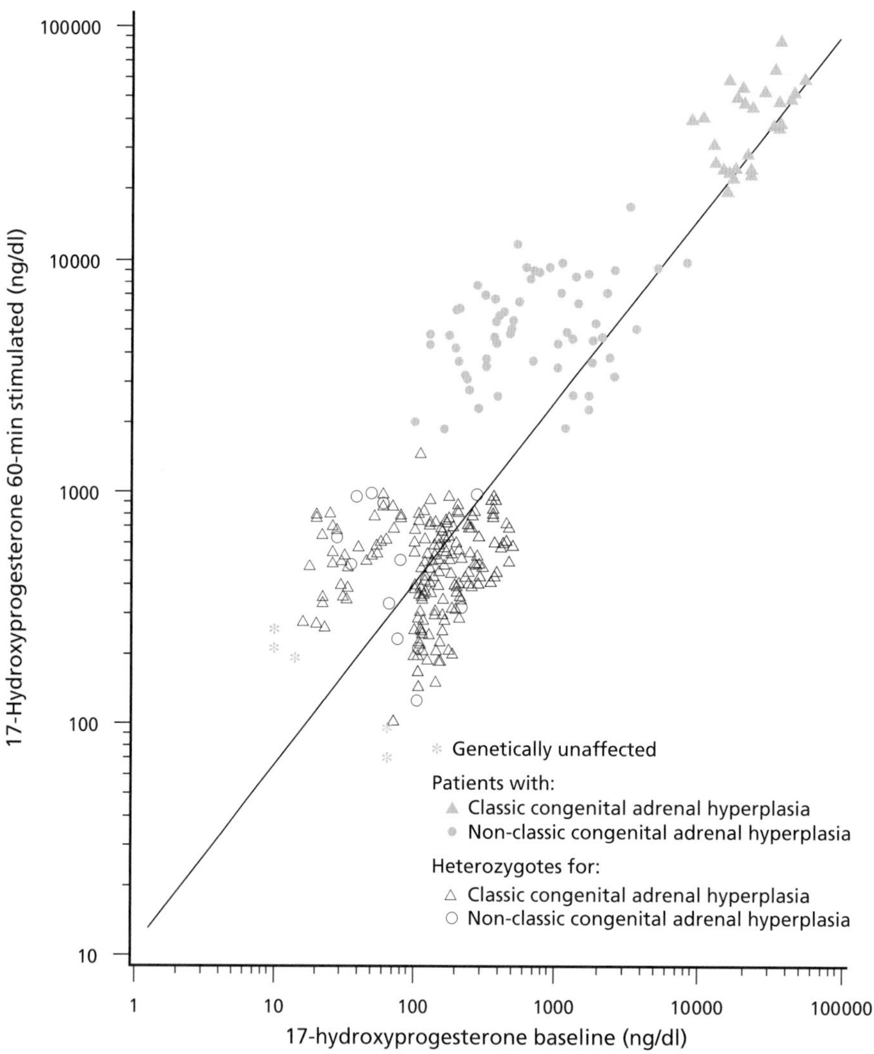

Figure 2 Nomogram relating baseline to adrenocorticotropic hormone (ACTH)-stimulated serum concentrations of 17-hydroxyprogesterone (17-OHP). The scales are logarithmic. A regression line for all data points is shown

grow normally, become much less prominent, so that surgery may not be required. When the clitoris is conspicuously enlarged or when the virilized genitalia interfere with parent–child bonding, surgical revision to correct the appearance of the clitoris may be performed by an experienced gynecological surgeon. The clitoral recession is performed in early childhood (preferably at 6–18 months of life); however, if a vaginoplasty is necessary, it is usually performed in late adolescence, because vaginal dilatations are often required to maintain a patent vagina.

Female internal genitalia (i.e. uterus, Fallopian tubes, upper vagina), which arise from Müllerian ducts, are normal in an affected CAH female because there are no Sertoli cells, the source of anti-Müllerian hormone. Females presenting with even extreme virilization from androgen overexposure will have normal development of their internal reproductive structures.

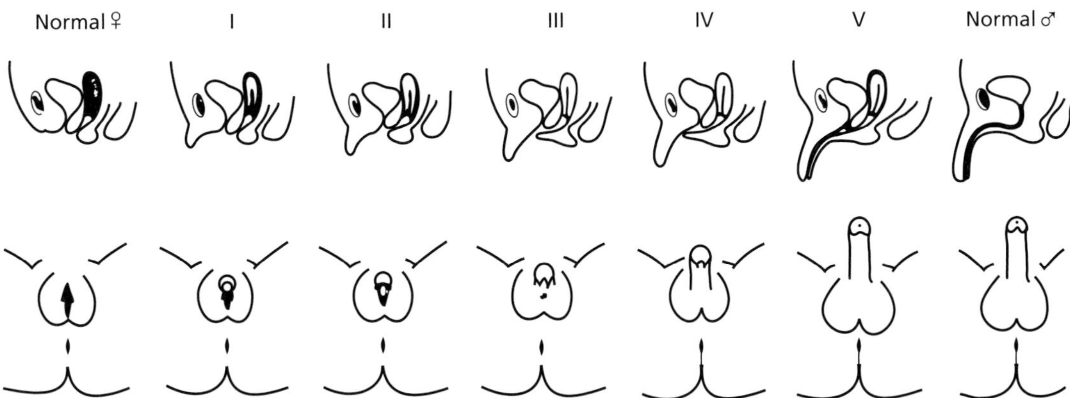

Figure 3 Diagrams representing five degrees of virilization affecting the urogenital sinus and external genitalia in females[7]

There have been several reports of adequately treated classic CAH patients who had successful, full-term pregnancies.[8–11] Menstrual irregularity and secondary amenorrhea, however, occur commonly in postmenarchal females who are untreated or are in poor hormonal control. Clinical experience suggests that excessive adrenal androgens disrupt gonadotropin secretion, leading to hypogonadism in 21-hydroxylase deficiency.[12–14] Reduced fertility in some CAH females may be attributed to inadequate introitus.[15]

APPLIED TECHNOLOGY

The gene for 21-hydroxylase, *CYP21*, is situated within the human leukocyte antigen (HLA) region on the short arm of chromosome 6. The enzyme is a heme protein belonging to the microsomal cytochrome P450 family, termed P450c21. *CYP21* sits in a tandem paired arrangement with serum complement components and a 21-hydroxylase pseudogene (*CYP21P*), in the order of *C4A–CYP21P–C4B–CYP21*. The *CYP21* and *CYP21P* genes are approximately 5 kb in length, and have ten exons each.[16] The pseudogene is non-functional owing to an 8-bp deletion in exon 3, causing a premature stop codon.[17] Additional frameshift and non-sense mutations that occur further downstream on the pseudogene have been confirmed to cause 21-hydroxylase deficiency.

A high degree of homology between *CYP21* and *CYP21P* permits two types of mutation causing recombination events: first, unequal crossing over during meiosis that results in complementary deletions/duplications of *CYP21P*;[17,18] and second, non-correspondences between the pseudogene and the coding gene that, if transferred by gene conversion, result in deleterious mutations.[19] Using *in vitro* expression studies, approximately 40 mutations in *CYP21* have been identified to cause CAH.[20,21] Specific mutations may be correlated with a given degree of enzymatic compromise and a clinical form of 21-hydroxylase deficiency.[22–24] The genotype for the classic form of CAH is predicted to be a severe mutation on both alleles at the 21-hydroxylase locus, with markedly decreased enzymatic activity generally associated with salt wasting. The point mutation A (or C) to G near the end of intron 2, which is the single most frequent mutation in classic 21-hydroxylase deficiency, causes premature splicing of the intron and a shift in the translational reading frame.[18,25] Most patients who are homozygous for this mutation have the salt-wasting form of the disorder.[23,26] Recent studies, however, have demonstrated that there is occasionally a divergence in phenotypes within mutation-identical groups, the reason for which requires further investigation.[26,27]

CURRENT CONCEPTS

Prenatal diagnosis and treatment

Females born with ambiguous genitalia not treated prenatally suffer consequences of the genital ambiguity even if the diagnosis is made at birth. Although 'corrective' surgery techniques for genital ambiguity have improved (i.e. preserving the sensitivity of the clitoris), there are insufficient data on long-term outcome, including the need for additional surgeries, the adequacy of sexual functioning and overall patient satisfaction. The possibility of eliminating the need for surgical reconstruction and frequent genital examinations in these female infants, and the accompanying reduction in emotional trauma to patient, parents and family, have generated great enthusiasm for prenatal treatment.

Since the report by Jeffcoate and colleagues in 1965 of the successful identification of an affected fetus by elevated concentrations of 17-ketosteroids and pregnanetriol in the amniotic fluid, investigators have carried out prenatal diagnosis for CAH by similar measurements of hormone levels in pregnancy.[28] Measurement of amniotic fluid 17-OHP levels is the most specific hormonal diagnostic test for 21-hydroxylase deficiency, with measurement of Δ^4-androstenedione employed as an adjunctive diagnostic assay.[29] Hormonal levels of 17-OHP are markedly elevated in salt-wasting cases; however, the test is not useful for detecting simple-virilizing and non-classic cases, whose amniotic 17-OHP levels may be in the normal range.[30]

When HLA was found to be linked to *CYP21*, diagnoses were made using HLA genetic linkage-marker analysis by serotyping cultured fetal cells from the amniotic fluid, when the HLA types of parents and of the family index case were known.[31] This method resulted in many diagnostic errors owing to recombination or haplotype sharing, and has now been superseded by direct molecular analysis of the 21-hydroxylase locus. With the advent of chorionic villus sampling (CVS), evaluation of the fetus at risk is now possible in the first trimester at 9–11 weeks' gestation. The fetal DNA is used for specific amplification of the *CYP21* gene utilizing polymerase chain reaction (PCR) and Southern blotting, which has the advantage of requiring small amounts of DNA.[32]

POTENTIAL CLINICAL RELEVANCE

Prenatal treatment with dexamethasone can be employed in pregnancies at risk for 21-hydroxylase deficiency. When properly administered, dexamethasone is effective in preventing ambiguous genitalia in the affected female. An algorithm has been developed for the diagnostic management of potentially affected pregnancies (Figure 4). The current recommendation is to treat the mother with a pregnancy at risk for 21-hydroxylase deficiency with dexamethasone in a dose of 20 μg/kg divided into three doses daily.

Institution of prenatal therapy must begin before 9 weeks' gestation to suppress effectively adrenal androgen production and allow normal separation of the vaginal and urethral orifices, in addition to preventing clitoromegaly. Because dexamethasone is to be administered at such an early date, treatment is blind to the status of the fetus. If the fetus is determined to be a male upon karyotype or an unaffected female upon DNA analysis, treatment is discontinued. Otherwise, treatment is continued to term.

Between 1986 and 1999, prenatal diagnosis and treatment of CAH due to 21-hydroxylase deficiency were carried out in over 400 pregnancies in the New York Presbyterian Hospital–Weill Medical College of Cornell University, in which 84 babies were affected with classic 21-hydroxylase deficiency.[34] Of these, 52 were females, 36 of whom were treated prenatally with dexamethasone. Dexamethasone administered at or before 10 weeks of gestation was effective in reducing virilization, thus avoiding postnatal genitoplasty.[34] There were no statistical differences in symptoms during pregnancy between those treated with dexamethasone and those not treated with dexamethasone, except for weight gain, which was greater in the treated group. No significant or enduring side-effects were noted in the fetuses, indicating that dexamethasone treatment is safe. Prenatally treated newborns did not differ in weight, head circumference or birth length from untreated,

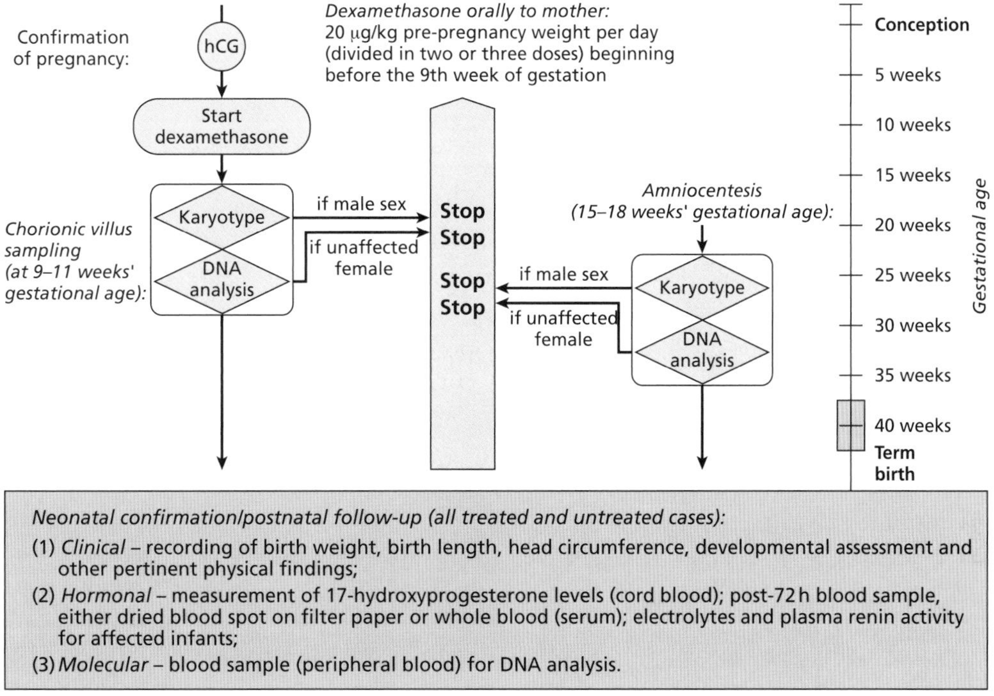

Figure 4 Algorithm depicting prenatal management of pregnancy in families at risk for a fetus affected with 21-hydroxylase deficiency. From reference 33 with permission

unaffected newborns, which is consistent with results of other large studies.[35,36]

The development of prenatal treatment for CAH may be viewed as an amalgamation of recent advancements in the fields of pediatric endocrinology, reproductive medicine and molecular genetics. While prenatal treatment for 21-hydroxylase deficiency represents a therapeutic milestone, research must continue into the long-term outcome of mothers and children. Large psychological studies are also under way to determine the effects of prenatal dexamethasone treatment on the behavior of affected females.

FUTURE PERSPECTIVES

Conventional medical therapy has proved disappointing, because finding a balance between providing sufficient glucocorticoid replacement to achieve adrenal suppression, and administering so much as to cause adverse side-effects, remains difficult in treating CAH patients. Because CAH is a monogenic inherited disorder, 21-hydroxylase deficiency has been proposed as a good candidate for gene therapy. In fact, mice with homozygous *CYP21* deletions have undergone successful genetic rescue by transgenesis.[37] It is yet to be seen whether this technique will serve as a good model system for humans. The problem of long-term expression of 21-hydroxylase will probably be the largest hurdle for future gene therapy research.

SUMMARY

Congenital adrenal hyperplasia due to 21-hydroxylase deficiency is the most common cause of ambiguous genitalia in the newborn, with an estimated frequency of 1 in 15 000 live births worldwide. CAH has served as a prototype for the examination of female pseudohermaphroditism and hormonal influence on the differentiation of the gonads. The pathophysiology can be traced to discrete, inherited defects in the genes encoding enzymes for adrenal steroidogenesis. Treatment of CAH is targeted to replace the hormones that are produced in insufficient quantity. With proper hormone replacement therapy, normal and healthy development may often be expected. Radioimmunoassay of serum and urinary steroid levels permits reliable diagnosis of the various forms of CAH. Prenatal therapy in 21-hydroxylase deficiency is used effectively to reduce or eliminate ambiguous genitalia in the female newborn.

REFERENCES

1. Pang SY, Wallace MA, Hofman L *et al.* Worldwide experience in newborn screening for classical congenital adrenal hyperplasia due to 21-hydroxylase deficiency. *Pediatrics* 1988 81: 866–874.
2. Speiser PW, Dupont B, Rubinstein P, Piazza A, Kastelan A, New IM. High frequency of nonclassical steroid 21-hydroxylase deficiency. *Am J Hum Genet* 1985 37: 650–667.
3. Alberti KGMM, Burger HG, Cohen RD *et al.* Genetic disorders of steroid hormone synthesis and metabolism. In Thakker RV, ed. *Genetic and Molecular Biological Aspects of Endocrine Disease. Ballière's Clinical Endocrinology and Metabolism. International Practice and Research, vol 9, no. 3.* London: Baillière Tindall, 1995: 526.
4. Tusie-Luna M, Traktman P, White PC. Determination of functional effects of mutations in the steroid 21-hydroxylase gene (*CYP21*) using recombinant vaccinia virus. *J Biol Chem* 1990 265: 20916–20922.
5. New MI, Lorenzen F, Lerner AJ *et al.* Genotyping steroid 21-hydroxylase deficiency: hormonal reference data. *J Clin Endocrinol Metab* 1983 57: 320–326.
6. Korth-Schutz S, Virdis R, Saenger P, Chow DM, Levine LS, New MI. Serum androgens as a continuing index of adequacy of treatment of congenital adrenal hyperplasia. *J Clin Endocrinol Metab* 1978 46: 452–458.
7. Prader A. Der genitalbefund beim pseudohermaphroditismus femininus der kengenitalen adrenoenitalen syndroms. *Helv Paediatr Acta* 1954 9: 231–248.
8. Mori M. Congenital adrenogenital syndrome and successful pregnancy: report of a case. *J Obstet Gynecol* 1970 35: 394.
9. Klingensmith G, Garcia S, Jones H, Migeon C, Blizzard R. Glucocorticoid treatment of girls with congenital adrenal hyperplasia: effects on height, sexual maturation, and fertility. *J Pediatr* 1977 90: 996–1004.
10. Riddick D, Hammond C. Adrenal virilism due to 21-hydroxylase deficiency in the postmenarchial female. *Obstet Gynecol* 1975 45: 21–24.
11. Lo J, Schwitzgebel V, Tyrrell J *et al.* Normal female infants born of mothers with classic congenital adrenal hyperplasia due to 21-hydroxylase deficiency. *J Clin Endocrinol Metab* 1999 84: 930–936.
12. Ghali I, David M, David L. Linear growth and pubertal development in treated congenital adrenal hyperplasia due to 21-hydroxylase deficiency. *Clin Endocrinol (Oxf)* 1977 6: 425–436.
13. Richards G, Styne D, Conte F, Kaplan S, Grumbach M. Plasma sex steroids and gonadotropins in pubertal girls with congenital adrenal hyperplasia: relationship to menstrual disorders. In Lee P, Plotnick L, Kowarski A, Migeon C, eds. *Congenital Adrenal Hyperplasia.* Baltimore: University Park Press, 1977: 233.
14. Hughes I, Read G. Menarche and subsequent ovarian function in girls with congenital adrenal hyperplasia. *Horm Res* 1982 16: 100–6.
15. Mulaikal RM, Migeon CJ, Rock JA. Fertility rates in female patients with congenital adrenal hyperplasia due to 21-hydroxylase deficiency. *N Engl J Med* 1987 316: 178–182.
16. White PC, Grossberger D, Onufer BJ *et al.* Two genes encoding steroid 21-hydroxylase are located near the genes encoding the fourth component of complement in man. *Proc Natl Acad Sci USA* 1985 82: 1089–1093.
17. White PC, New MI, Dupont B. Structure of the human steroid 21-hydroxylase genes. *Proc Natl Acad Sci USA* 1986 83: 5111–5115.
18. Higashi Y, Tanae A, Inoue H, Hiromasa T, Fujii-Kuriyama Y. Aberrant splicing and missense mutations cause steroid 21-hydroxylase (P-450(C21)) deficiency in humans: possible gene conversion products. *Proc Natl Acad Sci USA* 1988 85: 7486–7490.
19. Tusie-Luna M, White P. Gene conversions and unequal crossovers between *CYP21* (steroid 21-

hydroxylase gene) and *CYP21P* involve different mechanisms. *Proc Natl Acad Sci USA* 1995 92: 10796–10800.

20. Krawczak M, Cooper DN. The human gene mutation database. *Trends Genet* 1997 13: 121–122.
21. White P, Speiser P. Congenital adrenal hyperplasia due to 21-hydroxylase deficiency. *Endocr Rev* 2000 21: 245–291.
22. Werkmeister JW, New MI, Dupont B, White PC. Frequent deletion and duplication of the steroid 21-hydroxylase genes. *Am J Hum Genet* 1986 39: 461–469.
23. Speiser PW, Dupont J, Zhu D *et al.* Disease expression and molecular genotype in congenital adrenal hyperplasia due to 21-hydroxylase deficiency. *J Clin Invest* 1992 90: 584–595.
24. Wedell A, Ritzen EM, Haglund SB, Luthman H. Steroid 21-hydroxylase deficiency: three additional mutated alleles and establishment of phenotype–genotype relationships of common mutations. *Proc Natl Acad Sci USA* 1992 89: 7232–7236.
25. Higashi Y, Hiromasa T, Tanae A *et al.* Effects of individual mutations in the P-450(C21) pseudogene on the P-450(C21) activity and their distribution in the patient genomes of congenital steroid 21-hydroxylase deficiency. *J Biochem* 1991 10: 638–644.
26. Wilson RC, Mercado AB, Cheng KC, New MI. Steroid 21-hydroxylase deficiency: genotype may not predict phenotype. *J Clin Endocrinol Metab* 1995 80: 2322–2329.
27. Krone N, Braun A, Roscher A, Knorr D, Schwarz H. Predicting phenotype in steroid 21-hydroxylase deficiency? Comprehensive genotyping in 155 unrelated, well defined patients from southern Germany. *J Clin Endocrinol Metab* 2000 85: 1059–1065.
28. Jeffcoate T, Fleigner J, Russell S, Davis J, Wade A. Diagnosis of the adrenogenital syndrome before birth. *Lancet* 1965 2: 553.
29. Nagamani M, McDonough P, Ellegood J, Mahesh V. Maternal and amniotic fluid 17α-hydroxyprogesterone levels during pregnancy: diagnosis of congenital adrenal hyperplasia *in utero*. *Am J Obstet Gynecol* 1978 130: 791–794.
30. Pang S, Pollack MS, Loo M *et al.* Pitfalls of prenatal diagnosis of 21-hydroxylase deficiency congenital adrenal hyperplasia. *J Clin Endocrinol Metab* 1985 61: 89–97.
31. Pollack MS, Maurer D, Levine LS *et al.* Prenatal diagnosis of congenital adrenal hyperplasia (21-hydroxylase deficiency) by HLA typing. *Lancet* 1979 1: 1107–1108.
32. Wilson RC, Wei JQ, Cheng KC, Mercado AB, New MI. Rapid deoxyribonucleic acid analysis by allele-specific polymerase chain reaction for detection of mutations in the steroid 21-hydroxylase gene. *J Clin Endocrinol Metab* 1995 80: 1635–1640.
33. Mercado AB, Wilson RC, Cheng KC, Wei JQ, New MI. Extensive personal experience: prenatal treatment and diagnosis of congenital adrenal hyperplasia owing to steroid 21-hydroxylase deficiency. *J Clin Endocrinol Metab* 1995 80: 2014–2020.
34. Carlson AD, Obeid JS, Kanellopoulou N, Wilson RC, New MI. Congenital adrenal hyperplasia: update on prenatal diagnosis and treatment. *J Steroid Biochem Mol Biol* 1999 69: 19–29.
35. Forest M, Morel Y, David M. Prenatal treatment of congenital adrenal hyperplasia. *Trends Endocrinol Metab* 1998 9: 284–289.
36. Lajic S, Wedell A, Bui T, Ritzen E, Holst M. Long-term somatic follow-up of prenatally treated children with congenital adrenal hyperplasia. *J Clin Endocrinol Metab* 1998 83: 3872–3880.
37. Gotoh H, Kusakabe M, Shiroishi T, Moriwaki K. Survival of steroid 21-hydroxylase-deficient mice without endogenous corticosteroids after neonatal treatment and genetic rescue by transgenesis as a model system for treatment of congenital adrenal hyperplasia in humans. *Endocrinology* 1994 135: 1470–1476.

Chapter 31

Persistence of Müllerian derivatives in males

Nathalie Josso, Chrinne Belville and Jean-Yves Picard

INTRODUCTION

As demonstrated many years ago by Alfred Jost,[1] mammalian sex differentiation is driven, not by the nature of the fetal genotype, but by the hormonal capacity of the fetal testis (Figure 1). Androgens, produced by fetal Leydig cells, masculinize the external genitalia and urogenital sinus, promote the differentiation of Wolffian ducts into male efferent ducts and accessory organs, but fail to influence the fate of Müllerian ducts, which develop into the uterus, Fallopian tubes and upper vagina in normal and androgenized females alike.

Anti-Müllerian hormone (AMH), also known as Müllerian inhibiting substance (MIS) or factor (MIF), a member of the transforming growth factor-β (TGF-β) family, promotes regression of the Müllerian ducts by acting on receptors located on plasma membranes of peri-Müllerian mesenchymal cells. Fetal Sertoli cells begin to produce AMH immediately after testicular sex determination; it follows that Müllerian regression is the first sign of male sex differentiation, beginning at 9 fetal weeks in the human fetus and 16 days post-coitum in the fetal rat. Regression proceeds rapidly in a craniocaudal direction, provided that Müllerian ducts have been exposed to the hormone before the end of a critical 'window' of sensitivity, ending at 8 weeks in the human fetus[3] and 16 days post-coitum in the rat.[4]

BACKGROUND

The human anti-Müllerian hormone

AMH is a 560-amino-acid glycoprotein formed by two 70-kDa monomers linked by disulfide bonds. The hormone is cleaved at a proteolytic site 109 amino acids upstream of the C terminus, yielding a short bioactive C-terminal domain with homology to other members of the TGF-β family, and a long N terminus with no bioactivity of its own but which enhances the bioactivity of the C terminus.[5] AMH is synthesized by Sertoli cells immediately after testis determination and is produced at high levels

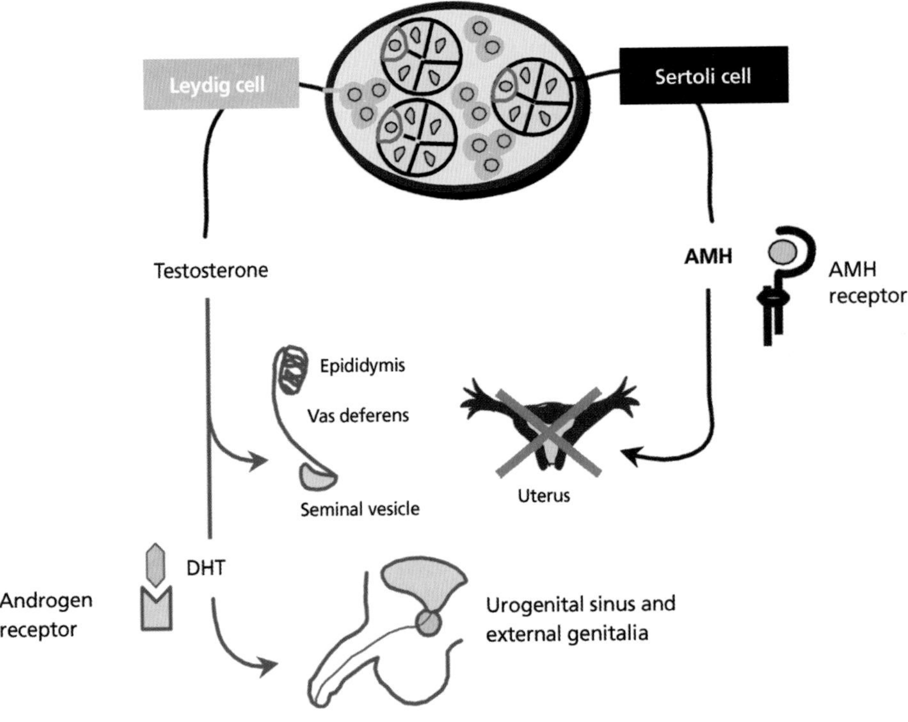

Figure 1 Hormonal control of male sex differentiation. AMH, anti-Müllerian hormone; DHT, dihydrotestosterone. From reference 2 with permission

up to puberty, at which time it is repressed by testicular androgens.[6] In the female, AMH is produced by ovarian granulosa cells from birth onwards.[7] AMH does not appear to play a major physiological role after birth, but has been used as a marker of gonadal function/tumorigenesis in both sexes. An assay kit is now commercially available for measurement of AMH in human serum (Immunotech Coulters, Marseilles, France).

The human *AMH* gene was cloned in 1986;[8] it is located on the tip of the short arm of chromosome 19, band p13.3.[9] Only 2.75 kbp in length, it contains five exons; the 3' end of the fifth exon is particularly GC-rich and encodes the bioactive C-terminal domain. The minimal promoter is only 200 bp[10] and is flanked by a housekeeping gene, encoding a spliceosome.[11] The *AMH* promoter contains binding sites for various transcription factors, namely steroidogenic factor-1 (SF-1), SRY-related HMG box-9 (SOX-9) and GATA-1 (reviewed in reference 12).

The anti-Müllerian hormone transduction cascade

Like other members of the TGF-β family, AMH signals through two distinct membrane-bound receptors, both serine/threonine kinases. Receptor type II binds to the ligand, and the type II–ligand complex recruits receptor type I which acts as a signal transducer by binding to and activating specific cytoplasmic substrates, the so-called Smad molecules.[13]

The gene for the AMH type II receptor (*AMHR-II*), located on chromosome 12q13,[14] is 8 kbp in length and divided into 11 exons. Exons 1–3 encode the signal sequence and extracellular domain, exon 4 most of the transmembrane domain and exons 5–11 the intracellular serine/threonine kinase domains. Exon 2 undergoes alternative splicing in the rabbit[15] and dog (Messika-Zeitoun and colleagues, unpublished data) but not in man[14] or the rat.[16]

AMHR-II is expressed in the mesenchymal cells that surround the Müllerian duct, and also on Sertoli, Leydig and granulosa cells. The mature form of the human receptor is expressed at the cell surface and has a mass of 82 kDa.[17] Binding of the ligand to the extracellular domain of AMHR-II successively triggers interaction with a bone morphogenetic protein (BMP) type I receptor, BMPR-IB/ALK6, phosphorylation of Smad1 and its entry into the nucleus.[18] *BMPR-IB* is duly expressed in developing Müllerian ducts but, to date, there is no evidence to suggest that its presence is required for AMH-mediated Müllerian regression, which follows a β-catenin-dependent pathway.[19] Smad1 could be activated by different type I receptors in different tissues.

APPLIED TECHNOLOGY

Many factors can interfere with Müllerian regression. If the testes do not produce enough hormone, if mutations of the AMH molecule destroy its bioactivity, if the AMH receptor does not appear on the membrane of target cells at the appointed time or if the transduction machinery is stalled by mutation of signalling molecules, the uterus and Fallopian tubes, which are the legacy of fetal Müllerian ducts, will be maintained in chromosomal and gonadal males.

Measurement of the level of circulating AMH is an important element for assessing testicular production of AMH, but only in prepubertal subjects, since AMH production by Sertoli cells is repressed by androgens and, thus, dramatically decreases at puberty in normal subjects.[20,21] An AMH enzyme-linked immunosorbent assay (ELISA) using two specific monoclonal antibodies is now commercially available (Immunotech Coulters, Marseilles, France). In prepubertal subjects with retained Müllerian ducts, a decreased AMH serum concentration shows that AMH is not produced in normal amounts, but does not differentiate between testicular dysgenesis on the one hand, affecting both androgen and AMH production, and persistent Müllerian duct syndrome (PMDS) on the other, in which the deficit is limited to AMH (Figure 1). A human chorionic gonadotropin (hCG) test followed by a serum testosterone assay will usually reveal Leydig cell insufficiency, characteristic of testicular dysgenesis. Isolated Müllerian duct persistence associated with a normal AMH concentration for age suggests an end-organ defect for AMH.

Gene sequencing is required to detect the molecular basis of PMDS. *AMH* mutations are detected by amplification of the gene by the polymerase chain reaction (PCR), followed by automatic sequencing. *AMHR-II* mutations are first explored by PCR, which allows the detection of an extremely frequent 27-bp deletion,[22] and then, in the event of negative results, by gene amplification followed by automatic sequencing.

CURRENT CONCEPTS

Clinical defects

The external phenotype of an individual with retained Müllerian ducts depends upon his potential for testicular androgen production. If the testis is dysgenetic, both Leydig and Sertoli cells are usually affected, leading to a combination of external undervirilization and Müllerian duct retention. This results in an intersex phenotype, which is not considered here. On the other hand, if androgen production and sensitivity are normal, the phenotype is unambiguously male, and the condition is known as PMDS, the subject of this chapter.

Affected subjects are genotypic and phenotypic males, with cryptorchidism sometimes associated with inguinal hernia. The hernia may appear incarcerated, despite the lack of symptoms of intestinal obstruction, but this discrepancy is usually not correctly interpreted. Sonography could be useful but is rarely performed, unless an older sibling has been diagnosed with the condition.

The presence of Müllerian derivatives is usually discovered at surgery. Two anatomical forms of PMDS have been described. The most common is characterized by the association of unilateral cryptorchidism and contralateral hernia. One testis has descended into the scrotum and the ipsilateral uterus and Fallopian tube either have entered the inguinal canal, a condition known as 'hernia uteri inguinalis', or alternatively can be dragged into it by gentle traction, pulling the contralateral testis and Fallopian tube in their wake. Often, no traction is necessary, because the contralateral testis is already in the hernial sac. Transverse testicular ectopia, as this condition is called, is extremely frequent in PMDS.[23] More rarely, PMDS presents as bilateral cryptorchidism; the uterus is fixed in the pelvis and both testes are embedded in the broad ligament (Figure 2). These clinical variants are not genetically determined and may occur within the same sibship.[25]

In PMDS the testes are abnormally mobile, because they are not anchored to the bottom of the processus vaginalis by a normal male gubernaculum but are connected instead to elongated, thin, round ligaments.[26] This hypermobility could favor testicular torsion and subsequent testicular degeneration, a high incidence of which has been reported in PMDS.[27]

Testes are normally differentiated and, in the absence of long-standing cryptorchidism, usually contain germ cells. However, they are often not properly connected to male excretory ducts, owing to aplasia of the epididymis and upper part of the vas deferens.[22] It is usually difficult to bring the testes down to a normal position; even after careful dissection, the free segment of the spermatic cord is very short because the vasa deferentia are

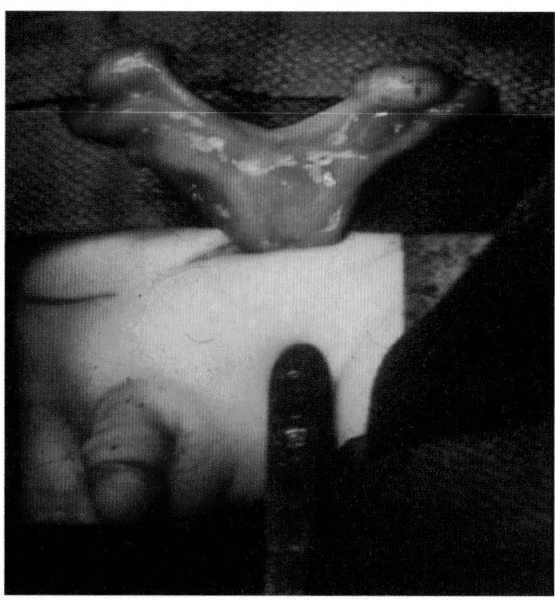

Figure 2 Operative view of patient with persistent Müllerian duct syndrome: the testes are tightly attached to the Müllerian derivatives. From reference 24, with permission

embedded in the mesosalpinx, lateral uterine wall and cervix. Lack of proper communication between the testis and excretory ducts, and difficulties at orchidopexy, probably explain why fertility is rare in PMDS patients: 11% according to a Kuwaiti study.[28]

Few associated clinical abnormalities have been described, apart from testicular tumors,[29,30] colon adenocarcinoma and medullary thyroid cancer, which developed in one patient at age 77 years.[22] Associated defects are more frequent in patients with unexplained PMDS (see below). Except in patients with testicular degeneration, serum testosterone levels and the response to chorionic gonadotropin are normal. In contrast, the level of circulating AMH is extremely variable, and depends on the molecular basis of the condition (Figure 3). Female relatives of AMH-resistant patients who share their genetic background are phenotypically normal and fertile.

Biological defects

Usually, patients with *AMH* mutations have very low or undetectable serum AMH concentrations, because *AMH* mutations prevent secretion from the cytoplasm of Sertoli cells (Belville and colleagues, unpublished data). We have observed three patients with *AMH* mutations who had a normal serum AMH concentration for age (Table 1); conceivably, these mutations could impair bioactivity but not secretion. Patients with defective *AMHR-II* have a normal AMH serum concentration for their age (Figure 3) unless the testes are regressed[27] or unless orchidopexy was carried out in the days preceding the assay[31] (Table 1).

Molecular basis of persistent Müllerian duct syndrome

Isolated persistence of Müllerian derivatives can be ascribed either to lack of production of AMH, usually because of mutations of the *AMH* gene, or to the resistance of target organs, a consequence of mutations of the components of the AMH transduction cascade. Our own experience is based upon 76 families screened for mutations of the *AMH* or *AMHR-II* gene from 1988 to 2001. Transmission is according to an autosomal recessive pattern.

Mutations of the AMH gene

AMH gene mutations were detected in 47% of PMDS families. Serum AMH levels were usually undetectable or extremely low for age (Figure 3). Thirty-five different mutations were found (Figure 4a). Sixty-one per cent were homozygous owing to a high proportion of patients from Arab or Mediterranean countries (Figure 5), characterized by a high rate of consanguinity.

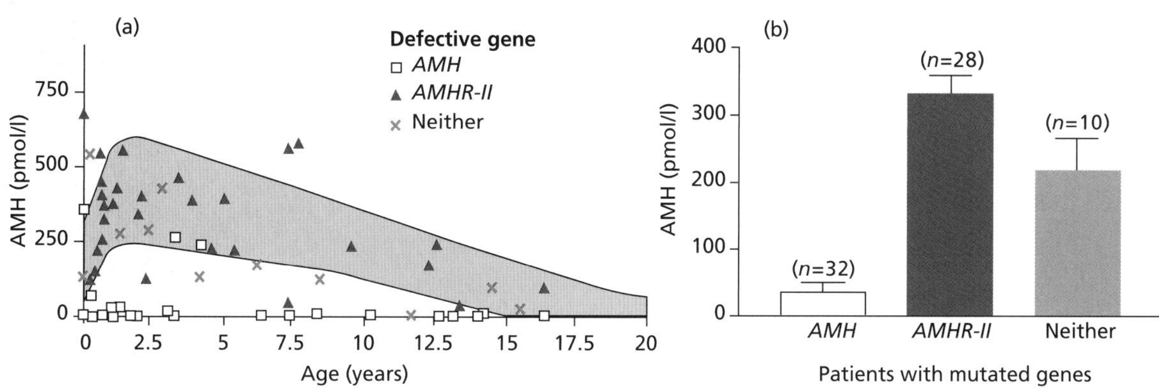

Figure 3 (a) Serum levels of anti-Müllerian hormone (AMH), measured by immunoassay, according to genotype. The normal range is shaded. (b) Mean and standard error of the mean (SEM) of serum AMH levels. Only patients under the age of 15 years are included, because serum AMH is not informative in older subjects

Table 1 Atypical cases of persistent Müllerian duct syndrome (PMDS)

Initials	*Age* (years)	*Serum AMH* (pmol/l)	*Normal value for age** (pmol/l)	*AMH mutations*†	*AMHR-II mutations*†
Unusual AMH serum concentration					
AA	3.25	263	430 ± 242	G101R, ex 1, homozygous	
ZD	0.05	357	495 ± 202	Q496H, ex 5, homozygous	
BB	4.15	241	430 ± 242	δ2501–2507, ex 5 (mother) A515V, ex 5 (mother)	
KK	0.22	117	495 ± 202		δ6331–6357, ex 10 (father) R504C, ex 11 (mother)
Mutation on a single allele					
BB	4.15	241	430 ± 242	δ2501–2507, ex 5 (mother) A515V, ex 5 (mother)	
MD	0.6	undetectable	495 ± 202	G101R, ex 1 (father)	
TP	0.28	65	495 ± 202	δ350–354 ex 1	
TB	13.5	31	132 ± 128		δ84–87, ex 1 (mother)
KC	14.5	NT	132 ± 128		δ6331–6357, ex 10 (mother)
Patients with mutations but no retained Müllerian derivatives					
CA	14			G101V, ex 1, homozygous	
AA	3.25	263	430 ± 242	G101R, ex 1, homozygous	

*Mean±SD; †the parent transmitting the mutation is indicated in parentheses; AMH, anti-Müllerian hormone; ex, exon; *AMHR-II,* gene for the AMH receptor type II; NT, not tested

Mutations, for the most part missense type, were present on the whole length of the gene, except exon 4. Mutations were particularly frequent in exon 1 and in the 3' end of exon 5, which encodes the bioactive C terminus. Nine mutations were recurrent: in exon 1 and in the 3' end of exon 5, representing 43% and 28% of total mutations, respectively (Figure 4a). As usual,[32] transitions were more frequent than transversions.

Mutations of the AMHR-II gene

AMHR-II mutations were present in 38% of PMDS families, and 52% were compound heterozygous (Figure 5). The ethnic origin of homozygous patients clearly differed from that of patients with *AMH* mutations, with a lower proportion of Arab and Mediterranean patients and a higher proportion of Asians (Turkish and Pakistani). In contrast, the ethnic origins of *AMH* and *AMHR-II* compound heterozygotes were similar, with a high proportion of French and Northern-European subjects. Serum AMH concentration was normal for age. Twenty-five different mutations of *AMHR-II* were spread over the whole length of the gene (Figure 4b). Most were missense, transitions being more frequent than transversions. Six mutations were recurrent (Figure 4b). The most frequent, δ6331–6357, a deletion of 27 bp in exon 10,[22] was present in 45% of families of this group, either in the homozygous state or coupled with missense mutations. An interesting mutation truncated the protein immediately after the transmembrane domain. This type of mutation is usually dominant negative in other members of the TGF-β family, but, in the PMDS family, there was no phenotype in heterozygotes. PMDS was present only in subjects bearing a mutation on the other *AMHR-II* allele.[33]

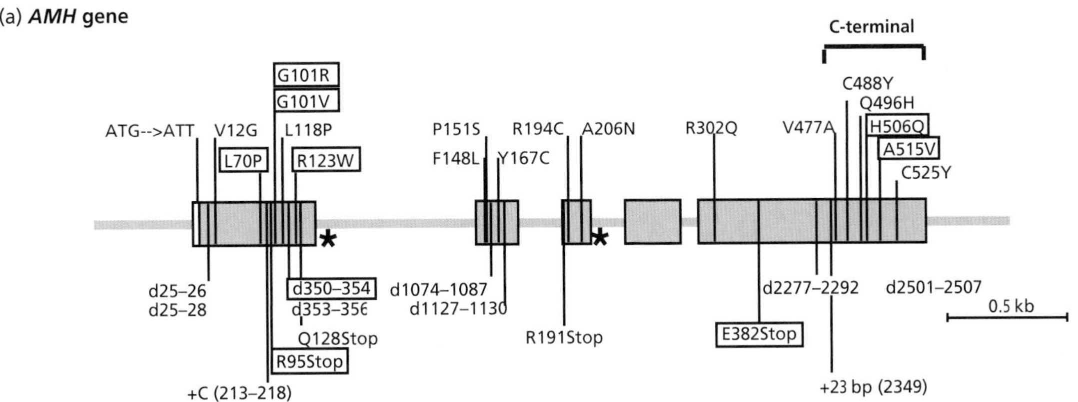

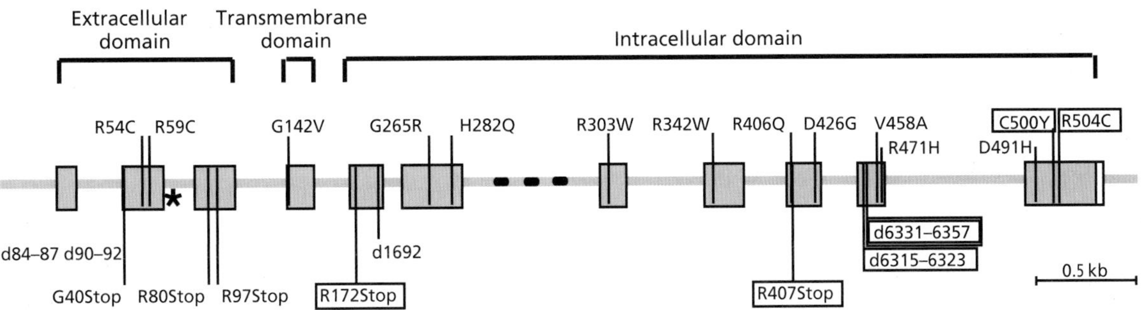

Figure 4 Mutations of the (a) anti-Müllerian hormone (*AMH*) and (b) anti-Müllerian hormone receptor type II (*AMHR-II*) genes. Exons are shaded. Recurrent mutations are boxed; the very frequent mutation d6331–6357 is surrounded by a double box. Asterisks represent splice mutations. The 3' end of exon 5 encodes the C-terminal, bioactive domain of the AMH protein

Unexplained persistent Müllerian duct syndrome

In 11 families, representing 15% of the total number, no mutation of either the *AMH* or the *AMHR-II* gene was detectable. No large DNA rearrangements were seen by Southern blotting. Associated diseases, such as jejunal atresia,[34] lipoatrophic diabetes and vitamin D-resistant rickets,[35] lymphangectasia[36] and other malformations, were present in approximately half the cases.

Unexplained PMDS may reflect mutations in the proteolytic enzyme involved in AMH processing, or in components downstream of the type II receptor in the AMH transduction cascade. The knock-out phenotype[37,38] of *BMPR-IB/ALK6*, a candidate type I receptor for AMH,[18] is characterized by severe chondrodystrophies, but cartilage defects have not been described in PMDS. Smad molecules are probably not involved either, because their mutations are usually associated with a high incidence of cancer, which is not observed in PMDS. In view of the high incidence of associated defects, unexplained PMDS may be part of complex malformative syndromes, with no relationship to the AMH pathway. Similarly, a high incidence of congenital malformations characterizes unexplained external male pseudohermaphroditism.[39]

Atypical cases

Cases with a mutation on only one allele Both *AMH* and *AMHR-II* mutations are transmitted as autosomal recessive conditions, owing to mutations on both alleles of the affected gene. However, in exceptional cases, extensive searches yielded

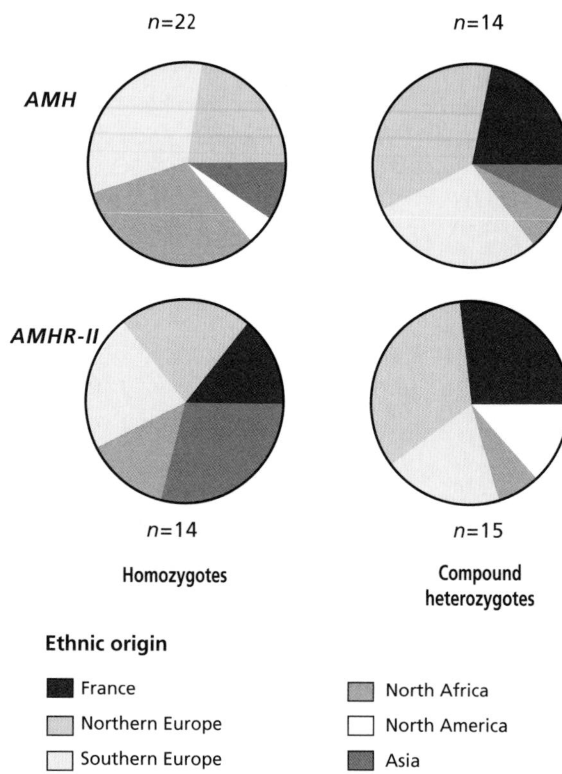

Figure 5 Ethnic origin and allelic status of patients with anti-Müllerian hormone (*AMH*) or its receptor type II (*AMHR-II*) gene mutations. The five patients in whom a mutation was detected on a single allele (Table 1) are included. *n* represents the number of families in each group

mutations on only one allele (Table 1). Possibly, the other allele bears a splicing mutation, which went undetected because testicular RNA was unavailable. Dominant negative activity of the two receptor mutations is unlikely, because one, δ6331–6357, is very common and does not produce a phenotype in other heterozygous males. The other receptor mutation is a deletion in the signal peptide; the resultant protein cannot be inserted into the plasma membrane and, thus, cannot bind to the wild-type protein to impair its activity.

Cases with AMH mutations but no uterus In rare cases (Table 1), no Müllerian derivatives were found in subjects bearing *AMH* gene mutations. One patient (CA) presented with transverse testicular ectopia, practically never seen in the absence of PMDS,[23] and the other (AA) had a brother with classical PMDS. Possibly, the Müllerian derivatives degenerated in late fetal life.

Animal models of persistent Müllerian duct syndrome

Male mice with persistence of Müllerian duct derivatives

Richard Behringer, in Houston, has become a specialist in the creation of *AMH/AMHR* transgenic lines. There is no phenotypic difference between male mice in whom either the *AMH*[40] or the *AMHR-II*[41] gene has been invalidated. In both instances, homozygous females are completely normal and fertile. Homozygous males have fully descended testes of normal size, normal male external genitalia and a combination of Wolffian and Müllerian duct derivatives, namely a uterus, oviducts and upper vagina superimposed upon the male reproductive system. A high proportion of males are infertile, apparently because the presence of Müllerian derivatives blocks sperm transport into the female reproductive tract.

Older *AMHR-II*-defective male mice develop Leydig cell hyperplasia and seminiferous tubule atrophy, leading to reduced spermatogenesis. Leydig cell hyperplasia in *AMH-AMHR-II* deficient mice is the mirror image of the Leydig cell hypoplasia observed in mice overexpressing the *AMH* gene under the control of the metallothionein promoter,[42] which is due to a block in the differentiation of the Leydig cell precursors into mature Leydig cells.[43] Leydig cell tumors have not been described in PMDS.

Persistent Müllerian duct syndrome in dogs

PMDS has been described in a strain of miniature schnauzer dogs[44] and, more recently, in German basset-hounds.[45] In both breeds, the condition is inherited as an autosomal recessive trait. Affected dogs present primarily with cystitis and pyometra due to communication between the vagina and urethra; cryptorchidism is relatively rare. PMDS dogs with scrotal testes are fertile.[44] AMH RNA and immunoreactive protein are present in embryonic testes during the critical period of Müllerian regression;[46] after birth, the concentration of serum AMH is similar in affected and normal animals (Messika-Zeitoun and colleagues, unpublished data). Taken together, these data suggest that PMDS in dogs is due to AMH resistance, not to an *AMH* defect. The dog *AMHR-II* has now been totally sequenced, and no mutation has been detected in the affected dogs, suggesting that the defect lies elsewhere in the transduction cascade (Messika-Zeitoun and colleagues, unpublished data).

Wnt-7a knock-out mice

Male mice in whom the *Wnt-7a* gene has been invalidated also have retained Müllerian ducts.[47] *Wnt* genes, the mammalian equivalent of the *Wingless* gene of *Drosophila*, play an important role in early development.[48] *Wnt-4* genes are required for normal organogenesis of Müllerian ducts in males and females alike,[49] and *Wnt-7* is required for normal expression of *AMHR-II* in the mesenchyme surrounding the Müllerian duct. In the absence of *AMHR-II*, Müllerian ducts become resistant to AMH and do not regress.

POTENTIAL PHYSIOLOGICAL RELEVANCE OF PERSISTENT MÜLLERIAN DUCT SYNDROME

Both human and animal models of PMDS have been very useful for understanding the impact of AMH on the developing organism. Several conclusions have been reached.

Anti-Müllerian hormone is not necessary for testicular development or descent

Because AMH is one of the first proteins expressed at the beginning of testicular differentiation, and because its promoter carries a binding site for a sex-determining gene, *SOX-9*,[12] it has been hypothesized that AMH participates in testicular organogenesis. Since both human and animal males with *AMH* or *AMHR-II* defects have normally developed testes, AMH cannot be a very important player in the game. Similarly, testicular descent is usually normal in PMDS mice or dogs, thus excluding a major role of AMH in the process of testicular descent.

Anti-Müllerian hormone is not required for normal ovarian function

There is biochemical and experimental evidence that AMH has an effect upon follicular development in the ovary. AMH is produced by granulosa cells during the period of reproductive activity,[20,50] and exerts a suppressive effect upon follicle recruitment[51] and maturation.[52] Nevertheless, ablation of the *AMH* or *AMHR-II* gene in human females does not impair ovarian function or fertility, suggesting that other growth factors can take over.

The transduction cascade of anti-Müllerian hormone is relatively simple

Compared with other members of the TGF-β family, which have several type II receptors, some of which bind to different ligands, the AMH transduction cascade appears to be relatively simple. Mutations of either the *AMH* or *AMHR-II* gene are associated with the same clinical symptoms, suggesting that AMH has only one type II receptor, AMHR-II, and that the latter binds to only one ligand, AMH. However, this simplicity does not apply to type I receptors, several of which could activate the Smad1 pathway, which controls the activation/repression of *AMH* target genes.[18]

SUMMARY AND FUTURE PERSPECTIVES

Persistent Müllerian duct syndrome (PMDS), characterized by the persistence of Müllerian ducts in genotypic and phenotypic males, is a striking illustration of the Jost theory of the duality of fetal testicular hormonal secretion. In PMDS, production and sensitivity to testosterone are normal, and thus external genital differentiation is normal, unlike the situation prevailing in gonadal dysgenesis, another cause of persistence of Müllerian duct derivatives. In most cases of PMDS, a mutation of either the *AMH* or *AMHR-II* gene can be detected. As in knock-out mice, the nature of the defective gene has no impact upon the clinical picture, suggesting that AMH is the only ligand of AMHR-II and has no other type II receptor.

In patients in whom both the *AMH* and *AMHR-II* genes are normal, the possibility that another element of the AMH transduction cascade is defective should be entertained. Mutations of candidate genes in patients with unexplained PMDS would be a strong argument in favor of their involvement in the AMH cascade. However, many patients in this group present with congenital associated malformations, indicating either that the unknown gene is also involved in other transduction pathways or that the retention of Müllerian ducts is part of a complex malformation syndrome unrelated to anti-Müllerian hormone. Further studies are required to elucidate this issue.

REFERENCES

1. Jost A. Problems of fetal endocrinology: the gonadal and hypophyseal hormones. *Rec Prog Horm Res* 1953 8: 379–418.
2. Josso N. Anatomy and endocrinology of sex differentiation. In deGroot LJ, Jameson JL, eds. *Endocrinology*. Philadelphia, PA: WB Saunders, 2000: 1947–1954
3. Josso N, Picard JY, Tran D. The anti-Müllerian hormone. *Rec Prog Horm Res* 1977 33: 117–160.
4. Picon R. Action du testicule foetal sur le développement *in vitro* des canaux de Müller chez le rat. *Arch Anat Microsc Morphol Exp* 1969 58: 1–19.
5. Wilson CA, di Clemente N, Ehrenfels C *et al.* Müllerian inhibiting substance requires its N-terminal domain for maintenance of biological activity, a novel finding within the TGF-β superfamily. *Mol Endocrinol* 1993 7: 247–257.
6. Al-Attar L, Noël K, Dutertre M *et al.* Hormonal and cellular regulation of Sertoli cell anti-Müllerian hormone production in the postnatal mouse. *J Clin Invest* 1997 100: 1335–1343.
7. Münsterberg A, Lovell-Badge R. Expression of the mouse anti-Müllerian hormone gene suggests a role in both male and female sex differentiation. *Development* 1991 113: 613–624.
8. Cate RL, Mattaliano RJ, Hession C *et al.* Isolation of the bovine and human genes for Müllerian inhibiting substance and expression of the human gene in animal cells. *Cell* 1986 45: 685–698.
9. Cohen-Haguenauer O, Picard JY, Mattei MG *et al.* Mapping of the gene for anti-Müllerian hormone to the short arm of human chromosome 19. *Cytogenet Cell Genet* 1987 44: 2–6.
10. Giuili G, Shen WH, Ingraham HA. The nuclear receptor SF-1 mediates sexually dimorphic expression of Müllerian inhibiting substance, *in vivo*. *Development* 1997 124: 1799–1807.
11. Dresser DW, Hacker A, Lovell-Badge R, Guerrier D. The genes for a spliceosome protein (SAP62) and the anti-Müllerian hormone (AMH) are contiguous. *Hum Mol Genet* 1995 4: 1613–1618.
12. Arango NA, Lovell-Badge R, Behringer RR. Targeted mutagenesis of the endogenous mouse *Mis* gene promoter: *in vivo* definition of genetic pathways of vertebrate sexual development. *Cell* 1999 99: 409–419.
13. Massagué J, Chen YG. Controlling TGF-β signaling. *Genes Dev* 2000 14: 627–644.
14. Imbeaud S, Faure E, Lamarre I *et al.* Insensitivity to anti-Müllerian hormone due to a spontaneous mutation in the human anti-Müllerian hormone receptor. *Nat Genet* 1995 11: 382–388.
15. di Clemente N, Wilson CA, Faure E *et al.* Cloning, expression and alternative splicing of the receptor for anti-Müllerian hormone. *Mol Endocrinol* 1994 8: 1006–1020.
16. Baarends WM, van Helmond MJL, Post M *et al.* A novel member of the transmembrane serine/threonine kinase receptor family is specifically expressed in the gonads and in mesenchymal cells adjacent to the Müllerian duct. *Development* 1994 120: 189–197.
17. Faure E, Gouédard L, Imbeaud S *et al.* Mutant isoforms of the anti-Müllerian hormone type II receptor are not expressed at the cell membrane. *J Biol Chem* 1996 271: 30571–30575.
18. Gouédard L, Chen YG, Thevenet L *et al.* Engagement of bone morphogenetic protein type IB receptor and Smad1 signaling by anti-Müllerian hormone and its type II receptor. *J Biol Chem* 2000 275: 27973–27978.
19. Allard S, Adin P, Gouédard L *et al.* Molecular mechanisms of hormone-mediated Müllerian duct regression: involvement of β-catenin. *Development* 2000 127: 3349–3360.

20. Lee MM, Donahoe PK, Hasegawa T *et al.* Müllerian inhibiting substance in humans: normal levels from infancy to adulthood. *J Clin Endocrinol Metab* 1996 81: 571–576.
21. Rey RA, Belville C, Nihoul-Fékété C *et al.* Evaluation of gonadal function in 107 intersex patients by means of serum anti-Müllerian hormone measurement. *J Clin Endocrinol Metab* 1999 84: 627–631.
22. Imbeaud S, Belville C, Messika-Zeitoun L *et al.* A 27 base-pair deletion of the anti-Müllerian type II receptor gene is the most common cause of the persistent Müllerian duct syndrome. *Hum Mol Genet* 1996 5: 1269–1279.
23. Thompson ST, Grillis MA, Wolkoff LH, Katzin WE. Transverse testicular ectopia in a man with persistent Müllerian duct syndrome. *Arch Pathol Lab Med* 1994 118: 752–755.
24. Carré-Eusèbe D, Imbeaud S, Harbison M, New MI, Josso N, Picard JY. Variants of the anti-Müllerian hormone gene in a compound heterozygote with the persistent Müllerian duct syndrome and his family. *Hum Genet* 1992 90: 389–394.
25. Guerrier D, Tran D, van der Winden JM *et al.* The persistent Müllerian duct syndrome: a molecular approach. *J Clin Endocrinol Metab* 1989 68: 46–52.
26. Hutson JM, Davidson PM, Reece L, Baker ML, Zhou B. Failure of gubernacular development in the persistent Müllerian duct syndrome allows herniation of the testes. *Pediatr Surg Int* 1994 9: 544–546.
27. Imbeaud S, Rey R, Berta P *et al.* Progressive testicular degeneration in the persistent Müllerian duct syndrome. *Eur J Pediatr* 1995 154 : 187–190.
28. Farag TI Familial persistent Müllerian duct syndrome in Kuwait and neighboring populations. *Am J Med Genet* 1993 47: 432–434.
29. Snow BW, Rowland RG, Seal GM, Williams SD. Testicular tumor in patient with persistent Müllerian duct syndrome. *Urology* 1985 26: 495–497.
30. Nishioka T, Kadowaki T, Miki T, Hanai J. Persistent Müllerian duct syndrome. *Hinyokika Kiyo* 1992 38: 89–92.
31. Korsch E, Rey R, Imbeaud S, Picard JY, Josso N. Persistent Müllerian duct syndrome: decrease of serum levels of anti-Müllerian hormone shortly after orchidopexy procedure. *Horm Res* 1997 48 (Suppl 2): 117 (abstr).
32. Cooper DN, Krawczak M. *Human Gene Mutations.* Eynsham, UK: Bios Scientific Publishers 1993.
33. Messika-Zeitoun L, Gouédard L, Belville C *et al.* Autosomal recessive segregation of a truncating mutation of anti-Müllerian hormone type II receptor in a family affected by the persistent Müllerian duct syndrome contrasts with its dominant negative activity *in vitro*. *J Clin Endocrinol Metab* 2001 86: 4390–4397.
34. Klosowski S, Abriak A, Morisot C *et al.* Atrésie jéjunale et syndrome de persistance des canaux de mülleriens. *Arch Pédiatr* 1997 4: 1264–1265.
35. Van Maldergem L, Bachy A, Feldman D *et al.* A syndrome exhibiting lipoatrophic diabetes, vitamin D resistant rickets, and persistent Müllerian ducts in a Turkish boy born to consanguineous parents. *Am J Med Genet* 1996 64: 506–513.
36. Tomboc M, Lee PA, Mitwally MF, Schneck FX, Bellinger M, Witchel SF. Insulin-like 3/relaxin-like factor gene mutations are associated with cryptorchidism. *J Clin Endocrinol Metab* 2000 85: 4013–4018.
37. Baur ST, Mai JJ, Dymecki SM. Combinatorial signaling through BMP receptor IB and GDF5: shaping of the distal mouse limb and the genetics of distal limb diversity. *Development* 2000 127: 605–619.
38. Yi SE, Daluiski A, Pederson R, Rosen V, Lyons KM. The type IBMP receptor BMPRIB is required for chondrogenesis in the mouse limb. *Development* 2000 127: 621–630.
39. Morel Y, Rey R, Teinturier C *et al.* Etiological diagnosis of male sex ambiguity: a collaborative study. *Eur J Pediatr* 2001 161: 49–59.
40. Behringer RR, Finegold MJ, Cate RL. Müllerian-inhibiting substance function during mammalian sexual development. *Cell* 1994 79: 415–425.
41. Mishina Y, Behringer RR The *in vivo* function of Müllerian inhibiting substance during mammalian sexual development. *Adv Dev Biol* 1996 4: 1–25.
42. Behringer RR, Cate RL, Froelick GJ, Palmiter RD, Brinster RL. Abnormal sexual development in transgenic mice chronically expressing Müllerian inhibiting substance. *Nature* 1990 345: 167–170.
43. Racine C, Rey R, Forest MG *et al.* Receptors for anti-Müllerian hormone on Leydig cells are responsible for its effects on steroidogenesis and

cell differentiation. *Proc Natl Acad Sci USA* 1998 95: 594–599.

44. Meyers-Wallen VN, Donahoe PK, Ueno S, Manganaro TF, Patterson DF. Müllerian inhibiting substance is present in testes of dogs with persistent Müllerian duct syndrome. *Biol Reprod* 1989 41: 881–888.
45. Jonen H, Nickel RF. Das Syndrom des persistierenden Müllerschen Gänge: eine erbliche Form von männlichem Pseudohermaphroditismus beim Bassethound. *Kleintierpraxis* 1996 41: 911–918.
46. Meyers-Wallen VN, Lee MM, Manganaro TF, Kuroda T, MacLaughlin D, Donahoe PK. Müllerian inhibiting substance is present in embryonic testes of dogs with persistent Müllerian duct syndrome. *Biol Reprod* 1993 48: 1410–1418.
47. Parr BA, McMahon AP. Sexually dimorphic development of the mammalian reproductive tract requires *Wnt-7a*. *Nature* 1998 395: 707–710.
48. McMahon AP. The *Wnt* family of developmental regulators. *Trends Genet* 1992 8: 236–242.
49. Vainio S, Heikkilä M, Kispert A, Chin N, McMahon AP. Female development in mammals is regulated by *Wnt-4* signaling. *Nature* 1999 397: 405–409.
50. Rey R, Lhommé C, Marcillac I *et al.* Anti-Müllerian hormone as a serum marker of granulosa-cell tumors of the ovary: comparative study with serum α-inhibin and estradiol. *Am J Obstet Gynecol* 1996 174: 958–965.
51. Durlinger ALL, Kramer P, Karels B *et al.* Control of primordial follicle recruitment by anti-Müllerian hormone in the mouse ovary. *Endocrinology* 1999 140: 5789–5796.
52. di Clemente N, Goxe B, Remy JJ *et al.* Inhibitory effect of AMH upon the expression of aromatase and LH receptors by cultured granulosa cells of rat and porcine immature ovaries. *Endocrine* 1994 2: 553–558.

Chapter 32

Premature ovarian failure

Gerard S. Conway and Sophie Christin-Maitre

INTRODUCTION

Physiological ovarian failure as measured by the last menstrual period, the menopause, occurs at an average age of 50.7 years, and conception is very unlikely after the age of 49 years.[1] This figure has been found to be constant across generations, unlike the age of menarche which has fallen, particularly over the first half of the 20th century. The age of menopause is known to be strongly inherited, as evidenced by a recent study of Australian twins,[2] and it is also affected by environmental factors such as smoking.[3] The distribution of the age of menopause is normal, with 1–2% of women menstruating after the age of 60 years and a similar proportion entering the menopause before the age of 40.[4] Menopause before the age of 40 years is most commonly taken to be the definition of 'premature ovarian failure'.

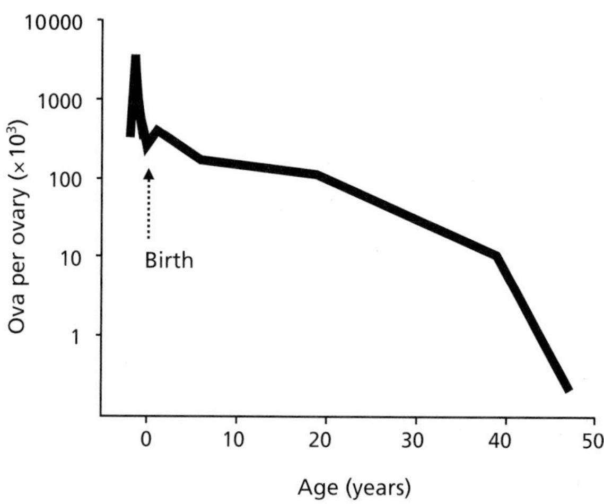

Figure 1 Number of germ cells per ovary throughout life, derived from postmortem studies[5,6]

BACKGROUND

The ovary comprises four cell types: germ cells, granulosa cells, theca cells and support cells. The most vital functions of the ovary, producing gametes and estrogen, are determined by germ cells and granulosa cells, respectively. Ovarian failure is usually considered to result from a defect in these components.

The life span of the ovary is often thought to be determined by the number of germ cells within it (Figure 1). In the 5-week human embryo, germ cells migrate from the yolk sac to the urogenital ridge, where they multiply by mitosis and form the primitive gonad. As the ovary develops oogonia, they are immediately surrounded by granulosa cells to form a stable primordial follicle. Early studies focused on counting germ cells in ovaries during fetal and postnatal life. Even though very few ovaries have been subjected to comprehensive germ cell quantification, from the available information it appears that there are at least four phases in the control of germ cell numbers throughout life. During the first 20 weeks of fetal life, there is a rapid multiplication of germ cell numbers resulting in 3.5 million per ovary.[5] In the last 3 months of fetal life, two-thirds of these are destroyed so that each ovary contains 1 million ova. There is then a fairly steady decline in ova, with only 300 000 left at puberty and 10 000 by age 40 years. This loss of 990 000 ova, presumably through apoptosis, compares with only 500 or so that might be used up through ovulation during reproductive life.[7] At about the age of 40, an acceleration of ovum destruction takes place, so that few are left at the time of the menopause at age 50.[6] The factors that control germ cell numbers are only now beginning to be characterized. For instance, disruption of ovarian apoptosis in *bax* knock-out mice can result in a delay of menopausal age.[8]

Variation in both timing and degree of ovarian failure determines the clinical picture and diagnostic label applied. Gonadal dysgenesis describes the failure of the ovary to develop, but is also loosely applied to the presentation of ovarian failure with primary amenorrhea and failure to develop secondary sexual characteristics. Premature ovarian failure covers the presentation between menarche and the age of 40 years, and this term has largely taken over from 'premature menopause', because the emphasis attached to the menopausal state only made a difficult diagnosis worse for young women. Premature ovarian failure should be distinguished from early menopause, which occurs between the ages of 40 and 45.[4] Partial ovarian failure is the usual explanation for the resistant ovary syndrome, the term originally applied to women with hypergonadotropic amenorrhea and intact ovarian apparatus on biopsy.[9] Occult ovarian failure describes a lesser form of partial ovarian failure whereby the serum follicle stimulating hormone (FSH) concentration is raised, higher than two standard deviations from the upper limit of the FSH normal range, but menstrual cycles persist.[10] The term hypergonadotropic hypogonadism has been applied to all of the above conditions.[11]

APPLIED TECHNOLOGY

Animal models of premature ovarian failure

In the next few years, there should be better knowledge of the etiology of premature ovarian failure, thanks to the description of different animal models. To date, more than 50 rodent models of gonadal deficiency have been described. As in human premature ovarian failure, mouse models can be divided into germ cell deficiency, increased follicular apoptosis and follicular growth blockade. Although no human equivalent of most animal models has been described so far, these examples are useful in understanding ovarian physiology. These mouse models of ovarian failure are listed in Table 1.

The c-*kit* knock-out mouse presents an absence of germ cells and no primordial follicle.[23] Mice null for *Wnt4* present the same phenotype. *Wnt4* has been identified as a factor necessary for ovarian germ migration.[24] Similarly, *DAXL* knock-outs have a depletion of germ cells, and, although this gene has been proposed as a candidate for premature ovarian failure, the fact that the human male phenotype is usually spared departs from the mouse model.[25] The same issue applies to the *ZFX* knock-out, in which both male and female mice have depleted germ cell numbers.[26]

Several genes have been implicated in follicular growth blockade. The first, *GDF-9*, belongs to the transforming growth factor family (growth differentiation factor) and is synthesized in the oocyte. It gives a signal to granulosa cells, so that they start to multiply. *GDF-9* knock-out mice are infertile because of follicular growth arrest.[27] Within the ovaries, only one layer of follicles can be identified. Mice lacking connexin-37 are infertile.[28] At birth, the ovary contains primordial follicles consisting of meiotically arrested oocytes surrounded by a single layer of supporting (granulosa) cells. Therefore, cell–cell signalling through intercellular channels made of connexins is required for successful oogenesis. Mice with overexpression of follistatin also have small ovaries, owing to a block in folliculogenesis at various stages.[29] Mice null for estrogen receptor α also have a blockade in follicular growth.[30]

Distinct from the murine model, a premature ovarian failure model has been described in Inverdale sheep. Sheep with a heterozygous *GDF-9B* (or *BMP15*) mutation, a new member of the

Table 1 Genes identified in mouse ovarian failure

	Gene	Reference(s)
Germ cell deficiency or increased follicular apoptosis	*c-kit*	12
	Wnt4	13
	Atm	14
	DAXL	15
	ZFX	16
	Bcl2	
Follicular growth blockade	*GDF-9* knock-out	17
	connexin-37 knock-out	18
	FSH receptor knock-out	19,20
	follistatin overexpression	21
	estrogen receptor α knock-out	22
	activin receptor type II knock-out	

FSH, follicle stimulating hormone

GDF-9 family, deliver two or three lambs. Sheep with a homozygous *GDF-9B* mutation develop premature ovarian failure.[31] Women with premature ovarian failure and polycystic ovary syndrome have been tested unsuccessfully for *GDF-9B* mutation.[32]

CURRENT CONCEPTS

In theory, mechanisms of primary ovarian failure might include defective germ cell migration to the gonad, impaired multiplication of germ cells or accelerated germ cell apoptosis, an abnormality of granulosa cell support of the germ cell or a blockade in follicular growth. In practice, the causes of primary ovarian failure can be classified into those that affect ovarian development, resulting in gonadal dysgenesis, and those that affect the developed ovary, presenting as ovarian failure (Table 2). The distinction between these categories is, however, not complete. Abnormalities in the X chromosome, for instance, may present as gonadal dysgenesis, as in most women with Turner's syndrome, or as secondary amenorrhea and premature ovarian failure in some Turner mosaics or in women with a smaller deletion of the long arm of the X chromosome.

POTENTIAL CLINICAL RELEVANCE

In clinical practice, ovarian failure can be considered under the headings of genetic, autoimmune and environmental causes. Despite recent inroads into our understanding of many of the pathogenic mechanisms leading to primary ovarian failure, the cause of premature ovarian failure remains idiopathic in the majority of women.[33]

Table 2 Various etiologies causing premature ovarian failure in women

Gonadal dysgenesis		Turner's syndrome: 45,X and variants X-chromosome translocation or deletion Perrault's syndrome 46,XX gonadal dysgenesis
Ovarian failure	genetic associations	X-chromosome deletions, translocations X, XX mosaicisms galactosemia enzyme defects: P450c17, StAR *FRAXA* premutations BPES small X-chromosome defects FSH receptor gene mutations ataxia–telangiectasia inhibin-α gene defects
	autoimmune	autoimmune polyendocrinopathy syndrome 1: APECED autoimmune polyendocrinopathy syndrome 2 association with various autoimmune diseases isolated autoimmune ovarian failure
	infection	viral oophoritis
	iatrogenic	chemotherapy radiotherapy pelvic surgery
	idiopathic	environmental toxins?

StAR, steroidogenic acute regulatory protein; BPES, blepharophimosis, ptosis and epicanthus syndrome; FSH, follicle stimulating hormone; APECED, autoimmune polyendocrinopathy, candidiasis and ectodermal dystrophy

GENETICS OF OVARIAN FAILURE

Genetic causes of ovarian failure could be involved in more than 40% of cases of premature ovarian failure. The best evidence for a genetic contribution comes from families in which more than one female has premature ovarian failure. Estimates vary as to the prevalence of familial premature ovarian failure from 5 to 30%, depending on the referral base of each clinic and the comprehensiveness of the family history.[34,35] The mode of inheritance can be X-linked or autosomal, dominant or recessive. The expression of gonadal failure in these families is primarily limited to females, and frequently no other organ is affected. The genetic associations of premature ovarian failure listed below and in Table 3 are growing each year, but still the mutations causing familial ovarian failure are unknown in the majority of pedigrees.

X-chromosome abnormalities

Turner's syndrome

Women with Turner's syndrome have a variable phenotype, in most cases the consistent features being short stature and ovarian dysgenesis. Other features include a wide carrying angle, a webbed neck and lymphedema.[36] Any defect of an X chromosome may cause Turner's syndrome, and the most common karyotype is 45,X, a complete absence of one X chromosome. Some of the variability in the Turner phenotype is the result of mosaicism, with only a proportion of cells in the body possessing the abnormal X chromosome. The very few women with Turner's syndrome enter puberty spontaneously, and even fewer escape primary amenorrhea. There are only a handful of pregnancies reported in the literature in women with Turner's syndrome. The ovaries are dysgenetic and appear as small fibrous streaks. In fact,

Table 3 Genes identified in human premature ovarian failure

Chromosome	*Gene*	*Disease*	*Reference(s)*
X chromosome	monosomy	Turner's syndrome	28
	deletions		29
	translocations		33
	mosaicism		36
	dia		35
	FRAXA premutation		37–39
Chromosome 2	FSH receptor mutations		40–43
Chromosome 2	inhibin-α		44
Chromosome 3	*FOXL2*	blepharophimosis	45
Chromosome 8	*StAR*		46
Chromosome 9	galactose-1-phosphate uridyl transferase	galactosemia	47
Chromosome 15	P450c17 hydroxylase		48
Chromosome 21	aromatase		49
Chromosome 21	*Atm*	ataxia–telangiectasia	
Chromosome 21	autoimmune polyendocrinopathy syndrome type 1	APECED: autoimmune polyendocrinopathy, candidiasis, ectodermal dystrophy	50
?	autoimmune polyendocrinopathy syndrome type 2		51
?	?	Perrault's syndrome	52

FSH, follicle stimulating hormone; StAR, steroidogenic acute regulatory protein

early ovarian development is relatively normal in Turner fetuses, but, in the third trimester of gestation, the ovary is rapidly destroyed so that few ova remain at birth.[37] It is presumed that accelerated germ cell destruction *in utero* is the result of unstable meiosis, which is disturbed by the absence of one X chromosome.

Small X-chromosome defects

While the deletion of one complete X chromosome causes Turner's syndrome, more minor defects of an X chromosome can cause less dramatic phenotypes.[38] In general, deletions of the short arm of the X chromosome result in a full Turner phenotype, with short stature and ovarian dysgenesis. The phenotype associated with partial deletions of the long arm of the X chromosome demonstrates features usually restricted to the ovary, with primary or secondary amenorrhea in both sporadic and familial forms.[39,48] Analysis of these small X-chromosome defects reveals critical regions that are likely loci for ovary-determining genes.[46] Balanced X;autosome translocations involving the interval of the X chromosome from Xq13 to Xq26 (but excluding part of Xq22) are often associated with premature ovarian failure.[49] Genes involved are yet unknown. Bione and colleagues have identified a translocation break-point in a woman with secondary amenorrhea.[47] The translocation disrupts *DIAPH2*, a human homolog of *Drosophila melanogaster diaphanous* (*dia*) involved in infertility in *Drosophila*. However, no other cases of *DIAPH2* defects have been identified in women. Prueitt and associates have mapped nine Xq translocation break-points, and identified *XPNPEP2* as a premature ovarian failure candidate gene.[53] Further studies are needed to dissect the different genes located on the long arm of chromosome X, involved in premature ovarian failure.

X mosaicism

Low-level 45,X/46,XX mosaicism has been shown to be more common in women with premature ovarian failure compared with age-matched controls.[54] However, the clinical significance of such mosaicism in the wider population of women with premature ovarian failure remains unknown.

Fragile-X premutations

The *FRAXA* fragile site is located in the untranslated exon 1 of the *FMR1* gene at Xq27.3. Normally, fewer than 60 trinucleotide (CCG) repeats occur in this exon. A *FRAXA* premutation is defined as between 60 and 200 repeats at this site, and the full mutation is defined as more than 200 repeats. The *FRAXA* premutation was thought to have no effect on the transcription of the *FMR1* gene or translation of its mRNA. The full *FRAXA* mutation is associated with methylation of the *FMR1* promoter and silencing of gene transcription with resulting mental impairment, most obvious in males. Recently, *FRAXA* premutations have been shown to be associated with premature ovarian failure, particularly in its familial form.[40,45,55] The mechanism of this association is unknown.

Enzyme defects

17-Hydroxylase

A deficiency of the enzyme 17-hydroxylase was originally described in genetic females presenting with primary amenorrhea and hypokalemic hypertension.[41] In fact, a single cytochrome, P450c17, catalyzes both 17-hydroxylase and 17,20-lyase enzyme activities, although clinical discrimination between these two activities is unclear. Autosomal recessive inheritance is transmitted through mutations of the *CYP17* gene in this syndrome. In this condition, the mineralocorticoid effect of raised deoxycorticosterone causes hypertension, with suppression of renin and aldosterone. An excess of corticosterone prevents glucocorticoid deficiency. Females fail to enter puberty because of an inability to synthesize estrogen, and ovaries are enlarged and cystic. Body hair is sparse because of the lack of androgens.

Steroidogenic acute regulatory protein

Mutations of the steroidogenic acute regulatory protein (StAR) gene cause congenital lipoid adrenal hyperplasia, which presents as a salt-wasting crisis in infancy.[56] In severe forms, no adrenal or gonadal steroids are synthesized.

Genetic females present with primary hypergonadotropic amenorrhea and lack of secondary sex characteristics. In some pedigrees, however, the ovary can be relatively spared, and spontaneous menarche occurs. Genetic males are phenotypic females.

Aromatase

Rare cases of aromatase mutations have been identified. Women have a defect in transforming androgens to estrogens. They present with primary amenorrhea and enlarged ovaries.[44]

Galactose-1-phosphate uridyl transferase

Galactosemia is inherited as an autosomal recessive disorder caused by mutations of the *GALT* gene mapped to locus 9p13. The resulting deficiency of galactose-1-phosphate uridyl transferase results in the presentation of failure to thrive, hypoglycemia, hepatomegaly and acidosis in the first few days of life. With early diagnosis and a strict galactose-restricted diet, children can be expected to thrive with little mental impairment. While dietary management can prevent liver and brain damage, it does not prevent ovarian failure. Female hypogonadism is frequent in affected girls, who usually fail to enter puberty and are found to have streak ovaries.[57] The risk of ovarian failure in galactosemia is greater in those with the least enzyme activity.[58] The mechanism of ovarian failure in galactosemia is unknown, but a toxic effect of galactose or its metabolites has been implicated. In contrast, males with galactosemia usually have normal gonadal function.

Blepharophimosis, ptosis and epicanthus inversus syndrome

In type I blepharophimosis, ptosis and epicanthus inversus syndrome (BPES), eyelid abnormalities are associated with ovarian failure. Type II BPES shows only the eyelid defects. Both BPES loci are on chromosome 3q23, as evidenced by deletions and translocations associated with the syndrome. BPES is transmitted as an autosomal dominant trait.[42] The characteristic appearance of the eyelids in type I BPES is accompanied by ovarian failure in one-half of pedigrees, leading some to suggest that the eyelid and the ovarian effects are the result of separate contiguous genes. Frequently, the ovarian failure is intermittent, with a clinical picture of 'resistant ovaries'.[43] The infertility is sex-limited, affected males being fertile. Recently, mutations of the forkhead transcription factor *FOXL2* have been identified.[19] These mutations produce a truncated protein. The target genes of *FOXL2* are yet unknown. *FOXL2* could interact with members of the transforming growth factor-β (TGF-β) family. Therefore, *FOXL2* might be involved in follicle apoptosis.

Follicle stimulating hormone receptor gene mutations

Follicle stimulating hormone is vital for normal ovarian development and function. It has a major antiapoptotic effect in the ovarian follicle. In Finland, an Ala189Val mutation affecting the extracellular domain of the FSH receptor gene was found to be associated with premature ovarian failure in 22 women from six families presenting with primary amenorrhea.[20] The phenotype of women homozygous for the Val189 mutation is particularly revealing. In those women for whom ovarian histology was obtained, all showed a reduced number of primordial follicles, rather than ovarian dysgenesis. A blockade in follicular growth was noted at the early preantral stage.[52] Conversely, male members of these families who shared the mutation had only a variable suppression of sperm count, and in some cases were fertile.[14] The Finnish Val189 mutation appears to be restricted to that geographically isolated population, as it has not been found in series from other countries, in Germany,[59] the United States,[60] Japan and England.[50]

Since the first Finnish mutation, three single cases have been reported. The phenotype of partial loss of function of the FSH receptor was demonstrated in a compound heterozygote who presented with secondary amenorrhea. Her ovaries were of normal size, containing small antral follicles.[61,62] Complete loss of function of the FSH receptor results in small hypoplastic ovaries, while partial loss of function results in secondary amenorrhea with normal ovarian morphology. Interestingly, two different mouse models of FSH receptor knock-out have been reported. In both models, a follicular growth was blocked at the early antral stage.[51,63] These human and mouse models have confirmed the role of FSH in final follicular growth.

Perrault's syndrome

Perrault's syndrome is the combination of congenital deafness, gonadal dysgenesis and short stature, and is inherited in an autosomal recessive manner. The genetic defect in Perrault's syndrome is yet to be defined.[64]

Ataxia–telangiectasia

Ataxia–telangiectasia is an autosomal recessive disorder with a wide variety of clinical manifestations. Patients usually present with cerebellar degeneration, oculomotor dysfunction, immunodeficiency, acute cancer predisposition and radiation sensitivity. Gonadal abnormalities are hypoplasia with germ cell deficiencies. The ataxia–telangiectasia locus was mapped to chromosome 11q22–23, and the gene called *ATM*. This gene is involved in cell regulation. A mouse model has been generated by disrupting the murine *Atm* locus.[65] Ovaries of mutant females are devoid of primordial and maturing follicles, as well as oocytes.

Inhibin gene defects

Using the approach of reproductive candidate gene screening, Shelling and colleagues have shown that a defect of the inhibin-α gene is more prevalent in women with premature ovarian failure than in controls.[66]

AUTOIMMUNITY AND THE OVARY

Evidence that an autoimmune process can damage the ovary comes from the association between ovarian failure and other autoimmune disease.[67] Between 10 and 30% of women with premature ovarian failure have a concurrent autoimmune disease, the most common of which is hypothyroidism. In addition to the syndromes described below, premature ovarian failure has been reported in association with myasthenia gravis, systemic lupus erythematosus, rheumatoid arthritis and Crohn's disease.

Autoimmune polyendocrinopathy syndrome type 1

Autoimmune polyendocrinopathy syndrome type 1 (APS1) is a rare autosomal recessive syndrome comprising mucocutaneous candidiasis, hypoparathyroidism and Addison's disease. Hypothyroidism, insulin-dependent diabetes mellitus (IDDM), pernicious anemia and ovarian failure are additional features of APS1, occurring with variable frequency.[68] Ovarian failure occurs in 60% of affected females, while the testis is spared autoimmune attack. This clinical picture of autoimmune polyendocrinopathy, candidiasis and ectodermal dystrophy is also known as APECED syndrome. APS1 has no human leukocyte antigen (HLA) association, and is caused by mutations in the *AIRE* gene, which encodes a putative transcription regulator.[69,70] To date, the variability of the clinical features has not been closely paralleled by the genotype.

Autoimmune polyendocrinopathy syndrome type 2: Schmidt's syndrome

Classic APS2 comprises Addison's disease, hypothyroidism, IDDM and ovarian failure. The most common accompaniment to ovarian failure is hypothyroidism, and this combination probably represents a form of APS2. The sequence of glands involved is variable, with up to 15 years spanning the first and last.[71] APS2 is associated with the HLA haplotypes HLA-B8, -DR3 and -DR4. The target antigens were originally recognized by steroid-producing cell antibodies, but are now known to be P450 cytochromes from the steroidogenic pathway: P450c17, P450c21 and P450scc.[71,72]

Autoimmunity in isolated ovarian failure

Further evidence of autoimmunity in women with premature ovarian failure comes from ovarian histology, and the detection of circulating antibodies directed against ovarian antigens.[12] Occasional reports of ovarian histology have noted the presence of a lymphocytic oophoritis in women with premature ovarian failure, but, because of small series, the significance of this finding is not known. A systematic series of biopsies from women with premature ovarian failure does not exist, and it is difficult to separate a pathological appearance from the normal inflammatory-like process that accompanies the growing follicle. The fact that the ovary and not the testis is normally populated with macrophages and other leukocytes is a possible explanation for the sparing of males from autoimmune hypogonadism.

Routine antiovarian immunofluorescence is rarely positive in women with isolated premature ovarian failure. Specific autoimmunity directed against the FSH receptor or the zona pellucida secreted by granulosa cells has been proposed as a pathogenic mechanism but has not yet found firm support. It is clear that antibodies directed against the cytochromes P450c17, P450c21 and P450scc, which are typical of APS2, are not in play in isolated premature ovarian failure. Recently, the enzyme 3β-hydroxysteroid dehydrogenase (3β-HSD) has been proposed as a marker for autoimmunity in women with premature ovarian failure.[13,15] The significance of anti-3β-HSD positivity in terms of pathogenesis is unclear. The struggle to find a sensitive marker for autoimmune ovarian failure is driven by occasional reports that treatment with glucocorticoids induces remission in some women with premature ovarian failure,[16] although a recent controlled trial has questioned the efficacy strategy.[17]

VIRAL OOPHORITIS

Not infrequently, women with premature ovarian failure recall a notable preceding infection, but a pathogenic link is only particularly strong for mumps–oophoritis.[18] Between 2 and 8% of women have been reported to have oophoritis after outbreaks of mumps infection, compared with 25% of males with orchitis. The clinical presentation of abdominal pain and enlarged ovaries in women may often go unnoticed, so these figures may be an underestimate of the prevalence of this complication. Moreover, the long-term sequelae of viral oophoritis and the mechanism of progression to subsequent ovarian failure are unknown. It may be that occult oophoritis underlies the majority of women in whom the etiology of premature ovarian failure is obscure.

Evidence for destructive oophoritis from other pathogens is slight. In immune-suppressed states, cytomegalovirus has been implicated. Varicella, malaria and shigella infections have also been associated with premature ovarian failure by review of clinical notes, but have not been systematically studied.

CHEMOTHERAPY/ RADIOTHERAPY

With the increasing success of treatment for childhood malignancies, more females are surviving to reproductive age. The cytotoxic treatments used in treating leukemia or Hodgkin's disease, in particular, cause ovarian failure in approximately 50% of females.[21] Interestingly, males exposed to these cytotoxic regimens experience a similar prevalence of testicular failure, in contrast to many of the genetic and autoimmune causes of ovarian failure listed above. Alkylating agents are particularly prone to cause gonadal damage, and this effect is related to both the dose and the duration of treatment. The same holds true for the use of cyclophosphamide in the treatment of systemic lupus erythematosus, where as many as one-half of the women treated experience early ovarian failure. Cytotoxic treatment-induced ovarian failure is notable for rare spontaneous remissions, which occur rather more often for other etiologies. Tilly and colleagues have shown, in the mouse treated by doxorubicin, that the ovarian toxicity occurs by increasing oocyte apoptosis.[22]

Women with hematological malignancies who receive total body irradiation almost universally experience ovarian failure, which is rarely reversible. In addition, total body irradiation damages the uterus, making successful pregnancy after egg donation a rare event. Radiation damage to the ovary in the mouse can be limited by disruption of the gene for acid sphingomyelinase, or by *in vivo* therapy with sphingosine-1-phosphate,[73] both suppressing apoptosis.

PELVIC SURGERY

The life span of ovaries preserved during hysterectomy is generally accepted to be reduced. The evidence on which this dogma is based is largely retrospective and anecdotal, so it is impossible to estimate the magnitude of this effect. There is even less information on the effect on the ovary of more distant pelvic or abdominal surgery, or of surgery upon one ovary.

IDIOPATHIC OVARIAN FAILURE

In a large proportion of women with primary ovarian failure, there is no identifiable cause. They are sporadic cases with a normal XX karyotype. Occult viral oophoritis or environmental toxins might account for some of these cases. While the effect of toxins on testicular function has been suggested as a cause of falling sperm counts over the last century, there is little comparable epidemiological data in women. There are, however, convincing examples of outbreaks of ovarian damage after exposure to industrial toxins. Further evidence that the ovary is vulnerable to damage by toxins is provided by the negative effect of smoking on the age of natural menopause. In heavy smokers, the menopause occurs 2 years earlier than in non-smokers.

SUMMARY

The duration of normal ovarian life is determined by apoptosis, controlling germ cell number. The mystery of premature ovarian failure lies in the observation that the etiology is unknown in the majority of cases. A focus on the genetic causes of premature ovarian failure has led to new understanding of the determinants of ovarian development and aging. For many women in whom the cause of ovarian failure is unknown, autoimmunity or viral damage may be the pathogenetic mechanism.

FUTURE PERSPECTIVES

The major drive in future research is to determine the 'unknown etiology group', which accounts for over half of all cases. In the past decade, most advances have been made in the definition of genetic causes. Linkage analysis was useful in identifying the FSH receptor mutations, but is not likely to be helpful for more genetically dispersed populations. Associated condition, such as fragile-X syndrome and blepharophimosis, ptosis and epicanthus inversus syndrome (BPES) have been useful in identifying genetic markers, but are only relevant for a minority of women. Furthermore, there are rather few similar conditions to offer new leads.

It seems likely that new genetic inroads will only be made by a consortium approach, to build a large repository of DNA, in particular from families, to allow a systematic gene screen.

Pursuit of an autoimmune link offers an exciting research opportunity, with the possibility that some cases of premature ovarian failure might be temporarily reversible with immune suppression. It is clear that current markers of immune damage to the ovary are not sufficiently sensitive to be useful in identifying an active process. A novel approach to this subject holds the greatest rewards in terms of treatment options.

Finally, the ovary as a model of the aging process will lead to greater insights into the control of apoptosis, which may in turn lead to new therapeutic opportunities.

REFERENCES

1. McKinlay SM, Brambilla DJ, Posner JG. The normal menopause transition. *Maturitas* 1992 14: 103–115.
2. Snieder H, MacGregor AJ, Spector TD. Genes control the cessation of a woman's reproductive life: a twin study of hysterectomy and age at menopause. *J Clin Endocrinol Metab* 1998 83: 1875–1880.
3. McKinlay SM. The normal menopause transition: an overview. *Maturitas* 1996 23: 137–145.
4. Torgerson DJ, Thomas RE, Reid DM. Mothers' and daughters' menopausal ages: is there a link? *Eur J Obstet Gynecol Reprod Biol* 1997 74 :63–66.
5. Baker TG. A quantitative and cytological study of germ cells in human ovaries. *Proc R Soc (London)* 1963 158 :417–433.
6. Richardson SJ, Senikas V, Nelson JF. Follicular depletion during the menopausal transition: evi-

dence for accelerated loss and ultimate exhaustion. *J Clin Endocrinol Metab* 1987 65 :1231–1237.

7. Gougeon A. Regulation of ovarian follicular development in primates: facts and hypotheses. *Endocr Rev* 1996 17: 121–155.
8. Perez GI, Robles R, Knudson CM *et al.* Prolongation of ovarian lifespan into advanced chronological age by *Bax*-deficiency. *Nat Genet* 1999 21 :200–203.
9. Jones GS, De Moraes-Ruehsen M. A new syndrome of amenorrhea in association with hypergonadotropism and apparently normal ovarian follicular apparatus. *Am J Obstet Gynecol* 1969 104: 597–600.
10. Cameron IT, O'Shea FC, Rolland JM *et al.* Occult ovarian failure: a syndrome of infertility, regular menses, and elevated follicle-stimulating hormone concentrations. *J Clin Endocrinol Metab* 1988 67: 1190–1194.
11. Rebar RW. Hypergonadotropic amenorrhea and premature ovarian failure: a review. *J Reprod Med* 1982 27: 179–186.
12. Hoek A, Schoemaker J, Drexhage HA. Premature ovarian failure and ovarian autoimmunity. *Endocr Rev* 1997 18: 107–134.
13. Arif S, Vallian S, Farzaneh F *et al.* Identification of 3β-hydroxysteroid dehydrogenase as a novel target of steroid cell autoantibodies: association of autoantibodies with endocrine autoimmune disease. *J Clin Endocrinol Metab* 1996 81: 4439–4445.
14. Tapanainen JS, Aittomaki K, Min J *et al.* Men homozygous for an inactivating mutation of the follicle-stimulating hormone (FSH) receptor gene present variable suppression of spermatogenesis and fertility. *Nat Genet* 1997 15: 205–206.
15. Reimand K, Peterson P, Hyoty H *et al.* 3β-Hydroxysteroid dehydrogenase autoantibodies are rare in premature ovarian failure. *J Clin Endocrinol Metab* 2000 85: 2324–2326.
16. Corenblum B, Rowe T, Taylor PJ. High-dose, short-term glucocorticoids for the treatment of infertility resulting from premature ovarian failure. *Fertil Steril* 1993 59: 988–991.
17. van Kasteren YM, Braat DD, Hemrika DJ *et al.* Corticosteroids do not influence ovarian responsiveness to gonadotropins in patients with premature ovarian failure: a randomized, placebo-controlled trial. *Fertil Steril* 1999 71: 90–95.
18. Wood C. Mumps and the menopause. *Br J Sex Med* 1975 2: 19.
19. Crisponi L, Deiana M, Loi A *et al.* The putative forkhead transcription factor *FOXL2* is mutated in blepharophimosis/ptosis/epicanthus inversus syndrome. *Nat Genet* 2001 27: 159–166.
20. Aittomaki K, Lucena JL, Pakarinen P *et al.* Mutation in the follicle-stimulating hormone receptor gene causes hereditary hypergonadotropic ovarian failure. *Cell* 1995 82: 959–968.
21. Clark ST, Radford JA, Crowther D *et al.* Gonadal function following chemotherapy for Hodgkin's disease: a comparative study of MVPP and a seven-drug hybrid regimen. *J Clin Oncol* 1995 13: 134–139.
22. Perez GI, Knudson CM, Leykin L *et al.* Apoptosis-associated signaling pathways are required for chemotherapy-mediated female germ cell destruction. *Nat Med* 1997 3: 1228–1232.
23. Manova K, Huang EJ, Angeles M *et al.* The expression pattern of the c-*kit* ligand in gonads of mice supports a role for the c-*kit* receptor in oocyte growth and in proliferation of spermatogonia. *Dev Biol* 1993 157 85–99.
24. Vainio S, Heikkila M, Kispert A *et al.* Female development in mammals is regulated by *Wnt-4* signalling. *Nature* 1999 397: 405–409.
25. Dorfman DM, Genest DR, Reijo Pera RA. Human *DAZL1* encodes a candidate fertility factor in women that localizes to the prenatal and postnatal germ cells. *Hum Reprod* 1999 14: 2531–2536.
26. Luoh SW, Bain PA, Polakiewicz RD *et al.* *Zfx* mutation results in small animal size and reduced germ cell number in male and female mice. *Development* 1997 124 :2275–2284.
27. Carabatsos MJ, Elvin J, Matzuk MM, Albertini DF. Characterization of oocyte and follicle development in growth differentiation factor-9-deficient mice. *Dev Biol* 1998 204: 373–384.
28. Simon AM, Goodenough DA, Li E, Paul DL. Female infertility in mice lacking connexin 37. *Nature* 1997 385: 525–529.
29. Guo Q, Kumar TR, Woodruff T *et al.* Overexpression of mouse follistatin causes reproductive defects in transgenic mice. *Mol Endocrinol* 1998 12: 96–106.
30. Korach KS. Estrogen receptor knock-out mice: molecular and endocrine phenotypes. *J Soc Gynecol Invest* 2000 7: S16–7.
31. Galloway SM, McNatty KP, Cambridge LM *et al.* Mutations in an oocyte-derived growth factor gene (*BMP15*) cause increased ovulation rate and

infertility in a dosage-sensitive manner. *Nat Genet* 2000 25: 279–283.
32. Takebayashi K, Takakura K, Wang H *et al.* Mutation analysis of the growth differentiation factor-9 and -9B genes in patients with premature ovarian failure and polycystic ovary syndrome. *Fertil Steril* 2000 74: 976–979.
33. Conway GS, Kaltsas G, Patel A *et al.* Characterization of idiopathic premature ovarian failure. *Fertil Steril* 1996 65: 337–341.
34. van Kasteren YM, Hundscheid RD, Smits AP *et al.* Familial idiopathic premature ovarian failure: an overrated and underestimated genetic disease? *Hum Reprod* 1999 14: 2455–2459.
35. Davis CJ, Davison RM, Payne NN *et al.* Female sex preponderance for idiopathic familial premature ovarian failure suggests an X chromosome defect: opinion. *Hum Reprod* 2000 15: 2418–2422.
36. Saenger P. Turner's syndrome. *N Engl J Med* 1996 335: 1749–1754.
37. Singh RP, Carr DH. The anatomy and histology of XO human embryos and fetuses. *Anat Rec* 1966 155: 369–383.
38. Simpson JL. Genes, chromosomes, and reproductive failure. *Fertil Steril* 1980 33: 107–116.
39. Powell CM, Taggart RT, Drumheller TC *et al.* Molecular and cytogenetic studies of an X;autosome translocation in a patient with premature ovarian failure and review of the literature. *Am J Med Genet* 1994 52: 19–26.
40. Hundscheid RD, Braat DD, Kiemeney LA *et al.* Increased serum FSH in female fragile X premutation carriers with either regular menstrual cycles or on oral contraceptives. *Hum Reprod* 2001 16 :457–462.
41. Kagimoto M, Winter JSD, Kagimoto K *et al.* Structural characterization of normal and mutant human steroid 17α-hydroxylase genes: molecular basis of one example of combined 17α-hydroxylase/17,20-lyase deficiency. *Mol Endocrinol* 1988 2: 564–570.
42. Amati P, Gasparini P, Zlotogora J *et al.* A gene for premature ovarian failure associated with eyelid malformation maps to chromosome 3q22–q23. *Am J Hum Genet* 1996 58: 1089–1092.
43. Morishima A, Grumbach MM, Simpson ER *et al.* Aromatase deficiency in male and female siblings caused by a novel mutation and the physiological role of estrogens. *J Clin Endocrinol Metab* 1995 80: 3689–3698.
44. Oley C, Baraitser M. Blepharophimosis, ptosis, epicanthus inversus syndrome (BPES syndrome). *J Med Genet* 1988 25: 47–51.
45. Vianna-Morgante AM, Costa SS, Pares AS, Verreschi IT. *FRAXA* premutation associated with premature ovarian failure. *Am J Med Genet* 1996 64: 373–375.
46. Sala C, Arrigo G, Torri G *et al.* Eleven X chromosome breakpoints associated with premature ovarian failure (POF) map to a 15-Mb YAC contig spanning Xq21. *Genomics* 1997 40: 123–131.
47. Bione S, Sala C, Manzini C *et al.* A human homologue of the *Drosophila melanogaster diaphanous* gene is disrupted in a patient with premature ovarian failure: evidence for conserved function in oogenesis and implications for human sterility. *Am J Hum Genet* 1998 62: 533–541.
48. Davison RM, Quilter CR, Webb J *et al.* A familial case of X chromosome deletion ascertained by cytogenetic screening of women with premature ovarian failure. *Hum Reprod* 1998 13: 3039–3041.
49. Therman E, Susman B. The similarity of phenotypic effects caused by Xp and Xq deletions in the human female: a hypothesis. *Hum Genet* 1990 85: 175–183.
50. Conway GS, Conway E, Walker C *et al.* Mutation screening and isoform prevalence of the follicle stimulating hormone receptor gene in women with premature ovarian failure, resistant ovary syndrome and polycystic ovary syndrome. *Clin Endocrinol (Oxf)* 1999 51: 97–99.
51. Abel MH, Wootton AN, Wilkins V *et al.* The effect of a null mutation in the follicle-stimulating hormone receptor gene on mouse reproduction. *Endocrinology* 2000 141: 1795–1803.
52. Aittomaki K, Herva R, Stenman UH *et al.* Clinical features of primary ovarian failure caused by a point mutation in the follicle-stimulating hormone receptor gene. *J Clin Endocrinol Metab* 1996 81: 3722–3726.
53. Prueitt RL, Ross JL, Zinn AR. Physical mapping of nine Xq translocation breakpoints and identification of *XPNPEP2* as a premature ovarian failure candidate gene. *Cytogenet Cell Genet* 2000 89: 44–50.
54. Devi AS, Metzger DA, Luciano AA, Benn PA. 45,X/46,XX mosaicism in patients with idiopathic premature ovarian failure. *Fertil Steril* 1998 70: 89–93.
55. Conway GS, Hettiarachchi S, Murray A, Jacobs PA. Fragile X premutations in familial premature ovarian failure. *Lancet* 1995 346: 309–310.

56. Bose HS, Sugawara T, Strauss JFI, Miller WL. The pathophysiology and genetics of congenital lipoid adrenal hyperplasia. *N Engl J Med* 1996 335: 1870–1878.
57. Kaufman FR, Kogut MD, Donnell GN *et al.* Hypergonadotropic hypogonadism in female patients with galactosemia. *N Engl J Med* 1981 304: 994–998.
58. Guerrero NV, Singh RH, Manatunga A *et al.* Risk factors for premature ovarian failure in females with galactosemia. *J Pediatr* 2000 137: 833–841.
59. Simoni M, Gromoll J, Nieschlag E. The follicle-stimulating hormone receptor: biochemistry, molecular biology, physiology, and pathophysiology. *Endocr Rev* 1997 18: 739–773.
60. Layman LC, Amde S, Cohen DP *et al.* The Finnish follicle-stimulating hormone receptor gene mutation is rare in North American women with 46,XX ovarian failure. *Fertil Steril* 1998 69: 300–302.
61. Beau I, Touraine P, Meduri G *et al.* A novel phenotype related to partial loss of function mutations of the follicle stimulating hormone receptor. *J Clin Invest* 1998 102: 1352–1359.
62. Touraine P, Beau I, Gougeon A *et al.* New natural inactivating mutations of the follicle-stimulating hormone receptor: correlations between receptor function and phenotype. *Mol Endocrinol* 1999 13: 1844–1854.
63. Danilovich N, Babu PS, Xing W *et al.* Estrogen deficiency, obesity, and skeletal abnormalities in follicle-stimulating hormone receptor knockout (FORKO) female mice. *Endocrinology* 2000 141: 4295–4308.
64. Meyers CM, Boughman JA, Rivas M *et al.* Gonadal (ovarian) dysgenesis in 46,XX individuals: frequency of the autosomal recessive form. *Am J Med Genet* 1996 63: 518–524.
65. Barlow C, Hirotsune S, Paylor R *et al.* *Atm*-deficient mice: a paradigm of ataxia telangiectasia. *Cell* 1996 86: 159–171.
66. Shelling AN, Burton KA, Chand AL *et al.* Inhibin: a candidate gene for premature ovarian failure. *Hum Reprod* 2000 15: 2644–2649.
67. Weetman AP. Autoimmunity to steroid-producing cells and familial polyendocrine autoimmunity. *Baillière's Clin Endocrinol Metab* 1995 9: 157–174.
68. Betterle C, Greggio NA, Volpato M. Autoimmune polyglandular syndrome type 1. *J Clin Endocrinol Metab* 1998 83: 1049–1055.
69. The Finnish–German APECED Consortium. An autoimmune disease, APECED, caused by mutations in a novel gene featuring two PHD-type zinc-finger domains. *Nat Genet* 1997 17: 399–403.
70. Nagamine K, Peterson P, Scott HS *et al.* Positional cloning of the APECED gene. *Nat Genet* 1997 17: 393–398.
71. Betterle C, Rossi A, Dalla Pria S *et al.* Premature ovarian failure: autoimmunity and natural history. *Clin Endocrinol (Oxf)* 1993 39: 35–43.
72. Chen S, Sawicka J, Betterle C *et al.* Autoantibodies to steroidogenic enzymes in autoimmune polyglandular syndrome, Addison's disease, and premature ovarian failure. *J Clin Endocrinol Metab* 1996 81: 1871–1876.
73. Morita Y, Perez GI, Paris F *et al.* Oocyte apoptosis is suppressed by disruption of the acid sphingomyelinase gene or by sphingosine-1-phosphate therapy. *Nat Med* 2000 6: 1109–1114.

Chapter 33

Polycystic ovary syndrome

Richard S. Legro and Jerome F. Strauss III

INTRODUCTION

Polycystic ovary syndrome (PCOS) is probably the most common but least understood endocrinopathy, affecting 5% of women in the developed world.[1] Unfortunately, there is no consensus as to the definition of PCOS, which confounds the selection of subjects for clinical studies. The diagnosis of PCOS has traditionally been based on the historical variables of oligomenorrhea and hirsutism; biochemical markers such as circulating total or bioavailable androgens and gonadotropin levels; or the ultrasound image of the ovaries. The criteria that emerged from the 1990 National Institutes of Health–National Institute of Child Health and Human Development (NIH–NICHD) conference – hyperandrogenism and/or hyperandrogenemia, oligo-ovulation in the absence of other endocrine disturbances including congenital adrenal hyperplasia, hyperprolactinemia, Cushing's syndrome, and androgen-secreting tumors – defined PCOS as unexplained hyperandrogenic chronic anovulation, making it in essence a diagnosis of exclusion.[2] This 'consensus' definition did not include the polycystic ovary morphology found on ultrasound scan consisting of multiple 2–8-mm subcapsular preantral follicles forming a 'black pearl necklace' sign.[3] The rationale for excluding ovarian morphology in the diagnostic criteria is that polycystic ovaries are distinct from polycystic ovary syndrome; up to 25% of an unsolicited population may have polycystic ovaries on ultrasound examination, and many of these women have normal androgen levels and regular menstrual cycles.[4]

Although usually characterized as a disturbance of the hypothalamic–pituitary–ovarian axis, PCOS clearly involves other organ systems (Figure 1). PCOS women appear to be uniquely insulin-resistant, which may be central to the pathophysiology of the disease.[5] The associated compensatory hyperinsulinemia may affect hypothalamic control of gonadotropin secretion and appetite, stimulate adrenal and ovarian androgen secretion, and suppress circulating levels of sex hormone-binding globulin, thus increasing the pool of bioavailable androgens. Despite the evidence implicating insulin as a key factor in PCOS, it is not included in any of the proposed diagnostic criteria.

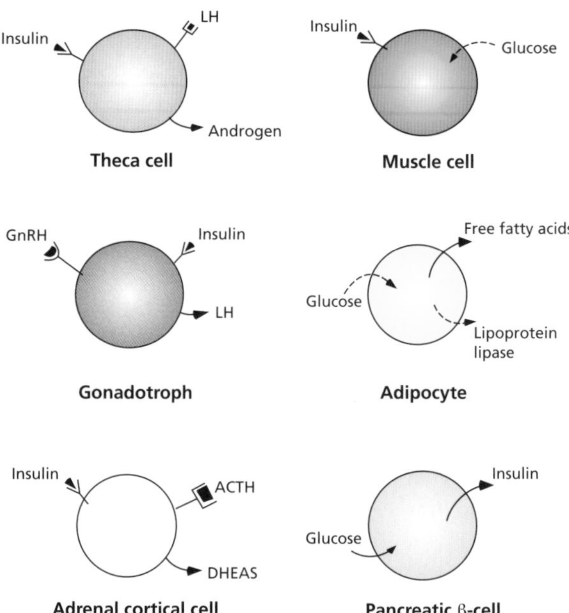

Figure 1 Metabolic defects associated with polycystic ovary syndrome (PCOS) in different cell types. Dotted lines represent reduced flux, thick solid lines increased secretion. LH, luteinizing hormone; GnRH, gonadotropin releasing hormone; ACTH, adrenocorticotropic hormone; DHEAS, dehydroepiandrosterone sulfate

PCOS usually becomes evident shortly after menarche, and lasts for most of reproductive life, although questions persist about its natural history during the reproductive years. It has been suggested that premature pubarche may be the earliest recognizable PCOS phenotype. Affected girls display hyperinsulinemia and elevated dehydroepiandrosterone sulfate (DHEAS) levels, and after menarche they become oligomenorrheic.[6] At the other end of the reproductive spectrum, both menstrual irregularity[7] and hyperandrogenemia[8] tend to normalize as PCOS women approach their late 30s and early 40s, erasing the reproductive symptoms for which most PCOS women seek medical attention.

BACKGROUND

The exploration of the genetics of PCOS has been plagued by debate over disease phenotypes and the lack of appropriate animal models which could be informative regarding the genetics of the disorder. The larger the number of distinct phenotypes within the affected category that need to be considered, the more complex is the analysis and the greater is the likelihood that investigators using different diagnostic criteria will arrive at different conclusions. In addition, the problem of assigning phenotypes to premenarchal girls and postmenopausal women, as well as the lack of a rigorously established male phenotype, make formal segregation analysis and genetic linkage studies more difficult. Unfortunately, the uncertainty of the male phenotype has not dissuaded some investigators from employing premature male balding as the 'male PCOS phenotype' in segregation and genetic studies. Instead of bringing greater clarity to the field, this may in fact have introduced more confusion. It is no wonder that some scholars have concluded that PCOS is a multigenic disorder beyond the comprehension of contemporary genetic analysis.

Family studies

The foundation of genetic studies is the evidence that disease clusters in families. None of the existing family studies of PCOS convincingly establishes a mode of inheritance, because the number of families studied was too small; the parental phenotypes could not be firmly established; and the male phenotype is uncertain.[9–18] Moreover, the diagnostic criteria used to assign affected status differed among the studies, as did the methods by which the status of first- and second-degree relatives was ascertained. By and large, ovarian morphology determined from tissue biopsy, direct visualization or diagnostic imaging, in association with menstrual disturbances and evidence for hyperandrogenism, have been used in most studies as the criteria for diagnosing PCOS in probands. When considered, the male phenotype has primarily been balding before the age of 30 years or increased pilosity. Earlier studies tended to characterize relatives on the basis of questionnaires, whereas later studies focused on more intensive phenotyping. Despite the heterogeneity in study design and the inability to obtain comprehensive phenotype information to permit a formal segregation analysis, collectively the existing literature strongly suggests the clustering of PCOS in families with a mode of inheritance that is not inconsistent with an autosomal dominant pattern (Table 1).

This conclusion has been supported by recent studies based on more intensive phenotyping of PCOS families. Carey and colleagues[14] reported that first-degree female relatives of PCOS women had a 51% chance of being affected. Accepting premature balding as the phenotype for male carriers, Carey and colleagues suggested that the segregation was consistent with autosomal dominant inheritance. Legro and colleagues[16] conducted a prospective study in which the probands were identified on the basis of oligomenorrhea (< six menses/year) and hyperandrogenemia, defined as a total or biologically available testosterone level that was > 2 standard deviations above the mean values of a control group of women. Twenty-two per cent of sisters of the affected women fulfilled the diagnostic criteria for PCOS, and an additional 24% were hyperandrogenemic but had > six menses per year. Moreover, the PCOS probands, their PCOS sisters and the hyperandrogenemic sisters had higher luteinizing hormone (LH) levels than control women. The Legro study suggested that testosterone levels in sisters of PCOS women had a bimodal distribution (Figure 2), which could be consistent with a monogenic trait controlled by two alleles at an autosomal locus. The idea that hyperandrogenemia alone is a marker linked to PCOS was supported by a subsequent report of Carmina and Lobo,[19] who noted that hyperandrogenic women display evidence of PCOS, including ultrasound polycystic ovarian morphology and an increased 17-hydroxyprogesterone response to a leuprolide acetate challenge despite normal menses. Further study of first-degree female relatives of PCOS probands by Kahsar-Miller and associates was also in agreement with these findings.[18] Govind and co-workers[17] found that sisters of the PCOS probands were more likely to have menstrual irregularities and higher serum androstenedione and DHEAS levels than sisters without polycystic ovaries. Additionally, men with premature balding were found to have higher serum testosterone levels than those without premature balding.

Table 1 Summary of diagnostic criteria for the proband in familial studies of polycystic ovary syndrome (PCOS) and proposed mode of inheritance

Author	*Diagnostic criteria for PCOS*	*Number studied*	*Mode of inheritance*
Cooper and Clayton[9]	oligomenorrhea, hirsutism, polycystic ovaries (by culdoscopy, gynecography or wedge resection)	18 PCOS women and their first-degree relatives and a control group	autosomal dominant with reduced penetrance
Givens[10]	oligomenorrhea, hirsutism and polycystic ovaries (by examination and surgery)	3 multigeneration kindreds	(X-linked?) dominant
Ferriman and Purdie[11]	hirsutism and/or oligomenorrhea, 60% with polycystic ovaries (by air contrast gynecography)	381 PCOS women and relatives and a control group	modified dominant
Lunde *et al.*[12]	clinical symptoms (menstrual irregularities, hirsutism, infertility and obesity) and multicystic ovaries on wedge resection	132 PCOS women and first- and second-degree relatives and a control group	unclear, most consistent with autosomal dominant
Hague *et al.*[13]	clinical symptoms (menstrual dysfunction, hyperandrogenism, obesity and infertility) and polycystic ovaries by transabdominal ultrasound	50 PCOS women and 17 women with CAH and a control group	segregation ratios exceeded autosomal dominant pattern
Carey *et al.*[14]	polycystic ovaries (by transabdominal ultrasound)	10 kindreds and 62 relatives	autosomal dominant with 90% penetrance
Norman *et al.*[15]	elevated androgens, decreased SHBG and polycystic ovaries on ultrasound	5 families with 24 females and 8 males	not stated
Legro *et al.*[16]	elevated testosterone levels combined with oligomenorrhea (≤ 6 menses/year)	80 PCOS probands and 115 sisters	hyperandrogenemia consistent with an autosomal dominant trait
Govind *et al.*[17]	polycystic ovary morphology on ultrasonography	29 families with 53 sisters and 18 brothers	autosomal dominant
Kahsar-Millar *et al.*[18]	oligomenorrhea and either hirsutism or elevated testosterone levels	90 PCOS probands, 50 sisters, 78 mothers	increased prevalence of symptoms in first-degree relatives suggesting genetic trait

SHBG, sex hormone-binding globulin; CAH, congenital adrenal hyperplasia

It is surprising that there have been only a few studies of PCOS monozygotic twins, given the power of this approach to disclose the genetic basis of disease. Case reports have described affected sets of female twins, yet the largest available twin study did not support a strong genetic component.[20] Both mono- and dizygotic twins were studied with ultrasound, as well as with clinical and biochemical parameters. From a starting population of 500 female–female twins who were contacted to participate, only 34 pairs were analyzed. There was an unusually high incidence of polycystic ovarian morphology on ultrasound scan, with 50% of the study population being affected. The authors found a high degree of discordance among the available twins for polycystic ovaries, and suggested that PCOS might have a more complex inheritance pattern than autosomal dominant, perhaps X-linked or polygenic. However, there was greater concordance among affected twins with respect to biochemical parameters, including fasting insulin levels and androgens.

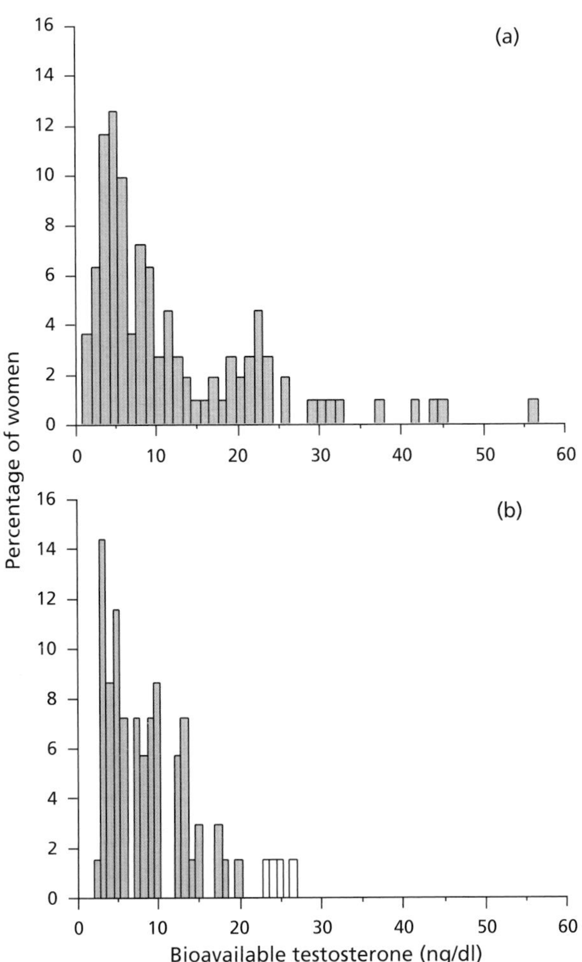

Figure 2 Frequency distribution of available testosterone levels in sisters (n = 113) of polycystic ovary syndrome (PCOS) probands (a) and control women (n = 70) (b). The distribution in the sisters appears to be bimodal. Modified from reference 16, with permission

CURRENT CONCEPTS

Polycystic ovary syndrome: a multisystem disorder

Elevated ovarian secretion of androgens, increased bioavailable testosterone and greater formation of 5α-dihydrotestosterone by the pilosebaceous unit result in hirsutism, balding and acne. Insulin resistance and defects in pancreatic β-cell function, both independent of obesity, have been well documented in PCOS.[21] Skeletal muscle in PCOS has reduced sensitivity to insulin for glucose transport. PCOS adipocytes display reduced *in vitro* responses to insulin with respect to glucose transport, GLUT4 expression and suppression of lipolysis. Moreover, they have been reported to show an impaired lipolytic response to adrenergic stimulation and contain reduced amounts of lipoprotein lipase. Insulin acting through its cognate receptor activates multiple signalling pathways. The insulin resistance of PCOS is likely to be the result of a yet to be identified 'post-receptor' defect that selectively impairs certain signalling arms. Thus, the hyperinsulinemia associated with insulin resistance of skeletal muscle and adipose tissue with respect to glucose disposal augments thecal androgen synthesis, reduces hepatic sex hormone-binding globulin production, and probably stimulates LH secretion because the latter processes are regulated by an insulin signalling pathway that is not affected.

The cardiovascular system also appears to be affected in PCOS. Recent studies based on surrogate risk factors (for example, increased carotid artery intimal thickness) indicate that PCOS women have evidence of atherosclerotic disease at an earlier and more advanced stage than appropriate control women.[22] There has yet to be confirmation of this notion based on actual cardiovascular events such as stroke or myocardial infarction. Gynecological cancers have been reported to be more common in women with PCOS, including ovarian, breast and endometrial cancer. Most of the risk factors for endometrial cancer, including obesity, chronic anovulation, hypertension and type II diabetes, overlap with the PCOS phenotype.[23]

The involvement of multiple organ systems in PCOS raises several important questions. Is PCOS many diseases, or do the factors that influence reproductive function also impact upon different cell types simultaneously, resulting in the multisystem PCOS phenotype? Are the metabolic abnormalities detected in different cells *in vitro* the result of a shared intrinsic defect, or are they the consequence of exposure to an altered endocrine state *in vivo*? The answers to these questions have profound significance with respect to the consideration of the genetics of PCOS.

APPLIED TECHNOLOGY

Karyotyping

There have been reports of X-chromosome aneuploidies and polyploidies as well as other cytogenetic abnormalities in women with PCOS. However, larger cytogenetic analyses have failed to identify common karyotypic abnormalities.[24] Thus, it appears that gross alterations in chromosome number or structure are not associated with PCOS.

Stable biochemical and molecular phenotype in PCOS cells studied *in vitro*

If PCOS is a genetic disease, it might be expected that cells collected from PCOS women would show persistent biochemical and molecular features that distinguish them from cells from normal women, studied under identical conditions. Indeed, studies conducted on theca cells collected from follicles from PCOS and normal ovaries, grown under conditions that support propagation through multiple population doublings, revealed that PCOS theca cells retained a phenotype characterized by increased production of progesterone, 17α-hydroxyprogesterone, and testosterone; increased expression of CYP11A, CYP17 and type II 3β-hydroxysteroid dehydrogenase, but not steroidogenic acute regulatory protein (StAR) mRNA (Figure 3);[25] and increased transcription of the *CYP17*, but not the *StAR* gene.[26] These observations parallel findings made on freshly isolated theca cells studied in short-term culture;[27] measurements of mRNAs encoding steroidogenic proteins in freshly isolated theca,[28] and ovarian responses to human chorionic gonadotropin (hCG) stimulation *in vivo*.[29]

Dunaif and colleagues[30] identified differential patterns of phosphorylation of the β-subunit of the insulin receptor in skeletal muscle cells and fibroblasts from women with PCOS. Approximately 50% of the sample population of PCOS women studied showed the increased insulin-receptor serine phosphorylation and reduced insulin-stimulated tyrosine phosphorylation. This finding was recently substantiated in ovarian tissue.[31] A stable alteration in skeletal muscle cells in culture has also

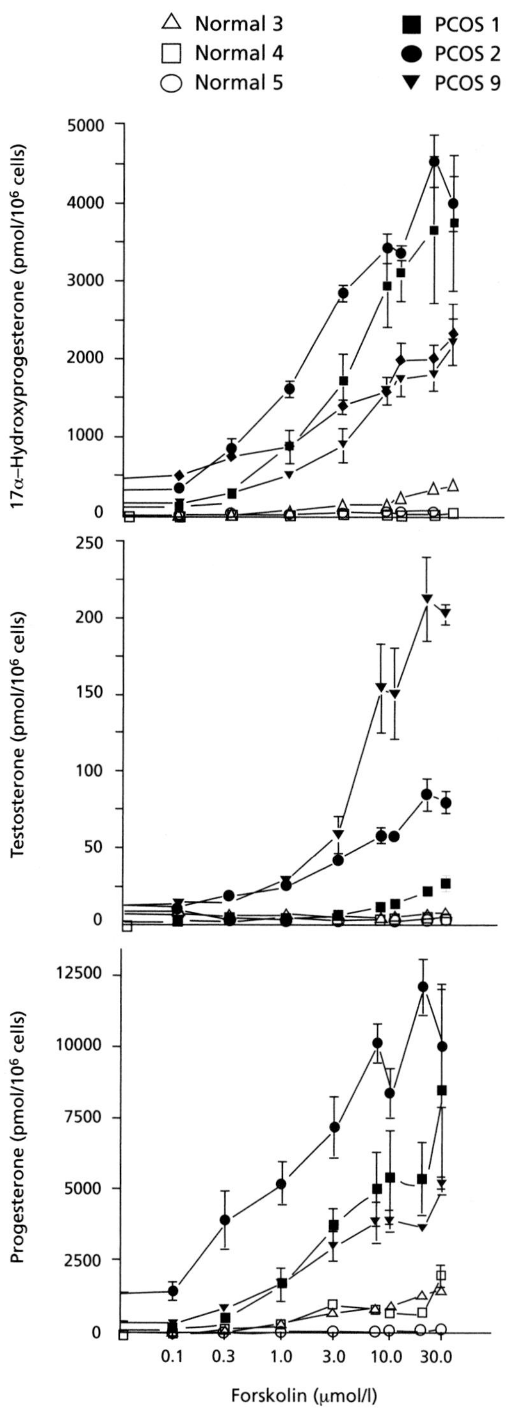

Figure 3 Accumulation of 17α-hydroxyprogesterone, progesterone and testosterone in the medium of normal and polycystic ovary syndrome (PCOS) theca cell cultures grown for 22–26 population doublings. Cells were exposed to increasing concentrations of the adenylate cyclase activator, forskolin, and steroid production was measured after 72 h. Modified from reference 25, with permission

been described by Jackson and associates,[32] who reported decreased insulin responsiveness in cells from non-diabetic insulin-resistant relatives from families with type II diabetes mellitus. Although the persistent biochemical/molecular alterations in PCOS cells maintained *in vitro* could possibly be explained by a stable metabolic imprint (epigenetic factor) obtained *in vivo*, the collective observations are more simply explained by genetic differences.

Association and linkage studies based on candidate genes

The identification of genes linked to complex diseases has been pursued using a number of strategies, including association studies, linkage analysis for candidate genes and genome-wide scanning. The last approach is ideal because it makes no assumptions regarding the gene(s) involved, the mode of inheritance, or disease penetrance or prevalence. The screen is based on the analysis of genomic DNA collected from a large number of families with one or more affected sibs, for informative markers spaced at 10–20 cM across the genome, so that identity by descent of alleles (inheritance of alleles from parents to affected sibs) can be established. Unfortunately, no investigative team has yet assembled a sufficient number of families with affected sib-pairs (of the order of several hundred) to perform a genome-wide screen. Consequently, the existing literature is based exclusively on association and linkage studies for candidate genes.[33–35] This reflects the difficulty in accruing families with affected sisters, especially when parental DNA is needed, despite the fact that PCOS is a common disorder.

Association studies test for over-representation of a specific allele of a candidate gene in a population of unrelated affected individuals, compared with a cohort of unrelated unaffected individuals. This type of case–control study can be used to study diseases in which the mode of inheritance and penetrance are not understood. These studies do not establish linkage of a disease phenotype with the allele, and they are subject to error because the test assumes that the frequencies of the alleles examined are similar in the populations studied. Studies incorporating mixed populations of ethnic groups in which allele frequencies vary may yield spurious results. Bias may also enter into these studies in the manner in which the control populations are selected. In addition, some investigators have studied the same populations for association with several different markers, and have failed to make appropriate compensation in their statistical analyses for multiple tests of significance.

Testing available family DNA sets for candidate genes is not an unreasonable exercise, because the candidate approach has been highly successful in identifying genes involved in a number of reproductive endocrine disorders. However, in those instances, the universe of possible candidates from which to choose was better defined than in the case of PCOS, for which multiple endocrine and metabolic abnormalities must be explained. Therefore, the number of wrong guesses for PCOS candidate genes is likely to be high until the pathophysiology is crystallized and a gene(s) becomes an 'obvious' candidate. Wrong guesses are not necessarily without value, since the exclusion of loci narrows the field of possibilities, and adjusts the level to which the gene products are emphasized in models of PCOS pathophysiology.

Polycystic ovary syndrome candidate genes

To date, the candidate genes that have been investigated include those involved in steroid hormone synthesis and action, genes involved in carbohydrate metabolism and fuel homeostasis, genes involved in gonadotropin action and regulation, and the major histocompatibility region. These candidates were selected because they could account for certain PCOS features. Given the recognition that PCOS affects multiple cell types, widely expressed genes involved in cell signalling would appear to be the most appropriate candidates. Negative results from candidate genes are summarized in Table 2; positive and/or mixed results are discussed in more detail in the text.

Genes involved in steroid hormone synthesis and action

Abnormalities in ovarian steroidogenesis, particularly androgen production, are a prominent feature

Table 2 Negative studies of candidate genes in polycystic ovary syndrome (PCOS)

Metabolic pathway	*Gene symbol*	*Gene name*	*Comments*
Steroid hormone synthesis and action	*CYP19*	aromatase	no linkage and/or association[34]
	HSD17B3	17β-hydroxysteroid dehydrogenase type 3 gene	no difference in allele frequency (a polymorphism in exon 11 results in substitution of a serine residue for a glycine at codon 289)[36]
	SF-1	steroidogenic factor-1, transcription factor	no mutation, association or linkage[34,37]
	StAR	steroidogenic acute regulatory protein	no mutation, association or linkage[34,37]
	DAX-1	dosage-sensitive sex reversal–adrenal hypoplasia gene	no mutation, association or linkage[34,37]
Carbohydrate metabolism/fuel homeostasis	*GYS*	glycogen synthetase	no association of an *Xba1* polymorphism with insulin sensitivity in PCOS[38]
	GCCR	glucocorticoid receptor	N362S (associated with insulin resistance) rare variant (< 6%) in a case–control study[39]
	TNFA	tumor necrosis factor-α (TNF-α)	no association between the –308 G/A *TNFA* polymorphism, which is implicated in increased TNF-α production, and PCOS[40]
	LEP	leptin	no mutations or association/ linkage[41]
	LEPR	leptin receptor	no mutations or association/ linkage[41]
Gonadotropin action and regulation	*LHB*	luteinizing hormone β-subunit	no association with PCOS[42]
	GNRHR	gonadotropin releasing hormone receptor	no mutations in PCOS[43]
	DRD2	dopamine receptor 2	no association with intron 5 polymorphism[44]

of PCOS. There has been intensive interest in the regulation of the enzyme that controls the gateway to androgen synthesis, 17α-hydroxylase/17,20-lyase, encoded by the *CYP17* gene. However, studies on isolated theca cells from the ovaries of women with PCOS revealed alterations in steroidogenesis that extend beyond 17α-hydroxylase/17,20-lyase activity, and include increased progesterone production.[27,45] These observations suggest abnormal activity or expression of a single steroidogenic enzyme gene cannot account for the PCOS-associated hyperandrogemia.

CYP11A Gharani and colleagues[46] examined 20 families with multiple affected women, and found evidence for weak linkage to the *CYP11A* locus. An association study conducted on 97 PCOS women and matched controls revealed significant association of a pentanucleotide repeat (TTTTA)*n* polymorphism at position –528 from the ATG initiation codon in the 5' region of the *CYP11A* gene, which encodes the cholesterol side-chain cleavage enzyme and total serum testosterone levels. Although no regulatory role has been assigned to this polymorphism in terms of *CYP11A* gene transcription, the investigators suggested that allelic variants of the *CYP11A* gene have a role in the hyperandrogenemia of PCOS. This conclusion was supported by another study that found an association between at least four repeats and PCOS, and within the PCOS group an association of this allele group with a lower testosterone level.[47] Urbanek and colleagues[34] found modestly increased sharing of alleles at the *CYP11A* locus among affected sib-pairs, but the sharing was not statistically significant. San Millan and associates[48] did not find a strong association between the pentanucleotide repeat and hirsutism. Based on the existing data, *CYP11A* still remains a potentially interesting PCOS candidate worthy of further scrutiny.

CYP17 A single nucleotide polymorphism (T–C) at base pair –34 in the *CYP17* promoter initially attracted attention because of the critical role this enzyme plays in governing androgen biosynthesis. Carey and colleagues[49] studied a small number of subjects, and suggested that the rarer allele increased the susceptibility for PCOS. In analyzing a larger population, these authors and others have failed to confirm a significantly different allele frequency in PCOS women, compared with their control women.[50] Moreover, no association with the phenotype of premature pubarche, hirsutism and oligomenorrhea in adolescence was found with *CYP17* variants/mutations.[51]

CYP21 Mutations in the *CYP21* gene, which encodes the 21-hydroxylase enzyme, are responsible for the major forms of congenital adrenal hyperplasia, a disorder that can mimic the PCOS phenotype. Recent studies have found a significant prevalence of *CYP21* mutations in the 'PCOS' population, including those with normal 17-hydroxyprogesterone responses after adrenocorticotropic hormone (ACTH) testing.[52,53] This observation raises important concerns regarding the phenotyping of subjects for genetic studies of PCOS as women with an identifiable genetic lesion could 'contaminate' the PCOS patient cohort.

Androgen receptor Using the transmission disequilibrium test, which tests for excess transmissions of alleles to affected individuals, Urbanek and colleagues[34] studied 150 families for the trinucleotide (CAG)*n* repeat polymorphism in the X-linked androgen receptor gene, and failed to find evidence for association with PCOS. However, this polymorphism has engendered continued interest because the number of repeats has been inversely associated with androgen action. Mifsud and coworkers[54] noted an association between short CAG repeat lengths and anovulation in women with polycystic ovaries and low testosterone levels, suggesting that the menstrual and ovarian abnormalities in this group of women were due to increased androgen action mediated by the short androgen receptor alleles.

Genes involved in carbohydrate metabolism and fuel homeostasis

The common occurrence of insulin resistance and pancreatic β-cell dysfunction in association with PCOS, and the increased risk for development of type II diabetes mellitus, are now well recognized. Moreover, insulin acting through its own receptor, and at high concentrations through the insulin-like growth factor-I (IGF-I) receptor, stimulates steroidogenesis.[55] This has led investigators to focus on insulin resistance as a potential central abnormality in PCOS.[5] In addition to mutations in the insulin receptor, the secretory products of adipose tissue including leptin, resistin and tumor necrosis factor-α (TNF-α) also promote insulin resistance, and have been considered as candidates for PCOS genes.[56]

Insulin receptor Mutations in the insulin receptor gene cause severe insulin resistance (the type A syndrome), associated with acanthosis nigricans and PCOS. Yet a number of studies have examined the insulin receptor gene sequence in PCOS women with and without insulin resistance, with negative results.[57] However, genes nearby the insulin receptor may contribute to PCOS. Two studies have found linkage or association between a marker (D19S884 at 19p13.3) located 1–2 Mb centromeric from the insulin receptor and

PCOS.[34,58] Because two independent studies suggested that this region near the insulin receptor is associated with PCOS, this locus emerges as a strong candidate region, deserving of detailed investigation for the identification of the putative PCOS gene.

Insulin gene Waterworth and colleagues[33] examined the role of a variable number tandem repeat (VNTR) in the 5' region of the insulin gene in PCOS. This locus has a bimodal distribution of repeats, which have been divided into class I alleles (averaging 40 repeats) and class III alleles with a larger number of repeats (average 157). Linkage of PCOS to this locus was examined in 17 families in which there were multiple individuals affected with polycystic ovaries on ultrasound scan, or premature male balding (before age 30), and the association of these alleles with polycystic ovaries was studied in two additional populations. The authors found that the class III allele was preferentially transmitted from heterozygous fathers, but not from mothers, to affected individuals. Linkage analysis suggested increased sharing of the locus in affected sibs. The authors concluded that they had discovered strong linkage and association between alleles at the insulin gene 5' VNTR and PCOS. However, others reviewing the study raised concerns regarding data analysis and use of the program to identify linkage.[59] Moreover, Urbanek and colleagues[34] found no evidence for linkage of the insulin gene in their analysis, or of association between the class III alleles of the insulin VNTR and hyperandrogenemia.

Genes involved in gonadotropin action and regulation

Abnormalities in gonadotropin secretion, particularly of LH, are characteristic of PCOS. Since LH plays a permissive role in driving thecal androgen production, there has been interest in exploring genes related to the regulation of LH secretion, LH bioactivity and LH action.[42]

Other gonadotropin/receptor genes An *AccI* polymorphism in exon 3 of the follicle stimulating hormone (FSH) β-subunit gene which does not alter the amino acid sequence has been reported in PCOS (and with obesity within the PCOS group).[60] However, screening of the FSH receptor gene[61] has failed to identify variants associated with PCOS.

Dopamine receptor genes Dopamine inhibits gonadotropin releasing hormone (GnRH) and prolactin secretion. Polymorphisms have been identified in the dopamine D2 and D3 receptor genes. Homozygosity for the rare allele (allele 2) of the D3 receptor has been associated with PCOS and clomiphene resistance in Hispanic women.[62] However, a subsequent case–control study carried out in non-Hispanic white women failed to show a significant association between the allele of the dopamine D3 receptor and PCOS.[63]

Follistatin Follistatin, an activin-binding protein, is widely expressed, as is activin, and could play an important role in PCOS. Activin stimulates pituitary FSH secretion and acts directly on the ovary to promote follicular maturation. Neutralization of activin would be expected to cause reductions in FSH levels and arrested follicular maturation, as occurs in transgenic mice overexpressing follistatin. In addition, activin inhibits thecal androgen biosynthesis, and its removal by binding to follistatin might lead to unrestrained thecal androgen synthesis driven by LH. The follistatin locus emerged as a promising candidate from the study of 39 affected sib-pairs,[34] although the strength of the link faded in a larger follow-up series that included detailed sequence analysis of the follistatin gene.[64] The follistatin gene was also not found to be linked to PCOS in other studies.[65,66]

Major histocompatibility region

Linkage of disorders to the major histocompatibility locus encoding human leukocyte antigens (HLAs) on chromosome 6 has been sought for numerous diseases, especially when an immunological basis has been entertained. Mandel and coworkers[65] reported four families in which at least two siblings had PCOS verified by elevated serum androgen and LH levels, and found no linkage between PCOS and alleles in the HLA region. Larger studies have yielded conflicting results. Hague and colleagues[66] found an increased frequency of DRW6 and a reduced frequency of DR7 in a study of 99 subjects with polycystic ovary morphology on ultrasound, compared with 110 control females. However, no linkage between HLA and PCOS was found in a study of 16 families.[30] Ober and associates[67] reported an association between DQA1*0501 and PCOS in 19 women affected with PCOS, compared with 46 controls. Homozygosity for this allele was also increased among the PCOS women, leading the authors to suggest that the PCOS susceptibility allele is linked to HLA, and that this allele is recessive.

POTENTIAL CLINICAL RELEVANCE

Advances in the understanding of the genetics of human disease rarely immediately translate into a clinical benefit. There is currently no genetic test or marker that is of clinical utility in the diagnosis or management of PCOS. Nonetheless, the elucidation of the genetic factors predisposing to PCOS may help identify individuals at risk of PCOS complications such as type II diabetes mellitus and to steer investigators towards novel therapeutic approaches.

SUMMARY

The existing literature provides a strong basis for arguing that PCOS clusters in families. In addition, studies of PCOS cells (theca, muscle and adipocytes) in culture have documented a persistent biochemical and molecular phenotype that distinguishes them from normal cells. Collectively, these findings are consistent with the concept that a gene or collection of genes is linked to PCOS susceptibility. However, the mode of inheritance of the disorder is still uncertain, although the majority of studies are consistent with an autosomal dominant pattern, modified perhaps by environmental factors.

Although several loci have been proposed as PCOS genes, including *CYP11A*, the insulin gene, and a region near the insulin receptor, the evidence supporting linkage is not overwhelming. The strongest case can be made for the region near the insulin receptor gene, as it has been identified in two separate studies. However, the responsible gene at chromosome 19p13.3 remains to be identified.

FUTURE PERSPECTIVES

The fact that genetic studies have not yet pinpointed a PCOS locus or mutation should not be disheartening news. The continued exploration of candidate genes will eventually shine light on loci that are worthy of intensive investigation, while also diminishing interest in other genes. The application of DNA microassay technology and proteomics to the characterization of the molecular and biochemical phenotypes of PCOS cells may provide new insight into potential candidate genes and the bases of the metabolic disturbance.

ACKNOWLEDGEMENTS

This work was supported by Public Health Service grant (PHS) K24 HD01476, the National Co-operative Program in Infertility Research (NCPIR) U54 HD34449 and a GCRC grant MO1 RR 10732 to Pennsylvania State University.

REFERENCES

1. Knochenhauer ES, Key TJ, Kahsar-Miller M, Waggoner W, Boots LR, Azziz R. Prevalence of the polycystic ovary syndrome in unselected black and white women of the Southeastern United States: a prospective study. *J Clin Endocrinol Metab* 1998 83: 3078–3082.
2. Dunaif A, Givens JR, Haseltine FP, Merriam GR, eds. *Polycystic Ovary Syndrome (Current Issues in Endocrinology and Metabolism)*, 1st edn. Boston: Blackwell Scientific Publications, 1992.
3. Polson DW, Adams J, Wadsworth J, Franks S. Polycystic ovaries – a common finding in normal women. *Lancet* 1988 1: 870–872.
4. Farquhar CM, Birdsall M, Manning P, Mitchell JM, France JT. The prevalence of polycystic ovaries on ultrasound scanning in a population of randomly selected women. *Aust NZ J Obstet Gynaecol* 1994 34: 67–72.
5. Dunaif A. Insulin resistance and the polycystic ovary syndrome: mechanism and implications for pathogenesis. *Endocr Rev* 1997 18: 774–800.
6. Ibanez L, de Zegher F, Potau N. Anovulation after precocious pubarche: early markers and time course in adolescence. *J Clin Endocrinol Metab* 1999 84: 2691–2695.

7. Elting MW, Korsen TJ, Rekers-Mombarg LT, Schoemaker J. Women with polycystic ovary syndrome gain regular menstrual cycles when ageing. *Hum Reprod* 2000 15: 24–28.
8. Winters SJ, Talbott E, Guzick DS, Zborowski J, McHugh KP. Serum testosterone levels decrease in middle age in women with the polycystic ovary syndrome. *Fertil Steril* 2000 73: 724–729.
9. Cooper DN, Clayton JF. DNA polymorphism and the study of disease associations. *Hum Genet* 1988 78: 299–312.
10. Givens JR. Familial polycystic ovarian disease. *Endocrinol Metab Clin North Am* 1988 17: 771–783.
11. Ferriman D, Purdie AW. The inheritance of polycystic ovarian disease and a possible relationship to premature balding. *Clin Endocrinol* 1979 11: 291–300.
12. Lunde O, Magnus P, Sandvik L, Hoglo S. Familial clustering in the polycystic ovarian syndrome. *Gynecol Obstet Invest* 1989 28: 23–30.
13. Hague WM, Adams J, Reeders ST, Peto TE, Jacobs HS. Familial polycystic ovaries: a genetic disease? *Clin Endocrinol* 1988 29: 593–605.
14. Carey AH, Chan KL, Short F, White D, Williamson R, Franks S. Evidence for a single gene effect causing polycystic ovaries and male pattern baldness. *Clin Endocrinol* 1993 38: 653–658.
15. Norman RJ, Masters S, Hague W. Hyperinsulinemia is common in family members of women with polycystic ovary syndrome. *Fertil Steril* 1996 66: 942–947.
16. Legro RS, Driscoll D, Strauss JF III, Fox J, Dunaif A. Evidence for a genetic basis for hyperandrogenemia in polycystic ovary syndrome. *Proc Natl Acad Sci USA* 1998 95: 14956–14960.
17. Govind A, Obhrai MS, Clayton RN. Polycystic ovaries are inherited as an autosomal dominant trait: analysis of 29 polycystic ovary syndrome and 10 control families. *J Clin Endocrinol Metab* 1999 84: 38–43.
18. Kahsar-Miller MD, Nixon C, Boots LR, Go RC, Azziz R. Prevalence of polycystic ovary syndrome (PCOS) in first-degree relatives of patients with PCOS. *Fertil Steril* 2001 75: 53–58.
19. Carmina E, Lobo RA. Do hyperandrogenic women with normal menses have polycystic ovary syndrome? *Fertil Steril* 1999 71: 319–322.
20. Jahanfar S, Eden JA, Warren P, Seppala M, Nguyen TV. A twin study of polycystic ovary syndrome. *Fertil Steril* 1995 63: 478–486.
21. O'Meara NM, Blackman JD, Ehrmann DA *et al.* Defects in β-cell function in functional ovarian hyperandrogenism. *J Clin Endocrinol Metab* 1993 76: 1241–1247.
22. Talbott EO, Guzick DS, Sutton-Tyrrell K *et al.* Evidence for association between polycystic ovary syndrome and premature carotid atherosclerosis in middle-aged women. *Arterioscler Thromb Vasc Biol [Online]* 2000 20: 2414–2421.
23. Dahlgren E, Friberg LG, Johansson S, Lindstrom B, Oden A, Samsioe G. Endometrial carcinoma; ovarian dysfunction – a risk factor in young women. *Eur J Obstet Gynecol Reprod Biol* 1991 41: 143–150.
24. Stenchever MA, Macintyre MN, Jarvis JA, Hempel JM. Cytogenetic evaluation of 41 patients with Stein–Leventhal syndrome. *Obstet Gynecol* 1968 32: 794–801.
25. Nelson VL, Legro RS, Strauss JF III, McAllister JM. Augmented androgen production is a stable steroidogenic phenotype of propagated theca cells from polycystic ovaries. *Mol Endocrinol* 1999 13: 946–957.
26. Wickenheisser JK, Quinn PG, Nelson VL, Legro RS, Strauss JF III, McAllister JM. Differential activity of the cytochrome P450 17α-hydroxylase and steroidogenic acute regulatory protein gene promoters in normal and polycystic ovary syndrome theca cells. *J Clin Endocrinol Metab* 2000; 85: 2304–2311.
27. Gilling-Smith C, Willis DS, Beard RW, Franks S. Hypersecretion of androstenedione by isolated thecal cells from polycystic ovaries. *J Clin Endocrinol Metab* 1994 79: 1158–1165.
28. Jakimiuk AJ, Weitsman SR, Navab A, Magoffin DA. Luteinizing hormone receptor, steroidogenesis acute regulatory protein, and steroidogenic enzyme messenger ribonucleic acids are overexpressed in thecal and granulosa cells from polycystic ovaries. *J Clin Endocrinol Metab* 2001 86: 1318–1323.
29. Gilling-Smith C, Story H, Rogers V, Franks S. Evidence for a primary abnormality of thecal cell steroidogenesis in the polycystic ovary syndrome. *Clin Endocrinol* 1997 47: 93–99.
30. Dunaif A, Xia J, Book CB, Schenker E, Tang Z. Excessive insulin receptor serine phosphorylation

in cultured fibroblasts and in skeletal muscle. A potential mechanism for insulin resistance in the polycystic ovary syndrome. *J Clin Invest* 1995 96: 801–810.

31. Moran C, Huerta R, Conway-Myers BA, Hines GA, Azziz R. Altered autophosphorylation of the insulin receptor in the ovary of a woman with polycystic ovary syndrome. *Fertil Steril* 2001 75: 625–628.
32. Jackson S, Bagstaff SM, Lynn S, Yeaman SJ, Turnbull DM, Walker M. Decreased insulin responsiveness of glucose uptake in cultured human skeletal muscle cells from insulin-resistant nondiabetic relatives of type 2 diabetic families. *Diabetes* 2000 49: 1169–1177.
33. Waterworth DM, Bennett ST, Gharani N *et al.* Linkage and association of insulin gene VNTR regulatory polymorphism with polycystic ovary syndrome. *Lancet* 1997 349: 986–990.
34. Urbanek M, Legro RS, Driscoll DA *et al.* Thirty-seven candidate genes for polycystic ovary syndrome: strongest evidence for linkage is with follistatin. *Proc Natl Acad Sci USA* 1999 96: 8573–8578.
35. Urbanek M, Wu X, Vickery KR *et al.* Allelic variants of the follistatin gene in polycystic ovary syndrome. *J Clin Endocrinol Metab* 2000 85: 4455–4461.
36. Moghrabi N, Hughes IA, Dunaif A, Andersson S. Deleterious missense mutations and silent polymorphism in the human 17β-hydroxysteroid dehydrogenase 3 gene (*HSD17B3*). *J Clin Endocrinol Metab* 1998 83: 2855–2860.
37. Calvo RM, Asuncion M, Telleria D, Sancho J, San Millan JL, Escobar-Morreale HF. Screening for mutations in the steroidogenic acute regulatory protein and steroidogenic factor-1 genes, and in *CYP11A* and dosage-sensitive sex reversal–adrenal hypoplasia gene on the X chromosome, gene-1 (*Dax-1*), in hyperandrogenic hirsute women. *J Clin Endocrinol Metab* 2001 86: 1746–1749.
38. Rajkhowa M, Talbot JA, Jones PW, Clayton RN. Polymorphism of glycogen synthetase gene in polycystic ovary syndrome. *Clin Endocrinol* 1996 44: 85–90.
39. Kahsar-Miller M, Azziz R, Feingold E, Witchel SF. A variant of the glucocorticoid receptor gene is not associated with adrenal androgen excess in women with polycystic ovary syndrome. *Fertil Steril* 2000 74: 1237–1240.
40. Milner CR, Craig JE, Hussey ND, Norman RJ. No association between the –308 polymorphism in the tumour necrosis factor α (TNFα) promoter region and polycystic ovaries. *Mol Hum Reprod* 1999 5: 5–9.
41. Oksanen L, Tiitinen A, Kaprio J, Koistinen HA, Karonen S, Kontula K. No evidence for mutations of the leptin or leptin receptor genes in women with polycystic ovary syndrome. *Mol Hum Reprod* 2000 6: 873–876.
42. Elter K, Erel CT, Cine N, Ozbek U, Hacihanefioglu B, Ertungealp E. Role of the mutations Trp8→Arg and Ile15→Thr of the human luteinizing hormone β-subunit in women with polycystic ovary syndrome. *Fertil Steril* 1999 71: 425–430.
43. Cohen DP, Stein EM, Li Z, Matulis CK, Ehrmann DA, Layman LC. Molecular analysis of the gonadotropin-releasing hormone receptor in patients with polycystic ovary syndrome. *Fertil Steril* 1999 72: 360–363.
44. Legro RS, Dietz GW, Comings DE, Lobo RA, Kovacs BW. Association of dopamine D2 receptor gene haplotypes with anovulation and fecundity in female Hispanics. *Hum Reprod* 1994 9: 1271–1275.
45. Almahbobi G, Misajon A, Hutchinson P, Lolatgis N, Trounson AO. Hyperexpression of epidermal growth factor receptors in granulosa cells from women with polycystic ovary syndrome. *Fertil Steril* 1998 70: 750–758.
46. Gharani N, Waterworth DM, Batty S *et al.* Association of the steroid synthesis gene *CYP11a* with polycystic ovary syndrome and hyperandrogenism. *Hum Mol Genet* 1997 6: 397–402.
47. Diamanti-Kandarakis E, Bartzis MI, Bergiele AT, Tsianateli TC, Kouli CR. Microsatellite polymorphism (TTTTA)(*n*) at –528 base pairs of gene *CYP11α* influences hyperandrogenemia in patients with polycystic ovary syndrome. *Fertil Steril* 2000 73: 735–741.
48. San Millan JL, Sancho J, Calvo RM, Escobar-Morreale HF. Role of the pentanucleotide (TTTTA)(*n*) polymorphism in the promoter of the *CYP11A* gene in the pathogenesis of hirsutism. *Fertil Steril* 2001 75: 797–802.
49. Carey AH, Waterworth D, Patel K *et al.* Polycystic ovaries and premature male pattern baldness are

associated with one allele of the steroid metabolism gene *CYP17*. *Hum Mol Genet* 1994 3: 1873–1876.

50. Gharani N, Waterworth DM, Williamson R, Franks S. 5' polymorphism of the *CYP17* gene is not associated with serum testosterone levels in women with polycystic ovaries. *J Clin Endocrinol Metab* 1996 81: 4174.

51. Witchel SF, Lee PA, Suda-Hartman M, Smith R, Hoffman EP. 17α-Hydroxylase/17,20-lyase dysregulation is not caused by mutations in the coding regions of CYP17. *J Pediatr Adolesc Gynecol* 1998 11: 133–137.

52. Escobar-Morreale HF, San Millan JL, Smith RR, Sancho J, Witchel SF. The presence of the 21-hydroxylase deficiency carrier status in hirsute women: phenotype–genotype correlations. *Fertil Steril* 1999 72: 629–638.

53. Witchel SF, Aston CE. The role of heterozygosity for *CYP21* in the polycystic ovary syndrome. *J Pediatr Endocrinol Metab* 2000 13 (Suppl 5): 1315–1317.

54. Mifsud A, Ramirez S, Yong EL. Androgen receptor gene CAG trinucleotide repeats in anovulatory infertility and polycystic ovaries. *J Clin Endocrinol Metab* 2000 85: 3484–3488.

55. Barbieri RL, Makris A, Randall RW, Daniels G, Kistner RW, Ryan KJ. Insulin stimulates androgen accumulation in incubations of ovarian stroma obtained from women with hyperandrogenism. *J Clin Endocrinol Metab* 1986 62: 904–910.

56. Moller DE. Potential role of TNF-α in the pathogenesis of insulin resistance and type 2 diabetes. *Trends Endocrinol Metab* 2000 11: 212–217.

57. Talbot JA, Bicknell EJ, Rajkhowa M, Krook A, O'Rahilly S, Clayton RN. Molecular scanning of the insulin receptor gene in women with polycystic ovarian syndrome. *J Clin Endocrinol Metab* 1996 81: 1979–1983.

58. Tucci S, Futterweit W, Concepcion ES *et al.* Evidence for association of polycystic ovary syndrome in Caucasian women with a marker at the insulin receptor gene locus. *J Clin Endocrinol Metab* 2001 86: 446–449.

59. McKeigue P, Wild S. Association of insulin gene VNTR polymorphism with polycystic ovary syndrome. *Lancet* 1997 349: 1771–1772.

60. Tong Y, Liao WX, Roy AC, Ng SC. Association of *ACCI* polymorphism in the follicle-stimulating hormone β gene with polycystic ovary syndrome. *Fertil Steril* 2000 74: 1233–1236.

61. Conway GS, Conway E, Walker C, Hoppner W, Gromoll J, Simoni M. Mutation screening and isoform prevalence of the follicle stimulating hormone receptor gene in women with premature ovarian failure, resistant ovary syndrome and polycystic ovary syndrome. *Clin Endocrinol* 1999 51: 97–99.

62. Legro RS, Muhleman DR, Comings DE, Lobo RA, Kovacs BW. A dopamine D3 receptor genotype is associated with hyperandrogenic chronic anovulation and resistant to ovulation induction with clomiphene citrate in female Hispanics. *Fertil Steril* 1995 63: 779–784.

63. Ciampelli M, Lanzone A. Insulin and polycystic ovary syndrome: a new look at an old subject. *Gynecol Endocrinol* 1998 12: 277–292.

64. Urbanek M, Wu X, Vickery KR *et al.* Allelic variants of the follistatin gene in polycystic ovary syndrome. *J Clin Endocrinol Metab* 2000 85: 4455–4461.

65. Mandel FP, Chang RJ, Dupont B *et al.* HLA genotyping in family members and patients with familial polycystic ovarian disease. *J Clin Endocrinol Metab* 1983 56: 862–864.

66. Hague WM, Adams J, Algar V, Drummond V, Schwarz G, Bottazzo GF. HLA associations in patients with polycystic ovaries and in patients with congenital adrenal hyperplasia caused by 21-hydroxylase deficiency. *Clin Endocrinol* 1990 32: 407–415.

67. Ober C, Weil S, Steck T, Billstrand C, Levrant S, Barnes R. Increased risk for polycystic ovary syndrome associated with human leukocyte antigen DQA1*0501. *Am J Obstet Gynecol* 1992 167: 1803–1806.

Chapter 34

Genetics of male reproductive dysfunction

Willy Lissens

INTRODUCTION

Infertility and subfertility affect up to 15% of all couples trying for a child.[1,2] In these couples, problems in the male are responsible for about a third of the cases, while, in a further third, male factors are a contributing cause. Several etiological factors in male infertility have been identified, and it is now clear that a proportion of these factors have a genetic basis. The purpose of this chapter is to review what is known of genetic factors contributing to male reproductive dysfunction in humans.

As a result of modern DNA technology, our knowledge of the molecular basis of male infertility is rapidly increasing. Moreover, improvements in artificial reproductive technology now offer many couples the possibility of having their own genetic children. The drawback to this technology is that the risk of transmitting infertility and genetic diseases to the offspring is considerably increased when the infertility is genetic in origin (see section on 'Clinical Relevance', below). This chapter therefore describes, with a view to the clinical relevance, diseases that are relevant to (male) reproductive medicine.

It is likely that many more genes involved in normal male reproduction will be identified in the next few years. Much will also probably be learnt from fertility genes identified and isolated in animal models. Homologous genes probably exist in humans, and may contribute to male infertility.[3] In addition, the study of RNA in ejaculated spermatozoa may further contribute to the isolation of genes involved in infertility.[4] Many infertility-associated genes therefore remain to be identified, and the clinician involved in reproductive medicine should be aware of this. On the other hand, knowledge of a given genetic defect in the male will help the couple concerned to make appropriate decisions in their situation, and, in an increasing number of cases, reproduction without disease transmission is becoming possible.

CLINICAL RELEVANCE

Before the introduction of intracytoplasmic sperm injection (ICSI) in reproductive centers, numerous couples, especially those suffering from andrological infertility, were unable to have their own genetic children by standard *in vitro* fertilization (IVF). In such couples, the fertilization rate was either very low or zero, or the number of motile sperm in the ejaculate was too low for the couple to be admitted to the IVF program. With the direct injection of a single spermatozoon into the cytoplasm of an oocyte, ICSI has dramatically changed this situation.[5,6] Although originally using only ejaculated sperm, ICSI has in recent years also used the injection of spermatozoa or spermatids recovered from the epididymis (through microsurgical epididymal sperm aspiration, MESA) or the testis (through testicular sperm extraction, TESE).[7]

ICSI is a more invasive method than standard IVF, since a sperm cell is injected through the membrane of the oocyte. Moreover, sperm cells are used that would never lead to fertilization in either natural or IVF conditions. The use of ICSI in assisted reproduction has therefore given cause for concern about its safety in terms of a possible increase in the incidence of major congenital malformations, chromosomal aberrations or developmental problems. So far, the most extensive studies conducted on this issue are fairly reassuring, and there seems to be no increased risk, compared with standard IVF, except perhaps for a slight increase in the number of sex chromosome abnormalities.[8] However, in some cases, transmission of defective genetic factors will not be readily recognized. Microdeletions of the Y chromosome, for instance, are found in about 10% of males with non-obstructive azoospermia, or with severe oligozoospermia, and are believed to cause infertility as a result of loss of the genes involved in spermatogenesis (see 'Microdeletions of the Y chromosome', below). While ICSI has been used for several years now, microdeletion detection was introduced only 4 or 5 years ago in infertility clinics. The transmission of a defective Y chromosome from an ICSI father to his sons will probably be recognized as a result of infertility only at the reproductive age of the sons, 20–30 years after the ICSI procedure. Similarly, other currently unknown genetic factors affecting only fertility may be transmitted in the same way. The clinical importance of screening ICSI candidates for known genetic disorders associated with infertility should therefore be stressed, as should that of long-term follow-up of children born after ICSI. X-linked recessive, autosomal dominant and autosomal recessive diseases will also be transmitted according to their mode of inheritance. X-linked diseases will be transmitted to female offspring who are healthy carriers; these females can then transmit the disease to their sons (for example, X-linked Kallmann syndrome). In autosomal dominant diseases, the risk of transmission is 50% for both male and female offspring, without reference to possible infertility (for example, Noonan syndrome). In autosomal recessive diseases, the risk to the offspring will depend on the incidence of the carrier frequency. If the female is also a carrier, then the risk of an affected child will be 1/2 (for example, congenital bilateral absence of the vas deferens linked to cystic fibrosis).

Screening for known genetic disorders involved in infertility is therefore mandatory for the prevention of transmission of genetic disease. The couple should be appropriately informed about their situation. In cases of negative screening, the possibility that a genetic factor might be involved in idiopathic infertility should be mentioned. When a genetic factor is the cause of the infertility, prenatal diagnosis can be proposed as a means of preventing transmission of the disease through ICSI. Alternatively, ICSI combined with preimplantation genetic diagnosis can be considered and proposed to the couple where this is technically feasible (see Chapter 27).

APPLIED TECHNOLOGY

The technology applied to the analysis of genes involved in the genetics of male reproductive dysfunction is the same as that used for the analysis of genetic disease in general. The molecular techniques used for this purpose are discussed in Chapters 1 and 3. Illustrations are given here for the application of the Southern blot technique and single-strand conformation polymorphism (SSCP) analysis in molecular genetic studies.

Localization of genes is often achieved by study of the segregation of polymorphic DNA markers with disease, or other particular phenotypes in families (linkage analysis). The human genome contains numerous polymorphic sites, and these can be used to follow genetic traits in families, as long as the markers analyzed are close enough to the gene involved. Depending on the mode of inheritance of the disease, all affected members of a family will then receive the same mutated gene or genes from their parent(s) and, hence, the same disease-linked polymorphic allele(s) of the DNA marker. All non-affected family members will either not receive the disease gene (in autosomal dominant disease or boys in X-linked disease) or will receive at most one disease allele (carriers in autosomal recessive disease or girls in X-linked disease), and they inherit the polymorphic markers according to their phenotype. When the polymorphic marker is located far from the disease gene, or is even on another chromosome, disease and marker will be transmitted in a random fashion. Until 1985, these DNA polymorphisms were mostly studied by means of the Southern blot technique. Application of the technique for autosomal dominant polycystic kidney disease (ADPKD) linked to the *PKD1* gene at chromosome 16p13.3 is shown in Figure 1. ADPKD is characterized by numerous cysts in the kidneys and in several other organs. In male patients, ADPKD is a rather uncommon cause of infertility due to obstructive azoospermia resulting from cysts in the seminal vesicles and ejaculatory ducts. The Southern blot technique can also be used for the detection of partial or complete deletions of a DNA region in the genomic DNA, or of an increase in fragment length. The latter is illustrated for the molecular diagnosis of myotonic dystrophy, which is due to an increase in the number of CTG repeats in the 3' end of the myotonic dystrophy-protein kinase gene (Figure 2). Although Southern blotting remains a valid method for molecular analysis today, it can often now be replaced by the use of polymerase chain reaction (PCR) amplification of the DNA region of interest, followed by analysis of the resulting fragment.

Several PCR-based screening methods have been developed to analyze large numbers of patients for mutations. Especially when large genes are involved, these screening methods have the advantage that the whole gene does not have to be sequenced for all patients, since sequencing is

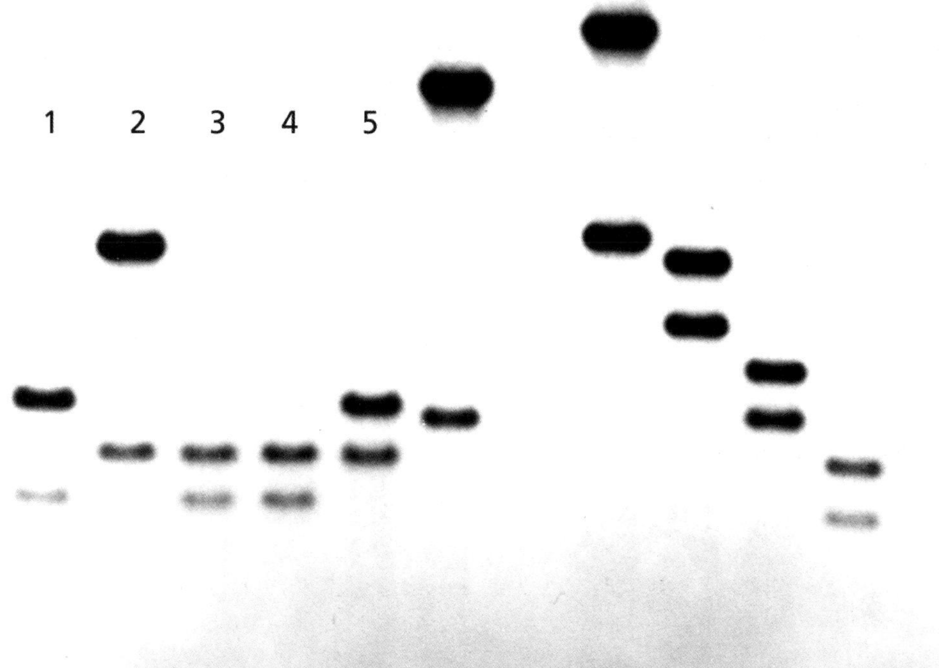

Figure 1 Linkage analysis for a family with adult polycystic kidney disease linked to the *PKD1* gene at chromosome 16p13.3. The highly polymorphic marker 3' hypervariable region (HVR), which is closely linked to the *PKD1* gene (4% of recombination), is studied by Southern blot analysis. The three affected children (lanes 3–5) inherit the same polymorphic band from their affected mother (lane 2). This band segregates with the mutant *PKD1* gene in this family. Lane 1: healthy father. Samples at the right are from five unrelated individuals; these samples are shown to demonstrate the highly polymorphic nature of the 3' HVR marker

both time-consuming and costly. A disadvantage of screening methods is that they will usually not detect all mutations present. This probably explains why sometimes no mutations are found in a clinically well-defined patient. In many cases, the conditions of analysis are very important. This is illustrated for the SSCP technique used for the analysis of samples from eight unrelated individuals for exon 15 of the cystic fibrosis gene (Figure 3). These samples were tested under the same conditions, except for temperature. The samples in Figure 3a were tested at 4°C, and those in Figure 3b at 20°C. Aberrant patterns are seen at both temperatures for samples 1 and 7. At 4°C, another two samples show mobility shifts (samples 5 and 6), but normal patterns are seen at 20°C. Nucleotide changes for samples 5 and 6 would not have been detected if the samples had been analyzed at 20°C only. Furthermore, normal SSCP patterns in the other patients do not exclude the presence of a nucleotide change.

CURRENT CONCEPTS

Introductory note

A growing number of genetic defects either specifically affecting male reproductive function, or affecting male reproductive function as part of a more complex disease or syndrome, have been identified in recent years. Normal structure and function of several of these genes, which, through mutation, are at the basis of these genetic defects, are described in detail in previous chapters (or in some detailed reviews).[9–11] These include genes or conditions involved in early sexual differentiation and development, the synthesis of testosterone and dihydrotestosterone, the androgen receptor, gonadotropin subunit genes and gonadotropin receptor genes, gonadotropin releasing hormone gene and receptor, persistent Müllerian duct syndrome, idiopathic hypogonadotropic hypogonadism and Kallmann syndrome. These topics are not repeated here.

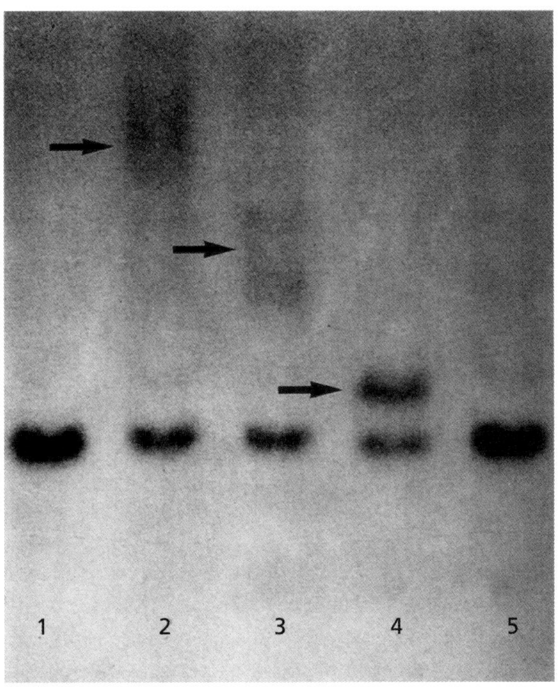

Figure 2 Molecular diagnosis of myotonic dystrophy by Southern blot analysis. Genomic DNA of five individuals was digested with restriction enzyme SacI and electrophoresed on a 0.8% agarose gel. After transfer of the DNA fragments to a nylon membrane, the 3' end of the myotonic dystrophy-protein kinase (*DMPK*) gene was probed with a ^{32}P-labelled probe from this region. In normal individuals (lanes 1 and 5), a DNA fragment of about 3.8 kb is detected. In myotonic dystrophy patients, a larger fragment is found because of an expansion of a CTG repeat in the 3' end of the *DMPK* gene of one allele (arrows). Individuals 2–4 are genetically related: lane 4, father of patient in lane 3; lane 3, mother of patient in lane 2 (boy). The length difference, in comparison with the 3.8-kb band, is about 300 bp (lane 4), 750 bp (lane 3) and 1800 bp (lane 2), which correspond to the presence of about 100, 250 and 600 CTG repeats, respectively. The bands in lanes 2 and 3 are rather diffuse, owing to somatic instability of the expanded CTG repeat. The number of repeats increases from generation to generation, and corresponds to the earlier appearance of the disease in successive generations with an increase of disease severity (anticipation)

Spinal and bulbar muscular atrophy

X-linked spinal and bulbar muscular atrophy or Kennedy disease is an adult-onset (30–50 years) form of motor neuron disease characterized by progressive muscle weakness and atrophy.[12] Affected males are often fertile in early adulthood, but many of them show signs of mild androgen insensitivity, often gynecomastia, after the third decade of life. Azoospermia, impotence and testicular atrophy may also develop. The genetic defect in Kennedy disease has been localized to exon 1 of the androgen receptor gene.[13] This exon contains a polymorphic (CAG)n repeat (n = 11–35 in normal individuals), and the number of repeats is almost doubled (n = 40–62) in Kennedy-disease patients. The severity of the disease correlates well with the size of the expansion.[14] The pathological defect caused by the triplet expansion is not known, but might be due to a reduced androgen-binding affinity of the receptor.[15]

Bardet–Biedl syndrome

Bardet–Biedl syndrome is an autosomal recessive disease characterized by obesity, polydactyly, mental retardation, renal malformations, retinal pigmentary dystrophy and hypogenitalism.[16] Hypogenitalism is most evident in males, and is associated with cryptorchidism, micropenis and various degrees of hypospadias. In females, vaginal anomalies and delayed puberty are observed. Other characteristics associated with Bardet–Biedl syndrome are deafness, diabetes mellitus, dental anomalies and small stature. The prevalence of Bardet–Biedl syndrome is less than 1 in 100 000.

At the molecular level, Bardet–Biedl syndrome is very heterogeneous. There is evidence for at least six different loci involved in this disorder, showing a broad phenotypic range of clinical manifestations within and between families. Six candidate loci have been mapped to chromosome regions 11q13 (*BBS1*), 16q21 (*BBS2*), 3p12 (*BBS3*), 15q22.3 – q23 (*BBS4*), 2q31 (*BBS5*) and 20p12 (*BBS6*). To date, defects in three genes have been implicated in Bardet–Biedl syndrome. Two are novel genes, called *BBS2* and *BBS4* according to their locus position, encoding proteins with as yet unknown function.[16,17] The third gene, *MKKS*, causes *BBS6* and had previously been shown to be involved in McKusick–Kaufman syndrome, a developmental anomaly syndrome.[18,19] This gene was considered a candidate for *BBS6* because of the mapping position of the *BBS6* locus and clinical similarities

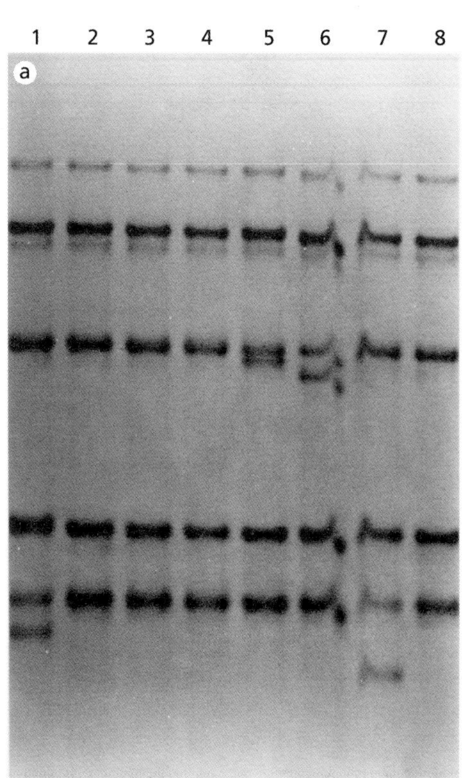

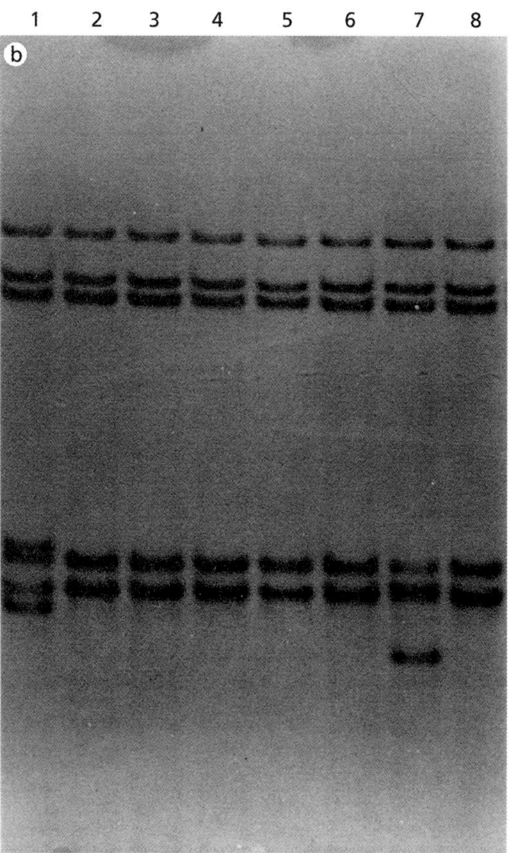

Figure 3 Single-strand conformation polymorphism (SSCP) analysis of exon 15 of the cystic fibrosis transmembrane conductance regulator (*CFTR*) gene at two temperatures. After polymerase chain reaction (PCR) amplification, the 485-bp fragment was cut into fragments of 267 bp and 218 bp by the restriction enzyme AccI. Eight samples were analyzed by SSCP at two different temperatures: (a) 4°C, (b) 20°C. Aberrant migration of single strands was present in both conditions for samples 1 and 7, but was seen for samples 5 and 6 only at 4°C. Lanes 1–7: cystic fibrosis patients; lane 8: normal control. Sequencing of this fragment for the four cystic fibrosis patients showing altered mobility led to the identification of two mutations and two polymorphisms (not shown)

between both syndromes.[20] The difference in phenotype between patients with McKusick–Kaufman syndrome and Bardet–Biedl syndrome is believed to result from different types of mutations: in McKusick–Kaufman syndrome, mutations lead to MKKS protein with retention of some function (mostly missense mutations), while, in Bardet–Biedl syndrome, mutations lead to complete loss of function of the MKKS protein (mostly non-sense and frameshift mutations). Although the MKKS protein shows similarities to the chaperonins, a family of proteins involved in protein processing, its function is still unknown.

Noonan syndrome

Noonan syndrome is a common (incidence 1/1000 to 1/2500 live births), multicongenital-anomalies syndrome characterized by short stature, typical facial dysmorphism and a congenital heart defect, which is mostly a pulmonary valve stenosis or hypertrophic cardiomyopathy. Most cases of Noonan syndrome are compatible with an autosomal dominant pattern of inheritance, with many new mutations (up to 50%); an autosomal recessive form might also exist.[21] Noonan syndrome has been found in association with neurofibromatosis type 1 (*NF1*; chromosome localization 17q11), and less frequently with 'deletion 22q11' syndromes. However, Noonan syndrome has been shown to segregate independently of both conditions. A first gene for autosomal dominant Noonan syndrome has been localized to chromosome 12q22–qter by linkage analysis.[22] It has recently been shown that mutations in the *PTPN11* gene in this region are responsible for Noonan syndrome in two pedigrees with the syndrome linked to this region, and account for more than 50% of the condition in a small sample of sporadic cases or small kindreds[23].

About half the males with Noonan syndrome are infertile. This infertility is strongly associated with bilateral cryptorchidism, resulting in azoospermia and oligozoospermia.[24] Males with normally descended testes are fertile. Fertility in affected females is not impaired.

Myotonic dystrophy

Myotonic dystrophy is the most common form of adult muscular dystrophy and is characterized by myotonia, muscle weakness and wasting, especially around the face and the neck, and a range of other symptoms including cardiac conduction disturbances, cataracts and, in males, testicular tubular atrophy. Symptoms usually manifest in the late 20s, but time of onset and clinical manifestations can vary considerably. A very severe, maternally transmitted form of the disease called congenital myotonic dystrophy can affect children, and is often associated with mental retardation or neonatal mortality. Milder phenotypes are often seen in older individuals. The disease is transmitted in an autosomal dominant fashion.

The gene involved in myotonic dystrophy was isolated in 1992, and encodes for a myotonic dystrophy-protein kinase (DMPK).[25–28] The disease is caused by an increased number of (CTG)*n* trinucleotide repeats in the 3' untranslated region of the *DMPK* gene, which is located on chromosome 19q13.3. In normal individuals, the gene contains between 5 and 35 CTG repeats. Affected individuals with a repeat number between 50 and 80 have mild disease, and often remain clinically undetected. Genes with this number of repeats are unstable, and expand in size upon transmission to the offspring. Severely affected patients may carry a few thousand CTG repeats in their *DMPK* gene (Figure 2). The progressively earlier appearance of the disease in successive generations with an increase of disease severity is called anticipation. In a few cases, the number of CTG repeats is reduced in the offspring. In some myotonic dystrophy patients, no expansion of the CTG repeats is found.[29] These patients either have other mutations in the *DMPK* gene or have myotonic dystrophy not linked to the *DMPK* gene. A second locus for myotonic dystrophy has indeed been identified at chromosome 3q.[30] Clinically, this form of the disease is very similar to classic myotonic dystrophy.

Male infertility in relation to cystic fibrosis

Cystic fibrosis (CF) is the most common autosomal recessive disease among populations of Caucasian descent.[31,32] The disorder affects about 1/2500 live births, and 1/25 individuals is an asymptomatic carrier of the disease. The typical form of the disease is characterized by chronic obstruction and infection of the respiratory tract, exocrine pancreatic insufficiency (in about 85% of patients) and elevated sweat chloride concentrations (over 60 mmol/l). Less common manifestations include neonatal meconium ileus, liver disease, pancreatitis and diabetes mellitus. More than 95% of adult male CF patients are infertile; infertility also occurs in female patients, although to a lesser extent. The clinical manifestations of CF result from defects (mutations) in the cystic fibrosis transmembrane conductance regulator (*CFTR*) gene, which encodes a 1480-amino-acid protein involved in chloride conduction across epithelial cell membranes. The gene contains 27 exons, spread over 250 kb of DNA, and is localized on chromosome 7q31.2. The most common mutation responsible for CF is the deletion of a phenylalanine amino acid at position 508 (ΔF508) of the protein; this mutation accounts for almost 70% of all CF chromosomes, although large population variations exist. Besides the ΔF508, more than 900 mutations, and a few hundred polymorphisms, have been identified in the *CFTR* gene.[33]

A wide variation in clinical presentation among CF patients has long been recognized. Since the identification of the *CFTR* gene, many reports have been published dealing with genotype (*CFTR* mutations)–phenotype (clinical presentation) correlations (for reviews, see references 31, 32 and 34). To date, only the pancreatic status of CF patients has been found to correlate well with the CF genotype. On the basis of pancreatic function, it was concluded that the presence of two severe mutations (i.e. mutations leaving no residual *CFTR* activity) in CF patients results in pancreatic insufficiency, while patients carrying one or two mild mutations (i.e. mutations retaining some residual *CFTR* activity) are likely to be pancreatic-sufficient. Other clinical manifestations for which no straight correlation was found between genotype and phenotype, such as pulmonary disease, may be influenced by non-genetic factors (such as environment), besides other genetic factors. In this context, it is interesting to mention the identification of a CF modifier locus for meconium ileus on chromosome 19q13.[35]

In recent years, mutation analysis of the *CFTR* gene has also provided ample evidence that defects in this gene are responsible for other clinical conditions, with some features in common with CF. An increased frequency of mutations was indeed found in some patients with idiopathic chronic and acute recurrent pancreatitis, chronic obstructive pulmonary disease, disseminated bronchiectasis, allergic bronchopulmonary aspergillosis, chronic bronchial hypersecretion, nasal polyposis, isolated hypotonic dehydration, neonatal transitory hypertrypsinemia and asthma.[34,36] It is thus becoming clear that *CFTR* is involved in a wide spectrum of clinical phenotypes, from typical CF to atypical mild disease involving isolated characteristics of CF.

A possible relationship between infertility in CF males due to congenital bilateral absence of the vas deferens (CBAVD) and infertility in otherwise healthy males with CBAVD was hypothesized as early as 1971.[37] CBAVD is responsible for up to 2% of male infertility and 6% of all obstructive azoospermias.[38] A high frequency of *CFTR* gene mutations was indeed found in males with isolated CBAVD.[36,39] About 20% of patients with CBAVD have mutations in both of their *CFTR* genes, and most carry a different mutation in each allele (they are compound heterozygotes). None of these patients has two severe mutations, but either a severe and a mild one or two mild ones. Fifty per cent of patients are carriers of one mutation, and the remaining 30% show no detectable mutation. However, the hypothesis of a common etiology of CF and isolated CBAVD implies that mutations in both *CFTR* genes should be present, and this is true for only about 20% of patients. So far, two reasons have been found for this discrepancy.

First, about 20% of CBAVD patients also have urinary tract malformations, and in these patients no mutations in the *CFTR* gene are found. CBAVD here might represent not a mild form of CF but a distinct clinical entity. In males, both the ureter and ducts of the epididymis and the vas deferens derive from the Wolffian or mesonephric duct. The Wolffian duct splits into a reproductive and a ureteral part at 7 weeks of gestation. As the ureteral ducts are normal in male CF patients, the Wolffian duct must be undamaged in early embryonic life. Defects in the genital ducts therefore probably occur after splitting of the ducts, and as a consequence of *CFTR* dysfunction. In patients with CBAVD combined with urinary tract malformations, the Wolffian duct is probably damaged before 7 weeks' gestation, and this might explain an etiology different from *CFTR* dysfunction.

Second, a partially penetrant *CFTR* splice site variant, called the '5T' allele, has been recognized as a mutation associated with the phenotype of CBAVD. This splice variant is located in intron 8, in front of exon 9, at the splice branch/acceptor site, and consists of 5, 7 or 9T nucleotides. The presence of 5T nucleotides is associated with a high degree of exon 9 loss (around 90%) upon transcription of the *CFTR* gene. Translation from this mRNA without exon 9 results in non-functional CFTR protein. Individuals carrying 7T or 9T alleles have, as a mean, about 25% or 15%, respectively, of mRNA without exon 9. Although the 5T, 7T and 9T alleles occur with frequencies of 5%, 85% and 10% in the general population, a fourfold increase of the incidence of the 5T allele was found in CBAVD patients.[40] Many patients with one *CFTR* mutation detected had a 5T on the other chromosome, and in a considerable proportion of patients with no mutation detected one or two 5T alleles were found. The 5T allele in combination with a *CFTR* mutation on the other chromosome, therefore, seems to be one of the major causes of CBAVD. Since the 5T allele also occurs on the normal chromosome of (fertile) fathers of CF patients, this allele is not expected to lead to CBAVD in all cases. On the basis of the (reduced) frequency of the 5T allele in fathers of CF patients, and compared with the frequency in mothers of CF patients, it can be calculated that the penetrance of the 5T allele is around 0.56. Recently, it has been shown that there is a variability in the splicing efficiency of the 5T allele among different individuals and between different tissues of the same individual. The proportion of mRNA without exon 9 was consistently higher in epididymal epithelium and in the vas deferens than in nasal epithelial cells.[41–43] These observations might explain both the partial penetrance of the 5T allele and the tissue specificity in CBAVD. Other frequent polymorphic *CFTR* variants might possibly also contribute to the CBAVD phenotype, and to other *CFTR*-associated atypical diseases.[44]

Nevertheless, between 10 and 30% of *CFTR* alleles cannot be identified even by excluding males with CBAVD and renal malformations. Several possibilities have been considered: failure to identify all mutations by the methodology used and the fraction of the gene studied, subtle forms of renal malformation that are not readily recognized, or genetic heterogeneity corresponding to a third clinical entity.

In patients with congenital unilateral absence of the vas deferens (CUAVD), an incidence of *CFTR* mutations comparable to that in CF-linked CBAVD is found when contralateral genitourinary anomalies are present and renal abnormalities are absent.[45] On the other hand, the condition of CUAVD is not related to CF when anatomically complete and patent vasa deferentia on the side of the palpable vas are present; in this group of patients, a high incidence of renal abnormalities at the ipsilateral side of the absent vas is found.

The etiology of Young's syndrome, a combination of obstructive azoospermia and chronic sinopulmonary disease, seems not to be related to *CFTR* dysfunction.[39,46] It has been suggested that the syndrome might be associated with childhood mercury poisoning.

In a study of patients with obstructive azoospermia of unknown etiology, a high incidence of *CFTR* mutations, including the 5T variant, has been described. In another study, dealing with patients with reduced sperm quality and quantity, but without CBAVD, an increased frequency of *CFTR* mutations was also found. On the other hand, this result could not be confirmed in Dutch patients with oligoasthenoteratozoospermia. Altogether, these data suggest that defects in *CFTR* might play a role in sperm production and maturation, but, since only a limited number of patients have been studied, these observations need to be confirmed.[36]

Primary ciliary dyskinesia

Primary ciliary dyskinesia or immotile cilia syndrome is a heterogeneous group of disorders resulting from severe or complete immotility of cilia.[47] It is characterized by recurrent pulmonary and upper respiratory tract infections, rhinitis and sinusitis, and in most affected males infertility due to poor mobility or immotility of sperm cells. Primary ciliary dyskinesia is inherited as an autosomal recessive trait in the majority of cases, and occurs with a prevalence of around 1/20 000. In about half of patients, primary ciliary dyskinesia is associated with situs inversus, a condition referred to as Kartagener syndrome.

Genetic heterogeneity of primary ciliary dyskinesia is expected on the basis of the complex structure of cilia. This heterogeneity is confirmed by the identification of several chromosomal regions suggestive of the presence of loci linked to the primary ciliary dyskinesia phenotype.[48] A first gene related to the *Chlamydomonas reinhardii* dynein gene *IC78* has been implicated in primary ciliary dyskinesia in one family.[49]

Chromosome anomalies

Chromosome anomalies represent an important contributory factor to sterility in azoospermic and severe oligozoospermic males. The existence of chromosomal abnormalities in patients attending a male fertility clinic was suspected as early as 1957. In ten of 91 males with azoospermia or severe oligozoospermia, a Barr body was found.[50] The presence of a Barr body was later shown to be due to a 47,XXY karyotype in these men with Klinefelter's syndrome.[51] Since then, several surveys have been conducted to determine the chromosome factor in male infertility. In the largest series reported, without the application of banding techniques, 1363 male partners from infertile couples were investigated cytogenetically.[52] In total, 70 males exhibited numerical sex chromosome abnormalities in their constitutional karyotype, 16 had a translocation and four had an extra marker chromosome. Moreover, this study also pointed to a relationship between declining sperm counts and increasing frequencies of chromosome abnormalities. On pooling data from seven surveys of 7876 infertile men, a 5.1% incidence of chromosome abnormalities was found: 3.8% involved the sex chromosomes (3.0% had a 47,XXY karyotype) and 1.3% the autosomes.[53] These frequencies were significantly higher than those reported in the newborn series considered, where the equivalent figures were 0.39% with chromosome anomalies, comprising 0.14% sex chromosome and 0.25% autosome aberrations (in this control group, only chromosome abnormalities compatible with a normal phenotype were considered). On averaging the overall incidence of chromosome anomalies in five surveys of oligozoospermic men and six studies of azoospermic men, frequencies of 4.6% and 13.7% were obtained, respectively.[53,54] It should be noted that, in the azoospermic group, sex chromosome abnormalities predominated (12.6%), whereas, in the oligozoospermic group, autosome anomalies were the most frequent (3.0%). These studies, and a survey recently published by Yoshida and colleagues, confirm the earlier observation by Kjessler: as mean sperm count increases, the number of chromosomally abnormal individuals declines.[52,54] In addition to the clinical importance of these chromosomal abnormalities, they may also lead to the mapping and the identification of genes involved in male infertility.[55]

Microdeletions of the Y chromosome

A first indication for the presence, on the Y chromosome, of a gene or genes involved in sperm production was found by Tiepolo and Zuffardi.[56] They described six azoospermic males with deletions on the long arm of this chromosome, detected by karyotype analysis. Each individual lacked both the distal fluorescent part of Yq (Yq12) and part of the flanking (non-fluorescent) euchromatic region (band Yq11). The deletions had occurred *de novo*, since the fertile fathers and brothers of these patients showed normal Y chromosomes. Since the fluorescent heterochromatic part of Yq was described as a natural variant in normal fertile males, it was postulated that the absence of one or more factors (azoospermia factor, *AZF*), normally present in Yq11, were responsible for the azoospermia observed in these six males. Many studies, at both the cytogenetic and molecular levels, have since confirmed the hypothesis of the presence of genes involved in spermatogenesis in Yq11. Moreover, the detection of non-overlapping microdeletions (defined as deletions not detectable by cytogenetic analysis) in patients with non-obstructive azoospermia and oligozoospermia has led to the subdivision of this region into three parts, *AZFa*, *AZFb* and *AZFc*, possibly containing different azoospermia factors.[57] Furthermore, analysis of testis sections of such males has shown the presence of different histopathologies ranging from complete absence of germ cells (Sertoli-cell-only syndrome) to the presence of spermatogonia and premeiotic spermatogenic cells in at least some seminiferous tubules (maturation arrest), but no postmeiotic cells. No clear correlation has been found between the size and location of the microdeletions and the severity of the spermatogenic defect. The existence of a possible fourth *AZF* locus, *AZFd*, associated with mild oligozoospermia or normal sperm cell count with abnormal morphology, has been suggested recently.

In the past few years, several candidate spermatogenesis genes or gene families have been isolated from the *AZFa*, *AZFb* and *AZFc* regions. A first family was isolated in 1993, and consists of 30–50 genes and pseudogenes. Although originally isolated from the *AZFb* region, these copies are spread over both arms of the Y chromosome.[58] The gene family is termed *RBMY* (RNA-binding motif, Y chromosome), and encodes proteins with an RNA recognition motif and an internal repeated sequence typical of proteins interacting with RNA or single-stranded DNA. A homolog on the X chromosome, *RBMX*, has recently been identified. The *RBMY* gene family is expressed at the RNA level in fetal, prepubertal and adult testes. The protein is present in the nuclei of germ cells, which suggests that it is involved in the nuclear metabolism of newly synthesized RNA. The functionally active gene or genes have been localized to *AZFb* in Yq11.23 by studying the expression of the protein in azoospermic patients with different Y deletions. In the absence of *RBMY* expression, germ cells are found that develop up to early meiosis but not beyond. *RBMY* function is therefore probably important for the completion of meiosis or for subsequent haploid stages.

A second gene family was found in *AZFc*, and is termed *DAZ/SPGY* (deleted in azoospermia/spermatogenesis gene on the Y; hereinafter referred to as *DAZ*). Several gene copies of this family are present in *AZFc* and encode, like *RBMY*, RNA-binding proteins.[58,59] *DAZ* genes are also conserved on the Y chromosome in old-world primates and apes; in other mammals, this gene family is present as a single copy gene localized on the autosomes. A functional human homolog of *DAZ* is also present on chromosome 3p24, and is termed *DAZL1* (for *DAZ*-like 1). The expression of *DAZ* and *DAZL1* is testis-specific, but *DAZL1* is possibly also expressed, although at a lower level, in prenatal and postnatal germ cells in females. Important questions remain as to the physiological roles of *DAZ* and *DAZL1*, and their roles in azoospermia and oligozoospermia. Neither is it clear whether *DAZ* and *DAZL1* have the same function. If they have the same function, *DAZL1* might compensate, at least partially, for the loss of *DAZ* copies in patients with deletions in *AZFc*. This might then possibly explain, in part, the phenotypic variability in histopathology observed in such patients. If they have a different function in spermatogenesis, then *DAZL1* might be involved in particular forms of autosomal recessive male infertility or even in hereditary female infertility. A possible role for *DAZ* and *DAZL1* in human spermatogenesis is further suggested by the study of homologous genes in *Drosophila* and in the mouse. In *Drosophila*, a *DAZ* homologous gene, termed *boule*, was isolated, and its absence was shown to result in azoospermia. Meiotic divisions are interrupted, but a limited spermatide differentiation is observed. The primary defect seems to occur at the level of the meiotic G_2/M transition stage. The mouse homolog of *DAZ*, *Dazl1*, is expressed in both sperm cells and oocytes, and the absence of Dazl1 protein leads to the complete absence of germ-cell production in both males and females. *Dazl1*, unlike *RBMY*, is localized in the cytoplasm, and is therefore unlikely to be a transcription or splicing factor, although it might be involved in translational control of mRNAs.

Another 12 genes were recently isolated from the long arm of the Y chromosome.[60] Some of these are expressed only in testes, while others are expressed ubiquitously and have functional homologs on the X chromosome. The function of these genes and their gene products is under study, and their possible role in spermatogenesis remains to be determined. Several of the Y-specific genes have multiple functional copies localized in *AZFb* and/or *AZFc*. It is therefore likely that these genes are also deleted in patients with deletions of these regions. Testicular phenotypes in these patients may therefore reflect the sum of deletion of several genes rather than the loss of a single gene.

Two genes have recently been localized to the *AZFa* region: *USP9Y* and *DBY*. In most cases, deletion of *AZFa* probably occurs through recombination between two *HERV15* provirus sequences, about 800 kb apart and localized on both sides of *AZFa*.[61] In patients with a complete *AZFa* deletion, no testicular germ cells are present. In contrast, patients lacking either *USP9Y* or *DBY* show severe hypospermatogenesis.[62,63] Both *USP9Y* and *DBY* are ubiquitously expressed and have functional homologs on the X chromosome. Their functional role in the process of spermatogenesis remains to be determined.

Numerous clinical studies looking for microdeletions of the Y chromosome in males with azoospermia and oligozoospermia have been conducted in recent years. In most of these studies, deletion detection usually depends on lack of PCR amplification of sequence-tagged sites (STSs), ordered on the physical map of the Y chromosome. Although the frequency of microdeletion detection varies greatly from study to study, the mean is around 10% (range from 1 to 55%). This variability probably reflects the different criteria used to select patients and the number of STSs studied.[64] Most microdeletions affect either *AZFc* or *AZFb* and *AZFc*, and are found in patients with fewer than 2×10^6 sperm cells/ml, but a few exceptions have been described. Moreover, several studies describe microdeletions in patients with cryptorchidism and varicocele. Although patients with greatly reduced numbers of sperm cells (fewer than 2×10^6/ml) are the primary target group for deletion detection, the precise definition of candidates remains difficult. Finally, natural transmission of *AZFc* deletions from fathers to their infertile sons has been reported in two families.[65,66]

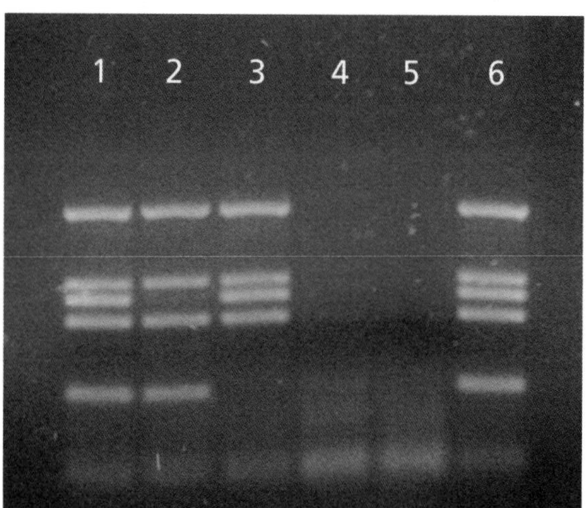

Figure 4 Y-chromosome microdeletion detection in azoospermic males by polymerase chain reaction (PCR) amplification of Y-specific sequence-tagged sites (STSs). The figure shown is part of a screening program consisting of five multiplex PCRs covering azoospermia factor (*AZF*) regions and other Y-chromosome sequences. STSs (from top to bottom): sY14 (*SRY*, Yp11), sY95 (region between *AZFa* and *AZFb*), sY127 (*AZFb*), sY109 (region between *AZFa* and *AZFb*) and sY149 (*AZFc*). Lanes 1–3: DNA from azoospermic males. Lanes 4 and 5: blank (no DNA) and female DNA, respectively. Lane 6: DNA from a normal fertile male. In patient 2, STS sY127 is absent, while in patient 3, STS sY149 is missing

An example of microdeletion screening is shown in Figure 4. The existence of male mosaics for normal and deleted Y chromosomes in white blood cells (and probably in the germ cells) has been suggested in recent studies of infertile fathers and their ICSI sons.[67] These mosaics will not be detected by the technology illustrated in Figure 4.

Other genetic diseases involved in male infertility

Several other genetic diseases are known to be involved in male infertility. A few examples are given below.

Prader–Willi syndrome and Beckwith–Wiedemann syndrome are two diseases that exhibit disturbances in parent-of-origin inheritance patterns, called imprinting defects. The term imprinting refers to the differential expression of genetic material depending on the parental origin of the material, thereby making the genetic region functionally hemizygous in the presence of both a paternal and a maternal contribution. A few regions in the human genome are known to be imprinted. One of these regions is the Prader–Willi syndrome region (and Angelman syndrome region, not described further here) at chromosome 15q11–q13. Prader–Willi syndrome is a neurobehavioral disease characterized by severe hypotonia in the newborn, hyperphagia and obesity in early childhood, hypogonadism, craniofacial dysmorphism and mental retardation. Several genes at 15q11–q13 are expressed only from the paternal chromosome, and in most cases Prader–Willi syndrome results from the absence of a paternal contribution owing to deletion of the 15q11–q13 region or to uniparental disomy of the maternal allele. Patients with mutations affecting an imprinting center have also been described. Beckwith–Wiedemann syndrome is a congenital overgrowth syndrome characterized by the association of gigantism, macroglossia and visceromegaly. Developmental anomalies further include anterior abdominal wall defects, hemihypertrophy, genitourinary abnormalities and mental retardation. Most cases of Beckwith–Wiedemann syndrome are sporadic; about 15% are familial cases showing an autosomal dominant heritance. The

Beckwith–Wiedemann syndrome gene has been mapped to chromosome 11p15.5, a region containing at least four imprinted genes: *H19*, p57^{KIP2} and *KVLQT1*, which are maternally expressed, and insulin-like growth factor-II (*IGF-II*), which is paternally expressed. Although these four genes might be impaired in Beckwith–Wiedemann syndrome, direct evidence has so far only been found for the p57^{KIP2} gene.

Some other examples of infertility-associated diseases are the Aarskog–Scott syndrome (X-linked recessive; cryptorchidism and sperm acrosome defects), adrenomyeloneuropathy (X-linked recessive; azoospermia or oligozoospermia), autoimmune polyglandular syndrome type 1 (autosomal recessive; hypogonadism in 10–20% of males) and chromosome-1 deletion q44 (Sertoli-cell-only syndrome).

SUMMARY

In recent years, an increasing number of genetic factors have been implicated in male reproductive dysfunction. Some of these factors have been well characterized by the identification of genes (and their corresponding gene products) that, when defective, are associated with male infertility. Others have been localized only to a specific region or regions on the chromosomes, while still others that are recognized as genetic in origin remain completely undefined. This chapter summarizes what is currently known about these genetic factors, especially those relevant to the clinician in reproductive medicine. It is important for the clinician to recognize these factors, especially in view of new developments in artificial reproductive technology. In several formerly untreatable conditions, reproduction has now become possible, but the risk of transmission of the genetic factors to the offspring is, in many cases, very high. Identification of these factors is therefore important, and will allow patients to be informed about their situation, so that they can make appropriate decisions.

ACKNOWLEDGEMENTS

I am indebted to many colleagues from the Center of Medical Genetics and the Center for Reproductive Medicine. I especially want to thank Professor I. Liebaers, Professor A. Van Steirteghem, Dr M. Bonduelle, Dr H. Tournaye, Dr S. Seneca, Mrs E. Van Assche and Mrs L. Van Landuyt. Frank Winter from the Language Education Center corrected the manuscript.

REFERENCES

1. Greenhall E, Vessey M. The prevalence of subfertility: a review of the current confusion and a report of two new studies. *Fertil Steril* 1990 54: 978–983.
2. Bhasin S, de Kretser DM, Baker HWG. Pathophysiology and natural history of male infertility. *J Clin Endocrinol Metab* 1994 79: 1525–1529.
3. Montgomery GW. Genome mapping in ruminants and map locations for genes influencing reproduction. *Rev Reprod* 2000 5: 25–37.
4. Miller D. RNA in the ejaculate spermatozoon: a window into molecular events in spermatogenesis and a record of the unusual requirements of haploid gene expression and post-meiotic equilibration. *Mol Hum Reprod* 1997 3: 669–676.
5. Palermo G, Joris H, Devroey P, Van Steirteghem AC. Pregnancies after intracytoplasmic injection of a single spermatozoon into an oocyte. *Lancet* 1992 340: 17–18.
6. Van Steirteghem AC, Nagy Z, Joris H *et al.* High fertilization and implantation rates after intracytoplasmic sperm injection. *Hum Reprod* 1993 8: 1061–1066.
7. Silber SJ, Nagy Z, Liu J *et al.* The use of epididymal and testicular spermatozoa for intracytoplasmic sperm injection: the genetic implications for male infertility. *Hum Reprod* 1995 10: 2031–2043.
8. Bonduelle M, Camus M, De Vos A *et al.* Seven years of intracytoplasmic sperm injection and follow-up of 1987 subsequent children. *Hum Reprod* 1999 14 (Suppl 1): 243–264.
9. Rugarli EI. Kallmann syndrome and the link between olfactory and reproductive development. *Am J Hum Genet* 1999 65: 943–948.
10. Layman LC. Mutations in human gonadotropin genes and their physiologic significance in puberty and reproduction. *Fertil Steril* 1999 71: 201–218.

11. Themmen APN, Huhtaniemi IT. Mutations of gonadotropins and gonadotropin receptors: elucidating the physiology and pathophysiology of pituitary–gonadal function. *Endocr Rev* 2000 21: 551–583.
12. Kennedy WR, Alter M, Sung JH. Progressive proximal spinal and bulbar muscular atrophy of late onset: a sex-linked recessive trait. *Neurology* 1968 18: 671–680.
13. La Spada AR, Wilson EM, Lubahn DB, Harding AE, Fischbeck KH. Androgen receptor gene mutations in X-linked spinal and bulbar muscular atrophy. *Nature* 1991 352: 77–79.
14. Igarashi S, Tanno Y, Onodera O *et al.* Strong correlation between the number of CAG repeats in androgen receptor genes and the clinical onset features of spinal and bulbar muscular atrophy. *Neurology* 1992 42: 2300–2302.
15. MacLean HE, Choi W-T, Rekaris G, Warne GL, Zajac JD. Abnormal androgen receptor binding affinity in subjects with Kenendy's disease (spinal and bulbar muscular atrophy). *J Clin Endocrinol Metab* 1995 80: 508–516.
16. Nishimura DY, Searby CC, Carmi R *et al.* Positional cloning of a novel gene on chromosome 16q causing Bardet–Biedl syndrome (BBS2). *Hum Mol Genet* 2001 10: 865–874.
17. Mykytyn K, Braun T, Carmi R *et al.* Identification of the gene that, when mutated, causes the human obesity syndrome BBS4. *Nat Genet* 2001 28: 188–191.
18. Slavotinek AM, Stone EM, Mykytyn K *et al.* Mutations in *MKKS* cause Bardet–Biedl syndrome. *Nat Genet* 2000 26: 15–16.
19. Katsanis N, Beales PL, Woods MO *et al.* Mutations in *MKKS* cause obesity, retinal dystrophy and renal malformations associated with Bardet–Biedl syndrome. *Nat Genet* 2000 26: 67–70.
20. Stone DL, Slavotinek A, Bouffard GG *et al.* Mutations of a gene encoding a putative chaperonin causes McKusick–Kaufman syndrome. *Nat Genet* 2000 25: 79–82.
21. van der Burgt I, Brunner H. Genetic heterogeneity in Noonan syndrome: evidence for an autosomal recessive form. *Am J Med Genet* 2000 94: 46–51.
22. Jamieson CR, van der Burgt I, Brady AF *et al.* Mapping a gene for Noonan syndrome to the long arm of chromosome 12. *Nat Genet* 1994 8: 357–360.
23. Tartaglia M, Mehler EL, Goldberg R *et al.* Mutations in *PTPN11*, encoding the protein tyrosine phosphatase SHP-2, cause Noonan syndrome. *Nat Genet* 2001 29: 465–468.
24. Elsawi MM, Pryor JP, Klufio G, Barnes C, Patton MA. Genital tract function in men with Noonan syndrome. *J Med Genet* 1994 31: 468–470.
25. Aslanidis C, Jansen G, Amemiya C *et al.* Cloning of the essential myotonic dystrophy region and mapping of the putative defect. *Nature* 1992 355: 548–551.
26. Brook DJ, McCurrach ME, Harley HG *et al.* Molecular basis of myotonic dystrophy: expansion of a trinucleotide (CTG) repeat at the 3' end of a transcript encoding a protein kinase family member. *Cell* 1992 68: 799–808.
27. Fu H-Y, Pizzuti A, Fenwick RG *et al.* An unstable triplet repeat in a gene related to myotonic muscular dystrophy. *Science* 1992 255: 1256–1258.
28. Mahadevan M, Tsilfidis C, Sabourin L *et al.* Myotonic dystrophy mutation: an unstable CTG repeat in the 3' untranslated region of the gene. *Science* 1992 255: 1253–1255.
29. Thornton CA, Griggs RC, Moxley RT. Myotonic dystrophy with no trinucleotide repeat expansion. *Ann Neurol* 1994 35: 269–272.
30. Ranum LPW, Rasmussen PF, Benzow KA, Koob MD, Day JW. Genetic mapping of a second myotonic dystrophy locus. *Nat Genet* 1998 19: 196–198.
31. Welsh MJ, Tsui L-C, Boat TF, Beaudet AL. Cystic fibrosis. In Scriver CR, Beaudet AL, Sly WS, Valle D, eds. *The Metabolic and Molecular Bases of Inherited Disease*. New York: McGraw-Hill, 1995: 3799–3876.
32. Zielenski J, Tsui L-C. Cystic fibrosis: genotypic and phenotypic variations. *Annu Rev Genet* 1995 29: 777–807.
33. Cystic Fibrosis Genetic Analysis Consortium. http://www.genet.sickkids.on.ca/cftr/.
34. Zielenski J. Genotype and phenotype in cystic fibrosis. *Respiration* 2000 67: 117–133.
35. Zielenski J, Corey M, Rozmahel R *et al.* Detection of a cystic fibrosis modifier locus for meconium ileus on human chromosome 19q13. *Nat Genet* 1999 22: 128–129.
36. Stuhrmann M, Dork T. *CFTR* gene mutations and male infertility. *Andrologia* 2000 32: 71–83.

37. Holsclaw DS, Lober B, Jockin H, Schwachman H. Genital abnormalities in male patients with cystic fibrosis. *J Urol* 1971 106: 568–574.
38. Dubin L, Amelar RD. Etiologic factors in 1294 consecutive cases of male infertility. *Fertil Steril* 1971 22: 469–474.
39. Lissens W, Mercier B, Tournaye H *et al.* Cystic fibrosis and infertility caused by congenital absence of the vas deferens and related clinical entities. *Hum Reprod* 1996 11 (Suppl 4): 55–80.
40. Chillon M, Casals T, Mercier B *et al.* Mutations in the cystic fibrosis gene in patients with congenital absence of the vas deferens. *N Engl J Med* 1995 332: 1475–1480.
41. Mak V, Jarvi KA, Zielenski J, Durie P, Tsui L-C. Higher proportion of intact exon 9 *CFTR* mRNA in nasal epithelium compared with vas deferens. *Hum Mol Genet* 1997 6: 2099–2107.
42. Rave-Harel N, Kerem E, Nissim-Rafinia M *et al.* The molecular basis of partial penetrance of splicing mutations in cystic fibrosis. *Am J Hum Genet* 1997 60: 87–94.
43. Teng H, Jorissen M, Van Poppel H *et al.* Increased proportion of exon 9 alternatively spliced *CFTR* transcripts in vas deferens compared with nasal epithelial cells. *Hum Mol Genet* 1997 6: 85–90.
44. Cuppens H, Lin W, Jaspers M *et al.* Polyvariant mutant cystic fibrosis transmembrane conductance regulator genes. The polymorphic *(Tg)m* locus explains the partial penetrance of the T5 polymorphism as a disease mutation. *J Clin Invest* 1998 101: 487–496.
45. Mickle J, Milunsky A, Amos JA, Oates RD. Congenital unilateral absence of the vas deferens: a heterogeneous disorder with two distinct subpopulations based upon aetiology and mutational status of the cystic fibrosis gene. *Hum Reprod* 1995 10: 1728–1735.
46. Friedman KJ, Teichtahl H, De Kretser DM *et al.* Screening Young syndrome patients for *CFTR* mutations. *Am J Respir Crit Care Med* 1995 152: 1353–1357.
47. Meeks M, Bush A. Primary ciliary dyskinesia (PCD). *Pediatr Pulmonol* 2000 29: 307–316.
48. Blouin J-L, Meeks M, Radhakrishna U *et al.* Primary ciliary dyskinesia: a genome-wide linkage analysis reveals extensive locus heterogeneity. *Eur J Hum Genet* 2000 8: 109–118.
49. Pennarun G, Escudier E, Chapelin C *et al.* Loss-of-function mutations in a human gene related to *Chlamydomonas reinhardii* dynein *IC78* result in primary ciliary dyskinesia. *Am J Hum Genet* 1999 65: 1508–1519.
50. Ferguson-Smith MA, Lennox B, Mack WS, Stewart JSS. Klinefelter's syndrome. Frequency and testicular morphology in relation to nuclear sex. *Lancet* 1957 2: 167–169.
51. Jacobs PA, Strong JA. A case of human intersexuality having a possible XXY sex determining mechanism. *Nature (London)* 1959 183: 302–303.
52. Kjessler B. Karyotype, meiosis and spermatogenesis in a sample of men attending an infertility clinic. *Monogr Hum Genet* 1966 2: 1–93.
53. Van Assche E, Bonduelle M, Tournaye H *et al.* Cytogenetics of infertile men. *Hum Reprod* 1996 11 (Suppl 4): 1–24.
54. Yoshida A, Miura K, Shirai M. Cytogenetic survey of 1007 infertile males. *Urol Int* 1996 58: 166–176.
55. Olesen C, Hansen C, Bendsen E *et al.* Identification of human candidate genes for male infertility by digital differential display. *Mol Hum Reprod* 2001 7: 11–20.
56. Tiepolo L, Zuffardi O. Localization of factors controlling spermatogenesis in the nonfluorescent position of the human Y chromosome long arm. *Hum Genet* 1976 34: 119–124.
57. Vogt PH. Human chromosome deletions in Yq11, *AZF* candidate genes and male infertility: history and update. *Mol Hum Reprod* 1998 4: 739–744.
58. Ma K, Mallidis C, Bhasin S. The role of Y chromosome deletions in male infertility. *Eur J Endocrinol* 2000 142: 418–430.
59. Reijo R, Lee T-Y, Salo P *et al.* Diverse spermatogenic defects in humans caused by Y chromosome deletions encompassing a novel RNA-binding protein gene. *Nat Genet* 1995 10: 383–393.
60. Lahn BT, Page DC. Functional coherence of the human Y chromosome. *Science* 1997 278: 675–680.
61. Sun C, Skaletsky H, Rozen S *et al.* Deletion of azoospermia factor a (*AZFa*) region of human Y chromosome caused by recombination between HERV15 proviruses. *Hum Mol Genet* 2000 9: 2291–2296.
62. Sun C, Skaletsky H, Birren B *et al.* An azoospermic man with a *de novo* point mutation in

the Y-chromosomal gene *USP9Y*. *Nat Genet* 1999 23: 429–432.

63. Foresta C, Moro E, Rossi A *et al.* Role of *AZFa* candidate genes in male infertility. *J Endocrinol Invest* 2000 23: 646–651.
64. McElreavey K, Krausz C. Male infertility and the Y chromosome. *Am J Hum Genet* 1999 64: 928–933.
65. Chang PL, Sauer MV, Brown S. Y chromosome microdeletion in a father and his four infertile sons. *Hum Reprod* 1999 14: 2689–2694.
66. Saut N, Terriou P, Navarro A, Lévy N, Mitchell MJ. The human Y chromosome genes *BPY2, CDY1* and *DAZ* are not essential for sustained fertility. *Mol Hum Reprod* 2000 6: 789–793.
67. Kent-First MG, Kol S, Muallem A *et al.* The incidence and possible relevance of Y-linked microdeletions in babies born after intracytoplasmic sperm injection and their infertile fathers. *Mol Hum Reprod* 1996 2: 943–950.

Chapter 35

Endocrine disruptors and male reproductive function

Marianne Jørgensen, Niels E. Skakkebaek and Henrik Leffers

INTRODUCTION

Increasing evidence suggests that the reproductive health of men in the Western world has declined during recent decades. Although controversial, the evidence suggests a decline in semen quality,[1,2] an increased frequency of malformations of the external genitalia such as hypospadias and cryptorchidism (undescended testis)[3] and an increased frequency of testicular cancer.[4] The reasons for the reproductive problems are unknown. Although genetic defects such as mutations in the androgen receptor[5] or Y-chromosome microdeletions[6] can lead to malformations and impaired spermatogenesis and an increased risk of testicular cancer, the rapid decline excludes a general genetic cause. Thus, the decline over the past few decades suggests that life-style and/or environmental factors are to blame.

There is a relationship between malformations and reduced semen quality whereby the former always results in the latter. This suggests that there is a common syndrome behind male reproductive problems, and that the development of the reproductive organs during embryogenesis, fetal life, and postnatally, is affected; this has led to the proposal of the 'testicular dysgenesis syndrome'.[7] Testicular dygenesis syndrome is not manifested as a single condition, but rather as a continuum of symptoms ranging from mildly decreased sperm counts to malformations of the genitalia and to complete absence of spermatogenesis, which all result in an increased risk of testicular cancer (Figure 1). Although hypothetical, both experimental and epidemiological evidence suggests that testicular dysgenesis syndrome can result from erroneous programming during gonadal development.

The development of the reproductive organs and the reproductive function in general are controlled by the endocrine hormone system, in a complex interplay between peptide and steroid hormones. Steroid hormones play a major role

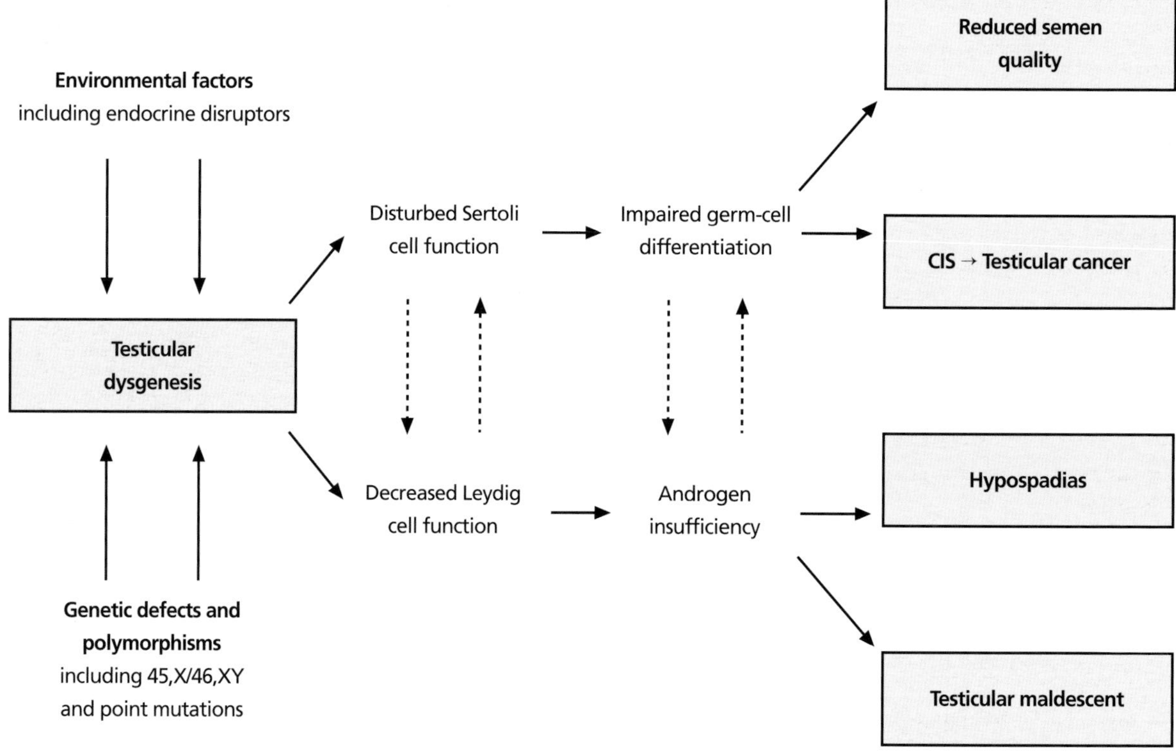

Figure 1 Schematic representation of pathogenic links between the components and clinical manifestations of testicular dysgenesis syndrome. CIS, carcinoma *in situ*. Reproduced from reference 7, with permission

during embryogenesis, where they control the development of internal and external genitalia, and in adults, where male reproductive function is dependent on the presence of testosterone. Thus, factors that affect the endocrine system can, potentially, strongly influence reproductive function. The endocrine disruptor hypothesis speculates that chemical compounds present in our food, drinking water or the environment can disturb the endocrine system.[8,9] The endocrine system can be affected by environmental factors in many ways, for example by chemical compounds that mimic the natural hormones leading to artifactual activation or repression of the corresponding receptors, or by compounds inhibiting the synthesis or metabolism of steroid hormones.[8–10] Because of the different targets, endocrine disruptors are defined broadly as 'exogenous substances that cause adverse health effects in an intact organism or its progeny, secondary to changes in the endocrine system'.[11]

During the past 40–50 years, the number of man-made chemicals that may eventually end up in our environment has increased dramatically. How many of these chemicals interfere with the endocrine system is uncertain, but the number could be considerable. Endocrine-disrupting ability has been demonstrated for a variety of chemicals used in industry, agriculture and private homes, and exposure routes can be multiple, including diet, drinking water, air and skin. Moreover, humans are exposed to a complex mixture of chemicals that have additive or even synergistic effects, and they act together with the endogenous hormones. This is further complicated by very different levels of natural hormones and different sensitivities to the hormones among the human population. Thus, it is difficult to estimate the actual exposure and the effect of the exposure on the individual.

BACKGROUND

Adverse trends in the human population

The adverse trends in humans include a possible decline in semen quality during the past 50 years.[1,2] However, the topic is controversial, and data are based on results from Western countries. Indeed, studies from developing countries have not shown any trends. In addition to the time trends, significant geographical gradients have been reported.[2,12] In this respect, data from Nordic countries are interesting, as several studies show that Finnish men, on average, have higher sperm counts than Danish men.[12] Furthermore, there seem to be birth-cohort effects, as a recent study shows that young men born around 1980 have significantly lower sperm counts than those of older men.[13]

Several studies have shown an increase in the frequency of malformations of the genitalia, such as cryptorchidism and hypospadias (Figure 2). The data suggest that the increase has occurred in multiple countries, including England and the USA, and the observations cannot be explained by improvement of the surveillance system or changes in detection or reporting.[14] Recently, a comparison of the frequencies in Denmark and Finland

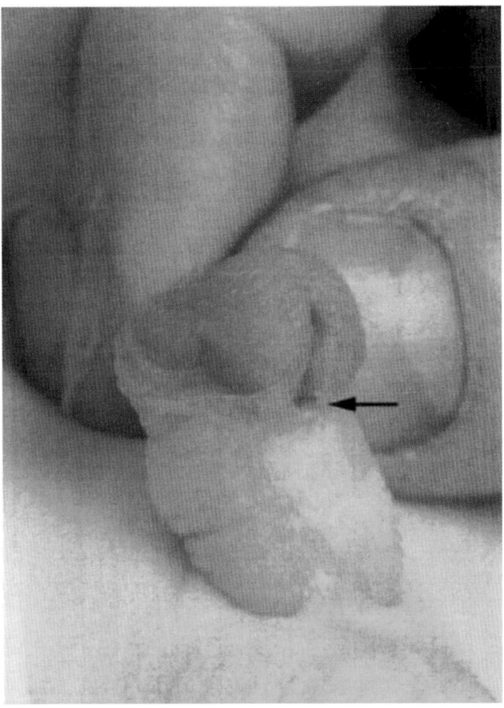

Figure 2 Example of hypospadias. Hypospadias is a congenital anomaly in which the meatus of the urethra (arrow) is situated on the ventral surface of the penis or in the perineum, owing to maldevelopment of the urethra. Hypospadias is classified according to the position of the meatus: glandular, coronal (as illustrated here), shaft, scrotal and perineal

revealed a much higher incidence of both cryptorchidism and hypospadias in Denmark.[3] This is in accordance with data on semen quality, showing that the situation in Denmark is almost the opposite of that in Finland with poor semen quality in Denmark, suggesting a common etiology.

The incidence of testicular germ cell tumors has increased in many countries during the past 40–50 years. The most common type of testicular cancer is derived from precursor cells, and is called carcinoma *in situ* (CIS).[15] Since CIS cells express a number of fetal germ-cell markers, they are thought to originate from fetal cells that did not differentiate properly. Later, when the levels of hormones increase during puberty, the CIS cells develop into cancer.[15] That testicular cancer may be caused by erroneous differentiation of fetal germ cells is supported by an increased risk if there are other malformations, such as cryptorchidism and hypospadias, or reduced semen quality. A comparison between Denmark and Finland, in accordance, revealed a five-fold higher incidence among Danish men than among Finns; almost 1% of young Danish men develop testicular cancer or CIS.[4]

Other adverse trends could potentially also be caused or influenced by exposure to endocrine disruptors. Among these is the significant increase in the incidence of endometrial and breast cancer in women and prostate cancer in men, observed during the past 40–50 years, in which a 'Western lifestyle' is clearly associated with a higher risk.[16] Furthermore, there are indications that other diseases, for example asthma and allergy, may be influenced or caused by exposure to endocrine disruptors.[17]

Are the environmental endocrine disruptors to be blamed for the apparently synchronized deterioration in reproductive function? Evidence obtained from studies of accidental exposure cases, whereby a population is exposed to many times higher concentrations of a potential endocrine disruptor compound than the general population, may shed light on how exogenous hormones affect humans. The most sinister example of 'accidental' exposure occurred from 1947 to 1971, when millions of pregnant women were treated with the potent estrogen diethylstilbestrol (DES) to prevent spontaneous abortion. Unfortunately, DES did not protect against abortion, but instead produced a number of serious side-effects, both in the mother and in the offspring, because of the potent estrogenicity of DES.[18] There was a clear dose–response relationship in the frequency and severity of the side-effects, and they also differed depending on the gestational age of the fetus at the time of exposure. The side-effects in male offspring included increased frequency of cryptorchidism and other malformations of the genitalia, and possibly reduced semen quality.[18] There are other examples of accidental exposure to hormonally active compounds, and most results seem to support the endocrine disruptor hypothesis.[9] Thus, some of the problems observed today in the general population are related to effects seen after accidental exposure to synthetic chemicals with potent hormone-like properties.

Evidence of endocrine disruption in wildlife

Substantial evidence of adverse developmental effects caused by endocrine disruptors comes from observations made in wildlife. For example, effluent from sewage-treatment plants entering British rivers contains estrogenic chemicals, including ethinylestradiol from birth-control pills, that may be responsible for the high incidence of intersex in fish, which has been observed in many rivers in the UK.[19] Feminization/demasculinization, hypospadias and altered sex hormone levels among alligators in Lake Apopka (Florida, USA) is another well-known example.[20] Lake Apopka is one of Florida's most polluted wetlands, which, after an accident, was extensively contaminated with dichlorodiphenyltrichloroethane (DDT), a weakly estrogenic compound, which is metabolized to an antiandrogen, dichlorodiphenylethylene (DDE). The effects on the alligators were consistent with exposure to an antiandrogen, since similar effects can be produced experimentally by exposure to flutamide, a potent antiandrogen.

Although most of the observed adverse effects in wildlife have been in heavily polluted areas, widespread occurrence of more subtle effects, such as imposex in marine snails (probably caused by tributyltin (Table 1) acting as an aromatase inhibitor in

Table 1 Chemical structures of potent estrogens and selected endocrine disruptors

Compound	*Chemical structure*	*Activity*	*Primary uses/sources*
17β-Estradiol	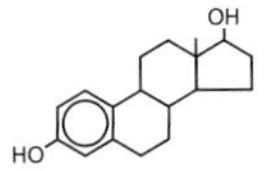	estrogenic	endogenous estrogen
DES	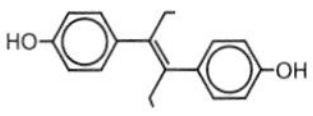	estrogenic	synthetic estrogen, used to prevent miscarriages in the USA and other countries from the 1940s until banned in 1971
Zeranol®	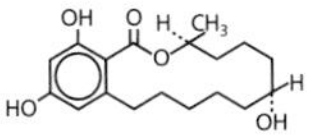	estrogenic	synthetic estrogen, anabolic growth promotion in cattle
Genistein	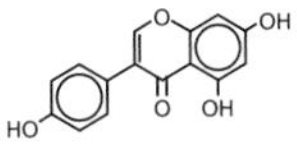	estrogenic	phytoestrogen, naturally occurring, found in many food products
Biochanin A	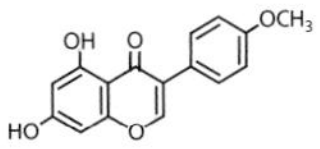	estrogenic, aromatase inhibitor	metabolite of genistein
Coumestrol	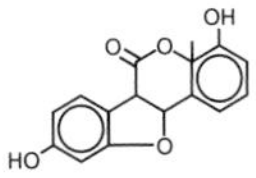	estrogenic, aromatase inhibitor	phytoestrogen, naturally occurring, found in many food products
Bisphenol A	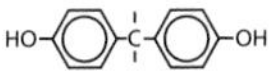	estrogenic, antiandrogenic	component of polycarbonates and of resins; found in dental sealants, metal food cans
4-*n*-Nonylphenol	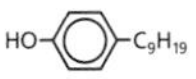	estrogenic	surfactant, present in detergents, paints, herbicides and pesticides
Dibutylphthalate	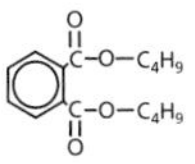	estrogenic, antiandrogenic	plasticizing agent, present in plastic, rubber, lacquers and glues
Tributyltin (TBT)	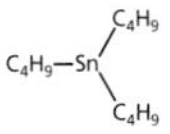	aromatase inhibitor in vertebrates	antifouling biocide, used in marine paints
Dieldrin	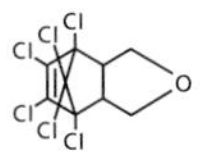	estrogenic	metabolite of the insecticide aldrin; restricted/ banned use
Endosulfan	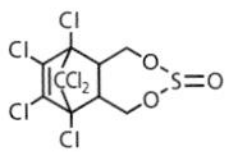	estrogenic	insecticide; chemical structure resembles aldrin/dieldrin
o,p'-DDT	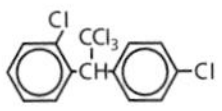	antiandrogenic	insecticide; banned in Western countries, still used in some developing countries
Vinclozolin	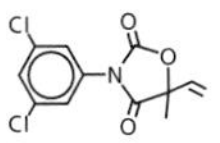	antiandrogenic	fungicide

DES, diethylstilbestrol; DDT, dichlorodiphenyltrichloroethane

invertebrates), thinning of egg shells in various bird species, production of female yolk proteins in livers of male fish and reproductive problems among polar bears, suggest that contamination with persistent endocrine disruptors may be a global problem.[21] This raises the question of how high the exposure must be to affect reproductive development. This was addressed by Sheehan and colleagues,[22] who studied sexual differentiation in reptile species and questioned the existence of a threshold below which no adverse outcomes are seen. In the red-eared slider turtle, female development requires endogenous estrogen produced at elevated temperatures. When turtle eggs were painted with 17β-estradiol, as little as 3.3 ng per egg resulted in sex reversal, and apparently there was no threshold dose.[22] Thus, any exposure may result in adverse effects, low exposures lead to few and small effects, while large exposures lead to widespread and severe effects; there may not exist a threshold below which exposure leads to no effects.

APPLIED TECHNOLOGY

Potential endocrine disruptors

A brief description follows of the main technologies targeted towards identifying potential endocrine disruptors. The current knowledge of the effects of exposure to endocrine disruptors is partly derived from animal studies (see below), and observations from some of these experimental animal models are described in the next section on 'current concepts'.

The number of chemicals that may be potential endocrine disruptors is unknown, and only a very small subset of the thousands of potential candidates has been assayed for effects on the endocrine system. Nevertheless, endocrine disrupting properties have been detected for many synthetic compounds. Examples of such industrial chemicals include phthalates, polychlorinated biphenyls (PCBs), organochlorine pesticides, alkylphenols, antioxidants and widely used preservatives such as parabenes (Table 1).[23] Some of the compounds are persistent and tend to bioaccumulate, with the result that they are present in the environment and in body fat many years after their use has ended. Recently, it has been suggested that potent synthetic androgens, estrogens and gestagens used for growth promotion in the livestock industry, particularly in cattle, should also be regarded as potential endocrine disruptors (Table 1). The use of anabolic growth-promoters in the cattle industry is banned in many countries, including all member states of the European Union, but they are used extensively and legally in the USA.

Food products can include naturally occurring estrogens, both in the form of natural hormones, which are found in large amounts in some animal and fish products, and from plants, the so-called phytoestrogens. Phytoestrogens are relatively weak estrogens compared with natural hormones, although many are more potent than the synthetic chemicals mentioned above. When ingested in large quantities, however, phytoestrogens can have profound biological effects; for example, ingestion of red clover, which contains high quantities of phytoestrogens, can cause infertility in sheep,[24] and phytoestrogens have estrogen-like effects in humans.[25] Phytoestrogens are not only estrogenic: they have a range of other properties, including tyrosine-kinase and aromatase inhibition, that make it difficult to estimate their effects. In contrast to other estrogens, phytoestrogens seem to protect against breast cancer,[25] emphasizing the complexity of the endocrine disruptor field.

Whether endocrine disruptors have an impact on the biological processes in the body depends presumably on the exposure level in relation to the endogenous production of hormones. Therefore, the segments of the population with the lowest daily production levels are probably the most sensitive to exogenous hormones. In the case of 17β-estradiol and progesterone, prepubertal boys synthesize the least; in the case of testosterone, the lowest levels are found in prepubertal girls.[26] Other specific sensitive periods may occur during fetal development and in women after the menopause. The consequences of exposure may vary according to the time of exposure. For example, *in utero* exposure to estrogens may affect testis development, and this may influence semen quality later in life, whereas estrogen exposure in adulthood may cause effects that are only temporary, such as gynecomastia in the adult man.

Identifying potential endocrine disruptors

The chemical structures of estrogenic compounds vary substantially, making it difficult to predict their activity from knowledge of their structures. Hence, there is a strong need for reliable short-term assays that can rapidly detect chemicals with endocrine-disrupting properties. The number of chemicals to be tested is considerable: the US Environmental Protection Agency (EPA) is considering more than 87 000 substances as potential candidates for testing. More information is available from the EPA web site (http://www.epa.gov). At present, there is no large-scale assay for antiandrogenicity; the current assay, the Hershberger assay,[27] is not suited to the testing of a large number of compounds. The assays described below test for estrogenic properties.

Tests for estrogenicity

The 'gold standard' *in vivo* estrogenicity assay measures uterine weight increase in ovariectomized or juvenile female rats after exposure to estrogenic compounds.[28] Another commonly used *in vivo* assay measures the level of the yolk protein vitellogenin in male fish.[29] However, *in vivo* tests for endocrine disruption should encompass at least two generations, and assay for effects on fertility and mating, embryonic and fetal development, sensitive neonatal growth and development, and transformation from the juvenile life-stage to sexual maturity.

The most simple *in vitro* assay for estrogenicity measures the binding of a compound to the estrogen receptors by competition with radiolabelled 17β-estradiol. However, estrogen receptor-binding assays do not define the ligand as an agonist or antagonist, and compounds that are metabolized to estrogenic compounds are not detected in cell-free binding assays. A widely used method is to assay the dose–response relationship between exposure to estrogens and proliferation of estrogen-dependent human breast cancer MCF7 cells.[30] Other *in vitro* screening assays include those for transcriptional activation, whereby the activity of a reporter gene is directly related to the transcriptional activation of estrogen receptors by the test compound.[31] All the end-points currently used to assay estrogenicity are derived from changes in gene expression. Hence, we have used the expression levels of endogenous estrogen-regulated genes in MCF7 cells as an indicator for estrogenicity. The results are in agreement with those of other assays, but the endogenous gene expression assay appears to be more sensitive, and may provide additional information in relation to the test compounds[32] (see below).

CURRENT CONCEPTS

Experimental models

Rodents are used as models for mammalian reproduction, because in many respects they resemble humans. Both mice and rats are used for reproductive studies. There may be an advantage in using rats, since they exhibit a short puberty-like period, whereas germ-cell differentiation leading to spermatogenesis is initiated a few days after birth in mice. The results discussed below include results from both species.

There are thousands of articles in the literature describing decreased reproductive function after exposure to chemical compounds, including many that have addressed the outcome of exposure to suspected or known endocrine disruptors. However, many studies have used doses that are very high, far higher than can be ingested from food or even via accidental exposure. Moreover, many deal with the effects of exposure to estrogens such as 17β-estradiol and DES, with potencies many orders of magnitude higher than those of potential endocrine disruptors (see below), which can make it difficult to extrapolate the results to other, much less potent estrogenic compounds. Nevertheless, there seem to be two classes of endocrine-modulating chemicals that can severely affect the development of the male reproductive apparatus: estrogens and antiandrogens.

Estrogens

Data on exposure to estrogens are mostly derived from studies of 17β-estradiol and DES, although there are a few reports about the effects of exposure to less potent estrogens. Results from studies of *in utero*, perinatal and postnatal exposure to DES showed that this compound dose-dependently altered the development of the reproductive tract in male mice. The affected tissues included hypoplastic testes, retainment of Müllerian tract-derived tissue and malformations of external genitalia, including hypospadias and microphallus.[33] In addition, there was evidence suggesting neoplastic transformations such as epididymal cysts, and tumors in the testes and prostate. The timing of the exposure was important. The worst effects were seen if the exposure occurred early during gestation; however, relatively high doses were required to induce the effects.[33] These results are in agreement with observations in the sons of DES-treated mothers. However, the results from a large 'benchmark' study using 17β-estradiol seem to contradict a general disrupting effect of potent estrogens.[34] In fact, treatment with 17β-estradiol in low and moderate doses did not significantly affect the morphology or function of the male reproductive system. The reason for the divergent results may be the different pharmacologies of the two compounds, and higher local concentrations of DES, caused by the lack of binding to carrier proteins. Results from estrogen receptor α and β knock-out mice (ERKO) suggest that estrogens do not play a crucial role during testis development and function.[35] Although the estrogen receptor α knock-out mice (αERKO) were infertile, this was caused not by primary malformations of the testes, but by failure of sperm maturation in the efferent ductules, where water resorption is dependent on the presence of estrogen receptor α.[36] High doses of potent estrogens, far higher than those attainable in a real-life exposure, can severely, but reversibly, impair spermatogenesis, probably through estrogen action at the hypothalamic–pituitary level. Thus, it is not clear whether, or how, low to moderate doses of potent estrogens affect male reproduction.

The action of estrogen-like compounds may diverge from that of actual estrogens in a number of ways. In particular, approximately 98–99% of the endogenous 17β-estradiol in plasma is not biologically active because it is bound to carrier proteins (mainly sex hormone-binding globulin (SHBG) and albumin).[37] Much less is known about the bioavailability of xenoestrogens, but studies indicate that these compounds do not bind to carrier proteins. Therefore, even though plasma concentrations of the xenoestrogens may be low, their biological activity may be higher than their concentrations suggest.[37] There are some conflicting results. Some studies showed altered testis and prostate development and reduced fertility after exposure to low doses of nonylphenol and bisphenol A.[38,39] However, the most comprehensive studies, with a large number of animals, failed to show any developmental alterations of the male reproductive system or reduced fertility, even when relatively high doses were used.[40,41]

The reason for the conflicting results may partly be genetic differences, since different mouse and rat strains have been used in the studies. To address possible genetic differences, Spearow and colleagues compared the effects of postnatal exposure to 17β-estradiol on sperm production and testis morphology in different mouse strains.[42] They showed that sensitivity to 17β-estradiol varied among mouse strains by more than 16-fold. Thus, genetic factors are very important for the response of mice to estrogens, and this is most probably also the case for other animals, including humans. This also indicates that it may not be possible to extrapolate from animal models to humans, because the sensitivities may be very different.

Antiandrogens

Development of the male reproductive system is dependent on androgens; in their absence, genetic males develop a female phenotype. This is exemplified by the effect of mutations in the androgen receptor, where inactivating mutations lead to a female phenotype of the external genitalia in genetic males.[5] Thus, any compound that impairs the function of the androgen receptor or the synthesis of androgens can severely affect male development. There are no examples of synthetic chemicals that act as androgen receptor agonists, whereas antiandrogenic properties have been detected in a range of compounds, including widely used chemicals such as phthalates, vinclozolin, methoxychlor (which is also an estrogen), linuron and persistent chemicals such as DDT and some of its metabolites (Table 1).[43] The most sensitive endpoints of *in utero* exposure to antiandrogens are retained nipples (Figure 3) and 'female-like' anogenital distance. At higher concentrations, hypospadias, cryptorchidism, cleft phallus, reduced sperm production and a general reduction in the size of androgen-sensitive organs, such as the prostate, are also observed.[43,44] The effects are dose-dependent, and the timing of exposure is very important. The worst effects are seen when exposure occurs at gestational days 16 and 17 (in rats), but antiandrogens can also affect adult animals, although much less severely. The effects are very similar to what is observed after exposure to pharmaceutical antiandrogens (flutamide).[45] It seems that the effects of exposure to antiandrogenic compounds in experimental animals are similar to the adverse trends observed in the human population. Much more research into the mechanisms and results of exposure to antiandrogens is required, and there is an urgent need to identify compounds with antiandrogenic properties.

Other compounds

Other chemicals affect the reproductive system at low doses; in particular, many organophosphorous compounds can have a large impact on male reproductive function.[46] Other compounds have also been shown to have an impact upon testis function; for example, exposure to the nematocide dibromochloropropane (DBCP) can completely abolish spermatogenesis via a presently unknown mechanism.[47] Thus, compounds that may not be endocrine disruptors as such can severely affect spermatogenesis, possibly through other mechanisms.

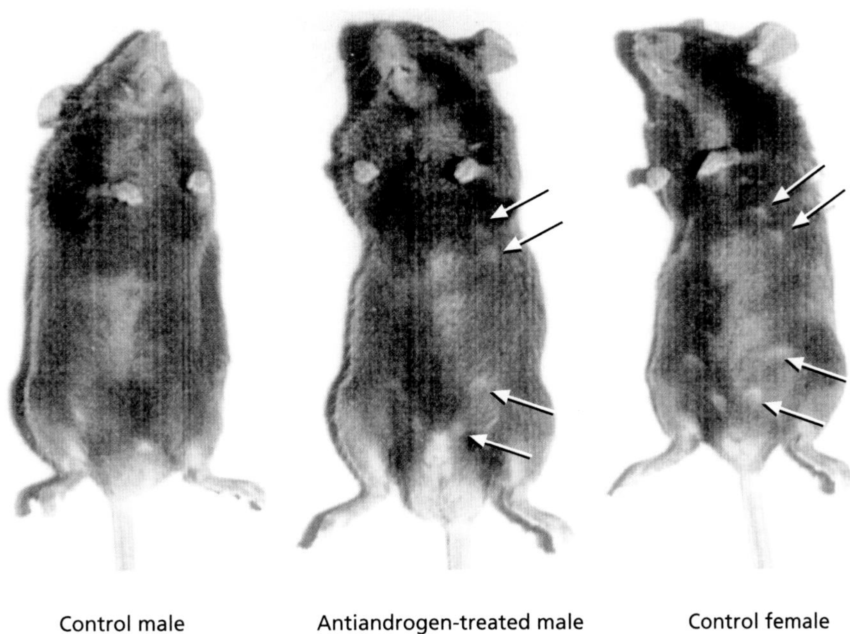

Figure 3 Nipples are formed in male mice, but they subsequently regress because androgens promote apoptosis in the cells constituting the nipples. When treated with antiandrogens, nipple retention is one of the most sensitive effects. In this case, it was caused by *in utero* exposure to the antiandrogen flutamide, but similar effects occur after exposure to endocrine disruptors with antiandrogenic properties

POTENTIAL CLINICAL RELEVANCE

Relative potencies and exposure levels

Chemicals with estrogenic properties vary greatly, which is reflected in their estrogenic potencies. Since their effects must be related to their potencies, it is important to know the relative potencies of the compounds in question. Many different estrogenicity assays have been applied (see above), but, in general, results from the various assays are reasonably similar.[48] We have used expression levels of the endogenous estrogen-regulated *pS2* and transforming growth factor-β_3 (TGF-β_3) genes in MCF7 cells as indicators of the estrogenic potencies of selected compounds, including potent hormones and synthetic chemicals (Figure 4). The most potent compounds by far were the natural 17β-estradiol, and the two synthetic hormones DES and Zeranol; next most potent was genistein, a phytoestrogen found in large amounts in soy products. The least potent compounds were the synthetic industrial chemicals, which included most of the synthetic chemicals tested.

The differences were large: 17β-estradiol, DES and Zeranol were at least five orders of magnitude more potent than nonylphenol and six orders more potent than the most estrogenic phthalate (dibutylphalate; Figure 4). Today, Zeranol is the only synthetic estrogen allowed for use in the cattle industry in the USA, and residual amounts have been found in finished consumer products.[32] The synthetic hormones used in meat production also include potent androgens (trenbolone acetate) and gestagens (melengestrol acetate), which perhaps should be considered as the most potent endocrine disruptors. Special emphasis must be given to elucidation of the effects that exposure may have on a human population, especially since the country where they are used in the largest quantities (USA) is also the country with the highest incidence of hormone-dependent cancers, such as breast and prostate cancer.

Except for phthalates, where it was recently shown that the urine of young women contained phthalate metabolites at levels suggesting exposures close to those resulting in adverse effects in laboratory animals,[43,44,49] the exposure levels of the synthetic chemicals in the general population are relatively low, in the microgram range. Together with their low potencies, this suggests that their putative adverse effects may not be directly related only to their estrogenic potencies, but also to other properties inherent in the compounds, especially antiandrogenicity.

SUMMARY

The evidence for a decline in male reproductive health in the Western world is mounting. There seems to be a correlation between the increase in a range of symptoms of reduced function of the reproductive organs and a probable decline in semen quality and increase in testicular cancer, suggesting that the observed symptoms may be due to maldevelopment of the reproductive tract, as hypothesized in the testicular dysgenesis syndrome. This occurs at a time when the exposure to chemicals that can affect the endocrine system is steadily increasing. Thus, there is sufficient evidence today to hypothesize that there could be a correlation between the observed reproductive problems and exposure to compounds that disturb the endocrine system. Much more research is needed, however, before we can draw any firm conclusion about the effects of the subclinical exposure of a whole population. Although the adverse trends and the exposure are currently not at levels to indicate that we should worry about the future of mankind, they are, nevertheless, at levels that should cause some concern. There is an urgent need for more research into the mechanisms whereby low-dose exposures may affect the endocrine system, a need for more epidemiological studies to determine the extent of the reproductive problems and, especially, a need to identify, phase out and replace compounds with potential endocrine disrupting properties.

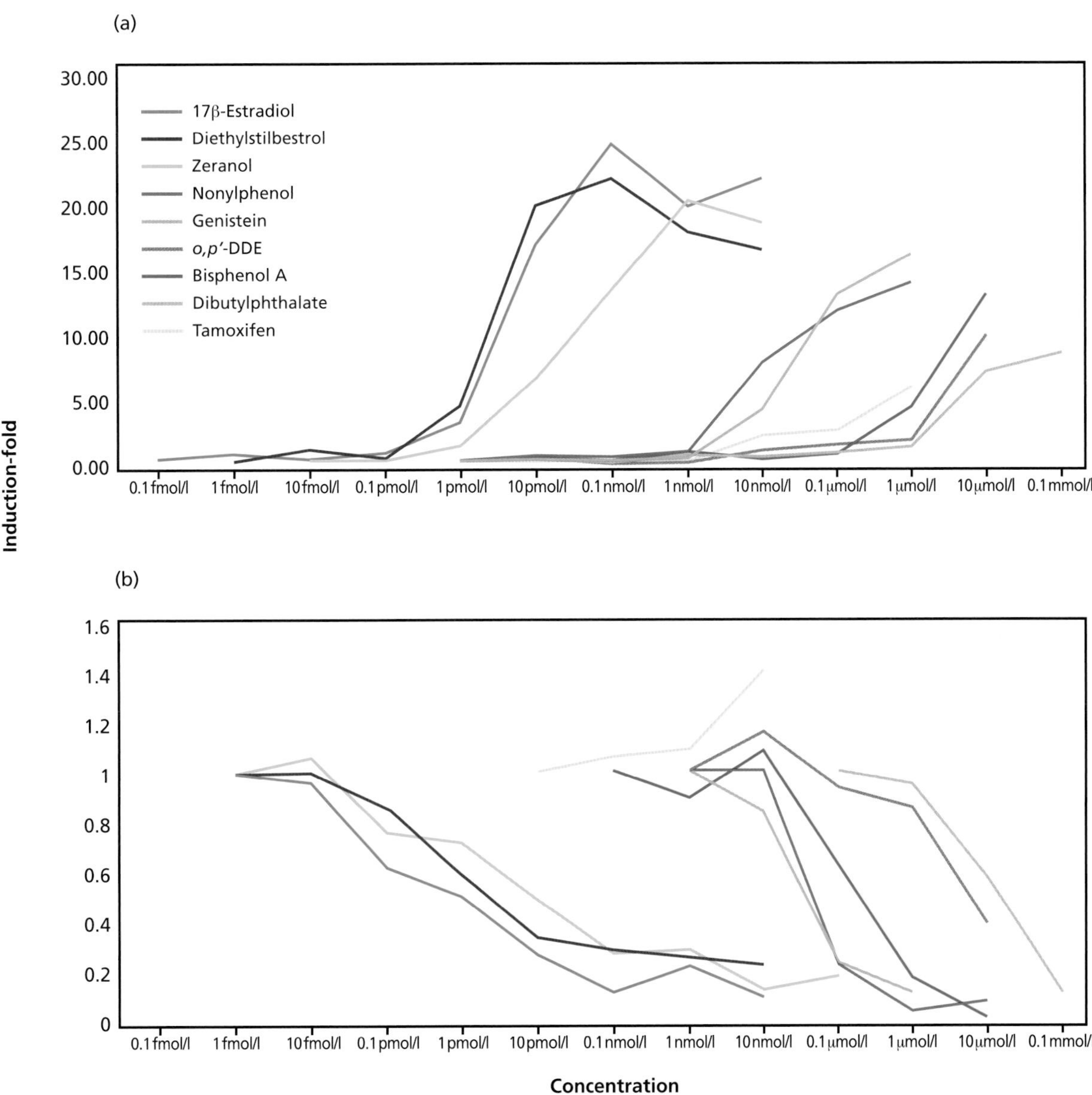

Figure 4 Relative potencies of selected estrogens. Human breast cancer MCF7 cells were incubated with increasing concentrations of the indicated compounds. After 24 h, the cells were harvested and total RNA prepared. The expression levels of two mRNAs (pS2) and transforming growth factor-β3 (TGF-β3) were subsequently determined by competitive polymerase chain reaction (PCR).[32] TGF-β_3 (b) is down-regulated, and pS2 (a) up-regulated by estrogens. Note that the antiestrogen tamoxifen, used in treatment of breast cancer, acted as an antiestrogen on TGF-β3, but as a weak estrogen on pS2. DDE, dichlorodiphenylethylene

REFERENCES

1. Swan SH, Elkin EP, Fenster L. The question of declining sperm density revisited: an analysis of 101 studies published 1934–1996. *Environ Health Perspect* 2000 108: 961–966.
2. Fisch H, Goluboff ET, Olson JH *et al.* Semen analyses in 1283 men from the United States over a 25-year period: no decline in quality. *Fertil Steril* 1996 65: 1009–1014.
3. Toppari J, Kaleva M, Virtanen HE. Trends in the incidence of cryptorchidism and hypospadias,

and methodological limitations of registry-based data. *Hum Reprod Update* 2001 7: 282–286.

4. Møller H. Trends in the incidence of testicular cancer and prostate cancer in Denmark. *Hum Reprod* 2001 16: 1007–1011.
5. Brinkmann AO, Jenster G, Ris-Stalpers C *et al.* Androgen receptor mutations. *J Steroid Biochem Mol Biol* 1995 53: 443–448.
6. McElreavey K, Krausz C, Bishop CE. The human Y chromosome and male infertility. *Results Probl Cell Differ* 2000 28: 211–232.
7. Skakkebaek NE, Rajpert-De Meyts E, Main KM. Testicular dysgenesis syndrome: an increasingly common developmental disorder with environmental aspects. *Hum Reprod* 2001 16: 972–978.
8. Sharpe RM, Skakkebaek NE. Are estrogens involved in falling sperm counts and disorders of the male reproductive tract? *Lancet* 1993 341: 1392–1395.
9. McLachlan JA. Environmental signaling: what embryos and evolution teach us about endocrine disrupting chemicals. *Endocr Rev* 2001 22: 319–341.
10. Walsh LP, McCormick C, Martin C, Stocco DM. Round up inhibits steroidogenesis by disrupting steroidogenic acute regulatory (StAR) protein expression. *Environ Health Perspect* 2000 108: 769–776.
11. Bergman A, Brandt I, Brouwer B *et al. European Workshop on the Impact of Endocrine Disrupters on Human Health and Wildlife. Reports of Proceedings.* Weybridge, UK: Environment and Climate Research Programme of DG XII of the European Commission, 1996.
12. Jorgensen N, Andersen AG, Eustache F *et al.* Regional differences in semen quality in Europe. *Hum Reprod* 2001 16: 1012–1019.
13. Andersen AG, Jensen TK, Carlsen E *et al.* High frequency of sub-optimal semen quality in an unselected population of young men. *Hum Reprod* 2000 15: 366–372.
14. Ansell PE, Bennett V, Bull D *et al.* Cryptorchidism: a prospective study of 7500 consecutive male births, 1984–8. *Arch Dis Child* 1992 67: 892–899.
15. Rajpert-De Meyts E, Skakkaek NE. Origin of germ cell cancer: biological and clinical significance of testicular carcinoma *in situ*. *Endocrinologist* 2000 10: 335–340.
16. Sasco AJ. Epidemiology of breast cancer: an environmental disease? *APMIS* 2001 109: 321–332.
17. Wjst M, Dold S. Is asthma an endocrine disease? *Pediatr Allergy Immunol* 1997 8: 200–204.
18. Swan SH. Intrauterine exposure to diethylstilbestrol: long-term effects in humans. *APMIS* 2000 108: 793–804.
19. Sumpter JP. Feminized responses in fish to environmental estrogens. *Toxicol Lett* 1995 82–83: 737–742.
20. Guillette LJ Jr, Gross TS, Masson GR *et al.* Developmental abnormalities of the gonad and abnormal sex hormone concentrations in juvenile alligators from contaminated and control lakes in Florida. *Environ Health Perspect* 1994 102: 680–688.
21. Vos JG, Dybing E, Greim HA *et al.* Health effects of endocrine-disrupting chemicals on wildlife, with special reference to the European situation. *Crit Rev Toxicol* 2000 30: 71–133.
22. Sheehan DM, Willingham E, Gaylor D, Bergeron JM, Crews D. No threshold dose for estradiol-induced sex reversal of turtle embryos: how little is too much? *Environ Health Perspect* 1999 107: 155–159.
23. Toppari J, Larsen JC, Christiansen P *et al.* Male reproductive health and environmental chemicals with estrogenic effects. *Environ Health Perspect* 1996 104 (Suppl 4): 741–803.
24. Bradbury RB, White DE. The chemistry of subterranean clover. Part I. Isolation of formononetin and genistein. *J Chem Soc* 1951 3447.
25. Tham DM, Gardner CD, Haskell WL. Potential health benefits of dietary phytoestrogens: a review of the clinical, epidemiological, and mechanistic evidence. *J Clin Endocrinol Metab* 1998 83: 2223–2235.
26. Andersson AM, Skakkebaek NE. Exposure to exogenous estrogens in food: possible impact on human development and health. *Eur J Endocrinol* 1999 140: 477–485.
27. Dorfman R. Androgens and anabolic agents. In Dorfman RI, ed. *Methods in Hormone Research.* New York: Academic Press, 1962: 275–313.
28. Lan NC, Katzenellenbogen BS. Temporal relationships between hormone receptor binding and biological responses in the uterus: studies with short- and long-acting derivatives of estriol. *Endocrinology* 1976 98: 220–227.

29. Sumpter JP, Jobling S. Vitellogenesis as a biomarker for estrogenic contamination of the aquatic environment. *Environ Health Perspect* 1995 103 (Suppl 7): 173–178.
30. Soto AM, Sonnenschein C, Chung KL *et al.* The E-SCREEN assay as a tool to identify estrogens: an update on estrogenic environmental pollutants. *Environ Health Perspect* 1995 103 (Suppl 7): 113–122.
31. Klein KO, Baron J, Colli MJ, McDonnell DP, Cutler GB Jr. Estrogen levels in childhood determined by an ultrasensitive recombinant cell bioassay. *J Clin Invest* 1994 94: 2475–2480.
32. Leffers H, Naesby M, Vendelbo B, Skakkebaek NE, Jorgensen M. Oestrogenic potencies of Zeranol, oestradiol, diethylstilboestrol, bisphenol-A and genistein: implications for exposure assessment of potential endocrine disrupters. *Hum Reprod* 2001 16: 1037–1045.
33. Newbold RR. Effects of developmental exposure to diethylstilbestrol (DES) in rodents: clues for other environmental estrogens. *APMIS* 2001 109 (Suppl 103): 261–271.
34. Cook JC, Johnson L, O'Connor JC *et al.* Effects of dietary 17β-estradiol exposure on serum hormone concentrations and testicular parameters in male Crl : CD BR rats. *Toxicol Sci* 1998 44: 155–168.
35. Couse JS, Korach KS. Estrogen receptor null mice: what have we learned and where will they lead us? *Endocr Rev* 1999 20: 358–417.
36. Mahato D, Goulding EH, Korach KS, Eddy EM. Spermatogenic cells do not require estrogen receptor-α for development or function. *Endocrinology* 2000 141: 1273–1276.
37. Nagel SC, vom Saal FS, Thayer KA *et al.* Relative binding affinity–serum modified access (RBA–SMA) assay predicts the relative *in vivo* bioactivity of the xenoestrogens bisphenol A and octylphenol. *Environ Health Perspect* 1997 105: 70–76.
38. Saunders PT, Majdic G, Parte P *et al.* Fetal and perinatal influence of xenoestrogens on testis gene expression. *Adv Exp Med Biol* 1997 424: 99–110.
39. Nagao T, Saito Y, Usumi K *et al.* Disruption of the reproductive system and reproductive performance by administration of nonylphenol to newborn rats. *Hum Exp Toxicol* 2000 19: 284–296.
40. Cagen SZ, Waechter JM, Dimond SS *et al.* Normal reproductive organ development in Wistar rats exposed to bisphenol A in the drinking water. *Regul Toxicol Pharmacol* 1999 30: 130–139.
41. Chapin RE, Delaney J, Wang Y *et al.* The effects of 4-nonylphenol in rats: a multigeneration reproduction study. *Toxicol Sci* 1999 52: 80–91.
42. Spearow JL, Doemeny P, Sera R, Leffler R, Barkley M. Genetic variation in susceptibility to endocrine disruption by estrogen in mice. *Science* 1999 285: 1259–1261.
43. Gray LE, Ostby J, Furr J *et al.* Effects of environmental antiandrogens on reproductive development in experimental animals. *Hum Reprod Update* 2001 7: 248–264.
44. Foster PM, Mylchreest E, Gaido KW, Sar M. Effects of phthalate esters on the developing reproductive tract of male rats. *Hum Reprod Update* 2001 7: 231–235.
45. McIntyre BS, Barlow NJ, Foster PM. Androgen-mediated development in male rat offspring exposed to flutamide *in utero*: permanence and correlation of early postnatal changes in anogenital distance and nipple retention with malformations in androgen-dependent tissues. *Toxicol Sci* 2001 62: 236–249.
46. Bustos-Obregón E. Adverse effects of exposure to agropesticides on male reproduction. *APMIS* 2001 109 (Suppl 103): 233–242.
47. Goldsmith JR. Dibromochloropropane: epidemiological findings and current questions. *Ann NY Acad Sci* 1997 837: 300–306.
48. Andersen HR, Andersson AM, Arnold SF *et al.* Comparison of short-term estrogenicity tests for identification of hormone-disrupting chemicals. *Environ Health Perspect* 1999 107 (Suppl 1): 89–108.
49. Blount BC, Silva MJ, Caudill SP *et al.* Levels of seven urinary phthalate metabolites in a human reference population. *Environ Health Perspect* 2000 108: 979–982.

Chapter 36

Ethics of assisted reproduction

Guido de Wert

INTRODUCTION

Assisted reproductive technologies can help prospective parents to reach their parental goals. Techniques that can circumvent subfertility or infertility are constantly being developed or improved. Furthermore, developments in (prenatal) diagnostic techniques make it increasingly possible to avoid the birth of a severely handicapped child. Although it is widely acknowledged that assisted reproductive technologies can contribute to human well-being, it is, at the same time, realized that the 'reproductive revolution' generates a wave of moral questions concerning the responsibilities of prospective parents, the scientists involved and society as a whole. While most, if not all, reproductive technologies involve some specific ethical issues, some more general ethical questions and issues can be identified. The following paragraphs concentrate on three of these broader issues, namely the ethics of embryo research, informed consent and the responsibility of doctors involved in medically assisted reproduction regarding the child-to-be. Furthermore, ethical aspects of some of the most discussed reproductive technologies, namely preimplantation genetic diagnosis and screening, and human cloning, are addressed. The final section presents some conclusions and recommendations.

RESEARCH WITH PREIMPLANTATION EMBRYOS

The introduction of new assisted reproductive technologies is closely related to preclinical research with human preimplantation embryos, aiming at investigating the safety, reliability and effectiveness of the respective techniques. This research is controversial.[1] Objections come from two different perspectives, one 'fetalist', the other feminist.[2] Fetalists, who focus on the moral status of preimplantation embryos, object that embryo research unjustifiably treats preimplantation embryos solely as a means. The fundamental controversy concerns whether preimplantation embryos have the same moral status as children or adults, who are to be protected from 'destructive' research. Most ethical theories hold that preimplantation embryos need not be respected as persons.[3] Most of the ethics committees who have studied the issue conclude that the preimplantation embryo has a relatively 'low' moral status. One of the arguments in favor of this position concerns the absence of developmental individuation: until the embryonic cells are differentiated and organized to become the primitive streak, there is no individual in any sense of the word, i.e. biological, legal or moral. Accordingly, the dominant view is that research with preimplantation embryos can be justified, within some ethical constraints, of course. Procedural conditions are the informed consent of the 'parents' (the providers of the gametes) (see below) and the approval of a (national) ethics committee. Material conditions relate, among other things, to the principle of proportionality – there is a strong consensus that the aim of such research should be to benefit human health – and to the time limit to be respected – most committees and regulations set a time limit of 14 days after fertilization.

A major point of controversy remains the origin or source of the preimplantation embryos to be used in research. The 'Convention on human rights and biomedicine' of the Council of Europe prohibits the creation of embryos solely for research purposes.[4] One may question, however, whether there is a fundamental moral difference between, on the one hand, using spare embryos ('left-overs') in research and, on the other hand, the generation of embryos for research purposes. Not only is the embryo used instrumentally in both cases, but also the moral status of the embryos is identical. So, if one accepts the instrumental use of spare embryos, one can hardly object to the creation of embryos for research purposes on the grounds that it involves instrumental use. Some commentators stress that the intention at the moment of fertilization is fundamentally different. This difference, however, although real, is only relative. It is a misunderstanding to think that, in the context of regular *in vitro* fertilization (IVF) treatment, every embryo is created as a 'goal in itself'. The goal of IVF is the solution of involuntary childlessness; the loss of some spare embryos is calculated beforehand.

From a feminist perspective, the issue becomes more complex. Some feminist critics question the acceptability of creating embryos for research out of concern for the autonomy and the interests of the women donating eggs for research. After all, women have to undergo hormone stimulation 'therapy' and invasive medical procedures, which carry some medical risks. Furthermore, so the argument runs, there is a serious risk of exploitation, in view of the temptation to withhold detailed information on risks, for fear of losing 'willing' candidate donors.[5] These feminist concerns are, of course, relevant, but they can hardly justify an absolute ban on creating embryos for research purposes. First, the interests of donors may be sufficiently protected by imposing the material condition that medical risks should be minimized, by limiting the number of hormone treatments as well as the dosage of hormones given to candidate donors. Of course, a valid informed consent presumes adequate information about potential residual risks. Second, the feminist objection is not an a priori objection, but a contingent one. After all, the creation of embryos for research purposes does not necessarily involve hormonal treatments and/or invasive interventions in donors to access the eggs. In clinical IVF/intracytoplasmic sperm injection (ICSI), it may happen that some oocytes have not yet reached the appropriate stage of nuclear maturity. If they mature further after 24–26 h of *in vitro* culture, these oocytes cannot be used clinically, but they may be a good source of research embryos. Furthermore, in the future, the *in vitro* maturation of oocytes might render hormonal treatments (for this purpose) obsolete; it could even imply access to altogether new 'sources' of research eggs, including aborted female fetuses and cadavers.[6]

Both from a feminist and from a fetalist perspective there is no categorical moral objection against the creation of embryos for reseach purposes. A complete prohibition would even be undesirable, as this would hamper an adequate preclinical risk assessment of some new, potentially valuable, techniques, such as the cryopreservation and *in vitro* maturation of oocytes.

Difference of opinion concerns the way in which the principle of proportionality should be made operational. In a number of countries, embryo research is restricted to that related to human reproduction. Internationally, this limitation is growingly, and rightly, disputed. The isolation of human embryonic stem cells for research into cell (replacement) therapy in this context operates as a catalyst (see below).[7]

INFORMED CONSENT

Information provision and consent procedures that enable applicants to make informed choices are crucial components of assisted reproduction services. It is important to discern the information component and the consent component of informed consent.[8] General information to be provided concerns, among others things, the nature of the particular reproductive technology at hand, the likelihood of success, the kind of decisions that may have to be made at a later stage in the treatment (for example, regarding surplus embryos) and the risks of the procedure(s), both for the woman and for the prospective child.

The extent, in terms of broadness and detail, of the information to be disclosed to patients/applicants depends on the standard of disclosure used by the professional involved. The professional practice standard holds that the criteria for adequate disclosure are determined by the customary practices of a professional community. This standard has been criticized, among other reasons, because it potentially undermines the individual autonomy of the patient. In order to satisfy this critique, the reasonable person standard was introduced. This standard requires that the applicant is informed about all the facts, probabilities and opinions that a reasonable man might be expected to consider before giving consent. This standard also met with criticism, as the information needs of a reasonable person depend partly upon the circumstances. This problem led to the third, subjective standard of disclosure, which requires the physician to disclose whatever information has relevance to the decision-making of a particular subject/applicant. Given the goal of informed consent – to enable individuals to make informed choices – the specific informational needs of the patient need to be taken into account; disclosure of information ought to be tailored to every individual applicant.

Informed consent should not be understood as an event, but as a process of communication, 'a process that includes ongoing shared information and developing choices as long as one is seeking medical assistance'.[9] The process model of informed consent is especially important in the context of treatments that do not require a single decision, but a chain of decisions, as in IVF/ICSI. It is necessary to point out the various moments in the course of a treatment at which the couple will be free to choose a different policy option from the procedure that generally applies at the center.[10] This involves choices such as whether or not to have all the oocytes fertilized, whether or not to have the maximum number of embryos transferred and whether or not to have any surplus embryos frozen. The center must make its policy on these points clear.

Three conditions need to be met before one can rightfully speak of informed consent: autonomy, competence and understanding. The first condition implies that informed consent should be given freely. Generally speaking, freedom in the context of informed consent implies the absence of undue influence upon the applicants. So-called radical feminists maintain that women cannot truly give voluntary consent for infertility treatments, because their choices are conditioned by the patriarchal power structure and the pronatalist ideology with which it is associated. It is argued that, if women were not conditioned to think that motherhood is their supreme fulfilment, they would not submit themselves to the risks and burdens of assisted reproductive technologies. Liberal feminists, such as Bonnie Steinbock, however, argue that to insist that the desire for children is merely a social construction is as untrue as insisting that women never freely choose to have an abortion.[11] Individuals must decide for themselves whether the benefits of assisted reproductive technologies outweigh their costs, risks and burdens.

Depending on the circumstances, the doctor may feel that testing the applicant for human immunodeficiency virus (HIV) or genetic testing is medically indicated. Clearly, the applicant's informed consent is a prerequisite for such testing. In some cases, doctors may argue that the applicant's consent is a precondition for medically assisted reproduction. The offer of testing, then, becomes a coercive offer, i.e. an offer patients can hardly refuse in view of the adverse consequences for access to assisted reproduction.[12,13] As discussed in the following paragraphs, this may be morally justified in the case of a conflict between the doctor's duty to respect the applicant's autonomy, on the one hand, and the doctor's responsibility to take into account the interests of the future child, on the other.

THE INTERESTS OF THE CHILD

The notion of 'the interests of the child' plays an important role in the ethics of medically assisted reproduction on two levels, or in two contexts. First, in the context of the evaluation of specific technologies, such as (experimental variants of) ICSI, surrogate motherhood or cloning (see below), appeals regarding the (presumed) interests of the resultant child often function as arguments to prohibit, discourage or, at least, regulate these reproductive technologies. Second, arguments regarding the 'interests of the child' are used to justify the screening and/or selection of applicants for reproductive technologies which are, in themselves, widely accepted. Here, the notion is used to justify counterindications for medical assistance regarding individual prospective parents. This second aspect is briefly discussed below.

What is at stake is the scope of responsibility of doctors involved in medically assisted reproduction. Should they honor any request for assisted reproduction? Is the doctor's sole responsibility to assess whether there is a medical indication for medically assisted reproduction, and, if patients are medically suited for treatment, to conduct assisted reproduction in conformity with the state of the art? Or do doctors also have a professional responsibility to take into account the interests of the future child?[14]

Some argue that it is not justified to deny access to medically assisted reproduction on grounds other than medical suitability. They claim this to be an unacceptable violation of the principle of respect for patient autonomy, and of the internationally proclaimed 'right to reproduce'. Furthermore, they fear ideological preoccupations as well as ungrounded prejudices of the physician. Also, it is often argued that people who can procreate 'normally' are not judged beforehand and do not need a 'license to procreate'. This criticism is unconvincing. The physician cannot deny responsibility for his actions; the resulting child would not have existed without the involvement of the doctor. The doctor, therefore, has a role-specific responsibility concerning the interests of the future child. This implies that respecting the autonomy of the prospective parents is an important prima facie principle, but not an absolute one. Those who object that the refusal to give unconditional access to 'high-risk' patients amounts to an invasion of their right to reproduce, wrongly interpret this negative claim right (or liberty right) as a positive claim right. Indeed, people's right to reproduce is the right to procreate without interference from the state or a third party. Infertile patients do not have an unqualified 'positive' right to reproduce, i.e. a right to assistance in reproduction, which would imply an unqualified duty of the physician to assist in reproduction.

While almost everybody agrees on the fundamental importance of the welfare of the future child in the context of deciding about access to medically assisted reproduction, there is no consensus on how to make this principle operational. What evaluation principle should guide the interpretation of the level or measure of welfare of the prospective child? In the literature, three guiding principles are distinguished:[14,15]

(1) The 'minimal risk' or 'maximum welfare' standard: doctors should refuse to give access to medically assisted reproduction if there is any risk for the well-being of the prospective child. This standard – '*in dubiis abstine*' is extremely strong; consistent application of this standard would imply that the large majority of applicants should be excluded from assisted reproduction.

(2) The 'minimum threshold', in particular the 'wrongful life' standard: this standard holds that assisted reproduction should be refused if there is a serious risk that the life of the child would be so miserable to him that no one would want to live such a life. In these cases, to be born is said to constitute a harm to the child. This is an extremely low standard, which leaves very little room for counterindications for assisted reproduction based on the welfare of the child. In practice, the doctor would be obliged to make decisions that many of us would consider to be counterintuitive and irresponsible. The fact that unconceived children could end up having lives that would be, on balance, worthwhile cannot function as an excuse for imposing grievous pain, suffering and deprivation on them.[16]

(3) The 'high risk of serious harm' or 'reasonable welfare' standard: this is intermediate between the others, and avoids their counter-intuitive implications.

While a further debate about these guiding principles is important, it would be naive to assume that consensus about the 'best guiding principle' will result in consensus about each individual case. Each of the principles needs interpretation (what is high risk? what is serious?, etc.), and may invite controversy if applied to individual cases. Individual infertility clinics have, at least to some degree, a discretionary authority regarding the interpretation of the professional's responsibility concerning the potential child. Clearly, the selection criteria should be transparent, and applied in a consistent way. Applicants who are denied access to medically assisted reproduction should be informed about the reason(s) for this denial, and be referred to another infertility center.

Traditionally, the debate about counterindications for medically assisted reproduction concentrated on (presumed) psychosocial risk factors for the welfare of the child, such as 'alternative' family structures (lesbian couples, single women). Obviously, medical risks, in particular the risk of transmitting HIV or genetic diseases, are equally relevant. Many clinics (routinely or selectively) offer HIV testing as a condition for medically assisted reproduction – an example of a coercive offer (see above). While HIV-seropositive applicants have traditionally been denied access to infertility treatment because of the high medical and psychosocial risks for the potential child, positive developments with regard to both reducing the risk of vertical transmission and combination therapy may justify a more permissive policy in individual cases.[17,18] As discussed below, in the case of genetic risks, preimplantation genetic diagnosis may be an option to prevent transmission of the disorder.

PREIMPLANTATION GENETIC DIAGNOSIS

IVF opened a 'new window' onto early human development, providing new options for intervention. One of these options is preimplantation genetic diagnosis (PGD), aiming at a selective transfer of preimplantation embryos. PGD was primarily developed in response to requests for help from people at high risk of passing on a serious genetic disorder to their children. In recent years, the number of PGD cycles as well as the number of centers involved in PGD have increased year by year.[19] At the same time, there is an ongoing debate about the ethics of PGD.[1] The following sections comment briefly on some of the objections presented by critics, and then focus on some of the main ethical issues regarding PGD.

Categorical objections?

Unjustified selection?

A fundamental objection to (post-conception) PGD holds that selecting preimplantation embryos is intrinsically wrong, given the 'sanctity of human life'; the physician has, according to this view, a moral obligation to transfer all embryos available.[20] This objection is highly problematic, for several reasons. First and foremost, the 'sanctity of life' doctrine is based on shaky premises: it presupposes that the embryo should be respected as a human person, i.e. has an absolute right to life (see above). A more moderate, and more reasonable, perspective is that the embryo, in view of its potentiality, deserves some respect, that one should not destroy embryos for trivial reasons. It is widely accepted that regular prenatal diagnosis and selective abortion can be morally justified as a means to prevent serious suffering for the future child and/or the parents/family. PGD can be justified as well: accepting selective abortion while categorically rejecting a selective transfer would, at least at first sight, be inconsistent.

To biopsy or not to biopsy?

Another objection to (invasive, post-conception) PGD concerns the preparatory biopsy. 'The individual blastomeres of the embryo are totipotent; they have exactly the same developmental potential as an embryo. Therefore, isolating a blastomere involves the creation of a second, duplicate, embryo, which will be destroyed during the diagnostic procedures'. For this reason, among others, post-conceptional PGD is prohibited by the German 'Embryo Protection Act'.[21] However, this objection is weak, for two reasons. First, it is unlikely that an individual blastomere isolated at the 6–10-cell stage is still totipotent.[22] Second, even if it were proven that (invasive) PGD inherently involves the creation of a second embryo, which will be destroyed during or after the diagnostic procedure, one could still justify this technique, in view of the relatively 'low' moral status of a preimplantation embryo, and the 'pros' of PGD in comparison with a selective abortion.

Disproportionately burdensome?

It is a major advantage of PGD that it can avoid traumatic (repeated) selective abortions. At the same time, however, PGD carries significant disadvantages. First, PGD presupposes medically assisted reproduction, in particular IVF/ICSI, which is not the most pleasant way to 'make' babies, carries various sorts of risk for women (infection, ovarian hyperstimulation syndrome, multifetal pregnancies, etc.) and has just a moderate 'take-home baby rate'. Second, PGD is still experimental: the reliability of the diagnostic tests as well as the safety of the preparatory biopsy (or combination of biopsies) remains to be proven. Paternalistic critics of PGD argue that PGD is too burdensome, whereas there are reasonable alternatives. They ignore that balancing the pros and cons of the respective options is highly subjective. For some couples, the so-called 'reasonable alternatives', in particular prenatal diagnosis and selective abortion, are a priori unacceptable. And for infertile couples at high risk of conceiving a handicapped child who opt for IVF/ICSI as an infertility treatment, PGD may be far more attractive than a non-selective transfer. These 'anti-paternalistic' reflections underscore the importance of the concept of informed choice: PGD requires the free and informed consent of the prospective parents (see above). It is of utmost importance that the couple understands the (practical) pros and cons of PGD, in comparison with regular prenatal diagnosis.

I conclude that there are no categorical objections to PGD. At the same time, PGD raises some complex questions, which need further scrutiny and debate, including: what indications should be accepted? How should the professional responsibility to take into account the interests of the future child be made operational?

Issues for further debate

Which indications?

One of the 'classical' questions concerning regular prenatal diagnosis is also relevant with regard to PGD: what genetic tests should be performed? Should a line be drawn, and, if so, where? There is a strong consensus (i.e. almost unanimity) that PGD, like regular prenatal testing, should be restricted to the identification of genetic anomalies, handicaps and disorders, allowing the prevention of the birth of affected children – the 'medical model'. According to some commentators, however, this model is too restrictive, as it limits the freedom of prospective parents to test and select for or against specific 'non-medical' traits. This model should, so these critics argue, be replaced by a 'parental autonomy' (or the 'designer') model.

The medical model The dominant moral justification for selective abortion, as well as a selective transfer, is the prevention of severe suffering related to genetic defects and disorders, not just suffering of the future affected child, but also suffering of the parents (family). The question then becomes: how severe does a defect or disorder have to be to qualify for prenatal testing or PGD? Most commentators agree that prenatal testing and PGD should not be performed for defects/conditions which cannot reasonably be considered to be a serious threat to the well-being of the future child and/or the family. The question of where precisely to draw the line is, however, difficult to answer. Three relevant factors should be simultaneously considered:[23] the severity of the defect/disorder, determined partly by the availability of treatment; the age of onset of the disorder; and the penetrance of the gene mutation (the probability that the genotype will affect the phenotype).

The following focuses on the ethics of PGD for two categories of late(r)-onset disorders, which is especially controversial: first, PGD for dominant, mid-life-onset, untreatable disorders, and second, PGD for mid-life-onset disorders that may be prevented or treated.

Huntington's disease is the paradigm example of the first category. Critics such as Post argue that a child carrying the Huntington's disease mutation will have 'many decades of good and unimpaired living. Moreover, the parents of the child are not immediately or even directly affected in the way they would be were the disease of early onset.'[24] Furthermore, so they argue, our desire not to bring suffering into the world must be tempered by the recognition that suffering is a part of life. Finally, we should acknowledge the moral ambiguity of the quest for 'perfect' babies ('perfectionism').

Do these considerations undermine the morality of (prenatal diagnosis or) PGD of Huntington's disease? No, they do not. Of course, it would be naive and misguided to fight against all contingencies of human life. The question is, however, whether carriers of Huntington's disease, who have a high (50%) risk of transmitting the mutation to their children, have the moral right to prevent this by making use of (prenatal diagnosis or) PGD. In view of the fact that Huntington's disease is severe, and that the penetrance of the mutation is complete, my answer is: 'Yes'. It is insensitive, if not an insult, to (dis)qualify preventive measures as symptomatic of 'genetic perfectionism'. Of course, carriers of Huntington's disease will be healthy during three to four decades. One needs to recognize, however, that the prospect of their eventual fate often imposes an extremely severe burden. Finally, Post completely ignores the perspective of the healthy relatives.

The ethical debate about PGD for disorders such as Huntington's disease should not be reduced to the 'fetalist' perspective (the status of the embryo) or to the 'humanist' reflections just mentioned. The request of a (presymptomatic) carrier of Huntington's disease (and even of a person at 50% risk of carrying Huntington's disease) to gain access to assisted reproduction may in itself pose an ethical dilemma for doctors involved.[25] After all, in view of the fact that a carrier will inevitably develop Huntington's disease, competence as a parent will be lost steadily, with increasing burdens placed on the other parent (partner). This may have adverse implications for any child born into this relationship. Furthermore, the child may well be confronted with the extreme suffering and death of the affected parent at a relatively young age. Do these psychosocial risks for the welfare of the child constitute an overriding argument not to assist in the reproduction of prospective parents carrying the Huntington's disease mutation (see above)?

Needless to say, the situation may be burdensome for the child. At the same time, however, it is well known that many children of Huntington's-disease families are able to cope reasonably well with the situation. Relevant variables include the coping skills of the parent not affected by Huntington's disease, the quality of the network of the family, the availability of social support and the age of onset of the disorder in the parent. Clearly, in many cases, the child will be an adolescent or even an adult when the parent who carries the mutation becomes symptomatic. In view of this, medically assisted reproduction may well be justified according to the third, intermediate, standard (see above). Another question that needs further discussion is whether it would be 'good clinical practice' to offer a neurological examination to an asymptomatic carrier, to obtain information about the progress of the process of neurodegeneration, and if so, under what conditions. While such an examination might be useful for estimating the number of years that the carrier could remain unaffected, and, thus, for quantifying the psychosocial risk for the potential child, (the result of) this examination could be very distressing and confrontational for the carrier, as it may result in an early diagnosis of Huntington's disease.

What if the prospective father or mother is already symptomatic? In this case, the psychosocial risks for the future child may be greater, as the child would be confronted with the agony of this progressive disorder 'right from the start'. I am, however, not convinced, as yet, that to assist in reproduction would be completely morally unjustified in all cases. Further discussion is needed to examine whether the variables just mentioned (and maybe other variables) would allow decisions to be made 'case by case'.

An example of the second above-mentioned category is hereditary breast/ovarian cancer. PGD of mutations in breast cancer genes (*BRCA1* and *-2*) is probably even more controversial then PGD for Huntington disease, because the penetrance of these mutations is incomplete (the life-time risk of a female carrier developing breast cancer is 50–85%; her risk of developing ovarian cancer is 20–65%), and preventive interventions may effectively reduce morbidity and mortality in carriers.[26] But again, we should resist premature conclusions with regard to preventing the birth of children predisposed to breast/ovarian cancer. Morally relevant questions concern the effectiveness of available preventive and/or therapeutic measures, and the burdens imposed by the respective medical interventions. Unfortunately, the effectiveness of medical surveillance is currently far from optimal. Although the effectiveness of prophylactic bilateral mastectomy appears to be high, longer follow-up studies and investigation of more carriers are necessary to establish definitively the protective value (and determine the long-term complications) of this procedure.[27] Furthermore, prophylactic surgery is irreversible, and may have major implications for a woman's quality of life. I would argue that the fear of prospective parents ('at risk') that their future children may inherit a *BRCA* gene mutation is far from unreasonable, and that prevention of the birth of daughters strongly predisposed to breast/ovarian cancer is not a trivial concern. A question that needs further discussion is whether PGD of mutations in *BRCA* should be restricted to female embryos because female carriers are at much higher risk of developing cancer than male carriers.

The parental autonomy model Some commentators argue that prospective parents should be free to use PGD for the selection of any offspring characteristic they prefer, as selection for medical as well as non-medical purposes is part of reproductive freedom.[28] The paradigm example is PGD for sex selection for social (non-medical) reasons.

Sex selection for social reasons is controversial.[29] Some of the objections are not convincing, at least in Western culture. Take, for example, the criticism that such selection is inherently sexist, or that it will result in a serious disturbance of the sex ratio. There are, however, some moral issues that need serious consideration. Most important, sex selection for non-medical reasons can set a precedent for other applications of so-called 'positive eugenics', i.e. selection for non-disease characteristics, such as intelligence, athletic genotype, etc.[30] One of the concerns is that this type of testing may selectively confer advantages upon the offspring, in particular the offspring of the economically advantaged who can afford the costs of such testing, thereby exacerbating existing socioeconomic differences. It would be symptomatic of a moral myopia to isolate the case of sex selection from this broader perspective of selecting embryos for other non-medical reasons. For the moment, I support Strong's conclusion, that we would at least need a greater confidence that the benefits would outweigh the harms before we would be morally justified in proceeding with preimplantation sex selection for non-medical reasons.[31]

According to some commentators, the fear that PGD will be used for 'trivial selection' (including 'positive-eugenic' purposes) is unfounded, as both the burdens and the only modest take-home baby rate of IVF, as well as the availability of just a few embryos, constitute a high barrier against such trivial use. This rebuttal is convincing and reassuring – for the moment. It is important, however, to acknowledge the potential implications of the combination of PGD with new technologies. If it becomes possible to create large numbers of embryos by fertilizing large numbers of eggs obtained by *in vitro* maturation, to perform multiplex genetic testing in the context of PGD, and perhaps even to multiply the embryo that proves to be 'the best', the 'designer baby' could well become a more realistic option in the future (see below).[32,33]

HLA typing: an intermediate case? Recently, Verlinsky and colleagues performed PGD including human leukocyte antigen (HLA) testing on embryos of a couple desiring not only to avoid the birth of an additional child affected with Fanconi anemia, but also to establish a pregnancy to provide an HLA match for an affected sibling requiring stem cell transplantation.[34] The transfer of a heterozygous, HLA-matched embryo resulted in the birth of a healthy child. The umbilical cord cells of the child were used for transfusion.

For the ethical analysis of this type of case, it is important to discern at least three ethical questions, the first two being preliminary.[17] First, can it be morally justified to use an incompetent child as a 'donor', or, more precisely, to use a child as a 'source' of a transplant, and, if so, under what conditions? Clearly, it will be crucial to evaluate the risk for the 'donor', the harm/probability ratio. The example of donating umbilical cord blood is the 'simple' case, as this does not entail any medical risk for the 'donor' child at all. Second, is it ethically justified to conceive a child in order to obtain a transplant? Critics, referring to the ethics of Kant, object that this would amount to treating the child as a means. Kant's proscription, however, is against using people solely as a means.[35,36] What matters, then, is whether the parents will value the child only as a transplant source or whether they will also love this child for itself.[37] Finally, can it be morally justified to perform PGD to (prevent the conception of an affected child and simultaneously) select an HLA-matched embryo for transfer? The fundamental ethical issue concerns the testing and selection of embryos for the required HLA type. On the one hand, this selection does not fit into the medical model for prenatal/preimplantation selection, as HLA types are not relevant for the health of the prospective child itself. On the other hand, the aim of the procedure is clearly health-related, as it serves a vital health interest of a sibling. Critics may argue that, even though this aim is not trivial, to allow exceptions to the medical model entails a slippery slope resulting in selecting embryos for whatever non-medical traits prospective parents prefer (athletic genotype, etc.). The slippery-slope argument, however, is questionable, as one can reasonably argue that the current case is a 'special' one in view of its 'intermediate' character, and should be evaluated on its own merits. One could object, finally, that 'mismatched' preimplantation embryos (which do not have the required HLA type) will be destroyed. This objection, however, is rather weak, in view of the low moral status of the early embryo and the 'life-saving' aim of the procedure.

It has not been convincingly argued that preimplantation testing for the required HLA type is necessarily morally wrong. At the same time, the risks and pitfalls involved should be taken seriously. First, the number of embryos which are both unaffected and a good HLA match will be small. As a consequence, the success rate of the procedure will be low. Furthermore, parents may feel pressurized to opt for PGD, although they cannot afford any more children, the affected child may die before the birth of the HLA-matched 'donor' and the transfusion may not save the child. To evaluate the risks and uncertainties case by case, and to provide adequate counselling are crucial prerequisites for PGD aiming at transferring HLA-matched embryos. Clearly, it would be misleading and even dangerous to present PGD as an easy way out of the parents' dilemma.

Autonomy or heteronomy?

In the context of regular prenatal diagnosis, the principle of respect for autonomy is of utmost importance. This means, first, that prospective parents at higher genetic risk should be free to decide whether or not to make use of prenatal diagnosis; the doctor should not put pressure on them to opt for testing. Second, in the case of a 'positive' result of prenatal diagnosis, the prospective parents should be free to decide whether or not to terminate the pregnancy; the doctor should support them, whatever decision they make. When applied to PGD, this normative principle, reasoning analogously, would imply that doctors should never put pressure on prospective parents to opt for PGD, and that prospective parents should be free to decide whether or not to accept (potentially) affected embryos for transfer. Clinical PGD practice, however, reveals that respect for autonomy is not unqualified. What, then, is 'good clinical practice'?

To clarify this issue, the following first comments on the locus of decision-making concerning PGD, and second on the locus of decision-making concerning the transfer. With regard to the first topic, the offering of PGD is discussed in a specific context, namely the 'treatment' of male subfertility.

In view of the higher levels of chromosomal and other genetic anomalies in subfertile males, the availability of genetic counselling and the (routine or selective) offering of genetic testing is a prerequisite for clinical ICSI. Genetic testing can take place before treatment (preconception testing for chromosome aberrations or gene mutations), during treatment (PGD) or during pregnancy. There is dissent with regard to the goal of such testing. The offering of genetic testing may have three different objectives. First, the testing may reveal the cause of subfertility and help patients to understand and handle their problem. Second, it benefits by enabling the couple in informed reproductive decision-making. If they have a higher risk of producing a handicapped child, they may – or may not – opt for preventive measures ('avoidance'). Third, genetic testing enables the doctor to take his own responsibility in preventing serious harm to children thus conceived. Sometimes, these goals may conflict. For example, if an infertile man refuses to be tested for chromosome aberrations or – in the case of congenital bilateral absence of the vas deferens (CBAVD) – for mutations in the cystic fibrosis gene, at least some clinics will not give access to ICSI. Should the man carry a balanced chromosome translocation involving a high risk for the future child, or should both partners be carriers of cystic fibrosis, then most centers will only offer ICSI if the couple intends to make use of prenatal diagnosis or PGD offered for 'avoidance'.[38] These examples illustrate that, at least in some cases, the offer of (preconception genetic testing and) PGD is, in fact, a 'coercive' offer.

Critics hold that a 'coercive offer' clashes with the traditional ethics of reproductive genetic counselling, in particular with the principle of nondirectiveness.[39,40] According to this principle, the doctor should respect the values and preferences of his clients and give unconditional support. This criticism is not convincing, because it ignores that doctors offering assisted reproductive technologies have their own responsibility to avoid serious harms to future children (see above). The real issue, then, is not whether it may be acceptable 'coercively' to offer PGD to infertile couples at high genetic risk who apply for medically assisted reproduction (i.e. to give access to those couples only on the condition that they opt for PGD), but, when this is acceptable, which criteria should be used. Again, even if we agree that medical assistance in procreation is morally unsound when there is a high risk of serious harm to the future child, there will always be a gray area, where the individual professional has discretionary authority (see above).

With regard to the above case, I tend to conclude that, when clients at high risk of conceiving a cystic fibrosis-affected child apply for ICSI, doctors may justifiably make access dependent upon the couple's willingness to make use of PGD (or prenatal diagnosis).

The second issue to be scrutinized is the decision-making authority when PGD has a 'positive' or inconclusive result. At first sight, this may seem a 'non-issue'. One may, after all, safely assume that doctors and prospective parents agree that severely defective embryos should not be transferred. Couples may, however, insist that embryos be transferred in the case of an inconclusive result of PGD, and/or when they consider the residual genetic risk for the future child to be acceptable, especially when there are no other, 'definitely healthy', embryos available for the transfer. It is sometimes argued that, given an adverse prognosis, the choice, of whether or not to have the embryo replaced, must lie with the potential mother.[41] In view of the physician's own professional responsibility to prevent serious harm to the prospective child, however, a (partial) shift with regard to the 'locus of decision-making' seems to be inevitable. At the same time, this shift raises complex moral questions: what standards should be used in overruling the preferences and autonomy of the couple?

In view of the potential dissent with regard to (un)acceptable 'harm/probability ratios', it is, again, imperative that the transfer policy of the IVF clinic be discussed with the couple in advance of PGD. Applicants should understand the Janus-faced character of PGD: while PGD may increase their reproductive options, it does not necessarily maximize their reproductive freedom.

PREIMPLANTATION SCREENING FOR ANEUPLOIDY

While PGD has been introduced as an alternative option for couples at very high (25–50%) risk of transmitting a genetic defect, it can also be routinely applied in the context of regular IVF. Of special interest are current trials regarding preimplantation genetic screening (PGS) for aneuploidy, especially, for example, in women aged 35 years and older (see Chapter 27). The potential 'pros and cons' of this screening have been neatly sketched by Reubinoff and Shushan.[42] First, if only chromosomally normal embryos (or at least embryos normal for the tested chromosomes) were to be transferred, there could be a significant increase in the take-home baby rate of IVF. Second, such selection could be reassuring, especially for older women. Currently, older women can opt for amniocentesis or chorionic villus sampling. These invasive procedures are associated with 0.5–1% pregnancy loss. Therefore, it is not uncommon that infertile women are reluctant to jeopardize their 'precious' conception by invasive prenatal testing, and risk the birth of an affected child. In addition, couples with moral or religious objections to pregnancy terminations could also avoid prenatal screening. Reubinoff and Shushan rightly stress that several difficulties must still be addressed before PGS can be routinely applied in women of advanced age opting for IVF, including the following. First, it is currently not known whether embryo biopsy might adversely affect implantation and pregnancy rates. Second, blastomere analysis will, at least in some cases, not be representative of the whole embryo, owing to the high frequency of chromosomal mosaicism. Mosaicism could lead to misdiagnosis. The biological significance of mosaicism is unclear. Some authors have claimed that abnormal cells in the early embryo are subsequently eliminated or diverted to the trophectoderm, and thus do not impair normal development. Therefore, discarding mosaic embryos might lead to the loss of potentially normal embryos, thereby impairing the take-home baby rate of IVF. According to some investigators, the high level of mosaicism makes PGS, on the basis of polar body analysis, more attractive.[43]

What about the ethics of PGS for aneuploidy? There are no valid, a priori, categorical, objections to this type of screening. After all, its primary aim, i.e. to increase the take-home baby rate of IVF, is unproblematic, if not commendable, while the means, namely to exclude embryos affected with serious chromosomal aberrations (which often lack viability) from transfer, is clearly morally acceptable. At the same time, PGS for aneuploidy raises ethical (and ethically relevant) issues and questions that need further debate. A first question concerns the 'timing' of the screening. Proponents of polar body screening claim that this strategy not only would avoid the complications concerning mosaicism, but also has a moral advantage; after all, (first) polar body diagnosis aims at a selective fertilization of normal oocytes, thus avoiding embryo selection. The leading investigators, however, have recommended that both the first and the second polar body should be studied.[44] Whereas they continue to qualify this strategy as preconception testing, this combined biopsy implies, in fact, a shift towards 'intraconception' screening; after all, the second polar body appears during the fertilization process. There is no consensus on the moral status of the oocyte during the fertilization process. In view of its potentiality, I would suggest that the oocyte '*in statu fertilisandi*' is (the moral equivalent of) an embryo, not (the moral equivalent of) an unfertilized egg. As a consequence, I doubt whether there is a fundamental difference between 'intraconception' and post-conception PGS from a 'fetalist' perspective. Second, PGS for aneuploidy is experimental. PGS should, therefore, be introduced in the context of a research project. Participation in this research presumes, of course, an informed consent (see above). An important aspect concerns the effectiveness of PGS. The preliminary findings of the first, controlled, randomized, prospective study suggest that selecting embryos with a normal chromosome complement has a positive impact on the implantation rate, although it does not reach the level of statistical difference.[45] Further data are clearly needed. Third, it may be even more difficult to answer the question of whether PGS for aneuploidy can be considered worthwhile, i.e. a justified investment in view of its presumed benefits, on the one hand, and the scarce resources available for health care, on the other. A fourth issue concerns, again, the 'locus of decision-making'. Who would have decision-making authority with regard to the actual use of such screening in an individual case: the doctor or the couple/woman? What if this screening would prove to be (cost-)effective and efficient? Will it, then, become the professional standard, part of 'good clinical practice', just like screening for the correct number of pronuclei? And who has decision-making authority with regard to selection of the chromosomes that would be included in the 'screening panel'?[46] What if patients do not want to be informed about mild chromosomal aberrations, such as XXY and XYY? And, finally, who has decision-making authority concerning the disposal of chromosomally 'abnormal' embryos? What if a couple insist on transferring an embryo with, for instance, an XXY or XYY karyotype (when there is no 'normal' embryo available for transfer)? I tend to argue that a doctor's categorical refusal to transfer embryos with these mild aberrations would be symptomatic of the disputable 'minimal risk' or 'maximum welfare' standard (see above).

HUMAN CLONING

Since the birth of Dolly, human cloning has re-emerged as a frontier biomedical and ethical issue. As it is important to be aware that there are, in principle, different sorts of human cloning, this section first sketches the various categories and applications of human cloning. Second, the ethical aspects of some of these potential applications are addressed.

Types of human cloning

At least two distinctions are relevant, namely between reproductive cloning and non-reproductive cloning, and between the cloning of adults/children and the cloning of preimplantation embryos (the term embryo is used to mean the early, preimplantation embryo). In view of these distinctions, there are at least four categories of human cloning:[33]

(1) Reproductive cloning of adults/children;

(2) Non-reproductive cloning of adults/children, including so-called 'therapeutic' cloning, i.e. cloning for transplantation purposes;

(3) Reproductive embryo cloning;

(4) Non-reproductive embryo cloning.

Reproductive cloning of adults/children

In theory, there are many reasons for would-be parents to resort to reproductive cloning. It may be useful to discern broadly two kinds of motivations, termed by Dena Davis as logistical and duplicative.[47] The logistical motivation concerns couples who have (for some reason) a difficult time procreating, and cloning offers them a unique opportunity. In these cases, the parents' goal is simply to have a (healthy) child, and they see cloning as the best option from an array of choices: adoption, childlessness or reproduction with the use of a third party. The duplicative element is, so to speak, a side-effect. As the term 'logistical' is somewhat confusing, and the most relevant examples of this kind of motivation relate to medical problems, namely infertility (for example, an infertile male who produces no sperm cells) or the presence of high genetic risks for progeny, I prefer the term medical reasons. In contrast, with duplication motives, the genetic replication itself is the attraction. Examples include parents who have lost a child and want to have a genetically identical child to replace the dead child, and parents who want to clone a genius in order to have an equally talented child.

Reproductive embryo cloning

Two methods may be used for reproductive embryo cloning: cell nucleus transfer and embryo splitting, more precisely, blastomere separation. (The suggestion that the latter technique does not constitute cloning is arbitrary.[33,47]) Reproductive embryo cloning might serve various purposes in different contexts. The first is the context of regular IVF. When there is only one embryo available, reproductive embryo cloning could increase the number of embryos for transfer in a single cycle, thereby increasing the take-home baby rate. Furthermore, for couples who produce enough embryos for one transfer, reproductive embryo cloning might provide additional embryos for a subsequent transfer, without them having to go through another oocyte retrieval cycle. This strategy could lessen the burdens and costs of IVF. Some experts seriously doubt whether 'blastomere separation' would increase the odds of women becoming pregnant.[48] Other experts, however, are less sceptical.[49] One may safely conclude that whether reproductive embryo cloning will be clinically feasible and valuable for regular IVF practice cannot be determined without further research. The second context is that of PGD, where reproductive embryo cloning may be particularly valuable when there is just one 'healthy' embryo available for transfer.[50] Again, the aim of reproductive embryo cloning would be to increase the take-home baby rate of medically assisted reproduction. The third context is that of genetic embryo therapy. In theory, one could prevent the inheritance of mitochondrial disorders by means of nuclear transplantation into an enucleated egg from a healthy donor, resulting in one or more embryos which have their nuclear DNA from the affected woman and her partner, and normal mitochondrial DNA from the egg donor. There are at least three options: to transplant the nucleus of the unfertilized egg or oocyte, both pronuclei of the zygote, or the nucleus of one or more blastomeres.[35] While the first and second methods would not involve the cloning (duplication) of an embryo, the last approach definitely would.

'Therapeutic' cloning

Human embryonic stem cells might be used for transplantation purposes in the future. Currently, research concentrates on human embryonic stem cells obtained from spare preimplantation embryos. From the outset, however, experts thought that the full therapeutic potential of human embryonic stem cells would depend on therapeutic cloning, aiming at the production of 'matched' transplants.[51] Therapeutic cloning could open the way towards a type of cell (or tissue) replacement therapy that avoids the problems of immune rejection.

Ethical exploration

As human cloning may have quite different applications, a differentiated ethical analysis is clearly required.[33]

Reproductive cloning of adults/children

In 1997, the (US) National Bioethics Advisory Commission concluded: 'At present, the use of this technique to create a child would be a premature experiment that would expose the fetus and the developing child to unacceptable risks'.[52] The data collected in recent years reinforce this view. As animal research using various species shows, current methods of cloning result in high numbers of spontaneous abortions and higher perinatal mortality and morbidity rates (poor kidneys, intestinal blockages, immune deficiencies, diabetes, etc.) (A remarkable exception is a recent publication which suggests that cloning cows may be relatively safe.[53]) Even clones that appear healthy may be 'ticking time-bombs'. The reproductive cloning of adults or children would almost certainly have similar adverse outcomes.[54] In view of this, there is a strong consensus that attempts to start reproductive cloning of humans at a time when the scientific issues have not been clarified, and the risks not substantially minimized, are completely irresponsible. Doctors who think otherwise disregard their professional responsibility to take the health interests of future cloned children seriously, and disqualify themselves as physicians performing good reproductive medicine. It has been suggested that the health risks of cloning could be adequately handled by PGS of the embryos thus conceived. This proposal is, however, a *testimonium paupertatis*, as it completely overplays the diagnostic potential of PGS and disregards the specific types of risks involved. The current health risks of human cloning are a compelling reason to impose a worldwide moratorium.

Let us assume, however, that, at sometime in the future, reproductive cloning would become substantially less risky, or even 'safe'. Could reproductive human cloning, then, be morally justified? The dominant view seems to be that reproductive human cloning (especially the cloning of adults and children) is categorically wrong. Many organizations, including the Council of Europe, and countries (want to) prohibit reproductive cloning altogether.[55] What are the (non-medical) objections, are these objections valid and, if so, do they apply to all cases of reproductive cloning?

Direct objections

'It is unnatural' The 'argument from nature', even though it is often used in debates regarding bioethical issues, is questionable. What matter morally are the goals, the means and the consequences of our actions. So, the 'argument from nature' is a *non sequitur*. Based on the fact that reproductive cloning by nuclear transfer is unnatural in the human, it does not follow that it is *per se* morally unjustified.

'It is an invasion of human dignity' (This is the main objection of the 'Additonal protocol' of the Council of Europe.[55]) It is not always clear what this objection exactly means: whose dignity is attacked – and how? According to Kantian ethics, respect for human dignity requires that a human being is never used exclusively as a means. Harris rightly states that critics of cloning do not convincingly demonstrate (or even try to demonstrate) that all sorts of reproductive cloning would necessarily involve a complete instrumentalization of the future child.[56] One can reasonably argue that the objection applies to some cases of cloning for duplicative reasons, most obviously when one would clone someone just to use the child for transplantation purposes. But, it is difficult to see how cloning for medical motivations is an invasion of human dignity too.

'It is a violation of the right to have a unique identity' This objection has been convincingly criticized by, among others, Brock.[57] The objection seems to

presume a crude, untenable, genetic determinism. It is important to acknowledge that the parent and his/her clone will grow up 'in another time, at another place': the further separated in time (the greater the 'asynchrony'), the more different the environment ('nurture'), and the more different the environmental influence on development, character and individuality.

Consequentialist objections

The main consequentialist objection, apart from that concerning the health risks involved, regards the psychosocial impact of cloning on the child.

'Cloning carries substantial psychosocial risks' The asynchrony between the person cloned ('the original') and the later twin could have adverse psychosocial consequences for the latter. After all, the later twin may perceive himself, and/or be perceived by others, as some sort of 'copy' of someone who already exists; he may feel that he lives 'in his shadow'.[58,59] Some commentators consider this type of criticism to be morally irrelevant. They argue that the value of being alive outweighs any disvalue associated with the putative psychosocial harms of cloning. The alternative to coming into existence as a clone of another is not to exist at all. Clearly, so they argue, for the clone, existence is to be preferred above non-existence. This rebuttal (the 'argument from non-existence'), however, is questionable.[57] It is not incoherent to assume that one should refrain from creating children if they would suffer greatly as a result of the procreative method used (see above). The argument from non-existence is not sufficiently compelling to disregard the psychosocial risks completely.

The real question is: do the psychological risks provide a compelling argument to refrain from reproductive cloning altogether? Obviously, the psychosocial risks to the clone are more or less speculative. It seems reasonable to assume that the psychological risks may well be smaller if the reasons for cloning are medical, and if the parents are carefully counselled.[47] Critics may argue that, as the psychological risk cannot be excluded, cloning for any reason is wrong. One might reply, however, that this view reflects a too restrictive standard for evaluating the psychological risks for the future child (see above), and that the psychosocial risks involved in cloning are not necessarily greater than the psychological risks of, for instance, the use of donor gametes or surrogate motherhood. If we accept the latter, it is not clear why we should categorically reject the former.

I conclude that, while currently a moratorium for the reproductive cloning of adults and children is justified in view of the health risks for the child, a definite ban is premature, and continuation and further differentiation of the ethical discussion are needed. This might well result in the conclusion that reproductive cloning of adults or children can be morally justified in some (mainly medical) cases. We should also reflect on some newer questions:

(1) When would cloning be 'safe enough' to enable a clinical trial to begin? Which 'standard for safety' should we accept?[60]

(2) How many clones (from one 'original') would be acceptable? We should acknowledge that the asynchrony between various 'identical' children sequentially produced by cloning may well entail greater psychosocial risks than the asynchrony between the 'original' and a single clone. After all, these children will not only share their (nuclear) DNA, but also live in the same environment, which may hamper the development of their own identity.

In the years to come, new reproductive technologies may be developed that could constitute an alternative option for some individuals who might now consider reproductive cloning as an infertility treatment. A first example is the artificial induction of meiosis in somatic cells, to generate haploid cells.[61,62] A second example would be the so-called interspecies transplantation of germ stem cells.[63,64] A comparative ethical analysis of these various options, scrutinizing the respective pros and cons, is beyond the scope of this chapter.

Reproductive embryo cloning

Preclinical human embryo research (non-reproductive embryo cloning) will be necessary to investigate (further) the feasibility, the safety and the potential clinical value of embryo cloning.

Whether this research can be justified depends partly upon the question of whether (clinical) reproductive embryo cloning could ever be acceptable. What, then, about the ethics of reproductive embryo cloning, in particular reproductive embryo cloning in the context of regular IVF?

According to the 'Additional protocol' regarding human cloning, 'Any intervention seeking to create a human being genetically identical to another human being, whether living or dead, is prohibited'.[55] This (first) article entails a categorical prohibition of reproductive embryo cloning. This prohibition is premature, in view of the fact that there has hardly been any public debate about the ethics of reproductive embryo cloning. Furthermore, one may wonder whether this first article, taken literally, applies to reproductive embryo cloning at all. After all, reproductive embryo cloning does not seek to create identical twins; its primary aim is to increase the success rate of IVF. Of course, the birth of a monozygotic twin may be the (unintended) consequence of reproductive embryo cloning but does that matter morally?

It may be useful to make a brief comparative analysis of reproductive adult cloning on the one hand and reproductive embryo cloning, on the other: do the objections to adult cloning also apply to reproductive embryo cloning?

The objection that cloning is unnatural is even less convincing if applied to reproductive embryo cloning. In view of the phenomenon of 'natural' twinning, one might even argue that reproductive embryo cloning is an imitation of nature, at least if one would make only one 'copy', and transfer the monozygotic embryos thus obtained in the same cycle – a strategy termed 'simple fresh-transfer twinning' by Bonnicksen.[65]

It is difficult to see how reproductive embryo cloning aimed at increasing the take-home baby rate of IVF could attack human dignity. In theory, reproductive embryo cloning could be combined with 'progeny testing', which would involve the storing of cloned embryos while a few are transferred to test their social and personal attributes.[66] The psychosocial and ethical problems inherent in this (hypothetical) scenario, however, do not constitute an objection to the applications of reproductive embryo cloning just described.

Regarding the 'medical risk' objection, reproductive embryo cloning by means of nuclear transfer will probably be less risky than reproductive adult cloning, because (this type of) embryo cloning would involve the transplantation of 'young' nuclei, which would not yet have accumulated DNA damage, or have shorter telomeres. It is premature, however, to suggest, as Edwards and Beard do, that cloning from embryonic nuclei is safe.[66] Reproductive embryo cloning by means of blastomere separation may carry fewer health risks for progeny than nuclear transfer. Clearly, (further) preclinical safety studies could contribute to a better understanding of the potential risks of both of these methods.

Finally, consider the 'psychosocial harm' objection. Psychosocial harms might easily be prevented by avoiding asynchrony between future monozygotic twins, (for instance) by a simultaneous transfer of monozygotic 'twin embryos'. Clearly, a sequential transfer resulting in asynchrony between identical twins could involve psychosocial harms. Bonnicksen points especially to the adverse effects of parental expectations or fears:[65] 'If a child is born as a spaced twin, pressures on the first born and the second born are possible, as parents wonder why the second one does not turn out like the first one (if the first one is successful) or worry that the second born will turn out like the first born (if the first born is unsuccessful).' Obviously, twins, especially the later twin, could also suffer if their parents did not have such expectations or fears. Even if the later twin does not believe in genetic determinism, he may feel that he lives constantly in the shadow of the older twin.

When sequential transfers would result in asynchrony between identical twins, reproductive embryo cloning would, paradoxically, be similar to reproductive adult cloning. The psychosocial risks of asynchrony between identical twins produced by reproductive embryo cloning might even be greater than the psychosocial risks of cloning adults, as, in the first case, nature as well as the environment ('nurture') would be identical, potentially resulting in greater problems, especially for the second (third, etc.) born twin, in developing his own identity.

With regard to reproductive embryo cloning in the context of regular IVF, my conclusions are that at least some of the moral objections to the reproductive cloning of adults/children do not, or not necessarily, apply to reproductive embryo cloning, and that there seem to be no valid categorical objections to 'simple fresh-transfer twinning' – the 'simple case' of reproductive embryo cloning. The real issue, then, is not whether reproductive embryo cloning can be morally justified, but what conditions should be imposed. I suspect that the main issue is whether or not we should allow sequential transfers of identical embryos, potentially resulting in asynchrony between monozygotic twins. On the one hand, the risk of psychosocial harm to the 'later' twin may be a good reason to avoid asynchrony between monozygotic twins. On the other hand, one should acknowledge that this policy could substantially decrease the potential clinical value of reproductive embryo cloning. After all, it would be at odds with one of the goals of reproductive embryo cloning, namely to minimize the burdens of IVF by cryopreserving a set of cloned embryos for transfer in a subsequent cycle.

The potential applications of reproductive embryo cloning in the context of genetic technologies would raise partly different issues. The first is that of selective reproductive embryo cloning to increase the success rate of IVF/PGD. No doubt, selective reproductive embryo cloning might be useful in individual cases. It is important, however, to anticipate its potential impact on the criteria for the selection of preimplantation embryos (see above). The second is that of nuclear transfer to prevent the transmission of mitochondrial disorders. One of the methods that could be used, namely the transplantation of nuclei isolated from blastomeres, would involve embryo cloning. It would, of course, be premature to conclude that, in order to avoid embryo cloning, one of the other methods should be investigated. Part of the ethical analysis needed is a comparative analysis of the respective pros and cons of the various methods (including their safety and effectiveness). A preliminary issue is, of course, whether genetic interventions in the germ line can be morally justified, and, if so, under what circumstances. This question is beyond the scope of this chapter.

Therapeutic cloning

The major ethical problem is that therapeutic cloning involves the production of embryos purely for instrumental use. (The view that the entity produced by cloning – asexual reproduction – is not an embryo, because an embryo is, by definition, the product of fertilization, is absurd. In the post-Dolly era, this definition is simply an anachronism.) In this respect, therapeutic cloning is similar to the issue of creating embryos for research purposes. In view of the relatively low moral status of the preimplantation embryo, it seems reasonable to argue that research using spare embryos, as well as creating embryos for research purposes, may be morally justified on some conditions (see above). Then, if there are no categorical moral objections to creating preimplantation embryos for instrumental use, can therapeutic cloning be morally acceptable? Objections (to be scrutinized) are connected to the principle of proportionality, the slippery-slope argument and the principle of subsidiarity.[67]

It is doubtful whether the principle of proportionality provides a convincing a priori objection to therapeutic cloning. After all, if it is considered acceptable to create embryos for research aiming at the improvement of assisted reproductive technologies (freezing of egg cells; *in vitro* maturation of egg cells, etc.), then it is inconsistent to reject therapeutic cloning beforehand as being disproportionate. A consequentialist objection (fashioned as a slippery-slope argument) is that therapeutic cloning will inevitably lead to reproductive cloning. This objection is not convincing; if reproductive cloning is categorically unacceptable, then it is reasonable to prohibit this specific type of cloning, and not to ban other, non-reproductive, applications of cloning. The presumed automatism in the slippery-slope argument is disputable.

A condition for the instrumental use of preimplantation embryos to be justified is that there are no suitable alternatives that may serve the goals of this (research) use – the principle of subsidiarity. Well, then, do suitable alternatives to therapeutic cloning exist? First, it is important to note that therapeutic cloning *strictu sensu* (starting with the first clinical trials) will not emerge soon. Much basic research is needed, among other things, about the question of whether it will be possible to control the differentiation of human embryonic stem cells *in vitro*. This research can and should be done with spare IVF embryos.

At the same time, research into potential 'embryo-saving' alternatives for therapeutic cloning should be conducted, including the use of adult stem cells, and the so-called 'direct reprogramming' of adult cells, i.e. to reprogram an adult cell to make it revert to its unspecialized state so that it can then be influenced to develop into a different type of tissue (this involves the development of undifferentiated cells without the need to create an embryo). No doubt, studies so far with adult stem cells are encouraging, and further research in this direction should be stimulated. It remains to be seen, however, whether adult stem cells can really fulfil the same potential as human embryonic stem cells. Critics of the use of human embryonic stem cells often argue that recent studies prove that adult stem cells have the same developmental potential equal to that of human embryonic stem cells. Alas, there is no such proof. At a recent workshop in the USA, both adult and embryonic stem cell researchers stated that hints of the versatility of adult stem cells have been overplayed and overinterpreted.[68] The most optimistic expectation concerning adult stem cells is that only in the long run may these cells appear to have equal plasticity to that of human embryonic stem cells and be as broadly applicable in the clinic.[69]

Direct reprogramming by some is seen as the ultimate future alternative for therapeutic cloning.[70] Its development, nevertheless, has a moral price; this requires nucleus transplantation in order to create embryos for basic research. This research can be morally justified (see above), but adequate animal research should be conducted first.

The question of whether therapeutic cloning should be allowed only becomes acute if research with spare IVF embryos suggests that usable transplants can be obtained *in vitro* from human embryonic stem cells, and if the (now) theoretical alternatives for therapeutic cloning are less promising or need more time for development than is currently hoped. In that case, therapeutic cloning could be justified on the basis of both the principle of proportionality and the principle of subsidiarity.

CONCLUSIONS AND RECOMMENDATIONS

(1) From both a fetalist and a feminist perspective there is no categorical moral objection to the creation of embryos for research purposes, either by *in vitro* fertilization (IVF) or by nuclear transfer.

(2) In the context of medically assisted reproduction, informed consent should not be understood as an event, but as a process.

(3) Reproductive doctors have a professional responsibility to take into account the interests of the future child. The question as to whether this responsibility can justify a 'coercive offer' of genetic testing of applicants for medically assisted reproduction and/or of PGD, and, if so, under what circumstances, needs further scrutiny.

(4) PGD for late-onset disorders cannot be categorically dismissed. Relevant variables include the penetrance of the mutation, the severity of the disorder and its age of onset.

(5) It would be symptomatic of a moral myopia to isolate the case of preimplantation sex selection for social reasons from the broader context of selecting embryos for other non-medical characteristics. The ethics of the 'designer' model of PGD should be given a high priority on the ethical agenda.

(6) PGD for the human leukocyte antigen (HLA) matching of siblings may well be morally justified.

(7) While there are no valid *a priori* objections to (experimental) preimplantation genetic screening (PGS) for aneuploidy, this screening raises complex ethical questions regarding decision-making authority and the just allocation of scarce resources.

(8) The current health risks of the reproductive cloning of adults/children are a compelling reason to impose a world-wide moratorium.

If human reproductive cloning became 'safe', however, it might well be justified in individual cases, especially to 'bypass' infertility or genetic risks.

(9) At least some of the objections to the reproductive cloning of adults/children do not, or not necessarily, apply to the reproductive cloning of embryos.

(10) If research with spare IVF embryos suggests that usable transplants can be obtained *in vitro* from human embryonic stem cells, and if the (now) theoretical alternatives for therapeutic cloning are less promising or need more time for development than is currently hoped, therapeutic cloning could be justified on the basis of both the principle of proportionality and the principle of subsidiarity.

(11) The ethics of the interspecies transplantation of human germ stem cells should be prospectively addressed.

REFERENCES

1. de Wert G. Ethics of assisted reproduction: the case of preimplantation genetic diagnosis. In Fauser BCJM, Rutherford AJ, Strauss JF III, Van Steirteghem A, eds. *Molecular Biology in Reproductive Medicine*. Carnforth, UK: Parthenon Publishing, 1999; 433–448.
2. Raymond J. Fetalists and feminists: they are not the same. In Spallone P, Steinberg DL, eds. *Made to Order: The Myth of Reproductive and Genetic Progress*. Oxford: Pergamon Press, 1987: 58–65.
3. Hursthouse R. *Beginning Lives*. London: Blackwell, 1987.
4. *Council of Europe. Convention for the protection of human rights and dignity of the human being with regard to the application of biology and medicine: Convention on Human Rights and Biomedicine*. Strasbourg: Council of Europe, 1996.
5. Gerrand N. Creating embryos for research. *J Appl Philos* 1993 10: 175–187.
6. Human Fertilisation and Embryology Authority. *Donated ovarian tissue in embryo research and assisted conception*. Report. London HFEA: 1994
7. De Wert G, Berghmans R, Boer G *et al.* Ethical guidance on human embryonic and fetal tissue transplantation: a European overview. *Med Health Care Philos* 2002 in press.
8. Beauchamp TL, Childress JF, eds. *Principles of Biomedical Ethics* (2nd edn). New York: Oxford University Press, 1983.
9. American College of Obstetricians and Gynecologists, Committee on Ethics. *Ethical dimensions of informed consent*. ACOG Committee Opinion no. 108. ACOG, 1992.
10. Health Council of The Netherlands. *In Vitro Fertilization*. Rijswijk: Health Council of the Netherlands, 1997.
11. Steinbock B. A philosopher looks at assisted reproduction. *J Assist Reprod Genet* 1995 12: 543–551.
12. de Wert G. Ethics of intracytoplasmic sperm injection: proceed with care. *Hum Reprod* 1998 13 (Suppl 1): 219–227.
13. Health Council of the Netherlands. *Assisted Fertilization: ICSI*. The Hague: Health Council of The Netherlands, 1996.
14. de Wert G. The postmenopause: playground for reproductive technology? Some ethical reflections. In Harris J, Holm S, eds. *The Future of Human Reproduction. Ethics, Choice and Regulation*. Oxford: Clarendon Press, 1998: 221–238.
15. Pennings G. The welfare of the child. *Hum Reprod* 1999 14: 1146–1150.
16. Arras JD. AIDS and reproductive decisions: having children in fear and trembling. *Milbank Q* 1990 68: 353–382.
17. de Wert G. *Reproductive Technologies, Genetics and Ethics*. Amsterdam: Thela Thesis, 1999.
18. Drapkin Lyerly A, Anderson J. Human immunodeficiency virus and assisted reproduction: reconsidering evidence, reframing ethics. *Fertil Steril* 2001 75: 843–858.
19. ESHRE Preimplantation Genetic Diagnosis (PGD) Consortium: data collection II (May 2000). *Hum Reprod* 2000 15: 2673–2683.
20. Iglesias T. *IVF and justice*. London: The Linacre Centre for Health Care Ethics, 1990.
21. Gesetz zum Schutz von Embryonen. (BGBI.I S.2746) December 1990.
22. McLaren A. A note on 'totipotency'. *Biomed Ethics* 1997 2: 7.
23. de Wert G. Prenatal diagnosis and ethics. In Ten Have H, Engberts DP, Kalkman-Bogerd LE *et al.*,

eds. *Ethiek en Recht in de Gezondheidszorg*. Dordrecht: Kluwer, 1990 XVI: 121–153.

24. Post S. Selective abortion and gene therapy: reflections on human limits. *Hum Gene Ther* 1991 2: 229–233.
25. Braude PR, de Wert GMWR, Evers-Kiebooms G *et al.* Non-disclosure preimplantation genetic diagnosis for Huntington's disease: practical and ethical dilemmas. *Prenat Diagn* 1998 18: 1422–1426.
26. Lancaster JM, Wiseman RW, Berchuck A. An inevitable dilemma: prenatal testing for mutations in *BRCA1* breast–ovarian cancer susceptibility gene. *Obstet Gynecol* 1996 87: 306–309.
27. Meijers-Heijboer H, Van Geel B, Van Putten W *et al.* Breast cancer after prophylactic bilateral mastectomy in women with *BRCA1* or *BRCA2* mutation. *N Engl J Med* 2001 345: 159–164.
28. Robertson JA. *Children of Choice. Freedom and the New Reproductive Technologies*. Princeton, NJ: Princeton University Press, 1994.
29. Warren MA. *Gendercide. The Implications of Sex Selection*. Totowa (NJ): Rowman & Allanheld, 1985.
30. Bouchard C, Malina R, Pérusse L. *Genetics of Fitness and Physical Performance*. Champain, IL: Human Kinetics, 1997.
31. Strong C. *Ethics in Reproductive and Perinatal Medicine. A New Framework*. New Haven & London: Yale University Press, 1997.
32. King DS. Preimplantation genetic diagnosis and the 'new' eugenics. *J Med Ethics* 1999 25: 176–182.
33. de Wert G. Human cloning: the case of the preimplantation embryo: an ethical exploration. In Gunning J, ed. *Assisted Conception. Research, Ethics and Law* Aldershot, UK: Ashgate/-Dartmouth, 2000: 83–97.
34. Verlinsky Y, Rechitsky S, Schoolcraft W *et al.* Preimplantation diagnosis for Fanconi anemia combined with HLA matching. *J Am Med Assoc* 2001 285: 3130–3133.
35. Kant I. *Foundations of the Metaphysics of Morals*. New York: Macmillan, 1959.
36. Davis N. Using persons and common sense. *Ethics* 1984 94: 387–406.
37. Kearney W, Caplan AL. Parity for donation of bone marrow: ethical and policy considerations. In Blank RH, Bonnicksen AL, eds. *Emerging Issues in Biomedical Policy*, Vol 1. *Genetic and Reproductive Technologies*. New York: Columbia University Press, 1992: 263–285.
38. Silber SJ, Nagy Z, Liu J *et al.* The use of epididymal and testicular spermatozoa for intracytoplasmic sperm injection: the genetic implications for male infertility. *Hum Reprod* 1995 10: 2031–2043.
39. Meschede D, De Geyter C, Nieschlag E, Horst J. Genetic risk in micromanipulative assisted reproduction. *Hum Reprod* 1995 10: 2880–2886.
40. Raeburn S. Genetic couselling. In Harper J, Delhanty JDA, Handyside AH, eds. *Preimplantation Genetic Diagnosis*. Chichester: J Wiley & Sons, 2001: 45–52.
41. Dunstan G. Ethics of gamete and embryo micromanipulation. In Fishel S, Symonds M, eds. *Gamete and Embryo Micromanipulation in Human Reproduction*. London: Edward Arnold, 1993: 212–218.
42. Reubinoff B, Shushan A. Preimplantation diagnosis in older patients. To biopsy or not to biopsy? *Hum Reprod* 1997 11: 2071–2078.
43. Strom CM. Mosaicism and aneuploidy in human pre-embryos. *J Assist Reprod Genet* 1996 13: 592–593.
44. Verlinsky Y, Cieslak J, Ivakhnenko V *et al.* Birth of healthy children after preimplantation diagnosis of common aneuploidies by polar body fluorescent *in situ* hybridization analysis. *Fertil Steril* 1996 66: 126–129.
45. Staessen C, Van Assche E, Platteau P *et al.* The impact of blastocyst transfer with or without preimplantation genetic diagnosis for aneuploidy screening with advanced maternal age. *Hum Reprod* 2001 16: Abstr Book 1, O-054: 23–24.
46. Egozcue J. Of course not. *Hum Reprod* 1996 11: 2077–2078.
47. Davis DS. *Genetic Dilemmas. Reproductive Technology, Parental Choices, and Children's Futures*. New York: Routledge, 2001: 107–128.
48. Jones HW Jr, Edwards RG, Seidel GE Jr. On attempts at cloning in the human. *Fertil Steril* 1994 61: 423–426.
49. Cohen J, Tomkin G. The science, fiction and reality of embryo cloning. *Kennedy Inst Ethics J* 1994 4: 193–203.
50. Wolf DP. Genetic manipulation by nuclear transfer. *J Assist Reprod Genet* 1997 14: 480 (abstr).
51. Gurdon JB, Colman A. The future of cloning. *Nature* 1999 402: 743–746.

52. National Bioethics Advisory Commission. Cloning human beings. Rockville, MD: NBAC, 1997.
53. Lanza RP, Cibelli JB, Faber D *et al.* Cloned cattle can be healthy and normal. *Science* 2001 294: 1893–1894.
54. Jaenisch R, Wilmut I. Don't clone humans! *Science* 2001 291: 2552.
55. Council of Europe. *Additional protocol to the Convention for the protection of human rights and dignity of the human being with regard to the application of biology and medicine on the prohibition of cloning human beings.* Strasbourg: Council of Europe, 1997.
56. Harris J. 'Goodbye Dolly?' The ethics of human cloning. *J Med Ethics* 1997 23: 353–360.
57. Brock DW. Cloning human beings: an assessment of the ethical issues pro and con. In Nussbaum MC, Sunstein CR, eds. *Clones and Clones. Facts and Fantasies about Human Cloning.* New York: Norton & Company, 1998: 141–164.
58. Jonas H. Lasst und einen Menschen klonieren. Betrachtungen zur Aussicht genetischer Versuche mit uns selbst. *Scheidewege* 1982 12: 462–489.
59. Holm S. A life in the shadow: one reason why we should not clone humans. *Camb Q Healthcare Ethics* 1998 7: 160–162.
60. Pence GE. *Recreating Medicine. Ethical Issues at the Frontiers of Medicine.* Lanham, UK: Rowman & Littlefield, 2000: 119–135.
61. Kaneko M, Takeuchi T, Veeck LL *et al.* Haploidization enhancement to manufacture human oocytes. *Hum Reprod* 2001 16, Abstr Book 1, O-010: 4–5.
62. Takeuchi T, Kaneko M, Veeck LL *et al.* Creation of viable human oocytes using diploid somatic nuclei. *Hum Reprod* 2001 16: Abstr Book 1, O-011: 5.
63. Dobrinski I, Avarbock MR, Brinster RL. Transplantation of germ cells from rabbits and dogs into mouse testes. *Biol Reprod* 1999 61: 1331–1339.
64. Ogawa T, Dobrinski I, Avarbock MR, Brinster RL. Transplantation of male germ line stem cells restores fertility in infertile mice. *Nat Med* 2000 6: 29–34.
65. Bonnicksen AL. Ethical and policy issues in human embryo twinning. *Camb Q Healthcare Ethics* 1995 4: 268–284.
66. Edwards RG, Beard HK. How identical would cloned children be? *Hum Reprod Update* 1998 4: 791–811.
67. de Wert G. The holy grail of human embryonic stem cells. An ethical reflection. *Filos Prakt* 2001 22: 34–56.
68. Vastag B. Many say adult stem cell reports overplayed. *J Am Med Assoc* 2001 286: 293.
69. Department of Health. *Stem cell research: medical progress with responsibility.* London: Department of Health, June 2000.
70. Aldhous P. Can they rebuild us? *Nature* 2001 410: 622–625.

Index